The Basic Practice of Statistics

Organizing a Statistical Problem: A Four-Step Process

STATE: What is the practical question, in the context of the real-world setting?

PLAN: What specific statistical operations does this problem call for?

SOLVE: Make the graphs and carry out the calculations needed for this problem.

CONCLUDE: Give your practical conclusion in the setting of the real-world problem.

Confidence Intervals: The Four-Step Process

STATE: What is the practical question that requires estimating a parameter?

PLAN: Identify the parameter, choose a level of confidence, and select the type of confidence interval that fits your situation.

SOLVE: Carry out the work in two phases:

1. **Check the conditions** for the interval you plan to use.
2. Calculate the **confidence interval**.

CONCLUDE: Return to the practical question to describe your results in this setting.

Tests of Significance: The Four-Step Process

STATE: What is the practical question that requires a statistical test?

PLAN: Identify the parameter, state null and alternative hypotheses, and choose the type of test that fits your situation.

SOLVE: Carry out the test in three phases:

1. **Check the conditions** for the test you plan to use.
2. Calculate the **test statistic**.
3. Find the ***P*-value**.

CONCLUDE: Return to the practical question to describe your results in this setting.

The Basic Practice of Statistics

Ninth Edition

DAVID S. MOORE
Purdue University

WILLIAM I. NOTZ
The Ohio State University

macmillan learning

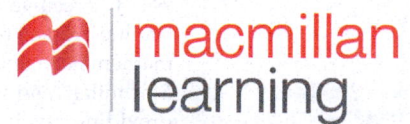

Austin • Boston • New York • Plymouth

Senior Vice President, STEM: Daryl Fox
Program Director, Math, Statistics, Earth Sciences, and Environmental Science: Andrew Dunaway
Program Manager: Sarah Seymour
Executive Content Development Manager, STEM: Debbie Hardin
Development Editor: David Dietz
Executive Project Manager, Content, STEM: Katrina Mangold
Director of Content, Math and Statistics: Daniel Lauve
Executive Media Editor: Catriona Kaplan
Lead Content Developer: Aaron Gladish
Content Development Manager: Ava Cas
Associate Editor: Andy Newton
Assistant Editor: Justin Jones
Marketing Manager: Leah Christians
Marketing Assistant: Morgan Psiuk
Director of Content Management Enhancement: Tracey Kuehn
Senior Managing Editor: Lisa Kinne
Senior Content Project Manager: Edward Dionne
Project Manager: Heidi Allgair, SPi Global
Senior Workflow Project Manager: Paul Rohloff
Production Supervisor: Robert Cherry
Director of Design, Content Management: Diana Blume
Design Services Manager: Natasha Wolfe
Cover Design Manager: John Callahan
Interior Design: Vicki Tomaselli
Art Manager: Matthew McAdams
Director of Digital Production: Keri deManigold
Media Project Manager: Hanna Squire
Executive Permissions Editor: Cecilia Varas
Rights and Billing Editor: Alexis Gargin
Composition: SPi Global
Printing and Binding: LSC Communications
Cover and Title Page Images: Gremlin/Getty Images

Library of Congress Control Number: 2020941936

Student Edition Paperback:
ISBN-13: 978-1-319-24437-8
ISBN-10: 1-319-24437-8

Student Edition Loose-leaf:
ISBN-13: 978-1-319-36523-3
ISBN-10: 1-319-36523-X

Printed in the United States of America

3 4 5 6 25 24 23 22

Macmillan Learning
One New York Plaza
Suite 4600
New York, NY 10004-1562
www.macmillanlearning.com

w.h.freeman
Macmillan Learning

In 1946, William Freeman founded W. H. Freeman and Company and published Linus Pauling's *General Chemistry*, which revolutionized the chemistry curriculum and established the prototype for a Freeman text. W. H. Freeman quickly became a publishing house where leading researchers can make significant contributions to mathematics and science. In 1996, W. H. Freeman joined Macmillan, and we have since proudly continued the legacy of providing revolutionary, quality educational tools for teaching and learning in STEM.

BRIEF CONTENTS

*Starred material is optional and can be skipped without loss of continuity.

CONTENTS

*Starred material is optional and can be skipped without loss of continuity.

WHY DID YOU DO THAT?

The Authors Answer Questions about The Basic Practice of Statistics

Welcome to the ninth edition of *The Basic Practice of Statistics*. As the title suggests, this text provides an introduction to the practice of statistics that aims to equip students to carry out common statistical procedures and to follow statistical reasoning in their fields of study and in their future employment.

There is no single best way to organize our presentation of statistics to beginners. That said, our choices reflect thinking about both content and pedagogy. Here are comments on several frequently asked questions about the order and selection of material in *The Basic Practice of Statistics*.

Why Did You Write *The Basic Practice of Statistics*?

Several factors influenced the writing of *The Basic Practice of Statistics*. Easy-to-use statistical software with graphical tools made it possible for students to explore and analyze data on their own. Statistics educators recognized that actually *doing* statistics—exploring data, analyzing data, thinking about what the data are telling us, and assessing the validity of the conclusions we make from data—is an effective way to *learn* statistics. Teachers also recognized the importance of using real data from actual studies to reinforce the fact that statistics is invaluable for answering real-world questions. Finally, an introductory course in statistics should expose students to how statistics is actually practiced by researchers. At the time of the writing of the first edition, few, if any, textbooks for courses intended for students with only college algebra as the mathematics prerequisite incorporated these ideas.

With this in mind, *The Basic Practice of Statistics* was designed to reflect the actual practice of statistics, where data analysis and design of data production join with probability-based inference to form a coherent science of data. *The Basic Practice of Statistics* was also designed to be accessible to college and university students with limited quantitative background—just "algebra" in the sense of being able to read and use simple equations.

Why Should I Use *The Basic Practice of Statistics* to Teach an Introductory Statistics Course?

The Basic Practice of Statistics is based on three principles: balanced content, experience with data, and the importance of ideas. These principles are widely accepted by statisticians concerned about teaching and are directly connected to and reflected by the themes of the College Report of the Guidelines in Assessment and Instruction for Statistics Education (GAISE) Project.

The GAISE guidelines include six recommendations for introductory statistics. The content, coverage, and features of *The Basic Practice of Statistics* are closely aligned to these recommendations:

1. *Teach statistical thinking.*
   - *Teach statistics as an investigative process of problem solving and decision making.* In *The Basic Practice of Statistics*, we present a four-step process for solving statistical problems. This begins by stating the practical question to be answered in the context of a real-world setting and ends with a practical conclusion, often a decision to be made, in the setting of the real-world problem. The process is illustrated in the text by revisiting data from a study in a series of examples or exercises. Different aspects of the data are investigated in different examples and exercises, with the ultimate goal of making some decision based on what has been learned.
   - *Give students experience with multivariable thinking. The Basic Practice of Statistics* exposes students to multivariate thinking early in the book. Chapters 4, 5, and 6 introduce students to methods for exploring bivariate data. In Chapter 7, we include online data with many variables, inviting students to explore aspects of these data. In Chapter 9, we discuss the importance of identifying the many variables that can affect a response and including them in the design of an experiment and the interpretation of the results. In Part V, we introduce students to formal methods of inference for bivariate data, and in the online supplemental chapters, we discuss multiple regression, two-way ANOVA, and statistical process control.
2. *Focus on conceptual understanding.* A first course in statistics introduces many skills, from making a stemplot and calculating a correlation to choosing and carrying out a significance test. In practice (even if not always in the course), calculations and graphs are automated. Moreover, anyone who makes serious use of statistics will need some specific procedures not taught in their college statistics course. *The Basic Practice of Statistics*, therefore, emphasizes conceptual understanding by making clear the larger patterns and big ideas of statistics—not in the abstract but in the context of learning specific skills and working with specific data. Many of the big ideas are summarized in graphical outlines. Three of the most useful of these appear opposite the title page. Formulas without guiding principles do students little good once the final exam is past, so it is worth the time to slow down a bit and explain the ideas.
3. *Integrate real data with a context and a purpose.* The study of statistics is supposed to help students work with data in their varied academic disciplines and in their unpredictable later employment. Students learn to work with data by working with data. *The Basic Practice of Statistics* is full of data from many fields of study and from everyday life. Data are more than mere numbers: they are numbers with a context that should play a role in making sense of the numbers and in stating conclusions. Examples and exercises in *The Basic Practice of Statistics*, though intended for beginners, use real data and give enough background to allow students to consider the meaning of their calculations.
4. *Foster active learning.* Fostering active learning is the business of the teacher, though an emphasis on working with data helps. To this end, we have created interactive applets to our specifications that are available online. These are designed primarily to help in learning statistics rather than in doing statistics. We suggest using selected applets for classroom demonstrations even if you do not ask students to work with them. The Correlation and Regression, Confidence Intervals, and *P*-Value of a Test of Significance applets, for example, convey core ideas more clearly than any amount of chalk and talk.

For each chapter (except the review chapters), web exercises are provided online. Our intent is to take advantage of the fact that most undergraduates are web savvy. These exercises require students to search the web for either data or statistical examples and then evaluate what they find. Teachers can use these as classroom activities or assign them as homework projects.

5. *Use technology to explore concepts and analyze data.* Automating calculations increases students' ability to complete problems, reduces their frustration, and helps them concentrate on ideas and problem recognition rather than mechanics. At a minimum, students should have a "two-variable statistics" calculator with functions for correlation and the least-squares regression line as well as for the mean and standard deviation.

   Many instructors will take advantage of more elaborate technology, as ASA/MAA and GAISE recommend. And many students who don't use technology in their college statistics course will find themselves using (for example) Excel on the job. *The Basic Practice of Statistics* does not assume or require use of software except in Part V, where the work is otherwise too tedious. It does accommodate software use and provides students with knowledge that will enable them to read and use output from almost any source. There are regular "Examples of Technology" sections throughout the text. Each of these sections displays and comments on output from the same three technologies, representing graphing calculators (the Texas Instruments TI-83 or TI-84), spreadsheets (Microsoft Excel), and statistical software (JMP, Minitab, R, and CrunchIt!). The output always concerns one of the main teaching examples so that students can compare text and output.

6. *Use assessments to improve and evaluate student learning.* Within chapters, a few "Apply Your Knowledge" exercises follow each new idea or skill for a quick check of basic mastery—and also to mark off digestible bites of material. Each of the first four parts of the book ends with a review chapter that includes a point-by-point outline of skills learned, problems students can use to test themselves, and several supplementary exercises. (Instructors can choose to cover any or none of the chapters in Part V, so each of these chapters includes a skills outline.) The review chapters present supplemental exercises without the "I just studied that" context, thus asking for another level of learning. We think it is helpful to assign some supplemental exercises. Many instructors will find that the review chapters appear at the right points for pre-exam review. Students can use the "Test Yourself" questions to review, self-assess, and prepare for exams. In addition, assessment materials in the form of a test bank and quizzes are available online.

Why Did You Choose to Order Topics as Listed in the Book?

There are good pedagogical reasons for beginning with data analysis (Chapters 1 through 7), then moving to data production (Chapters 8 through 11), and then to probability and inference (Chapters 12 through 27). In studying data analysis, students learn useful skills immediately and get over some of their fear of statistics. Data analytics is much in the media these days, and by discussing data analysis, instructors can link the course material to the current interest in data analytics. Data analysis is a necessary preliminary to inference in practice because inference requires clean data. Designed data production is the surest foundation for inference, and the deliberate use of chance in random sampling and randomized comparative experiments motivates the study of probability in a course that emphasizes data-oriented statistics. *The Basic Practice of Statistics* gives a full presentation of basic probability and inference (16 of the 27 chapters in the printed text) but places it in the context of statistics as a whole.

Why Does the Distinction between Population and Sample Not Appear in Part I?

There is more to statistics than inference. In fact, statistical inference is appropriate only in rather special circumstances. The chapters in Part I present tools and tactics for describing data—any data. These tools and tactics do not depend on the idea of inference from sample to population. Many data sets in these chapters (for example, the several sets of data about the 50 states) do not lend themselves to inference because they represent an entire population. Likewise, many modern big data sets are also viewed as information about an entire population, for which formal inference may not be appropriate. John Tukey of Bell Labs and Princeton, the philosopher of modern data analysis, insisted that the population–sample distinction be avoided when it is not relevant. He used the word *batch* for data sets in general. We see no need for a special word, but we think Tukey was right.

Why Not Begin with Data Production?

We prefer to begin with data exploration (Part I), as most students will use statistics mainly in settings other than planned research studies in their future employment. We place the design of data production (Part II) after data analysis to emphasize that data-analytic techniques apply to any data. However, it is equally reasonable to begin with data production; the natural flow of a planned study is from design to data analysis to inference. Because instructors have strong and differing opinions on this question, these two topics are now the first two parts of the book, with the text written so that it may be started with either Part I or Part II while maintaining the continuity of the material.

Another reason for beginning with data exploration is to give students experience exploring data and thinking about how to interpret what they discover. This experience provides a context for how data production affects the reliability of conclusions one might draw from data.

Why Do Normal Distributions Appear in Part I?

Density curves such as the Normal curves are just another tool to describe the distribution of a quantitative variable, along with stemplots, histograms, and boxplots. Professional statistical software offers to make density curves from data just as it offers histograms. We prefer not to suggest that this material is essentially tied to probability, as the traditional order does. And we find it helpful to break up the indigestible lump of probability that troubles students so much. Meeting Normal distributions early does this and strengthens the "probability distributions are like data distributions" way of approaching probability when we get there.

Why Not Delay Correlation and Regression Until Late in the Course, as Was Traditional?

The Basic Practice of Statistics begins by offering experience working with data and gives a conceptual structure for this nonmathematical, but essential, part of statistics. Students profit from more experience with data early and from seeing the conceptual structure worked out in relationships among variables as well as in describing single-variable data. Correlation and least-squares regression are very important descriptive tools and are often used in settings where there is no population–sample distinction, such as studies of all of a firm's employees. Perhaps most importantly, *The Basic Practice of Statistics* asks students to think about what kind of relationship lies behind the data (confounding, lurking variables, association not implying causation, and so on) without overwhelming them with the demands of formal inference methods. Inference in the correlation and regression setting is a bit complex, demands software, and often comes right at the end of the course. We find that delaying all mention of correlation and regression to that point often means that students don't master the basic uses and properties of these methods. We consider Chapters 4 and 5 (correlation and regression) essential and Chapter 26 (regression inference) optional.

Why Use the *z* Procedures for a Population Mean to Introduce the Reasoning of Inference?

This is a pedagogical issue, not a question of statistics in practice. The two most popular choices for introducing inference are z for a mean and z for a proportion. (Another option is resampling and permutation tests. We have included material on these topics but have not used them to introduce inference.)

We find z for means quite accessible to students. Positively, we can say up front that we are going to explore the reasoning of inference in the overly simple setting described in the box on page 367 titled "Simple Conditions for Inference about a Mean." As this box suggests, assuming an exactly Normal population and a true simple random sample are as unrealistic as known s. All the issues of practice—robustness against lack of Normality and application when the data aren't an SRS as well as the need to estimate s—are put off until, with the reasoning in hand, we discuss the practically useful t procedures. This separation of initial reasoning from messier practice works well.

Negatively, starting with inference for p introduces many side issues: no exact Normal sampling distribution but a Normal approximation to a discrete distribution; use of $\hat{p}$ in both the numerator and denominator of the test statistic to estimate both the parameter p and $\hat{p}$'s own standard deviation; loss of the direct link between test and confidence interval; and the need to avoid small and moderate sample sizes because the Normal approximation for the test is quite unreliable.

There are advantages to starting with inference for p. Starting with z for means takes a fair amount of time, and the ideas need to be rehashed with the introduction of the t procedures. Many instructors face pressure from client departments to cover a large amount of material in a single semester. Eliminating coverage of the "unrealistic" z for means with known variance enables instructors to cover additional, more realistic applications of inference. Also, many instructors believe that proportions are simpler and more familiar to students than means.

Why Didn't You Cover Topic X?

Introductory texts ought not to be encyclopedic. We chose topics on two grounds: they are the most commonly used in practice, and they are suitable vehicles for learning broader statistical ideas. Students who have completed the core of the book, Chapters 1 through 12 and 15 through 24, will have little difficulty moving on to more elaborate methods. Chapters 25 through 27 offer a choice of slightly more advanced topics, as do the optional supplemental chapters, available online.

Why Are Some Chapters and Sections Listed as Optional?

Many users have requested that we include the content listed as optional. However, as noted above, many instructors face pressure from client departments to cover many topics in a single semester. We have identified some material that can safely be omitted because it is not required for later parts of the book. Instructors can cover this optional content if they wish, but they can also omit it in order to cover topics that client departments have requested.

The content we designate as optional is not less important than other material in the book. For example, many instructors will want to cover Chapters 6 and 25 because they consider relationships between categorical variables an essential topic for their students.

We have enjoyed the opportunity to once again rethink how to help beginning students achieve a practical grasp of basic statistics. What students actually learn is not identical to what we teachers think we have "covered," so the virtues of concentrating on the essentials are considerable. We hope this new edition of *The Basic Practice of Statistics* offers a mix of concrete skills and clearly explained concepts that will help many teachers guide their students toward useful knowledge.

PREFACE

Empowering Problem Solving and Real-World Decision Making

Now available with Macmillan's new, ground-breaking online learning platform Achieve, the ninth edition of *The Basic Practice of Statistics* teaches statistical thinking through an investigative process of problem solving with pedagogy designed to help students of all levels. Examples and exercises from a wide variety of topic areas use current, real data to provide students with insight into how data is used to make decisions in the real world.

Achieve for *The Basic Practice of Statistics* connects the book's trusted Four-Step problem-solving approach and real-world examples to rich digital resources that foster understanding and facilitate the practice of statistics. The tools in Achieve support learning before, during, and after class for students and equip instructors with class performance analytics in an easy-to-use interface.

Overview of key features

Support for Learners on Every Page

- **Four-Step Problem-Solving** examples guide students through the Four-Step process for working through statistical problems: State, Plan, Solve, Conclude. Students are instructed to apply this process in designated exercises.
- **Apply Your Knowledge** exercises at the end of each section encourage students to read actively and to cement new concepts by applying them as they learn.
- **Examples of Technology**, located where most appropriate, display and comment on the output from popular statistical software applications (notably Excel, Minitab, JMP, and R) and TI 83/84 graphing calculators in the context of worked examples. Students learn to interpret output from any standard statistical package.
- **Definition and Theorem Boxes** in the text alert students to key concepts, terms, and procedures.
- **Caution Boxes** warn students of common mistakes or misconceptions.
- **Statistics in Your World** margin notes further connect statistics topics to the real world, highlighting interesting examples and applications from a variety of fields.
- The main themes of the text are strongly **aligned to the GAISE guidelines** (from the Guidelines in Assessment and Instruction in Statistics Education College Report).

New to the Ninth Edition

- Examples and exercises clearly emphasize **reaching conclusions and making decisions** based on data exploration and statistical inference.
- Chapter Summaries are in concise list form, and Skills Reviews (in Review Chapters) refer to relevant chapter sections to **help students check their knowledge and review for exams.**
- Data in examples and exercises have been **updated for relevance**, and new examples and exercises explore **contemporary issues** such as social media usage.
- Displayed in boldface type, **key terms are clearly defined** in running text or in the margin, to build understanding without focusing on vocabulary.

Highlight Four-Step Problem-Solving

Expanded in the Ninth Edition

Equips Students to Solve Complex Statistical Problems

Recognizing which approach to take and how to get started on a problem are often challenging for statistics students. David Moore and William Notz reduce student stress and support learning by using a problem-solving framework throughout *The Basic Practice of Statistics*.

Organizing a Statistical Problem: A Four-Step Process

STATE: What is the practical question, in the context of the real-world setting?

PLAN: What specific statistical operations does this problem call for?

SOLVE: Make the graphs and carry out the calculations needed for this problem.

CONCLUDE: Give your practical conclusion in the setting of the real-world problem.

EXAMPLE 4.3 State SAT Mathematics Scores

4step

MATHSAT

Figure 1.8 (page 26) reminded us that in some states, most high school graduates take the SAT to test their readiness for college, and in other states, most take the ACT. Who takes a test may influence the average score. Let's follow our four-step process (page 62) to examine this influence.[2]

STATE: The percentage of high school students who take the SAT varies from state to state. Does this fact help explain differences among the states in average SAT Mathematics score?

PLAN: Examine the relationship between percentage taking the SAT and state mean score on the Mathematics part of the SAT. Choose the explanatory and response variables. Make a *scatterplot* to display the relationship between the variables. Interpret the plot to understand the relationship.

SOLVE (make the plot): We suspect that "percentage taking" will help explain "mean score." So "percentage taking" is the explanatory variable, and "mean score" is the response variable. We want to see how mean score changes when percentage taking changes, so we put percentage taking (the explanatory variable) on the horizontal axis. Figure 4.1 is the scatterplot. Each point represents a single state. In Nevada, for example, 23% took the SAT, and their mean SAT Math score was 566. Find 23 on the x (horizontal) axis and 566 on the y (vertical) axis. Nevada appears as the point (23, 566) above 23 and to the right of 566.

CONCLUDE: We will explore conclusions in Example 4.4.

"*The Basic Practice of Statistics* has a great history and improves with each version. [It] always has a polished presentation with excellent explanations."

—Patricia Buchanan, *Professor, Pennsylvania State University*

Achieve

Achieve is the culmination of years of development work put toward creating the most powerful online learning tool for statistics students. It houses all of our renowned assessments, multimedia assets, e-books, and instructor resources in a powerful new platform.

Achieve supports educators and students throughout the full range of instruction, including assets suitable for pre-class preparation, in-class active learning, and post-class study and assessment. The pairing of a powerful new platform with outstanding statistics content provides an unrivaled learning experience.

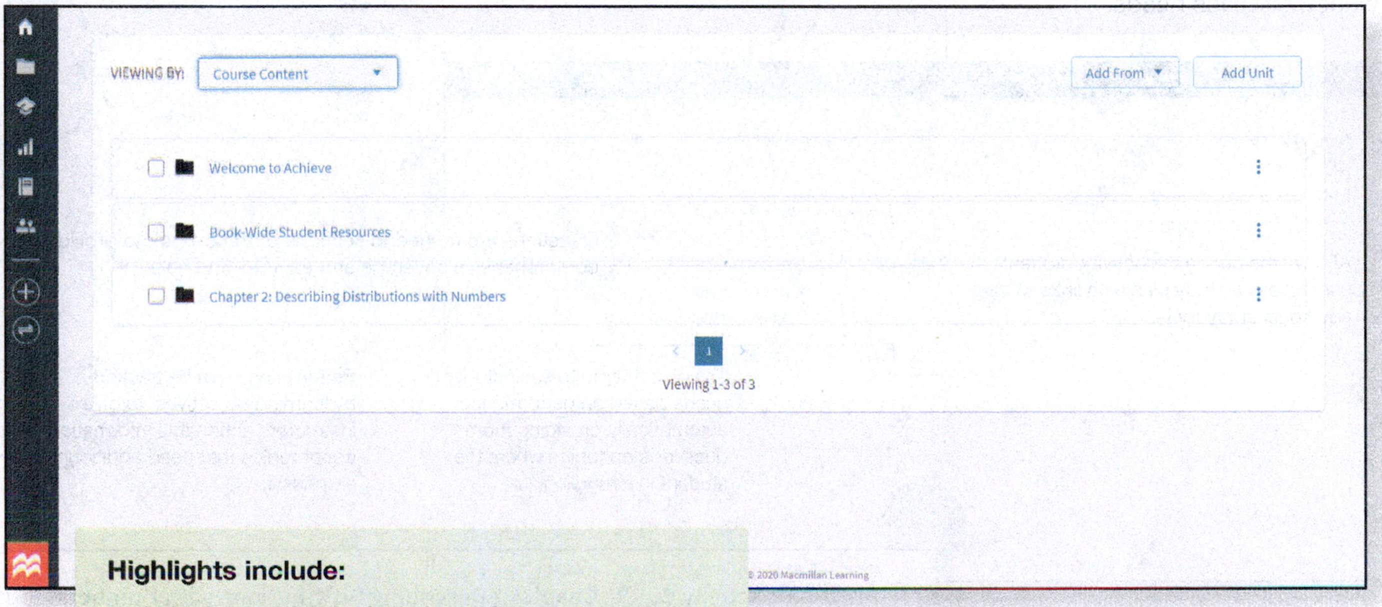

Highlights include:

- **A design guided by learning science research.** Co-designed through extensive collaboration and testing by both students and faculty including two levels of Institutional Review Board approval for every study of Achieve.
- **A learning path of powerful content** including pre-class, in-class, and post-class activities and assessments. A detailed gradebook with insights for just-in-time teaching and reporting on student achievement by learning objective.
- **Easy integration and gradebook sync** with iClicker classroom engagement solutions.
- **Simple integration** with your campus LMS and availability through **Inclusive Access** programs.

For more information or to sign up for a demonstration of Achieve, contact your local Macmillan representative or visit **macmillanlearning.com/achieve**

Robust tutorial-style assessment tools in Achieve help students develop problem-solving and statistical reasoning skills. Achieve contains more than 3,000 assessment questions designed for both pre-class foundational learning and post-class homework. For select questions, our formative assessment environment responds to students' incorrect answers with feedback to guide their study. This Socratic feedback mechanism emulates the office-hours experience, encouraging students to think critically about their identified misconceptions. Students make the most out of homework with Achieve's hallmark hints, detailed feedback, and fully worked solutions.

LearningCurve Adaptive Quizzing

LearningCurve's game-like quizzing motivates students to engage with the course content, and reporting tools help teachers get a handle on what their class needs.

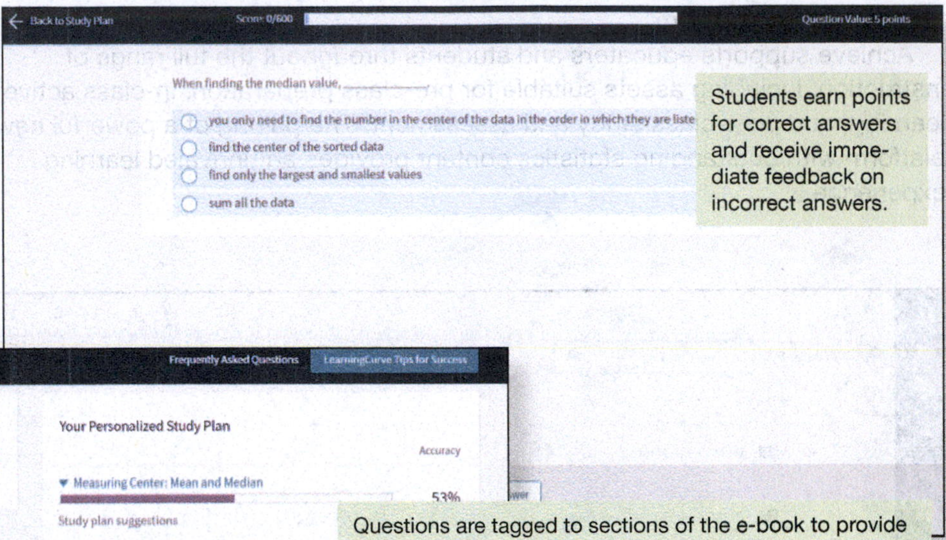

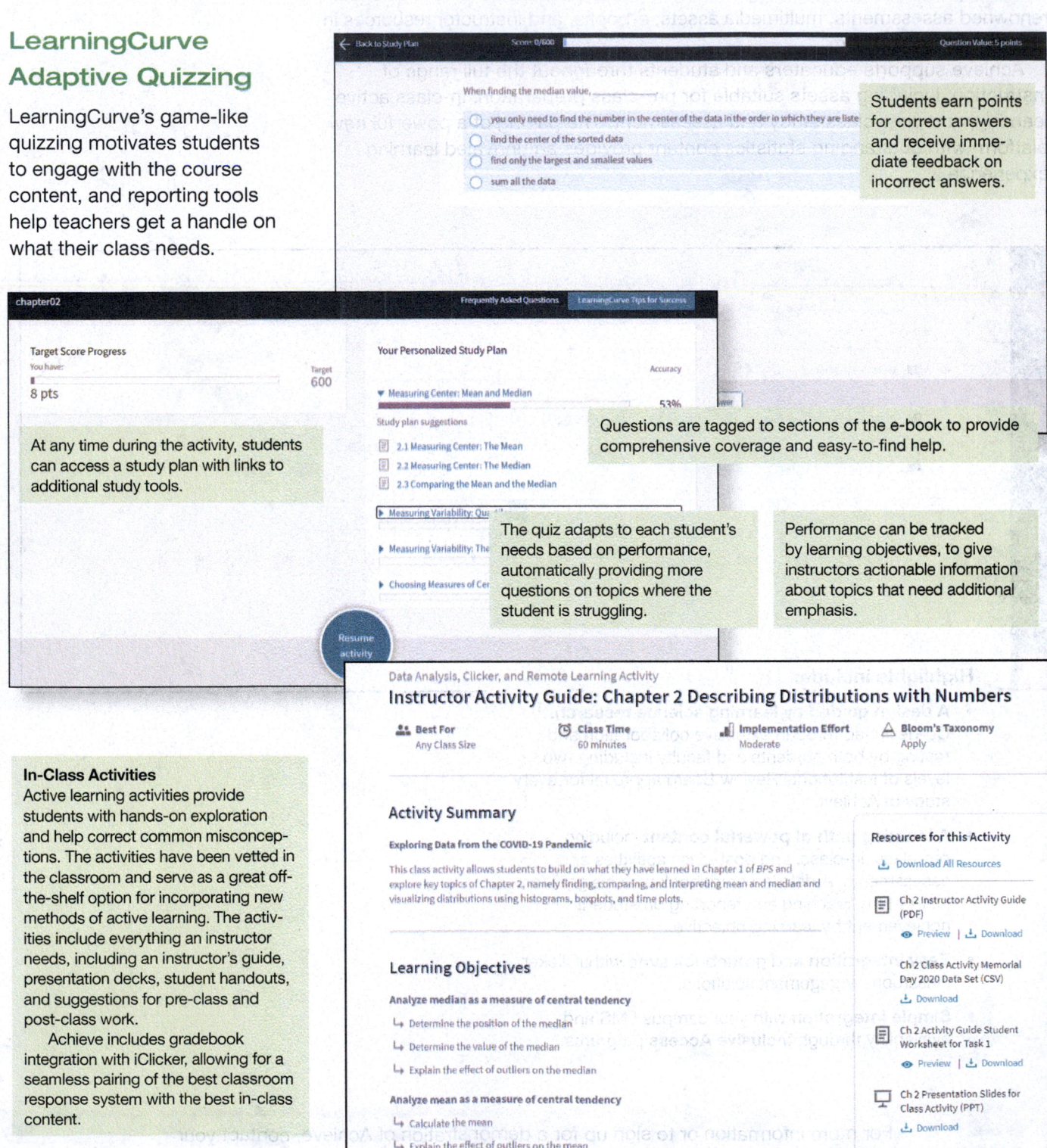

Students earn points for correct answers and receive immediate feedback on incorrect answers.

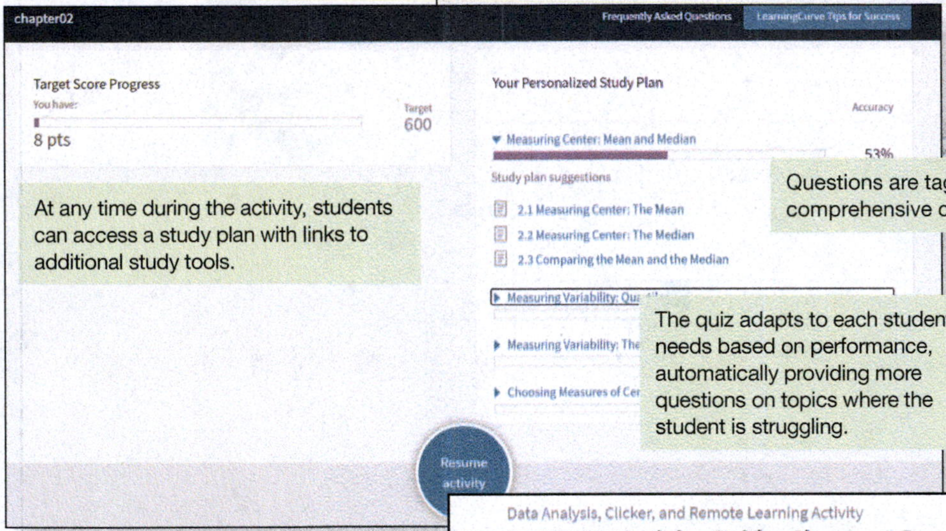

At any time during the activity, students can access a study plan with links to additional study tools.

Questions are tagged to sections of the e-book to provide comprehensive coverage and easy-to-find help.

The quiz adapts to each student's needs based on performance, automatically providing more questions on topics where the student is struggling.

Performance can be tracked by learning objectives, to give instructors actionable information about topics that need additional emphasis.

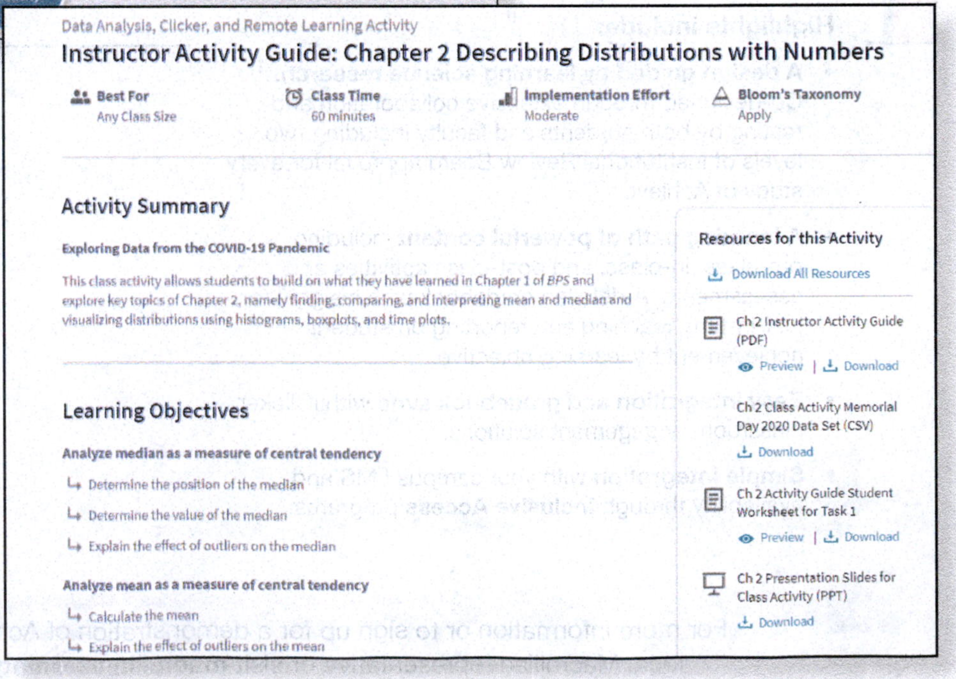

In-Class Activities
Active learning activities provide students with hands-on exploration and help correct common misconceptions. The activities have been vetted in the classroom and serve as a great off-the-shelf option for incorporating new methods of active learning. The activities include everything an instructor needs, including an instructor's guide, presentation decks, student handouts, and suggestions for pre-class and post-class work.

Achieve includes gradebook integration with iClicker, allowing for a seamless pairing of the best classroom response system with the best in-class content.

Using the Power of Computing to Work with Data

Students are provided with a wealth of multimedia resources and opportunities for practice, to coach them toward fuller conceptual understanding and proficiency in solving problems.

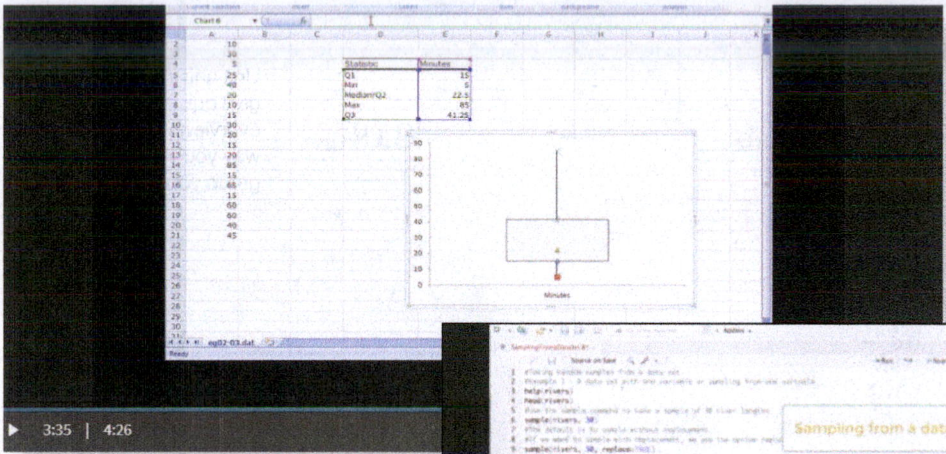

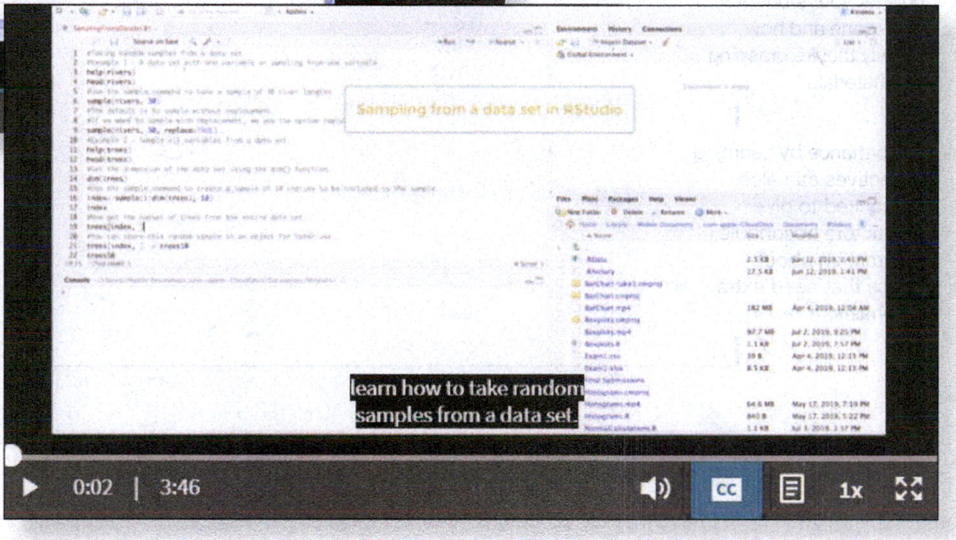

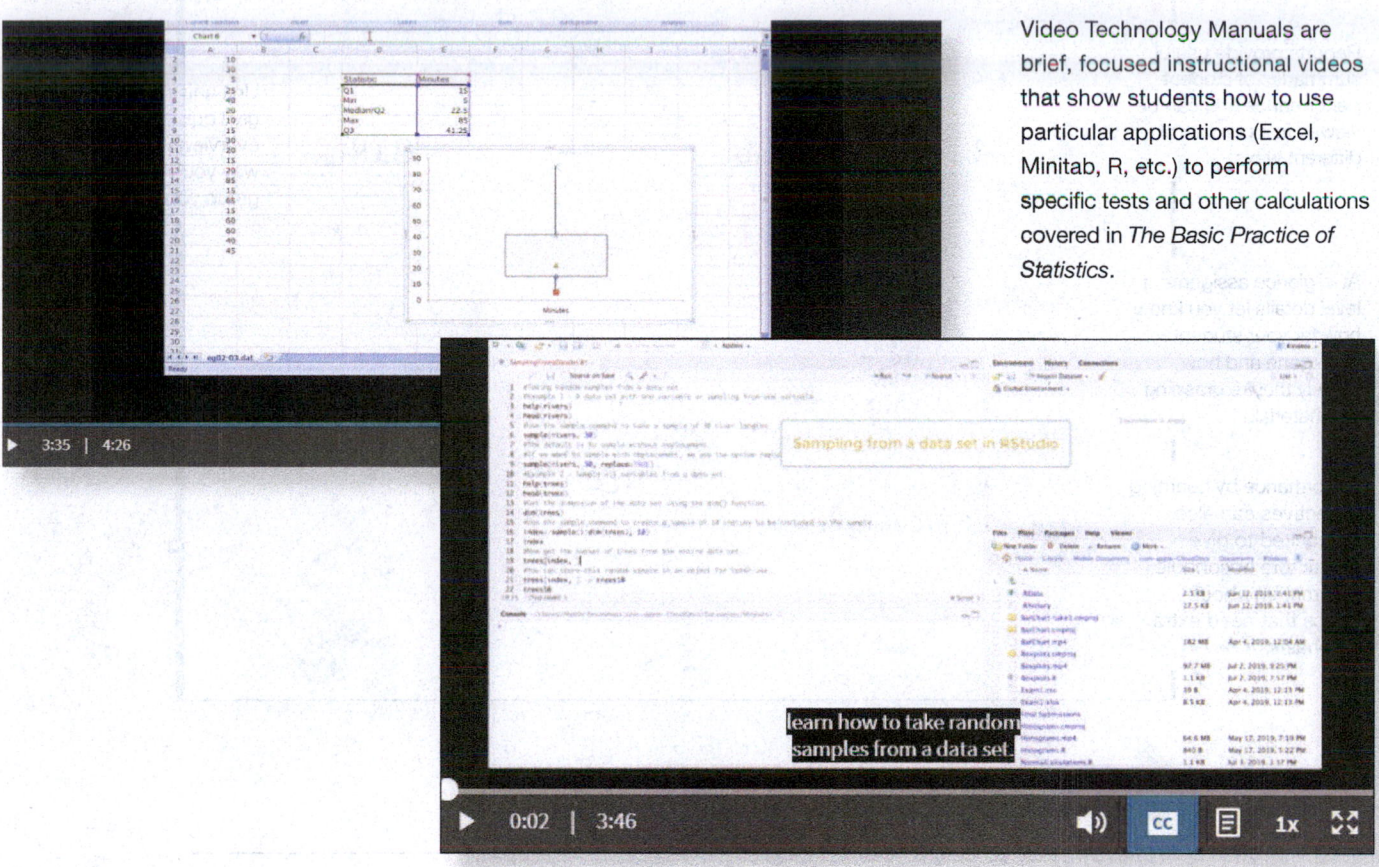

Video Technology Manuals are brief, focused instructional videos that show students how to use particular applications (Excel, Minitab, R, etc.) to perform specific tests and other calculations covered in *The Basic Practice of Statistics*.

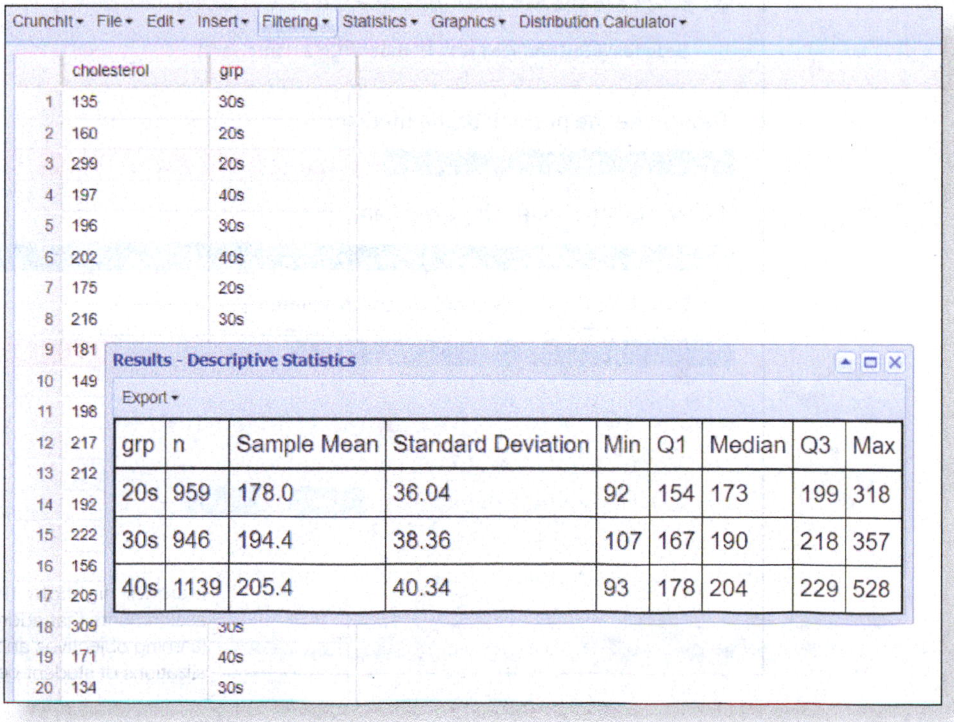

grp	n	Sample Mean	Standard Deviation	Min	Q1	Median	Q3	Max
20s	959	178.0	36.04	92	154	173	199	318
30s	946	194.4	38.36	107	167	190	218	357
40s	1139	205.4	40.34	93	178	204	229	528

Included with Achieve, **CrunchIt!**—Macmillan's proprietary online statistical software powered by R—handles every computation and graphing function an introductory statistics student needs. CrunchIt! is preloaded with the BPS data sets, and it allows editing and importing additional data.

Achieve includes data sets for homework problems using statistical software, including Excel, Minitab, TI Calculators, JMP, R, SPSS, CSV, and ASCII.

Powerful analytics, viewable in an elegant dashboard, offer instructors a window into student progress. Achieve gives you the insight to address students' weaknesses and misconceptions before they struggle on a test.

Reports provide useful summaries of student performance and can be viewed in a number of different ways.

At-a-glance assignment level details let you know how far your students have gone and how quickly they're grasping the material.

Performance by Learning Objectives can also be tracked to give instructors actionable information about topics that need extra emphasis.

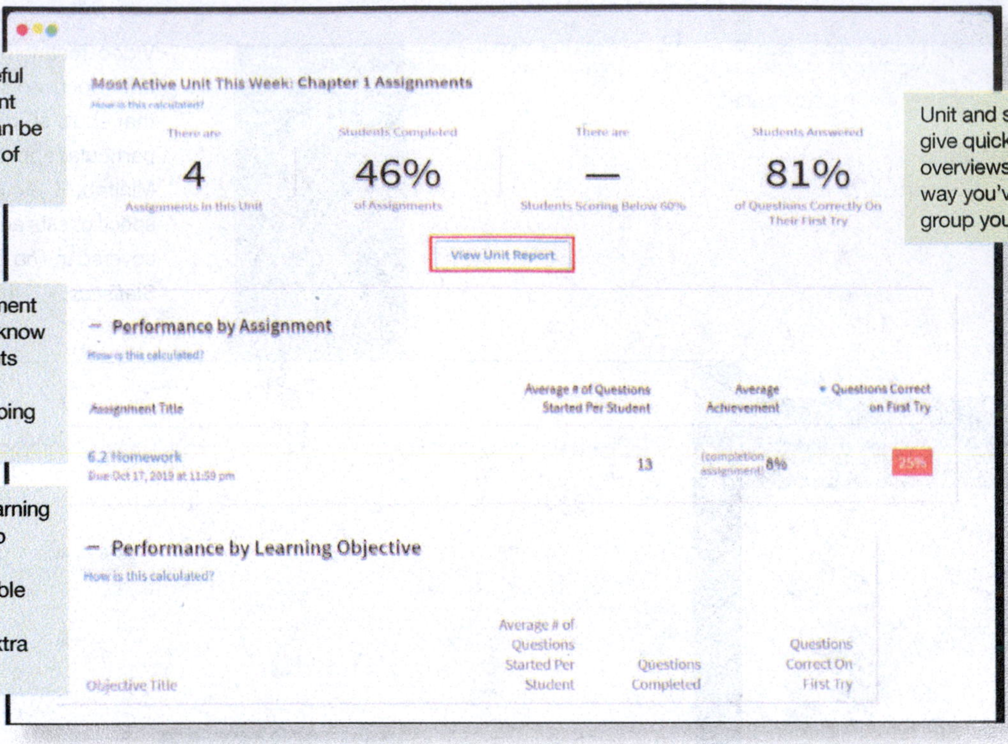

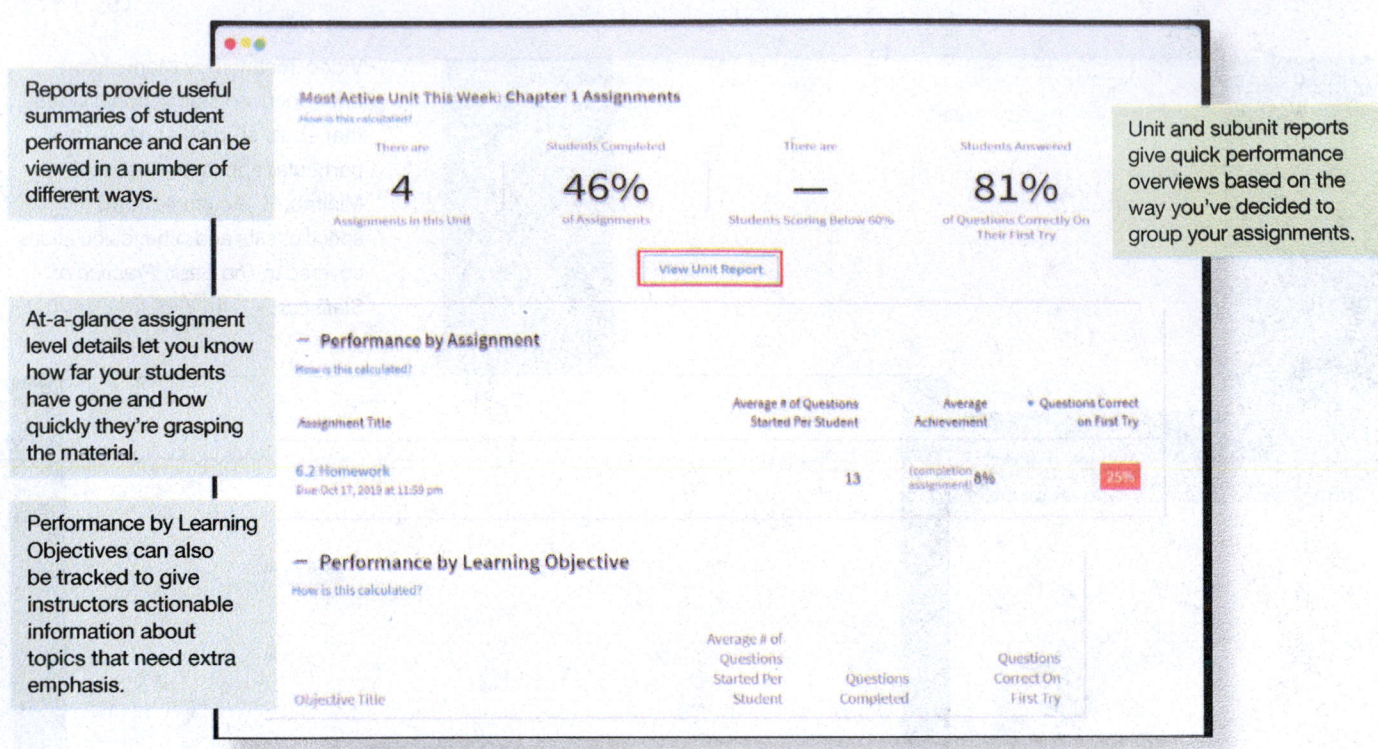

Unit and subunit reports give quick performance overviews based on the way you've decided to group your assignments.

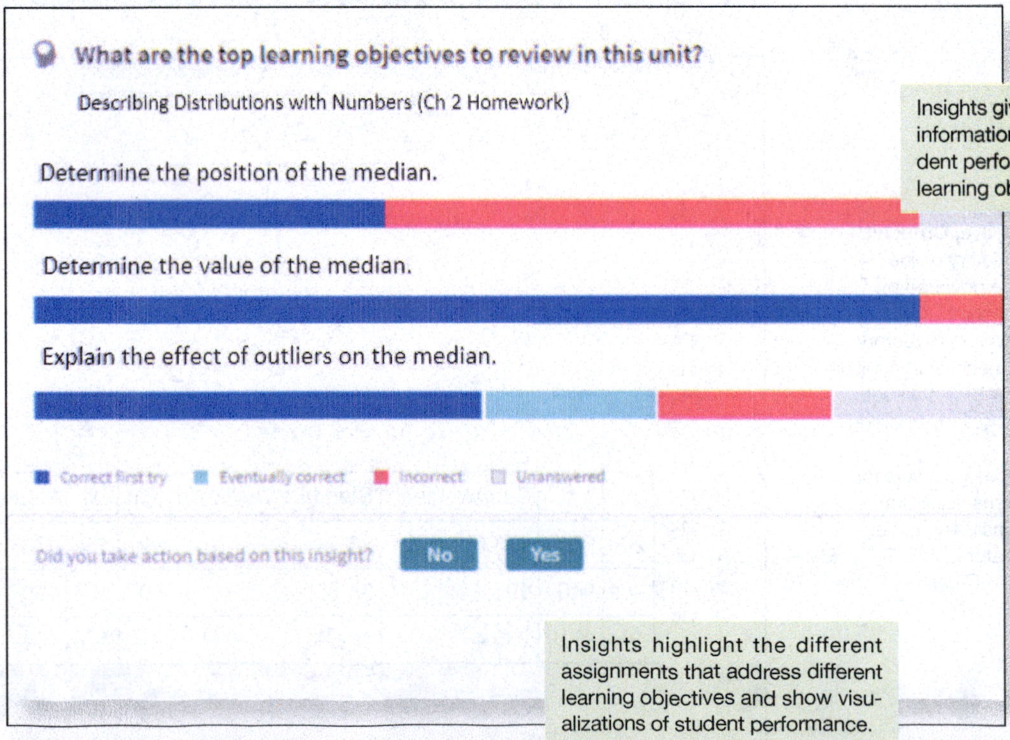

Insights give instructors information regarding student performance versus learning objectives.

Insights highlight the different assignments that address different learning objectives and show visualizations of student performance.

Achieve's Rich Digital Resources

Multimedia resources, such as interactives and videos, serve as an extension of the carefully constructed examples and exercises in the text.

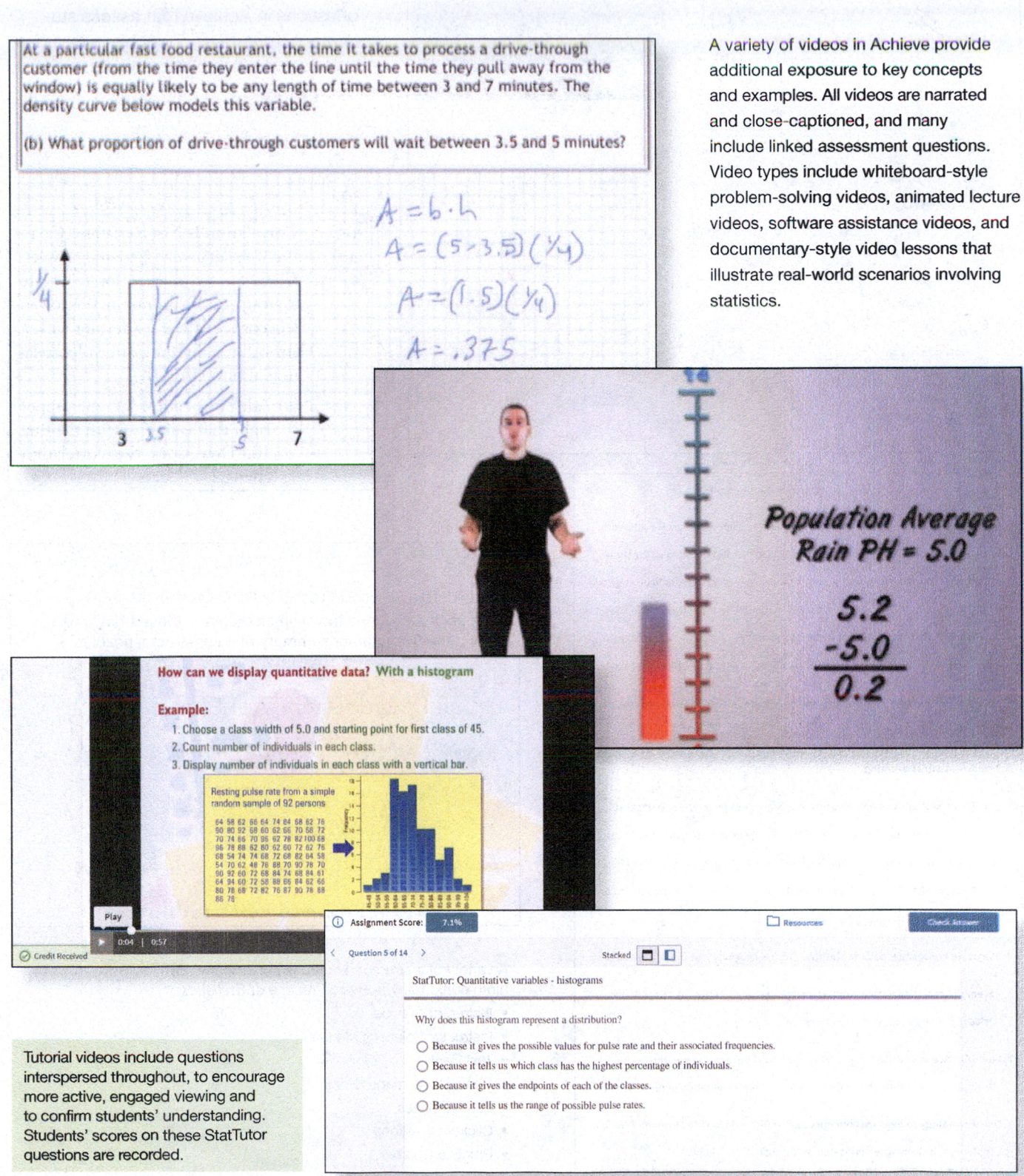

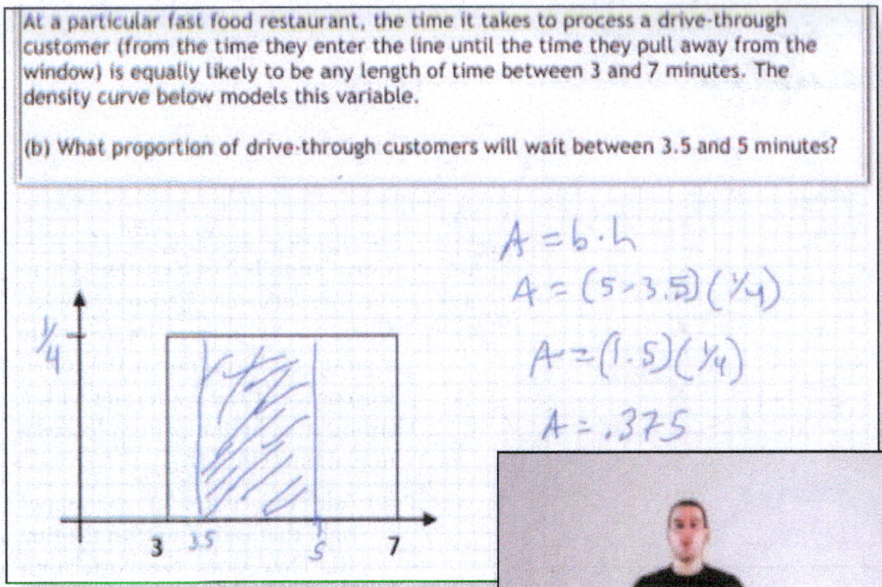

At a particular fast food restaurant, the time it takes to process a drive-through customer (from the time they enter the line until the time they pull away from the window) is equally likely to be any length of time between 3 and 7 minutes. The density curve below models this variable.
(b) What proportion of drive-through customers will wait between 3.5 and 5 minutes?
A=b·h
A=(5-3.5)(1/4)
A=(1.5)(1/4)
A=.375
1/4
3
3.5
5
7

A variety of videos in Achieve provide additional exposure to key concepts and examples. All videos are narrated and close-captioned, and many include linked assessment questions. Video types include whiteboard-style problem-solving videos, animated lecture videos, software assistance videos, and documentary-style video lessons that illustrate real-world scenarios involving statistics.

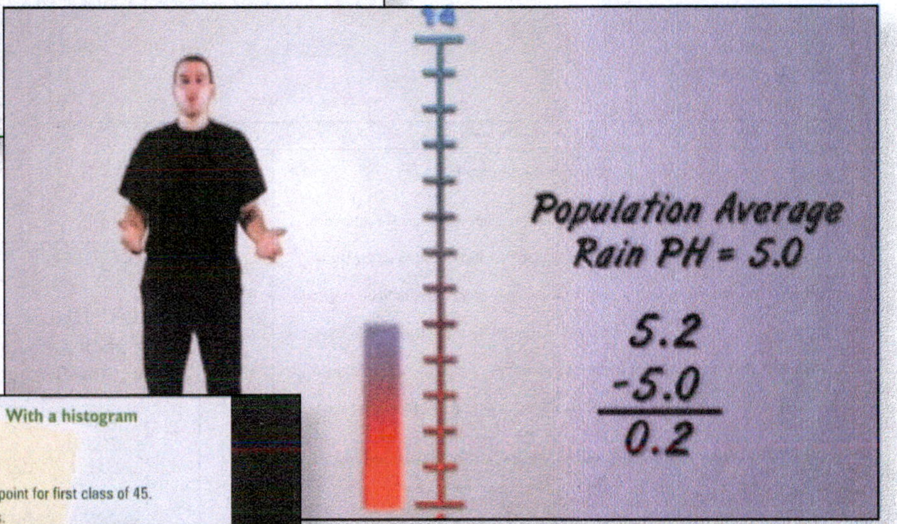

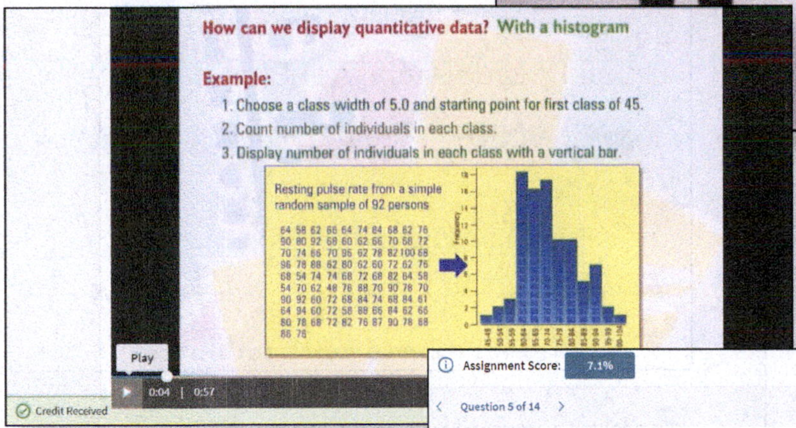

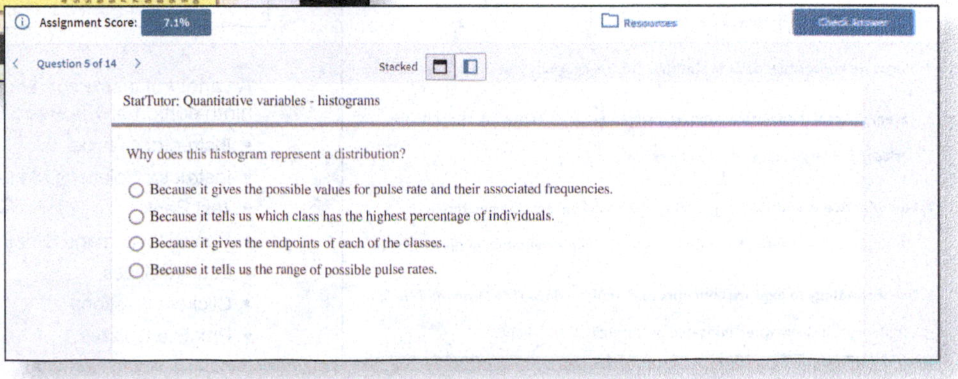

Population Average
Rain PH = 5.0

How can we display quantitative data? With a histogram

1. Choose a class width of 5.0 and starting point for first class of 45.

Credit Received

Assignment Score: 7.1%
Resources
Question 5 of 14
Stacked
StatTutor: Quantitative variables - histograms
Why does this histogram represent a distribution?
Because it gives the possible values for pulse rate and their associated frequencies.
Because it tells us which class has the highest percentage of individuals.
Because it gives the endpoints of each of the classes.
Because it tells us the range of possible pulse rates.

Tutorial videos include questions interspersed throughout, to encourage more active, engaged viewing and to confirm students' understanding. Students' scores on these StatTutor questions are recorded.

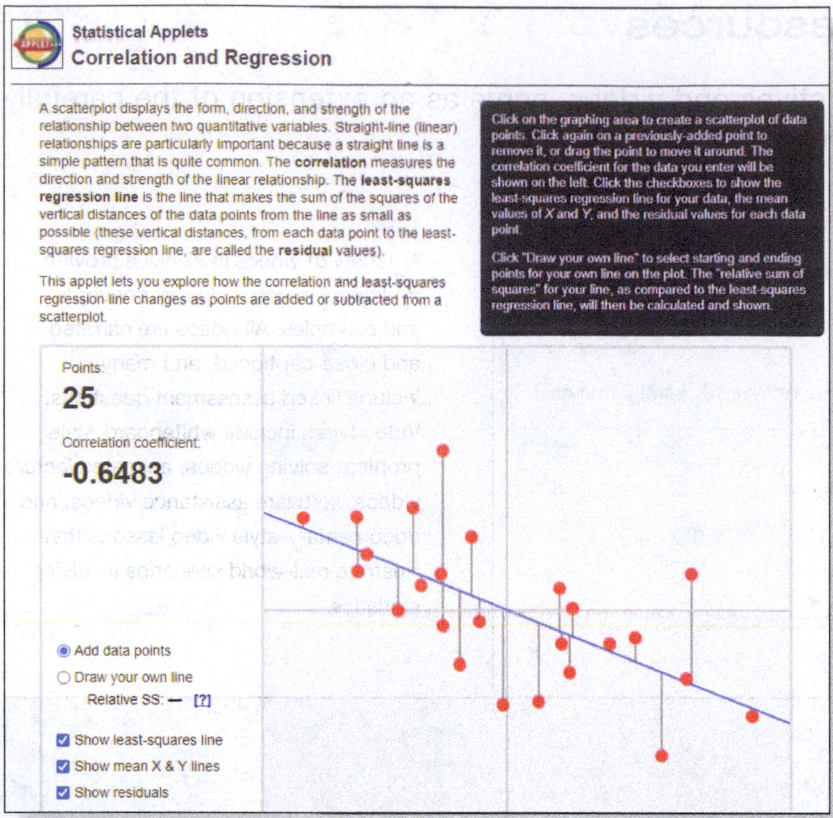
Statistical Applets
Correlation and Regression
A scatterplot displays the form, direction, and strength of the relationship between two quantitative variables. Straight-line (linear) relationships are particularly important because a straight line is a simple pattern that is quite common. The correlation measures the direction and strength of the linear relationship. The least-squares regression line is the line that makes the sum of the squares of the vertical distances of the data points from the line as small as possible (these vertical distances, from each data point to the least-squares regression line, are called the residual values).
This applet lets you explore how the correlation and least-squares regression line changes as points are added or subtracted from a scatterplot.
Click on the graphing area to create a scatterplot of data points. Click again on a previously-added point to remove it, or drag the point to move it around. The correlation coefficient for the data you enter will be shown on the left. Click the checkboxes to show the least-squares regression line for your data, the mean values of X and Y, and the residual values for each data point.
Click "Draw your own line" to select starting and ending points for your own line on the plot. The "relative sum of squares" for your line, as compared to the least-squares regression line, will then be calculated and shown.
Points:
25
Correlation coefficient:
-0.6483
Add data points
Draw your own line
Relative SS: — [?]
Show least-squares line
Show mean X & Y lines
Show residuals

Statistical Applets, referenced and displayed in the text, are visual interactives originally designed by David Moore. This feature enables students to manipulate data in calculations and see the results graphically. Applets are associated with questions in Achieve that assess students' comprehension. Achieve includes updated versions of the applets, powered by Desmos.

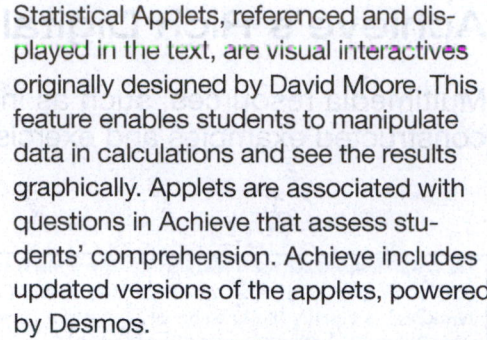

4.42 applet **Correlation Is Not Resistant.** Go to the *Correlation and Regression* applet. Click on the scatterplot to create a group of 10 points in the lower-left corner of the scatterplot with a strong straight-line pattern (correlation about 0.9).

(a) Add one point at the upper right that is in line with the first 10. How does the correlation change?

The Basic Practice of Statistics, 9e, Instructor's Guide, Chapter 2 — 2-7

5% raise pool, which is $20,000. We can give each teacher a 5% raise or give each teacher a $2000 raise. Have your students calculate the new salaries under these two scenarios. Examine both sets of new salaries. Your students should be able to see that, in addition to there being increased variability in the salary distribution, the top-paid teachers benefit more from the percent raise than do the lower-paid teachers (and the top-paid teachers have salaries that get further from the rest).

GAISE (Guidelines in Assessment and Instruction for Statistics Education, *revised May 2017*) **Guideline suggestions**:

1. **Teach statistical thinking.**
   a) **Teach statistics as an investigative process of problem-solving and decision-making.** This is covered thoroughly in section 2.10, "Organizing a Statistical Problem," culminating in the initial declaration of the *Four-Step Process:* STATE, PLAN, SOLVE, CONCLUDE. Exercises 2.43 through 2.49 are provided for practice.
2. **Focus on conceptual understanding.** This is especially true in Exercise 2.51.
3. **Integrate real data with a context and purpose.** Exercise 2.4, "New House Prices," can be given early to students.
4. **Foster active learning.** This guideline can easily be met using material throughout the chapter, but it depends on the implementation of the instructor.
5. **Use technology to explore concepts and analyze data.** The *Means and Medians* applet may be used when indicated or throughout the chapter.

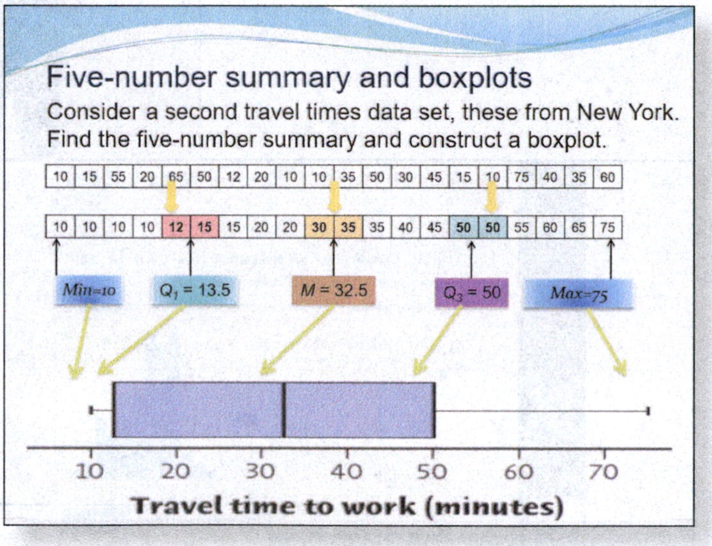
Five-number summary and boxplots
Consider a second travel times data set, these from New York. Find the five-number summary and construct a boxplot.
10 15 55 20 65 50 12 20 10 10 35 50 30 45 15 10 75 40 35 60
10 10 10 10 12 15 15 20 20 30 35 35 40 45 50 50 55 60 65 75
Min=10
Q1 = 13.5
M = 32.5
Q3 = 50
Max=75
10 20 30 40 50 60 70
Travel time to work (minutes)

A variety of instructor resources accompany the ninth edition of *The Basic Practice of Statistics*:

- Instructor's Guide
- Instructor Solutions Manual
- Test Bank
- PowerPoint Image Slides
- Lecture Slides
- Clicker Questions
- Practice Quizzes

ACKNOWLEDGMENTS

We are grateful to colleagues from two-year and four-year colleges and universities who reviewed and commented on the previous edition of *The Basic Practice of Statistics*, in preparation for this revision, or who commented on the ninth edition manuscript.

Todd Burus, Eastern Kentucky University
Rita Chattopadhyay, Eastern Michigan University
Elijah Dikong, Michigan State University
Kimberly Druschel, Saint Louis University
Cathy Frey, Norwich University
Petre Ghenciu, University of Wisconsin–Stout
Billie-Jo Grant, California State Polytechnic University
Mark Hardwidge, Danville Area Community College
Lisa Kay, Eastern Kentucky University
Michael Macon, Green River Community College
Connie Marberry, Kirkwood Community College
Andrew McDougall, Montclair State University
Juli Moore, Oregon State University
Roland Moore, Florida State University
Thomas Oliveri, University of Massachusetts–Lowell
Christina Pierre, Saint Mary's University of Minnesota
Mohammed Quasem, University of South Carolina
Joshua Roberts, Georgia Gwinnett College
N. Paul Schembari, East Stroudsburg University
Hilary Seagle, Southwestern Community College
Kristi Spittler-Brown, Arkansas Tech University
Tim Swartz, Simon Fraser University
Susan Toma, Madonna University
Carol Weideman, St. Petersburg College

We extend our appreciation to Sarah Seymour, Debbie Hardin, David Dietz, Katrina Mangold, Catriona Kaplan, Andy Newton, Justin Jones, Lisa Kinne, Edward Dionne, Paul Rohloff, Diana Blume, John Callahan, Vicki Tomaselli, and other publishing professionals who have contributed to the development, design, production, and cohesiveness of this book and its online resources.

Jack Miller and Mark McKibben are to be commended for their work on the full solutions manuals, as well as the back-of-book answers and exercise evaluations—offering their backgrounds in statistics and dedication to education. John Samons went above and beyond to ensure the accuracy, flow, and consistency of the presentation in the text, back-of-book answers, and full solutions. We thank all three of you for embracing the project in the spirit of teamwork and collaboration.

We also extend our gratitude to the following collaborators who contributed their expertise to the instructor and student resources for this edition:

- Nicole Dalzell, Wake Forest University, revised the Clicker Questions and the Test Bank.
- Terri Rizzo, Lakehead University, accuracy reviewed the Clicker Questions, Practice Quizzes, and Test Bank.
- Mark Gebert, University of Kentucky, revised the Lecture Slides and Practice Quizzes.
- Michelle Duda, Columbus State Community College, revised the Instructor's Guide.
- Karen Starin, Columbus State Community College, accuracy reviewed the Instructor's Guide.

We would like to thank Robert Wolf, University of San Francisco; Eugene Komaroff, Keiser University; Stephen Doty; and Aaron Gladish for pointing out errors in the previous edition and for helpful suggestions.

Finally, we are indebted to the many statistics teachers with whom we have discussed the teaching of our subject over many years; to people from diverse fields with whom we have worked to understand data; and, especially, to students whose compliments and complaints have changed and improved our teaching. Working with teachers, colleagues in other disciplines, and students constantly reminds us of the importance of hands-on experience with data and of statistical thinking in an era when computer routines quickly handle statistical details.

David S. Moore and William I. Notz

ABOUT THE AUTHORS

David S. Moore is Shanti S. Gupta Distinguished Professor of Statistics, Emeritus, at Purdue University and was 1998 president of the American Statistical Association. He received an AB from Princeton and a PhD from Cornell, both in mathematics. He has written many research papers in statistical theory and served on the editorial boards of several major journals. Professor Moore is an elected fellow of the American Statistical Association and of the Institute of Mathematical Statistics and an elected member of the International Statistical Institute. He has served as program director for statistics and probability at the National Science Foundation.

In recent years, Professor Moore has devoted his attention to the teaching of statistics. He was the content developer for the Annenberg/Corporation for Public Broadcasting college-level telecourse "Against All Odds: Inside Statistics" and for the series of video modules "Statistics: Decisions through Data," intended to aid the teaching of statistics in schools. He is the author of influential articles on statistics education and of several leading textbooks. Professor Moore has served as president of the International Association for Statistical Education and has received the Mathematical Association of America's national award for distinguished college or university teaching of mathematics.

William I. Notz is Professor Emeritus at The Ohio State University. He received a BS in physics from Johns Hopkins University and a PhD in mathematics from Cornell University. His first academic job was as assistant professor in the Department of Statistics at Purdue University. While there, he taught the introductory statistics concepts course with Professor Moore and developed an interest in statistical education. Professor Notz is a co-author of the *Electronic Encyclopedia of Statistical Examples and Exercises* and co-author of *Statistics: Concepts and Controversies*. His research interests have focused on experimental design and computer experiments. He is the author of several research papers and of a book on the design and analysis of computer experiments. William Notz is an elected fellow of the American Statistical Association and has served as the editor of the journal *Technometrics* and as editor of the *Journal of Statistics Education*. At The Ohio State University, he has served as the director of the Statistical Consulting Service, as acting chair of the Department of Statistics, and as an associate dean in the College of Mathematical and Physical Sciences. Professor Notz is a winner of Ohio State's Alumni Distinguished Teaching Award.

CHAPTER

0

When you complete this chapter, you will be able to:

0.1 Describe how the method used to collect data can affect the reliability of conclusions based on those data.

0.2 Recognize that looking at data by producing graphs and numerical summaries is the best way to begin most data analyses.

0.3 Describe the relationships among variation in data, uncertainty in the conclusions we can make from data, and statistics.

Getting Started

In this chapter we begin to think about how data can be used to answer practical questions. We raise some important issues about data and their use, which we will explore in detail in later chapters.

What's hot in popular music this week? SoundScan knows. SoundScan collects data electronically from the cash registers in more than 39,000 retail outlets around the world[1] and also collects data on download sales from websites. When you buy a CD or download a digital track, the checkout scanner or website is probably telling SoundScan what you bought. SoundScan provides this information to *Billboard* magazine, MTV, and VH1, as well as to record companies and artists' agents.

Should women take hormones such as estrogen after menopause, when natural production of these hormones ends? In 1992, several major medical organizations said "yes." In particular, women who took hormones seemed to reduce their risk of heart attack by 35% to 50%. The risks of taking hormones appeared small compared with the benefits. But in 2002, the National Institutes of Health declared these findings wrong. Use of hormones after menopause immediately plummeted. Both recommendations were based on extensive studies. What happened?

Is the global climate warming? Is it becoming more extreme? An overwhelming majority of scientists now agree that the earth is undergoing major changes in climate. Enormous quantities of data are continuously being collected from weather stations, satellites, and other sources to monitor factors such as the surface temperature on land and sea, precipitation, solar activity, and the chemical composition of air and water. Climate models incorporate this information to make projections of future climate change and can help us understand the effectiveness of proposed solutions.

Image Source RF/DreamPictures/Getty Images

SoundScan, medical studies, and climate research all produce data (numerical facts)—and lots of them. Using data effectively is a large and growing part of most professions, and reacting to data is part of everyday life. In fact, we define **statistics** as *the science of learning from data.*

Although data are numbers, they are not "just numbers." *Data are numbers with a context.* The number 8.5, for example, carries no information by itself. But if we hear that a friend's new baby weighed 8.5 pounds at birth, we congratulate her on the healthy size of the child. The context engages our background knowledge and allows us to make judgments. We know that a baby weighing 8.5 pounds is a little above average and that a human baby is unlikely to weigh 8.5 ounces or 8.5 kilograms (over 18 pounds). The context makes the number informative.

Data are used to answer some practical questions. To gain insight from data, we make graphs and do calculations. But graphs and calculations are guided by ways of thinking that amount to educated common sense. Let's begin our study of statistics with an informal look at some aspects of statistical thinking.[2]

0.1 How the Data Were Obtained Matters

Although data can be collected in a variety of ways, the type of conclusion that can be reached from data depends on how the data were obtained. *Observational studies* and *experiments* are two common methods for collecting data. Let's take a closer look at the hormone replacement data to understand the differences.

EXAMPLE 0.1 Hormone Replacement Therapy

What's behind the flip-flop in the advice offered to women about hormone replacement? The evidence in favor of hormone replacement came from a number of observational studies that compared women who were taking hormones with others who were not. But women who choose to take hormones are very different from women who do not: they tend to be better educated and more affluent. Because of this, as a group, they are more proactive about their health care and have the motivation and the means to seek preventive health care, including healthier diets and increased exercise. It isn't surprising that the group that takes better care of their health will have fewer heart attacks.

Large and careful observational studies are expensive, but they are easier to arrange than careful experiments. Experiments don't let women decide what to do. They assign women either to hormone replacement or to dummy pills that look and taste the same as the hormone pills. The assignment is done by a coin toss so that all kinds of women are equally likely to get either treatment. No longer will the women in the group receiving hormone therapy be better educated and more affluent than those who don't receive hormone therapy. Part of the difficulty of a good experiment is persuading women to accept the result—invisible to them—of the coin toss. By 2002, several experiments agreed that hormone replacement does *not* reduce the risk of heart attack, at least for older women. Faced with this better evidence, medical authorities changed their recommendations.[3]

Women who chose hormone replacement after menopause were, on average, better educated and more affluent than those who didn't. No wonder they had fewer heart attacks. We can't conclude that hormone replacement reduces heart attacks just because we see this relationship in data. In this example, education and affluence are background factors that help explain the relationship between hormone replacement and good health.

Children who play soccer do better in school (on the average) than children who don't play soccer. Does this mean that playing soccer increases school grades? Children who play soccer tend to have prosperous and well-educated parents. Once again, education and affluence are background factors that help explain the relationship between soccer and good grades.

Almost all relationships between two observed characteristics, or "variables," are influenced by other variables lurking in the background. To understand the relationship between two variables, you must often look at other variables. Careful statistical studies try to think of and measure possible *lurking variables* in order to correct for their influence. As the hormone saga illustrates, this doesn't always work well. News reports often just ignore possible lurking variables that might ruin a good headline like "Playing soccer can improve your grades." The habit of asking, "What might lie behind this relationship?" is part of thinking statistically.

Of course, observational studies are still quite useful. We can learn from observational studies how chimpanzees behave in the wild or which popular songs sold best last week or what percentage of workers were unemployed last month. SoundScan's data on popular music and the government's data on employment and unemployment come from *sample surveys,* an important kind of observational study that chooses a part (the sample) to represent a larger whole. Opinion polls interview perhaps 1000 of the 254 million adults in the United States to report the public's views on current issues. Can we trust the results? We'll see that this isn't a simple yes-or-no question. Let's just say that the government's unemployment rate is much more trustworthy than opinion poll results—and not just because the Bureau of Labor Statistics interviews 60,000 households rather than 1000. We can, however, say right away that some samples *can't* be trusted. Consider the following write-in poll.

EXAMPLE 0.2 Would You Have Children Again?

The advice columnist Ann Landers once asked her readers, "If you had it to do over again, would you have children?" A few weeks later, her column was headlined "70% OF PARENTS SAY KIDS NOT WORTH IT." Indeed, 70% of the nearly 10,000 parents who wrote in said they would not have children if they could make the choice again. Those 10,000 parents were upset enough with their children to write Ann Landers. Because of this, the views of these parents are not *representative* of parents in general. Most parents are happy with their kids and don't bother to write.

On August 24, 2011, Abigail Van Buren (the twin sister of Ann Landers) revisited this question in her column "Dear Abby." A reader asked, "I'm wondering when the information was collected and what the results of that inquiry were, and if you asked the same question today, what the majority of your readers would answer."

Ms. Van Buren responded, "The results were considered shocking at the time because the majority of responders said they would NOT have children if they had it to do over again. I'm printing your question because it will be interesting to see if feelings have changed over the intervening years."

In October 2011, Ms. Van Buren wrote that this time the majority of respondents would have children again. That is encouraging, but this was, again, a write-in poll.

Statistically designed samples, even opinion polls, don't let people choose themselves for the sample. They interview people selected by impersonal chance so that everyone has an equal opportunity to be in the sample. Such a poll showed that 91% of parents *would* have children again. *Where data come from matters a lot.* If you are careless about how you get your data, you may announce 70% No when the truth is close to 90% Yes. Understanding the importance of where data come from and their relationship to the conclusions that can be reached is an important part of learning to think statistically.

0.2 Always Look at the Data

Yogi Berra, the Hall of Fame New York Yankee, said it: "You can observe a lot by just watching." That's a motto for learning from data. A *few carefully chosen graphs are often more instructive than great piles of numbers.* Consider the outcome of the 2000 presidential election in Florida.

EXAMPLE 0.3 Palm Beach County

Elections don't come much closer: after much recounting, state officials declared that George Bush had carried Florida by 537 votes out of almost 6 million votes cast. Florida's vote decided the 2000 presidential election and made George Bush, rather than Al Gore, president. Let's look at some data. Figure 0.1 displays a graph that plots votes for the third-party candidate Pat Buchanan against votes for the Democratic candidate Al Gore in Florida's 67 counties.

What happened in Palm Beach County? The question leaps out from the graph. In this large and heavily Democratic county, a conservative third-party candidate did far better relative to the Democratic candidate than in any other county. The points for the other 66 counties show votes for both candidates increasing together in a roughly straight-line pattern. Both counts go up as county population goes up. Based on this pattern, we would expect Buchanan to receive around 800 votes in Palm Beach County. He actually received more than 3400 votes. That difference determined the election result in Florida and in the nation.

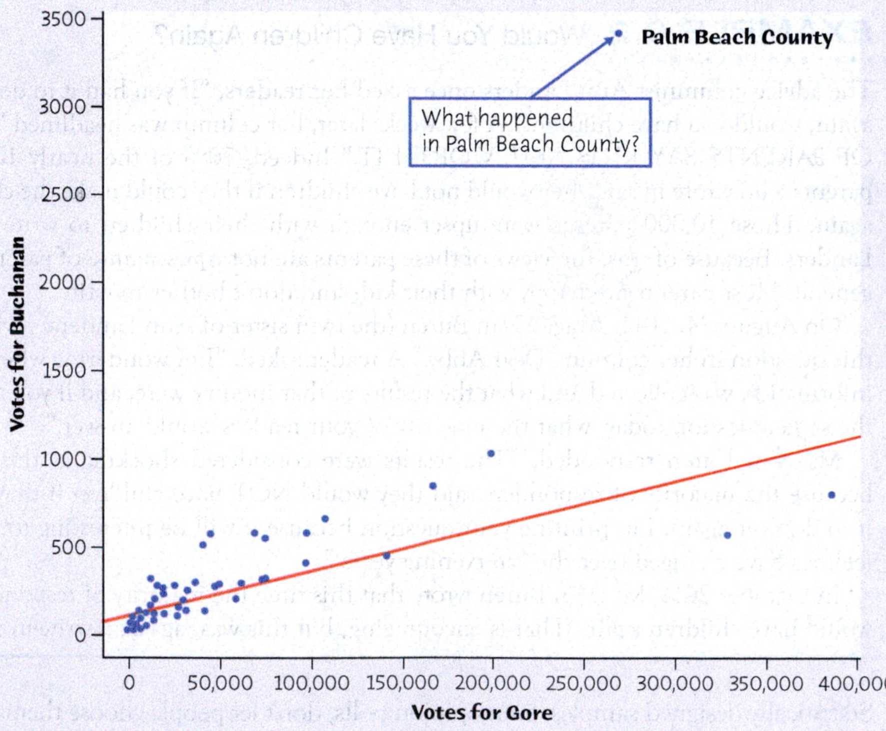

FIGURE 0.1
Votes in the 2000 presidential election for Al Gore and Patrick Buchanan in Florida's 67 counties. What happened in Palm Beach County?

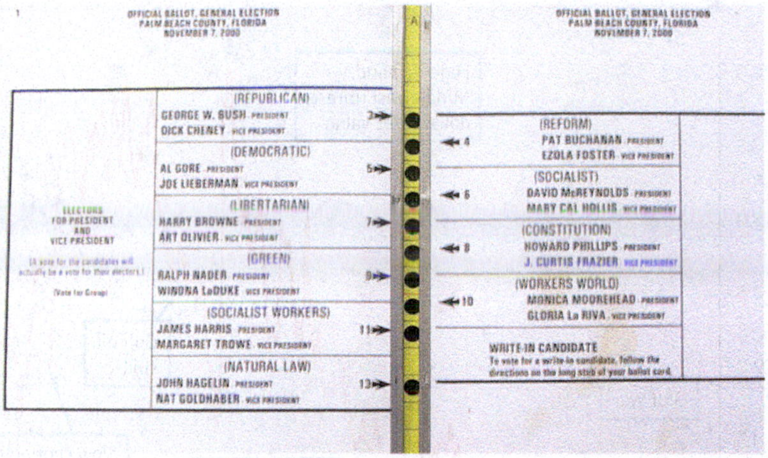

ROBERT DUYOS/Neuscom/Tribune News Service/PALM BEACH/FL/USA

The graph demands an explanation. It turns out that Palm Beach County used a confusing "butterfly" ballot, in which candidate names on both left and right pages led to a voting column in the center. It would be easy for a voter who intended to vote for Gore to in fact cast a vote for Buchanan. The graph is convincing evidence that this in fact happened.

Most statistical software will draw a variety of graphs with a few simple commands. Examining your data with appropriate graphs and numerical summaries is the correct place to begin most data analyses. These can often reveal important patterns or trends that will help you understand what your data have to say and, ultimately, help you answer the question that prompted you to examine the data.

0.3 Variation Is Everywhere

The company's sales reps file into their monthly meeting. The sales manager rises. "Congratulations! Our sales are up 2% this month, so we're all drinking champagne this morning. You remember that when sales were down 1% last month I fired half of our reps." This picture is only slightly exaggerated. Many managers overreact to small short-term variations in key figures. Here is Arthur Nielsen, former head of the country's largest market research firm, describing his experience:

> *Too many business people assign equal validity to all numbers printed on paper. They accept numbers as representing Truth and find it difficult to work with the concept of probability. They do not see a number as a kind of shorthand for a range that describes our actual knowledge of the underlying condition.*[4]

Business data such as sales and prices vary from month to month for reasons ranging from the weather to a customer's financial difficulties to the inevitable errors in gathering the data. The manager's challenge is to say when there is a real pattern behind the variation. We'll see that statistics provides tools for understanding variation and for seeking meaningful patterns behind the screen of variation. Let's look at some more data.

fotog/Getty Images

EXAMPLE 0.4 The Price of Gas

Figure 0.2 plots the average price of a gallon of regular unleaded gasoline each week from August 1990 to August 2019.[5] There certainly is variation! But a close look shows a yearly pattern: gas prices go up during the summer driving season, then down as demand drops in

FIGURE 0.2
Variation is everywhere: the average retail price of regular unleaded gasoline, mid-1990 to mid-2019.

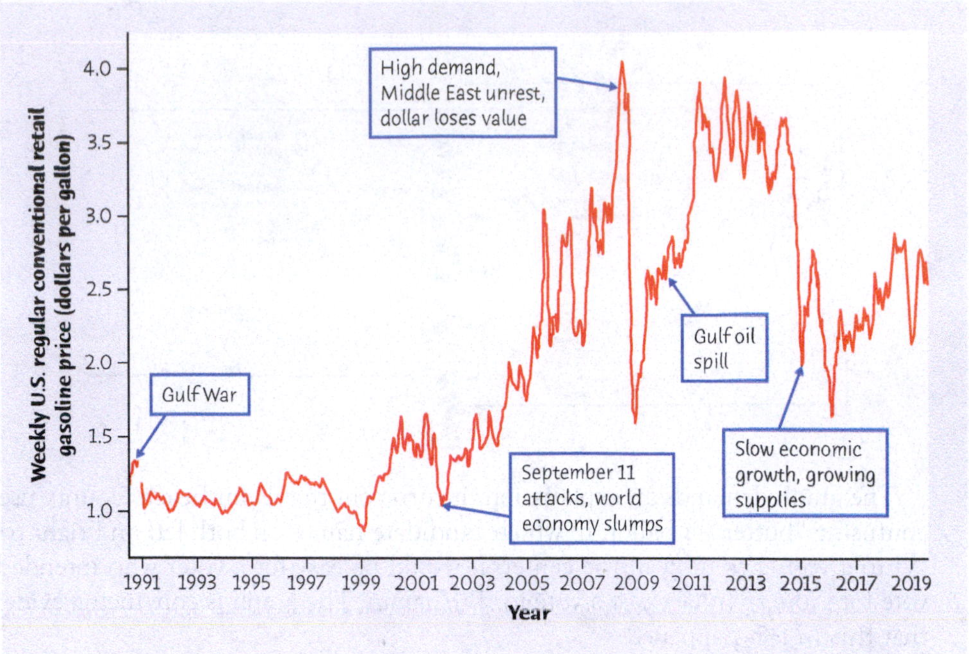

the fall. On top of this regular pattern, we see the effects of international events. For example, prices rose when the 1990 Gulf War threatened oil supplies and dropped when the world economy turned down after the September 11, 2001, terrorist attacks in the United States. The years 2007 and 2008 brought the perfect storm: the ability to produce oil and refine gasoline was overwhelmed by high demand from China and the United States and continued turmoil in the oil-producing areas of the Middle East and Nigeria. Add in a rapid fall in the value of the dollar, and prices at the pump skyrocketed to more than $4 per gallon. This increase was quickly followed by a downturn caused by the worldwide financial crisis of 2008. In 2010, the Gulf oil spill also affected supply and hence prices. In 2015 and 2016, slowing growth in emerging markets–and, most importantly, in China–along with rising oil supplies led to sharp drops in oil prices. The data carry an important message: because the United States imports much of its oil, we can't control the price we pay for gasoline.

Slowing growth in emerging markets, most importantly in China, has led to sharp drops in commodity prices almost across the board. Rising supply has been at least as important as falling demand.

Variation is everywhere. Individuals vary; repeated measurements on the same individual vary; almost everything varies over time. One reason we need to know some statistics is that it helps us deal with variation and describe the uncertainty in our conclusions. Let's look at another example to see how variation is incorporated into our conclusions.

EXAMPLE 0.5 The HPV Vaccine

Cervical cancer, once the leading cause of cancer deaths among women, is the easiest female cancer to prevent with regular screening tests and follow-up. Almost all cervical cancers are caused by human papillomavirus (HPV). The first vaccine to protect against the most common varieties of HPV became available in 2006. The Centers for Disease Control and Prevention recommends that all girls be vaccinated at age 11 or 12. In 2011, the CDC made the same recommendation for boys, to protect against anal and throat cancers caused by the HPV virus.

A natural question to ask is "How well does the vaccine work?" Doctors rely on experiments (called "clinical trials" in medicine) that give some women the new vaccine and others a dummy vaccine. (This is ethical when it is not yet known whether or not the vaccine is safe and effective.) The conclusion of the most important trial was that an estimated 98% of women up to age 26 who are vaccinated before they are infected with HPV will avoid cervical cancers over a three-year period.

Women who get the vaccine are much less likely to get cervical cancer. But because variation is everywhere, the results are different for different women. Some vaccinated women will get cancer, and many who are not vaccinated will escape. Statistical conclusions about the questions data are intended to answer are "on the average" statements only, and even these "on the average" statements have an element of uncertainty. Although we can't be 100% certain that the vaccine reduces risk on the average, statistics allows us to state how confident we are that this is the case.

Because variation is everywhere, conclusions are uncertain. Statistics gives us a language for talking about uncertainty that is used and understood by statistically literate people everywhere. In the case of HPV vaccine, the medical journal used that language to tell us: "Vaccine efficiency. . . was 98% (95 percent confidence interval 86% to 100%)."[6] That "98% effective" is, in Arthur Nielsen's words, "shorthand for a range that describes our actual knowledge of the underlying condition." The range is 86% to 100%, and we are 95 percent confident that the truth lies in that range. We will soon learn to understand this language. We can't escape variation and uncertainty. Learning statistics enables us to live more comfortably with these realities.

0.4 What Lies Ahead in This Book

The purpose of *Basic Practice of Statistics* is to give you a working knowledge of the ideas and tools of practical statistics. We will divide practical statistics into three main areas:

- **Data analysis** concerns methods and strategies for looking at data—for exploring, organizing, and describing data using graphs and numerical summaries. Your thoughtful exploration allows data to illuminate reality. Chapters 1 through 6 discuss data analysis.
- **Data production** provides methods for producing data that can give clear answers to specific questions. Where data come from matters and is often the most important limitation on their usefulness. Basic concepts about how to select samples and design experiments are some of the most influential ideas in statistics. These concepts are the subject of Chapters 8 and 9.
- **Statistical inference** moves beyond the data in hand to draw conclusions about some wider universe. Statistical conclusions aren't yes-or-no answers; they must take into account that variation is everywhere—variability among people, animals, or objects and uncertainty in data. To describe variation and uncertainty, inference uses the language of probability, introduced in Chapter 12. Because we are concerned with practice rather than theory, we need only a limited knowledge of probability. Chapters 13 and 14 offer more probability for those who want it. Chapters 15 through 18 discuss the reasoning of statistical inference. These chapters are the key to the rest of the book. Chapters 20 through 24 present inference as used in practice in the most common settings. Chapters 25 through 27 concern more advanced or specialized kinds of inference.

4step

Because data are numbers with a context, doing statistics means more than manipulating numbers. You must *state* a problem in its real-world context, *plan* your specific statistical work in detail, *solve* the problem by making the necessary graphs and calculations, and *conclude* by explaining what your findings say about the real-world setting. We'll make regular use of this four-step process to encourage good habits that go beyond graphs and calculations to ask, "What do the data tell me?"

Statistics does involve lots of calculating and graphing. The text presents the techniques you need, but you should use technology to automate calculations and graphs as much as possible. Because the big ideas of statistics don't depend on any particular level of access to technology, *Basic Practice of Statistics* does not require software or a graphing calculator until we reach the more advanced methods in Part V of the text. Even if you make little use of technology, you should look at the "Using Technology" sections throughout the book. You will see at once that you can read and apply the output from almost any technology used for statistical calculations. The ideas really are more important than the details of how to do the calculations.

Unless you have access to software or a graphing calculator, *you will need a basic calculator with some built-in statistical functions.* Specifically, your calculator should find means and standard deviations and calculate correlations and regression lines. Look for a calculator that claims to do "two-variable statistics" or mentions "regression."

Although ability to carry out statistical procedures is very useful in academics and employment, the most important asset you can gain from the study of statistics is an understanding of the big ideas about working with data. *Basic Practice of Statistics* tries to explain the most important ideas of statistics rather than just teach methods. Some examples of big ideas that you will meet (one from each of the three areas of statistics) are "always plot your data," "randomized comparative experiments," and "statistical significance."

You learn statistics by doing statistical problems. As you read, you will see several levels of exercises, arranged to help you learn. Short "Apply Your Knowledge" problem sets appear after each major idea. These are straightforward exercises that help you solidify the main points as you read. Be sure you can do these exercises before going on. The end-of-chapter exercises begin with multiple-choice "Check Your Skills" exercises (with odd-numbered answers in the back of the book). Use them to check your grasp of the basics. The regular "Chapter Exercises" help you combine all the ideas of a chapter. Finally, the four Part Review chapters (Chapters 7, 11, 19, and 24) look back over major blocks of learning, with many review exercises. At each step, you are given less advance knowledge of exactly what statistical ideas and skills the problems will require, so each type of exercise requires more understanding.

The key to learning is persistence. The main ideas of statistics, like the main ideas of any important subject, took a long time to discover and take some time to master. The gain will be worth the pain.

CHAPTER 0 EXERCISES

0.1 **Observational Studies and Experiments.** A study published in the *Journal of Epidemiology and Community Health* investigated the effect of vitamin C on health. A group of healthy men and women were followed for 16 years and their health tracked. Those people whose blood samples had the highest levels of vitamin C at the beginning of the study had significantly lower risks of dying at the end of the study.

An online article describing the study says that

> *While higher vitamin C levels are associated with people who practice healthier behavior patterns, this study nonetheless shows striking reductions in mortality rates in those with the highest blood levels of vitamin C.*[7]

(a) Reread Example 0.1 (page 2) and the comments following it. Explain why observational studies might suggest that vitamin C reduces the risk of dying by describing some lurking variables.

(b) A randomized controlled trial is a type of experiment. How does "higher vitamin C levels are associated with people who practice healthier behavior patterns" explain how people with higher levels of vitamin C could have lower risks of dying in observational studies but might not in experiments?

0.2 **The Price of Gas.** In Example 0.4 (page 5), we examined the variation in the price of gasoline from 1990 to 2019. We saw both a regular pattern and the effects of international events. Figure 0.3 plots the average annual retail price of gasoline from 1929 to 1990.[8] Prices are adjusted for inflation.

(a) What overall patterns do you observe? What departures from the overall patterns do you observe? To what international events do these departures correspond?

(b) Data are collected to answer real-world questions. What questions might one use these data to answer?

0.3 **Online Polls.** In 2016, MSNBC posted an online poll asking the question "What do you think? Has Europe let in too many refugees?"[9] Approximately 32% of those responding online said "no."

(a) This poll has some of the same problems as the Ann Landers poll of Example 0.2 (page 3). Do you think that the proportion of Americans who feel this way is higher, lower, or close to 32%? Explain.

(b) For this poll, 3000 people responded (as of August 26, 2019). Among those responding, 946 or 32% said "no." Do you think the results would have been more trustworthy if 30,000 people had responded instead of 3000? Explain.

0.4 **Traffic Fatalities and 9/11.** Figure 0.4 (page 10) provides information on the number of fatal traffic accidents by month for the years 1996–2001.[10] The vertical blue line above each month gives the lowest to highest numbers of fatal crashes for that month for the years 1996–2000. For example, in January the number of fatal crashes for the five years from 1996 through 2000 was between about 2600 and 2900. The blue dots give the number of fatal crashes for each month in 2001. The numbers of fatal crashes from January through September of 2001 follow the general pattern for the five preceding years, as we see the blue dots are well within the blue lines for each month.

(a) What happened in the last three months of 2001? The number of fatal crashes in October through December of 2001 are consistently at or above the values for the previous five years. How can you tell this from the graph?

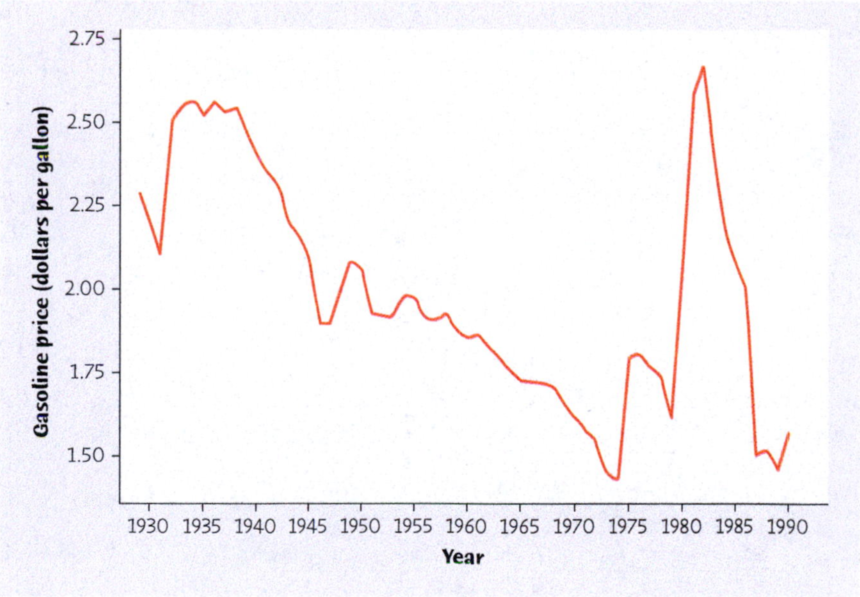

FIGURE 0.3
The average annual retail price of gasoline, 1929 to 1990. Prices are adjusted for inflation.

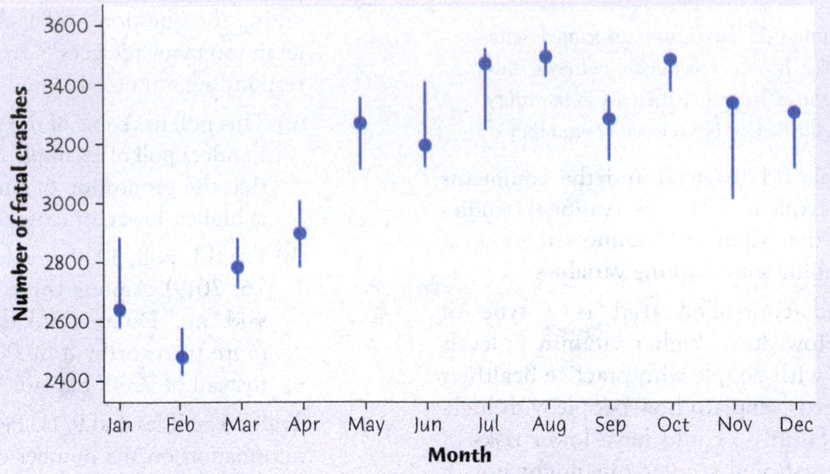

FIGURE 0.4
Number of fatal traffic accidents in the United States in 1996 through 2000 versus 2001. For each month, the blue lines represent the range of the number of fatal accidents from 1996 through 2000, and the blue dot gives the number of fatal accidents in 2001.

(b) On September 11, 2001, terrorists hijacked four U.S. airplanes and used them to strike various targets in the eastern United States. In part (a), we saw from the graph that fatal traffic accidents seemed to be unusually high in the three months following the attacks. Did the terrorists cause fatal crashes to increase? Can you give a simple explanation for the apparent increase in fatal crashes during these months? (Hint: Think about the effect of the September 11 attacks on air travel and hence the effect on alternative means of travel.)

VitalyEdush/Getty Images

PART I

Exploring Data

"What do the data say?" is the first question we ask in any statistical study. *Data analysis* answers this question through open-ended exploration of the data. The tools of data analysis are graphs such as histograms and scatterplots and numerical measures such as means and correlations. At least as important as the tools are principles that organize our thinking as we examine data. The seven chapters in Part I present the principles and tools of statistical data analysis. They equip you with skills that are immediately useful when you deal with numbers.

These chapters reflect the strong emphasis on exploring data that characterizes modern statistics. Sometimes, we hope to draw conclusions that apply to a setting that goes beyond the data in hand. This is *statistical inference*, the topic of much of the rest of the book. Data analysis is essential if we are to trust the results of inference, but data analysis isn't just preparation for inference. Roughly speaking, you can always do data analysis, but inference requires rather special conditions.

One of the organizing principles of data analysis is to first look at one thing at a time and then at relationships. Our presentation follows this principle. In Chapters 1, 2, and 3 you will study *variables and their distributions*. Chapters 4, 5, and 6 cover *relationships among variables*. Chapter 7 reviews this part of the text.

CHAPTER

1

When you complete this chapter, you will be able to:

1.1 Identify the individuals and variables in a given set of data and distinguish between variables that take on numerical values and variables that simply place individuals in different categories.

1.2 Explore data involving categorical variables using bar graphs and pie charts.

1.3 Draw a histogram to present the distribution of a variable that takes on numerical values.

1.4 Explore and understand data sets by interpreting histograms.

1.5 Make a stemplot to present the distribution of a variable that takes on numerical values.

1.6 Make a time plot to present variables whose values are varying over time.

Picturing Distributions with Graphs

Practical statistics uses data to draw conclusions about some broader universe. In our study of practical statistics, we will divide the subject into three main areas. In exploratory data analysis, graphs and numerical summaries are used for exploring, organizing, and describing data so that the patterns become apparent. Data production concerns where the data come from and helps us to understand whether what we learn from our data can be generalized to a wider universe. And statistical inference provides tools for generalizing what we learn to a wider universe.

In this chapter, we will begin to explore data analysis. A data set can consist of hundreds of observations on many variables. Even if we consider only one variable at a time, it is difficult to see what the data have to say by scanning a list containing many data values. A graph provides a visual tool for organizing and identifying patterns in data and is a good starting point in the exploration of the distribution of a variable.

Statistics is the science of data. The volume of data available to us is overwhelming. For example, the U.S. Census Bureau's American Community Survey collects data from about 3,000,000 housing units each year. Astronomers work with data on tens of millions of galaxies. The checkout scanners at Walmart's more than 11,000 stores in 27 countries record hundreds of millions of transactions every week, all saved to inform both Walmart and its suppliers. The first step in dealing with such a flood of data is to organize our thinking about data. Fortunately, we can do this without looking at millions of data points.

kali9/Getty Images

1.1 Individuals and Variables

Any set of data contains information about some group of *individuals*. The information is organized in *variables*.

Individuals and Variables

Individuals are the objects described by a set of data. Individuals may be people, but they may also be animals or things.

A **variable** is any characteristic of an individual. A variable can take different values for different individuals.

A college's student database, for example, includes data about every currently enrolled student. The students are the individuals described by the data set. For each individual, the data contain the values of variables such as date of birth, choice of major, and grade point average (GPA). In practice, any set of data is accompanied by background information that helps us understand the data. When you plan a statistical study or explore data from someone else's work, consider the following questions:

1. **Who?** What *individuals* do the data describe? *How many* individuals appear in the data?
2. **What?** How many *variables* do the data contain? What are the *exact definitions* of these variables? In what *unit of measurement* is each variable recorded? Weights, for example, might be recorded in pounds, in thousands of pounds, or in kilograms.
3. **Where?** Student GPAs and SAT scores (or lack of them) will vary from college to college, depending on many variables, including admissions "selectivity" for the college.
4. **When?** Students change from year to year, as do prices, salaries, and so forth.
5. **Why?** *What purpose* do the data have? Do we hope to answer some specific questions? Do we want answers for just these individuals or for some larger group that these individuals are supposed to represent? Are the individuals and variables suitable for the intended purpose?

STATISTICS IN YOUR WORLD

Data!

Despite the traditional image of lab-coated scientists peering through microscopes, experiments in modern particle physics produce digital data rather than directly observable phenomena. The documentary Particle Fever recreates the excitement of the Large Hadron Collider (LHC) experiment. When the first collisions of highly accelerated protons are recorded live in the film, American physicist Monica Dunford exclaims, "We have data. It's unbelievable how fantastic this data is."

Some variables, such as a person's sex or college major, simply place individuals into categories. Others, like height and GPA, take numerical values for which we can do arithmetic. It makes sense to give an average income for a company's employees, but it does not make sense to give an "average" sex. We can, however, count the numbers of female and male employees and do arithmetic with these counts.

Categorical and Quantitative Variables

A **categorical variable** places an individual into one of several groups or categories.

A **quantitative variable** takes numerical values for which arithmetic operations such as adding and averaging make sense. The values of a quantitative variable are usually recorded with a unit of measurement such as seconds or kilograms.

EXAMPLE 1.1 The American Community Survey

At the U.S. Census Bureau website, you can view the detailed data collected by the American Community Survey, though of course the identities of people and housing units are protected. If you choose the file of data on people, the *individuals* are the people living in the housing units contacted by the survey. More than 100 variables are recorded for each individual. Figure 1.1 displays a very small part of the data.

Excel

	A	B	C	D	E	F	G
1	SERIALNO	PWGTP	AGEP	JWMNP	SCHL	SEX	WAGP
2	283	187	66		6	1	24000
3	283	158	66		9	2	0
4	323	176	54	10	12	2	11900
5	346	339	37	10	11	1	6000
6	346	91	27	10	10	2	30000
7	370	234	53	10	13	1	83000
8	370	181	46	15	10	2	74000
9	370	155	18		9	2	0
10	487	233	26		14	2	800
11	487	146	23		12	2	8000
12	511	236	53		9	2	0
13	511	131	53		11	1	0
14	515	213	38		11	2	12500
15	515	194	40		9	1	800
16	515	221	18	20	9	1	2500
17	515	193	11		3	1	

Each row in the spreadsheet contains data on one individual.

FIGURE 1.1 A spreadsheet displaying data from the American Community Survey, for Example 1.1.

Each row records data on one individual. Each column contains the values of one *variable* for all the individuals. Translated from the U.S. Census Bureau's abbreviations, the variables are

SERIALNO	An identifying number for the household.
PWGTP	Weight in pounds.
AGEP	Age in years.
JWMNP	Travel time to work in minutes.
SCHL	Highest level of education. The numbers designate categories, *not* specific grades. For example, 9 = high school graduate, 10 = some college but no degree, and 13 = bachelor's degree.
SEX	Sex, designated by 1 = male and 2 = female.
WAGP	Wage and salary income last year, in dollars.

Look at the highlighted row in Figure 1.1. This individual is a 53-year-old man who weighs 234 pounds, travels 10 minutes to work, has a bachelor's degree, and earned $83,000 last year.

In addition to the household serial number, there are six variables. Education and sex are categorical variables. The values for education and sex are stored as numbers, but these numbers are just labels for the categories and have no units of measurement. The other four variables are quantitative. Their values do have units. These variables are weight in pounds, age in years, travel time in minutes, and income in dollars.

The *purpose* of the American Community Survey is to collect data that represent the entire nation to guide government policy and business decisions. To do this, the households contacted are chosen at random from all households in the country. We will see in Chapter 8 why choosing at random is a good idea.

Most data tables follow this format: each row is an individual, and each column is a variable. The data set in Figure 1.1 appears in a spreadsheet program that has rows and columns ready for your use. Spreadsheets are commonly used to enter and transmit data and to do simple calculations.

APPLY YOUR KNOWLEDGE

1.1 Fuel Economy. Here is a small part of a data set that describes the fuel economy in miles per gallon (mpg) of model year 2019 motor vehicles:

Make and Model	Vehicle Class	Transmission Type	Number of Cylinders	City mpg	Highway mpg	Annual Fuel Cost
⋮						
Subaru Impreza	Midsize	Manual	4	24	32	$1,450
Nissan Rogue Sport	Small station wagon	Automatic	4	25	32	$1,400
Hyundai Elantra	Midsize	Automatic	4	28	37	$1,200
Chevrolet Impala	Large	Automatic	6	19	28	$1,750
⋮						

The annual fuel cost is an estimate assuming 15,000 miles of travel a year (55% city and 45% highway) and an average fuel price.

(a) What are the individuals in this data set?

(b) For each individual, what variables are given? Which of these variables are categorical, and which are quantitative? In what units are the quantitative variables measured?

1.2 Students and Exercise. You are preparing to study the exercise habits of college students. Describe two categorical variables and two quantitative variables that you might measure for each student. Give the units of measurement for the quantitative variables.

1.2 Categorical Variables: Pie Charts and Bar Graphs

Statistical tools and ideas help us examine data in order to describe their main features. This examination is called **exploratory data analysis.** Like an explorer crossing unknown lands, we want first to simply describe what we see. Here are two principles that help us organize our exploration of a set of data.

exploratory data analysis
Use of graphs and numerical summaries to describe the variables in a data set and the relationships among them.

Exploring Data

1. Begin by examining each variable by itself. Then move on to study the relationships among the variables.
2. Begin with a graph or graphs. Then add numerical summaries of specific aspects of the data.

We will follow these principles in organizing our learning. Chapters 1 through 3 present methods for describing a single variable. We study relationships between two or more variables in Chapters 4 through 6. In each case, we begin with graphical displays and then add numerical summaries for more complete description.

The proper choice of graph depends on the nature of the variable. To examine a single variable, we usually want to display its *distribution.*

Distribution of a Variable

The **distribution** of a variable tells us what values it takes and how often it takes these values.

The values of a categorical variable are labels for the categories. The **distribution of a categorical variable** lists the categories and gives either the count or the percentage of individuals who fall in each category.

EXAMPLE 1.2 Which Major?

MAJORS

Approximately 1.5 million full-time, first-year students enrolled in colleges and universities in 2017. What did they plan to study? Here are data on the percentages of first-year students who planned to major in several discipline areas:[1]

Field of Study	Percentage of Students
Biological sciences	15.5
Business	13.8
Health professions	11.7
Engineering	11.5
Social sciences	11.0
Arts and humanities	8.8
Math and computer science	6.2
Education	4.4
Physical sciences	2.7
Other majors and undeclared	13.1
Total	98.7

It's a good idea to check data for consistency. The percentages should add to 100%. In fact, these data add to 98.7%. What happened? It is possible that there are errors in the data, but we believe there is another explanation. The data from which the table was created actually consist of the percentage of students who plan to major in a number of subfields. For example, for Education, the data consist of the percentage of students planning to major in Elementary Education, Music/Art Education, Physical Education/Recreation, Secondary School Teacher in a non-STEM subject, Special Education, or Other Education. The percentage listed under Education in the table is the sum of these percentages. Each of these percentages is rounded to the nearest tenth. The exact percentages for all of these subfields would add to 100, but the rounded percentages only come close. This is **roundoff error.** Roundoff errors don't point to mistakes in our work, just to the effect of rounding off results.

roundoff error The difference between the exact value of a number and its estimated value, occurring when results of calculations are altered by reducing the number of decimal places.

Columns of numbers take time to read. You can use a pie chart or a bar graph to display the distribution of a categorical variable more vividly. Figures 1.2 and 1.3 illustrate these displays for the distribution of intended college majors.

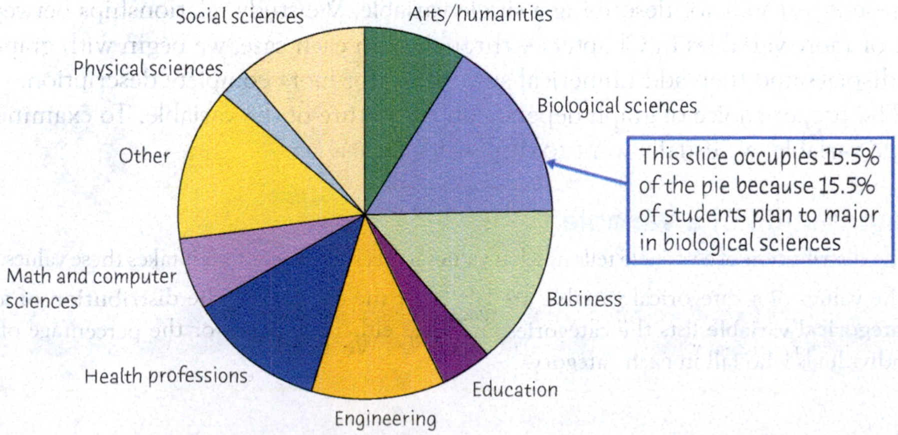

FIGURE 1.2
You can use a pie chart to display the distribution of a categorical variable. This pie chart, created with the JMP Pro 14 software package, shows the distribution of intended majors of students entering college. Statistical analysis relies heavily on statistical software, and JMP is one of the most popular software choices both in industry and in colleges and schools of business. Computer output from other statistical packages like Minitab, SPSS, and R is similar, so you can feel comfortable using any one of these packages. Spreadsheet programs such as Excel also can produce pie charts and other graphs.

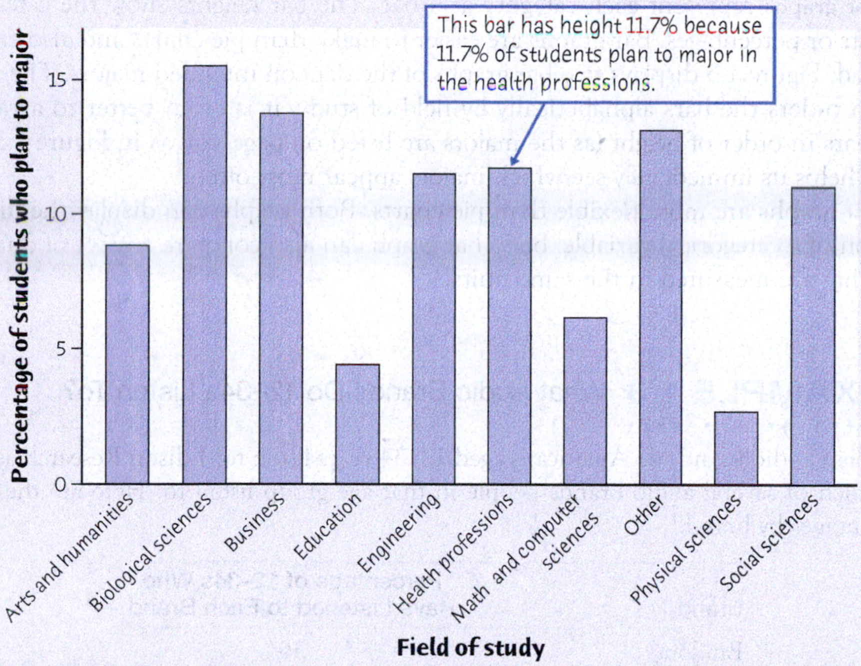

(a)

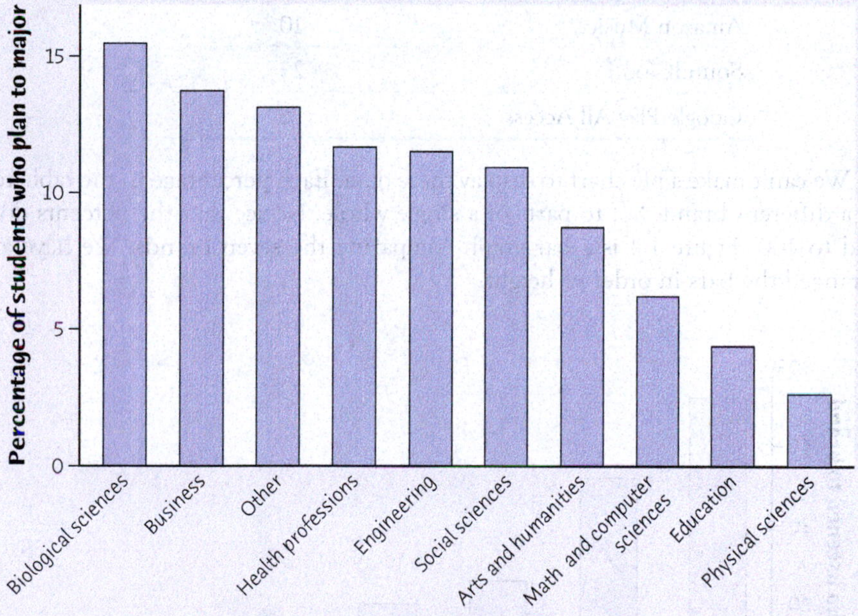

(b)

FIGURE 1.3
Bar graphs of the distribution of intended majors of students entering college. In part (a), the bars follow the alphabetical order of fields of study. In part (b), the same bars appear in order of height. These figures were created with the JMP Pro 14 software package.

Pie charts show the distribution of a categorical variable as a "pie" whose slices are sized by the counts or percentages for the categories. Pie charts are awkward to make by hand, but software will do the job for you. *A pie chart must include all the categories that make up a whole. Use a pie chart only when you want to emphasize each category's relationship to the whole.* We need the "Other majors and undeclared" category in Example 1.2 to complete the whole (all intended majors) and allow us to make the pie chart in Figure 1.2.

bar graph
A graph that shows the distribution of a categorical variable by representing each category as a bar, with the height of each bar representing the count or percentage for the corresponding category.

Bar graphs represent each category as a bar. The bar heights show the category counts or percentages. Bar graphs are easier to make than pie charts and also easier to read. Figure 1.3 displays two bar graphs of the data on intended majors. The first graph orders the bars alphabetically by field of study. It is often better to arrange the bars in order of height (as the majors are listed on page 16), as in Figure 1.3(b). This helps us immediately see which majors appear most often.

Bar graphs are more flexible than pie charts. Both graphs can display the distribution of a categorical variable, but a bar graph can also compare any set of quantities that are measured in the same units.

EXAMPLE 1.3 What Audio Brands Do 12–34s Listen To?

BRANDS

What audio brands do Americans aged 12–34 years listen to? Edison Research asked which of several audio brands people in that age group listen to. Here are the percentages by brand.[2]

Brand	Percentage of 12–34s Who Have Listened to Each Brand
Pandora	36
Spotify	46
iHeartRadio	14
Apple Music	20
Amazon Music	10
SoundCloud	23
Google Play All Access	8

We can't make a pie chart to display these data. Each percentage in the table refers to a different brand, not to parts of a single whole. Notice that the percents do not add to 100. Figure 1.4 is a bar graph comparing the seven brands. We have again arranged the bars in order of height.

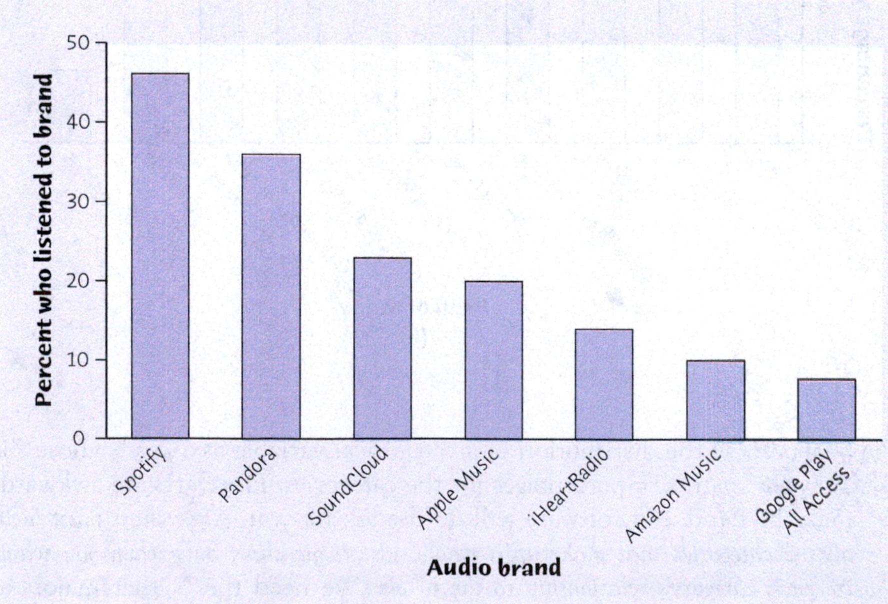

FIGURE 1.4
You can use a bar graph to compare quantities that are not part of a whole. This bar graph compares the percentages of Americans aged 12–34 who have listened to each audio brand, for Example 1.3.

Bar graphs and pie charts are mainly tools for presenting data: they help your audience grasp data quickly. Because it is easy to understand data on a single categorical variable without a graph, bar graphs and pie charts are of limited use for data analysis. We will move on to quantitative variables, where graphs are essential tools.

APPLY YOUR KNOWLEDGE

1.3 Social Media Preferences for Younger Audiences. By a small margin, Facebook remains the top choice of social media over all ages, with 29% using Facebook most often among those using social media sites. However, more visually oriented social networks such as Snapchat and Instagram continue to draw in younger audiences. When asked "Which one social networking brand do you use most often?" here are the top brands chosen by Americans aged 12–34 who currently use any social networking site or service:[3] SOCMEDIA

Social Media Site	Percentage That Use Most Often
Facebook	29
Snapchat	28
Instagram	26
Twitter	6
Pinterest	1

(a) What is the sum of the percentages for these top social media sites? What percentage of Americans aged 12–34 use other social media sites most often?

(b) Make a bar graph to display these data. Be sure to include an "Other social media brand" category.

(c) Would it be correct to display these data in a pie chart? Why or why not?

(d) Data are collected to answer real-world questions. What questions might one use these data to answer?

1.4 How Do Students Pay for College? The Higher Education Research Institute's Freshman Survey includes more than 120,000 first-time, full-time freshmen who entered college in 2017.[4] The survey reports the following data on the sources students use to pay for college expenses: EXPENSE

Source for College Expenses	Students
Family resources	69.9%
Student resources	55.4%
Aid—not to be repaid	70.2%
Aid—to be repaid	47.6%

(a) Explain why it is *not* correct to use a pie chart to display these data.

(b) Make a bar graph of the data. Notice that because the data contrast related groups such as family and student resources, it is better to keep these bars next to each other rather than to arrange the bars in order of height.

1.5 Never on Sunday? Births are not, as you might think, evenly distributed across the days of the week. Here are the average numbers of babies born on each day of the week in 2017:[5] BIRTHS

Day	Births
Sunday	7,164
Monday	11,008
Tuesday	11,943
Wednesday	11,949
Thursday	11,959
Friday	11,779
Saturday	8,203

Present these data in a well-labeled bar graph. Would it also be correct to make a pie chart? Suggest some possible reasons why there are fewer births on weekends.

1.3 Quantitative Variables: Histograms

histogram
A graph of the distribution of one quantitative variable. The horizontal axis is marked in the units of measurement for the variable. The range of the data is divided into classes of equal width. Bars on the graph each represent a class, with the base of each bar covering the class on the horizontal axis. The vertical axis displays the scale of counts, and each bar's height is the class count.

Quantitative variables often take many values. The distribution tells us what values the variable takes and how often it takes these values. A graph of the distribution is clearer if nearby values are grouped together. The most common graph of the distribution of one quantitative variable is a **histogram.**

EXAMPLE 1.4 Making a Histogram

GRADRATE

What percentage of your home state's high school students graduate within four years? Data about graduation rates can help answer the question "How successful is my state in graduating students and how does my state compare to others?" The Freshman Graduation Rate (FGR) counts the number of high school graduates in a given year for a state and divides this by the number of ninth-graders enrolled four years previously. The FGR neglects high school students moving into and out of a state and may include students who have repeated a grade. Several alternative measures are available, and states had been free to choose their own, but the resulting rates could differ by more than 10%. Federal law now requires all states to use a common, more rigorous computation, the *Adjusted Cohort Graduation Rate*, which tracks individual students. The use of the Adjusted Cohort Graduation Rate was first required for 2010–2011, and this finally allowed accurate comparisons of the graduation rates among states. Table 1.1 presents the data for 2016–2017.[6]

The *individuals* in this data set are the states. The *variable* is the percentage of a state's high school students who graduate within four years. The states vary quite a bit on this variable, from 71.1% in New Mexico to 91% in Iowa. It's much easier to see how your state compares with other states from a histogram than from the table. To make a histogram of the distribution of this variable, proceed as follows:

Step 1. Choose the classes. Divide the range of the data into classes of equal width. The data in Table 1.1 range from 71.1 to 91.0, so we decide to use these classes:

percentage on-time graduates between 70.0 and 72.5 (70.0 to $<$ 72.5)

percentage on-time graduates between 72.5 and 75.0 (72.5 to $<$ 75.0)

⋮

percentage on-time graduates between 90.0 and 92.5 (90.0 to $<$ 92.5)

It is important to specify the classes carefully so that each individual falls into exactly one class. Our notation 70.0 to $<$ 72.5 indicates that the first class includes states with

STATISTICS IN YOUR WORLD
What's That Number?

You might think that numbers, unlike words, are universal. Think again. A "billion" in the United States means 1,000,000,000 (nine zeros). In Europe, a "billion" is 1,000,000,000,000 (12 zeros). OK, those are words that describe numbers. But those commas in big numbers are periods in many other languages. This is so confusing that international standards call for spaces instead, so that an American billion is written 1 000 000 000. And the decimal point of the English-speaking world is the decimal comma in many other languages, so that 3.1416 in the United States becomes 3,1416 in Europe. So what is the number 10,642.389? It depends on where you are.

TABLE 1.1 Percent of state high school students graduating on time

State	Percent	Region	State	Percent	Region	State	Percent	Region
Alabama	89.3	S	Louisiana	78.1	S	Ohio	84.2	MW
Alaska	78.2	W	Maine	86.9	NE	Oklahoma	82.6	S
Arizona	78.0	W	Maryland	87.7	S	Oregon	76.7	W
Arkansas	88.0	S	Massachusetts	88.3	NE	Pennsylvania	86.6	NE
California	82.7	W	Michigan	80.2	MW	Rhode Island	84.1	NE
Colorado	79.1	W	Minnesota	82.7	MW	South Carolina	83.6	S
Connecticut	87.9	NE	Mississippi	83.0	S	South Dakota	83.7	MW
Delaware	86.9	S	Missouri	88.3	MW	Tennessee	89.8	S
Florida	82.3	S	Montana	85.8	W	Texas	89.7	S
Georgia	80.6	S	Nebraska	89.1	MW	Utah	86.0	W
Hawaii	82.7	W	Nevada	80.9	W	Vermont	89.1	NE
Idaho	79.7	W	New Hampshire	88.9	NE	Virginia	86.9	S
Illinois	87.0	MW	New Jersey	90.5	NE	Washington	79.4	W
Indiana	83.8	MW	New Mexico	71.1	W	West Virginia	89.4	S
Iowa	91.0	MW	New York	81.8	NE	Wisconsin	88.6	MW
Kansas	86.5	MW	North Carolina	86.6	S	Wyoming	86.2	W
Kentucky	89.7	S	North Dakota	87.2	MW	District of Columbia	73.2	S

graduation rates starting at 70.0% and up to, but not including, graduation rates of 72.5%. Thus, a state with an on-time graduation rate of 72.5% falls into the second class, whereas a state with an on-time graduation rate of 72.4% falls into the first class. It is equally correct to use classes 70.0 to < 72.0, 72.0 to < 74.0, and so forth. Just be sure to specify the classes precisely so that each individual falls into exactly one class.

Step 2. Count the individuals in each class. Here are the counts:

Class	Count	Class	Count
70.0 to < 72.5	1	82.5 to < 85.0	10
72.5 to < 75.0	1	85.0 to < 87.5	11
75.0 to < 77.5	1	87.5 to < 90.0	14
77.5 to < 80.0	6	90.0 to < 92.5	2
80.0 to < 82.5	5		

Check that the counts add to 51, the number of individuals in the data set (the 50 states and the District of Columbia).

Step 3. Draw the histogram. Mark the scale for the variable whose distribution you are displaying on the horizontal axis. That's the percentage of a state's high school students who graduate within four years. The scale runs from 70.0 to 92.5 because that is the span of the classes we chose. The vertical axis contains the scale of counts. Each bar represents a class. The base of the bar covers the class, and the bar height is the class count. Draw the bars with no horizontal space between them unless a class is empty so that its bar has height zero. Figure 1.5 is our histogram. Remember, an observation on the boundary of the bars—say, 75.0—is counted in the bar to its right.

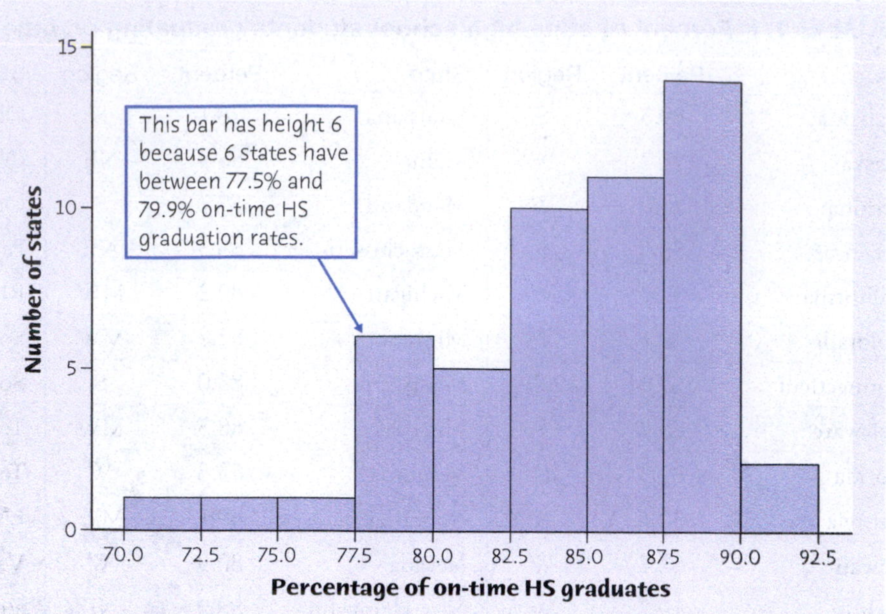

FIGURE 1.5 Histogram of the distribution of the percent of on-time high school graduates in 50 states and the District of Columbia, for Example 1.4. This figure was created with the JMP 14 software package.

Although histograms resemble bar graphs, their details and uses are different. A histogram displays the distribution of a quantitative variable. The horizontal axis of a histogram is marked in the units of measurement for the variable. A bar graph compares the sizes of different quantities. The horizontal axis of a bar graph simply identifies the quantities or categories being compared and need not have any measurement scale. These quantities may be the values of a categorical variable, but they may also be unrelated, like the sources used to learn about music in Example 1.3 (page 18). Draw bar graphs with blank space between the bars to separate the quantities being compared. Draw histograms with no space to indicate that all values of the variable are covered. A gap between bars in a histogram indicates that there are no values for that class.

Our eyes respond to the *area* of the bars in a histogram.[7] Because the classes are all the same width, area is determined by height, and all classes are fairly represented. There is no one right choice of the classes in a histogram. Too few classes will give a "skyscraper" graph, with all values in a few classes with tall bars. Too many will produce a "pancake" graph, with most classes having one or no observations. Neither choice will give a good picture of the shape of the distribution. You must use your judgment in choosing classes to display the shape. Statistics software will choose the classes for you. The software's choice is usually a good one, but you can change it if you want. The histogram function in the *Histograms* applet in Achieve allows you to change the number of classes by dragging with the mouse so that it is easy to see how the choice of classes affects the histogram.

applet

APPLY YOUR KNOWLEDGE

1.6 The Changing Face of America. In 1980, approximately 20% of adults aged 18–34 were considered minorities, reporting their ethnicity as other than non-Hispanic White. By the end of 2013, that percentage had more than doubled. How are minorities between the ages of 18 and 34 distributed in the United States? In the country as a whole, 42.8% of adults aged 18–34 are considered minorities, but the states vary from 8% in Maine and Vermont to 75% in Hawaii.

Table 1.2 presents the data for all 50 states and the District of Columbia.[8] Make a histogram of the percents using classes of width 10% starting at 0%. That is, the first bar covers 0% to < 10%, the second covers 10% to < 20%, and so on. (Make this histogram by hand, even if you have software, to be sure you understand the process. You may then want to compare your histogram with your software's choice.) MINORITY

TABLE 1.2 Percent of state population aged 18–34 who are minorities

State	Percent	State	Percent	State	Percent
Alabama	39	Louisiana	45	Ohio	23
Alaska	40	Maine	8	Oklahoma	37
Arizona	51	Maryland	52	Oregon	27
Arkansas	31	Massachusetts	31	Pennsylvania	26
California	67	Michigan	28	Rhode Island	31
Colorado	35	Minnesota	23	South Carolina	41
Connecticut	39	Mississippi	48	South Dakota	19
Delaware	23	Missouri	23	Tennessee	30
Florida	52	Montana	15	Texas	61
Georgia	51	Nebraska	23	Utah	22
Hawaii	75	Nevada	54	Vermont	8
Idaho	20	New Hampshire	11	Virginia	41
Illinois	42	New Jersey	51	Washington	34
Indiana	23	New Mexico	67	West Virginia	9
Iowa	16	New York	48	Wisconsin	22
Kansas	27	North Carolina	41	Wyoming	18
Kentucky	17	North Dakota	15	District of Columbia	53

1.7 Choosing Classes in a Histogram. The *Histograms* applet in this exercise includes a histogram of the data on the percentage of minorities between the ages of 18 and 34 in the states from Table 1.2. Choose these data and then click on the "Histogram" tab to see a histogram. MINORITY

applet

(a) How many classes does the applet choose to use? (You can see the count of classes on the toggle button below the histogram.)

(b) Click on the "Classes" toggle button and drag to the left. What is the smallest number of classes you can get? What are the lower and upper bounds of each class? (Click on the bar to find out.) Make a rough sketch of this histogram.

(c) Click and drag to the right. What is the greatest number of classes you can get? How many observations does the largest class have?

(d) You see that the choice of classes changes the appearance of a histogram. Drag back and forth until you get the histogram that you think best displays the distribution. How many classes did you use? Why do you think this is best?

1.4 Interpreting Histograms

Making a statistical graph is not an end in itself. *The purpose of graphs is to help us understand how the data help us answer some real-world question.* After you make a graph, always ask, "What do I see?" Once you have displayed a distribution, you can see its important features as follows.

Examining a Histogram

In any graph of data, look for the overall pattern and for striking deviations from that pattern.

You can describe the overall pattern of a histogram by its *shape, center,* and *variability.* You will sometimes see variability referred to as *spread.*

An important kind of deviation is an **outlier,** an individual value that falls outside the overall pattern.

One way to describe the center of a distribution is by its *midpoint,* the value with roughly half the observations taking smaller values and half taking larger values. In Chapter 2, we will formally define the midpoint as the *median.* To find the midpoint, order the observations from smallest to largest, making sure to include repeated observations as many times as they appear in the data. First cross off the largest and smallest observations, then the largest and smallest of those remaining, and continue this process. If you initially had an odd number of observations, you will be left with a single observation, which is the midpoint. If you initially had an even number of observations, you will be left with two observations, and their average is the midpoint.

For now, we will describe the variability of a distribution by giving the *smallest and largest values.* We will look at better ways to describe center and variability in Chapter 2. The overall shape of a distribution can often be described in terms of symmetry or skewness, defined as follows.

Symmetric and Skewed Distributions

A distribution is **symmetric** if the right and left sides of the histogram are approximately mirror images of each other.

A distribution is **skewed to the right** if the right side of the histogram (containing the half of the observations with larger values) extends much farther out than the left side. It is **skewed to the left** if the left side of the histogram extends much farther out than the right side.

EXAMPLE 1.5 Describing a Distribution

GRADRATE

Look again at the histogram in Figure 1.5. To describe the distribution, we want to look at its overall pattern and any deviations.

SHAPE: The distribution has a *single peak,* which represents states in which between 87.5% and 90.0% of students graduate high school on time. The distribution is *skewed to the left.* There is only one observation to the right of the peak, while to the left of the peak, most of the remaining states have graduation rates between 77.5% and 87.5%, but several states have much lower percents, so the graph extends quite far to the left of its peak.

CENTER: Arranging the observations from Table 1.1 in order of size shows that 86.0% is the midpoint of the distribution. There are a total of 51 observations, and if we cross off the 25 highest graduation rates and the 25 lowest graduation rates, we are left with a single graduation rate of 86.0%, which we take as the center of the distribution.

VARIABILITY: The graduation rates range from 71.1% to 91.0%, which shows considerable variability in graduation rates among the states.

OUTLIERS: Figure 1.5 shows no observations outside the overall single-peaked, left-skewed pattern of the distribution. Figure 1.6 is another histogram of the same distribution, with classes of width 2% rather than 2.5%. Now New Mexico at 71.1% and the District of Columbia at 73.2% are more clearly separated from the remaining states. Are New Mexico and the District of Columbia outliers or just the smallest

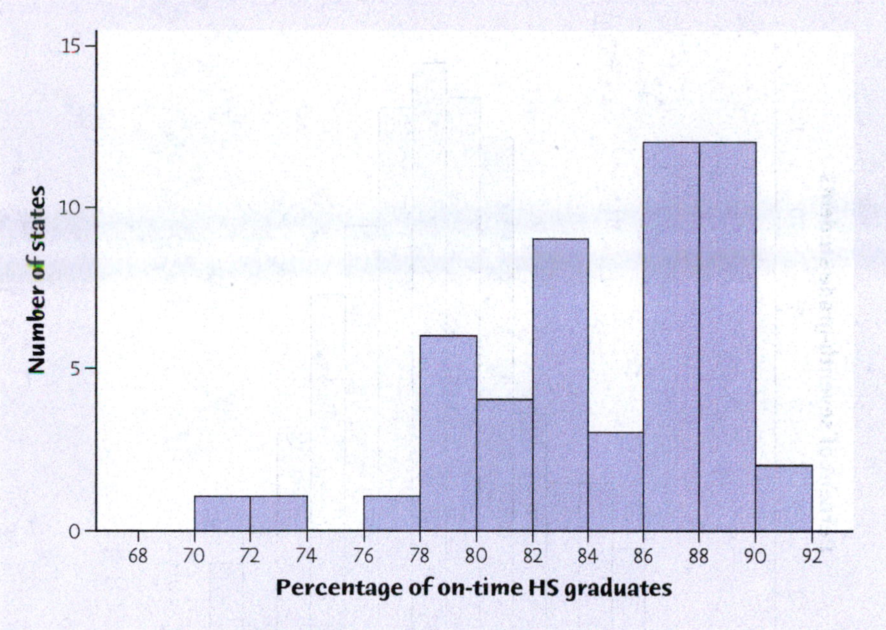

FIGURE 1.6 Another histogram of the distribution of the percentage of on-time high school graduates, with narrower class widths than in Figure 1.5. A histogram with more classes shows more detail but may have a less clear pattern.

observations in a strongly skewed distribution? Unfortunately, there is no rule. Let's agree to call attention to only strong outliers that suggest something special about an observation—or an error such as typing 10.1 as 101. Although the District of Columbia is often included with the other 50 states in data sets, for many variables it can differ markedly from the remaining states.

Figures 1.5 and 1.6 remind us that interpreting graphs calls for judgment. We also see that *the choice of classes in a histogram can influence the appearance of a distribution.* Because of this, and to avoid worrying about minor details, concentrate on the main features of a distribution that persist with several choices of class intervals. Look for major peaks, not for minor ups and downs, in the bars of the histogram. When you choose a larger number of class intervals, the histogram can become more jagged, leading to the appearance of multiple peaks that are close together. Always be sure to check for clear outliers, not just for the smallest and largest observations, and look for rough *symmetry* or clear *skewness.*

Here are more examples of describing the overall pattern of a histogram.

EXAMPLE 1.6 Iowa Tests Scores

IOWATEST

Figure 1.7 displays the scores of all 947 seventh-grade students in the public schools of Gary, Indiana, on the vocabulary part of the Iowa Test of Basic Skills.[9] The distribution is *single-peaked* and *symmetric.* Real data are almost never exactly symmetric. We are content to describe Figure 1.7 as symmetric. The center (half above, half below) is close to 7. This is seventh-grade reading level. The scores range from 2.1 (2nd-grade level) to 12.1 (12th-grade level).

Notice that the vertical scale in Figure 1.7 is not the *count* of students but the *percentage* of students in each histogram class. A histogram of percents rather than counts is convenient when we want to compare several distributions. To compare Gary with Los Angeles, a much bigger city, we would use percents so that both histograms have the same vertical scale.

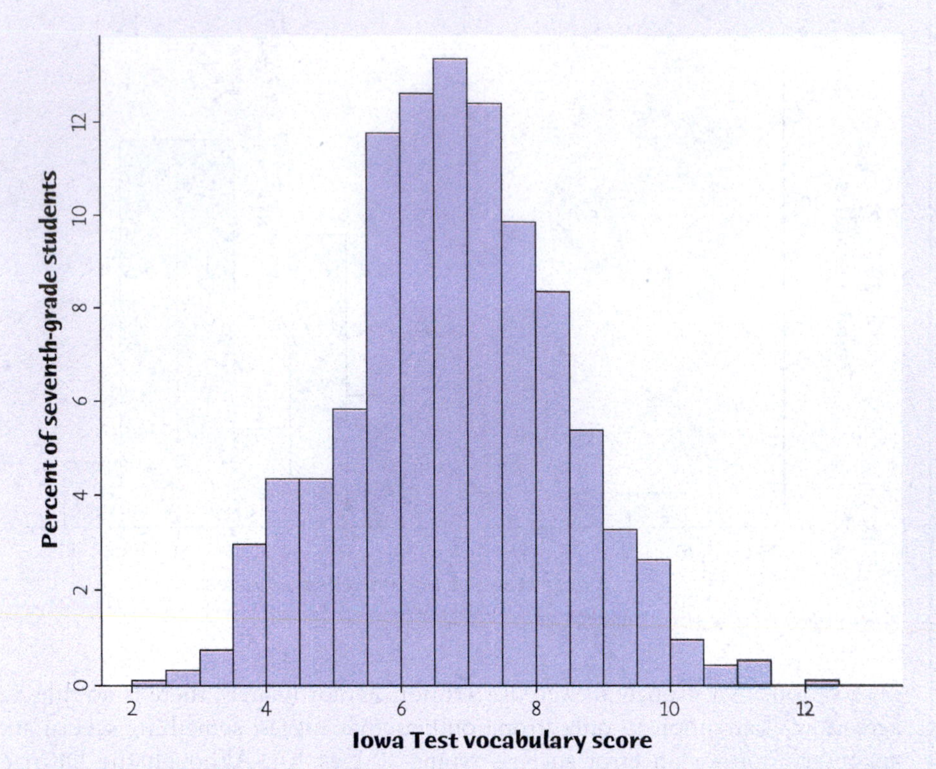

FIGURE 1.7
Histogram of the Iowa Test vocabulary scores of all seventh-grade students in Gary, Indiana, for Example 1.6. This distribution is single peaked and symmetric.

When describing the vertical scale of a histogram, you will sometimes see count referred to as *frequency* and percentage referred to as *relative frequency*, particularly when choosing an option for the vertical scale using software.

EXAMPLE 1.7 Who Takes the SAT?

SATPCT

Depending on where you went to high school, the answer to this question may be "almost everybody," "many but not everybody," or "almost nobody." Figure 1.8 is a histogram of the percentage of high school graduates in each state who took the SAT in 2018.[10]

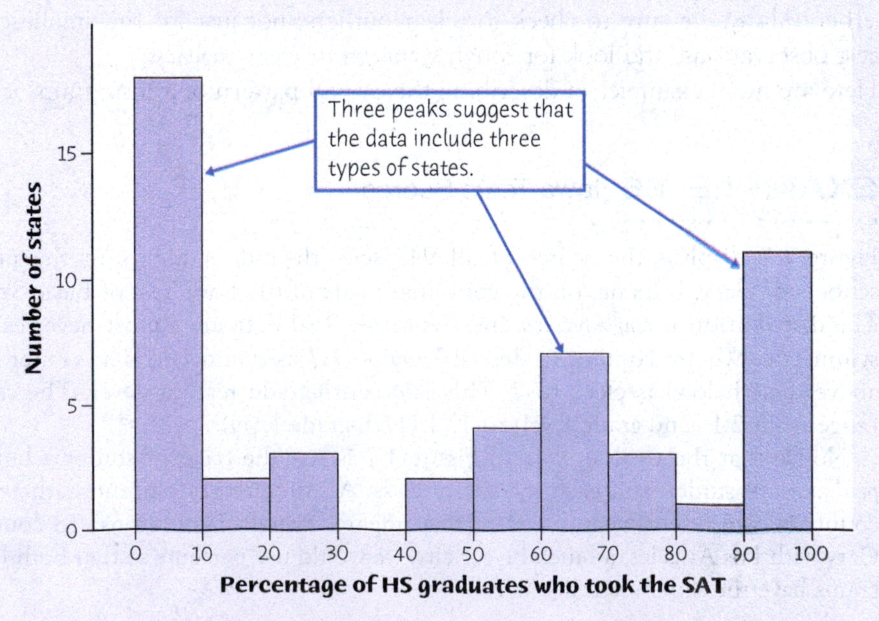

FIGURE 1.8
Histogram of the percentage of high school graduates in each state who took the SAT Reasoning test, for Example 1.7. The graph shows three groups of states: ACT states (where few students take the SAT) at the left, states where many take the SAT and some the ACT, and SAT states at the right.

To create the histogram with bars covering only the range 0% to 100%, we had to choose the class intervals wisely. A few states had 100% of high school students take the SAT. The lowest percentage is 2% for North Dakota. We would like to avoid having a class interval that begins at 100% and includes values larger than 100% (which is impossible) and would result in a histogram that suggests values larger than 100% in the data. We therefore used class intervals of 0.01% to < 10.01%, 10.01% to < 20.01%, etc. However, we have only displayed percents to the nearest integer on the horizontal scale.

The histogram shows three peaks: a high peak at the far left representing percents 10% and below, a lower peak in the 60.01% to < 70.01% class, and a high peak representing percents above 90%. The presence of more than one peak suggests that the distribution mixes several kinds of individuals. That is the case here. There are two major tests of readiness for college, the ACT and the SAT. Most states have a strong preference for one or the other. In some states, many students take the ACT exam and few take the SAT; these states form the peak on the left. In other states, some students take the ACT and some the SAT; these states are the bars near the peak in the 60.01% to < 70.01% class. In yet other states, almost all students take the SAT and very few choose the ACT; these states form the high peak at the far right.

Giving the center and variability of this distribution is not very useful. The midpoint falls in the 50.01% to < 60.01% class, between the peaks. The story told by the histogram is in the three peaks corresponding to primarily ACT states, states in which students take both, and primarily SAT states.

The overall shape of a distribution provides important information about a variable (although describing the shape of a bar graph for categorical data is not useful). Some variables have distributions with predictable shapes. Many biological measurements on specimens from the same species and sex–lengths of bird bills, heights of young women—have symmetric distributions. On the other hand, data on people's incomes are usually strongly skewed to the right. There are many moderate incomes, some large incomes, and a few enormous incomes. Many distributions have irregular shapes that are neither symmetric nor skewed. Some data show other patterns, such as the three peaks in Figure 1.8. Use your eyes, describe the pattern you see, and then try to explain the pattern.

APPLY YOUR KNOWLEDGE

1.8 The Changing Face of America. In Exercise 1.6 (page 22), you made a histogram of the percentage of minority residents aged 18–34 in each of the 50 states and the District of Columbia. These data are given in Table 1.2. Describe the shape of the distribution. Is it closer to symmetric or skewed? What is the center (midpoint) of the data? What is the variability in terms of the smallest and largest values? Are there any states with an unusually large or small percentage of minorities? MINORITY

Kallista Images/Superstock

1.9 Lyme Disease. Lyme disease is caused by a bacterion called *Borrelia burgdorferi* and is spread through the bite of an infected black legged tick, generally found in woods and grassy areas. There were 383,846 confirmed cases reported to the Centers for Disease Control and Prevention (CDC) between 2001 and 2017, and these are broken down by age and sex in Figure 1.9.[11]

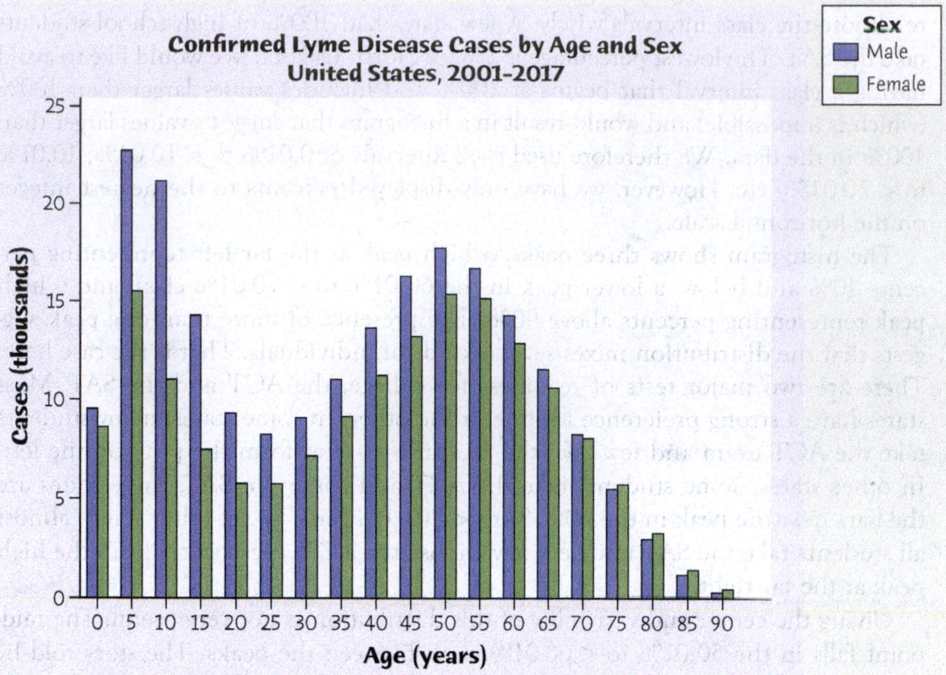

FIGURE 1.9
Histogram of the ages of individuals infected with Lyme disease for cases reported between 2001 and 2017 in the United States, for males and females, for Exercise 1.9.

Here is how Figure 1.9 relates to what we have been studying. The individuals are the 383,846 people with confirmed cases, and two of the variables measured on each individual are sex and age. Considering males and females separately, we could draw a histogram of the variable age using class intervals 0 to < 5 years old, 5 to < 10 years old, and so forth. Look at the leftmost two bars in Figure 1.9. The blue bar shows that almost 10,000 of the males were between 0 and 5 years old, and the green bar shows slightly fewer females were in this age range. If we were to take all the blue bars and put them side by side, we would have the histogram of age for males using the class intervals stated. Similarly, the green bars show females. Because we are trying to display both histograms in the same graph, the bars for males and females within each class interval have been placed alongside each other for ease of comparison, with the bars for different class intervals separated by small spaces.

(a) Describe the main features of the distribution of age for males. Why would describing this distribution in terms of only the center and variability be misleading?

(b) Suppose that different age groups of males spend differing amounts of time outdoors. How could this fact be used to explain the pattern that you found in part (a)? Remember to use your eyes to describe the pattern you see and then try to explain the pattern.

(c) A 45-year-old male friend of yours looks at the histogram and tells you that he is planning on giving up hiking because this graph suggests he is in a high-risk group for getting Lyme disease. He will resume hiking when he is 65, as he will be less likely to get Lyme disease at that age. Is this a correct interpretation of the histogram?

(d) Comparing the histograms for males and females, how are they similar? What is the main difference, and why do you think it occurs?

1.5 Quantitative Variables: Stemplots

Histograms are not the only graphical display of distributions. For small data sets, a **stemplot** is quicker to make and presents more detailed information.

stemplot
A graph of the distribution of a quantitative variable. Each observation is shown as a stem, consisting of all but the final (rightmost) digit, and a leaf, the final digit. The unique stems are stacked in a vertical column with the smallest at the top, and a vertical line appears to the right of the column. Each leaf for a stem appears in a row to the right of the stem.

Stemplot

To make a stemplot:

1. Separate each observation into a **stem,** consisting of all but the final (rightmost) digit, and a **leaf,** the final digit. Stems may have as many digits as needed, but each leaf contains only a single digit.
2. Write the stems in a vertical column with the smallest at the top and draw a vertical line at the right of this column. Be sure to include all the stems needed to span the data, even when some stems will have no leaves.
3. Write each leaf in the row to the right of its stem, in increasing order out from the stem.

EXAMPLE 1.8 Making a Stemplot

Table 1.2 (page 23) presents the percentage of adults aged 18–34 who were considered minorities in each of the states and the District of Columbia. To make a stemplot of these data, first write the percents 8 and 9 as 08 and 09 so that all the percents are two-digit numbers. Take the tens place (leftmost digit) of the percentage as the stem and the final digit (ones) as the leaf. Write stems from 0 for Maine, Vermont, and West Virginia to 7 for Hawaii. Now add leaves. Texas, 61%, has leaf 1 on the 6 stem. California and New Mexico, at 67%, each place a leaf of 7 on the same stem. These are the only observations on this stem. Arrange the leaves in order, so that 6 | 177 is one row in the stemplot. Figure 1.10 is the complete stemplot for the data in Table 1.2.

MINORITY

```
0 | 889
1 | 1556789
2 | 0223333336778
3 | 011145799
4 | 01112588
5 | 1112234
6 | 177
7 | 5
```

FIGURE 1.10
Stemplot of the percentages of minorities aged 18–34 in the states, for Example 1.8. The tens place of the percentage is the stem, and the ones place is the leaf.

A stemplot looks like a histogram turned on end, with the stems corresponding to the class intervals. The first stem in Figure 1.10 contains all states with percents between 0% and 10%. Examine the histogram in Figure 1.11, which is a histogram of the minority data using class intervals 0% to < 10%, 10% to < 20%, and so forth. Although Figures 1.10 and 1.11 display exactly the same pattern, the stemplot, unlike the histogram, preserves the actual value of each observation.

In a stemplot, the classes (the stems of a stemplot) are given to you. Histograms are more flexible than stemplots because you can choose the classes more easily. *Stemplots do not work well for large data sets, where each stem must hold a large number of leaves.* Don't try to make a stemplot of a large data set, such as the 947 Iowa Test scores in Figure 1.7.

Consider making a stemplot of the high school graduation rate data in Table 1.1 (page 21). If we use the tenths as the leaves, the neccesary stems begin at 71 for New Mexico and end at 91 for Iowa, requiring a total of 21 stems. When there are too many stems, as in this case, there are often no leaves or just one or two leaves on many of the stems. The number of stems can be reduced if we first round the data. In this example, we can round the data for each state to the nearest percent before drawing the stemplot. Here is the result:

```
7 | 13788899
8 | 00112233333444446667777777788888999999
9 | 00011
```

STATISTICS IN YOUR WORLD
The Vital Few
Skewed distributions can show us where to concentrate our efforts. Ten percent of the cars on the road account for half of all carbon dioxide emissions. A histogram of CO_2 emissions would show many cars with small or moderate values and a few with very high values. Cleaning up or replacing these cars would reduce pollution at a cost much lower than that of programs aimed at all cars. Statisticians who work at improving quality in industry make a principle of this: distinguish "the vital few" from "the trivial many."

FIGURE 1.11
Histogram of the percentages of minorities aged 18–34 in the states, for Example 1.8. The class widths have been chosen to agree with the widths of the stems in the stemplot in Figure 1.10.

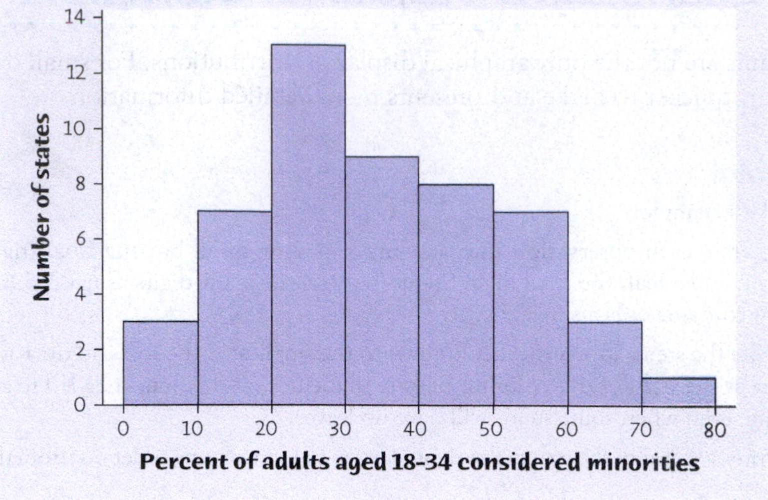

Now it seems that there are too few stems. You can also split stems in a stemplot to double the number of stems when all the leaves would otherwise fall on just a few stems, as occurred when we rounded to the nearest percent. Each stem then appears twice. Leaves 0 to 4 go on the upper stem, and leaves 5 to 9 go on the lower stem. If you split the stems with the data rounded to the nearest percent, the stemplot becomes:

```
7 | 13
7 | 788899
8 | 001122333334444
8 | 66677777777888889999999
9 | 00011
```

which makes the left skew pattern clearer. In fact, the stems in the previous stemplot correspond to the class intervals used in the histogram of Figure 1.5 (page 22), although there are minor differences in the histogram and the stemplot because the data were rounded for the stemplot but not for the histogram. When drawing stemplots, some data require rounding but don't require splitting stems, some require just splitting stems, and other data require both. Many statistical softwares, such as CrunchIt!, allow you to decide whether to split stems so that it is easy to see the effect.

A *dotplot* is another graphic for displaying small data sets. Dotplots are often used when the data have relatively few distinct values—for example, a small range of integer values such as 1 to 10. A dotplot is like a stemplot in which the leaves for each stem are replaced by dots. A dotplot is typically oriented like a histogram, with the stem values listed along the horizontal axis and the dots forming a vertical column above each stem. As such, dotplots resemble histograms. Figure 1.12 is a dotplot of the data in Table 1.1, corresponding to the split stemplot above. Dotplots, like stemplots, help us visualize the distribution of small data sets. We prefer stemplots to dotplots, and in the remainder of this book we will use stemplots rather than dotplots for small data sets.

Comparing Figures 1.11 (right-skewed) and 1.5 (left-skewed) reminds us that *the direction of skewness is the direction of the long tail, not the direction where most observations are clustered.*

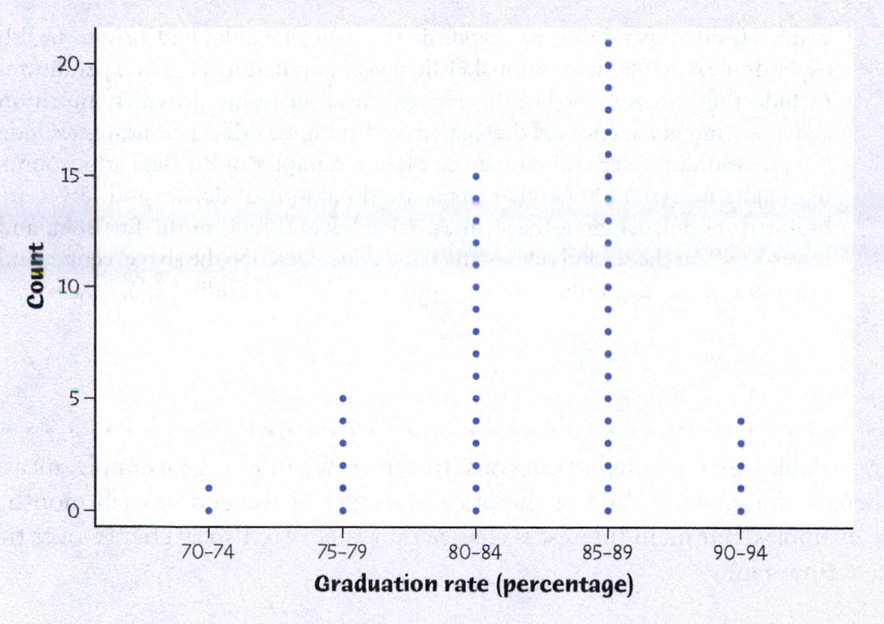

FIGURE 1.12
Dotplot of the data in Table 1.1.

APPLY YOUR KNOWLEDGE

1.10 The Changing Face of America. Figure 1.10 gives a stemplot of the percentages of adults aged 18–34 who were considered minorities in each of the states and the District of Columbia. MINORITY

(a) Make another stemplot of this data by splitting the stems, placing leaves 0 to 4 on the first stem and leaves 5 to 9 on the second stem of the same value. Does splitting the stems gives a different impression of the distribution? Explain.

(b) Draw a histogram of this data which uses class intervals that give the same pattern as the stemplot that you drew in part (a).

1.11 Health Care Spending. Table 1.3 shows the 2015 per capita total expenditure on health in 35 countries with the highest gross domestic product in that

TABLE 1.3 Per capita total expenditure on health (international dollars)

Country	Dollars	Country	Dollars	Country	Dollars
Argentina	1390	Indonesia	369	Saudi Arabia	3121
Australia	4492	Iran	1262	South Africa	1086
Austria	5138	Italy	3351	Spain	3183
Belgium	4782	Japan	4405	Sweden	5299
Brazil	1392	Korea, South	2556	Switzerland	7583
Canada	4600	Malaysia	1064	Thailand	610
China	762	Mexico	1009	Turkey	996
Colombia	853	Netherlands	5313	United Arab Emirates	2426
Denmark	5083	Nigeria	215	United Kingdom	4145
France	4542	Norway	6222	United States	9536
Germany	5357	Poland	1704	Venezuela	106
India	238	Russia	1414		

year.[12] Health expenditure per capita is the sum of public and private health expenditures (in PPP, international $) divided by population. Health expenditures include the provision of health services, family-planning activities, nutrition activities, and emergency aid designated for health; health expenditures exclude the provision of water and sanitation. Make a stemplot of the data after rounding to the nearest $100 (so that stems are thousands of dollars and leaves are hundreds of dollars). Split the stems, placing leaves 0 to 4 on the first stem and leaves 5 to 9 on the second stem of the same value. Describe the shape, center, and variability of the distribution. Which country is the high outlier? HEALTH

1.6 Time Plots

Many variables are measured at intervals over time. We might, for example, measure the height of a growing child or the price of a stock at the end of each month. In these examples, our main interest is change over time. To display change over time, make a **time plot.**

Time Plot

A **time plot** of a variable plots each observation against the time at which it was measured. Always put time on the horizontal scale of your plot and the variable you are measuring on the vertical scale. Connecting the data points by lines helps emphasize any change over time.

EXAMPLE 1.9 Water Levels in the Everglades

WATERLEV

Image courtesy of the U.S. Geological Survey

Water levels in Everglades National Park are critical to the survival of this unique region. Data on water levels can help us answer questions about threats to the survival of the region. The photo shows a water-monitoring station in Shark River Slough, the main path for surface water moving through the "river of grass" that is the Everglades. Each day the mean gauge height, the height in feet of the water surface above the gauge datum, is measured at the Shark River Slough monitoring station. (The gauge datum is a vertical control measure established in 1929 and is used as a reference for establishing varying elevations. It establishes a zero point from which to measure the gauge height.) Figure 1.13 is a time plot of mean daily gauge height at this station from January 1, 2000, to August 27, 2019.[13]

When you examine a time plot, look once again for an overall pattern and for strong deviations from the pattern. Figure 1.13 shows strong *cycles*—regular up-and-down movements in water level. The cycles show the effects of Florida's wet season (roughly June to November) and dry season (roughly December to May). Water levels are highest in late fall. If you look closely, you can see the year-to-year variation. The dry season in 2003 ended early, with the first-ever April tropical storm. In consequence, the dry-season water level in 2003 did not dip as low as in other years. The drought in the southeastern portion of the country in 2008 and 2009 shows up in the steep drop in the mean gauge height in 2009, and the lower peaks in 2006 and 2007 reflect lower water levels during the wet seasons in these years. Finally, in 2011, an extra-long dry season and a slow start to the 2011 rainy season compounded into the worst drought in the southwest Florida area in 80 years,

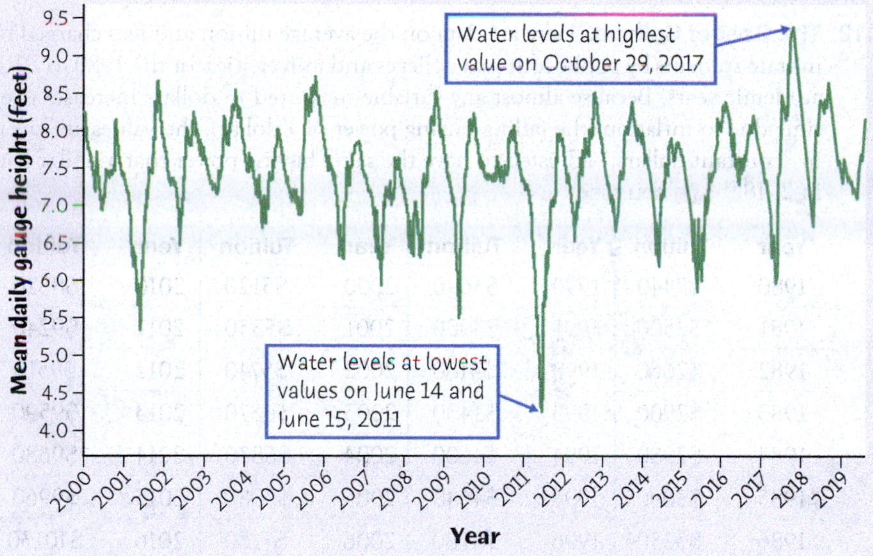

FIGURE 1.13
Time plot of average gauge height at a monitoring station in Everglades National Park over a 19-year period, for Example 1.9. The yearly cycles reflect Florida's wet and dry seasons. This figure was created with the JMP 14 software package.

which shows up as the steep drop in the mean gauge height in 2011. Hurricane Irma struck Florida in September 2017.

Another common overall pattern in a time plot is a *trend,* a long-term upward or downward movement over time. Many economic variables show an upward trend. Incomes, house prices, and (alas) college tuitions tend to move generally upward over time. Figure 1.14 plots the mean annual CO_2 concentration from 1959 to 2017, with a steady upward trend.

Histograms and time plots give different kinds of information about a variable. The time plot in Figure 1.13 presents **time series data** that show the change in water level at one location over time.

time series data
Measurements of a variable taken repeatedly, recording the time as well as the value of each measurement.

A histogram displays *cross-sectional* data, such as water levels at many locations in the Everglades at the same time.

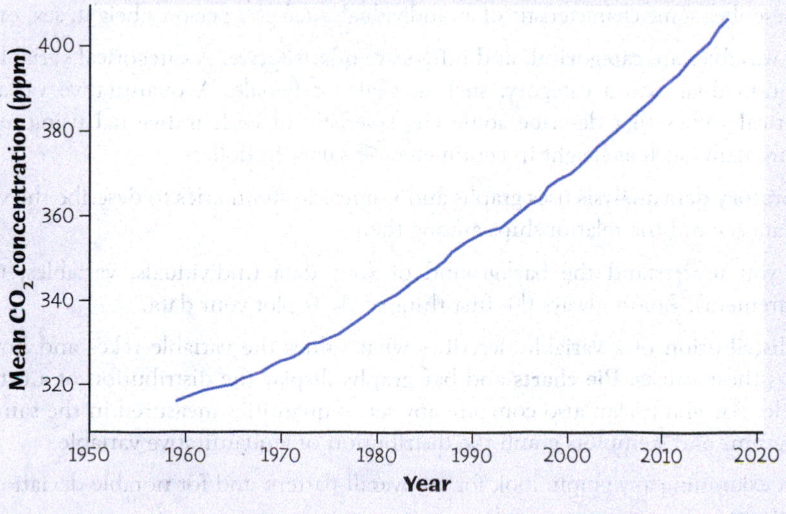

FIGURE 1.14
A time plot of the mean CO_2 concentration (in parts per million) in the atmosphere from 1959 to 2017. This figure was created with the JMP 14 software package.

APPLY YOUR KNOWLEDGE

1.12 The Cost of College. Here are data on the average tuition and fees charged to in-state students by public four-year colleges and universities for the 1980 to 2018 academic years. Because almost any variable measured in dollars increases over time due to inflation (the falling buying power of a dollar), the values are given in "constant dollars," adjusted to have the same buying power that a dollar had in 2018.[14] COLLEGEX

Year	Tuition	Year	Tuition	Year	Tuition	Year	Tuition
1980	$2440	1990	$3690	2000	$5120	2010	$8820
1981	$2500	1991	$3900	2001	$5350	2011	$9240
1982	$2660	1992	$4180	2002	$5740	2012	$9510
1983	$2900	1993	$4430	2003	$6370	2013	$9590
1984	$2980	1994	$4600	2004	$6830	2014	$9680
1985	$3090	1995	$4640	2005	$7080	2015	$9960
1986	$3250	1996	$4780	2006	$7180	2016	$10130
1987	$3300	1997	$4880	2007	$7490	2017	$10270
1988	$3360	1998	$5020	2008	$7560	2018	$10230
1989	$3440	1999	$5080	2009	$8270		

(a) Make a time plot of average tuition and fees.

(b) What overall pattern does your plot show?

(c) Some possible deviations from the overall pattern are outliers, periods when charges went down (in 2018 dollars) and periods of particularly rapid increase. Which are present in your plot, and during which years?

(d) In looking for patterns, do you think that it would be better to study a time series of the tuition for each year or the percentage increase for each year? Why? Think about what questions these data might answer.

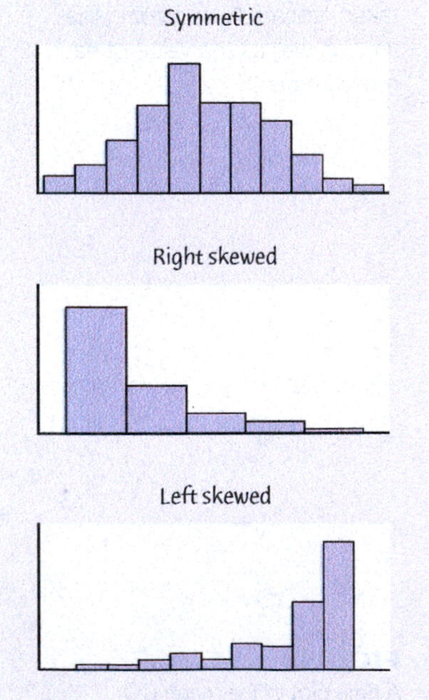

FIGURE 1.15
Symmetric, right-skewed, and left-skewed distributions.

CHAPTER 1 SUMMARY

Chapter Specifics

- A data set contains information on a number of **individuals.** Individuals may be people, animals, or things. For each individual, the data give values for one or more **variables.** A variable describes some characteristic of an individual, such as a person's height, sex, or salary.
- Some variables are **categorical,** and others are **quantitative.** A categorical variable places each individual into a category, such as male or female. A quantitative variable has numerical values that describe some characteristic of each individual using a unit of measurement, such as height in centimeters or salary in dollars.
- **Exploratory data analysis** uses graphs and numerical summaries to describe the variables in a data set and the relationships among them.
- After you understand the background of your data (individuals, variables, units of measurement), almost always the first thing to do is plot your data.
- The **distribution** of a variable describes what values the variable takes and how often it takes these values. **Pie charts** and **bar graphs** display the distribution of a categorical variable. Bar graphs can also compare any set of quantities measured in the same units. **Histograms** and **stemplots** graph the distribution of a quantitative variable.
- When examining any graph, look for an overall pattern and for notable deviations from the pattern.

- *Shape, center,* and *variability* describe the overall pattern of the distribution of a quantitative variable. Some distributions have simple shapes, such as **symmetric** or **skewed.** Not all distributions have a simple overall shape, especially when there are few observations.
- **Outliers** are observations that lie outside the overall pattern of a distribution. Always look for outliers and try to explain them.
- When observations on a variable are taken over time, make a **time plot** that graphs time horizontally and the values of the variable vertically. A time plot can reveal trends, cycles, or other changes over time.

CHECK YOUR SKILLS

The multiple-choice exercises in Check Your Skills ask straightforward questions about basic facts from the chapter. Answers to odd-numbered exercises appear in the back of the book. You should expect almost all your answers to be correct.

1.13 Here are the first lines of a professor's data set at the end of a statistics course:

Name	Major	Total Points	Grade
ADVANI, SURA	COMM	397	B
BARTON, DAVID	HIST	323	C
BROWN, ANNETTE	BIOL	446	A
CHIU, SUN	PSYC	405	B
CORTEZ, MARIA	PSYC	461	A

The individuals in these data are

(a) the students.

(b) the total points.

(c) the course grades.

1.14 According to the National Household Survey on Drug Use and Health, when asked in 2017, 31.0% of those aged 18 to 25 years said they had used cigarettes in the past year, 7.7% said they had used smokeless tobacco, 39.4% said they had used illicit drugs, and 7.8% said they had used pain relievers or sedatives.[15] To display this data, it would be correct to use

(a) either a pie chart or a bar graph.

(b) a pie chart, provided that a category for other is added to get to 100%.

(c) a bar graph but not a pie chart.

1.15 A description of different houses on the market includes the variables square footage of the house and the average monthly gas bill.

(a) Square footage and average monthly gas bill are both categorical variables.

(b) Square footage and average monthly gas bill are both quantitative variables.

(c) Square footage is a categorical variable, and average monthly gas bill is a quantitative variable.

1.16 A political party's data bank includes the zip codes of past donors, such as

47906 34236 53075 10010 90210 75204 30304 99709

Zip code is a

(a) quantitative variable.

(b) categorical variable.

(c) unit of measurement.

1.17 Figure 1.6 (page 25) is a histogram of the percentage of on-time high school graduates in each state. The rightmost bar in the histogram covers percents of on-time high school graduates ranging from about

(a) 85% to 90%.

(b) 88% to 92%.

(c) 90% to 92%.

1.18 Here are the exam scores of 10 students in a statistics class:

50 35 41 97 76 69 94 91 23 65

To make a stemplot of these data, you would use stems

(a) 2, 3, 4, 5, 6, 7, 9.

(b) 2, 3, 4, 5, 6, 7, 8, 9.

(c) 20, 30, 40, 50, 60, 70, 80, 90.

1.19 Where do students go to school? Although 78% of first-time first-year students attended college in the state in which they lived, this percentage varied considerably over the states. Here is a stemplot of the percentage of first-year students in each of the 50 states who were from the state where they enrolled. The stems are tens, and the leaves are ones. The stems have been split in the plot.[16] INSTATE

3	3
3	7
4	03
4	
5	022
5	5679
6	0134
6	789
7	0011333444
7	56667999
8	024
8	5677778
9	0224

The midpoint of this distribution is

(a) 60%. (b) 73.5%. (c) 80%.

1.20 The shape of the distribution in Exercise 1.19 is

(a) skewed to the left. (b) skewed upward.

(c) skewed to the right.

1.21 The state with the smallest percentage of first-year students enrolled in the state has

(a) 0.33% enrolled. (b) 3.3% enrolled.

(c) 33% enrolled.

1.22 You look at real estate ads for houses in Naples, Florida. There are many houses ranging from $200,000 to $500,000 in price. The few houses on the water, however, have prices up to $15 million. The distribution of house prices will be

(a) skewed to the left.

(b) roughly symmetric.

(c) skewed to the right.

CHAPTER 1 EXERCISES

1.23 **Medical Students.** Students who have finished medical school are assigned to residencies in hospitals to receive further training in a medical specialty. Here is part of a hypothetical database of students seeking residency positions. USMLE is the student's score on Step 1 of the national medical licensing examination.

Name	Medical School	Sex	Age	USMLE	Specialty Sought
Abrams, Laurie	Florida	F	28	238	Family medicine
Brown, Gordon	Meharry	M	25	205	Radiology
Cabrera, Maria	Tufts	F	26	191	Pediatrics
Ismael, Miranda	Indiana	F	32	245	Internal medicine

(a) What individuals does this data set describe?

(b) In addition to the student's name, how many variables does the data set contain? Which of these variables are categorical, and which are quantitative? If a variable is quantitative, what units is it measured in?

1.24 **Buying a Refrigerator.** *Consumer Reports* will have an article comparing refrigerators in the next issue. Some of the characteristics to be included in the report are the brand name and model; whether it has a top freezer, bottom freezer, or side-by-side layout; the estimated energy consumption per year (kilowatts); whether or not it is Energy Star compliant; the width, depth, and height in inches; and both the freezer and refrigerator net capacity, in cubic feet. Which of these variables are categorical, and which are quantitative? Give the units for the quantitative variables and the categories for the categorical variables. What are the individuals in the report?

1.25 **What Color Is Your Car?** The most popular colors for cars and light trucks vary with region and type of vehicle and over time. In North America, silver and white are the most popular choices for midsize cars, silver and black for convertibles and coupes, and white for light trucks. Despite this variation, overall white remains the top choice worldwide for the eighth consecutive year, increasing its lead by 2% over the previous year. Here is the distribution of the top colors for vehicles sold globally in 2018:[17] CARCOLOR

Color	Popularity
White	39%
Black	17%
Gray	12%
Silver	10%
Natural	7%
Red	7%
Blue	7%
Green	

Fill in the percentage of vehicles that are in green, Make a graph to display the distribution of color popularity.

1.26 **High School Tobacco Use.** Despite the intense anti-smoking campaigns sponsored by both federal and private agencies, smoking continues to be the single-biggest cause of preventable death in the United States. How has the tobacco use of high school students changed over the past few years? For each of several tobacco products, high school students were asked whether they had used each of them in the past 30 days. Here are some of the results:[18]

	Year							
Product	2011	2012	2013	2014	2015	2016	2017	2018
Any tobacco product	24.3	23.3	22.9	24.6	25.3	20.2	19.6	27.1
Cigarettes	15.8	14.0	12.7	9.2	9.3	8.0	7.6	8.1
Cigars	12.6	11.6	11.9	8.2	8.6	7.7	7.7	7.6
Pipes	4.5	4.0	4.1	1.5	1.0	1.4	0.8	1.1
Smokeless tobacco	7.3	6.4	5.7	5.5	6.0	5.8	5.5	5.9
E-cigarettes	1.5	2.8	4.5	13.4	16.0	11.3	11.7	20.8

The first row of the table gives the percentages of high school students who had used any tobacco product, including cigarettes, pipes, cigars, smokeless tobacco, e-cigarettes, hookahs, snus, bidis, or dissolvable tobacco, in the past 30 days for the years 2011–2018. The remaining rows give the percentage of high school students using the most common tobacco products in each of these years.

(a) Using the information in the first row of the table, draw a bar chart that shows the change in the use of any tobacco product between 2011 and 2018. How would you describe the pattern of change in this usage?

(b) Draw a bar chart that illustrates the change in usage in these years for the individual tobacco products. If your software allows it, give a single bar chart that contains the information for all products. Otherwise, provide a separate bar chart for each product.

(c) Using the bar charts in parts (a) and (b), give a simple description of the changes in the use of tobacco products by high school students between 2011 and 2018.

1.27 Deaths among Young People. Among persons aged 15–24 years in the United States, there were 32,025 deaths in 2017. The leading (distinct) causes of death and number of deaths were accidents, 13,441; suicide, 6252; homicide, 4905; cancer, 1374; heart disease, 1126; symptoms, signs, and abnormal clinical and laboratory findings, 501; congenital defects, 362.[19]

(a) Make a bar graph to display these data.

(b) Can you make a pie chart using the information given? Explain carefully why or why not.

1.28 Student Debt. At the end of 2016, the average outstanding student debt for bachelor's degree recipients at public and private nonprofit four-year institutions was $28,500. Figure 1.16 is a pie chart showing the distribution of outstanding education debt.[20] About what percentage of students had an outstanding debt between $20,000 and $49,999? $50,000 or more? You see that it is hard to determine numbers from a pie chart. Bar graphs are much easier to use. (Many agencies include the percents in their pie charts to aid in interpretation.)

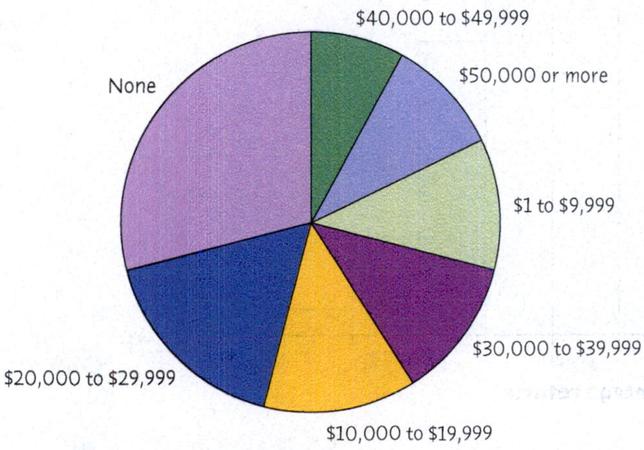

FIGURE 1.16
Pie chart of the distribution of outstanding education debt, for Exercise 1.28.

1.29 Mobile Apps We Cannot Go Without. Smartphone users have many apps on their phones. Which ones do they say they must have? Here is the percentage of smartphone users who say they cannot go without Amazon and without Google Search by age group:[21]

MOBILAPP

Age Group	Amazon	Google Search
18 to 34 years	35	11
35 to 54 years	30	21
55 years and over	24	39

(a) If your software allows it, draw a bar graph with adjacent bars for Amazon and Google Search percents for each of the three age categories, allowing easy comparison of these percents within each age category. If your software does not allow this, draw two bar charts, one for Amazon and the second for Google Search.

(b) Describe the main differences in the Amazon and Google Search percents by age group.

(c) Explain carefully why it is not correct to make a pie chart for the Amazon or the Google Search percents.

1.30 Do Adolescent Girls Eat Fruit? We all know that fruit is good for us. Many of us don't eat enough. Figure 1.17 is a histogram of the number of servings of fruit per day claimed by 74 17-year-old girls in a study in Pennsylvania.[22] Describe the shape, center, and variability of this distribution. Are there any outliers? What percentage of these girls ate six or more servings per day? How many of these girls ate fewer than two servings per day?

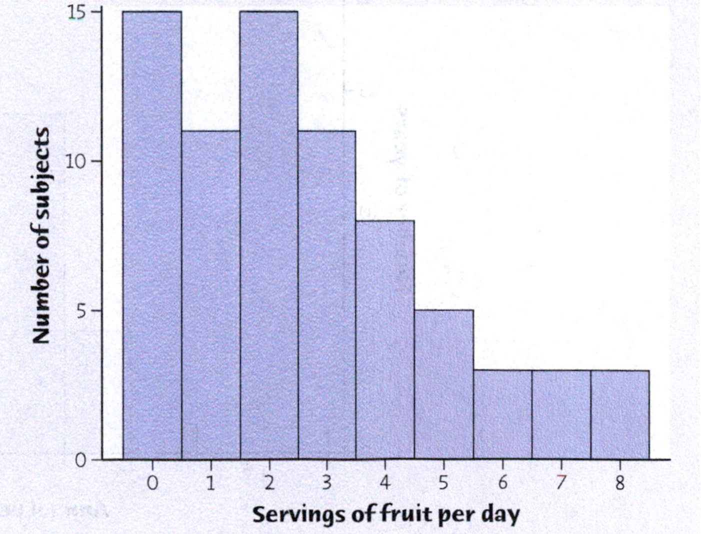

FIGURE 1.17
The distribution of fruit consumption in a sample of 74 17-year-old girls, for Exercise 1.30.

1.31 **IQ Test Scores.** Figure 1.18 is a stemplot of the IQ test scores of 78 seventh-grade students in a rural midwestern school.[23] IQ

(a) Four students had low scores that might be considered outliers. Ignoring these, describe the shape, center, and variability of the remainder of the distribution.

(b) We often read that IQ scores for large populations are centered at 100. What percentage of these 78 students have scores above 100?

7	24
7	79
8	
8	69
9	0133
9	6778
10	0022333344
10	555666777789
11	0000111122223334444
11	55688999
12	003344
12	677888
13	02
13	6

FIGURE 1.18

The distribution of IQ scores for 78 seventh-grade students, for Exercise 1.31.

1.32 **Returns on the S&P.** The return on a stock is the change in its market price plus any dividend payments made over some period. Total return is usually expressed as a percentage of the beginning price. Figure 1.19 is a histogram of the distribution of the annual combined returns for all stocks listed on the S&P 500 from 1928 to 2018 (91 years).[24] SPRET

(a) Describe the overall shape of the distribution of monthly returns.

(b) What is the approximate center of this distribution? (For now, take the center to be the value with roughly half the years having lower returns and half having higher returns.)

(c) Approximately what were the smallest and largest annual returns? (This is one way to describe the variability of the distribution.)

(d) A return less than zero means that stocks lost value in that year. About what percentage of all years had returns less than zero?

1.33 **Name That Variable.** A survey of a large college class asked the following questions:

1. Are you female or male? (In the data, male = 0, female = 1.)
2. Are you right-handed or left-handed? (In the data, right = 0, left = 1.)
3. What is your height, in inches?
4. How many minutes do you study on a typical weeknight?

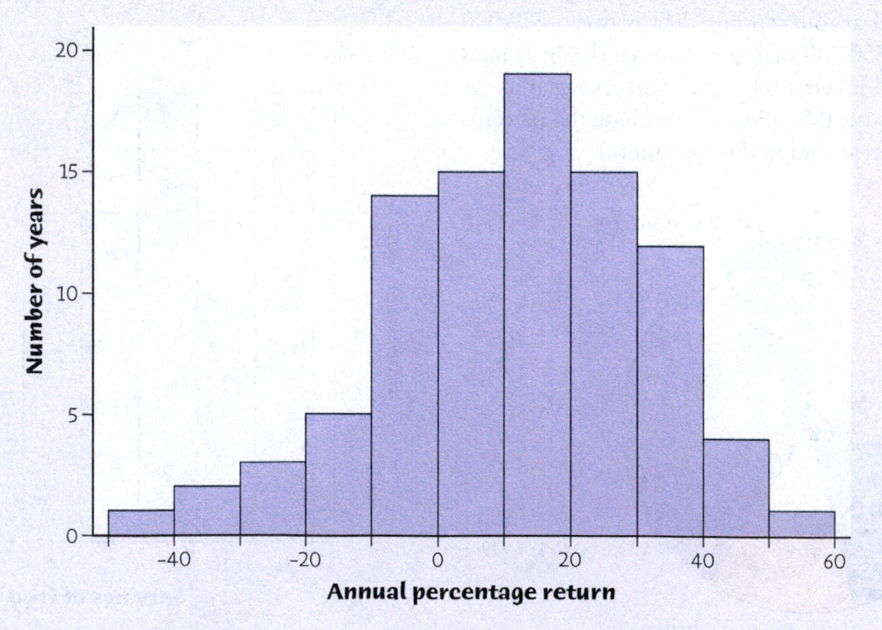

FIGURE 1.19

The distribution of annual percent returns on the S&P 500 from 1928 to 2018, for Exercise 1.32.

Figure 1.20 shows histograms of the student responses, in scrambled order and without scale markings. Which graph goes with each variable? Explain your reasoning.

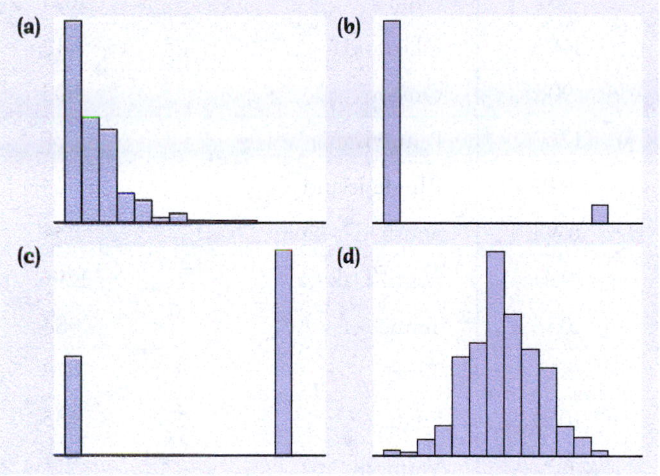

FIGURE 1.20
Histograms of four distributions, for Exercise 1.33.

1.34 **Food Oils and Health.** Fatty acids, despite their unpleasant name, are necessary for human health. Two types of essential fatty acids, called omega-3 and omega-6, are not produced by our bodies and so must be obtained from our food. Food oils, widely used in food processing and cooking, are major sources of these compounds. There is some evidence that a healthy diet should have more omega-3 than omega-6. Table 1.4 gives the ratio of omega-3 to omega-6 in some common food oils.[25] Values greater than 1 show that an oil has more omega-3 than omega-6. FOODOILS

(a) Make a histogram of these data, using classes bounded by the whole numbers from 0 to 6.

(b) What is the shape of the distribution? How many of the 30 food oils have more omega-3 than omega-6? What does this distribution suggest about the possible health effects of modern food oils?

(c) Table 1.4 contains entries for several fish oils (cod, herring, menhaden, salmon, sardine). How do these values help answer the question of whether eating fish is healthy?

1.35 **Where Are the Nurses?** Table 1.5 gives the number of active nurses per 100,000 people in each state.[26] NURSES

(a) Why is the number of nurses per 100,000 people a better measure of the availability of nurses than a simple count of the number of nurses in a state?

(b) Make a stemplot that displays the distribution of nurses per 100,000 people. The data will first need to be rounded (see page 29). What units are you going to use for the stems? The leaves? You should round the data to the units you are planning to use for the leaves before drawing the stemplot. Write a brief description of the distribution. Are there any outliers? If so, can you explain them?

(c) Do you think it would be useful to split the stems when drawing the stemplot for these data? Explain your reason.

TABLE 1.4 Omega-3 fatty acids as a fraction of omega-6 fatty acids in food oils

Oil	Ratio	Oil	Ratio
Perilla	5.33	Flaxseed	3.56
Walnut	0.20	Canola	0.46
Wheat germ	0.13	Soybean	0.13
Mustard	0.38	Grape seed	0.00
Sardine	2.16	Menhaden	1.96
Salmon	2.50	Herring	2.67
Mayonnaise	0.06	Soybean, hydrogenated	0.07
Cod liver	2.00	Rice bran	0.05
Shortening (household)	0.11	Butter	0.64
Shortening (industrial)	0.06	Sunflower	0.03
Margarine	0.05	Corn	0.01
Olive	0.08	Sesame	0.01
Shea nut	0.06	Cottonseed	0.00
Sunflower (oleic)	0.05	Palm	0.02
Sunflower (linoleic)	0.00	Cocoa butter	0.04

TABLE 1.5 Nurses per 100,000 people, by state

State	Nurses	State	Nurses	State	Nurses
Alabama	911	Louisiana	881	Ohio	1021
Alaska	717	Maine	1093	Oklahoma	742
Arizona	585	Maryland	906	Oregon	803
Arkansas	798	Massachusetts	1260	Pennsylvania	1030
California	630	Michigan	849	Rhode Island	1104
Colorado	831	Minnesota	1093	South Carolina	834
Connecticut	1017	Mississippi	950	South Dakota	1296
Delaware	1155	Missouri	1038	Tennessee	984
Florida	814	Montana	855	Texas	678
Georgia	665	Nebraska	1054	Utah	635
Hawaii	689	Nevada	609	Vermont	914
Idaho	682	New Hampshire	1006	Virginia	764
Illinois	901	New Jersey	858	Washington	814
Indiana	901	New Mexico	614	West Virginia	953
Iowa	1022	New York	848	Wisconsin	946
Kansas	934	North Carolina	940	Wyoming	864
Kentucky	1003	North Dakota	968	District of Columbia	1483

1.36 Child Mortality Rates. Although child mortality rates have dropped by more than 50% since 1990, it was still the case that 5.4 million children under five years old died in 2017. The mortality rates for children under five varied from 2.1 per 1000 in Slovenia to 127.2 per 1000 in Somalia. The data set is too large to print here, but here are the data for the first five countries:[27] MORTALTY

Country	Child Mortality Rate (per 1000)
Aruba	–
Afghanistan	67.9
Angola	81.1
Albania	8.8
Andorra	3.3

(a) Why do you think that mortality rates are measured as the number of deaths per 1000 children under age five rather than simply as the number of deaths?

(b) Make a histogram that displays the distribution of child mortality rates. Describe the shape, center, and variability of the distribution. Do any countries appear to be obvious outliers in the histogram?

1.37 Fur Seals on St. Paul Island. Every year, hundreds of thousands of northern fur seals return to their haul-outs in the Pribilof Islands in Alaska to breed, give birth, and teach their pups to swim, hunt, and survive in the Bering Sea. U.S. commercial fur sealing operations continued on St. Paul until 1984, but despite a reduction in harvest, the population of fur seals has continued to decline. Possible reasons include climate shifts in the North Pacific, changes in the availability of prey, and new or increased interaction with commercial fisheries that increase mortality. Here are data on the estimated number of fur seal pups born on St. Paul Island (in thousands) from 1979 to 2018, where a dash indicates a year in which no data were collected:[28] FURSEALS

Arco Images GmbH/Alamy

Year	Pups Born (thousands)	Year	Pups Born (thousands)	Year	Pups Born (thousands)	Year	Pups Born (thousands)
1979	245.93	1989	171.53	1999	–	2009	–
1980	203.82	1990	201.30	2000	158.74	2010	94.50
1981	179.44	1991	–	2001	–	2011	–
1982	203.58	1992	182.44	2002	145.72	2012	96.83
1983	165.94	1993	–	2003	–	2013	–
1984	173.27	1994	192.10	2004	122.82	2014	91.74
1985	182.26	1995	–	2005	–	2015	–
1986	167.66	1996	170.12	2006	109.96	2016	80.60
1987	171.61	1997	–	2007	–	2017	–
1988	202.23	1998	179.15	2008	102.67	2018	75.70

Make a stemplot to display the distribution of pups born per year. Describe the shape, center, and variability of the distribution. Are there any outliers?

1.38 Nintendo and Laparoscopic Skills. In laparoscopic surgery, a video camera and several thin instruments are inserted into the patient's abdominal cavity. The surgeon uses the image from the video camera positioned inside the patient's body to perform the procedure by manipulating the instruments that have been inserted. It has been found that the Nintendo Wii™ reproduces the movements required in laparoscopic surgery more closely than other video games with its motion-sensing interface. If training with a Nintendo Wii™ can improve laparoscopic skills, it can complement the more expensive training on a laparoscopic simulator. Forty-two medical residents were chosen, and all were tested on a set of basic laparoscopic skills. Twenty-one were selected at random to undergo systematic Nintendo Wii™ training for one hour a day, five days a week, for four weeks. The remaining 21 residents were given no Nintendo Wii™ training and asked to refrain from video games during this period. At the end of four weeks, all 42 residents were tested again on the same set of laparoscopic skills. One of the skills involved a virtual gall bladder removal, with several performance measures including time to complete the task recorded. Here are the improvement (before–after) times in seconds after four weeks for the two groups:[29] NINTENDO

Received Wii Training						No Wii Training					
281	134	186	128	84	243	21	66	54	85	229	92
212	121	134	221	59	244	43	27	77	−29	−14	88
79	333	−13	−16	71	−16	145	110	32	90	46	−81
71	77	144				68	61	44			

(a) In the context of this study, what do the negative values in the data set mean?

(b) *Back-to-back stemplots* can be used to compare the two samples. That is, use one set of stems with two sets of leaves, one to the right and one to the left of the stems. (Draw a line on either side of the stems to separate stems and leaves.) Order both sets of leaves from smallest at the stem to largest away from the stem. Complete the back-to-back stemplot started below. The data have been rounded to the nearest 10, with stems being 100s and leaves being 10s. The stems have been split. The first control observation corresponds to −80 and the next two to −30 and −10.

Received Wii Training		No Wii Training
	−0	8
1 2 2	−0	3 1
	0	
	0	
	1	
	1	
	2	
	2	
	3	

(c) Report the approximate midpoints of both groups. Does it appear that the treatment has resulted in a greater improvement in times than seen in the control group? (To better understand the magnitude of the improvements, note that the median time to complete the task on the first occasion was 11 minutes and 40 seconds, using the times of all 42 residents.)

1.39 **Fur Seals on St. Paul Island.** Make a time plot of the number of fur seals born per year from Exercise 1.37. What does the time plot show that your stemplot in Exercise 1.37 did not show? When you have data collected over time, a time plot is often needed to understand what is happening. FURSEALS

1.40 **Marijuana and Traffic Accidents.** Researchers in New Zealand interviewed 907 drivers at age 21. They had data on traffic accidents, and they asked the drivers about marijuana use. Here are data on the numbers of accidents caused by these drivers at age 19, broken down by marijuana use at the same age:[30]

	Marijuana Use per Year			
	Never	1–10 Times	11–50 Times	51+ Times
Accidents caused	59	36	15	50
Drivers	452	229	70	156

(a) Explain carefully why a useful graph must compare *rates* (accidents per driver) rather than *counts* of accidents in the four marijuana use classes.

(b) Compute the accident rates in the four marijuana use classes. After you have done this, make a graph that displays the accident rate for each class. What do you conclude? (You should not conclude that marijuana use *causes* accidents because risk takers are more likely both to drive aggressively and to use marijuana.)

1.41 **Mobile shopping in the United States, Canada, and the United Kingdom.** The use of mobile devices to perform financial transactions is steadily increasing as mobile shoppers become more and more comfortable with mobile finance. Here is the breakdown of the percentage of mobile shoppers performing mobile financial actions in a month in the United States, Canada, and the United Kingdom:[31] MOBILESHOP

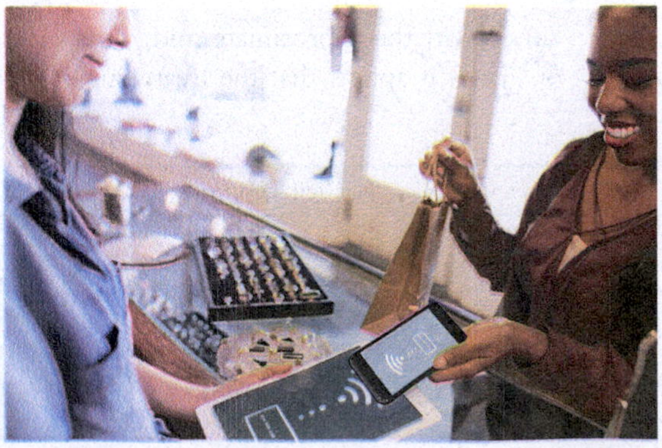

Hero/Media Bakery

Financial Action	Canada	U.S.	U.K.
Stock trading	54%	23%	11%
Bank account	45%	66%	68%
Credit card account	53%	50%	34%
e-payment	57%	52%	56%

(a) Draw a bar graph for the percentage of mobile shoppers performing mobile financial actions in a month for Canada. Do the same for the United States and United Kingdom, using the same scale for the percentage axis.

(b) Describe the most important difference in the percentage of mobile shoppers performing mobile financial actions in a month for the three countries. How does this difference show up in the bar graphs?

(c) Explain why it *is not* appropriate to use a pie chart to display any of these distributions.

1.42 **She Sounds Tall!** Presented with recordings of a pair of people of the same sex speaking the same phrase, can a listener determine which speaker is taller simply from the sound of their voice? Twenty-four young adults at Washington University listened to 100 pairs of speakers and, within each pair, were asked to indicate which of the two speakers was the taller. Here are the number correct (out of 100) for each of the 24 participants:[32] HEARING

65	61	67	59	58	62	56	67	61	67	63	53
68	49	66	58	69	70	65	56	68	56	58	70

Researchers believe that the key to correct discrimination is contained in a particular type of sound produced in the lungs, whose frequency is lower for taller people.

(a) Make two stemplots, with and without splitting the stems. Which plot do you prefer, and why?

(b) Describe the shape, center, and variability of the distribution. Are there any outliers?

(c) If the experimental subjects are just guessing which speaker is taller, they should correctly identify the taller person about 50% of the time. Does this data support the researchers' conjecture that there is information in a person's voice to help identify the taller person? Why or why not?

1.43 **Watch Those Scales!** Figures 1.21(a) and 1.21(b) both show time plots of tuition charged to in-state students from 1980 through 2018.[33]

(a) Which graph appears to show the biggest increase in tuition between 2000 and 2018?

(b) Read the graphs and compute the actual increase in tuition between 2000 and 2018 in each graph. Do you think these graphs are for the same or different data sets? Why?

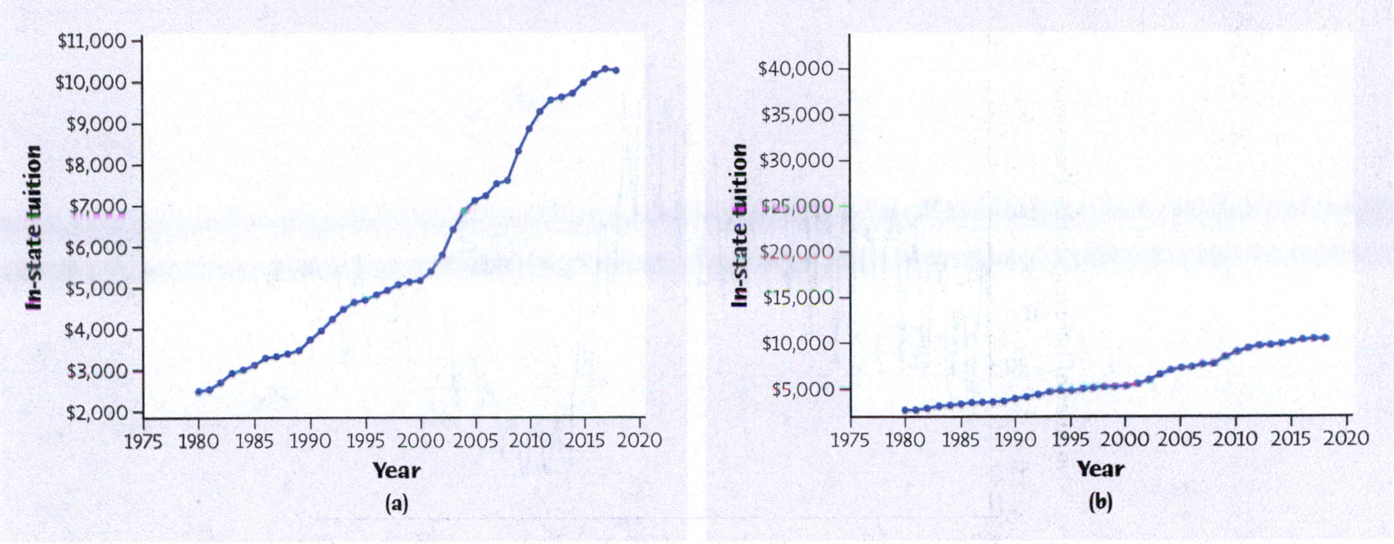

FIGURE 1.21

Time plots of in-state tuition between 1980 and 2018, for Exercise 1.43.

The impression that a time plot gives depends on the scales you use on the two axes. Changing the scales can make tuition appear to increase very rapidly or to have only a gentle increase. Also, truncating the vertical scale can lead to very different (and incorrect) conclusions. The moral of this exercise is: always pay close attention to the scales when you look at a time plot.

1.44 **The Value of a Four-Year Degree!** The big economic news of 2007 was a severe downturn in housing that began in mid-2006. This was followed by the financial crisis in 2008. And in recent years, the economy has been growing. How did these economic events affect the unemployment rate, and were all segments of the population affected similarly? The data are the monthly unemployment rates for those over 25 years of age with less education than a high school degree and those over 25 years of age with a bachelor's degree, from January 1992 through August 2019. The data set is too large to print here, but here are the data for the unemployment rates for both groups for the first five months:[34] UNEMPLOY

Month	Four-Year College Degree	Less Than High School Degree
January 1992	3.1	10.8
February 1992	3.2	11.0
March 1992	2.9	11.2
April 1992	3.2	10.8
May 1992	3.2	12.2

(a) Make a time plot of the monthly unemployment rates for those over 25 years of age without a high school diploma and no college and those over 25 years of age with a four-year college degree. If your software allows it, make both time plots on the same set of axes. Otherwise, make separate time plots for each group but use the same scale for both plots for ease of comparison. Are the patterns in the two time plots similar? What is the primary difference between the two time plots?

(b) How are economic events described reflected in the time plots of the unemployment rates? Since the end of 2009, how would you describe the behavior of the unemployment rates for both groups?

(c) Are there any other periods during which there were patterns in the unemployment rate? Describe them.

1.45 **Housing Starts.** Figure 1.22 is a time plot of the number of single-family homes started by builders each month from January 1990 through July 2019.[35] The counts are in thousands of homes. HOUSING

(a) The most notable pattern in this time plot is yearly up-and-down cycles. At what season of the year are housing starts highest? Lowest? The cycles are explained by the weather in the northern part of the country.

(b) Is there a longer-term trend visible in addition to the cycles? If so, describe it.

(c) The big economic news of 2007 was a severe downturn in housing that began in mid-2006. This was followed by the financial crisis in 2008. How are these economic events reflected in the time plot?

(d) How would you describe the behavior of the time plot since January 2011?

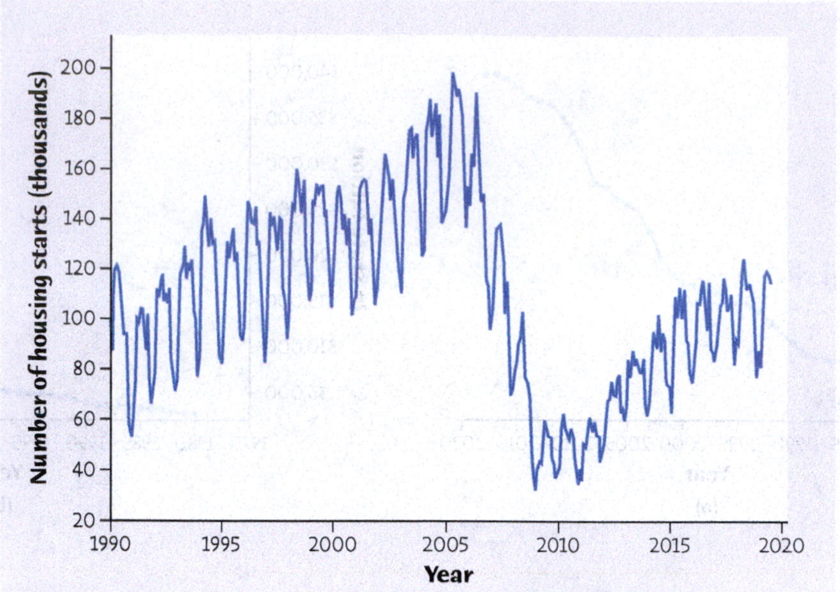

FIGURE 1.22
Time plot of the monthly count of new single-family houses started (in thousands) between January 1990 and July 2019, for Exercise 1.45.

1.46 Choosing Class Intervals. applet Student engineers learn that, although handbooks give the strength of a material as a single number, in fact the strength varies from piece to piece. A vital lesson in all fields of study is that "variation is everywhere." Here are data from a typical student laboratory exercise: the load in pounds needed to pull apart pieces of Douglas fir 4 inches long and 1.5 inches square:

33,190	31,860	32,590	26,520	33,280
32,320	33,020	32,030	30,460	32,700
23,040	30,930	32,720	33,650	32,340
24,050	30,170	31,300	28,730	31,920

The data sets in the *Histograms* applet in Achieve include the "pulling wood apart" data given in this exercise. How many class intervals does the applet choose when drawing the histogram? Use the applet to make several histograms with a larger number of class intervals. Are there any important features of the data that are revealed using a larger number of class intervals? Which histogram do you prefer? Explain your choice.

Brasil2/Getty Images

CHAPTER

2

Describing Distributions with Numbers

When you complete this chapter, you will be able to:

2.1 Find the mean of a set of observations and interpret it as the center of the set.

2.2 Find the median of a set of observations and interpret it as the center of the set.

2.3 Compare the mean and median values of a data set and distinguish their meanings.

2.4 Calculate and use quartiles to describe the variation of a data set.

2.5 Use the five-number summary (the minimum, the maximum, the quartiles, and the median) and boxplot to characterize a distribution.

2.6 Use the $1.5 \times IQR$ rule to identify outliers in a data set.

2.7 Calculate and use the standard deviation to describe the variation of a data set.

2.8 Discriminate between the five-number summary and the mean and standard deviation for describing the distribution of data, depending on features of the data set.

2.9 Interpret presentations of descriptive statistics output by graphing calculators and computer programs.

2.10 Apply four-step state, plan, solve, and conclude process to examine data based on context.

In Chapter 1 we began to learn about data analysis using graphs to provide a visual tool for organizing and identifying patterns in data. Pie charts and bar graphs can summarize the information in a categorical variable by giving us the percentage of the distribution in the various categories. Histograms and stemplots are graphical tools for summarizing the information provided by a quantitative variable. The overall pattern in a histogram or stemplot illustrates some of the important features of the distribution of a variable. The center of the histogram tells us about the value of a "typical" observation on this variable, whereas the variability gives us a sense of how close most of the observations are to this value. Other interesting features are the presence of outliers and the general shape of the plot. For data collected over time, time plots can show patterns such as seasonal variation and trends in the variable. In the next chapter, we will see how the information about the distribution of a variable can also be described using numerical summaries.

In this chapter, we continue our study of exploratory data analysis. A graph is an important visual tool for organizing and identifying patterns in data. It gives a fairly complete description of a distribution, although for many problems, the important information in the data can be described by using a few numbers. A graph is a numerical summary that can be useful for describing a single distribution as well as for comparing the distributions from several groups of observations.

We saw in Chapter 1 (page 13) that the American Community Survey asks, among much else, about workers' travel times to work. Here are the travel times in minutes for 15 workers in North Carolina, chosen at random by the U.S. Census Bureau:[1]

20 35 8 70 5 15 25 30 40 35 10 12 40 15 20

We aren't surprised that most people estimate their travel time in multiples of five minutes. Here is a stemplot of these data:

0	58
1	0255
2	005
3	055
4	00
5	
6	
7	0

The distribution is single peaked and right-skewed. The longest travel time (70 minutes) may be an outlier. Our goal in this chapter is to describe with numbers the center and variability of this and other distributions.

2.1 Measuring Center: The Mean

The most common measure of center is the ordinary arithmetic average, or *mean*.

The Mean $\bar{x}$

To find the **mean** of a set of observations, add their values and divide by the number of observations. If the n observations are $x_1, x_2, \ldots, x_n$, their mean is

$$\bar{x} = \frac{x_1 + x_2 + \cdots + x_n}{n}$$

or, in more compact notation,

$$\bar{x} = \frac{1}{n}\sum x_i$$

The Σ (capital Greek sigma) in the formula for the mean is short for "add them all up." The subscripts on the observations x_i are just a way of keeping the n observations distinct. They do not necessarily indicate order or any other special facts about the data. The bar over the x indicates the mean of all the x-values. Pronounce the mean $\bar{x}$ as "x-bar." This notation is very common. When writers who are discussing data use $\bar{x}$ or $\bar{y}$, they are talking about a mean.

EXAMPLE 2.1 Travel Times to Work

NCTRAVEL

The mean travel time of our 15 North Carolina workers is

$$\bar{x} = \frac{x_1 + x_2 + \cdots + x_n}{n} = \frac{20 + 35 + \cdots + 20}{15} = \frac{380}{15} = 25.3 \text{ minutes}$$

In practice, you can enter the data into your calculator and ask for the mean. You don't have to actually add and divide. But you should know that this is what the calculator is doing.

Notice that only 6 of the 15 travel times are larger than the mean. If we leave out the longest single travel time, 70 minutes, the mean for the remaining 14 people is 22.1 minutes. That one 70-minute observation raises the mean by 3.2 minutes.

Example 2.1 illustrates an important fact about the mean as a measure of center: it is sensitive to the influence of a few extreme observations. These may be outliers, but a skewed distribution that has no outliers will also pull the mean toward its long tail. Because the mean cannot resist the influence of extreme observations, we say that it is not a *resistant measure* of center.

STATISTICS IN YOUR WORLD

Don't Hide the Outliers

Data from an airliner's control surfaces, such as the vertical tail rudder, go to cockpit instruments and then to the "black box" flight data recorder. To avoid confusing the pilots, short erratic movements in the data are "smoothed" so that the instruments show overall patterns. When a crash killed 260 people, investigators suspected a catastrophic movement of the tail rudder. But the black box contained only the smoothed data. Sometimes, outliers are more important than the overall pattern.

Resistant Measure

A **resistant measure** is a statistical measure that is relatively unaffected by large changes in numerical values of a small proportion of the observations in the distribution that the measure describes.

APPLY YOUR KNOWLEDGE

2.1 *E. Coli* in Swimming Areas. To investigate water quality, the *Columbus Dispatch* took water specimens at 16 Ohio State Park swimming areas in central Ohio. Those specimens were taken to laboratories and tested for *E. coli*, which are bacteria that can cause serious gastrointestinal problems. For reference, if a 100-milliliter specimen (about 3.3 ounces) of water contains more than 130 *E. coli* bacteria, it is considered unsafe. Here are the *E. coli* levels per 100 milliliters found by the laboratories:[2] ECOLI

291.0	10.9	47.0	86.0	44.0	18.9	1.0	50.0
190.4	45.7	28.5	18.9	16.0	34.0	8.6	9.6

Find the mean *E. coli* level. How many of the lakes have *E. coli* levels greater than the mean? What feature of the data explains the fact that the mean is greater than most of the observations?

2.2 Health Care Spending. Table 1.3 (page 31) gives the 2015 health care expenditure per capita in 35 countries with the highest gross domestic products in 2015. The United States, at 9536 international dollars per person, is a high outlier. Find the mean health care spending in these nations with and without the United States. How much does the one outlier increase the mean? HEALTH

2.2 Measuring Center: The Median

In Chapter 1, we used the midpoint of a distribution as an informal measure of center and gave a method for its computation. The *median* is the formal version of the midpoint, and we now provide a more detailed rule for its calculation.

The Median *M*

The **median *M*** is the midpoint of a distribution, the number such that half the observations are smaller and the other half are larger. To find the median of a distribution:

1. Arrange all observations in order of size, from smallest to largest.
2. If the number of observations n is odd, the median M is the center observation in the ordered list. If the number of observations n is even, the median M is midway between the two center observations in the ordered list.
3. You can always locate the median in the ordered list of observations by counting up $(n+1)/2$ observations from the start of the list.

Note that the formula $(n+1)/2$ does not give the median; it just gives the location of the median in the ordered list. Medians require little arithmetic, so they are easy to find by hand for small sets of data. Arranging even a moderate number of observations in order is very tedious, however, so finding the median by hand for larger sets of data is unpleasant. Even simple calculators have an $\bar{x}$ button, but you will need to use software or a graphing calculator to automate finding the median.

EXAMPLE 2.2 Finding the Median: Odd *n*

NCTRAVEL

What is the median travel time for our 15 North Carolina workers? Here are the data arranged in order:

5 8 10 12 15 15 20 **20** 25 30 35 35 40 40 70

The count of observations, $n = 15$, is odd. The bold **20** is the center observation in the ordered list, with seven observations to its left and seven to its right. This is the median, $M = 20$ minutes.

Because $n = 15$, our rule for the location of the median gives

$$\text{location of } M = \frac{n+1}{2} = \frac{16}{2} = 8$$

That is, the median is the eighth observation in the ordered list. It is more reliable and faster to use this rule than to locate the center by eye.

EXAMPLE 2.3 Finding the Median: Even *n*

NYTRAVEL

Travel times to work in New York State are (on the average) longer than in North Carolina. Here are the travel times in minutes of 20 randomly chosen New York workers:

10 15 55 20 65 50 12 20 10 10 35 50 30 45 15 10 75 40 35 60

A stemplot not only displays the distribution but also makes finding the median easy because it arranges the observations in order:

0	
1	0000255
2	00
3	055
4	05
5	005
6	05
7	5

The distribution is single peaked and right-skewed, with several travel times of an hour or more. There is no center observation, but there is a center pair. These are the bold **30** and **35** in the stemplot, which have nine observations before them in the ordered list and nine after them. The median is midway between these two observations:

$$M = \frac{30 + 35}{2} = 32.5 \text{ minutes}$$

With $n = 20$, the rule for locating the median in the list gives

$$\text{location of } M = \frac{n+1}{2} = \frac{21}{2} = 10.5$$

The location 10.5 means "halfway between the 10th and 11th observations in the ordered list." That agrees with what we found by eye.

2.3 Comparing the Mean and the Median

Examples 2.1 and 2.2 illustrate an important difference between the mean and the median. The median travel time (the midpoint of the distribution) is 20 minutes. The mean travel time is higher, 25.3 minutes. The mean is pulled toward the right tail of this right-skewed distribution. The median, unlike the mean, is *resistant.* If the longest travel time were 700 minutes rather than 70 minutes, the mean would increase to 67.3 minutes, but the median would not change at all. The outlier just counts as one observation above the center, no matter how far above the center it lies. The mean uses the actual value of each observation and so will chase a single large observation upward. Using the *Mean and Median* applet is an excellent way to compare the resistance of M and $\bar{x}$.

applet

Comparing the Mean and the Median

The mean and median of a roughly symmetric distribution are close together. If the distribution is exactly symmetric, the mean and median are exactly the same. In a skewed distribution, the mean is usually farther out in the long tail than the median.[3]

Many economic variables have distributions that are skewed to the right. For example, the median endowment of colleges and universities in the United States and Canada in 2018 was about $142 million—but the mean endowment was over $770 million. Most institutions have modest endowments, but a few are very wealthy. Harvard's endowment was more than $38 billion.[4] The few wealthy institutions pull the mean up but do not affect the median. Reports about incomes and other strongly skewed distributions usually give the median ("midpoint") rather than the mean ("arithmetic average"). However, a county that is about to impose a tax of 1% on the incomes of its residents cares about the mean income, not the median. The tax revenue will be 1% of total income, and because the total income is the mean income times the number of residents, the tax revenue can be computed easily from the mean. The mean and median measure center in different ways, and both are useful. *Don't confuse the "average" value of a variable (the mean) with its "middle" value, which we might describe by the median, or with its "typical" value, which we might describe by the* mode.

The **mode** of a set of values is the most frequently occurring value. The mode can be calculated for both numerical and categorical data. In Example 1.3, the mode of the audio brands is Spotify because the highest percentage of 12- to 34-years-olds have listened to this brand. For the 20 New York travel times, the mode is 10 minutes because this time occurred most often in the 20 travel times. There can be multiple modes; for example, in the 15 North Carolina travel times, 15, 20, 35, and 40 minutes are all modes because all these numbers are tied for occurring most often.

APPLY YOUR KNOWLEDGE

2.3 New York Travel Times. Find the mean of the travel times to work for the 20 New York workers in Example 2.3. Compare the mean and median for these data. What general fact does your comparison illustrate? NYTRAVEL

2.4 New House Prices. The mean and median sales prices of new homes sold in the United States in July 2019 were \$312,800 and \$388,000.[5] Which of these numbers is the mean, and which is the median? Explain how you know.

2.5 Carbon Dioxide Emissions. Burning fuels in power plants and motor vehicles emits carbon dioxide (CO_2), which contributes to global warming. The CO_2 emissions (metric tons per capita) for countries varies from 0.04 in Burundi to 43.86 in Qatar. Although the data set includes 203 countries, the CO_2 emissions of 14 countries are not available on the World Bank database. The data set is too large to print here, but here are the data for the first 5 countries:[6] CO2EMISS

Albert Valles/Getty Images

Country	CO_2 Emissions (metric tons per capita)
Aruba	8.41
Afghanistan	0.29
Angola	1.29
Albania	1.98
Andorra	5.83

Find the mean and the median for the full data set (included among the data sets available for this chapter). Make a histogram of the data. What features of the distribution explain why the mean is larger than the median?

2.4 Measuring Variability: The Quartiles

The mean and median provide two different measures of the center of a distribution. But a measure of center alone can be misleading. The U.S. Census Bureau reports that in 2017 the median income of American households was \$61,372. Half of all households had incomes below \$61,372, and half had higher incomes. The mean was much higher, \$86,220, because the distribution of incomes is skewed to the right. But the median and mean don't tell the whole story. The bottom 20% of households had incomes less than \$24,638, and households in the top 5% took in more than \$237,034.[7] We are interested in the *variability* of incomes as well as their center. *The simplest useful numerical description of a distribution requires both a measure of center and a measure of variability.*

One way to measure variability is to give the smallest and largest observations. For example, the travel times of our 15 North Carolina workers range from 5 minutes to 70 minutes. These single observations show the full variability of the data, but they may be outliers. We can improve our description of variability by also looking at the variability of the middle half of the data. The *quartiles* mark out the middle half. Count up the ordered list of observations, starting from the smallest. The *first quartile* lies one-quarter of the way up the list. The *third quartile* lies three-quarters of the way up the list. In other words, the first quartile is larger than 25% of the observations, and the third quartile is larger than 75% of the observations. The second quartile is the median, which is larger than 50% of the observations. That is the idea of quartiles. We need a rule to make the idea exact. The rule for calculating the quartiles uses the rule for the median.

The Quartiles Q_1 and Q_3

To calculate the **quartiles:**

1. Arrange the observations in increasing order and locate the median, M, in the ordered list of observations.
2. The **first quartile,** Q_1, is the median of the observations whose position in the ordered list is to the left of the location of the overall median.
3. The **third quartile,** Q_3, is the median of the observations whose position in the ordered list is to the right of the location of the overall median.

The following examples show how the rules for the quartiles work for both odd and even numbers of observations.

EXAMPLE 2.4 Finding the Quartiles: Odd *n*

NCTRAVEL

Our North Carolina sample of 15 workers' travel times, arranged in increasing order, is

5 8 10 12 15 15 20 **20** 25 30 35 35 40 40 70

There is an odd number of observations, so the median is the middle one, the bold **20** in the list. The first quartile is the median of the seven observations to the left of the median. This is the fourth of these seven observations, so $Q_1 = 12$ minutes. If you want, you can use the rule for the location of the median with $n = 7$:

$$\text{location of } Q_1 = \frac{n+1}{2} = \frac{7+1}{2} = 4$$

The third quartile is the median of the seven observations to the right of the median, $Q_3 = 35$ minutes. *When there is an odd number of observations, leave out the overall median when you locate the quartiles in the ordered list.*

The quartiles are *resistant* because they are not affected by a few extreme observations. For example, Q_3 would still be 35 if the outlier were 700 rather than 70.

EXAMPLE 2.5 Finding the Quartiles: Even *n*

NYTRAVEL

Here are the travel times to work of the 20 New Yorker workers from Example 2.3, arranged in increasing order:

10 10 10 10 12 15 15 20 20 30 | 35 35 40 45 50 50 55 60 65 75

There is an even number of observations, so the median lies midway between the middle pair, the 10th and 11th in the list. Its value is $M = 32.5$ minutes. We have marked the location of the median by |. The first quartile is the median of the first 10 observations because these are the observations to the left of the location of the median. Check that $Q_1 = 13.5$ minutes and $Q_3 = 50$ minutes. *When the number of observations is even, include all the observations when you locate the quartiles.*

Be careful when, as in these examples, several observations take the same numerical value. Write down all the observations, arrange them in order, and apply the rules just as if they all had distinct values.

There are several rules for finding the quartiles. Some calculators and software use rules that give results that differ from ours for some sets of data (see Example 2.8). Our rule is the simplest for hand calculation, with the results from the various rules generally being close to each other.

2.5 The Five-Number Summary and Boxplots

The smallest and largest observations tell us little about a distribution as a whole, but they give information about the tails of the distribution that is missing if we know only the median and the quartiles. To get a quick summary of both center and variability, combine all five numbers.

The Five-Number Summary

The **five-number summary** of a distribution consists of the smallest observation, the first quartile, the median, the third quartile, and the largest observation, written in order from smallest to largest. In symbols, the five-number summary is

Minimum Q_1 M Q_3 Maximum

These five numbers offer a reasonably complete description of center and variability. The five-number summaries of travel times to work from Examples 2.4 and 2.5 are

North Carolina	5	12	20	35	70
New York	10	13.5	32.5	50	75

The five-number summary of a distribution leads to a new graph, the *boxplot*. Figure 2.1 shows boxplots comparing travel times to work in North Carolina and New York.

Boxplot

A **boxplot** is a graph of the five-number summary:

- A central box spans the quartiles Q_1 and Q_3.
- A line in the box marks the median M.
- Lines extend from the box out to the smallest and largest observations.

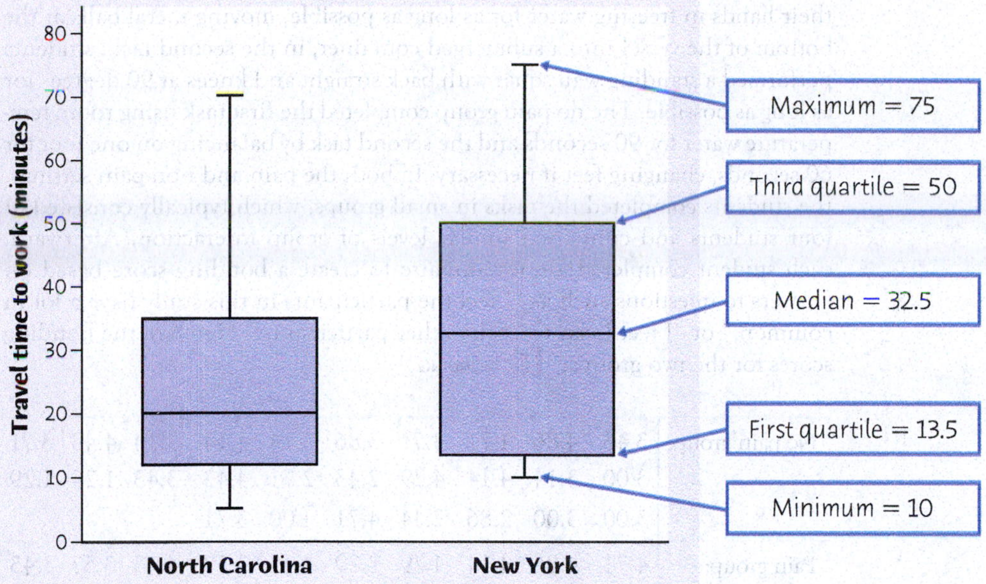

FIGURE 2.1
Boxplots comparing the travel times to work of samples of workers in North Carolina and New York.

Because boxplots show less detail than histograms or stemplots, they are best used for side-by-side comparison of more than one distribution, as in Figure 2.1. Be sure to include a numerical scale in the graph. When you look at a boxplot, first locate the median, which marks the center of the distribution. Then look at the variability. The span of the central box shows the variability of the middle half of the data, and the extremes (the smallest and largest observations) show the variability of the entire data set. We see from Figure 2.1 that travel times to work are in general a bit longer in New York than in North Carolina. The median, both quartiles, the minimum, and the maximum are all larger in New York. New York travel times are also more variable, as shown by the span of the box. Note that the boxes with arrows in Figure 2.1 that indicate the location of the five-number summary are *not* part of the boxplot but are included purely for illustration.

Finally, the North Carolina data are more strongly right-skewed. In a symmetric distribution, the first and third quartiles are equally distant from the median. In most distributions that are skewed to the right, on the other hand, the third quartile is farther above the median than the first quartile is below it. The extremes behave the same way, but remember that they are just single observations and may say little about the distribution as a whole.

APPLY YOUR KNOWLEDGE

2.6 Shared Pain and Bonding. Although painful experiences are involved in social rituals in many parts of the world, little is known about the social effects of pain. Will sharing painful experiences in a small group lead to greater bonding of group members than sharing a similar non-painful experience? Fifty-four university students in South Wales were divided at random into a pain group containing 27 students, with the remaining students in the no-pain group. Pain was induced by two tasks. In the first task, students submerged their hands in freezing water for as long as possible, moving metal balls at the bottom of the vessel into a submerged container; in the second task, students performed a standing wall squat with back straight and knees at 90 degrees for as long as possible. The no-pain group completed the first task using room temperature water for 90 seconds and the second task by balancing on one foot for 60 seconds, changing feet if necessary. In both the pain and non-pain settings, the students completed the tasks in small groups, which typically consisted of four students and contained similar levels of group interaction. Afterward, each student completed a questionnaire to create a bonding score based on answers to questions such as "I feel the participants in this study have a lot in common," or "I feel I can trust the other participants." Here are the bonding scores for the two groups:[8] BONDING

No-pain group:	3.43	4.86	1.71	1.71	3.86	3.14	4.14	3.14	4.43	3.71
	3.00	3.14	4.14	4.29	2.43	2.71	4.43	3.43	1.29	1.29
	3.00	3.00	2.86	2.14	4.71	1.00	3.71			
Pain group:	4.71	4.86	4.14	1.29	2.29	4.43	3.57	4.43	3.57	3.43
	4.14	3.86	4.57	4.57	4.29	1.43	4.29	3.57	3.57	3.43
	2.29	4.00	4.43	4.71	4.71	2.14	3.57			

(a) Find the five-number summaries for the pain and no-pain groups.

(b) Construct a comparative boxplot for the two groups following the model of Figure 2.1. It doesn't matter if your boxplots are horizontal or vertical, but they should be drawn on the same set of axes.

(c) Which group tends to have higher bonding scores? Is the variability in the two groups similar, or does one of the groups tend to have less variable bonding scores? Does either group contain one or more clear outliers?

2.7 Fuel Economy for Midsize Cars. The Department of Energy provides fuel economy ratings for all cars and light trucks sold in the United States. Here are the estimated miles per gallon for combined city and highway driving for the 189 cars classified as midsize in 2019, arranged in increasing order:[9] MIDCARS

12	14	16	16	16	17	17	17	18	18	18	18	19	19	19	19	19	19
20	20	20	20	20	20	20	20	21	21	21	21	21	21	22	22	22	22
22	23	23	23	23	23	23	23	23	23	23	23	23	23	24	24	24	24
24	24	24	24	24	24	24	24	24	24	24	24	24	25	25	25	25	25
25	25	25	25	25	25	25	25	25	25	25	25	26	26	26	26	26	26
26	26	26	26	26	26	26	26	26	26	26	26	26	26	27	27	27	27
27	27	27	27	27	27	27	27	27	27	27	27	27	28	28	28	28	28
29	29	29	29	29	29	29	29	29	29	29	30	30	30	30	30	31	31
31	31	32	32	32	32	32	32	32	32	32	32	32	33	33	33	33	33
33	34	34	34	34	35	35	35	35	36	41	41	41	41	42	42	43	44
44	46	46	48	50	52	52	52	56									

(a) Give the five-number summary of this distribution.

(b) Draw a boxplot of these data. What is the shape of the distribution shown by the boxplot? Which features of the boxplot led you to this conclusion? Are any observations unusually small or large?

2.6 Spotting Suspected Outliers and Modified Boxplots*

Look again at the stemplot of travel times to work in North Carolina in Example 2.3. The five-number summary for this distribution is

$$5 \quad 12 \quad 20 \quad 35 \quad 70$$

How shall we describe the variability of this distribution? The smallest and largest observations are extremes that don't describe the variability of the majority of the data. The distance between the quartiles (the range of the center half of the data) is a more resistant measure of variability. This distance is called the *interquartile range.*

The Interquartile Range (*IQR*)

The **interquartile range (*IQR*)** is the distance between the first and third quartiles,

$$IQR = Q_3 - Q_1$$

For our data on North Carolina travel times, $IQR = 35 - 12 = 23$ minutes. However, *no single numerical measure of variability, such as IQR, is very useful for describing skewed distributions.* The two sides of a skewed distribution have different variability, so one number can't summarize them. That's why we give the full five-number summary. The interquartile range is mainly used as the basis for a rule of thumb for identifying suspected outliers.

The 1.5 × *IQR* Rule for Outliers

Call an observation a suspected outlier if it falls more than $1.5 \times IQR$ above the third quartile or below the first quartile.

STATISTICS IN YOUR WORLD

How Much Is That House Worth?

The town of Manhattan, Kansas, is sometimes called "the Little Apple" to distinguish it from that other Manhattan, "the Big Apple." A few years ago, a house there appeared in the county appraiser's records valued at $200,059,000. That would be quite a house, even on Manhattan Island. As you might guess, the entry was wrong: the true value was $59,500. But before the error was discovered, the county, the city, and the school board had based their budgets on the total appraised value of real estate, which the one outlier jacked up by 6.5%. It can pay to spot outliers before you trust your data.

EXAMPLE 2.6 Using the 1.5 × *IQR* Rule

NCTRAVEL

For the North Carolina travel time data, $IQR = 23$ and

$$1.5 \times IQR = 1.5 \times 23 = 34.5$$

Any values not falling between

$$Q_1 - (1.5 \times IQR) = 12.0 - 34.5 = -22.5 \quad \text{and}$$

$$Q_3 + (1.5 \times IQR) = 35 + 34.5 = 69.5$$

are flagged as suspected outliers. Look again at the stemplot in Example 2.3 (page 48): the only suspected outlier is the longest travel time, 70 minutes. The $1.5 \times IQR$ rule suggests that the two next-longest travel times of 40 minutes are just part of the long right tail of this skewed distribution.

In a modified boxplot, which is provided by many software packages, the suspected outliers are identified in the boxplot with a special plotting symbol such as a dot (•). Comparing Figure 2.2 with Figure 2.1, we see that the largest observation from North Carolina is flagged as an outlier. The line

*This short section is optional.

FIGURE 2.2 Horizontal modified boxplots comparing the travel times to work of samples of workers in North Carolina and New York.

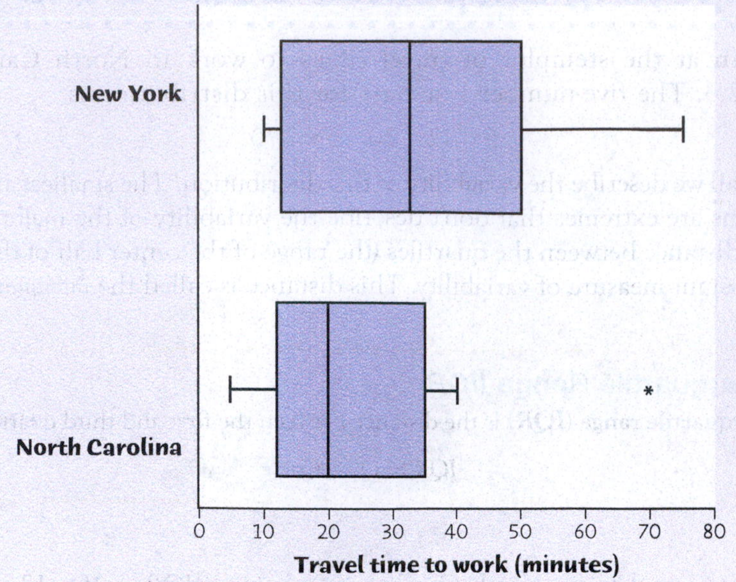

beginning at the third quartile no longer extends to the maximum but now ends at 40, which is the largest observation from North Carolina that is not identified as an outlier. Figure 2.2 also displays the modified boxplots horizontally rather than vertically, an option available in some software packages that does not change the interpretation of the plot. Finally, the $1.5 \times IQR$ rule is not a replacement for looking at the data. It is most useful when large volumes of data are processed automatically.

APPLY YOUR KNOWLEDGE

Jecapix/Getty Images

2.8 Travel Time to Work. In Example 2.3 (page 48), there is one long travel time of 75 minutes in our sample of 20 New York travel times. Does the $1.5 \times IQR$ rule identify this travel time as a suspected outlier? NYTRAVEL

2.9 Fuel Economy for Midsize Cars. Exercise 2.7 (page 54) gives the estimated miles per gallon (mpg) for city driving for the 189 cars classified as midsize in 2019. Are any of the larger values outliers by the $1.5 \times IQR$ rule? Although outliers can be produced by errors or incorrectly recorded observations, they are often observations that differ from the others in some particular way. In this case, the cars producing the high outliers share a common feature. What do you think that is? MIDCARS

2.7 Measuring Variability: The Standard Deviation

The five-number summary is not the most common numerical description of a distribution. That distinction belongs to the combination of the mean to measure center and the *standard deviation* to measure variability. The standard deviation and its close relative the *variance* measure variability by looking at how far the observations are from their mean.

The Standard Deviation s

The **variance** s^2 of a set of observations is an average of the squares of the deviations of the observations from their mean. In symbols, the variance of n observations $x_1, x_2, \ldots, x_n$ is

$$s^2 = \frac{(x_1 - \bar{x})^2 + (x_2 - \bar{x})^2 + \cdots + (x_n - \bar{x})^2}{n - 1}$$

or, more compactly,

$$s^2 = \frac{1}{n-1}\sum (x_i - \bar{x})^2$$

The **standard deviation** s is the square root of the variance s^2:

$$s = \sqrt{\frac{1}{n-1}\sum (x_i - \bar{x})^2}$$

In practice, use software or your calculator to obtain the standard deviation from keyed-in data. Doing an example step-by-step will help you understand how the variance and standard deviation work, however.

EXAMPLE 2.7 Calculating the Standard Deviation

SATMATH

Georgia Southern University had 2786 students with regular admission in its freshman class of 2015. For each student, data are available on their SAT and ACT scores (if taken), high school GPA, and the college within the university to which they were admitted.[10] In Exercise 3.49 (page 96), the full data set for the SAT Mathematics scores will be examined. Here are the first five observations from that data set:

490 580 450 570 650

We will compute $\bar{x}$ and s for these students. First find the mean:

$$\bar{x} = \frac{490 + 580 + 450 + 570 + 650}{5} = \frac{2740}{5} = 548$$

Figure 2.3 displays the data as points above the number line, with their mean marked by an asterisk (*). The arrows mark two of the deviations from the mean. The deviations show how variable the data are about their mean. They are the starting point for calculating the variance and the standard deviation.

Observations x_i	**Deviations** $x_i - \bar{x}$	**Squared Deviations** $(x_i - \bar{x})^2$
490	$490 - 548 = -58$	$(-58)^2 = 3{,}364$
580	$580 - 548 = 32$	$32^2 = 1{,}024$
450	$450 - 548 = -98$	$(-98)^2 = 9{,}604$
570	$570 - 548 = 22$	$22^2 = 484$
650	$650 - 548 = 102$	$102^2 = 10{,}404$
	sum = 0	sum = 24,880

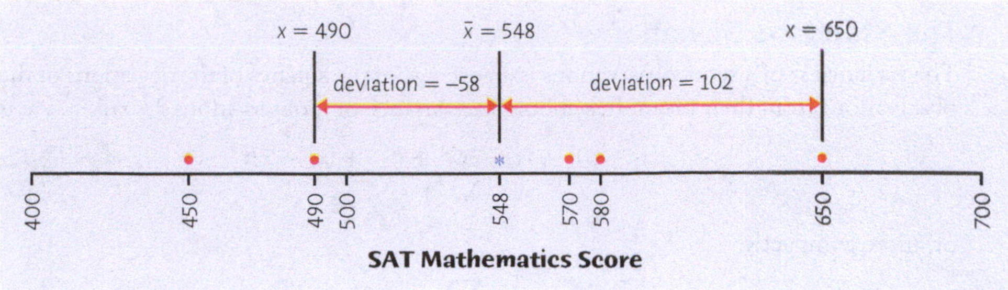

FIGURE 2.3
SAT Mathematics scores for five students, with their mean (*) and the deviations of two observations from the mean shown, for Example 2.7.

The variance is the sum of the squared deviations divided by one less than the number of observations:

$$s^2 = \frac{1}{n-1}\sum (x_i - \overline{x})^2 = \frac{24{,}880}{4} = 6220$$

The standard deviation is the square root of the variance:

$$s = \sqrt{6220} = 78.87$$

Notice that the "average" in the variance s^2 divides the sum by one fewer than the number of observations, that is, $n - 1$ rather than n. The reason is that the deviations $x_i - \overline{x}$ always sum to exactly 0 so that knowing $n - 1$ of them determines the last one. Only $n - 1$ of the squared deviations can vary freely, and we average by dividing the total by $n - 1$. The number $n - 1$ is called the *degrees of freedom*[11] of the variance or standard deviation. Some calculators offer a choice between dividing by n and dividing by $n - 1$, so be sure to use $n - 1$.

More important than the details of hand calculation are the properties that determine the usefulness of the standard deviation:

- s measures *variability about the mean* and should be used only when the mean is chosen as the measure of center.
- s is *always zero or greater than zero.* $s = 0$ only when there is no variability. This happens only when all observations have the same value. Otherwise, $s > 0$. As the observations become more variable about their mean, s gets larger.
- s has the *same units of measurement as the original observations.* For example, if you measure weight in kilograms, both the mean $\overline{x}$ and the standard deviation s are also in kilograms. This is one reason to prefer s to the variance s^2, which would be in squared kilograms.
- Like the mean $\overline{x}$, s is *not resistant.* A few outliers can make s very large.

The use of squared deviations renders s even more sensitive than $\overline{x}$ to a few extreme observations. For example, the standard deviation of the travel times for the 15 North Carolina workers in Example 2.1 is 16.97 minutes. (Use your calculator or software to verify this.) If we omit the high outlier, the standard deviation drops to 12.07 minutes.

If you feel that the importance of the standard deviation is not yet clear, you are right. We will see in Chapter 3 that the standard deviation is the natural measure of variability for a very important class of symmetric distributions, the Normal distributions. The usefulness of many statistical procedures is tied to distributions of particular shapes. This is certainly true of the standard deviation.

2.8 Choosing Measures of Center and Variability

We now have a choice between two descriptions of the center and variability of a distribution: the five-number summary or $\bar{x}$ and s. Because $\bar{x}$ and s are sensitive to extreme observations, they can be misleading when a distribution is strongly skewed or has outliers. In fact, because the two sides of a skewed distribution have different variability, no single number describes the variability well. The five-number summary, with its two quartiles and two extremes, does a better job.

Choosing a Summary

- The five-number summary is usually better than the mean and standard deviation for describing a skewed distribution or a distribution with strong outliers.
- Use $\bar{x}$ and s only for reasonably symmetric distributions that are free of outliers.

Outliers can greatly affect the values of the mean $\bar{x}$ and the standard deviation s, the most common measures of center and variability. Many more elaborate statistical procedures also can't be trusted when outliers are present. *Whenever you find outliers in your data, try to find an explanation for them.* Sometimes the explanation is as simple as a typing error, such as typing 10.1 as 101; if this is the case, correct the typing error. Sometimes a measuring device broke down or a subject gave a frivolous response, like the student in a class survey who claimed to study 30,000 minutes per night. (Yes, that really happened.) In all these cases, you can simply remove the outlier from your data. When outliers are "real data," like the long travel times of some New York workers, you should choose statistical methods that are not greatly disturbed by the outliers. For example, use the five-number summary rather than $\bar{x}$ and s to describe a distribution with extreme outliers. We will meet other examples later in the book.

Remember that a graph gives the best overall picture of a distribution. If data have been entered into a calculator or statistical program, it is very simple and quick to create several graphs to see all the different features of a distribution. Numerical measures of center and variability report specific facts about a distribution, but they do not describe its entire shape. Numerical summaries do not disclose the presence of multiple peaks or clusters, for example. Exercise 2.11 shows how misleading numerical summaries can be. *Always plot your data.*

APPLY YOUR KNOWLEDGE

2.10 $\bar{x}$ and s by Hand. Radon is a naturally occurring gas and is the second leading cause of lung cancer in the United States.[12] It comes from the natural breakdown of uranium in the soil and enters buildings through cracks and other holes in foundations. Radon is found throughout the United States, but levels vary considerably from state to state. Several methods can reduce the levels of radon in a home, and the Environmental Protection Agency recommends using one of them if the measured level in a home is above 4 picocuries per liter. Four readings from Franklin County, Ohio, where the county average is 8.2 picocuries per liter, were 3.8, 1.9, 12.1, and 14.4.

(a) Find the mean step-by-step. That is, find the sum of the four observations and divide by 4.

(b) Find the standard deviation step-by-step. That is, find the deviation of each observation from the mean, square the deviations, and obtain the variance and the standard deviation. Example 2.7 (page 57) shows the method.

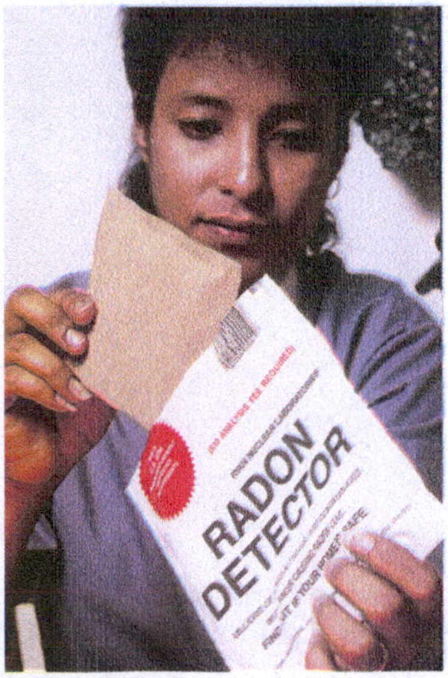

Doug Martin/Science Source

(c) Now enter the data into your calculator and use the mean and standard deviation buttons to obtain $\bar{x}$ and s. Do the results agree with your hand calculations?

2.11 $\bar{x}$ and s Are Not Enough. The mean $\bar{x}$ and standard deviation s measure center and variability but are not a complete description of a distribution. Data sets with different shapes can have the same mean and standard deviation. To demonstrate this fact, use your calculator to find $\bar{x}$ and s for these two small data sets. Then make a stemplot of each and comment on the shape of each distribution. DATASET2

Data A:	9.14	8.14	8.74	8.77	9.26	8.10	6.13	3.10	9.13	7.26	4.74
Data B:	6.58	5.76	7.71	8.84	8.47	7.04	5.25	6.89	5.56	7.91	12.50

2.12 Choose a Summary. The shape of a distribution is a rough guide to whether the mean and standard deviation are a helpful summary of center and variability. For which of the following distributions would $\bar{x}$ and s be useful? In each case, give a reason for your decision.

(a) Percentages of high school graduates in the states taking the SAT, Figure 1.8 (page 26)

(b) Iowa Tests scores, Figure 1.7 (page 26)

(c) New York travel times, Example 2.3 (page 48)

2.9 Examples of Technology

Although a calculator with "two-variable statistics" functions will do many of the basic calculations we need throughout the text, more elaborate technology is helpful. Graphing calculators and computer software will do calculations and make graphs as you command, freeing you to concentrate on choosing the right methods and interpret your results. Figure 2.4 displays outputs from three technology tools for describing the travel times to work of 20 people in New York State (Example 2.3, page 48). Can you find $\bar{x}$, s, and the five-number summary in each output? The big message is: *once you know what to look for, you can read output from any technological tool.*

The displays in Figure 2.4 come from a Texas Instruments graphing calculator, the Microsoft Excel spreadsheet program, and JMP statistical software. JMP allows you to choose what descriptive measures you want, whereas the descriptive measures in Excel and the calculator give some things you don't need. Just ignore the extras. Because Excel's "Descriptive Statistics" menu item doesn't give the quartiles, we used the spreadsheet's separate quartile function to get Q_1 and Q_3.

EXAMPLE 2.8 What Is the Third Quartile?

In Example 2.5, we saw that the quartiles of the New York travel times are $Q_1 = 13.5$ and $Q_3 = 50$. Look at the output displays in Figure 2.4. The calculator agrees with our work, while Excel says $Q_1 = 14.25$ and JMP says that $Q_1 = 12.75$. What happened? *There are several rules for finding the quartiles. Some calculators and software use rules that give results different from ours for some sets of data.* This is true of JMP and Excel. Results from the various rules are generally close to each other, so the differences are not important in practice. Our rule is the simplest for hand calculation.

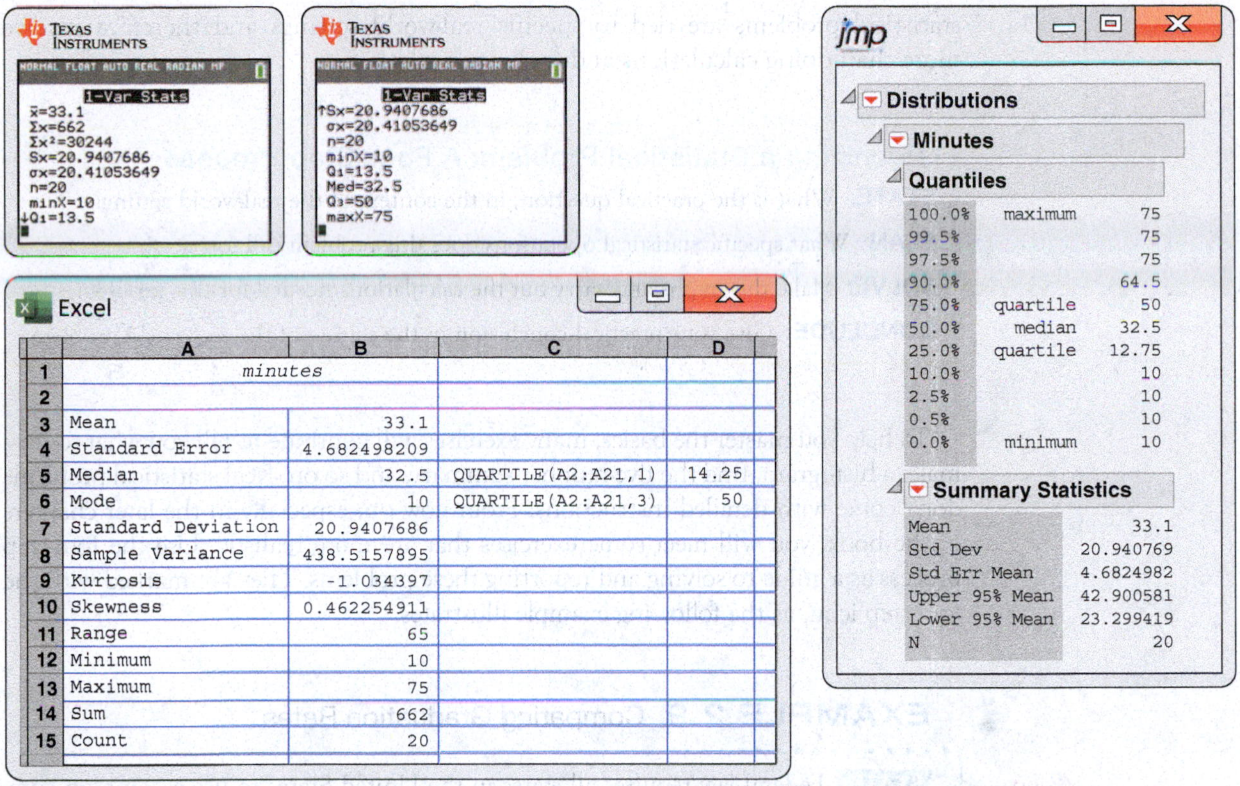

FIGURE 2.4

Output from a graphing calculator, a spreadsheet program, and a statistical software package describing the data on travel times to work in New York State.

2.10 Organizing a Statistical Problem

Most of our examples and exercises have aimed to help you learn basic tools (graphs and calculations) for describing and comparing distributions. You have also learned principles that guide use of these tools, such as "start with a graph" and "look for the overall pattern and striking deviations from the pattern." The data you work with are not just numbers; they describe specific settings such as water depth in the Everglades or travel time to work. Because data come from a specific setting, the final step in examining data is *coming to a conclusion for that setting*. Water depth in the Everglades has a yearly cycle that reflects Florida's wet and dry seasons. Travel times to work are generally longer in New York than in North Carolina.

Let's return to the on-time high school graduation rates discussed in Example 1.4 (page 20). We know from the example that the on-time graduation rates vary from 71.1% in New Mexico to 91% in Iowa, with a median of 86%. State graduation rates are related to many factors, and in a statistical problem, we often try to explain the differences or variation in a variable such as graduation rate by some of these factors. For example, do states with lower household incomes tend to have lower high school graduation rates? Or, do the states in some regions of the country tend to have lower high school graduation rates than the states in other regions?

As you learn more statistical tools and principles, you will face more complex statistical problems. Although no framework accommodates all the varied issues that arise in applying statistics to real settings, we find the following four-step thought process gives useful guidance. In particular, the first and last steps emphasize that statistical problems are tied to specific real-world settings and therefore involve more than doing calculations and making graphs.

Organizing a Statistical Problem: A Four-Step Process

STATE: What is the practical question, in the context of the real-world setting?

PLAN: What specific statistical operations does this problem call for?

SOLVE: Make the graphs and carry out the calculations needed for this problem.

CONCLUDE: Give your practical conclusion in the setting of the real-world problem.

To help you master the basics, many exercises will continue to tell you what to do—make a histogram, find the five-number summary, and so on. Real statistical problems don't come with detailed instructions. From now on, especially in the later chapters of the book, you will meet some exercises that are more realistic. Use the four-step process as a guide to solving and reporting these problems. They are marked with the four-step icon, as the following example illustrates.

EXAMPLE 2.9 Comparing Graduation Rates

GRADRATE

STATE: Federal law requires all states in the United States to use a common computation of on-time high school graduation rates beginning with the 2010–11 school year. Previously, states chose one of several computation methods that gave answers that could differ by more than 10%. This common computation allows for meaningful comparison of graduation rates between the states.

We know from Table 1.1 (page 21), that the on-time high school graduation rates in the 2016–17 school year varied from 71.1% in New Mexico to 91% in Iowa. The U.S. Census Bureau divides the 50 states and the District of Columbia into four geographical regions: the Northeast (NE), Midwest (MW), South (S), and West (W). The region for each state is included in Table 1.1. Do the states in the four regions of the country display distinct distributions of graduation rates? How do the mean graduation rates of the states in each of these regions compare?

PLAN: Use graphs and numerical descriptions to describe and compare the distributions of on-time high school graduation rates of the states in the four regions of the United States.

SOLVE: We might use boxplots to compare the distributions, but stemplots preserve more detail and work well for data sets of these sizes. Figure 2.5 displays the stemplots with the stems lined up for easy comparison. The stems have been split to better display the distributions, and the data have been rounded to the nearest percentage (with no decimal places). The stemplots overlap, and some care is needed when comparing the four stemplots because the sample sizes differ, with some stemplots having more leaves than others. The states in the Northeast and Midwest have distributions that are similar to each other. The South, with the most observations, has one low observation corresponding to the District of Columbia that stands apart from the others and some skewness to the left. With little skewness and no serious outliers, we report $\bar{x}$ and s as our summary measures of center and variability of the distribution of the on-time graduation rates of the states in each region. Because the District of Columbia is not a state, although often included with state data, we have reported summary statistics for the South with and without this observation.

Region	*n*	Mean	Standard Deviation
Midwest	12	86.03	3.12
Northeast	9	87.12	2.70
South (including DC)	17	85.14	4.72
South (excluding DC)	16	85.89	3.69
West	13	80.5	4.27

Midwest		Northeast		South		West	
7		7		7	3	7	1
7		7		7	8	7	688999
8	02334	8	14	8	02233	8	0225
8	677889	8	667889	8	666789999	8	66
9	1	9	0	9		9	

FIGURE 2.5 Stemplots comparing the distributions of graduation rates for the four census regions from Table 1.1, for Example 2.9.

CONCLUDE: The table of summary statistics and the stemplots lead to similar conclusions. The states in the Midwest and Northeast are most similar to each other, with the South, excluding the District of Columbia, having a slightly lower mean and higher standard deviation. The states in the West have a lower mean graduation rate than the other three regions, with a standard deviation similar to that of the South but higher than those of the Midwest or Northeast.

It is important to remember that the individuals in Example 2.9 are the states. For example, the mean of 87.12 is the mean of the on-time graduation rates for the nine Northeastern states, and the standard deviation tells us how much these state rates vary about this mean. However, the mean of these nine states *is not* the same as the graduation rate for all high school students in the Northeast, unless the states have the same number of high school graduates. The graduation rate for all high school students in the Northeast would be a *weighted* average of the state rates, with the larger states receiving more weight. For example, because New York is the most populous state in the Northeast and also has the lowest graduation rate, we would expect the graduation rate of all high school students in the Northeast to be lower than 87.12 because New York would pull down the overall graduation rate. See Exercise 2.37 (page 68) for a similar example.

APPLY YOUR KNOWLEDGE

2.13 Logging in the Rain Forest. "Conservationists have despaired over destruction of tropical rain forest by logging, clearing, and burning." These words begin a report on a statistical study of the effects of logging in Borneo.[13] Charles Cannon of Duke University and his coworkers compared forest plots that had never been logged (Group 1) with similar plots nearby that had been logged one year earlier (Group 2) and eight years earlier (Group 3). Each plot was 0.1 hectare in area. Here are the counts of trees for plots in each group: LOGGING

AustralianCamera/Shutterstock

Group 1:	27	22	29	21	19	33	16	20	24	27	28	19
Group 2:	12	12	15	9	20	18	17	14	14	2	17	19
Group 3:	18	4	22	15	18	19	22	12	12			

To what extent has logging affected the count of trees? Follow the four-step process in reporting your work.

4step

2.14 Worldwide Child Mortality. Although child mortality rates worldwide have dropped by more than 50% since 1990, in 2017 it was still the case that 16,000 children under five years old died each day. The mortality rates for children under five varied from 2.1 per 1000 in Iceland to 127.2 per 1000 in Somalia. In Exercise 1.36 (page 40), you were asked to draw a histogram of these data. In this exercise, you will explore the relationship between child mortality and a measure of a country's economic wealth. One measure used by the World Bank is the gross national income (GNI) per capita, the dollar value of a country's final income in a year divided by its population. It reflects the average income of a country's citizens, and the World Bank uses GNI per capita to classify countries into low-income, lower-middle-income, upper-middle-income, and high-income economies. Although the data set includes 214 countries, the child mortality rates of 21 countries are not available in the World Health Organization database. Because the data set is too large to print here, we give the data for the first five countries:[14] MORTALTY

Country	Child Mortality Rate (per 1000)	Economy Classification
Aruba	—	High
Afghanistan	67.9	High
Angola	81.1	Low
Albania	8.8	Upper-middle
Andorra	3.3	Upper-middle

Give a full description of the distribution of child mortality rates for the countries in each of the four economic classifications and identify any high outliers. Compare the four groups. Does the economic classification used by the World Bank do a good job of explaining the differences in child mortality rates among the countries?

CHAPTER 2 SUMMARY

- A numerical summary of a distribution should report at least its center and its variability.
- The **mean** $\bar{x}$ and the **median M** describe the center of a distribution in different ways. The mean is the arithmetic average of the observations, and the median is the midpoint of the values.
- When you use the median to indicate the center of the distribution, describe its variability by giving the **quartiles**. The **first quartile**, Q_1, has one-fourth of the observations below it, and the **third quartile**, Q_3, has three-fourths of the observations below it.
- The **five-number summary** consisting of the median, the quartiles, and the smallest and largest individual observations provides a quick overall description of a distribution. The median describes the center, and the quartiles and extremes show the variability.
- **Boxplots** based on the five-number summary are useful for comparing several distributions. The box spans the quartiles and shows the variability of the central half of the distribution. The median is marked within the box. Lines extend from the box to the extremes and show the full variability of the data.
- The **variance** s^2 and especially its square root, the **standard deviation** s, are common measures of variability about the mean as center. The standard deviation s is zero when there is no variability and gets larger as the variability increases.

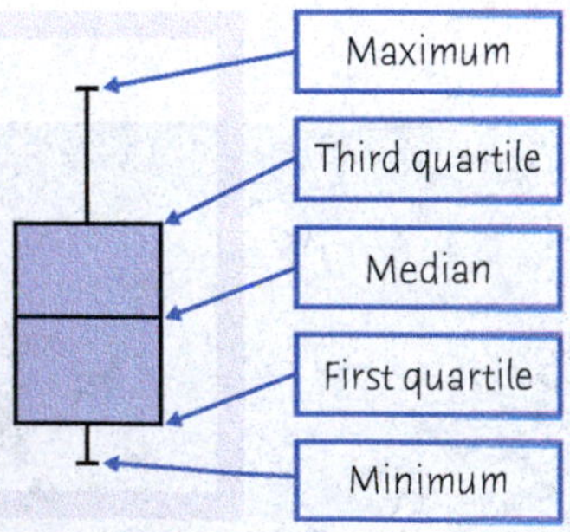

FIGURE 2.6
Boxplot showing maximum, third quartile, median, first quartile, and minimum.

- A **resistant measure** of any aspect of a distribution is relatively unaffected by changes in the numerical value of a small proportion of the total number of observations, no matter how large these changes are. The median and quartiles are resistant, but the mean and the standard deviation are not.
- The mean and standard deviation are good descriptions for symmetric distributions without outliers. They are most useful for the Normal distributions introduced in the next chapter. The five-number summary is a better description for skewed distributions.
- Numerical summaries do not fully describe the shape of a distribution. Always plot your data.
- A statistical problem has a real-world setting. You can organize many problems by using the following four steps: *state, plan, solve,* and *conclude.*

CHECK YOUR SKILLS

2.15 The 2019–20 roster of the New England Patriots, winners of the 2019 NFL Super Bowl, included 10 defensive linemen. The weights in pounds of the 10 defensive linemen were LINEMEN

275 300 300 315 345 260 250 275 280 250

The mean of these data is

(a) 277.50. (b) 285.00. (c) 300.25.

2.16 The median of the data in Exercise 2.15 is LINEMEN

(a) 277.50. (b) 285.00. (c) 300.25.

2.17 The first quartile of the data in Exercise 2.15 is LINEMEN

(a) 260.00. (b) 300.00. (c) 303.75.

2.18 If a distribution is skewed to the left,

(a) the mean is less than the median.

(b) the mean and median are equal.

(c) the mean is greater than the median.

2.19 What percentage of the observations in a distribution are greater than the first quartile?

(a) 25% (b) 50% (c) 75%

2.20 To make a boxplot of a distribution, you must know

(a) all the individual observations.

(b) the mean and the standard deviation.

(c) the five-number summary.

2.21 The standard deviation of the 10 weights in Exercise 2.15 (use your calculator) is about LINEMEN

(a) 28.72. (b) 30.28. (c) 46.25.

2.22 What are all the values that a standard deviation s can possibly take?

(a) $0 \leq s$ (b) $0 \leq s \leq 1$ (c) $-1 \leq s \leq 1$

2.23 The correct units for the standard deviation in Exercise 2.21 are

(a) no units—it's just a number.

(b) pounds.

(c) pounds squared.

2.24 Which of the following is most affected if an extreme high outlier is added to the data?

(a) The median

(b) The mean

(c) The first quartile

CHAPTER 2 EXERCISES

2.25 **Incomes of College Grads.** According to the U.S. Census Bureau's Current Population Survey, the mean and median 2018 income of people aged 25–34 years who had a bachelor's degree but no higher degree were \$50,350 and \$60,178.[15] Which of these numbers is the mean, and which is the median? Explain your reasoning.

2.26 **Household Assets.** Once every three years, the Board of Governors of the Federal Reserve System collects data on household assets and liabilities through the Survey of Consumer Finances (SCF).[16] Here are some results from the 2016 survey.

(a) Transaction accounts, which include checking, savings, and money market accounts, are the most commonly held type of financial asset. The mean value of transaction accounts per household for those holding transaction accounts was \$40,200, and the median value was \$4,500. What explains the differences between the two measures of center?

(b) The median value of cash value life insurance per household was \$0. What does a median of \$0 say about the percentage of households with cash value life insurance?

2.27 University Endowments. The National Association of College and University Business Officers collects data on college endowments. In 2018, its report included the endowment values of 809 colleges and universities in the United States and Canada. When the endowment values are arranged in order, what are the locations of the median and the quartiles in this ordered list?

2.28 Pulling Apart Wood. Exercise 1.46 (page 44) gives the breaking strengths in pounds of 20 pieces of Douglas fir. WOOD

(a) Give the five-number summary of the distribution of breaking strengths.

(b) Here is a stemplot of the data rounded to the nearest hundred pounds. The stems are thousands of pounds, and the leaves are hundreds of pounds.

23	0
24	1
25	
26	5
27	
28	7
29	
30	259
31	399
32	033677
33	0237

The stemplot shows that the distribution is skewed to the left. Does the five-number summary show the skew? Remember that only a graph gives a clear picture of the shape of a distribution.

2.29 Comparing Graduation Rates. An alternative presentation to compare the graduation rates in Table 1.1 (page 21) by region of the country reports five-number summaries and uses boxplots to display the distributions. Calculate the five-number summaries and make the boxplots. Do the boxplots fail to reveal any important information visible in the stemplots of Figure 2.5 (page 63)? Which plots make it simpler to compare the regions? Why? GRADRATE

2.30 How Much Fruit Do Adolescent Girls Eat? Figure 1.17 (page 37) is a histogram of the number of servings of fruit per day claimed by 74 17-year-old girls.

(a) With a little care, you can find the median and the quartiles from the histogram. What are these numbers? How did you find them?

(b) You can also find the mean number of servings of fruit claimed per day from the histogram. First use the information in the histogram to compute the sum of the 74 observations, and then use this to compute the mean. What is the relationship between the mean and median? Is this what you expected?

(c) In general, you cannot find the exact values of the median, quartiles, or mean from the histogram. What is special about the histogram of the number of servings of fruit that allows you to do this?

2.31 Guinea Pig Survival Times. Here are the survival times, in days, of 72 guinea pigs after they were injected with infectious bacteria in a medical experiment.[17] Survival times, whether of machines under stress or cancer patients after treatment, usually have distributions that are skewed to the right. GUINPIGS

43	45	53	56	56	57	58	66	67	73	74	79
80	80	81	81	81	82	83	83	84	88	89	91
91	92	92	97	99	99	100	100	101	102	102	102
103	104	107	108	109	113	114	118	121	123	126	128
137	138	139	144	145	147	156	162	174	178	179	184
191	198	211	214	243	249	329	380	403	511	522	598

(a) Graph the distribution and describe its main features. Does it show the expected right-skew?

(b) Which numerical summary would you choose for these data? Calculate your chosen summary. How does it reflect the skewness of the distribution?

2.32 Maternal Age at Childbirth. How old are women when they have their first child? Here is the distribution of the age of the mother for all firstborn children in the United States in 2017:[18]

Age	Count	Age	Count
10–14 years	1,892	30–34 years	329,623
15–19 years	162,536	35–39 years	124,637
20–24 years	395,927	40–44 years	24,049
25–29 years	417,162	45–49 years	2,377

The number of firstborn children to mothers under 10 or over 50 years of age represent a negligible percentage of all first births and are not included in the table.

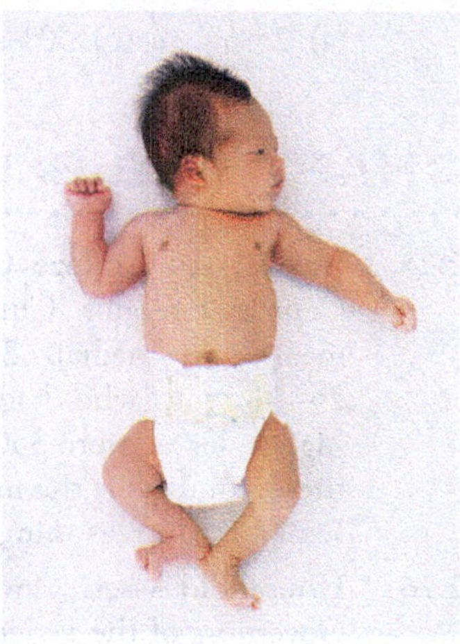
Tom Merton/Getty Images

(a) For comparison with other years and with other countries, we prefer a histogram of the *percentages* in each age class rather than the counts. Explain why.

(b) How many babies were there?

(c) Make a histogram of the distribution, using percentages on the vertical scale. Using this histogram, describe the distribution of the age at which women have their first child.

(d) What are the locations of the median and quartiles in the ordered list of all maternal ages? In which age classes do the median and quartiles fall?

2.33 More on Nintendo and Laparoscopic Surgery. In Exercise 1.38 (page 41), you examined the improvement in times to complete a virtual gall bladder removal for those with and without four weeks of Nintendo Wii™ training. The most common methods for formal comparison of two groups use $\bar{x}$ and s to summarize the data. NINTENDO

(a) What kinds of distributions are best summarized by $\bar{x}$ and s? Do you think these summary measures are appropriate in this case?

(b) In the control group, one subject improved his/her time by 229 seconds. How much does removing this observation change $\bar{x}$ and s for the control group? You will need to compute $\bar{x}$ and s for the control group, both with and without the high outlier.

(c) Compute the median for the control group with and without the high outlier. What does this show about the resistance of the median and $\bar{x}$?

2.34 Making Resistance Visible. applet In the *Mean and Median* applet, place three observations on the line by dragging movable points from the Data Bank one at a time to a position on the line: two close together near the center of the line and one somewhat to the right of these two.

(a) Pull the single rightmost observation out to the right. (Place the cursor on the point, hold down a mouse button, and drag the point.) How does the mean behave? How does the median behave? Explain briefly why each measure acts as it does.

(b) Now drag the single rightmost point to the left as far as you can. What happens to the mean? What happens to the median as you drag this point past the other two? (Watch carefully.)

2.35 Behavior of the Median. applet Place five observations on the line in the *Mean and Median* applet by dragging movable points from the Data Bank one at a time to a position on the line.

(a) Add one additional observation *without changing the median*. Where is your new point?

(b) Use the applet to convince yourself that when you add yet another observation (there are now seven in all), the median does not change, no matter where you put the seventh point. Explain why this must be true.

2.36 Never on Sunday: Also in Canada? Exercise 1.5 (page 19) gives the number of births in the United States on each day of the week during an entire year. The boxplots in Figure 2.7 are based on more detailed data from Toronto, Canada: the number of births on each of the 365 days in a year, grouped by day of the week.[19] Based on these plots, compare the day-of-the-week distributions using shape, center, and variability. Summarize your findings.

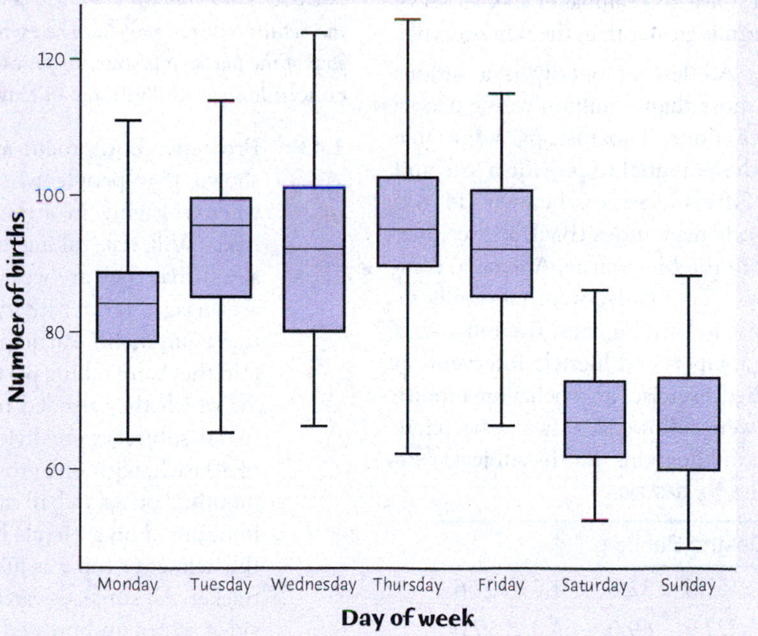

FIGURE 2.7
Boxplots of the distributions of numbers of births in Toronto, Canada, on each day of the week during a year, for Exercise 2.36.

2.37 Thinking about Means. Table 1.2 (page 23) gives the percentage of minority residents in each of the states. For the nation as a whole, 42.8% of residents are minorities. Find the mean of the 51 entries in Table 1.2. It is *not* 42.8%. Explain carefully why this happens. (*Hint:* The states with the largest populations are California, Texas, New York, and Florida. Look at their entries in Table 1.2.) MINORITY

2.38 Thinking about Means and Medians. In 2018, approximately 2.1% of hourly rate workers were being paid at the federal minimum wage level or lower. Would federal legislation to increase the minimum wage have a greater effect on the mean or the median income of *all* workers? Explain your answer.

2.39 A Standard Deviation Contest. You are to choose four numbers from the whole numbers 0 to 10, with repeats allowed.

(a) Choose four numbers that have the smallest possible standard deviation.

(b) Choose four numbers that have the largest possible standard deviation.

(c) Is more than one choice possible in either part (a) or (b)? Explain.

2.40 You Create the Data. Create a set of seven numbers (repeats allowed) that have the five-number summary

$$\text{Minimum} = 4 \quad Q_1 = 8 \quad M = 12 \quad Q_3 = 15 \quad \text{Maximum} = 19$$

There is more than one set of seven numbers with this five-number summary. What must be true about the seven numbers to have this five-number summary?

2.41 You Create the Data. Give an example of a small set of data for which the mean is greater than the third quartile.

2.42 Adolescent Obesity. Adolescent obesity is a serious health risk affecting more than 5 million young people in the United States alone. Laparoscopic adjustable gastric banding has the potential to provide a safe and effective treatment. Fifty adolescents between 14 and 18 years old with a body mass index (BMI) higher than 35 were recruited from the Melbourne, Australia, community for the study.[20] Twenty-five were randomly selected to undergo gastric banding, and the remaining 25 were assigned to a supervised lifestyle intervention program involving diet, exercise, and behavior modification. All subjects were followed for two years. Here are the weight losses, in kilograms, for the subjects who completed the study: GASTRIC

Gastric Banding					
35.6	81.4	57.6	32.8	31.0	37.6
36.5	−5.4	27.9	49.0	64.8	39.0
43.0	33.9	29.7	20.2	15.2	41.7
53.4	13.4	24.8	19.4	32.3	22.0

Lifestyle Intervention					
6.0	2.0	−3.0	20.6	11.6	15.5
−17.0	1.4	4.0	−4.6	15.8	34.6
6.0	−3.1	−4.3	−16.7	−1.8	−12.8

(a) In the context of this study, what do the negative values in the data set mean?

(b) Give a graphical comparison of the weight loss distributions for the two groups, using side-by-side boxplots. Provide appropriate numerical summaries for the two distributions and identify any high outliers in either group. What can you say about the effects of gastric banding versus lifestyle intervention on weight loss for the subjects in this study?

(c) The measured variable was weight loss, in kilograms. Would two subjects with the same weight loss always have similar benefits from a weight-reduction program? Does it depend on their initial weights? Other variables considered in this study were the percentage of excess weight lost and the reduction in BMI. Do you see any advantages to either of these variables when comparing weight loss for two groups?

(d) One subject from the gastric-banding group dropped out of the study, and seven subjects from the lifestyle group dropped out. Of the seven dropouts in the lifestyle group, six had gained weight at the time they dropped out. If all subjects had completed the study, how do you think it would have affected the comparison between the two groups?

Exercises 2.43 through 2.49 ask you to analyze data without having the details outlined for you. The exercise statements give you the ***state*** *step of the four-step process. In your work, follow the* ***plan, solve,*** *and* ***conclude*** *steps as illustrated in Example 2.9 (page 62).*

2.43 Protective Equipment and Risk Taking. 4step Studies have shown that people who are using safety equipment when engaging in an activity tend to take increased risks. Will risk taking increase when people are not aware that they are wearing protective equipment and are engaged in an activity that cannot be made safer by this equipment? Participants in the study were falsely told they were taking part in an eye-tracking experiment for which they needed to wear an eye-tracking device. Eighty subjects were divided at random into two groups of 40 each, with one group wearing the tracking device mounted on a baseball cap and the other group wearing it mounted on a bicycle helmet. Subjects were told that the helmet or cap was just being used to mount the eye tracker. All subjects watched an animated balloon on a video screen and pressed a button to inflate it. The balloon was programmed to burst at a random point, but until that point, each press of the button inflated the balloon further and increased the amount of fictional currency a subject would earn. Subjects were free to stop pumping at any point and keep their earnings, knowing that if the balloon burst, they would lose all earnings for that round. The score was the average number of pumps on the trials, with lower scores corresponding to less risk taking and more conservative play. Here are the first 10 observations from each group:[21] HELMET

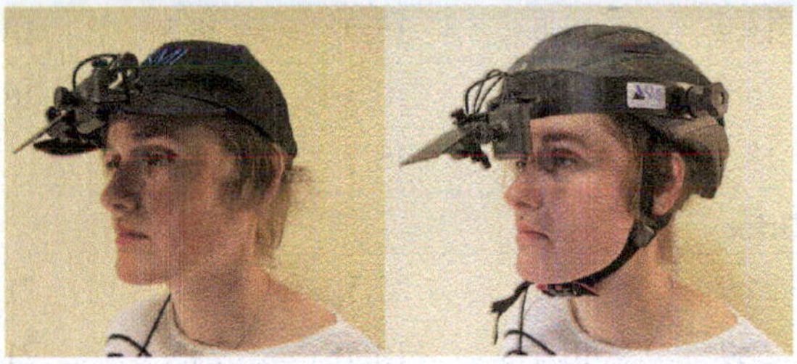

Tim Gamble and Ian Walker, "Wearing a bicycle helmet can increase risk taking and sensation seeking in adults," Psychological Science, 27 (2016), pp. 289–294: https://doi.org/10.1177/0956797615620784 (http://www.creativecommons.org/licenses/by/3.0/).

Helmet:	3.67	36.50	29.28	30.50	24.08
	32.10	50.67	26.26	41.05	20.56
Baseball Cap:	29.38	42.50	41.57	47.77	32.45
	30.65	7.04	2.68	22.04	25.86

Compare the distributions for the two groups. How is wearing of a helmet related to the measure of risk behavior?

2.44 **Athletes' Salaries.** The Montreal Canadiens were founded in 1909 and are the longest continuously operating professional ice hockey team. The team has won 24 Stanley Cups, making them one of the most successful professional sports teams of the traditional four major sports of Canada and the United States. Table 2.1 gives the salaries of the 2019–20 roster prior to the start of the 2019–20 season.[22] Provide the team owner with a full description of the distribution of salaries and a brief summary of its most important features. HOCKEY

2.45 **Returns on Stocks.** How well have stocks done over the past generation? The Wilshire 5000 index describes the average performance of all U.S. stocks. The average is weighted by the total market value of each company's stock, so think of the index as measuring the performance of the average investor. Shown below are the percentage returns on the Wilshire 5000 index for the years 1971–2018: What can you say about the distribution of yearly returns on stocks?[23] WILSHIRE

Year	Return	Year	Return	Year	Return
1971	17.68	1987	2.27	2003	31.64
1972	17.98	1988	17.94	2004	12.62
1973	−18.52	1989	29.17	2005	6.32
1974	−28.39	1990	−6.18	2006	15.88
1975	38.47	1991	34.20	2007	5.73
1976	26.59	1992	8.97	2008	−37.34
1977	−2.64	1993	11.28	2009	29.42
1978	9.27	1994	−0.06	2010	17.87
1979	25.56	1995	36.45	2011	0.59
1980	33.67	1996	21.21	2012	16.12
1981	−3.75	1997	31.29	2013	34.02
1982	18.71	1998	23.43	2014	12.07
1983	23.47	1999	23.56	2015	−0.24
1984	3.05	2000	−10.89	2016	13.04
1985	32.56	2001	−10.97	2017	21.00
1986	16.09	2002	−20.86	2018	−5.29

TABLE 2.1 Salaries for the 2019–20 Montreal Canadiens

Player	Salary	Player	Salary	Player	Salary
Carey Price	$15,000,000	Phillip Danault	$3,000,000	Christian Folin	$800,000
Shea Weber	$6,000,000	Brett Kulak	$1,950,000	Victor Mete	$750,000
Jonathan Drouin	$5,500,000	Dale Weise	$1,750,000	Charlie Lindgren	$750,000
Tomas Tatar	$4,981,132	Jordan Weal	$1,300,000		
Karl Alzner	$4,625,000	Matthew Peca	$1,300,000		
Paul Byron	$4,000,000	Nate Thompson	$1,000,000		
Jeff Petry	$4,000,000	Nicolas Deslauriers	$950,000		
Brendan Gallagher	$4,000,000	Jesperi Kotkaniemi	$925,000		
Andrew Shaw	$3,250,000	Ryan Poehling	$925,000		
Max Domi	$3,150,000	Noah Juulsen	$832,000		

TABLE 2.2 Amount spent (euros) by customers in a restaurant when exposed to odors

No Odor									
15.9	18.5	15.9	18.5	18.5	21.9	15.9	15.9	15.9	15.9
15.9	18.5	18.5	18.5	20.5	18.5	18.5	15.9	15.9	15.9
18.5	18.5	15.9	18.5	15.9	18.5	15.9	25.5	12.9	15.9
Lemon Odor									
18.5	15.9	18.5	18.5	18.5	15.9	18.5	15.9	18.5	18.5
15.9	18.5	21.5	15.9	21.9	15.9	18.5	18.5	18.5	18.5
25.9	15.9	15.9	15.9	18.5	18.5	18.5	18.5		
Lavender Odor									
21.9	18.5	22.3	21.9	18.5	24.9	18.5	22.5	21.5	21.9
21.5	18.5	25.5	18.5	18.5	21.9	18.5	18.5	24.9	21.9
25.9	21.9	18.5	18.5	22.8	18.5	21.9	20.7	21.9	22.5

2.46 **Do Good Smells Bring Good Business?** Businesses know that customers often respond to background music. Do they also respond to odors? Nicolas Guéguen and his colleagues studied this question in a small pizza restaurant in France on Saturday evenings in May. On one evening, a relaxing lavender odor was spread through the restaurant; on another evening, a stimulating lemon odor; a third evening served as a control, with no odor. Table 2.2 shows the amounts (in euros) that customers spent on each of these evenings.[24] Compare the three distributions. Were both odors associated with increased customer spending? ODORS

2.47 **Policy Justification: Pragmatic vs. Moral.** How does a leader's justification of his/her organization's policy affect support for the policy? A study compared a moral, pragmatic, and ambiguous justification for three policy proposals: a politician's plan to fund a retirement planning agency, a state governor's plan to repave state highways, and a president's plan to outlaw child labor in a developing country. For example, for the retirement agency proposal, the moral justification was the importance of retirees "to live with dignity and comfort," the pragmatic was "to not drain public funds," and the ambiguous was "to have sufficient funds." Three hundred seventy-four volunteer subjects were assigned at random to read all three proposals: 122 subjects read the three proposals with a moral justification, 126 subjects read the three proposals with a pragmatic justification, and 126 subjects read the three proposals with an ambiguous justification. Several questions measuring support for each policy proposal were answered by each subject to create a support score for each proposal, and their scores for the three proposals were then averaged to create an index of policy support for each subject, higher values indicating greater support.[25] Here are the first five observations: JUSTIFY

Justification:	Pragmatic	Ambiguous	Pragmatic	Moral	Ambiguous
Policy support index:	5	7	4.75	7	5.75

The first individual read the proposals with a pragmatic justification, with a policy support index of 5, the second with an ambiguous justification and a policy support index of 7, and so forth. Compare the three distributions. How does the support index vary with the type of justification?

2.48 **Does Playing Video Games Improve Surgical Skill?** In laparoscopic surgery, a video camera and several thin instruments are inserted into the patient's abdominal cavity. The surgeon uses the image from the video camera positioned inside the patient's body to perform the procedure by manipulating the instruments that have been inserted. The Top Gun Laparoscopic Skills and Suturing Program was developed to help surgeons develop the skill set necessary for laparoscopic surgery. Because of the similarity in many of the skills involved in video games and laparoscopic surgery, it was hypothesized that surgeons with greater prior video game experience might acquire the skills required in

yacobchuk/Getty Images

laparoscopic surgery more easily. Thirty-three surgeons participated in the study and were classified into the three categories—never used, under three hours per day, and more than three hours per day—depending on the number of hours they played video games at the height of their video game use. They also performed Top Gun drills and received a score based on the time to complete the drill and the number of errors made, with lower scores indicating better performance. Here are the Top Gun scores and video game categories for the 33 participants:[26] TOPGUN

Never played:	9379	8302	5489	5334	4605	4789	9185	7216	9930
	4828	5655	4623	7778	8837	5947			
Under three hours:	5540	6259	5163	6149	4398	3968	7367	4217	5716
Three or more hours:	7288	4010	4859	4432	4845	5394	2703	5797	3758

Compare the distributions for the three groups. How is prior video game experience related to Top Gun scores?

2.49 **Cholesterol Levels and Age.** The National Health and Nutrition Examination Survey (NHANES) is a unique survey that combines interviews and physical examinations.[27] It includes basic demographic information; questions about topics such as diet, physical activity, and prescription medications; and results of a physical examination measuring a variety of variables, including blood pressure and cholesterol levels. The program began in the early 1960s, and the survey currently examines a nationally representative sample of about 5000 persons each year. You will work with the total cholesterol measurements (mg/dL) obtained from participants in the survey in 2009–2010. CHOLEST

To examine changes in cholesterol with age, we consider only the 3044 participants between 20 and 50 years of age and have classified them into the three age categories: 20s, 30s, and 40s. The full data set is too large to print here, but here are the first 10 individuals:

Age category:	30s	20s	20s	40s	30s	40s	20s	30s	30s	20s
Total cholesterol:	135	160	299	197	196	202	175	216	181	149

The first individual is in the 30s with total cholesterol of 135, the second in the 20s with total cholesterol of 160, and so forth.

(a) Use graphical and numerical summaries to compare the three distributions. How does cholesterol change with age?

(b) The ideal range of total cholesterol is below 200 mg/dL. For individuals with elevated cholesterol levels, prescription drugs are often recommended to reduce levels. Among the 3044 participants between 20 and 50 years of age, 4 individuals in their 20s, 24 individuals in their 30s, and 117 individuals in their 40s were taking prescription medications to reduce their cholesterol levels. How do you think your comparison of the distribution would be changed if none of the individuals were taking medication? Explain.

Exercises 2.50 through 2.53 make use of the optional material on the 1.5 × IQR rule for suspected outliers.

2.50 **The Changing Face of America.** Figure 1.10 (page 29) gives a stemplot of the percentage of minority residents aged 18–34 in each of the 50 states and the District of Columbia. These data are given in Table 1.2 (page 23). MINORITY

(a) Give the five-number summary of this distribution.

(b) Although there do not appear to be any outliers in Figure 1.10, when you split the stems for the data in Exercise 1.10, Texas, California, New Mexico, and Hawaii are separated from the remaining states. Are these four states outliers or just the largest observations in a strongly skewed distribution? What does the 1.5 × *IQR* rule say?

2.51 **Shared Pain and Bonding.** In Exercise 2.6, you should have noticed some low outliers in the pain group. BONDING

(a) Compute the mean and the median of the bonding scores for the pain group, both with and without the two smallest scores. Do they have more of an effect on the mean or the median? Explain why.

(b) Does the 1.5 × *IQR* rule identify these two low bonding scores as suspected outliers?

(c) Unusual observations are not necessarily mistakes. Suppose a small percentage of subjects would experience little bonding regardless of whether they were in the no-pain group or the pain group. Explain how the randomization of the students to the two groups could have led to these "outliers."

2.52 **The *Fortune* Global 500.** The *Fortune* Global 500, also known as the Global 500, is an annual ranking by *Fortune* magazine of the top 500 corporations worldwide as measured by revenue. In total, the Global 500 generated $32.7 trillion in revenues in 2018. Table 2.3 provides a list of the 30 companies with the highest revenues (in billions of dollars) in 2018.[28] A stemplot or histogram shows that the distribution is strongly skewed to the right. GLOBE500

(a) Give the five-number summary. Explain why this summary suggests that the distribution is right-skewed.

(b) Which companies are outliers according to the $1.5 \times IQR$ rule? Make a stemplot of the data. Do you agree with the rule's suggestions about which companies are and are not outliers?

(c) If you consider *all* 500 companies, the 30 companies in Table 2.3 each represent a high outlier among all Global 500 companies. Is there a common feature shared by many of the 30 companies in the table? What proportion of the total of the Global 500 revenues is accounted for by these 30 companies?

2.53 Cholesterol for People in Their 20s. Exercise 2.49 contains the cholesterol levels of individuals in their 20s from the NHANES survey in 2009–10. The cholesterol levels are right-skewed, with a few large cholesterol levels. Which cholesterol levels are suspected outliers by the $1.5 \times IQR$ rule? CHOLES20

TABLE 2.3 Revenues for the top Global 500 companies in 2018

Company Name	Revenues ($b)	Company Name	Revenues ($b)
Wal-Mart Stores	514.4	Glencore	219.8
Sinopec Group	414.6	McKesson	214.3
Royal Duch Shell	396.6	Daimler	197.5
China National Petroleum	393.0	Total	184.1
State Grid	387.1	China State Construction Engineering	181.5
Saudi Aramco	355.9	Trafigura Group	180.7
BP	303.7	Hon Hai Precision Industry	175.6
Exxon Mobil	290.2	EXOR Group	175.0
Volkswagen	278.3	AT&T	170.8
Toyota Motor	272.6	Industrial & Commercial Bank of China	169.0
Apple	265.6	AmerisourceBergen	167.9
Berkshire Hathaway	247.8	Chevron	166.3
Amazon.com	232.9	Ping An Insurance	163.6
UnitedHealth Group	226.2	Ford Motor	160.3
Samsung Electronics	221.6	China Construction Bank	151.1

CHAPTER

3

The Normal Distributions

In Chapters 1 and 2 we discussed methods for summarizing a large data set. These included graphical displays, such as bar charts and histograms, and numerical summaries, such as the mean, median, quartiles, and standard deviation. In this chapter we will look at another way to summarize a large data set using a smooth curve to summarize the general shape of its distribution.

We now have a toolbox of graphical and numerical methods for describing distributions. What is more, we have a clear strategy for exploring data on a single quantitative variable.

Exploring a Distribution

1. Always plot your data: make a graph, usually a histogram or a stemplot.
2. Look for the overall pattern (shape, center, variability) and for striking deviations, such as outliers.
3. Calculate a numerical summary to briefly describe center and variability.

In this chapter, we add one more step to this strategy:

4. Sometimes the overall pattern of a large number of observations is so regular that we can describe it by a smooth curve.

spxChrome/Getty Images

When you complete this chapter, you will be able to:

3.1 Understand the properties of density curves.

3.2 Use density curves to infer characteristics of a data set.

3.3 State the properties of Normal distributions and understand their role.

3.4 Use the 68–95–99.7 rule to estimate proportions of a Normal distribution above or below a given value or between two values.

3.5 Use technology and an understanding of cumulative proportions to find proportions of a Normal distribution above or below a given value or between two values.

3.6 Use a table of standard Normal cumulative proportions to find proportions of a Normal distribution above or below a given value or between two values.

3.7 Given the proportion of observations above or below a particular observed value in a distribution, find the observed value.

3.1 Density Curves

Figure 3.1 is a histogram of the scores of all 947 seventh-grade students in Gary, Indiana, on the vocabulary part of the Iowa Test of Basic Skills.[1] Scores of many students on this national test have a quite regular distribution. The histogram is symmetric, and both tails fall off smoothly from a single center peak. There are no large gaps or obvious outliers. The smooth curve drawn closely approximates the tops of the histogram bars in Figure 3.1 and provides a good description of the overall pattern of the data.

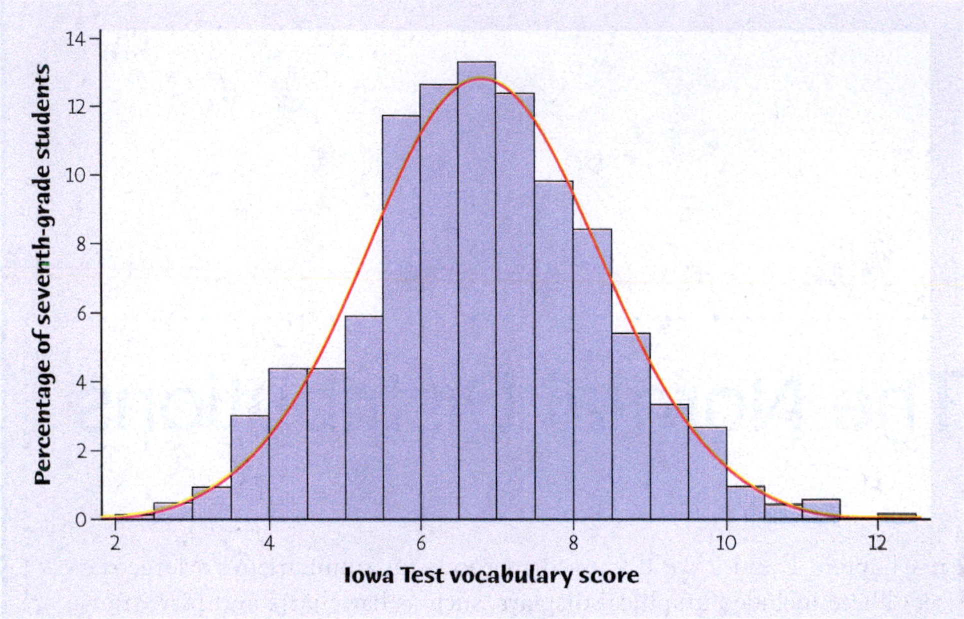

FIGURE 3.1
Histogram of the Iowa Test vocabulary scores of all seventh-grade students in Gary, Indiana. The smooth curve shows the overall shape of the distribution.

EXAMPLE 3.1 From Histogram to Density Curve

IOWATEST

Our eyes respond to the *areas* of the bars in a histogram. The bar areas represent proportions of the observations. Figure 3.2(a) is a copy of Figure 3.1 with the leftmost bars shaded. The area of the shaded bars in Figure 3.2(a) represents the students with vocabulary scores of 6.0 or lower. There are 287 such students, who make up the proportion $287/947 = 0.303$ of all Gary seventh-graders.

Now look at the curve drawn through the bars. In Figure 3.2(b), the area under the curve to the left of 6.0 is shaded. We can draw histogram bars taller or shorter by adjusting the vertical scale. In moving from histogram bars to a smooth curve, we make a specific choice: adjust the scale of the graph so that *the total area under the curve is exactly 1*. The total area represents the proportion 1—that is, all the observations. We can then interpret areas under the curve as proportions of the observations. The curve is now a *density curve*. The shaded area under the density curve in Figure 3.2(b) represents the proportion of students with score 6.0 or lower. This area is 0.293, only 0.010 away from the actual proportion 0.303. The method for finding this area will be presented shortly. For now, note that the areas under the density curve give quite good approximations to the actual distribution of the 947 test scores.

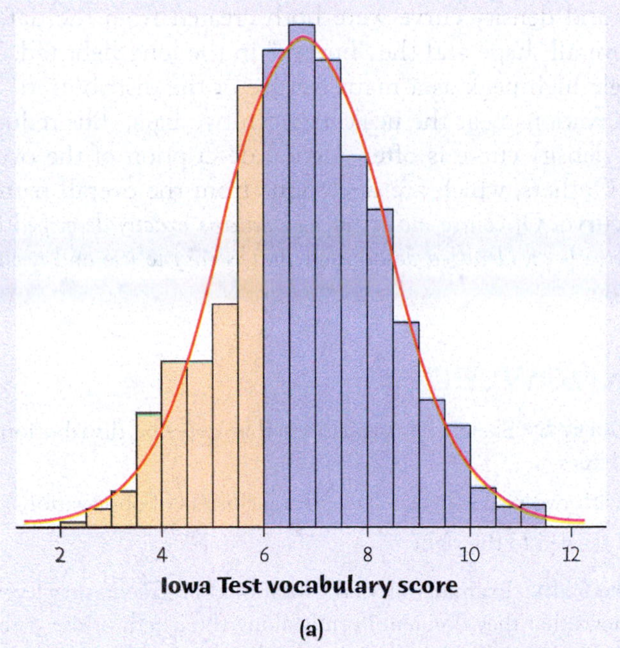

FIGURE 3.2(a) The proportion of scores less than or equal to 6.0 in the actual data is 0.303. The vertical scale is adjusted so that the total area under the curve equals 1.

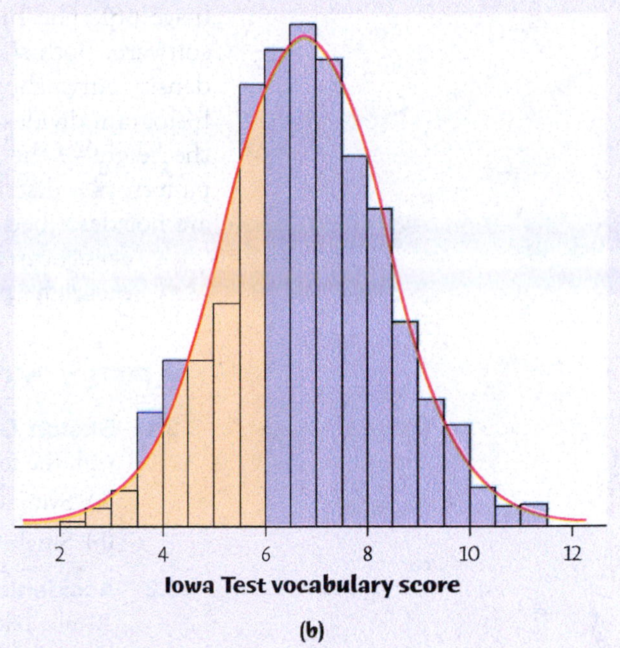

FIGURE 3.2(b) The proportion of scores less than or equal to 6.0 from the density curve is 0.293. The density curve is a good approximation to the distribution of the data.

Density Curve

A **density curve** is a curve that

- Is always on or above the horizontal axis.
- Has area exactly 1 underneath it.

A density curve describes the overall pattern of a distribution. The area under the curve and above any range of values is the proportion of all observations that fall in that range.

Density curves, like distributions, come in many shapes. Figure 3.3 shows a strongly skewed distribution, the survival times of 72 guinea pigs from Exercise 2.31

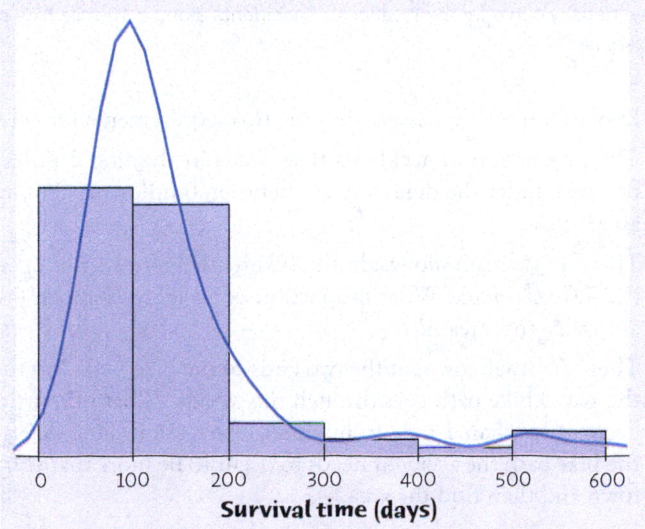

FIGURE 3.3
A right-skewed distribution pictured using both a histogram and a density curve.

(page 66). The histogram and density curve were both created from the data by software. Both show the overall shape and the "bumps" in the long right tail. The density curve shows a single high peak as a main feature of the distribution. The histogram divides the observations near the peak between two bars, thus reducing the height of the peak. A density curve is often a good description of the overall pattern of a distribution. Outliers, which are deviations from the overall pattern, are not described by the curve. *Of course, no set of real data is exactly described by a density curve. The curve is an idealized description that is easy to use and accurate enough for practical use.*

APPLY YOUR KNOWLEDGE

3.1 Sketch Density Curves. Sketch density curves that describe distributions with the following shapes:

(a) Symmetric but with two peaks (that is, two strong clusters of observations)

(b) Single peak and skewed to the right

3.2 Accidents on a Bike Path. Examination of the locations of accidents on a level, 10-mile bike path shows that they occur uniformly along the length of the path. Figure 3.4 displays the density curve that describes the distribution of accidents.

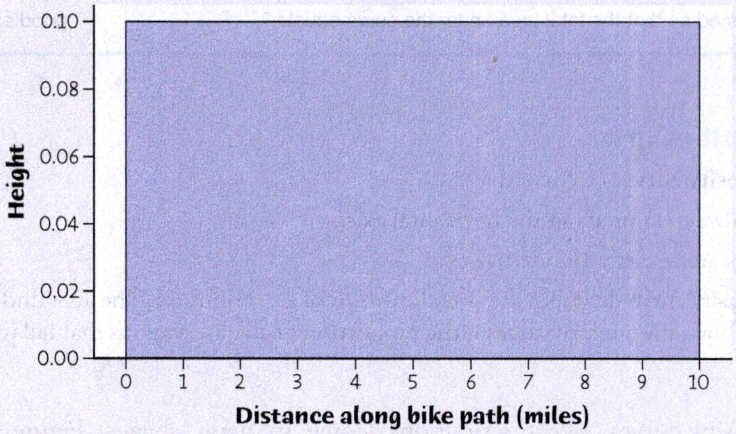

FIGURE 3.4
The density curve for the locations of accidents along a 10-mile bike path, for Exercise 3.2.

(a) Explain why this curve satisfies the two requirements for a density curve.

(b) The proportion of accidents that occur in the first 2 miles of the path is the area under the density curve between 0 miles and 2 miles. What is this area?

(c) There is a stream alongside the bike path between the 2.5-mile mark and the 4.0-mile mark. What proportion of accidents happen on the bike path alongside the stream?

(d) There are small towns at the two ends of the bike path, but the remainder of the paved bike path goes through the woods. What proportion of accidents occur more than 1 mile from either town? (*Hint*: First determine where on the bike path the accident needs to occur to be more than 1 mile from either town and then find the area.)

3.2 Describing Density Curves

Our measures of center and variability apply to density curves as well as to actual sets of observations. The median and quartiles are easy. Areas under a density curve represent proportions of the total number of observations. The median is the point with half the observations on either side. So *the median of a density curve is the equal-areas point,* the point with half the area under the curve to its left and the remaining half of the area to its right. The quartiles divide the area under the curve into quarters. One-fourth of the area under the curve is to the left of the first quartile, and three-fourths of the area is to the left of the third quartile. You can roughly locate the median and quartiles of any density curve by eye by dividing the area under the curve into four equal parts.

Because density curves are idealized patterns, a symmetric density curve is exactly symmetric. The median of a symmetric density curve is therefore at its center. Figure 3.5(a) shows a symmetric density curve with the median marked. It isn't so easy to spot the equal-areas point on a skewed curve. There are mathematical ways of finding the median for any density curve. That's how we marked the median on the skewed curve in Figure 3.5(b).

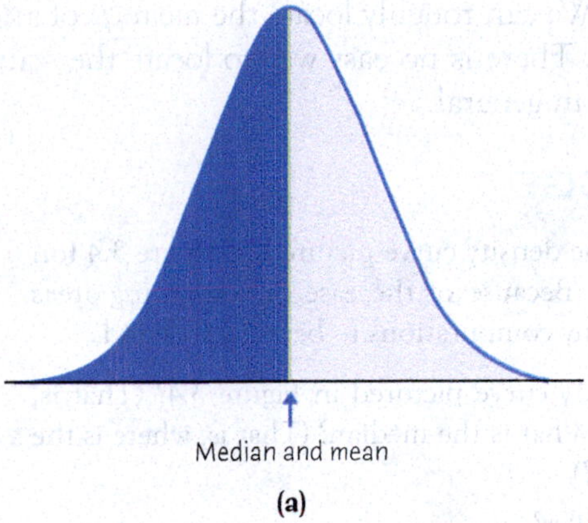
Median and mean
(a)

FIGURE 3.5(a) The median and mean of a symmetric density curve both lie at the center of symmetry.

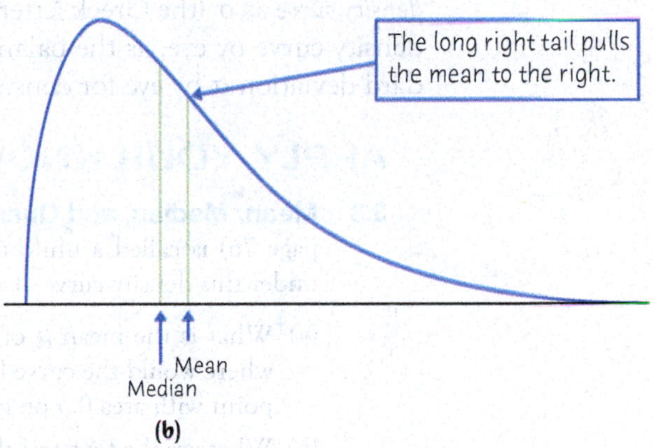
The long right tail pulls the mean to the right.
Mean
Median
(b)

FIGURE 3.5(b) The median and mean of a right-skewed density curve. The mean is pulled away from the median toward the long tail.

What about the mean? The mean of a set of observations is their arithmetic average. If we think of the observations as weights strung out along a thin rod, the mean is the point at which the rod would balance. This fact is also true of density curves. *The mean is the point at which the curve would balance if made of solid material.* Figure 3.6 illustrates this fact about the mean. A symmetric curve balances at its center because the two sides are identical. *The mean and median of a symmetric density curve are equal,* as in Figure 3.5(a). We know that the mean of a skewed distribution is pulled toward the long tail. Figure 3.5(b) shows how the mean of a skewed density curve is pulled toward the long tail more than is the median. It's hard to locate the

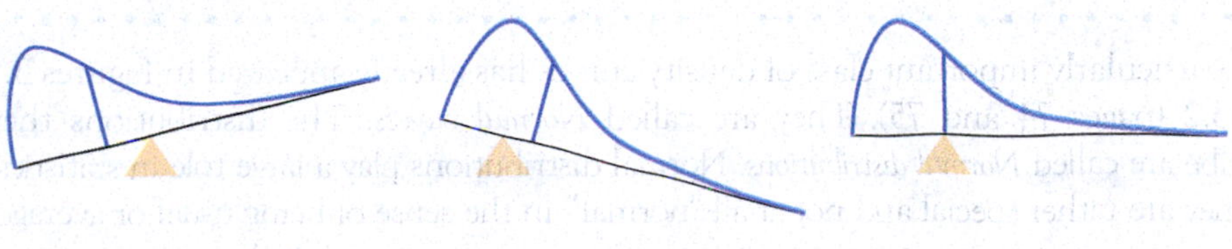

FIGURE 3.6 The mean is the balance point of a density curve.

balance point by eye on a skewed curve. There are mathematical ways of calculating the mean for any density curve, so we are able to mark the mean as well as the median in Figure 3.5(b).

Median and Mean of a Density Curve

The **median of a density curve** is the equal-areas point, the point that divides the area under the curve in half.

The **mean of a density curve** is the balance point at which the curve would balance if made of solid material. The usual notation for the mean of a density curve is μ (the Greek letter mu).

The median and mean are the same for a symmetric density curve. They both lie at the center of the curve. The mean of a skewed curve is pulled away from the median in the direction of the long tail.

Because a density curve is an idealized description of a distribution of data, we need to distinguish between the mean and standard deviation of the density curve and the mean $\bar{x}$ and standard deviation s computed from the actual observations. We use μ for the mean of a density curve and we write the *standard deviation of a density curve* as σ (the Greek letter sigma). We can roughly locate the mean μ of any density curve by eye, as the balance point. There is no easy way to locate the standard deviation σ by eye for density curves in general.

APPLY YOUR KNOWLEDGE

3.3 Mean, Median, and Quartiles. The density curve pictured in Figure 3.4 (on page 76) is called a uniform density. Because of the ease of computing areas under this density curve, it allows many computations to be done by hand.

(a) What is the mean μ of the density curve pictured in Figure 3.4? (That is, where would the curve balance?) What is the median? (That is, where is the point with area 0.5 on either side?)

(b) What are the first and third quartiles?

3.4 Mean and Median. Figure 3.7 displays three density curves, each with three points marked on it. At which of these points on each curve do the mean and the median fall?

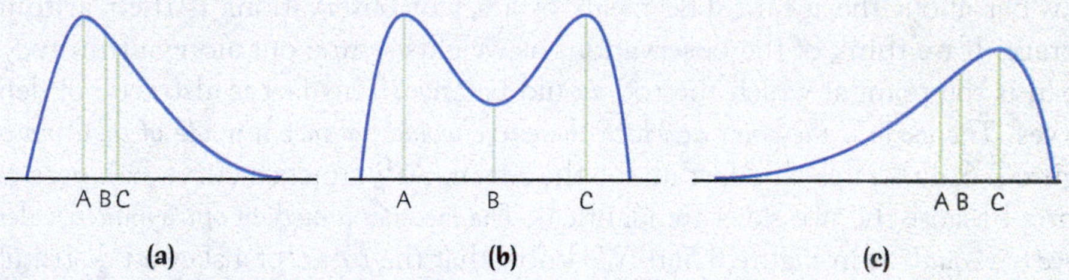

FIGURE 3.7
Three density curves, for Exercise 3.4.

3.3 Normal Distributions

One particularly important class of density curves has already appeared in Figures 3.1 and 3.2 (pages 74 and 75). They are called *Normal curves*. The distributions they describe are called *Normal distributions*. Normal distributions play a large role in statistics, but they are rather special and not at all "normal" in the sense of being usual or average.

FIGURE 3.8
Two Normal curves, showing the mean μ and standard deviation σ.

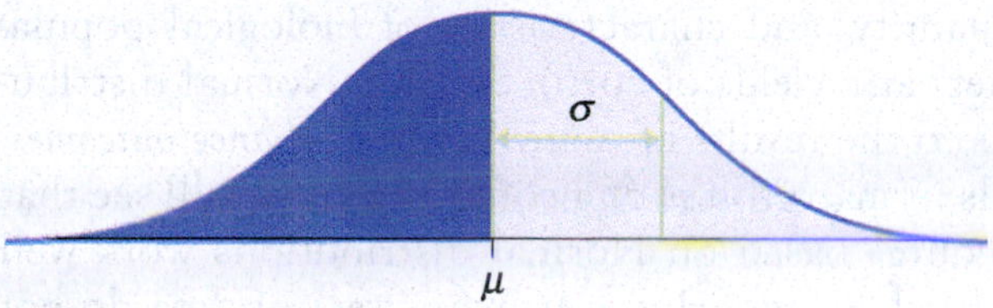

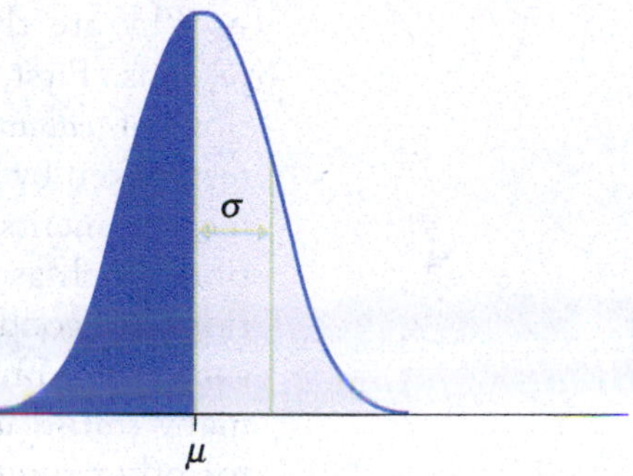

We capitalize Normal to remind you that these curves are special. Look at the two Normal curves in Figure 3.8. They illustrate several important facts:

- All Normal curves have the same overall shape: symmetric, single-peaked, bell-shaped.
- Any specific Normal curve is completely described by giving its mean μ and its standard deviation σ.
- The mean is located at the center of the symmetric curve and is the same as the median. Changing μ without changing σ moves the Normal curve along the horizontal axis without changing its variability.
- The standard deviation σ controls the variability of a Normal curve. When the standard deviation is larger, the area under the normal curve is less concentrated about the mean.

The standard deviation σ is the natural measure of variability for Normal distributions. Not only do μ and σ completely determine the shape of a Normal curve, but we can also locate σ by eye on a Normal curve. Here's how. Imagine that you are skiing down a mountain that has the shape of a Normal curve. At first, you descend at an ever-steeper angle as you go out from the peak:

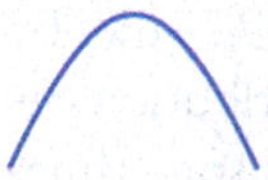

Fortunately, before you find yourself going straight down, the slope begins to grow flatter rather than steeper as you go out and down:

The points at which this change of curvature takes place are located at distance σ on either side of the mean μ. You can feel the change as you run a pencil along a Normal curve, and by doing that find (roughly) the standard deviation. Remember that *μ and σ alone do not specify the shape of most distributions* and that the shape of density curves in general does not reveal σ. These are special properties of Normal distributions.

Normal Distributions

A **Normal distribution** is described by a Normal density curve. Any particular Normal distribution is completely specified by two numbers: its mean μ and standard deviation σ.

The mean of a Normal distribution is at the center of the symmetric Normal curve. The standard deviation is the distance from the center to the change-of-curvature points on either side.

Why are the Normal distributions important in statistics? Here are three reasons. First, Normal distributions are good descriptions for some distributions of *real data*. Distributions that are often close to Normal include scores on tests taken by many people (such as Iowa Test and SAT exams), repeated careful measurements of the same quantity, and characteristics of biological populations (such as lengths of crickets and yields of corn). Second, Normal distributions are good approximations to the results of many kinds of *chance outcomes*, such as the proportion of heads in many tosses of a coin. Third, we will see that many *statistical inference* procedures based on Normal distributions work well for other roughly symmetric distributions. However, many sets of data do not follow a Normal distribution. Most income distributions, for example, are skewed to the right and so are not Normal. Non-Normal data, like non-normal people, not only are common but also are sometimes more interesting than their Normal counterparts.

3.4 The 68–95–99.7 Rule

Although there are many Normal curves, they all have common properties. In particular, all Normal distributions obey the following rule.

The 68–95–99.7 Rule

In the Normal distribution with mean μ and standard deviation σ:

- Approximately 68% of the observations fall within σ of the mean μ.
- Approximately 95% of the observations fall within 2σ of μ.
- Approximately 99.7% of the observations fall within 3σ of μ.

Figure 3.9 illustrates the 68–95–99.7 rule. By remembering these three numbers, you can think about Normal distributions without constantly making detailed calculations. The 68–95–99.7 rule is sometimes called the *empirical rule.*

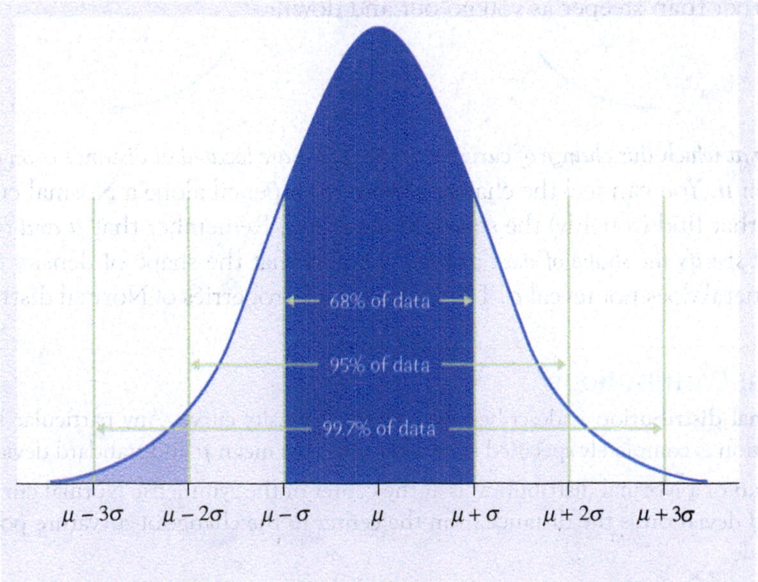

FIGURE 3.9
The 68–95–99.7 rule for Normal distributions.

EXAMPLE 3.2 Iowa Test Scores

IOWATEST

Figures 3.1 and 3.2 (see pages 74 and 75) show that the distribution of *Iowa Test* vocabulary scores for seventh-grade students in Gary, Indiana, is close to Normal. Suppose that the distribution is exactly Normal with mean $\mu = 6.84$ and standard deviation $\sigma = 1.55$. (These are the mean and standard deviation of the 947 actual scores.)

Figure 3.10 applies the 68–95–99.7 rule to the Iowa Test scores. The 95 part of the rule says that approximately 95% of all scores are between

$$\mu - 2\sigma = 6.84 - (2)(1.55) = 6.84 - 3.10 = 3.74$$

and

$$\mu + 2\sigma = 6.84 + (2)(1.55) = 6.84 + 3.10 = 9.94$$

The other 5% of scores are outside this range. Because Normal distributions are symmetric, half of these scores are lower than 3.74, and half are higher than 9.94. That is, about 2.5% of the scores are below 3.74, and 2.5% are above 9.94.

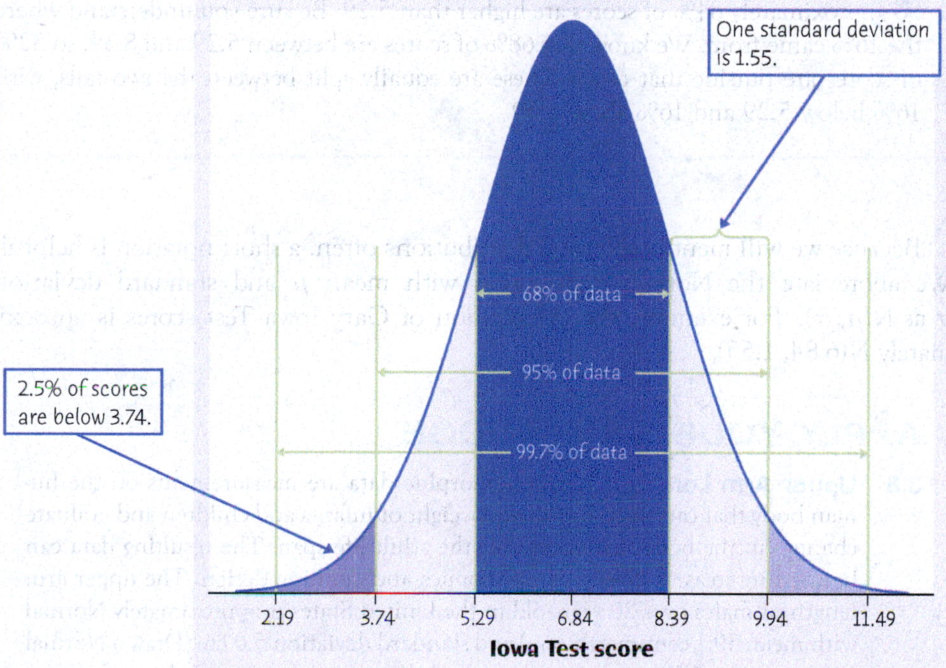

FIGURE 3.10 The 68–95–99.7 rule applied to the distribution of Iowa Test scores for seventh-grade students in Gary, Indiana, for Example 3.2. The mean and standard deviation are $\mu = 6.84$ and $\sigma = 1.55$.

The 68–95–99.7 rule describes distributions that are exactly Normal. Real data such as the actual Gary scores are never exactly Normal. For one thing, Iowa Test scores are reported only to the nearest tenth. A score can be 9.9 or 10.0 but not 9.94. We use a Normal distribution because it's a good approximation and because we think the knowledge that the test measures is continuous rather than stopping at tenths.

How well does our work in Example 3.2 describe the actual Iowa Test scores? Well, 900 of the 947 scores are between 3.74 and 9.94. That's 95.04%, very accurate indeed. Of the remaining 47 scores, 20 are below 3.74, and 27 are above 9.94. The tails of the actual data are not quite equal, as they would be in an exactly Normal distribution. Normal distributions often describe real data better in the center of the distribution than in the extreme high and low tails.

EXAMPLE 3.3 Iowa Test Scores

Look again at Figure 3.10. A score of 5.29 is one standard deviation below the mean. What percentage of scores are higher than 5.29? Find the answer by adding areas in the figure. Here is the calculation in pictures:

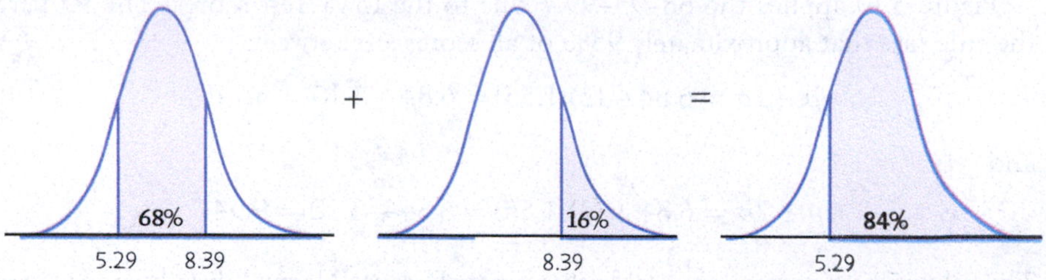

percentage between 5.29 and 8.39 + percentage above 8.39 = percentage above 5.29

68% + 16% = 84%

So approximately 84% of scores are higher than 5.29. Be sure you understand where the 16% came from. We know that 68% of scores are between 5.29 and 8.39, so 32% of scores are outside that range. These are equally split between the two tails, with 16% below 5.29 and 16% above 8.39.

Because we will mention Normal distributions often, a short notation is helpful. We abbreviate the Normal distribution with mean μ and standard deviation σ as $N(\mu, \sigma)$. For example, the distribution of Gary Iowa Test scores is approximately $N(6.84, 1.55)$.

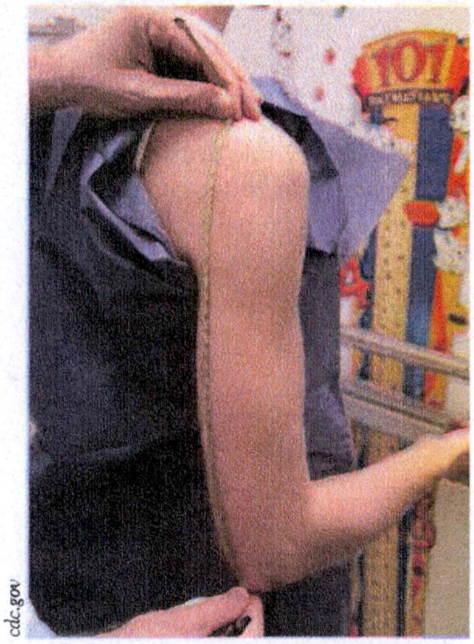

FIGURE 3.11 Correct tape placement when measuring upper arm length, for Exercise 3.5. The upper arm length is measured from the acromion process, the highest point of the shoulder, down the posterior surface of the arm to the tip of the olecranon process, the bony part of the mid-elbow.

APPLY YOUR KNOWLEDGE

3.5 Upper Arm Lengths. Anthropomorphic data are measurements on the human body that can track growth and weight of infants and children and evaluate changes in the body that occur over the adult life span. The resulting data can be used in areas as diverse as ergonomics and clothing design. The upper arm length of males over 20 years old in the United States is approximately Normal with mean 39.1 centimeters (cm) and standard deviation 5.0 cm. Draw a Normal curve on which this mean and standard deviation are correctly located. (*Hint:* Draw an unlabeled Normal curve, locate the points where the curvature changes, then add number labels on the horizontal axis.)

Figure 3.11 shows how arm length is measured.[2]

3.6 Upper Arm Lengths. The upper arm length of males over 20 years old in the United States is approximately Normal with mean 39.1 centimeters (cm) and standard deviation 5.0 cm. Use the 68–95–99.7 rule to answer the following questions. (Start by making a sketch like Figure 3.10.)

(a) What range of lengths covers the middle 99.7% of this distribution?

(b) What percentage of men over 20 have upper arm lengths greater than 44.1 cm?

3.7 Monsoon Rains. The summer monsoon rains bring 80% of India's rainfall and are essential for the country's agriculture. Records going back more than a century show that the amount of monsoon rainfall varies from year to year according to a distribution that is approximately Normal, with mean

852 millimeters (mm) and standard deviation 82 mm.[3] Use the 68–95–99.7 rule to answer the following questions.

(a) Between what values do the monsoon rains fall in the middle 95% of all years?

(b) How small are the monsoon rains in the driest 2.5% of all years?

3.5 The Standard Normal Distribution

As the 68–95–99.7 rule suggests, all Normal distributions share many properties. In fact, all Normal distributions are the same if we measure in units of size σ about the mean μ as center. Changing to these units is called *standardizing*. To standardize a value, subtract the mean of the distribution and then divide by the standard deviation.

Standardizing and z-Scores

If x is an observation from a distribution that has mean μ and standard deviation σ, the **standardized value** of x is

$$z = \frac{x - \mu}{\sigma}$$

A standardized value is often called a **z-score.**

STATISTICS IN YOUR WORLD

He Said, She Said.

Height, weight, and body mass distributions in this book come from actual measurements by a government survey. That is a good thing. When *asked* their weight in a recent Gallup Poll, 38% of respondents said they were overweight. However, a body mass index of over 25 is considered overweight, and the actual measurements from the government survey indicate that well over 50% are overweight.

A z-score tells us how many standard deviations the original observation falls away from the mean and in which direction. Observations larger than the mean are positive when standardized, and observations smaller than the mean are negative.

EXAMPLE 3.4 Standardizing Women's Heights

The heights of women aged 20–29 in the United States are approximately Normal, with $\mu = 64.1$ inches and $\sigma = 3.7$ inches.[4] The standardized height is

$$z = \frac{\text{height} - 64.1}{3.7}$$

A woman's standardized height is the number of standard deviations by which her height differs from the mean height of all women aged 20–29. A woman 70 inches tall, for example, has standardized height

$$z = \frac{70 - 64.1}{3.7} = 1.59$$

or 1.59 standard deviations above the mean. Similarly, a woman 5 feet (60 inches) tall has standardized height

$$z = \frac{60 - 64.1}{3.7} = -1.11$$

or 1.11 standard deviations less than the mean height.

We often standardize observations from symmetric distributions to express them in a common scale. We might, for example, compare the heights of two children of different ages by calculating their *z*-scores. The standardized heights tell us where each child stands in the distribution for his or her age group.

If the variable we standardize has a Normal distribution, standardizing does more than give a common scale. It makes all Normal distributions into a single distribution, and this distribution is still Normal. Standardizing a variable that has any Normal distribution produces a new variable that has the *standard Normal distribution*.

Standard Normal Distribution

The **standard Normal distribution** is the Normal distribution $N(0, 1)$ with mean 0 and standard deviation 1.

If a variable x has any Normal distribution $N(\mu, \sigma)$ with mean μ and standard deviation σ, then the **standardized variable**

$$z = \frac{x - \mu}{\sigma}$$

has the standard Normal distribution.

APPLY YOUR KNOWLEDGE

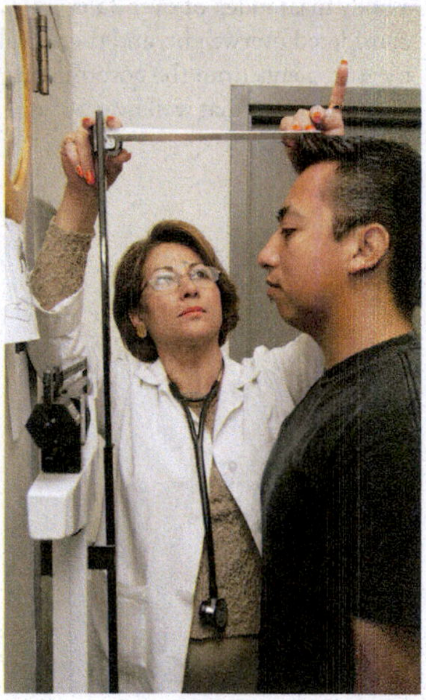

3.8 SAT versus ACT. In 2018, when she was a high school senior, Linda scored 680 on the mathematics part of the SAT.[5] The distribution of SAT math scores in 2018 was Normal with mean 528 and standard deviation 117. Jack took the ACT and scored 26 on the mathematics portion. ACT math scores for 2018 were Normally distributed with mean 20.5 and standard deviation 5.5. Find the standardized scores for both students. Assuming that both tests measure the same kind of ability, who had the higher score?

3.9 Men's and Women's Heights. The heights of women aged 20–29 in the United States are approximately Normal, with mean 64.1 inches and standard deviation 3.7 inches. Men the same age have mean height 69.4 inches, with standard deviation 3.1 inches.[6] What are the *z*-scores for a woman 5.5 feet tall and a man 5.5 feet tall? Say in simple language what information the *z*-scores give that the original nonstandardized heights do not.

3.6 Finding Normal Proportions

Areas under a Normal curve represent proportions of observations from that Normal distribution. There is no formula for areas under a Normal curve. Calculations use either software that calculates areas or a table of areas. Most tables and software calculate one kind of area, *cumulative proportions*. The idea of "cumulative" is "everything that came before." Here is the exact statement.

Cumulative Proportions

The **cumulative proportion** for a value *x* in a distribution is the proportion of observations in the distribution that are less than or equal to *x*.

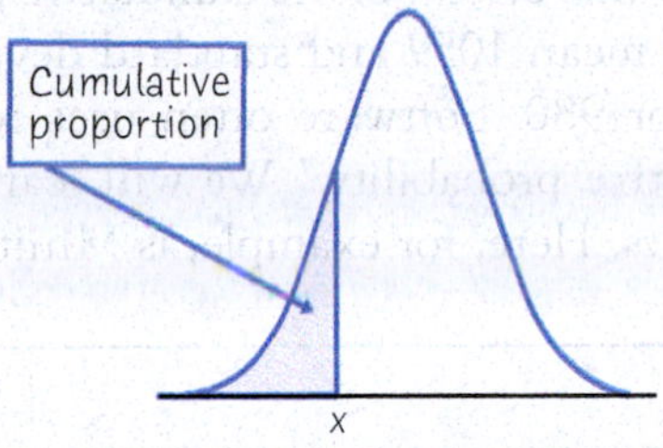

The key to calculating Normal proportions is to match the area you want with areas that represent cumulative proportions. If you make a sketch of the area you want, you will almost never go wrong. Find areas for cumulative proportions either from software or (with an extra step) from a table. The following example shows the method in a picture.

EXAMPLE 3.5 Who Qualifies for College Sports?

The National Collegiate Athletic Association (NCAA) uses a sliding scale for eligibility for Division I athletes.[7] Those students with a lower high school GPA must score higher on the combined mathematics and critical reading parts of the SAT (or the composite ACT) to compete in their first college year. Beginning in August 2016, first-year eligibility requires a minimum 2.3 GPA in high school core courses. Those students with a 2.3 high school core GPA who take the SAT are required to score at least 980 on the combined mathematics and evidence-based reading and writing parts of the SAT to be eligible. The combined scores of the almost 2.2 million high school seniors taking the SAT in 2018 were approximately Normal with mean 1059 and standard deviation 210. What proportion of high school seniors meet this SAT requirement of a combined score of 980 or better?

Here is the calculation in a picture: the proportion of scores above 980 is the area under the Normal curve to the right of 980. We know that 980 is to the left of the center of the graph because $980 < \mu$. The area to the right of 980 is the total area under the curve (which is always 1) minus the cumulative proportion up to 980.

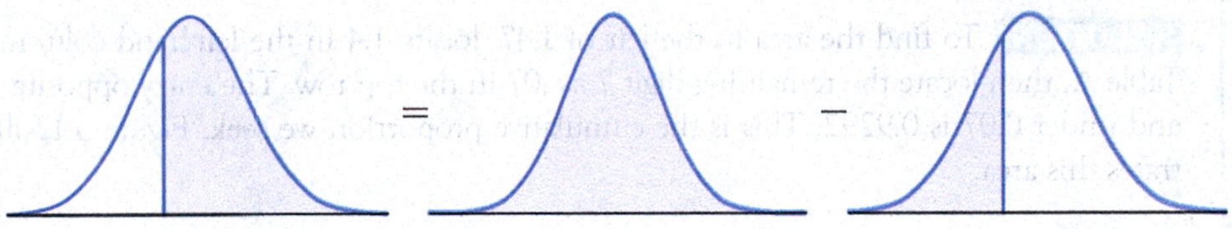

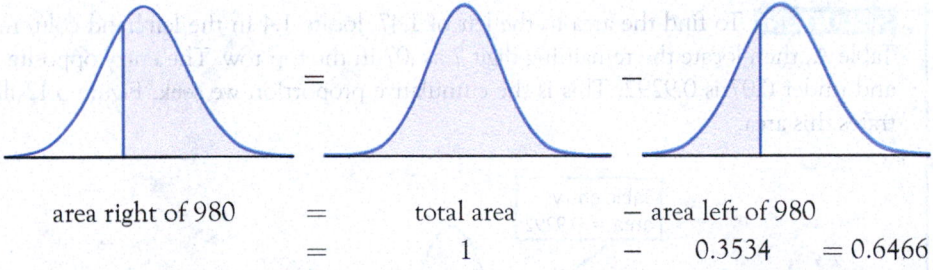

area right of 980	=	total area	− area left of 980	
	=	1	− 0.3534	= 0.6466

The area to the left of 980 was found to be 0.3534 using software. The use of software to find areas under a Normal curve will be discussed further at the end of this section. To conclude the example, about 65% of all high school seniors meet this SAT requirement of a combined math and reading score of 980 or higher.

There is *no* area under a smooth curve and exactly over the point 980. Consequently, the area to the right of 980 (the proportion of scores > 980) is the same as the area at or to the right of this point (the proportion of scores ≥ 980).

The actual data may contain a student who scored exactly 980 on the SAT. That the proportion of scores exactly equal to 980 is 0 for a Normal distribution is a consequence of the idealized smoothing of Normal distributions for data.

To find the numerical value 0.3534 of the cumulative proportion in Example 3.5 using software, enter the mean 1059 and standard deviation 210 and ask for the cumulative proportion for 980. Software often uses terms such as "cumulative distribution" or "cumulative probability." We will learn in Chapter 11 why the language of probability fits. Here, for example, is Minitab's output:

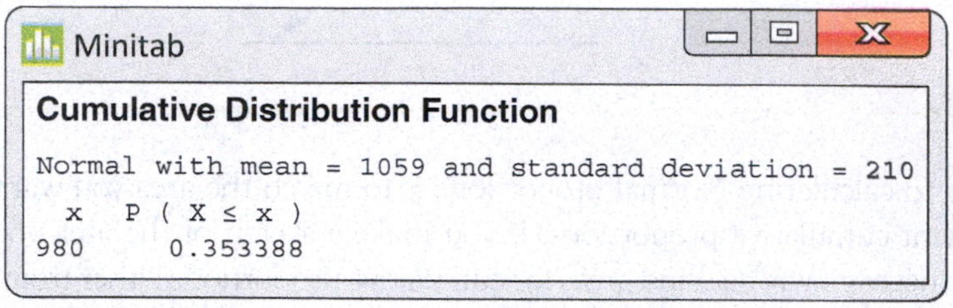

Minitab

Cumulative Distribution Function

```
Normal with mean = 1059 and standard deviation = 210
  x    P ( X ≤ x )
980     0.353388
```

applet

The *P* in the output stands for "probability," but we can read it as "proportion of the observations." The *Normal Density Curve* applet is even handier because it draws pictures as well as finding areas. If you are not using software, you can find cumulative proportions for Normal curves from a table. This requires an extra step.

3.7 Using the Standard Normal Table

The extra step in finding cumulative proportions from a table is that we must first standardize to express the problem in the standard scale of *z*-scores. This allows us to get by with just one table, a table of *standard Normal cumulative proportions*. Table A in the back of the book gives cumulative proportions for the standard Normal distribution. The pictures at the top of the table remind us that the entries are cumulative proportions, areas under the curve to the left of a value *z*.

EXAMPLE 3.6 The Standard Normal Table

What proportion of observations on a standard Normal variable *z* take values less than 1.47?

SOLUTION: To find the area to the left of 1.47, locate 1.4 in the left-hand column of Table A, then locate the remaining digit 7 as .07 in the top row. The entry opposite 1.4 and under 0.07 is 0.9292. This is the cumulative proportion we seek. Figure 3.12 illustrates this area.

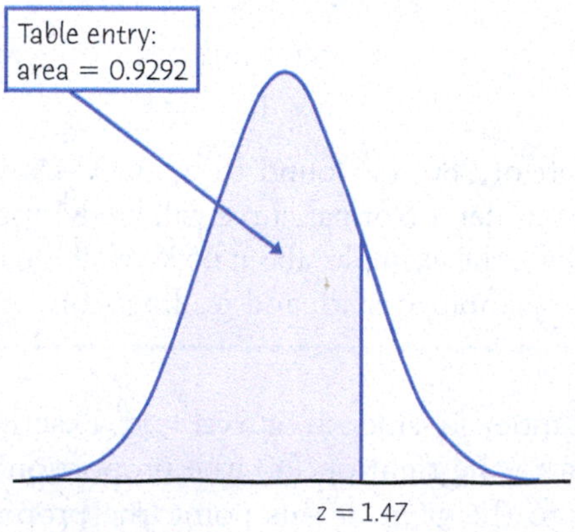

FIGURE 3.12 The area under a standard Normal curve to the left of the point $z = 1.47$ is 0.9292. Table A gives areas under the standard Normal curve.

Now that you see how Table A works, let's redo Example 3.5 using the table. We can break Normal calculations using the table into three steps.

EXAMPLE 3.7 Who Qualifies for College Sports?

Scores of high school seniors on the SAT follow the Normal distribution, with mean $\mu = 1059$ and standard deviation $\sigma = 210$. What percentage of seniors score at least 980?

Step 1. Draw a picture. The picture is exactly as in Example 3.5. It shows that

$$\text{area to the right of } 980 = 1 - \text{area to the left of } 980$$

Step 2. Standardize. Call the SAT score x. Subtract the mean and then divide by the standard deviation to transform the problem about x into a problem about a standard Normal z:

$$x \geq 980$$
$$\frac{x - 1059}{210} \geq \frac{980 - 1059}{210}$$
$$z \geq -0.38$$

Step 3. Use the table. The picture shows that we need the cumulative proportion for $x = 980$. Step 2 says this is the same as the cumulative proportion for $z = -0.38$. The Table A entry for $z = -0.38$ says that this cumulative proportion is 0.3520. The area to the right of -0.38 is therefore $1 - 0.3520 = 0.6480$.

The area from the table in Example 3.7 (0.6480) is slightly less accurate than the area from software in Example 3.5 (0.6466) because we must round z to two decimal places when we use Table A. The difference is rarely important in practice. Here's the method in outline form.

Using Table A to Find Normal Proportions

Step 1. State the problem in terms of the observed variable x. Draw a picture that shows the proportion you want in terms of cumulative proportions.

Step 2. Standardize x to restate the problem in terms of a standard Normal variable z.

Step 3. Use Table A to find areas to the left of z. The fact that the total area under the curve is 1 may also be necessary to find the required area.

EXAMPLE 3.8 Who Qualifies for College Sports?

Recall that the NCAA uses a sliding scale for eligibility for Division I athletics. Students with the minimum GPA of 2.3 must have a combined SAT score of 980 or higher to be eligible. Students with higher GPAs can have a lower SAT score and still be eligible. For example, students with a 2.75 GPA are only required to have a combined SAT score that is at least 810. What proportion of all students who take the SAT would meet an SAT requirement of at least 810 but not 980?

Step 1. State the problem and draw a picture. Call the SAT score x. The variable x has the $N(1059, 210)$ distribution. What proportion of SAT scores fall between 810 and 980? Here is the picture:

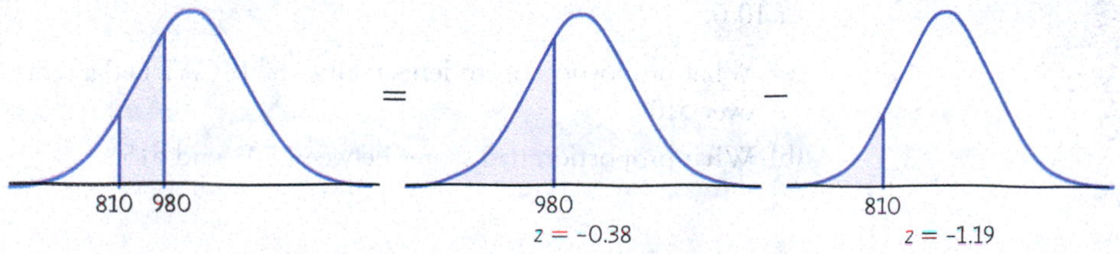

Step 2. Standardize. Subtract the mean and then divide by the standard deviation to turn x into a standard Normal z:

$$810 \leq x < 980$$

$$\frac{810-1059}{210} \leq \frac{x-1059}{210} < \frac{980-1059}{210}$$

$$-1.19 \leq z < -0.38$$

Step 3. Use the table. Follow the picture (we added the z-scores to the picture label to help you):

$$\text{area between } -1.19 \text{ and } -0.38 = (\text{area left of } -0.38) - (\text{area left of } -1.19)$$
$$= 0.3520 - 0.1170 = 0.2350$$

About 24% of high school seniors have SAT scores between 810 and 980.

Sometimes we encounter a value of z more extreme than those appearing in Table A. For example, the area to the left of $z = -4$ is not given directly in the table. The z-values in Table A leave only area 0.0002 in each tail unaccounted for. For practical purposes, we can act as if there is approximately zero area outside the range of Table A. Specifically, we act as if there is approximately zero area below $z = -3.5$ and approximately zero area above $z = 3.5$. While saying the area above $z = 3.5$ is 0.0002 versus saying the area is approximately 0 makes little difference in statistical practice, conceptually it is important to remember that these areas are not actually zero.

APPLY YOUR KNOWLEDGE

3.10 Use the Normal Table. Use Table A to find the proportion of observations from a standard Normal distribution that satisfies each of the following statements. In each case, sketch a standard Normal curve and shade the area under the curve that is the answer to the question.

(a) $z < -0.42$ (b) $z > -1.58$ (c) $z < 2.12$ (d) $-0.42 < z < 2.12$

3.11 Monsoon Rains. The summer monsoon rains in India follow approximately a Normal distribution with mean 852 millimeters (mm) of rainfall and standard deviation 82 mm.

roberthardimg/Alamy Stock Photo

(a) In the drought year 1987, 697 mm of rain fell. In what percentage of all years will India have 697 mm or less of monsoon rain?

(b) "Normal rainfall" means within 20% of the long-term average, or between 682 mm and 1022 mm. In what percentage of all years is the rainfall normal?

3.12 The Medical College Admissions Test. Almost all medical schools in the United States require students to take the Medical College Admission Test (MCAT).[8] The total score of the four sections on the test ranges from 472 to 528. In spring of 2019, the mean score was 500.9, with a standard deviation of 10.6.

(a) What proportion of students taking the MCAT had a score over 510?

(b) What proportion had scores between 505 and 515?

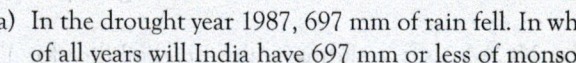

3.8 Finding a Value Given a Proportion

Examples 3.5 through 3.8 illustrated the use of software or Table A to find what proportion of the observations satisfies some condition, such as "SAT score above 810." We may instead want to find the observed value with a given proportion of the observations above or below it. Statistical software can do this directly.

EXAMPLE 3.9 Find the Top 10% Using Software

Scores on the SAT Evidence-Based Reading and Writing (ERW) test in 2018 follow approximately the $N(531, 104)$ distribution. How high must a student score to place in the top 10% of all students taking the SAT?

We want to find the SAT score x with area 0.1 to its *right* under the Normal curve with mean $\mu = 531$ and standard deviation $\sigma = 104$. That's the same as finding the SAT score x with area 0.9 to its *left*. Figure 3.13 poses the question in graphical form. Most software will tell you x when you enter mean 531, standard deviation 104, and cumulative proportion 0.9. Here is Minitab's output:

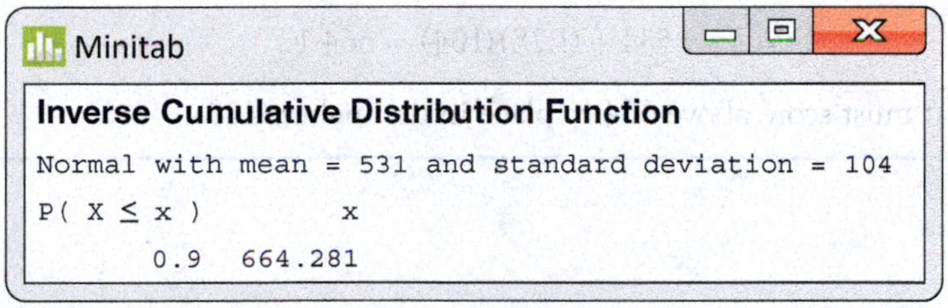

Minitab gives $x = 664.281$. So scores above 664 are in the top 10%. (Round up because SAT scores can only be whole numbers.)

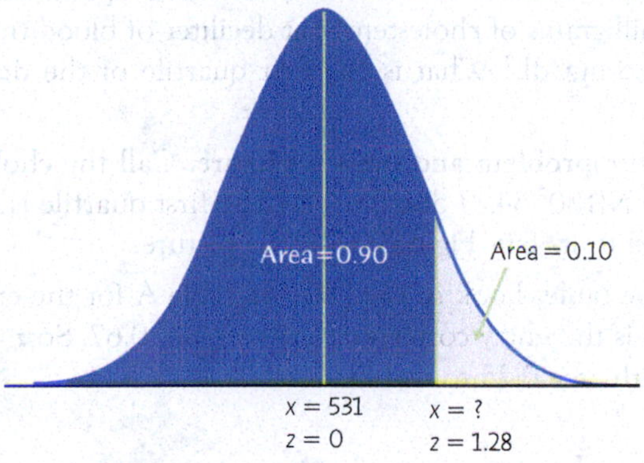

FIGURE 3.13 Locating the point on a Normal curve with area 0.10 to its right, for Examples 3.9 and 3.10.

Without software, use Table A backward. Find the given proportion in the body of the table and then read the corresponding z from the left column and top row. There are again three steps.

Unstandardize a z-Score

Step 1. State the problem in terms of the observed variable x. Draw a picture that shows the proportion you want in terms of cumulative proportions.

Step 2. Use Table A to find the entry in the body of the table that is closest to the cumulative proportion you want. Determine the corresponding value of z in the table.

Step 3. Unstandardize to transform z back to the original x scale using $x = \mu + z \times \sigma$. (Note that the value of z in Step 2 may be a negative number.)

EXAMPLE 3.10 Find the Top 10% Using Table A

Scores on the SAT Critical Reading test in 2018 follow approximately the $N(531, 104)$ distribution. How high must a student score in order to place in the top 10% of all students taking the SAT?

Step 1. State the problem and draw a picture. This step is exactly as in Example 3.9. The picture is Figure 3.13. The *x*-value that puts a student in the top 10% is the same as the *x*-value for which 90% of the area is to the left of *x*.

Step 2. Use the table. Look in the body of Table A for the entry closest to 0.9. It is 0.8997. This is the entry corresponding to $z = 1.28$. So $z = 1.28$ is the standardized value with area 0.9 to its left.

Step 3. Unstandardize to transform z back to the original x scale. We know that the standardized value of the unknown x is $z = 1.28$. This means that x itself lies 1.28 standard deviations above the mean on this particular Normal curve. That is,

$$x = \text{mean} + (1.28)(\text{standard deviation})$$
$$= 531 + (1.28)(104) = 664.12$$

A student must score above 664 to place in the highest 10%.

EXAMPLE 3.11 Find the First Quartile

High levels of cholesterol in the blood increase the risk of heart disease. For males aged 20–24, the distribution of blood cholesterol is approximately Normal, with mean $\mu = 180$ milligrams of cholesterol per deciliter of blood (mg/dl) and standard deviation $\sigma = 34.8$ mg/dl.[9] What is the first quartile of the distribution of blood cholesterol?

Step 1. State the problem and draw a picture. Call the cholesterol level x. The variable x has the $N(180, 34.8)$ distribution. The first quartile is the value with 25% of the distribution to its left. Figure 3.14 is the picture.

Step 2. Use the table. Look in the body of Table A for the entry closest to 0.25. It is 0.2514. This is the entry corresponding to $z = -0.67$. So $z = -0.67$ is the standardized value with area 0.25 to its left.

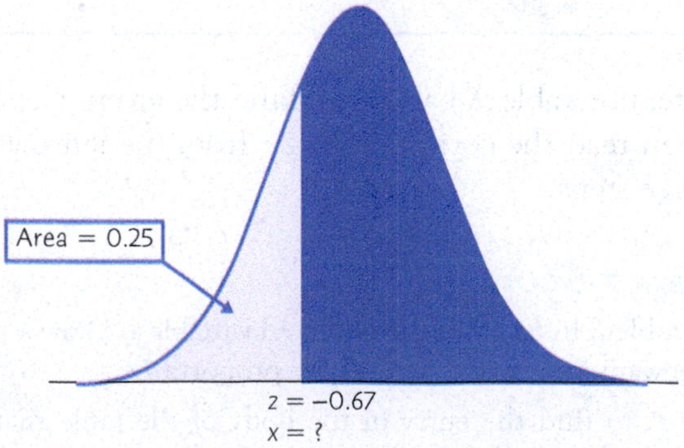

FIGURE 3.14
Locating the first quartile of a Normal curve, for Example 3.11.

Step 3. Unstandardize. The cholesterol level corresponding to $z = -0.67$ lies 0.67 standard deviation below the mean, so

$$x = \text{mean} + (0.67)(\text{standard deviation})$$
$$= 180 - (0.67)(34.8) = 156.7$$

The first quartile of blood cholesterol levels in males aged 20–24 is about 157 mg/dl.

APPLY YOUR KNOWLEDGE

3.13 Table A. Use Table A to find the value z of a standard Normal variable that satisfies each of the following conditions. (Use the value of z from Table A that comes closest to satisfying the condition.) In each case, sketch a standard Normal curve with your value of z marked on the axis.

(a) The point z with 75% of the observations falling below it

(b) The point z with 15% of the observations falling above it

(c) The point z with 15% of the observations falling below it

3.14 The Medical College Admissions Test. Almost all medical schools in the United States require students to take the Medical College Admission Test (MCAT). The total score of the four sections on the test ranges from 472 to 528. In spring of 2019, the mean score was 500.9, with a standard deviation of 10.6.

(a) What are the median and the first and third quartiles of the MCAT scores? What is the interquartile range?

(b) Give the interval that contains the central 80% of the MCAT scores.

STATISTICS IN YOUR WORLD

The Bell Curve?

Does the distribution of human intelligence follow the "bell curve" of a Normal distribution? Scores on IQ tests do roughly follow a Normal distribution. That is because a test score is calculated from a person's answers in a way that is designed to produce a Normal distribution. To conclude that intelligence follows a bell curve, we must agree that the test scores directly measure intelligence. Many psychologists don't think there is one human characteristic that we can call "intelligence" and can measure using a single test score.

CHAPTER 3 SUMMARY

- We can sometimes describe the overall pattern of a distribution by using a **density curve.** A density curve has total area 1 underneath it. An area under a density curve gives the proportion of observations that fall in a range of values.
- A density curve is an idealized description of the overall pattern of a large number of observations that smooths out the irregularities in the actual data. We write the mean of a density curve as μ and the standard deviation of a density curve as σ to distinguish them from the mean $\bar{x}$ and standard deviation s of the actual data.
- The mean, the median, and the quartiles of a density curve can be located by eye. The **mean** μ is the balance point of the curve. The **median** divides the area under the curve in half. The quartiles and the median divide the area under the curve into quarters. The *standard deviation* σ cannot be located by eye on most density curves.
- The mean and median are equal for symmetric density curves. The mean of a skewed curve is located farther toward the long tail than is the median.
- The **Normal distributions** are described by a special family of bell-shaped, symmetric density curves, called *Normal curves.* The mean μ and standard deviation σ completely specify a Normal distribution $N(\mu, \sigma)$. The mean is the center of the curve, and σ is the distance from μ to the change-of-curvature points on either side.

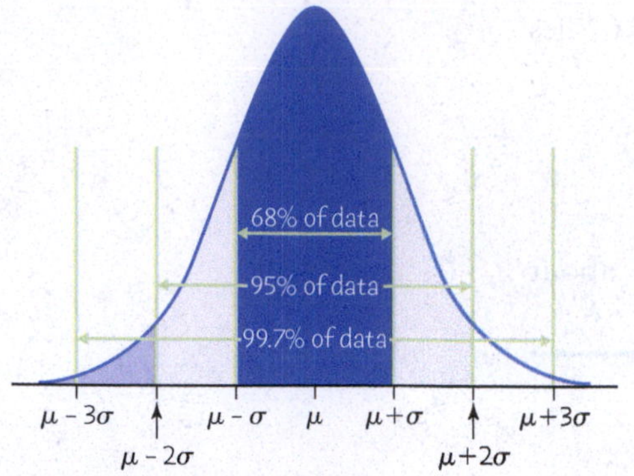

FIGURE 3.15
The 68–95–99.7 rule for Normal distributions.

- To **standardize** any observation x, subtract the mean of the distribution and then divide by the standard deviation. The resulting ***z*-score**

$$z = \frac{x - \mu}{\sigma}$$

says how many standard deviations x lies from the distribution mean.
- All Normal distributions are the same when measurements are transformed to the standardized scale. In particular, all Normal distributions satisfy the **68–95–99.7 rule,** which describes what percentage of observations lie within one, two, and three standard deviations of the mean, respectively.
- If x has the $N(\mu, \sigma)$ distribution, then the **standardized variable** $z = (x - \mu)/\sigma$ has the **standard Normal distribution** $N(0, 1)$ with mean 0 and standard deviation 1. Table A gives the cumulative proportions of standard Normal observations that are less than z for many values of z. By standardizing, we can use Table A for any Normal distribution.

CHECK YOUR SKILLS

3.15 Which of these variables is most likely to have a Normal distribution?

(a) Income per person for 150 different countries

(b) Sale prices of 200 homes in Santa Barbara, California

(c) Lengths of 100 newborns in Connecticut

3.16 To completely specify the shape of a Normal distribution, you must give

(a) the mean and the standard deviation.

(b) the five-number summary.

(c) the median and the quartiles.

3.17 Figure 3.16 shows a Normal curve. The mean of this distribution is

(a) 0.

(b) 2.

(c) 3.

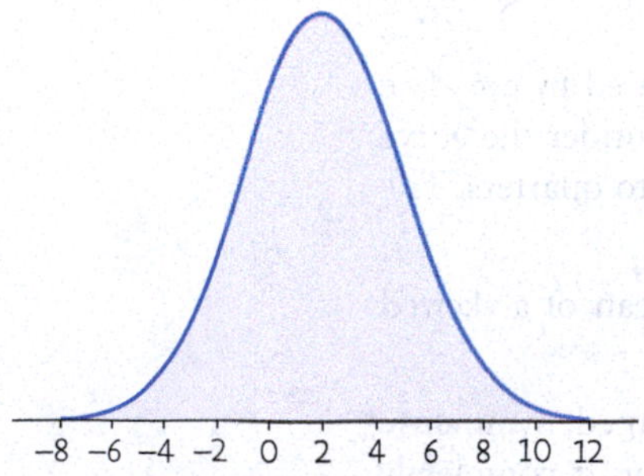

FIGURE 3.16
A Normal curve, for Exercises 3.17 and 3.18.

3.18 The standard deviation of the Normal distribution in Figure 3.16 is

(a) 2. (b) 3. (c) 5.

3.19 The distribution of hours of sleep per weeknight among college students is found to be Normally distributed, with a mean of 6.5 hours and a standard deviation of 1 hour. What range contains the middle 95% of hours slept per weeknight by college students?

(a) 5.5 and 7.5 hours per weeknight

(b) 4.5 and 7.5 hours per weeknight

(c) 4.5 and 8.5 hours per weeknight

3.20 The distribution of hours of sleep per weeknight among college students is found to be Normally distributed, with a mean of 6.5 hours and a standard deviation of 1 hour. The percentage of college students that sleep at least eight hours per weeknight is about

(a) 95%. (b) 6.7%. (c) 2.5%.

3.21 The scores of adults on an IQ test are approximately Normally distributed, with mean 100 and standard deviation 15. Alysha scores 135 on such a test. Her *z*-score is about

(a) 1.33. (b) 2.33. (c) 6.33.

3.22 The proportion of observations from a standard Normal distribution that take values greater than 1.78 is about

(a) 0.9554.

(b) 0.0446.

(c) 0.0375.

3.23 The proportion of observations from a standard Normal distribution that take values between 1 and 2 is about

(a) 0.025.

(b) 0.135.

(c) 0.160.

3.24 The scores of adults on an IQ test are approximately Normally distributed, with mean 100 and standard deviation 15. Alysha scores 135 on such a test. She scores higher than what percentage of all adults?

(a) About 5% (b) About 95% (c) About 99%

CHAPTER 3 EXERCISES

3.25 Understanding Density Curves. Remember that it is areas under a density curve, not the height of the curve, that give proportions in a distribution. To illustrate this, sketch a density curve that has a tall, thin peak at 0 on the horizontal axis but has most of its area close to 1 on the horizontal axis without a high peak at 1.

3.26 Daily Activity. It appears that people who are mildly obese are less active than leaner people. One study looked at the average number of minutes per day that people spend standing or walking.[10] Among mildly obese people, minutes of activity varied according to the $N(373, 67)$ distribution. Minutes of activity for lean people had the $N(526, 107)$ distribution. Within what limits do the active minutes for about 95% of the people in each group fall? Use the 68–95–99.7 rule.

3.27 Cholesterol. Low density lipoprotein, or LDL, is the main source of cholesterol buildup and blockage in the arteries. This is why LDL is known as "bad cholesterol." LDL is measured in milligrams per deciliter of blood, or mg/dL. In a population of adults at risk for cardiovascular problems, the distribution of LDL levels is Normal, with a mean of 123 mg/dL and a standard deviation of 41 mg/dL. If an individual's LDL is at least 1 standard deviation or more above the mean, he or she will be monitored carefully by a doctor. What percentage of individuals from this population will have LDL levels 1 or more standard deviations above the mean? Use the 68–95–99.7 rule.

3.28 Standard Normal Drill. Use Table A to find the proportion of observations from a standard Normal distribution that falls in each of the following regions. In each case, sketch a standard Normal curve and shade the area representing the region.

(a) $z \leq -1.63$ (b) $z \geq -1.63$

(c) $z > 0.92$ (d) $-1.63 < z < 0.92$

3.29 Standard Normal Drill.

(a) Find the number z such that the proportion of observations that are less than z in a standard Normal distribution is 0.2.

(b) Find the number z such that 40% of all observations from a standard Normal distribution are greater than z.

3.30 Fruit Flies. The common fruit fly *Drosophila melanogaster* is the most studied organism in genetic research because it is small, is easy to grow, and reproduces rapidly. The length of the thorax (where the wings and legs attach) in a population of male fruit flies is approximately Normal with mean 0.800 millimeter (mm) and standard deviation 0.078 mm.

(a) What proportion of flies have thorax length less than 0.7 mm?

(b) What proportion have thorax length greater than 1.0 mm?

(c) What proportion have thorax length between 0.7 mm and 1.0 mm?

© Plastique1 | Dreamstime.com

3.31 Acid Rain? Emissions of sulfur dioxide by industry set off chemical changes in the atmosphere that result in "acid rain." The acidity of liquids is measured by pH on a scale of 0 to 14. Distilled water has pH 7.0, and lower pH values indicate acidity. Normal rain is somewhat acidic, so acid rain is sometimes defined as rainfall with a pH below 5.0. The pH of rain at one location varies among rainy days according to a Normal distribution with mean 5.43 and standard deviation 0.54. What proportion of rainy days have rainfall with pH below 5.0?

3.32 **Runners.** In a study of exercise, a large group of male runners walk on a treadmill for 6 minutes. Their heart rates in beats per minute at the end vary from runner to runner according to the $N(104, 12.5)$ distribution. The heart rates for male nonrunners after the same exercise have the $N(130, 17)$ distribution.

(a) What percentage of the runners have heart rates above 140?

(b) What percentage of the nonrunners have heart rates above 140?

3.33 **Are We Getting Smarter?** When the Stanford-Binet IQ test came into use in 1932, it was adjusted so that scores for each age group of children followed roughly the Normal distribution with mean 100 and standard deviation 15. The test is readjusted from time to time to keep the mean at 100. If present-day American children took the 1932 Stanford-Binet test, their mean score would be about 120. The reasons for the increase in IQ over time are not known but probably include better childhood nutrition and more experience in taking tests.[11]

(a) IQ scores above 130 are often called "very superior." What percentage of children had very superior scores in 1932?

(b) If present-day children took the 1932 test, what percentage would have very superior scores? (Assume that the standard deviation 15 does not change.)

3.34 **Body Mass Index.** Your body mass index (BMI) is your weight in kilograms divided by the square of your height in meters. Many online BMI calculators allow you to enter weight in pounds and height in inches. High BMI is a common but controversial indicator of overweight or obesity. A study by the National Center for Health Statistics found that the BMI of 2-year-old American male children is approximately Normally distributed, with mean 16.8 and standard deviation 1.9.[12]

(a) What percentage of 2-year-old American male children have a BMI less than 15.0?

(b) What percentage of 2-year-old American male children have a BMI less than 18.5?

Miles per Gallon. *In its* Fuel Economy Guide *for 2019 model vehicles, the Environmental Protection Agency gives data on 1259 vehicles. There are a number of high outliers, mainly hybrid gas–electric vehicles. If we ignore the vehicles identified as outliers, however, the combined city and highway gas mileage of the other 1231 vehicles is approximately Normal with mean 22.8 miles per gallon (mpg) and standard deviation 4.8 mpg. Exercises 3.35 through 3.38 concern this distribution.*

3.35 **I Love My Bug!** The 2019 Volkswagen Beetle with a four-cylinder 2.0-L engine and automatic transmission has combined gas mileage of 29 mpg. What percentage of all vehicles have better gas mileage than the Beetle?

3.36 **The Top 15%.** How high must a 2019 vehicle's gas mileage be to fall in the top 15% of all vehicles?

3.37 **The Middle Half.** The quartiles of any distribution are the values with cumulative proportions 0.25 and 0.75. They span the middle half of the distribution. What are the quartiles of the distribution of gas mileage?

3.38 **Quintiles.** The quintiles of any distribution are the values with cumulative proportions 0.20, 0.40, 0.60, and 0.80. What are the quintiles of the distribution of gas mileage?

3.39 **What's Your Percentile?** Reports on a student's test score such as the SAT or a child's height or weight usually give the *percentile* as well as the actual value of the variable. The percentile is just the cumulative proportion stated as a percentage: the percentage of all values of the variable that were lower than this one. The upper arm lengths of females 20 years and over in the United States are approximately Normally distributed, with mean 35.9 cm and standard deviation 5.1 cm, and those for males 20 years and over are approximately Normally distributed, with mean 39.1 cm and standard deviation 5.0 cm.

(a) Cecile, a 73-year-old female in the United States, has an upper arm length of 33.9 cm. What is her percentile?

(b) Measure your upper arm length to the nearest tenth of a centimeter, referring to Figure 3.11 (page 82) for the measurement instructions. What is your arm length, in centimeters? What is your percentile?

3.40 **Perfect SAT Scores.** It is possible to score higher than 1600 on the combined mathematics and evidence-based reading and writing portions of the SAT, but scores 1600 and above are reported as 1600. The distribution of SAT scores (combining Mathematics and Reading) in 2019 was close to Normal, with mean 1059 and standard deviation 210. What proportion of SAT scores for these two parts were reported as 1600? (That is, what proportion of SAT scores were actually 1600 or higher?)

3.41 **Heights of Men and Women.** The heights of women aged 20–29 follow approximately the $N(64.1, 3.7)$ distribution. Men the same age have heights distributed as $N(69.4, 3.1)$. What percentage of men aged 20–29 are taller than the mean height of women aged 20–29?

3.42 **Weights Aren't Normal.** The heights of people of the same sex and similar ages follow a Normal distribution reasonably closely. Weights, on the other hand, are not Normally distributed. The weights of men aged 20–29 in the United States have mean 186.8 pounds and median 177.8 pounds. The first and third quartiles are 152.9 pounds and 208.5 pounds, respectively. In addition, the bottom 10% have weights less than or equal to 137.6 pounds while the top 10% have weights greater than or equal to 247.2. What can you say about the shape of the weight distribution? Why?

3.43 A Surprising Calculation. Changing the mean and standard deviation of a Normal distribution by a moderate amount can greatly change the percentage of observations in the tails. Suppose a college is looking for applicants with either SAT Math or Evidence-Based Reading and Writing (ERW) scores 780 and above.

(a) In 2018, the scores on the math SAT followed the $N(528, 117)$ distribution. What percentage scored 780 or better?

(b) The ERW scores that year had the $N(531, 104)$distribution. What percentage scored 780 or better? You see that the percentage of students with math SAT scores above 780 is almost two times the percentage of students with such high ERW scores.

3.44 Grading Managers. Some companies "grade on a bell curve" to compare the performance of their managers and professional workers. This forces the use of some low performance ratings so that not all workers are listed as "above average." Ford Motor Company's "performance management process" assigned 10% A grades, 80% B grades, and 10% C grades to the company's managers. Suppose Ford's performance scores really are Normally distributed. This year, managers with scores less than 25 received C grades, and those with scores above 475 received A grades. What are the mean and standard deviation of the scores?

3.45 Osteoporosis. Osteoporosis is a condition in which the bones become brittle due to loss of minerals. To diagnose osteoporosis, an elaborate apparatus measures bone mineral density (BMD). BMD is usually reported in standardized form. The standardization is based on a population of healthy young adults. The World Health Organization (WHO) criterion for osteoporosis is a BMD lower than 2.5 standard deviations below the mean for healthy young adults. BMD measurements in a population of people similar in age and sex roughly follow a Normal distribution.

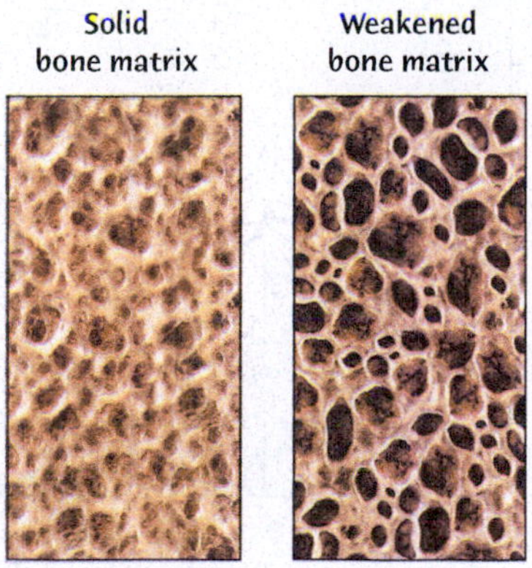

Nucleus Medical Media Inc/Alamy Stock Photo

(a) What percentage of healthy young adults have osteoporosis by the WHO criterion?

(b) Women aged 70–79 are, of course, not young adults. The mean BMD in this age is about −2 on the standard scale for young adults. Suppose the standard deviation is the same as for young adults. What percentage of this older population have osteoporosis?

In later chapters, we will meet many statistical procedures that work well when the data are "close enough to Normal." Exercises 3.46 through 3.50 concern data that are mostly close enough to Normal for statistical work, whereas Exercise 3.51 concerns data for which the data are not close to Normal. These exercises ask you to do data analysis and Normal calculations to investigate how close to Normal real data are.

3.46 Normal Is Only Approximate: ACT Scores. Composite scores on the ACT for the 2019 high school graduating class had mean 20.8 and standard deviation 5.8. In all, 1,914,817 students in this class took the test. Of these, 227,221 had scores higher than 28, and another 54,848 had scores exactly 28. ACT scores are always whole numbers. The exactly Normal $N(20.8, 5.8)$ distribution can include any value, not just whole numbers. What is more, there is *no* area exactly above 28 under the smooth Normal curve. So ACT scores can be only approximately Normal. To illustrate this fact, find

(a) the percentage of 2019 ACT scores greater than 28, using the actual counts reported.

(b) the percentage of 2019 ACT scores greater than or equal to 28, using the actual counts reported.

(c) the percentage of observations that are greater than 28 using the $N(20.8, 5.8)$ distribution. (The percentage greater than or equal to 28 is the same because there is no area exactly over 28.)

3.47 Are the Data Normal? Returns on Stocks. The return on a stock is the change in its market price plus any dividend payments made. Total return is usually expressed as a percentage of the beginning price. Exercise 1.32 provides a histogram of the distribution of the annual returns for all stocks listed on the S&P 500 from 1928 to 2018 (91 years). The average return is 11.36%, with a standard deviation of 19.58%.[13] SPRET

(a) If the distribution of the 91 years were exactly $N(11.36, 19.58)$, what would be the proportion of months with returns greater than 0? Greater than 30%?

(b) What proportion of the actual returns are greater than 0? Greater than 30%? Do these results suggest that the $N(11.36, 19.58)$ provides a good approximation to the distribution of returns over this period?

3.48 Are the Data Normal? Acidity of Rainfall. Exercise 3.31 (page 93) concerns the acidity (measured by pH) of rainfall. A sample of 105 rainwater specimens had mean pH 5.43, standard deviation 0.54, and five-number summary 4.33, 5.05, 5.44, 5.79, 6.81.[14]

(a) Compare the mean and median and also the distances of the two quartiles from the median. Does it appear that the distribution is quite symmetric? Why?

(b) If the distribution is really $N(5.43, 0.54)$, what proportion of observations would be less than 5.05? Less than 5.79? Do these proportions suggest that the distribution is close to Normal? Why?

3.49 Are the Data Normal? SAT Mathematics Scores. Georgia Southern University (GSU) had 2786 students with regular admission in its freshman class of 2015. For each student, data are available on his/her SAT and ACT scores, if taken; high school GPA; and the college within the university to which admitted.[15] Here are the first 20 SAT Mathematics scores from that data set: SATMATH

490	580	450	570	650	420	560	410	480	510
540	530	620	380	440	460	460	600	640	450

The complete data is in the file SATMATH, which contains both the original and ordered mathematics scores.

(a) Make a histogram of the distribution (if your software allows it, superimpose a normal curve over the histogram, as in Figure 3.1). Although the resulting histogram depends a bit on your choice of classes, the distribution appears roughly symmetric, with no outliers.

(b) Find the mean, median, standard deviation, and quartiles for these data. Comparing the mean and the median and comparing the distances of the two quartiles from the median suggest that the distribution is fairly symmetric. Why?

(c) In 2015, the mean score on the mathematics portion of the SAT for all college-bound seniors was 511. If the distribution were exactly Normal, with the mean and standard deviation you found in part (b), what proportion of GSU freshmen scored above the mean for all college-bound seniors?

(d) Compute the exact proportion of GSU freshmen who scored above the mean for all college-bound seniors. It will be simplest to use the ordered scores in the SATMATH file to calculate this. How does this percentage compare with the percentage calculated in part (c)? Despite the discrepancy, this distribution is "close enough to Normal" for statistical work in later chapters.

3.50 Are the Data Normal? Monsoon Rains. Here are the amounts of summer monsoon rainfall (millimeters) for India in the 100 years 1901 to 2000:[16] MONSOON

722.4	792.2	861.3	750.6	716.8	885.5	777.9	897.5	889.6	935.4
736.8	806.4	784.8	898.5	781.0	951.1	1004.7	651.2	885.0	719.4
866.2	869.4	823.5	863.0	804.0	903.1	853.5	768.2	821.5	804.9
877.6	803.8	976.2	913.8	843.9	908.7	842.4	908.6	789.9	853.6
728.7	958.1	868.6	920.8	911.3	904.0	945.9	874.3	904.2	877.3
739.2	793.3	923.4	885.8	930.5	983.6	789.0	889.6	944.3	839.9
1020.5	810.0	858.1	922.8	709.6	740.2	860.3	754.8	831.3	940.0
887.0	653.1	913.6	748.3	963.0	857.0	883.4	909.5	708.0	882.9
852.4	735.6	955.9	836.9	760.0	743.2	697.4	961.7	866.9	908.8
784.7	785.0	896.6	938.4	826.4	857.3	870.5	873.8	827.0	770.2

(a) Make a histogram of these rainfall amounts. Find the mean and the median.

(b) Although the distribution is reasonably Normal, your work shows some departure from Normality. In what way are the data not Normal?

3.51 Are the Data Normal? Weight of Females in Their Twenties. Many body measurements of people of the same sex and similar ages such as height and upper arm length follow a Normal distribution reasonably closely. Weights, on the other hand, are not Normally distributed. The NHANES survey of 2009–2010[17] includes the weights of a representative sample of 548 females in the United States aged 20–29. The mean of the weights was 161.58 pounds, and the standard deviation was 48.96 pounds. Figure 3.17 gives a histogram of the data along with a smooth curve representing an $N(161.58, 48.96)$ distribution. From the figure, the Normal curve does not appear to follow the pattern in the histogram that

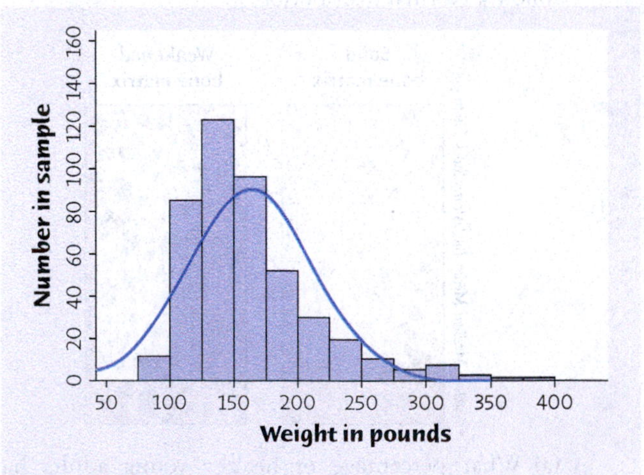

FIGURE 3.17
Histogram of the weights of 548 females aged 20–29 in the 2009–2010 NHANES survey, with a Normal curve superimposed, for Exercise 3.51.

closely. Because of this, the use of areas under the Normal curve may not provide a good approximation to weights in various intervals. FEMWEIGH

(a) Using the data file on the text website, what proportion of females aged 20–29 weighed under 100 pounds? What percentage of the $N(161.58,\ 48.96)$ distribution is below 100?

(b) What proportion of females aged 20–29 weighed over 250 pounds? What percentage of the $N(161.58,\ 48.96)$ distribution is above 250?

(c) Based on your answers in parts (a) and (b), do you think it is a good idea to summarize the distribution of weights by an $N(161.58,\ 48.96)$ distribution?

The Normal Density Curve *applet allows you to do Normal calculations quickly. It is somewhat limited by the number of pixels available for use, so it can't hit every value exactly. In the following exercises, use the closest available values. In each case, make a sketch of (or use software to capture) the curve from the applet marked with the values you used to answer the questions.*

3.52 applet **How Accurate Is 68–95–99.7?** The 68–95–99.7 rule for Normal distributions is a useful approximation. To see how accurate the rule is, note that the area under the curve between the two points on the applet is displayed on the screen.

(a) Place the points 1 standard deviation on either side of the mean. What is the area between these two values? What does the 68–95–99.7 rule say this area is?

(b) Repeat for locations 2 and 3 standard deviations on either side of the mean. Again compare the 68–95–99.7 rule with the area given by the applet.

3.53 applet **Where Are the Quartiles?** How many standard deviations above and below the mean do the quartiles of any Normal distribution lie? (Use the standard Normal distribution to answer this question.)

3.54 applet **Grading Managers.** In Exercise 3.44, we saw that Ford Motor Company once graded its managers in such a way that the top 10% received an A grade, the bottom 10% a C, and the middle 80% a B. Let's suppose that performance scores follow a Normal distribution. How many standard deviations above and below the mean do the A/B and B/C cutoffs lie? (Use the standard Normal distribution to answer this question.)

CHAPTER

4

When you complete this chapter, you will be able to:

4.1 Given two related variables, identify the explanatory variable and the response variable.

4.2 Given two related variables, create a scatterplot.

4.3 Use the characteristics of a scatterplot to make statements about the displayed data.

4.4 Create a scatterplot that distinguishes values of a categorical variable by using a different symbol for each category.

4.5 Given two related variables, calculate their correlation r with and without technology.

4.6 Use the correlation r to describe the relationship between two variables.

Scatterplots and Correlation

In Chapters 1 through 3, we focused on exploring features of a single variable. In this chapter, we continue our study of exploratory data analysis but for the purpose of examining relationships *between* variables. We introduce a useful graphical tool for exploring the relationship between two variables and a numerical measure of the strength of the relationship between two variables.

A medical study finds that short women are more likely to have heart attacks than women of average height, and tall women have the fewest heart attacks. An insurance group reports that heavier cars have fewer deaths per 10,000 vehicles registered than do lighter cars. These and many other statistical studies look at the *relationship between two variables.* Statistical relationships are overall tendencies, not ironclad rules. They allow individual exceptions. Although smokers on the average die younger than nonsmokers, some people live to 90 while smoking three packs a day.

To understand a statistical relationship between two variables, we measure both variables on the same individuals. Often, we must examine other variables as well. To conclude that shorter women have higher risk of heart attack, for example, the researchers had to eliminate the effects of other variables, such as weight and exercise habits. In this chapter and the following chapter, we study relationships between variables. One of our main themes is that the relationship between two variables can be strongly influenced by other variables that are lurking in the background.

Michael Wood/Stocktrek Images/Getty Images

4.1 Explanatory and Response Variables

We think that car weight helps explain accident deaths and that smoking influences life expectancy. In each of these relationships, the two variables play different roles: one explains or influences the other.

Response Variable, Explanatory Variable

A **response variable** measures an outcome of a study. An **explanatory variable** may explain or influence changes in a response variable.

You will often find explanatory variables called *independent variables* and response variables called *dependent variables*. The idea behind this language is that the response variable depends on the explanatory variable. Because "independent" and "dependent" have other meanings in statistics that are unrelated to the explanatory-response distinction, we prefer to avoid those words.

We will sometimes refer to explanatory variables as *predictor variables*. The idea here is that in many applications, the explanatory-response relationship is exploited to predict a response from the value of the explanatory variable.

It is easiest to identify explanatory and response variables when we actually set values of one variable to see how it affects another variable.

EXAMPLE 4.1 Beer and Blood Alcohol

How does drinking beer affect the level of alcohol in our blood? The legal limit for driving in all states is 0.08%. Student volunteers at The Ohio State University drank different numbers of cans of beer. Thirty minutes later, a police officer measured their blood alcohol content. Number of beers consumed is the explanatory variable, and percentage of alcohol in the blood is the response variable.

When we don't set the values of either variable but just observe both variables, there may or may not be explanatory and response variables. Whether there are depends on how we plan to use the data.

EXAMPLE 4.2 College Debts

A college student aid officer looks at the findings of the National Student Loan Survey. She notes data on the amount of debt of recent graduates, their current income, and how stressed they feel about college debt. She isn't interested in predictions but is simply trying to understand the situation of recent college graduates. The distinction between explanatory and response variables does not apply.

A sociologist looks at the same data with an eye to using amount of debt and income, along with other variables, to explain the stress caused by college debt. Now amount of debt and income are explanatory variables, and stress level is the response variable.

In many studies, the goal is to show that changes in one or more explanatory variables actually *cause* changes in a response variable. Other explanatory-response relationships do not involve direct causation. Nations with more television sets per person have greater life expectancies, but shipping many television sets to Botswana won't *cause* life expectancy to increase. Even when *direct* causation is not present, explanatory variables can be used to predict response variables. For example, for many years colleges have used SAT scores to predict success in college. And in 2008

STATISTICS IN YOUR WORLD

After You Plot Your Data, Think!

The statistician Abraham Wald (1902–1950) worked on war problems during World War II. Wald invented some statistical methods that were military secrets until the war ended. Here is one of his simpler ideas. Asked where extra armor should be added to airplanes, Wald studied the location of enemy bullet holes in planes returning from combat. He plotted the locations on an outline of the plane. As data accumulated, most of the outline filled up. Put the armor in the few spots with no bullet holes, said Wald. That's where bullets hit the planes that didn't make it back.

Google used data on Internet searches to accurately predict the spread of influenza. (We doubt that Internet searches caused influenza, but it is plausible that as people developed symptoms, they searched online for information about influenza.)

Most statistical studies examine data on more than one variable. Fortunately, statistical analysis of several-variable data builds on the tools we used to examine individual variables. The principles that guide our work also remain the same:

- Plot your data. Look for overall patterns and deviations from those patterns.
- Based on what your plot shows, choose numerical summaries for some aspects of the data.

APPLY YOUR KNOWLEDGE

4.1 Explanatory and Response Variables? You have data on a large group of college students. Here are four pairs of variables measured on these students. For each pair, is it more reasonable to simply explore the relationship between the two variables or to view one of the variables as an explanatory variable and the other as a response variable? In the latter case, which is the explanatory variable, and which is the response variable?

(a) Number of times a student accessed the course website for your statistics course and grade on the final exam for the course

(b) Number of hours per week spent exercising and calories burned per week

(c) Hours per week spent online using social media and grade point average

(d) Hours per week spent online using social media and IQ

4.2 Coral Reefs. How sensitive to changes in water temperature are coral reefs? To find out, scientists examined data on sea surface temperatures and coral growth per year at locations in the Gulf of Mexico and the Caribbean Sea.[1] What are the explanatory and response variables? Are they categorical or quantitative?

4.3 Predicting Life Expectancy. Identifying variables that can be used to predict life expectancy is important for insurance companies, economists, and policymakers. Several researchers have investigated the extent to which poverty level can be used to predict life expectancy. Name two other variables that could be used to predict life expectancy.

4.2 Displaying Relationships: Scatterplots

The most useful graph for displaying the relationship between two quantitative variables is a *scatterplot*.

Scatterplot

A **scatterplot** is a plot that displays the relationship between two quantitative variables measured on the same individuals. The values of one variable appear on the horizontal axis, and the values of the other appear on the vertical axis. Each individual in the data appears as the point in the plot fixed by the values of both variables for that individual.

Always plot the explanatory variable, if there is one, on the horizontal axis (the x axis) of a scatterplot. As a reminder, we usually call the explanatory variable x and the response variable y. If there is no explanatory/response distinction, either variable can go on the horizontal axis.

EXAMPLE 4.3 State SAT Mathematics Scores

Figure 1.8 (page 26) reminded us that in some states, most high school graduates take the SAT to test their readiness for college, and in other states, most take the ACT. Who takes a test may influence the average score. Let's follow our four-step process (page 62) to examine this influence.[2]

4step

MATHSAT

STATE: The percentage of high school students who take the SAT varies from state to state. Does this fact help explain differences among the states in average SAT Mathematics score?

PLAN: Examine the relationship between percentage taking the SAT and state mean score on the Mathematics part of the SAT. Choose the explanatory and response variables. Make a *scatterplot* to display the relationship between the variables. Interpret the plot to understand the relationship.

SOLVE (make the plot): We suspect that "percentage taking" will help explain "mean score." So "percentage taking" is the explanatory variable, and "mean score" is the response variable. We want to see how mean score changes when percentage taking changes, so we put percentage taking (the explanatory variable) on the horizontal axis. Figure 4.1 is the scatterplot. Each point represents a single state. In Nevada, for example, 23% took the SAT, and their mean SAT Math score was 566. Find 23 on the *x* (horizontal) axis and 566 on the *y* (vertical) axis. Nevada appears as the point (23, 566) above 23 and to the right of 566.

CONCLUDE: We will explore conclusions in Example 4.4.

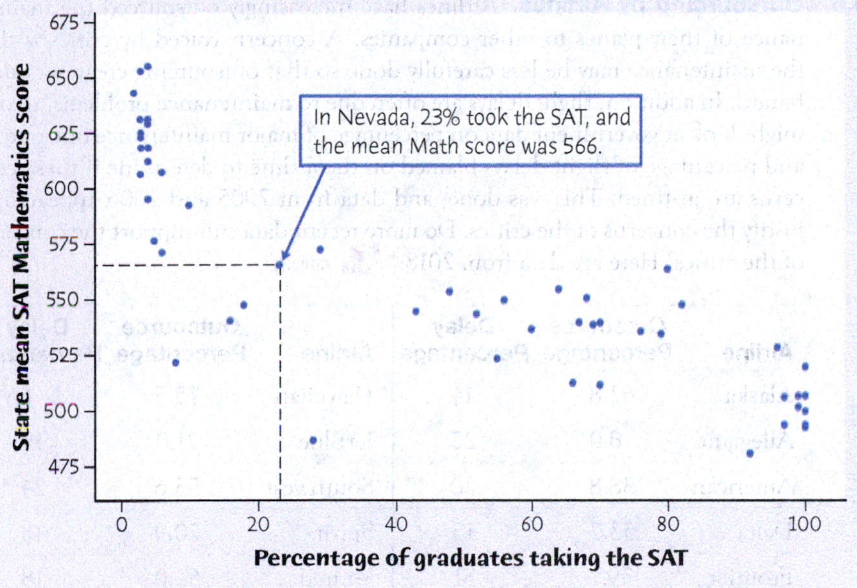

FIGURE 4.1 Scatterplot of the mean SAT Mathematics score in each state against the percentage of that state's high school graduates who take the SAT, for Example 4.3. The dotted lines intersect at the point (23, 566), the data point for Nevada.

APPLY YOUR KNOWLEDGE

4.4 Homicide and Suicide. Preventing suicide is an important issue facing mental health workers. Predicting geographic regions where the risk of suicide is high could help people decide where to increase or improve mental health resources and care. Some psychiatrists have argued that homicide and suicide may have some causes in common. If so, one would expect homicide and suicide rates to be correlated. And if this is true, areas with high rates of homicide might be predicted to have high rates of suicide and therefore deserving of increased mental health resources. Research has had mixed results, including some evidence that there is a positive correlation in certain European countries but not in the United States. Here are data from 2015 for the 11 counties in Ohio with sufficient data for homicides and suicides to allow for estimating rates for both.[3] Rates are per 100,000 people. DEATH

County	Homicide Rate	Suicide Rate	County	Homicide Rate	Suicide Rate
Butler	4.0	11.2	Lucas	6.0	12.6
Clark	10.8	15.3	Mahoning	11.7	15.2
Cuyahoga	12.2	11.4	Montgomery	8.9	15.7
Franklin	8.7	12.3	Stark	5.8	16.1
Hamilton	10.2	11.0	Summit	7.1	17.9
Lorain	3.3	14.3			

Make a scatterplot to examine whether homicide and suicide rates are correlated. For these data, we are simply interested in exploring the relationship between the two variables, so neither variable is an obvious choice for the explanatory variable. For convenience, use homicide rate as the explanatory variable and suicide rate as the response. (The *Two-Variable Statistical Calculator* applet provides an easy way to make scatterplots.)

applet

4.5 Outsourcing by Airlines. Airlines have increasingly outsourced the maintenance of their planes to other companies. A concern voiced by critics is that the maintenance may be less carefully done so that outsourcing creates a safety hazard. In addition, flight delays are often due to maintenance problems, so one might look at government data on percentage of major maintenance outsourced and percentage of flight delays blamed on the airline to determine if these concerns are justified. This was done, and data from 2005 and 2006 appeared to justify the concerns of the critics. Do more recent data still support the concerns of the critics? Here are data from 2018:[4] AIRLINE

Airline	Outsource Percentage	Delay Percentage	Airline	Outsource Percentage	Delay Percentage
Alaska	62.8	14	Hawaiian	75.5	8
Allegiant	8.0	22	JetBlue	71.0	19
American	38.8	20	Southwest	53.6	24
Delta	53.7	15	Spirit	20.9	18
Frontier	39.1	31	United	52.0	18

Make a scatterplot that shows the relation between delays and outsourcing.

4.3 Interpreting Scatterplots

To interpret a scatterplot, adapt the strategies of data analysis learned in Chapters 1 and 2 to the new two-variable setting. Describe the overall pattern of the plot by its direction, form, and strength. *Direction* is fairly simple and indicates whether the overall pattern moves from lower left to upper right, from upper left to lower right, or neither.

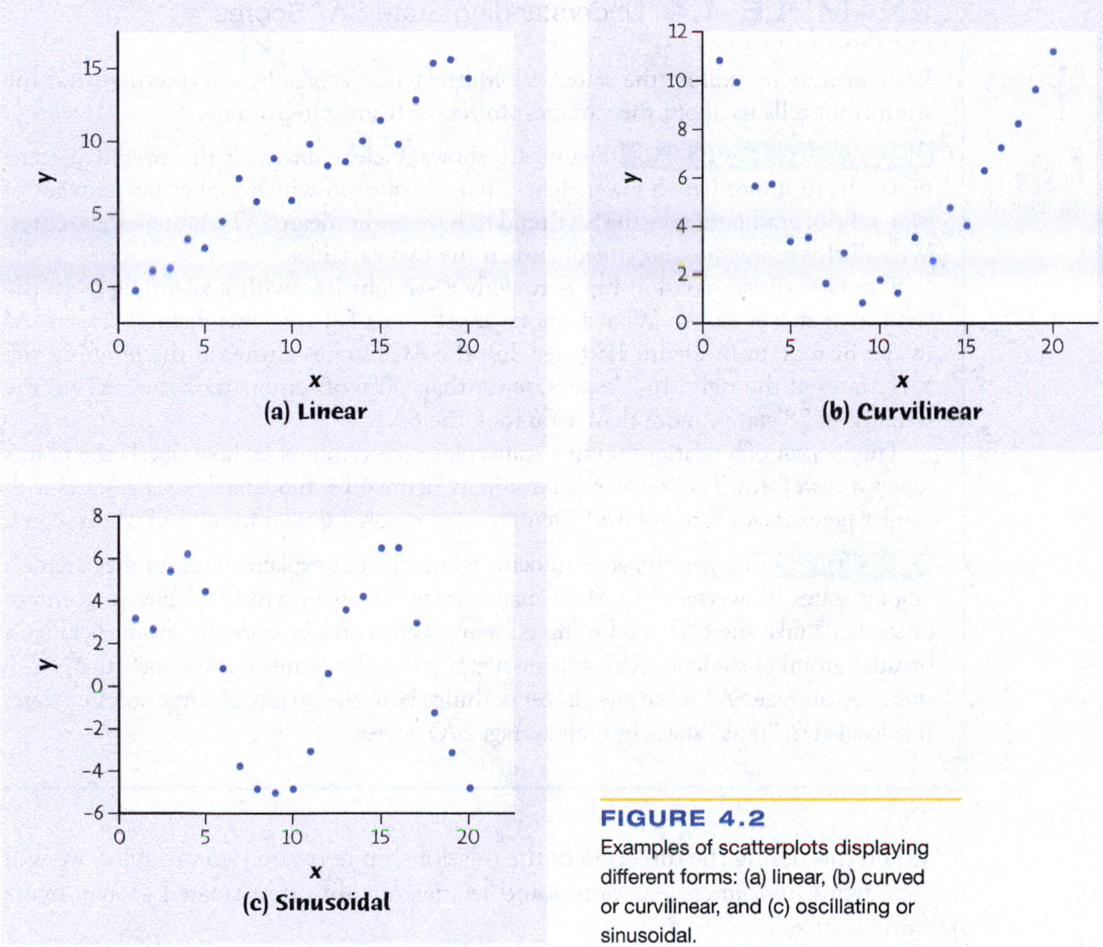

FIGURE 4.2
Examples of scatterplots displaying different forms: (a) linear, (b) curved or curvilinear, and (c) oscillating or sinusoidal.

Form refers to the approximate functional form. For example, is it roughly a straight line, is it curved, or does it oscillate in some way? Figure 4.2 displays three different forms. Figure 4.2(a) is a scatterplot in which the form would be described as *linear*. The form in Figure 4.2(b) is curved or *curvilinear*, and the form in Figure 4.2(c) oscillates and might be described as *sinusoidal*.

Strength refers to how closely the points in the plot follow the form. If they fall almost perfectly on a straight line, we say there is a strong straight-line relationship. If they are widely scattered around a straight line, we say the relationship is weak.

strength
A characteristic of a relationship between two variables determined by how close the points in the scatterplot lie to a simple form such as a line.

Examining a Scatterplot

In any graph of data, look for the *overall pattern* and for striking *deviations* from that pattern.

You can describe the overall pattern of a scatterplot by the *direction, form,* and *strength* of the relationship.

An important kind of deviation is an *outlier,* an individual value that falls outside the overall pattern of the relationship.

Be careful not to confuse ways we describe patterns for distributions of a single variable, such as symmetric or skewed, with ways we describe patterns in scatterplots.

EXAMPLE 4.4 Understanding State SAT Scores

4step

MATHSAT

We continue to explore the state SAT Mathematics scores by interpreting what the scatterplot tells us about the variation in scores from state to state.

SOLVE (interpret the plot): Figure 4.1 shows a clear *direction:* the overall pattern moves from upper left to lower right. That is, states in which higher percentages of high school graduates take the SAT tend to have lower mean SAT Mathematics scores. We call this a *negative association* between the two variables.

The *form* of the relationship is roughly a straight line with a slight curve to the right as it moves down. What is more, most states fall into two distinct *clusters.* As in the histogram in Figure 1.8 (page 26), the ACT states cluster at the left, and the SAT states at the right. In 23 states, fewer than 30% of seniors took the SAT; in the remaining 28 states, more than 40% took the SAT.

The *strength* of a relationship in a scatterplot is determined by how closely the points follow a clear form. The overall relationship in Figure 4.1 is moderately strong: states with similar percentages taking the SAT tend to have roughly similar mean SAT Math scores.

CONCLUDE: The percentage of students taking the test explains much of the variation among states in average SAT Mathematics score. States in which a higher percentage of students take the SAT tend to have lower mean scores because the mean includes a broader group of students. SAT states as a group have lower mean SAT scores than ACT states. So average SAT score says almost nothing about the quality of education in a state. It is foolish to "rank" states by their average SAT scores.

When discussing the direction of the relationship between two variables, we will use the word *association*. *Association* and *relationship* are often treated as synonyms by statisticians.

Positive Association, Negative Association

Two variables are **positively associated** when above-average values of one tend to accompany above-average values of the other, and below-average values also tend to occur together.

Two variables are **negatively associated** when above-average values of one tend to accompany below-average values of the other and vice versa.

Of course, not all relationships have a clear direction that we can describe as positive association or negative association. Exercise 4.8 gives an example that does not have a single direction. Here is an example of a strong positive association with a simple and important form.

EXAMPLE 4.5 The Endangered Manatee

4step

STATE: Manatees are large, gentle, slow-moving creatures found along the coast of Florida. Many manatees are injured or killed by boats. Table 4.1 contains data on the number of boats registered in Florida (in thousands) and the number of manatees killed by boats for the years between 1977 and 2018.[5] Examine the relationship. Is it plausible that restricting the number of boats would help protect manatees?

MANATEE

PLAN: Make a scatterplot with "boats registered" as the explanatory variable and "manatees killed" as the response variable. Describe the form, direction, and strength of the relationship.

TABLE 4.1 Florida boat registrations (thousands) and manatees killed by boats

Year	Boats	Manatees	Year	Boats	Manatees	Year	Boats	Manatees
1977	447	13	1992	679	38	2007	1027	73
1978	460	21	1993	678	35	2008	1010	90
1979	481	24	1994	696	49	2009	982	97
1980	498	16	1995	713	42	2010	942	83
1981	513	24	1996	732	60	2011	922	88
1982	512	20	1997	755	54	2012	902	82
1983	526	15	1998	809	66	2013	897	73
1984	559	34	1999	830	82	2014	900	69
1985	585	33	2000	880	78	2015	916	86
1986	614	33	2001	944	81	2016	931	106
1987	645	39	2002	962	95	2017	944	111
1988	675	43	2003	978	73	2018	951	124
1989	711	50	2004	983	69			
1990	719	47	2005	1010	79			
1991	681	53	2006	1024	92			

SOLVE: Figure 4.3 is the scatterplot. There is a positive association: more boats goes with more manatees killed. This form is a **linear relationship**. That is, the overall pattern follows a straight line from lower left to upper right. The relationship is strong because the points don't deviate greatly from a line.

linear relationship An important form of a relationship between two variables, in which the points on a scatterplot show a straight-line pattern.

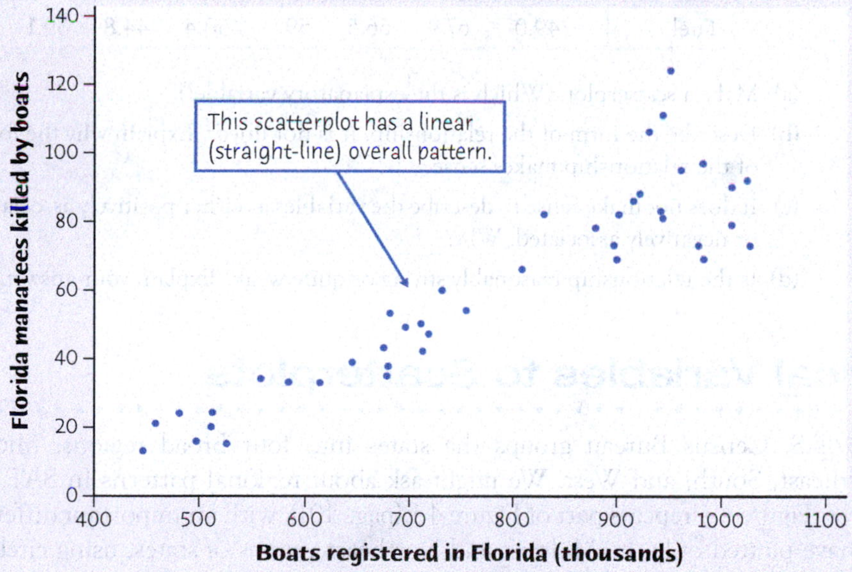

FIGURE 4.3
Scatterplot of the number of Florida manatees killed by boats in the years 1977 to 2018 against the number of boats registered in Florida that year, for Example 4.5. There is a strong linear (straight-line) pattern.

CONCLUDE: As more boats are registered, the number of manatees killed by boats goes up linearly. Data from the Florida Wildlife Commission indicate that in 2018, boats accounted for 15.0% of all manatee deaths (both those for which the cause could be determined and those for which the cause could not be determined) based on the statistical association shown above, and 21.6% of those deaths whose causes could be determined. Although many manatees die from other causes, it appears that fewer boats would mean fewer manatee deaths.

As the following chapter will emphasize, *it is wise to always ask what other variables lurking in the background might contribute to the relationship displayed in a scatterplot.* Because both boats registered and manatees killed are recorded year by year, any change in conditions over time might affect the relationship. For example, if boats in Florida have tended to go faster over the years, that might result in more manatees being killed by the same number of boats.

APPLY YOUR KNOWLEDGE

4.6 Homicide and Suicide. Describe the direction, form, and strength of the relationship between homicide rate and suicide rate, as displayed in your plot for Exercise 4.4. Are there any deviations from the overall pattern? DEATH

4.7 Outsourcing by Airlines. Does your plot for Exercise 4.5 show a positive association, a negative association, or no association between maintenance outsourcing and delays caused by the airline? If it shows association, is the relationship very strong? Are there any outliers? AIRLINE

4.8 Does Fast Driving Waste Fuel? How does the fuel consumption of a car change as its speed increases? Here are data for a 2013 Volkswagen Jetta Diesel. Speed is measured in miles per hour, and fuel consumption is measured in miles per gallon:[6] FASTDR

Speed	20	30	40	50	60	70	80
Fuel	49.0	67.9	66.5	59	50.4	44.8	39.1

(a) Make a scatterplot. (Which is the explanatory variable?)

(b) Describe the form of the relationship. It is not linear. Explain why the form of the relationship makes sense.

(c) It does not make sense to describe the variables as either positively associated or negatively associated. Why?

(d) Is the relationship reasonably strong or quite weak? Explain your answer.

4.4 Adding Categorical Variables to Scatterplots

The U.S. Census Bureau groups the states into four broad regions: Midwest, Northeast, South, and West. We might ask about regional patterns in SAT exam scores. Figure 4.4 repeats part of Figure 4.1 (page 101), with an important difference: we have plotted only the Midwest and Northeast groups of states, using circles for the Midwest states and plus signs for the Northeast states.

The regional comparison is striking. The nine Northeast states are all SAT states—at least 63% of high school graduates in each of these states take the SAT. The 13 Midwest states are mostly ACT states. In 9 of these states, fewer than 5% of high school graduates take the SAT. Three states in the Midwest region are clearly

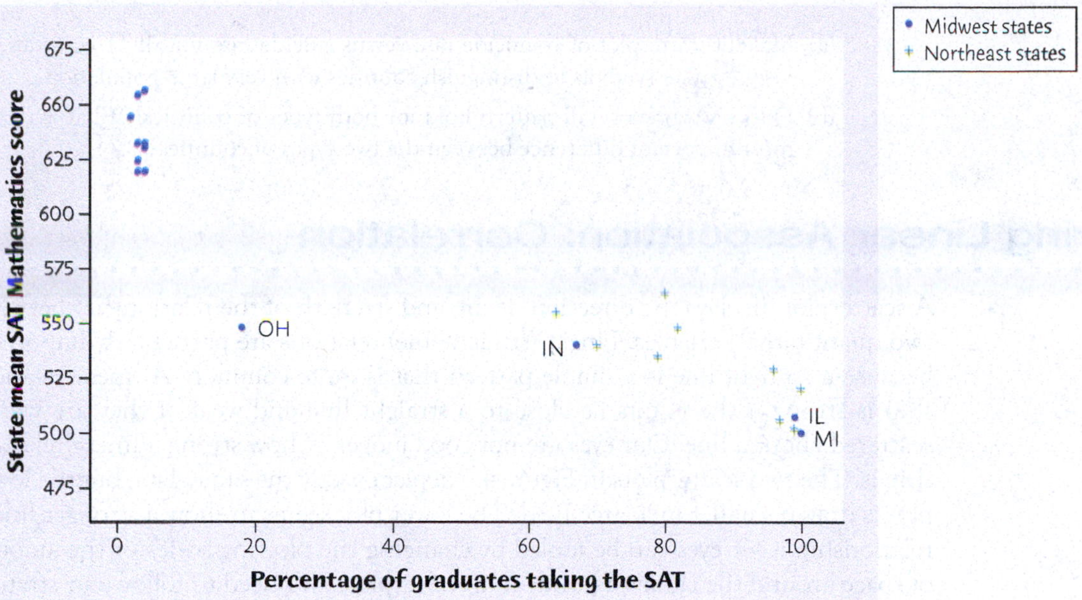

FIGURE 4.4
Mean SAT Mathematics score and percentage of high school graduates who take the test for only the Midwest (o) and Northeast (+) states.

outliers: Indiana, Illinois, and Michigan are SAT states (67%, 99%, and 100%, respectively, take the SAT) whose mean scores fall within the Northeast cluster on the scatterplot. Mean scores for the Midwest state of Ohio, where 18% take the SAT, also lie outside the Midwest cluster.

Dividing the states into regions introduces a third variable into the scatterplot. "Region" is a categorical variable that has four values, although we plotted data from only two of the four regions. The two regions are identified by the two different plotting symbols.

Categorical Variables in Scatterplots

To add a categorical variable to a scatterplot, use a different plot color or symbol for each category.

APPLY YOUR KNOWLEDGE

4.9 Homicide and Suicide. The data described in Exercise 4.4 includes counties of varying sizes. Those with very large populations (greater than 800,000) are indicated in the table: DEATH2

County	Homicide Rate	Suicide Rate	Population Above 800,000	County	Homicide Rate	Suicide Rate	Population Above 800,000
Butler	4.0	11.2	N	Lucas	6.0	12.6	N
Clark	10.8	15.3	N	Mahoning	11.7	15.2	N
Cuyahoga	12.2	11.4	Y	Montgomery	8.9	15.7	N
Franklin	8.7	12.3	Y	Stark	5.8	16.1	N
Hamilton	10.2	11.0	Y	Summit	7.1	17.9	N
Lorain	3.3	14.3	N				

(a) Make a scatterplot of homicide rate versus suicide rate for all 11 counties. Use separate symbols to distinguish counties with very large populations.

(b) Does the same overall pattern hold for both types of counties? What is the most important difference between the two types of counties?

4.5 Measuring Linear Association: Correlation

A scatterplot displays the direction, form, and strength of the relationship between two quantitative variables. Linear (straight-line) relations are particularly important because a straight line is a simple pattern that is quite common. A linear relationship is strong if the points lie close to a straight line and weak if they are widely scattered about a line. Our eyes are not good judges of how strong a linear relationship is. The two scatterplots in Figure 4.5 depict exactly the same data, but the lower plot is drawn smaller in a large field. The lower plot seems to show a stronger linear relationship. Our eyes can be fooled by changing the plotting scales or the amount of space around the cloud of points in a scatterplot.[7] We need to follow our strategy for data analysis by using a numerical measure to supplement the graph. *Correlation* is the measure we use.

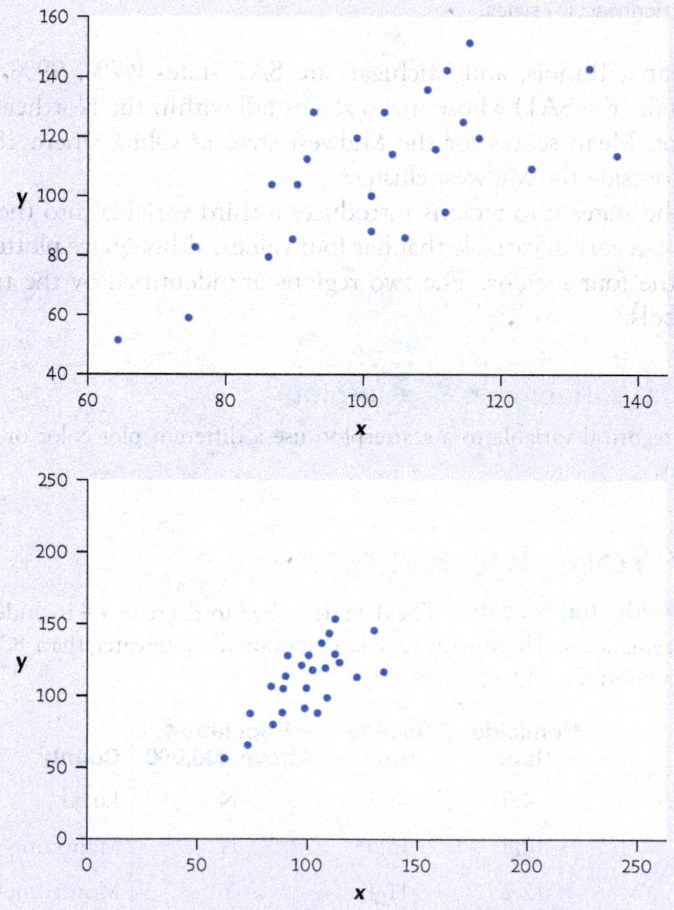

FIGURE 4.5
Two scatterplots of the same data. The straight-line pattern in the lower plot appears stronger because of the surrounding space.

Correlation

The **correlation** measures the direction and strength of the linear relationship between two quantitative variables. Correlation is usually written as r.

Suppose that we have data on variables x and y for n individuals. The values for the first individual are x_1 and y_1, the values for the second individual are x_2 and y_2, and so on. The means and standard deviations of the two variables are $\bar{x}$ and s_x for the x-values, and $\bar{y}$ and s_y for the y-values. The correlation r between x and y is

$$r = \frac{1}{n-1}\left[\left(\frac{x_1-\bar{x}}{s_x}\right)\left(\frac{y_1-\bar{y}}{s_y}\right)+\left(\frac{x_2-\bar{x}}{s_x}\right)\left(\frac{y_2-\bar{y}}{s_y}\right)+\cdots+\left(\frac{x_n-\bar{x}}{s_x}\right)\left(\frac{y_n-\bar{y}}{s_y}\right)\right]$$

or, more compactly,

$$r = \frac{1}{n-1}\sum\left(\frac{x_i-\bar{x}}{s_x}\right)\left(\frac{y_i-\bar{y}}{s_y}\right)$$

The formula for the correlation r is a bit complex. It helps us see what correlation is, but in practice you should use software or a calculator that finds r from keyed-in values of two variables, x and y. Exercise 4.10 asks you to calculate a correlation step-by-step from the definition to solidify its meaning.

The formula for r begins by standardizing the observations. Suppose, for example, that x is height in centimeters and y is weight in kilograms and that we have height and weight measurements for n people. Then $\bar{x}$ and s_x are the mean and standard deviation of the n heights, both in centimeters. The value

$$\frac{x_i-\bar{x}}{s_x}$$

is the standardized height of the ith person, from Chapter 3. The standardized height says how many standard deviations above or below the mean a person's height lies. Standardized values have no units; in this example, they are no longer measured in centimeters. Standardize the weights also. The correlation r is an average of the products of the standardized height and the standardized weight for all the individuals. Just as in the case of the standard deviation s, the "average" here divides by one fewer than the number of individuals.

STATISTICS IN YOUR WORLD

Death from Superstition?

Is there a relationship between superstitious beliefs and bad things happening? Apparently there is. Chinese and Japanese people think that the number 4 is unlucky because when pronounced it sounds like the word for "death." Sociologists looked at 15 years' worth of death certificates for Chinese and Japanese Americans and for white Americans. Deaths from heart disease were notably higher on the fourth day of the month among Chinese and Japanese but not among whites. The sociologists think the explanation is increased stress on "unlucky days."

APPLY YOUR KNOWLEDGE

4.10 Coral Reefs. Exercise 4.2 discusses a study in which scientists examined data on mean sea surface temperatures (in degrees Celsius) and mean coral growth (in centimeters per year) over a several-year period at locations in the Gulf of Mexico and the Caribbean. Here are the data for the Gulf of Mexico:[8]

Sea surface temperature	26.7	26.6	26.6	26.5	26.3	26.1
Growth	0.85	0.85	0.79	0.86	0.89	0.92

Georgie Holland/AGE Fotostock

(a) Make a scatterplot. Which is the explanatory variable? The plot shows a negative linear pattern.

(b) Find the correlation r step by step. You may wish to round off to two decimal places in each step. First, find the mean and standard deviation of each variable. Then, find the six standardized values for each variable. Finally, use the formula for r. Explain how your value for r matches the direction of the linear pattern in your graph in part (a).

CORAL

(c) Enter these data into your calculator or software and use the correlation function to find r. Check that you get the same result as in part (b), up to roundoff error.

4.11 Brain Size and Intelligence. For centuries, people have associated intelligence with brain size. A recent study used magnetic resonance imaging to measure the brain sizes of several individuals. The IQs and brain sizes (in units of 10,000 pixels) of six individuals are as follows:

Brain size:	100	90	95	92	88	106
IQ:	140	90	100	135	80	103

(a) Make a scatterplot. Which is the explanatory variable? The plot shows a positive linear pattern.

(b) Find the correlation r step by step. You may wish to round off to two decimal places in each step. First, find the mean and standard deviation of each variable. Then, find the six standardized values for each variable. Finally, use the formula for r. Explain how your value for r matches the direction of the linear pattern in your graph in part (a).

BRAIN

(c) Enter these data into your calculator or software, and use the correlation function to find r. Check that you get the same result as in part (b), up to roundoff error.

4.6 Facts about Correlation

The formula for correlation helps us see that r is positive when there is a positive association between the variables. Height and weight, for example, have a positive association. People who are above average in height tend also to be above average in weight. Both the standardized height and the standardized weight are positive. People who are below average in height tend also to have below-average weight. Then both standardized height and standardized weight are negative. In both cases, the products in the formula for r are mostly positive, and so r is positive. In the same way, we can see that r is negative when the association between x and y is negative. More detailed study of the formula gives more detailed properties of r. Here is what you need to know to interpret correlation:

1. *Correlation makes no distinction between explanatory and response variables.* It makes no difference which variable you call x and which you call y in calculating the correlation.
2. Because r uses the standardized values of the observations, *r does not change when we change the units of measurement of x, y, or both.* Measuring height in inches rather than centimeters and weight in pounds rather than kilograms does not change the correlation between height and weight. The correlation r itself has no unit of measurement; it is just a number.
3. *Positive r indicates positive association between the variables, and negative r indicates negative association.*

4. *The correlation r is always a number between -1 and 1.* Values of r near 0 indicate a very weak linear relationship. The strength of the linear relationship increases as r moves away from 0 toward either -1 or 1. Values of r close to -1 or 1 indicate that the points in a scatterplot lie close to a straight line. The extreme values $r = -1$ and $r = 1$ occur only in the case of a perfect linear relationship, when the points lie exactly along a straight line.

EXAMPLE 4.6 From Scatterplot to Correlation

The scatterplots in Figure 4.6 illustrate how values of r closer to 1 or -1 correspond to stronger linear relationships. To make the meaning of r clearer, the standard deviations of both variables in these plots are equal, and the horizontal and vertical scales are the same. In general, it is not so easy to guess the value of r from the appearance of a scatterplot. Remember that changing the plotting scales in a scatterplot may mislead our eyes, but it does not change the correlation.

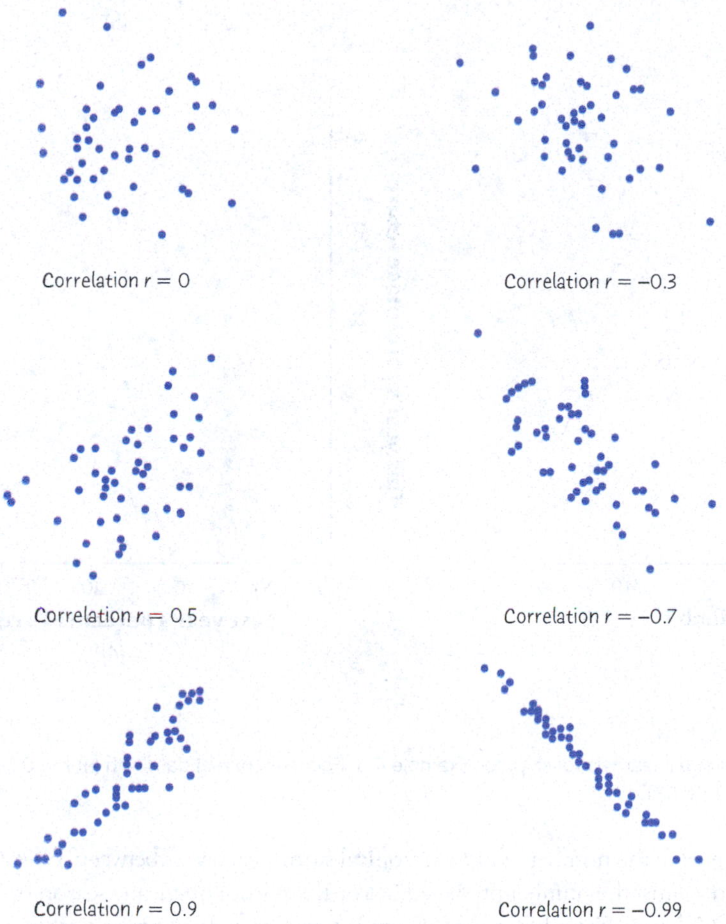

FIGURE 4.6
How correlation measures the strength of a linear relationship, for Example 4.6. Patterns closer to a straight line have correlations closer to 1 or -1.

The scatterplots in Figure 4.7 show four sets of real data. The patterns are less regular than those in Figure 4.6, but they also illustrate how correlation measures the strength of linear relationships:[9]

(a) This repeats the manatee plot in Figure 4.3. There is a strong positive linear relationship, $r = 0.919$.

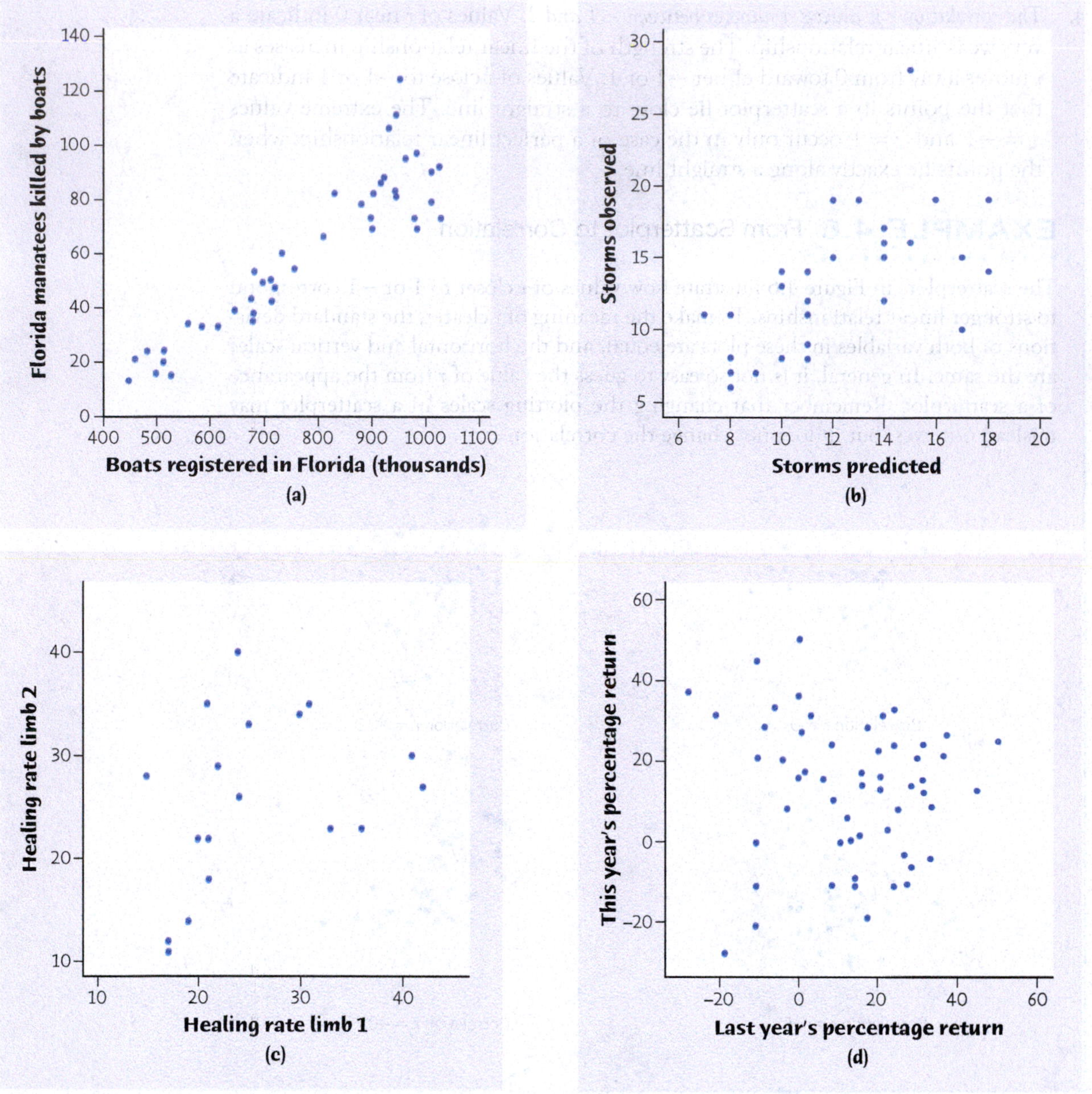

FIGURE 4.7

How correlation measures the strength of a linear relationship, for Example 4.6. Four sets of real data with (a) $r = 0.919$, (b) $r = 0.628$, (c) $r = 0.358$, and (d) $r = -0.081$.

(b) Here are the number of named tropical storms each year between 1984 and 2018 plotted against the number predicted before the start of hurricane season by William Gray of Colorado State University. There is a moderate linear relationship, $r = 0.628$.

(c) These data come from an experiment that studied how quickly cuts in the limbs of newts heal. Each point represents the healing rate in micrometers (millionths of a meter) per hour for the two front limbs of the same newt. This relationship is weaker than those in parts (a) and (b), with $r = 0.358$.

(d) Does last year's stock market performance help predict how stocks will do this year? No. The correlation between last year's percentage return and this year's percentage return over a 56-year period is only $r = -0.081$. The scatterplot shows a cloud of points with no visible linear pattern.

Describing the relationship between two variables is a more complex task than describing the distribution of one variable. Here are some more facts about correlation, cautions to keep in mind when you use r:

1. *Correlation requires that both variables be quantitative, so it makes sense to do the arithmetic indicated by the formula for r.* We cannot calculate a correlation between the incomes of a group of people and what city they live in because city is a categorical variable.
2. Correlation measures the strength of only the linear relationship between two variables. *Correlation does not describe curved relationships between variables, no matter how strong they are.* Exercise 4.14 illustrates this important fact.
3. *Like the mean and standard deviation, the correlation is not resistant: r is strongly affected by a few outlying observations.* Use r with caution when outliers appear in the scatterplot. Reporting the correlation both with the outliers included and with the outliers removed is informative.
4. *Correlation is not a complete summary of two-variable data,* even when the relationship between the variables is linear. You should give the means and standard deviations of both x and y along with the correlation.

Because the formula for correlation uses the means and standard deviations, these measures are the proper choice to accompany a correlation. Here is an example in which understanding requires both means and correlation.

EXAMPLE 4.7 Scoring *American Idol* at Home

One website recommended that fans of the television show *American Idol* score contestants at home on a scale from 1 to 10, with higher scores indicating a better performance. Two friends, Angela and Elizabeth, decide to follow this advice and score contestants over the course of the final season in 2016. How well did they agree? We calculate that the correlation between their scores is $r = 0.9$, suggesting that they agreed. But the mean of Angela's scores was 0.8 point lower than Elizabeth's mean. Does this suggest that the two friends disagreed?

These facts do not contradict each other. They are simply different kinds of information. The mean scores show that Angela awarded lower scores than Elizabeth. But because Angela gave *every* contestant a score about 0.8 point lower than Elizabeth, the correlation remains high. Adding the same number to all values of either x or y does not change the correlation. Angela and Elizabeth actually scored consistently because they agreed on which performances were better. The high r shows their agreement.

Nagaiets/Getty Images

Of course, even giving means, standard deviations, and the correlation for state SAT scores and percent taking will not point out the clusters in Figure 4.1. Numerical summaries complement plots of data, but they don't replace them.

APPLY YOUR KNOWLEDGE

4.12 Changing the Units. The sea surface temperatures in Exercise 4.10 are measured in degrees Celsius and growth in centimeters per year. The correlation between sea surface temperature and coral growth is $r = -0.8111$. If the measurements were made in degrees Fahrenheit and inches per year, would the correlation change? Explain your answer.

4.13 Changing the Correlation. Use your calculator or software to demonstrate how outliers can affect correlation.

(a) What is the correlation between homicide rate and suicide rate for the 11 counties in Exercise 4.4? DEATH

(b) Make a scatterplot of the data with one new point added. Point A: homicide rate 30, suicide rate 30. Find the new correlation for the original data plus Point A. DEATH3

(c) By looking at your plot, explain why adding Point A makes the correlation stronger (closer to 1).

4.14 Strong Association but No Correlation. The gas mileage of an automobile first increases and then decreases as the speed increases. Suppose this relationship is very regular, as shown by the following data on speed (miles per hour) and mileage (miles per gallon): MPG

Speed	20	30	40	50	60	70	80
Mileage	21	26	29	30	29	26	21

Make a scatterplot of mileage versus speed. Show that the correlation between speed and mileage is $r = 0$. Explain why the correlation is 0 even though there is a strong relationship between speed and mileage.

CHAPTER 4 SUMMARY

- To study relationships between variables, we must measure the variables on the same group of individuals.
- If we think that a variable x may explain or even cause changes in another variable y, we call x an **explanatory variable** and y a **response variable.**
- A **scatterplot** displays the relationship between two quantitative variables measured on the same individuals. Mark values of one variable on the horizontal **axis** (**x axis**) and values of the other variable on the vertical axis (**y axis**). Plot each individual's data as a point on the graph. Always plot the explanatory variable, if there is one, on the x axis of a scatterplot.
- Plot points with different colors or symbols to see the effect of a categorical variable in a scatterplot.
- In examining a scatterplot, look for an overall pattern showing the *direction, form,* and *strength* of the relationship and then for *outliers* or other deviations from this pattern.
- Direction: If the relationship has a clear direction, we speak of either **positive association** (high values of the two variables tend to occur together) or **negative association** (high values of one variable tend to occur with low values of the other variable).
- Form: **Linear relationships,** where the points show a straight-line pattern, are an important form of relationship between two variables. Curved relationships and *clusters* are other forms to watch for.
- Strength: The **strength** of a relationship is determined by how close the points in the scatterplot lie to a simple form such as a line.
- The **correlation r** measures the direction and strength of the linear association between two quantitative variables x and y. Although you can calculate a correlation for any scatterplot, r measures only straight-line relationships.

- Correlation indicates the direction of a linear relationship by its sign: $r > 0$ for a positive association and $r < 0$ for a negative association. Correlation always satisfies $-1 \leq r \leq 1$ and indicates the strength of a relationship by how close it is to -1 or 1. Perfect correlation, $r = \pm 1$, occurs only when the points on a scatterplot lie exactly on a straight line.
- Correlation ignores the distinction between explanatory and response variables. The value of r is not affected by changes in the unit of measurement of either variable. Correlation is not resistant, so outliers can greatly change the value of r.

CHECK YOUR SKILLS

4.15 The Department of Energy website contains data on 1259 model year 2019 cars and SUVs.[10] Included in the data are the engine size (as measured by engine displacement in liters) and combined city and highway gas mileage (in miles per gallon). When you make a scatterplot to predict gas mileage from engine size, the explanatory variable on the x axis

(a) is the gas mileage.

(b) is the engine size.

(c) can be either gas mileage or engine size.

4.16 Examining the data in the previous exercise, one finds that cars with bigger engines tend to have lower gas mileages. In a scatterplot of the engine size and the gas mileage, you expect to see

(a) a positive association.

(b) very little association.

(c) a negative association.

4.17 Figure 4.8 is a scatterplot of the price of a hot dog against the price of beer (per ounce) at 30 major league ballparks in 2019.[11] There are two low outliers in the plot. The prices of a hot dog for these two ballparks are approximately

(a) \$0.28 and \$0.33.

(b) \$1.50 and \$2.00.

(c) \$0.72 and \$5.25.

4.18 If we leave out the two low outliers, the correlation for the remaining 28 points in Figure 4.8 is closest to

(a) 0.8. (b) −0.8. (c) 0.2.

4.19 What are all the values that a correlation r can possibly take?

(a) $r \geq 0$

(b) $0 \leq r \leq 1$

(c) $-1 \leq r \leq 1$

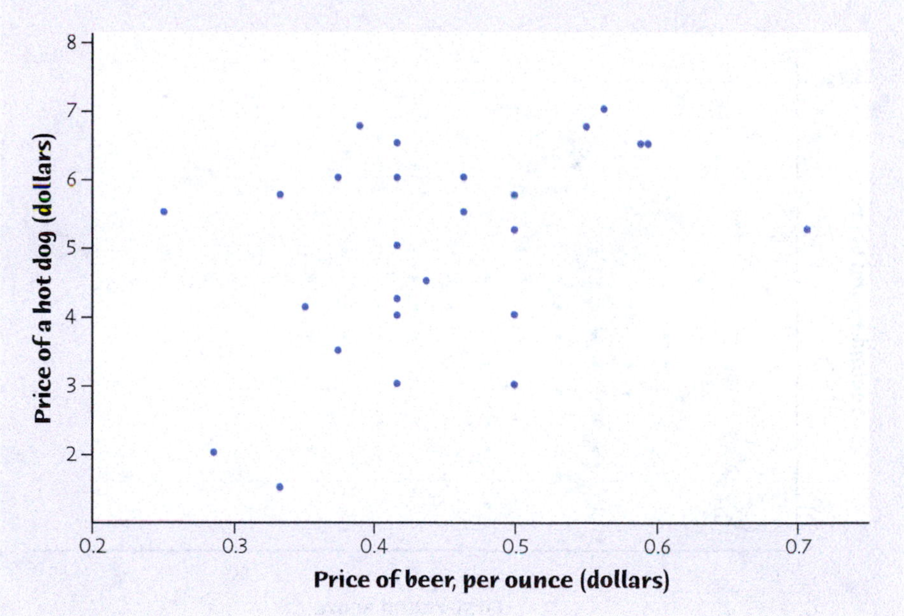

FIGURE 4.8
Scatterplot of the price of a hot dog against the price of beer (per ounce) for 24 major league ballparks, for Exercises 4.17 and 4.18.

4.20 If the correlation between two variables is close to 0, you can conclude that a scatterplot would show

(a) a strong straight-line pattern.

(b) a cloud of points with no visible pattern.

(c) no straight-line pattern, but there might be a strong pattern of another form.

4.21 The points on a scatterplot lie very close to a straight line. The correlation between x and y is close to

(a) -1.

(b) 1.

(c) either -1 or 1, depending on the direction.

4.22 A statistics professor warns her class that her second exam is always harder than the first. She tells her class that students always score 10 points worse on the second exam compared to their score on the first exam. This means that the correlation between students' scores on the first and second exam is

(a) 1.

(b) -1.

(c) Can't tell without seeing the data.

4.23 Researchers asked mothers how much soda (in ounces) their kids drank in a typical day. They also asked these mothers to rate how aggressive their kids were on a scale of 1 to 10, with larger values corresponding to a greater degree of aggression.[12] The correlation between amount of soda consumed and aggression rating was found to be $r = 0.3$. If the researchers had measured amount of soda consumed in liters instead of ounces, what would be the correlation? (There are 35 ounces in a liter.)

(a) $0.3/35 = 0.009$

(b) 0.3

(c) $(0.3)(35) = 10.5$

4.24 Researchers measured the percentage body fat and the preferred amount of salt (percent weight/volume) for several children. Here are data for seven children:[13] SALT

Preferred amount of salt x	0.2	0.3	0.4	0.5	0.6	0.8	1.1
Percentage body fat y	20	30	22	30	38	23	30

Use your calculator or software: The correlation between percentage body fat and preferred amount of salt is about

(a) $r = 0.08$. (b) $r = 0.3$. (c) $r = 0.8$.

CHAPTER 4 EXERCISES

4.25 Scores at the Masters. The Masters is one of the four major golf tournaments. Figure 4.9 is a scatterplot of the scores for the first two rounds of the 2019 Masters for all the golfers entered. The plot has a grid pattern because golf scores must be whole numbers.[14] MASTERS19

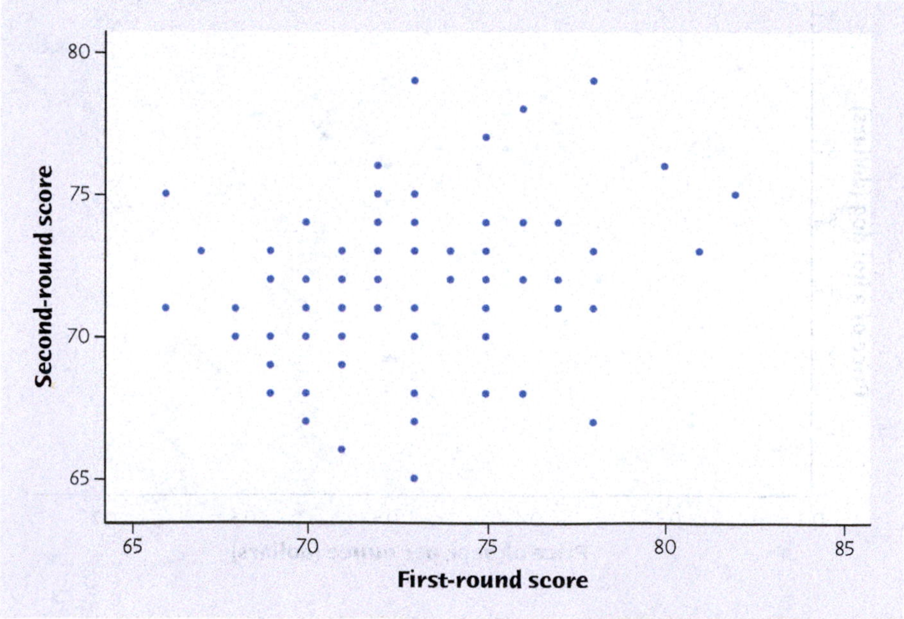

FIGURE 4.9
Scatterplot of the scores in the first two rounds of the 2019 Masters Tournament, for Exercise 4.25.

(a) Read the graph: What was the lowest score in the first round of play? How many golfers had this low score? For each golfer with the lowest score in the first round, what score did he have in the second round?

(b) Read the graph: Jose Maria Olazabal and Jovan Rebula had the highest score in the second round. What was this score? What were their scores in the first round?

(c) Is the correlation between first-round scores and second-round scores closest to $r = 0.01$, $r = 0.25$, $r = 0.75$, or $r = 0.99$? Explain your choice. Does the graph suggest that knowing a professional golfer's score for one round is much help in predicting his score for another round on the same course?

4.26 **Happy States.** Human happiness or well-being can be assessed either subjectively or objectively. Subjective assessment can be accomplished by listening to what people say. Objective assessment can be made from data related to well-being such as income, climate, availability of entertainment, housing prices, lack of traffic congestion, and so on. Do subjective and objective assessments agree? To study this, investigators made both subjective and objective assessments of happiness for each of the 50 states. The subjective measurement was the mean score on a life-satisfaction question found on the Behavioral Risk Factor Surveillance System (BRFSS), which is a state-based system of health surveys. Lower scores indicate a greater degree of happiness. To objectively assess happiness, the investigators computed a mean well-being score (called the compensating-differentials score) for each state, based on objective measures that have been found to be related to happiness or well-being. The states were then ranked according to this score (Rank 1 being the happiest). Figure 4.10 is a scatterplot of mean BRFSS scores (response) against the rank based on the compensating differentials (explanatory).[15] HAPPY

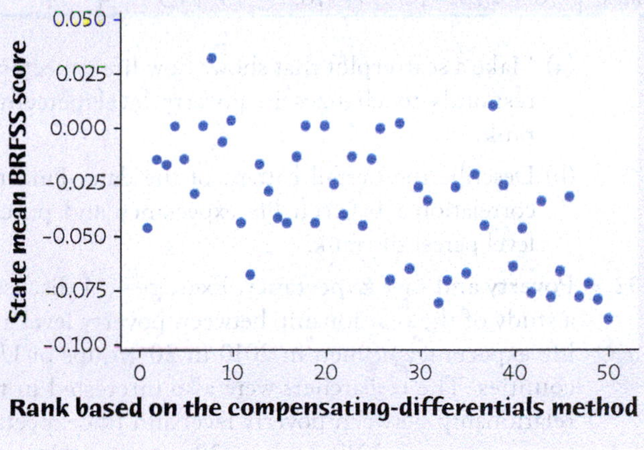

FIGURE 4.10
Scatterplot of mean BRFSS score in each state against each state's well-being rank, for Exercise 4.26.

(a) Is there an overall positive association or an overall negative association between mean BRFSS score and rank based on the compensating-differentials method?

(b) Does the overall association indicate agreement or disagreement between the mean subjective BRFSS score and the ranking based on objective data used in the compensating-differentials method?

(c) Are there any outliers? If so, what are the BRFSS scores corresponding to these outliers?

4.27 **Wine and Cancer in Women.** Some studies have suggested that a nightly glass of wine may not only take the edge off a day but also improve health. Is wine good for your health? A study of nearly 1.3 million middle-aged British women examined wine consumption and the risk of breast cancer. The researchers were interested in how risk changed as wine consumption increased. Risk is based on breast cancer rates in drinkers relative to breast cancer rates in nondrinkers in the study, with higher values indicating greater risk. In particular, a value greater than 1 indicates a greater breast cancer rate than that of nondrinkers. Wine intake is the mean wine intake, in grams of alcohol per day (where one glass of wine is approximately 10 grams of alcohol), of groups of women in the study who drank approximately the same amount of wine per week. Here are the data (for drinkers only):[16] CANCER

Wine intake x (grams of alcohol per day)	2.5	8.5	15.5	26.5
Relative risk y	1.00	1.08	1.15	1.22

(a) Make a scatterplot of these data. Based on the scatterplot, do you expect the correlation to be positive or negative? Near ± 1 or not?

(b) Find the correlation r between wine intake and relative risk. Do the data show that women who consume more wine tend to have higher relative risks of breast cancer?

(c) Can you conclude a causal relationship between wine consumption and breast cancer in women? Explain.

4.28 **Ebola and Gorillas.** The deadly Ebola virus is a threat to both people and gorillas in Central Africa. An outbreak in 2002 and 2003 killed 91 of the 95 gorillas in seven home ranges in the Congo. To study the spread of the virus, measure "distance" by the number of home ranges separating a group of gorillas from the first group infected. Here are data on distance and time in number of days until deaths began in each later group:[17] EBOLA

Distance	1	3	4	4	4	5
Time	4	21	33	41	43	46

(a) Make a scatterplot. Which is the explanatory variable? What kind of pattern does your plot show?

(b) Find the correlation r between distance and time.

(c) If time in days were replaced by time in number of weeks until death began in each later group (fractions allowed so that four days becomes 4/7 week), would the correlation between distance and time change? Explain your answer.

4.29 **Sparrowhawk Colonies.** One of nature's patterns connects the percentage of adult birds in a colony that return from the previous year and the number of new adults that join the colony. Here are data for 13 colonies of sparrowhawks:[18] SPARROW

Age fotostock/Superstock

Percentage returned	74	66	81	52	73	62	52	45	62	46	60	46	38
New adults	5	6	8	11	12	15	16	17	18	18	19	20	20

(a) Plot the count of new adults (response) against the percentage of returning birds (explanatory). Describe the direction and form of the relationship. Is the correlation r an appropriate measure of the strength of this relationship? If so, find r.

(b) For short-lived birds, the association between these variables is positive: changes in weather and food supply drive the populations of new and returning birds up or down together. For long-lived territorial birds, on the other hand, the association is negative because returning birds claim their territories in the colony and don't leave room for new recruits. Which type of species is the sparrowhawk?

4.30 **Our Brains Don't Like Losses.** Most people dislike losses more than they like gains. In money terms, people are about as sensitive to a loss of \$10 as to a gain of \$20. To discover what parts of the brain are active in decisions about gain and loss, psychologists presented subjects with a series of gambles with different odds and different amounts of winnings and losses. From a subject's choices, they constructed a measure of "behavioral loss aversion." Higher scores show greater sensitivity to losses. Observing brain activity while subjects made their decisions pointed to specific brain regions. Here are data for 16 subjects on behavioral loss aversion and "neural loss aversion," a measure of activity in one region of the brain:[19] LOSSES

Neural	−50.0	−39.1	−25.9	−26.7	−28.6	−19.8	−17.6	5.5
Behavioral	0.08	0.81	0.01	0.12	0.68	0.11	0.36	0.34
Neural	2.6	20.7	12.1	15.5	28.8	41.7	55.3	155.2
Behavioral	0.53	0.68	0.99	1.04	0.66	0.86	1.29	1.94

(a) Make a scatterplot that shows how behavior responds to brain activity.

(b) Describe the overall pattern of the data. There is one clear outlier. What is the behavioral score associated with this outlier?

(c) Find the correlation r between neural and behavioral loss aversion both with and without the outlier. Does the outlier have a strong influence on the value of r? By looking at your plot, explain why adding the outlier to the other data points causes r to increase.

4.31 **Poverty and Life Expectancy.** Do poorer people tend to have shorter lives than richer people? Two researchers ranked all counties in the United States by their poverty level and then divided them into 20 groups, each representing approximately 5% of the overall U.S. population. The bottom 5% were the least impoverished (wealthiest), and the top 5% (the 100th percentile) were the most impoverished. Life expectancies at birth for each of the 20 groups were calculated based on life tables. Here are the data for poverty percentiles (with higher percentiles corresponding to greater poverty) and life expectancies at birth (in years) in 2010 for males in the 20 groups:[20] LIFEEXP

Percentile rank	5	10	15	20	25	30	35	40	45	50
Life expectancy of males	79	79	78	77.5	77.5	77.5	77	76.5	77	77
Percentile rank	55	60	65	70	75	80	85	90	95	100
Life expectancy of males	76.5	76	76	77	76.5	75	75.5	74	74	73

(a) Make a scatterplot that shows how life expectancy responds to changes in poverty level percentile rank.

(b) Describe the overall pattern of the data. Find the correlation r between life expectancy and poverty level percentile rank.

4.32 **Poverty and Life Expectancy.** Exercise 4.31 discusses a study of the relationship between poverty level and life expectancy of men in 2010 in 20 groups of U.S. counties. The researchers were also interested in the relationship between poverty level and life expectancy of women in 2010 in these 20 counties. Here are the data for both men and women in the 20 groups of counties: LIFEEXP2

Percentile rank	5	10	15	20	25	30	35	40	45	50
Life expectancy of males	79	79	78	77.5	77.5	77.5	77	76.5	77	77
Percentile rank	55	60	65	70	75	80	85	90	95	100
Life expectancy of males	76.5	76	76	77	76.5	75	75.5	74	74	73

Percentile rank	5	10	15	20	25	30	35	40	45	50
Life expectancy of females	83	83	82.5	82.5	82.5	82.5	82	81.5	82	82
Percentile rank	55	60	65	70	75	80	85	90	95	100
Life expectancy of females	81	81	81	82	81	80	80.5	79	79.5	79

(a) Make a scatterplot of the life expectancy versus poverty level percentile rank, using separate symbols for men and women.

(b) What does your plot show about the pattern of life expectancy? What does it show about the effect of sex on life expectancy?

4.33 Feed the Birds. Canaries provide more food to their babies when the babies beg more intensely. Researchers wondered if begging was the main factor determining how much food baby canaries receive or if parents also take into account whether the babies are theirs or not. To investigate, researchers conducted an experiment allowing canary parents to raise two broods: one of their own and one fostered from a different pair of parents. If begging determines how much food babies receive, then differences in the "begging intensities" of the broods should be strongly associated with differences in the amount of food the broods receive. The researchers decided to use the relative growth rates (that is, the growth rates of the foster babies relative to those of the natural babies, with values greater than 1 indicating that the foster babies grew more rapidly than the natural babies) as a measure of the difference in the amount of food received. They recorded the difference in begging intensities (that is, the begging intensity of the foster babies minus that of the natural babies) and relative growth rates. Here are data from the experiment:[21] CANARY

Difference in begging intensity	−14.0	−12.5	−12.0	−8.0	−8.0	−6.5	−5.5
Relative growth rate	0.85	1.00	1.33	0.85	0.90	1.15	1.00
Difference in begging intensity	−3.5	−3.0	−2.0	−1.5	−1.5	0.0	0.0
Relative growth rate	1.30	1.33	1.03	0.95	1.15	1.13	1.00
Difference in begging intensity	2.00	2.00	3.00	4.50	7.00	8.00	8.50
Relative growth rate	1.07	1.14	1.00	0.83	1.15	0.93	0.70

(a) Make a scatterplot that shows how relative growth rate responds to the difference in begging intensity.

(b) Describe the overall pattern of the relationship. Is it linear? Is there a positive or negative association, or neither? Find the correlation r. Is r a helpful description of this relationship?

(c) If begging intensity is the main factor determining food received, with higher intensity leading to more food, one would expect the relative growth rate to increase as the difference in begging intensity increases. However, if both begging intensity and a preference for their own babies determine the amount of food received (and hence the relative growth rate), we might expect growth rate to increase initially as begging intensity increases but then to level off (or even decrease) as the parents begin to ignore increases in begging by the foster babies. Which of these theories do the data appear to support? Explain your answer.

4.34 Good Weather and Tipping. Favorable weather has been shown to be associated with increased tipping of servers in restaurants.

Jupiterimages/Getty Images

Will just the belief that future weather will be favorable lead to higher tips? Researchers gave 60 index cards to a waitress at an Italian restaurant in New Jersey. Before delivering the bill to each customer, the waitress randomly selected a card and wrote on the bill the same message that was printed on the index card. Twenty of the cards had the message "The weather is supposed to be really good tomorrow. I hope you enjoy the day!" Another 20 cards contained the message "The weather is supposed to be not so good tomorrow. I hope you enjoy the day anyway!" The remaining 20 cards were blank, indicating that the waitress was not supposed to write any message. Choosing a card at random ensured that there was a random assignment of the diners to the three experimental conditions. Here are the percentage tips for the three messages:[22] TIPPING

Weather Report	Tip Percentage									
Good	20.8	18.7	19.9	20.6	21.9	23.4	22.8	24.9	22.2	20.3
	24.9	22.3	27.0	20.5	22.2	24.0	21.2	22.1	22.0	22.7
Bad	18.0	19.1	19.2	18.8	18.4	19.0	18.5	16.1	16.8	14.0
	17.0	13.6	17.5	20.0	20.2	18.8	18.0	23.2	18.2	19.4
None	19.9	16.0	15.0	20.1	19.3	19.2	18.0	19.2	21.2	18.8
	18.5	19.3	19.3	19.4	10.8	19.1	19.7	19.9	21.3	20.6

(a) Make a plot of tip percentage against the weather report on the bill (spacing the three weather reports equally on the horizontal axis). Which weather report appears to lead to the best tips?

(b) Does it make sense to speak of a positive or negative association between weather report and tip percentage? Why? Is correlation r a helpful description of the relationship? Why?

4.35 Thinking about Correlation. Exercise 4.27 presents data on wine intake and the relative risk of breast cancer in women.

(a) If wine intake is measured in ounces of alcohol per day rather than grams per day, how would the correlation change? (There is 0.035 ounce in a gram.)

(b) How would r change if all the relative risks were 0.25 less than the values given in the table? Does the correlation tell us that among women who drink, those who drink more wine tend to have a greater relative risk of cancer than women who don't drink at all?

(c) If drinking an additional gram of alcohol each day raised the relative risk of breast cancer by exactly 0.01, what would be the correlation between alcohol in wine intake and relative risk of breast cancer? (*Hint:* Draw a scatterplot for several values of alcohol in wine intake.)

4.36 The Effect of Changing Units. Changing the units of measurement can dramatically alter the appearance of a scatterplot. Return to the data on percentage body fat and preferred amount of salt in Exercise 4.24: SALT2

Preferred amount of salt x	0.2	0.3	0.4	0.5	0.6	0.8	1.1
Percentage body fat y	20	30	22	30	38	23	30

In calculating the preferred amount of salt, the weight of the salt was in milligrams. A mad scientist decides to measure weight in tenths of milligrams. The same data in these units are

Preferred amount of salt x	2	3	4	5	6	8	11
Percentage body fat y	20	30	22	30	38	23	30

(a) Make a plot with the x axis extending from 0 to 12 and the y axis from 15 to 40. Plot the original data on these axes. Then plot the new data using a different color or symbol. The two plots look very different.

(b) The correlation is exactly the same for the two sets of measurements. Why do you know that this is true without doing any calculations? Find the two correlations to verify that they are the same.

4.37 Statistics for Investing. Investment reports now often include correlations. Following a table of correlations among mutual funds, a report adds: "Two funds can have perfect correlation yet different levels of risk. For example, Fund A and Fund B may be perfectly correlated, yet Fund A moves 20% whenever Fund B moves 10%." Write a brief explanation, for someone who knows no statistics, of how this can happen. Include a sketch to illustrate your explanation.

4.38 Statistics for Investing. A mutual funds company's newsletter says "A well-diversified portfolio includes assets with low correlations." The newsletter includes a table of correlations between the returns on various classes of investments. For example, the correlation between municipal bonds and large-cap stocks is 0.50, and the correlation between municipal bonds and small-cap stocks is 0.21.

(a) Rachel invests heavily in municipal bonds. She wants to diversify by adding an investment whose returns do not closely follow the returns on her bonds. Should she choose large-cap stocks or small-cap stocks for this purpose? Explain your answer.

(b) If Rachel wants an investment that tends to increase when the return on her bonds drops, what kind of correlation should she look for?

4.39 Teaching and Research. A college newspaper interviews a psychologist about student ratings of the teaching of faculty members. The psychologist says, "The evidence indicates that the correlation between the research productivity and teaching rating of faculty members is close to zero." The paper reports this as "Professor McDaniel said that good researchers tend to be poor teachers, and vice versa." Explain why the paper's report is wrong. Write a statement in plain language (without using the word *correlation*) to explain the psychologist's meaning.

4.40 Sloppy Writing about Correlation. Each of the following statements contains a blunder. Explain in each case what is wrong.

(a) "There is a high correlation between the gender of an adult and and their political affiliation."

(b) "We found a strong negative correlation ($r = -1.09$) between the amount of time spent on social media and the number of books read in the last year."

(c) "The correlation between height and weight of the subjects was $r = 0.63$ centimeter per kilogram."

4.41 More about Scatterplots and Correlation. Here are two sets of data:

Data Set A				
x	1	2	3	4
y	1	1.5	0.5	4

Data Set B								
x	1	1	1	2	3	4	4	4
y	1	1	1	1.5	0.5	4	4	4

(a) Make a scatterplot of both sets of data. Comment on any differences you see in the two plots.

(b) Compute the correlation for both sets of data. Comment on any differences in the two values. Are these differences what you would expect from the plots in part (a)?

4.42 Correlation Is Not Resistant. Go to the *Correlation and Regression* applet. Click and drag points from the Data Bank to the graph to create a group of 10 points in the lower-left corner of the scatterplot with a strong straight-line pattern (correlation about 0.9).

(a) Add one point at the upper right that is in line with the first 10. How does the correlation change?

(b) Drag this last point down until it is opposite the group of 10 points. How small can you make the correlation? Can you make the correlation negative? You see that a single outlier can greatly strengthen or weaken a correlation. Always plot your data to check for outlying points.

4.43 Match the Correlation. You are going to use the *Correlation and Regression* applet to make scatterplots with 10 points that have correlation close to 0.7. The lesson is that many patterns can have the same correlation. Always plot your data before you trust a correlation.

(a) Click and drag points from the Data Bank to the graph to add the first two points. What is the value of the correlation? Why does it have this value?

(b) Make a lower-left to upper-right pattern of 10 points with correlation about $r = 0.7$. (You can drag points up or down to adjust r after you have 10 points.) Make a rough sketch of your scatterplot.

(c) Make another scatterplot with nine points in a vertical stack at the left of the plot. Add one point far to the right and move it until the correlation is close to 0.7. Make a rough sketch of your scatterplot.

(d) Make yet another scatterplot with 10 points in a curved pattern that starts at the lower left, rises to the right, then falls again at the far right. Adjust the points up or down until you have a fairly smooth curve with correlation close to 0.7. Make a rough sketch of this scatterplot also.

The following exercises ask you to answer questions from data without having the details outlined for you. The exercise statements give you the ***State*** *step of the four-step process. In your work, follow the* ***Plan***, ***Solve***, *and* ***Conclude*** *steps of the process, described on page 62.*

4.44 Global Warming. Have average global temperatures been increasing in recent years? Here are annual average global temperatures for the past 25 years, in degrees Celsius:[23] GTEMPS

Year	1994	1995	1996	1997	1998	1999	2000	2001
Temperature	14.25	14.37	14.23	14.42	14.56	14.34	14.33	14.47
Year	2002	2003	2004	2005	2006	2007	2008	2009
Temperature	14.52	14.54	14.49	14.57	14.54	14.52	14.45	14.55
Year	2010	2011	2012	2013	2014	2015	2016	2017
Temperature	14.63	14.48	14.54	14.58	14.64	14.83	14.90	14.81
Year	2018							
Temperature	14.73							

Discuss what the data show about change in average global temperatures over time.

4.45 Will Women Outrun Men? Does the physiology of women make them better suited than men to long-distance running? Will women eventually outperform men in long-distance races? In 1992, researchers examined data on world record times (in seconds) for men and women in the marathon. Here are data for women:[24] RUNNING

Year	1926	1964	1967	1970	1971	1974	1975
Time	13,222.0	11,973.0	11,246.0	10,973.0	9990.0	9834.5	9499.0
Year	1977	1980	1981	1982	1983	1985	
Time	9287.5	9027.0	8806.0	8771.0	8563.0	8466.0	

Here are data for men:

Year	1908	1909	1913	1920	1925	1935	1947
Time	10,518.4	9751.0	9366.6	9155.8	8941.8	8802.0	8739.0
Year	1952	1953	1954	1958	1960	1963	1964
Time	8442.2	8314.8	8259.4	8117.0	8116.2	8068.0	7931.2
Year	1965	1967	1969	1981	1984	1985	1988
Time	7920.0	7776.4	7713.6	7698.0	7685.0	7632.0	7610.0

(a) What do the data show about women's and men's times in the marathon? (Start by plotting both sets of data on the same plot, using two different plotting symbols.)

(b) Based on these data, researchers (in 1992) predicted that women would outrun men in the marathon in 1998. How do you think they arrived at this date? Was their prediction accurate? (You may want to look on the web; try doing a Google search on "women's world record marathon times.")

4.46 **Toucan's Beak.** 4step The toco toucan, the largest member of the toucan family, possesses the largest beak relative to body size of all birds. This exaggerated feature has received various interpretations, such as being a refined adaptation for feeding. However, the large surface area may also be an important mechanism for radiating heat (and hence cooling the bird) as outdoor temperature increases. Here are data for beak heat loss, as a percentage of total body heat loss, at various temperatures in degrees Celsius:[25] TOUCAN

Fedor Selivanov/Shutterstock

Temperature (°C)	15	16	17	18	19	20	21	22
Percentage heat loss from beak	32	34	35	33	37	46	55	51
Temperature (°C)	23	24	25	26	27	28	29	30
Percentage heat loss from beak	43	52	45	53	58	60	62	62

Investigate the relationship between outdoor temperature and beak heat loss, as a percentage of total body heat loss.

4.47 **Does Social Rejection Hurt?** 4step We often describe our emotional reaction to social rejection as "pain." Does social rejection cause activity in areas of the brain that are known to be activated by physical pain? If it does, we really do experience social and physical pain in similar ways. Psychologists first included and then deliberately excluded individuals from a social activity while they measured changes in brain activity. After each activity, the subjects filled out questionnaires that assessed how excluded they felt. Here are data for 13 subjects:[26] REJECT

Subject	Social Distress	Brain Activity	Subject	Social Distress	Brain Activity
1	1.26	−0.055	8	2.18	0.025
2	1.85	−0.040	9	2.58	0.027
3	1.10	−0.026	10	2.75	0.033
4	2.50	−0.017	11	2.75	0.064
5	2.17	−0.017	12	3.33	0.077
6	2.67	0.017	13	3.65	0.124
7	2.01	0.021			

The explanatory variable is change in brain activity in a region of the brain that is activated by physical pain. Negative values show a decrease in activity, suggesting less physical distress. The response variable is "social distress," measured by each subject's questionnaire score after exclusion relative to the score after inclusion. (So values greater than 1 show the degree of distress caused by exclusion.) Discuss what the data show about the relationship between social distress and brain activity.

4.48 **Yukon Squirrels.** 4step The population density of North American red squirrels in Yukon, Canada, fluctuates annually. Researchers believe one reason for the fluctuation may be the availability of white spruce cones in the spring, a significant source of food for the squirrels. To explore this, researchers measured red squirrel population density in the spring and spruce cone production the previous autumn over a 23-year period. The data for one study area appear in Table 4.2.[27] Squirrel population density is measured in squirrels per hectare. Spruce cone production is an index on a logarithmic scale, with larger values indicating larger spruce cone production. Discuss whether the data support the idea that higher spruce cone production in the autumn leads to a higher squirrel population density the following spring. SQRLCO

TABLE 4.2 Red squirrel population density and spruce cone production in Yukon, Canada

Squirrel Density (Squirrels per ha)	Cone Production (Index)	Squirrel Density (Squirrels per ha)	Cone Production (Index)
1.0	0.1	1.3	2.3
1.5	0.3	2.3	2.5
1.4	0.4	2.0	2.9
1.5	0.5	1.0	3.1
0.7	0.5	1.4	3.3
1.2	0.8	0.9	3.6
0.9	1.2	1.3	3.8
1.0	1.4	1.9	4.2
1.2	1.8	1.8	5.1
1.4	2.0	1.9	5.2
1.5	2.1	3.4	5.3
0.8	2.2		

4.49 4step **Teacher Salaries.** For each of the 50 states and the District of Columbia, average Mathematics SAT scores and average high school teacher salaries for 2018 are available.[28] Discuss whether the data support the idea that higher teacher salaries lead to higher Mathematics SAT scores. TCHSAL

CHAPTER

5

When you complete this chapter, you will be able to:

5.1 Use a regression line to predict values of a response variable and interpret the slope and intercept of a regression line.

5.2 Understand the meaning of the least-squares regression line and its equation.

5.3 Identify the least-squares regression line equation in the outputs of calculators and computers.

5.4 Based on the properties of the least-squares regression line, judge its usefulness in specific situations.

5.5 Use residuals and plots of residuals to reach conclusions about the relationship between two quantitative variables.

5.6 Describe the impact of outliers and influential observations on least-squares regression lines and formulas.

5.7 Consider the impact of ecological correlation, extrapolation, and lurking variables when making predictions based on regression.

5.8 Distinguish between correlation and causation when interpreting associations between variables.

5.9 Recognize the limitations of reaching conclusions about associations between variables based on patterns in very large data sets.

Regression

In this chapter we continue the exploration of the relationships between two variables that we began in Chapter 4. In Chapter 4 we saw that a useful tool for exploring the relationship between two variables is the scatterplot. When the relationship is linear, correlation is a numerical measure of the strength of the linear relationship.

In this chapter we introduce a method for summarizing the straight-line relationship between two variables when it exists. When there is a clear explanatory variable and a strong straight-line relationship, it is tempting to assume that large values of the correlation imply that there is a cause-and-effect relationship between the explanatory variable and the response variable. This need not be true. Correlation does not imply causation! In this chapter we explore this issue more carefully.

Linear (straight-line) relationships between two quantitative variables are easy to understand and quite common. In Chapter 4, we found linear relationships in settings as varied as Florida manatee deaths, the risk of cancer, and predicting tropical storms. Correlation measures the direction and strength of these linear relationships. When a scatterplot shows a linear relationship, we would like to summarize the overall pattern by drawing a line of best fit (called a regression line) on the scatterplot.

5.1 Regression Lines

A *regression line* summarizes the relationship between two variables—but only in a specific setting: when one of the variables helps explain or predict the other. That is, regression describes a relationship between an explanatory variable and a response variable.

iStock/Getty Images

Regression Line

A **regression line** is a straight line that describes how a response variable y changes as an explanatory variable x changes. We often use a regression line to predict the value of y for a given value of x when we believe the relationship between y and x is linear.

EXAMPLE 5.1 Does Fidgeting Keep You Slim?

4step

FATGAIN

Why is it that some people find it easy to stay slim? Here, following our four-step process (page 62), is an account of a study that sheds some light on gaining weight.

STATE: Some people don't gain weight even when they overeat. Perhaps fidgeting and other "nonexercise activity" (NEA) explains why. In fact, some people may spontaneously increase nonexercise activity when fed more, thus reducing the amount of weight they gain from overeating. To investigate the effect of NEA on fat gain, researchers deliberately overfed 16 healthy young adults for eight weeks. They measured fat gain (in kilograms) and, as an explanatory variable, change in energy use (in calories) from activity other than deliberate exercise–fidgeting, daily living, and the like. Change in energy use was energy use measured the last day of the eight-week period minus energy use measured the day before the overfeeding began. Here are the data:[1]

NEA change (cal)	−94	−57	−29	135	143	151	245	355
Fat gain (kg)	4.2	3.0	3.7	2.7	3.2	3.6	2.4	1.3
NEA change (cal)	392	473	486	535	571	580	620	690
Fat gain (kg)	3.8	1.7	1.6	2.2	1.0	0.4	2.3	1.1

Do people with larger increases in NEA tend to gain less fat?

PLAN: Make a scatterplot of the data and examine the pattern. If it is linear, use correlation to measure its strength and draw a regression line on the scatterplot to predict fat gain from change in NEA.

SOLVE: Figure 5.1 is a scatterplot of these data. The plot shows a moderately strong negative linear association with no outliers. The correlation is $r = -0.7786$.

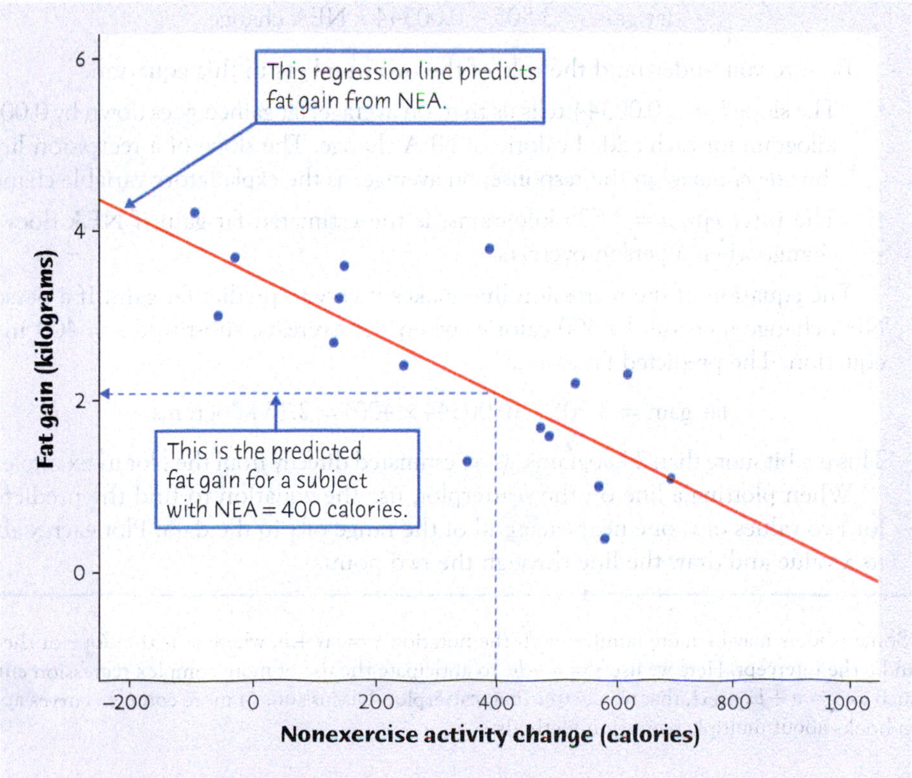

FIGURE 5.1 Fat gain after eight weeks of overeating, plotted against NEA (increase in nonexercise activity) over the same period, for Example 5.1.

The line on the plot is a regression line for predicting fat gain from change in NEA.

CONCLUDE: People with larger increases in NEA do indeed gain less fat. To add to this conclusion, we must study regression lines in more detail.

We can, however, already use the regression line to predict fat gain from NEA. Suppose that an individual's NEA increases by 400 calories when she overeats. Go "up and over" on the graph in Figure 5.1. From 400 calories on the x axis, go up to the regression line and then over to the y axis. The graph shows that the predicted gain in fat is a bit more than 2 kilograms.

Many calculators and software programs will give you the equation of a regression line from keyed-in data. Understanding and using the line are more important than the details of where the equation comes from.

STATISTICS IN YOUR WORLD

Regression Toward the Mean

To "regress" means to go backward. Why are statistical methods for predicting a response from an explanatory variable called "regression"? Sir Francis Galton (1822–1911), who was the first to apply regression to biological and psychological data, looked at examples such as the heights of children versus the heights of their parents. He found that taller-than-average parents tended to have children who were also taller than average but not as tall as their parents. Galton called this fact "regression toward the mean," and the name came to be applied to the statistical method.

Review of Straight Lines

Suppose that y is a response variable (plotted on the vertical axis) and x is an explanatory variable (plotted on the horizontal axis). A straight line relating y to x has an equation of the form*

$$y = a + bx$$

In this equation, b is the **slope,** the amount by which y changes when x increases by one unit. The number a is the **intercept,** the value of y when $x = 0$.

EXAMPLE 5.2 Using a Regression Line

Any straight line describing the NEA data has the form

$$\text{fat gain} = a + (b \times \text{NEA change})$$

The line in Figure 5.1 is the regression line with the equation

$$\text{fat gain} = 3.505 - 0.00344 \times \text{NEA change}$$

Be sure you understand the role of the two numbers in this equation:

- The slope $b = -0.00344$ tells us that, on average, fat gained goes down by 0.00344 kilogram for each added calorie of NEA change. The slope of a regression line is the *rate of change* in the response, on average, as the explanatory variable changes.
- The intercept, $a = 3.505$ kilograms, is the estimated fat gain if NEA does not change when a person overeats.

The equation of the regression line makes it easy to predict fat gain. If a person's NEA change increases by 400 calories when she overeats, substitute $x = 400$ in the equation. The predicted fat gain is

$$\text{fat gain} = 3.505 - (0.00344 \times 400) = 2.13 \text{ kilograms}$$

This is a bit more than 2 kilograms, as we estimated directly from the plot in Example 5.1.

When plotting a line on the scatterplot, use the equation to find the predicted y for two values of x, one near each end of the range of x in the data. Plot each y above its x value and draw the line through the two points.

*Some readers may be more familiar with the notation $y = mx + b$, where m is the slope of the line and b the intercept. Here we use $y = a + bx$ to anticipate the use of more complex regression curves, such as $y = a + bx + cx^2$, that may better fit a scatterplot. Discussions of more complex curves appear in books about multiple regression methods.

The slope of a regression line is an important numerical description of the relationship between the two variables. Although we need the value of the intercept to draw the line, this value is statistically meaningful only when, as in Example 5.2, the explanatory variable can actually take values close to zero. The slope $b = -0.00344$ in Example 5.2 is small. This does *not* mean that change in NEA has little effect on fat gain. The size of the slope depends on the units in which we measure the two variables. In this example, the slope is the change in fat gain in kilograms when NEA increases by 1 calorie. There are 1000 grams in a kilogram. If we measured fat gain in grams, the slope would be 1000 times larger, $b = -3.44$. *You can't say how important a relationship is by looking at the size of the slope of the regression line.*

APPLY YOUR KNOWLEDGE

5.1 City Mileage, Highway Mileage. We expect a car's highway gas mileage to be related to its city gas mileage (in miles per gallon, mpg). Data for all 1259 vehicles in the government's *2019 Fuel Economy Guide* give the regression line

$$\text{highway mpg} = 8.720 + (0.914 \times \text{city mpg})$$

for predicting highway mileage from city mileage.

(a) What is the slope of this line? Say in words what the numerical value of the slope tells you.

(b) What is the intercept? Explain why the value of the intercept is not statistically meaningful.

(c) Find the predicted highway mileage for a car that gets 16 mpg in the city. Do the same for a car with city mileage of 28 mpg.

(d) Draw a graph of the regression line for city mileages between 10 and 50 mpg. (Be sure to show the scales for the x and y axes.)

5.2 What's the Line? An online article suggested that for each additional person who took up regular running for exercise, the number of cigarettes smoked daily would decrease by 0.178.[2] If we assume that 48 million cigarettes would be smoked per day if nobody ran, what is the equation of the regression line for predicting number of cigarettes smoked per day from the number of people who regularly run for exercise?

5.3 Shrinking Forests. Scientists measured the annual forest loss (in square kilometers) in Indonesia from 2000 to 2012.[3] They found the regression line

$$\text{forest loss} = 7500 + (1021 \times \text{year since 2000})$$

for predicting forest loss, in square kilometers, from years since 2000.

(a) What is the slope of this line? Say in words what the numerical value of the slope tells you.

(b) If we measured forest loss in meters^2 per year, what would the slope be? Note that there are 10^6 square meters in a square kilometer.

(c) If we measured forest loss in thousands of square kilometers per year, what would the slope be?

5.2 The Least-Squares Regression Line

In most cases, no line will pass exactly through all the points in a scatterplot. Different people will draw different lines by eye. We need a way to draw a regression line that doesn't depend on our guess of where the line should go. Because we use the line to predict y from x, the prediction errors we make are errors in y, the vertical direction in the scatterplot. A *good regression line makes the vertical distances of the points from the line as small as possible.*

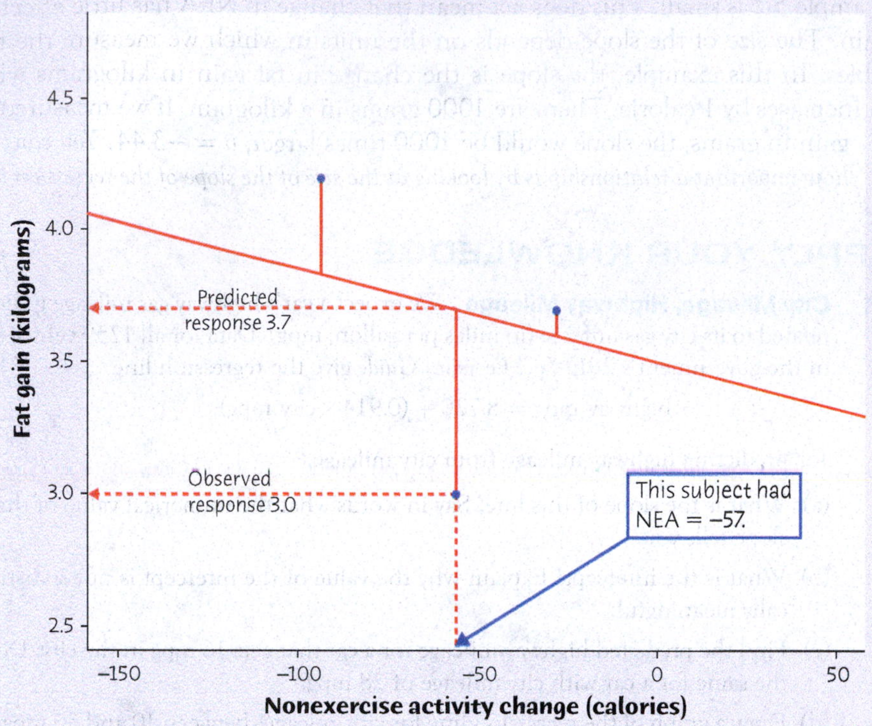

FIGURE 5.2
The least-squares idea. For each observation, find the vertical distance of each point on the scatterplot from a regression line. The least-squares regression line makes the sum of the squares of these distances as small as possible.

Figure 5.2 illustrates the idea. This plot shows three of the points from Figure 5.1, along with the line, on an expanded scale. The line passes above one of the points and below two of them. The three prediction errors appear as vertical line segments. For example, one subject had $x = -57$, a decrease of 57 calories in NEA. The line predicts a fat gain of 3.7 kilograms, but the actual fat gain for this subject was 3.0 kilograms. The prediction error is

$$\text{error} = \text{observed response} - \text{predicted response}$$
$$= 3.0 - 3.7 = -0.7 \text{ kilogram}$$

There are many ways to make the collection of vertical distances "as small as possible." The most common is the *least-squares* method.

Least-Squares Regression Line

The **least-squares regression line** of y on x is the line that makes the sum of the squares of the vertical distances of the data points from the line as small as possible.

One reason for the popularity of the least-squares regression line is that the problem of finding the line has a simple answer. We can give the equation for the least-squares line in terms of the means and standard deviations of the two variables and the correlation between them.

Equation of The Least-Squares Regression Line

We have data on an explanatory variable x and a response variable y for n individuals. From the data, calculate the means $\bar{x}$ and $\bar{y}$ and the standard deviations s_x and s_y of the two variables and their correlation r. The least-squares regression line is the line

$$\hat{y} = a + bx$$

with slope

$$b = r\frac{s_y}{s_x}$$

and intercept

$$a = \bar{y} - b\bar{x}$$

We write $\hat{y}$ (read "y hat") in the equation of the regression line to emphasize that the line gives a *predicted* response $\hat{y}$ for any x. Because of the scatter of points about the line, the predicted response will usually not be exactly the same as the actually *observed* response y. In practice, you don't need to calculate the means, standard deviations, and correlation first. Software or your calculator will give the slope b and intercept a of the least-squares line from the values of the variables x and y. You can then concentrate on understanding and using the regression line.

5.3 Examples of Technology

Least-squares regression is one of the most common statistical procedures. Any technology you use for statistical calculations will give you the least-squares line and related information. Figure 5.3 displays the regression output for the data of

Texas Instruments

```
LinReg
 y=a+bx
 a=3.505122916
 b=-.003441487
 r²=.6061492049
 r=-.7785558457
```

CRUNCHIT! CRUNCHIT!

Export

```
Fitted Equation:
Fat = 3.505 - 0.003441 * NEA

              Estimate    Std. Error   t value    Pr(>|t|)
(Intercept)     3.505       0.3036      11.54      <0.0001
NEA          -0.003441    0.0007414    -4.642    0.0003810

r-Squared: 0.6061  Adjusted r-Squared: 0.5780 sigma: 0.7399
```

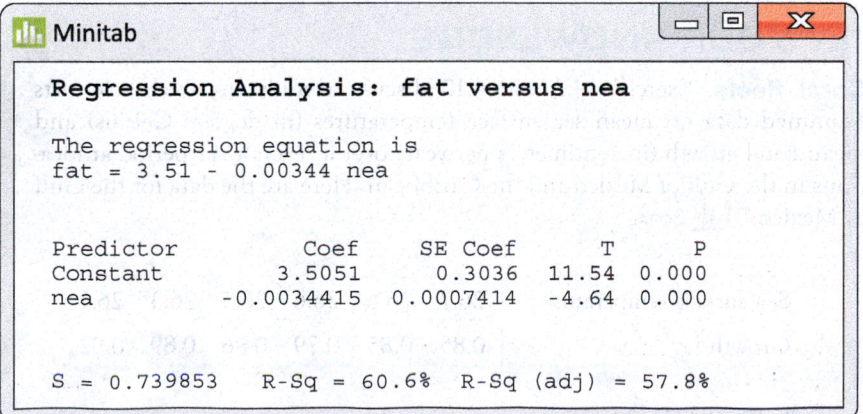

Minitab

```
Regression Analysis: fat versus nea

The regression equation is
fat = 3.51 - 0.00344 nea


Predictor         Coef     SE Coef      T      P
Constant        3.5051      0.3036  11.54  0.000
nea         -0.0034415   0.0007414  -4.64  0.000


S = 0.739853   R-Sq = 60.6%   R-Sq (adj) = 57.8%
```

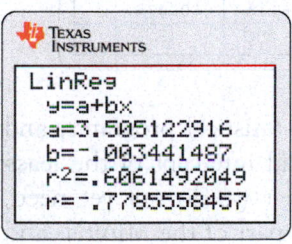

FIGURE 5.3 Least-squares regression for the nonexercise activity data: output from a graphing calculator, three statistical programs, and a spreadsheet program.

FIGURE 5.3
(Continued)

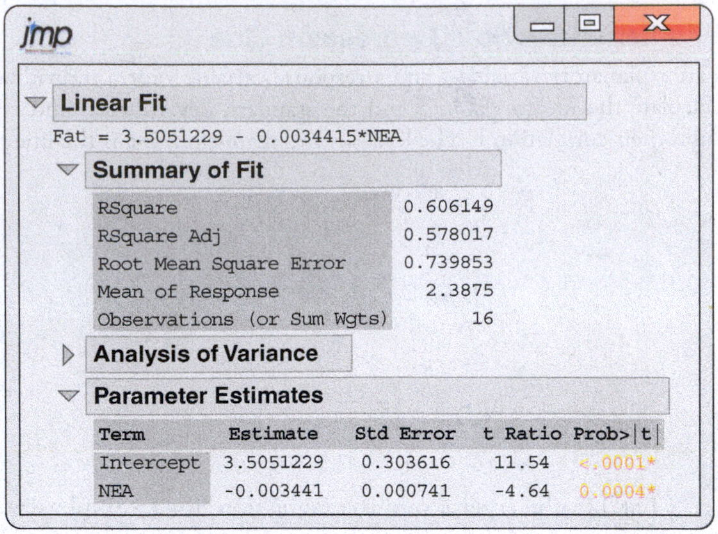

jmp

Linear Fit

Fat = 3.5051229 - 0.0034415*NEA

Summary of Fit

RSquare	0.606149
RSquare Adj	0.578017
Root Mean Square Error	0.739853
Mean of Response	2.3875
Observations (or Sum Wgts)	16

Analysis of Variance

Parameter Estimates

Term	Estimate	Std Error	t Ratio	Prob>|t|
Intercept	3.5051229	0.303616	11.54	<.0001*
NEA	-0.003441	0.000741	-4.64	0.0004*

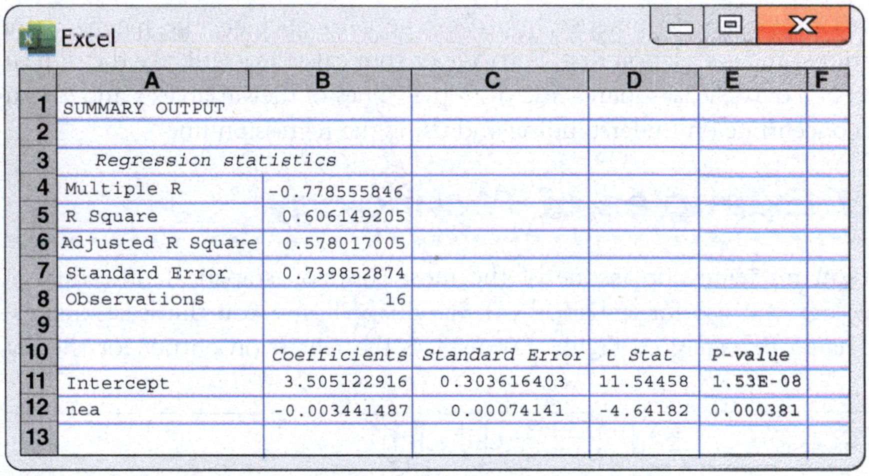

Excel

	A	B	C	D	E	F
1	SUMMARY OUTPUT					
2						
3	*Regression statistics*					
4	Multiple R	-0.778555846				
5	R Square	0.606149205				
6	Adjusted R Square	0.578017005				
7	Standard Error	0.739852874				
8	Observations	16				
9						
10		*Coefficients*	*Standard Error*	*t Stat*	*P-value*	
11	Intercept	3.505122916	0.303616403	11.54458	1.53E-08	
12	nea	-0.003441487	0.00074141	-4.64182	0.000381	
13						

Examples 5.1 and 5.2 from a graphing calculator, three statistical programs, and a spreadsheet program. Each output records the slope and intercept of the least-squares line. The software also provides information that we do not yet need, although we will use much of it later. (In fact, we left out part of the Minitab and Excel outputs.) Be sure that you can locate the slope and intercept on all four outputs. *Once you understand the statistical ideas, you can read and work with almost any software output.*

APPLY YOUR KNOWLEDGE

5.4 Coral Reefs. Exercises 4.2 and 4.10 discuss a study in which scientists examined data on mean sea surface temperatures (in degrees Celsius) and mean coral growth (in centimeters per year) over a several-year period at locations in the Gulf of Mexico and the Caribbean. Here are the data for the Gulf of Mexico:[4] CORAL

Sea surface temperature	26.7	26.6	26.6	26.5	26.3	26.1
Growth	0.85	0.85	0.79	0.86	0.89	0.92

(a) Use your calculator to find the mean and standard deviation of both sea surface temperature x and growth y and the correlation r between x and y. Use these basic measures to find the equation of the least-squares line for predicting y from x.

(b) Enter the data into your software or calculator and use the regression function to find the least-squares line. The result should agree with your work in part (a) up to roundoff error.

(c) Say in words what the numerical value of the slope tells you.

5.5 Homicide and Suicide. Preventing suicide is a important issue facing mental health workers. Predicting geographic regions where the risk of suicide is high could help people decide where to increase or improve mental health resources and care. Some psychiatrists have argued that homicide and suicide may have some causes in common. If so, one would expect homicide and suicide rates to be correlated. And if this is true, areas with high rates of homicide might be predicted to have high rates of suicide and therefore be in need of increased mental health resources. Research has had mixed results, including some evidence that there is a positive correlation in certain European countries but not in the United States. Here are data from 2015 for the 11 counties in Ohio with sufficient data for homicides and suicides to allow for estimating rates for both.[5] Rates are per 100,000 people. DEATH

County	Homicide Rate	Suicide Rate	County	Homicide Rate	Suicide Rate
Butler	4.0	11.2	Lucas	6.0	12.6
Clark	10.8	15.3	Mahoning	11.7	15.2
Cuyahoga	12.2	11.4	Montgomery	8.9	15.7
Franklin	8.7	12.3	Stark	5.8	16.1
Hamilton	10.2	11.0	Summit	7.1	17.9
Lorain	3.3	14.3			

(a) Make a scatterplot that shows how suicide rate can be predicted from homicide rate. There is a weak linear relationship, with correlation $r = -0.0645$.

(b) Find the least-squares regression line for predicting suicide rate from homicide rate. Add this line to your scatterplot.

(c) Explain in words what the slope of the regression line tells us.

(d) Another Ohio county has a homicide rate of 8.0 per 100,000 people. What is the county's predicted suicide rate?

5.4 Facts About Least-Squares Regression

One reason for the popularity of least-squares regression lines is that they have many convenient properties. The least-squares regression line is the line that makes the vertical distances of the data points from the line as small as possible. Here are some additional facts about least-squares regression lines.

Fact 1. *The distinction between explanatory and response variables is essential in regression.* Least-squares regression makes the distances of the data points from the line small only in the y direction. If we reverse the roles of the two variables, we get a different least-squares regression line.

Fact 2. *There is a close connection between correlation and the slope of the least-squares line.* The slope is

$$b = r\frac{s_y}{s_x}$$

EXAMPLE 5.3 Predicting Fat Gain, Predicting Change in NEA

Figure 5.4 repeats the scatterplot of the NEA data in Figure 5.1 (page 125) but with *two* least-squares regression lines. The solid line is the regression line for predicting fat gain from change in NEA. This is the line that appeared in Figure 5.1. We might also use the data on these 16 subjects to predict the change in NEA for another subject based on that subject's fat gain when overfed for eight weeks. Now the roles of the variables are reversed: fat gain is the explanatory variable, and change in NEA is the response variable. The dashed line in Figure 5.4 is the least-squares line for predicting NEA change from fat gain. The two regression lines are not the same. *In the regression setting, you must know clearly which variable is explanatory.*

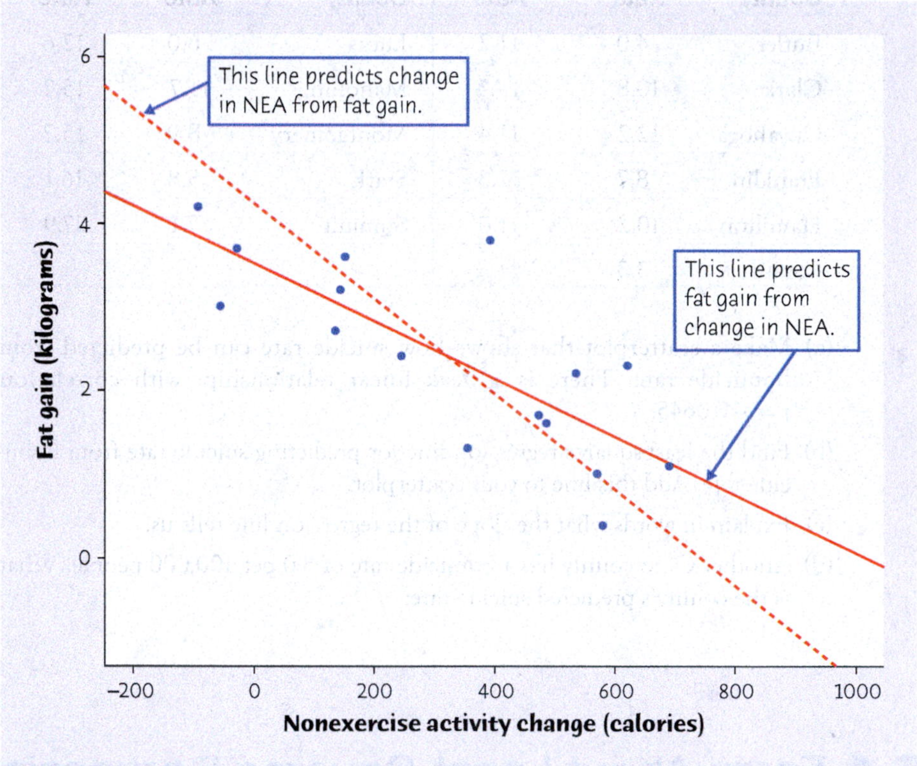

FIGURE 5.4
Two least-squares regression lines for the nonexercise activity data, for Example 5.3. The solid line predicts fat gain from change in nonexercise activity. The equation for this line is fat = 3.505 − 0.00344 NEA. The dashed line predicts change in nonexercise activity from fat gain. The equation for this line is NEA = 745.3 − 176 fat (or, after rearranging terms, fat = 4.23 − 0.00568 NEA).

You see that the slope and the correlation always have the same sign. For example, if a scatterplot shows a positive association, then both b and r are positive. The formula for the slope b says more: along the regression line, a change of one standard deviation in x corresponds to a change of r standard deviations in y. When the variables are perfectly correlated ($r = 1$ or $r = -1$), the change in the predicted response $\hat{y}$ is the same (in standard deviation units) as the change in x. Otherwise, because $-1 \leq r \leq 1$, the change in $\hat{y}$ (in standard deviation units) is less than the change in x (in standard deviation units). As the correlation grows less strong, the prediction $\hat{y}$ moves less in response to changes in x.

Fact 3. *The least-squares regression line always passes through the point* $(\bar{x}, \bar{y})$ *on the graph of* y *against* x. This is a consequence of the equation of the least-squares regression line (box on page 129). In Exercise 5.55, we ask you to confirm this.

Fact 4. *The correlation* r *describes the strength of a straight-line relationship.* In the regression setting, this description takes a specific form: the square of the correlation, r^2, is the fraction of the variation in the values of y that is explained by the least-squares regression of y on x.

The idea is that when there is a linear relationship, some of the variation in y is accounted for by the fact that as x changes, y changes along with it. Look again at Figure 5.1 (page 125), the scatterplot of the NEA data. The variation in y appears as the spread of fat gains from 0.4 to 4.2 kg. Some of this variation is explained by the fact that x (change in NEA) varies from a loss of 94 calories to a gain of 690 calories. As x changes from -94 to 690, y changes along the line. You would predict a smaller fat gain for a subject whose NEA increased by 600 calories than for someone with 0 change in NEA. But the straight-line tie of y to x doesn't explain *all* of the variation in y. The remaining variation appears as the scatter of points above and below the line.

Although we won't do the algebra, it is possible to break the variation in the observed values of y into two parts. One part measures the variation in $\hat{y}$ along the least-squares regression line as x varies. The other measures the vertical scatter of the data points above and below the line. The squared correlation r^2 is the first of these as a fraction of the whole:

$$r^2 = \frac{\text{variation in } \hat{y} \text{ along the regression line as } x \text{ varies}}{\text{total variation in observed values of } y}$$

EXAMPLE 5.4 Using r^2

For the NEA data, $r = -0.7786$ and $r^2 = (-0.7786)^2 = 0.6062$. About 61% of the variation in fat gained is accounted for by the linear relationship with change in NEA. The other 39% is individual variation among subjects that is not explained by the linear relationship.

Figure 4.3 (page 105) shows a stronger linear relationship between boat registrations in Florida and manatees killed by boats. The correlation is $r = 0.919$ and $r^2 = (0.919)^2 = 0.845$. About 85% of the year-to-year variation in number of manatees killed by boats is explained by regression on number of boats registered. Only about 15% is variation among years with similar numbers of boats registered.

You can find a regression line for any relationship between two quantitative variables, but the usefulness of the line for prediction depends on the strength of the linear relationship. So r^2 is almost as important as the equation of the line in reporting a regression. All the outputs in Figure 5.3 (pages 129 and 130) include r^2, either in decimal form or as a percentage. When you see a correlation, square it to get a better feel for the strength of the association. Perfect correlation ($r = -1$ or $r = 1$) means the points lie exactly on a line. Then $r^2 = 1$, and all the variation in one variable is accounted for by the linear relationship with the other variable. If $r = -0.7$ or $r = 0.7$, $r^2 = 0.49$ and about half the variation is accounted for by the linear relationship. In the r^2 scale, correlation ± 0.7 is about halfway between 0 and ± 1.

Facts 2, 3, and 4 are special properties of least-squares regression. They are not true for other methods of fitting a line to data that are discussed in more advanced courses.

APPLY YOUR KNOWLEDGE

5.6 How Useful Is Regression? Figure 4.9 (page 116) displays the relationship between golfers' scores on the first and second rounds of the 2019 Masters Tournament. The correlation is $r = 0.283$. Figure 4.3 (page 105) gives data on number of boats registered in Florida and the number of manatees killed by boats for the years 1977 to 2018. The correlation is $r = 0.919$. Explain in simple language why knowing only these correlations enables you to say that prediction of manatee deaths from number of boats registered by a regression line will be much more accurate than prediction of a golfer's second-round score based on that golfer's first-round score.

Arco Images GmbH/Alamy Stock Photo

5.7 Feed the Birds. Exercise 4.33 (page 119) gives data from a study in which canary parents cared for both their own babies and those of other parents. Investigators looked at how the growth rate of the foster babies relative to the growth rate of the natural babies changed as the begging intensity for food by the foster babies increased over the begging intensity of the natural babies. If begging intensity is the main factor determining food received, with higher intensity leading to more food, one would expect the relative growth rate to increase as the difference in begging intensity increases. However, if both begging intensity and a preference for parents' own babies determine the amount of food received (and hence the relative growth rate), we might expect growth rate to increase initially as begging intensity increases but then to level off (or even decrease) as the parents begin to ignore further increases in begging by the foster babies. CANARY

(a) Make a scatterplot of the data. Find the least-squares regression line for predicting relative growth rate of the foster brood from the difference in begging intensity between the foster brood and the actual babies of the parents and add this line to your plot. Should we *not* use the regression line for prediction in this setting?

(b) What is r^2? What does this value say about the success of the regression line in predicting relative growth rate?

5.5 Residuals

One of the first principles of data analysis is to look for an overall pattern and also for striking deviations from the pattern. A regression line describes the overall pattern of a linear relationship between an explanatory variable and a response variable. We see deviations from this pattern by looking at the scatter of the data points about the regression line. The vertical distances from the points to the least-squares regression line are as small as possible, in the sense that they have the smallest possible sum of squares. Because they represent "leftover" variation in the response after fitting the regression line, these distances are called *residuals*.

Residuals

A **residual** is the difference between an observed value of the response variable and the value predicted by the regression line. That is, a residual is the prediction error that remains after we have chosen the regression line:

$$\begin{aligned}\text{residual} &= \text{observed } y - \text{predicted } y \\ &= y - \hat{y}\end{aligned}$$

EXAMPLE 5.5 More Exercise, More Weight Loss?

EXERC

Lee Torrens/Shutterstock

The more calories you burn, the more weight you lose. Wearable technology such as Fitbits and smart watches allows wearers to track daily activity such as walking, running, and stair climbing. As such, wearable technology is viewed as a useful tool for those seeking to lose weight. Increasing the amount you exercise should increase the number of calories you burn, leading to weight loss. But is it true that increasing the amount you exercise always increases the number of calories you burn? To explore this, researchers measured the physical activity of 332 subjects, in mean counts per minute per day (CPM/d), using wearable accelerometers. They also measured the total energy expenditure in kilocalories per day (kcal/d) for each subject. At low to moderate levels of exercise, increases in physical activity as measured by CPM/d were positively associated with increases in actual energy expenditure. But at high levels of physical activity, the relationship was less clear. Here are the data for the 34 subjects with the highest levels (top 10%) of physical activity:[6]

Subject	1	2	3	4	5	6	7	8	9
CPM/d	700	640	590	550	510	510	500	500	490
Energy expenditure (kcal/d)	2800	4500	2600	2700	2400	2600	2200	3300	3500
Subject	10	11	12	13	14	15	16	17	18
CPM/d	450	430	425	420	410	410	405	380	375
Energy expenditure (kcal/d)	2800	2500	3200	2800	3150	3500	2850	3600	2900
Subject	19	20	21	22	23	24	25	26	27
CPM/d	370	370	360	360	350	350	350	350	345
Energy expenditure (kcal/d)	3700	2500	3100	2450	3200	2700	2300	1950	3150
Subject	28	29	30	31	32	33	34		
CPM/d	340	330	330	330	325	321	320		
Energy expenditure (kcal/d)	2000	3650	2700	2400	3100	2500	2950		

Figure 5.5 is a scatterplot, with CPM/d as the explanatory variable x and energy expenditure as the response variable y. The plot shows a weak positive association. That is, subjects with higher CPM/d do have higher energy expenditures. The overall pattern is moderately linear, with correlation $r = 0.181$. The line on the plot

FIGURE 5.5 Scatterplot of energy expenditure (kcal/d) versus physical activity (CPM/d) for Example 5.5. The line is the least-squares regression line.

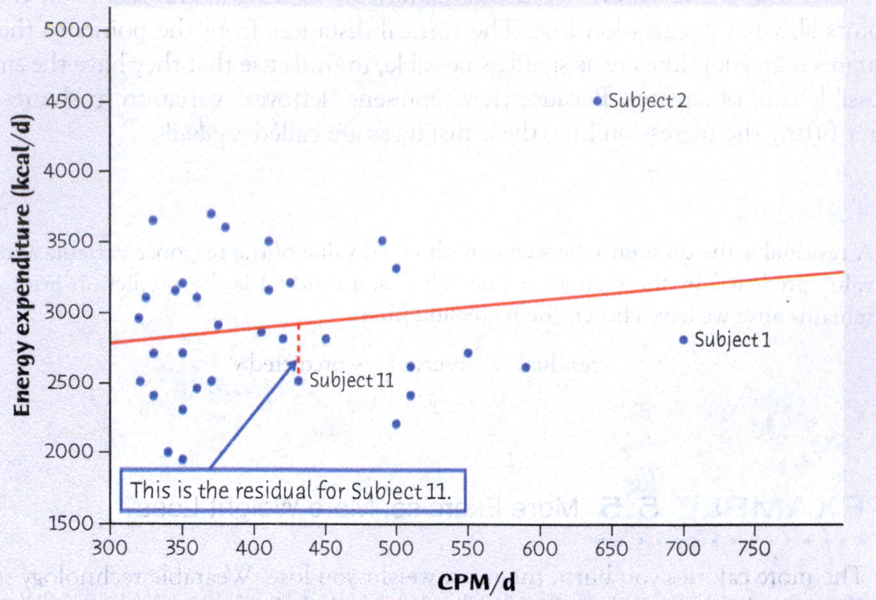

is the least-squares regression line of energy expenditure on CPM/d. Its equation (rounding to two decimal places) is

$$\text{energy expenditure} = 2467.55 + 1.01(\text{CPM/d})$$

For Subject 11, with CPM/d 430, we predict

$$\text{energy expenditure} = 2467.55 + (1.01)(430) = 2901.85$$

This subject's actual energy expenditure was 2500. The residual is

$$\begin{aligned}\text{residual} &= \text{observed energy expenditure} - \text{predicted energy expenditure} \\ &= 2500 - 2901.85 = -401.85\end{aligned}$$

The residual is negative because the data point lies below the regression line. The dashed line segment in Figure 5.5 shows the size of the residual.

There is a residual for each data point. Finding the residuals is a bit unpleasant because you must first find the predicted response for every x. Software or a graphing calculator gives you the residuals all at once. Here are the 34 residuals for the physical activity study data, from software:

Subject	1	2	3	4	5	6	7	8	9
Residual	−375.307	1385.358	−464.088	−323.645	−583.201	−383.201	−773.090	326.910	537.020
Subject	10	11	12	13	14	15	16	17	18
Residual	−122.536	−402.315	302.741	−92.204	267.907	617.907	−27.038	748.239	53.295
Subject	19	20	21	22	23	24	25	26	27
Residual	858.350	−341.650	268.461	−381.539	378.572	−121.428	−521.428	−871.428	333.627
Subject	28	29	30	31	32	33	34		
Residual	−811.317	848.794	−101.206	−401.206	303.849	−292.107	158.904		

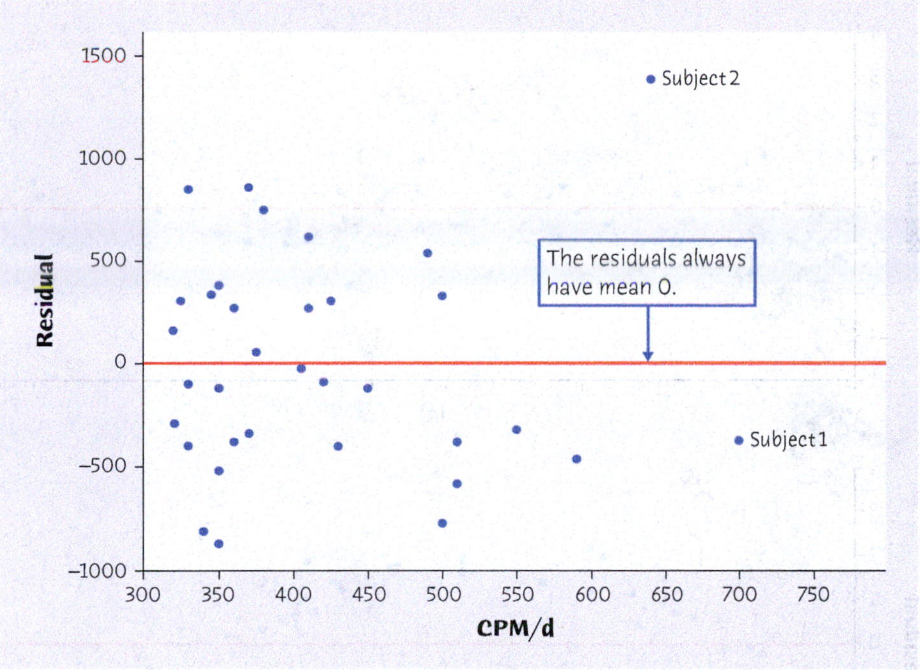

FIGURE 5.6 Residual plot for the data shown in Figure 5.5. The horizontal line at zero residual corresponds to the regression line in Figure 5.5.

Because the residuals show how far the data fall from our regression line, examining the residuals helps us assess how well the line describes the data. Although residuals can be calculated from any curve or line fitted to the data, the residuals from the least-squares line have a special property: the mean of the least-squares residuals is always zero.

Compare the scatterplot in Figure 5.5 with the *residual plot* for the same data in Figure 5.6. The horizontal line at zero in Figure 5.6 helps orient us. This "residual = 0" line corresponds to the regression line in Figure 5.5.

Residual Plots

A **residual plot** is a scatterplot of the regression residuals against the explanatory variable. Residual plots help us assess how well a regression line fits the data.

A residual plot in effect turns the regression line horizontal. It magnifies the deviations of the points from the line and makes it easier to see unusual observations and patterns.

Figure 5.7 shows the overall pattern of some typical residual plots in simplified form. The residuals are plotted in the vertical direction against the corresponding values of the explanatory variable in the horizontal direction. If our assumptions hold, the pattern of this plot will be an unstructured horizontal band centered at 0 (the mean of the residuals) and symmetric about 0, as in Figure 5.7(a). A curved pattern, like the one in Figure 5.7(b), indicates that the relationship between the response and explanatory variable is curved rather than linear. A straight line is not a good description of such a relationship. A fan-shaped pattern like the one in Figure 5.7(c) shows that the variation of the response about the least-squares line increases as the explanatory variable increases. Predictions of the response will be more precise for smaller values of the explanatory variable, where the response shows less variability about the line.

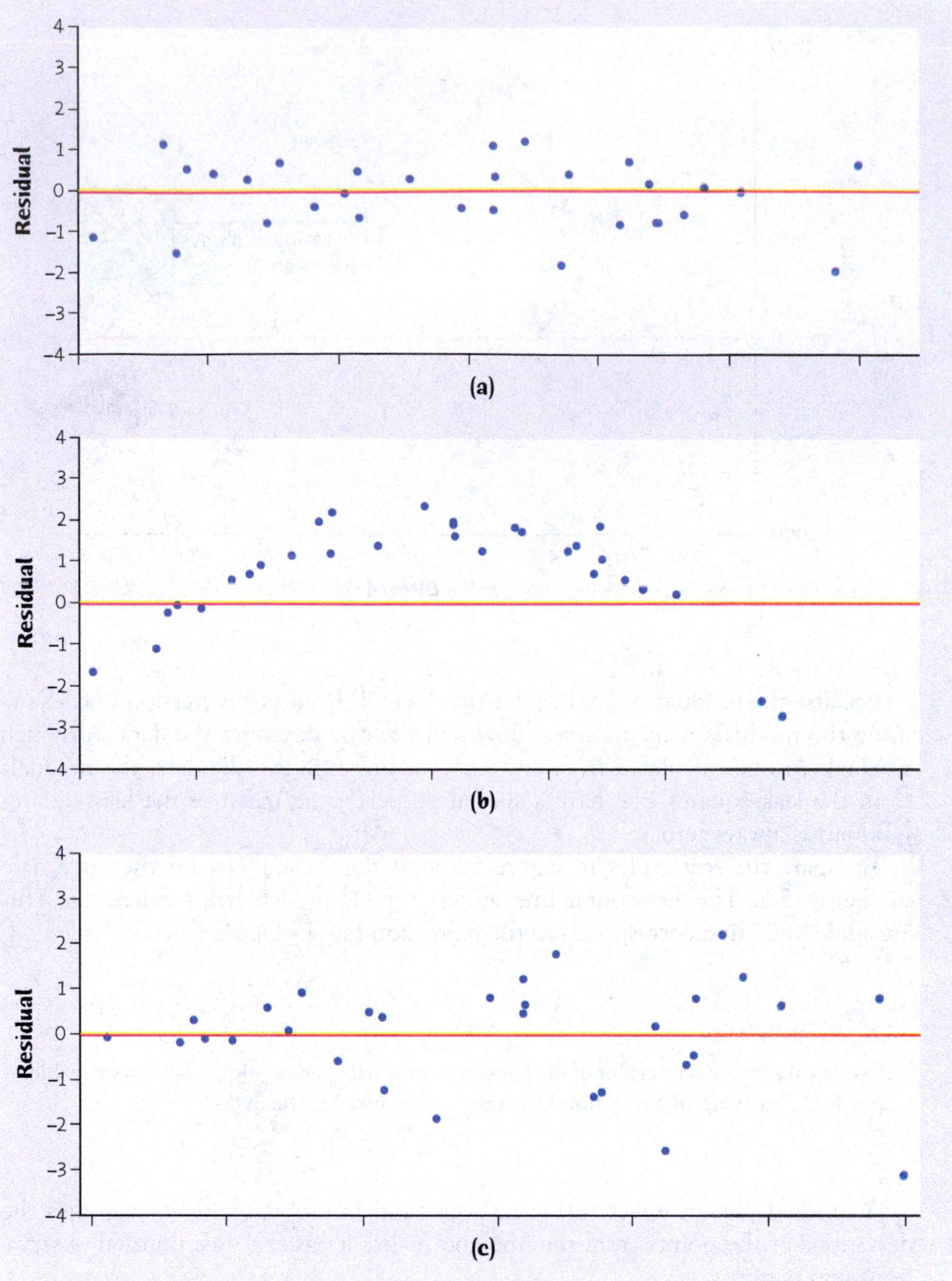

FIGURE 5.7 Some typical residual plots in simplified form.

APPLY YOUR KNOWLEDGE

5.8 Residuals by Hand. In Exercise 5.4 (page 130), you found the equation of the least-squares line for predicting coral growth y from mean sea surface temperature x.

(a) Use the equation to obtain the seven residuals step by step. That is, find the prediction $\hat{y}$ for each observation and then find the residual $y - \hat{y}$.

(b) Check that (up to roundoff error) the residuals add to 0.

(c) The residuals are the part of the response y left over after the straight-line tie between y and x is removed. Show that the correlation between the residuals and x is 0 (up to roundoff error). That this correlation is always 0 is another special property of least-squares regression.

5.9 Does Fast Driving Waste Fuel? Exercise 4.8 (page 106) gives data on the fuel consumption y of a car at various speeds x. Fuel consumption is measured in miles per gallon, and speed is measured in miles per hour. Software tells us that the equation of the least-squares regression line is FASTDR2

$$\hat{y} = 70.243 - 0.329x$$

Using this equation, we can add the residuals to the original data:

Speed	20	30	40	50	60	70	80
Fuel	49.0	67.9	66.5	59.0	50.4	44.8	39.1
Residual	−14.67	7.51	9.40	5.19	−0.13	−2.44	−4.86

(a) Make a scatterplot of the observations and draw the regression line on your plot.

(b) Would you use the regression line to predict y from x? Explain your answer.

(c) Verify the value of the first residual for $x = 20$. Verify that the residuals have sum zero (up to roundoff error).

(d) Make a plot of the residuals against the values of x. Draw a horizontal line at height zero on your plot. How does the pattern of the residuals about this line compare with the pattern of the data points about the regression line in your scatterplot from part (a)?

5.10 Not Obvious to the Naked Eye. The data set RESIDS contains the values of a response y, an explanatory variable x, and the residuals from the least-squares regression line for predicting y from x. RESIDS

(a) Make a scatterplot of the observations and draw the regression line on your plot.

(b) Would you use the regression line to predict y from x? Explain your answer.

(c) Make a plot of the residuals against the values of x. Draw a horizontal line at height zero on your plot. How does the pattern of the residuals about this line compare with the pattern of the data points about the regression line in your scatterplot from part (a)?

5.6 Influential Observations

Figures 5.5 and 5.6 show two unusual observations: Subjects 1 and 2. Subject 2 is an outlier in both the x and y directions, with CPM/d 50 counts higher than all subjects except Subject 1 and energy expenditure 850 kcal/d higher than all other subjects. Because of its extreme position on both the CPM/d and energy expenditure scales, Subject 2 has a strong influence on the correlation. Dropping Subject 2 reduces the correlation from $r = 0.181$ (a weak positive association) to $r = -0.043$ (a weak negative association). Reporting that there is a weak positive association between exercise and energy expenditure is qualitatively different from reporting that there is a weak negative association.

We say that Subject 2 is *influential* for calculating the correlation.

Influential Observations

An observation is **influential** for a statistical calculation if removing it would markedly change the result of the calculation.

The result of a statistical calculation may be of little practical use if it depends strongly on a few influential observations.

Points that are outliers in either the x or the y direction of a scatterplot are often influential for the correlation. Points that are outliers in the x direction are often influential for the least-squares regression line.

What constitutes a "marked" change? This is somewhat subjective. Changes in a calculation that are the same size as roundoff error are often not evidence that an observation is influential. A change in a calculation that differs by a factor of 1.5 or more is often evidence that an observation is influential. A change in the direction of an association is often evidence that an observation is influential. If the least-squares regression line computed after removing an observation still fits the original data in the scatterplot, the observation is probably not influential. However, one can find exceptions to these guidelines, and statisticians may disagree as to whether an observation should be considered influential.

If an observation is influential, or if there is some doubt as to whether an observation is influential, it can be informative to report statistical calculations with both the observation included and the observation removed. This provides readers with the ability to assess the effect of the observation.

EXAMPLE 5.6 An Influential Observation?

In Figures 5.5 and 5.6, Subject 1 is an outlier in the x direction, with CPM/d 60 counts higher than Subject 2 and 110 higher than all the other subjects. Is it influential for the correlation? We can explore this by seeing what happens when we remove Subject 1 from the data. Dropping Subject 1 increases the correlation from 0.181 to 0.229, an increase by a factor of 1.27. One might not consider this a marked change. Figure 5.8 shows that this observation is not influential for the least-squares line.

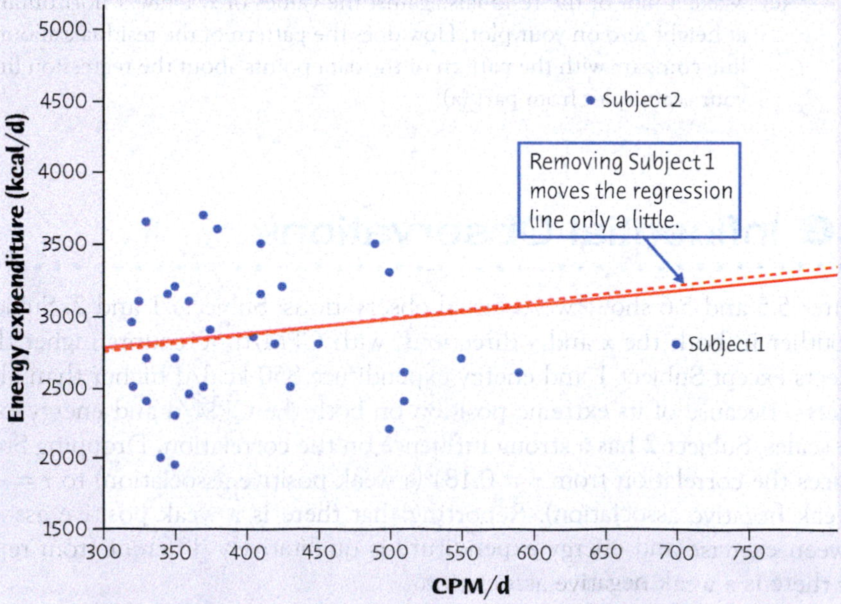

FIGURE 5.8 Subject 1 is an outlier in the x direction. The outlier is not influential for least-squares regression because removing it moves the regression line only a little.

The regression line calculated without Subject 1 (dashed) differs little from the line that uses all the observations (solid). The reason that the outlier has little influence on the regression line is that it lies reasonably close to the dashed regression line calculated from the other observations.

Subject 2 is an outlier in both the x and y directions. We investigate whether Subject 2 is influential by seeing what happens when we remove Subject 2 from the data. Subject 2 in Example 5.5 (page 135) is influential for the correlation between CPM/d and energy expenditure because removing it reduces r from 0.181 to -0.043. A change in the sign of the correlation would usually be considered a marked change, so Subject 2 is influential for the correlation. This suggests that $r = 0.181$ is not a very useful description of the data because the value depends so strongly on just 1 of the 34 subjects.

Is this observation also influential for the least-squares line? Figure 5.9 shows that it is. The regression line calculated without Subject 2 (dashed) differs from the line that uses all the observations (solid). The dashed line's slightly downward slope indicates that the association between CPM/d and energy expenditure is (weakly) negative, while the positive slope of the solid line indicates a weak positive association. The reason that the outlier has influence on the regression line is that it lies well above the dashed regression line calculated from the other observations.

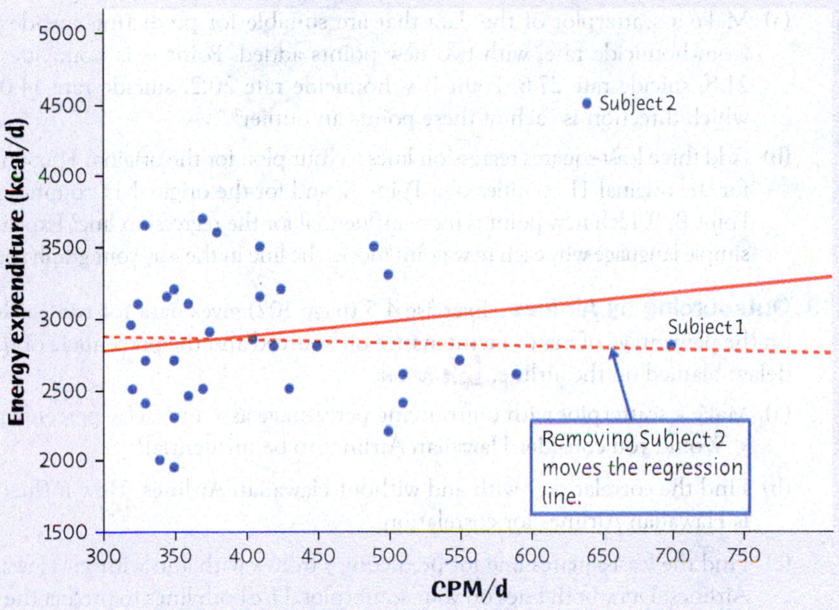

FIGURE 5.9 Subject 2 is an outlier in the x and y directions. The outlier is influential for least-squares regression because removing it moves the regression line from having a positive slope to having a negative slope.

Because the positive association disappears when Subject 2 is dropped, the researchers concluded that energy expenditure appears to level off for large amounts of exercise (CPM/d). In other words, increasing one's exercise from moderate to extreme levels may produce little or no additional calories burned and hence little or no additional weight loss.

applet

The *Correlation and Regression* applet allows you to experiment with the effect of outliers in the x direction (see Exercise 5.11). *An outlier in x pulls the least-squares line toward itself. If the outlier does not lie close to the line calculated from the other observations, it will be influential.*

We did not need the distinction between outliers and influential observations in Chapter 2. A single high salary that pulls up the mean salary $\bar{x}$ for a group of workers is an outlier because it lies far above the other salaries. It is also influential because the mean changes when it is removed. In the regression setting, however, not all outliers are influential.

APPLY YOUR KNOWLEDGE

applet

5.11 Influence in Regression. The *Correlation and Regression* applet allows you to animate Figure 5.9. Click to and drag a group of 10 points from the Data Bank to the graph in the lower-left corner of the scatterplot with a strong straight-line pattern (correlation about 0.9). Toggle the "Show regression line" button to display the least-squares regression line.

(a) Add one point at the upper right that is far from the other 10 points but exactly on the regression line. Why does this outlier have no effect on the line, even though it changes the correlation?

(b) Now use the mouse to drag this last point straight down. You see that one end of the least-squares line chases this single point, whereas the other end remains near the middle of the original group of 10. What makes the last point so influential?

5.12 Homicide and Suicide. Return to the data of Exercise 5.5 (page 131) on homicide rate and suicide rate. We will use these data to illustrate influence. DEATH4

(a) Make a scatterplot of the data that are suitable for predicting suicide rate from homicide rate, with two new points added. Point A is homicide rate 21.8, suicide rate 27.6. Point B is homicide rate 20.2, suicide rate 14.0. In which direction is each of these points an outlier?

(b) Add three least-squares regression lines to your plot: for the original 11 counties, for the original 11 counties plus Point A, and for the original 11 counties plus Point B. Which new point is more influential for the regression line? Explain in simple language why each new point moves the line in the way your graph shows.

5.13 Outsourcing by Airlines. Exercise 4.5 (page 102) gives data for nine airlines on the percentage of major maintenance outsourced and the percentage of flight delays blamed on the airline. AIRLINE

(a) Make a scatterplot with outsourcing percentage as x and delay percentage as y. Would you consider Hawaiian Airlines to be influential?

(b) Find the correlation r with and without Hawaiian Airlines. How influential is Hawaiian Airlines for correlation?

(c) Find the least-squares line for predicting y from x with and without Hawaiian Airlines. Draw both lines on your scatterplot. Use both lines to predict the percentage of delays blamed on an airline that has outsourced 78.4% of its major maintenance. How influential is Hawaiian Airlines for the least-squares line?

5.7 Cautions About Correlation and Regression

Correlation and regression are powerful tools for describing the relationship between two variables. When you use these tools, you must be aware of their limitations. You already know that

- *Correlation and regression lines describe only linear relationships.* You can do the calculations for any relationship between two quantitative variables, but the results are useful only if the scatterplot shows a linear pattern.
- *Correlation and least-squares regression lines are not resistant.* Always plot your data and look for observations that may be influential.

Here are three more things to keep in mind when you use correlation and regression.

Beware of ecological correlation. There is a large positive correlation between *average* income and number of years of education. The correlation is smaller if we compare the incomes of *individuals* with number of years of education. The correlation based on average income ignores the large variation in the incomes of individuals having the same amount of education. The variation from individual to individual increases the scatter in a scatterplot, reducing the correlation. The correlation between average income and education overstates the strength of the relationship between the incomes of individuals and number of years of education. *Correlations based on averages can be misleading if they are interpreted to be about individuals.*

Ecological Correlation

A correlation based on averages rather than on individuals is called an **ecological correlation.**

Beware of extrapolation. Suppose that you have data on a child's growth between 3 and 8 years of age. You find a strong linear relationship between age x and height y. If you fit a regression line to these data and use it to predict height at age 25 years, you may well predict that the child will be 8 feet tall. Growth slows down and then stops at maturity, so extending the straight line to adult ages is foolish. *Few relationships are linear for all values of x. Don't make predictions far outside the range of x that actually appears in your data.*

Extrapolation

Extrapolation is the use of a regression line for prediction far outside the range of values of the explanatory variable x that you used to obtain the line. Such predictions are often not accurate.

Beware of the lurking variable. Another caution is even more important: *the relationship between two variables can often be understood only by taking other variables into account. Lurking variables* can make a correlation or regression misleading.

Lurking Variable

A **lurking variable** is a variable that is not among the explanatory or response variables in a study and yet may influence the interpretation of relationships among those variables.

You should always think about possible lurking variables before you draw conclusions based on correlation or regression.

EXAMPLE 5.7 Magic Mozart?

The Kalamazoo (Michigan) Symphony once advertised a "Mozart for Minors" program with this statement: "Question: Which students scored 51 points higher in verbal skills and 39 points higher in math? Answer: Students who had experience in music."[7]

We could as well answer "Students who played soccer." Why? Children with prosperous and well-educated parents are more likely than poorer children to have

STATISTICS IN YOUR WORLD

Do Left-handers Die Early?

Yes, said a study of 1000 deaths in California. Left-handed people died at an average age of 66 years; right-handers died at 75 years of age. Should left-handed people fear an early death? No. Here the lurking variable has struck again. Older people grew up in an era when many natural left-handers were forced to use their right hands. So right-handers are more common among older people, and left-handers are more common among the young. When we look at deaths, the left-handers who die are younger on the average because left-handers in general are younger. Mystery solved.

experience with music and also to play soccer. They are also likely to attend good schools, get good health care, and be encouraged to study hard. These advantages lead to high test scores. Family background is a lurking variable that explains why test scores are related to experience with music.

APPLY YOUR KNOWLEDGE

5.14 SAT Scores. The correlation between mean 2018 Math SAT scores and mean 2018 ERW (Evidence-Based Reading and Writing) SAT scores for all 50 states and the District of Columbia is 0.985. Would you expect the correlation between the mean state SAT scores for these two tests to be lower, about the same, or higher than the correlation between the scores of individuals on these two tests? Explain your answer.

PhotoDisc/Getty Images

5.15 The Endangered Manatee. Table 4.1 (page 105) shows 42 years of data on boats registered in Florida and manatees killed by boats. Figure 4.3 shows a strong positive linear relationship. The correlation is $r = 0.919$. MANATEE

(a) Find the equation of the least-squares line for predicting manatees killed from thousands of boats registered. Because the linear pattern is so strong, we expect predictions from this line to be quite accurate—but only if conditions in Florida remain similar to those of the past 42 years.

(b) Suppose we expect that the number of boats registered in Florida to be 950,000 in 2019. What would you predict the number of manatees killed by boats to be if there are 950,000 boats registered? Explain why we can trust this prediction.

(c) Predict manatee deaths if there were *no* boats registered in Florida. Explain why the predicted count of deaths is impossible. (We use $x = 0$ to find the intercept of the regression line, but unless the explanatory variable x actually takes values near 0, prediction for $x = 0$ is an example of extrapolation.)

5.16 Is Math the Key to Success in College? A College Board study of 15,941 high school graduates found a strong correlation between how much math minority students took in high school and their later success in college. News articles quoted the head of the College Board as saying that "math is the gatekeeper for success in college."[8] Maybe so, but we should also think about lurking variables. What might lead minority students to take more or fewer high school math courses? Would these same factors influence success in college?

5.8 Association Does Not Imply Causation

Thinking about lurking variables leads to the most important caution about correlation and regression. When we study the relationship between two variables, we often hope to show that changes in the explanatory variable *cause* changes in the response variable. *A strong association between two variables is not enough to draw conclusions about cause and effect.* A least-squares regression line that fits the data well and gives accurate predictions is also not enough to draw conclusions about cause and effect. Sometimes an observed association really does reflect cause and effect. A household that heats with natural gas uses more gas in colder months because more gas must be burned in cold weather to stay warm. In other cases, an association is explained by lurking variables, and the conclusion that x causes y is either wrong or not proved.

EXAMPLE 5.8 Does Having More Cars Make You Live Longer?

A serious study once found that people with two cars live longer than people who own only one car.[9] Owning three cars is even better, and so on. There is a substantial positive correlation between number of cars x and length of life y.

The basic meaning of causation is that by changing x, we can bring about a change in y. Could we lengthen our lives by buying more cars? No. The study used number of cars as a quick indicator of affluence. Well-off people tend to have more cars. They also tend to live longer, probably because they are better educated, take better care of themselves, and get better medical care. The cars have nothing to do with it. There is no cause-and-effect tie between number of cars and length of life.

America/Alamy Stock Photo

Correlations such as that in Example 5.8 are sometimes called "nonsense correlations." The correlation is real. What is nonsense is the conclusion that changing one of the variables causes changes in the other. A lurking variable—such as personal affluence in Example 5.8—that influences both x and y can create a high correlation even though there is no direct connection between x and y.

Association Does Not Imply Causation

An association between an explanatory variable x and a response variable y, even if it is very strong, is not by itself good evidence that changes in x actually cause changes in y.

STATISTICS IN YOUR WORLD

The Super Bowl Effect

The Super Bowl is the most-watched TV broadcast in the United States. Data show that, on Super Bowl Sunday, we consume 3 times as many potato chips as on an average day and 17 times as much beer. What's more, the number of fatal traffic accidents goes up in the hours after the game ends. Could that be celebration? Or catching up with tasks left undone? Or maybe it's the beer.

EXAMPLE 5.9 SAT Scores and Teacher Salaries

Exercise 4.49 asks you to explore the relationship between Mathematics SAT scores and average teacher salaries using 2018 data for each of the 50 states and the District of Columbia. Figure 5.10 is a scatterplot of these data, which shows that SAT scores and salaries are negatively correlated ($r = -0.266$).

Average scores on the Mathematics SAT are partly determined by the percentage of test takers. States that don't require the SAT have fewer test takers. Typically, only better students take the SAT exam in these states—hence average scores are

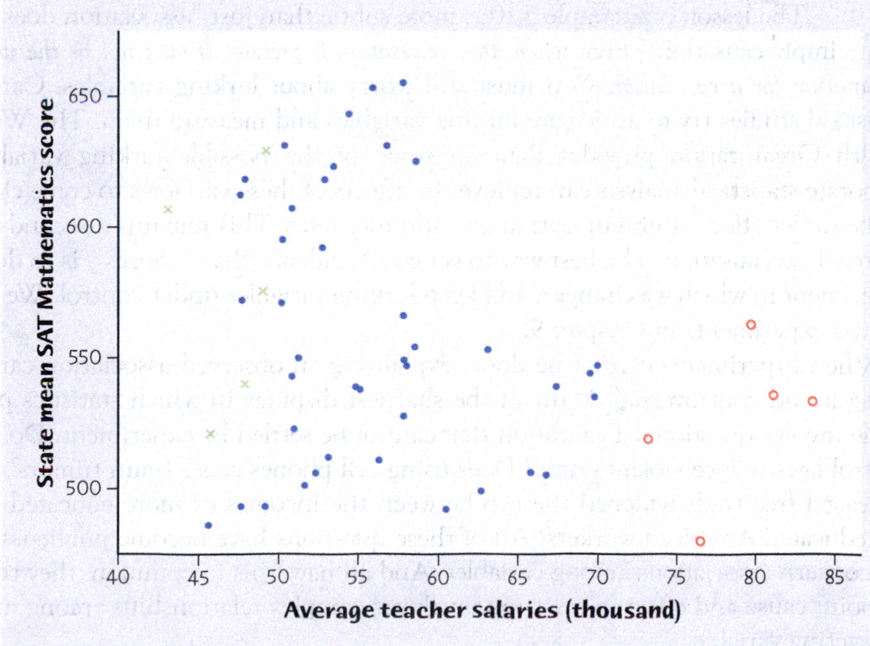

FIGURE 5.10 Scatterplot of state mean SAT Mathematics scores versus average teacher salaries for the 50 states and the District of Columbia. The five red circles are California, Connecticut, the District of Columbia, Massachusetts, and New York, where the cost of living is high. The five green x's are Arkansas, Mississippi, Missouri, New Mexico, and Oklahoma, where the cost of living is low.

high. States that do require the SAT tend to be states with higher costs of living and hence higher average teacher salaries. The points represented by red circles in Figure 5.10 are California, Connecticut, the District of Columbia, Massachusetts, and New York, all of which have a high cost of living. The points represented by green x's are Arkansas, Mississippi, Missouri, New Mexico, and Oklahoma, all of which have a low cost of living and don't require the SAT. These lurking variables explain the negative correlation, and you will explore this in Exercises 5.50 and 5.51. In fact, you will discover that, after accounting for the percentage of test takers, the association actually reverses!

EXAMPLE 5.10 Immunization and Mortality Rates

Measles is a highly contagious and serious disease. Before the introduction of the measles vaccine in 1963 and widespread vaccination, major epidemics occurred approximately every 2 to 3 years, and measles caused an estimated 2.6 million deaths each year, including many children under the age of 5. Data from the World Health Organization give the percentage of children ages 12–23 months worldwide who were immunized for measles and the worldwide mortality rate per 1000 live births for children under 5 from 1990 to 2017. The correlation between the percentage immunized for measles and the mortality rate for children under 5 during this periods is $r = -0.953$.[10]

The mortality rate for children under 5 is partly due to deaths from measles. There is a direct cause-and-effect link between immunization for measles, which prevents deaths, and mortality rate. The negative correlation reflects the fact that immunization reduces mortality. But there are many other causes of death in children under 5, such as other diseases and malnourishment, and these causes may be more prevalent in regions with low vaccination rates. There are also treatments for measles, which, if administered early, can prevent the disease or reduce the seriousness of the symptoms. All these factors contribute to mortality rates and obscure the actual effect of immunization.

The lesson of Example 5.9 is that association does not imply causation. A lurking variable may help explain the observed association, and once accounted for, it may even reverse the observed association. In Chapter 6, we refer to this as "Simpson's paradox."

The lesson of Example 5.10 is more subtle than just "association does not imply causation." *Even when direct causation is present, it may not be the whole explanation for a correlation.* You must still worry about lurking variables. Careful statistical studies try to anticipate lurking variables and measure them. The World Health Organization provides data on some of the possible lurking variables. Elaborate statistical analysis can remove the effects of these variables to come closer to the direct effect of immunization on mortality rates. This remains a second-best approach to causation. The best way to get good evidence that x causes y is to do an experiment in which we change x and keep lurking variables under control. We will discuss experiments in Chapter 9.

When experiments cannot be done, explaining an observed association can be difficult and controversial. Many of the sharpest disputes in which statistics plays a role involve questions of causation that cannot be settled by experiment. Do gun control laws reduce violent crime? Does using cell phones cause brain tumors? Has increased free trade widened the gap between the incomes of more educated and less educated American workers? All of these questions have become public issues. All concern associations among variables. And all have this in common: they try to pinpoint cause and effect in a setting involving complex relationships among many interacting variables.

EXAMPLE 5.11 Does Smoking Cause Lung Cancer?

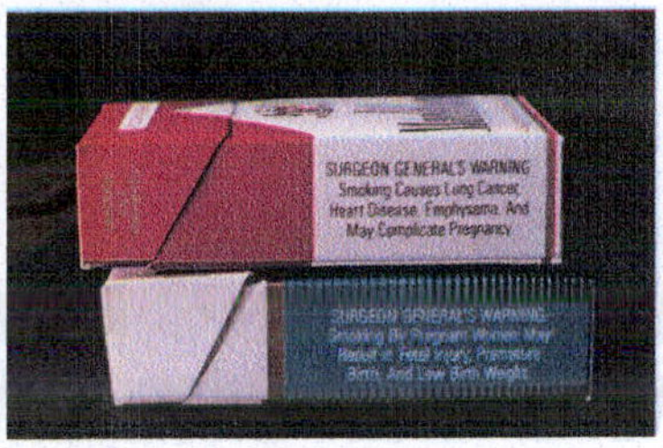

James Leynse/Corbis via Getty Images

Despite the difficulties, it is sometimes possible to build a strong case for causation in the absence of experiments. The evidence that smoking causes lung cancer is about as strong as nonexperimental evidence can be.

Doctors had long observed that most lung cancer patients were smokers. Comparison of smokers and "similar" nonsmokers showed a very strong association between smoking and death from lung cancer. Could the association be explained by lurking variables? Might there be, for example, a genetic factor that predisposes people both to nicotine addiction and to lung cancer? Smoking and lung cancer would then be positively associated, even if smoking had no direct effect on the lungs. How were these objections overcome?

Let's answer this question in general terms: What are the criteria for establishing causation when we cannot do an experiment?

- *The association is strong.* The association between smoking and lung cancer is very strong.
- *The association is consistent.* Many studies of different kinds of people in many countries link smoking to lung cancer. That reduces the chance that a lurking variable specific to one group or one study explains the association.
- *Higher doses are associated with stronger responses.* People who smoke more cigarettes per day or who smoke over a longer period get lung cancer more often. People who stop smoking reduce their risk.
- *The alleged cause precedes the effect in time.* Lung cancer develops after years of smoking. The number of men dying of lung cancer rose as smoking became more common, with a lag of about 30 years. Lung cancer kills more men than any other form of cancer. Lung cancer was rare among women until women began to smoke. Lung cancer in women rose along with smoking, again with a lag of about 30 years, and has now passed breast cancer as the leading cause of cancer death among women.
- *The alleged cause is plausible.* Experiments with animals show that tars from cigarette smoke do cause cancer.

Medical authorities do not hesitate to say that smoking causes lung cancer. The U.S. Surgeon General has long stated that cigarette smoking is "the largest avoidable cause of death and disability in the United States."[11] The evidence for causation is overwhelming—but it is not as strong as the evidence provided by well-designed experiments.

APPLY YOUR KNOWLEDGE

5.17 Another Reason Not to Smoke? A research article reports that children at age five of women who smoked 10 or more cigarettes per day during pregnancy had IQs four points lower, on average, than children of nonsmokers.[12] Suggest some lurking variables that may help explain the association between smoking during pregnancy and children's later test scores. The association by itself is not good evidence that mothers' smoking *causes* lower scores.

5.18 Education and Income. There is a strong positive association between workers' education and their income. For example, the U.S. Census Bureau reported in 2018 that the mean income of young adults (ages 25–34) who worked full-time year round increased from $28,511 for those with less than a ninth-grade education, to $35,327 for high school graduates, to $60,178 for holders of a bachelor's degree, and on up for yet more education. In part, this association reflects causation: education helps people qualify for better jobs. Suggest several lurking variables that also contribute. (Ask yourself what kinds of people tend to get more education.)

5.19 To Earn More, Get Married? Data show that men who are married, and also divorced or widowed men, earn quite a bit more than men the same age who have never been married. This does not mean that a man can raise his income by getting married because men who have never been married are different from married men in many ways other than marital status. Suggest several lurking variables that might help explain the association between marital status and income.

5.9 Correlation, Prediction, and Big Data*

In 2008, researchers at Google were able to track the spread of influenza across the United States much faster than the Centers for Disease Control and Prevention (CDC). By using computer algorithms to explore millions of Internet searches, the researchers discovered a correlation between what people searched for online and whether they had flu symptoms. The researchers used this correlation to make their surprisingly accurate predictions.

Massive databases, or "big data," that are collected by Google, Facebook, credit card companies, and others contain petabytes, or 10^{15} bytes, of data and continue to grow in size. Big data allow researchers, businesses, and industry to search for correlations and patterns in data that will enable them to make accurate predictions about public health, economic trends, or consumer behavior. Using big data to make predictions is increasingly common. Big data explored with clever algorithms opens exciting possibilities. Will the experience of Google become the norm?

Proponents of big data often make the following claims for its value. First, there is no need to worry about causation because correlations are all we need to know for making accurate predictions. Second, scientific and statistical theory is unnecessary because, with enough data, the numbers speak for themselves.

Are these claims correct? It is true that correlation can be exploited for purposes of prediction even if there is no causal relation between explanatory and response variables. However, if you have no idea what is behind a correlation, you have no idea what might cause a prediction to fail, especially when one exploits the correlation to extrapolate to new situations. For a few winters after their success in 2008, Google Flu Trends continued to accurately track the spread of influenza using the correlations they discovered. But during the 2012–2013 flu season, data from the CDC showed that Google's estimate of the spread of flu-like illnesses was overstated by almost a factor of two. A possible explanation was that the news was full of stories about the flu, and this provoked Internet searches by people who were otherwise healthy. The failure to understand why search terms were correlated with the spread of flu resulted in incorrectly assuming that previous correlations extrapolated into the future.

Bias (systematic departures from what is true about a particular group because data are not representative of the group) is another source of error and is not eliminated because of the large amount of data. Big data are often enormous data sets, the result of recording huge numbers of web searches, credit card purchases, or mobile phones pinging the nearest phone tower. This is not equivalent to having good information about the group of interest. For example, in principle, it is possible to record every message on Twitter and use these data to draw conclusions about public opinion. However, Twitter users are not representative of the public as a whole. According to the Pew Research Internet Project, in 2013, U.S.-based users were disproportionally young, urban or suburban, and Black. In other words, the large amount of data generated by Twitter users is biased when the goal is to draw

*This material is optional. Some instructors may prefer to defer this section until they have discussed the material in Chapters 8 and 9.

conclusions about public opinion of all adults in the United States. (See Chapter 8 for a discussion of bias in samples.)

Adding to the perception of the infallibility of big data are news reports touting successes, with few reports of the failures. The claim that theory is unnecessary because the numbers speak for themselves is misleading when all the numbers concerning successes and failures of big data are not reported. Statistical theory has much to say that can prevent data analysts from making serious errors. Providing examples of where mistakes have been made and explaining how, with proper statistical understanding and tools, those mistakes could have been avoided is an important contribution.

The era of big data is exciting and challenging, and it has opened incredible opportunities for researchers, businesses, and industry. But simply being big does not exempt big data from statistical pitfalls such as bias and extrapolation.[13]

CHAPTER 5 SUMMARY

- A **regression line** is a straight line that describes how a response variable y changes as an explanatory variable x changes. You can use a regression line to predict the value of y for any value of x by substituting this x into the equation of the line.
- The **slope** b of a regression line $\hat{y} = a + bx$ is the rate at which the predicted response $\hat{y}$ changes along the line as the explanatory variable x changes. Specifically, b is the change in $\hat{y}$ when x increases by 1.
- The **intercept** a of a regression line $\hat{y} = a + bx$ is the predicted response $\hat{y}$ when the explanatory variable $x = 0$. This prediction is of no statistical interest unless x can actually take values near 0.
- The most common method of fitting a line to a scatterplot is least squares. The **least-squares regression line** is the straight line $\hat{y} = a + bx$ that minimizes the sum of the squares of the vertical distances of the observed points from the line.
- The least-squares regression line of y on x is the line with slope $b = rs_y/s_x$ and intercept $a = \bar{y} - b\bar{x}$. This line always passes through the point $(\bar{x}, \bar{y})$.
- The least-squares regression line depends on the choice of explanatory and response variables.
- Correlation and regression are closely connected. The correlation r is the slope of the least-squares regression line when we measure both x and y in standardized units. The square of the correlation r^2 is the fraction of the variation in one variable that is explained by least-squares regression on the other variable.
- Correlation and regression must be interpreted with caution. Plot the data to be sure the relationship is roughly linear and to detect outliers and influential observations. A plot of the **residuals** makes these effects easier to see.
- Look for **influential observations,** individual points that substantially change the correlation or the regression line. Outliers in the x direction are often influential for the regression line. It may be helpful to report statistical calculations both including and excluding influential observations.
- Be aware of **ecological correlation,** the tendency for correlations based on averages to be stronger than correlations based on individuals. Be careful not to misinterpret correlations based on averages as applying to individuals.
- Avoid **extrapolation,** the use of a regression line for prediction for values of the explanatory variable far outside the range of the data from which the line was calculated.

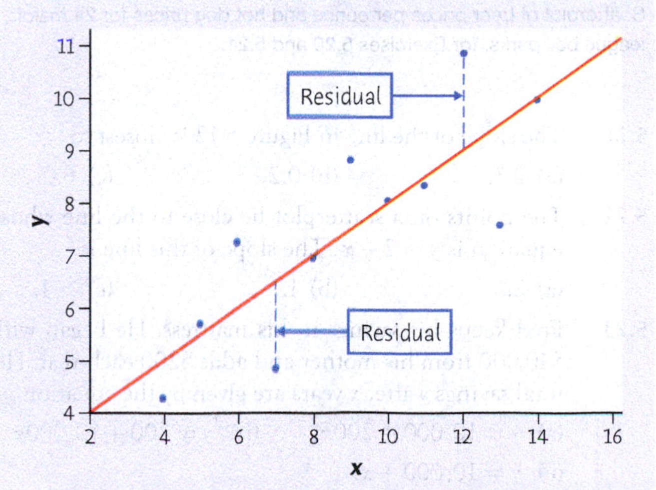

FIGURE 5.11
Scatterplot showing least-squares line and residuals.

- **Lurking variables** may explain the relationship between the explanatory and response variables. Correlation and regression can be misleading if you ignore important lurking variables.
- Most of all, be careful not to conclude that there is a cause-and-effect relationship between two variables just because they are strongly associated. *High correlation does not imply causation.* The best evidence that an association is due to causation comes from an experiment in which the explanatory variable is directly changed and other influences on the response are controlled.

CHECK YOUR SKILLS

5.20 Figure 5.12 is a scatterplot of the price of a hot dog against the price of beer (per ounce) at 30 major league ballparks in 2019.[14] The line is the least-squares regression line for predicting the price of a hot dog from the price of beer. If another ballpark charges $0.60 per ounce for beer, you predict the price of a hot dog to be close to

(a) $3.50. (b) $6.00. (c) $7.00.

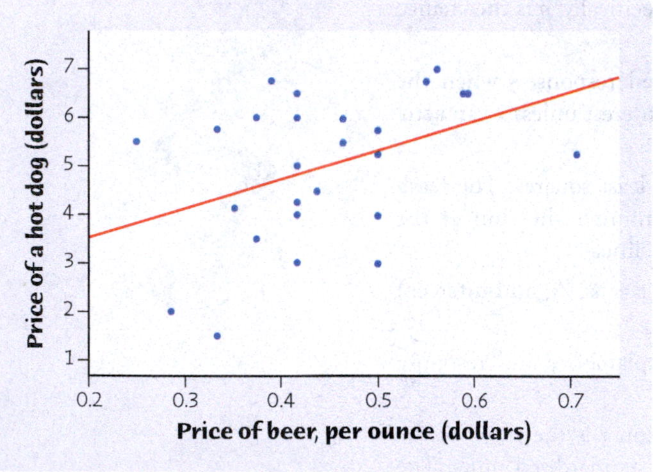

FIGURE 5.12
Scatterplot of beer prices per ounce and hot dog prices for 24 major league ball parks, for Exercises 5.20 and 5.21.

5.21 The slope of the line in Figure 5.12 is closest to

(a) 2.3. (b) 0.2. (c) 6.0.

5.22 The points on a scatterplot lie close to the line whose equation is $y = 2 - x$. The slope of this line is

(a) 2. (b) 1. (c) -1.

5.23 Fred keeps his savings in his mattress. He began with $10,000 from his mother and adds $200 each year. His total savings y after x years are given by the equation

(a) $y = 10{,}000 + 200x$. (b) $y = 200 + 10{,}000x$.

(c) $y = 10{,}000 + x$.

5.24 Smokers don't live as long (on the average) as nonsmokers, and heavy smokers don't live as long as light smokers. You regress the age at death of a group of male smokers on the number of packs per day they smoked. The slope of your regression line

(a) will be greater than 0.

(b) will be less than 0.

(c) can't be determined without seeing the data.

5.25 An owner of a home in the Midwest installed solar panels to reduce heating costs. After installing the solar panels, he measured the amount of natural gas used y (in cubic feet) to heat the home and outside temperature x (in degree-days, where a day's degree-days are the number of degrees its average temperature falls below 65°F) over a 23-month period. He then computed the least-squares regression line for predicting y from x and found it to be[15]

$$\hat{y} = 85 + 16x$$

How much, on average, does gas used increase for each additional degree-day?

(a) 23 cubic feet (b) 85 cubic feet

(c) 16 cubic feet

5.26 According to the regression line in Question 5.25, the predicted amount of gas used when the outside temperature is 20 degree-days is about

(a) 405 cubic feet. (b) 320 cubic feet.

(c) 105 cubic feet.

5.27 By looking at the equation of the least-squares regression line in Question 5.25, you can see that the correlation between amount of gas used and degree-days

(a) is greater than zero. (b) is less than zero.

(c) can't be determined without seeing the data.

5.28 The software used to compute the least-squares regression line in Question 5.25 says that $r^2 = 0.98$. This suggests that

(a) although degree-days and gas used are correlated, degree-days do not predict gas used very accurately.

(b) gas used increases by $\sqrt{0.98} = 0.99$ cubic feet for each additional degree-day.

(c) prediction of gas used from degree-days will be quite accurate.

5.29 Researchers measured the percentage of body fat and the preferred amount of salt (percentage weight/volume) for several children. Here are data for seven children:[16] SALT

Preferred amount of salt x	0.2	0.3	0.4	0.5	0.6	0.8	1.1
Percentage of body fat y	20	30	22	30	38	23	30

Using your calculator or software, what is the equation of the least-squares regression line for predicting percentage of body fat based on the preferred amount of salt?

(a) $\hat{y} = 24.2 + 6.0x$

(b) $\hat{y} = 0.15 + 0.01x$

(c) $\hat{y} = 6.0 + 24.2x$

CHAPTER 5 EXERCISES

5.30 Penguins Diving. A study of king penguins looked for a relationship between how deep the penguins dive to seek food and how long they stay underwater.[17] For all but the shallowest dives, there is a linear relationship that is different for different penguins. The study report gives a scatterplot for one penguin titled "The relation of dive duration (DD) to depth (D)." Duration DD is measured in minutes, and depth D is in meters. The report then says, "The regression equation for this bird is, $DD = 2.69 + 0.0138D$."

Paul A. Souders/Getty Images

(a) What is the slope of the regression line? Explain in specific language what this slope says about this penguin's dives.

(b) According to the regression line, how long does a typical dive to a depth of 200 meters last?

(c) The dives varied from 40 meters to 300 meters in depth. Use the regression equation to determine DD for $D = 40$ and $D = 300$ and then plot the regression line from $D = 40$ to $D = 300$.

5.31 The Price of Diamond Rings. Online advertisements contained pictures of diamond rings and listed their prices, diamond weights (in carats), and gold purity. Based on data for only the 18-carat gold ladies' rings in the advertisements, the least-squares regression line for predicting price (in dollars) from the weight of the diamond (in carats) is[18]

$$\text{price} = -6047.75 + 11975.14 \text{ carats}$$

(a) What does the slope of this line say about the relationship between price and number of carats?

(b) What is the predicted price when number of carats = 0? How would you interpret this price?

5.32 Does Social Rejection Hurt? Exercise 4.47 (page 122) gives data from a study that shows that social exclusion causes "real pain." That is, activity in an area of the brain that responds to physical pain goes up as distress from social exclusion goes up. A scatterplot shows a moderately strong linear relationship. Figure 5.13 shows JMP regression output for these data. REJECT

(a) What is the equation of the least-squares regression line for predicting social distress score from brain activity? Interpret the slope in the context of the problem. Use the equation to predict social distress score for a brain activity of 0.020.

(b) What percentage of the variation in social distress score among these subjects is explained by the straight-line relationship with brain activity?

(c) Use the information in Figure 5.13 to find the correlation r between brain activity and social distress score. How do you know whether the sign of r is positive or negative?

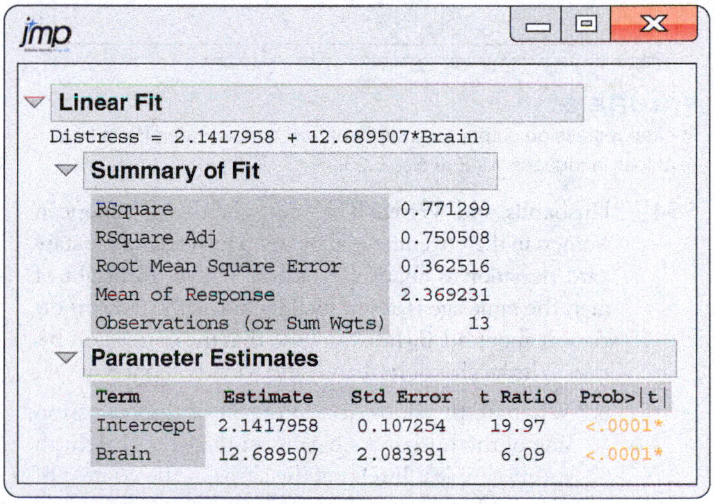

jmp

Linear Fit

Distress = 2.1417958 + 12.689507*Brain

Summary of Fit

RSquare	0.771299
RSquare Adj	0.750508
Root Mean Square Error	0.362516
Mean of Response	2.369231
Observations (or Sum Wgts)	13

Parameter Estimates

Term	Estimate	Std Error	t Ratio	Prob>|t|
Intercept	2.1417958	0.107254	19.97	<.0001*
Brain	12.689507	2.083391	6.09	<.0001*

FIGURE 5.13
JMP regression output for a study of the effects of social rejection on brain activity, for Exercise 5.32.

5.33 Toucan's Beak. Exercise 4.46 (page 122) gives data on beak heat loss, as a percentage of total body heat loss from all sources, at various temperatures. The data show that beak heat loss is higher at higher temperatures and that the relationship is roughly linear. Figure 5.14 shows Minitab regression output for these data. TOUCAN

(a) What is the equation of the least-squares regression line for predicting beak heat loss, as a percentage of total body heat loss from all sources, from temperature? Explain in specific language what the slope of this line says about the relationship between beak heat loss and temperature.

(b) Use the equation of the least-squares regression line to predict beak heat loss, as a percentage of total body heat loss from all sources, at a temperature of 25°C.

(c) What percentage of the variation in beak heat loss is explained by the straight-line relationship with temperature?

(d) Use the information in Figure 5.14 to find the correlation r between beak heat loss and temperature. How do you know whether the sign of r is positive or negative?

```
Minitab

Regression Analysis: Percent heat loss versus Temperature

The regression equation is
Percent heat loss = 0.92 + 2.06 Temperature


Predictor       Coef     SE Coef      T       P
Constant        0.919     5.613      0.16   0.872
Temperatures   2.0647    0.2444      8.45   0.000


S = 4.50655    R-Sq = 83.6%   R-Sq(adj) = 82.4%
```

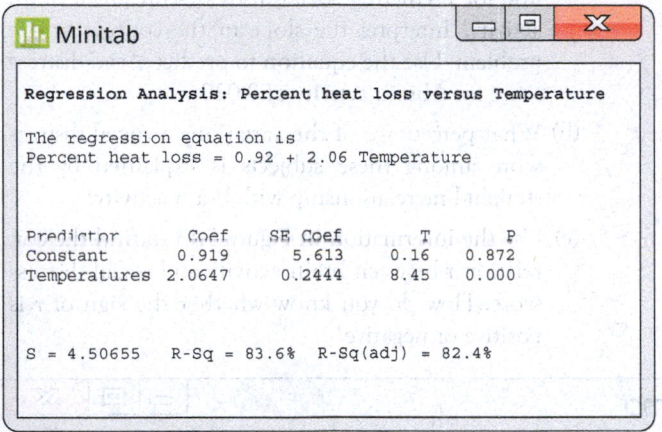

FIGURE 5.14
Minitab regression output for a study of how temperature affects beak heat loss in toucans, for Exercise 5.33.

5.34 **Husbands and Wives.** The mean height of American women in their twenties is about 64.3 inches, and the standard deviation is about 2.7 inches. The mean height of men the same age is about 69.9 inches, with standard deviation about 3.1 inches. Suppose that the correlation between the heights of husbands and wives is about $r = 0.5$.

(a) What are the slope and intercept of the regression line of the husband's height on the wife's height in young couples? Interpret the slope in the context of the problem.

(b) Draw a graph of this regression line for heights of wives between 56 and 72 inches. Predict the height of the husband of a woman who is 67 inches tall and plot the wife's height and predicted husband's height on your graph.

(c) You don't expect this prediction for a single couple to be very accurate. Why not?

5.35 **What's My Grade?** In Professor Krugman's economics course, the correlation between the students' total scores prior to the final examination and their final-examination scores is $r = 0.5$. The pre-exam totals for all students in the course have mean 280 and standard deviation 40. The final-exam scores have mean 75 and standard deviation 8. Professor Krugman has lost Julie's final exam but knows that her total before the exam was 300. He decides to predict her final-exam score from her pre-exam total.

(a) What is the slope of the least-squares regression line of final-exam scores on pre-exam total scores in this course? What is the intercept? Interpret the slope in the context of the problem.

(b) Use the regression line to predict Julie's final-exam score.

(c) Julie doesn't think this method accurately predicts how well she did on the final exam. Use r^2 to argue that her actual score could have been much higher (or much lower) than the predicted value.

5.36 **More Exercise, More Weight Loss.** In the study described in Example 5.5 (page 135), the researchers found that, in general, subjects who engaged in more physical activity had higher total energy expenditures. In particular, they found that physical activity explained 3.3% of the variation in total energy expenditure. What is the numerical value of the correlation between physical activity and total energy expenditure?

5.37 **Sisters and Brothers.** How strongly do physical characteristics of sisters and brothers correlate? Here are data on the heights (in inches) of 12 adult pairs:[19] BROSIS

Brother	71	68	66	67	70	71	70	73	72	65	66	70
Sister	69	64	65	63	65	62	65	64	66	59	62	64

(a) Use your calculator or software to find the correlation and the equation of the least-squares line for predicting sister's height from brother's height. Make a scatterplot of the data and add the regression line to your plot.

(b) Damien is 70 inches tall. Predict the height of his sister Tonya. Based on the scatterplot and the correlation r, do you expect your prediction to be very accurate? Why?

5.38 **Keeping Water Clean.** Keeping water supplies clean requires regular measurement of levels of pollutants. The measurements are indirect: a typical analysis involves forming a dye by a chemical reaction with the dissolved pollutant and then passing light through the solution and measuring its "absorbence." To calibrate such measurements, the laboratory measures known standard solutions and uses regression to relate absorbence and pollutant concentration. This is usually done every day. Here is one series of data on the absorbence for different levels of nitrates. Nitrates are measured in milligrams per liter of water:[20] NITRATES

Nitrates	50	50	100	200	400	800	1200	1600	2000	2000
Absorbence	7.0	7.5	12.8	24.0	47.0	93.0	138.0	183.0	230.0	226.0

(a) Chemical theory says that these data should lie on a straight line. If the correlation is not at least 0.997, something went wrong, and the calibration procedure is repeated. Plot the data and find the correlation. Must the calibration be done again?

(b) The calibration process sets nitrate level and measures absorbence. The linear relationship that results is used to estimate the nitrate level in water from a measurement of absorbence. What is the equation of the line used to estimate nitrate level? What does the slope of this line say about the relationship between nitrate level and absorbence? What is the estimated nitrate level in a water specimen with absorbence 40?

(c) Do you expect estimates of nitrate level from absorbence to be quite accurate? Why?

5.39 Sparrowhawk Colonies. One of nature's patterns connects the percentage of adult birds in a colony that return from the previous year and the number of new adults that join the colony. Here are data for 13 colonies of sparrowhawks:[21] SPARROW

Percent return x	74	66	81	52	73	62	52	45	62	46	60	46	38
New adults y	5	6	8	11	12	15	16	17	18	18	19	20	20

You saw in Exercise 4.29 that there is a moderately strong linear relationship, with correlation $r = 0.748$.

(a) Find the least-squares regression line for predicting y from x. Make a scatterplot and draw your line on the plot.

(b) Explain in words what the slope of the regression line tells us.

(c) An ecologist uses the line, based on 13 colonies, to predict how many new birds will join another colony, to which 60% of the adults from the previous year return. What is the prediction?

5.40 Global Warming. Exercise 4.44 (page 121) gives data on annual average global temperature anomalies (the difference between the average global temperature for a given year and the average global temperature between 1901 and 2000, which was 13.9 degrees Celsius) from 1994 to 2018, in degrees Celsius. GTEMPS

(a) Find the least-squares regression line for predicting average global temperature anomaly from year. Make a scatterplot and draw your line on the plot.

(b) Explain in words what the slope of the regression line tells us.

(c) An environmentalist uses the line, based on the 25 years, to predict average global temperature anomaly in 2050. What is the prediction? How reliable do you think this prediction is?

5.41 Our Brains Don't Like Losses. Exercise 4.30 (page 118) describes an experiment that showed a linear relationship between how sensitive people are to monetary losses ("behavioral loss aversion") and activity in one part of their brains ("neural loss aversion"). LOSSES

(a) Make a scatterplot with neural loss aversion as x and behavioral loss aversion as y. One point is a high outlier in both the x and y directions.

(b) Find the least-squares line for predicting y from x, *leaving out the outlier*, and add the line to your plot.

(c) The outlier lies very close to your regression line. Looking at the plot, you now expect that adding the outlier will increase the correlation but will have little effect on the least-squares line. Explain why.

(d) Find the correlation with and without the outlier. Your results verify the expectations from part (c).

5.42 Always Plot Your Data! Table 5.1 presents four sets of data prepared by the statistician Frank Anscombe to illustrate the dangers of calculating without first plotting the data.[22] ANSCOMBE A, B, C, D

TABLE 5.1 Four data sets for exploring correlation and regression

Data Set A											
x	10	8	13	9	11	14	6	4	12	7	5
y	8.04	6.95	7.58	8.81	8.33	9.96	7.24	4.26	10.84	4.82	5.68
Data Set B											
x	10	8	13	9	11	14	6	4	12	7	5
y	9.14	8.14	8.74	8.77	9.26	8.10	6.13	3.10	9.13	7.26	4.74
Data Set C											
x	10	8	13	9	11	14	6	4	12	7	5
y	7.46	6.77	12.74	7.11	7.81	8.84	6.08	5.39	8.15	6.42	5.73
Data Set D											
x	8	8	8	8	8	8	8	8	8	8	19
y	6.58	5.76	7.71	8.84	8.47	7.04	5.25	5.56	7.91	6.89	12.50

TABLE 5.2 Two measures of glucose level in diabetics

Subject	HbA (%)	FPG (mg/mL)	Subject	HbA (%)	FPG (mg/mL)	Subject	HbA (%)	FPG (mg/mL)
1	6.1	141	7	7.5	96	13	10.6	103
2	6.3	158	8	7.7	78	14	10.7	172
3	6.4	112	9	7.9	148	15	10.7	359
4	6.8	153	10	8.7	172	16	11.2	145
5	7.0	134	11	9.4	200	17	13.7	147
6	7.1	95	12	10.4	271	18	19.3	255

(a) Without making scatterplots, find the correlation and the least-squares regression line for all four data sets. What do you notice? Use the regression line to predict y for $x = 10$.

(b) Make a scatterplot for each of the data sets and add the regression line to each plot.

(c) In which of the four cases would you be willing to use the regression line to describe the dependence of y on x? Explain your answer in each case.

5.43 **Managing Diabetes.** People with diabetes must manage their blood sugar levels carefully. They measure their fasting plasma glucose (FPG) several times a day with a glucose meter. Another measurement, made at regular medical checkups, is called HbA. This is roughly the percentage of red blood cells that have a glucose molecule attached. It measures average exposure to glucose over a period of several months. Table 5.2 gives data on both HbA and FPG for 18 diabetics five months after they had completed a diabetes education class.[23] DIABETES

Glow Wellness/Alamy Stock Photo

(a) Make a scatterplot with HbA as the explanatory variable. There is a positive linear relationship, but it is surprisingly weak.

(b) Subject 15 is an outlier in the y direction. Subject 18 is an outlier in the x direction. Find the correlation for all 18 subjects, for all except Subject 15, and for all except Subject 18. Are either or both of these subjects influential for the correlation? Explain in simple language why r changes in opposite directions when we remove each of these points.

5.44 **The Effect of Changing Units.** The equation of a regression line, unlike the correlation, depends on the units we use to measure the explanatory and response variables. Return to the data on beer and hot dog prices in major league ballparks in Exercise 5.20: BEER2

Team	Hot Dog	Beer
Angels	5.00	0.42
Astros	6.00	0.46
Athletics	5.75	0.50
Blue Jays	4.13	0.35
Braves	4.25	0.42
Brewers	6.00	0.42
Cardinals	5.00	0.42
Cubs	6.50	0.59
Diamondbacks	2.00	0.29
Dodgers	6.75	0.39
Giants	6.50	0.59
Indians	4.25	0.42
Mariners	6.50	0.42
Marlins	3.00	0.42
Mets	6.75	0.55
Nationals	7.00	0.56
Orioles	1.50	0.33
Padres	5.25	0.50
Phillies	4.00	0.50
Pirates	3.50	0.38
Rangers	6.00	0.38
Rays	5.00	0.42
Reds	5.50	0.46
Red Sox	5.25	0.71
Rockies	5.50	0.25
Royals	5.75	0.33
Tigers	5.00	0.42
Twins	4.00	0.42
White Sox	4.50	0.44
Yankees	3.00	0.50

Beer prices are measured in dollars per ounce (rounded to two decimal places).

(a) Find the equation of the regression line for predicting hot dog price from beer price, when beer price is in dollars per ounce.

(b) A mad economist decides to measure beer price in dollars per pound. The same data in these units are:

Team	Hot Dog	Beer
Angels	5.00	6.67
Astros	6.00	7.43
Athletics	5.75	8.00
Blue Jays	4.13	5.63
Braves	4.25	6.67
Brewers	6.00	6.67
Cardinals	5.00	6.67
Cubs	6.50	9.50
Diamondbacks	2.00	4.57
Dodgers	6.75	6.25
Giants	6.50	9.43
Indians	4.25	6.67
Mariners	6.50	6.67
Marlins	3.00	6.67
Mets	6.75	8.80
Nationals	7.00	9.00
Orioles	1.50	5.33
Padres	5.25	8.00
Phillies	4.00	8.00
Pirates	3.50	6.00
Rangers	6.00	6.00
Rays	5.00	6.67
Reds	5.50	7.43
Red Sox	5.25	11.33
Rockies	5.50	4.00
Royals	5.75	5.33
Tigers	5.00	6.67
Twins	4.00	6.67
White Sox	4.50	7.00
Yankees	3.00	8.00

Find the equation of the regression line for predicting hot dog price from beer price when beer price is in dollars per pound.

(c) Use both lines to predict hot dog price from beer price when beer price is \$0.60 per ounce, which is the same as \$9.60 when beer price is measured in dollars per pound. Are the two predictions the same (up to any roundoff error)?

5.45 **Managing Diabetes (continued).** Add three regression lines for predicting FPG from HbA to your scatterplot from Exercise 5.43: for all 18 subjects, for all except Subject 15, and for all except Subject 18. Is either Subject 15 or Subject 18 strongly influential for the least-squares line? Explain in simple language what features of the scatterplot explain the degree of influence. DIABETES

5.46 **Are You Happy?** Exercise 4.26 (page 117) discusses a study in which the mean BRFSS life-satisfaction score of individuals in each state was compared with the mean of an objective measure of well-being (based on the "compensating-differentials method") for each state. Suppose that instead of the means for the states, the BRFSS life-satisfaction scores for individuals were compared with the corresponding measure of well-being (based on the compensating-differentials method) for these individuals. Would you expect the correlation between the mean state scores on these two measures to be lower, about the same, or higher than the correlation between the scores of individuals on these two measures? Explain your answer.

5.47 **A Few More Dollars, One More Year.** Data on the *average* income of all men who died in the past year in several U.S. counties showed a positive correlation with *average* age of death of men who died in the past year in the counties. Would the correlation be greater, smaller, or about the same if you calculated the correlation between the incomes of individual men who died in the past year and their age at death? Explain your answer.

5.48 **Does Diet Soda Cause Weight Gain?** Researchers analyzed data from more than 5000 adults and found that the more diet sodas a person drank, the greater the person's weight gain.[24] Does this mean that drinking diet soda causes weight gain? Give a more plausible explanation for this association.

5.49 **Learning Online.** Many colleges offer online versions of courses that are also taught in the classroom. It often happens that the students who enroll in the online version do better than the classroom students on the course exams. This does not show that online instruction is more effective than classroom teaching because the people who sign up for online courses are often quite different from the classroom students. Suggest some differences between online and classroom students that might explain why online students do better.

5.50 **SAT Scores and Teacher Salaries.** The data set TCHSAL gives the mean Mathematics SAT score and mean salary of teachers in each of the 50 states and the District of Columbia in 2018.[25] The correlation between mean Mathematics SAT score and mean teacher salary is $r = -0.266$. TCHSAL

(a) Find the least-squares line for predicting mean Mathematics SAT score from mean teacher salary. Interpret the slope in the context of the problem.

(b) Is it reasonable to conclude that states can increase the mean Mathematics SAT score in their state by reducing teacher salaries? (*Hint*: Which states have the highest and lowest mean Mathematics SAT scores? Which states have the highest and lowest cost of living?)

5.51 SAT Scores and Teacher Salaries (continued). The data set TCHSAL2 gives the mean Mathematics SAT score and mean salary of teachers in each of the 50 states and the District of Columbia in 2018. It also includes a categorical variable, pct. taking, that indicates whether the percentage taking is above 35% (Y) or below 35% (N). TCHSAL2

(a) Find the least-squares line for predicting mean Mathematics SAT score from mean teacher salary for only the cases where the percentage taking is above 35%. Interpret the slope in the context of the problem.

(b) Find the least-squares line for predicting mean Mathematics SAT score from mean teacher salary for only the cases where the percentage taking is below 35%. Interpret the slope in the context of the problem.

(c) If you did Exercise 5.50, compare your results with those in part (a) of Exercise 5.50. What do you conclude?

5.52 Correlation and Causation. The data set DJI gives the Dow Jones Industrial Average (at the end of the year), the mean global CO_2 level in parts per million (ppm), and the mean global average temperature deviation from the 1901–2000 average, in degrees Celsius, for the years 1984–2017.[26] DJI

(a) Compute the correlation between mean CO_2 level and temperature deviation.

(b) Compute the correlation between temperature deviation and the Dow Jones Industrial Average.

(c) In each case, do you think the correlation is due to causation? That is, are increasing CO_2 levels causing increases in temperature, and are increasing temperatures causing the increases in the Dow Jones Industrial Average? Explain the reasons for your answers.

5.53 Grade Inflation and the SAT. The effect of a lurking variable can be surprising when individuals are divided into groups. In recent years, the mean SAT score of all high school seniors has increased. But the mean SAT score has decreased for students at each level of high school grade averages (A, B, C, and so on). Explain how grade inflation in high school (the lurking variable) can account for this pattern.

5.54 Workers' Incomes. Here is another example of the group effect cautioned about in the previous exercise. Explain how, as a nation's population grows older, median income can go down for workers in each age group yet still go up for all workers.

5.55 Some Regression Math. Use the equation of the least-squares regression line (box on page 129) to show that the regression line for predicting y from x always passes through the point $\bar{x}, \bar{y}$. That is, when $x = \bar{x}$, the equation gives $\hat{y} = \bar{y}$.

5.56 Regression to the Mean. Figure 4.9 (page 116) displays the relationship between golfers' scores on the first and second rounds of the 2019 Masters Tournament. The least-squares line for predicting second-round scores (y) from first-round scores (x) has equation $\hat{y} = 62.91 + 0.164x$. Find the predicted second-round scores for a player who shot 80 in the first round and for a player who shot 70. The mean second-round score for all players was 75.02. So, a player who does well in the first round is predicted to do less well, but still better than average, in the second round. In addition, a player who does poorly in the first is predicted to do better, but still worse than average, in the second.

(*Comment*: This is regression to the mean. If you select individuals with extreme scores on some measure, they tend to have less extreme scores when measured again. That's because their extreme position is partly merit and partly luck, and the luck will be different next time. Regression to the mean contributes to lots of "effects." The rookie of the year often doesn't do as well the next year; the best player in an orchestral audition may play less well once hired than the runners-up; a student who feels she needs coaching after taking the SAT often does better on the next try without coaching.)

5.57 Regression to the Mean. We expect that students who do well on the midterm exam in a course will usually also do well on the final exam. Gary Smith of Pomona College looked at the exam scores of all 346 students who took his statistics class over a 10-year period.[27] The least-squares line for predicting final exam score from midterm-exam score was $\hat{y} = 46.6 + 0.41x$. (Both exams have a 100-point scale.)

Octavio scores 10 points above the class mean on the midterm. How many points above the class mean do you predict that he will score on the final? (*Hint*: Use the fact that the least-squares line passes through the point $\bar{x}, \bar{y}$ and the fact that Octavio's midterm score is $\bar{x} + 10$.) This is another example of regression to the mean: students who do well on the midterm will, in general, do less well, but still above average, on the final.

5.58 Is Regression Useful? applet In Exercise 4.43 (page 121), you used the *Correlation and Regression* applet to create three scatterplots having correlation about $r = 0.7$ between the horizontal variable x and the vertical variable y. Create three similar scatterplots again and toggle the "Show regression line" button to display the least-squares regression lines. Correlation $r = 0.7$ is considered reasonably strong in many areas of work. Because there is a reasonably strong correlation, we might use a regression line to predict y from x. In which of your three scatterplots does it make sense to use a straight line for prediction?

5.59 applet **Guessing a Regression Line.** In the *Correlation and Regression* applet, click on the scatterplot to create a group of 15 to 20 points from lower left to upper right with a clear positive straight-line pattern (correlation around 0.7). Click the "Draw your own line" button and use the mouse (click at two locations on the scatterplot to create a line connecting these two locations) to draw a line through the middle of the cloud of points from lower left to upper right. The "Relative SS" entry below the "Draw your own line" button shows how well the line you draw fits the data relative to the fit of the least-squares line. Values are greater than or equal to 1, with 1 indicating your line fits as well as the least-squares line.

(a) You drew a line by eye through the middle of the pattern, yet the relative SS is greater than 1. What does that tell you?

(b) Now click the "Show least-squares line" box. Is the slope of the least-squares line smaller (new line is less steep) or larger (new line is steeper) than that of your line? If you repeat this exercise several times, you will consistently get the same result. The least-squares line minimizes the vertical distances of the points from the line. It is not the line through the "middle" of the cloud of points. This is one reason it is hard to draw a good regression line by eye.

The following exercises ask you to answer questions from data without having the details outlined for you. The exercise statements give you the **State** *step of the four-step process. In your work, follow the* **Plan, Solve,** *and* **Conclude** *steps of the process, described on page 62.*

5.60 4step **Beavers and Beetles.** Do beavers benefit beetles? Researchers laid out 23 circular plots, each 4 meters in diameter, in an area where beavers were cutting down cottonwood trees. In each plot, they counted the number of stumps from trees cut by beavers and the number of clusters of beetle larvae. Ecologists think that the new sprouts from stumps are more tender than other cottonwood growth, so that beetles prefer them. If so, more stumps should produce more beetle larvae. Here are the data:[28] BEAVERS

Stumps	2	2	1	3	3	4	3	1	2	5	1	3
Beetle larvae	10	30	12	24	36	40	43	11	27	56	18	40
Stumps	2	1	2	2	1	1	4	1	2	1	4	
Beetle larvae	25	8	21	14	16	6	54	9	13	14	50	

Analyze these data to see if they support the "beavers benefit beetles" idea.

5.61 4step **A Computer Game.** A multimedia statistics learning system includes a test of skill in using the computer's mouse. The software displays a circle at a random location on the computer screen. The subject clicks in the circle with the mouse as quickly as possible. A new circle appears as soon as the subject clicks in the old one. Table 5.3 gives data for one subject's trials, 20 with each hand. Distance is the distance from the cursor location to the center of the new circle, in units whose actual size depends on the size of the screen. Time is the time required to click in the new circle, in milliseconds.[29] We suspect that time depends on distance. We also suspect that performance will not be the same with the right and left hands. Analyze the data with a view to predicting performance separately for the two hands. COMGAME

5.62 4step **Predicting Tropical Storms.** William Gray heads the Tropical Meteorology Project at Colorado State University (well away from the hurricane belt). His forecasts before each year's hurricane season attract lots of attention. Here are data on the number of named Atlantic tropical storms predicted by Dr. Gray and the actual number of storms for the years 1984–2018:[30]

NASA/Goddard Space Flight Center/ Scientific Visualization Studio

Year	Forecast	Actual
1984	10	13
1985	11	11
1986	8	6
1987	8	7
1988	11	12
1989	7	11
1990	11	14
1991	8	8
1992	8	7
1993	11	8
1994	9	7
1995	12	19
1996	10	13
1997	11	8
1998	10	14
1999	14	12
2000	12	15
2001	12	15
2002	11	12
2003	14	16
2004	14	15
2005	15	28
2006	17	10
2007	17	15

Year	Forecast	Actual
2008	15	16
2009	11	9
2010	18	19
2011	16	19
2012	13	19
2013	18	14
2014	10	8
2015	8	11
2016	14	15
2017	14	17
2018	14	15

Analyze these data. How accurate are Dr. Gray's forecasts? How many tropical storms would you expect in a year when his preseason forecast calls for 16 storms? What is the effect of the disastrous 2005 season on your answers? STORMS

5.63 4step **Great Arctic Rivers.** One effect of global warming is to increase the flow of water into the Arctic Ocean from rivers. Such an increase may have major effects on the world's climate. Six rivers (Yenisey, Lena, Ob, Pechora, Kolyma, and Severnaya Dvina) drain two-thirds of the Arctic in Europe and Asia. Several of these are among the largest rivers on earth. Table 5.4 presents the total discharge from these rivers each year from 1936 to 2014.[31] Discharge is measured in cubic kilometers of water. Analyze these data to uncover the nature and strength of the trend in total discharge over time. ARCTIC

5.64 4step **Will Women Outrun Men?** Does the physiology of women make them better suited than men to long-distance running? Will women eventually outperform men in long-distance races? Researchers examined data on world record times (in seconds) for men and women in the marathon. Based on these data, researchers (in 1992) attempted to predict when women would outrun men in the marathon. Here are data for women:[32] MARATHON

Year	1926	1964	1967	1970	1971	1974	1975
Time	13,222.0	11,973.0	11,246.0	10,973.0	9990.0	9834.5	9499.0
Year	1977	1980	1981	1982	1983	1985	
Time	9287.5	9027.0	8806.0	8771.0	8563.0	8466.0	

TABLE 5.3 Reaction times (in milliseconds) in a computer game

Time	Distance	Hand	Time	Distance	Hand
115	190.70	right	240	190.70	left
96	138.52	right	190	138.52	left
110	165.08	right	170	165.08	left
100	126.19	right	125	126.19	left
111	163.19	right	315	163.19	left
101	305.66	right	240	305.66	left
111	176.15	right	141	176.15	left
106	162.78	right	210	162.78	left
96	147.87	right	200	147.87	left
96	271.46	right	401	271.46	left
95	40.25	right	320	40.25	left
96	24.76	right	113	24.76	left
96	104.80	right	176	104.80	left
106	136.80	right	211	136.80	left
100	308.60	right	238	308.60	left
113	279.80	right	316	279.80	left
123	125.51	right	176	125.51	left
111	329.80	right	173	329.80	left
95	51.66	right	210	51.66	left
108	201.95	right	170	201.95	left

TABLE 5.4 **Arctic river discharge (in cubic kilometers), 1936 to 2017**

Year	Discharge	Year	Discharge	Year	Discharge	Year	Discharge
1936	1721	1957	1762	1978	2008	1999	1970
1937	1713	1958	1936	1979	1970	2000	1905
1938	1860	1959	1906	1980	1758	2001	1890
1939	1739	1960	1736	1981	1774	2002	2085
1940	1615	1961	1970	1982	1728	2003	1780
1941	1838	1962	1849	1983	1920	2004	1900
1942	1762	1963	1774	1984	1823	2005	1930
1943	1709	1964	1606	1985	1822	2006	1910
1944	1921	1965	1735	1986	1860	2007	2270
1945	1581	1966	1883	1987	1732	2008	2078
1946	1834	1967	1642	1988	1906	2009	1900
1947	1890	1968	1713	1989	1932	2010	1813
1948	1898	1969	1742	1990	1861	2011	1859
1949	1958	1970	1751	1991	1801	2012	1702
1950	1830	1971	1879	1992	1793	2013	1758
1951	1864	1972	1736	1993	1845	2014	1988
1952	1829	1973	1861	1994	1902	2015	2042
1953	1652	1974	2000	1995	1842	2016	1863
1954	1589	1975	1928	1996	1849	2017	1988
1955	1656	1976	1653	1997	2007		
1956	1721	1977	1698	1998	1903		

Here are data for men:

Year	1908	1909	1913	1920	1925	1935	1947
Time	10,518.4	9751.0	9366.6	9155.8	8941.8	8802.0	8739.0
Year	1952	1953	1954	1958	1960	1963	1964
Time	8442.2	8314.8	8259.4	8117.0	8116.2	8068.0	7931.2
Year	1965	1967	1969	1981	1984	1985	1988
Time	7920.0	7776.4	7713.6	7698.0	7685.0	7632.0	7610.0

Analyze these data using least-squares regression to estimate when men's and women's record times will be equal. How reliable is your estimate? (You may wish to look online to find the current men's and women's record times.)

5.65 Will Women Outrun Men? (continued). 4step The data set RUN2 contains all the world record times (in seconds) for men and women in the marathon up to 2019. Use these data to repeat the analysis in Exercise 5.64; that is, use least-squares regression to estimate when men's and women's record times will be equal. How reliable is this estimate? Explain your answer. RUN2

CHAPTER

6

When you complete this chapter, you will be able to:

6.1 Compute and interpret marginal distributions in two-way tables.

6.2 Compute and interpret conditional distributions in two-way tables.

6.3 Recognize and explain Simpson's paradox.

Two-Way Tables*

In Chapters 4 and 5, we considered relationships between two quantitative variables. In this chapter, we use two-way tables to describe relationships between two categorical variables. Some variables—such as sex, race, and occupation—are categorical by nature. Other categorical variables are created by grouping values of a quantitative variable into classes.

To explore relationships between two categorical variables, we use the counts or percentages of individuals that fall into various categories. As with quantitative variables, we must be alert to the influence of lurking variables and be careful not to assume that the patterns we observe would continue to hold for additional data or in a broader setting.

EXAMPLE 6.1 Who Earns Academic Degrees?

In 2017, the National Center for Education Statistics projected the number of academic degrees to be awarded in 2020–2021 for men and women. Table 6.1 shows their projections.[1] This is a **two-way table** because it describes two categorical variables. One is the sex of an individual. The other is the academic degree earned. Sex is the *row variable* because each row in the table describes the sex of an individual. Academic degree conferred is the *column variable* because each column describes a degree. Because academic degree conferred has a natural order from "Associate" to "Professional/doctorate," the columns are in this order. The entries in the table are the counts of individuals (in thousands) in

*This material is important in statistics, but it is needed later in this book only for Chapter 25. You may omit it if you do not plan to read Chapter 25 or may delay reading it until you reach Chapter 25.

miodrag ignjatovic/Getty Images

TABLE 6.1 Academic degrees by sex

	Degrees Conferred (thousands)				
Sex	Associate's	Bachelor's	Master's	Professional/Doctorate	Total
Women	639	1087	460	97	2283
Men	402	804	329	87	1622
Total	1041	1891	789	184	3905

each sex-by-academic-degree class. The entries in the right margin are the total of the row entries, the entries in the bottom margin are the total of the column entries, and the entry at the bottom right is the total of all students projected to receive an academic degree in 2020–2021.

two-way table
A table of counts used to organize data about two categorical variables. Values of each row variable run across the table, and values of each column variable run vertically down the table. Entries in the table are the counts of how often each combination of the corresponding row and column variables occur. Two-way tables are often used to summarize large amounts of information by grouping observations into categories.

6.1 Marginal Distributions

How can we best grasp the information contained in Table 6.1? First, *look at the distribution of each variable separately.* The distribution of a categorical variable says how often each outcome occurred. The "Total" column at the right of the table contains the totals for the rows. These row totals give the distribution of sex in the entire group of 3905 thousand students: 2283 thousand are women, and 1622 thousand are men.

If the row and column totals are missing, the first thing to do in studying a two-way table is to calculate them. The distributions of sex alone and degree conferred alone are called *marginal distributions* because they appear at the right and bottom margins of the two-way table.

Marginal Distributions

The **marginal distribution** of one of the categorical variables in a two-way table of counts is the distribution of values of that variable among all individuals described by the table.

Percentages are often more informative than counts. We can display the marginal distribution of sex in percentages by dividing each row total by the table total and converting to a percentage.

EXAMPLE 6.2 Calculating a Marginal Distribution

The percentage of these students in Table 6.1 who are women is

$$\frac{\text{women total}}{\text{table total}} = \frac{2283}{3905} = 0.585 = 58.5\%$$

Repeat such a calculation to obtain the marginal distribution of men (or subtract 58.5% from 100%). Here is the complete distribution:

Response	Percentage
Women	$\frac{2283}{3905} = 58.5\%$
Men	$\frac{1622}{3905} = 41.5\%$

More women are projected to receive degrees than men. The total is 100% because everyone belongs to one of the four sex and education classes.

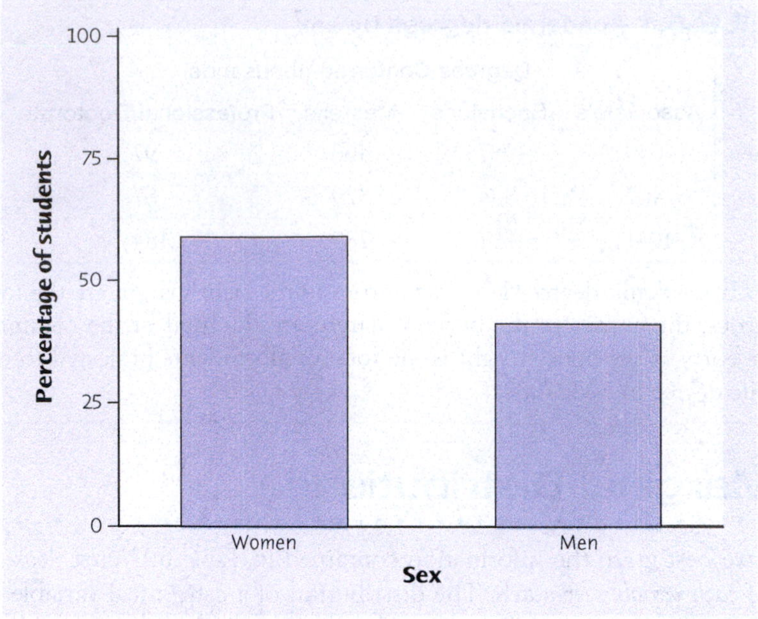

FIGURE 6.1
Bar graph of the distribution of sex of adults who participated in the survey. This is one of the marginal distributions for Table 6.1.

Each marginal distribution from a two-way table is a distribution for a single categorical variable. As we saw in Chapter 1, we can use a bar graph or a pie chart to display such a distribution. Figure 6.1 is a bar graph of the distribution of sex among students in the sample.

In working with two-way tables, you must calculate lots of percentages. Here's a tip to help you decide what fraction gives the percentage you want: Ask, "What group represents the total of which I want a percentage?" The count for that group is the denominator of the fraction that leads to the percentage. In Example 6.2, we want a percentage "of students," so the count of students (the table total) is the denominator.

APPLY YOUR KNOWLEDGE

6.1 Video Gaming and Grades. The popularity of computer, video, online, and virtual reality games has raised concerns that they might negatively affect youth. The data in this exercise are based on a recent survey of 14- to 18-year-olds in Connecticut high schools. Here are the grade distributions of boys who have and have not played video games:[2] GAMING

	Grade Average		
	A's and B's	C's	D's and F's
Played games	736	450	193
Never played games	205	144	80

(a) How many people does this table describe? How many of these have played video games?

(b) Give the marginal distribution of the grades. What percentage of the boys represented in the table received a grade of C or lower?

6.2 Ages of College Students. Here is a two-way table of U.S. Census Bureau data describing the age and sex of all American students enrolled in college. The table entries are counts in thousands of students:[3] AGES

Age Group	Female	Male
15 to 19 years	2348	1831
20 to 24 years	4280	3713
25 to 34 years	2166	1714
35 years or older	1492	853

(a) How many college students are there?

(b) Find the marginal distribution of age group. What percentage of undergraduates are in the 20- to 24-year-old college age group?

Dirk Kruell/Laif/Redux

6.2 Conditional Distributions

Table 6.1 contains much more information than the two marginal distributions of sex alone and degree alone confer. *Marginal distributions tell us nothing about the relationship between two variables.* To describe a relationship between two categorical variables, we must calculate some well-chosen percentages from the counts given in the body of the table.

Let's say we want to compare the proportions of women and men who receive a professional/doctorate degree. To do this, compare percentages for each sex category. To study women, we look only at the "Women" row in Table 6.1. To find the percentage *of women* who receive a professional/doctorate degree, divide the count of such women by the total number of women (the row total):

$$\frac{\text{women who receive a professional/doctorate degree}}{\text{row total}} = \frac{97}{2283} = 0.042 = 4.2\%$$

Doing this for all four entries in the "Women" row gives the *conditional distribution* of degrees conferred among women. We use the term *conditional* because this distribution describes only students who satisfy the condition that they are women.

Conditional Distributions

A **conditional distribution** of a variable is the distribution of values of that variable among only individuals who have a given value of the other variable. There is a separate conditional distribution for each value of the other variable.

STATISTICS IN YOUR WORLD

Smiling Faces

Women smile more than men. The same data that produce this fact allow us to link smiling to other variables in two-way tables. For example, as the second variable, add whether or not the person thinks he or she is being observed. If yes, that's when women smile more. If no, there's no difference between women and men. Next, take the second variable to be the person's social role (for example, is he or she the boss in an office?). Within each role, there is very little difference in smiling between women and men.

EXAMPLE 6.3 Comparing Women and Men

4step

STATE: How do women and men differ in terms of the degrees they are projected to receive in 2020–2021?

PLAN: Make a two-way table of response by sex category. Find the conditional distribution of response for each sex category. Compare these two distributions.

SOLVE: Table 6.1 is the two-way table we need. Look first at just the "Women" row to find the conditional distribution for women, then at just the "Men" row to find the conditional distribution for men. Here are the calculations and the two conditional distributions:

Response	Associate's	Bachelor's	Master's	Professional/ Doctorate
Women	$\frac{639}{2283} = 28.0\%$	$\frac{1087}{2283} = 47.6\%$	$\frac{460}{2283} = 20.1\%$	$\frac{97}{2283} = 4.2\%$
Men	$\frac{402}{1622} = 24.8\%$	$\frac{804}{1622} = 49.6\%$	$\frac{329}{1622} = 20.3\%$	$\frac{87}{1622} = 5.4\%$

The percentages in each row should be 100% because for each sex category, everyone receives one of the four degrees. However, in general, the percentages may not add exactly to 100% because we round off to a fixed number of decimal places. This is *roundoff error*, and we see that there is roundoff error here.

CONCLUDE: The percentage of women projected to receive an associate's degree is higher than the percentage of men projected to receive an associate's degree, while the percentage of men projected to receive each of the degrees other than an associate's degree is slightly higher than the percentage of women.

Roundoff Error

The **roundoff error** is the small difference between a rounded decimal number and its precise value before rounding.

Software can do these calculations for you. Most programs allow you to choose which conditional distributions you want to compare. The output in Figure 6.2

Minitab

Tabulated statistics: sex, degree

Rows: Sex Columns: Degree

	Associate's	Bachelor's	Master's	Professional/Doctorate	All
Men	402	804	329	87	1622
	24.78	49.57	20.28	5.36	100.00
Women	639	1087	460	97	2283
	27.99	47.61	20.15	4.25	100.00
All	1041	1891	789	184	3905
	26.66	48.43	20.20	4.71	100.00

Cell contents: Count
% of row

jmp

Contingency Table

Degree

Sex	Count row %	Associate's	Bachelor's	Master's	Professional/ Doctorate	Total
	Men	402	804	329	87	1622
		24.78	49.57	20.28	5.36	
	Women	639	1087	460	97	2283
		27.99	47.61	20.15	4.25	
	Total	1041	1891	789	184	3905

FIGURE 6.2
Minitab and JMP output for the two-way table of adults by sex and education. Each entry in the Minitab output includes the count and percentage of its row total. The "Men" and "Women" rows give the conditional distributions of responses for each sex category, and the "All" row shows the marginal distribution of responses for all these adults. Notice that Minitab and JMP order variables in the table alphabetically. Each entry in the JMP output includes the count and percentage of its row total. The second entry in each cell gives the conditional distribution of responses for the different sex categories. The "Total" row and column show the corresponding marginal totals of responses for all these adults.

presents the two conditional distributions of degrees conferred, one for each sex category, and also the marginal distribution of degrees conferred for all the students. The distributions agree (up to roundoff) with the results in Examples 6.2 and 6.3.

Remember that there are two sets of conditional distributions for any two-way table. Example 6.3 looked at the conditional distributions of degrees conferred for the two sex categories. Figure 6.3(a) makes this comparison in a bar graph with separate bars for men and women side by side for each category of degree. In this graph, the total of the four blue bars is 100%, and the total of the green bars is also 100%. We could also examine the four conditional distributions of sex, one for each of the four degrees conferred, by looking separately at the four columns in Table 6.1 (page 161). Figure 6.3(b) makes this comparison in a bar graph, again with

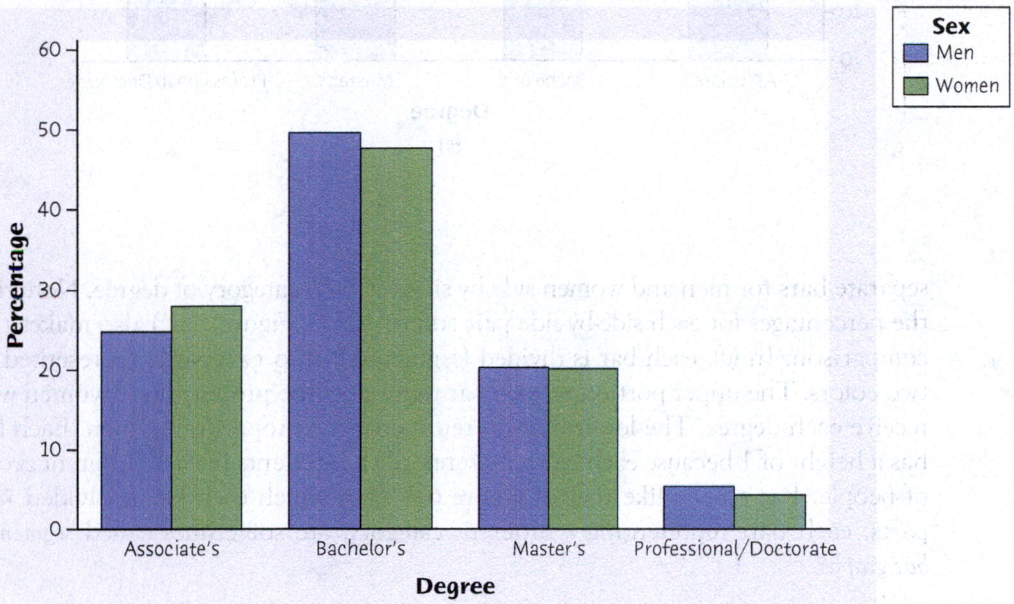

(a)

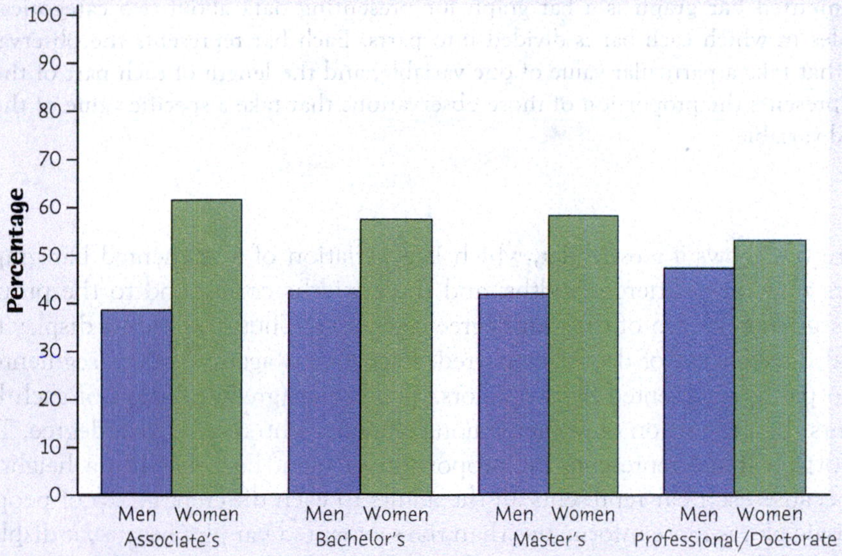

(b)

FIGURE 6.3 (a) Side-by-side bar graph comparing the percentages of women and men among those in each sex category, for each degree conferred. (b) Side-by-side bar graph comparing the percentages of women and men among those in each degree-conferred category. (c) Segmented bar graph comparing the percentages of women (green) and men (blue) among those in each degree-conferred category.

FIGURE 6.3
(Continued)

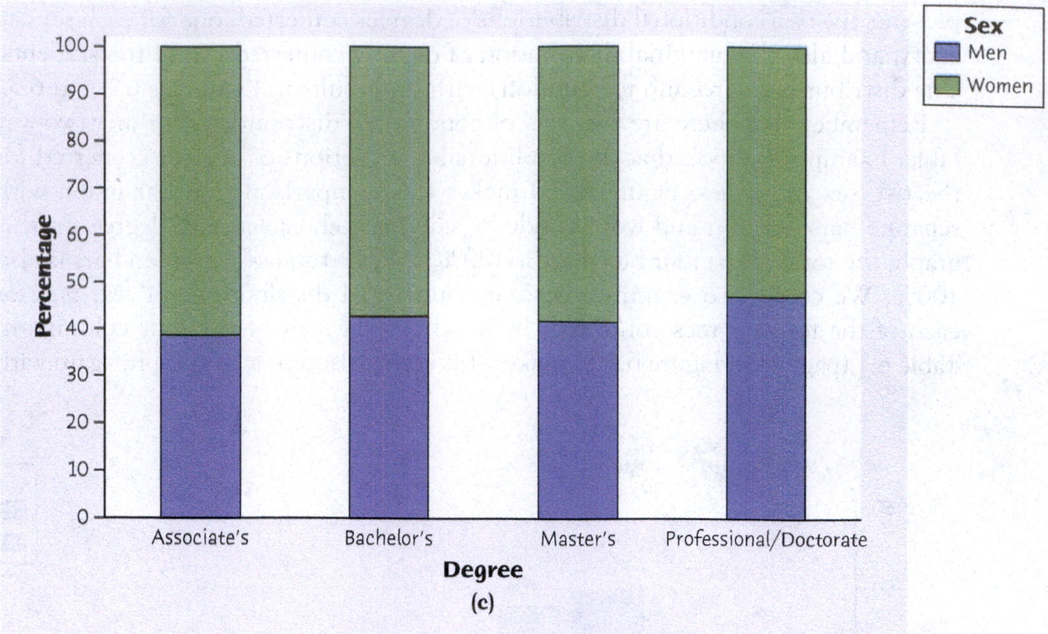

separate bars for men and women side by side for each category of degree. Note that the percentages for each side-by-side pair add to 100%. Figure 6.3(c) also makes this comparison. In (c), each bar is divided (segmented) into two parts, represented by two colors. The upper portion of each bar represents the proportion of women who receive each degree. The lower portion represents the proportion of men. Each bar has a height of 1 because each bar represents all the students in each different group of people. Bar graphs like that in Figure 6.3(c), in which each bar is divided into parts, each part representing a different category, are sometimes called *segmented bar graphs.*

Segmented Bar Graph

A **segmented bar graph** is a bar graph for presenting data about two categorical variables in which each bar is divided into parts. Each bar represents the observations that take a particular value of one variable, and the length of each part of the bar represents the proportion of those observations that take a specific value of the second variable.

mosaic plot
A segmented bar graph in which the width of each bar represents the proportion of all observations that fall into the category that the bar represents.

Figure 6.4 shows a **mosaic plot,** which is a variation of a segmented bar graph. The bars now have different widths, and these widths correspond to the proportion of students in each of the four degree categories. Thus, the widths display the marginal distribution of degree conferred. Each bar is again divided (segmented) into two parts, represented by two colors. The upper (green) portion of each bar represents the proportion of women among students who receive each degree. The other portion (blue) represents the proportion of men. Each bar has a height of 100% because each bar represents all the adults in each different group of people. The mosaic plot is more informative than the segmented bar plot because it displays the marginal distribution of degree conferred as well as the conditional distribution of sex, given degree conferred.

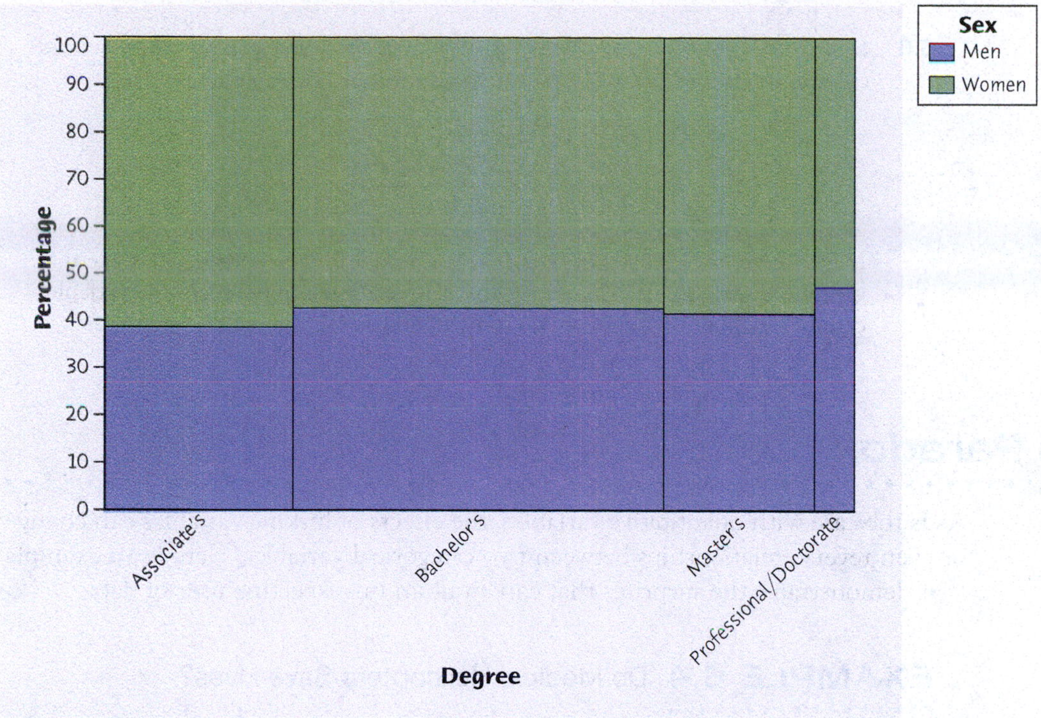

FIGURE 6.4 Mosaic plot comparing the proportions of women (green) and men (blue) among those in each degree-conferred category.

Figure 6.4 displays only one of the two sets of conditional distributions. We would need another graph to display the other (the conditional distribution of degree conferred, given sex). Also, the graphs in Figures 6.3 and 6.4 only indicate percentages or proportions, not total counts.

No single graph portrays the form of the relationship between categorical variables (as a scatterplot does for quantitative variables). No single numerical measure (such as the correlation) summarizes the strength of the association. Bar graphs are flexible enough to be helpful, but you must think about what comparisons you want to display. For numerical measures, we rely on well-chosen percentages. You must decide which percentages you need. Here is a hint: *If there is an explanatory-response relationship, compare the conditional distributions of the response variable for the separate values of the explanatory variable.* If you think that sex influences degree conferred, compare the conditional distributions of degree conferred for each sex category, as in Example 6.3.

APPLY YOUR KNOWLEDGE

6.3 Video Gaming and Grades. Exercise 6.1 (page 162) gives data on the grade distribution of boys who have and have not played video games. To see the relationship between grades and game-playing experience, find the conditional distributions of grades (the response variable) for players and nonplayers. What do you conclude? GAMING

6.4 Ages of College Students Exercise 6.2 (page 163) gives U.S. Census Bureau data describing the age and sex of all American students enrolled in college. We suspect that the percentage of women is higher among students in the 25- to 34-year-old age group than in the 20- to 24-year-old age group. Do the data support this suspicion? Follow the four-step process, as illustrated in Example 6.3. AGES

4step

6.5 Marginal Distributions Aren't the Whole Story. Here are the row and column totals for a two-way table with two rows and two columns:

a	b	50
c	d	50
60	40	100

Make up *two different* sets of counts a, b, c, and d for the body of the table that give these same totals. This shows that the relationship between two variables cannot be obtained from the two individual distributions of the variables.

6.3 Simpson's Paradox

As is the case with quantitative variables, the effects of lurking variables can change or even reverse relationships between two categorical variables. Here is an example that demonstrates the surprises that can await an unsuspecting user of data.

EXAMPLE 6.4 Do Medical Helicopters Save Lives?

© Ashley Cooper/Getty Images

Accident victims are sometimes taken by helicopter from the accident scene to a hospital. Helicopters save time. Do they also save lives? Let's compare the percentages of accident victims who die with helicopter evacuation and with the usual transport to a hospital by road. Here are hypothetical data that illustrate a practical difficulty:[4]

	Helicopter	Road
Victim died	64	260
Victim survived	136	840
Total	200	1100

We see that 32% (64 out of 200) of helicopter patients died, but only 24% (260 out of 1100) of the others did. That seems discouraging.

The explanation is that the helicopter is sent mostly to serious accidents, so that the victims transported by helicopter are more often seriously injured. They are more likely to die with or without helicopter evacuation. Here are the same data broken down by the seriousness of the accident:

Serious Accidents		
	Helicopter	Road
Died	48	60
Survived	52	40
Total	100	100

Less Serious Accidents		
	Helicopter	Road
Died	16	200
Survived	84	800
Total	100	1000

Inspect these tables to convince yourself that they describe the same 1300 accident victims as the original two-way table. For example, 200 $(= 100 + 100)$ were moved by helicopter, and $64 (= 48 + 16)$ of these died.

Among victims of serious accidents, the helicopter saves 52% (52 out of 100) compared with 40% for road transport. If we look only at less serious accidents, 84% of those transported by helicopter survive, versus 80% of those transported by road. Both groups of victims have a higher survival rate when evacuated by helicopter.

How can it happen that the helicopter does better for both groups of victims but worse when all victims are lumped together? Examining the data makes the explanation clear. Half the helicopter transport patients are from serious accidents, compared with only 100 of the 1100 road transport patients. So the helicopter carries patients who are more likely to die. The seriousness of the accident was a lurking variable that, until we uncovered it, hid the true relationship between survival and mode of transport to a hospital. Example 6.4 illustrates *Simpson's paradox.*

Simpson's Paradox

An association or comparison that holds for all of several groups can reverse direction when the data are combined to form a single group. This reversal is called **Simpson's paradox.**

The lurking variable in Simpson's paradox is categorical. That is, it breaks the individuals into groups, as when accident victims are classified as injured in a "serious accident" or a "less serious accident." Simpson's paradox is just an extreme form of the fact that observed associations can be misleading when there are lurking variables.

APPLY YOUR KNOWLEDGE

6.6 Field Goal Shooting. Here are data on field goal shooting for two members of the University of Michigan 2017–2018 men's basketball team:[5] BSKTBLL

	Charles Matthews		Duncan Robinson	
	Made	Missed	Made	Missed
Two-pointers	171	136	44	30
Three-pointers	34	73	78	125

(a) What percentage of all field goal attempts did Charles Matthews make? What percentage of all field goal attempts did Duncan Robinson make?

(b) Now find the percentage of all two-point field goals and all three-point field goals that Charles made. Do the same for Duncan.

(c) Charles had a lower percentage for *both* types of field goals, but he had a better overall percentage. That sounds impossible. Explain carefully, referring to the data, how this can happen.

6.7 Bias in the Jury Pool? The New Zealand Department of Justice did a study of the composition of juries in court cases. Of interest was whether Maori, the indigenous people of New Zealand, were adequately represented in jury pools. Here are the results for two districts, Rotorua and Nelson, in New Zealand (similar results were found in all districts):[6] JURY

Rotorua	Maori	Non-Maori
In jury pool	79	258
Not in jury pool	8810	23,751
Total	8889	24,009

Nelson	Maori	Non-Maori
In jury pool	1	56
Not in jury pool	1328	32,602
Total	1329	32,658

Molly Marshall/Alamy Stock Photo

(a) Compare percentages to show that the percentage of all Maori in the jury pool in each district is less than the percentage of non-Maori in the jury pool.

(b) Combine the data into a single two-way table of outcome ("in jury pool" or "not in jury pool") by ethnicity (Maori or non-Maori). The original study only reported such an overall rate. Which ethnic group has a higher percentage of its people in the jury pool?

(c) Explain from the data, in language that a reporter can understand, how Maori can have a higher percentage overall even though non-Maori have higher percentages for both districts.

CHAPTER 6 SUMMARY

- A **two-way table** of counts organizes data about two categorical variables. Values of the **row variable** label the rows that run across the table, and values of the **column variable** label the columns that run down the table. Two-way tables are often used to summarize large amounts of information by grouping outcomes into categories.
- The **row totals** and **column totals** in a two-way table give the **marginal distributions** of the two individual variables. It is clearer to present these distributions as percentages of the table total. Marginal distributions tell us nothing about the relationship between the variables.
- There are two sets of **conditional distributions** for a two-way table: the distributions of the row variable for each fixed value of the column variable and the distributions of the column variable for each fixed value of the row variable. Comparing one set of conditional distributions is one way to describe the association between the row and the column variables.
- To find the **conditional distribution** of the row variable for one specific value of the column variable, look only at that one column in the table. Find each entry in the column as a percentage of the column total.
- Bar graphs are flexible means of presenting categorical data. There is no single best way to describe an association between two categorical variables.
- A comparison between two variables that holds for each individual value of a third variable can be changed or even reversed when the data for all values of the third variable are combined. This is **Simpson's paradox.** Simpson's paradox is an example of the effect of lurking variables on an observed association.

CHECK YOUR SKILLS

The Pew Research Center conducts an annual Internet Project, which includes research related to social networking. The following two-way table about the percentage of U.S. adults surveyed who use at least one social media site, broken down by age, is based on data reported by Pew as of February 2019:[7]

Use Social Media	Yes	No
Age 18–29	212	24
Age 30–49	324	71
Age 50–64	293	131
Age 65+	156	235

Questions 6.8 through 6.16 are based on this table. MEDIA

6.8 How many individuals are described by this table?

(a) 1446

(b) 985

(c) Need more information

6.9 How many individuals ages 18–29 were among the respondents?

(a) 212

(b) 236

(c) Need more information

6.10 The percentage of individuals ages 18–29 who were among the respondents was

(a) about 22%.

(b) about 16%.

(c) about 90%.

6.11 Your percentage from the previous question is part of

(a) the marginal distribution of age.

(b) the marginal distribution of whether one uses social media.

(c) the conditional distribution of whether one uses social media among individuals ages 18–29.

6.12 What percentage of individuals who use social media are ages 18–29?

(a) about 15%

(b) about 22%

(c) about 90%

6.13 Your percentage from the previous question is part of

(a) the marginal distribution of whether one uses social media.

(b) the conditional distribution of whether one uses social media among individuals ages 18–29.

(c) the conditional distribution of age among those who use social media.

6.14 What percentage of individuals ages 18–29 use social media?

(a) about 15%

(b) about 22%

(c) about 90%

6.15 Your percentage from the previous question is part of

(a) the marginal distribution of age.

(b) the conditional distribution of whether one uses social media among individuals ages 18–29.

(c) the conditional distribution of age among those who use social media.

6.16 A bar graph showing the conditional distribution of whether one uses social media among age groups would have

(a) 2 bars. (b) 4 bars. (c) 8 bars.

6.17 A college looks at the grade point average (GPA) of its full-time and part-time students. Grades in science courses are generally lower than grades in other courses. There are few science majors among part-time students but many science majors among full-time students. The college finds that full-time students who are science majors have higher GPAs than part-time students who are science majors. Full-time students who are not science majors also have higher GPAs than part-time students who are not science majors. Yet part-time students as a group have higher GPAs than full-time students. This finding is

(a) not possible: if both science and other majors who are full-time have higher GPAs than those who are part-time, then all full-time students together must have higher GPAs than all part-time students together.

(b) an example of Simpson's paradox: full-time students do better in both kinds of courses but worse overall because they take more science courses.

(c) due to comparing two conditional distributions that should not be compared.

CHAPTER 6 EXERCISES

6.18 **Is Astrology Scientific?** The University of Chicago's General Social Survey (GSS) is the nation's most important social science sample survey. The GSS asked a random sample of adults their opinion about whether astrology is very scientific, sort of scientific, or not at all scientific. Here is a two-way table of counts for people in the sample who had three levels of higher education degrees:[8] ASTRLGY

	Degree Held		
	Junior College	Bachelor	Graduate
Not at all scientific	47	181	113
Very or sort of scientific	36	43	13

Find the two conditional distributions of degree held: one for those who hold the opinion that astrology is not at all scientific and one for those who say astrology is very or sort of scientific. Based on your calculations, describe with a graph and in words the differences between those who say astrology is not at all scientific and those who say it is very or sort of scientific.

6.19 **Weight-Lifting Injuries.** Resistance training is a popular form of conditioning aimed at enhancing sports performance and is widely used among high school, college, and professional athletes, although its use for younger athletes is controversial. A random sample of 4111 patients between the ages of 8 and 30 admitted to U.S. emergency rooms with the injury code "weightlifting" was obtained. These injuries were classified as "accidental" if caused by dropped weight or improper equipment use. The patients were also classified into the four age categories 8–13 years, 14–18, 19–22, and 23–30. Here is a two-way table of the results:[9] LIFTING

Age	Accidental	Not Accidental
8–13	295	102
14–18	655	916
19–22	239	533
23–30	363	1008

Compare the distributions of ages for accidental and nonaccidental injuries. Use percentages and draw a bar graph. What do you conclude?

Marital Status and Income. *We sometimes hear that getting married is good for your career. Table 6.2 presents data from the U.S. Census Bureau that classifies men ages 18 and over according to marital status and annual income in 2018. We include only data on men.*[10] *Exercises 6.20 through 6.24 are based on these data.* STATUS

6.20 Marginal Distributions. Give (in percentages) the two marginal distributions, for marital status and for income. Do each of your two sets of percentages add to exactly 100%? If not, why not?

6.21 Percentages. What percentage of single men have no income? What percentage of men with no income are single men?

6.22 Conditional Distribution. Give (in percentages) the conditional distribution of income level among single men. Should your percentages add to 100% (up to roundoff error)? Explain your reasoning.

6.23 Marital Status and Income. One way to see the relationship is to look at who has no income.

(a) There are 1,933,000 married men with no income and 5,514,000 single men with no income. Explain why these counts by themselves don't describe the relationship between marital status and income.

(b) Among men who have no income, find the percentages for marital status. Do the same for men who have an income of \$100,000 and over. What do these percentages say about the relationship?

6.24 Association Is Not Causation. The data in Table 6.2 show that single men are more likely to hold lower-income jobs than are married men. We should not conclude that single men can increase their income by getting married. What lurking variables might help explain the association between marital status and income?

6.25 Race and the Death Penalty. Whether a convicted murderer gets the death penalty seems to be influenced by the race of the victim. Several researchers studied this issue in the 1970s and 1980s, resulting in several landmark, oft-cited, and controversial papers. Here are data from one of these studies on 326 cases in which the defendant was convicted of murder:[11] DISCRIM

White Defendant		
	White Victim	Black Victim
Death	19	0
Not	132	9

Black Defendant		
	White Victim	Black Victim
Death	11	6
Not	52	97

(a) Use these data to make a two-way table of defendant's race (White or Black) versus death penalty (yes or no).

(b) Show that Simpson's paradox holds: a higher percentage of White defendants are sentenced to death overall, but for both Black and White victims, a higher percentage of Black defendants are sentenced to death.

(c) Use the data to explain why the paradox holds, in language that a judge could understand.

6.26 Obesity and Health. To estimate the health risks of obesity, we might compare how long obese and nonobese people live. Smoking is a lurking variable that may reduce the gap between the two groups because smoking tends to both reduce weight and lead to earlier death. So if we ignore smoking, we may underestimate the health risks of obesity. Illustrate Simpson's paradox with a simplified version of this situation: make up two-way tables of obese (yes or no) by early death (yes or no) separately for smokers and nonsmokers such that

TABLE 6.2 Marital status and salary level (thousands of men)

	Marital Status				
Income	Single (Never Married)	Married	Divorced	Widowed	Total
No income	5514	1933	552	136	8135
\$1–\$49,999	24,827	30,609	5917	2437	63,790
\$50,000–\$99,999	6359	21,546	2895	629	31,429
\$100,000 and over	2076	14,332	1276	265	17,949
Total	39,776	68,420	10,640	3467	121,303

- Obese smokers and obese nonsmokers are both more likely to die earlier than those who are not obese.
- But when smokers and nonsmokers are combined into a two-way table of obese by early death, persons who are not obese are more likely to die earlier because more of them are smokers.

The following exercises ask you to answer questions from data without having the details outlined for you. The exercise statements give you the **State** *step of the four-step process. In your work, follow the* **Plan, Solve,** *and* **Conclude** *steps of the process, as illustrated in Example 6.3 (page 163).*

6.27 Smoking Cessation. A large randomized trial was conducted to assess the efficacy of Chantix for smoking cessation compared with bupropion (more commonly known as Wellbutrin or Zyban) and a placebo. Chantix is different from most other quit-smoking products in that it targets nicotine receptors in the brain, attaches to them, and blocks nicotine from reaching them, whereas bupropion is an antidepressant often used to help people stop smoking. Generally healthy smokers who smoked at least 10 cigarettes per day were assigned at random to take Chantix ($n = 352$), bupropion ($n = 329$), or a placebo ($n = 344$). The response measure is continuous cessation from smoking for weeks 9 through 12 of the study. Here is a two-way table of the results:[12] SMOKE

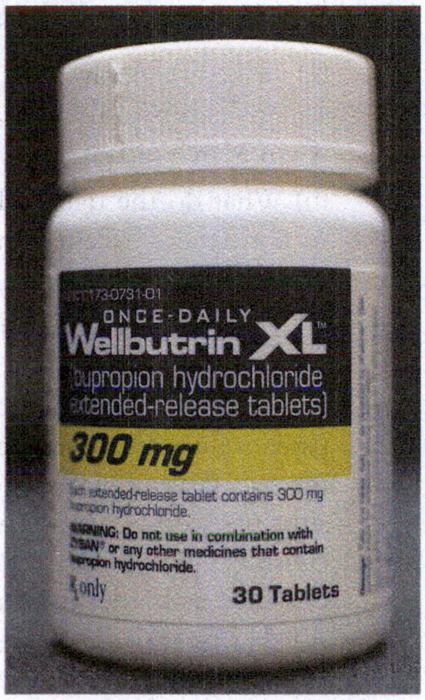

Joe Raedle/Getty Images

	Treatment		
	Chantix	Bupropion	Placebo
No smoking in weeks 9–12	155	97	61
Smoked in weeks 9–12	197	232	283

How does whether a subject smoked in weeks 9–12 depend on the treatment received?

6.28 Animal Testing. "It is right to use animals for medical testing if it might save human lives." The General Social Survey asked 1152 adults to react to this statement. Here is the two-way table of their responses: ANTEST

Response	Male	Female
Strongly agree	76	59
Agree	270	247
Neither agree nor disagree	87	139
Disagree	61	123
Strongly disagree	22	68

How do the distributions of opinion differ between men and women?

6.29 College Degrees. "Colleges and universities across the country are grappling with the case of the mysteriously vanishing male." So said an article in the *Washington Post*. Here are projections of the numbers of degrees that will be earned in 2023–2024, as projected by the National Center for Education Statistics. The table entries are counts of degrees, in thousands:[13] DEGREES

Degree	Female	Male
Associate's	644	405
Bachelor's	1092	806
Master's	467	335
Professional or doctorate	99	89

Briefly contrast the counts and distributions of men and women in earning degrees. Are men projected to be "vanishing" from colleges and universities across the country?

6.30 Complications of Bariatric Surgery. Bariatric surgery, or weight-loss surgery, includes a variety of procedures performed on people who are obese. Weight loss is achieved by reducing the size of the stomach with an implanted medical device (gastric banding), by removing a portion of the stomach (sleeve gastrectomy), or by resecting and rerouting the small intestines to a small stomach pouch (gastric bypass surgery). Because there can be complications using any of these methods, the National Institutes of Health recommends bariatric surgery only for obese people with a body mass index (BMI) of at least 40 or with BMI 35 and serious coexisting medical conditions, such as diabetes. Serious complications include potentially life-threatening, permanently disabling, and fatal outcomes. Here is a two-way table for data collected in Michigan over several years, giving counts of non-life-threatening complications, serious complications, and no complications for these three types of surgeries:[14] BARI

	Type of Complication			
	Non-Life-Threatening	Serious	None	Total
Gastric banding	81	46	5253	5380
Sleeve gastrectomy	31	19	804	854
Gastric bypass	606	325	8110	9041

What do the data say about differences in complications for the three types of surgeries?

6.31 4step **Smokers Rate Their Health.** The University of Michigan Health and Retirement Study (HRS) surveys more than 22,000 Americans over the age of 50 every two years. A subsample of the HRS participated in a 2009 Internet-based survey that collected information on a number of topical areas, including health (physical and mental health behaviors), psychosocial items, economics (income, assets, expectations, and consumption), and retirement.[15] Two of the questions asked were, "Would you say your health is excellent, very good, good, fair, or poor?" and "Do you smoke cigarettes now?" The two-way table summarizes the answers on these two questions. SMRATE

	Current Smoker	
Health	Yes	No
Excellent	25	484
Very good	115	1557
Good	145	1309
Fair	90	545
Poor	29	11

What do the data say about differences in self-evaluation of health for current smokers and nonsmokers?

6.32 4step **Punxsutawney Phil.** In the United States, Groundhog Day is celebrated on February 2. On February 2, Punxsutawney Phil, the legendary groundhog in Punxsutawney, Pennsylvania, emerges from his home, and if he sees his shadow and returns to his hole, he has predicted six more weeks of winter-like weather. How accurate is Phil? We have data for 119 years (up to 2019) indicating whether Phil saw his shadow and average temperature in March. It is not clear what constitutes "six more weeks of winter-like weather," so here we define it to occur if the average temperature for March is not above the historical average. Here we only use temperature data for Pennsylvania, Phil's home state. The results are 51 years where Phil saw his shadow and the average March temperature was not above the historical average, 51 years where Phil saw his shadow and the average March temperature was above the historical average, 7 years where Phil did not see his shadow and the average March temperature was not above the historical average, and 10 years where Phil did not see his shadow and the average March temperature was above the historical average.[16]

(a) Make a two-way table of "Phil saw his shadow or not" against "above historical average temperatures in March or not."

(b) What do the data tell us about Phil as a weather forecaster for Pennsylvania?

6.33 **Sleep Quality.** A random sample of 871 students ages 20–24 at a large midwestern university completed a survey including questions about their sleep quality, moods, academic performance, physical health, and psychoactive drug use. Sleep quality was measured using the Pittsburgh Sleep Quality Index (PSQI), with students scoring less than or equal to 5 on the index classified as optimal sleepers, those scoring a 6 or 7 classified as borderline, and those scoring over 7 classified as poor sleepers. The following table looks at the relationship between sleep quality classification and the use of over-the-counter (OTC) or prescription (Rx) stimulant medication more than once a month to help keep awake.[17] SLEEPQ

Use of OTC/Rx Meds to Wake > 1x/Month	Sleep Quality on PSQI		
	Optimal	Borderline	Poor
Yes	37	53	84
No	266	186	245

What do the data say about differences in sleep quality for those who use over-the-counter or prescription stimulant medication more than once a month to keep awake and those who don't?

CHAPTER

7

Exploring Data: Part I Review

IN THIS CHAPTER, WE COVER...

Data analysis is the art of describing data using graphs and numerical summaries. The purpose of exploratory data analysis is to help us see and understand the most important features of a set of data. Chapter 1 described graphs used to display distributions: pie charts and bar graphs for categorical variables, histograms and stemplots for quantitative variables. In addition, time plots show how a quantitative variable changes over time. Chapter 2 presented numerical tools for describing the center and spread of the distribution of one variable. Chapter 3 discussed density curves for describing the overall pattern of a distribution, with emphasis on the Normal distributions.

The first STATISTICS IN SUMMARY figure on the next page organizes the big ideas for exploring a quantitative variable. Plot your data, then describe their center and spread using either the mean and standard deviation or the five-number summary. The last step, which makes sense only for data whose distribution is single-peaked and roughly symmetric, is to summarize the data in compact form by using a Normal curve as a description of the overall pattern. The question marks at the last two stages remind us that the usefulness of numerical summaries and Normal distributions depends on what we find when we examine graphs of our data. No short summary does justice to irregular shapes or to data with several distinct clusters.

VitalyEdush/Getty Images

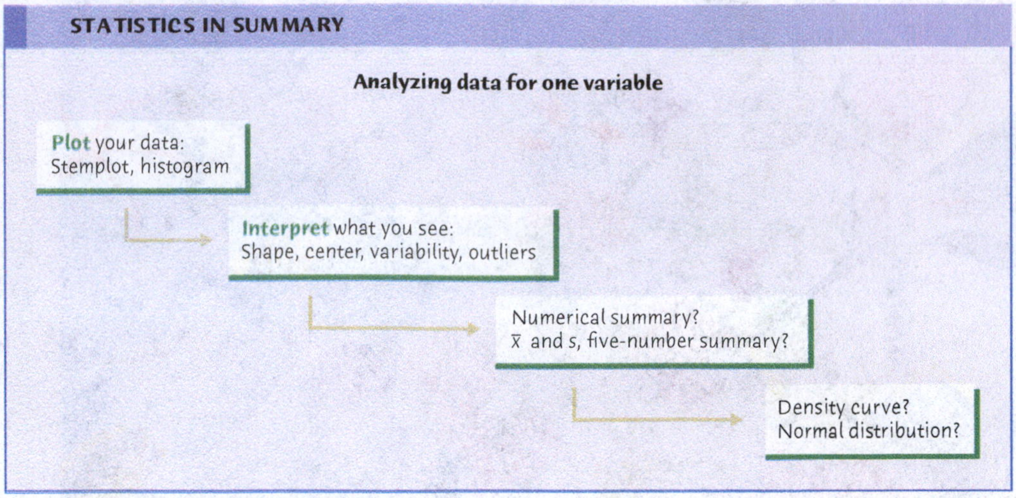

Chapters 4 and 5 applied the same ideas to relationships between two quantitative variables. The second STATISTICS IN SUMMARY figure retraces the big ideas, with details that fit the new setting. Always begin by making graphs of your data. In the case of a scatterplot, we have learned a numerical summary only for data that show a roughly linear pattern on the scatterplot. The summary is then the means and standard deviations of the two variables and their correlation. A regression line drawn on the plot gives a compact description of the overall pattern that we can use for prediction. Once again, there are question marks at the last two stages to remind us that correlation and regression describe only straight-line relationships. Chapter 6 shows how to understand relationships between two categorical variables; comparing well-chosen percentages is the key.

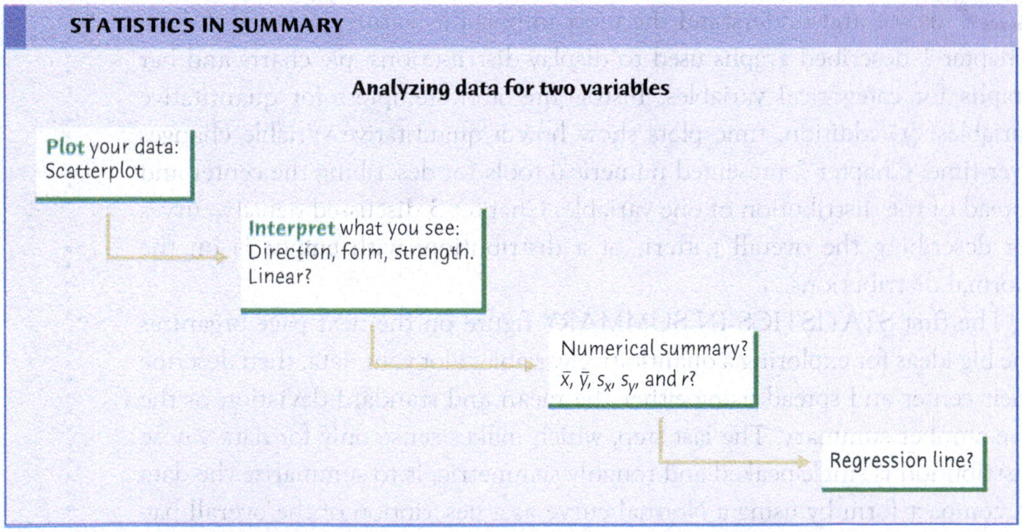

You can organize your work in any open-ended data analysis setting by following the four-step **State, Plan, Solve,** and **Conclude** process first introduced in Chapter 2. After we have mastered the extra background needed for statistical inference, this process will also guide practical work on inference later in the book.

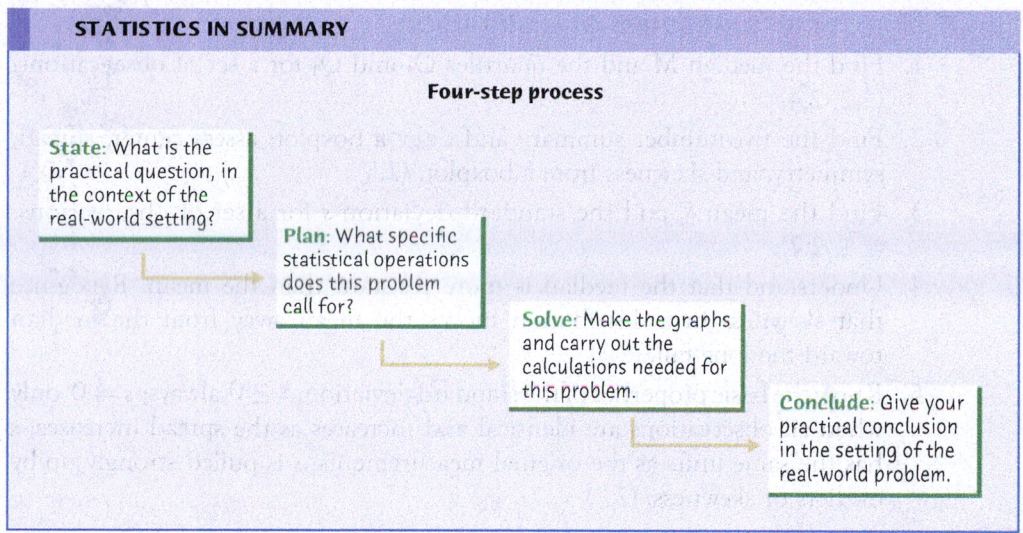

STATISTICS IN SUMMARY

Four-step process

State: What is the practical question, in the context of the real-world setting?

Plan: What specific statistical operations does this problem call for?

Solve: Make the graphs and carry out the calculations needed for this problem.

Conclude: Give your practical conclusion in the setting of the real-world problem.

Part I Skills Review

Here are the most important skills you should have acquired from reading Chapters 1 through 6. Following each skill in parentheses is the section of the text where the topic is introduced.

A. Data

1. Identify the individuals and variables in a set of data. (1.1)
2. Identify each variable as categorical or quantitative. Identify the units in which each quantitative variable is measured. (1.1)
3. Identify the explanatory and response variables in situations where one variable explains or influences another. (4.1)

B. Displaying Distributions

1. Recognize when pie charts can and cannot be used. (1.2)
2. Make a bar graph of the distribution of a categorical variable or, in general, to compare related quantities. (1.2)
3. Interpret pie charts and bar graphs. (1.2)
4. Make a histogram of the distribution of a quantitative variable. (1.3)
5. Make a stemplot of the distribution of a small set of observations. Round leaves or split stems as needed to make an effective stemplot. (1.5)
6. Make a time plot of a quantitative variable over time. Recognize patterns such as trends and cycles in time plots. (1.6)

C. Describing Distributions (Quantitative Variable)

1. Look for the overall pattern and for major deviations from the pattern. (1.4)
2. Assess from a histogram or stemplot whether the shape of a distribution is roughly symmetric, distinctly skewed, or neither. Assess whether the distribution has one or more major peaks. (1.4, 1.5)
3. Describe the overall pattern by giving numerical measures of center and spread in addition to a verbal description of shape. (2.1–2.8)
4. Decide which measures of center and spread are more appropriate: the mean and standard deviation (especially for symmetric distributions) or the five-number summary (especially for skewed distributions). (2.8)
5. Recognize outliers and give plausible explanations for them. (2.4–2.8)

D. Numerical Summaries of Distributions

1. Find the median M and the quartiles Q_1 and Q_3 for a set of observations. (2.2, 2.4)
2. Find the five-number summary and draw a boxplot; assess center, spread, symmetry, and skewness from a boxplot. (2.5)
3. Find the mean $\overline{x}$ and the standard deviation s for a set of observations. (2.1, 2.7)
4. Understand that the median is more resistant than the mean. Recognize that skewness in a distribution moves the mean away from the median toward the long tail. (2.3)
5. Know the basic properties of the standard deviation: $s \geq 0$ always; $s = 0$ only when all observations are identical and increases as the spread increases; s has the same units as the original measurements; s is pulled strongly up by outliers or skewness. (2.7)

E. Density Curves and Normal Distributions

1. Know that areas under a density curve represent proportions of all observations and that the total area under a density curve is 1. (3.1)
2. Approximately locate the median (equal-areas point) and the mean (balance point) on a density curve. (3.2)
3. Know that the mean and median both lie at the center of a symmetric density curve and that the mean moves farther toward the long tail of a skewed curve. (3.2)
4. Recognize the shape of Normal curves and estimate by eye both the mean and standard deviation from such a curve. (3.3)
5. Use the 68–95–99.7 rule and symmetry to state what percentage of the observations from a Normal distribution fall between two points when both points lie at the mean or 1, 2, or 3 standard deviations on either side of the mean. (3.4)
6. Find the standardized value (z-score) of an observation. Interpret z-scores and understand that any Normal distribution becomes the standard Normal $N(0, 1)$ distribution when standardized. (3.5)
7. Given that a variable has a Normal distribution with a stated mean μ and standard deviation σ, calculate the proportion of values above a stated number, below a stated number, or between two stated numbers. (3.6)
8. Given that a variable has a Normal distribution with a stated mean μ and standard deviation σ, calculate the point having a stated proportion of all values above it or below it. (3.6)

F. Scatterplots and Correlation

1. Make a scatterplot to display the relationship between two quantitative variables measured on the same subjects. Place the explanatory variable (if any) on the horizontal scale of the plot. (4.2)
2. Add a categorical variable to a scatterplot by using a different plotting symbol or color. (4.4)
3. Describe the direction, form, and strength of the overall pattern of a scatterplot. In particular, recognize positive or negative association and linear (straight-line) patterns. Recognize outliers in a scatterplot. (4.3)
4. Judge whether it is appropriate to use correlation to describe the relationship between two quantitative variables. Find the correlation r. (4.5)

5. Know the basic properties of correlation: r measures the direction and strength of only straight-line relationships. (4.6)
   - r is always a number between -1 and 1.
   - $r > 0$ for positive associations and $r < 0$ for negative associations.
   - $r = \pm 1$ only for perfect straight-line relationships.
   - r moves away from 0 toward ± 1 as the straight-line relationship gets stronger.

G. Regression Lines

1. Understand that regression requires an explanatory variable and a response variable. Correctly identifying which variable is the explanatory variable and which is the response variable is important. Switching them will result in different regression lines. Use a calculator or software to find the least-squares regression line of a response variable y on an explanatory variable x from data. (5.1)
2. Explain what the slope b and the intercept a mean in the equation $\hat{y} = a + bx$ of a regression line. (5.2)
3. Draw a graph of a regression line when you are given its equation. (5.3)
4. Use a regression line to predict y for a given x. Recognize extrapolation and be aware of its dangers. (5.1, 5.7)
5. Find the slope and intercept of the least-squares regression line from the means and standard deviations of x and y and their correlation. (5.2)
6. Use r^2, the square of the correlation, to describe how much of the variation in one variable can be accounted for by a straight-line relationship with another variable. (5.4)
7. Recognize outliers and potentially influential observations from a scatterplot with the regression line drawn on it. (5.6)
8. Calculate the residuals and plot them against the explanatory variable x. Recognize that a residual plot magnifies the pattern of the scatterplot of y versus x and helps us assess how well a regression line fits the data. (5.5)

H. Cautions about Correlation and Regression

1. Understand that both r and the least-squares regression line can be strongly influenced by a few extreme observations. (5.6)
2. Recognize possible lurking variables that may explain the observed association between two variables x and y. (5.7)
3. Understand that even a strong correlation does not mean that there is a cause-and-effect relationship between x and y. (5.8)
4. Give plausible explanations for an observed association between two variables: direct cause and effect, the influence of lurking variables, or both. (5.8)

I. Categorical Data

1. From a two-way table of counts, find the marginal distributions of both variables by obtaining the row sums and column sums. (6.1)
2. Express any distribution in percentages by dividing the category counts by their total. (6.2)
3. Describe the relationship between two categorical variables by computing and comparing percentages. Often this involves comparing the conditional distributions of one variable for the different categories of the other variable. (6.2)
4. Recognize Simpson's paradox and be able to explain it. (6.3)

STATISTICS IN YOUR WORLD

Driving in Canada

Canada is a civilized and restrained nation, at least in the eyes of Americans. A survey sponsored by the Canada Safety Council suggests that driving in Canada may be more adventurous than expected. Of the Canadian drivers surveyed, 88% admitted to aggressive driving in the past year, and 76% said that sleep-deprived drivers are common on Canadian roads. What really alarms us is the name of the survey: the Nerves of Steel Aggressive Driving Study.

TEST YOURSELF

The questions below include multiple-choice questions, calculations, and short-answer questions. They will help you review the basic ideas and skills presented in Chapters 1 through 6.

7.1 As part of a database on new births at a hospital, some variables recorded are the age of the mother, marital status of the mother (single, married, divorced, other), weight of the baby, and sex of the baby. Of these variables

(a) age, marital status, and weight are quantitative variables.

(b) age and weight are categorical variables.

(c) sex and marital status are categorical variables.

(d) sex, marital status, and age are categorical variables.

7.2 You are interested in obtaining information about the performance of students in your statistics class and seeing how this performance is affected by several factors, such as age. To do this, you are going to give a questionnaire to all students in the class. Give two questions for which the response is categorical and two questions for which the response is quantitative. For the categorical variables give the possible values and for the quantitative variables give the unit of measurement.

Nigel Cattlin/Science Source

Weeds Among the Corn. *Velvetleaf is a particularly annoying weed in corn fields. It produces lots of seeds, and the seeds wait in the soil for years until conditions are right. How many seeds do velvetleaf plants produce? Figure 7.1 is a histogram of the number of seeds produced from 28 velvetleaf plants that came up in a corn field when no herbicide was used.[1] Use this histogram to answer Questions 7.3 through 7.5.*

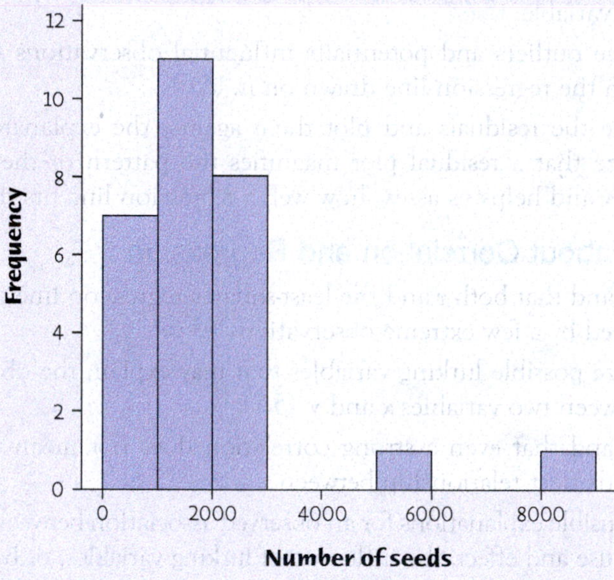

FIGURE 7.1
Histogram of the number of seeds produced by velvetleaf plants when no herbicide was used, for Questions 7.3 through 7.5.

7.3 The histogram

(a) is skewed right.

(b) has outliers.

(c) is asymmetric.

(d) is all of the above.

7.4 The median number of seeds produced is

(a) under 1000.

(b) between 1000 and 2000.

(c) between 2000 and 3000.

(d) over 3000.

7.5 The *percentage* of plants that produced fewer than 2000 seeds is

(a) about 55%.

(b) about 65%.

(c) about 75%.

(d) about 85%.

7.6 A reporter wishes to portray women professional soccer players as underpaid. Which measure of center should he report as the average salary of women professional soccer players?

(a) The mean

(b) The median

(c) Either the mean or the median. It doesn't matter since they will be equal.

(d) Neither the mean nor the median. Both will be much lower than the actual average salary.

Bachelor's Degrees. *The Organisation for Economic Cooperation and Development (OECD) began in 1961 to stimulate economic progress and world trade. It originally consisted of European countries, the United States, and Canada but has now grown to include 37 countries spanning the globe. Here is a stemplot of the percentage of the population between 25 and 64 years of age with a bachelor's degree or equivalent. The stems are 10s, the leaves are 1s, and the stems have been split.*[2] *Use the stemplot to answer Questions 7.7 through 7.9.* BACHELORS

0	344
0	6677
1	00333
1	555677889
2	222333334
2	66779
3	11

7.7 The shape of the distribution is

(a) slightly skewed to the right.

(b) almost exactly symmetric.

(c) slighlty skewed to the left.

7.8 The median percentage of those with a post secondary education in these countries is

(a) 15%.

(b) 17%.

(c) 17.5%.

(d) 18%.

7.9 What is the third quartile for these data?

(a) 3%

(b) 10%

(c) 23%

(d) 31%

7.10 A report gives the mean and median credit card debt per American household as of October 2019. The two values that are reported are \$2300 and \$5700. Which of these is the mean? Explain how you know this.

7.11 A biology project examines the effect on growth of the type of music that a plant is exposed to. At the end of the experiment, you measure the height, in centimeters, and weight, in grams, of the plants. What units of measurement do each of the following have?

(a) The mean weight of the plants

(b) The first quartile of the heights of the plants

(c) The standard deviation of the heights of the plants

(d) The variance of the weights of the plants

El Niño and the Monsoon. *The earth is interconnected. For example, it appears that El Niño, the periodic warming of the Pacific Ocean west of South America, affects the monsoon rains that are essential for agriculture in India. Figure 7.2 is a boxplot of the monsoon rains (in millimeters) for the 23 strong El Niño years between 1871 and 2004. Use the figure to answer Questions 7.12 and 7.13.*[3]

MONSOON

7.12 What is the interquartile range for the amount of rainfall in the strong El Niño years?

(a) 90 millimeters (b) 130 millimeters

(c) 200 millimeters (d) 784 millimeters

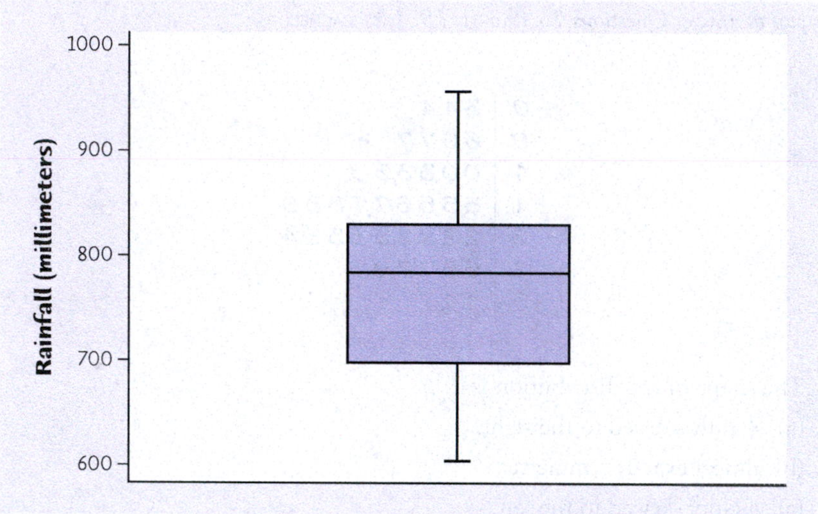

FIGURE 7.2
Boxplot for the monsoon rains (in millimeters) in the strong El Niño years, for Questions 7.12 and 7.13.

7.13 The average monsoon rainfall for all years from 1871 to 2004 is about 850 millimeters. What effect does El Niño appear to have on monsoon rains?

(a) Strong El Niño years tend to have higher monsoon rainfalls than other years.

(b) Strong El Niño years tend to have the same monsoon rainfalls as other years.

(c) Strong El Niño years tend to have lower monsoon rainfalls than other years.

(d) None of the above.

Do You Listen to Adult Contemporary Radio? *The rating service Arbitron places U.S. radio stations into more than 50 categories that describe the kinds of programs they broadcast. Which formats attract the largest audiences? The bar graph in Figure 7.3 gives the percentages of the listening audience (ages 12 and over) at a given time for the most popular formats.*[4] *Use this bar graph to help answer Questions 7.14 and 7.15.*

7.14 Approximately what percentage of the audience listens to Country?

(a) 3% (b) 8%

(c) 13% (d) 20%

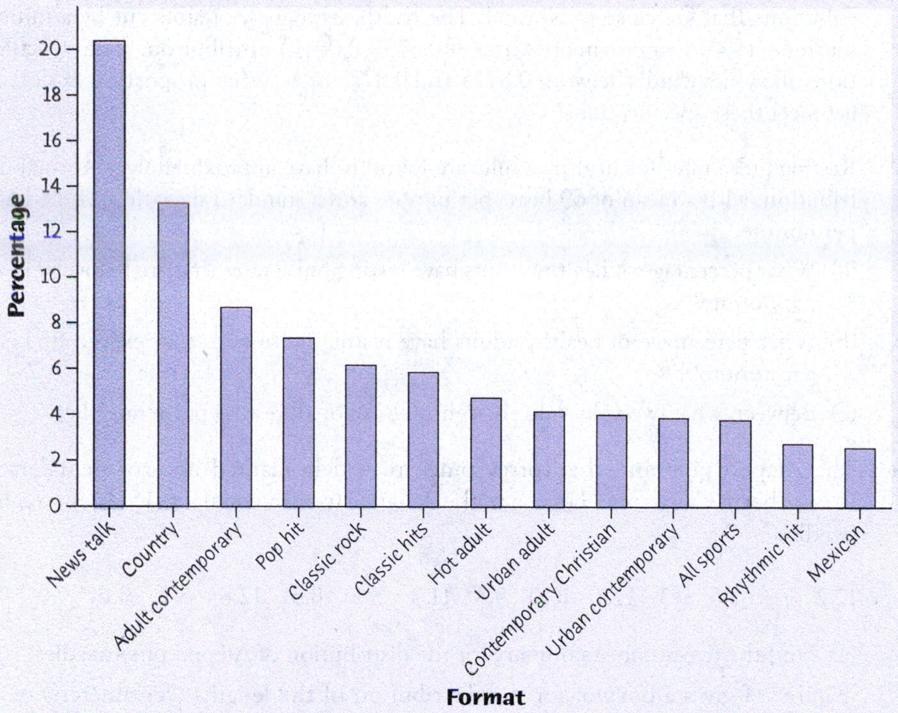

FIGURE 7.3 Bar graph of the distribution of audience share for the most popular radio formats in 2019, for Questions 7.14 and 7.15.

7.15 Approximately what percentage of the audience listens to formats other than those listed in the bar graph?

(a) 12% (b) 25%

(c) 75% (d) 88%

7.16 Reports on a student's ACT, SAT, or MCAT usually give the percentile as well as the actual score. The percentile is the cumulative proportion stated as a percentage: the percentage of all scores that were lower than this one. In 2019, the scores on the Mathematics portion of the SAT were close to Normal, with mean 528 and standard deviation 117.[5]

(a) Find the 85th percentile for the scores on the Mathematics portion of the SAT.

(b) Joseph scored 451. What was his percentile?

(c) Find the first quartile for the scores on the Mathematics portion of the SAT.

7.17 The heights of male five-year-olds have a Normal distribution, with a mean of 44.8 inches and a standard deviation of 2.1 inches.[6]

(a) What percentage of male five-year-olds have heights between 40 and 50 inches?

(b) What range of heights covers the central 95% of this distribution?

(c) You are informed by your doctor that your five-year-old boy's height is at the 70th percentile of heights. How tall is your child?

7.18 The length of human pregnancies from conception to birth varies according to a distribution that is approximately Normal, with mean 266 days and standard deviation 16 days. Use the 68–95–99.7 rule to answer the following questions.

(a) What range of pregnancy lengths covers almost all (99.7%) of this distribution?

(b) What percentage of pregnancies last longer than 282 days?

Emilija Manevska/Getty Images

STATISTICS IN YOUR WORLD

Beer in South Dakota

Take a break from doing exercises to apply your math to beer cans in South Dakota. A newspaper there reported that, every year, an average of 650 beer cans per mile are tossed onto the state's highways. South Dakota has about 83,000 miles of roads. How many beer cans is that in all? The U.S. Census Bureau says that there are about 882,235 people in South Dakota. Assuming the cans are tossed by South Dakotans, how many beer cans does each man, woman, and child in the state toss on the road each year? That's pretty impressive. Maybe the paper got its numbers wrong.

7.19 Automated manufacturing operations are quite precise but still vary, often with distributions that are close to Normal. The width, in inches, of slots cut by a milling machine follows approximately the $N(0.8750, 0.0012)$ distribution. The specifications allow slot widths between 0.8725 and 0.8775 inch. What proportion of slots do *not* meet these specifications?

7.20 Resting pulse rates for healthy adults are found to have approximately a Normal distribution, with a mean of 69 beats per minute and a standard deviation of 8.5 beats per minute.

(a) What percentage of healthy adults have resting pulse rates that are below 50 beats per minute?

(b) What percentage of healthy adults have resting pulse rates that exceed 85 beats per minute?

(c) Between what two values do the central 80% of all resting pulse rates lie?

7.21 The Aleppo pine and the Torrey pine are widely planted as ornamental trees in southern California. Here are the lengths (centimeters) of 15 Aleppo pine needles:[7]

10.2 7.2 7.6 9.3 12.1 10.9 9.4 11.3 8.5 8.5 12.8 8.7 9.0 9.0 9.4

(a) Find the five-number summary for the distribution of Aleppo pine needles.

Figure 7.4 gives a boxplot for the distribution of the lengths (centimeters) of 18 Torrey pine needles. Use this information to help answer the remainder of this question.

(b) The median of the distribution of Torrey pine needles is closest to which of the following values?

24 25 27 30

(c) Twenty-five percent of the Torrey pine needles exceed what value?

(d) Given only the length of a needle, do you think you could say which pine species it comes from? Explain briefly.

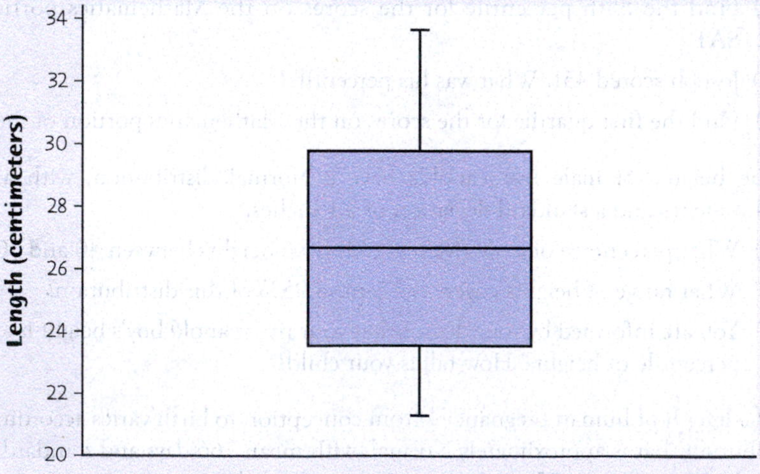

FIGURE 7.4

Boxplot for the distribution of the lengths (centimeters) of 18 Torrey pine needles, for Question 7.21.

NHL Salaries. *One can find online*[8] *the amount each NHL team spent on players (in millions of dollars) for the 2018–2019 NHL season and the total points each team earned by the end of the season. Questions 7.22 through 7.25 are based on the NHL data set.* NHLSP

Team	Spending	Points	Team	Spending	Points	Team	Spending	Points
Anaheim Ducks	66.3	80	Edmonton Oilers	73.9	79	Pittsburgh Penguins	72.0	100
Arizona Coyotes	51.2	86	Florida Panthers	64.2	86	San Jose Sharks	84.0	101
Boston Bruins	64.9	107	Los Angeles Kings	71.7	71	St. Louis Blues	83.2	99
Buffalo Sabres	64.5	76	Minnesota Wild	55.0	83	Tampa Bay Lightning	79.1	128
Calgary Flames	75.3	107	Montreal Canadiens	64.8	96	Toronto Maple Leafs	69.0	100
Carolina Hurricanes	55.0	99	Nashville Predators	76.5	100	Vancouver Canucks	60.0	81
Chicago Blackhawks	74.2	84	New Jersey Devils	44.9	72	Vegas Golden Knights	61.7	93
Colorado Avalanche	64.5	90	New York Islanders	71.6	103	Washington Capitals	90.3	104
Columbus Blue Jackets	69.4	98	New York Rangers	58.5	78	Winnipeg Jets	74.9	99
Dallas Stars	71.2	93	Ottawa Senators	38.3	64			
Detroit Red Wings	56.4	74	Philadelphia Flyers	59.3	82			

7.22 Figure 7.5 is a scatterplot of points earned against spending. How would you describe the overall pattern?

(a) Sharply curved

(b) Two distinct clusters that are widely separated

(c) A negative association

(d) A positive association

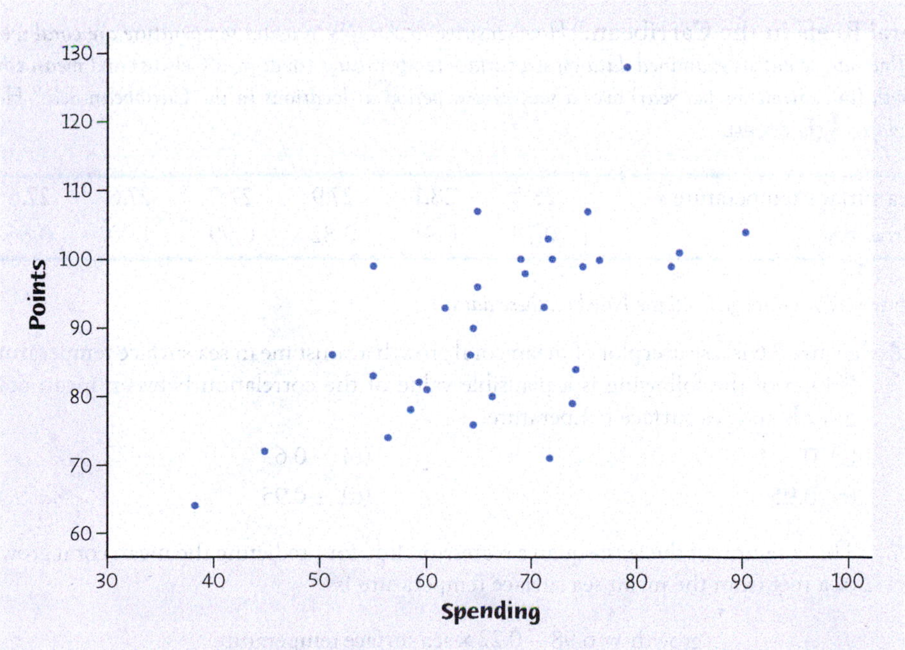

FIGURE 7.5 Scatterplot of the points earned in the 2018–2019 NHL regular season against spending on player salaries, for Question 7.22.

7.23 The equation of the least-squares regression line for predicting points earned from spending is

$$\text{points} = 39.64 + 0.77 \times \text{spending}$$

What does this tell us about points gained for each additional million dollars spent?

(a) A team gains about 0.77 point per million dollars spent.

(b) A team gains about 0.3964 point per million dollars spent.

(c) A team gains about 39.64 points per million dollars spent.

(d) A team gains about 40.41 points per million dollars spent.

7.24 The equation of the least-squares regression line for predicting points earned from spending is

$$\text{points} = 39.64 + 0.77 \times \text{spending}$$

Use the regression equation to predict the points earned if a team spent $60 million.

(a) 46.2
(b) 81.0
(c) 85.8
(d) 99.6

7.25 The equation of the least-squares regression line for predicting points earned from spending is

$$\text{points} = 39.64 + 0.77 \times \text{spending}$$

We use the regression equation to predict the points a team would earn if they spent no money. We conclude that

(a) the team will earn 39.64 points.

(b) the prediction is not sensible because the prediction is far outside the range of values of the response variable.

(c) the prediction is not sensible because no money is far outside the range of values of the explanatory variable.

(d) the prediction is not sensible because of the outlier present in the data.

Coral Reefs in the Carribean. *How sensitive to changes in water temperature are coral reefs? To find out, scientists examined data on sea surface temperatures (in degrees Celsius) and mean coral growth (in centimeters per year) over a several-year period at locations in the Caribbean Sea.*[9] *Here are data:* CCORAL

Sea surface temperature x	28.2	28.1	27.9	27.6	27.6	27.6
Growth y	0.78	0.91	0.82	0.99	1.00	0.86

Questions 7.26 through 7.28 are based on these data.

7.26 Figure 7.6 is a scatterplot of mean coral growth against mean sea surface temperature. Which of the following is a plausible value of the correlation between mean coral growth and sea surface temperature?

(a) 0
(b) −0.6
(c) 0.95
(d) −0.95

7.27 The equation of the least-squares regression line for predicting the mean coral growth of a reef from the mean sea surface temperature is

$$\text{growth} = 6.98 - 0.22 \times \text{sea surface temperature}$$

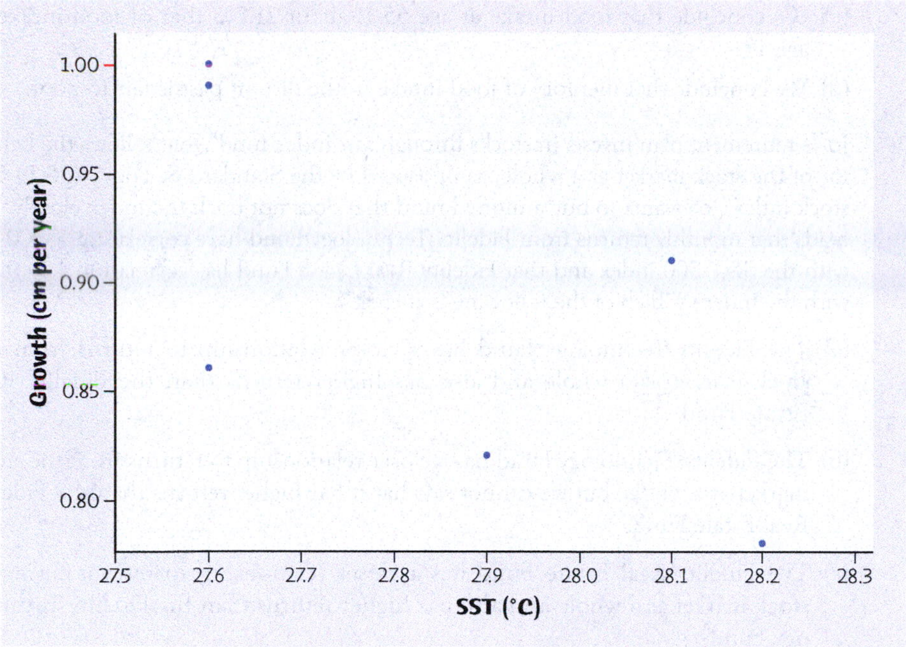

FIGURE 7.6
Scatterplot of the mean coral growth against mean sea surface temperature, for Question 7.26.

What does the slope −0.22 tell us?

(a) The mean coral growth of reefs in the study is decreasing 0.22 centimeter per year.

(b) The predicted mean coral growth of reefs in the study is 0.22 centimeter per degree of mean sea surface temperature.

(c) The predicted mean coral growth of a reef in the study when the mean sea surface temperature is 0 degrees is 6.98 centimeters.

(d) For each degree increase in mean sea surface temperature, the predicted mean coral growth of a reef decreases by 0.22 centimeter.

7.28 The equation of the least-squares regression line for predicting the mean coral growth of a reef from the mean sea surface temperature is

$$\text{growth} = 6.98 - 0.22 \times \text{sea surface temperature}$$

Use this to predict the mean coral growth of a reef in the Carribean Sea with a mean sea surface temperature of 28.

(a) −6.16

(b) −0.82

(c) 0.82

(d) 6.16

7.29 How well do people remember their past diet? Data are available for 91 people who were asked about their diet when they were 18 years old. Researchers asked them at about age 55 to describe their eating habits at age 18. For each subject, the researchers calculated the correlation between actual intakes of many foods at age 18 and the intakes the subjects now remember. The median of the 91 correlations was $r = 0.217$.[10] Which of the following conclusions is consistent with this correlation?

(a) We conclude that subjects remember approximately 21.7% of their food intakes at age 18.

(b) We conclude that subjects remember approximately $r^2 = 0.217^2 = 0.047$ of their food intakes at age 18.

(c) We conclude that food intake at age 55 is about 21.7% that of food intake at age 18.

(d) We conclude that memory of food intake in the distant past is fair to poor.

7.30 Joe's retirement plan invests in stocks through an "index fund" that follows the behavior of the stock market as a whole, as measured by the Standard & Poor's (S&P) 500 stock index. Joe wants to buy a mutual fund that does not track the index closely. He reads that monthly returns from Fidelity Technology Fund have correlation $r = 0.77$ with the S&P 500 index and that Fidelity Real Estate Fund has correlation $r = 0.37$ with the index. Which of the following is correct?

(a) The Fidelity Technology Fund has a closer relationship to returns from the stock market as a whole and also has higher returns than the Fidelity Real Estate Fund.

(b) The Fidelity Technology Fund has a closer relationship to returns from the stock market as a whole, but we cannot say that it has higher returns than the Fidelity Real Estate Fund.

(c) The Fidelity Real Estate Fund has a closer relationship to returns from the stock market as a whole and also has higher returns than the Fidelity Technology Fund.

(d) The Fidelity Real Estate Fund has a closer relationship to returns from the stock market as a whole, but we cannot say that it has higher returns than the Fidelity Technology Fund.

JMx Images/Shutterstock

Monkey Calls. *The usual way to study the brain's response to sounds is to have subjects listen to "pure tones." The response to recognizable sounds may differ. To compare responses, researchers anesthetized macaque monkeys. They fed pure tones and also monkey calls directly to their brains by inserting electrodes. Response to the stimulus was measured by the firing rate (electrical spikes per second) of neurons in various areas of the brain. Table 7.1 contains the responses for 37 neurons.*[11] *Figure 7.7 is a scatterplot of monkey call response against pure-tone response (explanatory variable). Questions 7.31 and 7.32 refer to these data and the scatterplot.* MONKEY

TABLE 7.1 Neuron response (electrical firing rate per second) to pure tones and monkey calls

Neuron	Tone	Call	Neuron	Tone	Call	Neuron	Tone	Call
1	474	500	14	145	42	26	71	134
2	256	138	15	141	241	27	68	65
3	241	485	16	129	194	28	59	182
4	226	338	17	113	123	29	59	97
5	185	194	18	112	182	30	57	318
6	174	159	19	102	141	31	56	201
7	176	341	20	100	118	32	47	279
8	168	85	21	74	62	33	46	62
9	161	303	22	72	112	34	41	84
10	150	208	23	20	193	35	26	203
11	19	66	24	21	129	36	28	192
12	20	54	25	26	135	37	31	70
13	35	103						

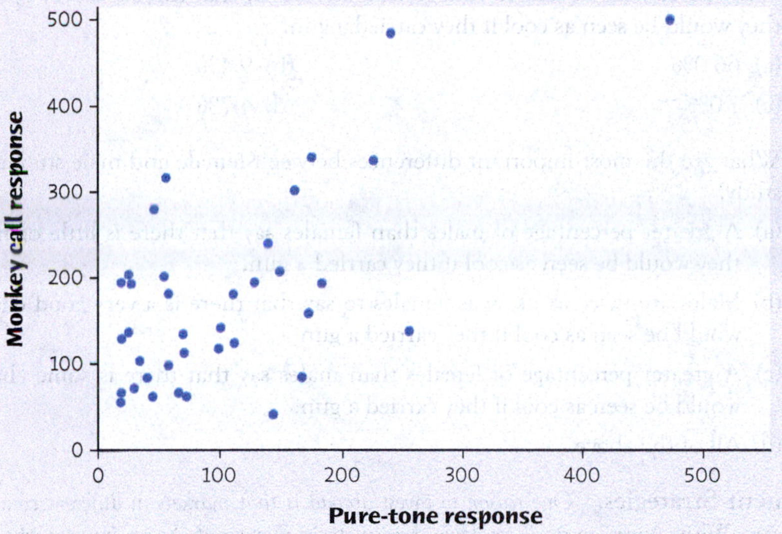

FIGURE 7.7
Scatterplot of monkey tone response against pure-tone response, for Questions 7.31 and 7.32.

7.31 We might expect some neurons to have strong responses to any stimulus and others to have consistently weak responses. There would then be a strong relationship between tone response and call response. From the scatterplot of monkey call response against pure-tone response in Figure 7.7, what would you estimate the correlation r to be?

(a) −0.6 (b) −0.1
(c) 0.1 (d) 0.6

7.32 Which of the following statements about the scatterplot in Figure 7.7 is correct?

(a) There is moderate evidence that pure-tone response causes monkey call response.
(b) There is moderate evidence that monkey call response causes pure-tone response.
(c) There are one or two outliers, and at least one of them may also be influential.
(d) None of the above.

Is Carrying a Gun Cool? *The Indiana Youth Risk Behavior Survey asked Indiana High School students, "What are the chances you would be seen as cool if you carried a gun?"*[12] *Here are the counts for 12th-grade males and females. Use these counts to answer Questions 7.33 through 7.35.* GUN

Chance	Female	Male
Very good	276	540
Pretty good	224	298
Some	447	580
Little	729	862
Very little or none	4206	3474
Total	5882	5754

7.33 What percentage of all 12th-grade students say that there is a very good chance they would be seen as cool if they carried a gun?

(a) 66.0% (b) 9.4%
(c) 7.0% (d) 4.7%

7.34 What percentage of female 12th-grade students say that there is a very good chance they would be seen as cool if they carried a gun?

(a) 66.0%
(b) 9.4%
(c) 7.0%
(d) 4.7%

7.35 What are the most important differences between female and male students in this study?

(a) A greater percentage of males than females say that there is little or no chance they would be seen as cool if they carried a gun.
(b) Males are twice as likely as females to say that there is a very good chance they would be seen as cool if they carried a gun.
(c) A greater percentage of females than males say that there is some chance they would be seen as cool if they carried a gun.
(d) All of the above.

Investment Strategies. *One reason to invest abroad is that markets in different countries don't move in step. When American stocks go down, foreign stocks may go up. So an investor who holds both bears less risk. That's the theory. But then we read in a magazine article that the correlation between changes in American and European stock prices rose from 0.4 in the mid-1990s to 0.8 in 2000.*[13] *Questions 7.36 and 7.37 refer to this article.*

7.36 Explain to an investor who knows no statistics why the fact stated in this article reduces the protection provided by buying European stocks.

7.37 The same article that claims that the correlation between changes in stock prices in Europe and the United States is 0.8 goes on to say: "Crudely, that means that movements on Wall Street can explain 80% of price movements in Europe."

(a) Is this true?
(b) What is the correct percentage of price movements explained if $r = 0.8$?

7.38 Researchers wished to determine whether individual differences in introspective ability are reflected in the anatomy of brain regions responsible for this function. They measured introspective ability (using a score on a test of introspective ability, with larger values indicating greater introspective ability) and gray-matter volume in milliliters (the Brodmann area) in the anterior prefrontal cortex of the brains of 29 subjects. Here are the data: INTROSP

Volume	0.55	0.58	0.59	0.59	0.59	0.61	0.62	0.63	0.63	0.63
Introspective ability	59	62	43	63	83	61	55	57	57	67
Volume	0.63	0.64	0.65	0.65	0.65	0.65	0.65	0.66	0.66	0.67
Introspective ability	72	62	58	62	65	70	75	60	63	71
Volume	0.67	0.67	0.68	0.69	0.70	0.70	0.71	0.72	0.75	
Introspective ability	71	80	68	72	66	73	61	80	75	

The researchers wished to determine the equation of the least-squares regression line for predicting introspective ability (y) from gray-matter volume (x). To do this, they calculated the following summary statistics:

$$\bar{x} = 0.649,\ s_x = 0.045$$
$$\bar{y} = 65.897,\ s_y = 8.69$$
$$r = 0.448$$

(a) Use this information to calculate the equation of the least-squares regression line. State clearly what the slope of the least-squares regression line means in this setting.

(b) Based on the least-squares regression line, what would you predict introspective ability to be for someone with gray-matter volume 0.60?

(c) Based on the least-squares regression line, what would you predict introspective ability to be for someone with gray-matter volume 0.99? How reliable do you think this prediction is? Explain your answer.

7.39 Animals and people that take in more energy than they expend will add body fat. Here are data on 12 rhesus monkeys: six lean monkeys (4% to 9% body fat) and six obese monkeys (13% to 44% body fat). The following data report the energy expended in 24 hours (kilojoules per minute) and the lean body mass (kilograms, leaving out fat) for each monkey:[14] THINFAT

Lean		Obese	
Mass	Energy	Mass	Energy
6.6	1.17	7.9	0.93
7.8	1.02	9.4	1.39
8.9	1.46	10.7	1.19
9.8	1.68	12.2	1.49
9.7	1.06	12.1	1.29
9.3	1.16	10.8	1.31

(a) Compute the mean lean body mass of the lean monkeys.

(b) Compute the mean lean body mass of the obese monkeys.

(c) The goal of the study is to compare the energy expended in 24 hours by the lean monkeys with that of the obese monkeys. However, animals with higher lean mass usually expend more energy. Based on your calculations in parts (a) and (b), would it make sense to simply compute the mean energy expended by lean and obese monkeys and compare the means? Explain.

(d) To investigate how energy expended is related to body mass, make a scatterplot of energy versus mass, using different plot symbols for lean and obese monkeys.

(e) What do the trends in your scatterplot suggest about the monkeys?

7.40 The number of adult Americans who smoke continues to drop. Here are estimates of the percentages of adults (ages 18 and over) who were smokers in the years between 1965 and 2017:[15] SMOKERS

Year x	1965	1970	1974	1978	1980	1983	1985	1987	1990	1993	1995
Smokers y	41.9	37.4	37.1	34.1	33.2	32.1	30.1	28.8	25.5	25.0	24.7
Year x	1997	1999	2001	2002	2004	2006	2008	2010	2012	2014	2017
Smokers y	24.7	23.5	22.8	22.5	20.9	20.8	20.6	19.3	18.1	16.8	14.0

(a) Make a scatterplot of these data.

(b) Describe the direction, form, and strength of the relationship between percentage of smokers and year. Are there any outliers?

(c) Here are the means and standard deviations for both variables and the correlation between percentage of smokers and year:

$$\bar{x} = 1994.1,\ s_x = 14.9$$
$$\bar{y} = 26.1,\ s_y = 7.4$$
$$r = -0.99$$

Use this information to find the least-squares regression line for predicting percentage of smokers from year and add the line to your plot.

(d) According to your regression line, how much did smoking decline per year during this period, on the average?

(e) What percentage of the observed variation in percentage of adults who smoke can be explained by linear change over time?

(f) Use your regression line to predict the percentage of adults who will smoke in 2030.

(g) Use your regression line to predict the percentage of adults who will smoke in 2075. Why is your result impossible? Why was it foolish to use the regression line for this prediction?

7.41 People who get angry easily tend to have more heart disease. That's the conclusion of a study that followed a random sample of 12,986 people from three locations for about four years. All subjects were free of heart disease at the beginning of the study. The subjects took the Spielberger Trait Anger Scale test, which measures how prone a person is to sudden anger. Here are data for the 8474 people in the sample who had normal blood pressure. CHD stands for "coronary heart disease." This includes people who had heart attacks and those who needed medical treatment for heart disease.[16] ANGER

	Low Anger	Moderate Anger	High Anger	Total
CHD	53	110	27	190
No CHD	3057	4621	606	8284
Total	3110	4731	633	8474

(a) What percentage of all 8474 people with normal blood pressure had CHD?

(b) What percentage of all 8474 people were classified as having high anger?

(c) What percentage of those classified as having high anger had CHD?

(d) What percentage of those with no CHD were classified as having moderate anger?

(e) Do these data provide any evidence that as anger score increases, the percentage who suffer CHD increases? Explain.

SUPPLEMENTARY EXERCISES

Supplementary exercises apply the skills you have learned in ways that require more thought or more elaborate use of technology. Some of these exercises ask you to follow the ***Plan, Solve,*** *and* ***Conclude*** *steps of the four-step process introduced on page 62.*

7.42 **Ozone Hole.** The ozone hole is a region in the stratosphere over the Antarctic with exceptionally depleted ozone. The size of the hole is not constant over the year but is largest at the beginning of the Southern Hemisphere spring (August–October). The increase in the size of the ozone hole led to the Montreal Protocol in 1987, an international treaty designed to protect the ozone layer by phasing out the production of substances, such as chlorofluorocarbons (CFCs), believed to be responsible for ozone depletion. The following table gives the average ozone hole size for the period September 7 to October 13 for each of the years from 1979 to 2019 (with no data acquired in 1995).[17] To get a better feel for the magnitude of the numbers, the area of North America is approximately 24.5 million square kilometers (km^2). OZONE

Year	Area (millions of km²)	Year	Area (millions of km²)	Year	Area (millions of km²)
1979	0.1	1993	24.2	2007	22.0
1980	1.4	1994	23.6	2008	25.2
1981	0.6	1995	–	2009	22.0
1982	4.8	1996	22.7	2010	19.4
1983	7.9	1997	22.1	2011	24.7
1984	10.1	1998	25.9	2012	17.8
1985	14.2	1999	23.2	2013	21.0
1986	11.3	2000	24.8	2014	20.9
1987	19.3	2001	25.0	2015	25.6
1988	10.0	2002	12.0	2016	20.7
1989	18.7	2003	25.8	2017	17.4
1990	19.2	2004	19.5	2018	22.9
1991	18.8	2005	24.4	2019	9.3
1992	22.3	2006	26.6		

(a) Make a graph of the distribution of the size of the ozone hole. Describe the overall shape of the distribution and any outliers.

(b) Based on the shape of the distribution, do you expect the mean to be close to the median, clearly less than the median, or clearly greater than the median? Why? Find the mean and the median to check your answer.

7.43 **More on the Ozone Hole.** The data in Exercise 7.42 are a time series. The severity of the ozone hole will vary from year to year, depending on the meteorology of the atmosphere above Antarctica. Make a time plot that shows how the size of the ozone hole changed between 1979 and 2019. Does the time plot illustrate only year-to-year variation, or are other patterns apparent? Specifically, is there a trend over any period of years? What about cyclical fluctuation? Explain in words the change in the average size of the ozone hole over this 41-year period. It is a good idea to always make a time plot of time series data because a histogram cannot show changes over time. OZONE

4step

Falling Through the Ice. *The Nenana Ice Classic is an annual contest to guess the exact time in the spring thaw when a tripod erected on the frozen Tanana River near Nenana, Alaska, will fall through the ice. The 2019 jackpot prize was $311,652. The contest has been run since 1917. Table 7.2 gives simplified data that record only the date on which the tripod fell each year. The earliest date so far is April 14. To make the data easier to use, the table gives the date each year in days starting with April 14. That is, April 14 is 1, April 15 is 2, and so on. Exercises 7.44 through 7.46 concern these data.*[18]

7.44 **When Does the Ice Break Up?** We have 103 years of data on the date of ice breakup on the Tanana River. Describe the distribution of the breakup date with both a graph or graphs and appropriate numerical summaries. What is the median date (month and day) for ice breakup? TANANA

7.45 **Global Warming?** Because of the high stakes, the falling of the tripod has been carefully observed for many years. If the date the tripod falls has been getting earlier, that may be evidence for the effects of global warming. TANANA

(a) Make a time plot of the date the tripod falls against year.

(b) There is a great deal of year-to-year variation. Fitting a regression line to the data may help us see the trend. Fit the least-squares line and add it to your time plot. What do you conclude?

(c) There is much variation about the line. Give a numerical description of how much of the year-to-year variation in ice breakup time is accounted for by the time trend represented by the regression line. (This simple example is typical of more complex evidence for the effects of global warming: large year-to-year variation requires many years of data to see a trend.)

TABLE 7.2 Days from April 14 for the Tanana River tripod to fall

Year	Day	Year	Day	Year	Day	Year	Day	Year	Day	Year	Day
1917	17	1935	32	1953	16	1971	25	1989	18	2007	14
1918	28	1936	17	1954	23	1972	27	1990	11	2008	22
1919	20	1937	29	1955	26	1973	21	1991	18	2009	18
1920	28	1938	23	1956	18	1974	23	1992	31	2010	16
1921	28	1939	16	1957	22	1975	27	1993	10	2011	20
1922	29	1940	7	1958	16	1976	19	1994	16	2012	9
1923	26	1941	20	1959	25	1977	23	1995	13	2013	37
1924	28	1942	17	1960	19	1978	17	1996	22	2014	12
1925	22	1943	15	1961	22	1979	17	1997	17	2015	11
1926	13	1944	21	1962	29	1980	16	1998	7	2016	10
1927	29	1945	33	1963	22	1981	17	1999	16	2017	18
1928	23	1946	22	1964	37	1982	27	2000	18	2018	18
1929	22	1947	20	1965	24	1983	16	2001	25	2019	1
1930	25	1948	30	1966	25	1984	26	2002	24		
1931	27	1949	31	1967	21	1985	29	2003	16		
1932	18	1950	23	1968	25	1986	25	2004	11		
1933	25	1951	17	1969	15	1987	22	2005	15		
1934	17	1952	29	1970	21	1988	14	2006	19		

7.46 More on Global Warming. Side-by-side boxplots offer a different look at the data. Group the data into periods of roughly equal length: 1917–1942, 1943–1968, 1969–1994, and 1995–2019. Make boxplots to compare ice breakup dates in these four time periods. Write a brief description of what the plots show. TANANA

7.47 4step **Diplomatic Scofflaws.** Until Congress allowed some enforcement in 2002, the thousands of foreign diplomats in New York City could freely violate parking laws. Two economists looked at the number of unpaid parking tickets per diplomat over a five-year period ending when enforcement reduced the problem.[19] They concluded that large numbers of unpaid tickets indicated a "culture of corruption" in a country and lined up well with more elaborate measures of corruption. The data set for 145 countries is too large to print here, but look at the data file on the text website. The first 32 countries in the list (Australia to Trinidad and Tobago) are classified by the World Bank as "developed." The remaining countries (Albania to Zimbabwe) are "developing." The World Bank classification is based only on national income and does not take into account measures of social development. SCOFFLAW

Give a full description of the distribution of unpaid tickets for both groups of countries and identify any high outliers. Compare the two groups. Does national income alone do a good job of distinguishing countries whose diplomats do and do not obey parking laws?

7.48 4step **Cicadas as Fertilizer?** Every 17 years, swarms of cicadas emerge from the ground in the eastern United States, live for about six weeks, then die. (There are several "broods," so we experience cicada eruptions more often than every 17 years.) There are so many cicadas that their dead bodies can serve as fertilizer and increase plant growth. In an experiment, a researcher added 10 cicadas under some plants in a natural plot of American bellflowers in a forest, leaving other plants undisturbed. One of the response variables was the size of seeds produced by the plants. Here are data (seed mass in milligrams) for 39 cicada plants and 33 undisturbed (control) plants:[20] CICADA

Alastair Shay/Getty Images

Cicada Plants				Control Plants			
0.237	0.277	0.241	0.142	0.212	0.188	0.263	0.253
0.109	0.209	0.238	0.277	0.261	0.265	0.135	0.170
0.261	0.227	0.171	0.235	0.203	0.241	0.257	0.155
0.276	0.234	0.255	0.296	0.215	0.285	0.198	0.266
0.239	0.266	0.296	0.217	0.178	0.244	0.190	0.212
0.238	0.210	0.295	0.193	0.290	0.253	0.249	0.253
0.218	0.263	0.305	0.257	0.268	0.190	0.196	0.220
0.351	0.245	0.226	0.276	0.246	0.145	0.247	0.140
0.317	0.310	0.223	0.229	0.241			
0.192	0.201	0.211					

Describe and compare the two distributions. Do the data support the idea that dead cicadas can serve as fertilizer?

7.49 **A Big-Toe Problem.** *Hallux abducto valgus* (call it HAV) is a deformation of the big toe that is not common in youth and often requires surgery. Doctors used X-rays to measure the angle (in degrees) of deformity in 38 consecutive patients under the age of 21 who came to a medical center for surgery to correct HAV.[21] The angle is a measure of the seriousness of the deformity. The data appear in Table 7.3 as "HAV Angle." (The "MA Angle" data in this table are used in Exercise 7.51.) Describe the distribution of the angle of deformity among young patients needing surgery for this condition. BIGTOE

TABLE 7.3 **Angle of deformity (degrees) for two types of foot deformity**

HAV Angle	MA Angle	HAV Angle	MA Angle	HAV Angle	MA Angle
28	18	21	15	16	10
32	16	17	16	30	12
25	22	16	10	30	10
34	17	21	7	20	10
38	33	23	11	50	12
26	10	14	15	25	25
25	18	32	12	26	30
18	13	25	16	28	22
30	19	21	16	31	24
26	10	22	18	38	20
28	17	20	10	32	37
13	14	18	15	21	23
20	20	26	16		

7.50 **Prey Attract Predators.** Here is one way in which nature regulates the size of animal populations: high population density attracts predators, who remove a higher proportion of the population than when the density of the prey is low. One study looked at kelp perch and their common predator, the kelp bass. The researcher set up four similarly sized circular pens on sandy ocean bottom in southern California. He chose young perch at random from a large group and placed 10, 20, 40, and 60 perch in the four pens. Then he dropped the nets protecting the pens, allowing bass to swarm in, and counted the perch left after two hours. Here are data on the proportions of perch eaten in four repetitions of this setup:[22] PREY

Perch	Proportion Killed			
10	0.0	0.1	0.3	0.3
20	0.2	0.3	0.3	0.6
40	0.075	0.3	0.6	0.725
60	0.517	0.55	0.7	0.817

Do the data support the principle that "more prey attract more predators, who drive down the number of prey"?

7.51 **Predicting Foot Problems.** *Metatarsus adductus* (call it MA) is a turning in of the front part of the foot that is common in adolescents and usually corrects itself. Table 7.3 gives the severity of MA ("MA Angle"). Doctors speculate that the severity of MA can help predict the severity of HAV. Describe the relationship between MA and HAV. Do you think the data confirm the doctors' speculation? Why or why not? BIGTOE

7.52 **Change in the Serengeti.** Long-term records from the Serengeti National Park in Tanzania show interesting ecological relationships. When wildebeest are more abundant, they graze the grass more heavily, so there are fewer fires, and more trees grow. Lions feed more successfully when there are more trees, so the lion population increases. Here are data on one part of this cycle, wildebeest abundance (in thousands of animals) and the percentage of the grass area that burned in the same year:[23] SERENG

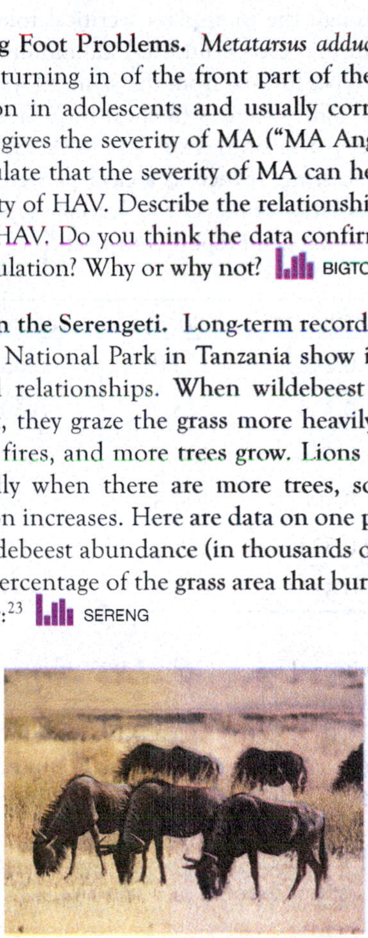

Gallo Images-Anthony Bannister/ Getty Images

Wildebeest (1000s)	Percentage Burned	Wildebeest (1000s)	Percentage Burned
396	56	622	60
476	50	600	56
698	25	902	45
1049	16	1440	21
1178	7	1147	32
1200	5	1173	31
1302	7	1178	24
360	88	1253	24
444	88	1249	53
524	75		

To what extent do these data support the claim that more wildebeest reduce the percentage of grasslands that burn? How rapidly does burned area decrease as the number of wildebeest increases? Include a graph and suitable calculations.

7.53 Casting Aluminum. 4step In casting metal parts, molten metal flows through a "gate" into a die that shapes the part. The gate velocity (the speed at which metal is forced through the gate) plays a critical role in die casting. A firm that casts cylindrical aluminum pistons examined 12 types formed from the same alloy. How does the piston wall thickness (inches) influence the gate velocity (feet per second) chosen by the skilled workers who do the casting? If there is a clear pattern, it can be used to direct new workers or to automate the process. Analyze these data and report your findings:[24] ALUM

Thickness	Velocity	Thickness	Velocity
0.248	123.8	0.628	326.2
0.359	223.9	0.697	302.4
0.366	180.9	0.697	145.2
0.400	104.8	0.752	263.1
0.524	228.6	0.806	302.4
0.552	223.8	0.821	302.4

7.54 Twitter Users. 4step Only 22% of American adults use Twitter. In late 2018, the Pew Research Center conducted a poll to learn about Twitter users, how they compare to American adults in general, and how the most frequent users of Twitter differ from other Twitter users. Here are the survey results comparing the sex of the top 10% of Twitter users with the bottom 90% of Twitter users:[25] TWITTER

Gender	Top 10%	Bottom 90%
Female	181	1206
Male	98	1306
Total	279	2512

Write a brief analysis of these results that focuses on the relationship between sex and Twitter use.

7.55 Influence: Hot Sector Funds? Investment advertisements always warn that "past performance does not guarantee future results." Here is an example that shows why you should pay attention to this warning. Stocks fell sharply in 2002, then rose sharply in 2003. The following table gives the percentage returns from 23 Fidelity Investments "sector funds" in these two years. Sector funds invest in narrow segments of the stock market. They often rise and fall faster than the market as a whole: MFUNDS

2002 Return	2003 Return	2002 Return	2003 Return	2002 Return	2003 Return
−17.1	23.9	−0.7	36.9	−37.8	59.4
−6.7	14.1	−5.6	27.5	−11.5	22.9
−21.1	41.8	−26.9	26.1	−0.7	36.9
−12.8	43.9	−42.0	62.7	64.3	32.1
−18.9	31.1	−47.8	68.1	−9.6	28.7
−7.7	32.3	−50.5	71.9	−11.7	29.5
−17.2	36.5	−49.5	57.0	−2.3	19.1
−11.4	30.6	−23.4	35.0		

(a) Make a scatterplot of 2003 return (response) against 2002 return (explanatory). The funds with the best performance in 2002 tend to have the worst performance in 2003. Fidelity Gold Fund, the only fund with a positive return in both years, is an extreme outlier.

(b) To demonstrate that correlation is not resistant, find r for all 23 funds and then find r for the 22 funds other than Gold. Explain from Gold's position in your plot why omitting this point makes r more negative.

(c) Find the equations of two least-squares lines for predicting 2003 return from 2002 return, one for all 23 funds and one omitting Fidelity Gold Fund. Add both lines to your scatterplot. Starting with the least-squares idea, explain why adding Fidelity Gold Fund to the other 22 funds moves the line in the direction that your graph shows.

7.56 Influence: Monkey Calls. Table 7.1 (page 188) contains data on the responses of 37 monkey neurons to pure tones and to monkey calls. Figure 7.7 (page 189) is a scatterplot of these data. MONKEY

(a) Find the least-squares line for predicting a neuron's call response from its pure tone response. Add the line to your scatterplot. Mark on your plot the point (call it A) with the largest residual (either positive or negative) and also the point (call it B) that is an outlier in the x direction.

(b) How influential are each of these points for the correlation r?

(c) How influential are each of these points for the regression line?

7.57 Influence: Bushmeat. Table 7.4 gives data on fish catches in a region of West Africa and the percentage change in the biomass (total weight) of 41 animals in nature reserves. It appears that years with smaller fish catches see greater declines in animals, probably because local people turn to "bushmeat" when other sources of protein are not available. The next year (1999) had a fish catch of 23.0 kilograms per person and animal biomass change of −22.9%. BUSHMEAT

(a) Make a scatterplot that shows how change in animal biomass depends on fish catch. Be sure to include the additional data point. Describe the overall pattern. The added point is a low outlier in the y direction.

(b) Find the correlation between fish catch and change in animal biomass both with and without the outlier. The outlier is influential for correlation. Explain from your plot why adding the outlier makes the correlation smaller.

(c) Find the least-squares line for predicting change in animal biomass from fish catch both with and without the additional data point for 1999. Add both lines to your scatterplot from part (a). The outlier is not influential for the least-squares line. Explain from your plot why this is true.

7.58 Python Eggs. How is the hatching of water python eggs influenced by the temperature of the snake's nest? Researchers placed 104 newly laid eggs in a hot environment, 56 in a neutral environment, and 27 in a cold environment. Hot duplicates the warmth provided by the mother python. Neutral and cold are cooler, as when the mother is absent. The results: 75 of the hot eggs hatched, along with 38 of the neutral eggs and 16 of the cold eggs.[26]

(a) Make a two-way table of "environment temperature" against "hatched or not."

(b) The researchers anticipated that eggs would hatch less well at cooler temperatures. Do the data support that anticipation?

TABLE 7.4 Fish supply and wildlife decline in West Africa

Year	Fish Supply (kilograms per person)	Biomass Change (percent)	Year	Fish Supply (kilograms per person)	Biomass Change (percent)
1971	34.7	2.9	1985	21.3	−5.5
1972	39.3	3.1	1986	24.3	−0.7
1973	32.4	−1.2	1987	27.4	−5.1
1974	31.8	−1.1	1988	24.5	−7.1
1975	32.8	−3.3	1989	25.2	−4.2
1976	38.4	3.7	1990	25.9	0.9
1977	33.2	1.9	1991	23.0	−6.1
1978	29.7	−0.3	1992	27.1	−4.1
1979	25.0	−5.9	1993	23.4	−4.8
1980	21.8	−7.9	1994	18.9	−11.3
1981	20.8	−5.5	1995	19.6	−9.3
1982	19.7	−7.2	1996	25.3	−10.7
1983	20.8	−4.1	1997	22.0	−1.8
1984	21.1	−8.6	1998	21.0	−7.4

7.59 Normal Is Only Approximate: IQ Test Scores. Here are the IQ test scores of 31 seventh-grade girls in a midwestern school district:[27] IQ

114	100	104	89	102	91	114	114	103	105	
108	130	120	132	111	128	118	119	86	72	
111	103	74	112	107	103	98	96	112	112	93

(a) We expect IQ scores to be approximately Normally distributed. Make a stemplot to check that there are no major departures from Normality.

(b) Proportions calculated from a Normal distribution are not always very accurate for small numbers of observations. Find the mean $\bar{x}$ and standard deviation s for these IQ scores. What proportions of the scores are within 1 standard deviation and within 2 standard deviations of the mean? What would these proportions be in an exactly Normal distribution?

Online Data for Additional Analyses

1. Data from the Ohio Department of Health website are available in the PDF 2013OHH Detail Tables. This is a source of many tables that can be used for further analyses using methods discussed in Chapter 6. HEALTH

2. SAT, ACT, and teacher salaries for 2018 for each of the 50 states and the District of Columbia are available in the data set SATACT. One could use these data to carry out analyses for ACT scores similar to those for the SAT scores in Chapters 5 and 6. SATACT

3. The data set MLB contains hitting, pitching, fielding, salary, and win–loss performance data from the 2019 season for all major league baseball teams. These data can be used to determine the correlation between payroll and winning percentage. One can also explore what variables are most highly correlated with winning percentage and whether variables that measure pitching performance are more highly correlated with winning percentage than variables that measure hitting performance. These data are from http://www.baseball-reference.com/. Visit this website for definitions of several of the variables in the data set. MLB

4. Historical temperature data and whether Punxsutawney Phil saw his shadow are available in the data set PHIL. One can use various measures of whether spring arrived early based on how warm February and March were compared to the historical average to assess the performance of Phil. PHIL

5. The data set WHAT contains three variables and 3848 observations on each. At one time, this was considered a large data set and difficult to explore with software. Students can use various exploratory methods available in software packages such as JMP and Minitab to find the "hidden pattern" in these data. WHAT

David Paynter/AGE Fotostock

PART II

Producing Data

The purpose of statistics is to gain understanding from data. We can seek understanding in different ways, depending on the circumstances. We have studied one approach to data, *exploratory data analysis,* in some detail. Now we begin the move from data analysis toward *statistical inference.* Both types of reasoning are essential to effective work with data. Here is a brief sketch of the differences between them:

Exploratory Data Analysis	Statistical Inference
Purpose is unrestricted exploration of the data, searching for interesting patterns.	Purpose is to answer specific questions, posed before the data were produced.
Conclusions apply only to the individuals and circumstances for which we have data in hand.	Conclusions apply to a larger group of individuals or a broader class of circumstances.
Conclusions are informal, based on what we see in the data.	Conclusions are formal, backed by a statement of our confidence in them.

Our journey toward inference begins in Chapters 8 through 10, which describe statistical designs for *producing data* by samples and experiments, and the ethical issues involved. The important lesson of Part II is that the quality of the inferences we make from data depends heavily on how the data are produced.

CHAPTER

8

When you complete this chapter, you will be able to:

8.1 Identify the population and the sample in an observational statistical study.

8.2 Recognize and avoid possible sources of bias in sampling design.

8.3 Construct a simple random sample from a list of study subjects using software or a table of random digits.

8.4 Use facts about random samples, including sample size, to make judgments about the reliability of conclusions about populations based on those samples.

8.5 Describe the methods used to construct stratified random samples and multistage samples.

8.6 Recognize and avoid undercoverage, nonresponse, response bias, and question wording effects as sources of bias, especially in studies of human subjects using surveys.

8.7 Describe the advantages and limitations of random digit dialing and Internet-based surveys, and how the viability of these survey methods is changing.

Producing Data: Sampling

Statistics, the science of data, provides ideas and tools that we can use in many settings. Sometimes we have data that describe a group of individuals and want to learn what the data say. That's the job of exploratory data analysis, and the methods of Chapters 1 through 6 can be used. Sometimes we have specific questions but no data to answer them. To get sound answers, we must produce data in a way that is designed to answer our questions. This chapter introduces one way to produce data: using samples. In Chapter 9, we explore statistical designs for experiments, a quite different means of producing data.

Suppose our question is, "What percentage of college students think that people should not obey laws that violate their personal values?" To answer the question, we interview undergraduate college students. We can't afford to ask all students, so we put the question to a *sample* chosen to represent the entire student *population*. How shall we choose a sample that truly represents the opinions of the entire population? Statistical designs for choosing samples are the topic of this chapter. We will see that

- A sound statistical design is necessary if we are to trust data from a sample for drawing sound conclusions about the population.
- When sampling from large human populations, even with a sound design, there are still many practical difficulties that arise.
- The impact of technology (particularly cell phones and the Internet) is making it harder to produce trustworthy national data by sampling.

sturti/Getty Images

8.1 Population Versus Sample

A political scientist wants to know what percentage of college-age adults consider themselves conservatives. An automaker hires a market research firm to learn what percentage of adults aged 18–35 recall seeing television advertisements for a new gas/electric hybrid car. Government economists inquire about average household income. In all these cases, we want to gather information about a large group of individuals. Time, cost, and inconvenience preclude contacting every individual. So we gather information about only part of the group to draw conclusions about the whole.

Population, Sample, and Sampling Design

The **population** in a statistical study is the entire group of individuals about which we want information.

A **sample** is a part of the population from which we actually collect information. We use a sample to draw conclusions about the entire population.

A **sampling design** describes exactly how to choose a sample from the population.

Pay careful attention to the details of the definitions of *population* and *sample*. Exercise 8.1 will help you check your understanding.

We often draw conclusions about a whole on the basis of a sample. Everyone has tasted a sample of ice cream and ordered a cone on the basis of that taste. But ice cream is uniform so that the single taste represents the whole. Choosing a representative sample from a large and varied population is not so easy. The first step in planning a **sample survey** is to say exactly *what population* we want to describe.

sample survey
A survey conducted on a sample from the population of all individuals about which we desire information. We base conclusions about the population on data from the sample.

The second step is to say exactly *what we want to measure*—that is, to give exact definitions of our variables. These preliminary steps can be complicated, as the following example illustrates.

EXAMPLE 8.1 The Current Population Survey

The most important government sample survey in the United States is the monthly Current Population Survey (CPS) conducted by the Bureau of the Census for the Bureau of Labor Statistics. The CPS contacts about 60,000 households each month. It produces the monthly unemployment rate and much other economic and social information. (See Figure 8.1.) To measure unemployment, we must first specify the population we want to describe. Which age groups will we include? Will we include undocumented immigrants or people in prisons? The CPS defines its population as all U.S. residents (legal or not) 16 years of age and over who are civilians and are not in an institution such as a prison. The unemployment rate announced in the news refers to this specific population.

The second question is harder: What does it mean to be "unemployed"? Someone who is not looking for work—for example, a full-time student—should not be called unemployed just because she is not working for pay. If you are chosen for the CPS sample, the interviewer first asks whether you are available to work and whether you actually looked for work in the past four weeks. If not, you are neither employed nor unemployed; you are not in the labor force. So discouraged workers who haven't looked for a job in four weeks are excluded from the count.

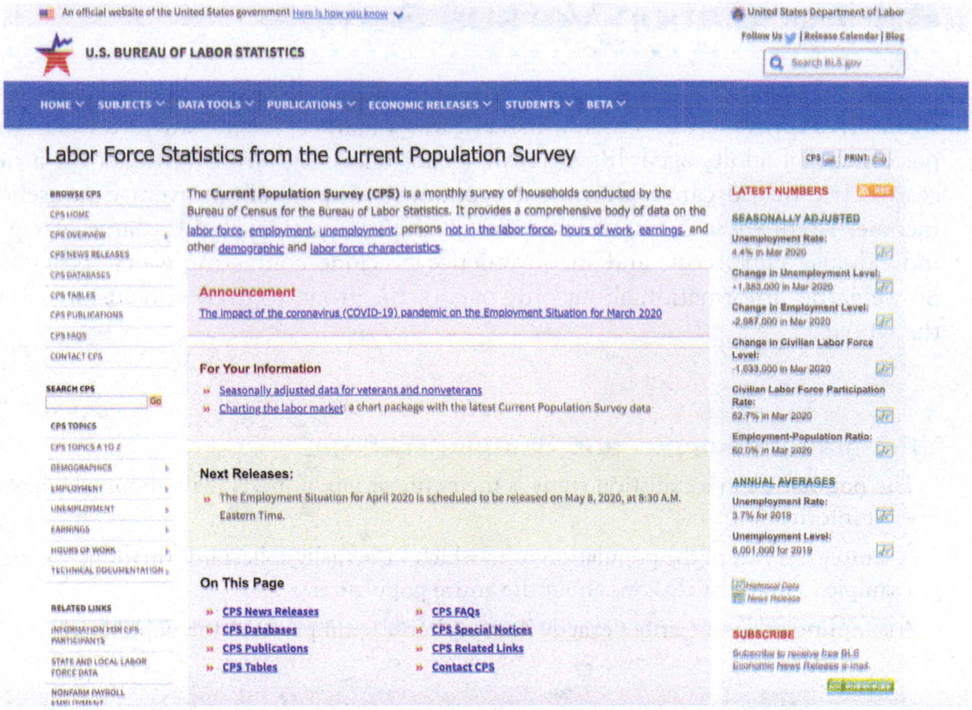

U.S. Bureau of Labor Statistics

FIGURE 8.1
The home page of the Current Population Survey at the Bureau of Labor Statistics.

If you are in the labor force, the interviewer goes on to ask about employment. If you did any work for pay or in your own business during the week of the survey, you are employed. If you worked at least 15 hours in a family business without pay, you are employed. You are also employed if you have a job but didn't work during the week of the survey because of a vacation, being on strike, or for another good reason. An unemployment rate of 6.7% means that 6.7% of the sample was unemployed, using the exact CPS definitions of both *labor force* and *unemployed*.

The final step in planning a sample survey is the sampling design. We will now introduce basic statistical designs for sampling.

APPLY YOUR KNOWLEDGE

8.1 Off-Campus Housing. A university's housing and residence office wants to know how much students pay per month for rent in off-campus housing. The university does not have enough on-campus housing for students, and this information will be used in a brochure about student housing. The housing office obtains a list of the 12,304 students who live in off-campus housing and have not yet graduated and mails a questionnaire to a randomly selected group of 200 of these students. Only 78 questionaires are returned.

(a) What is the population in this study? Be careful: about what group does the office *want information?*

(b) What is the sample? Be careful: from what group does the office *actually obtain information?*

The important message in this problem is that the sample can redefine the population about which information is obtained.

8.2 Student Archaeologists. An archaeological dig turns up large numbers of pottery shards, broken stone implements, and other artifacts. Students working on the project classify each artifact and assign it a number. The counts in different categories are important for understanding the site, so the project director chooses 2% of the artifacts at random and checks the students' work. What are the population and the sample here?

John Elk III/Alamy

8.3 Software Survey. A statistical software company is planning on updating Version 8.1 of its software and wants to know what features are most important to users. The company's managers have the email addresses of 1100 individuals, mostly faculty at universities, for whom they have supplied free courtesy copies of Version 8.1. They email these 1100 individuals and ask them to complete a survey online. A total of 186 of these individuals complete the survey.

(a) What is the population of interest to the software company? Do you think the 1100 individuals contacted are representative of the population? Explain your reasons.

(b) What is the sample? From what group is information actually obtained?

8.2 How to Sample Badly

How can we choose a sample that we can trust to represent the population? A sampling design is a specific method for choosing a sample from the population. The easiest—but not the best—design just chooses individuals close at hand. If we are interested in finding out how many people have jobs, for example, we might go to a shopping mall and ask people passing by if they are employed. A sample composed of the members of the population who are easiest to reach is called a **convenience sample**. Convenience samples often produce unrepresentative data.

convenience sample
A sample composed of members of the population who are selected by the researcher because they are easy to reach.

EXAMPLE 8.2 Sampling at the Mall

Finding a sample of mall shoppers is fast and cheap. But people at shopping malls tend to be more prosperous than typical Americans. They are also especially likely to be teenagers or retired. Moreover, unless interviewers are carefully trained, they tend to question well-dressed, respectable-looking people and avoid poorly dressed or tough-looking individuals. The types of people at the mall also vary by time of day and day of week. In short, a mall interview does not contact a sample that is representative of the entire population.

Interviews at shopping malls will almost surely overrepresent middle-class and retired people and underrepresent the poor. This will happen almost every time we take such a sample. That is, it is a systematic error caused by a bad sampling design, not just bad luck on one sample. This is *bias:* the outcomes of mall surveys will repeatedly miss the truth about the population in the same ways.

Bias

The design of a statistical study is **biased** if it systematically favors certain outcomes.

EXAMPLE 8.3 Online Polls

In June 2019, KGAB AM 650 in Cheyenne, Wyoming, posted an online poll at its website. This "pot poll" asked whether Wyoming should consider legalizing marijuana as a way to improve state revenues in light of a budget shortfall that some economists were estimating at $1 billion over the next few years. Of more than 1100 respondents, 57% said "Yes, it should not be illegal anyway," 24% said "Let's cut spending and legalize recreational cannabis," 16% said "Absolutely not, it will create more problems than it solves," 3% said "Let's just cut spending," and 0% said "I would rather impose state income taxes."[1] These results would seem to indicate strong (81%) support for legalizing marijuana in Wyoming.

The kgab.com poll was biased because people chose whether or not to participate. *People who take the trouble to respond to an open invitation are usually not representative of any clearly defined population.* That's true of the people who bother to respond to write-in, call-in, or online polls in general. Polls like these are examples of *voluntary response sampling.*

Voluntary Response Sample

A **voluntary response sample** consists of people who choose themselves by responding to a broad appeal. Voluntary response samples are biased because people with strong opinions are most likely to respond.

We don't know whether those responding to the KGAB poll in Example 8.3 are likely voters or even registered voters. In fact, a telephone survey of Wyoming residents conducted by the Wyoming Survey and Analysis Center at the University of Wyoming in October 2018 showed only 49% favoring legalization.

APPLY YOUR KNOWLEDGE

8.4 Sampling on Campus. You would like to start a club for psychology majors on campus, and you are interested in finding out what proportion of psychology majors would join. The dues would be $35 and used to pay for speakers to come to campus. You ask five psychology majors from your senior psychology honors seminar whether they would be interested in joining this club and find that four of the five students questioned are interested. Is this sampling method biased, and if so, what is the likely direction of bias?

8.5 Airport Shuttle. Blue Ribbon taxis offers shuttle service to the nearest airport. You look up the online reviews for Blue Ribbon taxis and find that there are 17 reviews, 6 of which report that the taxi never showed up. Is this a biased sampling method for obtaining customer opinion on the taxi service? If so, what is the likely direction of bias? Explain your reasoning carefully.

8.3 Simple Random Samples

Random sampling, the use of chance to select a sample, is the essential principle of statistical sampling. In a voluntary response sample, people choose whether to respond. In a convenience sample, the interviewer makes the choice. In both cases, personal choice produces bias. The statistician's remedy is to allow impersonal chance to choose the sample. A sample chosen by chance rules out both favoritism by the sampler and self-selection by respondents. Choosing a sample by chance attacks bias by giving all individuals an equal chance to be chosen. Rich and poor, young and old, liberal and conservative—all have the same chance to be in the sample.

The simplest way to use chance to select a sample is to place names in a hat (the population) and draw out a handful (the sample). This is the idea of *simple random sampling.* Although the idea of drawing names from a hat is a good way to conceptualize a simple random sample, it is generally *not* practical for large populations.

Simple Random Sample

A **simple random sample** (SRS) of size n consists of n individuals from the population chosen in such a way that every set of n individuals has an equal chance to be the sample actually selected.

An SRS not only gives each individual an equal chance to be chosen but also gives every possible sample an equal chance to be chosen. There are other random sampling designs that give each individual, but not each sample, an equal chance. Exercise 8.43 (page 221) describes one such design.

When you think of an SRS, you can still picture the conceptual situation of drawing names from a hat to remind yourself that an SRS doesn't favor any part of the population. That's why an SRS is a better method of choosing samples than convenience or voluntary response sampling. However, in practice, most samplers use software to obtain an SRS. The use of software or the *Simple Random Sample* applet makes choosing an SRS very fast. If you don't use the applet or other software, you can randomize by using a *table of random digits.* In fact, software for choosing samples starts by generating random digits, so using a table just does by hand what the software does more quickly.

applet

Random Digits

A **table of random digits** is a long string of the digits 0, 1, 2, 3, 4, 5, 6, 7, 8, and 9 with these two properties:

1. Each entry in the table is equally likely to be any of the 10 digits 0 through 9.
2. The entries are independent of each other. That is, knowledge of one part of the table gives no information about any other part.

Table B at the back of the book is a table of random digits. Table B begins with the digits 19223950340575628713. To make the table easier to read, the digits appear in groups of five and in numbered rows. The groups and rows have no meaning; the table is just a long list of randomly chosen digits. There are two steps in using the table to choose a simple random sample.

Using Table B to Choose an SRS

Label: Give each member of the population a numerical label of the *same length.*

Table: To choose an SRS, read from Table B successive groups of digits of the length you used as labels. Your sample contains the individuals whose labels you find in the table.

You can label up to 100 items with two digits: 01, 02, . . . , 99, 00. Up to 1000 items can be labeled with three digits, and so on. Always use the shortest labels that will cover your population. As standard practice, we recommend that you begin with label 1 (or 01 or 001, as needed). Reading groups of digits from the table gives all individuals the same chance to be chosen because all labels of the same length

STATISTICS IN YOUR WORLD

Are These Random Digits Really Random?

Not a chance. The random digits in Table B were produced by a computer program. Computer programs do exactly what you tell them to do. Give the program the same input, and it will produce exactly the same "random" digits. Of course, clever people have devised computer programs that produce output that *looks* like random digits. These are called "pseudo-random numbers," and that's what Table B contains. Pseudo-random numbers work fine for statistical randomizing, but they have hidden nonrandom patterns that can mess up more refined uses.

have the same chance to be found in the table. For example, any pair of digits in the table is equally likely to be any of the 100 possible labels 01, 02, . . . , 99, 00. Ignore any group of digits that was not used as a label or that duplicates a label already in the sample. You can read digits from Table B in any order—across a row, down a column, and so on—because the table has no order. As standard practice, we recommend reading across rows.

EXAMPLE 8.4 Sampling Spring Break Resorts

A campus newspaper plans a major article on spring break destinations. The reporters intend to call four randomly chosen resorts at each destination to ask about their attitudes toward groups of students as guests. Here are the resorts listed in one city:

01 Aloha Kai	08 Captiva	15 Palm Tree	22 Sea Shell
02 Anchor Down	09 Casa del Mar	16 Radisson	23 Silver Beach
03 Banana Bay	10 Coconuts	17 Ramada	24 Sunset Beach
04 Banyan Tree	11 Diplomat	18 Sandpiper	25 Tradewinds
05 Beach Castle	12 Holiday Inn	19 Sea Castle	26 Tropical Breeze
06 Best Western	13 Lime Tree	20 Sea Club	27 Tropical Shores
07 Cabana	14 Outrigger	21 Sea Grape	28 Veranda

LABEL: Because two digits are needed to label the 28 resorts, every label will have two digits. We have added labels 01 to 28 in the list of resorts. (If you wanted a sample from a major vacation area containing 1240 resorts, you would label the resorts 0001, 0002, . . . , 1239, 1240.) Always say how you labeled the members of the population.

applet

SELECT SAMPLE: To use the *Simple Random Sample* applet, move the "Population" slider to a size of 28 and move the "Sample size" slider to n = 4. Then move the "Generate a sample" toggle to obtain your sample. Figure 8.2 shows the result of one sample that contains the resorts labeled 20, 16, 21, and 10. These are Sea Club, Radisson, Sea Grape, and Coconuts.

To use Table B, read two-digit groups until you have chosen four resorts. Starting at line 130 (any line will do), we find

69051 64817 87174 09517 84534 06489 87201 97245

Because the labels are two digits long, read successive two-digit groups from the table: the first three two-digit groups here are 69, 05, and 16. Ignore groups not used as labels, like the initial 69. Also ignore any repeated labels, like the second and third 17s in this row, because you can't choose the same resort twice. Your sample contains the resorts labeled 05, 16, 17, and 20. These are Beach Castle, Radisson, Ramada, and Sea Club.

Most statistical software packages will produce an SRS if you enter the size of the sample and the size of the population.

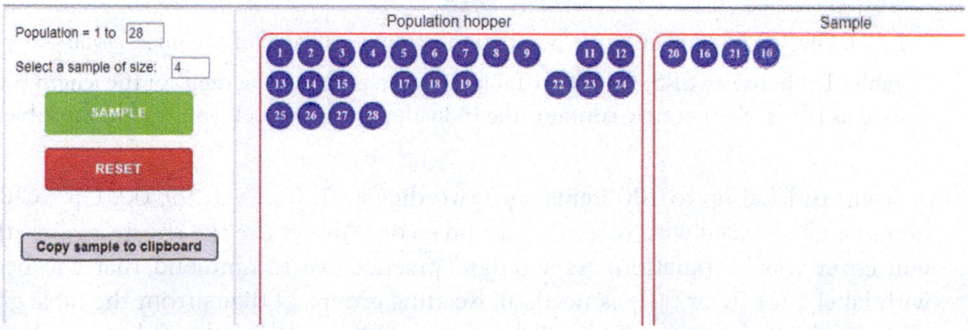

FIGURE 8.2
The *Simple Random Sample* applet used to choose an SRS of size $n = 4$ from a population of size 28.

We can trust results from an SRS, as well as from other types of random samples that we will meet later, because the use of impersonal chance avoids bias. Online polls and mall interviews also produce samples. We can't trust results from these samples because they are chosen in ways that invite bias. *The first question to ask about any sample is whether it was chosen at random.*

EXAMPLE 8.5 Sale of Guns

"Would you support or oppose requiring background checks on all potential gun buyers, including private sales and gun shows?" When ABC News and the *Washington Post* asked this question of 1003 adults in September 2019, 89% said support, 9% said oppose, and the rest said no opinion. Can we trust the opinions of this sample to fairly represent the opinions of all adults? Here are some of the key features of the poll methodology provided by the *Times:*[2]

- For this survey, the ABC News/*Washington Post* poll used telephone interviews of 1003 adults throughout the United States taken September 2–5, 2019.
- For landline interviews, a sample of landline households in the continental United States is selected via random digit dialing, in which all landline telephone numbers, listed and unlisted, have an equal probability of selection. Landline numbers are drawn proportionate to their estimated distribution in the country's nine census divisions. The database consists of all listed landline telephone numbers, updated on a four- to six-week rolling basis, 25% of listings at a time.
- Cell phone numbers were generated by a similar random process. The two samples were then combined and adjusted to ensure the proper ratios of landine-only, cell-phone-only, and dual-phone users.

The selection of landline phone numbers is a good description of a common method for choosing national samples, called *random digit dialing.* Although the information about the conduct of the poll provides the reader with some basic details, data collection and analysis of a national survey is a great deal more complex than this short description suggests. We'll come back to random digit dialing and its problems later (see Exercise 8.16, page 216), but this description of the conduct of the poll does contain important information. We know the size of the sample, when the poll was taken, and that a random process was used in the selection of the sample.

APPLY YOUR KNOWLEDGE

8.6 Apartment Living. You are planning a report on apartment living in a college town. You decide to select four apartment complexes at random for in-depth interviews with residents. Use the *Simple Random Sample* applet, other software, or Table B to select a simple random sample of four of the following apartment complexes. If you use Table B, start at line 133.

Ashley Oaks	Country View	Mayfair Village
Bay Pointe	Country Villa	Nobb Hill
Beau Jardin	Crestview	Pemberly Courts
Bluffs	Del-Lynn	Peppermill
Brandon Place	Fairington	Pheasant Run
Briarwood	Fairway Knolls	River Walk
Brownstone	Fowler	Sagamore Ridge
Burberry Place	Franklin Park	Salem Courthouse
Cambridge	Georgetown	Village Square

8.7 Minority Managers. A firm wants to understand the attitudes of its minority managers toward its system for assessing management performance. Following is a list of all the firm's managers who are members of minority groups. Use the *Simple Random Sample* applet, other software, or Table B at line 127 to choose three managers to be interviewed in detail about the performance appraisal system.

Adelaja	Draguljic	Huo	Modur
Ahmadiani	Fernandez	Ippolito	Rettiganti
Barnes	Fox	Jiang	Rodriguez
Bonds	Gao	Jung	Sanchez
Burke	Gemayel	Mani	Sgambellone
Deis	Gupta	Mazzeo	Yajima

The Photo Works

8.8 Sampling Gravestones. The local genealogical society in Coles County, Illinois, has compiled records on all 55,914 gravestones in cemeteries in the county for the years 1825 through 1985. Historians plan to use these records to learn about African Americans in Coles County's history. They first choose an SRS of 395 records to check their accuracy by visiting the actual gravestones.[3]

(a) How would you label the 55,914 records?

(b) Use software, the *Simple Random Sample* applet, or Table B (starting at line 141) to choose the first six records for the SRS.

8.4 Trustworthiness of Inference from Samples

The purpose of a sample is to get information about a larger population. The process of drawing conclusions about a population on the basis of sample data is called *inference* because we *infer* information about the population from what we *know* about the sample.

Inference from convenience samples or voluntary response samples would be misleading because these methods of choosing a sample are biased. We are almost certain that the sample does *not* fairly represent the population. *The first reason to rely on random sampling is to eliminate bias in selecting samples from the list of available individuals.*

Nonetheless, it is unlikely that results from a random sample are exactly the same as for the entire population. Sample results, like the unemployment rate obtained from the monthly Current Population Survey, are only estimates of the truth about the population. If we select two samples at random from the same population, we will almost certainly draw different individuals. So the sample results will differ somewhat, just by chance. Properly designed samples avoid systematic bias, but their results are rarely exactly correct, and they vary from sample to sample.

Why can we trust random samples? The big idea is that the results of random sampling don't change haphazardly from sample to sample. Because we deliberately use chance, the results obey the laws of probability that govern chance behavior. These laws allow us to say how likely it is that sample results are close to the truth about the population. *The second reason to use random sampling is that the laws of probability allow trustworthy inference about the population.* Results from random samples come with a margin of error that sets bounds on the size of the likely error. How to do this is part of the technique of statistical inference. We will describe the reasoning in Chapter 16 and present details throughout the rest of the book.

One point is worth making now: *larger random samples give more accurate results than smaller random samples.* By taking a very large *random* sample, you can be confident that the sample result is very close to the truth about the population. The Current Population Survey contacts about 60,000 households, so it estimates the national unemployment rate very accurately. Opinion polls that contact 1000 or 1500 people give less accurate results.

It is a common misconception that larger sample sizes *always* give more accurate results. After the Democratic debate in July 2019, a New Jersey online poll listed Bernie Sanders as winning the debate, capturing 53% of the 13,468 votes in the poll. However, a July 29 Quinnipiac University Poll of a random sample of 807 Democrat voters found only 8% chose Sanders as the winner. Other polls found similar results.

When reading the results of a survey, don't assume because the sample size is large that the survey is accurate. You should pay more attention to how the sample was selected. Biased sampling techniques continue to yield biased results no matter how large the sample.

APPLY YOUR KNOWLEDGE

8.9 Ask More People. In the 2016 presidential pre-election surveys, ABC/Post sampled 740 likely voters during October 10–13, 2016, and asked if they were planning to vote for Clinton, and then asked the same question of a sample of 1135 likely voters taken from October 22–25, 2016. However, in their last survey, taken November 3–6, 2016, just before the election held on November 8, 2016, they asked this question of a sample of 2220 likely voters. Why do you think ABC/Post did this?

8.10 How Accurate Is the Poll? A Pew Research Center survey called *Teens, Social Media & Technology* in the spring of 2018 included 743 teens, of which 355 were White, non-Hispanic; 129 were Black, non-Hispanic; 202 were Hispanic; and 57 were other races or ethnic groups. Each teen sampled was asked about technology usage, including access to mobile devices, online platform usage, views on social media, and video game playing. The margin of error (we will give more detail in later chapters) was reported as ±5.0% for the entire sample. When considering technology usage of only the Hispanic teens, the margin of error was reported as ±9.5%.[4] What do you think explains the fact that estimates for Hispanic teens were less precise than for the entire sample?

8.5 Other Sampling Designs

Designs for random sampling from large populations spread out over a wide area are usually more complex than an SRS. For example, it is common to sample important groups within the population separately and then combine these samples. This is the idea of a *stratified random sample.*

Stratified Random Sample

To select a **stratified random sample,** first classify the population into groups of similar individuals, called *strata.* Then, choose a separate SRS in each stratum and combine these SRSs to form the full sample.

Choose the strata based on facts known before the sample is taken. For example, a population of election districts might be divided into urban, suburban, and rural strata. A stratified design can produce more precise information than an SRS of the same size by taking advantage of the fact that individuals in the same stratum are similar to one another.

STATISTICS IN YOUR WORLD

Golfing at Random

Random drawings give everyone the same chance to be chosen, so they offer a fair way to decide who gets a scarce good—like a round of golf. Lots of golfers want to play the famous Old Course at St. Andrews, Scotland. Some can reserve in advance, at considerable expense. Most must hope that chance favors them in the daily random drawing for tee times. At the height of the summer season, only one in six wins the right to pay $250 for a round.

EXAMPLE 8.6 Seat Belt Use in Alaska

Ryan McVey/Getty Images

Each state conducts an annual survey of seat belt use by drivers, following guidelines set by the federal government. The guidelines require random sampling, with seat belt use observed at randomly chosen road locations at random times during daylight hours.

In the Alaskan survey, the locations are not an SRS of all locations in the state but rather a stratified sample using the state's boroughs as strata. The seat belt survey sample consists of 256 randomly selected road locations in the five boroughs (the strata) that account for 85% of crash-related fatalities from 2005 to 2009: a random sample of 112 in Anchorage, a random sample of 56 in Matanuska-Susitna, a random sample of 40 in Fairbanks North Star, a random sample of 24 in Kenai Peninsula, and a random sample of 24 in Juneau. The sample sizes in the boroughs are proportional to the populations of the boroughs.[5]

multistage sample A method of sampling in which groups are randomly selected from within larger groups so that the groups are smaller at each stage until individuals are sampled from the smallest groups.

Most large-scale sample surveys use a **multistage sample**. For example, the opinion poll described in Example 8.5 has three stages: choose a random sample of telephone exchanges (stratified by region of the country), then an SRS of household telephone numbers within each exchange, and then a random adult in each household.

Analysis of data from sampling designs more complex than an SRS takes us beyond basic statistics. But the SRS is the building block of more elaborate designs, and analysis of other designs differs more in complexity of detail than in fundamental concepts.

APPLY YOUR KNOWLEDGE

8.11 Sampling Metro Chicago. Cook County, Illinois, has the second-largest population of any county in the United States (after Los Angeles County, California). Cook County has 30 suburban townships and an additional 8 townships that make up the city of Chicago. The suburban townships are

Barrington	Elk Grove	Maine	Orland	Riverside
Berwyn	Evanston	New Trier	Palatine	Schaumburg
Bloom	Hanover	Niles	Palos	Stickney
Bremen	Lemont	Northfield	Proviso	Thornton
Calumet	Leyden	Norwood Park	Rich	Wheeling
Cicero	Lyons	Oak Park	River Forest	Worth

The Chicago townships are

Hyde Park	Lake	North Chicago	South Chicago
Jefferson	Lake View	Rogers Park	West Chicago

Because city and suburban areas may differ, the first stage of a multistage sample is to choose a stratified sample of four suburban townships and two of the Chicago townships, which are more heavily populated. Use software, the *Simple Random Sample* applet, or Table B to choose this sample. (If you use Table B, assign labels in alphabetical order and start at line 118 for the suburbs and at line 127 for Chicago.)

8.12 Academic Dishonesty. A study of academic dishonesty among college students used a two-stage sampling design. The first stage involved choosing a sample of 30 colleges and universities. Then, the study authors mailed questionnaires to a stratified sample of 200 seniors, 100 juniors, and 100 sophomores at each school.[6] One of the schools chosen had 1127 freshmen, 989 sophomores,

943 juniors, and 895 seniors. You have alphabetical lists of the students in each class. Explain how you would assign labels for stratified sampling. Then use software, the *Simple Random Sample* applet, or Table B, starting at line 138, to select the first four students in the sample from each stratum. After selecting four students for a stratum, continue to select the students for the next stratum. If you are using Table B, to select the students in the next stratum, continue along in the table where you left off.

8.6 Cautions About Sample Surveys

Random selection eliminates bias in the choice of a sample from a list of the population. When the population consists of human beings, however, accurate information from a sample requires more than a good sampling design.

To begin, we need an accurate and complete list of the population. Because such a list is rarely available, most samples suffer from some degree of *undercoverage.* A sample survey of households, for example, will miss not only homeless people but also prison inmates and students in dormitories. An opinion poll conducted by calling landline telephone numbers will miss households that have only cell phones as well as households without a phone. The results of national sample surveys, therefore, have some bias if the people not covered differ from the rest of the population.

A more serious source of bias in most sample surveys is *nonresponse,* which occurs when a selected individual cannot be contacted or refuses to cooperate. Nonresponse to sample surveys often exceeds 50%, even with careful planning and several callbacks. Because nonresponse is especially high in urban areas, most sample surveys substitute other people in the same area to avoid favoring rural areas in the final sample. If the people contacted differ from those who are rarely at home or who refuse to answer questions, some bias remains.

Undercoverage and Nonresponse

Undercoverage occurs when some groups in the population are left out of the process of choosing the sample.

Nonresponse occurs when an individual chosen for a sample can't be contacted or refuses to participate.

EXAMPLE 8.7 How Bad Is Nonresponse?

The U.S. Census Bureau's American Community Survey (ACS) has the lowest nonresponse rate of any poll we know: in 2018, only about 8.0% of the households in the sample refused to respond; the overall nonresponse rate, including "never at home" and other causes, was just 3.3%.[7] This monthly survey of almost 300,000 housing units replaces the "long form" that in the past was sent to some households in the every-10-years national census. Participation in the ACS is mandatory, and the U.S. Census Bureau follows up by telephone and then in person if a household fails to return the mail questionnaire.

The University of Chicago's General Social Survey (GSS) is the nation's most important social science survey. (See Figure 8.3.) The GSS contacts its sample in person, and it is run by a university. Its response rate in 2016 was 61.3%, among the highest in the world.

FIGURE 8.3
The home page of the General Social Survey at the University of Chicago's National Opinion Research Center. The GSS has tracked opinions about a wide variety of issues since 1972.

GSS General Social Survey

What about opinion polls by news media and opinion-polling firms? We don't know the rates of nonresponse for many of these surveys because they don't say, which in itself is a bad sign. In a January 2014 survey on book readers and ebook readers, the Pew Research Center provided a full disposition of the sampled phone numbers. Here are the details. Initially, 26,388 landline and 16,000 cell phone numbers were dialed, with 7,767 of the landlines and 9,654 of the cell numbers being working numbers. Of these working numbers, Pew was able to contact 3,839 landline numbers and 4,747 cell phone numbers, or about 50%, from each group as a large portion of the calls went to voicemail. Among those contacted, 13.6% of the landline numbers and 18.6% of the cell numbers cooperated. Of those cooperating, some numbers were ineligible due to language barriers or contacting a child's cell phone, and some calls were eventually broken off without being completed. To sum it up, the 17,421 working numbers dialed resulted in a final sample of 500 landline numbers and 505 cell phone numbers, giving a response rate of about 6% for landline numbers and 5% for cell phone numbers, the fraction of all *eligible* respondents in the sample who were ultimately interviewed.[8] We don't know whether those who were not interviewed differed in systematic ways from those who were. It is unlikely that the two groups are similar, and no evidence is provided that they are. If they are not, bias (perhaps substantial) would be present, rendering the results unreliable as indicators of what might be true about the population in general. It would be unfortunate if personal or public policy decisions were based on such results.

In addition, the behavior of the respondent or of the interviewer can cause *response bias* in sample results People know that they should take the trouble to vote, for example, so many who didn't vote in the last election will tell an interviewer that they did. The race or sex of the interviewer can influence responses to questions about race relations or attitudes toward feminism. Answers to questions that ask respondents to recall past events are often inaccurate because of faulty memory. For example, many people "telescope" events in the past, bringing them forward in memory to more recent time periods. "Have you visited a dentist in the last six months?" will often draw a Yes from someone who last visited a dentist eight months ago.[9] Careful training of interviewers and careful supervision to avoid variation among the interviewers can reduce response bias. Good interviewing technique is another aspect of a well-done sample survey.

Response Bias

Response bias occurs when survey respondents answer falsely, either deliberately or by mistake.

In our discussion of bias, it is important to distinguish between two ways bias can occur in a survey. Bias can result from the selection of the sample through voluntary sampling or undercoverage. However, even with a well-selected sample, bias can still occur through nonresponse or untruthful answers to the question. Don't confuse voluntary sampling with nonresponse. Voluntary sampling relates to the selection of the sample, while nonresponse relates to a choice made by possible participants to not respond after the sample has been selected.

The **wording effects** are the most important influence on the answers given to a sample survey. Confusing or leading questions can introduce strong bias, and variations in wording can greatly change a survey's outcome. Even the order in which questions are asked matters. Here are some examples.[10]

wording effects
The influence of question wording on survey responses, in which different phrasings of the same question can evoke different responses.

EXAMPLE 8.8 What Was That Question?

How do Americans feel about conditions in immigration detention centers? "Do you think that the conditions in immigration detention centers are a serious problem, or don't you think so?" Asked this question in an opinion poll, 42% of Republicans said it was a serious problem. But when the very same sample was asked "Do you think that the conditions in immigration detention centers are inhumane, or don't you think so?" only 13% of Republicans said conditions were inhumane. Different questions give quite different impressions of attitudes toward conditions in immigration detention centers.

What about electing a gay president? Only 36% think "we" (Americans) are ready to elect a gay president, but 70% say they personally are open to electing a gay president.

EXAMPLE 8.9 Are You Happy?

Ask a sample of college students these two questions:

How happy are you with your life in general? (Answers on a scale of 1 to 5)

How many dates did you have last month?

When asked in this order, it appears that dating and happiness are not associated or that dating has little to do with happiness. Reverse the order of the questions, however, and there is a strong association between the answers, with greater happiness associated with more dates. Asking a question that brings dating to mind makes dating success a big factor in happiness.

Don't trust the results of a sample survey until you have read the exact questions asked. The amount of nonresponse and the date of the survey are also important. Good statistical design is a part—but only a part—of a trustworthy survey.

APPLY YOUR KNOWLEDGE

8.13 A Survey of 100,000 Physicians. In 2010, the Physicians Foundation conducted a survey of physicians' attitudes about health care reform, saying that the report was "a survey of 100,000 physicians." The survey was sent to 100,000 randomly selected physicians practicing in the United States: 40,000 via postal mail and 60,000 via email. A total of 2,379 completed surveys were received.[11]

(a) State carefully what population is sampled in this survey and what is the sample size. Could you draw conclusions from this study about all physicians practicing in the United States?

(b) What is the rate of nonresponse for this survey? How might this affect the credibility of the survey results?

(c) Why is it misleading to call the report "a survey of 100,000 physicians"?

8.14 Universal Health Care. In 2019, a Monmouth University poll and an NBC News/*Wall Street Journal* poll each asked a nationwide sample about their views on universal health care.[12] Here are the two questions:

Question A: *Do you favor or oppose creating a universal health care system in America?*

Question B: *Would you favor or oppose a single payer health care system in which all Americans would get their health insurance from one government plan that is financed in part by taxes?*

One of these questions had 58% responding favor, and the other question had only 44% responding favor. Which wording is slanted toward a more negative response on universal health care? Why?

8.7 The Impact of Technology

A few national sample surveys, including the General Social Survey and the U.S. government's American Community Survey and Current Population Survey, interview some or all of their subjects in person. This is expensive and time-consuming, so most national surveys contact subjects by telephone using the random digit dialing (RDD) method described in Example 8.5 (page 207). Technology, especially the spread of cell phones, is making traditional RDD methods outdated.

First, *call screening* is now common. A large majority of American households have answering machines, voicemail, or caller ID, and many use these methods to screen their calls. Calls from polling organizations are rarely returned.

More seriously, the percentage of *cell-phone-only households* is increasing rapidly. By the end of 2009, 25% of American households had a cell phone but no landline phone; by the end of 2014, that percentage had increased to 45%, and by the middle of 2017, the percentage was 53.9%. It's clear from these numbers that RDD reaching only landline numbers is in trouble. Can surveys just add cell phone numbers? Not easily. Federal regulations prohibit automated dialing to cell phones, which rules out computerized RDD sampling and requires hand dialing of cell phone numbers, which is expensive. A cell phone can be anywhere, and many people keep their cell number despite moving, so stratifying by location becomes difficult. And a cell phone user may be driving or otherwise unable to talk safely.

People who screen calls and people who have only a cell phone tend to be younger than the general population. By mid-2017, 73.3% of adults aged 25 to 29 years lived in households with no landline phone. So RDD surveys using only landlines may be biased (see Exercise 8.16, page 216). Careful surveys weight their responses to reduce

STATISTICS IN YOUR WORLD

Do Not Call!

People who do sample surveys hate telemarketing. We all get so many unwanted sales pitches by phone that many people hang up before learning that the caller is conducting a survey rather than selling vinyl siding. You can eliminate calls from commercial telemarketers by placing your phone number on the National Do Not Call Registry. Just go to www.donotcall.gov to sign up. Political organization, charity, and telephone research calls are not covered by the National Do Not Call Registry. But if they conduct research and also offer to sell goods or services, they must comply with the rules.

bias. For example, if a sample contains too few young adults, the responses of the young adults who do respond are given extra weight. This might be accomplished by multiplying the number of responses by some number greater than 1 based on the known proportion of young adults in the general population.

But with response rates steadily dropping and cell-phone-only use steadily growing, the future of RDD landline telephone surveys is not promising. Many polling organizations now include a minimum quota of cell phone users in their samples to help adjust for bias[13] (see Example 8.5, page 207, and Exercise 8.45, page 222).

One alternative is to use *web surveys*, an increasingly popular survey method, rather than telephone surveys. Web surveys have several advantages over more traditional survey methods. It is possible to collect large amounts of survey data at lower costs than traditional methods allow. Anyone can put survey questions on dedicated sites offering free services; thus, large-scale data collection is available to almost every person with access to the Internet. Furthermore, web surveys allow delivery of multimedia survey content to respondents, opening up new realms of survey possibilities that would be extremely difficult to implement using traditional methods. Some argue that eventually web surveys will replace traditional survey methods.

Although web surveys are easy to do, they are not easy to do well. Three major problems are voluntary response, undercoverage, and nonresponse. Voluntary response appears in several forms in online surveys. Example 8.3 (page 204) is a survey that invited individuals to a particular website to participate in a poll. Other web surveys solicit participation through announcements in news groups, email invitations, and banner ads on high-traffic sites.[14]

EXAMPLE 8.10 Back-to-School Shopping

NerdWallet, a financial company, reports that 52% of parents say they feel pressured by their children to buy back-to-school items they want, even if they cost more than parents would normally want to spend.[15] How did NerdWallet come up with this number? It is based on an online survey of 2010 U.S. adults, conducted by Harris Interactive during May 30–June 3, 2019. This is a more sophisticated example of voluntary response. It occurs when the polling organization, in this case Harris Interactive, maintains a panel of volunteers. The panel is recruited to fill out questionnaires on the Internet, often in exchange for points redeemable for cash and gifts. Panelists join by going to the polling company website and filling out some personal information that is later used to select them for specific surveys. Harris Interactive uses an online research panel of more than 6 million volunteers worldwide, from which it selected the sample for the NerdWallet poll. In its report on online panels,[16] the American Association of Public Opinion Research warns against the use of nonprobability online panels (panels that are not random samples) when the objective is to accurately estimate population values. It also recommends against the reporting of a margin of sampling error from this type of sample because the result can be misleading. In the methodology section of their poll, NerdWallet does state, "This online survey is not based on a probability sample, and therefore, no estimate of theoretical sampling error can be calculated," although the issue of potential bias in the estimate is not addressed.

Undercoverage is still a serious problem for even careful web surveys because as of 2019, about 10% of Americans lacked Internet access, and only about 73% had broadband access. People without Internet access are more likely to be poor, elderly, minority, or rural than the overall population, so the potential for bias in a web survey is clear. There is no easy way to choose a random sample even from people with web access because there is no technology that generates personal email addresses at random in the way that RDD generates residential telephone numbers, and individuals may have several email addresses. Even if such technology existed, etiquette and regulations aimed at spammers would prevent mass emailing. For the present, web surveys work well only for restricted populations, such as for surveying students at your university using the school's list of student email addresses. Here is an example of a successful web survey.

EXAMPLE 8.11 Doctors and Placebos

A placebo is a dummy treatment such as a pill that has no direct physiological effect on a patient but may bring about a response because patients expect it to. Do academic physicians who maintain private practices sometimes give their patients placebos? A web survey of doctors in internal medicine departments at Chicago-area medical schools was possible because almost all the doctors had listed email addresses.

An email was sent to each doctor, explaining the purpose of the study, promising anonymity, and giving an individual web link for response. In all, 231 of 443 doctors responded. The response rate was helped by the fact that the email came from a team at a medical school. Result: 45% said they sometimes used placebos in their clinical practice.[17]

APPLY YOUR KNOWLEDGE

8.15 SurveyMonkey. In 2019, the *New York Times* conducted an online poll using SurveyMonkey to determine how people felt about their financial situation. SurveyMonkey is a free online survey development service and also provides a "pro" option with fees based on additional features. The survey was conducted October 7–13, and one question was "Now looking ahead - do you think that a year from now you and your family will be better off financially, worse off financially, or just about the same as now?" A total of 2701 people answered the survey, and 85% answered the same or better.

(a) Here is what the *New York Times* says about the survey methodology: "This SurveyMonkey online poll was conducted October 7 through 13, 2019 among a national sample of 2701 adults. Respondents for this survey were selected from the more than 2 million people who take surveys on the SurveyMonkey platform each day. Data were weighted for age, race, sex, education, and geography using the Census Bureau's American Community Survey to reflect the demographic composition of the United States."[18] What concerns do you have about whether the results of this survey represent the opinions of all U.S. adults?

(b) What groups of U.S. adults are likely to be underrepresented by this survey?

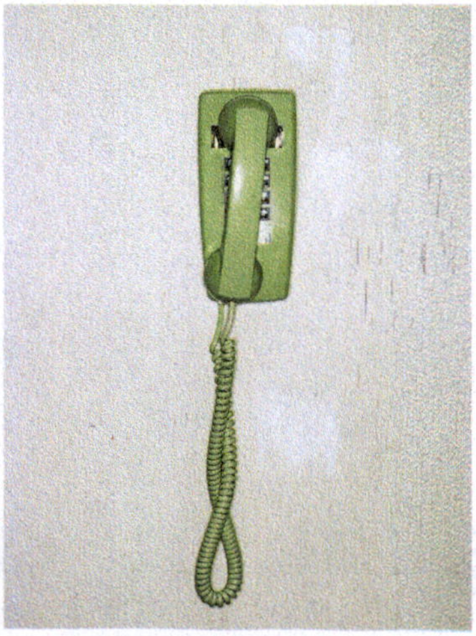

Steven Puetzer/Getty Images

8.16 More on Random Digit Dialing. By mid-2017, about 53.9% of adults lived in households with a cell phone and no landline phone. Among adults aged 25–29, this percentage was about 73.3%, while among adults over 65, the percentage was only 23.9%.[19]

(a) Write a survey question for which the opinions of adults with landline phones only are likely to differ from the opinions of adults with cell phones only. Give the direction of the difference of opinion.

(b) For the survey question in part (a), suppose a survey were conducted using random digit dialing of landline phones only. Would the results be biased? What would be the direction of bias?

(c) Most surveys now supplement the landline sample contacted by RDD with a second sample of respondents reached through random dialing of cell phone numbers. The landline respondents are weighted to take account of household size and number of telephone lines into the residence, whereas the cell phone respondents are weighted according to whether they were reachable only by cell phone or also by landline. Explain why it is important to include both a landline sample and a cell phone sample. Why is the number of telephone lines into the residence important? (*Hint*: How does the number of telephone lines into the residence affect the chance of the household being included in the RDD sample?)

CHAPTER 8 SUMMARY

- A **sample survey** selects a sample from the population of all individuals about which we desire information. We base conclusions about the population on data from the sample. It is important to specify exactly what population you are interested in and what variables you will measure.
- The design of a sample describes the method used to select the sample from the population. Random sampling designs use chance to select a sample.
- The basic random sampling design is a **simple random sample** (SRS). An SRS gives every possible sample of a given size the same chance to be chosen.
- Choose an SRS by labeling the members of the population and using **random digits** to select the sample. Software can automate this process.
- To choose a **stratified random sample,** classify the population into **strata,** groups of individuals that are similar in some way that is important to the response. Then choose a separate SRS from each stratum.
- Failure to use random sampling often results in **bias,** or systematic errors in the way the sample represents the population. **Voluntary response samples,** in which the respondents choose themselves, are particularly prone to large bias.
- In human populations, even random samples can suffer from bias due to **undercoverage** or **nonresponse,** from **response bias,** or from misleading results due to poorly worded questions. Sample surveys must deal expertly with these potential problems in addition to using a random sampling design.
- Most national sample surveys are carried out by telephone, using *random digit dialing* to choose residential telephone numbers at random. Because call screening is increasing nonresponse to such surveys, and the rise of cell-phone-only households is increasing undercoverage, many surveys include a minimum quota of cell phone users in their samples to help adjust for bias. *Web surveys* are becoming more frequent, but many suffer from volunteer response, undercoverage, and nonresponse.

CHECK YOUR SKILLS

8.17 An online store contacts 1000 customers from its list of customers who have purchased in the past year. In all, 696 of the 1000 say that they are very satisfied with the store's website. The population in this setting is

(a) all customers who have purchased something in the past year.

(b) the 1000 customers contacted.

(c) the 696 customers who were very satisfied with the store's website.

8.18 An opinion poll calls 2000 randomly chosen residential telephone numbers in Portland and asks to speak with an adult member of the household. The interviewer asks, "How many movies have you watched in a movie theater in the past 12 months?" In all, 831 people respond. The sample in this study is

(a) all adults living in Portland.

(b) the 2000 residential phone numbers called.

(c) the 831 people who responded.

8.19 The rate (percentage) of nonresponse in the previous question is

(a) 8.31%.

(b) 41.5%.

(c) 58.5%.

8.20 The National Health and Nutrition Examination Study (NHANES) had a random sample of 9317 participants recall their diet over the past 24 hours. The information in this sample was used in a recent study that found that, on average, 57.9% of the calories eaten by participants were obtained from ultra-processed foods that include substances not used in culinary preparations, such as flavors, colors, sweeteners, emulsifiers, and other additives. One of the limitations of the study reported by the authors was the dependence on the dietary recall of individuals.[20] The authors were concerned with

(a) response bias.

(b) undercoverage.

(c) overstratification.

8.21 Archaeologists plan to examine a sample of 2-meter-square plots near an ancient Greek city for artifacts visible in the ground. They choose separate samples of plots from floodplain, coast, foothills, and high hills. What kind of sample is this?

(a) A simple random sample

(b) A stratified random sample

(c) A voluntary response sample

8.22 You must choose an SRS of 10 of the 440 retail outlets in New York that sell your company's products. How would you label this population to select a simple random sample?

(a) 001, 002, 003, . . . , 439, 440

(b) 000, 001, 002, . . . , 439, 440

(c) 1, 2, . . . , 439, 440

8.23 You are using the table of random digits to choose a simple random sample of six students from a class of 30 students. You label the students 01 to 30, in alphabetical order, and then select a simple random sample. Which of the following is a possible sample that could be obtained?

(a) 45, 74, 04, 18, 07, 65

(b) 04, 18, 07, 13, 02, 07

(c) 04, 18, 07, 13, 02, 05

8.24 An online store takes a sample, selected at random, from a list of all people who have purchased an item from the store in the last year. The store sends each person selected an email inviting the person to take a brief online survey about his or her experience shopping at the online store. The store is only interested in the population of all its customers in the last year. The sample will certainly suffer from

(a) nonresponse.

(b) undercoverage.

(c) poor question wording.

8.25 The Pew Research Center Report titled "Libraries 2016," released September 9, 2016, asked a random sample of 1601 Americans aged 16 and over, "Have you personally ever visited a public library or used a public library bookmobile in person in the last 12 months?" In the entire sample, 48% said Yes. But only 40% of those in the sample over 65 years of age said Yes. Which of these two sample percentages will be more accurate as an estimate of the truth about the population?

(a) The result for those over 65 is more accurate because it is easier to estimate a proportion for a small group of people.

(b) The result for the entire sample is more accurate because it comes from a larger sample.

(c) Both are equally accurate because both come from the same sample.

CHAPTER 8 EXERCISES

In all exercises asking for an SRS, you may use software, the Simple Random Sample *applet, or Table B.*

8.26 **Importance of Childhood Vaccinations.** The debate about childhood vaccinations continued to be a major news issue in 2019, with the CDC reporting that in the period January 1–October 1, 2019, the United States experienced a record number of cases of measles, with 1249 cases from 31 states reported. This is the greatest number of cases reported in a single year since 1992. (Measles elimination was documented in the United States in 2000.) According to the same report by the CDC, "1107 (89%) [of measles cases] were in patients who were unvaccinated or had an unknown vaccination status." A Gallup World Poll asked the following question: "Do you agree, disagree, or neither agree nor disagree with the following statement? Vaccines are important for children to have." According to the survey methods section, the results of the poll are based on telephone interviews conducted July 12–August 23, 2018, with a random sample of 1006 people, aged 15 and older, living in the United States.[21]

(a) What is the population for this sample survey?

(b) What is the sample?

8.27 Racial Profiling and Traffic Stops. The Denver Police Department wants to know if Hispanic residents of Denver believe that the police use racial profiling when making traffic stops. A sociologist prepares several questions about the police. The police department chooses an SRS of 200 mailing addresses in predominantly Hispanic neighborhoods and sends a uniformed Hispanic police officer to each address to ask the questions of an adult living there.

(a) What are the population and the sample?

(b) Why are the results likely to be biased even though the sample is an SRS?

8.28 Race Relations. A 2018 Gallup poll showed that most Black Americans rate race relations with White people as bad. The poll is based on telephone interviews conducted November 19–December 22, 2018, with a random sample of 6502 adults, aged 18 and older, living in all 50 U.S. states and the District of Columbia. Landlines and cell phones were selected using random digit dialing.[22]

(a) The survey wants the opinion of an individual adult, but a landline phone reaches a household in which several adults may live. In that case, the survey interviewed the adult with the most recent birthday. Why is this preferable to simply interviewing the person who answers the phone?

(b) What is the population that this survey wants to describe? Why do you think it is important to include both landline and cell phones in your sample?

8.29 Sampling Greenville County. The rails to trails program involves the conversion of old rail corridors into multipurpose trails for recreation and transportation. Researchers were interested in obtaining information on characteristics of users and nonusers of a 10-mile-long paved greenway trail in Greenville, South Carolina, that connects residential areas to both a university campus and the commercial downtown area of the city. Random digit dialing of residential numbers was done using a database of exchanges. A total of 2461 persons were contacted, and 726 of them completed the survey. When a household was reached, surveyors asked to speak to the adult over 18 with the next birthday. No cell phone numbers were included in the sample.[23]

(a) What is the population of interest? What is the response rate for the survey?

(b) The following table gives the number of adults between 18 and 64 and over 65 in both the sample and the county (only 689 of the 726 respondents to the survey provided data on their ages). The county counts were obtained from the U.S. Census Bureau:

	In the Sample	In Greenville County
Age 18–64	436	356,123
Age 65 or older	253	105,176
Total	689	461,299

What percentage of the sample is between 18 and 64? What percentage of the population is between 18 and 64? Does this difference surprise you, given the sampling method described? Explain briefly.

(c) Among the 726 respondents, 181, or 24.9%, reported having used the trail in the past six months. Do you think this sampling method gives biased information about the percentage of Greenville County adults who have used the trail in the past six months? What is the likely direction of bias? Explain briefly.

8.30 Race Relations, Continued. The sample for the survey described in Exercise 8.28 included 4578 non-Hispanic White adults, 701 non-Hispanic Blacks, and 760 Hispanics. Results for the entire sample are considered to be accurate within ±2%. The margin of error for non-Hispanic Whites is ±2%, and for non-Hispanic Blacks and Hispanics, the margin of error is ±5%.

(a) Why is the accuracy greater for non-Hispanic Whites than for non-Hispanic Blacks and Hispanics?

(b) The total sample size was 6502. Why do you think such a large sample size was used? If we want to compare the responses of many different ethnic groups, explain briefly the need for a large overall sample size.

8.31 Paying Taxes. In April 2019, a Gallup Poll asked two questions about the amount one pays in federal income taxes.[24] Here are the two questions:

> Question A: *Do you regard the income tax which you will have to pay this year as fair?*
>
> Question B: *Do you consider the amount of federal income tax you have to pay as too high, about right, or too low?*

One of these drew 57% saying the amount was fair or about right; the other, 48%. Which wording produced the lower percentage? Why?

8.32 Online News Polls. Following the October 15, 2019, fourth Democratic debate, *Drudge Report* ran an online poll on its website and asked readers who won the debate. A total of 208,001 votes were cast, with Gabbard receiving 79,881 votes (38%), Yang receiving 37,944 votes (18%), Buttigieg receiving 25,337 votes (12%), Warren receiving 14,405 votes (7%), Kloubachar receiving 13,142 votes (6%), Biden receiving 11,246 votes (5%), Steyers receiving 7,791 votes (4%), Sanders receiving 7,672 votes (4%), and the remaining candidates each receiving 1% or fewer votes.[25] The sample size for this poll is much larger than is typical for polls such as the Gallup Poll. Explain why the poll may give unreliable information about the voting population, even with such a large sample size.

8.33 **Wording of Questions.** In August 2019, Gallup asked a random sample of 2291 American adults the question

> *Are you for or against a law which would make it illegal to manufacture, sell or possess semi-automatic guns known as assault rifles?*

47% said for. At the same time, Gallup also asked the question,

> *Do you think there should or should not be a ban on the manufacture, possession and sale of semi-automatic guns, known as assault rifles?*

61% said there should be.[26] Why do you think the results for the two questions differ by so much?

8.34 **Nonresponse.** Exercise 8.10 discusses the Pew Research Center survey *Teens, Social Media & Technology* conducted in the spring of 2018. The report mentions that 743 teens completed the survey and that the response rate for teens was 18%.[27] Approximately how many teens must have been recruited for the survey for a response rate of 18%?

8.35 **Running Red Lights.** A survey about driving habits by academic researchers produced a list of 5024 licensed drivers.[28] The investigators chose an SRS of 880 of these drivers to answer questions about their driving habits.

(a) How would you assign labels to the 5024 drivers? Choose the first 5 drivers in the sample. If you use Table B, start at line 118.

(b) One question asked was, "Recalling the last ten traffic lights you drove through, how many of them were red when you entered the intersections?" Of the 880 respondents, 171 admitted that at least one light had been red. A practical problem with this survey is that people may not give truthful answers. What is the likely direction of the bias: Do you think more or fewer than 171 of the 880 respondents really ran a red light? Why?

8.36 **Seat Belt Use.** A study in El Paso, Texas, looked at seat belt use by drivers. Drivers were observed at randomly chosen convenience stores. After they left their cars, they were invited to answer questions that included questions about seat belt use. In all, 75% said they always used seat belts, yet only 61.5% were wearing seat belts when they pulled into the store parking lots.[29] Explain the reason for the bias observed in responses to the survey. Do you expect bias in the same direction in most surveys about seat belt use?

8.37 **Student Opinions.** A university has 30,000 undergraduate and 10,000 graduate students. A survey of student opinion concerning health care benefits for domestic partners of students selects 300 of the 30,000 undergraduate students at random and then separately selects 100 of the 10,000 graduate students at random. The 400 students chosen make up the sample.

(a) What is the probability that any of the 30,000 undergraduates is in your random sample of 300 undergraduates selected? What is the probability that any of the 10,000 graduate students is in your random sample of 100 graduate students selected?

(b) If you have done the calculations correctly in part (a), the probability of any student at the university being selected is the same. Why is your sample of 400 students from the university not an SRS of students? Explain.

8.38 **Ring-no-answer.** A common form of nonresponse in telephone surveys is "ring-no-answer." That is, a call is made to an active number, but no one answers. The Italian National Statistical Institute looked at nonresponse to a government survey of households in Italy during the periods January 1 to Easter and July 1 to August 31. All calls were made between 7 and 10 P.M., but 21.4% gave "ring-no-answer" in one period versus 41.5% "ring-no-answer" in the other period.[30] Which period do you think had the higher rate of no answers? Why? Explain why a high rate of nonresponse makes sample results less reliable.

8.39 **Retweeters.** Twitter and Compete, a marketing services company, conducted a survey to investigate some of the characteristics of those who retweet (that is repost someone else's tweet). Among other findings, it was found the Twitter users who retweet are demographically similar to those who don't, use Twitter more often during the day, and are more likely to use Twitter on a mobile phone. Here is the methodology section contained with the survey results:

> The findings are based on data from surveys fielded in the United States during 2012. Twitter and Compete worked together to build a questionnaire that asked respondents about their propensity to use Twitter and other services as well as the when, where, how and why of their usage patterns. Compete interviewed 655 Internet users in the U.S. for this study.[31]

(a) Explain in simple language why it is important to know how the sample was selected when drawing conclusions about a survey.

(b) Do you feel the methodology section adequately explains how this sample was selected? Explain why or why not. If not, what information is lacking, and why is it important?

8.40 Sampling Pharmacists. All pharmacists in the Canadian province of Ontario are required to be members of the Ontario College of Pharmacists. At the end of 2018, here are the types of practices that they had:[32]

Practice Type	Number of Pharmacists
Community pharmacy	11,323
Hospital or other health care facility	2,664
Academia or government	333
Industry	500
Corporate office, professional practice, or clinic	167

Suppose the college is interested in obtaining members' views and understanding of the 2012 Expanded Scope of Practice Regulations, which authorizes pharmacists to provide additional services, including prescribing drug products for smoking cessation and administering the publicly funded influenza vaccine. To be sure that the opinions of all practice types are represented, you choose a stratified random sample of 10 pharmacists from each practice type. Explain how you will assign labels within each practice type and then give the label numbers for the 10 pharmacists in each of community pharmacy and industry who will be part of your sample. Use software, the *Simple Random Sample* applet, or Table B. If you use Table B, start at line 125 for community pharmacies and at line 133 for industry.

8.41 Sampling Amazon Forests. Stratified samples are widely used to study large areas of forest. Based on satellite images, a forest area in the Amazon basin is divided into 14 types. Foresters studied the four most commercially valuable types: alluvial climax forests of quality levels 1, 2, and 3, and mature secondary forest. They divided the area of each type into large parcels, chose parcels of each type at random, and counted tree species in a 20- by 25-meter rectangle randomly placed within each parcel selected. Here is some detail:

age fotostock/Superstock

Forest Type	Total Parcels	Sample Size
Climax 1	36	4
Climax 2	72	7
Climax 3	31	3
Secondary	42	4

Choose the stratified sample of 18 parcels. Be sure to explain how you assigned labels to parcels. If you use Table B, start at line 112.

8.42 Canadian Health Care Survey. The Thirteenth Annual Health Care in Canada Survey, conducted by POLLARA Research in May and June 2018, is a survey of the opinions of the Canadian public and health care providers on a variety of health care issues, including quality of health care, access to health care, health and the environment, and so forth. According to POLLARA, the survey was based on telephone interviews and included

> nationally representative samples of 1,500 members of the Canadian public, 100 doctors, 100 nurses, 100 pharmacists and 100 health managers. Public results are considered to be accurate within ±2.5%, while the margin of error for results for doctors, nurses, pharmacists and managers is ±9.8%.[33]

(a) Why is the accuracy greater for the public than for health care providers and managers?

(b) Why do you think the researchers sampled the public as well as health care providers and managers?

8.43 Systematic Random Samples. A *systematic random sample* goes through a list of the population at fixed intervals from a randomly chosen starting point. For example, a study of dating among college students chose a systematic sample of 200 single male students at a university as follows.[34] Start with a list of all 9000 single male students. Because $9000/200 = 45$, choose one of the first 45 names on the list at random and then every 45th name after that. For example, if the first name chosen is at position 23, the systematic sample consists of the names at positions 23, 68, 113, 158, and so on up to 8978.

(a) Choose a systematic random sample of 5 names from a list of 200. If you use Table B, enter the table at line 128.

(b) Like an SRS, a systematic sample gives all individuals the same chance to be chosen. Explain why this is true and explain carefully why a systematic sample is nonetheless *not* an SRS.

8.44 More on Systematic Sampling. Foresters were interested in studying a remote-sensing measure of standing timber as an alternative to taking measurements on the ground. The study area was a 1200-acre pine forest in Louisiana on which the U.S. Forest Service first created a grid of 1410 equally spaced circular plots of 0.05 acre in size over a map of the forest. The ground survey then visited every 10th plot and took measurements of tree volume.[35]

(a) Assuming that the plots are numbered from 1 to 1410, use the information in the previous exercise to explain how you will select a systematic sample of every 10th plot.

(b) Now choose a systematic random sample of 141 plots for the ground survey. If you use Table B, start at line 133.

8.45 **Why Random Digit Dialing Is Common.** The list of individuals from which a sample is actually selected is called the *sampling frame.* Ideally, the frame should list every individual in the population, but in practice, this is often difficult. A frame that leaves out part of the population is a common source of undercoverage.

(a) Suppose that a sample of households in a community is selected at random from the telephone directory. What households are omitted from this frame? What types of people do you think are likely to live in these households? These people will probably be underrepresented in the sample.

(b) It is usual in telephone surveys to use random digit dialing equipment that selects the last four digits of a telephone number at random after being given the exchange (the first three digits), as described in Example 8.5 (page 207). Which of the households that you mentioned in your answer to part (a) will be included in the sampling frame by random digit dialing?

8.46 **A Twitter Poll.** In September 2019, ProgressPolls asked the question "Do you think the Democrats have a valid reason for wanting to impeach Trump?" Of the 1437 votes cast, 91% said no. This poll was a Twitter poll. A CNN random digit dialing telephone poll of 1003 respondents in September 2019 asked the question "In your view, why do most Democrats in Congress support impeachment of Donald Trump?" Thirty eight percent responded "Because they are out to get Donald Trump at all costs." Explain to someone who knows no statistics why the two polls can give such widely differing results and which poll is likely to be more reliable.

8.47 **Wording Survey Questions.** Comment on each of the following as a potential sample survey question. Is the question sufficiently clear? Is it slanted toward a desired response?

(a) "In light of increasing threats from climate change, we should decrease our dependence on fossil fuels. Do you agree or disagree?"

(b) "Do you agree that a national system of health insurance should be favored because it would provide health insurance for everyone and would reduce administrative costs?"

(c) "In view of the negative externalities in parent labor force participation and pediatric evidence associating increased group size with morbidity of children in day care, do you support government subsidies for day care programs?"

8.48 **Your Own Bad Questions.** Write your own examples of bad sample survey questions.

(a) Write a biased question designed to get one answer rather than another.

(b) Write the "same question" in two different ways to get different responses.

(c) Write a question to which many people may not give truthful answers.

8.49 **The Canadian Census.** The Canadian government's decision to eliminate the mandatory long-form version of the census and to move these questions to an optional survey has many concerned. Many members of the business community and economists stressed the importance of the census data for crafting public policy. The minister of industry was given the task of defending the government's decision. In response to an argument that making the long form of the census voluntary would skew the data by eliminating the statistical randomness of the survey, the minister replied: "Wrong. Statisticians can ensure validity with a larger sample size."[36] Is the minister correct? If not, explain in simple terms the error in his statement.

8.50 **Election Polls.** In response to the question "If the 2016 presidential elections were being held today, would you vote for Hilary Clinton or Donald Trump?" the *New York Times* reported the result as 43% for Hilary Clinton and 39% for Donald J. Trump on July 7, 2016. This result was described as a "National Polling Average." Here are some details on how the average was computed:

> The New York Times polling averages use all polls currently listed in The Huffington Post's polling database. Polls conducted more recently and polls with a larger sample size are given greater weight in computing the averages, and polls with partisan sponsors are excluded.[37]

(a) Why do you think the surveyors gave greater weight to polls with larger sample sizes?

(b) Why should more recent polls be given greater weight? What population were the surveyors interested in on July 7, 2016, and how does that population continue to change over the election period?

(c) Why were polls with partisan sponsors excluded?

8.51 **Cluster Sampling.** *Cluster sampling* begins by dividing the population into separate groups, or clusters. An SRS of the clusters is selected, and individuals in the cluster are sampled. If all the individuals in a cluster are sampled, this is called one-stage cluster sampling. If a random sample of individuals in a cluster is sampled, this is called two-stage sampling. Cluster sampling can be convenient when the individuals in a cluster are easily sampled as a group, such as all people in a neighborhood for a door-to-door survey. Here is a simple example of one-stage cluster sampling. All students at a small college are required to live in dormitories. There are 25 such dormitories on campus, each with 30 students.

(a) To select a cluster sample of 150 students, do the following. Label the dormitories from 01 to 25. Choose an SRS of 5 dormitories from the list of the 25. If you use Table B, enter the table at line 121 and indicate which dormitories you selected. Your cluster sample is the 150 students in these dormitories.

(b) How many dormitories would you have to sample if you wanted a sample of 100 students?

CHAPTER

9

When you complete this chapter, you will be able to:

9.1 Distinguish observational studies from experiments.

9.2 Identify subjects, factors, and treatments in a statistical study.

9.3 Recognize how confounding lurking variables can invalidate uncontrolled experiments.

9.4 Use software or a table of random digits to randomly assign subjects to treatment groups in an experiment.

9.5 Explain how the principles of experimental design—control, randomization, and replication—help researchers to obtain results that are statistically significant and indicate a cause-and-effect relationship.

9.6 Explain the role of placebo treatment groups and blinding in randomized comparative experiments.

9.7 Design experiments using matched pairs and other block designs.

Producing Data: Experiments

A sample survey aims to gather information about a population without disturbing the population in the process. Sample surveys are one kind of *observational study*. Other observational studies observe the behavior of animals in the wild or the interactions between teacher and students in the classroom. This chapter is about statistical designs for *experiments*, a quite different means of producing data.

Experiments are used when the situation calls for a conclusion about whether a treatment causes a change in a response. The distinction between observational studies and experiments will be important when stating your conclusions in later chapters. Only well-designed experiments provide a sound basis for concluding cause-and-effect relationships.

9.1 Observation Versus Experiment

In contrast to observational studies, experiments don't just observe individuals or ask them questions. They actively impose some treatment to observe the response. Experiments can answer questions such as "Does aspirin reduce the chance of heart attack?" and "Do a majority of college students prefer Pepsi to Coke when they taste both without knowing which they are drinking?"

UPI/Alamy Stock Photo

Observation Versus Experiment

An **observational study** observes individuals and measures variables of interest but does not attempt to influence the responses. The purpose of an observational study is to describe some group or situation.

An **experiment,** on the other hand, deliberately imposes some treatment on individuals to observe their responses. The purpose of an experiment is to study whether the treatment causes a change in the response.

An observational study, even one based on a statistical sample, is not the best way to gauge the effect of a treatment. To see the response to a change, we must actually impose the change. *When our goal is to understand cause and effect, experiments are the preferred source for fully convincing data.* For this reason, the distinction between observational studies and experiments is one of the most important in statistics.

STATISTICS IN YOUR WORLD

You Just Don't Understand

A sample survey of journalists and scientists found quite a communications gap. Journalists think that scientists are arrogant, and scientists think that journalists are ignorant. We won't take sides, but here is one interesting result from the survey: 82% of the scientists agree that the "media do not understand statistics well enough to explain new findings" in medicine and other fields.

EXAMPLE 9.1 Does Playing Video Games Improve Surgical Skills?

In laparoscopic surgery, a video camera and several thin instruments are inserted into the patient's abdominal cavity. The surgeon uses the image from the video camera positioned inside the patient's body to perform the procedure by manipulating the instruments that have been inserted. Because of the similarity in many of the skills involved in video games and laparoscopic surgery, it was hypothesized that surgeons with greater prior video game experience might acquire the skills required in laparoscopic surgery more easily.

Thirty-three surgeons participated in the study and were classified into the three categories—never used, under three hours per day, and more than three hours per day—depending on the number of hours they played video games at the height of their video game use. It was found that those who reported more video game playing did better in a simulator program measuring laparoscopic skills. However, in their conclusions, the authors correctly point out, "This is a correlational (observational) study and, therefore, causality cannot be definitely determined."[1]

Although the data showed a clear association between prior video game experience and improved scores in the simulator program, we cannot conclude that more video game playing caused the improvement. People who play more video games may be different from those who don't, in terms of both their interest and the natural skills that are required in video games. Those who played more video games may have scored better simply because they have more ability in areas such as fine motor skills, eye-hand coordination, and depth perception that are required in both video games and laparoscopic surgery. The game players may have had these greater skills before they ever played video games.

It is easy to imagine an experiment that would settle the issue of whether playing video games really causes improvement in laparoscopic skills. Choose half of a group of surgeons at random to be the "treatment" group. The remaining half becomes the "control" group. Require the treatment group to play video games on a regular basis for several weeks and require the control group to abstain from video games. This experiment isolates the effect of playing video games. See Exercise 9.8 (page 233) for a description of such an experiment.

The point of Example 9.1 is the contrast between observing people who chose for themselves how many hours of video games to play and an experiment that requires some people play video games and others to abstain. When we simply observe people's video game choices, the effect of choosing to play more video games is confounded with (mixed up with) the characteristics of people who choose to play

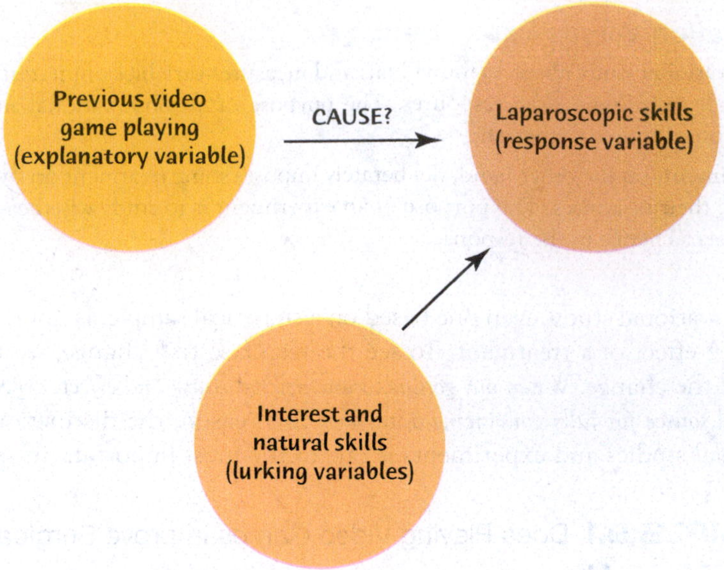

FIGURE 9.1
Confounding: we can't distinguish the effects of previous video game experience from the effects of interest and natural skills.

more. These characteristics are lurking variables (see page 143) that make it hard to see the true relationship between the explanatory and response variables. Figure 9.1 shows the confounding in picture form.

Confounding

Two variables (explanatory variables or lurking variables) are **confounded** when their effects on a response variable cannot be distinguished from each other.

Observational studies of the effect of one variable on another often fail because the explanatory variable is confounded with lurking variables. Well-designed experiments take steps to prevent confounding.

APPLY YOUR KNOWLEDGE

9.1 More Education Improves Driving? Although traffic fatalities have been decreasing for years, this decrease has not been experienced equally in all segments of the population. In fact, although the overall rate of traffic fatalities has been decreasing, the rate has declined the most for those with more education and has actually gone up for those without high school degrees. A recent study shows that among those over 25, as education level increased from less than high school, to high school grad, to some college, to college grad, the rate of motor vehicle crash deaths decreased.[2]

(a) What are the explanatory and response variables?

(b) Those with less education tend to drive cars that are older, have poorer crash test ratings, and have fewer safety features such as side airbags. Are the variables age of car, crash test rating, and presence of safety features explanatory variables, response variables, or lurking variables? Explain your reason.

(c) Is the *association* between traffic fatalities and education level good reason to think that a higher level of education actually *causes* an individual to be a safer driver? Explain why or why not.

9.2 The Font Matters! In general, when trying to change your behavior, if the effort required is perceived as high, this will be an impediment to change, whether it is modifying your diet or your study habits. Researchers divided 40 students into two groups of 20. The first group reads instructions for an exercise program printed in an easy-to-read font (Arial, 12 point), and the second group reads identical instructions in a difficult-to-read font (Brush, 12 point). Each subject estimates how many minutes the program will take and also uses a seven-point rating scale to report whether they are likely to include the exercise program as part of their daily routine (7 = very likely). The researchers hypothesized that those reading about the exercise program in the more difficult-to-read font would estimate that the program would take longer and, they would be less likely to make the exercise program part of their regular routine.[3] Is this an experiment? Why or why not? What are the explanatory and response variables?

9.3 Quitting Smoking and Risk for Type 2 Diabetes. Researchers studied a group of 10,892 middle-aged adults over a period of nine years. They found that smokers who quit had a higher risk of diabetes within three years of quitting than either nonsmokers or continuing smokers.[4] Does this show that stopping smoking causes the short-term risk for Type 2 diabetes to increase? (Weight gain has been shown to be a major risk factor for developing Type 2 diabetes and is often a side effect of quitting smoking. Smokers also often quit due to health reasons.) Based on this research, should you tell a middle-aged adult who smokes that stopping smoking can *cause* diabetes and advise him or her to continue smoking? Carefully explain your answers to both questions.

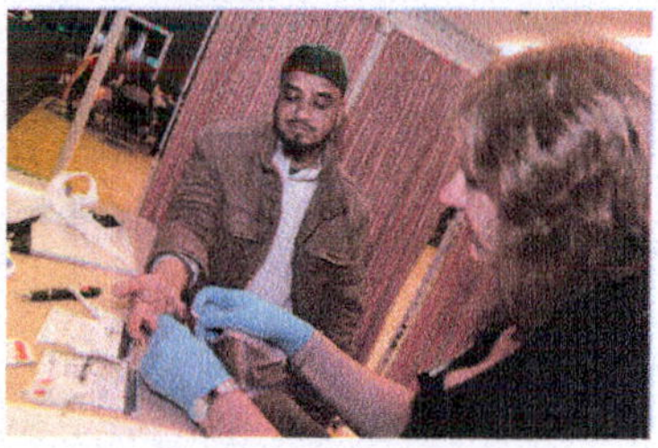
Paula Solloway/Alamy

9.2 Subjects, Factors, and Treatments

A study is an experiment when we actually do something to people, animals, or objects to observe the response. Because the purpose of an experiment is to reveal the response of one variable to changes in other variables, the distinction between explanatory and response variables is essential. Here is the basic vocabulary of experiments.

Subjects, Factors, and Treatments

The individuals studied in an experiment are often called **subjects,** particularly when they are people.

The explanatory variables in an experiment are often called **factors.**

A **treatment** is any specific experimental condition applied to the subjects. If an experiment has more than one factor, a treatment is a combination of specific values of each factor.

EXAMPLE 9.2 Foster Care Versus Orphanages

Do abandoned children placed in foster homes do better than similar children placed in an institution? The Bucharest Early Intervention Project found that the answer is a clear Yes. The *subjects* were 136 young children abandoned at birth and living in orphanages in Bucharest, Romania. Half of the children, chosen at random, were placed in foster homes. The other half remained in the orphanages. The experiment compared these two *treatments.* There is a single *factor,* "type of care," with two values, foster care and institutional care. When there is only one factor, the levels or values of the factor correspond to the treatments. The *response variables* included measures of mental and physical development.[5] (Foster care was not easily available in Romania at the time and so was paid for by the study. See Exercise 10.24 on page 261 in Chapter 10 for ethical questions concerning this experiment.)

EXAMPLE 9.3 Policy Justification: Pragmatic Versus Moral

How does a leader's justification of his or her organization's policy affect support for the policy? A study compared a moral, pragmatic, and ambiguous justification for both public and private policies using a sample of 300 *subjects*. As an example of a public policy, subjects read a politician's proposal to fund a retirement planning agency. The moral justification was the importance of retirees "to live with dignity and comfort," the pragmatic was "to not drain public funds," and the ambiguous was "to have sufficient funds." For a private policy, subjects read about a CEO's plan to provide healthy meals for the employees. The moral justification was increased access to meals "should improve our employees' well-being," the pragmatic was "to improve our employees' productivity," and the ambiguous was "to improve the status-quo."[6]

This experiment has two *factors:* type of justification used for the proposal with three values and public versus private policies with two values. The six combinations of one value of each factor form the six *treatments.* Figure 9.2 shows the layout of the treatments. The subjects were divided into six groups of size 50, each assigned one of the treatments. For example, those in group 3 read a public policy statement with an ambiguous justification. After reading the proposals for one of the treatment conditions, the subjects answered questions to measure their support of the policy, the moral character of the leader, and the policy ethicality. These are the *response variables.*

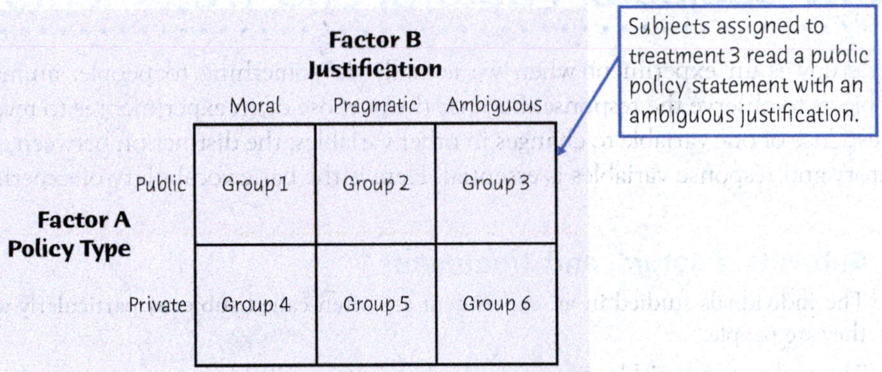

FIGURE 9.2 The treatments in the experimental design of Example 9.3. Combinations of values of the two factors form six treatments.

Examples 9.2 and 9.3 illustrate the advantages of experiments over observational studies. In an experiment, we can study the effects of the specific treatments we are interested in. By assigning subjects to treatments randomly, we can avoid confounding. For example, observational studies of the effects of foster homes versus institutions on the development of children have often been biased because healthier or more alert children tend to be placed in homes. The random assignment in Example 9.2 eliminates bias in placing the children. Moreover, we can control the environment of the subjects to hold constant factors that are of no interest to us, such as the specific proposals read in Example 9.3.

Another advantage of experiments is that we can study the combined effects of several factors simultaneously. The interaction of several factors can produce effects that could not be predicted from looking at the effect of each factor alone. Perhaps a moral justification improves support for public policies but not for private policies. The two-factor experiment in Example 9.3 will help us find out.

APPLY YOUR KNOWLEDGE

For the experiments described in Exercises 9.4 and 9.5, identify the subjects, the factors, the treatments, and the response variables.

9.4 Adolescent Obesity. Adolescent obesity is a serious health risk affecting more than 5 million young people in the United States alone. Laparoscopic adjustable gastric banding has the potential to provide a safe and effective treatment. Fifty adolescents between 14 and 18 years old with a body mass index higher than 35 were recruited from the Melbourne, Australia, community for the study. Twenty-five were randomly selected to undergo gastric banding, and the remaining 25 were assigned to a supervised lifestyle intervention program involving diet, exercise, and behavior modification. All subjects were followed for two years, and their weight loss was recorded.[7]

9.5 Sounds Big. Does the lower pitch of a voice in an ad lead consumers to envision a bigger product? To test this, researchers had students listen to a radio advertisement for the new Southwest Turkey Club Sandwich at a fictitious sandwich chain, Cosmo. Half the students were randomly assigned to hear the ad spoken at a high pitch and the other half at a low pitch. In all other respects, the ads were identical, and no clues were given as to the size of the sandwich. After hearing the ad, students were asked to rate the perceived size of the sandwich on a 7-point scale, ranging from −3 (much smaller than average) to +3 (much larger than average).[8]

9.6 Ripening Mangoes. The mango is considered the "king of fruits" in many parts of the world. Mangoes are generally harvested at the mature green stage and ripen up during the marketing process of transport, storage, and so on. During this process, about 30% of the fruit is wasted. Because of this, the impact of harvest stage and storage conditions on the postharvest quality are of interest. In one experiment, the fruit was harvested at 80, 95, or 110 days after the fruit setting (the transition from flower to fruit) and then stored at temperatures of 20°C, 30°C, or 40°C. For each harvest time and storage temperature, a random sample of mangoes was selected, and the time to ripening was measured.[9]

(a) What are the factors, the treatments, and the response variables? Use a diagram like Figure 9.2 to display the factors and treatments.

(b) For simplicity, the experimenter thought about selecting nine mango trees and randomly assigning one tree to each of the nine treatments. All the mangoes on a tree would then receive the same treatment. Do you think this would be a good way to assign the mangoes to the treatments? Explain your reasoning in simple English.

9.3 How to Experiment Badly

Experiments are the preferred method for examining the effect of one variable on another. By imposing the specific treatment of interest and controlling other influences, we can pin down cause and effect. Statistical designs are often essential for effective experiments. To see why, let's look at an example in which an experiment suffers from confounding just as observational studies do.

EXAMPLE 9.4 An Uncontrolled Experiment

A college regularly offers a review course to prepare candidates for the Graduate Management Admission Test (GMAT), which is required by most graduate business schools. This year, it offers only an online version of the course. The average GMAT score of students in the online course is 10% higher than the longtime average for those who took the classroom review course. Is the online course more effective?

This experiment has a very simple design. A group of subjects (the students) were exposed to a treatment (the online course), and the outcome (GMAT scores) was observed. Here is the design:

Subjects → Online course → GMAT scores

A closer look at the GMAT review course showed that the students in the online review course were quite different from the students who in past years had taken the classroom course. In particular, they were older and more likely to be employed. An online course appeals to these mature people, but we can't compare their performance with that of the undergraduates who previously dominated the course. The online course might even be less effective than the classroom version. The effect of online versus in-class instruction is confounded with the effect of lurking variables. As a result of confounding, the experiment is biased in favor of the online course.

The simple design of this experiment failed to control for possible differences between the students who took the online course and students in the past who took the classroom course. Would the situation have been different if both the online and the classroom courses had been given this year? If students still chose the course they wanted, with older students tending to sign up for the online course and younger students tending to sign up for the classroom course, then the effect of type of course would still be confounded with the lurking variable age. Again, such an experiment fails to control for possible differences between those who choose to sign up for the online course compared to those who choose the classroom course. The solution to such uncontrolled experiments will be described in the next section.

Many laboratory experiments use a design like that in Example 9.4:

Subjects → Treatment → Measure response

In the tightly monitored environment of the laboratory with inanimate objects as the individuals, simple designs often work well. Field experiments and experiments with living subjects are exposed to more variable conditions and deal with more variable subjects. *Outside the laboratory, uncontrolled experiments often yield worthless results because of confounding with lurking variables.*

APPLY YOUR KNOWLEDGE

9.7 Unhappy Marriage, Unhappy Gut. To investigate how an unhappy marriage can affect an individual's health, scientists recruited 43 healthy couples between 24 and 61 years old who had been married for at least three years to take part in an experiment.[10] The researchers asked couples to discuss touchy topics likely to spark disagreement, such as money or in-laws, and taped the conversations. They used this footage to analyze verbal and nonverbal modes of conflict, including eye rolls. The team also took blood samples from the couples before and after arguing, and found that those who were most hostile toward their spouses had higher levels of LPS-binding protein, a biomarker for a leaky gut. Couples choose to argue and engage in hostile behavior when discussing touchy subjects. And anger and unhappiness that can lead to fighting may be symptoms of a physiological or mental health problem. Explain why these facts make any conclusion about cause and effect untrustworthy. Use the language of lurking variables and confounding in your explanation.

9.4 Randomized Comparative Experiments

The remedy for the confounding in Example 9.4 is to be sure that we do a *comparative experiment* in which some students are taught in the classroom and other, similar students take the course online. In Example 9.4, the classroom group is called a "control group." A **control group** receives either a standard treatment (which may be no treatment at all) or, in some cases, a sham treatment, and provides a basis for comparison with the other treatment groups. Although, in many experiments, one of the treatments is a control treatment, a comparative experiment does not require this. In Example 9.4, a comparative experiment could simply compare two newly developed online courses without including a control or classroom treatment. Most well-designed experiments compare two or more treatments (one of which may be a sham treatment or no treatment at all). Part of the design of an experiment is a description of the factors (explanatory variables) and the layout of the treatments, with comparison as the leading principle.

However, as discussed at the end of Example 9.4, comparison alone isn't enough to produce results we can trust. If the treatments are given to groups that differ markedly when the experiment begins, bias will result. If we allow students to select online or classroom instruction, students who are older and employed are likely to sign up for the online course. Personal choice will bias our results in the same way that volunteers bias the results of online opinion polls. The solution to the problem of bias in sampling is random selection, and the same is true in experiments. The subjects assigned to any treatment should be chosen at random from the available subjects.

Randomized Comparative Experiment

An experiment that uses both comparison of two or more treatments and random assignment of subjects to treatments is a **randomized comparative experiment**.

EXAMPLE 9.5 Classroom Versus Online

The college decides to compare the progress of 25 on-campus students taught in the classroom with that of 25 students taught the same material online. Select the students who will be taught online by taking a simple random sample of size 25 from the 50 available subjects. The remaining 25 students form the control group. They will receive classroom instruction. The result is a randomized comparative experiment with two groups. Figure 9.3 outlines the design in graphical form.

The selection procedure is exactly the same as it is for sampling.

LABEL: Label the 50 students 01 to 50.

TABLE: Go to Table B and read successive two-digit groups. The first 25 labels encountered select the online group. As usual, ignore repeated labels and groups of

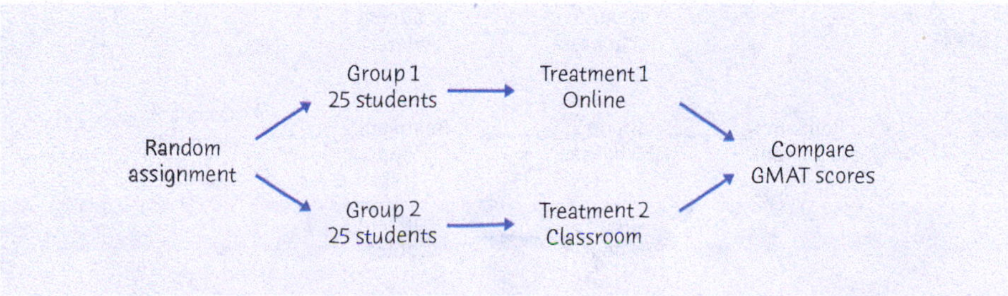

FIGURE 9.3 Outline of a randomized comparative experiment to compare online and classroom instruction, for Example 9.5.

digits not used as labels. For example, if you begin at line 125 in Table B, the first five students chosen are those labeled 21, 49, 37, 18, and 44. Software such as the *Simple Random Sample* applet makes it particularly easy to choose treatment groups at random.

applet

The design in Example 9.5 is *comparative* because it compares two treatments (the two instructional settings). It is *randomized* because the subjects are assigned to the treatments by chance. This flowchart outline in Figure 9.3 presents all the essentials: randomization, the sizes of the groups and which treatment they receive, and the response variable. There are, as we will see later, statistical reasons for generally using treatment groups that are about equal in size. We call designs like that in Figure 9.3 *completely randomized.*

Completely Randomized Design

In a **completely randomized** experimental design, all the subjects are allocated at random among all the treatments.

Completely randomized designs can compare any number of treatments. Here is an example that compares three treatments.

EXAMPLE 9.6 Conserving Energy

Maskot/Getty Images

Many utility companies have introduced programs to encourage energy conservation among their customers. An electric company considers placing electronic meters in households to show what the cost would be if the electricity use at that moment continued for a month. Will meters reduce electricity use? Would cheaper methods work almost as well? The company decides to design an experiment.

One cheaper approach is to give customers an app and information about using the app to monitor their electricity use. The experiment compares these two approaches (meter, app) and also a control. The control group of customers receives information about energy conservation but no help in monitoring electricity use. The response variable is total electricity used in a year. The company finds 60 single-family residences in the same city willing to participate, so it assigns 20 residences at random to each of the three treatments. Figure 9.4 outlines the design.

applet

To use the *Simple Random Sample* applet, set the population labels as 1 to 60 by moving the "Population size" toggle and the sample size to 20 by moving the "Sample size" toggle and toggle the "Generate a sample" button. The 20 households chosen receive the displays. Continue toggling the "Generate a sample" button until you get 20 more unique values, separate from the original 20. This next sample of 20 households will receive charts. The 20 remaining households form the control group.

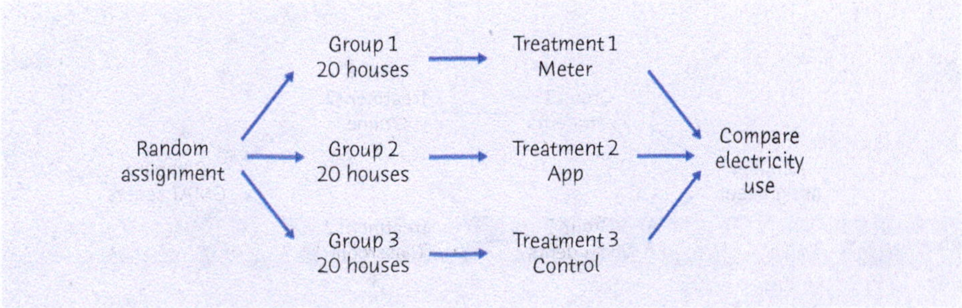

FIGURE 9.4
Outline of a completely randomized design comparing three energy-saving programs, for Example 9.6.

To use Table B, label the 60 households 01 to 60. Enter the table to select an SRS of 20 to receive the displays. Continue in Table B, selecting 20 more to receive charts. The remaining 20 form the control group.

Examples 9.5 and 9.6 describe completely randomized designs that compare values of a single factor. In Example 9.5, the factor is the type of instruction. In Example 9.6, it is the method used to encourage energy conservation. Completely randomized designs can have more than one factor. The policy justification experiment of Example 9.3 (page 228) has two factors: type of justification and public versus private policy. Their combinations form the six treatments outlined in Figure 9.2 (page 228). A completely randomized design assigns subjects at random to these six treatments. Once the layout of treatments is set, the randomization needed for a completely randomized design is tedious but straightforward.

APPLY YOUR KNOWLEDGE

9.8 Does Playing Video Games Improve Surgical Skills? Another Look. In laparoscopic surgery, a video camera and several thin instruments are inserted into the patient's abdominal cavity. The surgeon uses the image from the video camera positioned inside the patient's body to perform the procedure by manipulating the instruments that have been inserted. It has been found that the Nintendo Wii, with its motion-sensing interface, reproduces the movements required in laparoscopic surgery more closely than do other video games. If training with a Nintendo Wii can improve laparoscopic skills, it can complement the more expensive training on a laparoscopic simulator. Forty-two medical residents were chosen, and all were tested on a set of basic laparoscopic skills. Twenty-one were selected at random to undergo systematic Nintendo Wii training for one hour a day, five days a week, for four weeks. The remaining 21 residents were the control group and received no Nintendo Wii training and were asked to refrain from video games during this period. At the end of four weeks, all 42 residents were tested again on the same set of laparoscopic skills. The difference in time (before − after) to complete a virtual gall bladder removal on the simulator was measured.[11]

(a) Compare the study described here with the study in Example 9.1 (page 225). What are the important differences between the two studies? Ignoring the fact that different video games and measures of laparoscopic skill were used in the two studies, explain in simple language which study gives stronger evidence that playing video games is helpful in improving laparoscopic skills and why.

(b) Outline the design of this experiment, following the model of Figure 9.3 (page 231). What is the response variable?

(c) Carry out the random assignment of 21 residents to the Nintendo training group, using the *Simple Random Sample* applet, other software, or Table B, starting at line 130.

9.9 Shared Pain and Bonding. Although painful experiences are involved in social rituals in many parts of the world, little is known about the social effects of pain. Will sharing painful experiences in a small group lead to greater bonding of group members than sharing a similar nonpainful experience? Twenty-seven of 54 university students in New South Wales, Australia, were assigned at random into a pain group, with the remaining students in the no-pain group. Pain was induced by two tasks. In the first task, students submerged their hands in freezing water for as long as possible, moving metal balls at the bottom of the vessel into a submerged container; in the second task, students performed a standing wall squat with back straight and knees at 90 degrees for as long as possible. The no-pain group completed the first task using room temperature water for 90 seconds, and the second task by balancing on one foot for 60 seconds, changing feet if necessary. In both the pain and no-pain settings, the students completed the tasks in small groups, which typically consisted of four students and contained similar levels of group interaction. Afterward, each student completed a questionnaire to create a bonding score based on answers to questions such as "I feel the participants in this study have a lot in common" or "I feel I can trust the other participants."[12]

(a) Outline the design of the experiment, following the model of Figure 9.4.

(b) Explain how you will randomly assign the subjects at random to the two groups and then carry out this randomization using software, the *Simple Random Sample* applet, or Table B, beginning at line 125.

(c) Why do you think the experimenter had students in the no-pain group complete similar pain-free tasks in small groups? Do you think this is important for the type of conclusion that can be reached? Explain.

9.10 Policy Justification. Figure 9.2 (page 228) displays the six treatments for the two-factor experiment on policy justification. The 30 volunteers named here will serve as subjects. Outline the design and randomly assign the subjects to the six treatments, an equal number of subjects to each treatment. If you use Table B, start at line 133.

Aaron	Alexander	Banks	Campanella	Duffy	Foxx	Greenberg
Herman	Hubbell	Johnson	Kiner	Lajoie	Lemon	Lombardi
Mize	Newhouser	Ott	Paige	Palmer	Robinson	Ruffing
Sisler	Speaker	Terry	Traynor	Vance	Wagner	Waner
Williams	Young					

9.5 The Logic of Randomized Comparative Experiments

Randomized comparative experiments are designed to give good evidence that differences in the treatments actually *cause* the differences we see in the response. The logic is as follows:

- Random assignment of subjects forms groups that should be similar in all respects before the treatments are applied. Exercise 9.52 (page 249) demonstrates this.
- A comparative experiment with randomization ensures that influences other than the experimental treatments operate equally on all groups.
- Therefore, differences in average response must be due either to the treatments or to the play of chance in the random assignment of subjects to the treatments.

That "either-or" deserves more thought. In Example 9.5, we cannot say that *any* difference between the average GMAT scores of students enrolled online and in the classroom must be caused by a difference in the effectiveness of the two types of instruction. There would be some difference even if both groups received the same instruction because of variation among students in background and study habits. Chance assigns students to one group or the other, and this creates a chance difference between the groups. We would not trust an experiment with just one student in each group, for example. The results would depend too much on which group got lucky and received the stronger student. If we replicate the experiment by assigning many subjects to each group, the effects of chance will average out, and there will be little difference in the average responses in the two groups unless the treatments themselves cause a difference. *Replication* or the use of enough subjects to reduce chance variation, is the third big idea of statistical design of experiments.

Principles of Experimental Design

The basic principles of statistical design of experiments are

1. **Control**—restrict the effects of lurking variables on the response, most simply by comparing two or more treatments.
2. **Randomization**—use chance to assign subjects to treatments.
3. **Replication**—use enough subjects in each group to reduce chance variation in the results.

We hope to see a difference in the responses between the two groups so large that it is unlikely to happen just because of chance variation. We can use the laws of probability, which describe chance behavior, to learn if the treatment effects are larger than we would expect to see if only chance were operating. If they are, we call them *statistically significant.*

Statistical Significance

An observed effect so large that it would rarely occur by chance is called **statistically significant.**

If we observe statistically significant differences among the groups in a randomized comparative experiment, we have good evidence that the treatments actually caused these differences. You will often see the phrase "statistically significant" in reports of investigations in many fields of study. The great advantage of randomized comparative experiments is that they can produce data that give good evidence for a cause-and-effect relationship between the explanatory and response variables. We know that in general a strong association does not imply causation. A statistically significant association in data from a well-designed experiment *does* imply causation.

APPLY YOUR KNOWLEDGE

9.11 Prayer and Meditation. You read in a magazine that "nonphysical treatments such as meditation and prayer have been shown to be effective in controlled scientific studies for such ailments as high blood pressure, insomnia, ulcers, and asthma." Explain in simple language what the article means by "controlled scientific studies." Why can such studies in principle provide good evidence that, for example, meditation is an effective treatment for high blood pressure?

9.12 Conserving Energy. Example 9.6 (page 232) describes an experiment to learn whether providing households with digital displays or charts will reduce their electricity consumption. An executive of the electric company objects to including a control group. He says: "It would be simpler to just compare electricity use last year (before the meter or app was provided) with consumption in the same period this year. If households use less electricity this year, the display or chart must be working." Explain clearly why this design is inferior to that in Example 9.6.

STATISTICS IN YOUR WORLD
What's News?

Randomized comparative experiments provide the best evidence for medical advances. Do newspapers care? Maybe not. University researchers looked at 1192 articles in medical journals, of which 7% were turned into stories by the two newspapers examined. Of the journal articles, 37% concerned observational studies, and 25% described randomized experiments. Among the articles publicized by the newspapers, 58% were observational studies, and only 6% were randomized experiments. Conclusion: the newspapers want exciting stories, especially bad-news stories, whether or not the evidence is good.

STATISTICS IN YOUR WORLD
Correlation = Causation?

Randomized comparative experiments are the best way to determine that differences in treatments actually *cause* the differences we see in the response. Observational studies that find an association between differences in treatments and differences observed in a response do not provide a sound basis for concluding treatments cause the differences in the response. Do researchers recognize this? A 2019 study in the journal *Psychology of Aesthetics, Creativity, and the Arts* evaluated 114 published articles in which observational studies found an association between music training and a nonmusical ability, brain structure, or brain function. Of these, 72 attributed the observed association to causation. Apparently, even researchers fall prey to the temptation to conclude correlation = causation.

9.13 Healthy Diet and Cataracts. The relationship between healthy diet and prevalence of cataracts was assessed using a sample of 1808 participants from the Women's Health Initiative Observational Study. Having a high Healthy Eating Index score was the strongest predictor of a reduced risk of cataracts, among modifiable behaviors considered. The Healthy Eating Index score, created by the U.S. Department of Agriculture, measures how well a person's diet conforms to recommended healthy eating patterns. The report concludes: "These data add to the body of evidence suggesting that eating foods rich in a variety of vitamins and minerals may contribute to postponing the occurrence of the most common type of cataract in the United States."[13]

(a) Explain why this is an observational study rather than an experiment.

(b) Although the result was statistically significant, the authors did not use strong language in stating their conclusions, instead using words such as *suggesting* and *may.* Do you think that their language is appropriate, given the nature of the study? Why?

9.6 Cautions About Experimentation

The logic of a randomized comparative experiment depends on our ability to treat all the subjects identically in every way except for the actual treatments being compared. Good experiments therefore require careful attention to detail to ensure that all subjects really are treated identically.

STATISTICS IN YOUR WORLD

Scratch My Furry Ears

Rats and rabbits, specially bred to be uniform in their inherited characteristics, are the subjects in many experiments. Animals, like people, are quite sensitive to how they are treated. This can create opportunities for hidden bias. For example, human affection can change the cholesterol level of rabbits. Choose some rabbits at random and regularly remove them from their cages to have their heads scratched by friendly people. Leave other rabbits unloved. All the rabbits eat the same diet, but the rabbits that receive affection have lower cholesterol.

If some subjects in a medical experiment take a pill each day and a control group takes no pill, the subjects are not treated identically. Many medical experiments are therefore *placebo controlled.* A study of the effects of taking vitamin E on heart disease is a good example. All the subjects receive the same medical attention during the several years of the experiment. All take a pill every day: vitamin E in the treatment group and a placebo in the control group. A **placebo** is a dummy treatment that is as similar to the treatment as possible but contains no active ingredient. In this experiment, the placebo would be a pill that looked like the vitamin E pill but contained no active ingredient. As a second example, a study compared arthroscopic surgery versus no-surgery on several recovery outcomes for a partial meniscus tear.[14] Patients randomly assigned to the no-surgery group received a sham surgery in which the surgeon asked for all instruments, manipulated the knee as in surgery, and kept the sham surgery sufficiently realistic so that afterward patients were unaware of whether they had received the actual surgery or the sham surgery.

Many patients respond favorably to any treatment, even a placebo, perhaps because they trust the doctor or they believe that the treatment will work. The favorable response to a placebo treatment or a treatment with no therapeutic value is called the *placebo effect.* If the control group did not take any pills, the effect of vitamin E in the treatment group would be confounded with the placebo effect, the effect of simply taking pills. That is, if the vitamin E group improved, we wouldn't know whether it was due to simply taking a pill or the actual vitamin E content of the pill. Having a placebo group for comparison allows us to see if the effect of taking a vitamin E pill in the treatment group is larger than the effect of simply taking a pill in the placebo group.

In addition, such studies are usually *double-blind.* The subjects don't know whether they are taking vitamin E or a placebo. Neither do the medical personnel who work with them. The double-blind method avoids unconscious bias by, for example, a doctor who is convinced that a vitamin must be better than a placebo. In many medical studies, only the statistician who does the randomization knows which treatment each patient is receiving.

Double-Blind Experiments

In a **double-blind** experiment, neither the subjects nor the people who interact with them know which treatment each subject is receiving.

When testing a drug against a placebo, we indicated that the placebo contains no active ingredient, but the situation can be more complex than this would suggest. Since many drugs being tested have known side effects, such as dry mouth, an *active placebo* may be used. An active placebo contains ingredients designed to mimic the side effects of the drug but does not treat the particular disease.[15] This is important in blinding as it can prevent the subject from being aware of whether he or she received the drug or the placebo. Drug companies create these placebo drugs, and many do not provide information regarding the contents of an active placebo. Unfortunately, this then allows drug companies to make advertising claims such as that the occurrence of dry mouth was similar in the drug and the placebo group.

Placebo controls and the double-blind method are more ways to eliminate possible confounding. But even well-designed experiments often face another problem: *lack of realism.* Practical constraints may mean that the subjects or treatments or setting of an experiment don't realistically duplicate the conditions we really want to study. Here are two examples.

EXAMPLE 9.7 Studying Frustration

A psychologist wants to study the effects of failure and frustration on the relationships among members of a work team. She forms a team of students, brings them to the psychology laboratory, and has them play a game that requires teamwork. The game is rigged so that they lose regularly. The psychologist observes the students through a one-way window and notes the changes in their behavior during an evening of game playing.

Playing a game in a laboratory for small stakes, knowing that the session will soon be over, is a long way from working for months developing a new product that never works right and is finally abandoned by your company. Does the behavior of the students in the lab tell us much about the behavior of the team whose product failed? Many behavioral science experiments use as subjects students or other volunteers who know they are subjects in an experiment. That's often not a realistic setting.

EXAMPLE 9.8 Center Brake Lights

Do those high center brake lights, required on all cars sold in the United States since 1986, really reduce rear-end collisions? Randomized comparative experiments with fleets of rental and business cars done before the lights were required showed that the third brake light reduced rear-end collisions by as much as 50%. Alas, requiring the third light in all cars led to only a 5% drop.

What happened? Most cars did not have the extra brake light when the experiments were carried out, so it caught the eye of following drivers. Now that almost all cars have the third light, these lights no longer capture attention.

Lack of realism can limit our ability to apply the conclusions of an experiment to the settings of greatest interest. Most experimenters want to generalize their conclusions to some setting wider than that of the actual experiment. *Statistical analysis of an experiment cannot tell us how far the results will generalize.* Nonetheless, the randomized comparative experiment, because of its ability to give convincing evidence for causation, is one of the most important ideas in statistics.

APPLY YOUR KNOWLEDGE

9.14 Prayer and Healing. To study the effect of prayer on healing, patients with health problems are randomly divided into two groups. In one group, intercessors pray for the health of the patients. In the other group, patients are not prayed for. Patients do not know that they are being prayed for, and the persons who are praying do not come in contact with the patients for whom they pray. Medical outcomes in the two groups of patients are compared. Finally, the medical treatment team is also blind to the prayer group status of individual patients.[16] Is this experiment double-blind? What are the treatments? Is one of the treatments a placebo? Explain your answers.

9.15 Eggs and Cholesterol. An article in a medical journal reports on an experiment to see the effect on cholesterol levels of eating three whole eggs per day compared to eating the equivalent of a yolk-free egg substitute. The article describes the experiment as a randomized, single-blinded experiment of 37 subjects with metabolic syndrome.[17] What do you think "single-blinded" means here? Why isn't a double-blind experiment possible?

9.7 Matched Pairs and Other Block Designs

Completely randomized designs are the simplest statistical designs for experiments. They illustrate clearly the principles of control, randomization, and adequate number of subjects. However, completely randomized designs are often inferior to more elaborate statistical designs. In particular, matching the subjects in various ways can produce more precise results than simple randomization.

One common design that combines matching with randomization is the **matched pairs design.** A matched pairs design compares just two treatments. Choose pairs of subjects that are as closely matched as possible. Use chance to decide which subject in a pair gets the first treatment. The other subject in that pair gets the other treatment. That is, the random assignment of subjects to treatments is done within each matched pair, not for all subjects at once. Sometimes each "pair" in a matched pairs design consists of just one subject, who gets both treatments one after the other. Each subject serves as his or her own control. The *order* of the treatments can influence the subject's response, so we randomize the order for each subject.

matched pairs design Design of an experiment that compares two treatments. Each subject receives both treatments in random order, or the subjects are matched in pairs that are similar in some way that is expected to affect the response, and one subject in each pair receives each treatment.

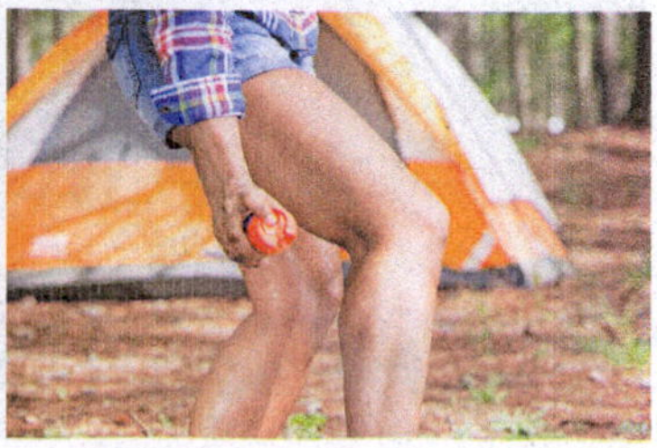
poweroffforever/Getty Images

EXAMPLE 9.9 Testing Insect Repellants

Consumer Reports describes a method for comparing the effectiveness of two insect repellants. The active ingredient in one is 15% Deet. The active ingredient in the other is oil of lemon eucalyptus. Beginning 30 minutes after applying the repellant, once every hour, volunteer subjects put an arm in an 8-cubic-foot cage containing 200 disease-free female mosquitoes in need of a blood meal to lay their eggs. Subjects leave their arm in the cage for five minutes. The repellant is considered to have failed if a subject is bitten two or more times in a five-minute session. The response is the number of one-hour sessions until a repellant fails. Let's compare two designs for this experiment.

In design 1, the *completely randomized design,* all subjects are assigned at random to one of the two repellants: half to the repellant with 15% Deet and the other half to the repellant with oil of lemon eucalyptus. In design 2, the *matched pairs design* that was actually used, all subjects use both repellants. For each subject, the left arm is sprayed with one of the repellants and the right arm with the other. To guard against the possibility that responses may depend on which arm is sprayed, which arm receives which repellant is determined randomly.[18]

For reasons that are mostly genetic, some subjects are more susceptible to mosquito bites than others. The completely randomized design relies on chance to distribute the more susceptible subjects roughly evenly between the two groups. The matched pairs design compares each subject's response using both types of repellant. This makes it easier to see the difference in the effects of the two repellants.

Matched pairs designs use the principles of comparison of treatments and randomization. However, the randomization is not complete: we do not randomly assign all the subjects at once to the two treatments; instead we randomize only within each matched pair. This allows matching to reduce the effect of variation among the subjects. Matched pairs are one kind of *block design,* with each pair forming a *block.*

Block Design

A **block** is a group of individuals that are known before an experiment to be similar in some way that is expected to affect the response to the treatments.

In a **block design,** the random assignment of individuals to treatments is carried out separately within each block.

A block design combines the idea of creating equivalent treatment groups by matching with the principle of forming treatment groups at random. Blocks are another form of *control.* They control the effects of some outside variables by bringing those variables into the experiment to form the blocks. Here are some typical examples of block designs.

EXAMPLE 9.10 Men, Women, and Advertising

Women and men respond differently to advertising. An experiment to compare the effectiveness of three advertisements for the same product will want to look separately at the reactions of men and women, as well as to assess the overall response to the ads.

A *completely randomized design* considers all subjects, both men and women, as a single pool. The randomization assigns subjects to three treatment groups without regard to their sex. This ignores the differences between men and women. A *block design* considers women and men separately. Randomly assign the women to three groups, with one to view each advertisement. Then separately assign the men at random to three groups. Figure 9.5 outlines this improved design. Notice the assignment of subjects to blocks is not random.

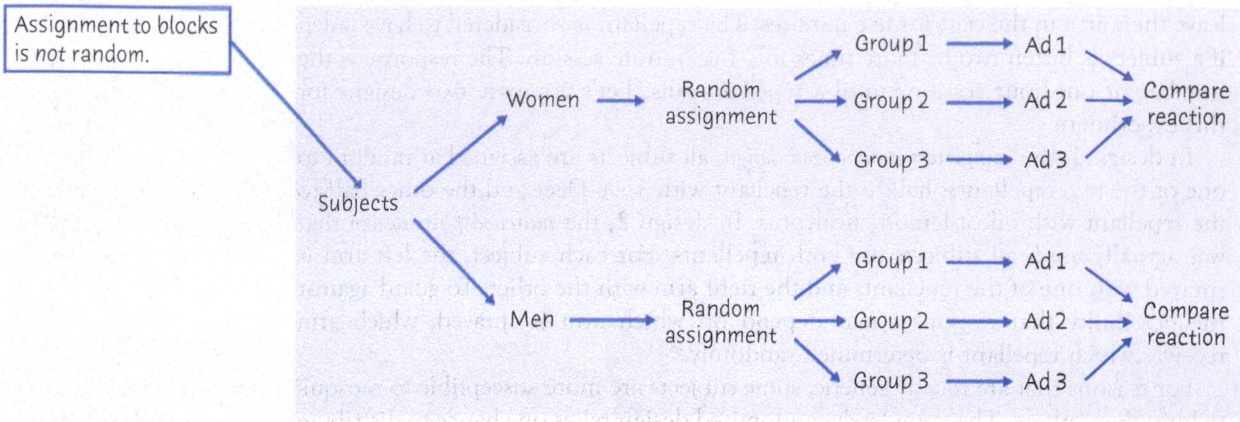

FIGURE 9.5
Outline of a block design, for Example 9.10. The blocks consist of male and female subjects. The treatments are three advertisements for the same product.

EXAMPLE 9.11 Comparing Welfare Policies

A social policy experiment will assess the effect on family income of several proposed new welfare systems and compare them with the present welfare system. Because the future income of a family is strongly related to its present income, the families who agree to participate are divided into blocks of similar income levels. The families in each block are then allocated at random among the welfare systems.

A block design allows us to draw separate conclusions about each block—for example, about men and women in Example 9.10. Blocking also allows more precise overall conclusions because the systematic differences between men and women can be removed when we study the overall effects of the three advertisements. The idea of blocking is an important additional principle of statistical design of experiments. A wise experimenter will form blocks based on the most important unavoidable sources of variability among the subjects. Randomization will then average out the effects of the remaining variation and allow an unbiased comparison of the treatments.

Like the design of samples, the design of complex experiments is a job for experts. Now that we have seen a bit of what is involved, we will concentrate for the most part on completely randomized experiments.

APPLY YOUR KNOWLEDGE

9.16 Comparing Breathing Frequencies in Swimming. Researchers from the United Kingdom studied the effect of two breathing frequencies on both performance times and several physiological parameters in front crawl swimming.[19] The breathing frequencies were one breath every second stroke (B2) and one breath every fourth stroke (B4). Subjects were 10 male collegiate swimmers. Each subject swam 200 meters, once with breathing frequency B2 and once on a different day with breathing frequency B4.

(a) Describe the design of this matched pairs experiment, including the randomization required by this design. Explain how you might carry out the randomization.

(b) Could this experiment be conducted using a completely randomized design? How would the design differ from the matched pairs experiment?

(c) Suppose we allow each swimmer to choose a breathing frequency and then swim 200 meters using the selected frequency. Are there any problems with then comparing the performance of the two breathing frequencies?

9.17 Athletes Take Oxygen. We often see players on the sidelines of a football game inhaling oxygen. Their coaches think this will speed their recovery. We might measure recovery from intense exertion as follows: Have a football player run 100 yards three times in quick succession. Then allow three minutes to rest before running 100 yards again. Time the final run. Describe the design of two experiments to investigate the effect of inhaling oxygen during the rest period. One of the experiments is to be a completely randomized design and the other a matched pairs design in which each athlete serves as their own control. Suppose you have 20 football players available as subjects. For both experiments, carry out the randomization of the 20 football players to treatments as required by the design.

9.18 Technology for Teaching Statistics. The Brigham Young University statistics department is performing randomized comparative experiments to compare teaching methods. Response variables include students' final-exam scores and a measure of their attitude toward statistics. One study compares two levels of technology for large lectures: standard (overhead projectors and chalk) and multimedia. The individuals in the study are the eight lectures in a basic statistics course. There are four instructors, each of whom teaches two lectures. Because the lecturers differ, their lectures form four blocks.[20] Suppose the lectures and lecturers are as follows:

Lecture	Lecturer	Lecture	Lecturer
1	Grimshaw	5	Tolley
2	Hilton	6	Grimshaw
3	Reese	7	Tolley
4	Reese	8	Hilton

Outline a block design and do the randomization that your design requires.

CHAPTER 9 SUMMARY

- We can produce data intended to answer specific questions by **observational studies** or **experiments.** Sample surveys that select a part of a population of interest to represent the whole are one type of observational study. Experiments, unlike observational studies, actively impose some treatment on the subjects of the experiment.
- Variables are **confounded** when their effects on a response can't be distinguished from each other. Observational studies and uncontrolled experiments often fail to show that changes in an explanatory variable actually cause changes in a response variable because the explanatory variable is confounded with lurking variables.
- In an experiment, we impose one or more **treatments** on individuals, often called **subjects.** Each treatment is a combination of values of the explanatory variables, which we call **factors.**

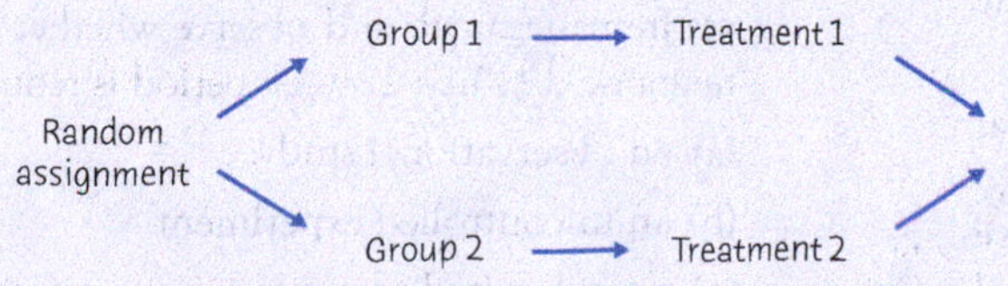

FIGURE 9.6 Outline of a randomized comparative experiment to compare two treatements.

- The design of an experiment describes the choice of treatments and the manner in which the subjects are assigned to the treatments.
- The basic principles of statistical design of experiments are **control** and **randomization** to combat bias and **replication** (use enough subjects) to reduce chance variation.
- The simplest form of **control** is comparison. Experiments should compare two or more treatments to avoid confounding of the effect of a treatment with other influences, such as lurking variables.
- **Randomization** uses chance to assign subjects to the treatments. Randomization creates treatment groups that are similar (except for chance variation) before the treatments are applied. Randomization and comparison together prevent bias, or systematic favoritism, in experiments.
- You can carry out randomization by giving numerical labels to the subjects and using software or a table of random digits to choose treatment groups.
- Applying each treatment to many subjects reduces the role of chance variation and makes the experiment more sensitive to differences among the treatments.
- Good experiments also require attention to detail as well as good statistical design. Many behavioral and medical experiments are **double-blind.** Some give a **placebo** to a control group.
- Lack of realism in an experiment can prevent us from generalizing its results.
- In addition to comparison, a second form of control is to restrict randomization by forming **blocks** of individuals that are similar in some way that is important to the response. Randomization is then carried out separately within each block.
- **Matched pairs** are a common form of blocking for comparing just two treatments. In some matched pairs designs, each subject receives both treatments in a random order. In others, the subjects are matched in pairs as closely as possible, and each subject in a pair receives one of the treatments.

CHECK YOUR SKILLS

9.19 Does peer victimization during adolescence have an impact on depression in early adulthood? A study in the United Kingdom examined data on 3898 participants for which the researchers had information on both victimization by peers at age 13 and the presence of depression at age 18. The study found more than a two-fold increase in the odds of depression between children who were not victimized and those who were frequently victimized.[21] This is an example of

(a) an observational study.

(b) a randomized comparative experiment.

(c) a block design, with level of victimization as the blocks.

9.20 The study described in Question 9.19 also found that adolescents who were victimized by peers at age 13 were more likely to be female and more likely to have displayed higher levels of emotional and behavioral problems before being bullied. Sex and previous emotional and behavioral problems are examples of

(a) blocking variables.

(b) explanatory variables.

(c) lurking variables.

9.21 Migraine is a prevalent disease characterized by headaches that are often severe and throbbing and accompanied by associated symptoms, such as nausea, vomiting, vertigo, and cognitive dysfunction. A drug, fremanezumab, may be an effective preventive treatment for migraine. To investigate this, researchers give 20 migraine sufferers fremanezumab and observe whether the number of migraine days in a 12-week period is reduced. This is

(a) an observational study.

(b) an uncontrolled experiment.

(c) a randomized comparative experiment.

9.22 What is the effect of a salesperson's demeanor on a customer? When purchasing clothing, it was hypothesized that for a more luxurious brand, such as Louis Vuitton, consumers would aspire more toward the brand if the salesperson were condescending, while for a mass market brand such as American Eagle, the opposite would be true. Participants in the study read the following hypothetical scenario:[22]

> *Imagine that you're out shopping for some new clothes. You decide to go to* (**Louis Vuitton**) *because you've always liked the clothing there. As you are browsing the store, you encounter a saleswoman. She greets you and* (**condescendingly**) *asks you if she can help you find what you're looking for.*

For the mass market brand, Louis Vuitton is replaced by American Eagle, and for the salesperson demeanor, the word *condescending* is omitted for the neutral condition. Three hundred sixty participants were assigned at random to one of the four conditions: (luxury brand, condescending), (luxury brand, neutral), (mass market brand, condescending), or (mass market brand, neutral). An aspiration measure toward the product was computed based on the responses to questions about liking the product, distinctiveness, fashionability, and desire to be seen wearing the product. This is an

(a) experiment with four factors corresponding to the four conditions.

(b) experiment with factors corresponding to the different brands.

(c) experiment with two factors, luxury/mass market and condescending/neutral.

9.23 In Question 9.22, generalizing results to consumers beyond this experiment may be

(a) limited because of small sample sizes.

(b) limited because of lack of blocking.

(c) limited because of lack of realism.

9.24 In Question 9.22, the response is

(a) the condescension level of the salesperson.

(b) the luxury level of the product.

(c) the aspiration measure toward the product.

9.25 Does exposure to aircraft noise increase the risk of hospitalization for cardiovascular disease in older people (≥ 65 years) residing near airports? Selecting a random sample of approximately 650,000 Medicare claims, it was found that about 75,000 people had zip codes near airports, and the remaining 575,000 did not. The proportions of hospital admissions related to cardiovascular disease were computed for those with zip codes near airports and those who did not have zip codes near airports. A larger proportion of admissions for cardiovascular disease was found for older people living in zip codes near airports. Which of the following statements is correct?

(a) Since this is an observational study, living in a zip code near an airport may or may not be causing the increase in the proportions of admissions for cardiovascular disease.

(b) Because of the large sample sizes from each group, we can claim that living in a zip code near an airport is causing the increase in the proportion of admissions for cardiovascular disease.

(c) Because this is an experiment, although not a randomized experiment, we can still conclude that living in a zip code near an airport is causing the increase in the proportions of admissions for cardiovascular disease.

9.26 The Community Intervention Trial for Smoking Cessation asked whether a community-wide advertising campaign would reduce smoking. The researchers located 11 pairs of communities, with each pair similar in location, size, economic status, and so on. One community in each pair was chosen at random to participate in the advertising campaign and the other was not. This is

(a) an observational study.

(b) a matched pairs experiment.

(c) a completely randomized experiment.

9.27 To decide which community in each pair in the previous question should get the advertising campaign, it is best to

(a) toss a coin.

(b) choose the community that will help pay for the campaign.

(c) choose the community with a mayor who will participate.

9.28 Researchers recruited 60 undergraduate students, in exchange for course credit, for a study on the effect of recycling on how much wrapping paper subjects used to wrap a gift. The subjects were randomly assigned to one of two rooms. In one room there was a large recycling bin and in the other a large trash bin. Subjects were asked to wrap a gift. Unknown to the students, the researchers were interested in how much paper the students used. The researchers found that students in the room with the recycling bin used (statistically) significantly more paper than those in the room with a trash bin. The researchers had hypothesized that people in general would rather recycle than throw things in the trash and hence would use less of a disposable resource when recycling is not available. Which of the following is an important weakness of this study?

(a) The study should have used a matched pairs design instead of a completely randomized design.

(b) Because undergraduate students were used as subjects, the results may not generalize to all adults and all situations involving disposable items.

(c) This is an observational study, not an experiment.

CHAPTER 9 EXERCISES

applet *In exercises that ask you to outline the experiment, use a figure similar to Figure 9.3, 9.4 or 9.5, depending on the setting. In all exercises that require randomization, you may use software, the Simple Random Sample applet, or Table B. See Example 9.6 (page 232) for directions on using the applet for more than two treatment groups.*

9.29 **Red Meat and Mortality.** Many studies have found an association between red meat consumption and an increased risk of chronic diseases. What is the relationship between red meat consumption and mortality? A large study followed 120,000 men and women who were free of coronary heart disease and cancer at the beginning of the study. Participants were asked detailed questions about their eating habits every 4 years, and the study spanned almost 30 years. It was found that the risk of dying at an early age–from heart disease, cancer, or any other cause, rises with the amount of red meat consumed.[23]

(a) Is this an observational study or an experiment? What are the explanatory and response variables?

(b) The authors noted that "Men and women with higher intake of red meat were less likely to be physically active and were more likely to be current smokers, to drink alcohol, and to have a higher body mass index." Explain carefully why differences in these variables make it more difficult to conclude that higher intake of red meat explains the increased death rate. What are the variables physical activity, smoking status, drinking behavior, and body mass index called?

(c) Suggest at least one lurking variable related to diet that may be confounded with higher intake of red meat. Explain why you chose the variable or variables you chose.

9.30 **Price Change and Fairness.** A marketing researcher wishes to study what factors affect the perceived fairness of a change in the price of an item from its advertised price. In particular, does the type of change in price (an increase or decrease) and the source of the information about the change (from a store clerk or from the price tag on the item) affect the perceived fairness? In an experiment, 20 subjects interested in purchasing a new rug are recruited. They are told that the price of a rug in a certain store was advertised at \$500. Subjects are sent, one at a time, to the store, where they learn that the price has changed. Five subjects are told by a store clerk that the price has *increased* to \$550. Five subjects learn that the price has *increased* to \$550 from the price tag on the rug. Five subjects are told by a store clerk that the price has *decreased* to \$450. Five subjects learn that the price has *decreased* to \$450 from the price tag on the rug. After learning about the change in price, each subject is asked to rate the fairness of the change on a 10-point scale, from 1 = "very unfair" to 10 = "very fair".

(a) What are the explanatory variables and the response variables for this experiment?

(b) Make a diagram like Figure 9.1 (page 226) to describe the treatments. How many treatments are there?

(c) In the experiment, the first five subjects learn from a store clerk that the price has increased to \$550, the next five learn that the price has increased to \$550 from the price tag on the rug, and so on. Would it be better to determine the order in which subjects are sent to the store and which scenario they will encounter (type of change and source of information about the change) randomly? Explain your answer.

9.31 **Don't Stop Exercising!** An investigation of the effect of different levels of physical activity was conducted on identical male twins in Finland. For each twin pair, although the twins had maintained the same activity level for most of their lives, one twin had significantly reduced activity over the last few years due to work or family pressures. For each pair of twins, body fat percentage, endurance levels, and insulin sensitivity were measured. For the less active twin, the results showed greater body fat, worse endurance, and levels of insulin sensitivity indicating early signs of metabolic disease.[24]

(a) What type of design is being used in this investigation? Give the explanatory and response variables.

(b) Is this an experiment or an observational study? Why?

(c) The article reports that the measurements were carried out blind. Explain what this means and why it is important.

9.32 **Running and Sleep.** Sufficient sleep is important for adolescents for both their neural and psychological development. Despite this, daytime sleepiness and poor physical and psychological functioning related to chronic sleep disturbances are common. A growing body of evidence suggests that exercise is associated with both better sleep and improved psychological functioning. Sixty participants were recruited from a high school in northwestern Switzerland. They were randomly assigned to either a running group or a control group, 30 to each group. The running group ran every morning for a little over 30 minutes on weekdays for a three-week period. All participants used a sleep log for subjective evaluation of sleep, and sleep was also objectively assessed at the beginning and end of the study using a sleep electroencephalographic device that measured quantities such as sleep efficiency and time spent in the four different sleep phases. Running was found to positively impact both objective and subjective measures of sleep functioning.[25]

(a) What are the explanatory variable(s) and the response variable(s)?

(b) Outline the design of the experiment.

(c) Here are some more details on the treatment and control groups. All participants arrived at school at 7 A.M., and the running group did two laps on the track and then ran cross country in groups of least four people for 30 minutes. The control group remained seated at the track, worked on homework, and interacted with each other. When the runners returned, all participants prepared for school and ate a breakfast that was provided. Why do you think the experimenters had the control group arrive at 7 A.M., interact with classmates, and have breakfast together? Explain. Do you think having the control group do these activities is important for the types of conclusions that can be reached? How?

(d) Time to sleep onset was measured before the beginning of the study and again at the end of the study for participants in both groups. Can this be considered a randomized controlled experiment with time to sleep onset as the response and four treatments (runners before, runners after, controls before, and controls after)? Explain why or why not.

9.33 Observation Versus Experiment. Researchers at the University of Pennsylvania found that patients who were divorced, separated, or widowed had approximately a 40% greater chance of dying or developing a new functional disability in the first two years following cardiac surgery than their married peers. The data included 1576 subjects who underwent cardiac surgery, of whom 65% were married; 33% were divorced, separated, or widowed; and 2% had never been married. The findings were reported to be statistically significant.[26]

(a) Without reading any further details of this study, how do you know that this was an observational study?

(b) Suggest some variables that might differ between the subjects in the study who were married versus those who were divorced, separated, or widowed. Are any of these possible confounding variables? Explain.

(c) Summarize briefly the limitations of this study. Despite these limitations, explain why this study still furnishes useful information in formulating a recovery plan for those undergoing cardiac surgery.

9.34 Attitudes Toward Homeless People. Negative attitudes toward poor people are common. Are attitudes more negative when a person is homeless? To find out, a description of a poor person is read to subjects. There are two versions of this description. One begins

> *Jim is a 30-year-old single man. He is currently living in a small single-room apartment.*

The other description begins

> *Jim is a 30-year-old single man. He is currently homeless and lives in a shelter for homeless people.*

Otherwise, the descriptions are the same. After reading the description, you ask subjects what they believe about Jim and what they think should be done to help him. The subjects are 544 adults interviewed by telephone.[27] Outline the design of this experiment.

9.35 Whole Grains and Metabolism. The *American Journal of Clinical Nutrition* published a research study to investigate the effect of a diet rich in whole grains on metabolism. Fifty adults with metabolic syndrome were randomly divided into two groups. Both groups had a diet of reduced calories, but for one of the groups, all of their grains were whole grains (brown rice, whole wheat bread, etc.), and the other group got all their grains as refined grains (white bread, white rice, etc.). Both groups lost weight at the end of 12 weeks, but the refined-grain group lost 11 pounds, and the whole-grain group lost 8 pounds. However, the whole-grain group lost more body fat from their abdomens and had other health benefits.[28]

(a) Outline the design of this experiment.

(b) Label the adults and choose the first 10 adults for the treatment (whole grains) group. If you use Table B, start at line 107.

9.36 The Effect of Product Density and Ambient Scent on Consumer Anxiety. Retail stores overflowing with merchandise can make consumers anxious, and minimally stocked spaces can have the same effect. Researchers investigated whether the use of ambient scents can reduce anxiety by creating feelings of openness in a crowded environment or coziness in a minimally stocked environment. Participants were invited to a lab that simulated a retail environment that was either jam-packed or nearly empty. For each of these two product densities, the lab was infused with one of three scents: (1) a scent associated with spaciousness, such as the seashore, (2) a scent associated with an enclosed space, like the smell of firewood, and (3) no scent at all. Consumers evaluated several products, and their level of anxiety was measured.[29]

(a) Use a diagram like Figure 9.2 (page 228) to display the treatments in a design with two factors: "product density" and "ambient scent." Then outline the design of a completely randomized experiment, to compare these treatments.

(b) There are 30 subjects available for the experiment, and they are to be randomly assigned to the treatments, an equal number of subjects in each treatment. Explain how you would number subjects and then randomly assign the subjects to the treatments. If you use the *Simple Random Sample* applet or other software, assign all the subjects. If you use Table B, start at line 133 and assign subjets to only the first treatment group.

9.37 Let Them Eat Chocolate. There is some evidence that cocoa has beneficial effects on heart health. To study this, researchers decide to give subjects either a cocoa pill or a placebo daily for a two-year period. Measurements of the subjects' heart health, based on a questionnaire, before and after the two-year period, are to be compared.[30]

(a) Outline the design of this experiment, using 20 subjects, with 10 assigned to each group

(b) Here are the names of the 20 subjects. Use software or Table B at line 129 to carry out the randomization your design requires.

Abel	DeVore	Kennedy	Reichert	Stout
Aeffner	Fleming	Lamone	Riddle	Williams
Birkel	Fritz	Mani	Sawant	Wilson
Bower	Giriunas	Mattos	Scannell	Worbis

(c) Do you think this can be run as a double-blind experiment? Explain.

9.38 **Handwriting Versus Keyboard Writing.** Do people who write by hand have a better memory of what they write than those who write using a keyboard? To test this, researchers had 36 participants in a study write down a long list of words read out loud to them. They were then asked to put aside their list and try to recall as many of the words as possible. Two methods of writing down words were used. One was using a blue-ink regular ball point pen and a notepad. The other was using a laptop equipped with a full-size keyboard. The number of words correctly recalled was the response.[31]

(a) Outline a completely randomized design to learn the effect of method of writing words on number of words correctly recalled.

(b) Describe in detail the design of a matched pairs experiment, using the same 36 subjects, in which each subject serves as his or her own control.

(c) The researchers reported that word recall was better when the words were written down on a notepad, and the result was statistically significant. What does statistically significant mean in describing the outcome of this study?

9.39 **Can Low-fat Food Labels Lead to Obesity?** What are the effects of low-fat food labels on food consumption? Do people eat more of a snack food when the food is labeled as low-fat? The answer may depend both on whether the snack food is labeled low-fat and whether the label includes serving-size information. An experiment investigated this question using university staff, graduate students, and undergraduate students at a large university as subjects. Subjects were asked to evaluate a pilot episode for an upcoming TV show in a theater on campus and were given a cold 24-ounce bottle of water and a bag of granola from a respected campus restaurant called The Spice Box. They were told to enjoy as much or as little of the granola as they wanted. Depending on the condition randomly assigned to the subjects, the granola was labeled as either "Regular Rocky Mountain Granola" or "Low-Fat Rocky Mountain Granola." Below this, the label indicated "Contains 1 Serving" or "Contains 2 Servings," or it provided no serving-size information.[32] Twenty subjects were assigned to each treatment, and their granola bags were weighed at the end of the session to determine how much granola was eaten.

(a) What are the factors and the treatments? How many subjects does the experiment require?

(b) Outline a completely randomized design for this experiment. (You need not actually do the randomization.)

9.40 **Store Window Creativity and Shopper Behavior.** Do more creative store-window displays affect shopper behavior? Six main-street retailers selling everyday fashion items were used in the study. Pretests with shoppers showed the six stores to be comparable on brands and consumer perceptions of value for the money. Three of the retailers had more creative windows, in terms of displaying items in a more innovative and artistic manner versus the less creative windows, which had a more concrete focus on the items on display. All display windows were of similar dimensions. Observers, in close proximity but out of sight of shoppers, watched their behavior as they passed the display windows, and for each shopper the observers recorded whether the shopper looked at the window or entered the store. A total of 863 shoppers passed the more creative windows and 971 passed the less creative windows. The study found that a higher percentage of shoppers looked at and entered the stores with the more creative windows, and the differences in shopper behavior between the more/less creative windows were statistically significant.[33]

More creative

Creative Lab/Shutterstock

Less creative

Creative Lab/Shutterstock

(a) Is this an observational study or an experiment? What are the explanatory and response variables?

(b) Explain what statistical significance means in describing the outcome of this study.

(c) Despite the results being statistically significant, the authors state:

The field study did not support an examination of why more creative store windows led consumers to enter the stores . . . . The use of actual retailers' real store windows meant that the level of creativity was not the only variable that differed among the retailers and their windows.

Using the language of this chapter, explain the authors' concerns and suggest at least one variable that might differ among the retailers and their windows.

9.41 Treating Sinus Infections. Sinus infections are common, and doctors commonly treat them with antibiotics. Another treatment is to spray a steroid solution into the nose. A well-designed clinical trial found that these treatments, alone or in combination, do not reduce the severity or the length of sinus infections.[34] The clinical trial was a completely randomized experiment that assigned 240 patients at random among four treatments, as follows:

	Antibiotic Pill	Placebo Pill
Steroid spray	53	64
Placebo spray	60	63

(a) The report of this study in the *Journal of the American Medical Association* describes it as a "double-blind, randomized, placebo-controlled factorial trial." "Factorial" means that the treatments are formed from more than one factor. What are the factors? What do "double-blind" and "placebo-controlled" mean?

(b) If the random assignment of patients to treatments did a good job of eliminating bias, possible lurking variables such as smoking history, asthma, and hay fever should be similar in all four groups. After recording and comparing many such variables, the investigators said that "all showed no significant difference between groups." Explain to someone who knows no statistics what "no significant difference" means. Does it mean that the presence of all these variables was exactly the same in all four treatment groups?

9.42 Store Window Creativity and Shopper Behavior (continued). In their paper, the authors of the study in Exercise 9.40 also reported the results of a second study to compare more/less creative window displays. In this second study, the authors used a single retailer and displayed the same merchandise in exactly the same way for both the more and less creative window displays. The differences between the window displays only involved the design surrounding the merchandise being more or less creative, not the content. Subjects recruited from the retailer's customer database were randomly assigned to view an *image* of one of the two window displays. After viewing the image, subjects answered questions about whether the products in the display made them want to enter the store.

(a) Is this an observational study or an experiment? What are the explanatory and response variables.

(b) Exercise 9.40 considered some of the drawbacks of the first study. Explain how this second study addresses those drawbacks. Does either study suffer from a lack of realism? Explain briefly.

9.43 Liquid Water Enhancers? Bottled water, flavored and plain, is expected to become the largest segment of the liquid refreshment market by the end of this decade, surpassing traditional carbonated soft drinks.[35] Kraft's MiO, a liquid water enhancer, comes in a variety of flavors, and a few drops added to water creates a zero-calorie flavored water drink. You wonder if those who drink flavored water like the taste of MiO as well as they like the taste of a competing flavored water product that comes ready to drink.

(a) Describe a matched pairs experiment to answer this question. Be sure to include proper blinding of your subjects. What is your response variable going to be?

(b) You have 20 people on hand who prefer to drink flavored water. Use the *Simple Random Sample* applet, software, or Table B, at line 138, to do the randomization that your design requires.

9.44 Algal Blooms. Algal blooms have become a recurring problem on many American lakes. Among other things, they can cause damage to a person's liver, kidneys, and nervous system. Phosphorus runoff from farms is one factor that contributes to algal blooms. Will inserting fertilizer into soil rather than spreading it across the surface help reduce runoff? To study this, researchers compare the effects of these two methods of fertilizing fields on the amount of phosphorus in runoff. Specific features of a field, such as slope of the ground and nature of the soil, can affect runoff, so the researchers divide each of four fields into two plots of equal size in such a way that the runoff from each plot can be measured separately. They use a matched pairs design, with the two plots in the same field as the matched pairs.

(a) Draw a sketch of the four fields, displaying each as a rectangle. Divide each field (rectangle) in half, each half representing one of the two plots. Label the two plots for each field as Plot 1 and Plot 2.

(b) Do the randomization required by the matched pairs design. That is, randomly assign the two treatments to the two plots in each field. Mark on your sketch which treatment is used in each plot.

9.45 **Save Money, the Environment, or Both!** Many consumers have both monetary and environmental reasons for saving energy. A study was designed to see how different types of advertising affect a consumer's willingness to enroll in an energy-saving program. The energy-saving programs were of two types: either reducing overall usage (Conservation program) or reducing consumption when demand is highest (Peak Saving program). For each of the two types of energy-saving programs, the advertisement described the program's benefits as saving money, saving energy, or saving both money and energy. For the study, 1406 participants were recruited and randomly assigned to the combinations of energy-saving program and program benefits. After reading the advertisement, a subject then indicated willingness to enroll in the program on a scale of 1 (definitely not) to 8 (definitely yes). Here is some of the wording used for the conservation program emphasizing saving money:

> We are offering a new program that will help you reduce your electricity bill . . . . To help you SAVE MONEY, you will get a free display that shows how much electricity you are using . . . . For example, you can set your thermostat higher in the summer, turn off your air conditioner, ...

The Peak Saving program advertisement emphasizing saving money was similar, but the free display was designed to track the price and use of electricity in the consumer's area to allow the individual to use electricity during cheaper, off-peak times.[36]

(a) Identify the subjects, the factors, the treatments, and the response variables.

(b) Use a diagram like Figure 9.2 (page 228) to display the factors and the treatments.

9.46 **Political Polarization and Social Media.** On September 1, 2018, *The Columbus Dispatch* reported on a study about political polarization and social media. In this study, 901 Democrats and 751 Republicans were recruited. The Democrats were randomly divided into two groups. All were asked to follow an automated Twitter account (Twitter bot) each day for one month. One group received tweets with a liberal point-of-view, and the other tweets with a conservative point-of-view. Likewise, Republicans were randomly divided into two groups and received the same two treatments (liberal or conservative tweets). All subjects were given a test, both before and after the experiment, that scored them on a liberal/conservative scale. Changes in scores were the response variable.[37]

(a) Is the political affiliation of the subjects (Democrat or Republican) a treatment variable or a block? Why?

(b) Is the type of tweet (liberal or conservative) a treatment variable or a block? Why?

(c) Use a diagram to outline the design.

9.47 **Better Sleep?** Is the number of times you awaken during the night affected by whether you have a glass of wine before bed and whether you have a snack before you go to bed? Describe briefly the design of an experiment with two explanatory variables—whether or not you have a glass of wine and whether or not you have a snack before going to bed—to investigate this question. Be sure to specify what the response variable will be. Also tell how you will handle lurking variables such as amount of sleep the previous night.

9.48 **Better Sleep?** Sleep habits of men and women may differ. We can improve the completely randomized design of Exercise 9.47 by using women and men as blocks. Your 300 subjects include 120 women and 180 men. Outline a block design for comparing the effect on sleep of whether or not you have a glass of wine and whether or not you have a snack before going to bed. Be sure to say how many subjects you will put in each group in your design.

9.49 **Quick Randomizing.** Here's a quick and easy way to randomize. You have 100 subjects: 50 adults under the age of 65 and 50 who are 65 or older. Toss a coin. If it's heads, assign all the adults under the age of 65 to the treatment group and all those 65 and over to the control group. If the coin comes up tails, assign all those 65 and over to treatment and all those under the age of 65 to the control group. This gives every individual subject a 50–50 chance of being assigned to treatment or control. Why isn't this a good way to randomly assign subjects to treatment groups?

9.50 **Do Antioxidants Prevent Cancer?** People who eat lots of fruits and vegetables have lower rates of colon cancer than those who eat little of these foods. Fruits and vegetables are rich in antioxidants such as vitamins A, C, and E. Will taking antioxidants help prevent colon cancer? A medical experiment studied this question with 864 people who were at risk of colon cancer. The subjects were divided into four groups: daily beta-carotene, daily vitamins C and E, all three vitamins every day, or daily placebo. After four years, the researchers were surprised to find no significant difference in colon cancer among the groups.[38]

(a) What are the explanatory and response variables in this experiment?

(b) Outline the design of the experiment. Use your judgment in choosing the group sizes.

(c) The study was double-blind. What does this mean?

(d) What does "no significant difference" mean in describing the outcome of the study?

(e) Suggest some lurking variables that could explain why people who eat lots of fruits and vegetables have lower rates of colon cancer. The experiment suggests that these lurking variables or other properties of fruits and vegetables, rather than the antioxidants, may be responsible for the observed benefits of fruits and vegetables.

9.51 SAMe for Depression? S-adenosyl methionine (SAMe), a naturally occurring molecule found throughout the body, has been used as an antidepressant with some success. It has been available commercially in Europe since the late 1970s and is now available over-the-counter in the United States. Participants in the current study were 73 individuals with major depressive disorder who had not responded to a standard treatment using serotonin reuptake inhibitors (SRI) to relieve their symptoms. The effect of augmenting their SRI treatment with SAMe was investigated.[39]

(a) The study was a *randomized, double-blind* trial conducted over six weeks, with 34 participants receiving a placebo (dummy pills) and the remaining 39 receiving pills containing SAMe (a trial is a medical experiment using actual patients as subjects). Explain why it is important to have a placebo group rather than having all participants receive pills containing SAMe. What is the purpose of the two italicized terms in the context of this study?

(b) A 50% reduction in the Hamilton Rating scale for depression over the treatment period was considered a positive response to treatment. It was found that 36.1% of the SAMe group had a positive response versus 17.6% in the placebo group, a statistically significant difference. Explain what statistical significance means in the context of this trial.

(c) From the information given, use a diagram to outline the design of this trial.

9.52 Randomization at Work. To demonstrate how randomization reduces confounding, consider the following situation. A nutrition experimenter intends to compare the weight gain of prematurely born infants fed Diet A with those fed Diet B. To do this, she will feed each diet to 40 prematurely born infants whose parents have enrolled them in the study. She has available 20 baby girls and 20 baby boys. The researcher is concerned that baby boys may respond more favorably to the diets, so if all the baby boys were fed diet A, the experiment would be biased in favor of Diet A.

(a) Label the infants 01, 02, . . . , 40. Use Table B to assign 20 infants to Diet A. Or, if you have access to statistical software, use it to assign 20 infants to Diet A. Do this 10 times, using different parts of the table (or different runs of your software) and write down the 10 groups assigned to Diet A.

(b) The infants labeled 21, 22, . . . , 40 are the 20 baby boys. How many of these infants were in each of the 10 Diet A groups that you generated?

(c) You see that there is considerable chance variation in the number of baby boys assigned to Diet A. Draw a stem-and-leaf plot of the number of baby boys assigned to Diet A. Do you see any systematic bias in favor of one or the other diet being assigned the baby boys? Larger samples from a larger population will, on the average, do an even better job of creating two similar groups.

CHAPTER

10

When you complete this chapter, you will be able to:

10.1 Describe the purposes, processes, and effects of institutional review boards.

10.2 Explain standards of informed consent in statistical studies involving human subjects.

10.3 Evaluate confidentiality practices of observational studies and experiments and distinguish between confidentiality and anonymity.

10.4 Explain the ethical issues associated with clinical trials.

10.5 Discuss the differences between ethical principals underlying studies in behavioral and social science and those in medicine.

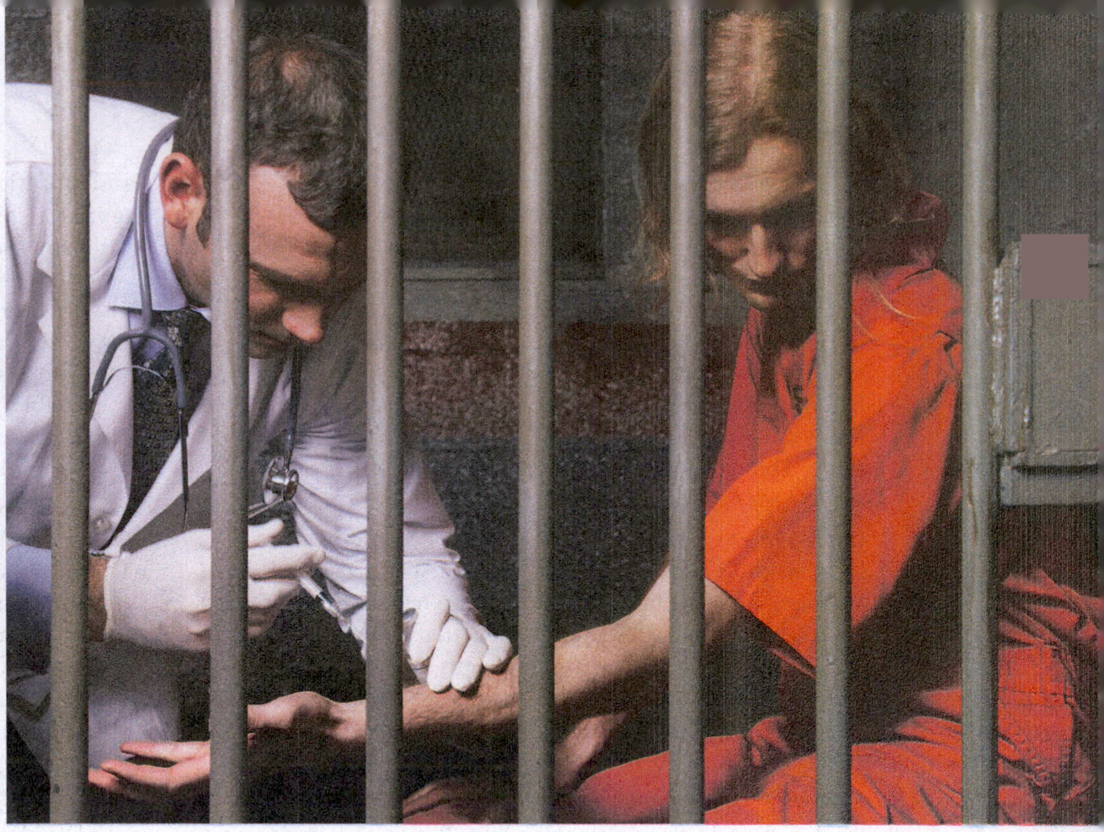

Radius Images/Alamy Stock Photo

Data Ethics*

Chapters 8 and 9 discuss methods for producing data. Applying these methods in practice raises ethical questions. In this chapter, we present some of these ethical issues.

We won't discuss the telemarketer who begins a telephone sales pitch with, "I'm conducting a survey." Such deception is clearly unethical. It enrages legitimate survey organizations, which find the public less willing to talk with them. Neither will we discuss those few researchers who, in the pursuit of professional advancement, publish fake data. There is no ethical question here: faking data to advance your career is just wrong.[1] It will end your career when uncovered. But just how honest must researchers be about real, unfaked data? Here is an example suggesting that the answer is "More honest than they often are."

EXAMPLE 10.1 The Whole Truth?

Papers reporting scientific research are supposed to be short, with no extra baggage. Brevity, however, can allow researchers to avoid complete honesty about their data. Did they choose their subjects in a biased way? Did they report data on only some of their subjects? Did they try several statistical analyses and report only the ones that looked best? The statistician John Bailar screened more than 4000 medical papers in more than a decade as consultant to the *New England Journal of Medicine*. He says, "When it came to the statistical

*This short chapter concerns a very important topic, but the material is not needed to read the rest of the book.

review, it was often clear that critical information was lacking, and the gaps nearly always had the practical effect of making the authors' conclusions look stronger than they should have."[2] The situation is no doubt worse in fields that screen published work less carefully. This problem continues to grow with the proliferation of open-access online journals that "will print seemingly anything for a fee" and provide little or no peer review.[3]

The most complex issues of data ethics arise when we collect data from people (but research with animals also raises ethical issues—see Web Exercise 10.28, available online). The ethical difficulties are more severe for experiments that impose some treatment on people than for sample surveys that simply gather information. Trials of new medical treatments, for example, can do harm as well as good to their subjects. Here are some basic standards of data ethics that must be obeyed by all studies that gather data from human subjects, both observational studies and experiments.

Basic Data Ethics for Human Subjects

All planned studies must be reviewed in advance by an *institutional review board* charged with protecting the safety and well-being of the subjects.

All individuals who are subjects in a study must give their *informed consent* before data are collected.

All individual data must be kept *confidential*. Only statistical summaries for groups of subjects may be made public.

If subjects are children, then their consent is needed in addition to that of the parents or guardians.

Many journals have a formal requirement of explicitly addressing human subjects issues if the study is classified as human subjects research. For example, here is a statement from the instructions for authors for *JAMA* (the *Journal of the American Medical Association*):[4]

> For all manuscripts reporting data from studies involving human participants or animals, formal review and approval, or formal review and waiver, by an appropriate institutional review board or ethics committee is required and should be described in the Methods section.

For situations in which a formal ethics review committee does not exist, the *Journal of the American Medical Association* instructs investigators to follow the principles outlined in the 1964 Helsinki Declaration of the World Medical Association. When human subjects are involved, investigators are to state in their paper's "Methods" section the manner in which informed consent was obtained from the study participants (that is, oral or written). Also, the law requires that studies carried out or funded by the federal government obey these principles.[5] But neither the law nor the consensus of experts is completely clear about the details of their application.

10.1 Institutional Review Boards

The purpose of an **institutional review board** is not to decide whether a proposed study will produce valuable information or whether it is statistically sound. The board's purpose is, in the words of one university's board, "to protect the rights and welfare of human subjects or patients recruited to participate in research activities."[6] The board reviews the plan of the study and can require changes. It reviews the consent

institutional review board
A panel of experts, associated with a university or other research organization, that reviews all planned studies in order to protect the safety and well-being of the subjects.

form to ensure that subjects are informed about the nature of the study and about any potential risks. Once research begins, the board monitors the study's progress at least once a year.

The most pressing issue concerning institutional review boards is whether their workload has become so large that their effectiveness in protecting subjects drops. When the government temporarily stopped human subject research at Duke University Medical Center in 1999 due to inadequate protection of subjects, more than 2000 studies were going on. That's a lot of review work. There are shorter review procedures for projects that involve only minimal risks to subjects, such as most sample surveys. When a board is overloaded, there is a temptation to put more proposals in the minimal-risk category to speed the work.

APPLY YOUR KNOWLEDGE

10.1 Minimal Risk? You are a member of your college's institutional review board. You must decide whether several research proposals qualify for less rigorous review because they involve only minimal risk to subjects. Federal regulations say that "minimal risk" means the risks are no greater than "those ordinarily encountered in daily life or during the performance of routine physical or psychological examinations or tests." That's vague. Which of these do you think qualifies as "minimal risk"?

(a) Give subjects an experimental drug that may produce temporary dizziness as a side effect, and warn the subjects about this risk.

(b) Give subjects an experimental drug that may produce episodes of depression as a side effect.

(c) Recruit women for a study on physical abuse from spouses or partners by putting up posters in the surrounding community. The posters instruct any interested women who have experienced abuse to call the lab phone number and leave a message with their name and phone numbers.

(d) Recruit women for a study on physical abuse from spouses or partners by visiting shelters for victims of domestic violence and asking for volunteers.

(e) Asking women if they have had an abortion in a country where it is illegal.

(f) Asking women if they have had an abortion in a country in which it is legal but the issue is fraught with religious and political controversy.

10.2 Does This Really Need to Be Reviewed? A college professor would like to conduct a taste test of a new breakfast bar that contains only wholesome ingredients, such as whole grains, dried fruit, and honey without additives. He plans to ask students in his class if any would like to volunteer to serve as taste testers. Should he seek institutional review board approval before proceeding? Discuss.

10.2 Informed Consent

informed consent
The requirement that subjects must be told in advance about the nature and purpose of a study and any risk of harm it may bring. Subjects must then give approval of their participation in writing before the study begins.

Both words in the phrase **informed consent** are important, and both can be controversial. Subjects must be *informed* in advance about the nature of a study and any risk of harm it may bring. In the case of a sample survey, physical harm is not possible. The subjects should be told what kinds of questions the survey will ask and about how much of their time it will take. Experimenters must tell subjects the nature and purpose of the study and outline possible risks. Subjects must then *consent* in writing.

EXAMPLE 10.2 Who Can Consent?

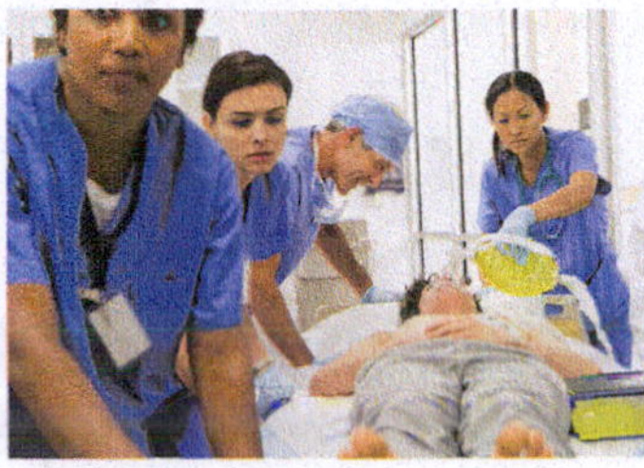
Ariel Skelley/Getty Images

Are there some subjects who can't give informed consent? It was once common, for example, to test new vaccines on prison inmates who gave their consent in return for good-behavior credit. Now we worry that prisoners are not really free to refuse, and the law forbids almost all medical research in prisons.

Children can't give fully informed consent, so the usual procedure is to ask their parents. A study of new ways to teach reading is about to start at a local elementary school, so the study team sends consent forms home to parents. Many parents don't return the forms. Can their children take part in the study because the parents did not say No, or should we allow only children whose parents returned the form and said Yes?

What about research into new medical treatments for people with mental disorders? What about studies of new ways to help emergency room patients who may be unconscious? In most cases, there is not time to get the consent of the family. Does the principle of informed consent bar realistic trials of new treatments for unconscious patients?

These are questions without clear answers. Reasonable people differ strongly on all of them. There is nothing simple about informed consent.[7]

The difficulties of informed consent do not vanish even for capable subjects. Some researchers, especially in medical trials, regard consent as a barrier to getting patients to participate in research. They may not explain all possible risks; they may not point out that there are other therapies that might be better than those being studied; they may be too optimistic in talking with patients, even when the consent form has all the right details. On the other hand, mentioning every possible risk leads to very long consent forms that really are barriers. "They are like rental car contracts," one lawyer said. Some subjects don't read forms that run five or six printed pages. Others are frightened by the large number of possible (but unlikely) disasters that might happen and so refuse to participate. Of course, unlikely disasters sometimes happen. When they do, lawsuits follow, and the consent forms become yet longer and more detailed.

APPLY YOUR KNOWLEDGE

10.3 Coercion? U.S. Department of Health and Human Services regulations for informed consent state that "an investigator shall seek such consent only under circumstances that provide the prospective subject or the representative sufficient opportunity to consider whether or not to participate and that minimize the possibility of coercion or undue influence."[8] Coercion occurs when an overt or implicit threat of harm is intentionally presented by one person to another in order to obtain compliance. Which of the following circumstances do you believe constitutes coercion? Discuss.

(a) A researcher has developed a vaccine against a new virus. The researcher is recruiting healthy adult volunteers from an inner city to determine if the vaccine is safe in humans. Volunteers will be paid for their participation. One participant tells one of the research nurses that he would not have enrolled in the study, but he recently lost his job and needs the money. He claims that he feels as though he has no alternative but to participate.

(b) A research nurse is asked to consent and provide samples for three minimal risk studies during her first week on the job. She is told "everyone working here is enrolled in these studies."

10.4 Undue Influence? Undue influence in obtaining informed consent often occurs through an offer of an excessive or inappropriate reward or other overture in order to obtain compliance. Which of the following circumstances do you believe constitutes undue influence? Discuss.

(a) The students in a professor's class are told they will be given extra credit if they participate in a research study she is conducting. An alternative means of obtaining extra credit is available for students not wishing to participate.

(b) The students in a professor's class are told they will be given extra credit if they participate in a research study she is conducting. Extra credit is only available for students who choose to participate but will be awarded even if a student drops out of the study before it is completed.

(c) The students in a professor's class are told they will be given extra credit if they participate in a research study she is conducting. The extra credit will only be awarded to those students who continue in the study until it is finished.

10.3 Confidentiality

Ethical problems do not disappear once a study has been cleared by the review board, has obtained consent from its subjects, and has actually collected data about the subjects. It is important to protect the subjects' privacy. Privacy refers to a person's interest in controlling the access of others to himself or herself, including information about himself or herself. One way this is done is by keeping all data about individuals confidential. **Confidentiality** refers to the agreement between the investigator and participant about how data will be managed and used. The report of an opinion poll may say what percentage of the 1200 respondents felt that legal immigration should be reduced. It may not report what *you* said about this or any other issue. However, the investigator who collected the data will know what you said about this or other issues in the poll.

Confidentiality is not the same as *anonymity*. Anonymity means that subjects are anonymous—their names are not known even to the director of the study. Anonymity provides a high degree of privacy, but anonymity is rare in statistical studies. Even where it is possible (mainly in surveys conducted by mail), anonymity prevents any follow-up to improve nonresponse or inform subjects of results.

Any breach of confidentiality is a serious violation of data ethics. The best practice is to separate the identity of the subjects from the rest of the data at once. Sample surveys, for example, use the identification only to check on who did or did not respond. In an era of advanced technology, however, it is no longer enough to be sure that each individual set of data protects people's privacy. The government, for example, maintains a vast amount of information about citizens in many separate databases—census responses, tax returns, Social Security information, data from surveys such as the Current Population Survey, and so on. Many of these databases can be searched by computers for statistical studies. A clever computer search of several databases might be able, by combining information, to identify you and learn a great deal about you, even if your name and other identification have been removed from the data available for search. A colleague from Germany once remarked that "female full professor of statistics with a PhD from the United States" was enough to identify her among all the 83 million residents of Germany. Privacy and confidentiality of data are hot issues among statisticians in the computer age. Computer

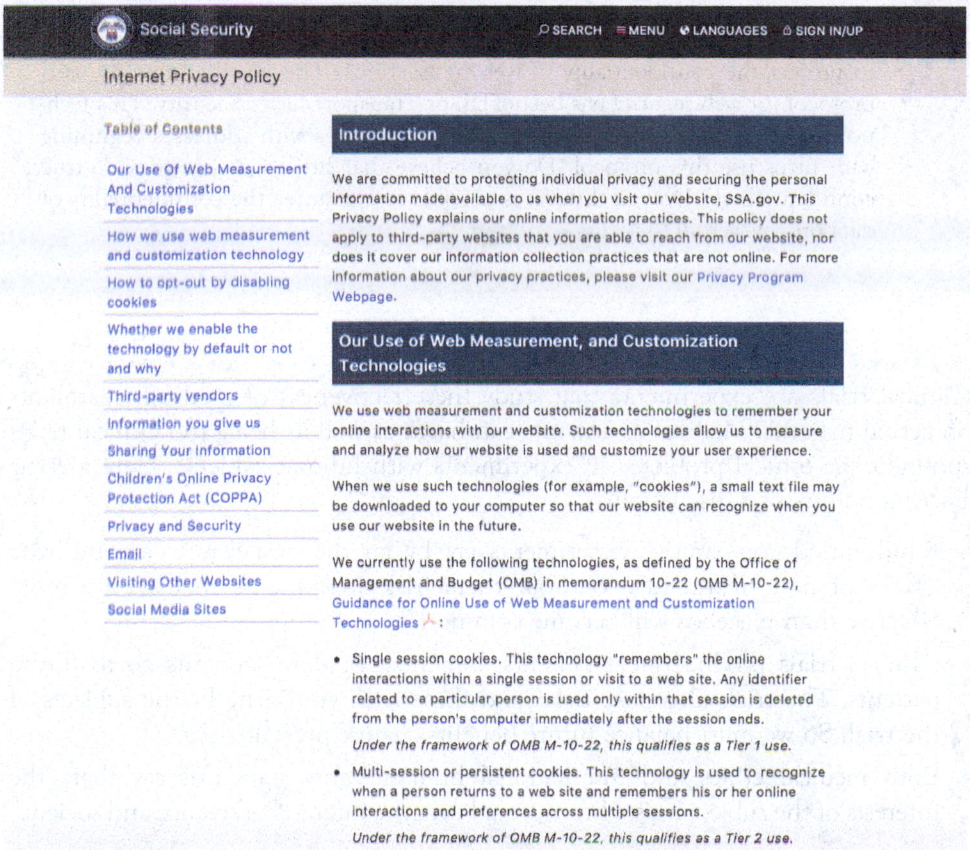
Social Security
SEARCH MENU LANGUAGES SIGN IN/UP

Internet Privacy Policy

Table of Contents

Introduction

We are committed to protecting individual privacy and securing the personal information made available to us when you visit our website, SSA.gov. This Privacy Policy explains our online information practices. This policy does not apply to third-party websites that you are able to reach from our website, nor does it cover our information collection practices that are not online. For more information about our privacy practices, please visit our Privacy Program Webpage.

Our Use of Web Measurement, and Customization Technologies

We use web measurement and customization technologies to remember your online interactions with our website. Such technologies allow us to measure and analyze how our website is used and customize your user experience.

When we use such technologies (for example, "cookies"), a small text file may be downloaded to your computer so that our website can recognize when you use our website in the future.

We currently use the following technologies, as defined by the Office of Management and Budget (OMB) in memorandum 10-22 (OMB M-10-22), Guidance for Online Use of Web Measurement and Customization Technologies:

- Single session cookies. This technology "remembers" the online interactions within a single session or visit to a web site. Any identifier related to a particular person is used only within that session is deleted from the person's computer immediately after the session ends.
  Under the framework of OMB M-10-22, this qualifies as a Tier 1 use.
- Multi-session or persistent cookies. This technology is used to recognize when a person returns to a web site and remembers his or her online interactions and preferences across multiple sessions.
  Under the framework of OMB M-10-22, this qualifies as a Tier 2 use.

Social Security Administration

FIGURE 10.1 The privacy policy of the government's Social Security Administration website, www.ssa.gov/agency/privacy.html.

hacking and thefts of laptops containing data add to the difficulties. Is it even possible to guarantee confidentiality of data stored in databases that can be hacked or stolen? The U.S. Social Security Administration has devised a comprehensive Internet privacy policy, part of which can be seen in Figure 10.1.

EXAMPLE 10.3 Uncle Sam Knows

Citizens are required to give information to the government. Think of tax returns and Social Security contributions. The government needs these data for administrative purposes—to see if you paid the right amount of tax and how large a Social Security benefit you are owed when you retire. Some people feel that individuals should be able to forbid any other use of their data, even with all identification removed. This would prevent using government records to study, say, the ages, incomes, and household sizes of Social Security recipients. Such a study could well be vital to debates on reforming Social Security.

APPLY YOUR KNOWLEDGE

10.5 Sunshine Laws. All states in the United States have open records laws, sometimes known as "Sunshine Laws," that give citizens access to government meetings and records.[9] This includes, for example, reports of crimes and recordings of 911 calls. A crime report includes the name of anyone accused of the crime. Suppose a 10-year-old juvenile is accused of committing a crime. A reporter from the local newspaper asks for a copy of the crime report. The sheriff refuses to provide the report because the accused is a juvenile, and he believes the name of the accused should be confidential. Is this an issue of confidentiality? Discuss.

10.6 https. Generally, secure websites use encryption and authentication standards to protect the confidentiality of web transactions. The most commonly used protocol for web security has been TLS, or Transport Layer Security. This technology is still commonly referred to as SSL. Websites with addresses beginning with https use this protocol. Do you believe that https websites provide true confidentiality?[10] Do you think it is possible to guarantee the confidentiality of data on any website? Discuss.

10.4 Clinical Trials

Clinical trials are experiments that study the effectiveness of medical treatments on actual patients. Medical treatments can harm as well as heal, and clinical trials spotlight the ethical problems of experiments with human subjects. Here are the starting points for a discussion:

- Randomized comparative experiments are by far the best way to see the true effects of new treatments. Without them, risky treatments that are no more effective than placebos will become common.[11]
- Clinical trials produce great benefits, but most of these benefits go to future patients. The trials also pose risks, and these risks are borne by the subjects of the trial. So we must balance future benefits against present risks.
- Both medical ethics and international human rights standards say that "the interests of the subject must always prevail over the interests of science and society."

The quoted words are from the 1964 Helsinki Declaration of the World Medical Association, the most respected international standard. The most outrageous examples of unethical experiments are those that ignore the interests of the subjects.

EXAMPLE 10.4 The Tuskegee Study

In the 1930s, syphilis was common among black men in the rural South, a group that had almost no access to medical care. The Public Health Service Tuskegee study recruited 399 poor black sharecroppers with syphilis and 201 others without the disease to observe how syphilis progressed when no treatment was given. Beginning in 1943, penicillin became available to treat syphilis. The study subjects were not treated. In fact, the Public Health Service prevented any treatment until word leaked out and forced an end to the study in the 1970s.

The Tuskegee study is an extreme example of investigators following their own interests and ignoring the well-being of their subjects. A 1996 review said, "It has come to symbolize racism in medicine, ethical misconduct in human research, paternalism by physicians, and government abuse of vulnerable people." In 1997, President Clinton formally apologized to the surviving participants in a White House ceremony.[12]

The Tuskegee study may help explain the reluctance of many Black Americans to take part in clinical trials: "From a historical perspective, the Tuskegee syphilis study is widely recognized as a reason for mistrust because of the extent and duration of deception and mistreatment and the study's impact on human subject review and approval."[13] Unfortunately, Black Americans have been impacted by more than the Tuskegee study. Enough studies have taken advantage of Black Americans for Harriet S. Washington to write the 528-page book *Medical Apartheid: The Dark History of Medical Experimentation on Black Americans From Colonial Times to the Present.* Washington documents the mistreatment that Black Americans have experienced in the name of research.

Because "the interests of the subject must always prevail," medical treatments can be tested in clinical trials only when there is reason to hope that they will help the patients who are subjects in the trials. Future benefits aren't enough to justify experiments with human subjects. Of course, if there is already strong evidence that a treatment works and is safe, it is unethical *not* to give it. Dr. Charles Hennekens of the Harvard Medical School, who directed the large clinical trial that showed that aspirin reduces the risk of heart attacks in men, discussed the issue of when to do or not to do a randomized trial. Here are his words:

> On the one hand, there must be sufficient belief in the agent's potential to justify exposing half the subjects to it. On the other hand, there must be sufficient doubt about its efficacy to justify withholding it from the other half of subjects who might be assigned to placebos.[14]

Why is it ethical to give a control group of patients a placebo? Well, we know that placebos often work. Moreover, placebos have no harmful side effects. So in the state of balanced doubt described by Dr. Hennekens, the placebo group may be getting a better treatment than the drug group. If we *knew* which treatment was better, we would give it to everyone. When we don't know, it is ethical to try both and compare them.[15]

APPLY YOUR KNOWLEDGE

10.7 Ethics and Scientific Validity. The authors of a paper on clinical research and ethics[16] stated the following:

> . . . the research must have a clear scientific objective; be designed using accepted principles, methods, and reliable practices; have sufficient power to definitively test the objective; and offer a plausible data analysis plan. In addition, it must be possible to execute the proposed study.

Do you think this rules out observational studies as "ethical"? Discuss. You might wish to refer to the discussion on correlation and causation in Section 5.8. (page 146).

10.5 Behavioral and Social Science Experiments

When we move from medicine to the behavioral and social sciences, the direct risks to experimental subjects are less acute, but so are the possible benefits to the subjects. Consider, for example, the experiments conducted by psychologists in their study of human behavior.

EXAMPLE 10.5 Psychologists in the Men's Room

Psychologists observe that people have a "personal space" and are uneasy if others come too close to them. We don't like strangers to sit at our table in a coffee shop if other tables are available, and we see people move apart in elevators if there is room to do so. Americans tend to require more personal space than people in most other cultures. Can violations of personal space have physical, as well as emotional, effects?

Investigators set up shop in a men's public restroom. They blocked off urinals to force men walking in to use either a urinal next to an experimenter (treatment group) or a urinal separated from the experimenter (control group). Another experimenter, using a periscope from a toilet stall, measured how long the subject took to start urinating and how long he continued.[17]

© David Pollack/Corbis/Getty Images

This personal space experiment illustrates the difficulties facing those who plan and review behavioral studies:

- There is no risk of harm to the subjects, although they would certainly object to being watched through a periscope. What should we protect subjects from when physical harm is unlikely? Possible emotional harm? Undignified situations? Invasion of privacy?
- What about informed consent? The subjects did not even know they were participating in an experiment. Many behavioral experiments rely on hiding the true purpose of the study. The subjects would change their behavior if told in advance what the investigators were looking for. Subjects are asked to consent on the basis of vague information. They receive full information only after the experiment.

The "Ethical Principles" of the American Psychological Association require consent unless a study merely observes behavior in a public place. They allow deception only when it is necessary to the study, does not hide information that might influence a subject's willingness to participate, and is explained to subjects as soon as possible. The personal space study just described was conducted in the 1970s; it does not meet current ethical standards.

We see that the basic requirement for informed consent is understood differently in medicine and psychology. Here is an example of another setting with yet another interpretation of what is ethical. The subjects get no information and give no consent. They don't even know that an experiment may be sending them to jail for the night.

EXAMPLE 10.6 Reducing Domestic Violence

How should police respond to domestic violence calls? In the past, the usual practice was to remove male offenders and order them to stay out of the household overnight. Police were reluctant to make arrests because the victims rarely pressed charges. Women's groups argued that arresting offenders would help prevent future violence even if no charges were filed. Is there evidence that arrest will reduce future offenses? That's a question that experiments have tried to answer.

A typical domestic violence experiment compares two treatments: arrest the suspect and hold him overnight or warn the suspect and release him. When police officers reach the scene of a domestic violence call, they calm the participants and investigate. Weapons or death threats require an arrest. If the facts permit an arrest but do not require it, an officer radios headquarters for instructions. The person on duty opens the next envelope in a file prepared in advance by a statistician. The envelopes contain the treatments in random order. The police either arrest the suspect or warn and release him, depending on the contents of the envelope. The researchers then watch police records and visit the victim to see if the domestic violence recurs.

Such experiments show that arresting domestic violence suspects does reduce their future violent behavior.[18] As a result of this evidence, arrest has become the common police response to domestic violence.

The domestic violence experiments shed light on an important issue of public policy. Because there is no informed consent, the ethical rules that govern clinical trials and most social science studies would forbid these experiments. They were cleared by review boards because, in the words of one domestic violence researcher, "These people became subjects by committing acts that allow the police to arrest them. You don't need consent to arrest someone."

APPLY YOUR KNOWLEDGE

10.8 Deceiving Subjects. Researchers are interested in assessing the "Good Samaritan" behavior of unsuspecting travelers in a subway train. An actor, either apparently drunk or carrying a cane, would collapse, and the number of helpful interventions by travelers would be observed and recorded. The results of the experiment determined that people were generally very helpful, although they were a little more reluctant to help a drunk. Do you think this study is ethically okay? Discuss.

CHAPTER 10 SUMMARY

- All planned studies must be reviewed in advance by an **institutional review board** charged with protecting the safety and well-being of the subjects.
- All individuals who are subjects in a study must give their **informed consent** before data are collected.
- All individual data must be kept **confidential.** Only statistical summaries for groups of subjects may be made public. The goal is to protect subjects' privacy.

CHAPTER 10 EXERCISES

Most of these exercises pose issues for discussion. There are no right or wrong answers, but there are more and less thoughtful answers.

10.9 Who Reviews? Government regulations require that an institutional review board consist of at least five people, including at least one scientist, one nonscientist, and one person from outside the institution. Most boards are larger, but many contain just one outsider.

(a) Why should review boards contain people who are not scientists?

(b) Do you think that one outside member is enough? How would you choose that member? (For example, would you prefer a medical doctor? A member of the clergy? An activist for patients' rights?)

10.10 Informed Consent. A researcher suspects that people who are abused as children tend to be more prone to severe depression as young adults. She prepares a questionnaire that measures depression and that also asks many personal questions about childhood experiences. Write a description of the purpose of this research to be read by subjects in order to obtain their informed consent. You must balance the conflicting goals of not deceiving the subjects about what the questionnaire will tell about them and not biasing the sample by scaring off people with painful childhood experiences.

In January 2012, Facebook performed an experiment on more than 689,000 users without informing them even after the experiment was over. Facebook adjusted people's newsfeeds so that half of these individuals saw only happy posts from their friends and the other half saw only sad posts from their friends. Facebook then determined the mood of the user by judging the quality of his or her own posts. Why would Facebook want to learn how to manipulate emotions? It is well known that sad people tend to shop more, and Facebook sells advertisement space on its site. Questions 10.11 to 10.14 refer to this experiment.

10.11 Review Board Approval. Any organization that receives federal funding must receive review board approval for research with humans. Facebook does not receive federal funding. Facebook partnered with Cornell University to write the article and analyze the data after the experiment was already performed. The researcher at Cornell consulted his institutional review board to get approval for his part of this work, but since his involvement started after the experiment was already completed, his review board said that he did not need approval from them. What do you think about this experiment happening without a review board?

10.12 Confidentiality. Facebook did take an unusual step for a business by publishing the results from this experiment in the *Proceedings of the National Academy of Sciences*, a prestigious journal.[19] Facebook knew who all of the individuals were and what they had posted, but Facebook did not publish any individual information in the article. Did Facebook provide confidentiality? Explain your answer.

10.13 Informed Consent. Facebook claims that its data privacy policy covered this experiment because it included this line: "For example, in addition to helping people see and find things that you do and share, we may use the information we receive about you... for internal operations, including troubleshooting, data analysis, testing, research and service improvement." Do you agree that this policy does enough to count as informed consent? Discuss your reasoning.

10.14 **Informed Consent, Continued.** Sometimes exceptions can be made to the informed consent process. Examples include education research studies with normal classroom activities posing no unusual risks (like trying a lecture versus an active learning activity to teach a new concept) or behavioral studies in a public place. These ethical guidelines were written in the middle of the twentieth century, well before the Internet and social media existed. Do you believe that Facebook and other social media sites count as "public places"? If so, does that change your answer to whether informed consent was necessary for this experiment?

10.15 **Anonymous? Confidential?** One of the most important nongovernment surveys in the United States is the National Opinion Research Center's General Social Survey (GSS). The GSS regularly monitors public opinion on a wide variety of political and social issues. Interviews are conducted in person in the subject's home. Are a subject's responses to GSS questions anonymous, confidential, or both? Explain your answer.

10.16 **Anonymous or Confidential?** The website for STDcheck.com contains the following information about HIV testing: "We offer 100% private testing. You are not required to show your ID at the lab, you're given a unique code which allows the lab to perform testing without your ID, and your results are uploaded to your private online account . . . . We encrypt our data with industry standard 128-bit encryption. All communication and transactions between you and our website are secure." Does this practice offer anonymity or confidentiality or both? Explain your answer.

10.17 **Political Polls.** The presidential election campaign is in full swing, and the candidates have hired polling organizations to take sample surveys to find out what the voters think about the issues. What information should the pollsters be required to give out?

(a) What does the standard of informed consent require the pollsters to tell potential respondents?

(b) The standards accepted by polling organizations also require giving respondents the name and address of the organization that carries out the poll. Why do you think this is required?

(c) The polling organization usually has a professional name, such as "Samples Incorporated," so respondents don't know that the poll is being paid for by a political party or candidate. Would revealing the sponsor to respondents bias the poll? Should the sponsor always be announced whenever poll results are made public?

10.18 **Charging for Data?** Data produced by the government are often available free or at low cost to private users. For example, satellite weather data produced by the U.S. National Weather Service are available free to TV stations for their weather reports and to anyone on the web. *Opinion 1: Government data should be available to everyone at minimal cost.* European governments, on the other hand, charge TV stations for weather data. *Opinion 2: The satellites are expensive, and the TV stations are making a profit from their weather services, so they should share the cost.* Which opinion do you support, and why?

10.19 **Undue Influence?** An investigator wants to conduct a funded study of the safety of a vaccine to prevent hepatitis C involving prisoners as subjects. Prisoners will receive either vaccine or placebo and then be asked to complete surveys and undergo physical exams to assess for adverse effects. In order to ensure that subjects will report side effects and cooperate with exams, prisoners who are judged by the guards to be most compliant and well behaved are nonrandomly assigned to the experimental arm; others are assigned to the control (placebo) arm. To encourage participation, prisoners are offered better meals and the opportunity for better-paying jobs in the prison. Are there any aspects of this study that you object to? Why?

10.20 **The Willowbrook Hepatitis Studies.** In the 1960s, children entering the Willowbrook State School, an institution for the intellectually disabled, were deliberately infected with hepatitis. The researchers argued that almost all children in the institution quickly became infected anyway. The studies showed for the first time the existence of two strains of hepatitis. This finding contributed to the development of effective vaccines. Despite these valuable results, the Willowbrook studies are now considered an example of unethical research. Explain why, according to current ethical standards, useful results are not enough to allow a study.

10.21 **Unequal Benefits.** Researchers on depression proposed to investigate the effect of supplemental therapy and counseling on the quality of life of adults with depression. Eligible patients on the rolls of a large medical clinic were to be randomly assigned to treatment and control groups. The treatment group would be offered dental care, vision testing, transportation, and other services not available without charge to the control group. The review board felt that providing these services to some but not other persons in the same institution raised ethical questions. Do you agree? Explain your answer.

10.22 **Immortal Cells.** In 1951 Henrietta Lacks died at the Johns Hopkins Hospital from complications due to cervical cancer. Some of her cells were taken without her permission. It was subsequently discovered that these were "immortal cells," cells that do not die after a set number of cell divisions. These were the first human cells grown in a lab that were naturally immortal, making them invaluable for research. For example, in medical experiments if the cells died, they could simply be discarded and the experiment attempted again on fresh cells from the culture. Henrietta's "immortal" cells became the He-La cell line and have been used to develop the polio vaccine and flu treatments and in HIV/AIDS, leukemia, tuberculosis, and Parkinson's disease research, just to name a few applications. The research from He-La cells has saved hundreds of thousands, if not millions, of people. Does the benefit society received from the cells of Henrietta Lacks outweigh the ethics of failing to receive permission to use the cells from anyone in the Lacks family, including Henrietta herself? Explain your reasoning.

10.23 **AIDS Trials in Africa.** The drug programs that treat AIDS in rich countries are very expensive, and some African nations cannot afford to give them to large numbers of people. Yet AIDS is more common in parts of Africa than anywhere else. "Short-course" drug programs that are much less expensive might help, for example, in preventing infected pregnant women from passing the infection to their unborn children. Is it ethical to compare a short-course program with a placebo in a clinical trial? Some say No: this is a double standard because, in rich countries, the full drug program would be the control treatment. Others say Yes: the intent is to find treatments that are practical in Africa, and the trial does not withhold any treatment that subjects would otherwise receive. What do you think?

10.24 **Abandoned Children in Romania.** The study described in Example 9.2 (page 227) randomly assigned abandoned children in Romanian orphanages to move to foster homes or to remain in an orphanage. All the children would otherwise have remained in an orphanage. The foster care was paid for by the study. There was no informed consent because the children had been abandoned and had no adult to speak for them. The experiment was considered ethical because "people who cannot consent can be protected by enrolling them only in minimal-risk research, whose risks do not exceed those of everyday life" and because the study "aimed to produce results that would primarily benefit abandoned, institutionalized children." Do you agree?

10.25 **Asking Teens about Vaping.** A survey of more than 44,000 teenagers asked the subjects if they had used vaping devices in the past 12 months. In a follow-up question, subjects were asked what they vaped. Should consent of parents be required to ask minors about drug use and other such issues, or is consent of the minors themselves enough? Give reasons for your opinion.

10.26 **Deceiving Subjects.** Students sign up to be subjects in a psychology experiment. When they arrive, they are placed in a room and assigned a task. During the task, the subject hears a loud thud from an adjacent room and then a piercing cry for help. Some subjects are placed in a room by themselves. Others are placed in a room with "confederates" (a research methods term for accomplices) who have been instructed by the researcher to look up upon hearing the cry and then return to their task. The treatments being compared are whether the subject is alone in the room or in the room with confederates, will the subject ignore the cry for help?

The students had agreed to take part in an unspecified study, and the true nature of the experiment is explained to them afterward. Do you think this study is therefore ethically acceptable?

10.27 **Show Me the Data.** In Example 10.1 (page 250), we mentioned that researchers are not always completely honest about their data. In the interests of transparency, some suggest that researchers should be required to publish their data as part of any publication whose findings are based on these data.

In defense of keeping data private, researchers expend considerable time and effort in collecting data. The data may be part of ongoing research and used in future publications. Making it public before researchers have the opportunity to complete all their research involving the data allows others to exploit the data for their own advantage (perhaps preempting publications by the researchers who collected the data) without having to expend any effort in collecting the data themselves. This seems unfair.

Should researchers be required to publish their data along with their findings based on the data? If not, can you suggest how one might confirm the accuracy of any findings based on the data while preventing others from using the data to their advantage?

CHAPTER

11

IN THIS CHAPTER, WE COVER . . .

Producing Data: Part II Review

In Part I of this book, you mastered *data analysis,* the use of graphs and numerical summaries to organize and explore any set of data. Part II has introduced designs for data production. Part III will discuss basic probability and the foundations of inference, and Parts IV and V will deal in detail with statistical inference.

Designs for producing data are essential if the data are intended to represent some wider population or process. Figures 11.1 and 11.2 display the big ideas visually. Random sampling and randomized comparative experiments are perhaps the most important statistical inventions of the twentieth century. Both were slow to gain acceptance, and you will still see many voluntary response samples and uncontrolled experiments. You should now understand good designs for producing data and also why bad designs often produce data that are worthless for inference. The deliberate use of chance in producing data is a central idea in statistics. It not only reduces bias but also allows us to use probability, the mathematics of chance, as the basis for inference.

David Paynter/AGE Fotostock

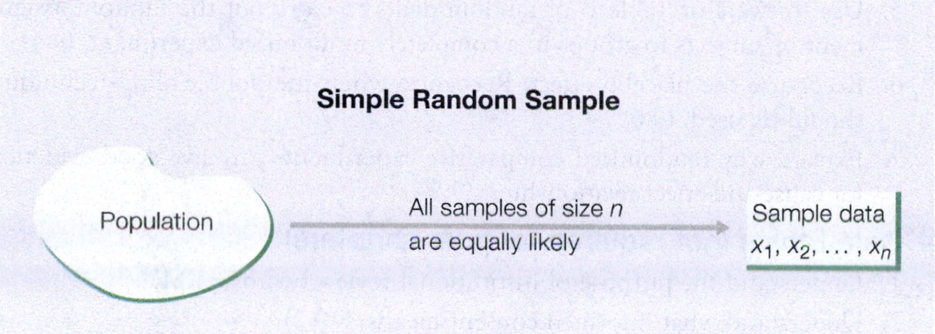

FIGURE 11.1 Statistics in summary.

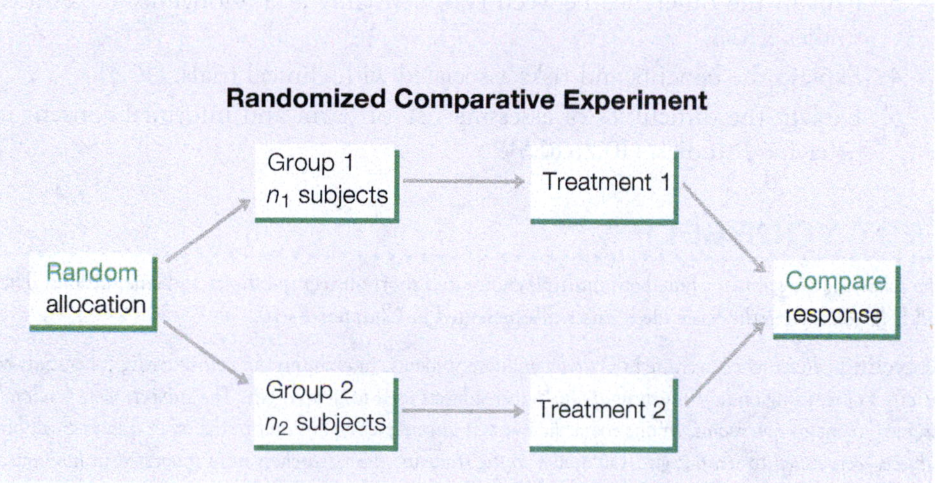

FIGURE 11.2 Statistics in summary.

Part II Skills Review

Here are the most important skills you should have acquired from reading Chapters 8 to 10. Following each sill in parentheses is the section of the text where the topic is introduced.

A. Sampling

1. Identify the population in a sampling situation. (8.1)
2. Recognize bias due to voluntary response samples and other inferior sampling methods. (8.2)
3. Use software or Table B of random digits to select a simple random sample (SRS) from a population. (8.3)
4. Recognize the presence of undercoverage and nonresponse as sources of error in a sample survey. Recognize the effect of the wording of questions on the responses. (8.6)
5. Use software or Table B of random digits to select a stratified random sample from a population when the strata are identified. (8.5)

B. Experiments

1. Recognize whether a study is an observational study or an experiment. (9.1)
2. Recognize bias due to confounding of explanatory variables with lurking variables in either an observational study or an experiment. (9.3)
3. Identify the factors (explanatory variables), treatments, response variables, and individuals or subjects in an experiment. (9.2)
4. Outline the design of a completely randomized experiment using a diagram like that in Figure 9.3 (page 231). The diagram in a specific case should show the sizes of the groups, the specific treatments, and the response variable. (9.4)

STATISTICS IN YOUR WORLD

Icing the Kicker

The football team lines up for what they hope will be the winning field goal . . . and the other team calls time-out. "Make the kicker think about it" is their motto. Does "icing the kicker" really work? That is, does the probability of making a field goal go down when the kicker must wait around during the time-out? This isn't a simple question. A detailed statistical study considered the distance, the weather, the kicker's skill, and so on. The conclusion is cheering to coaches: yes, icing the kicker does reduce the probability of success.

5. Use software or Table B of random digits to carry out the random assignment of subjects to groups in a completely randomized experiment. (9.4)
6. Recognize the placebo effect. Recognize when the double-blind technique should be used. (9.6)
7. Explain why randomized comparative experiments can give good evidence for cause-and-effect relationships. (9.5)

C. Data Ethics (not required for later chapters)

1. Understand the purpose of institutional review boards. (10.1)
2. Understand what informed consent means. (10.2)
3. Explain the difference between confidentiality and anonymity in research studies. (10.3)
4. Explain the benefits and risks associated with clinical trials. (10.3)
5. Explain the difficulties of assessing risk of harm and informed consent in behavioral studies. (10.2, 10.5)

TEST YOURSELF

The following questions include both multiple-choice and short-answer questions and calculations. They will help you review the basic ideas and skills presented in Chapters 8–10.

Recycling. *Researchers recruited 60 undergraduate students, in exchange for course credit, for a study on the effect of recycling on how much wrapping paper subjects used to wrap a gift. The subjects were randomly assigned to one of two rooms. In one room there was a large recycling bin and in the other a large trash bin. Subjects were asked to wrap a gift. Unknown to the students, the researchers were interested in how much paper the students used. The researchers found that students in the room with the recycling bin used (statistically) significantly more paper than those in the room with a trash bin.[1] The researchers had hypothesized that people in general would rather recycle than throw things in the trash, and hence would use less of a disposable resource when recycling is not available. Use this information to answer Questions 11.1 and 11.2.*

11.1 This experiment has

(a) two factors: recycling bin and trash bin.

(b) matched pairs.

(c) one treatment.

(d) stratification by gift.

11.2 The response in this experiment is

(a) the amount of paper used.

(b) the type of bin available.

(c) the desire to recycle.

(d) the 60 undergraduate students.

11.3 **Do You Trust the Internet?** You want to ask a sample of college students, "How much do you trust information about politics that you find on the Internet—a great deal, somewhat, not much, or not at all?" You try out this and other questions on a pilot group of five students chosen from your class. The class members are

Allenby	Drake	Kelbick	Rumsey
Bach	Ding	Kim	Scott
Baker	Drake	Lee	Smith
Chen	Farmer	Linder	Stewart
Collins	Hans	Miner	Verducci
Critchlow	Howell	O'Neill	Wolfe
Davis	Jeter	Paul	
Dean	Jones	Richards	

(a) Describe how you will label the students to select the sample.

(b) Use the *Simple Random Sample* applet, other software, or Table B, beginning at line 122, to select the five students in the sample.

(c) What is the response variable in this study?

American Community Survey. *Each month the U.S. Census Bureau's American Community Survey mails survey forms to 300,000 households, asking questions about demographic, social, economic, and housing characteristics such as mortgage and utility costs. Telephone calls are made to households that don't return the form. In one month, responses were obtained from 295,000 of the households contacted. Use this information to answer Questions 11.4 to 11.6.*

11.4 The sample is

(a) the 300,000 households initially contacted.

(b) the 295,000 households that responded.

(c) the 5,000 households that did not respond.

(d) all U.S. households.

11.5 The population of interest is

(a) all households with mortgages.

(b) the 300,000 households contacted.

(c) only U.S. households with phones.

(d) all U.S. households.

11.6 A source of bias in this survey is

(a) voluntary response.

(b) nonresponse.

(c) the fact that the survey was not double-blind.

(d) only U.S. households were contacted.

11.7 Traumatic brain injury (TBI) can have serious long-term consequences, including psychiatric disorders. To determine if there is a relationship between TBI and the risk of suicide, researchers examined the medical records of 7,418,391 individuals living in Denmark from 1980 to 2014. The researchers found that suicide rates were statistically significantly higher in those individuals who had medical contact for TBI compared to those with no evidence of TBI. However, the medical records did not contain information about TBI suffered prior to 1977, nor did the records indicate what treatment patients with TBI received.[2] This is an example of

(a) an observational study.

(b) a randomized comparative experiment.

(c) a block design, with blocks being whether or not there was information prior to 1977.

11.8 In the study described in the Question 11.7, we can conclude that

(a) having TBI leads to a greater risk for suicide. We can reach this conclusion because we have actual medical records over a long period of time.

(b) having TBI leads to a greater risk for suicide. We can reach this conclusion because we have a very large sample.

(c) individuals who had medical contact for TBI tended to have higher suicide rates than those who did not have medical contact for TBI.

11.9 A common definition of "binge drinking" is five or more drinks at one setting for men and four or more for women. An observational study finds that students who binge have lower average GPA than those who don't. Suggest two lurking variables that may be confounded with binge drinking and be sure to give a reason you have chosen each of these variables. The possibility of confounding means that we can't conclude that binge drinking *causes* lower GPA.

11.10 Many polling organizations conduct surveys about public opinion concerning the death penalty on a regular basis. Here are the questions asked in two polls conducted in 2018.

Question A: Which punishment do you prefer for people convicted of murder: the death penalty or life in prison with no chance of parole?

Question B: Are you in favor of the death penalty for a person convicted of murder?

One of these questions drew 56% favoring/preferring the death penalty. The other drew only 37%. Which wording pulls respondents favoring/preferring the death penalty? Why?

Reactions to Simulated News Reports. *A sample of University of Colorado students each viewed one of two simulated news reports about a terrorist bombing against the United States by a fictitious country. One report showed the bombing attack on a military target and the other on a cultural/educational site. In addition, before viewing the news report, each student read one of two "primes." The first was a prime for forgiveness based on the biblical saying "Love thy enemy," and the second was a retaliatory prime based on the biblical saying "An eye for an eye, and a tooth for a tooth." After viewing the news report, the students were asked to rate on a scale of 1 to 12 what the U.S. reaction should be, with the lowest score (1) corresponding to the United States sending a special ambassador to the country and the highest score (12) corresponding to an all-out nuclear attack against the country.[3] Use this information to answer Questions 11.11 and 11.12.*

11.11 Which of the following is correct?

(a) There are two factors in this experiment.

(b) There are four treatments in this experiment.

(c) Both (a) and (b) are correct.

11.12 The response in this experiment is

(a) the type of prime.

(b) the rating of U.S. reaction.

(c) the type of report.

11.13 In July 2018, C-SPAN ran a poll on Twitter that asked, "Do you SUPPORT or OPPOSE the nomination of Judge Brett Kavanaugh to the Supreme Court?" The final result was 54% SUPPORT, 39% OPPOSE, and 7% UNDECIDED. The number of votes received was 42,145. Explain why these sample results are almost certainly biased.

11.14 A study attempts to determine whether a football filled with helium travels farther when kicked than one filled with air. Each subject kicks twice: once with a football filled with helium and once with a football filled with air. The order of the type of football kicked is randomized. This is an example of

(a) a matched pairs experiment.

(b) a randomized controlled experiment.

(c) a stratified experiment.

(d) the placebo effect.

11.15 A *Columbus Dispatch* article reported that researchers at the Columbia University Department of Medicine examined records for an incredible 1.75 million patients born between 1900 and 2000 who had been treated at Columbia University Medical Center. Using statistical analysis, the researchers found that for cardiovascular disease, those born in the fall (September through December) were more protected, while those born in winter and spring (January to June) had higher risk. And because so many lives are cut short due to cardiovascular diseases, being born in the autumn was actually associated with living longer than being born in the spring.[4] This is as an example of

(a) a two-factor study with factors cardiovascular disease and length of life.

(b) an observational study.

(c) a single-blind experiment as the patients were unaware that their medical records were being studied.

(d) a matched pairs experiment, with each subject born in the fall paired with a subject born in winter or spring.

11.16 **(Optional topic)** A researcher is conducting a written survey about people's attitudes toward walking as an exercise option at a large shopping mall that supports a walking program. The survey is anonymous (without codes, names, or other information), and volunteers may complete the survey and place it in a box at a shopping mall exit. Which of the following is the most important ethical issue that the researcher addressed in planning the research?

(a) Providing a large sample size

(b) Confidentiality of the individual subject's responses

(c) Minimizing the risk of emotional distress from the questions themselves

11.17 **(Optional topic)** The purpose of informed consent is

(a) to obtain a signature from a study subject in order to protect the investigator, the study staff, and the institution.

(b) to obtain a signature from a study subject in order to document his or her agreement to participate in research.

(c) to provide a potential subject with appropriate information in an appropriate manner and allow that person to make an informed decision about participation in research.

SUPPLEMENTARY EXERCISES

Supplementary exercises apply the skills you have learned in ways that require more thought or more elaborate use of technology.

11.18 **Sampling Students.** You want to investigate the attitudes of students at your school toward the school's policy on sexual harassment. You have a grant that will pay the costs of contacting about 500 students.

Elinor Jones/REX/Shutterstock

(a) Specify the exact population for your study. For example, will you include part-time students?

(b) Describe your sample design. Will you use a stratified sample?

(c) Briefly discuss the practical difficulties that you anticipate. For example, how will you contact the students in your sample?

11.19 **Reducing Nonresponse.** How can we reduce the rate of refusals in telephone surveys? Most people who answer at all listen to the interviewer's introductory remarks and then decide whether to continue. One study made telephone calls to randomly selected households to ask opinions about the next election. In some calls, the interviewer gave her name, in others she identified the university she was representing, and in still others she identified both herself and the university. The study recorded what percentage of each group of interviews was completed. Is this an observational study or an experiment? Why? What are the explanatory and response variables?

11.20 Choosing Controls. The requirement that human subjects give their informed consent to participate in an experiment can greatly reduce the number of available subjects. For example, a study of new teaching methods asks the consent of parents for their children to be taught using either a new method or the standard method. Many parents do not return the forms, so their children must continue to follow the standard curriculum. Why is it not correct to consider these children as part of the control group along with children who are randomly assigned to the standard method?

11.21 Fixing Health Care. The cost of health care and health insurance is the biggest health concern among Americans, even ahead of cancer and other diseases. Changing to a national government health insurance system is controversial. An opinion poll will give different results depending on the wording of the question asked. For each of the following claims, indicate whether including it in the question would *increase* or *decrease* the percentage of a poll sample indicating support for a government health insurance system.

(a) A national system would mean that everybody has health insurance.

(b) A national system would probably require an increase in taxes.

(c) Eliminating private insurance companies and their profits would reduce insurance costs.

(d) A national system would limit the medical treatments available to contain costs.

11.22 Market Research. Stores advertise price reductions to attract customers. What type of price cut is most attractive? Market researchers prepared ads for athletic shoes announcing different levels of discounts (20%, 40%, or 60%). The student subjects who read the ads were also given "inside information" about the fraction of shoes on sale (50% or 100%). Each subject then rated the attractiveness of the sale on a scale of 1 to 7.[5]

(a) There are two factors. Make a sketch like Figure 9.2 (page 228) that displays the treatments formed by all combinations of levels of the factors.

(b) Outline a completely randomized design using 60 student subjects. Use software or Table B, at line 111, to choose the subjects for the first treatment.

11.23 The Safest Level of Drinking Is None. The news site *Vox* reported on a study in the journal *Lancet*. Researchers looked at the results of 700 studies from around the world, involving millions of people, and concluded that "the safest level of drinking is none." The study found that the more people drank around the globe, the greater their risk of cancer rose. In their paper, the researchers stated, "Alcohol use is a leading risk factor for global disease burden and causes substantial health loss," and "the level of consumption that minimizes health loss is zero."[6]

(a) What are the explanatory and response variables?

(b) The article in *Vox* goes on to say that "the data in the paper do not support a zero drinks recommendation." Why do you think *Vox* makes this statement?

11.24 Typeface Naturalness and Perceived Healthiness. Can the typeface used on a packaged product affect our perception of the product's healthiness? It was hypothesized that use of a "natural" typeface, which looks more handwritten and tends to be more slanted and curved, would lead to a higher perception of product healthiness than an unnatural typeface. Two typefaces, Impact and Sketchflow Print, were used. These typefaces had been shown in a previous study to differ in their perceived naturalness, but otherwise they were rated similarly on factors such as readability and likability. Images of two identical packages differing only in the typeface used were presented to the subjects. Participants read statements such as "This product is healthy natural/wholesome/organic," and they rated how much they agreed with the statement on a seven-point scale from 1 = strong agreement to 7 = strong disagreement. The subjects' responses were combined to create a perceived healthiness score.[7] The researcher had 100 students available to serve as subjects.

(a) What are the response and explanatory variables?

(b) Describe the design of a completely randomized experiment to learn the effect of typeface naturalness on perceived healthiness.

(c) Describe the design of a matched pairs experiment using the same 100 subjects.

Alysha Fligner

Alysha Fligner

11.25 Observation Versus Experiment. An *LA Times* article reported that the artery walls of people living within 100 meters of a highway thicken more than twice as fast as the average person's.[8] Researchers used ultrasound to measure the carotid artery wall thickness of 1483 people living near freeways in the Los Angeles area. The artery wall thickness among those living within 100 meters of a highway increased by 5.5 micrometers (roughly 1/20 the thickness of a human hair) each year during the three-year study, which is more than twice the progression observed in participants who did not live within this distance of a highway.

(a) The study compared artery thickening of subjects in the study who were living within 100 meters of a highway with those who were not. Without reading any further details of this study, how do you know that this was an observational study?

(b) Suggest some variables that might differ between the subjects in the study living within 100 meters of a highway and those who were farther away. Are any of these possible confounding variables? Explain. (Think about whether living very close to a highway indicates a desirable neighborhood.)

(c) Could this study be conducted as a randomized comparative experiment? What would be the difficulties?

11.26 Flu Shots. A *New York Times* article reported a study that investigated whether giving flu shots to schoolchildren protects a whole community from the disease. Researchers in Canada recruited 49 remote Hutterite farming colonies in western Canada for the study. In 25 of the colonies, all children aged 3 to 15 received flu shots in late 2008; in the 24 other colonies, all children aged 3 to 15 received a placebo. Which colonies received flu shots and which received the placebo was determined by randomization, and the colonies did not know whether they received the flu shots or the placebo. The researchers recorded the percentage of all children and adults in each colony who had laboratory-confirmed flu over the ensuing winter and spring.[9]

(a) Outline the design of this experiment. You do not need to do the randomization that your design requires.

(b) The placebo was actually the hepatitis A vaccine. The researchers stated that "hepatitis was not studied, but to keep the investigators from knowing which colonies received flu vaccine, they had to offer placebo shots, and hepatitis shots do some good while sterile water injections do not." In addition, the article mentions that the colonies were studied "without the investigators being subconsciously biased by knowing which received the placebo." Why was it important that investigators not be subconsciously biased by knowing which received the placebo?

(c) By June 2009, more than 10% of all the adults and children in colonies that received the placebo had laboratory-confirmed seasonal flu. Fewer than 5% of those in the colonies that received flu shots had. This difference was statistically significant. Explain to someone who knows no statistics what "statistically significant" means in this context.

fotografixx/Getty Images

PART III

From Data Production to Inference

Armed with designs for producing trustworthy data, we continue our journey toward *statistical inference*. Exploratory data analysis equips us to examine the data obtained from sampling or experiments, but simply describing or looking for patterns in the data at hand is often not the primary goal. Usually, data are used to answer specific questions, posed before the data are collected. If the sample has been selected using the principles presented in Chapter 8, the sample can tell us about important aspects of the population from which it was obtained. In a comparative experiment, the data can indicate how strong the evidence is that our treatment would be superior to the placebo for a broader class of circumstances.

Generalizing the results of sampling or experiments to a larger group of individuals or a broader class of circumstances is one goal of statistical inference. The conclusions of inference use the language of *probability*, the mathematics of chance. Chapters 12 and 15 present the ideas we need, and the optional Chapters 13 and 14 add more detail. Armed with designs for producing trustworthy data, data analysis to examine the data, and the language of probability, we are prepared to understand the big ideas of inference in Chapters 16, 17, and 18. These chapters are the foundation for the discussion of inference in practice that occupies the rest of the book.

*This material is not required for later parts of the text.

Christine Korten/EyeEm/Getty Images

CHAPTER

12

Introducing Probability

When you complete this chapter, you will be able to:

- 12.1 Understand probability as a way of quantifying the long-run behavior of random phenomena.
- 12.2 Interpret the concept of randomness in the physical world and in computer simulations.
- 12.3 Given facts about a situation or process that involves random outcomes, construct a probability model.
- 12.4 Use basic probability rules to assess the legitimacy of a given probability model, and to determine probabilities of specific events.
- 12.5 Use a finite probability model to calculate probabilities of events.
- 12.6 Given a continuous probability model, calculate probabilities represented by areas under the density curve.
- 12.7 Distinguish finite and continuous random variables and apply the concept of a random variable to real-world scenarios.
- 12.8 Interpret judgments about the likelihood of events as personal probabilities.

Why is probability, the mathematics of chance behavior, needed to understand statistics, the science of data? We collect data to provide insight into the larger population from which the data come. However, deciding what is true about this population by observing only a portion of it involves some degree of uncertainty. This is where the mathematics of chance behavior, or uncertainty, can help. To see how, let's look at a typical sample survey.

EXAMPLE 12.1 Does Anyone in Your Household Own a Gun?

What proportion of all U.S. adults say someone in their household owns a gun? We don't know, but we do have results from a *Washington Post*/ABC News poll.[1] The poll took a random sample of 1003 adults. The poll found that 461 of the people in the sample said that someone in their household owned a gun. The proportion who said that someone in their household owned a gun is

$$\text{sample proportion} = \frac{461}{1003} = 0.46 \text{ (that is, 46\%)}$$

If the sample was a simple random sample of all adults,[2] then, as discussed in Chapter 8 (page 205), all adults had the same chance to be among the chosen 1003. It would be reasonable to use this 46% as an estimate of the unknown proportion in the population. It's a *fact* that 46% of the sample said that someone in their household owned a gun; we know because the poll asked them. We don't know what percentage of all adults would say that someone in their household owned a gun, but we *estimate* that about 46% did at the time of the poll. This is a basic move in statistics: use a result from a sample to estimate something about a population.

What if the *Washington Post*/ABC News poll took a second random sample of 1003 adults? The new sample would have different people in it. It is almost certain that there would not be exactly 461 positive responses. That is, the *Washington Post*/ABC News poll's estimate of the proportion of adults who would say that someone in their household owns a gun will vary from sample to sample. Could it happen that one random sample finds that 46% of adults say that someone in their household owned a gun and another random sample taken at the same time finds that 64% have a gun in their home? *Random samples eliminate bias from the act of choosing a sample, but they can still fail to perfectly agree with the true population proportion because of the variability that results when we choose at random.* If the variation when we take repeated samples from the same population is too great, we can't trust the results of any one sample.

This is where we need facts about probability to make progress in statistics. When a poll uses chance to choose its samples, the laws of probability govern the behavior of the samples. The *Washington Post*/ABC News poll says that the probability is 0.95 that an estimate from one of its samples comes within ±3.5 percentage points of the truth about the population of all adults. The first step toward understanding this statement is to understand what "probability is 0.95" means. Our purpose in this chapter is to understand the language of probability—but without going into the full mathematics of probability theory.

12.1 The Idea of Probability

To understand why we can trust random samples and randomized comparative experiments, we must look closely at chance behavior. The big fact that emerges is this: *chance behavior is unpredictable in the short run but has a regular and predictable pattern in the long run.*

Toss a coin or choose a random sample. The result can't be predicted in advance because the result will vary when you toss the coin or choose the sample repeatedly. But there is a regular pattern in the results, a pattern that emerges clearly only after many repetitions. This remarkable fact is the basis for the idea of probability.

EXAMPLE 12.2 Coin Tossing

When you toss a coin, there are only two possible outcomes: heads or tails. Figure 12.1 shows the results of tossing a coin 5000 times, twice. For each number of tosses from 1 to 5000, we have plotted the proportion of those tosses that gave a head. Trial A (solid red line) begins tail, head, tail, tail. You can see that the proportion of heads for Trial A starts at 0 on the first toss, rises to 0.5 when the second toss gives a head, then falls to 0.33 and 0.25 as we get two more tails. Trial B, on the other hand, starts with five straight heads, so the proportion of heads is 1 until the sixth toss.

The proportion of tosses that produce heads is quite variable at first. Trial A starts low, and Trial B starts high. As we make more and more tosses, however, the proportion of heads for both trials gets close to 0.5 and stays there. If we made yet a third trial at tossing the coin a great many times, the proportion of heads would again settle down to 0.5 in the long run. This is the intuitive idea of probability. Probability 0.5 means "occurs half the time in a very large number of trials." The probability 0.5 appears as a horizontal line on the graph.

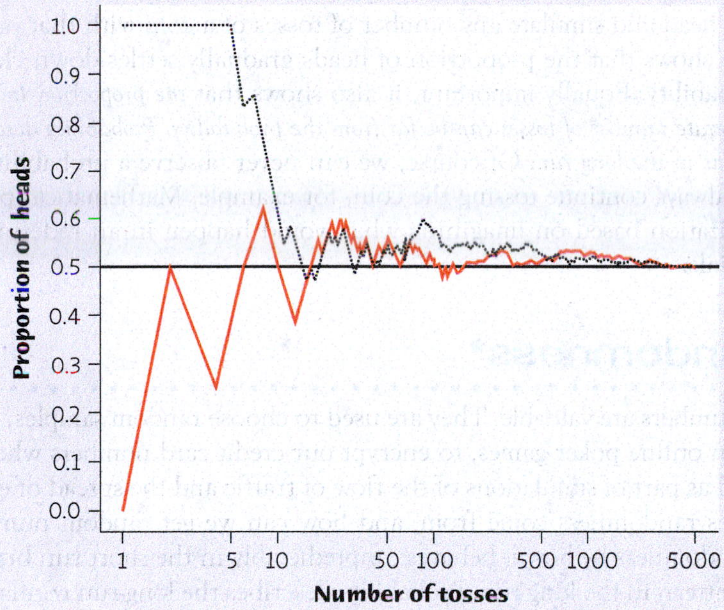

FIGURE 12.1 The proportion of tosses of a coin that give a head changes as we make more tosses. Eventually, however, the proportion approaches 0.5, the probability of a head. This figure shows the results of two trials of 5000 tosses each.

We might suspect that a coin has probability 0.5 of coming up heads just because the coin has two sides. But we can't be sure. In fact, spinning a penny on a flat surface, rather than tossing the coin, gives heads probability about 0.45 rather than 0.5.[3] The idea of probability is empirical. That is, it is based on observation rather than theory. Probability describes what happens in very many trials, and we must actually observe many trials to pin down a probability. In the case of tossing a coin, some diligent people have, in fact, made thousands of tosses.

EXAMPLE 12.3 Some Coin Tossers

The French naturalist Count Buffon (1707–1788) tossed a coin 4040 times. Result: 2048 heads, or proportion 2048/4040 = 0.5069 for heads.

Around 1900, the English statistician Karl Pearson heroically tossed a coin 24,000 times. Result: 12,012 heads, a proportion of 0.5005.

While imprisoned by the Germans during World War II, the South African mathematician John Kerrich tossed a coin 10,000 times. Result: 5067 heads, a proportion of 0.5067.

Randomness and Probability

We call a phenomenon **random** if individual outcomes are uncertain but there is nonetheless a regular pattern or distribution of outcomes in a large number of repetitions.

The **probability** of any outcome of a random phenomenon is the proportion of times the outcome would occur in a very long series of repetitions.

The best way to understand randomness is to observe random behavior, as in Figure 12.1. You can do this with physical devices like coins, but computer simulations (imitations) of random behavior allow faster exploration. The *Probability* applet

applet

is a computer simulation that animates Figure 12.1. It allows you to choose the probability of a head and simulate any number of tosses of a coin with that probability. Experience shows that the proportion of heads gradually settles down close to the probability. Equally important, it also shows that *the proportion in a small or moderate number of tosses can be far from the probability. Probability describes* only *what happens in the long run.* Of course, we can never observe a probability exactly. We could always continue tossing the coin, for example. Mathematical probability is an idealization based on imagining what would happen in an indefinitely long series of trials.

12.2 The Search for Randomness*

STATISTICS IN YOUR WORLD

Does God Play Dice?

Few things in the world are truly random in the sense that no amount of information will allow us to predict the outcome. But, according to the branch of physics called quantum mechanics, randomness does rule events inside individual atoms. Although Albert Einstein helped quantum theory get started, he always insisted that physical events are determined by observable causes, not just probabilities. "I shall never believe that God plays dice with the world," said the great scientist. A century after Einstein's first work on quantum theory, it appears that he was wrong.

Random numbers are valuable. They are used to choose random samples, to shuffle the cards in online poker games, to encrypt our credit card numbers when we buy online, and as part of simulations of the flow of traffic and the spread of epidemics. Where does randomness come from, and how can we get random numbers? We defined randomness by how it behaves: unpredictably in the short run but showing a regular pattern in the long run. Probability describes the long-run regular pattern. That many things are random in this sense is an observed fact about the world. Not all these things are really random. Here's a quick tour of how to find random behavior and get random numbers.

The easiest way to get random numbers is from a *computer program.* Of course, a computer program just does what it is told to do. Run the program again, and you get exactly the same result. The random numbers in Table B, the outcomes of the *Probability* applet, and the random numbers that shuffle cards for online poker come from computer programs, so they aren't really random. Clever computer programs produce outcomes that look random even though they really aren't. These *pseudo-random numbers* are more than good enough for choosing samples and shuffling cards. But they may have hidden patterns that can distort scientific simulations.

You might think that *physical devices such as coins and dice* produce really random outcomes. But a tossed coin obeys the laws of physics. If we knew all the inputs of the toss (forces, angles, and so on), then we could say in advance whether the outcome will be heads or tails. The outcome of a toss is predictable rather than random. Why do the results of tossing a coin *look* random? The outcomes are extremely sensitive to the inputs, so that very small changes in the forces you apply when you toss a coin change the outcome from heads to tails and back again. In practice, the outcomes are not predictable. Probability is a lot more useful than physics for describing coin tosses.

We call a phenomenon with "small changes in, big changes out" behavior *chaotic.* If we can feed chaotic behavior into a computer, we can do better than pseudo-random numbers. Coins and dice are awkward, but you can go to the website `www.random.org` to get random numbers from radio noise in the atmosphere, a chaotic phenomenon that is easy to feed to a computer.

Is anything really random? As far as current science can say, behavior inside atoms really is random—that is, there isn't any way to predict behavior in advance, no matter how much information we have. It was this "really, truly random" idea that Einstein disliked as he watched the new science of quantum mechanics emerge. Really, truly random numbers generated from the radioactive decay of atoms is available at the HotBits website, `www.fourmilab.ch/hotbits`.

*This short discussion is optional.

APPLY YOUR KNOWLEDGE

12.1 A Flush. You read online that the probability of being dealt a flush (all five cards of the same suit) in a five-card poker hand is 1/508. Explain carefully what this means. In particular, explain why it does *not* mean that if you are dealt 508 five-card poker hands, one will be a flush.

12.2 Probability Says . . . Probability is a measure of how likely an event is to occur. Match one of the probabilities that follow with each statement of likelihood given. (The probability is usually a more exact measure of likelihood than is the verbal statement.)

0 0.05 0.45 0.50 0.55 0.95 1

(a) This event is impossible. It can never occur.

(b) This event is just as likely to occur as it is to not occur.

(c) This event is very likely, but it will not occur once in a while in a long sequence of trials.

(d) This event will occur slightly less often than not.

12.3 Random Digits. The table of random digits (Table B) was produced by a random mechanism that gives each digit probability 0.1 of being a 0.

applet

(a) What proportion of the first 200 digits (those in the first five lines) in the table are 0s? This proportion is an estimate, based on 200 repetitions, of the true probability, which we know is 0.1.

(b) The *Probability* applet can imitate random digits. Set the probability of heads in the applet to 0.1. The horizontal green line on the graph represents the true probability. A head stands for a 0 in the random digit table, and a tail stands for any other digit. Simulate 200 digits (set "Number of Tosses" to 200 and click on "Toss"). What was the result of your 200 tosses?

12.4 The Long Run but Not the Short Run. Our intuition about chance behavior is not very accurate. In particular, we tend to expect that the long-run pattern described by probability will show up in the short run as well. For example, we tend to think that tossing a coin 10 times will give close to five heads.

applet

(a) Set the probability of heads in the *Probability* applet to 0.5 and the number of tosses to 10 to simulate 10 tosses of a balanced coin. What was the proportion of heads?

(b) Toggle the "Re-Toss" button to toss again. The simulation is fast, so do it 25 times and keep a record of the proportion of heads in each set of 10 tosses. Make a stemplot of your results. You see that the result of tossing a coin 10 times is quite variable and need not be very close to the probability 0.5 of heads.

12.3 Probability Models

Gamblers have known for centuries that coins, cards, and dice yield clear patterns in the long run. The idea of probability rests on the observed fact that the average result of many thousands of chance outcomes can be known with near certainty. How can we give a mathematical description of long-run regularity?

To see how to proceed, think first about a very simple random phenomenon: tossing a coin once. When we toss a coin, we cannot know the outcome in advance. What *do* we know? We are willing to say that the outcome will be either heads or tails. We believe that each of these outcomes has probability 1/2. This description of coin tossing has two parts:

- A list of possible outcomes
- A probability for each outcome

Such a description is the basis for all *probability models*. Here is the basic vocabulary we use.

Probability Models

The **sample space S** of a random phenomenon is the set of all possible outcomes.

An **event** is an outcome or a set of outcomes of a random phenomenon. That is, an event is a subset of the sample space.

A **probability model** is a mathematical description of a random phenomenon consisting of two parts: a sample space S and a way of assigning probabilities to events.

A sample space S can be very simple or very complex. When we toss a coin once, there are only two outcomes: heads and tails. The sample space is S = {H,T}. When the *Washington Post*/ABC News poll draws a random sample of 1003 adults, the sample space contains all possible choices of 1003 of the 254 million adults in the United States. This S is extremely large. Each member of S is a possible sample, so S is the collection, or "space," of all possible samples. This explains the term *sample space*.

EXAMPLE 12.4 Rolling Dice

Rolling two dice is a common way to lose money in casinos. There are 36 possible outcomes when we roll two dice and record the up-faces in order (first die, second die). Figure 12.2 displays these outcomes. They make up the sample space S. "Roll a sum of 5" is an event, call it A, that contains four of these 36 outcomes:

A =
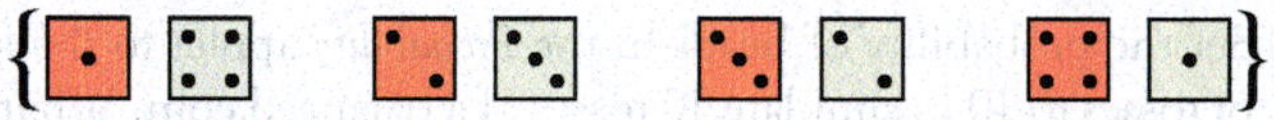

How can we assign probabilities to this sample space? We can find the actual probabilities for two specific dice only by actually tossing the dice many times and, even then, only approximately. So we will give a probability model that assumes ideal, perfectly balanced dice. This model will be quite accurate for carefully made casino dice and less accurate for the cheap dice that come with a board game.

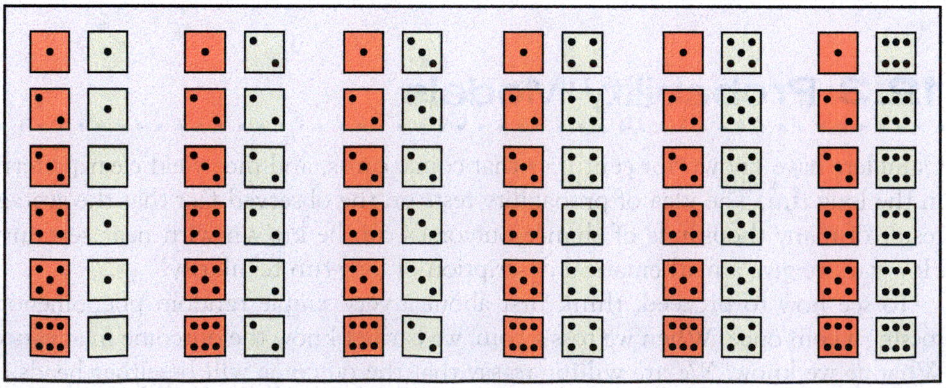

FIGURE 12.2
The 36 possible outcomes in rolling two dice. If the dice are carefully made, all these outcomes have the same probability.

If the dice are perfectly balanced, all 36 outcomes in Figure 12.2 will be *equally likely*. That is, each of the 36 outcomes will come up on one-thirty-sixth of all rolls in the long run. So each outcome has probability 1/36. There are four outcomes in the event A ("roll a sum of 5"), so this event has probability 4/36. In this way, we can assign a probability to any event. So we have a complete probability model.

In general, if all outcomes in a sample space are equally likely, we find the probability of any event by

$$\frac{\text{number of ways the event could occur}}{\text{total number of outcomes in the sample space}}$$

EXAMPLE 12.5 Rolling Dice and Counting the Spots

Gamblers care only about the total number of spots on the up-faces of the dice. The sample space for rolling two dice and counting the spots is

$$S = \{2, 3, 4, 5, 6, 7, 8, 9, 10, 11, 12\}$$

Comparing this S with Figure 12.2 reminds us that *we can change S by changing the detailed description of the random phenomenon we are describing.*

What are the probabilities for this new sample space? The 11 possible outcomes are *not* equally likely because there are six ways to roll a sum of 7, and there is only one way to roll a sum of 2 or a sum of 12. That's the key: each outcome in Figure 12.2 has probability 1/36. So "roll a sum of 7" has probability 6/36 because this event contains six of the 36 outcomes. Similarly, "roll a sum of 2" has probability 1/36, and "roll a sum of 5" (four outcomes from Figure 12.2) has probability 4/36. Here is the complete probability model:

Total spots	2	3	4	5	6	7	8	9	10	11	12
Probability	1/36	2/36	3/36	4/36	5/36	6/36	5/36	4/36	3/36	2/36	1/36

APPLY YOUR KNOWLEDGE

12.5 Sample Space. Choose a student at random from a large statistics class. Describe a sample space S for each of the following. (In some cases, you may have some freedom in specifying S.)

(a) Does the student have a pet or not?

(b) What is the student's height, in meters?

(c) What are the last three digits of the student's cell phone number?

(d) What is the student's birth month?

12.6 Role-Playing Games. Computer games in which the players take the roles of characters are very popular. They go back to earlier tabletop games such as Dungeons & Dragons. These games use many different types of dice. A four-sided die has faces with one of the numbers 1, 2, 3, or 4 appearing at the bottom of each visible face.

(a) What is the sample space for rolling a four-sided die twice (numbers on first and second rolls)? Follow the example of Figure 12.2.

(b) What is the assignment of probabilities to outcomes in this sample space? Assume that the die is perfectly balanced and follow the method of Example 12.4.

12.7 Role-Playing Games. Suppose the strength of a character in a game is determined by rolling the four-sided die twice and adding 2 to the sum of the numbers. Start with your work in the previous exercise to give a probability model (sample space and probabilities of outcomes) for the character's strength. Follow the method of Example 12.5.

12.4 Probability Rules

In Examples 12.4 and 12.5, we found probabilities for tossing dice. As random phenomena go, dice are pretty simple. Even so, we had to assume idealized, perfectly balanced dice. In most situations, it isn't easy to give a "correct" probability model. We can make progress by listing some facts that must be true for *any* assignment of probabilities. These facts follow from the idea of probability as "the long-run proportion of repetitions on which an event occurs":

1. *Any probability is a number between 0 and 1, inclusive.* Any proportion is a number between 0 and 1, so any probability is also a number between 0 and 1. An event with probability 0 never occurs, and an event with probability 1 occurs in every trial. An event with probability 0.5 occurs in half the trials in the long run.
2. *All possible outcomes together must have probability 1.* Because some outcome must occur on every trial, the sum of the probabilities for all possible outcomes must be exactly 1.
3. *If two events have no outcomes in common, the probability that one or the other occurs is the sum of their individual probabilities.* If one event occurs in 40% of all trials, a different event occurs in 25% of all trials, and the two can never occur together, then one or the other occurs on 65% of all trials because $40\% + 25\% = 65\%$.
4. *The probability that an event does not occur is 1 minus the probability that the event does occur.* If an event occurs in (say) 70% of all trials, it fails to occur in the other 30%. The probability that an event occurs and the probability that it does not occur always add to 100%, or 1.

We can use mathematical notation to state Facts 1 to 4 more concisely. Capital letters near the beginning of the alphabet denote events. If A is any event, we write its probability as $P(A)$. Here are our probability facts in formal language. As you apply these rules, remember that they are just another form of intuitively true facts about long-run proportions.

STATISTICS IN YOUR WORLD

Equally Likely?

A game of bridge begins by dealing all 52 cards in the deck to the four players, 13 to each. If the deck is well shuffled, all the immense number of possible hands will be equally likely. But don't expect the hands that appear in newspaper bridge columns to reflect the equally likely probability model. Writers on bridge choose "interesting" hands, especially those that lead to high bids that are rare in actual play.

Probability Rules

Rule 1. The probability $P(A)$ of any event A satisfies $0 \leq P(A) \leq 1$.

Rule 2. If S is the sample space in a probability model, then $P(S) = 1$.

Rule 3. Two events A and B are **disjoint** if they have no outcomes in common and so can never occur together. If A and B are disjoint,

$$P(A \text{ or } B) = P(A) + P(B)$$

This is the **addition rule for disjoint events.**

Rule 4. For any event A,

$$P(A \text{ does not occur}) = 1 - P(A)$$

The addition rule extends to more than two events that are disjoint in the sense that no two have any outcomes in common. If events A, B, and C are disjoint, the probability that one of these events occurs is $P(A) + P(B) + P(C)$.

EXAMPLE 12.6 Using the Probability Rules

We already used the addition rule for disjoint events, without calling it by that name, to find the probabilities in Example 12.5. The event "roll a sum of 5" contains the four disjoint outcomes displayed in Example 12.4, so the addition rule (Rule 3) says that its probability is

$$P(\text{roll a sum of 5}) = P(⚀\,⚃) + P(⚁\,⚂) + P(⚂\,⚁) + P(⚃\,⚀)$$

$$= \frac{1}{36} + \frac{1}{36} + \frac{1}{36} + \frac{1}{36}$$

$$= \frac{4}{36} = 0.111$$

Check that the probabilities in Example 12.5, found using the addition rule, are all between 0 and 1 and add to exactly 1. That is, this probability model obeys Rules 1 and 2.

What is the probability of rolling anything other than a sum of 5? By Rule 4,

$$P(\text{roll does not give a sum of 5}) = 1 - P(\text{roll a sum of 5})$$

$$= 1 - 0.111 = 0.889$$

Our model assigns probabilities to individual outcomes. To find the probability of an event, just add the probabilities of the outcomes that make up the event. For example:

$$P(\text{outcome is odd}) = P(3) + P(5) + P(7) + P(9) + P(11)$$

$$= \frac{2}{36} + \frac{4}{36} + \frac{6}{36} + \frac{4}{36} + \frac{2}{36}$$

$$= \frac{18}{36} = \frac{1}{2}$$

Image Source Plus/Alamy Stock Photo

APPLY YOUR KNOWLEDGE

12.8 Who Takes the GMAT? In many settings, the "rules of probability" are just basic facts about percentages. The Graduate Management Admission Test (GMAT) website provides the following information about the geographic region of citizenship of those who took the test in 2018: 1.9% were from Africa; 0.3% were from Australia and the Pacific Islands; 2.4% were from Canada; 14.3% were from Central and South Asia; 36.1% were from East and Southeast Asia; 1.7% were from Eastern Europe; 3.2% were from Mexico, the Caribbean, and Latin America; 2.2% were from the Middle East; 30.3% were from the United States; and 7.6% were from Western Europe.[4]

(a) What percentage of those who took the test in 2018 were from the Americas (either Canada, the United States, Mexico, the Caribbean, or Latin America)? Which rule of probability did you use to find the answer?

(b) What percentage of those who took the test in 2018 were from some other region than the United States? Which rule of probability did you use to find the answer?

12.9 Overweight? Although the rules of probability are just basic facts about percentages or proportions, we need to be able to use the language of events and their probabilities. Choose an American adult aged 20 years or over at random. Define two events:

$$A = \text{the person chosen is obese}$$

$$B = \text{the person chosen is overweight but not obese}$$

According to the National Center for Health Statistics, $P(A) = 0.40$ and $P(B) = 0.32$.

(a) Explain why events A and B are disjoint.

(b) Say in plain language what the event "A or B" is. What is $P(A \text{ or } B)$?

(c) If C is the event that the person chosen has normal weight or less, what is $P(C)$?

12.10 Languages in Canada. Canada has two official languages: English and French. Choose a Canadian at random and ask, "What is your mother tongue?" Here is the distribution of responses, combining many separate languages:[5]

Language	English	French	Other
Probability	0.57	0.21	?

(a) What is the probability that a Canadian's mother tongue is either English or French?

(b) What probability should replace "?" in the distribution?

(c) What is the probability that a Canadian's mother tongue is not English?

12.11 Are They Disjoint? Which of the following pairs of events, A and B, are disjoint? Explain your answers.

(a) A person is selected at random. A is the event "the person selected is less than age 18"; B is the event "the person selected is age 18 or over."

(b) A person is selected at random. A is the event "the person selected earns more than \$100,000 per year"; B is the event "the person selected earns more than \$250,000 per year."

(c) A pair of dice are tossed. A is the event "one of the dice is a 3"; B is the event "the sum of the two dice is 3."

12.5 Finite Probability Models

Examples 12.4, 12.5, and 12.6 illustrate one way to assign probabilities to events: assign a probability to every individual outcome and then add these probabilities to find the probability of any event. This idea works well when there are only a finite (fixed and limited) number of outcomes.

Finite Probability Model

A probability model with a finite sample space is called **finite**.

To assign probabilities in a finite model, list the probabilities of all the individual outcomes. These probabilities must be numbers between 0 and 1 that add to exactly 1. The probability of any event is the sum of the probabilities of the outcomes making up the event.

Statisticians often refer to finite probability models as discrete probability models.[6] In this book, we will sometimes refer to finite probability models as discrete.

EXAMPLE 12.7 Faking Data?

Faked numbers in tax returns, invoices, or expense account claims often display patterns that aren't present in legitimate records. Some patterns, such as too many round numbers, are obvious and are easily avoided by a clever crook. Others are more subtle. It is a striking fact that the first digits of numbers in legitimate records often follow a model known as Benford's law.[7] Call the first digit of a randomly chosen record X for short. Benford's law gives this probability model for X (note that a first digit can't be 0):

First digit X	1	2	3	4	5	6	7	8	9
Probability	0.301	0.176	0.125	0.097	0.079	0.067	0.058	0.051	0.046

Check that the probabilities of the outcomes sum to exactly 1. This is, therefore, a valid finite (or discrete) probability model. With these probabilities, investigators can detect fraud by comparing the first digits in records such as invoices paid by a business.

The probability that a first digit is equal to or greater than 6 is

$$P(X \geq 6) = P(X = 6) + P(X = 7) + P(X = 8) + P(X = 9)$$
$$= 0.067 + 0.058 + 0.051 + 0.046 = 0.222$$

This is less than the probability that a record has first digit 1,

$$P(X = 1) = 0.301$$

Fraudulent records tend to have too few 1s and too many higher first digits.

Note that the probability that a first digit is greater than or equal to 6 is not the same as the probability that a first digit is strictly greater than 6. The latter probability is

$$P(X > 6) = 0.058 + 0.051 + 0.046 = 0.155$$

The outcome $X = 6$ is included in "greater than or equal to" and is not included in "strictly greater than."

EXAMPLE 12.8 A Completely Randomized Design

In Chapter 9 (page 232) we discussed completely randomized experimental designs. Suppose you have three men–Ari, Luis, and Troy–and three women–Ana, Deb, and Hui–for an experiment. Three of the six subjects are to be assigned completely at random to a new experimental weight loss treatment and three to a placebo. Here are all 20 possible ways of selecting three of these subjects for the treatment group. (The remaining three are in the placebo group.)

Treatment Group	Treatment Group
Ari, Luis, Troy	Luis, Troy, Ana
Ari, Luis, Ana	Luis, Troy, Deb
Ari, Luis, Deb	Luis, Troy, Hui
Ari, Luis, Hui	Luis, Ana, Deb
Ari, Troy, Ana	Luis, Ana, Hui
Ari, Troy, Deb	Luis, Deb, Hui
Ari, Troy, Hui	Troy. Ana, Deb
Ari, Ana, Deb	Troy, Ana, Hui
Ari, Ana, Hui	Troy, Deb, Hui
Ari, Deb, Hui	Ana, Deb, Hui

These 20 possible treatment groups are the outcomes of assigning three of the six subjects to the treatment, and these outcomes constitute a sample space. With a completely randomized design, each of these 20 possible treatment groups (outcomes) is equally likely; thus each has probability 1/20 of being the actual group assigned to the treatment. Notice that the chance that all the men are assigned to the treatment group is 1/20, and the chance that the treatment group consists of either all men or all women is 2/20.

APPLY YOUR KNOWLEDGE

12.12 Rolling a Die. Figure 12.3 displays several possible finite probability models for rolling a die. We can learn which model is actually *accurate* for a particular die only by rolling the die many times. However, some of the models are not *valid*. That is, they do not obey the rules. Which are valid and which are not? In the case of the invalid models, explain what is wrong.

	Probability			
Outcome	Model 1	Model 2	Model 3	Model 4
⚀	1/7	1/3	1/3	1
⚁	1/7	1/6	1/6	1
⚂	1/7	1/6	1/6	2
⚃	1/7	0	1/6	1
⚄	1/7	1/6	1/6	1
⚅	1/7	1/6	1/6	2

FIGURE 12.3
Four assignments of probabilities to the six faces of a die, for Exercise 12.12.

12.13 Benford's Law. The first digit of a randomly chosen expense account claim follows Benford's law (Example 12.7). Consider the events

$$A = \{\text{first digit is 4 or greater}\}$$
$$B = \{\text{first digit is even}\}$$

(a) What outcomes make up the event *A*? What is $P(A)$?

(b) What outcomes make up the event *B*? What is $P(B)$?

(c) What outcomes make up the event "*A* or *B*"? What is $P(A \text{ or } B)$? Why is this probability not equal to $P(A) + P(B)$?

12.14 How Many Cups of Coffee? Choose an adult age 18 or over in the United States at random and ask, "How many cups of coffee do you drink on average per day?" Call the response *X* for short. Based on a large sample survey, here is a probability model for the answer you will get:[8]

Number	0	1	2	3	4 or more
Probability	0.36	0.26	0.19	0.08	0.11

(a) Verify that this is a valid finite probability model.

(b) Describe the event $X < 4$ in words. What is $P(X < 4)$?

(c) Express the event "have at least one cup of coffee on an average day" in terms of *X*. What is the probability of this event?

12.6 Continuous Probability Models

When we use the table of random digits to select a digit between 0 and 9, the finite probability model assigns probability 1/10 to each of the 10 possible outcomes. Suppose that we want to choose a number at random between 0 and 1, allowing *any* number between 0 and 1 as the outcome. Software random number generators will do this. For example, here is the result of asking software to produce five random numbers between 0 and 1 (with the results rounded to seven decimal places):[9]

0.2893511 0.3213787 0.5816462 0.9787920 0.4475373

Ignoring the fact that the results must be rounded in order to display them, the sample space is now an entire interval of numbers:

$$S = \{\text{all numbers between 0 and 1}\}$$

Call the outcome of the random number generator Y for short. How can we assign probabilities to events such as $\{0.3 \leq Y \leq 0.7\}$? As in the case of selecting a random digit, we would like all possible outcomes to be equally likely. But we cannot assign probabilities to each individual value of Y and then add them because there is an infinite continuum of possible values. In fact, we cannot even make a list of the individual values of Y. For example, what is the next largest value of Y after 0.3?

We use a new way of assigning probabilities directly to events—as *areas under a density curve*. Any density curve has area exactly 1 underneath it, corresponding to total probability 1. We met density curves as models for data in Chapter 3 (page 77).

STATISTICS IN YOUR WORLD

Really Random Digits

For purists, the RAND Corporation long ago published a book titled *One Million Random Digits*. The book lists 1,000,000 digits that were produced by a very elaborate physical randomization and really are random. An employee of RAND once said that this is not the most boring book that RAND has ever published.

Continuous Probability Model

A **continuous probability model** assigns probabilities as areas under a density curve. The area under the curve and between any range of values is the probability of an outcome in that range.

EXAMPLE 12.9 Random Numbers

The random number generator will spread its output uniformly across the entire interval from 0 to 1 as we allow it to generate a long sequence of numbers. Figure 12.4 is a histogram of the result of generating 10,000 random numbers. They are quite uniform but not exactly so. The bar heights would all be exactly equal (1,000 numbers for each bar) if the 10,000 numbers were exactly uniform. In fact, the counts vary from a low of 978 to a high of 1,060.

As in Chapter 3, we have adjusted the histogram scale so that the total area of the bars is exactly 1. Now we can add the density curve that describes the distribution of perfectly random numbers. This density curve also appears in Figure 12.4. It has height 1 over the interval from 0 to 1. This is the density curve of a *uniform distribution*. It is the continuous probability model for the results of generating very many random numbers. Like the probability models for perfectly balanced coins and dice, the density curve is an idealized description of the outcomes of a perfectly uniform random number generator. It is a good approximation for software outcomes, but even 10,000 tries isn't enough for actual outcomes to look exactly like the idealized model.

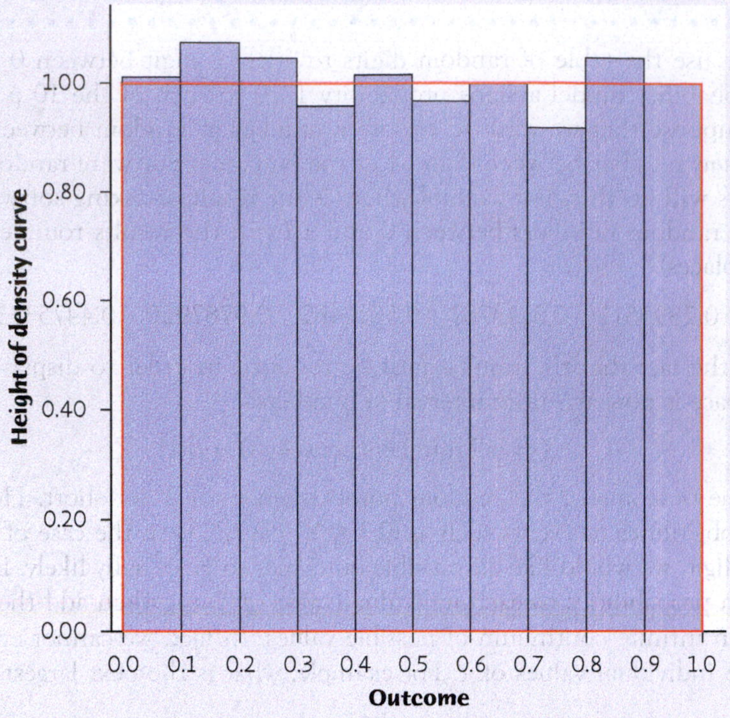

FIGURE 12.4 The probability model for the outcomes of a software random number generator, for Example 12.9. Compare the histogram of 10,000 actual outcomes with the uniform density curve, in red, that spreads probability evenly between 0 and 1.

Uniform Distribution

A **uniform distribution** is a continuous probability distribution that assigns equal probability to every number (or to every interval of a given length) in the range of values over which it is defined.

The uniform density curve has height 1 over the interval from 0 to 1. The area under the curve is 1, and the probability of any event is the area under the curve and above the interval that corresponds to the event in question. Figure 12.5 illustrates finding probabilities as areas under the density curve. The probability that the random number generator produces a number between 0.3 and 0.7 is

$$P(0.3 \le Y \le 0.7) = 0.4$$

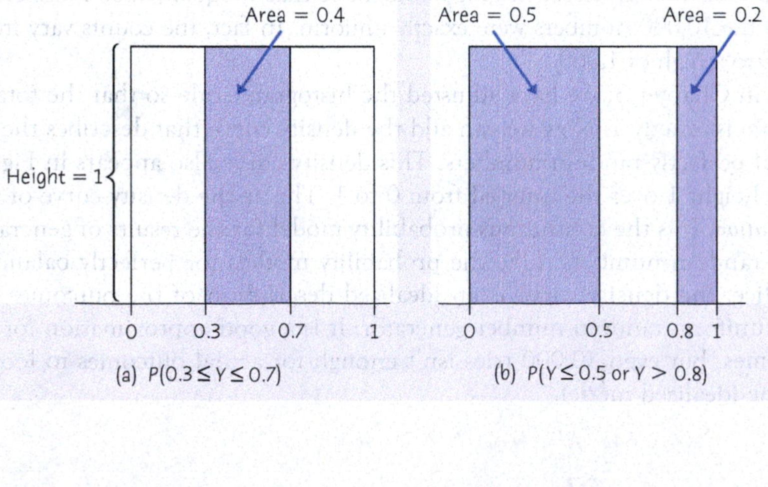

FIGURE 12.5 Probability as area under a density curve. The uniform density curve spreads probability evenly between 0 and 1.

because the area under the density curve and above the interval from 0.3 to 0.7 is 0.4. The height of the curve is 1, and the area of a rectangle is the product of height and length, so the probability of any interval of outcomes is just the length of the interval. Similarly,

$$P(Y \le 0.5) = 0.5$$
$$P(Y > 0.8) = 0.2$$
$$P(Y \le 0.5 \text{ or } Y > 0.8) = 0.7$$

The last event consists of two nonoverlapping intervals, so the total area above the event is found by adding two areas, as illustrated by Figure 12.5(b). This assignment of probabilities obeys all our rules for probability.

Continuous probability models assign probabilities to intervals of outcomes rather than to individual outcomes. In fact, *all continuous probability models assign probability 0 to every individual outcome.* Only intervals of values have positive probability. To see that this is true, consider a specific outcome such as $P(Y = 0.8)$. The probability of any interval is the same as its length. The point 0.8 has no length, so its probability is 0. Put another way, $P(Y > 0.8)$ and $P(Y \ge 0.8)$ are both 0.2 because that is the area in Figure 12.5(b) between 0.8 and 1.

We can use any density curve to assign probabilities. We discussed density curves in Chapter 3, and the density curves that are most familiar to us are the Normal curves. In Chapter 3, Normal curves were used to describe the distribution of data, and we used them to answer questions about the proportion of the data taking values between two numbers. *Normal distributions are continuous probability models* as well as descriptions of data. We use them to answer questions about the probability of a random variable taking on a value between two numbers. There is a close connection between a Normal distribution as an idealized description for data and a Normal probability model. If we look at the heights of all young women, we find that they closely follow the Normal distribution with mean $\mu = 64.1$ inches and standard deviation $\sigma = 3.7$ inches. This is a distribution for a large set of data. Now choose one young woman at random. Call her height X. If we repeat the random choice very many times, the distribution of values of X is the same Normal distribution that describes the heights of all young women.

EXAMPLE 12.10 The Heights of Young Women

What is the probability that a randomly chosen woman aged 20–29 has height between 68 and 70 inches? The height X of the woman we choose has the $N(64.1, 3.7)$ distribution. We want $P(68 \le X \le 70)$. This is the area under the Normal curve in Figure 12.6. Software or the *Normal Density Curve* applet will give us the answer at once: $P(68 \le X \le 70) = 0.0905$.

applet

We can also find the probability by standardizing, as discussed in Section 3.5, and using Table A, the table of standard Normal probabilities. We will reserve capital Z for a standard Normal variable.

Peathegee Inc/Getty Images

$$\begin{aligned} P(68 \le X \le 70) &= P\left(\frac{68 - 64.1}{3.7} \le \frac{X - 64.1}{3.7} \le \frac{70 - 64.1}{3.7}\right) \\ &= P(1.05 \le Z \le 1.59) \\ &= P(Z \le 1.59) - P(Z \le 1.05) \\ &= 0.9441 - 0.8531 = 0.0910 \end{aligned}$$

The calculation is the same as those we did in Chapter 3. Only the language of probability is new. In Chapter 3, we asked what proportion or percentage of the values for a population fall in a given interval. Now we ask: What is the probability that a randomly selected value falls within the interval? The probability equals the proportion.

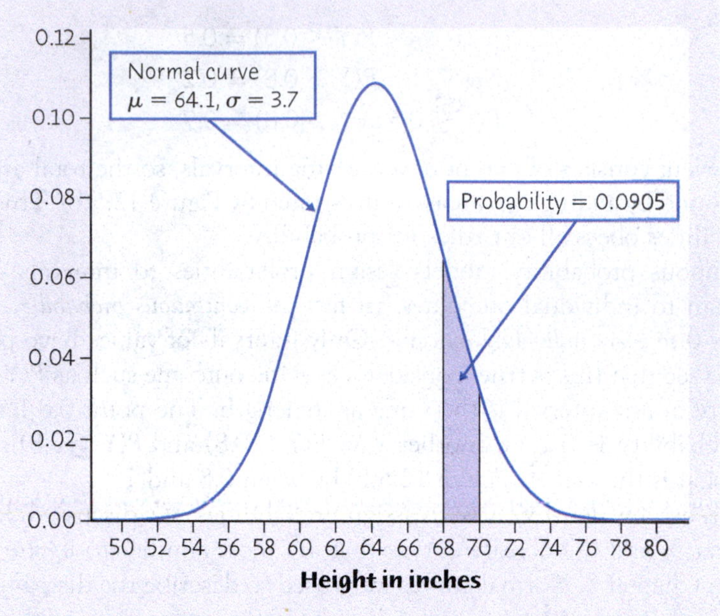

FIGURE 12.6
The probability in Example 12.10 as an area under a Normal curve.

APPLY YOUR KNOWLEDGE

12.15 Random Numbers. Let Y be a random number between 0 and 1 produced by the idealized random number generator described in Example 12.9 and Figure 12.4. Find the following probabilities:

(a) $P(Y \leq 0.6)$

(b) $P(Y < 0.6)$

(c) $P(0.4 \leq Y \leq 0.8)$

(d) $P(0.4 < Y \leq 0.8)$

12.16 Adding Random Numbers. Generate two random numbers between 0 and 1 and take X to be their sum. The sum X can take any value between 0 and 2. The density curve of X is the triangle shown in Figure 12.7.

(a) Verify by using geometry that the area under this curve is 1.

(b) What is the probability that X is less than 1? Sketch the density curve, shade the area that represents the probability, and then find that area. Do this for part (c) also.

(c) What is the probability that X is less than 0.5?

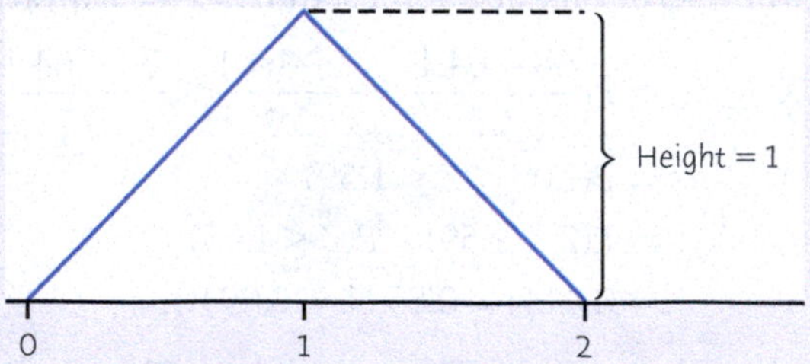

FIGURE 12.7
The density curve for the sum of two random numbers, for Exercise 12.16. This density curve spreads probability between 0 and 2.

12.17 The Medical College Admission Test. The Normal distribution with mean $\mu = 500.9$ and standard deviation $\sigma = 10.6$ is a good description of the total score on the Medical College Admission Test (MCAT).[10] This is a continuous probability model for the score of a randomly chosen student. Call the score of a randomly chosen student X for short.

(a) Write the event "the student chosen has a score of 510 or higher" in terms of X.

(b) Find the probability of this event.

12.7 Random Variables

Examples 12.7, 12.9, and 12.10 use a shorthand notation that is often convenient. In Example 12.10, we let X stand for the result of choosing a woman at random and measuring her height. X takes numerical values. We know that X would take a different numerical value if we made another random choice. Because its value changes from one random choice to another, we call the height X a *random variable.*

Random Variable

A **random variable** is a variable whose value is a numerical outcome of a random phenomenon.

The **probability distribution** of a random variable X tells us what values X can take and how to assign probabilities to those values.

We usually denote random variables by using capital letters near the end of the alphabet, such as X or Y. Of course, the random variables of greatest interest to us are outcomes such as the mean $\bar{x}$ of a random sample, for which we will keep the familiar notation. There are two main types of random variables, corresponding to two types of probability models: *finite* (or *discrete*) and *continuous*.

EXAMPLE 12.11 Finite and Continuous Random Variables

The first digit X in Example 12.7 (page 281) is a random variable whose possible values are the whole numbers $\{1, 2, 3, 4, 5, 6, 7, 8, 9\}$. The distribution of X assigns a probability to each of these outcomes. A **finite random variable** has a finite list of possible outcomes.

finite random variable

Compare the output Y of the random number generator in Example 12.8. The values of Y fill the entire interval of numbers between 0 and 1. The probability distribution of Y is given by its density curve, shown in Figure 12.4. A **continuous random variable** can take on any value in an interval, with probabilities given as areas under a density curve.

continuous random variable

We have defined a random variable to be a variable whose value is a *numerical* outcome of a random phenomenon. However, there are situations in which the outcomes of a random phenomenon are not numerical. In Example 12.2 (page 272), the possible outcomes of a coin toss are heads and tails. We can represent these outcomes with numbers, letting 1 represent heads and 0 represent tails. Using numbers to represent the nonnumerical outcomes of a random phenomenon is a common practice in statistics. For mathematical purposes, it is convenient to restrict random variables to take on numerical values, even if this means representing nonnumerical outcomes by numbers.

APPLY YOUR KNOWLEDGE

12.18 Grades in a Business Course. Indiana University posts the grade distributions for its courses online.[11] Students in Business 100 in the spring 2019 semester received these grades: 9% A+, 15% A, 13% A−, 10% B+, 13% B, 8% B−, 7% C+, 11% C, 0% C−, 2% D+, 4% D, 0% D−, and 8% F. Choose a Business 100 student at random. To "choose at random" means to give every student the same chance to be chosen. The student's grade on a four-point scale (with A+ = 4.3, A = 4, A− = 3.7, B+ = 3.3, B = 3.0, B− = 2.7, C+ = 2.3, C = 2.0, C− = 1.7, D+ = 1.3, D = 1.0, D− = 0.7, and F = 0.0) is a random variable X with this probability distribution:

Value of X	0.0	0.7	1.0	1.3	1.7	2.0	2.3	2.7	3.0	3.3	3.7	4.0	4.3
Probability	0.08	0.00	0.04	0.02	0.00	0.11	0.07	0.08	0.13	0.10	0.13	0.15	0.09

(a) Is X a finite or continuous random variable? Explain your answer.

(b) Say in words what the meaning of $P(X \geq 3.0)$ is. What is this probability?

(c) Write the event "the student got a grade poorer than B−" in terms of values of the random variable X. What is the probability of this event?

12.19 Running a Mile. A study of 12,000 able-bodied male students at the University of Illinois found that their times for the mile run were approximately Normal with mean 7.11 minutes and standard deviation 0.74 minute.[12] Choose a student at random from this group and call his time for the mile Y.

(a) Is Y a finite or continuous random variable? Explain your answer.

(b) Say in words what the meaning of $P(Y \geq 8)$ is. What is this probability?

(c) Write the event "the student could run a mile in less than six minutes" in terms of values of the random variable Y. What is the probability of this event?

12.8 Personal Probability*

STATISTICS IN YOUR WORLD

What Are the Odds?

Gamblers often express chance in terms of *odds* rather than probability. Odds of *A* to *B* against an outcome means that the probability of that outcome is $B/(A + B)$. So "odds of 5 to 1" is another way of saying "probability 1/6." A probability is always between 0 and 1, but odds range from 0 to infinity. Although odds are mainly used in gambling, they give us a way to make very small probabilities clearer. "Odds of 999 to 1" may be easier to understand than "probability 0.001."

We began our discussion of probability with one idea: the probability of an outcome of a random phenomenon is the proportion of times that outcome would occur in a very long series of repetitions. This idea ties probability to actual outcomes. It allows us, for example, to estimate probabilities by simulating random phenomena. Yet we often meet another, quite different, idea of probability.

EXAMPLE 12.12 Joe and the Cleveland Indians

Joe sits staring into his beer as his favorite baseball team, the Cleveland Indians, just miss in the playoffs. The Indians have some talented players, so let's ask Joe, "What's the chance that the Indians will go to the playoffs next year?" Joe brightens up. "Oh, about 60%," he says.

Does Joe assign probability 0.60 to the Indians appearing in the playoffs? The outcome of next year's season is certainly unpredictable, but we can't reasonably ask what would happen in many repetitions. Next year's baseball season will happen only once and will differ from all other seasons in terms of players, weather, and many other factors. If probability measures "what would happen if we did this many times," Joe's 0.60 is not a probability. Probability is based on data about many repetitions of the same random phenomenon. Joe is giving us something else: his personal judgment.

*This short section is optional.

Although Joe's 0.60 isn't a probability in our usual sense, it gives useful information about Joe's opinion. More seriously, a company asking, "How likely is it that building this plant will pay off within five years?" can't employ an idea of probability based on many repetitions of the same thing. The opinions of company officers and advisers are nonetheless useful information, and these opinions can be expressed in the language of probability. These are *personal probabilities*.

Personal Probability

A **personal probability** of an outcome is a number between 0 and 1 that expresses an individual's judgment of how likely the outcome is.

Rachel's opinion about the Indians may differ from Joe's, and the opinions of several company officers about the new plant may differ. These opinions may be based on careful analysis of past data, established theory, and logical arguments by experts, but they are still opinions. Personal probabilities are indeed personal: they vary from person to person. Moreover, if two people assign different personal probabilities to an event, it may be difficult or impossible to determine who is more correct. If we say, "In the long run, this coin will come up heads 60% of the time," we can find out if we are right by actually tossing the coin several thousand times. If Joe says, "I think the Indians have a 60% chance of going to the playoffs next year," that's just Joe's opinion. Why think of personal probabilities as probabilities? Because *any set of personal probabilities that makes sense obeys the same basic Rules 1 through 4 that describe any legitimate assignment of probabilities to events.* If Joe thinks there's a 60% chance that the Indians will go to the playoffs, he must also think that there's a 40% chance that they won't go. There is just one set of rules of probability, even though we now have two interpretations of what probability means.

APPLY YOUR KNOWLEDGE

12.20 Will You Be in a Crash? The probability that a randomly chosen driver will be involved in a car crash in the next year is about 0.051.[13] This is based on the proportion of millions of drivers who have crashes.

(a) What do you think is your own probability of being in a crash in the next year? This is a personal probability.

(b) Give some reasons why your personal probability might be a more accurate prediction of your "true chance" of being in a crash than the probability for a random driver.

(c) Almost everyone says their personal probability is lower than the random driver probability. Why do you think this is true?

12.21 Winning the ACC Tournament. The annual Atlantic Coast Conference men's basketball tournament has temporarily taken Joe's mind off of the Cleveland Indians. He says to himself, "I think that Louisville has probability 0.05 of winning. North Carolina's probability is twice Louisville's, and Duke's probability is four times Louisville's."

(a) What are Joe's personal probabilities for North Carolina and Duke?

(b) What is Joe's personal probability that one of the 15 teams other than Louisville, North Carolina, and Duke will win the tournament?

CHAPTER 12 SUMMARY

- A **random phenomenon** (chance experiment) has outcomes that we cannot predict but that nonetheless have a regular distribution in very many repetitions.
- The **probability** of an outcome is the proportion of times the outcome occurs in many repeated trials of a random phenomenon.
- A **probability model** for a random phenomenon consists of a sample space S and an assignment of probabilities P.
- The **sample space S** is the set of all possible outcomes of the random phenomenon. Sets of outcomes are called **events**. P assigns a number P(A) to an event A as its **probability**.
- Any assignment of probability must obey the rules that state the basic properties of probability:
  1. $0 \leq P(A) \leq 1$ for any event A.
  2. $P(S) = 1$.
  3. **Addition rule for disjoint events:** Events A and B are **disjoint** if they have no outcomes in common. If A and B are disjoint, then $P(A \text{ or } B) = P(A) + P(B)$.
  4. For any event A, $P(A \text{ does not occur}) = 1 - P(A)$.
- When a sample space S contains finitely many possible outcomes, a **finite probability model** assigns each of these outcomes a probability between 0 and 1 such that the sum of all the probabilities is exactly 1. The probability of any event is the sum of the probabilities of all the outcomes that make up the event. Finite probability models are also referred to as discrete probability models.
- A sample space can contain as outcomes all values in some interval of numbers. A **continuous probability model** assigns probabilities as areas under a density curve. The probability of any event is the area under the curve above the values that make up the event.
- A **random variable** is a variable taking numerical values determined by the outcome of a random phenomenon. The **probability distribution** of a random variable X tells us what the possible values of X are and how probabilities are assigned to those values.
- A random variable X and its distribution can be *discrete* or *continuous*. The distribution of a **discrete random variable** with finitely many possible values gives the probability of each value. A **continuous random variable** takes all values in some interval of numbers. A density curve describes the probability distribution of a continuous random variable.

CHECK YOUR SKILLS

12.22 You read in a book on poker that the probability of being dealt a straight in a five-card poker hand is 1/255. This means that

(a) if you deal millions of poker hands, the fraction of them that contain a straight will be very close to 1/255.

(b) if you deal 255 poker hands, exactly one of them will contain a straight.

(c) if you deal 25,500 poker hands, exactly 100 of them will contain a straight.

12.23 A basketball player shoots six free throws during a game. The sample space for counting the number she makes is

(a) S = any number between 0 and 1.

(b) S = whole numbers 0 to 6.

(c) S = all sequences of six hits or misses, like HMMHHH.

Here is the probability model for the political affiliation of a randomly chosen adult in the United States.[14] Questions 12.24 through 12.27 use this information.

Political affiliation	Republican	Independent	Democrat	Other
Probability	0.30	0.38	0.31	?

12.24 This probability model is

(a) finite. (b) continuous. (c) equally likely.

12.25 The probability that a randomly chosen American adult's political affiliation is "Other" must be

(a) any number between 0 and 1.

(b) 0.01.

(c) 0.1.

12.26 What is the probability that a randomly chosen American adult is a member of one of the two major political parties (Republicans and Democrats)?

(a) 0.38

(b) 0.61

(c) 0.99

12.27 What is the probability that a randomly chosen American adult is not a Republican?

(a) 0.39

(b) 0.70

(c) 0.01

12.28 In a table of random digits such as Table B, each digit is equally likely to be any of 0, 1, 2, 3, 4, 5, 6, 7, 8, or 9. What is the probability that a digit in the table is a 7?

(a) 1/9

(b) 1/10

(c) 9/10

12.29 In a table of random digits such as Table B, each digit is equally likely to be any of 0, 1, 2, 3, 4, 5, 6, 7, 8, or 9. What is the probability that a digit in the table is 7 or greater?

(a) 7/10 (b) 4/10 (c) 3/10

12.30 Choose an American household at random and let the random variable X be the number of cars (including SUVs and light trucks) the residents own. Here is the probability model if we ignore the few households that own more than seven cars:[15]

Number of cars X	0	1	2	3	4	5	6	7
Probability	0.004	0.247	0.383	0.212	0.097	0.037	0.011	0.009

A housing company builds houses with two-car garages. What percentage of households have more cars than the garage can hold?

(a) 21.2% (b) 25.1% (c) 36.6%

12.31 Choose a common fruit fly *Drosophila melanogaster* at random. Call the length of the thorax (where the wings and legs attach) Y. The random variable Y has the Normal distribution with mean $\mu = 0.800$ millimeter (mm) and standard deviation $\sigma = 0.078$ mm. The probability $P(Y > 1)$ that the fly you choose has a thorax more than 1 mm long is about

(a) 0.995. (b) 0.5. (c) 0.005.

CHAPTER 12 EXERCISES

12.32 **Sample Space.** In each of the following situations, describe a sample space S for the random phenomenon.

(a) A basketball player shoots four free throws. You record the sequence of hits and misses.

(b) A basketball player shoots four free throws. You record the number of baskets she makes.

Darrell Walker/HWMS/Icon SMI 945/Icon Sportswire/Athens GA United States

12.33 **Probability Models?** In each of the following situations, state whether or not the given assignment of probabilities to individual outcomes is legitimate—that is, satisfies the rules of probability. Remember, a legitimate model need not be a practically reasonable model. If the assignment of probabilities is not legitimate, give specific reasons for your answer.

(a) Roll a six-sided die and record the count of spots on the up-face:

$P(1) = 0$ $P(2) = 1/6$ $P(3) = 1/3$
$P(4) = 1/3$ $P(5) = 1/6$ $P(6) = 0$

(b) Deal a card from a shuffled deck:

$P(\text{clubs}) = 12/52$ $P(\text{diamonds}) = 12/52$
$P(\text{hearts}) = 12/52$ $P(\text{spades}) = 16/52$

(c) Choose a college student at random and record sex and enrollment status:

$P(\text{female full-time}) = 0.56$ $P(\text{male full-time}) = 0.44$
$P(\text{female part-time}) = 0.24$ $P(\text{male part-time}) = 0.17$

12.34 **Education Among Young Adults.** Choose a young adult (aged 25–29) at random. The probability is 0.07 that the person chosen did not complete high school, 0.46 that the person has a high school diploma but no further education, and 0.37 that the person has at least a bachelor's degree.

(a) What must be the probability that a randomly chosen young adult has some education beyond high school but does not have a bachelor's degree?

(b) What is the probability that a randomly chosen young adult has at least a high school education?

12.35 **Land in Canada.** Canada's national statistics agency, Statistics Canada, says that the land area of Canada is 9,094,000 square kilometers. Of this land, 4,176,000 square kilometers are forested. Choose a square kilometer of land in Canada at random. (Assume a selected square is classified as either forested or not forested.)

(a) What is the probability that the area you chose is forested?

(b) What is the probability that it is not forested?

12.36 Foreign-language Study. Choose a student in a U.S. public high school at random and ask if he or she is studying a language other than English. Here is the distribution of results:

Language	Spanish	French	German	All others	None
Probability	0.30	0.08	0.02	0.03	0.57

(a) Explain why this is a legitimate probability model.

(b) What is the probability that a randomly chosen student is studying a language other than English?

(c) What is the probability that a randomly chosen student is studying French, German, or Spanish?

12.37 Car Colors. See Exercise 1.25. Choose a new car or light truck at random and note its color. Here are the probabilities of the most popular colors for vehicles sold globally in 2018:[16]

Color	White	Black	Gray	Silver	Natural	Red	Blue
Probability	0.39	0.17	0.12	0.10	0.07	0.07	0.07

(a) What is the probability that the vehicle you chose has any color other than those listed?

(b) What is the probability that a randomly chosen vehicle is neither white nor silver?

12.38 Drawing Cards. You are about to draw a card at random (that is, all choices have the same probability) from a set of seven cards. Although you can't see the cards, here they are:

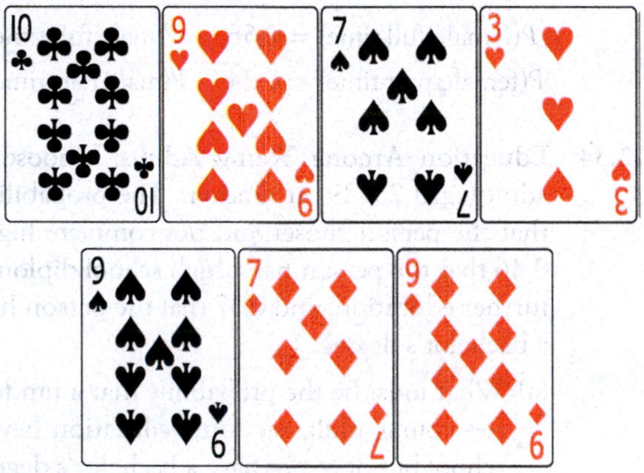

(a) What is the probability that you draw a 9?

(b) What is the probability that you draw a red 9?

(c) What is the probability that you do not draw a 7?

12.39 Loaded Dice. There are many ways to produce crooked dice. To *load* a die so that 6 comes up too often and 1 (which is opposite 6) comes up too seldom, add a bit of lead to the filling of the spot on the 1 face. If a die is loaded so that 6 comes up with probability 0.2 and the probabilities of the 2, 3, 4, and 5 faces are not affected, what is the assignment of probabilities to the six faces?

12.40 A Door Prize. A party host gives a door prize to one guest chosen at random. There are 48 men and 42 women at the party. What is the probability that the prize goes to a woman? Explain how you arrived at your answer.

12.41 Race and Ethnicity. The U.S. Census Bureau allows each person to choose from a long list of races. That is, in the eyes of the U.S. Census Bureau, you belong to whatever race you say you belong to. "Hispanic/Latino" is a separate category; Hispanics may be of any race. If we choose a resident of the United States at random, the U.S. Census Bureau gives these probabilities:[17]

	Hispanic	Not Hispanic
Asian	0.003	0.064
Black	0.011	0.131
White	0.161	0.605
Other	0.009	0.016

(a) Verify that this is a legitimate assignment of probabilities.

(b) What is the probability that a randomly chosen American is Hispanic?

(c) Non-Hispanic Whites are the historical majority in the United States. What is the probability that a randomly chosen American is not a member of this group?

Choose at random a person 15 years of age or older. Ask their gender and marital status (never married, married, or widowed/divorced/separated). Here is the probability model for 8 possible answers:[18]

	Gender	
	Men	Women
Never married	0.171	0.152
Married	0.259	0.261
Divorced	0.042	0.057
Widowed	0.013	0.045

Exercises 12.42 through 12.44 use this probability model.

12.42 Marital Status.

(a) Why is this a legitimate finite probability model?

(b) What is the probability that the person chosen is a woman who is married?

(c) What is the probability that the person chosen is a woman?

(d) What is the probability that the person chosen is married?

12.43 Marital Status, Continued.

(a) List the outcomes that make up the event

A = {The person chosen is *either* a woman *or* is married}

(b) What is $P(A)$? Explain carefully why $P(A)$ is not the sum of the probabilities you found in parts (c) and (d) of the previous exercise.

12.44 Marital Status, Continued.

(a) What is the probability that the person chosen is a man?

(b) What is the probability that the person chosen is or has been married?

12.45 Spelling Errors. Spell-checking software catches "nonword errors" that result in a string of letters that is not a word, as when "the" is typed as "teh." When undergraduates are asked to type a 250-word essay (without spell-checking), the number X of nonword errors has the following distribution:

Value of X	0	1	2	3	4
Probability	0.1	0.2	0.3	0.3	0.1

(a) Is the random variable X discrete or continuous? Why?

(b) Write the event "at least one nonword error" in terms of X. What is the probability of this event?

(c) Describe the event $X \leq 2$ in words. What is its probability? What is the probability that $X < 2$?

12.46 First Digits Again. A crook who never heard of Benford's law might choose the first digits of his faked invoices so that all of 1, 2, 3, 4, 5, 6, 7, 8, and 9 are equally likely. Call the first digit of a randomly chosen fake invoice W for short.

(a) Write the probability distribution for the random variable W.

(b) Find $P(W \geq 6)$ and compare your result with the Benford's law probability from Example 12.7 (page 281).

12.47 Who Gets Interviewed? Abby, Deborah, Mei-Ling, Sam, and Roberto are students in a small seminar course. Their professor decides to choose two of them to interview about the course. To avoid unfairness, the choice will be made by drawing two names from a hat. (This is an SRS of size 2.)

(a) Write down all possible choices of two of the five names. This is the sample space.

(b) The random drawing makes all choices equally likely. What is the probability of each choice?

(c) What is the probability that Mei-Ling is chosen?

(d) Abby, Deborah, and Mei-Ling liked the course. Sam and Roberto did not like the course. What is the probability that both people selected liked the course?

12.48 Birth Order. A couple plans to have three children. There are eight possible arrangements of girls and boys. For example, GGB means the first two children are girls and the third child is a boy. All eight arrangements are (approximately) equally likely.

Jaren Jai Wicklund/Shutterstock

(a) Write down all eight arrangements of the sexes of three children. What is the probability of any one of these arrangements?

(b) Let X be the number of girls the couple has. What is the probability that $X = 2$?

(c) Starting from your work in part (a), find the distribution of X. That is, what values can X take, and what are the probabilities for each value?

12.49 Unusual Dice. Nonstandard dice can produce interesting distributions of outcomes. You have two balanced, six-sided dice. One is a standard die, with faces having 1, 2, 3, 4, 5, and 6 spots. The other die has three faces with 0 spots and three faces with 6 spots. Find the probability distribution for the total number of spots Y on the up-faces when you roll these two dice. *Hint:* Start with a picture like Figure 12.2 (page 276) for the possible up-faces. Label the three 0 faces on the second die 0a, 0b, 0c in your picture and similarly distinguish the three 6 faces.)

12.50 A Taste Test. A tea-drinking Canadian friend of yours claims to have a very refined palate. She tells you that she can tell if, in preparing a cup of tea, milk is first added to the cup and then hot tea poured into the cup or the hot tea is first poured into the cup and then the milk is added.[19] To test her claims, you prepare six cups of tea. Three have the milk added first and the other three the tea first. In a blind taste test, your friend tastes all six cups and is asked to identify the three that had the milk added first.

(a) How many different ways are there to select three of the six cups? (*Hint:* See Example 12.8, page 281.)

(b) If your friend is just guessing, what is the probability that she correctly identifies the three cups with the milk added first?

12.51 Random Numbers. Many random number generators allow users to specify the range of the random numbers to be produced. Suppose you specify that the random number Y can take any value between 0 and 2. Then the density curve of the outcomes has constant height between 0 and 2 and height 0 elsewhere.

(a) Is the random variable Y discrete or continuous? Why?

(b) What is the height of the density curve between 0 and 2? Draw a graph of the density curve.

(c) Use your graph from part (b) and the fact that probability is area under the curve to find $P(Y \leq 1)$.

12.52 More Random Numbers. Find these probabilities as areas under the density curve you sketched in Exercise 12.51.

(a) $P(0.5 < Y < 1.3)$ (b) $P(Y \geq 0.8)$

12.53 Survey Accuracy. A sample survey contacted an SRS of 2220 registered voters shortly before the 2016 presidential election and asked respondents whom they planned to vote for. Election results show that 46% of registered voters voted for Donald Trump. The proportion of the sample who voted for Trump varies, depending on which

2220 voters are in the sample. We will see later that in this situation, if we consider *all possible samples* of 2220 voters, the proportion of voters in each sample who planned to vote for Trump (call it V) has approximately the Normal distribution with mean $\mu = 0.46$ and standard deviation $\sigma = 0.011$.

(a) If the respondents answer truthfully, what is $P(0.44 \leq V \leq 0.48)$? This is the probability that the sample proportion V estimates the population proportion 0.46 within plus or minus 0.02.

(b) In fact, 43% of the respondents in the actual sample said they planned to vote for Donald Trump. If respondents answer truthfully, what is $P(V \geq 0.43)$?

12.54 Friends. How many close friends do you have? Suppose that the number of close friends adults claim to have varies from person to person with mean $\mu = 9$ and standard deviation $\sigma = 2.5$. An opinion poll asks this question of an SRS of 1100 adults. We will see later, in Chapter 19, that in this situation, the sample mean response $\bar{x}$ has approximately the Normal distribution with mean 9 and standard deviation 0.075. What is $P(8.9 \leq \bar{x} \leq 9.1)$, the probability that the sample result $\bar{x}$ estimates the population truth $\mu = 9$ to within ± 0.1?

12.55 Playing Pick 4. The Pick 4 games in many state lotteries announce a four-digit winning number each day. Each of the 10,000 possible numbers 0000 to 9999 has the same chance of winning. You win if your choice matches the winning digits. Suppose your chosen number is 5974.

(a) What is the probability that the winning number matches your number exactly?

(b) What is the probability that the winning number has the same digits as your number *in any order*?

12.56 What Type of Probability? (optional topic) The NASA website on global climate change says that "The current warming trend is of particular significance because most of it is extremely likely (greater than 95% probability) to be the result of human activity since the mid-20th century."[20] This probability is based on satellite data collecting many different types of information, historical data, scientific theory, and sophisticated computer models implementing the latest theory. What type of probability is this? Is it a probability based on the proportion of times an outcome would occur in a very long series of repetitions or a personal probability?

12.57 What Probability Doesn't Say. applet The idea of probability is that the *proportion* of heads in many tosses of a balanced coin eventually gets close to 0.5. But does the actual *count* of heads get close to one-half the number of tosses? Let's find out. Set the "Probability of heads" in the *Probability* applet to 0.5 and the number of tosses to 50. You can increase the number of tosses by moving the "Number of Tosses" slider or toggling the "Re-Toss" button to toss 50 more and keep track of the cumulative totals.

(a) After 50 tosses, what is the proportion of heads? What is the count of heads? What is the difference between the count of heads and 25 (one-half the number of tosses)?

(b) Now make 150 tosses by changing the number of tosses to 150 and toggling the "Re-Toss" button. Again record the proportion and count of heads and the difference between the count and 75 (half the number of tosses).

(c) Keep going. Make 250 tosses and then 500 tosses by moving the Number of Tosses slider and toggling the "Re-Toss" button. Record the same facts. The laws of probability say that the proportion of heads will always get close to 0.5 and also that the difference between the count of heads and half the number of tosses will always grow without limit.

12.58 LeBron's Free Throws. applet LeBron James makes about 70% of his free throws over an entire season. Use the *Probability* applet or statistical software to simulate 100 free throws shot by a player who has probability 0.70 of making each shot. (In most software, the key phrase to look for is "Bernoulli trials." This is the technical term for independent trials with Yes/No outcomes. Our outcomes here are "Hit" and "Miss.")

(a) What percentage of the 100 shots did he hit?

(b) Examine the sequence of hits and misses. How long was the longest run of shots made? Of shots missed? (Sequences of random outcomes often show runs longer than our intuition thinks likely.)

12.59 Simulating an Opinion Poll. applet A 2019 Gallup Poll showed that about 34% of the American public have very little or no confidence in big business. Suppose that this is exactly true of the population. Choosing a person at random then has probability 0.34 of getting one who has very little or no confidence in big business. Use the *Probability* applet or statistical software to simulate choosing many people at random. (In most software, the key phrase to look for is "Bernoulli trials." This is the technical term for independent trials with Yes/No outcomes. Our outcomes here are "Favorable" or "Not Favorable.")

(a) Simulate drawing 50 people, then 100 people, then 400. What proportion have very little or no confidence in big business in each case? We expect (but because of chance variation we can't be sure) that the proportion will be closer to 0.34 with larger samples.

(b) Simulate drawing 50 people 10 times and record the percentages in each sample who have very little or no confidence in big business. Drag the "Re-Toss" toggle button on the applet to re-draw a sample of the same size. Then simulate drawing 400 people 10 times and record the 10 percentages. Which set of 10 is less variable? We expect the results of samples of size 400 to be more predictable (less variable) than the results of samples of size 50. That is "long-run regularity" showing itself.

CHAPTER

13

General Rules of Probability*

When you complete this chapter, you will be able to:

- 13.1 Use Venn diagrams and the general addition rule to find probabilities involving events that share outcomes (events that are not disjoint).
- 13.2 Use the multiplication rule to find the probability of two or more independent events occurring together.
- 13.3 Calculate conditional probabilities.
- 13.4 Use the general multiplication rule to find the probability of two or more events occurring together.
- 13.5 Determine whether two events A and B are independent by testing whether $P(B) = P(B \mid A)$.
- 13.6 Draw and interpret tree diagrams to organize probability models in multiple stages and to find conditional probabilities.
- 13.7 Apply Bayes' rule to solve problems involving conditional probabilities.

Probability models can describe the flow of traffic through a highway system, a telephone interchange, or a computer processor; the genetic makeup of populations; the energy states of subatomic particles; the spread of epidemics or rumors; and the rate of return on risky investments. Although we are interested in probability mainly because it is the foundation for statistical inference, the mathematics of chance is important in many fields of study. Our introduction to probability in Chapter 12 concentrated on basic ideas and facts. Now we look at some further details. With more probability at our command, we can model more complex random phenomena.

Although we won't emphasize the math, everything in this chapter (and much more) follows from the four rules we met in Chapter 12. Here they are again.

*This chapter introduces some of the mathematics of probability. The material is not needed to understand the rest of the book.

praetorianphoto/Getty Images

Probability Rules

- **Rule 1.** For any event A, $0 \leq P(A) \leq 1$.
- **Rule 2.** If S is the sample space, $P(S) = 1$.
- **Rule 3. Addition rule for disjoint events:** If A and B are disjoint events,

$$P(A \text{ or } B) = P(A) + P(B)$$

- **Rule 4.** For any event A,

$$P(A \text{ does not occur}) = 1 - P(A)$$

13.1 The General Addition Rule

venn diagram In probability, a picture used to display relationships among events, where the sample space is shown as a rectangular area and events are shown as areas within circles or closed curves. The areas of overlap among circles represent outcomes common to the events represented by the circles.

We know that if A and B are disjoint events then, using Rule 3, $P(A \text{ or } B) = P(A) + P(B)$. We would like to be able to compute $P(A \text{ or } B)$ when the events A and B are *not* disjoint and therefore have some outcomes in common. You may find it helpful to distinguish between these two situations by using a **Venn diagram**—a picture that will help you to visualize relationships among several events.

The Venn diagram in Figure 13.1 shows the sample space S as a rectangular area and the events A and B as areas within S. The events A and B in Figure 13.1 are disjoint because they do not overlap—that is, they have no outcomes in common. Contrast this with the Venn diagram in Figure 13.2, which illustrates two events that are not disjoint. The event $\{A \text{ and } B\}$ appears as the overlapping area that contains the outcomes that are common to both A and B.

When two events are not disjoint, the probability that one or the other occurs is *less* than $P(A) + P(B)$. As Figure 13.3 illustrates, the outcomes that are common to both A and B are counted twice when we add these two probabilities, so we must subtract $P(A \text{ and } B)$ from the sum to avoid this double counting. Here is the addition rule for any two events, disjoint or not.

Addition Rule For Any Two Events

For any two events A and B,

$$P(A \text{ or } B) = P(A) + P(B) - P(A \text{ and } B)$$

If A and B are disjoint, the event $\{A \text{ and } B\}$ that both occur contains no outcomes and therefore has probability 0. Because the general addition rule includes Rule 3, the addition rule for disjoint events, we can always use the general addition rule to find $P(A \text{ or } B)$.

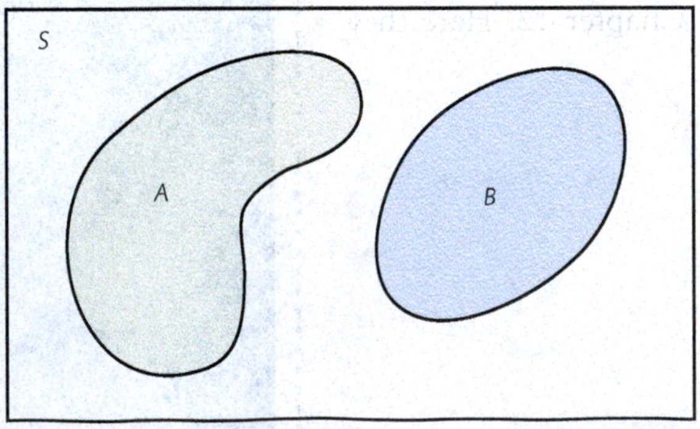

FIGURE 13.1
Venn diagram showing disjoint events *A* and *B*.

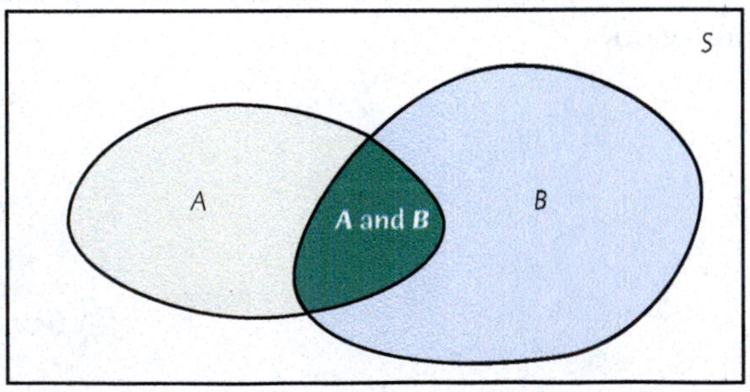

FIGURE 13.2
Venn diagram showing events *A* and *B* that are not disjoint. The event {*A* and *B*} consists of outcomes common to *A* and *B*.

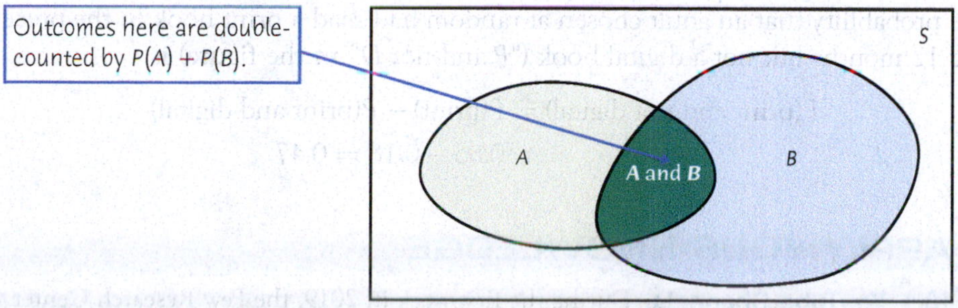

FIGURE 13.3
The general addition rule: for any events A and B, $P(A \text{ or } B) = P(A) + P(B) - P(A \text{ and } B)$.

EXAMPLE 13.1 Reading: Print Versus Digital Formats

Although an increasing share of Americans are reading e-books on tablets and smartphones rather than dedicated e-readers, print books continue to be much more popular than books in *digital format*. (Digital format includes both e-books and audio books.) A 2019 survey found that 65% of adults had read a print book in the preceding 12 months, 25% had read a book in digital format, and 18% had read both a print book and a book in digital format.[1] Choose an adult at random. Then

$$P(\text{print or digital}) = P(\text{print}) + P(\text{digital}) - P(\text{print and digital})$$
$$= 0.65 + 0.25 - 0.18 = 0.72.$$

That is, 72% of adults had read either a print book, a digital book, or both in the preceding 12 months. "Nonreaders" had read *neither* a print book nor a digital book in the preceding 12 months. So

$$P(\text{nonreader}) = 1 - 0.72 = 0.28$$

Venn diagrams clarify events and their probabilities because you can just think of adding and subtracting areas. Figure 13.4 shows all the events formed from "print" and "digital" in Example 13.1. The four probabilities that appear in the figure add to 1 because they refer to four disjoint events that make up the entire sample space. All of these probabilities come from the information in Example 13.1. For example,

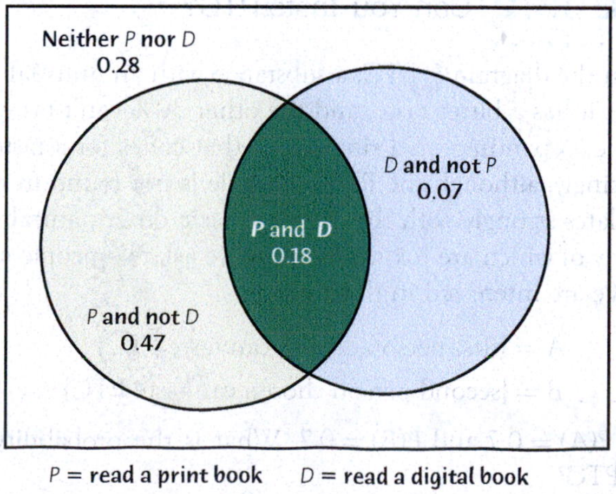

FIGURE 13.4
Venn diagram and probabilities for print versus digital format, for Example 13.1.

the probability that an adult chosen at random had read a print book in the preceding 12 months but not a digital book ("*P* and not *D*" in the figure) is

$$P(\text{print and not digital}) = P(\text{print}) - P(\text{print and digital})$$
$$= 0.65 - 0.18 = 0.47$$

APPLY YOUR KNOWLEDGE

13.1 YouTube Channels. During the first week in 2019, the Pew Research Center tracked the types of English language YouTube videos posted. Let *A* be the event that the video posted involved games of some sort, including sports. Let *B* be the event that the video posted involved hobbies and skills, including sports. Pew Research Center finds that $P(A) = 0.30$, $P(B) = 0.13$, and $P(A \text{ or } B) = 0.34$.[2]

(a) Make a Venn diagram similar to Figure 13.4 showing the events {*A* and *B*}, {*A* and not *B*}, {*B* and not *A*}, and {neither *A* nor *B*}.

(b) Describe each of these events in words.

(c) Find the probabilities of all four events and add the probabilities to your Venn diagram. The four probabilities you have found should add to 1.

13.2 College Degrees. Of all postsecondary degrees awarded in the United States, including master's and doctorate degrees, 21% are associate's degrees, 58% are earned by people whose race is White, and 12% are associate's degrees earned by Whites.[3] Make a Venn diagram and use it to answer these questions.

(a) What percentage of all degrees are associate's degrees earned by non-Whites?

(b) What percentage of non-Whites earn a college degree other than an associate's degree?

(c) What percentage of all degrees are earned by non-Whites?

13.2 Independence and the Multiplication Rule

When can we find the probability $P(A \text{ and } B)$ that both events occur if we know the individual probabilities $P(A)$ and $P(B)$? In this section, we show how this can be done in the special situation when the events *A* and *B* are *independent*. The more general rule for finding $P(A \text{ and } B)$, which we return to in Section 13.4, requires conditional probabilities, which are covered in Section 13.3.

EXAMPLE 13.2 Can You Taste PTC?

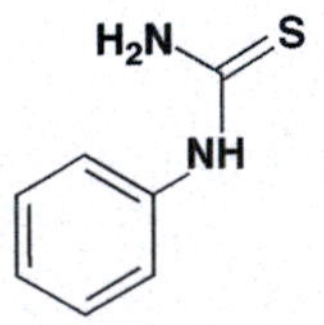

The molecule in the diagram is PTC, a substance with an unusual property: 70% of people find that it has a bitter taste, and the other 30% can't taste it at all. The difference is genetic, depending on a single gene that codes for a taste receptor on the tongue. Interestingly, although the PTC molecule is not found in nature, the ability to taste it correlates strongly with the ability to taste other naturally occurring bitter substances, many of which are toxins. Suppose we ask two people chosen at random to taste PTC. We are interested in the events

$$A = \{\text{first person chosen can taste PTC}\}$$
$$B = \{\text{second person chosen can taste PTC}\}$$

We know that $P(A) = 0.7$ and $P(B) = 0.7$. What is the probability $P(A \text{ and } B)$ that both can taste PTC?

We can think our way to the answer. The first person chosen can taste PTC in 70% of all samples, and then the second person chosen can taste it in 70% of those samples. We will get two tasters in 70% of 70% of all samples. That is $P(A \text{ and } B) = 0.7 \times 0.7 = 0.49$.

The argument in Example 13.2 works because knowing that the first person can taste PTC tells us nothing about the second person. The probability is still 0.7 that the second person can taste PTC, whether or not the first person can. We say that the events "first person can taste PTC" and "second person can taste PTC" are independent. Now we have another rule of probability.

Multiplication Rule For Independent Events

Two events *A* and *B* are **independent** if knowing that one occurs does not change the probability that the other occurs. If *A* and *B* are independent,

$$P(A \text{ and } B) = P(A)P(B)$$

EXAMPLE 13.3 Independent or Not?

To use the multiplication rule for independent events, we must decide whether events are independent. Here are some examples to help you recognize when you can assume events are independent.

In Example 13.2, we think that the ability of one randomly chosen person to taste PTC tells us nothing about whether or not a second person, also randomly chosen, can taste PTC. That's independence. But if the two people are members of the same family, the fact that ability to taste PTC is inherited warns us that they are not independent.

Independence is clearly recognized in artificial settings such as games of chance. Because a coin has no memory and most coin tossers cannot influence the fall of the coin, it is safe to assume that successive coin tosses are independent, so that the probability of three heads in succession is $0.5 \times 0.5 \times 0.5 = 0.125$.

On the other hand, the colors of successive cards dealt from the same deck are not independent. A standard 52-card deck contains 26 red cards and 26 black cards. For the first card dealt from a shuffled deck, the probability of a red card is $26/52 = 0.50$. Once we see that the first card is red, we know that there are only 25 reds among the remaining 51 cards. The probability that the second card is red is therefore only $25/51 = 0.49$. Knowing the outcome of the first deal changes the probability for the second.

STATISTICS IN YOUR WORLD

Condemned by Independence

Assuming independence when it isn't true can lead to disaster. Several mothers in England were convicted of murder simply because two of their children had died in their cribs with no visible cause. An "expert witness" for the prosecution said that the probability of an unexplained crib death in a nonsmoking middle-class family is 1/8500. He then multiplied 1/8500 by 1/8500 to claim that there is only a 1 in 73 million chance that two children in the same family could have died naturally. This is nonsense: it assumes that crib deaths are independent, and data suggest that they are not. Some common genetic or environmental cause, not murder, probably explains the deaths.

The multiplication rule extends to collections of more than two events, provided that all are independent. Independence of events *A*, *B*, and *C* means that no information about any one or any two can change the probability of the remaining events. Independence is often assumed in setting up a probability model when the events we are describing seem to have no connection.

If two events *A* and *B* are independent, the event that *A* does not occur is also independent of *B*, and so on. For example, choose two people at random and ask if they can taste PTC. Because 70% can taste PTC and 30% cannot, the probability that the first person is a taster and the second is not is $(0.7)(0.3) = 0.21$.

EXAMPLE 13.4 Surviving?

Historical/Getty Images

During World War II, the British found that the probability of a bomber being lost through enemy action on a mission over occupied Europe was 0.05. The probability of the bomber returning safely from a mission was therefore 0.95. It is reasonable to assume that missions were independent. Take A_i to be the event that a bomber survived its ith mission. The probability of surviving two missions is

$$\begin{aligned} P(A_1 \text{ and } A_2) &= P(A_1)P(A_2) \\ &= (0.95)(0.95) = 0.9025 \end{aligned}$$

The multiplication rule also applies to more than two independent events, so the probability of surviving three missions is

$$\begin{aligned} P(A_1 \text{ and } A_2 \text{ and } A_3) &= P(A_1)P(A_2)P(A_3) \\ &= (0.95)(0.95)(0.95) = 0.8574 \end{aligned}$$

In 1941, the tour of duty for an airman was established as 30 missions. The probability of surviving 30 missions was only

$$\begin{aligned} P(A_1 \text{ and } A_2 \text{ and} \ldots \text{and } A_{30}) &= P(A_1)P(A_2)\cdots P(A_{30}) \\ &= (0.95)(0.95)\cdots(0.95) \\ &= (0.95)^{30} = 0.2146 \end{aligned}$$

The probability of surviving two tours of duty was much smaller.

Here is another example of using the multiplication rule for independent events to compute probabilities.

EXAMPLE 13.5 Rapid HIV Testing

4step

STATE: Many people who come to clinics to be tested for HIV, the virus that causes AIDS, don't come back to learn the test results. Clinics now use "rapid HIV tests" that give a result while the client waits. In a clinic in Malawi, for example, use of rapid tests increased the percentage of clients who learned their test results from 69% to 99.7%.

The trade-off for fast results is that rapid tests are less accurate than slower laboratory tests. Applied to people who have no HIV antibodies, one rapid test has probability about 0.004 of producing a false positive (that is, of falsely indicating that antibodies are present).[4] If a clinic tests 200 people who are free of HIV antibodies, what is the chance that at least one false positive will occur?

PLAN: It is reasonable to assume that the test results for different individuals are independent. We have 200 independent events, each with probability 0.004. What is the probability that at least one of these events occurs?

SOLVE: The event "At least one positive" combines many outcomes. In this situation, the use of Rule 4 (page 296), which says that, for any event A,

$$P(A \text{ does not occur}) = 1 - P(A)$$

leads to a more direct solution. For our setting, the event A corresponds to "At least one positive," and the event A does not occur when there are "No positives." To solve our problem, it is simplest to use Rule 4

$$P(\text{at least one positive}) = 1 - P(\text{no positives})$$

and find P (no positives) first.

The probability of a negative result for any one person is $1 - 0.004 = 0.996$. To find the probability that all 200 people tested have negative results, use the multiplication rule:

$$\begin{aligned} P(\text{no positives}) &= P(\text{all 200 negative}) \\ &= (0.996)(0.996) \cdots (0.996) \\ &= 0.996^{200} = 0.4486 \end{aligned}$$

The probability we want is, therefore,

$$P(\text{at least one positive}) = 1 - 0.4486 = 0.5514$$

CONCLUDE: The probability is greater than 1/2 that at least one of the 200 people will test positive for HIV, even though no one has the virus.

In everyday speech, *independent* and *disjoint* are often interpreted as referring to events that are unrelated, separated in some way, or not connected. But they have quite different meanings in probability. If I flip a fair coin once, it can come up either heads or tails. Both events have probability 1/2. But the event "the coin comes up heads and comes up tails" is impossible and has probability 0. The events "the coin comes up heads" and "the coin comes up tails" are disjoint. If they were independent, the probability of the event "the coin comes up heads and comes up tails" would have probability $1/2 \times 1/2 = 1/4$.

You must be careful not to confuse disjointness and independence. If A and B are disjoint, then the fact that A occurs tells us that B cannot occur; look again at Figure 13.1 (page 296). So disjoint events are not independent. We cannot depict independence in a Venn diagram because it involves the probabilities of the events rather than just the outcomes that make up the events.

You must also remember that the special multiplication rule $P(A \text{ and } B) = P(A)P(B)$ *holds if* A *and* B *are independent but not otherwise.* Resist the temptation to use this simple rule when the circumstances that justify it are not present. The next three sections provide more details to help you determine when two events are independent and also introduce the general multiplication rule, which can be used for events that are not independent.

APPLY YOUR KNOWLEDGE

13.3 Older College Students. Government data show that 4% of adults are full-time college students and that 37% of adults are aged 55 or older. Nonetheless, we can't conclude that because $(0.04)(0.37) = 0.015$, about 1.5% of adults are college students 55 or older. Why not?

13.4 Common Names. The U.S. Census Bureau says that the 10 most common names in the United States are (in order) Smith, Johnson, Williams, Brown, Jones, Miller, Davis, Garcia, Rodriguez, and Wilson. These names account for 9.6% of all U.S. residents. Out of curiosity, you look at the authors of the textbooks for your current courses. There are 9 authors in all. Would you be surprised if none of the names of these authors were among the 10 most common? (Assume that authors' names are independent and follow the same probability distribution as the names of all residents.)

13.5 Lost Internet References. Internet sites often vanish or move so that references to them can't be followed. In fact, 47% of Internet sites referenced in major medical journals are lost.[5] If a paper contains seven Internet references, what is the probability that all seven are still good? What specific assumptions did you make to calculate this probability?

13.3 Conditional Probability

The probability we assign to an event can change if we know that some other event has occurred. This idea is the key to many applications of probability, and it is simplest to understand in the context of a two-way table of counts, such as those we discussed in Chapter 6.

EXAMPLE 13.6 Imported Motor Vehicles

Imported motor vehicles sold in the United States are classified as either cars, light trucks, medium trucks, or heavy trucks, and they can be classified as either from NAFTA or from other countries. "Light trucks" include SUVs and minivans. "NAFTA" means made in Canada or Mexico. Here is a two-way table giving the counts of vehicles sold in 2018, classified by light truck or car/medium or heavy truck and NAFTA/Other:[6]

	NAFTA	Other	Total
Light truck/car	4,337,091	3,881,650	8,218,741
Medium/heavy truck	189,722	40,995	230,717
Total	4,526,813	3,922,645	8,449,458

The entries in the body of the table are quantities of vehicles with two specific classifications. For instance, the entry in the row for "Light truck/car" and the column for "NAFTA" tells us that 4,337,091 vehicles sold were light trucks or cars from the NAFTA countries Canada and Mexico. The entries in the right column are the totals for each row (8,218,741 vehicles sold were light trucks or cars), and the entries in the bottom row are the totals for each column (that is, 3,922,645 vehicles sold were imported from countries other than Canada or Mexico). Finally, the entry 8,449,458 in the bottom right is the total number of vehicles sold.

If we select a motor vehicle sold at random, what is the probability that it is from Canada or Mexico? Because there were 8,449,458 vehicles sold and 4,526,813 of them were from Canada or Mexico, the probability a randomly selected vehicle is from Canada or Mexico is just the proportion of "NAFTA" vehicles,

$$P(\text{NAFTA}) = \frac{4,526,813}{8,449,458} = 0.536$$

Similarly, the probability that a randomly selected vehicle is imported from Canada or Mexico and a medium/heavy truck is the proportion of "NAFTA" medium/heavy trucks sold,

$$P(\text{NAFTA and medium/heavy truck}) = \frac{189,722}{8,449,458} = 0.022$$

Suppose we are told that the vehicle chosen is a medium/heavy truck. That is, it is one of the 230,717 vehicles in the "medium/heavy truck." row of the table. The probability that a vehicle is from Canada or Mexico, *given the information that it is a medium/heavy truck,* is the proportion of NAFTA vehicles in the "Medium/heavy truck" row,

$$P(\text{NAFTA} \mid \text{medium/heavy truck}) = \frac{189,722}{230,717} = 0.822.$$

This is a **conditional probability**. You can read the bar | as "given the information that."

conditional probability The probability of one event B, given that another event A occurs.

Although 53.6% of all vehicles sold are from Canada or Mexico, 82.2% of medium/heavy trucks are imported from Canada or Mexico. It's common sense that knowing that one event (the vehicle is a medium/heavy truck) occurs often changes the probability of another event (the vehicle is from Canada or Mexico). Although conditional probabilities have been presented here with a two-way table of counts, a conditional probability can be expressed in terms of the original probabilities by

$$P(\text{NAFTA} \mid \text{medium/heavy truck}) = \frac{189{,}722}{230{,}717}$$

$$= \frac{\frac{189{,}722}{8{,}449{,}458}}{\frac{230{,}717}{8{,}449{,}458}}$$

$$= \frac{P(\text{NAFTA and medium/heavy truck})}{P(\text{medium/heavy truck})}$$

The idea of a conditional probability $P(B \mid A)$ of one event B, given that another event A occurs, is the proportion *of all occurrences of* A for which B also occurs.

Conditional Probability Formula

When $P(A) > 0$, the **conditional probability** of B, given A is

$$P(B \mid A) = \frac{P(A \text{ and } B)}{P(A)}$$

Figure 13.5 is a Venn diagram illustrating the conditional probability of B, given A.

The conditional probability $P(B \mid A)$ makes no sense if the event A can never occur, so we require that $P(A) > 0$ whenever we talk about $P(B \mid A)$. *Be sure to keep in mind the distinct roles of the events* A and B in $P(B \mid A)$. Event A represents the information we are given, and B is the event whose probability we are calculating. Here is an example that emphasizes this distinction.

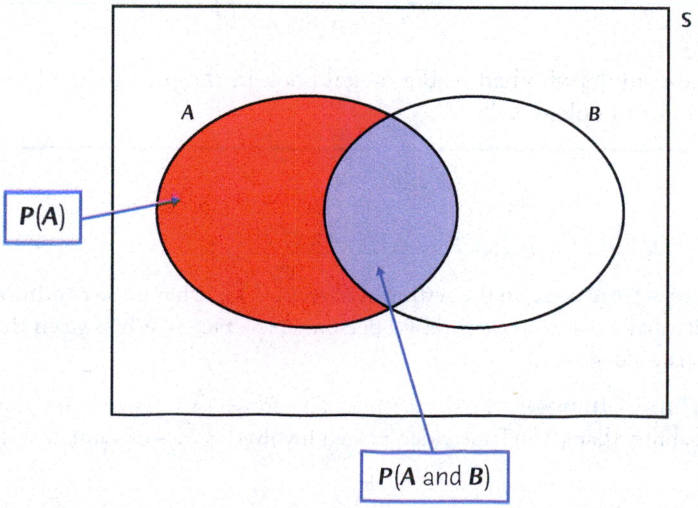

FIGURE 13.5 The probability of B given A is the proportion of the outcomes in the red- and blue-shaded oval (event A) that are in the blue-shaded area (event A and B). That is, $P(B|A) = \frac{P(A \text{ and } B)}{P(A)}$.

EXAMPLE 13.7 Trucks Among NAFTA

What is the conditional probability that a randomly chosen vehicle is a medium/heavy truck, *given the information that it is from Canada or Mexico?* Using the definition of conditional probability,

$$P(\text{medium/heavy truck} \mid \text{NAFTA}) = \frac{P(\text{medium/heavy truck and NAFTA})}{P(\text{NAFTA})}$$

$$= \frac{\frac{189{,}722}{8{,}449{,}458}}{\frac{4{,}526{,}813}{8{,}449{,}458}}$$

$$= \frac{0.022}{0.536} = 0.041$$

Only 4.1% of vehicles from Canada or Mexico sold are medium/heavy trucks.

Be careful not to confuse the two different conditional probabilities

$$P(\text{NAFTA} \mid \text{medium/heavy truck}) = 0.822$$

$$P(\text{medium/heavy truck} \mid \text{NAFTA}) = 0.041$$

The first answers the question, "What proportion of medium/heavy trucks are from Canada or Mexico?" The second answers, "What proportion of vehicles from Canada or Mexico are medium/heavy trucks?"

In many applications of conditional probability, we are given the probabilities of several events rather than being provided a table of counts. Here is an example of using the conditional probability formula in this setting.

EXAMPLE 13.8 Reading: Print Versus Digital Formats, Continued

Returning to Example 13.1 (page 297), a 2019 survey found 65% of adults had read a print book in the preceding 12 months, 25% had read a book in digital format, and 18% had read both a print book and a book in digital format. If we choose an adult at random, the conditional probability that he or she had read a print book, given that he or she had read a digital book, is

$$P(\text{print} \mid \text{digital}) = \frac{P(\text{print and digital})}{P(\text{digital})}$$

$$= \frac{0.18}{0.25} = 0.72$$

Among those adults who had read a digital book in the preceding 12 months, 72% had read a print book as well.

APPLY YOUR KNOWLEDGE

13.6 College Degrees. In the setting of Exercise 13.2, what is the conditional probability that a degree is earned by a person whose race is White given that it is an associate's degree?

13.7 YouTube Channels. In the setting of Exercise 13.1, what is the conditional probability that a YouTube video posted involved games of some sort, including sports, given that the YouTube video posted involved hobbies or skills, including sports? (*Hint*: First find the probability that a YouTube video posted involved games of some sort, including sports, *and* involved hobbies or skills, including sports. This can be found using the addition rule for general events directly, or you may find it simpler to first summarize the information in a Venn diagram, as in Exercise 13.1.)

13.8 Computer Games. Here is the distribution of computer games sold by type of game:[7]

Game Type	Probability
Action	0.269
Shooter	0.209
Role Play	0.113
Sport	0.111
Adventure	0.079
Fighting	0.078
Racing	0.058
Strategy	0.037
Other	0.046

What is the conditional probability that a computer game is a role-playing game, given that it is not an action game?

13.4 The General Multiplication Rule

The definition of conditional probabilities leads to a more general version of the multiplication rule that allows us to calculate the probability that several events occur simultaneously, even when they are not independent. More importantly, in Section 13.5, conditional probabilities are used to provide a formal definition of the independence of two events.

Multiplication Rule for Any Two Events

The probability that both of two events A and B happen together can be found by

$$P(A \text{ and } B) = P(A)P(B \mid A)$$

Here, $P(B \mid A)$ is the conditional probability that B occurs, given the information that A occurs.

In words, this rule says that for both of two events to occur, first one must occur and then, given that the first event has occurred, the second must occur. As the next two examples illustrate, the use of the general multiplication rule is straightforward once you have taken the information given and expressed it in the language of probability.

STATISTICS IN YOUR WORLD

Winning the Lottery Twice

In 1986, Evelyn Marie Adams won the New Jersey lottery for the second time, adding $1.5 million to her previous $3.9 million jackpot. The *New York Times* claimed that the odds of one person winning the big prize twice were 1 in 17 trillion. Nonsense, said two statisticians in a letter to the *Times*. The chance that Evelyn Marie Adams would win twice is indeed tiny, but it is almost certain that *someone* among the millions of lottery players would win two jackpots. Sure enough, Robert Humphries won his second Pennsylvania lottery jackpot ($6.8 million total) in 1988, and more recently, Ernest Pullen of St. Louis won the Missouri lottery in June 2010 and then again in September 2010 ($3 million total). When commenting on the double win, Ernest Pullen said he considers himself to be a "lucky guy."

EXAMPLE 13.9 Gen X and the Internet

The Pew Internet and Technology Project finds that 91% of gen X-ers (adults born between 1965 and 1980) use the Internet and that 89% of online gen X-ers say their use of the Internet is a good thing for them personally.[8] What percentage of gen X-ers are online *and* say the Internet is a good thing for them personally?

Use the multiplication rule:

$$P(\text{online}) = 0.91$$

$$P(\text{say their use of the Internet is a good thing for them personally} \mid \text{online}) = 0.89$$

$$\begin{aligned} P(\text{online and say their use of the Internet is a good thing for them personally}) &= P(\text{online}) \times P(\text{say their use of the Internet is a good thing for them personally} \mid \text{online}) \\ &= (0.91)(0.89) \\ &= 0.8099 \end{aligned}$$

That is, about 81% of all gen X-ers are online and say their use of the Internet is a good thing for them personally.

You should think your way through this: if 91% of gen X-ers are online and 89% *of these* say their use of the Internet is a good thing for them personally, then 89% of 91% are both online and say their use of the Internet is a good thing for them personally.

It is important to remember that the conditional probability of an event generally depends on the event we condition on. Although we have seen

$$P(\text{say their use of the Internet is a good thing for them personally} \mid \text{online}) = 0.89$$

since someone who is not online cannot say their use of the Internet is a good thing for them personally, we have

$$P(\text{their use of the Internet is a good thing for them personally} \mid \text{not online}) = 0$$

We can extend the multiplication rule to find the probability that all of several events occur. The key is to condition each event on the occurrence of *all* of the preceding events. So for any three events A, B, and C,

$$P(A \text{ and } B \text{ and } C) = P(A)P(B \mid A)P(C \mid \text{both } A \text{ and } B)$$

Here is an example of the extended multiplication rule.

EXAMPLE 13.10 Fundraising by Telephone

STATE: A charity raises funds by calling a list of prospective donors to ask for pledges. It is able to talk with 40% of the people on its list. Of those the charity reaches, 30% make a pledge. But only half of those who pledge actually make a contribution. What percentage of the donor list contributes?

PLAN: Express the information we are given in terms of events and their probabilities:

If A = {the charity reaches a prospect}	then	$P(A) = 0.4$
If B = {the prospect makes a pledge}	then	$P(B \mid A) = 0.3$
If C = {the prospect makes a contribution}	then	$P(C \mid \text{both } A \text{ and } B) = 0.5$

We want to find $P(A \text{ and } B \text{ and } C)$.

SOLVE: Use the multiplication rule:

$$P(A \text{ and } B \text{ and } C) = P(A)P(B \mid A)P(C \mid \text{both } A \text{ and } B)$$
$$= 0.4 \times 0.3 \times 0.5 = 0.06$$

CONCLUDE: Only 6% of the prospective donors make contributions.

RuslanDashinsky/Getty Images

As the examples in this section illustrate, formulating a problem in the language of probability is often the key to success in applying probability ideas.

APPLY YOUR KNOWLEDGE

13.9 At the Gym. Suppose that 8% of adults belong to health clubs, and 45% of these health club members go to the club at least twice a week. What percentage of all adults go to a health club at least twice a week? Write the information given in terms of probabilities and use the general multiplication rule.

13.10 Gen X and the Internet. We saw in Example 13.9 that 91% of gen X-ers are online and that 89% of online gen X-ers say their use of the Internet is a good thing for them personally. Gen X-ers make up 20.3% of the U.S. adult population. What percentage of all U.S. adults are gen X-ers, are online, and say their use of the Internet is a good thing for them personally? Define events and probabilities and follow the pattern of Example 13.10.

4step

13.5 Showing That Events Are Independent

The conditional probability $P(B \mid A)$ is generally not equal to the unconditional probability $P(B)$. That's because the occurrence of event A generally gives us some additional information about whether or not event B occurs. If knowing that A occurs gives no additional information about B, then A and B are independent events. The precise definition of independence is expressed in terms of conditional probability.

Independent Events

Two events A and B that both have positive probability are **independent** if

$$P(B \mid A) = P(B)$$

The converse is also true: if events A and B are independent, then $P(B \mid A) = P(B)$. This gives us a way to check whether two events are independent.

EXAMPLE 13.11 Reading: Print Versus Digital Formats, Conclusion

Returning to Example 13.1, a 2019 survey found 65% of adults had read a print book in the preceding 12 months, 25% had read a book in digital format, and 18% had read both a print book and a book in digital format. If we choose an adult at random, does the probability that he or she had read a print book depend on whether he or she had read a digital book? In the language of probability, we are asking if the event that the selected adult had read a print book is independent of the event that he or she had read a digital book. From Example 13.8 (page 304), we know the conditional probability of having read a print book, given that he or she had read a digital book is

$$P(\text{print} \mid \text{digital}) = \frac{P(\text{print and digital})}{P(\text{digital})}$$
$$= \frac{0.18}{0.25} = 0.72$$
$$\neq P(\text{print})$$

The events "had read a print book" and "had read a digital book" are *not* independent. While 65% of adults had read a print book, among those who had read a digital book, the proportion who had read a print book is 72%.

From the definition of independence, we now see that the multiplication rule for independent events, $P(A \text{ and } B) = P(A)P(B)$, is a special case of the general multiplication rule because if A and B are independent

$$P(A \text{ and } B) = P(A)P(B \mid A) = P(A)P(B)$$

In Example 13.11, we could also show that the events "had read a print book" and "had read a digital book" are not independent by verifying directly that

$$P(\text{print and digital}) \neq P(\text{print})P(\text{digital})$$

Although we have concentrated in this chapter on understanding the meaning of independence and the use of the definition to determine if events are independent, independence is often an assumption that we will be making about observations in a data set such as the assumed independence of successive coin tosses.

APPLY YOUR KNOWLEDGE

13.11 Independent? The 2017 update to the *Report on the UC Berkeley Faculty Salary Equity Study* shows that 94 of the university's 253 assistant professors were women, along with 134 of the 314 associate professors and 244 of the 949 full professors. Note that the study only classified faculty members as a man or a woman.

(a) What is the probability that a randomly chosen Berkeley professor (of any rank) is a woman?

(b) What is the conditional probability that a randomly chosen professor is a woman, given that the person chosen is a full professor?

(c) Are the rank and sex of Berkeley professors independent? How do you know?

13.6 Tree Diagrams

Probability models often have several stages, with probabilities at each stage conditional on the outcomes of earlier stages. These models require us to combine several of the basic rules into a more elaborate calculation. Here is an example.

EXAMPLE 13.12 Who Dates Online

STATE: The number of American adults who have used online dating or mobile dating apps continues to grow, with the largest increase occuring for young adults ages 18–24. Looking only at adults under 65, about 17% are 18–24 years old, another 41% are 25–44 years old, and the remaining 42% are 45–64 years old. Pew Research reports that 27% of those aged 18–24 have used online dating sites, along with 22% of those aged 25–44 and 13% of those aged 45–64.[9] What percentage of American adults under 65 have used an online dating site?

PLAN: To use the tools of probability, restate all these percentages as probabilities. If we choose an adult under 65 at random,

$$P(\text{aged 18–24}) = 0.17$$
$$P(\text{aged 25–44}) = 0.41$$
$$P(\text{aged 45–64}) = 0.42$$

These three probabilities add to 1 because all adults under 65 are in one of the three age groups. The percentages of each group that have used online dating sites are *conditional* probabilities:

$$P(\text{online date yes} \mid \text{aged 18–24}) = 0.27$$
$$P(\text{online date yes} \mid \text{aged 25–44}) = 0.22$$
$$P(\text{online date yes} \mid \text{aged 45–64}) = 0.13$$

We want to find the unconditional probability $P(\text{online date yes})$.

SOLVE: The *tree diagram* in Figure 13.6 organizes this information. Each segment in the tree is one stage of the problem. Each complete branch shows a path through the two stages. The probability written on each segment is the conditional probability of an adult following that segment, given that he or she has reached the node from which it branches.

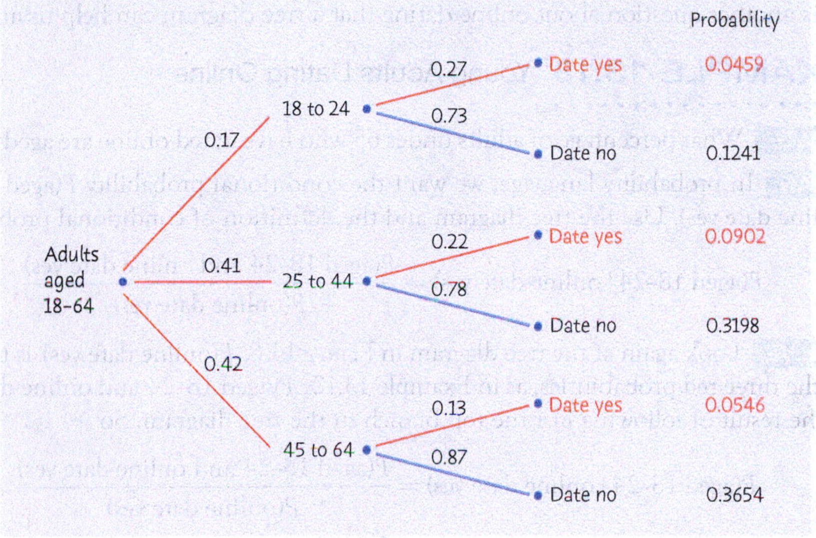

FIGURE 13.6 Tree diagram for who dates online, for Example 13.12. The three disjoint paths to the outcome that an adult aged 18–34 has used online dating or a mobile dating app are colored red.

Starting at the left, an adult under 65 falls into one of the three age groups. The probabilities of these groups mark the leftmost segments in the tree. Look at age 18–24, the top branch. The two segments going out from the "18–24" branch point carry the conditional probabilities

$$P(\text{online date yes} \mid \text{aged 18–24}) = 0.27$$
$$P(\text{online date no} \mid \text{aged 18–24}) = 0.73$$

The full tree shows the probabilities for all three age groups.

Now use the multiplication rule. The probability that a randomly chosen adult under 65 is an 18- to 24-year-old and has used an online dating site is

$$P(\text{aged 18–24 and online date yes}) = P(\text{aged 18–24})P(\text{online date yes} \mid \text{aged 18–24})$$
$$= (0.17)(0.27) = 0.0459$$

This probability appears at the end of the topmost branch. The multiplication rule says that the probability of any complete branch in the tree is the product of the probabilities of the segments in that branch.

There are three disjoint paths to "online date yes," one for each of the three age groups. These paths are colored red in Figure 13.6. Because the three paths are disjoint, the probability that an adult under 65 has used an online dating site is the sum of their probabilities:

$$P(\text{online date yes}) = (0.17)(0.27) + (0.41)(0.22) + (0.42)(0.13)$$
$$= 0.0459 + 0.0902 + 0.0546 = 0.1907$$

CONCLUDE: Slightly under 20% of American adults under 65 have used an online dating site.

Tree Diagram

A **tree diagram** is a graphical way to organize a probability model that has several stages, in which leftmost branch segments represent probabilities of outcomes, and branches to the right represent conditional probabilities, given those outcomes.

It takes longer to explain a tree diagram than it does to use one. Once you understand a problem well enough to draw the tree, the rest is easy. An important property of a tree diagram is that the events in the last branch are all disjoint and include all outcomes in the sample space, so the probabilities in the last branch of the tree must sum to 1. Here is another question about online dating that a tree diagram can help us answer.

EXAMPLE 13.13 Young Adults Dating Online

4step

STATE: What percentage of adults under 65 who have dated online are aged 18–24?

PLAN: In probability language, we want the conditional probability $P(\text{aged 18–24} \mid \text{online date yes})$. Use the tree diagram and the definition of conditional probability:

$$P(\text{aged 18–24} \mid \text{online date yes}) = \frac{P(\text{aged 18–24 and online date yes})}{P(\text{online date yes})}$$

SOLVE: Look again at the tree diagram in Figure 13.6. $P(\text{online date yes})$ is the sum of the three red probabilities, as in Example 13.12. $P(\text{aged 18–24 and online date yes})$ is the result of following just the top branch in the tree diagram. So

$$P(\text{aged 18–24} \mid \text{online date yes}) = \frac{P(\text{aged 18–24 and online date yes})}{P(\text{online date yes})}$$
$$= \frac{0.0459}{0.1907} = 0.2407$$

CONCLUDE: About 24% of adults under 65 who have used an online dating site are aged 18–24. Compare this conditional probability with the original information (unconditional) that 17% of adults under 65 are aged 18–24. Knowing that a person is using an online dating site increases the probability that he or she is young.

Examples 13.12 and 13.13 illustrate a common setting for tree diagrams. Some outcome (such as dating online) has several sources (such as the three age groups). Starting from

- the probability of each source and
- the conditional probability of the outcome, given each source,

the tree diagram leads to the overall probability of the outcome. Example 13.12 does this. You can then use the probability of the outcome and the definition of conditional probability to find the conditional probability of one of the sources, given that the outcome occurred. Example 13.13 shows how.

APPLY YOUR KNOWLEDGE

13.12 White Cats and Deafness. Although cats generally possess an acute sense of hearing, due to an anomaly in their genetic makeup, deafness among white cats with blue eyes is quite common. Approximately 95% of the general cat population are non-white cats (i.e., not pure white), and congenital deafness is extremely rare in non-white cats. However, among white cats, approximately 75% with two blue eyes are deaf, 40% with one blue eye are deaf, and only 19% with eyes of other colors are deaf. In addition, among white cats, approximately 23% have two blue eyes, 4% have one blue eye, and the remainder have eyes of other colors.[10]

(a) Draw a tree diagram for selecting a white cat (outcomes: one blue eye, two blue eyes, or eyes of other colors) and deafness (outcomes: deaf or not deaf).

(b) What is the probability that a randomly chosen white cat is deaf?

13.13 More on Online Dating and Age. Example 13.12 provides the conditional probabilities for using an online dating site, given three age groups, as well as the probabilities of an adult being in each of the three age groups. Use that information to complete this exercise.

(a) Find the conditional probabilities for each of the three age groups, given that the adult has *not* used an online dating site. This is the conditional distribution for age, given "online date no."

(b) Compare this conditional distribution to the unconditional distribution for the three age groups given in the example. Is this what you would expect? Explain.

13.14 White Cats and Deafness, Continued. Continue your work from Exercise 13.12, using the probabilities given in the exercise.

(a) Among white cats that are deaf, what is the probability that a randomly selected cat will have two blue eyes? One blue eye? Eyes of other colors? Verify that these three probabilities add to one. These three probabilities are the *conditional distribution* of eye color, given that a white cat is deaf.

(b) Find the probability that a white cat has two blue eyes and is deaf. What can you say about the events two blue eyes and deafness among white cats?

jkitan/Getty Images

STATISTICS IN YOUR WORLD

Politically Correct

In 1950, the Soviet mathematician B. V. Gnedenko (1912–1995) wrote *The Theory of Probability*, a text that was popular around the world. The introduction contains a mystifying paragraph that begins, "We note that the entire development of probability theory shows evidence of how its concepts and ideas were crystallized in a severe struggle between materialistic and idealistic conceptions." It turns out that "materialistic" is jargon for "Marxist-Leninist." It was good for the health of Soviet scientists in the Stalin era to add such statements to their books.

13.7 Bayes' Rule*

This optional section provides coverage of Bayes' rule. Although the exercises in this section can be solved using the material on conditional probability and tree diagrams covered in Section 13.6, Bayes' rule provides a unifying structure and is a basic result in an important area of statistics known as *Bayesian statistics*. The notation in Bayes' rule can be daunting at first, but utilizing the connection to tree diagrams will help in understanding the ideas and notation in the theorem.

The prostate-specific antigen (PSA) test is a simple blood test to screen for prostate cancer. It has been used on men over 50 as a routine part of a physical exam, with levels above 4 ng/mL indicating possible prostate cancer. The test result is

*This section is not needed to understand the main ideas of this chapter.

not always correct, sometimes indicating prostate cancer when it is not present and often missing prostate cancer that is present. Here are the approximate conditional probabilities of a positive test result (above 4 ng/mL) and negative test result, given that cancer is present or cancer is absent:[11]

	Test Result	
	Positive	Negative
Cancer present	0.21	0.79
Cancer absent	0.06	0.94

These probabilities are properties of the screening test and are the same whether we screen for prostate cancer in men 30–40 years old, for whom prostate cancer is relatively rare, or in men over 50, for whom prostate cancer is much more common.

For men over 50, the screening population of interest, it is found that about 6.3% of the population has prostate cancer. Figure 13.7 is a tree diagram for selecting a person from this population (outcomes: cancer present or absent) and testing his blood (outcomes: test positive or negative).

false-positive rate
Given a test for a disease or some other condition, the probability that a subject who tests positive does not actually have the condition.

The conditional probability that a person does not have prostate cancer, given that the PSA test is positive, is called the **false-positive rate**. The false-positive rate depends on both the properties of the diagnostic test and the incidence of disease in the population. Example 13.14 uses the information in the tree diagram to compute the false-positive rate of the PSA test.

EXAMPLE 13.14 False Positives

4step

STATE: The PSA test has conditional probability 0.21 of giving a positive test result when there is cancer and conditional probability 0.06 of giving a positive test result when cancer is absent. Approximately 6.3% of the screened population has prostate cancer. What is the false-positive rate when the PSA test is used to screen for prostate cancer on this population?

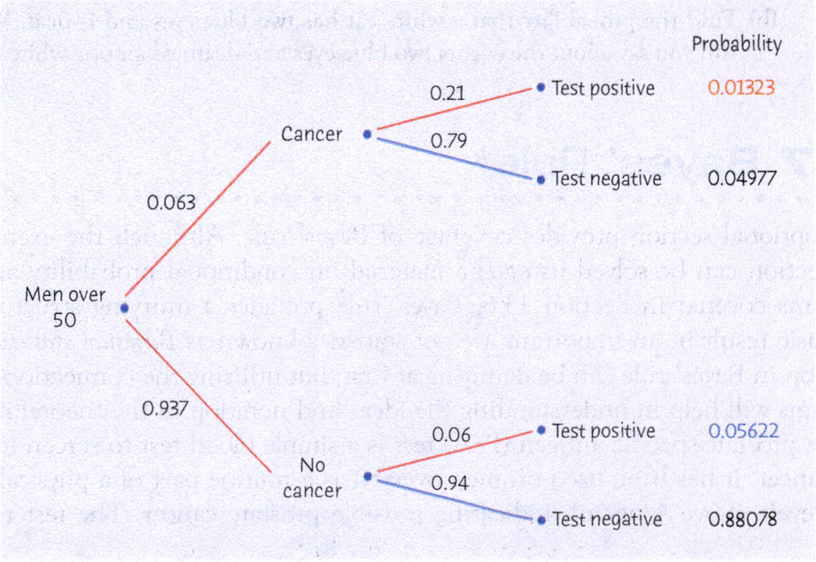

FIGURE 13.7
Tree diagram for the PSA test, for Example 13.14. The red probability is the probability of having cancer and a positive test result, and the blue probability is the probability of not having cancer and having a positive test result.

PLAN: Express the information given in terms of events and their probabilities:

$$\begin{aligned} &\text{If } B_1 = \{\text{cancer is present}\} && \text{then} && P(B_1) = 0.063 \\ &\text{If } B_2 = \{\text{cancer is absent}\} && \text{then} && P(B_2) = 0.937 \\ &\text{If } A = \{\text{test is positive}\} && \text{then} && P(A \mid B_1) = 0.21 \\ & && \text{and} && P(A \mid B_2) = 0.06 \end{aligned}$$

We want to find $P(B_2 \mid A)$.

SOLVE: From the definition of conditional probability,

$$P(B_2 \mid A) = \frac{P(B_2 \text{ and } A)}{P(A)}$$

Both probabilities on the right-hand side of the equation follow easily from the information in the tree diagram. Using the multiplication rule,

$$\begin{aligned} P(B_2 \text{ and } A) &= P(B_2) \times P(A \mid B_2) \\ &= (0.937)(0.06) = 0.05622 \end{aligned}$$

is the probability in blue in the tree diagram of Figure 13.7. In the tree diagram, there are two disjoint paths to "test positive," one for each disease status. Because the two paths are disjoint, the probability of A = {test positive} is the sum of the probabilities:

$$\begin{aligned} P(A) &= P(A \text{ and } B_1) + P(A \text{ and } B_2) \\ &= P(B_1) \times P(A \mid B_1) + P(B_2) \times P(A \mid B_2) \\ &= (0.063)(0.21) + (0.937)(0.06) \\ &= 0.01323 + 0.05622 = 0.06945 \end{aligned}$$

Thus, $P(A)$ is the sum of the red and blue probabilities in the tree diagram of Figure 13.7. Combining these answers, the probability of a false positive is

$$\begin{aligned} P(B_2 \mid A) &= \frac{P(B_2 \text{ and } A)}{P(A)} \\ &= \frac{0.05622}{0.06945} = 0.81 \end{aligned}$$

CONCLUDE: The probability of a false positive when using the PSA test as a routine screen for men over 50 is approximately 81%.

Treatment for prostate cancer can have serious side effects, including incontinence and impotence, and many men diagnosed with prostate cancer would eventually die of another cause if the cancer were left untreated. In fact, it has been found in some large clinical trials that screening did not reduce overall mortality.[12] In October 2011, the U.S. Preventive Services Task Force (USPSTF) released a draft report in which it recommended against using the PSA test to screen for prostate cancer in the general population. The false-positive rate is quite high, showing that in routine screening, about four-fifths of the positive test results occur for a person without cancer, an important fact in the USPSTF decision.

In Example 13.14, we are given the conditional probabilities of the test result, given the disease status, but we are interested in the conditional probability of the disease status, given the test result—the reverse condition. The relationship between these conditional probabilities is given formally in Bayes' rule.

Bayes' Rule

Suppose $B_1, B_2, \ldots, B_n$ are disjoint events with positive probabilities and that the sum of these probabilities is 1. If A is any event with probability greater than 0, then

$$P(B_i \mid A) = \frac{P(A \mid B_i)P(B_i)}{P(A \mid B_1)P(B_1) + P(A \mid B_2)P(B_2) + \cdots + P(A \mid B_n)P(B_n)}$$

In Example 13.14, we have intentionally used notation to match that used in Bayes' rule when there are only two events B_1 and B_2. In the general theorem, the events B_1, B_2, . . . , Bn correspond to the leftmost segments in the tree diagram, with every outcome belonging to exactly one of the B_i. Each event B_i then has a segment going to A with conditional probability $P(A \mid B_i)$. The probability of the branch beginning at B_i and continuing through A is $P(B_i) \times P(A \mid B_i)$. Bayes' rule states that the $P(B_i \mid A)$ is the probability of the branch going from B_i through A divided by the sum of the probabilities of the n branches going from B_1, B_2, . . . , B_n through A. Here is another example.

EXAMPLE 13.15 Who Uses the Internet?

STATE: Internet usage varies among millennials (adults born between 1981 and 1986), gen X-ers (adults born between 1965 and 1980), boomers (adults born between 1946 and 1964), and the silent generation (adults born before 1946). Among all millennials, gen X-ers, boomers, and the silent generation, millennials comprise 30%, gen X-ers 27%, boomers 32%, and the silent generation 11%. Also, 100% of millennials use the Internet, 91% of gen X-ers use the Internet, 85% of boomers use the Internet, and 78% of the silent generation use the Internet.[13] What percentage of Internet users from these four groups are millennials? Are gen X-ers? Are boomers? Are the silent generation?

PLAN: Express the information given in terms of events and their probabilities:

If B_1 = (millennial)	then	$P(B_1) = 0.30$
If B_2 = (gen X-er)	then	$P(B_2) = 0.27$
If B_3 = (boomer)	then	$P(B_3) = 0.32$
If B_4 = (silent)	then	$P(B_4) = 0.11$
If A = (use Internet)	then	$P(A \mid B_1) = 1.00$
	and	$P(A \mid B_2) = 0.91$
	and	$P(A \mid B_3) = 0.85$
	and	$P(A \mid B_4) = 0.78$

We want to find $P(B_1 \mid A)$, $P(B_2 \mid A)$, $P(B_3 \mid A)$, and $P(B_4 \mid A)$.

SOLVE: Using Bayes' rule,

$$P(B_1 \mid A) = \frac{P(A \mid B_1)P(B_1)}{P(A \mid B_1)P(B_1) + P(A \mid B_2)P(B_2) + P(A \mid B_3)P(B_3) + P(A \mid B_4)P(B_4)}$$

$$= \frac{(1.00)(0.30)}{(1.00)(0.30) + (0.91)(0.27) + (0.85)(0.32) + (0.78)(0.11)}$$

$$= \frac{0.30}{0.9035} = 0.3320$$

Similar calculations show $P(B_2 \mid A) = 0.2719$, $P(B_3 \mid A) = 0.3011$, and $P(B_4 \mid A) = 0.0950$.

CONCLUDE: About 33% of Internet users born before 1987 are millennials, 27% are gen X-ers, 30% are boomers, and the remaining 10% are members of the silent generation.

$P(B_1)$, $P(B_2)$, $P(B_3)$, and $P(B_4)$ are the *prior probabilities* for the four age categories. If we know the person has used the Internet, we have used Bayes' rule to compute the conditional probabilities $P(B_1 \mid A)$, $P(B_2 \mid A)$, $P(B_3 \mid A)$, and $P(B_4 \mid A)$. These are referred to as the *posterior probabilities* for the four age categories, given the information that the person has used the Internet. As we might expect, the posterior probability is greater than the prior probability for millennials and smaller than the prior probability for people who are members of the silent generation.

APPLY YOUR KNOWLEDGE

13.15 False-Negative Rate. In a diagnostic test for a particular disease, the conditional probability that a person has the disease, given that the test is negative is called the false-negative rate. Suppose the PSA screening test with the properties described in Example 13.14 is used to screen for prostate cancer in the population of men over 50, for which 6.3% have prostate cancer. Compute the false-negative rate using Bayes' rule.

13.16 More on False-Positives. The false-positive rate for a diagnostic test depends on the properties of the diagnostic test as well as the rate of the disease in the population. Suppose we are going to use the PSA screening test to screen a population in which only 3% of the population has prostate cancer, with the properties given in Example 13.14.

(a) Compute the false-positive rate of the PSA test if it were used to screen for prostate cancer in this population. How does this compare with the false-positive rate computed in Example 13.14, in which 6.3% of the population had prostate cancer? Explain simply why the false-positive rate has changed in this direction.

(b) What is the relationship between the false-positive rate and the rate of disease in the population? What does this say about screening for a very rare disease?

13.17 Eye Color and Hair Color. A large study of children of Caucasian descent in Germany looked at the effect of eye color, hair color, and freckles on the reported extent of burning from sun exposure.[14] The population's distribution of hair color and eye color is given in the tree diagram of Figure 13.8. (You will not need to use the rightmost column in the tree for this exercise.)

(a) What are the prior probabilities of black, brown, blonde, and red hair for a child of Caucasian descent in Germany?

(b) Find the posterior probabilities of black, brown, blonde, and red hair, given that the child has blue eyes. Is the relationship between the prior and posterior probabilities what you would expect, given the population distribution of hair color and eye color? Explain.

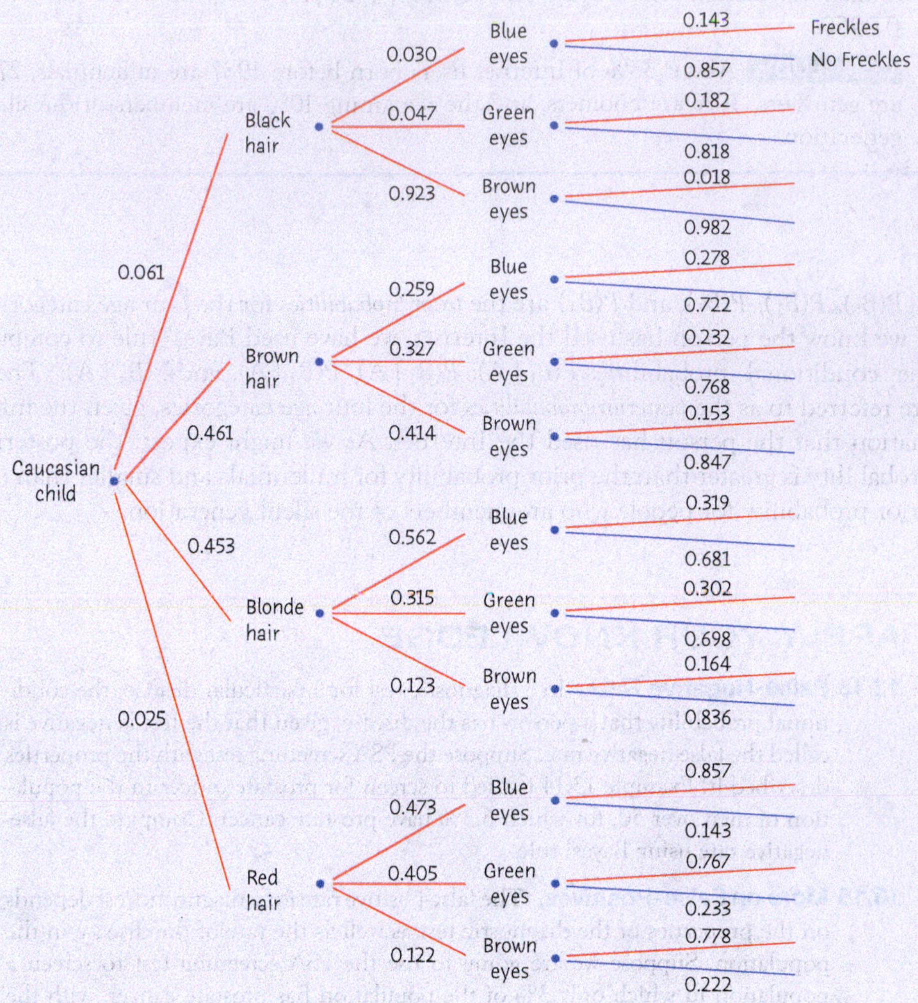

FIGURE 13.8

Tree diagram of the features of children of Caucasian descent in Germany, for Exercise 13.17. The three stages are hair color, eye color, and whether or not the child has freckles.

CHAPTER 13 SUMMARY

- Events A and B are *disjoint* if they have no outcomes in common. In that case, $P(A \text{ or } B) = P(A) + P(B)$.
- The **conditional probability** $P(B \mid A)$ of an event B, given an event A, is defined by

$$P(B \mid A) = \frac{P(A \text{ and } B)}{P(A)}$$

  when $P(A) > 0$. In practice, we most often find conditional probabilities from directly available information rather than from the definition.
- Events A and B are **independent** if knowing that one of the events occurs does not change the probability we would assign to the other event; that is, $P(B \mid A) = P(B)$. In that case, $P(A \text{ and } B) = P(A)P(B)$.
- Any assignment of probability obeys these rules:

  Addition rule for disjoint events: If events A, B, C, . . . are all disjoint in pairs, then

$$P(A \text{ or } B \text{ or } C \text{ or} \ldots) = P(A) + P(B) + P(C) + \cdots$$

Multiplication rule for independent events: If events A, B, C, . . . are independent, then

$$P(\text{all of these events occur}) = P(A)P(B)P(C)\ldots$$

General addition rule: For any two events A and B,

$$P(A \text{ or } B) = P(A) + P(B) - P(A \text{ and } B)$$

General multiplication rule: For any two events A and B,

$$P(A \text{ and } B) = P(A)P(B \mid A)$$

- **Tree diagrams** organize probability models that have several stages.

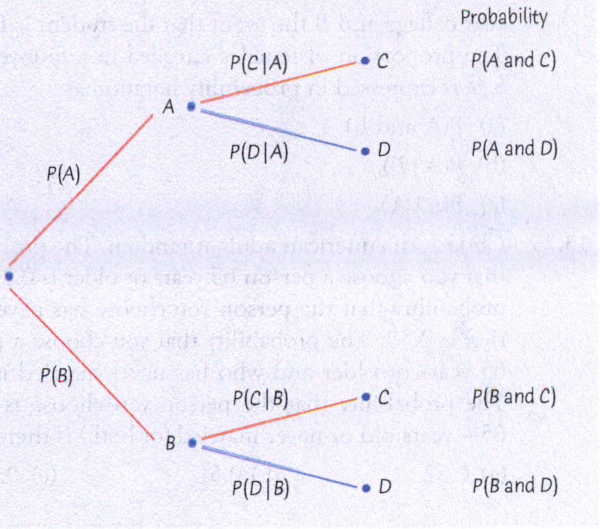

FIGURE 13.9
A generalized tree diagram.

CHECK YOUR SKILLS

13.18 An instant lottery game gives you probability 0.02 of winning on any one play. Plays are independent of each other. If you play three times, the probability that you win on none of your plays is about

(a) 0.98.

(b) 0.94.

(c) 0.000008.

13.19 The probability that you win on one or more of your three plays of the game in the previous exercise is about

(a) 0.02.

(b) 0.06.

(c) 0.999992.

13.20 An athlete suspected of having used steroids is given two tests that operate independently of each other. Test A has probability 0.9 of being positive if steroids have been used. Test B has probability 0.8 of being positive if steroids have been used. What is the probability that *at least one* test is positive if steroids have been used?

(a) 0.98

(b) 0.72

(c) 0.28

What do students do after completing high school? Here are the counts of enrollments in two-year and four-year colleges for males and females completing high school in 2017. High school completers include individuals ages 16–24 who graduated from high school or completed a GED. All counts in the table are in thousands:[15]

	Male	Female	Total
Enrolled in two-year college	321	326	647
Enrolled in four-year college	500	767	1267
Other	524	432	956
Total	1345	1525	2870

Questions 13.21 through 13.24 are based on this table.

13.21 Choose a student completing high school at random from this group. The probability that the student does not attend either a two-year or four-year college is

(a) 0.28.

(b) 0.33.

(c) 0.44.

13.22 The conditional probability that the student is enrolled in a two-year college, given that the student is male, is about

(a) 0.11.

(b) 0.24.

(c) 0.50.

13.23 The conditional probability that the student is female, given that the student is enrolled in a four-year college, is about

(a) 0.27.

(b) 0.50.

(c) 0.61.

13.24 Let A be the event that the student is enrolled in a four-year college and B the event that the student is female. The proportion of females enrolled in a four-year college is expressed in probability notation as

(a) $P(A \text{ and } B)$.

(b) $P(A \mid B)$.

(c) $P(B \mid A)$.

13.25 Choose an American adult at random. The probability that you choose a person 65 years or older is 0.20. The probability that the person you choose has never married is 0.32. The probability that you choose a person 65 years or older and who has never married is 0.01. The probability that the person you choose is either 65+ years old or never married (or both) is therefore

(a) 0.52. (b) 0.51. (c) 0.06.

13.26 Of people who died in the United States in recent years, 78% were non-Hispanic White, 12% were non-Hispanic Black, 7% were Hispanic, and 3% were Asian. (This ignores a small number of deaths among other races.) Diabetes caused 2.5% of deaths among non-Hispanic Whites, 4.4% among non-Hispanic Blacks, 4.7% among Hispanics, and 4.2% among Asians. The probability that a randomly chosen death is a non-Hispanic White person and died of diabetes is about

(a) 0.81. (b) 0.025. (c) 0.020.

13.27 Using the information in Question 13.26, the probability that a randomly chosen death was due to diabetes is about

(a) 0.158. (b) 0.029. (c) 0.019.

CHAPTER 13 EXERCISES

13.28 Playing the Lottery. New York State's "Quick Draw" lottery moves right along. Players choose between 1 and 10 numbers from the range 1 to 80; 20 winning numbers are displayed on a screen every four minutes. If you choose just one number, your probability of winning is 20/80, or 0.25. Lester plays one number eight times as he sits in a bar. What is the probability that all eight bets lose?

13.29 Universal Blood Donors. People with type O-negative blood are referred to as universal donors, although if you give type O-negative blood to any patient, you run the risk of a transfusion reaction due to certain antibodies present in the blood. However, any patient can receive a transfusion of O-negative *red blood cells*. Only 7.2% of the American population have O-negative blood. If 10 people appear at random to give blood, what is the probability that at least one of them is a universal donor?

13.30 Playing the Slots. Slot machines are now video games, with outcomes determined by random number generators. Old slot machines work like this: you pull the lever to spin three wheels; each wheel has 20 symbols, all equally likely to show when the wheel stops spinning; the three wheels are independent of each other. Suppose that the middle wheel has nine cherries among its 20 symbols, and the left and right wheels have one cherry each.

Peter Dazeley/Getty Images

(a) You win the jackpot if all three wheels show cherries. What is the probability of winning the jackpot?

(b) There are three ways that the three wheels can show two cherries and one symbol other than a cherry. Find the probability of each of these ways.

(c) What is the probability that the wheels stop with exactly two cherries showing among them?

13.31 Tendon Surgery. (4step) You have torn a tendon and are facing surgery to repair it. The surgeon explains the risks to you: infection occurs in 3% of such operations, the repair fails in 14%, and both infection and failure occur together in 1%. What percentage of these operations succeed and are free from infection? Follow the four-step process in your answer.

13.32 Screening Job Applicants. A company retains a psychologist to assess whether job applicants are suited for assembly-line work. The psychologist classifies applicants as one of A (well suited), B (marginal), or C (not suited). The company is concerned about the event D that an employee leaves the company within a year of being hired. Data on all people hired in the past five years give these probabilities:

$$P(A) = 0.4 \quad P(B) = 0.3 \quad P(C) = 0.3$$
$$P(A \text{ and } D) = 0.1 \quad P(B \text{ and } D) = 0.1 \quad P(C \text{ and } D) = 0.2$$

Sketch a Venn diagram of the events A, B, C, and D and mark on your diagram the probabilities of all combinations of psychological assessment and leaving (or not) within a year. What is $P(D)$, the probability that an employee leaves within a year?

13.33 Tendon Surgery (continued). (4step) You have torn a tendon and are facing surgery to repair it. The surgeon explains the risks to you: infection occurs in 3% of such operations, the repair fails in 14%, and both infection and failure occur together in 1%. What is the probability of infection, given that the repair is successful? Follow the four-step process in your answer.

13.34 Cancer-Detecting Dogs. Research has shown that specific biochemical markers are found exclusively in the breath of patients with lung cancer. However, no lab test can currently distinguish the breath of lung cancer patients from that of other subjects. Could dogs be trained to identify these markers in specimens of human breath, as they can be to detect illegal substances or to follow a person's scent? An experiment trained dogs to distinguish breath specimens of lung cancer patients from breath specimens of control individuals by using a food-reward training method. After the training was complete, the dogs were tested on new breath specimens without any reward or clue using a double-blind, completely randomized design. Here are the results for a random sample of 1286 breath specimens:[16]

	Breath Specimen		
Dog Test Result	Control Subject	Cancer Subject	Total
Negative	708	10	718
Positive	4	564	568
Total	712	574	1286

(a) The *sensitivity* of a diagnostic test is its ability to correctly give a positive result when a person tested has the disease, or P(positive test | disease). Find the sensitivity of the dog cancer-detection test for lung cancer.

(b) The *specificity* of a diagnostic test is the conditional probability that the subject tested doesn't have the disease, given that the test has come up negative. Find the specificity of the dog cancer-detection test for lung cancer.

13.35 Student Debt. At the end of 2016, the average outstanding student debt for bachelor's degree recipients was $28,500. Here is the distribution of outstanding education debt (in thousands of dollars):[17]

Debt	< 10	10 to < 20	20 to < 30	30 to < 40	40 to < 50	≥ 50
Probability	0.40	0.13	0.17	0.12	0.08	0.10

(a) What is the probability that a randomly chosen student has an outstanding debt of $20,000 or more?

(b) Given that a student has an outstanding debt of at least $20,000, what is the conditional probability that the debt is at least $50,000?

13.36 A Probability Teaser. Let's assume it is safe to say that people are either male or female at birth[18] and that each child born is equally likely to be a boy or a girl and that the sexes of successive children are independent. If we let BG mean that the older child is a boy and the younger child is a girl, then each of the combinations BB, BG, GB, and GG has probability 0.25. Ashley and Brianna each have two children.

(a) You know that at least one of Ashley's children is a boy. What is the conditional probability that she has two boys?

(b) You know that Brianna's older child is a boy. What is the conditional probability that she has two boys?

13.37 College Degrees. A striking trend in higher education is that more women than men reach each level of attainment. The National Center for Education Statistics (which classifies all subjects as male or female) provides projections for the number of degrees earned, classified by level and by the sex of the degree recipient. Here are the projected number of earned degrees (in thousands) in the United States for the 2027–28 academic year:[19]

Barry Austin Photography/Getty Images

	Associate's	Bachelor's	Master's	Doctorate	Total
Female	653	1106	473	101	2333
Male	411	816	340	90	1657
Total	1064	1922	813	191	3990

(a) If you choose a degree recipient at random, what is the probability that the person you choose is a man?

(b) What is the conditional probability that you choose a man, given that the person chosen received a master's?

(c) Are the events "choose a man" and "choose a master's degree recipient" independent? How do you know?

13.38 Eye Color, Hair Color, and Freckles. Tree diagrams can organize problems having more than two stages. A large study of children of Caucasian descent in Germany looked at the effect of eye color, hair color, and freckles on the reported extent of burning from sun exposure. The population's distribution of hair color, eye color, and freckles is shown in the tree diagram of Figure 13.8 (see page 316). Find the following probabilities and describe them in plain English:

(a) P(blue eyes | red hair) and P(blue eyes and red hair).

(b) P(freckles | red hair and blue eyes) and P(freckles and red hair and blue eyes).

13.39 College Degrees (continued). Exercise 13.37 gives the projected counts (in thousands) of earned degrees in the United States in the 2027–28 academic year. Use these data to answer the following questions:

(a) What is the probability that a randomly chosen degree recipient is a woman?

(b) What is the conditional probability that the person chosen received an associate's degree, given that she is a woman?

(c) Use the general multiplication rule to find the probability of choosing a female associate's degree recipient. Check your result by finding this probability directly from the table of counts.

13.40 Eye Color, Hair Color, and Freckles (continued). Continue your work from Exercise 13.38 using the tree diagram of Figure 13.8. Find the following probabilities and describe them in plain English:

(a) $P(\text{red hair})$, $P(\text{blue eyes})$, and $P(\text{freckles})$.

(b) $P(\text{freckles and red hair})$ and $P(\text{freckles} \mid \text{red hair})$.

(c) What can you say about the events "freckles" and "red hair" in this population?

13.41 The Probability of a Flush. A poker player holds a flush when all five cards in the hand belong to the same suit (clubs, diamonds, hearts, or spades). We will find the probability of a flush when five cards are drawn in succession from the top of the deck. Remember that a deck contains 52 cards, 13 of each suit, and that when the deck is well shuffled, each card drawn is equally likely to be any of those that remain in the deck.

(a) Concentrate on spades. What is the probability that the first card drawn is a spade? What is the conditional probability that the second card drawn is a spade, given that the first is a spade? (*Hint*: How many cards remain? How many of these are spades?)

(b) Continue to count the remaining cards to find the conditional probabilities of a spade for the third, the fourth, and the fifth card drawn, given in each case that all previous cards are spades.

(c) The probability of drawing five spades in succession from the top of the deck is the product of the five probabilities you have found. Why? What is this probability?

(d) The probability of drawing five hearts or five diamonds or five clubs is the same as the probability of drawing five spades. What is the probability that the five cards drawn all belong to the same suit?

13.42 Deer and Pine Seedlings. As suburban gardeners know, deer will eat almost anything green. In a study of pine seedlings at an environmental center in Ohio, researchers noted how deer damage varied with how much of the seedling was covered by thorny undergrowth:[20]

	Deer Damage	
Thorny Cover	Yes	No
None	60	151
$< 1/3$	76	158
1/3 to 2/3	44	177
$> 2/3$	29	176

Peter Skinner/Science Source

(a) What is the probability that a randomly selected seedling was damaged by deer?

(b) What are the conditional probabilities that a randomly selected seedling was damaged, given each level of cover?

(c) Does knowing about the amount of thorny cover on a seedling change the probability of deer damage? If so, cover and damage are not independent.

13.43 Deer and Pine Seedlings, Continued. In the setting of Exercise 13.42, what percentage of the trees that were not damaged by deer were more than two-thirds covered by thorny plants?

13.44 Deer and Pine Seedlings, Continued. In the setting of Exercise 13.42, what percentage of the trees that were damaged by deer were less than one-third covered by thorny plants?

Although cigarette smoking has declined among U.S. youth in recent years, the use of some other tobacco products has increased. When high school students were asked which of several tobacco products they had used in the past 30 days, more than 40% of those who had used any tobacco product had used multiple tobacco products. Let A, B, and C be the events corresponding to the use of the following types of tobacco products in the past 30 days:

A = cigarettes

B = electronic cigarettes

C = other tobacco products including cigars, pipes, smokeless tobacco, and hookahs

Here are the probabilities that a randomly selected high school student used these different tobacco products:[21]

$P(A) = 0.08$ $\quad P(B) = 0.21$ $\quad P(C) = 0.19$

$P(A \text{ and } B) = 0.06$ $\quad P(A \text{ and } C) = 0.03$ $\quad P(B \text{ and } C) = 0.06$

$P(A \text{ and } B \text{ and } C) = 0.02$

Make a Venn diagram of the events A, B, and C. As in Figure 13.4 (page 297), mark the probabilities of every intersection involving these events. Use this diagram for Exercises 13.45 through 13.47.

13.45 Do You Use Tobacco Products? What is the probability that a randomly selected high school student did not use any tobacco product?

13.46 Do You Just Use Electronic Cigarettes? What is the probability that a randomly selected high school student used electronic cigarettes but no other tobacco products?

13.47 Conditional Probabilities. If a student smokes electronic cigarettes, what is the conditional probability that he or she also smokes cigarettes? If a student smokes cigarettes, what is the conditional probability that he or she also smokes electronic cigarettes?

13.48 The Geometric Distributions. You are rolling a pair of balanced dice in a board game. Rolls are independent. You land in a danger zone that requires you to roll doubles (both faces show the same number of spots) before you are allowed to play again. How long will you wait to play again?

(a) What is the probability of rolling doubles on a single toss of the dice? (For a review, see the possible outcomes in Figure 12.2 on page 276. All 36 outcomes are equally likely.)

(b) What is the probability that you do not roll doubles on the first toss but do on the second toss?

(c) What is the probability that the first two tosses are not doubles and the third toss is doubles? This is the probability that the first doubles occurs on the third toss.

(d) Now you see the pattern. What is the probability that the first doubles occurs on the fourth toss? On the fifth toss? Give the general result, what is the probability that the first doubles occurs on the kth toss?

(e) What is the probability that you get to go again within three turns? (*Comment:* The distribution of the number of trials to the first success is called a *geometric distribution*. In this problem, you have found geometric distribution probabilities when the probability of a success on each trial is 1/6. The same idea works for any probability of success.)

13.49 Winning at Tennis. A player serving in tennis has two chances to get a serve into play. If the first serve is out, the player serves again. If the second serve is also out, the player loses the point. Here are probabilities based on four years of the Wimbledon Championship:[22]

$$P(\text{1st serve in}) = 0.59$$

$$P(\text{win point} \mid \text{1st serve in}) = 0.73$$

$$P(\text{2nd serve in} \mid \text{1st serve out}) = 0.86$$

$$P(\text{win point} \mid \text{1st serve out and 2nd serve in}) = 0.59$$

Make a tree diagram for the results of the two serves and the outcome (win or lose) of the point. (The branches in your tree have different numbers of stages, depending on the outcome of the first serve.) What is the probability that the serving player wins the point?

13.50 Peanut Allergies Among Children. About 2% of children in the United States are allergic to peanuts.[23] Choose three children at random and let the random variable X be the number in this sample who are allergic to peanuts. The possible values X can take are 0, 1, 2, and 3. Make a three-stage tree diagram of the outcomes (allergic or not allergic) for the three individuals and use it to find the probability distribution of X.

4step

13.51 Winning at Tennis, Continued. Use your work in Exercise 13.49 to answer the following. Considering only points that the server won, in what percentage of those points was the first serve in? (Write this as a conditional probability and use the definition of conditional probability.)

13.52 Peanut Allergies Among Children, Continued. Continue your work from Exercise 13.50. What is the conditional probability that exactly two of the children will be allergic to peanuts, given that at least one of the three children suffers from this allergy?

13.53 Lactose Intolerance. Lactose intolerance causes difficulty digesting dairy products that contain lactose (milk sugar). It is particularly common among people of African and Asian ancestry. In the United States (ignoring other groups and people who consider themselves to belong to more than one race), 82% of the population is White, 14% is Black, and 4% is Asian. Moreover, 15% of Whites, 70% of Blacks, and 90% of Asians are lactose intolerant.[24]

(a) What percentage of the entire population is lactose intolerant?

(b) What percentage of people who are lactose intolerant are Asian?

13.54 (Optional Topic) Should the Government Help the Poor? In the 2014 General Social Survey, 32% of those sampled thought of themselves as Democrats, 45% as Independents, 21% as Republicans, and 2% as Other.[25] When asked, "Should the government in Washington do everything possible to improve the standard of living of all poor Americans?" 23% of the Democrats, 18% of the Independents, 4% of the Republicans, and 15% of Others agreed. Given that a person agrees that the government in Washington should do everything possible to improve the standard of living of all poor Americans, use Bayes' rule to find the probability that the person thinks of him- or herself as a Democrat.

13.55 (Optional Topic) College Degrees and Race. According to the National Center for Education Statistics, in 2016, 26% of college degrees were associate's degrees, 49% were bachelor's degrees, 20% were master's degrees, and 5% were doctorates or other advanced degrees (including professional degrees such as MD, DDS, and law degrees). Asians/Pacific Islanders earned 7% of the associate's degrees, 6% of the bachelor's degrees, 6% of the master's degrees, and 11% of the doctorates and other advanced degrees.

(a) What are the prior probabilities for the four types of degrees?

(b) Find the posterior probabilities for the type of degree, given that the recipient is Asian or a Pacific Islander. Is the relationship between the prior and posterior probabilities what you would expect? Explain briefly.

DNA Forensics. *When a suspect's DNA is compared to a sample of DNA collected at a crime scene, the comparison is made between certain sections of the DNA called loci. Each locus has two alleles (gene forms), one inherited from the mother and the other from the father. Suppose there are two alleles for a particular locus called A and B. These alleles can be present at the locus in three combinations. A person could have both alleles at the locus be A, one allele could be A and the other B, or both alleles could be B, giving the three combinations (A and A), (A and B), and (B and B). Here's how the math works. If the proportion of the population with allele A as at least one of the alleles at the locus is a, and the proportion of the population with allele B as at least one of the alleles at the locus is b, then the proportion of the population with the three combinations of these allele types at the locus follows:*

Alleles at the Locus	Population Proportion with Allele Combination
A and A	a^2
A and B	$2ab$
B and B	b^2

Use this information in Exercises 13.56 and 13.57. The numbers used in the exercises are from the FBI database.[26]

13.56 Suppose the locus D21S11 has two alleles, called 29 and 31. The proportion of the Caucasian population with allele 29 is 0.181 and with allele 31 is 0.071. What proportion of the Caucasian population has the combination (29, 31) at the locus D21S11? What proportion has the combination (29, 29)?

13.57 Suppose the locus D3S1358 has two alleles, called 16 and 17. The proportion of the Caucasian population with allele 16 is 0.232 and with allele 17 is 0.213. What proportion of the Caucasian population has the combination (16, 17) at the locus D3S1358?

One important fact regarding the loci evaluated in such forensic tests is that the allele combinations at each locus have been shown to be independent. Use this information in Exercise 13.58.

13.58 What proportion of the Caucasian population has the combination (29, 31) at the locus D21S11 and combination (16, 17) at the locus D3S1358? As we specify the alleles present at more loci, what will happen to the proportion of the Caucasian population that matches the allele combinations at all the loci?

A defendant in Ohio was indicted on December 17, 2009, on the charges of Aggravated Burglary and Assault (State of Ohio vs. Myers, Case No. 09 CR 666). In this case, a hair found at the crime scene was tested at six loci and demonstrated a specific combination of alleles found in a proportion of about one in 1.6 million individuals in the population. Comparison of the DNA profile found on the hair to a database of convicted felons revealed a match between the allelic profile found on the hair and an individual in the database (the defendant). Defense attorneys in the case requested that the state perform additional DNA testing because several previously untested loci were available to test. The results of this testing revealed that the defendant did not match at some of these newly tested markers, indicating that the DNA from the hair was not the defendant's, and the charges were dropped. Use this information in Exercise 13.59.

13.59 If the DNA profile (or combination of alleles) found on the hair is possessed by one in 1.6 million individuals, and the database of convicted felons contains 4.5 million individuals, approximately how many individuals in the database would demonstrate a match between their DNA and that found on the hair?

13.60 Are They Independent? For each of the following pair of events A and B, do you think the events are independent? Explain your reasoning.

(a) You select an adult U.S. citizen at random. Event A is "the person is a registered democrat," and event B is "the person is opposed to the death penalty."

(b) You select an adult U.S. citizen at random. Event A is "the person is a baby boomer" (baby boomers are people born 1946–1964), and event B is "the person favors legalization of marijuana."

(c) You draw a card at random from a deck of playing cards. Event A is "the card is the king of hearts," and event B is "the card is the queen of diamonds."

(d) You select a student at your college at random. Event A is "the student is taking a Spanish class," and event B is "the student has visited Mexico."

CHAPTER

14

Binomial Distributions*

When you complete this chapter, you will be able to:

14.1 Recognize whether a given situation is a binomial setting, in which the count of successes has a binomial distribution.

14.2 Determine whether the proportion of successes in a simple random sample has approximately a binomial distribution.

14.3 Use the binomial probability formula to find probabilities of events in binomial settings.

14.4 Use technology to perform binomial calculations.

14.5 Find the mean and standard deviation of a random variable that has a binomial distribution.

14.6 Recognize when a Normal approximation to a binomial distribution is valid and use the Normal distribution to approximate binomial distributions.

We began our study of probability in Chapter 12. We introduced probability rules, probability models, and probability distributions as tools for describing and predicting the long-run behavior of random phenomena. In this chapter we study a simple but very useful probability distribution for the proportion of times an event with only two possible outcomes can occur in several independent trials. In Chapter 22 we will use this distribution to make inferences about the proportion of some outcome in a population, and in Chapter 23 we will use it to compare the proportion of some outcome in two populations.

14.1 The Binomial Setting and Binomial Distributions

A basketball player shoots five free throws; how many does she make? A sample survey dials 1200 landline phone numbers at random; how many of the numbers dialed correspond to working residential numbers? You plant 10 dogwood trees; how many live through the winter? In all these situations, we want a probability model for a *count* of successful outcomes out of a fixed (known) number of trials. These scenarios share the following attributes.

*This chapter concerns a special topic in probability. The material is not needed to understand the rest of the book.

The Binomial Setting

1. There are a fixed number of observations, n.
2. The n observations are all **independent.** That is, knowing the result of one observation does not change the probabilities we assign to other observations.
3. Each observation falls into one of just two categories, which for convenience we call "success" and "failure."
4. The probability of a success, call it p, is the same for each observation.

STATISTICS IN YOUR WORLD

Was He Good or Was He Lucky?

When a baseball player hits .300, everyone applauds. A .300 hitter gets a hit in 30% of times at bat. Could a .300 year just be luck? Typical Major League players bat about 500 times a season and hit about .260. A hitter's successive tries seem to be independent (i.e., a Major League hitter in the midst of a "streak" is *not* more likely than usual to get a hit), so we have a binomial setting. From this model, we can calculate or simulate the probability of hitting .300. It is about 0.025. Out of 100 run-of-the-mill Major League hitters, two or three each year will bat .300 because they were lucky.

Think of tossing a coin n times as an example of the binomial setting. Each toss gives either heads or tails. Knowing the outcome of one toss doesn't change the probability of a head on any other toss, so the tosses are independent. If we call heads a success, then p is the probability of a head and remains the same as long as we toss the same coin. For tossing a coin, p is close to 0.5. If we spin the coin on a flat surface rather than toss it, p is not equal to 0.5. The number of heads we count is a discrete random variable X. The distribution of X is called a *binomial distribution.*

Binomial Distribution

The count X of successes in the binomial setting has the **binomial distribution** with parameters n and p. The parameter n is the number of observations, and p is the probability of a success on any one observation. The possible values of X are the whole numbers from 0 to n.

Binomial distributions are an important class of discrete probability models. *Pay attention to the binomial setting because not all counts have binomial distributions.*

EXAMPLE 14.1 Blood Types

Genetics says that children receive genes from their parents independently. Each child of a particular pair of parents has probability 0.25 of having type O blood. If these parents have five children, the number who have type O blood is the count X of successes in five independent observations with probability 0.25 of a success on each observation. So X has the binomial distribution with $n = 5$ and $p = 0.25$.

EXAMPLE 14.2 Counting Boys

Here is a set of genetic examples that require more thought.

Choose two births at random from last year's births at a large hospital and count the number of boys (0, 1, or 2). The sexes of children born to different mothers are surely independent. The probability that a randomly chosen birth in Canada and the United States is a boy is about 0.52. (Why it is not 0.5 is something of a mystery.) So the count of boys has a binomial distribution with $n = 2$ and $p = 0.52$.

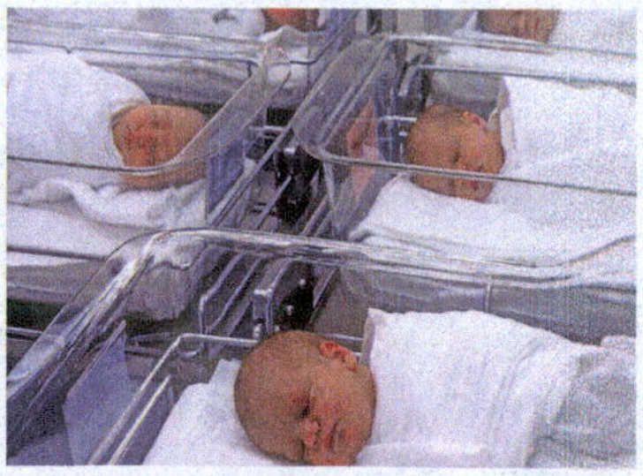

Randy Duchaine/Alamy Stock Photo

Next, observe successive births at a large hospital and let X be the number of births until the first boy is born. Births are independent, and each has probability 0.52 of being a boy. However, X is *not* binomial because there is no fixed number of observations. "Count observations until the first success" is a different setting than "count the number of successes in a fixed number of observations."

Finally, choose at random a family with exactly two children and count the number of boys. Careful study of such families shows that the count of boys is *not* binomial: the probability of exactly 1 boy is too high.[1] Families are less likely to have a third child if the first two are a boy and a girl, so when we look at families that stopped at two children, "one of each" is more common than if we look at randomly chosen births. The sexes of successive children in two-child families are *not independent* because the parents' choices interfere with the genetics.

14.2 Binomial Distributions in Statistical Sampling

The binomial distributions are important in statistics when we wish to make inferences about the proportion p of "successes" in a population. Here is a typical example.

EXAMPLE 14.3 Choosing an SRS of Tomatoes

Physical damage to tomatoes, which can occur throughout the distribution system from field to consumer, has a major impact on market loss of fresh tomatoes. At the packing house, a distributor inspects an SRS of 10 tomatoes from a shipment of 10,000 tomatoes. Suppose that (unknown to the distributor) 11% of tomatoes in the shipment can be considered unmarketable due to physical damage, most generally bruising.[2] Count the number X of unmarketable tomatoes in the sample.

This is not quite a binomial setting. The binomial distribution assumes that the SRS of size 10 is a series of 10 independent selections of a single tomato and that the probability of selecting an unmarketable tomato is the same for each selection. Removing one tomato changes the proportion of unmarketable tomatoes remaining in the shipment. So the probability that the second tomato chosen is unmarketable changes when we know whether the first is unmarketable or not. (Notice that this also violates the assumption of independence because the probability that a tomato selected at random is independent depends on the outcomes of the tomatoes previously selected.) However, removing one tomato from a shipment of 10,000 changes the makeup of the remaining 9999 tomatoes very little. In practice, the distribution of X is very close to the binomial distribution with $n = 10$ and $p = 0.11$.

Example 14.3 shows how we can use the binomial distributions in the statistical setting of selecting an SRS. When the population is much larger than the sample, a count of successes in an SRS of size n has approximately the binomial distribution with n equal to the sample size and p equal to the proportion of successes in the population.

Sampling Distribution of a Count

Choose an SRS of size n from a population with proportion p of successes. When the sample size is less than 5% of the population size, the count X of successes in the sample has approximately the binomial distribution with parameters n and p.

In Example 14.3, the sample size of 10 is much less than 5% of the population size of 10,000, so we can safely act as if the number of unmarketable tomatoes in our sample has approximately a binomial distribution. If you are taking a sample of 20 students from your statistics class of 100 students to estimate the proportion in your class who live off campus, the count X of the number of students in your sample who live off campus does not have approximately the binomial distribution as you are sampling more than 5% of the population.

APPLY YOUR KNOWLEDGE

In each of Exercises 14.1 to 14.3, X is a count. Does X have a binomial distribution? Give your reasons in each case.

14.1 Response Rates for Random Digit Dialing. When an opinion poll uses random digit dialing to select respondents for polls, the response rate (the percentage who actually provide a usable response to the poll) is approximately 10% for people contacted by cell phone.[3] A pollster dials 20 cell phone numbers. X is the number that respond to the pollster.

14.2 Response Rates for Random Digit Dialing. When an opinion poll uses random digit dialing to select respondents for polls, the response rate is approximately 10% for people contacted by cell phone. You watch a pollster dial cell numbers that have been selected in this manner. X is the number of calls dialed before the pollster obtains a second response to the poll.

14.3 Boxes of Tiles Boxes of six-inch slate flooring tile contain 40 tiles per box. The count X is the number of cracked tiles in a box. You have noticed that most boxes contain no cracked tiles, but if there are cracked tiles in a box, then there are usually several.

14.4 Disabled Adults in Canada. Statistics Canada reports that 22.3% of adult Canadians (aged 15 and older) report being limited in their daily activities due to a disability.[4] If you take an SRS of 4000 Canadians aged 15 and over, what is the approximate distribution of the number in your sample who report being limited in their daily activities due to a disability? Explain why the approximation is valid in this situation.

14.3 Binomial Probabilities

We can find a formula for the probability that a binomial random variable takes any value by adding probabilities for the different ways of getting exactly that many successes in n observations. Here is an example that illustrates the idea.

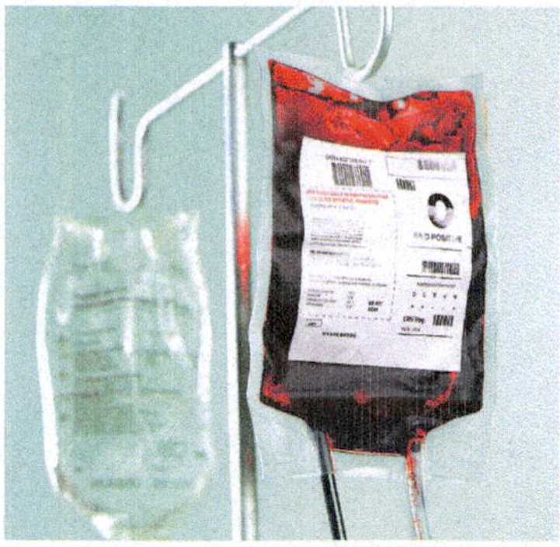

Peter Dazeley/Getty Images

EXAMPLE 14.4 Inheriting Blood Type

The blood types of successive children born to the same parents are independent and have fixed probabilities that depend on the genetic makeup of the parents. Each child born to a particular set of parents has probability 0.25 of having blood type O. If these parents have five children, what is the probability that exactly two of them have type O blood?

The count of children with type O blood is a binomial random variable X with $n = 5$ tries and probability $p = 0.25$ of a success on each try. We want $P(X = 2)$.

Because the method doesn't depend on the specific example, let's use "S" for success and "F" for failure for short, with a success representing type O blood. Do the work in two steps.

Step 1. Find the probability that a specific two of the five tries—say, the first and the third—give successes. This is the outcome SFSFF. Because tries are independent, the multiplication rule for independent events applies. The probability we want is

$$\begin{aligned} P(\text{SFSFF}) &= P(S)P(F)P(S)P(F)P(F) \\ &= (0.25)(0.75)(0.25)(0.75)(0.75) \\ &= (0.25)^2(0.75)^3 \end{aligned}$$

Step 2. Observe that *any one arrangement* of two Ss and three Fs has this same probability. This is true because we multiply together 0.25 twice and 0.75 three times whenever we have two Ss and three Fs. The probability that $X = 2$ is the probability of getting two Ss and three Fs in any arrangement whatsoever. Here are all the possible arrangements:

SSFFF SFSFF SFFSF SFFFS FSSFF

FSFSF FSFFS FFSSF FFSFS FFFSS

There are 10 of them, all with the same probability. The overall probability of two successes is therefore

$$P(X = 2) = 10(0.25)^2(0.75)^3 = 0.2637$$

The pattern of this calculation works for any binomial random variable. To use it, we must count the number of arrangements of k successes in n observations. We use the following fact to do the counting without actually listing all the arrangements.

STATISTICS IN YOUR WORLD

What Looks Random?

Toss a coin six times and record heads (H) or tails (T) on each toss. Which of these outcomes is more probable: HTHTTH or TTTHHH? Almost everyone says that HTHTTH is more probable because TTTHHH does not "look random." In fact, both are equally probable. That heads has probability 0.5 says that about half of a very long sequence of tosses will be heads. It doesn't say that heads and tails must come close to alternating in the short run. The coin doesn't know what past outcomes were, and it can't try to create a balanced sequence.

Binomial Coefficient

The number of ways of arranging k successes among n observations is given by the **binomial coefficient**

$$\binom{n}{k} = \frac{n!}{k!(n-k)!}$$

for $k = 0, 1, 2, \ldots, n$.

The formula for binomial coefficients uses the *factorial* notation. For any positive whole number n, its **factorial $n!$** is

$$n! = n \times (n-1) \times (n-2) \times \cdots \times 3 \times 2 \times 1$$

In addition, we define $0! = 1$.

The larger of the two factorials in the denominator of a binomial coefficient will cancel much of the $n!$ in the numerator. For example, the binomial coefficient we need for Example 14.4 is

$$\begin{aligned} \binom{5}{2} &= \frac{5!}{(2!)(3!)} \\ &= \frac{(5)(4)(3)(2)(1)}{(2)(1) \times (3)(2)(1)} \\ &= \frac{(5)(4)}{(2)(1)} = \frac{20}{2} = 10 \end{aligned}$$

The binomial coefficient $\binom{5}{2}$ *is not related to the fraction* $\frac{5}{2}$. A helpful way to remember its meaning is to read it as "5 choose 2." Binomial coefficients have many uses, but we are interested in them only as an aid to finding binomial probabilities. The binomial coefficient $\binom{n}{k}$ counts the number of different ways in which k successes can be arranged among n observations. The *binomial probability* $P(X = k)$ is this count multiplied by the probability of any one specific arrangement of the k successes. Here is the result we seek.

Binomial Probability

If X has the binomial distribution with n observations and probability p of success on each observation, the possible values of X are $0, 1, 2, \ldots, n$. If k is any one of these values,

$$P(X = k) = \binom{n}{k} p^k (1-p)^{n-k}$$

EXAMPLE 14.5 Inspecting Tomatoes

The number X of unmarketable tomatoes in Example 14.3 has approximately the binomial distribution with $n = 10$ and $p = 0.11$.

The probability that the sample contains no more than one unmarketable tomato is

$$\begin{aligned} P(X \le 1) &= P(X = 1) + P(X = 0) \\ &= \binom{10}{1}(0.11)^1(0.89)^9 + \binom{10}{0}(0.11)^0(0.89)^{10} \\ &= \frac{10!}{(1!)(9!)}(0.11)(0.3504) + \frac{10!}{(0!)(10!)}(1)(0.3118) \\ &= (10)(0.11)(0.3504) + (1)(1)(0.3118) \\ &= 0.3854 + 0.3118 = 0.6972 \end{aligned}$$

This calculation uses the facts that $0! = 1$ and that $a^0 = 1$ for any number a other than 0. We see that about 70% of all samples will contain no more than one unmarketable tomato. In fact, about 31% of the samples will contain no unmarketable tomatoes. A sample of size 10 cannot be trusted to alert the distributor to the presence of unacceptable tomatoes in the shipment.

The complement rule described in Chapter 12 can make the computation of certain binomial probabilities simpler. For example, the probability that the sample contains at least one unmarketable tomato is

$$\begin{aligned} P(X \ge 1) &= P(X = 1) + P(X = 2) + \cdots + P(X = 10) \\ &= 1 - P(X = 0) \\ &= 1 - 0.3118 = 0.6882 \end{aligned}$$

When computing binomial probabilities by hand, it is useful to keep the complement rule in mind.

14.4 Examples of Technology

The binomial probability formula is awkward to use unless the number of observations n is quite small. You can find tables of binomial probabilities $P(X = k)$ and cumulative probabilities $P(X \le k)$ for selected values of n and p, but the most efficient way to do binomial calculations is to use technology. Figure 14.1 shows output for the calculation in Example 14.5 from a graphing calculator, two statistical programs, and a spreadsheet program. We asked all four to give cumulative probabilities. The calculator, Minitab, and CrunchIt! have menu entries for binomial cumulative probabilities. Excel has no menu entry, but the worksheet function BINOM.DIST is available. All the outputs agree with the result 0.6972 for Example 14.5.

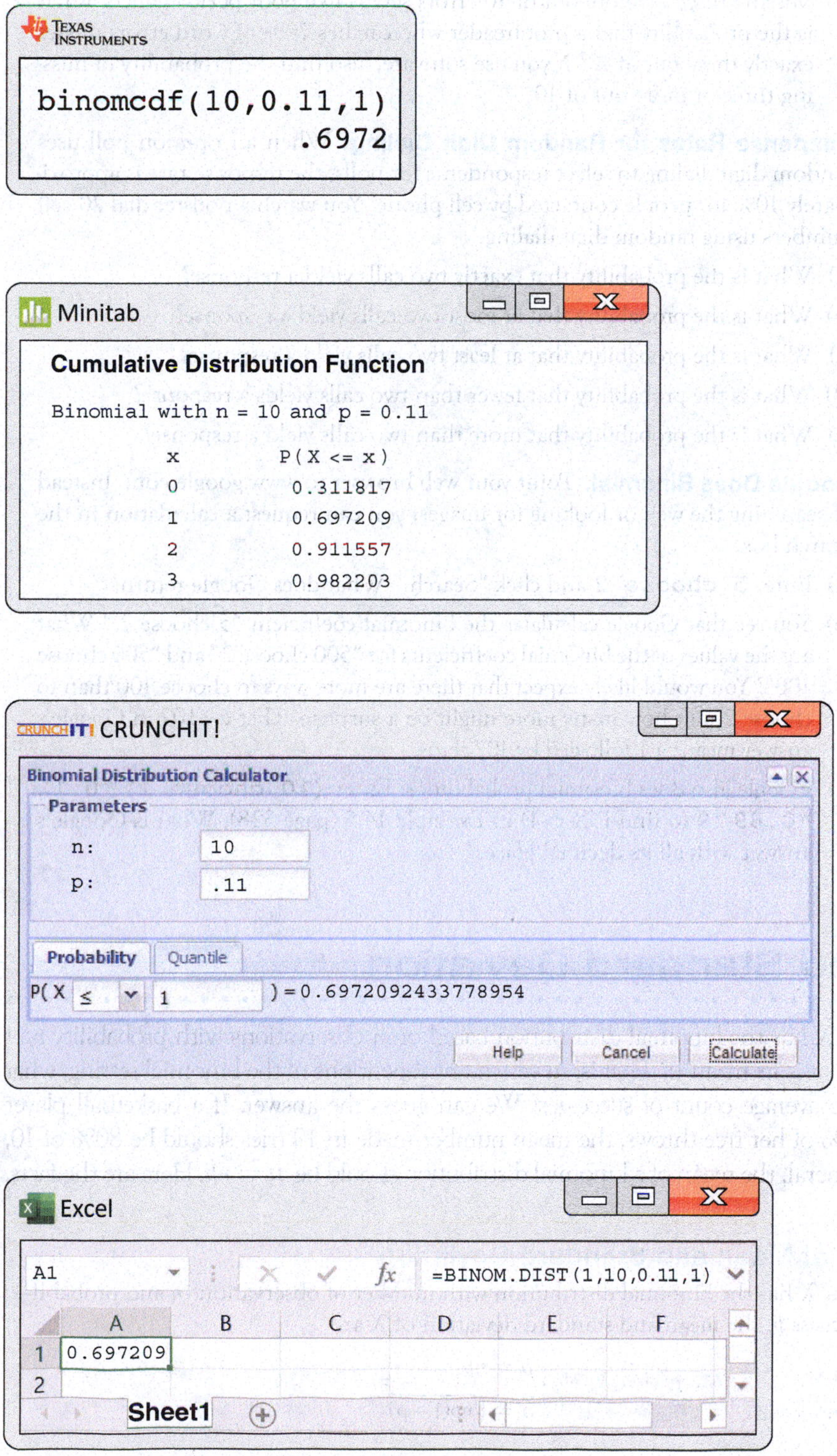

FIGURE 14.1
The binomial probability $P(X \leq 1)$ for Example 14.5: output from a graphing calculator, two statistical programs, and a spreadsheet program.

APPLY YOUR KNOWLEDGE

14.5 Proofreading. Typing errors in text are either nonword errors (as when "the" is typed as "teh") or word errors that result in real but incorrect words. Spell-checking software will catch nonword errors but not word errors. Human proofreaders catch 70% of word errors. You ask a fellow student to proofread an essay in which you have deliberately made 10 word errors.

(a) If the student matches the usual 70% rate, what is the distribution of the number of errors caught? What is the distribution of the number of errors missed?

(b) Missing three or more out of 10 errors seems like poor performance. What is the probability that a proofreader who catches 70% of word errors misses exactly three out of 10? If you use software, also find the probability of missing three or more out of 10.

14.6 Response Rates for Random Digit Dialing. When an opinion poll uses random digit dialing to select respondents for polls, the response rate is approximately 10% for people contacted by cell phone. You watch a pollster dial 20 cell numbers using random digit dialing.

(a) What is the probability that exactly two calls yield a response?

(b) What is the probability that at most two calls yield a response?

(c) What is the probability that at least two calls yield a response?

(d) What is the probability that fewer than two calls yields a response?

(e) What is the probability that more than two calls yield a response?

14.7 Google Does Binomial. Point your web browser to www.google.com. Instead of searching the web or looking for images, you can request a calculation in the Search box.

(a) Enter `5 choose 2` and click "Search." What does Google return?

(b) You see that Google calculates the binomial coefficient "5 choose 2." What are the values of the binomial coefficients for "500 choose 2" and "500 choose 100"? You would likely expect that there are more ways to choose 100 than to choose 2, but how many more might be a surprise. That e+107 in Google's answer means a 1 followed by 107 zeros.

(c) Google also does binomial probabilities. Enter `(10 choose 1)*0.11*0.89^9` to find $P(X = 1)$ in Example 14.5 (page 328). What is Google's answer with all its decimal places?

14.5 Binomial Mean and Standard Deviation

If a count X has the binomial distribution based on n observations with probability p of success, what is its mean μ? That is, in very many repetitions of the binomial setting, what will be the average count of successes? We can guess the answer. If a basketball player makes 80% of her free throws, the mean number made in 10 tries should be 80% of 10, or 8. In general, the mean of a binomial distribution should be $\mu = np$. Here are the facts.

Binomial Mean and Standard Deviation

If a count X has the binomial distribution with number of observations n and probability of success p, the mean and standard deviation of X are

$$\mu = np$$
$$\sigma = \sqrt{np(1-p)}$$

Remember that these short formulas are good only for binomial distributions. They can't be used for other distributions.

EXAMPLE 14.6 Inspecting Tomatoes

Continuing Example 14.5 (page 328), the count X of unmarketable tomatoes is binomial with $n = 10$ and $p = 0.11$. The histogram in Figure 14.2 displays this probability distribution. (Because probabilities are long-run proportions, using probabilities as the heights of the bars shows what the distribution of X would be in very many repetitions.) The distribution is strongly right-skewed. Although X can take any whole-number value from zero to 10, the probabilities of values larger than five are so small that they do not appear in the histogram.

The mean and standard deviation of the binomial distribution in Figure 14.2 are

$$\mu = np$$
$$= (10)(0.11) = 1.1$$
$$\sigma = \sqrt{np(1-p)}$$
$$= \sqrt{(10)(0.11)(0.89)} = \sqrt{0.979} = 0.9894$$

The mean is marked on the probability histogram in Figure 14.2.

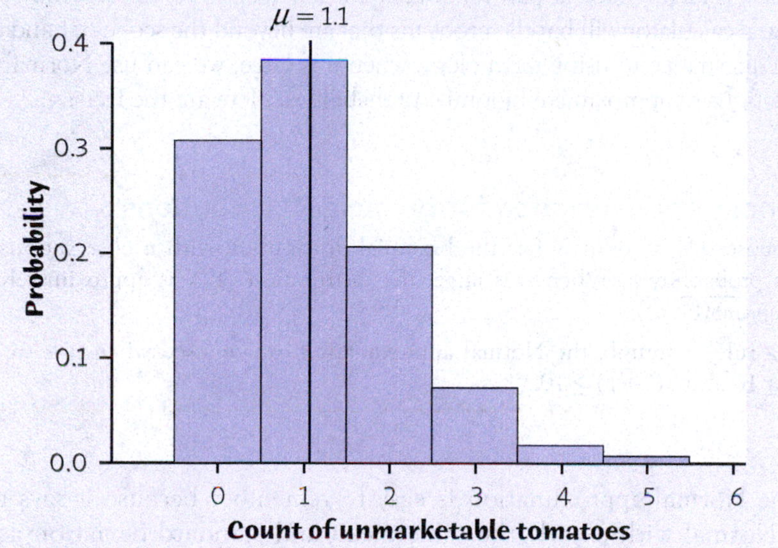

FIGURE 14.2
Probability histogram for the binomial distribution with $n = 10$ and $p = 0.11$, for Example 14.6.

STATISTICS IN YOUR WORLD

Randomness Turns Silver to Bronze

After many charges of favoritism by judges, the rules for scoring international figure skating competitions changed in 2004. The big change was that 12 judges scored all performances, then scores from three judges chosen at random were dropped for each part of the program. So there are $\binom{12}{9} = 220$ possible panels of nine judges for (say) the "Free Skate," and these panels would have slightly different scores. Result: at the 2006 World Figure Skating Championships, the Russian pair Maria Petrova and Alexei Tikhonov received the bronze medal when the consensus of all 12 judges would have given them the silver medal. The system was changed again in 2008 and now uses a panel of nine judges and eliminates the random dropping of scores. The high and low scores among the nine judges are dropped, and the performance score is simply the average of the remaining seven judges.

APPLY YOUR KNOWLEDGE

14.8 Response Rates for Random Digit Dialing. When an opinion poll uses random digit dialing to select respondents for polls, the response rate is approximately 10% for people contacted by cell phone. You watch a pollster dial 20 cell numbers many times using random digit dialing.

(a) What is the mean number of calls that yield a response?

(b) What is the standard deviation σ of the count of calls that yield a response?

(c) Suppose that the probability of getting a response were $p = 0.05$. How does this new p affect the standard deviation? What would be the standard deviation if $p = 0.01$? What does your work show about the behavior of the standard deviation of a binomial distribution as the probability of success gets closer to zero?

14.9 Proofreading. Return to the proofreading setting of Exercise 14.5 (page 330).

(a) If X is the number of word errors missed, what is the distribution of X? If Y is the number of word errors caught, what is the distribution of Y?

(b) What is the mean number of errors caught? What is the mean number of errors missed? The mean counts of successes and of failures always add to n, the number of observations.

(c) What is the standard deviation of the number of errors caught? What is the standard deviation of the number of errors missed? The standard deviations of the count of successes and the count of failures are always the same.

14.6 The Normal Approximation to Binomial Distributions

It isn't practical to use the formula for binomial probabilities when the number of observations n is large. (Look at part (b) of Exercise 14.7 (page 330) to see why.) Software or a graphing calculator will handle problems that are beyond the scope of hand calculation. As an alternative to using technology, when n is large, we can use Normal probability calculations to approximate binomial probabilities. Here are the facts:

Normal Approximation for Binomial Distributions

Suppose that a count X has the binomial distribution with n observations and success probability p. When n is large, the distribution of X is approximately Normal, $N(np, \sqrt{np(1-p)})$.

As a rule of thumb, the Normal approximation can be used when n is so large that $np \geq 10$ and $n(1-p) \geq 10$.

The Normal approximation is easy to remember because it says to act as if X is Normal with exactly the same mean and standard deviation as the binomial. The accuracy of the Normal approximation improves as the sample size n increases. It is most accurate for any fixed n when p is close to 1/2 and least accurate when p is near 0 or 1. This is why the rule of thumb in the box depends on p as well as n.

EXAMPLE 14.7 Living with Parents

Beginning in 2016 and for the first time since 1880, young adults aged 18 to 34 are more likely to be living with a parent than they are to be living with a romantic partner in their own home. Although approximately 32% of young adults now live with a parent, this varies by gender with 34% of men living with a parent versus 29% of women.[5] In a nationwide random sample of 1200 young adults, what is the probability that 400 or more live with a parent?

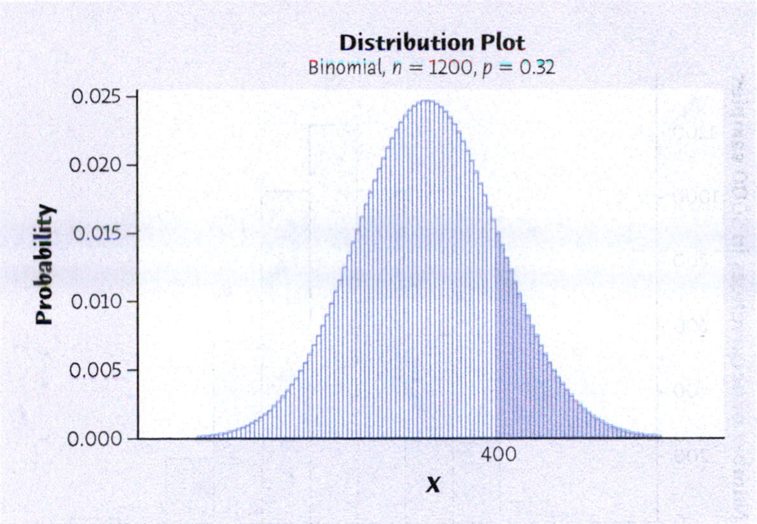

FIGURE 14.3 Probability histogram for the binomial distribution with $n = 1200$, $p = 0.32$. The bars at and above 400 are shaded to highlight the probability of getting at least 400 successes. The shape of this binomial probability distribution closely resembles a Normal curve.

Because there are over 60 million young adults in the United States, the sample size of 1200 is much less than 5% of the population. So the number in our sample who live with a parent is a random variable X having the binomial distribution with $n = 1200$ and $p = 0.32$. To find the probability $P(X \geq 400)$ that at least 400 of the young adults in the sample are living with a parent, we must add the binomial probabilities of all outcomes from $X = 400$ to $X = 1200$. Figure 14.3 is a probability histogram of this binomial distribution, from Minitab. As the Normal approximation suggests, the shape of the distribution looks Normal. The probability we want is the sum of the heights of the shaded bars. Here are three ways to find this probability:

1. Use technology. Statistical software can find the exact binomial probability. In most cases, software finds cumulative probabilities $P(X \leq x)$. So start by writing

$$P(X \geq 400) = 1 - P(X \leq 399)$$

Here is Minitab's answer for $P(X \leq 399)$:

```
Binomial with n = 1200 and p = 0.32

  x  p(X<=x)
399  0.831350
```

The probability we want is $1 - 0.831350 = 0.168650$, correct to six decimal places.

2. Simulate a large number of samples. Figure 14.4 displays a histogram of the counts X from 5000 samples of size 1200 when the truth about the population is $p = 0.32$. The simulated distribution, like the exact distribution in Figure 14.3, looks Normal. Because 832 of these 5000 samples have X at least 400, the probability estimated from the simulation is

$$P(X \geq 400) = \frac{832}{5000} = 0.1664$$

This estimate misses the true probability by about 0.002. The law of large numbers says that the results of such simulations always get closer to the true probability as we simulate more and more samples.

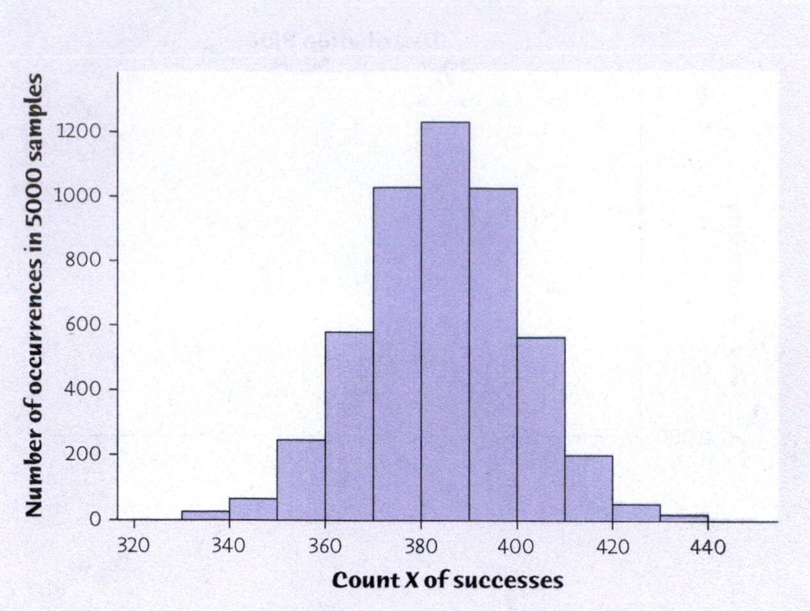

FIGURE 14.4
Histogram of 5000 simulated binomial counts ($n = 1200$, $p = 0.32$).

3. Both of the previous methods require software. We can avoid the need for software by using the Normal approximation.

EXAMPLE 14.8 Normal Approximation of a Binomial Probability

Approximate the count X in Example 14.7 by using the Normal distribution with the same mean and standard deviation as the binomial distribution:

$$\mu = np = (1200)(0.32) = 384$$

$$\sigma = \sqrt{np(1-p)} = \sqrt{(1200)(0.32)(0.68)} = 16.159$$

Standardizing X gives a standard Normal variable Z. The probability we want is

$$\begin{aligned} P(X \geq 400) &= P\left(\frac{X - 384}{16.159} \geq \frac{400 - 384}{16.159}\right) \\ &= P(Z \geq 0.99) \\ &= 1 - 0.8389 = 0.1611 \end{aligned}$$

The Normal approximation 0.1611 misses the true probability calculated in Example 14.7 by about 0.007. The accuracy of the Normal approximation can often be improved through the use of a *continuity correction* (see Exercise 14.43, page 341).

With the availability of technology for binomial calculations, including online binomial calculators (see Web Exercise 14.45, available online), there is little reason not to use exact calculations for binomial probabilities. The close agreement of simulation and the Normal approximation to the exact calculation does not imply that we should use these when exact calculations are available. The reason for describing them in this simple case is that simulation and the Normal approximation are both useful in other settings when exact calculations become too complex, even with technology. Chapter 32 describes simulation-based statistical methods, and Chapters 22, 23, and 28 use the Normal distribution extensively as an approximation to the distributions encountered in these chapters.

The *Normal Approximation to Binomial* applet shows in visual form how well the Normal approximation fits the binomial distribution for any n and p. You can slide n and watch the approximation get better. Whether or not the Normal approximation is satisfactory depends on how accurate your calculations need to be. For most statistical purposes, great accuracy is not required. Our rule of thumb for use of the Normal approximation reflects this judgment.

applet

APPLY YOUR KNOWLEDGE

14.10 Using Benford's Law. According to Benford's law (Example 12.7, page 281) the probability that the first digit of the amount of a randomly chosen invoice is a 1 or a 2 is 0.477. You examine 90 invoices from a vendor and find that 29 have first digits 1 or 2. If Benford's law holds, the count of 1s and 2s will have the binomial distribution with $n = 90$ and $p = 0.477$. Too few 1s and 2s suggests fraud. What is the approximate probability of 29 or fewer 1s and 2s if the invoices follow Benford's law? Do you suspect that the invoice amounts are not genuine?

14.11 College Admissions. A small liberal arts college in Ohio would like to have an entering class of 500 students next year. Past experience shows that about 40% of the students admitted will decide to attend. The college is planning to admit 1250 students. Suppose that students make their decisions independently and that the probability is 0.40 that a randomly chosen student will accept the offer of admission.

(a) What are the mean and standard deviation of the number of students who accept the admissions offer from this college?

(b) Using the Normal approximation, what is the probability that the college gets more students than it wants? Check that you can safely use the approximation.

(c) Use software or an online binomial calculator to compute the exact probability that the college gets more students than it wants. How good is the approximation in part (b)?

(d) To decrease the probability of getting more students than are wanted, does the college need to increase or decrease the number of students it admits? Using software or an online binomial calculator, what is the largest number of students that the college can admit if administrators want the exact probability of getting more students than they want to be no larger than 5%?

14.12 Checking for Survey Errors. One way of checking the effect of undercoverage, nonresponse, and other sources of error in a sample survey is to compare the sample with known facts about the population. About 24% of the Canadian population is first generation—that is, they were born outside Canada.[6] The number X of first-generation Canadians in random samples of 1500 persons should therefore vary with the binomial ($n = 1500$, $p = 0.24$) distribution.

(a) What are the mean and standard deviation of X?

(b) Use the Normal approximation to find the probability that a sample will contain between 340 and 390 first-generation Canadians. Check that you can safely use the approximation.

CHAPTER 14 SUMMARY

- A count X of successes has a **binomial distribution** under the conditions of the **binomial setting:** there are n observations, the observations are independent of each other, each observation results in a success or a failure, and each observation has the same probability p of a success.
- The binomial distribution with n observations and probability p of success gives a good approximation to the sampling distribution of the count of successes in an SRS of size n from a large population containing proportion p of successes.
- If X has the binomial distribution with parameters n and p, the possible values of X are the whole numbers 0, 1, 2, . . . , n. The **binomial probability** that X takes any of these values is

$$P(X = k) = \binom{n}{k} p^k (1-p)^{n-k}$$

- Binomial probabilities in practice are best found using software.
- The **binomial coefficient**

$$\binom{n}{k} = \frac{n!}{k!(n-k)!}$$

counts the number of ways k successes can be arranged among n observations. Here the **factorial** $n!$ is

$$n! = n \times (n-1) \times (n-2) \times \cdots \times 3 \times 2 \times 1$$

for positive whole numbers n, and $0! = 1$.
- The mean and standard deviation of a binomial count X are

$$\mu = np$$
$$\sigma = \sqrt{np(1-p)}$$

- The **Normal approximation to the binomial distribution** says that if X is a count having the binomial distribution with parameters n and p, then when n is large, X is approximately $N(np, \sqrt{np(1-p)})$. Use this approximation only when $np \geq 10$ and $n(1-p) \geq 10$.

CHECK YOUR SKILLS

14.13 Larry reads that half of all super jumbo eggs contain double yolks. So he always buys super jumbo eggs and uses two whenever he cooks. If eggs do or don't contain two yolks independently of each other, the number of eggs with double yolks when Larry uses two chosen at random has the distribution

(a) binomial with $n = 2$ and $p = 1/2$.

(b) binomial with $n = 2$ and $p = 1/3$.

(c) binomial with $n = 3$ and $p = 1/2$.

14.14 In the previous exercise, the probability that at least one of Larry's two eggs contains double yolks is about

(a) 0.75. (b) 0.50. (c) 0.33.

14.15 In a group of 10 college students, three are psychology majors. You choose three of the 10 students at random and ask their major. The distribution of the number of psychology majors you choose is

(a) binomial with $n = 10$ and $p = 0.3$.

(b) binomial with $n = 3$ and $p = 0.3$.

(c) not binomial.

In a test for ESP (extrasensory perception), a subject is told that cards that the experimenter can see, but that the subject cannot see, contain either a star, a circle, a wave, or a square. As the experimenter looks at each of 4 cards in turn, the subject names the shape on the card. A subject who is just guessing has probability 0.25 of guessing correctly on each card. Questions 14.16 to 14.18 use this information.

14.16 If the subject guesses two shapes correctly and two incorrectly, in how many ways can you arrange the sequence of correct and incorrect guesses?

(a) $\binom{3}{2} = 3$ (b) $\binom{4}{2} = 12$ (c) $\binom{4}{2} = 6$

14.17 Assume the subject's guesses are independent of each other. The probability that the subject guesses the shape correctly on the first and last cards but incorrectly on the other two cards is about

(a) 0.211. (b) 0.045. (c) 0.035.

14.18 Assume the subject's guesses are independent of each other. The probability that the subject guesses the shape correctly exactly half the time is about

(a) 0.250. (b) 0.211. (c) 0.035.

Each entry in a table of random digits like Table B has probability 0.1 of being any of the 10 digits 0 to 9, and digits are independent of each other. Questions 14.19 to 14.21 use this setting.

14.19 The probability of an entry being either a 0, a 1, or a 2 is

(a) 0.1. (b) 0.2. (c) 0.3.

14.20 Each line in Table B has 40 digits. The number of times a 0, a 1, or a 2 occurs in *two lines* of the table is

(a) binomial with $n = 80$ and $p = 0.3$.

(b) binomial with $n = 40$ and $p = 0.3$.

(c) binomial with $n = 30$ and $p = 0.1$.

14.21 The mean number of times a 0, a 1, or a 2 occurs in *two lines* of the table is

(a) 24. (b) 12. (c) 3.

CHAPTER 14 EXERCISES

14.22 **Binomial Setting?** In each of the following situations, is it reasonable to use a binomial distribution for the random variable X? Give reasons for your answer in each case.

(a) An auto manufacturer chooses one car from each hour's production for a detailed quality inspection. One variable recorded is the count X of finish defects (dimples, ripples, etc.) in the car's paint.

(b) The pool of potential jurors for a murder case contains 100 persons chosen at random from the adult population of a large city. Each person in the pool is asked whether he or she opposes the death penalty; X is the number who say Yes.

(c) Joe buys a ticket in his state's Pick 3 lottery game every week; X is the number of times in a year that he wins a prize.

14.23 **Binomial Setting?** A binomial distribution will be approximately correct as a model for one of these two sports settings and not for the other. Explain why by briefly discussing both settings.

(a) A National Football League kicker has made 90% of his field goal attempts in the past. This season he attempts 20 field goals. The attempts differ widely in distance, angle, wind, and so on.

Lachlan Cunningham/Getty Images

(b) A National Basketball Association player has made 90% of his free-throw attempts in the past. This season he takes 150 free throws. Basketball free throws are always attempted from 15 feet away from the basket, with no interference from other players.

14.24 **Antibiotic Resistance.** According to CDC estimates, at least 2.8 million people in the United States are sickened each year with antibiotic-resistant infections, and at least 35,000 die as a result. Antibiotic resistance occurs when disease-causing microbes become resistant to antibiotic drug therapy. Because this resistance is typically genetic and transferred to the next generations of microbes, it is a very serious public health problem. Of the infections considered most serious by the CDC, gonorrhea has an estimated 1.14 million new cases occurring annually, and approximately 50% of those cases are resistant to any antibiotic.[7] A public health clinic in California sees eight patients with gonorrhea in a given week.

(a) What is the distribution of X, the number of these eight cases that are resistant to any antibiotic?

(b) What are the mean and standard deviation of X?

(c) Find the probability that exactly one of the cases is resistant to any antibiotic. What is the probability that at least one case is resistant to any antibiotic? (*Hint:* It is easier to first find the probability that exactly zero of the eight cases were resistant.)

14.25 **Hot Spot.** Hot Spot is a California lottery game. Players pick 1 to 10 Spots (sets of numbers, each from 1 to 80) that they want to play per draw. For example, if you select a 4 Spot, you play four numbers. The lottery draws 20 numbers, each from 1 to 80. Your prize is based on how many of the numbers you picked match one of those selected by the lottery. The odds of winning depend on the number of Spots you choose to play. For example, the overall odds of winning some prize in 4 Spot is approximately 0.256.

You decide to play the 4 Spot game and buy 5 tickets. Let X be the number of tickets that win some prize.

(a) X has a binomial distribution. What are n and p?

(b) What are the possible values that X can take?

(c) Find the probability of each value of X. Draw a probability histogram for the distribution of X. (See Figure 14.2 on page 331 for an example of a probability histogram.)

(d) What are the mean and standard deviation of this distribution? Mark the location of the mean on your histogram.

14.26 Roulette—Betting on Red. A roulette wheel has 38 slots, numbered 0, 00, and 1 to 36. The slots 0 and 00 are colored green, 18 of the others are red, and 18 are black. The dealer spins the wheel and at the same time rolls a small ball along the wheel in the opposite direction. The wheel is carefully balanced so that the ball is equally likely to land in any slot when the wheel slows. Gamblers can bet on various combinations of numbers and colors.

(a) If you bet on "red," you win if the ball lands in a red slot. What is the probability of winning with a bet on red in a single play of roulette?

(b) You decide to play roulette four times, each time betting on red. What is the distribution of X, the number of times you win?

(c) If you bet the same amount on each play and win on exactly two of the four plays, then you will "break even." What is the probability that you will break even?

(d) If you win on *fewer than* two of the four plays, then you will lose money. What is the probability that you will lose money?

14.27 Aerosolized Vaccine for Measles. An aerosolized vaccine for measles was developed in Mexico and has been used on over 4 million children since 1980. Aerosolized vaccines have the advantages of being able to be administered by people without clinical training and do not cause injection-associated infections. The percentage of children developing an immune response to measles after receiving the subcutaneous injection of the vaccine is 95%, and for those receiving the aerosolized vaccine it is 85%.[8] There are 20 children to be vaccinated for measles using the aerosolized vaccine in a small rural village in India. We are going to count the number who developed an immune response to measles after vaccination.

(a) Explain why this is a binomial setting.

(b) What is the probability that at least one child does not develop an immune response to measles after receiving the aerosolized vaccine? What would be the probability that at least one child does not develop an immune response to measles if all children were vaccinated using the subcutaneous injection of the vaccine?

14.28 Roulette (continued). You decide to play roulette 200 times, each time betting the same amount on red. You will lose money if you win on fewer than 100 of the plays. Based on the information in Exercise 14.26, what is the probability that you will lose money? (Check that the Normal approximation is permissible and use it to find this probability. If your software allows, find the exact binomial probability and compare the two results.) In general, if you bet the same amount on red every time, you will lose money if you win on fewer than half of the plays. What do you think happens to the probability of making money the longer you continue to play?

14.29 Aerosolized Vaccine for Measles (continued). The aerosolized measles vaccine is to be used on a random sample of 100 children.

(a) Based on the information about the effectiveness of the aerosolized vaccine in Exercise 14.27, what is the probability that at least 90 of the children develop an immune response to measles after vaccination? Check that the Normal approximation is permissible and use it to find this probability. If your software allows, find the exact binomial probability and compare the two results.

(b) If the 100 children receive the subcutaneous injection, we can't use the Normal approximation to find this probability using the information about the effectiveness of the subcutaneous injection described in Exercise 14.27. Why not?

14.30 Does the Wall Street Culture Corrupt Bankers? Bank employees from a large international bank were recruited, and 67 were randomly assigned to a control group and the remaining 61 to a treatment group. All subjects first completed a short online survey. After answering some general filler questions, the members of the treatment group were asked seven questions about their professional background, such as "At which bank are you currently employed?" or "What is your function at this bank?" These are referred to as "identity priming" questions. The members of the control group were asked seven innocuous questions unrelated to their profession, such as "How many hours a week, on average, do you watch television?" After the survey, all subjects performed a coin tossing task that required tossing any coin 10 times and reporting the results online. They were told they would win $20 for each head tossed, for a maximum payoff of $200. Subjects were unobserved during the task, making it impossible to tell if a particular subject cheated. If the banking culture favors dishonest behavior, it was conjectured that it should be possible to trigger this behavior by reminding subjects of their profession.[9] Here are the results. The first line gives the possible number of heads on 10 tosses, and the next two lines give the number of subjects that reported tossing this number of heads for the control and treatment groups, respectively (for example, 16 control subjects reported getting 4 heads).

Number of heads	0	1	2	3	4	5	6	7	8	9	10
Control group	0	0	1	8	16	17	14	6	2	1	2
Treatment group	0	0	2	4	8	14	15	7	6	0	5

(a) Suppose that a subject tosses a fair coin and truthfully reports the number of heads. What is the distribution of the number of heads reported? What is the probability of collecting $160 or more?

(b) For a truthful subject, what would be the probability of doing better than chance—that is, tossing 6 or more heads? What proportion of subjects in the two groups reported tossing 6 or more heads?

(c) What does your result in (b) suggest about cheating in the two groups? (In Chapter 23 we will return to this example with more formal tools for comparing the groups.)

14.31 Genetics. According to genetic theory, the blossom color in the second generation of a certain cross of sweet peas should be red or white in a 3:1 ratio. That is, each plant has probability 3/4 of having red blossoms, and the blossom colors of separate plants are independent.

blickwinkel/Alamy Stock Photo

(a) What is the probability that exactly three out of four of these plants have red blossoms?

(b) What is the mean number of red-blossomed plants when 60 plants of this type are grown from seeds?

(c) What is the probability of obtaining at least 45 red-blossomed plants when 60 plants are grown from seeds? Use the Normal approximation. If your software allows, find the exact binomial probability and compare the two results.

14.32 False Positives in Testing for HIV. A rapid test for the presence in the blood of antibodies to HIV, the virus that causes AIDS, gives a positive result with probability about 0.004 when a person who is free of HIV antibodies is tested. A clinic tests 1000 people who are all free of HIV antibodies.

(a) What is the distribution of the number of positive tests?

(b) What is the mean number of positive tests?

(c) You cannot safely use the Normal approximation for this distribution. Explain why.

14.33 Hyundai Sales in 2018. Hyundai Motor America sold 677,946 vehicles in the United States in 2018, with the U.S.-built Elantra leading sales, with 200,415 cars sold. The other top-selling nameplates in 2018 were the Tucson, with 135,348 sold, the Santa Fe, with 123,989 sold, and the Sonata, with 105,118.[10] The company wants to undertake a survey of 2018 Hyundai buyers to ask them about satisfaction with their purchase.

(a) What proportion of the Hyundais sold in 2018 were Elantras?

(b) If Hyundai plans to survey a simple random sample of 1000 Hyundai buyers, what are the expected number and standard deviation of the number of Elantra buyers in the sample?

(c) What is the probability Hyundai will get fewer than 300 Elantra buyers in the sample?

14.34 Preference for the Middle? When choosing an item from a group, researchers have shown that an important factor influencing choice is the item's location. This occurs in varied situations, such as shelf positions when shopping, when filling out a questionnaire, and even when choosing a preferred candidate during a presidential debate. Experimenters displayed five identical pairs of white socks by attaching them vertically to a blue background, which was then mounted on an easel for viewing. One hundred participants from the University of Chester were used as subjects and asked to choose their preferred pair of socks.[11]

(a) Suppose each subject selects a preferred pair of socks at random. What is the probability that a subject would choose the pair of socks in the center position? Assuming that the subjects make their choices independently, what is the distribution of X, the number of subjects among the 100 who would choose the pair of socks in the center position?

(b) What is the mean of the number of subjects who would choose the pair of socks in the center position? What is the standard deviation?

(c) In choice situations of this type, subjects often exhibit the "center stage effect," which is a tendency to choose the item in the center. In this experiment, 34 subjects chose the pair of socks in the center. What is the probability that 34 or more subjects would choose the item in the center if each subject were selecting the preferred pair of socks at random? Use the Normal approximation. If your software allows, find the exact binomial probability and compare the two results.

(d) Do you feel that this experiment supports the "center stage effect"? Explain briefly.

14.35 Multiple-Choice Tests. Here is a simple probability model for multiple-choice tests. Suppose each student has probability p of correctly answering a question chosen at random from a universe of possible questions. (A strong student has a higher p than a weak student.) Answers to different questions are independent.

(a) Stacey is a good student for whom $p = 0.75$. Use the Normal approximation to find the probability that Stacey scores between 70% and 80% on a 100-question test.

(b) If the test contains 250 questions, what is the probability that Stacey will score between 70% and 80%? You see that Stacey's score on the longer test is more likely to be close to her "true score."

14.36 Is This Coin Balanced? While he was a prisoner of war during World War II, John Kerrich tossed a coin 10,000 times. He got 5067 heads. If the coin is perfectly balanced, the probability of a head is 0.5. Is there reason to think that Kerrich's coin was not balanced? To answer this question, find the probability that tossing a balanced coin 10,000 times would give a count of heads at least this far from 5000 (that is, at least 5067 heads or no more than 4933 heads).

14.37 Binomial Variation. Never forget that probability describes only what happens in the long run. Example 14.5 (page 328) concerns the count of unmarketable tomatoes in inspection samples of size 10. The count has the binomial distribution with $n = 10$ and $p = 0.11$. The *Probability* applet simulates inspecting an SRS of 10 tomatoes if you set the probability of heads to 0.11, toss 10 times, and let each head stand for an unmarketable tomato.

(a) The mean number of unmarketable tomatoes in a sample is 1.1. Drag the "Re-Toss" toggle button repeatedly to simulate 20 samples of size 10. How many unmarketable tomatoes did you find in each sample? How close to the mean 1.1 is the average number of unmarketable tomatoes in these samples?

(b) Example 14.5 shows that the probability of exactly one unmarketable tomato is 0.3854. How close to the probability is the proportion of the 20 samples that have exactly one unmarketable tomato?

Whooping Cough. *Whooping cough (pertussis) is a highly contagious bacterial infection that was a major cause of childhood deaths before the development of vaccines. About 80% of unvaccinated children who are exposed to whooping cough will develop the infection, as opposed to only about 5% of vaccinated children. Exercises 14.38 to 14.41 are based on this information.*

14.38 Vaccination at Work. A group of 20 children at a nursery school are exposed to whooping cough by playing with an infected child.

(a) If all 20 have been vaccinated, what is the mean number of new infections? What is the probability that no more than two of the 20 children develop infections?

(b) If none of the 20 has been vaccinated, what is the mean number of new infections? What is the probability that 18 or more of the 20 children develop infections?

14.39 A Whooping Cough Outbreak. In 2007, Bob Jones University in Greenville, South Carolina, ended its fall semester a week early because of a whooping cough outbreak; 158 students were isolated and another 1200 given antibiotics as a precaution.[12] Authorities react strongly to whooping cough outbreaks because the disease is very contagious. Because the effect of childhood vaccination often wears off by late adolescence, treat the Bob Jones students as if they were unvaccinated. It appears that about 1400 students were exposed. What is the probability of at least 75% of these students developing infections if not treated? (Fortunately, whooping cough is much less serious after infancy than during infancy.)

14.40 A Mixed Group: Means. Twenty children at a nursery school are exposed to whooping cough by playing with an infected child. Of these children, 17 have been vaccinated and three have not.

(a) What is the distribution of the number of new infections among the 17 vaccinated children? What is the mean number of new infections?

(b) What is the distribution of the number of new infections among the three unvaccinated children? What is the mean number of new infections?

(c) Add your means from parts (a) and (b). This is the mean number of new infections among all 20 exposed children.

14.41 A Mixed Group: Probabilities. We would like to find the probability that exactly two of the 20 exposed children in the previous exercise develop whooping cough.

(a) One way to get two infections is to get one among the 17 vaccinated children and one among the three unvaccinated children. Find the probability of exactly one infection among the 17 vaccinated children. Find the probability of exactly one infection among the three unvaccinated children. These events are independent. What is the probability of exactly one infection in each group?

(b) Write down all the ways in which two infections can be divided between the two groups of children. Follow the pattern of part (a) to find the probability of each of these possibilities. Add all of your results (including the result of part (a)) to obtain the probability of exactly two infections among the 20 children.

14.42 Estimating π from Random Numbers. Kenyon College student Eric Newman used basic geometry to evaluate software random number generators as part of a summer research project. He generated 2000 independent random points (X, Y) in the unit square. (That is, X and Y are independent random numbers between 0 and 1, each having the density function illustrated in Figure 12.5 (page 284). The probability that (X, Y) falls in any region within the unit square is the area of the region.)[13]

(a) Sketch the unit square, the region of possible values for the point (X, Y).

(b) The set of points (X, Y) where $X^2 + Y^2 < 1$ describes a circle of radius 1. Add this circle to your sketch in part (a) and label the intersection of the two regions A.

(c) Let T be the total number of the 2000 points that fall into the region A. T follows a binomial distribution. Identify n and p. (*Hint:* Recall that the area of a circle is πr^2.)

(d) What are the mean and standard deviation of T?

(e) Explain how Eric used a random number generator and the facts given here to estimate π.

14.43 The Continuity Correction. One reason the Normal approximation may fail to give accurate estimates of binomial probabilities is that binomial distributions are discrete, and Normal distributions are continuous. That is, counts take only whole number values, but Normal variables can take any value. We can improve the Normal approximation by treating each whole number count as if it occupied the interval from 0.5 below the number to 0.5 above the number. For example, approximate a binomial probability $P(X \geq 10)$ by finding the Normal probability $P(X \geq 9.5)$. Be careful: binomial $P(X > 10)$ is approximated by Normal $P(X \geq 10.5)$.

We saw in Exercise 14.24 that 50% of gonorrhea cases are resistant to any antibiotic. Suppose a local health clinic sees 20 cases. The exact binomial probability that 13 or more cases are resistant to any antibiotic is 0.1316.

(a) Show that this setting satisfies the rule of thumb for use of the Normal approximation (just barely).

(b) What is the Normal approximation to $P(X \geq 13)$?

(c) What is the Normal approximation using the continuity correction? That's a lot closer to the true binomial probability.

CHAPTER

15

When you complete this chapter, you will be able to:

15.1 Given a statistical measure and its context, determine whether the measure is a parameter or a statistic.

15.2 Use the law of large numbers to describe the behavior of the mean of a set of observed values from a population as the number of observed values increases.

15.3 Interpret the distribution of all possible values of a statistic in a given situation as a sampling distribution and distinguish that from a population distribution.

15.4 Recognize that $\bar{x}$ is an unbiased estimator of the population mean and that the variability of a sampling distribution decreases as the size of the sample increases.

15.5 Use the central limit theorem to calculate probabilities related to random samples from a population whose parameters are known.

15.6 Assess the statistical significance of a result by calculating the probability of the result under the assumption that no actual effect is present.

Sampling Distributions

As we mentioned in Chapter 12, probability is a tool we can use to generalize from data produced by random samples and randomized comparative experiments to some wider population. In this chapter, we begin to formalize this process. More specifically, we begin to think about how the mean of a sample can provide information about the mean of the population from which the sample was taken.

Each spring, the U.S. government's Current Population Survey asks detailed questions about income. The 128,579 households included in 2018 had a mean "total money income" of \$90,021.[1] (The median income was of course lower, \$63,179.) That \$90,021 describes the sample, but we use it to estimate the mean income of all households. This is an example of statistical inference: we use information from a sample to infer something about a wider population.

Because the results of random samples and randomized comparative experiments include an element of chance, we can't guarantee that our inferences are always accurate. What we can guarantee is that our methods usually give accurate answers. The reasoning of statistical inference rests on determining the answer to the question "How often would this method give an accurate answer if we used it very many times?" If our data come from random sampling or randomized comparative experiments, the laws of probability answer the question "What would happen if we did this many times?" This chapter presents some facts about probability that help answer this question.

kokouu/Getty Images

15.1 Parameters and Statistics

As we begin to use sample data to draw conclusions about a wider population, we must take care to keep straight whether a number describes a sample or a population. Here is the vocabulary we use.

Parameter, Statistic

A **parameter** is a number that describes the population. In practice, the value of a parameter is not known because we can rarely examine the entire population.

A **statistic** is a number that can be computed from the sample data without making use of any unknown parameters. In practice, we often use a statistic to estimate an unknown parameter.

EXAMPLE 15.1 Household Earnings

The mean income of the sample of 128,579 households included in the 2018 Current Population Survey was $\bar{x} = \$90,021$. The number \$90,021 is a *statistic* because it describes this one Current Population Survey sample. The population that the poll wants to draw conclusions about is all 128 million U.S. households. The *parameter* of interest is the mean income of all these households. We don't know the value of this parameter.

Remember *s* and *p*: statistics come from samples, and parameters come from populations. When we are just doing data analysis, searching for patterns or summarizing features of our data, the distinction between population and sample is not important. As we begin to understand what our data (sample) tell us about a population, the distinction is essential, and the notation we use must reflect this distinction. We write μ (the Greek letter mu) for the *population mean* and σ (the Greek letter sigma) for the *population standard deviation*. These are fixed parameters that are unknown when we use a sample for inference. The *sample mean* is the familiar $\bar{x}$, the average of the observations in the sample. The *sample standard deviation* is denoted by s, the standard deviation of the observations in the sample. These are statistics that would almost certainly take different values if we chose another sample from the same population. The sample mean $\bar{x}$ and sample standard deviation s from a sample or an experiment are estimates of the mean μ and standard deviation σ of the underlying population.

Parameter and Statistics: Notation

Denote the **population mean** by μ, the **population standard deviation** by σ, the **sample mean** by $\bar{x}$, and the **sample standard deviation** by s.

APPLY YOUR KNOWLEDGE

15.1 Genetic Engineering. Here's an idea for treating advanced melanoma, the most serious kind of skin cancer: genetically engineer white blood cells to better recognize and destroy cancer cells; then infuse these cells into patients. The subjects in a small initial study of this approach were 11 patients whose melanoma had not responded to existing treatments. One outcome of this experiment was measured by a test for the presence of cells that trigger an immune response in the body and so may help fight cancer. The mean counts of active cells per 100,000 cells for the 11 subjects were **3.8** before infusion and **160.2** after infusion. Is each of the boldface numbers a parameter or a statistic?

15.2 Florida Voters. Florida has played a key role in recent presidential elections. Voter registration records in September 2019 show that **37%** of Florida voters are registered as Democrats and **35%** as Republicans. (Most of the others did not choose a party.) To test a random digit dialing device that you plan to use to poll voters for the 2020 presidential elections, you use it to call 250 randomly chosen residential telephones in Florida. Of the registered voters contacted, **35%** are registered Democrats. Is each of the boldface numbers a parameter or a statistic?

15.3 Guns in School. Researchers surveyed 14,765 American high school students (grades 9–12) and found that **27.3%** of those surveyed were in grade 9. The percentage of all American high school students who are are in grade 9 is **26.5%**. The percentage of those surveyed who were in grade 9 and had carried a gun to school was **4.4%**. Is each of the boldface numbers a parameter or a statistic?

15.2 Statistical Estimation and the Law of Large Numbers

Statistical inference uses sample data to draw conclusions about the entire population. Because good samples are chosen randomly, statistics such as $\overline{x}$ computed from these samples are random variables. We can describe the behavior of a sample statistic by using a probability model that answers the question "What would happen if we did this many times?" Here is an example that will lead us toward the probability ideas that are most important for statistical inference.

STATISTICS IN YOUR WORLD

High-Tech Gambling

There are twice as many slot machines as bank ATMs in the United States. Once upon a time, you put in a coin and pulled the lever to spin three wheels, each with 20 symbols. No longer. Now the machines are video games with flashy graphics and outcomes produced by random number generators. Machines can accept many coins at once, can pay off on a bewildering variety of outcomes, and can be networked to allow common jackpots. Gamblers still search for systems, but in the long run, the law of large numbers guarantees the house its 5% profit.

Hudzilla/Getty Images

EXAMPLE 15.2 Does This Wine Smell Bad?

One of the reasons that winemaking is referred to as an art is because so many things can go wrong during production. Wine is chemically delicate and must be carefully supervised and nurtured. Sulfur compounds such as dimethyl sulfide (DMS) are formed naturally in the process of making wine. DMS is present in all wines. At low levels, it contributes to roundness, fruitiness, and the complexity of wine. Unfortunately, at higher levels, it can contribute a vegetative, cooked cabbage, onion-like, or sulfur smell. In restaurants, when you order a bottle of wine, you are often presented with a small sample of the newly opened bottle to confirm that there are no unpleasant odors.

Winemakers need to know the "odor threshold"—the lowest concentration of DMS that the human nose can detect. People vary in their ability to detect DMS, and it is important to understand this variation.

Because different people have different thresholds, we start by asking about the mean threshold μ in the population of all adults. The number μ is a parameter that describes this population.

To estimate μ, we present tasters with both natural wine and the same wine spiked with DMS at different concentrations to find the lowest concentration at which they identify the spiked wine. Here are the odor thresholds (measured in micrograms of DMS per liter of wine) for 10 randomly chosen subjects:

28 40 28 33 20 31 29 27 17 21

The mean threshold for these subjects is $\overline{x} = 27.4$. It seems reasonable to use the sample result $\overline{x} = 27.4$ to estimate the unknown μ. An SRS should fairly represent the population, so the mean $\bar{x}$ of the sample should be somewhere near the mean μ of the population. Of course, we don't expect $\bar{x}$ to be exactly equal to μ. We realize that if we choose another SRS, the luck of the draw will probably produce a different $\bar{x}$.

If $\bar{x}$ is rarely exactly right and varies from sample to sample, why is it nonetheless a reasonable estimate of the population mean μ? Here is one answer: *if we keep taking larger and larger samples, the statistic $\bar{x}$ is guaranteed to get closer and closer to the parameter μ.* We have the comfort of knowing that if we can afford to keep measuring more subjects, eventually we will estimate the mean odor threshold of all adults very accurately. This remarkable fact is called the *law of large numbers.* It is remarkable because it holds for *any* population, not just for some special class such as Normal distributions.

Law of Large Numbers

If we draw observations at random from any population with finite mean μ, as the number of observations drawn increases, the mean $\bar{x}$ of the observed values tends to get closer and closer to the mean μ of the population.[2]

The law of large numbers can be proven mathematically starting from the basic laws of probability. The behavior of $\bar{x}$ is similar to the idea of probability. In the long run, the *proportion* of outcomes taking any value gets close to the probability of that value, and the *average* outcome gets close to the population mean. Figure 12.1 (page 273) shows how proportions approach probability in one example. Here is an example of how sample means approach the population mean.

EXAMPLE 15.3 The Law of Large Numbers in Action

Suppose the distribution of odor thresholds among all adults has mean 25. The mean $\mu = 25$ is the true value of the parameter we seek to estimate. Figure 15.1 shows how the sample mean $\bar{x}$ of an SRS drawn from this population changes as we add more subjects to our sample.

The first subject in Example 15.2 had threshold 28, so the line in Figure 15.1 starts there. The mean for the first two subjects is

$$\bar{x} = \frac{28 + 40}{2} = 34$$

This is the second point on the graph. At first, the graph shows that the mean of the sample changes as we take more observations. Eventually, however, the mean of the observations gets close to the population mean $\mu = 25$ and settles down at that value.

If we started over, again choosing people at random from the population, we would get a different path from left to right in Figure 15.1. The law of large numbers says that whatever path we get will always settle down at 25 as we draw more and more people.

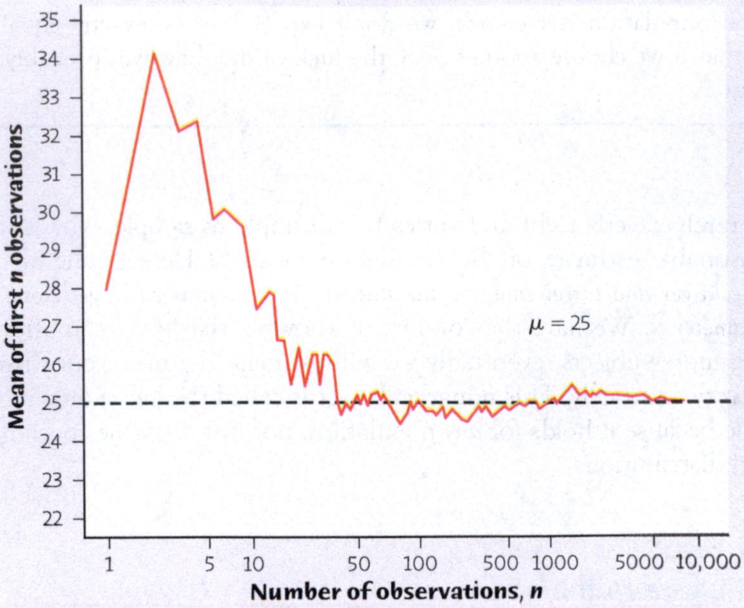

FIGURE 15.1
The law of large numbers in action: as we take more observations, the sample mean $\bar{x}$ always approaches the mean μ of the population.

applet

The *Law of Large Numbers* applet animates Figure 15.1 in a different setting. You can use the applet to watch $\bar{x}$ change as you average more observations until it eventually settles down at the mean μ.

The law of large numbers is the foundation of such business enterprises as gambling casinos and insurance companies. The winnings (or losses) of a gambler on a few plays are uncertain—which is why some people find gambling exciting. In Figure 15.1, the mean of 100 observations has not settled down to μ. It is only *in the long run* that the mean outcome is predictable. The house plays tens of thousands of times. So the house, unlike individual gamblers, can count on the long-run regularity described by the law of large numbers. The average winnings of the house on tens of thousands of plays will be very close to the mean of the distribution of winnings determined by the probabilities of the games. Needless to say, this mean guarantees the house a profit. That's why gambling can be a business.

APPLY YOUR KNOWLEDGE

applet

15.4 The Law of Large Numbers Made Visible. Roll two balanced dice and count the total spots on the up-faces. The probability model appears in Example 12.5 (page 277). You can see that this distribution is symmetric with 7 as its center, so it's no surprise that the mean is $\mu = 7$. This is the population mean for the idealized population that contains the results of rolling two dice forever. The law of large numbers says that the average $\bar{x}$ from a finite number of rolls tends to get closer and closer to 7 as we do more and more rolls.

(a) Click "More dice" once in the *Law of Large Numbers* applet to get two dice. Click "Show μ_x" to see the mean 7 on the graph. Leaving the number of rolls at 1, click "Roll dice" three times, recording each roll. How many spots did each roll produce? What is the average for the three rolls? You see that the graph displays at each point the average number of spots for all rolls up to the last one. This is exactly like Figure 15.1.

(b) Click "Reset" to start over. Set the number of rolls to 100 and click "Roll dice." The applet rolls the two dice 100 times. The graph shows how the average count of spots changes as we make more rolls. That is, the graph shows $\bar{x}$ as we continue to roll the dice. Sketch (or print out) the final graph.

(c) Repeat your work from part (b). Click "Reset" to start over, then roll two dice 100 times. Make a sketch of the final graph of the mean $\bar{x}$ against the number of rolls. Your two graphs will often look very different. What they have in common is that the average eventually gets close to the population mean $\mu = 7$. The law of large numbers says that this will *always* happen if you keep on rolling the dice.

15.5 Insurance. The idea of insurance is that we all face risks that are unlikely but carry high cost. Think of a fire or flood destroying your apartment. Insurance spreads the risk: we all pay a small amount, and the insurance policy pays a large amount to those few of us whose apartments are damaged. An insurance company looks at the records for millions of apartment owners and sees that the mean loss from apartment damage in a year is $\mu = \$150$ per person. (Most of us have no loss, but a few lose most of their possessions. The \$150 is the average loss.) The company plans to sell renters' insurance for \$150 plus enough to cover its costs and profit. Explain clearly why it would be unwise to sell only 10 policies. Then explain why selling thousands of such policies is a safe business.

15.3 Sampling Distributions

The law of large numbers assures us that if we measure enough randomly selected subjects, the statistic $\bar{x}$ will eventually get very close to the unknown parameter μ. But the odor threshold study in Example 15.2 (page 344) had just 10 subjects. What can we say about estimating μ by $\bar{x}$ from a sample of 10 subjects? Put this one sample in the context of all such samples by considering the question "What would happen if we took many samples of 10 subjects from this population?" Here's how to answer this question:

- Take a large number of samples of size 10 from the population.
- Calculate the sample mean $\bar{x}$ for each sample.
- Make a histogram of the values of $\bar{x}$.
- Examine the shape, center, and variability of the distribution displayed in the histogram.

In practice, it is too expensive to take many samples from a large population such as all adult U.S. residents. But we can imitate many samples by using software. Using software to imitate chance behavior is called **simulation**.

EXAMPLE 15.4 What Would Happen in Many Samples?

Extensive studies have found that the DMS odor threshold of adults follows roughly a Normal distribution with mean $\mu = 25$ micrograms per liter and standard deviation $\sigma = 7$ micrograms per liter. We call this the *population distribution* of odor threshold.

Figure 15.2 illustrates the process of choosing many samples and finding the sample mean threshold $\bar{x}$ for each one. Follow the flow of the figure from the population at the left, to choosing an SRS and finding the $\bar{x}$ for this sample, to collecting together the $\bar{x}$'s from many samples. The first sample has $\bar{x} = 26.42$. The second sample contains a different 10 people, with $\bar{x} = 24.28$, and so on. The histogram at the right of the figure shows the distribution of the values of $\bar{x}$ from 1000 separate SRSs of size 10. This histogram displays the *sampling distribution* of the statistic $\bar{x}$.

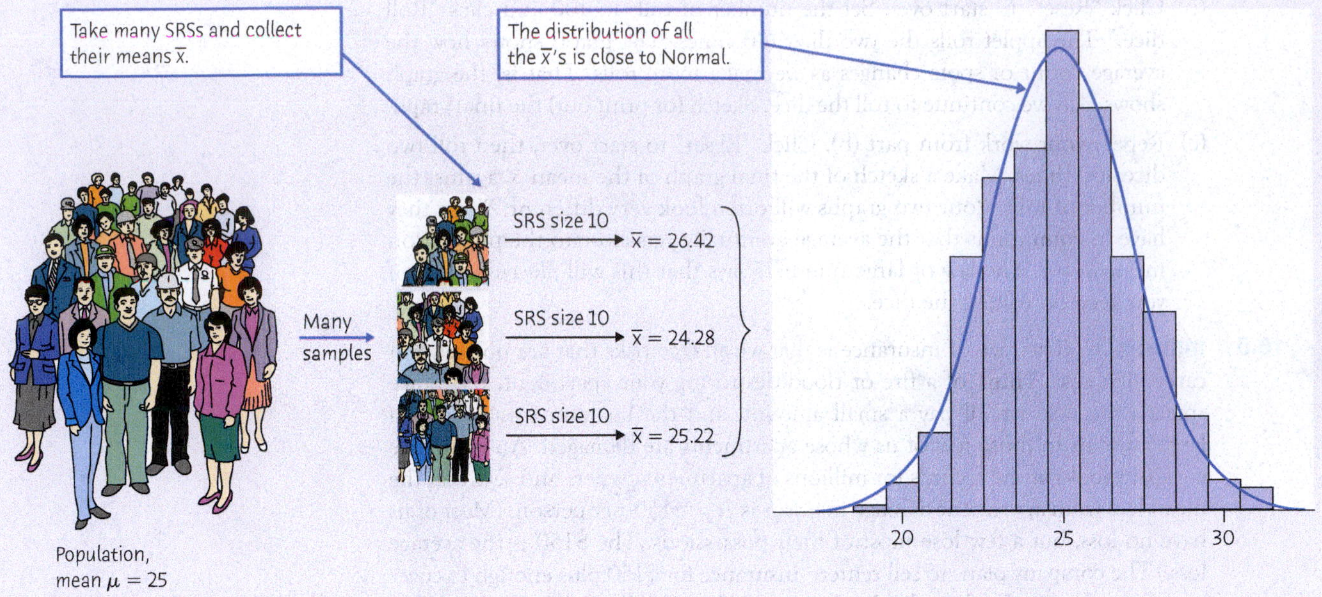

FIGURE 15.2
The idea of a sampling distribution: take many samples from the same population, collect the $\bar{x}$'s from all the samples, and display the distribution of the $\bar{x}$'s. The histogram shows the results of 1000 samples.

Population Distribution, Sampling Distribution

The **population distribution** of a variable is the distribution of values of the variable among all the individuals in the population.

The **sampling distribution** of a statistic is the distribution of values taken by the statistic in all possible samples of the same size from the same population.

Be careful: the population distribution describes the *individuals* who make up the population. A sampling distribution describes how a *statistic* varies in many samples from the population.

Strictly speaking, the sampling distribution is the ideal pattern that would emerge if we looked at all possible samples of size 10 from our population. A distribution obtained from a fixed number of trials, like the 1000 trials in Figure 15.2, is only an approximation to the sampling distribution. One of the uses of probability theory in statistics is to obtain sampling distributions without simulation. The interpretation of a sampling distribution is the same, however, whether we obtain it by simulation or by the mathematics of probability.

We can use the tools of data analysis to describe any distribution. Let's apply those tools to Figure 15.2. What can we say about the shape, center, and variability of this distribution?

- *Shape:* It looks Normal! Detailed examination confirms that the distribution of $\bar{x}$ from many samples is very close to Normal.
- *Center:* The mean of the 1000 $\bar{x}$'s is 24.95. That is, the distribution is centered very close to the population mean $\mu = 25$.
- *Variability:* The standard deviation of the 1000 $\bar{x}$'s is 2.214, notably smaller than the standard deviation $\sigma = 7$ of the population of individual subjects.

Although these results describe just one simulation of a sampling distribution, they reflect facts that are true whenever we use random sampling.

APPLY YOUR KNOWLEDGE

15.6 Sampling Distribution Versus Population Distribution. The 2018 American Time Use Survey contains data on how many minutes of sleep per night each of 9600 survey participants estimated they get.[3] The times follow the Normal distribution with mean 529.2 minutes and standard deviation 135.6 minutes. An SRS of 100 of the participants has a mean time of $\bar{x} = 514.4$ minutes. A second SRS of size 100 has mean $\bar{x} = 539.3$ minutes. After many SRSs, the many values of the sample mean $\bar{x}$ follow the Normal distribution with mean 529.9 minutes and standard deviation 13.56 minutes.

(a) What is the population? What values does the population distribution describe? What is this distribution?

(b) What values does the sampling distribution of $\bar{x}$ describe? What is the sampling distribution?

15.7 Generating a Sampling Distribution. Let's illustrate the idea of a sampling distribution in the case of a very small sample from a very small population. The population is the scores of 10 students on an exam:

Student	1	2	3	4	5	6	7	8	9	10
Score	98	63	91	75	72	84	65	51	88	69

The parameter of interest is the mean score μ in this population. The sample is an SRS of size $n = 4$ drawn from the population. The *Simple Random Sample* applet can be used to select simple random samples of four numbers between 1 and 10, corresponding to the students.

(a) Make a histogram of these 10 scores.

(b) Find the mean of the 10 scores in the population. This is the population mean μ.

(c) Use the *Simple Random Sample* applet to draw an SRS of size 4 from this population. Set the population size to 10 and the sample size to 4. What are the four scores in your sample? What is their mean $\bar{x}$? This statistic is an estimate of μ. (If you prefer not to use applets, use Table B, beginning at line 121, to chose an SRS of size 4 from this population.)

(d) Repeat this process nine more times, using the applet (or Table B, continuing on line 121, if you are not using applets). Calculate and record the mean for each of the nine additional samples. Make a histogram of the 10 values of $\bar{x}$. You are constructing the sampling distribution of $\bar{x}$. Is the center of your histogram close to μ? How does the shape of this histogram compare with the histogram you made in part (a)?

15.4 The Sampling Distribution of $\bar{x}$

Figure 15.2 suggests that when we choose many SRSs from a population, the sampling distribution of the sample means is centered at the mean of the original population and is less variable (spread out) than the distribution of individual observations. Here are the facts.

Mean and Standard Deviation of a Sample Mean[4]

Suppose that $\bar{x}$ is the mean of an SRS of size n drawn from a population with mean μ and standard deviation σ. Then the sampling distribution of $\bar{x}$ has mean μ and standard deviation $\sigma/\sqrt{n}$.

These facts assume n is not too big a fraction of the population size—say, at most no more that 5% of the size of the population.

These facts about the mean and the standard deviation of the sampling distribution of $\bar{x}$ are true for *any* population, not just for some special class, such as when the population has a Normal distribution. They have important implications for statistical inference:

- The mean of the statistic $\bar{x}$ is always equal to the mean μ of the population. That is, the sampling distribution of $\bar{x}$ is centered at μ. In repeated sampling, $\bar{x}$ will sometimes fall above the true value of the parameter μ and sometimes below, but there is no systematic tendency to overestimate or underestimate the parameter. This makes the idea of lack of bias in the sense of "no favoritism" more precise. Because the mean of $\bar{x}$ is equal to μ, we say that the statistic $\bar{x}$ is an *unbiased estimator* of the parameter μ.
- An unbiased estimator is "correct on the average"[5] in many samples. How close the estimator falls to the parameter in most samples is determined by the variability of the sampling distribution. If individual observations have standard deviation σ, then sample means $\bar{x}$ from samples of size n have standard deviation $\sigma/\sqrt{n}$. That is, *averages are less variable than individual observations.*
- Not only is the standard deviation of the distribution of $\bar{x}$ smaller than the standard deviation of individual observations, but it gets smaller as we take larger samples. *The results of large samples are less variable than the results of small samples.*

STATISTICS IN YOUR WORLD

Sample Size Matters

A recent trend in baseball is using statistics to evaluate players, with new measures of performance to help decide which players are worth the high salaries they demand. This method challenges traditional subjective evaluation of young players and the usefulness of traditional measures such as batting average. But success has led many Major League teams to hire statisticians. The statisticians say that sample size matters in baseball also: the 162-game regular season is long enough for the better teams to come out on top, but five-game and seven-game playoff series are so short that luck has a lot to do with who wins.

Unbiased Estimator

An **unbiased estimator** is a statistic used to estimate a parameter where the mean of the statistic's sampling distribution equals the true value of the population parameter being estimated.

The upshot of all this is that we can trust the sample mean from a large random sample to estimate the population mean accurately. If the sample size n is large, the standard deviation of $\bar{x}$ is small, and almost all samples will give values of $\bar{x}$ that lie very close to the true parameter μ. *However, the standard deviation of the sampling distribution gets smaller only at the rate $\sqrt{n}$. To cut the standard deviation of $\bar{x}$ in half, we must take four times as many observations, not just twice as many.* So very precise estimates (estimates with very small standard deviation) may be expensive.

We have described the center and variability of the sampling distribution of a sample mean $\bar{x}$ but not its shape. The shape of the sampling distribution depends on the shape of the population distribution. In one important case, there is a simple relationship between the two distributions: if the population distribution is Normal, then so is the sampling distribution of the sample mean.

Sampling Distribution of a Sample Mean for Normally Distributed Populations

If individual observations have the $N(\mu, \sigma)$ distribution, then the sample mean $\bar{x}$ of an SRS of size n has the $N(\mu, \sigma/\sqrt{n})$ distribution.

Notice that if the population distribution is Normal, then the sampling distribution of the sample mean is Normal, regardless of the sample size n.

EXAMPLE 15.5 Population Distribution, Sampling Distribution

If we measure the DMS odor thresholds of individual adults (as described in Example 15.2, page 344), the values follow the Normal distribution with mean $\mu = 25$ micrograms per liter and standard deviation $\sigma = 7$ micrograms per liter. This is the population distribution of odor threshold.

Take many SRSs of size 10 from this population and find the sample mean $\bar{x}$ for each sample, as in Figure 15.2 (page 348). The sampling distribution describes how the values of $\bar{x}$ vary among samples. That sampling distribution is also Normal, with mean $\mu = 25$ and standard deviation

$$\frac{\sigma}{\sqrt{n}} = \frac{7}{\sqrt{10}} = 2.2136$$

Figure 15.3 contrasts these two Normal distributions. Both are centered at the population mean, but sample means are much less variable than individual observations.

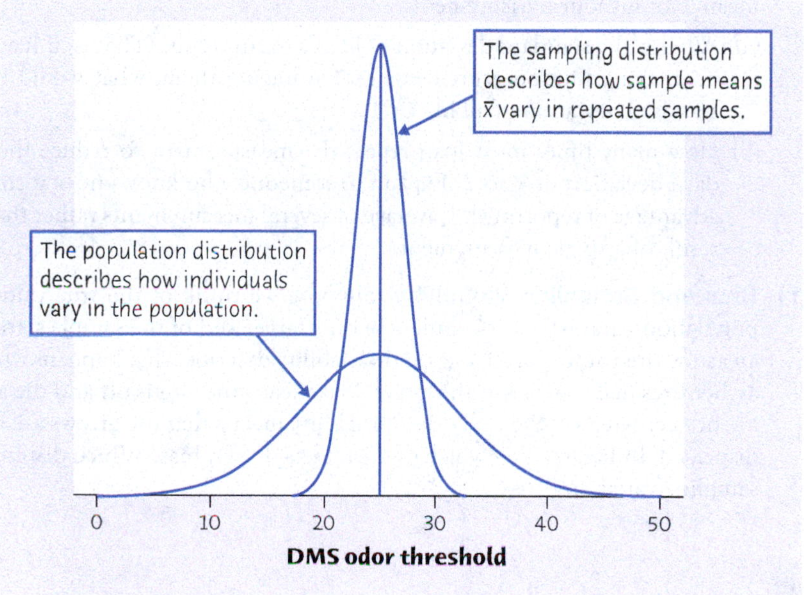
The sampling distribution describes how sample means $\bar{x}$ vary in repeated samples.
The population distribution describes how individuals vary in the population.
0
10
20
30
40
50
DMS odor threshold

FIGURE 15.3 The distribution of single observations (the population distribution) compared with the sampling distribution of the means $\bar{x}$ of 10 observations taken many times, for Example 15.5. Both have the same mean, but averages are less variable than individual observations.

The smaller variation of sample means shows up in probability calculations. You can show (using software or standardizing and using Table A) that about 52% of all adults have odor thresholds between 20 and 30. But almost 98% of means of samples of size 10 lie in this range.

APPLY YOUR KNOWLEDGE

15.8 A Sample of Young Men. A government sample survey plans to measure the mean total cholesterol level of an SRS of men aged 20–34. The researchers will report the mean $\bar{x}$ from their sample as an estimate of the mean total cholesterol level μ in this population.

(a) Explain to someone who knows no statistics what it means to say that $\bar{x}$ is an "unbiased" estimator of μ.

(b) The sample result $\bar{x}$ is an unbiased estimator of the population truth μ no matter what size SRS the study uses. Explain to someone who knows no statistics why a large sample gives more trustworthy results than a small sample.

15.9 **Larger Sample, More Accurate Estimate.** Suppose that, in fact, the total cholesterol level of all men aged 20–34 follows the Normal distribution with mean $\mu = 182$ milligrams per deciliter (mg/dL) and standard deviation $\sigma = 37$ mg/dL.

(a) Choose an SRS of 100 men from this population. What is the sampling distribution of $\bar{x}$? What is the probability that $\bar{x}$ takes a value between 180 and 184 mg/dL? This is the probability that $\bar{x}$ estimates μ within ± 2 mg/dL.

(b) Choose an SRS of 1000 men from this population. Now what is the probability that $\bar{x}$ falls within ± 2 mg/dL of μ? The larger sample is much more likely to give an accurate estimate of μ.

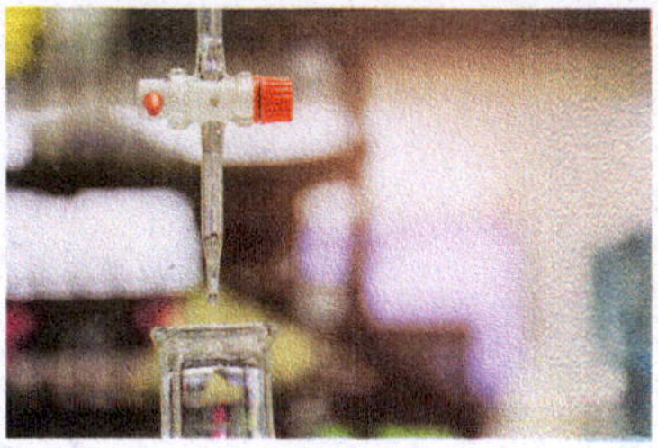

Ropisme/Shutterstock

15.10 **Measurements in the Lab.** Juan makes a measurement in a chemistry laboratory and records the result in his lab report. Suppose that if Juan makes this measurement repeatedly, the standard deviation of his measurements will be $\sigma = 12$ milligrams. Juan repeats the measurement nine times and records the mean $\bar{x}$ of his four measurements.

(a) What is the standard deviation of Juan's mean result? (That is, if Juan kept on making nine measurements and averaging them, what would be the standard deviation of all his $\bar{x}$'s?)

(b) How many times must Juan repeat the measurement to reduce the standard deviation of $\bar{x}$ to 2? Explain to someone who knows no statistics the advantage of reporting the average of several measurements rather than the result of a single measurement.

15.11 **Bias and Sampling Variability.** Suppose we think of the true value of a population parameter as the bull's-eye on a target and of the sample statistic as an arrow fired at the target. Bias and variability describe what happens when an archer fires many arrows at the target. Bias means the aim is off and the arrows are not centered on the bull's-eye. Variability means that the arrows are widely dispersed. In Figure 15.4, which of the targets display bias? Which display large sampling variability?

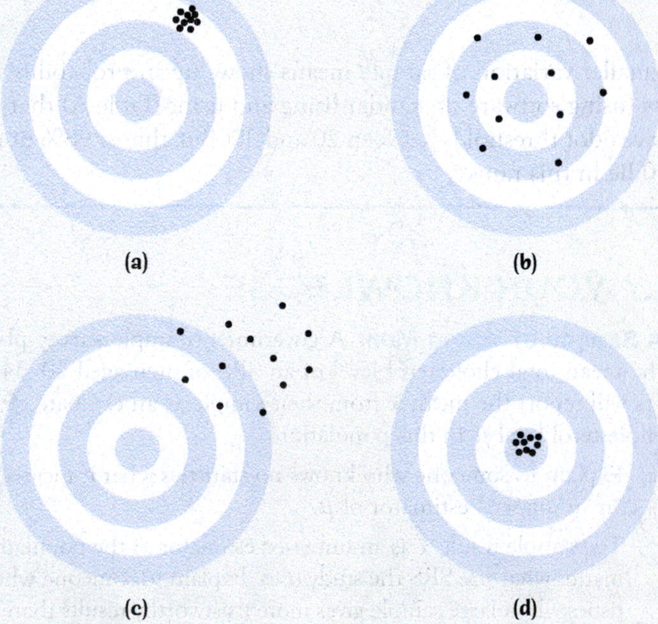

FIGURE 15.4
Four targets with the patterns of several arrows fired at the target displayed as dots, for Exercise 15.11.

15.5 The Central Limit Theorem

The facts about the mean and standard deviation of $\bar{x}$ are true no matter what the shape of the population distribution may be. But what is the shape of the sampling distribution when the population distribution is not Normal? *It is a remarkable fact that as the sample size increases, the distribution of $\bar{x}$ changes shape: it looks less like that of the population and more like a Normal distribution.* When the sample is large enough, the distribution of $\bar{x}$ is very close to Normal. This is true no matter what shape the population distribution has, as long as the population has a finite standard deviation σ. This famous fact of probability theory is called the *central limit theorem*. It is much more useful than the fact that the distribution of $\bar{x}$ is exactly Normal if the population is exactly Normal.

STATISTICS IN YOUR WORLD

What Was That Probability Again?

In real-world applications, probability calculations can be very complex and are usually based on assumptions that may not be accurate. Wall Street uses sophisticated mathematics to predict the probabilities that fancy investments will go wrong. The estimated probabilities are always too low—sometimes because something was assumed to be Normal but was not. The results can be devastating to investors.

Central Limit Theorem

Draw an SRS of size n from any population with mean μ and finite standard deviation σ. The **central limit theorem** says that when n is large, the sampling distribution of the sample mean $\bar{x}$ is approximately Normal:

$$\bar{x} \text{ is approximately } N\left(\mu, \frac{\sigma}{\sqrt{n}}\right)$$

The central limit theorem allows us to use Normal probability calculations to answer questions about sample means from many observations even when the population distribution is not Normal.

More general versions of the central limit theorem say that the distribution of any sum or average of many small random quantities is close to Normal. This is true even if the quantities are correlated with each other (as long as they are not too highly correlated) and even if they have different distributions (as long as no one random quantity is so large that it dominates the others). The central limit theorem suggests why the Normal distributions are common models for observed data. Any variable that is an average of many small influences will have approximately a Normal distribution.

How large a sample size n is needed for $\bar{x}$ to be close to Normal depends on the population distribution. More observations are required if the shape of the population distribution is far from Normal. Here are two examples in which the population is far from Normal.

EXAMPLE 15.6 The Central Limit Theorem in Action

In 2019, the Annual Social and Economic Supplement to the Current Population Survey obtained data on the personal income of 141,251 individuals. Figure 15.5(a) is a histogram of the total personal income of these individuals.[6] As we expect, the distribution of total incomes is strongly skewed to the right and very spread out. Notice that some of the incomes are negative. The right tail of the distribution is even longer than the histogram shows because there are too few high incomes for their bars to be visible on this scale. In fact, we cut off the earnings scale at \$400,000 to save space; a few individuals had incomes of even more than \$400,000. The mean income for the 141,251 individuals was \$43,663.

Regard these 141,251 individuals as a population with mean $\mu = \$43{,}663$. Take an SRS of 100 households. Suppose the mean earnings in this sample is $\bar{x} = \$46{,}279$. That's higher than the mean of the population. Take another SRS of size 100. The mean for this sample is $\bar{x} = \$41{,}266$. That's less than the mean of the population. *What would happen if we did this many times?* Figure 15.5(b) is a histogram of the mean

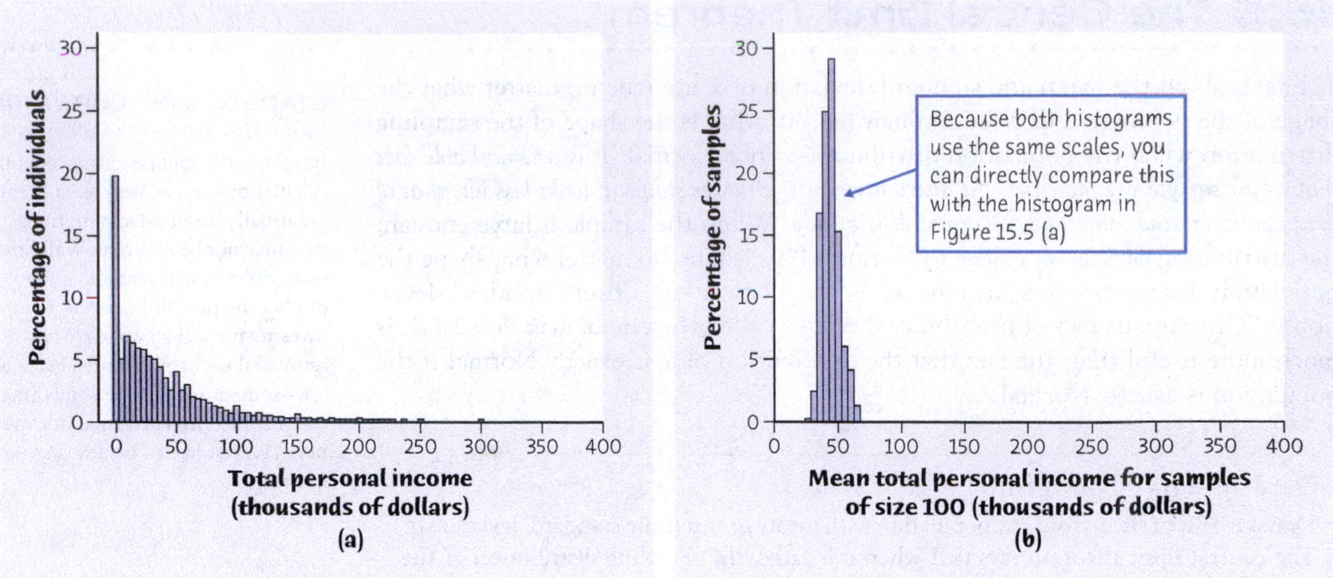

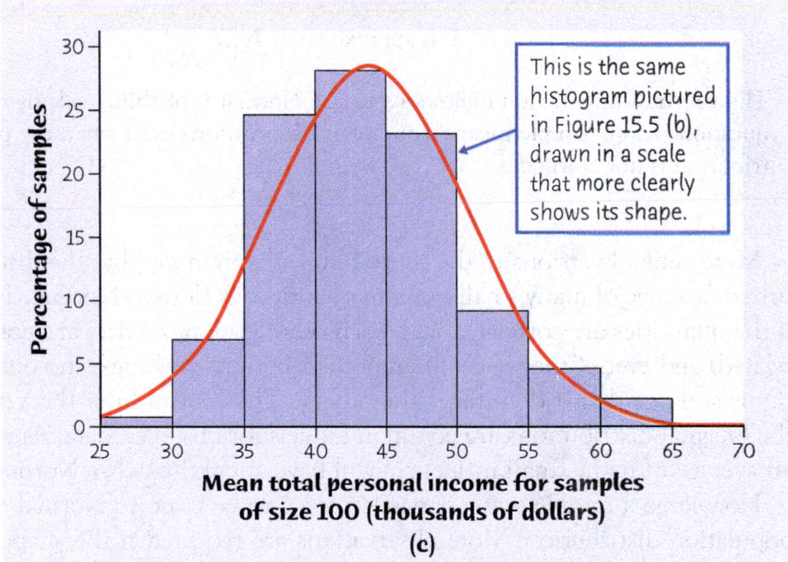

FIGURE 15.5
The central limit theorem in action, for Example 15.6. (a) The distribution of total personal income in a population of 141,251 individuals. (b) The distribution of the mean earnings for 500 SRSs of 100 individuals each from this population. (c) The distribution of the sample means in more detail: the shape is close to Normal.

earnings for 500 samples, each of size 100. The scales in Figures 15.5(a) and 15.5(b) are the same for easy comparison. Although the distribution of individual earnings is skewed and with large variability, the distribution of sample means is roughly symmetric and shows much less variability.

Figure 15.5(c) zooms in on the center part of the histogram in Figure 15.5(b) to more clearly show its shape. Although $n = 100$ is not a very large sample size and the population distribution is extremely skewed, we can see that the distribution of sample means is close to Normal.

Comparing Figure 15.5(a) with Figures 15.5(b) and 15.5(c) illustrates the two most important ideas of this chapter.

Thinking about Sample Means

Means of random samples are *less variable* than individual observations.

Means of random samples are *more Normal* than individual observations.

EXAMPLE 15.7 The Central Limit Theorem in Action

Exponential distributions are used as models for the lifetime of electronic components and for the time required to serve a customer or repair a machine. Figure 15.6(a) shows the exponential population distribution–that is, the density curve–of a single observation. This distribution is strongly right-skewed, and the most probable outcomes are near 0. The mean μ of this distribution is 1, and its standard deviation σ is also 1.

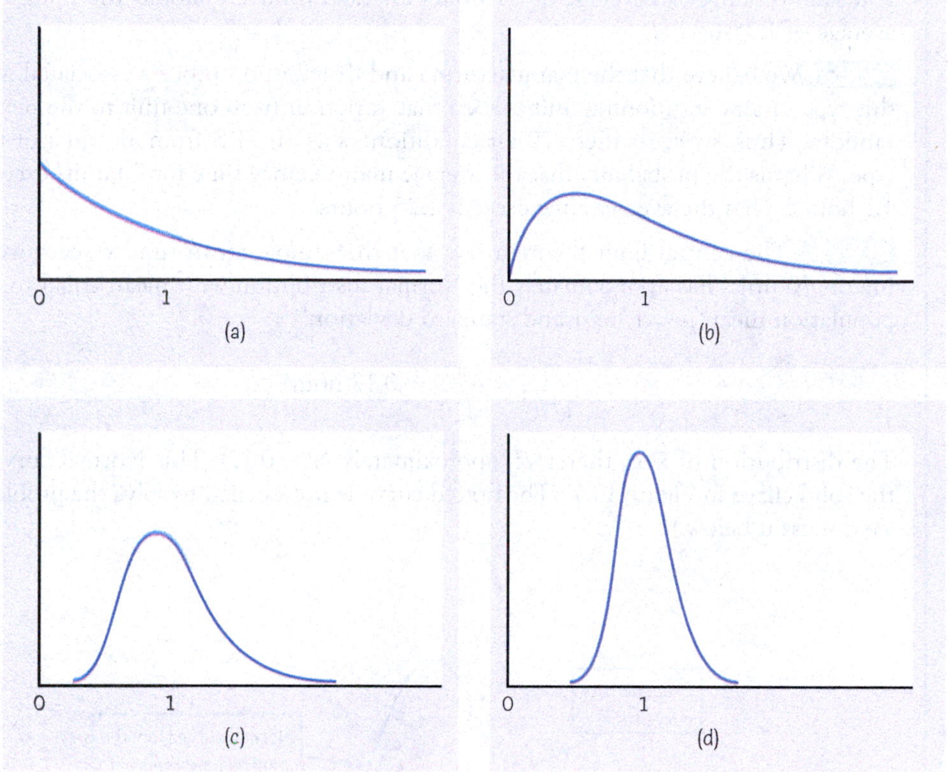

FIGURE 15.6 The central limit theorem in action, for Example 15.7. The distribution of sample means $\bar{x}$ from a strongly non-Normal population becomes more Normal as the sample size increases. (a) The distribution of one observation (population distribution). (b) The distribution of $\bar{x}$ for two observations. (c) The distribution of $\bar{x}$ for 10 observations. (d) The distribution of $\bar{x}$ for 25 observations.

Mathematics can be used to derive the theoretical sampling distribution of $\bar{x}$ when sampling from an exponential distribution. Figures 15.6(b), 15.6(c), and 15.6(d) are the theoretical density curves of the sample means of samples of size two, 10, and 25 from this population. As n increases, the shape becomes more Normal. The mean remains at $\mu = 1$, and the standard deviation decreases, taking the value $1/\sqrt{n}$. The density curve for 10 observations is still somewhat skewed to the right but already resembles a Normal curve having $\mu = 1$ and $\sigma = 1/\sqrt{10} = 0.32$. The density curve for $n = 25$ is yet more Normal. The contrast between the shapes of the population distribution and of the distribution of the mean of 10 or 25 observations is striking.

The *Central Limit Theorem* applet allows you to watch the central limit theorem in action. The applet simulates the sampling distribution of $\bar{x}$ for several population distributions, and you can see how the sampling distribution changes shape as the sample size increases.

applet

Let's use Normal calculations based on the central limit theorem to answer a question about the very non-Normal distribution in Figure 15.6(a).

EXAMPLE 15.8 Maintaining Air Conditioners

4step

STATE: The time (in hours) that a technician requires to perform preventive maintenance on an air-conditioning unit is governed by the exponential distribution whose density curve appears in Figure 15.6(a). The exponential distribution arises in many engineering and industrial problems, such as time until failure of a machine or time until a success. The mean time is $\mu = 1$ hour, and the standard deviation is $\sigma = 1$ hour. Your company has a contract to maintain 70 of these units in an apartment building. You must schedule technicians' time for a visit to this building. Is it safe to budget an average of 1.1 hours for each unit? Or should you budget an average of 1.25 hours?

PLAN: We believe that the manufacturing and distribution process associated with this type of air-conditioning unit is such that variation from one unit to the next is random. Thus, we treat these 70 air conditioners as an SRS from all units of this type. What is the probability that the average maintenance time for 70 units exceeds 1.1 hours? That the average time exceeds 1.25 hours?

SOLVE: The central limit theorem says that the sample mean time $\bar{x}$ spent working on 70 units has approximately the Normal distribution with mean equal to the population mean $\mu = 1$ hour and standard deviation

$$\frac{\sigma}{\sqrt{70}} = \frac{1}{\sqrt{70}} = 0.12 \text{ hour}$$

The distribution of $\bar{x}$ is, therefore, approximately $N(1, 0.12)$. This Normal curve is the solid curve in Figure 15.7. (The dotted curve is not needed to solve the problem; we discuss it below.)

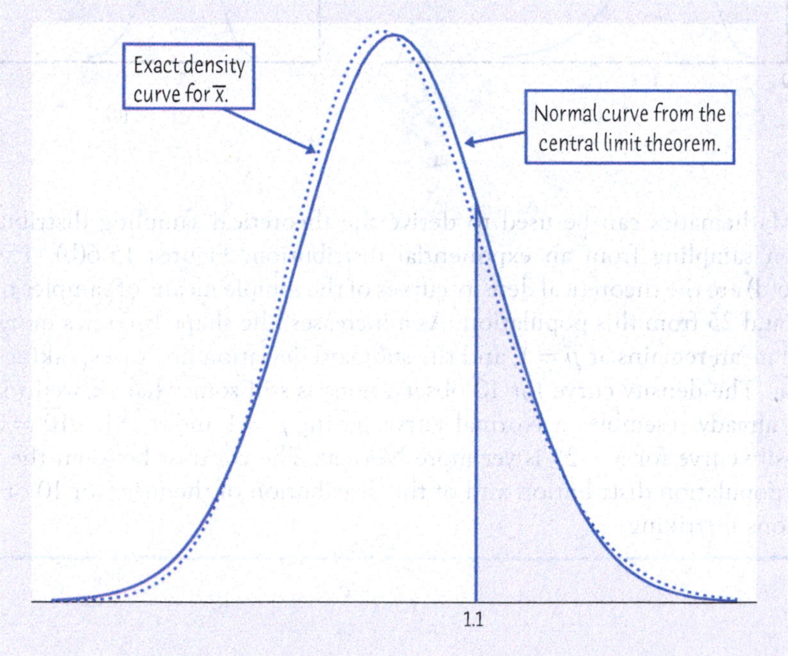

FIGURE 15.7 The exact distribution (dotted) and the Normal approximation from the central limit theorem (solid) for the average time needed to maintain an air conditioner, for Example 15.8. The probability we want is the area to the right of 1.1.

Using this Normal distribution, the probabilities we want are

$$P(\bar{x} > 1.10 \text{ hours}) = 0.2014$$
$$P(\bar{x} > 1.25 \text{ hours}) = 0.0182$$

Software gives these probabilities immediately, or you can standardize and use Table A. For example,

$$P(\overline{x} > 1.10) = P\left(\frac{\overline{x}-1}{0.12} > \frac{1.10-1}{0.12}\right)$$
$$= P(Z > 0.83) = 1 - 0.7967 = 0.2033$$

with the usual roundoff error. Don't forget to use standard deviation 0.12 in your software or when you standardize $\overline{x}$.

CONCLUDE: If you budget 1.1 hours per unit, there is a 20% chance that the technicians will not complete the work in the building within the budgeted time. This chance drops to 2% if you budget 1.25 hours. You should therefore budget 1.25 hours per unit.

Using more mathematics, we can start with the exponential distribution and find the actual density curve of $\overline{x}$ for 70 observations. This is the dotted curve in Figure 15.7. You can see that the solid Normal curve is a good approximation. The exactly correct probability for 1.1 hours is an area to the right of 1.1 under the dotted density curve. It is 0.1977. The central limit theorem Normal approximation 0.2014 is off by less than 0.004.

APPLY YOUR KNOWLEDGE

15.12 What Does the Central Limit Theorem Say? Asked what the central limit theorem says, a student replies, "As you take larger and larger samples from a population, the histograms of the sample values look more and more Normal." Is the student right? Explain your answer.

15.13 Detecting the Emerald Ash Borer. The emerald ash borer is a serious threat to ash trees. A state agriculture department places traps throughout the state to detect the emerald ash borer. When traps are checked periodically, the mean number of ash borers trapped is only 2.2, but some traps have many ash borers. The distribution of ash borer counts is finite and strongly skewed, with standard deviation 3.9.

Herman Wong HM/Shutterstock

(a) What are the mean and standard deviation of the average number of ash borers $\overline{x}$ in 50 traps?

(b) Use the central limit theorem to find the probability that the average number of ash borers in 50 traps is greater than 3.0.

15.14 More on Insurance. An insurance company knows that in the entire population of millions of apartment owners, the mean annual loss from damage is $\mu = \$150$, and the standard deviation of the loss is $\sigma = \$300$. The distribution of losses is strongly right-skewed: most policies have \$0 loss, but a few have large losses. If the company sells 10,000 policies, can it safely base its rates on the assumption that its average loss will be no greater than \$160? Follow the four-step process, as illustrated in Example 15.8.

15.6 Sampling Distributions and Statistical Significance*

We have looked carefully at the sampling distribution of a sample mean. However, any statistic we can calculate from a sample will have a sampling distribution.

*This material is optional. It is more challenging than the material in the rest of the chapter, but it makes three points. First, it presents sampling distributions for statistics other than means. Second, it clarifies the concept of statistical significance first introduced in Chapter 9. Third, it introduces the type of reasoning that is the basis for hypothesis testing discussed in Chapter 17.

EXAMPLE 15.9 Median, Variance, and Standard Deviation

In Example 15.5, we took 1000 SRSs of size 10 from a Normal population with mean $\mu = 25$ micrograms per liter and standard deviation $\sigma = 7$ micrograms per liter. This Normal distribution is the distribution of the DMS odor thresholds of all adults. Figure 15.3 (page 351) is a histogram of the distribution of the sample means.

Now take 1000 SRSs of size 5 from a Normal population with mean $\mu = 25$ and standard deviation $\sigma = 7$. For each sample, compute the sample median, variance, and standard deviation. Figure 15.8 displays histograms of the 1000 sample results. These histograms show the sampling distributions of the three statistics. The sampling distribution of the sample median is symmetric, centered at 25, and approximately Normal. The sampling distribution of the sample variance is strongly skewed to the right. The sampling distribution of the sample standard deviation is very slightly skewed to the right.

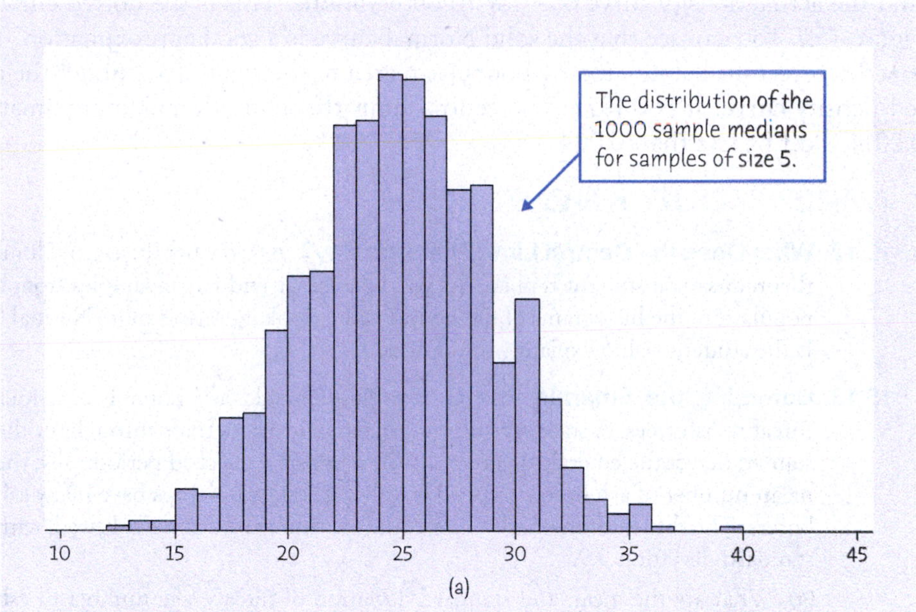
The distribution of the 1000 sample medians for samples of size 5.
10
15
20
25
30
35
40
45
(a)

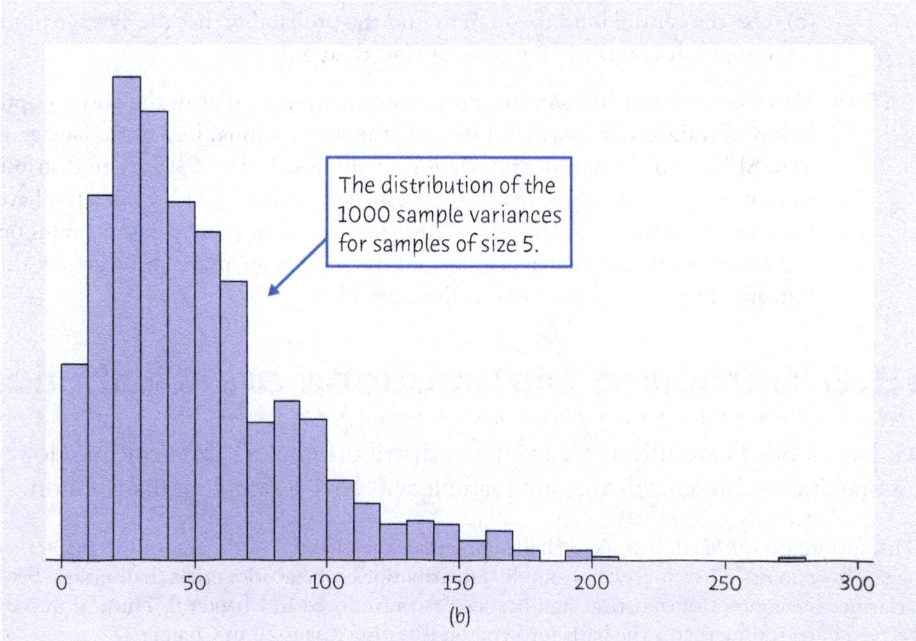
The distribution of the 1000 sample variances for samples of size 5.
0
50
100
150
200
250
300
(b)

FIGURE 15.8
(a) The distribution of 1000 sample medians for samples of size 5 from a Normal population with $\mu = 25$ and $\sigma = 7$. (b) The distribution of 1000 sample variances for samples of size 5 from a Normal population with $\mu = 25$ and $\sigma = 7$. (c) The distribution of 1000 sample standard deviations for samples of size 5 from a Normal population with $\mu = 25$ and $\sigma = 7$.

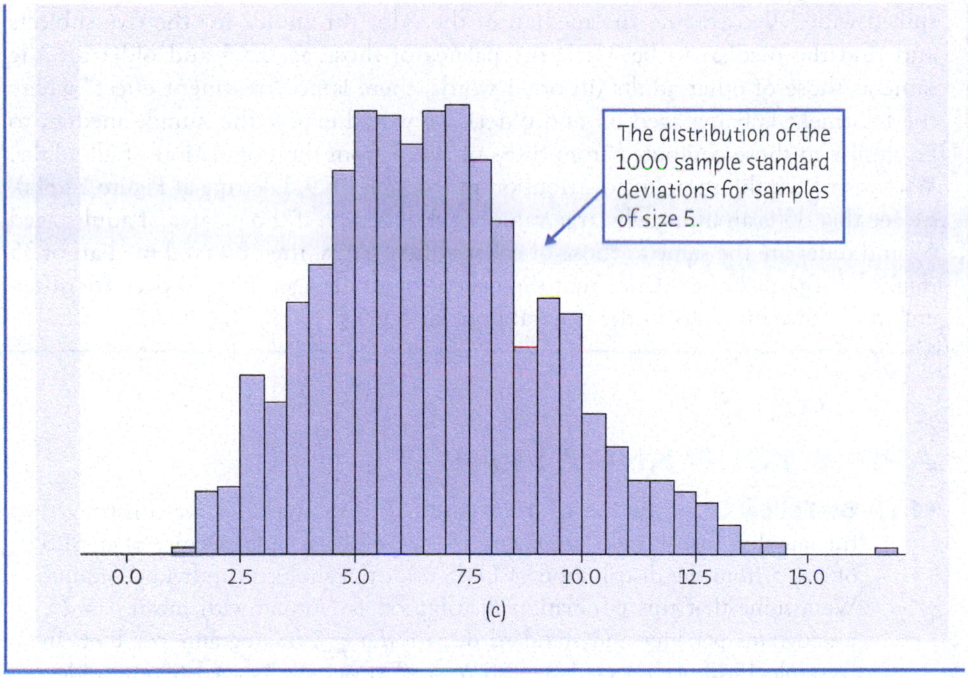
The distribution of the 1000 sample standard deviations for samples of size 5.
0.0
2.5
5.0
7.5
10.0
12.5
15.0
(c)

FIGURE 15.8
(Continued)

The sampling distribution of a sample statistic is determined by the particular sample statistic we are interested in, the distribution of the population of individual values from which the sample statistic is computed, and the method by which samples are selected from the population. The sampling distribution allows us to determine the probability of observing any particular value of the sample statistic in another such sample from the population. Let's look at one application of this idea to a concept we first introduced in Chapter 9.

In Chapter 9 (page 235), we stated that we can use the laws of probability to learn if an observed treatment effect is larger than we would expect if only chance were operating. We said that an observed effect that is so large that it would rarely occur by chance is called *statistically significant*. How does one determine whether an observed effect would rarely occur if only chance were operating? By "only chance operating" we mean under the assumption that there is no treatment effect.

Let the observed treatment effect be represented by the value of a sample statistic—for example, the mean of the responses of those in a treatment group minus the mean of those in a control group. Then we can determine whether the observed effect is statistically significant by considering the sampling distribution of the sample statistic, under the assumption that no effect is present. Use this sampling distribution to determine the probability that we would observe values as extreme as we did observe, if the treatment actually had no effect. The next example illustrates this calculation.

EXAMPLE 15.10 A Sampling Distribution and Statistical Significance

Do adults aged 65 and older have palates similar to those of other adults? Are they just as capable as younger adults of appreciating a fine wine? Suppose we take an SRS of five adults from the population of all adults aged 65 and older. We present them with both natural wine and the same wine spiked with DMS at different concentrations to find the lowest concentration (odor threshold) at which they identify the spiked wine. We compute the median of the odor thresholds for the five subjects and find the median to be 35. If the palates of those aged 65 and older are the same as those of other adults (in other words, there is no "treatment effect" where the treatment is being aged 65 and older), we would expect the sample median to be similar to those computed from SRSs of size 5 from the population of all adults. We constructed this sampling distribution in Example 15.9. Looking at Figure 15.8(a), we see that 35 is an unusually large value for the median if the palates of adults aged 65 and older are the same as those of other adults. Thus, the observed median of 35 might be regarded as evidence that the palates of adults aged 65 and over are different from those of adults in the general population.

APPLY YOUR KNOWLEDGE

15.15 Statistical Significance of a Variance. In Example 15.9, we constructed the sampling distribution [see Figure 15.8(b)] of the sample variance of an SRS of size 5 from the distribution of DMS odor thresholds of individual adults. We assume that this population distribution is Normal with mean $\mu = 25$ micrograms per liter and standard deviation $\sigma = 7$ micrograms per liter. In Example 15.10, we selected an SRS of size 5 from adults aged 65 and older. Suppose the variance of this SRS is 9.2. Would you consider this value to be statistically significant if the palates of adults aged 65 and older are no different than those of all other adults? Explain your answer.

applet

15.16 Statistical Significance from a Sampling Distribution. In Exercise 15.7, you generated 10 samples of size 4 from the population of 10 students, calculated $\bar{x}$ for each sample, and constructed a histogram of these 10 values of $\bar{x}$.

(a) Use the *Simple Random Sample* applet to generate 25 new samples of size 4, calculate $\bar{x}$ for each, and construct a histogram of the 25 values. Once again, you are constructing the sampling distribution of $\bar{x}$.

(b) Based on your histogram in part (a), what would you estimate to be the chance of obtaining a simple random sample of four students with $\bar{x} \geq 78$?

(c) Suppose you learn that students 1, 3, 5, and 7 are honors students. Would you regard their mean score as being "statistically significant"?

CHAPTER 15 SUMMARY

- A **parameter** in a statistical problem is a number that describes a population, such as the population mean μ. To estimate an unknown parameter, use a **statistic** calculated from a sample, such as the sample mean $\bar{x}$.
- The **law of large numbers** states that the actual observed mean outcome $\bar{x}$ must approach the mean μ of the population as the number of observations increases.
- The **population distribution** of a variable describes the values of the variable for all individuals in a population.
- The **sampling distribution** of a statistic describes the values of the statistic in all possible samples of the same size from the same population.
- When the sample is an SRS from the population, the mean of the sampling distribution of the sample mean $\bar{x}$ is the same as the population mean μ. That is, $\bar{x}$ is an **unbiased estimator** of μ.

- The standard deviation of the sampling distribution of $\bar{x}$ is $\sigma/\sqrt{n}$ for an SRS of size n if the population has standard deviation σ. That is, averages are less variable than individual observations.
- When the sample is an SRS from a population that has a Normal distribution, the sample mean $\bar{x}$ also has a Normal distribution.
- Choose an SRS of size n from any population with mean μ and finite standard deviation σ. The **central limit theorem** states that when n is large, the sampling distribution of $\bar{x}$ is approximately Normal. That is, averages are more Normal than individual observations. We can use the $N(\mu, \sigma/\sqrt{n})$ distribution to calculate approximate probabilities for events involving $\bar{x}$.

CHECK YOUR SKILLS

15.17 The Bureau of Labor Statistics announces that last month it interviewed all members of the labor force in a sample of 60,000 households; **3.5%** of the people interviewed were unemployed. The boldface number is a

(a) sampling distribution.

(b) statistic.

(c) parameter.

15.18 An October 20, 2019, poll of Canadian adults who were registered voters found that 31.6% said they would vote conservative in the October 2019 elections. Election records show that **34.4%** actually voted conservative. The boldface number is a

(a) sampling distribution.

(b) statistic.

(c) parameter.

15.19 Annual returns on stocks vary a lot. The long-term mean return on stocks in the S&P 500 is 9.8%, and the long-term standard deviation of returns is 16.8%. The law of large numbers says that

(a) you can get an average return higher than the mean 9.8% by investing in a large number of the S&P stocks.

(b) as you invest in more and more stocks chosen at random, your long-term average return on these stocks gets ever closer to 9.8%.

(c) if you invest in a large number of stocks chosen at random, your long-term average return will have approximately a Normal distribution.

15.20 Scores on the Evidence-Based Reading part of the SAT exam in a recent year were roughly Normal with mean 536 and standard deviation 102. You choose an SRS of 100 students and average their SAT Evidence-Based Reading scores. If you do this many times, the mean of the average scores you get will be close to

(a) 536.

(b) $536/100 = 5.36$.

(c) $536/\sqrt{100} = 53.6$.

15.21 Scores on the Evidence-Based Reading part of the SAT exam in a recent year were roughly Normal with mean 536 and standard deviation 102. You choose an SRS of 100 students and average their SAT Evidence-Based Reading scores. If you do this many times, the standard deviation of the average scores you get will be close to

(a) 102.

(b) $102/100 = 1.02$.

(c) $102/\sqrt{100} = 10.2$.

15.22 A newborn baby has extremely low birth weight (ELBW) if it weighs less than 1000 grams. A study of the health of such children in later years examined a random sample of 219 children who had been born with ELBW. Their mean weight at birth was $\bar{x} = 810$ grams. This sample mean is an *unbiased estimator* of the mean weight μ in the population of all ELBW babies. This means that

(a) in many samples from this population, the mean of the many values of $\bar{x}$ will be equal to μ.

(b) as we take larger and larger samples from this population, $\bar{x}$ will get closer and closer to μ.

(c) in many samples from this population, the many values of $\bar{x}$ will have a distribution that is close to Normal.

15.23 The number of hours a battery lasts before failing varies from battery to battery. The distribution of failure times follows an exponential distribution (see Example 15.7, page 355), which is strongly skewed to the right. The central limit theorem says that

(a) as we look at more and more batteries, their average failure time gets ever closer to the mean μ for all batteries of this type.

(b) the average failure time of a large number of batteries has a distribution of the same shape (strongly skewed) as the distribution for individual batteries.

(c) the average failure time of a large number of batteries has a distribution that is close to Normal.

15.24 The length of human pregnancies from conception to birth varies according to a distribution that is approximately Normal with mean 266 days and standard deviation 16 days. The probability that the average pregnancy length for six randomly chosen women exceeds 270 days is about

(a) 0.40.

(b) 0.27.

(c) 0.07.

15.25 Figure 15.9 shows the behavior of a sample statistic in many samples in four situations. The true value of the population parameter is marked on each graph. The graphs for which the sample statistic is unbiased are

(a) a, b, and c.

(b) b and c.

(c) only b.

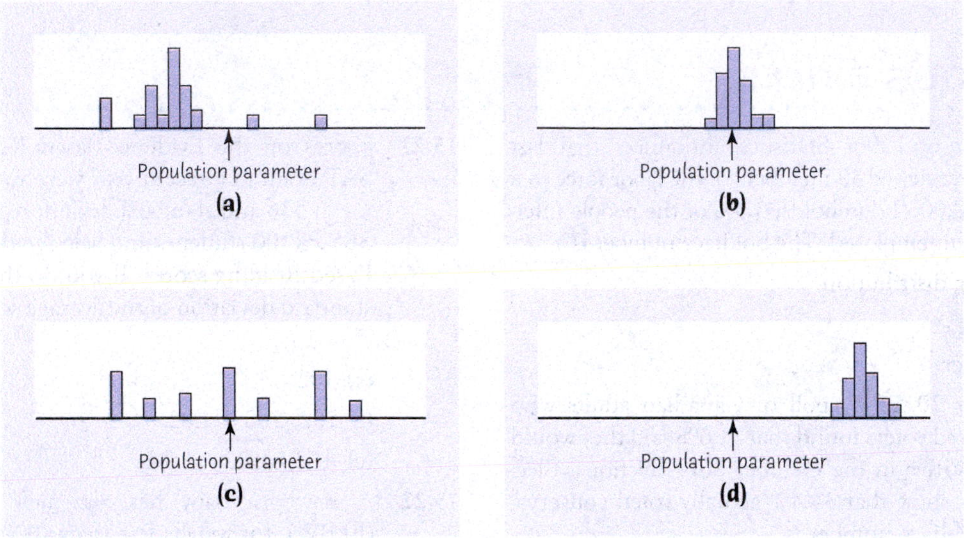

FIGURE 15.9
Take many samples from the same population and make a histogram of the values taken by a sample statistic. Here are four different sampling methods for Question 15.25.

CHAPTER 15 EXERCISES

15.26 **Testing Glass.** How well materials conduct heat matters when designing houses. As a test of a new measurement process, 10 measurements are made on pieces of glass known to have conductivity **1**. The average of the 10 measurements is **1.07**. For each of the boldface numbers, indicate whether it is a parameter or a statistic. Explain your answer.

15.27 **Statistics Anxiety.** What can teachers do to alleviate statistics anxiety in their students? To explore this question, statistics anxiety for students in two classes was compared. In one class, the instructor lectured in a formal manner, including dressing formally. In the other, the instructor was less formal, dressed casually, was more personal, used humor, and called on students by their first names. Anxiety was measured using a questionnaire. Higher scores indicate a greater level of anxiety. The mean anxiety score for students in the formal lecture class was **25.40**; in the informal class, the mean was **20.41**. For each of the boldface numbers, indicate whether it is a parameter or a statistic. Explain your answer.

15.28 **Roulette.** A roulette wheel has 38 slots, of which 18 are black, 18 are red, and two are green. When the wheel is spun, the ball is equally likely to come to rest in any of the slots. One of the simplest wagers is to choose red or black. A bet of $1 on red returns $2 if the ball lands in a red slot. Otherwise, the player loses the dollar. When gamblers bet on red or black, the two green slots result in losses. Because the probability of winning $2 is 18/38, the mean payoff from a $1 bet is twice 18/38, or 94.7 cents. Explain what the law of large numbers tells us about what will happen if a gambler makes very many bets on red.

MagMos/Getty Images

15.29 The Medical College Admission Test. Almost all medical schools in the United States require students to take the Medical College Admission Test (MCAT). To estimate the mean score μ of those who took the MCAT on your campus, you will obtain the scores of an SRS of students. The scores follow a Normal distribution, and from published information, you know that the standard deviation of scores for all MCAT takers is 10.6. Suppose that (unknown to you) the mean score of those taking the MCAT on your campus is 500.0.

(a) If you choose one student at random, what is the probability that the student's score is between 495 and 505?

(b) You sample 25 students. What is the sampling distribution of their average score $\bar{x}$?

(c) What is the probability that the mean score of your sample is between 495 and 505?

15.30 Glucose Testing. Shelia's doctor is concerned that she may suffer from gestational diabetes (high blood glucose levels during pregnancy). There is variation both in the actual glucose level and in the blood test that measures the level. In a test to screen for gestational diabetes, a patient is classified as needing further testing for gestational diabetes if the glucose level is above 130 milligrams per deciliter (mg/dL) one hour after having a sugary drink. Shelia's measured glucose level one hour after the sugary drink varies according to the Normal distribution with $\mu = 122$ mg/dL and $\sigma = 12$ mg/dL.

(a) If a single glucose measurement is made, what is the probability of Shelia being diagnosed as needing further testing for gestational diabetes?

(b) If measurements are made on four separate days and the mean result is compared with the criterion 130 mg/dL, what is the probability that Shelia is diagnosed as needing further testing for gestational diabetes?

15.31 Daily Activity. It appears that people who are mildly obese are less active than leaner people. One study looked at the average number of minutes per day that people spend standing or walking.[7] Among mildly obese people, the mean number of minutes of daily activity (standing or walking) is approximately Normally distributed with mean 373 minutes and standard deviation 67 minutes. The mean number of minutes of daily activity for lean people is approximately Normally distributed with mean 526 minutes and standard deviation 107 minutes. A researcher records the minutes of activity for an SRS of five mildly obese people and an SRS of five lean people.

(a) What is the probability that the mean number of minutes of daily activity of the five mildly obese people exceeds 420 minutes?

(b) What is the probability that the mean number of minutes of daily activity of the five lean people exceeds 420 minutes?

15.32 Glucose Testing (continued). Shelia's measured glucose level one hour after having a sugary drink varies according to the Normal distribution with $\mu = 122$ mg/dL and $\sigma = 12$ mg/dL. What is the level L such that there is probability only 0.05 that the mean glucose level of four test results falls above L? (*Hint:* This requires a backward Normal calculation. See page 89 in Chapter 3 if you need a review.)

15.33 Pollutants in Auto Exhausts. In 2017, the entire fleet of light-duty vehicles sold in the United States by each manufacturer was required to emit an average of no more than 86 milligrams per mile (mg/mi) of nitrogen oxides (NOX) and nonmethane organic gas (NMOG) over the useful life (150,000 miles of driving) of the vehicle. NOX + NMOG emissions over the useful life for one car model vary Normally with mean 80 mg/mi and standard deviation 4 mg/mi.

(a) What is the probability that a single car of this model emits more than 86 mg/mi of NOX + NMOG?

(b) A company has 25 cars of this model in its fleet. What is the probability that the average NOX + NMOG level $\bar{x}$ of these cars is above 86 mg/mi?

15.34 Runners. In a study of exercise, a large group of male runners walk on a treadmill for six minutes. After this exercise, their heart rates vary with mean 8.8 beats per five seconds and standard deviation 1.0 beats per five seconds. The researcher records the number of heartbeats per five seconds for each runner over a period of time. This distribution takes only whole-number values, so it is certainly not Normal.

(a) Let $\bar{x}$ be the mean number of beats per five seconds after measuring heart rate for 24 five-second intervals (two minutes). What is the approximate distribution of $\bar{x}$ according to the central limit theorem?

(b) What is the approximate probability that $\bar{x}$ is less than 8?

(c) What is the approximate probability that the heart rate of a runner is less than 100 beats per minute? (*Hint:* Restate this event in terms of $\bar{x}$.)

15.35 Pollutants in Auto Exhausts (continued). The level of nitrogen oxides (NOX) and nonmethane organic gas (NMOG) in the exhaust over the useful life (150,000 miles of driving) of cars of a particular model varies Normally with mean 80 mg/mi and standard deviation 4 mg/mi. A company has 25 cars of this model in its fleet. What is the level L such that the probability that the average NOX + NMOG level $\bar{x}$ for the fleet is greater than L is only 0.01? (*Hint:* This requires a backward Normal calculation. See page 89 in Chapter 3 if you need a review.)

15.36 Returns on Stocks. Andrew plans to retire in 40 years. He plans to invest part of his retirement funds in stocks, so he seeks out information on past returns. He learns that from 1969 to 2018, the annual returns on the S&P 500 had mean 9.8% and standard deviation 16.8%.[8] The mean return over even a moderate number of years is close to Normal. What is the probability (assuming that the past pattern of variation continues) that the mean annual return on common stocks over the next 40 years will exceed 10%? What is the probability that the mean return will be less than 5%? Follow the four-step process as illustrated in Example 15.8.

15.37 Airline Passengers Get Heavier. In response to the increasing weight of airline passengers, the Federal Aviation Administration (FAA) in 2003 told airlines to assume that passengers average 195 pounds in the winter, including clothing and carry-on baggage. But passengers vary, and the FAA did not specify a standard deviation. A reasonable standard deviation is 35 pounds. Weights are not Normally distributed, especially when the population includes both men and women, but they are not very non-Normal. A commuter plane carries 22 passengers. What is the approximate probability that the total weight of the passengers exceeds 4500 pounds? Use the four-step process to guide your work. (*Hint:* To apply the central limit theorem, restate the problem in terms of the mean weight.)

anandoart/Shutterstock

15.38 Sampling Students. To estimate the mean score μ of those who took the Medical College Admission Test on your campus, you will obtain the scores of an SRS of students. From published information, you know that the scores are approximately Normal, with standard deviation about 10.6. How large an SRS must you take to reduce the standard deviation of the sample mean score to 1?

15.39 Sampling Students, Continued. To estimate the mean score μ of those who took the Medical College Admission Test on your campus, you will obtain the scores of an SRS of students. From published information, you know that the scores are approximately Normal, with standard deviation about 10.6. You want your sample mean $\bar{x}$ to estimate μ with an error of no more than one point in either direction.

(a) What standard deviation must $\bar{x}$ have so that 99.7% of all samples give an $\bar{x}$ within one point of μ? (Use the 68–95–99.7 rule.)

(b) How large an SRS do you need in order to reduce the standard deviation of $\bar{x}$ to the value you found in part (a)?

15.40 Playing the Numbers. The numbers racket is a well-entrenched illegal gambling operation in most large cities. One version works as follows: you choose one of the 1000 three-digit numbers 000 to 999 and pay your local numbers runner a dollar to enter your bet. Each day, one three-digit number is chosen at random and pays off $600. The mean payoff for the population of thousands of bets is $\mu = 60$ cents. Joe makes one bet every day for many years. Explain what the law of large numbers says about Joe's results as he keeps on betting.

15.41 Playing the Numbers: A Gambler Gets Chance Outcomes. The law of large numbers tells us what happens in the long run. Like many games of chance, the numbers racket described in the previous exercise has outcomes that vary considerably—one three-digit number wins $600 and all others win nothing—that gamblers never reach "the long run." Even after many bets, their average winnings may not be close to the mean. For the numbers racket, the mean payout for single bets is $0.60 (60 cents), and the standard deviation of payouts is about $18.96. If Joe plays 350 days a year for 40 years, he makes 14,000 bets.

(a) What are the mean and standard deviation of the average payout $\bar{x}$ that Joe receives from his 14,000 bets?

(b) The central limit theorem says that his average payout is approximately Normal with the mean and standard deviation you found in part (a). What is the approximate probability that Joe's average payout per bet is between $0.50 and $0.70? You see that Joe's average may not be very close to the mean $0.60 even after 14,000 bets.

15.42 Playing the Numbers: The House Has a Business. Unlike Joe (see the previous exercise), the operators of the numbers racket can rely on the law of large numbers. It is said that the New York City mobster Casper Holstein took as many as 25,000 bets per day in the Prohibition era. That's 150,000 bets in a week if he takes Sunday off. Casper's mean winnings per bet are $0.40 (he pays out an average of 60 cents per dollar bet to people like Joe and keeps the other 40 cents). His standard deviation for single bets is about $18.96, the same as Joe's.

New York Daily News Archive/Getty Images

(a) What are the mean and standard deviation of Casper's average winnings $\bar{x}$ on his 150,000 bets?

(b) According to the central limit theorem, what is the approximate probability that Casper's average winnings per bet are between \$0.30 and \$0.50? After only a week, Casper can be pretty confident that his winnings will be quite close to \$0.40 per bet.

15.43 Can We Trust the Central Limit Theorem? The central limit theorem says that "when n is large," we can act as if the distribution of a sample mean $\bar{x}$ is close to Normal. How large a sample we need depends on how far the population distribution is from being Normal. Example 15.8 shows that we can trust this Normal approximation for quite moderate sample sizes even when the population has a strongly skewed continuous distribution.

The central limit theorem requires much larger samples for Joe's bets with his local numbers racket (see the previous exercises on "playing the numbers"). The population of individual bets has a finite distribution with only two possible outcomes: \$600 (probability 0.001) and \$0 (probability 0.999). This distribution has mean $\mu = 0.6$ and standard deviation about $\sigma = 18.96$. With more math and good software, we can find exact probabilities for Joe's average winnings.

(a) If Joe makes 14,000 bets, the exact probability $P(0.5 \leq \bar{x} \leq 0.7) = 0.4961$. How accurate was your Normal approximation from part (b) of Exercise 15.41?

(b) If Joe makes only 3500 bets, $P(0.5 \leq \bar{x} \leq 0.7) = 0.4048$. How accurate is the Normal approximation for this probability?

(c) If Joe and his buddies make 150,000 bets, $P(0.5 \leq \bar{x} \leq 0.7) = 0.9629$. How accurate is the Normal approximation?

15.44 What's the Mean? (applet) Suppose that you roll three balanced dice. We wonder what the mean number of spots on the up-faces of the three dice is. The law of large numbers says that we can find out by experience: roll three dice many times, and the average number of spots will eventually approach the true mean. Set up the *Law of Large Numbers* applet to roll three dice. Don't click "Show mean" yet. Roll the dice until you are confident you know the mean quite closely; then click "Show mean" to verify your discovery. What is the mean? Make a rough sketch of the path the averages $\bar{x}$ followed as you kept adding more rolls.

15.45 Statistical Significance? Look again at Exercise 12.50 (page 293). If your Canadian friend correctly identifies all three cups with the milk added first, would you regard the result as statistically significant?

CHAPTER

16

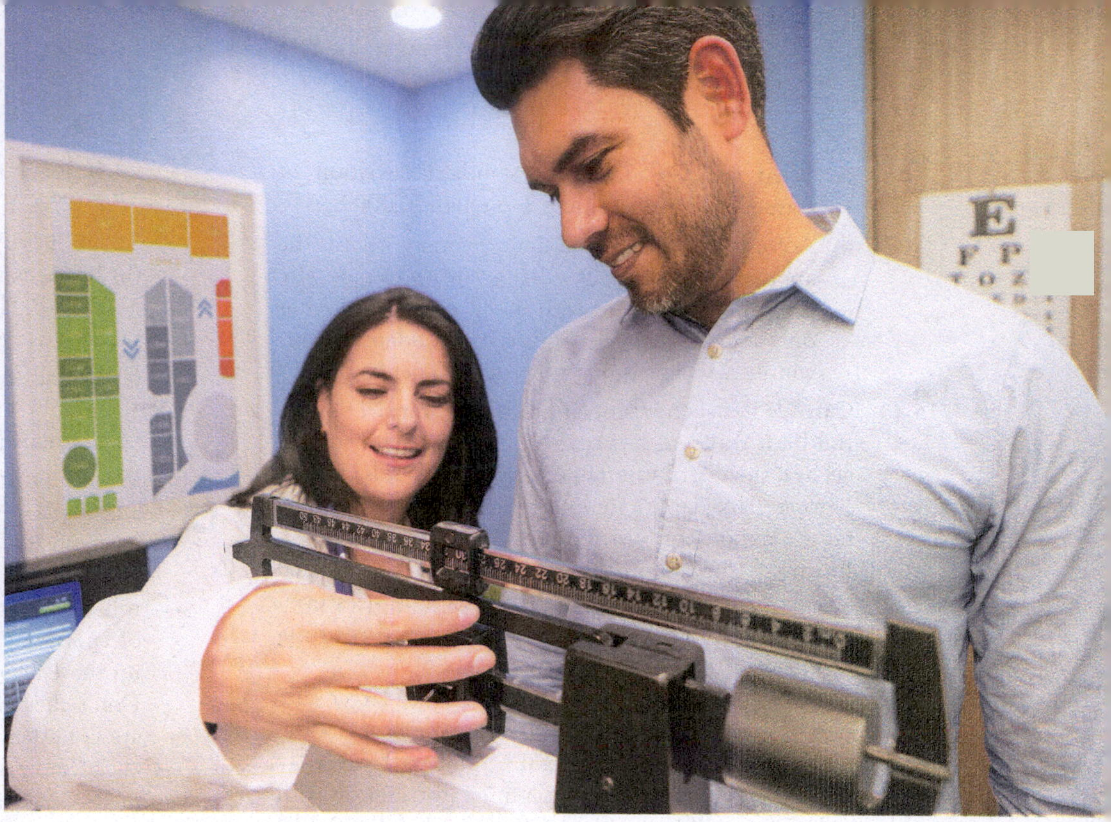

When you complete this chapter, you will be able to:

16.1 Use the principles of statistical inference and estimation to interpret confidence intervals.

16.2 Articulate the meaning of statements involving levels of confidence and margins of error.

16.3 Compute confidence intervals for means after confirming that necessary conditions are met.

16.4 Understand how the margin of error changes with sample size and level of confidence.

Confidence Intervals: The Basics

Chapters 8 and 9 say that the way we produce data (sampling, experimental design) affects whether we have a good basis for generalizing to some wider population. Chapters 12, 13, and 14 discuss probability, the mathematical tool that determines the nature of the inferences we make. Chapter 15 discusses sampling distributions, which tell us how repeated SRSs behave and what a statistic (in particular, a sample mean) computed from our sample is likely to tell us about the corresponding parameter of the population from which the sample was selected. In this chapter, we discuss the basic reasoning of statistical estimation, with emphasis on estimating a population mean.

After we have selected a sample, we know the responses of the individuals in the sample. The usual reason for taking a sample is not to learn about the individuals in the sample but to *infer* from the sample data some conclusion about the wider population that the sample represents.

Statistical Inference

Statistical inference provides methods for drawing conclusions about a population from sample data.

Because a different sample might lead to different conclusions, we can't be certain that our conclusions are correct. Statistical inference uses the language

andresr/Getty Images

of probability to say how trustworthy our conclusions are. This chapter introduces one of the two most common types of inference, *confidence intervals* for estimating the value of a population parameter. The next chapter discusses the other common type of inference, *tests of significance* for assessing the evidence for a claim about a population parameter. Both types of inference are based on the sampling distributions of statistics. That is, both use probability to say what would happen if we applied the inference method many times.

This chapter presents the basic reasoning of statistical inference. To make the reasoning as clear as possible, we start with a setting that is too simple to be realistic. Here is the setting for our work in this chapter.

Simple Conditions for Inference About a Mean

1. We have a simple random sample (SRS) from the population of interest. There is no nonresponse or other practical difficulty. The population is large compared to the size of the sample.
2. The variable we measure has an exactly Normal distribution $N(\mu, \sigma)$ in the population.
3. We don't know the population mean μ. But we do know the population standard deviation σ.

The condition that the population is large relative to the size of the sample will be adequately satisfied if the population is, say, at least 20 times as large.[1] *The conditions that we have a perfect SRS, that the population is exactly Normal, and that we know the population* σ *are all unrealistic.* Chapter 18 begins to move from the "simple conditions" toward the reality of statistical practice. Later chapters deal with inference in fully realistic settings.

If these "simple conditions" are unrealistic, why study them? One reason is that under these simple conditions, we can apply what we have learned in previous chapters about the Normal distribution and the sampling distribution of a sample mean to develop step-by-step methods for inference about a mean. The reasoning used under simple conditions applies to more realistic settings with more complicated mathematics.

Although we never know whether a population is exactly Normal and we never know the population σ, the methods we discuss in this and the next two chapters are approximately correct for sufficiently large sample sizes, provided we treat the sample standard deviation as though it were the population σ. Thus, there are situations (admittedly rare) where these methods can be used in practice.

16.1 The Reasoning of Statistical Estimation

Body mass index (BMI) is used to screen for possible weight problems. It is calculated as weight divided by the square of height, measuring weight in kilograms and height in meters. Many online BMI calculators allow you to enter weight in pounds and height in inches. Adults with BMI less than 18.5 kg/m^2 are considered underweight, and those with BMI greater than 25 kg/m^2 may be overweight. For data about BMI, we turn to the National Health and Nutrition Examination Survey (NHANES), a continuing government sample survey that monitors the health of the American population.

EXAMPLE 16.1 Body Mass Index of Young Men

An NHANES report gives data for 936 men aged 20–29 years.[2] The mean BMI of these 936 men was $\bar{x} = 27.2$. On the basis of this sample, we want to estimate the mean BMI μ in the population of all 23.2 million American men in this age group.

To match the "simple conditions," we will treat the NHANES sample as an SRS from a Normal population, and we will assume that we know that the standard deviation $\sigma = 11.6$. (The sample standard deviation for these 936 men is 11.63 kg/m^2. For purposes of the example, we round this to 11.6 and act as though this is the population standard deviation σ.)

Here is the reasoning of statistical estimation in a nutshell:

1. To estimate the unknown population mean BMI μ, use the mean $\bar{x} = 27.2$ of the random sample. We don't expect $\bar{x}$ to be exactly equal to μ, so we want to say how accurate this estimate is.

2. We know the sampling distribution of $\bar{x}$. In repeated samples, $\bar{x}$ has the Normal distribution with mean μ and standard deviation $\sigma/\sqrt{n}$. So the average BMI $\bar{x}$ of an SRS of 936 young men has standard deviation

$$\frac{\sigma}{\sqrt{n}} = \frac{11.6}{\sqrt{936}} = 0.4 \quad \text{(rounded off)}$$

3. The 95 part of the 68–95–99.7 rule for Normal distributions says that $\bar{x}$ is within 2 standard deviations of the mean μ in 95% of all samples. The standard deviation is 0.4, so 2 standard deviations is 0.8. Thus, for 95% of all samples of size 936, the distance between the sample mean $\bar{x}$ and the population mean μ is less than 0.8. If we estimate that μ lies somewhere in the interval from $\bar{x} - 0.8$ to $\bar{x} + 0.8$, we'll be right for 95% of all possible samples. For this particular sample, this interval is

$$\bar{x} - 0.8 = 27.2 - 0.8 = 26.4$$

to

$$\bar{x} + 0.8 = 27.2 + 0.8 = 28.0$$

4. Because we got the interval 26.4 to 28.0 from a method that captures the population mean for 95% of all possible samples, we say that we are *95% confident* that the mean BMI μ of all young men is some value in that interval—no lower than 26.4 and no higher than 28.0.

STATISTICS IN YOUR WORLD

Ranges Are for Statistics?

Many people like to think that statistical estimates are exact. The Nobel Prize-winning economist Daniel McFadden tells a story of his time on the Council of Economic Advisers. Presented with a range of forecasts for economic growth, President Lyndon Johnson replied: "Ranges are for cattle; give me one number."

The big idea is that the sampling distribution of $\bar{x}$ tells us how close to μ the sample mean $\bar{x}$ is likely to be. Statistical estimation just turns that information around to say how close to $\bar{x}$ the unknown population mean μ is likely to be. We call the interval of numbers between the values $\bar{x} \pm 0.8$ a *95% confidence interval* for μ.

APPLY YOUR KNOWLEDGE

16.1 Number Skills of Eighth-Graders. The National Assessment of Educational Progress (NAEP) includes a mathematics test for eighth-grade students.[3] Scores on the test range from 0 to 500. Demonstrating the ability to use the mean to solve a problem is an example of the skills and knowledge associated with performance at the Basic level. An example of the knowledge and skills associated with the Proficient level is being able to read and interpret a stem-and-leaf plot.

In 2019, 147,400 eighth-graders were in the NAEP sample for the mathematics test. The mean mathematics score was $\overline{x} = 282$. We want to estimate the mean score μ in the population of all eighth-graders. Consider the NAEP sample as an SRS from a Normal population with standard deviation $\sigma = 40$.

(a) If we take many samples, the sample mean $\overline{x}$ varies from sample to sample according to a Normal distribution with mean equal to the unknown mean score μ in the population. What is the standard deviation of this sampling distribution?

(b) According to the 95 part of the 68–95–99.7 rule, 95% of all values of $\overline{x}$ fall within ____ on either side of the unknown mean μ. What is the missing number?

(c) What is the 95% confidence interval for the population mean score μ based on this one sample?

16.2 Retaking the SAT. An SRS of 400 high school seniors gained an average of $\overline{x} = 40$ points in their second attempt at the SAT Mathematics exam. Assume that the change in score has a Normal distribution with standard deviation $\sigma = 25$. We want to estimate the mean change in score μ in the population of all high school seniors.

(a) Give a 95% confidence interval for μ based on this sample.

(b) Based on your confidence interval in part (a), how certain are you that the mean change in score μ in the population of all high school seniors is greater than 0? (*Hint:* Does the interval in part (a) include 0?)

16.2 Margin of Error and Confidence Level

The 95% confidence interval for the mean BMI of young men, based on the NHANES sample, is $\overline{x} \pm 0.8$. Once we have the sample results in hand, we know that for this sample, $\overline{x} = 27.2$, so that our confidence interval is 27.2 ± 0.8. Most confidence intervals have a form similar to this:

$$\text{estimate} \pm \text{margin of error}$$

The estimate ($\overline{x} = 27.2$ in our example) is our guess for the value of the unknown parameter. The *margin of error* ± 0.8 shows how accurate we believe our guess is, based on the variability of the estimate. We have a 95% confidence interval because the interval $\overline{x} \pm 0.8$ catches the unknown parameter in 95% of all possible samples.

Margin of Error

The **margin of error** is a number that is added to, and subtracted from, a statistical estimate to define a confidence interval at a given confidence level.

Confidence Interval

A **level C confidence interval** for a parameter has two parts:

- An interval calculated from the data, usually of the form

$$\text{estimate} \pm \text{margin of error}$$

- A **confidence level C,** which gives the probability that the interval will capture the true parameter value in repeated samples. That is, the confidence level is the success rate for the method.

This form for a confidence interval and its interpretation apply to most parameters we will consider in this book, including means and proportions.

Users can choose the confidence level, usually 90% or higher because we usually want to be quite sure of our conclusions. The most common confidence level is 95%.

Interpreting a Confidence Level

The confidence level is the success rate of the method that produces the interval. We don't know whether the 95% confidence interval from a particular sample is one of the 95% that capture μ or one of the unlucky 5% that miss.

To say that we are 95% **confident** that the unknown μ lies between 26.4 and 28.0 is shorthand for "We got these numbers using a method that gives correct results 95% of the time."

EXAMPLE 16.2 Statistical Estimation in Pictures

Figures 16.1 and 16.2 illustrate the behavior of confidence intervals. Study these figures carefully. If you understand what they say, you have mastered one of the big ideas of statistics.

Figure 16.1 illustrates the behavior of the interval $\bar{x} \pm 0.8$ for the mean BMI of young men. Starting with the population, imagine taking many SRSs of 936 young men. The first sample has $\bar{x} = 27.2$, the second has $\bar{x} = 27.4$, the third has $\bar{x} = 26.8$, and so on. The sample mean varies from sample to sample, but when we use the formula $\bar{x} \pm 0.8$ to get an interval based on each sample, *95% of these intervals capture the unknown population mean* μ. Notice that we use the same margin of error, 0.8, for each interval because the sample size and standard deviation σ are the same.

Figure 16.2 illustrates the idea of a 95% confidence interval in a different form. It shows the result of drawing many SRSs from the same population and calculating a 95% confidence interval from each sample. The center of each interval is at $\bar{x}$ and therefore varies from sample to sample. The sampling distribution of $\bar{x}$ appears at the top of the figure to show the long-term pattern of this variation. The population mean μ is at the center of the sampling distribution. The 95% confidence intervals from 25 SRSs appear underneath. The center $\bar{x}$ of each interval is marked by a dot. The arrows on either side of the dot span the confidence interval. All except one of these 25 intervals capture the true value of μ. If we take a very large number of samples, 95% of the confidence intervals will contain μ.

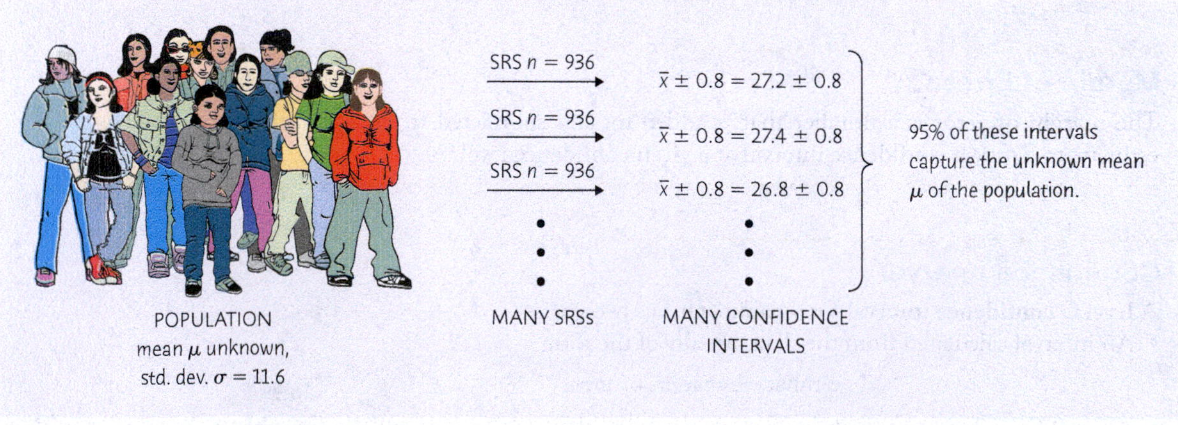

FIGURE 16.1
To say that $\bar{x} \pm 0.8$ is a 95% confidence interval for the population mean μ is to say that, in repeated samples, 95% of these intervals capture μ.

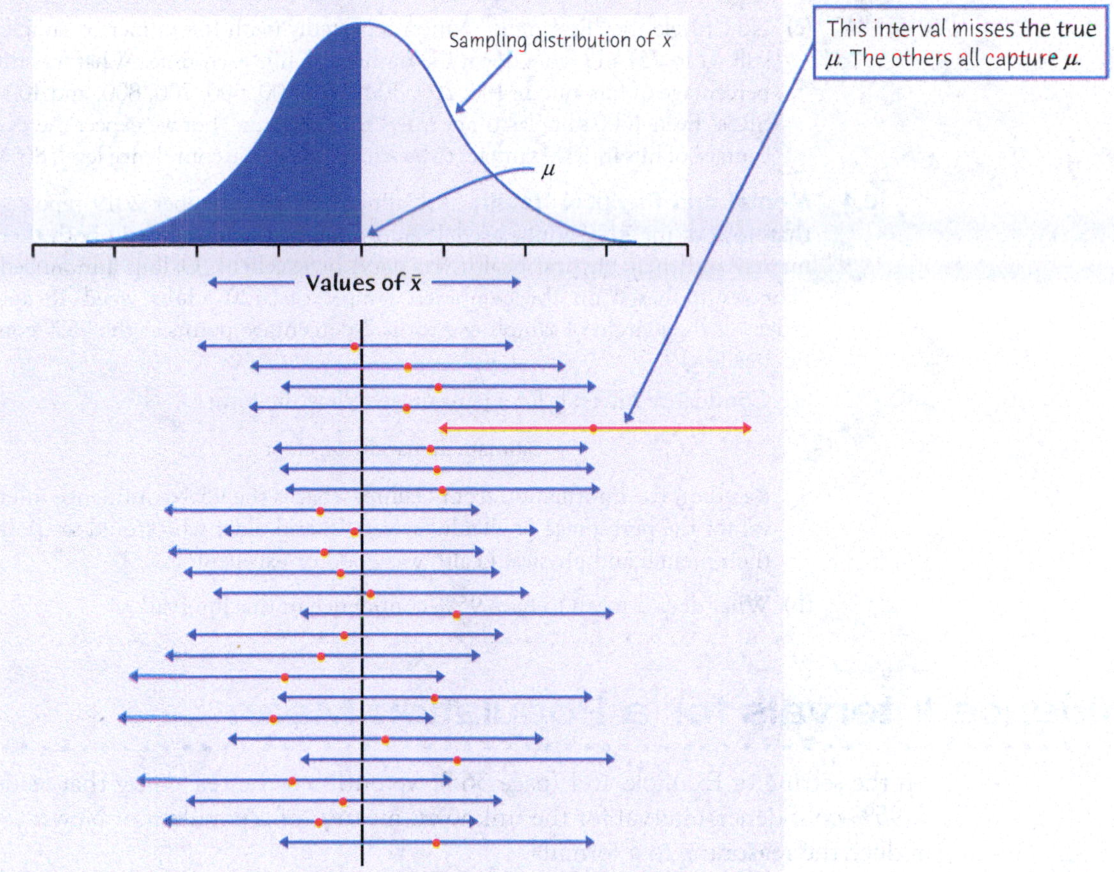

FIGURE 16.2
Twenty-five samples from the same population gave these 95% confidence intervals. In the long run, 95% of all samples give an interval that contains the population mean μ.

The *Confidence Intervals* applet creates figures similar to Figure 16.2. You can use the applet to watch confidence intervals from one sample after another capture or fail to capture the true parameter.

applet

APPLY YOUR KNOWLEDGE

16.3 Confidence Intervals in Action. The idea of an 80% confidence interval is that in 80% of all samples, the method produces an interval that captures the true parameter value. That's not high enough confidence for practical use, but 80% hits and 20% misses make it easy to see how a confidence interval behaves in repeated samples from the same population. Go to the *Confidence Intervals* applet.

applet

(a) Set the confidence level to 80% and the sample size to 50. Click "Sample" to choose an SRS and calculate the confidence interval. Do this 10 times to simulate 10 SRSs with their 10 confidence intervals. How many of the 10 intervals captured the true mean μ? How many missed?

(b) You see that we can't predict whether the next sample will hit or miss. The confidence level, however, tells us what percentage will hit in the long run. Now move the "Number of Samples" slider to "25" to get the confidence intervals from 25 SRSs. How many hit?

(c) Now toggle the "Re-sample" button repeatedly (with the numer of samples still set to 25) and write down the number of hits each time. What was the percentage of hits among 100, 200, 300, 400, 500, 600, 700, 800, and 1000 SRSs? Even 1000 samples is not truly "the long run," but we expect the percentage of hits in 1000 samples to be fairly close to the confidence level, 80%.

16.4 Mental and Physical Health. A Gallup Poll in December 2019 reported that 46% of the 5120 adults aged 18 and older in the sample said both their mental and their physical health was good or excellent. Gallup announced, "For results based on the combined sample of 5120 adults, aged 18 and older . . . the margin of sampling error is 2 percentage points at the 95% confidence level."

(a) Confidence intervals for a percentage follow the form

$$\text{estimate} \pm \text{margin of error}$$

Based on the information from Gallup, what is the 95% confidence interval for the percentage of all adults aged 18 and older who would say both their mental and physical health was good or excellent?

(b) What does it mean to have 95% confidence in this interval?

16.3 Confidence Intervals for a Population Mean

In the setting of Example 16.1 (page 368), we outlined the reasoning that leads to a 95% confidence interval for the unknown mean μ of a population. Now we will reduce the reasoning to a formula.

In Example 16.1, what role did 95% play in determining the confidence interval? To find a 95% confidence interval for the mean BMI of young men, we first caught the central 95% of the Normal sampling distribution by going out *two* standard deviations in both directions from the mean. The value 95% determined how many standard deviations we go out in both directions from the mean to capture this central 95%. To find a level C confidence interval, we first capture the central area C under the Normal sampling distribution. How many standard deviations must we go out in both directions from the mean to capture this central area C? Because all Normal distributions are the same in the standard scale, we can obtain everything we need from the standard Normal curve.

Figure 16.3 shows how the central area C under a standard Normal curve is marked off by two points z^* and $-z^*$. Numbers like z^* that mark off specified areas are called *critical values* of the standard Normal distribution.

Critical Value

The **critical value** is a number z^* chosen so that the standard Normal curve has a prespecified area C between z^* and $-z^*$.

Values of z^* for many choices of C appear at the bottom of Table C in the back of the book, in the row labeled z^*. Here are the entries for the most common confidence levels:

Confidence level C	90%	95%	99%
Critical value z^*	1.645	1.960	2.576

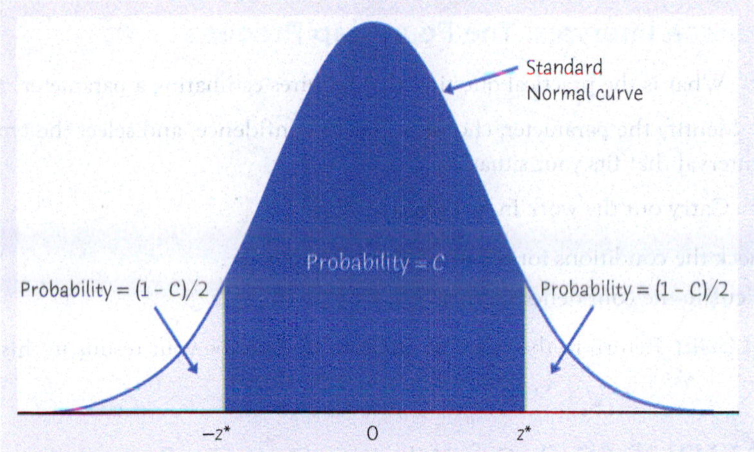

FIGURE 16.3
The critical value z^* is the number that catches central probability C under a standard Normal curve between $-z^*$ and z^*.

You see that, for C = 95%, the table gives $z^* = 1.960$. This is a bit more precise than the approximate value $z^* = 2$ based on the 68–95–99.7 rule. You can, of course, use software to find critical values z^*, as well as the entire confidence interval.

Figure 16.3 shows that there is area C under the standard Normal curve between $-z^*$ and z^*. So *any* Normal curve has area C within z^* standard deviations on either side of its mean. That is, a proportion of C values lie between $\mu - (z^* \times \sigma)$ and $\mu + (z^* \times \sigma)$.

The Normal sampling distribution of $\bar{x}$ has area C within $z^* \times \sigma/\sqrt{n}$ on either side of the population mean μ because it has mean μ and standard deviation $\sigma/\sqrt{n}$. If we start at $\bar{x}$ and go out $z^*\sigma/\sqrt{n}$ in both directions, we get an interval that contains the population mean μ in a proportion C of all samples. This interval is

$$\text{from } \bar{x} - z^*\frac{\sigma}{\sqrt{n}} \text{ to } \bar{x} + z^*\frac{\sigma}{\sqrt{n}}$$

or

$$\bar{x} \pm z^*\frac{\sigma}{\sqrt{n}}$$

It is a level C confidence interval for μ.

Confidence Interval for the Mean of a Normal Population

Draw an SRS of size n from a Normal population having unknown mean μ and known standard deviation σ. A level C **confidence interval for** μ is

$$\bar{x} \pm z^*\frac{\sigma}{\sqrt{n}}$$

The critical value z^* corresponding to the confidence level C is illustrated in Figure 16.3 and found at the bottom of Table C.

The steps in finding a confidence interval mirror the overall four-step process for organizing statistical problems.

Confidence Intervals: The Four-Step Process

STATE: What is the practical question that requires estimating a parameter?

PLAN: Identify the parameter, choose a level of confidence, and select the type of confidence interval that fits your situation.

SOLVE: Carry out the work in two phases:

1. Check the conditions for the interval you plan to use.
2. Calculate the confidence interval.

CONCLUDE: Return to the practical question to describe your results in this setting.

EXAMPLE 16.3 Good Weather, Good Tips?

4step

Ariel Skelley/AGE Fotostock

STATE: Does the expectation of good weather lead to more generous behavior? Psychologists studied the size of the tip in a restaurant when a message indicating that the next day's weather would be good was written on the bill. Here are tips from 20 patrons, measured in percentage of the total bill:[4] TIP2

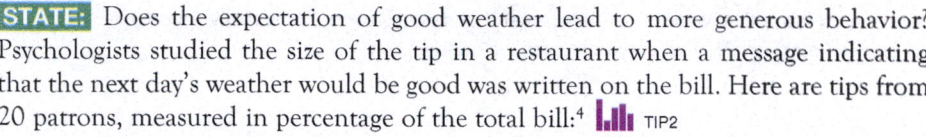

20.8	18.7	19.9	20.6	21.9	23.4	22.8	24.9	22.2	20.3
24.9	22.3	27.0	20.4	22.2	24.0	21.1	22.1	22.0	22.7

This is one of three sets of measurements made, the others being tips received when the message on the bill said that the next day's weather would not be good or there was no message on the bill. We want to estimate the mean tip for comparison with tips under the other conditions. As part of the "simple conditions," suppose that, from past experience with patrons of this restaurant, we know that the standard deviation of percentage tip is $\sigma = 2$.

PLAN: We will estimate the mean percentage tip μ for all patrons of this restaurant when they receive a message on their bill indicating that the next day's weather will be good by giving a 95% confidence interval. The confidence interval just introduced fits this situation.

SOLVE: We should start by checking the conditions for inference. For this example, we will first find the interval and then discuss how statistical practice deals with conditions that are never perfectly satisfied.

The mean percentage tip of the sample is $\bar{x} = 22.21$. For 95% confidence, the critical value is $z^* = 1.960$. A 95% confidence interval for μ is therefore

$$\begin{aligned} \bar{x} \pm z^* \frac{\sigma}{\sqrt{n}} &= 22.21 \pm 1.960 \frac{2}{\sqrt{20}} \\ &= 22.21 \pm 0.88 \\ &= 21.33 \text{ to } 23.09 \end{aligned}$$

CONCLUDE: We are 95% confident that the mean percentage tip for all patrons of this restaurant when their bill contains a message that the next day's weather will be good is between 21.33 and 23.09.

In practice, the first part of the "Solve" step is to check the conditions for inference. The "simple conditions" are as follows:

1. **SRS:** We don't have an actual SRS from the population of all patrons of this restaurant. Scientists often act as if subjects are SRSs if there is nothing special about how the subjects were obtained. But it is always better to have an actual SRS because

otherwise we can never be sure that hidden biases aren't present. This study was actually a randomized comparative experiment in which these 20 patrons were assigned at random from a larger group of patrons to get one of the treatments being compared.

2. **Normal distribution:** The psychologists expect from past experience that measurements like this on patrons of the same restaurant under the same conditions will follow approximately a Normal distribution. We can't look at the population, but we can examine the sample. Figure 16.4 is a stemplot that is roughly bell-shaped, with perhaps a modest outlier but no strong skewness. Shapes like this often occur in small samples from Normal populations, so we have no reason to doubt that the population distribution is Normal.

3. **Known σ:** It really is unrealistic to suppose that we know that $\sigma = 2$. We will see in Chapter 20 that it is easy to do away with the need to know σ.

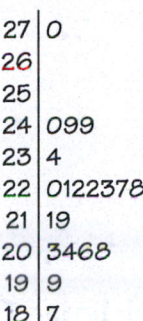

27	0
26	
25	
24	099
23	4
22	0122378
21	19
20	3468
19	9
18	7

FIGURE 16.4
Stemplot of the percentage tips for Example 16.3.

As this discussion suggests, inference methods are often used when conditions like SRS and Normal population are not exactly satisfied. In this introductory chapter, we act as though the "simple conditions" are satisfied. In reality, wise use of inference requires judgment. Chapter 17 and the later chapters on each inference method will give you better bases for judgment.

APPLY YOUR KNOWLEDGE

16.5 Find a Critical Value. The critical value z^* for confidence level 75% is not in Table C. Use software or Table A of standard Normal probabilities to find z^*. Include in your answer a sketch like Figure 16.3 with $C = 0.75$ and your critical value z^* marked on the axis.

16.6 Measuring Melting Point. The National Institute of Standards and Technology (NIST) supplies "standard materials" whose physical properties are supposed to be known. For example, you can buy from NIST a copper sample whose melting point is certified to be 1084.80°C. Of course, no measurement is exactly correct. NIST knows the variability of its measurements very well, so it is quite realistic to assume that the population of all measurements of the same sample has the Normal distribution with mean μ equal to the true melting point and standard deviation $\sigma = 0.25$°C. Here are six measurements on the same copper sample, which is supposed to have melting point 1084.80°C: MELT

1084.55 1084.89 1085.02 1084.79 1084.69 1084.86

NIST wants to give the buyer of this copper sample a 90% confidence interval for its true melting point. What is this interval? Follow the four-step process as illustrated in Example 16.3.

16.7 IQ Test Scores. Here are the IQ test scores of 74 seventh-grade girls in a midwestern school district:[5] MWSTIQ

115	111	97	112	104	106	113	109	113	128	128
118	113	124	127	136	106	123	124	126	116	127
119	97	102	110	120	103	115	93	123	119	110
110	107	105	105	110	90	114	106	114	100	104
89	102	91	114	114	103	105	111	108	130	120
132	111	128	118	119	86	107	100	111	103	107
112	107	103	98	96	112	112	93			

(a) These 74 girls are an SRS of all seventh-grade girls in the school district. Suppose that the standard deviation of IQ scores in this population is known to be $\sigma = 11$. We expect the distribution of IQ scores to be close to Normal. Make a stemplot of the distribution of these 74 scores (split the stems) to verify that there are no major departures from Normality. You have now checked the "simple conditions" to the extent possible.

(b) Estimate the mean IQ score for all seventh-grade girls in the school district, using a 99% confidence interval. Follow the four-step process as illustrated in Example 16.3.

(c) Does 99% confidence mean that your interval in part (b) is guaranteed to include 99% of all IQ test scores of seventh-grade girls in the Midwestern school district? If not, how should you interpret the interval?

16.4 How Confidence Intervals Behave

The z confidence interval $\bar{x} \pm z^* \sigma/\sqrt{n}$ for the mean of a Normal population illustrates several important properties that are shared by all confidence intervals in common use. The user chooses the confidence level, and the margin of error follows from this choice. We would like high confidence and also a small margin of error. High confidence says that our method almost always gives correct answers. A small margin of error says that we have pinned down the parameter quite precisely. The factors that influence the margin of error of the z confidence interval are typical of most confidence intervals.

How do we get a small margin of error? The margin of error for the z confidence interval is

$$\text{margin of error} = z^* \frac{\sigma}{\sqrt{n}}$$

This expression has z^* and σ in the numerator and $\sqrt{n}$ in the denominator. Therefore, the margin of error gets smaller when

- z^* gets smaller. Smaller z^* is the same as lower confidence level C. (Look again at Figure 16.3 on page 373.) *There is a trade-off between the confidence level and the margin of error. To obtain a smaller margin of error from the same data, you must be willing to accept lower confidence.*
- σ is smaller. The standard deviation σ measures the variation in the population. You can think of the variation among individuals in the population as noise that obscures the average value μ. It is easier to pin down μ when σ is small.
- n gets larger. Increasing the sample size n reduces the margin of error for any confidence level. Larger samples thus allow more precise estimates. However, *because n appears under a square root sign, we must take four times as many observations to cut the margin of error in half.*

In practice, we can control the confidence level and sample size, but we can't control σ.

EXAMPLE 16.4 Changing the Margin of Error

In Example 16.3, psychologists recorded the size of the tip of 20 patrons in a restaurant when a message indicating that the next day's weather would be good was written on their bill. The data gave the mean size of the tip, as a percentage of the total bill, as $\bar{x} = 22.21$, and we know that $\sigma = 2$. The 95% confidence interval for the mean percentage tip for all patrons of the restaurant when their bill contains a message that the next day's weather will be good is

$$\bar{x} \pm z^* \frac{\sigma}{\sqrt{n}} = 22.21 \pm 1.960 \frac{2}{\sqrt{20}}$$
$$= 22.21 \pm 0.88$$

The 90% confidence interval based on the same data replaces the 95% critical value $z^* = 1.960$ by the 90% critical value $z^* = 1.645$. This interval is

$$\begin{aligned}\bar{x} \pm z^* \frac{\sigma}{\sqrt{n}} &= 22.21 \pm 1.645 \frac{2}{\sqrt{20}} \\ &= 22.21 \pm 0.74\end{aligned}$$

Lower confidence results in a smaller margin of error, ±0.74 in place of ±0.88. You can calculate that the margin of error for 99% confidence is larger, ±1.15. Figure 16.5 compares these three confidence intervals.

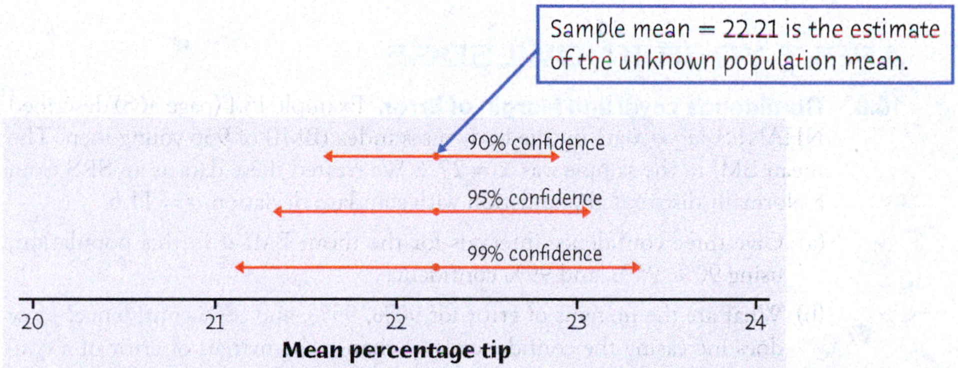

FIGURE 16.5
The lengths of three confidence intervals, for Example 16.4. All three are centered at the estimate $\bar{x} = 22.21$. When the data and the sample size remain the same, higher confidence results in a larger margin of error.

If we had a sample of only 10 patrons, you can check that the margin of error for 95% confidence increases from ±0.88 to ±1.24. Cutting the sample size in half does *not* double the margin of error because the sample size n appears under a square root sign.

EXAMPLE 16.5 Interpreting Our Confidence Interval

In Example 16.3 we found that a 95% confidence interval for the mean percentage tip for all patrons of the restaurant, when their bill contains a message that the next day's weather will be good, is the interval 21.33 to 23.09. The confidence level refers to the success rate of the method for computing the interval. Thus, the probability is 95% that our method will produce an interval that will capture the true mean percentage tip for all patrons of the restaurant when their bill contains a message that the next day's weather will be good.

Confidence intervals are often misinterpreted. Here are some incorrect interpretations of this interval:

- The probability is 95% that the true mean percentage tip is between 21.33 and 23.09 for all patrons of the restaurant when their bill contains a message that the next day's weather will be good. *The confidence level is not the probability that a given interval contains the true parameter value. The true parameter value is a fixed number, and a particular confidence interval either contains it or doesn't. To say that we are 95% confident that the true mean percentage tip is between 21.33 and 23.09 is shorthand for "We got these numbers using a method that gives correct results 95% of the time."*
- The probability is 95% that the mean percentage tip is between 21.33 and 23.09 for our sample of 20 patrons of the restaurant when their bill contains a message that the next day's weather will be good. *The confidence level is not the probability that a given interval contains the value of the statistic computed from the sample.*

- 95% of the tips, measured in percentage of the total bill, will be between 21.33 and 23.09 for any sample of 20 patrons receiving a bill containing a message that the next day's weather will be good. *The confidence level is not the proportion of the sample values that are within the limits of the confidence interval.*
- 95% of the tips, measured in percentage of the total bill, will be between 21.33 and 23.09 for all patrons of the restaurant when their bill contains a message that the next day's weather will be good. *The confidence level is not the proportion of the population value that are within the limits of the confidence interval.*

APPLY YOUR KNOWLEDGE

16.8 Confidence Level and Margin of Error. Example 16.1 (page 368) described NHANES survey data on the body mass index (BMI) of 936 young men. The mean BMI in the sample was $\bar{x} = 27.2$. We treated these data as an SRS from a Normally distributed population with standard deviation $\sigma = 11.6$.

(a) Give three confidence intervals for the mean BMI μ in this population, using 90%, 95%, and 99% confidence.

(b) What are the margins of error for 90%, 95%, and 99% confidence? How does increasing the confidence level change the margin of error of a confidence interval when the sample size and population standard deviation remain the same?

16.9 Sample Size and Margin of Error. Example 16.1 (page 368) described NHANES survey data on the body mass index (BMI) of 936 young men. The mean BMI in the sample was $\bar{x} = 27.2$. We treated these data as an SRS from a Normally distributed population with standard deviation $\sigma = 11.6$.

(a) Suppose that we had an SRS of just 100 young men. What would be the margin of error for 95% confidence?

(b) Find the margins of error for 95% confidence based on SRSs of 400 young men and 1600 young men.

(c) Compare the three margins of error. How does increasing the sample size change the margin of error of a confidence interval when the confidence level and population standard deviation remain the same?

16.10 Retaking the SAT. In Exercise 16.2 (page 371), we saw that an SRS of 400 high school seniors gained an average of $\bar{x} = 40$ points in their second attempt at the SAT Mathematics exam. Assuming that the change in score has a Normal distribution with standard deviation $\sigma = 25$, we computed a 95% confidence interval for the mean change in score μ in the population of all high school seniors.

(a) Find a 90% confidence interval for μ based on this sample.

(b) What is the margin of error for a confidence level of 90%? How does decreasing the confidence level change the margin of error of a confidence interval when the sample size and population standard deviation remain the same?

(c) Suppose we had an SRS of just 100 high school seniors. What would be the margin of error for 95% confidence?

(d) How does decreasing the sample size change the margin of error of a confidence interval when the confidence level and population standard deviation remain the same?

CHAPTER 16 SUMMARY

- A **confidence interval** uses sample data to estimate an unknown population parameter with an indication of how accurate the estimate is and of how confident we are that the result is correct.
- Any confidence interval has two parts: an interval calculated from the data and a confidence level C. The **confidence interval** often has the form

estimate ± **margin of error**

- The **confidence level** is the success rate of the method that produces the interval. That is, C is the probability that the method will give a correct answer. If you use 95% confidence intervals often, in the long run 95% of your intervals will contain the true parameter value. You do not know whether or not a 95% confidence interval calculated from a particular set of data contains the true parameter value.
- A level C **confidence interval for the mean** μ of a Normal population with known standard deviation σ, based on an SRS of size n, is given by

$$\bar{x} \pm z^* \frac{\sigma}{\sqrt{n}}$$

- The **critical value** z^* is chosen so that the standard Normal curve has area C between $-z^*$ and z^*.

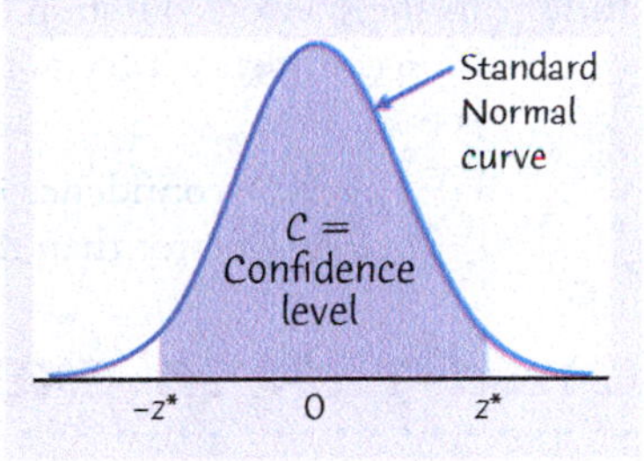

FIGURE 16.6
The confidence level *C* corresponds to the area between $-z^*$ and z^* under a standard Normal curve.

Other things being equal, the margin of error of a confidence interval gets smaller as

- The confidence level C decreases
- The population standard deviation σ decreases
- The sample size n increases

CHECK YOUR SKILLS

16.11 To give a 99.9% confidence interval for a population mean μ, you would use the critical value

(a) $z^* = 1.960$.

(b) $z^* = 2.576$.

(c) $z^* = 3.291$.

Use the following information for Questions 16.12 through 16.14. A laboratory scale is known to have a standard deviation of $\sigma = 0.001$ gram in repeated weighings. Scale readings in repeated weighings are Normally distributed, with mean equal to the true weight of the specimen. Three weighings of a specimen on this scale give 3.412, 3.416, and 3.414 grams.

16.12 A 95% confidence interval for the true weight of this specimen is WEIGHTS

(a) 3.414 ± 0.00113.

(b) 3.414 ± 0.00065.

(c) 3.414 ± 0.00196.

16.13 You want a 99% confidence interval for the true weight of this specimen. The margin of error for this interval will be

(a) smaller than the margin of error for 95% confidence.

(b) greater than the margin of error for 95% confidence.

(c) about the same as the margin of error for 95% confidence.

16.14 Another specimen is weighed eight times on this scale. The average weight is 4.1602 grams. A 99% confidence interval for the true weight of this specimen is

(a) 4.1602 ± 0.00032.

(b) 4.1602 ± 0.00069.

(c) 4.1602 ± 0.00091.

Use the following information for Questions 16.15 through 16.18. The National Assessment of Educational Progress (NAEP) includes a mathematics test for eighth-grade students. Scores on the test range from 0 to 500. Suppose that you give the NAEP test to an SRS of 2500 eighth-graders from a large population in which the scores have mean $\mu = 282$ and standard deviation $\sigma = 40$. The mean $\bar{x}$ will vary if you take repeated samples.

16.15 The sampling distribution of $\bar{x}$ is approximately Normal. It has mean $\mu = 282$. What is its standard deviation?

(a) 40 (b) 0.8 (c) 0.016

16.16 Suppose that an SRS of 2500 eighth-graders has $\bar{x} = 285$. Based on this sample, a 95% confidence interval for μ is

(a) 1.57 ± 0.031. (b) 285 ± 1.57.

(c) 282 ± 1.57.

16.17 In the previous question, suppose that we computed a 90% confidence interval for μ. Which of the following is true?

(a) This 90% confidence interval would have a smaller margin of error than the 95% confidence interval.

(b) This 90% confidence interval would have a larger margin of error than the 95% confidence interval.

(c) This 90% confidence interval could have either a smaller or a larger margin of error than the 95% confidence interval. This varies from sample to sample.

16.18 Suppose that we took an SRS of 1600 eighth-graders and found $\bar{x} = 285$. Compared with an SRS of 2500 eighth-graders, the margin of error for a 95% confidence interval for μ is

(a) smaller.

(b) larger.

(c) either smaller or larger, but we can't say which.

CHAPTER 16 EXERCISES

16.19 Student Study Times. A class survey in a large class for first-year college students asked, "About how many hours do you study during a typical week?" The mean response of the 463 students was $\bar{x} = 13.7$ hours.[6] Suppose that we know that the study time follows a Normal distribution with standard deviation $\sigma = 7.4$ hours in the population of all first-year students at this university.

(a) Use the survey result to give a 99% confidence interval for the mean study time of all first-year students.

(b) What condition not yet mentioned must be met for your confidence interval to be valid?

16.20 I Want More Muscle. Many young men in North America and Europe (but not in Asia) tend to think they need more muscle to be attractive. One study presented 200 young American men with 100 images of men with various levels of muscle.[7] Researchers measure level of muscle in kilograms of fat-free body mass per square meter of body surface area (kg/m^2). Typical young men have about 20 kg/m^2. Each subject chose two images, one that represented his own level of body muscle and one that he thought represented "what women prefer." The mean gap between self-image and "what women prefer" was 2.35 kg/m^2.

© Rubberball/age fotostock

Suppose that the "muscle gap" in the population of all young men has a Normal distribution with standard deviation 2.5 kg/m^2. Give a 90% confidence interval for the mean amount of muscle young men think they should add to be attractive to women. (They are wrong: women actually prefer a level close to that of typical men.)

16.21 An Outlier Strikes. There were actually 464 responses to the class survey in Exercise 16.19. One student claimed to study 10,000 hours per week (10,000 is more than the number of hours in a year). We know that student is joking, so we left out this value. If we did a calculation without looking at the data, we would get $\bar{x} = 35.2$ hours for all 464 students. Now what is the 99% confidence interval for the population mean? (Continue to use $\sigma = 7.4$.) Compare the new interval with that in Exercise 16.19. The message is clear: always look at your data because outliers can greatly change your result.

16.22 Explaining Confidence. A student reads that a recent poll finds that a 95% confidence interval for the mean "ideal" weight given by adult American men is 183 ± 4.4 pounds. Asked to explain the meaning of this interval, the student says, "95% of all adult American men would say that their ideal weight is between 178.6 and 187.4 pounds." Is the student right? Explain your answer. (*Hint:* See the box "Interpreting a Confidence Level" on page 370.)

16.23 Explaining Confidence. You ask another student to explain the confidence interval for mean ideal weight described in the previous exercise. The student answers, "We can be 95% confident that future samples of adult American men will say that their mean ideal weight is between 178.6 and 187.4 pounds." Is this explanation correct? Explain your answer. (*Hint:* See the box "Interpreting a Confidence Level" on page 370.)

16.24 Explaining Confidence. Here is an explanation from the Associated Press concerning one of its opinion polls. Explain briefly but clearly in what way this explanation is incorrect.

For a poll of 1,600 adults, the variation due to sampling error is no more than three percentage points either way. The error margin is said to be valid at the 95 percent confidence level. This means that, if the same questions were repeated in 20 polls, the results of at least 19 surveys would be within three percentage points of the results of this survey.

(*Hint:* See the box "Interpreting a Confidence Level" on page 370.)

Exercises 16.25 through 16.27 ask you to answer questions from data. Assume that the "simple conditions" hold in each case. The exercise statements give you the **State** *step of the four-step process. In your work, follow the* **Plan, Solve,** *and* **Conclude** *steps as illustrated in Example 16.3 (page 374) for a confidence interval.*

16.25 Pulling Apart Wood. How heavy a load (in pounds) is needed to pull apart pieces of Douglas fir 4 inches long and 1.5 inches square? Here are data from students doing a laboratory exercise: WOOD

33,190	31,860	32,590	26,520	33,280
32,320	33,020	32,030	30,460	32,700
23,040	30,930	32,720	33,650	32,340
24,050	30,170	31,300	28,730	31,920

(a) We are willing to regard the wood pieces prepared for the lab session as an SRS of all similar pieces of Douglas fir. Engineers also commonly assume that characteristics of materials vary Normally. Make a graph to show the shape of the distribution for these data. Does it appear safe to assume that the Normality condition is satisfied? Suppose that the strength of pieces of wood like these follows a Normal distribution with standard deviation 3000 pounds.

(b) Give a 95% confidence interval for the mean load required to pull apart the wood.

(c) One of the students stated that the confidence interval computed in part (b) will contain at least 95% of any additional measurements of the load needed to pull apart pieces of Douglas fir 4 inches long and 1.5 inches square. Is the student correct? Explain your answer.

16.26 Bone Loss by Nursing Mothers. Breast-feeding mothers secrete calcium into their milk. Some of the calcium may come from their bones, so mothers may lose bone mineral. Researchers measured the percentage change in mineral content of the spines of 47 mothers during three months of breast-feeding.[8] Here are the data: BONELS

−4.7	−2.5	−4.9	−2.7	−0.8	−5.3	−8.3	−2.1	−6.8	−4.3
2.2	−7.8	−3.1	−1.0	−6.5	−1.8	−5.2	−5.7	−7.0	−2.2
−6.5	−1.0	−3.0	−3.6	−5.2	−2.0	−2.1	−5.6	−4.4	−3.3
−4.0	−4.9	−4.7	−3.8	−5.9	−2.5	−0.3	−6.2	−6.8	1.7
0.3	−2.3	0.4	−5.3	0.2	−2.2	−5.1			

(a) The researchers are willing to consider these 47 women to be an SRS from the population of all nursing mothers. Suppose that the percentage change in this population has standard deviation $\sigma = 2.5\%$. Make a stemplot of the data to verify that the data follow a Normal distribution quite closely. (Don't forget that you need both a 0 and a −0 stem because there are both positive and negative values.)

Blend Images/ Superstock

(b) Use a 99% confidence interval to estimate the mean percentage change in the population.

(c) Would it be correct to say that the probability is 99% that the mean percentage change in the population lies in the interval you computed in part (b)? Explain your answer.

16.27 This Wine Stinks. Sulfur compounds cause "off odors" in wine, and winemakers want to know the odor threshold—the lowest concentration of a compound that the human nose can detect. The odor threshold for dimethyl sulfide (DMS) in trained wine tasters is about 25 micrograms per liter of wine (μg/L). The untrained noses of consumers may be less sensitive, however. Here are the DMS odor thresholds for 10 untrained students: WINE2

30 30 42 35 22 33 31 29 19 23

(a) Assume that the standard deviation of the odor threshold for untrained noses is known to be $\sigma = 7\ \mu\text{g/L}$. Briefly discuss the other two "simple conditions" and create a stemplot to verify that the distribution is roughly symmetric with no outliers.

(b) Give a 95% confidence interval for the mean DMS odor threshold among all students. Use the four-step process.

16.28 Why Are Larger Samples Better? Statisticians prefer large samples. Describe briefly the effect of increasing the size of a sample on the margin of error of a 95% confidence interval.

16.29 The Margins of Error in a Poll. A Gallup Poll reported that 89% of Republicans in the poll approved of the job President Trump was doing, 37% of Independents in the poll approved, and 6% of Democrats in the poll approved. At the end of the poll, the section on the survey methods states that the poll was based on a random sample of 1108 adults aged 18 and older living in all 50 states and the District of Columbia. The methods section also states that for results based on the total sample, the margin of sampling error is ±4 percentage points at the 95% confidence level. Can we conclude that a 95% confidence interval for the percentage of the population of all adults aged 18 and older living in all 50 states and the District of Columbia who, at the time of the poll, were Independents and who approved of the job President Trump was doing is 37% ± 4%? Explain your answer.

CHAPTER

17

When you complete this chapter, you will be able to:

17.1 Use the reasoning of statistical tests to state whether sample data support a claim about a population.

17.2 State the null and alternative hypotheses when testing a claim about the mean of a population.

17.3 Find and interpret *P*-values and state whether a test result is statistically significant at a given level.

17.4 Calculate one-sample *z* test statistics for both one-sided and two-sided tests of a population mean and draw conclusions from the results.

17.5 Use a table to look up approximate *P*-values based on a *z* statistic and state whether the result is statistically significant.

Tests of Significance: The Basics

Confidence intervals are one of the two most common types of statistical inference. In this chapter, we discuss tests of significance, the second type of statistical inference. The mathematics of probability—in particular, the sampling distributions discussed in Chapter 15—provides the formal basis for a test of significance. Here we will apply the reasoning of tests of significance for the mean of a population that has a Normal distribution in a simple and artificial setting (where we assume that we know the population standard deviation). We will use the same logic in future chapters to construct tests of significance for population parameters in more realistic settings.

Use a confidence interval when your goal is to estimate a population parameter. Tests of significance have a different goal: to assess the evidence provided by data about some prior claim concerning a population parameter. Here is the reasoning of statistical tests in a nutshell.

EXAMPLE 17.1 I'm a Good Free-Throw Shooter

I claim that I make 80% of my basketball free throws. To test my claim, you ask me to shoot 20 free throws. I make only eight of the 20. "Aha!" you say. "Someone who makes 80% of his free throws would almost never make only eight out of 20. So I don't believe your claim."

Your reasoning is based on asking what would happen if my claim were true and we repeated the sample of 20 free throws many times: I would almost never

WILLIAM ARCHIE/Tribune News Service/Detroit/MI/USA

make eight or fewer. The eight out of 20 outcome is so unlikely that it gives strong evidence that my claim is not true.

You can say how strong the evidence against my claim is by giving the probability that I would make eight or fewer out of 20 free throws if I really make 80% in the long run. This probability is 0.0001; as described in Chapter 14, this is computed using the binomial distribution. Thus, I would make eight or fewer of 20 only one time in 10,000 tries in the long run—where each "try" is 20 attempted free throws—if my claim to make 80% were true. The small probability convinces you that my claim is false.

The *Reasoning of a Statistical Test* applet animates Example 17.1. You can ask a player to shoot free throws until the data do (or don't) convince you that he makes fewer than 80%. Tests of significance, also called significance tests, use an elaborate vocabulary, but the basic idea is simple: *an outcome that would rarely happen if a claim were true is good evidence that the claim is not true.*

applet

17.1 The Reasoning of Tests of Significance

The reasoning of statistical tests of significance, like that of confidence intervals, is based on asking what would happen if we repeated the sample or experiment many times. We will act as if the "simple conditions" listed on page 367 are true: we have a perfect SRS from an exactly Normal population with standard deviation σ known to us. Here is an example we will explore.

EXAMPLE 17.2 Sweetening Colas

COLA

Diet colas use artificial sweeteners to avoid sugar. These sweeteners gradually lose their sweetness over time. Manufacturers therefore test new colas for loss of sweetness before marketing them. Trained tasters sip the cola along with drinks of standard sweetness and score the cola on a "sweetness score" of 1 to 10, with larger scores corresponding to greater sweetness. The cola is then stored for a month at high temperature to imitate the effect of four months' storage at room temperature. Each taster scores the cola again after storage. This is a matched pairs experiment. Our data are the differences (score before storage minus score after storage) in the tasters' scores. The bigger these positive differences (differences > 0), the bigger the loss of sweetness.

Suppose we know that, for any cola, the sweetness loss scores vary from taster to taster according to a Normal distribution, with standard deviation $\sigma = 1$. The mean μ for all tasters measures loss of sweetness and is different for different colas.

Here are the sweetness losses for a cola currently on the market, as measured by 10 trained tasters:

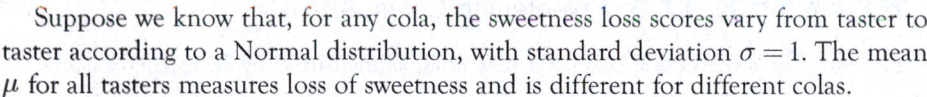

1.6 0.4 0.5 −2.0 1.5 −1.1 1.3 −0.1 −0.3 1.2

The average sweetness loss is given by the sample mean $\bar{x} = 0.3$ so that, on average, the 10 tasters found a small loss of sweetness. Also, more than half (six) of the tasters found a loss of sweetness. Are these data good evidence that the cola lost sweetness in storage?

The reasoning is the same as in Example 17.1. We make a claim and ask if the data give evidence *against* it. We seek evidence that there *is* a sweetness loss, so the claim we test is that there *is not* a loss. In that case, the mean loss for the population of all trained testers would be $\mu = 0$.

FIGURE 17.1
If the cola does not lose sweetness in storage, the mean score $\bar{x}$ for 10 tasters will have this sampling distribution. The actual result for one SRS for the cola was $\bar{x} = 0.3$. That could easily happen just by chance. A sample of sweetness losses for another cola had $\bar{x} = 1.02$. That's so far out on the Normal curve that it is good evidence that this cola did lose sweetness.

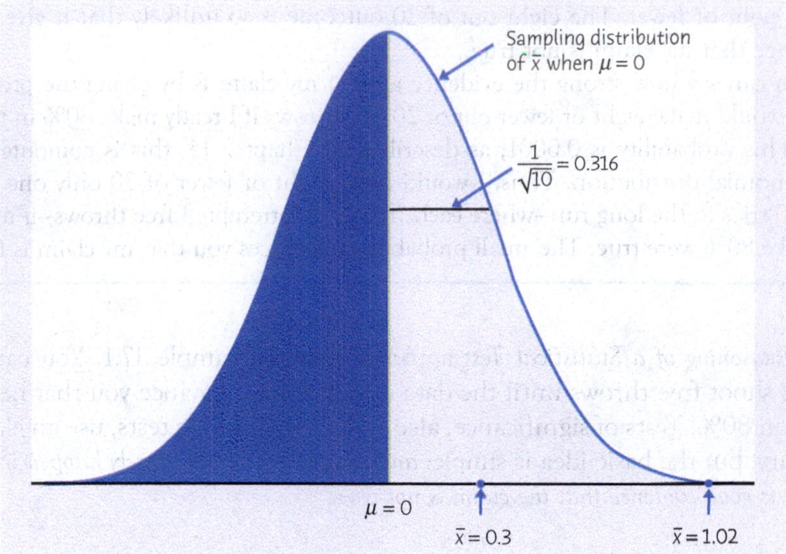

- If the claim that $\mu = 0$ is true, the sampling distribution of $\bar{x}$ from 10 tasters is Normal with mean $\mu = 0$ and standard deviation

$$\frac{\sigma}{\sqrt{n}} = \frac{1}{\sqrt{10}} = 0.316$$

  This is just like the calculations we did in Chapter 15 (see Example 15.5 on page 351) and Chapter 16 (see Example 16.1 on page 368). Figure 17.1 shows this sampling distribution. We can judge whether any observed $\bar{x}$ is surprising by locating it on this distribution.
- For this cola, 10 tasters had mean loss $\bar{x} = 0.3$. It is clear from Figure 17.1 that an $\bar{x}$ this large is not particularly surprising. It could easily occur just by chance when the population mean is $\mu = 0$. That 10 tasters found $\bar{x} = 0.3$ is not *strong* evidence that this cola loses sweetness.

EXAMPLE 17.3 Sweetening Colas, Again

COLA2

Here are the sweetness losses for a new cola, as measured by 10 trained tasters:

2.0	0.4	0.7	2.0	−0.4	2.2	−1.3	1.2	1.1	2.3

The average sweetness loss is given by the sample mean $\bar{x} = 1.02$. Most scores are positive. That is, most tasters found a loss of sweetness. But the losses are small, and two tasters (the negative scores) thought the cola gained sweetness. Are these data good evidence that the cola lost sweetness in storage?

The taste test for the new cola produced $\bar{x} = 1.02$. That's way out in the tail of the Normal curve in Figure 17.1—so far out that *an observed value this large would rarely occur just by chance if the true μ were* 0. This observed value is good evidence that the true μ is in fact greater than 0—that is, that the cola lost sweetness. The manufacturer must reformulate the new cola and try again.

APPLY YOUR KNOWLEDGE

17.1 The GMAT. The Graduate Management Admission Test (GMAT) is taken by individuals interested in pursuing graduate management education. GMAT scores are used as part of the admissions process for more than 6000 graduate management programs worldwide. The mean score for all test takers is 563, with a standard deviation of 118.[1] A researcher in the Philippines is concerned about the performance of undergraduates in the Philippines on the GMAT. She believes that the mean scores for this year's college seniors in the Philippines who are interested in pursuing graduate management education will be less than 563. She has a random sample of 250 college seniors in the Philippines interested in pursuing graduate management education take the GMAT. Suppose we know that GMAT scores are Normally distributed with standard deviation $\sigma = 118$.

(a) We seek evidence *against* the claim that $\mu = 563$. What is the sampling distribution of the mean score $\bar{x}$ of a sample of 250 students if the claim is true? Draw the density curve of this distribution. (Sketch a Normal curve and then mark on the axis the values of the mean and 1, 2, and 3 standard deviations of the sampling distribution on either side of the mean.)

(b) Suppose that the sample data give $\bar{x} = 555$. Mark this point on the axis of your sketch.

(c) Suppose that the sample data give $\bar{x} = 540$. Mark this point on your sketch. Using your sketch, explain in simple language why one result is good evidence that the mean score of all college seniors in the Philippines interested in pursuing graduate management education who plan to take the GMAT would be less than 563 and why the other outcome is not.

17.2 Inspecting Weights of Boxes of Cookies. The National Institute of Standards and Technology (NIST) publishes procedures for inspecting the advertised net contents of packaged goods.[2] The discussion of these procedures includes an example of inspecting the weight of a sample of boxes of thin mint cookies from a particular company that are labeled as containing a net weight of 1 lb. of cookies. An SRS of 12 boxes is weighed according to the NIST procedures, and the net weights are recorded. Assume that the net weights for the population of all boxes of thin mint cookies from this company come from a Normal population with mean net weight μ and standard deviation $\sigma = 0.01$ lb.

(a) We seek evidence *against* the claim that $\mu = 1.000$. What is the sampling distribution of the mean $\bar{x}$ in many samples of 12 boxes if the claim is true? Make a sketch of the Normal curve for this distribution. (Draw a Normal curve and then mark on the axis the values of the mean and 1, 2, and 3 standard deviations on either side of the mean.)

(b) Suppose that the sample mean is $\bar{x} = 0.998$. Mark this value on the axis of your sketch. Another sample of 12 boxes has $\bar{x} = 1.010$ for 12 measurements. Mark this value on the axis as well. Explain in simple language why one result is good evidence that the population mean net weight of boxes of cookies differs from 1.000 and why the other result gives no reason to doubt that 1.000 is correct.

17.2 Stating Hypotheses

A statistical test of significance starts with a careful statement of the claims we want to compare. In Example 17.3, we saw that the taste test data are not plausible if the new cola loses no sweetness. Because the reasoning of tests of significance looks for evidence *against* a claim, we start with the claim we seek evidence against, such as "no loss of sweetness."

Null and Alternative Hypotheses

The claim tested by a statistical test of significance is called the **null hypothesis.** The test is designed to assess the strength of the evidence *against* the null hypothesis. Usually the null hypothesis is a statement of "no effect" or "no difference."

The claim about the population that we are trying to find evidence *for* is the **alternative hypothesis.** The alternative hypothesis is **one-sided** if it states that a parameter is *larger than* or *smaller than* the null hypothesis value. It is **two-sided** if it states that the parameter is *different from* the null value. (It could be either smaller or larger.)

We abbreviate the null hypothesis as H_0 and the alternative hypothesis as H_a. *Hypotheses always refer to a population parameter, not to a particular sample outcome. Be sure to state H_0 and H_a in terms of population parameters.* Because H_a expresses the effect that we hope to find evidence *for,* it is sometimes easier to begin by stating H_a and then set up H_0 as the statement that the hoped-for effect is not present. H_0 usually will include "equals."

In Examples 17.2 and 17.3, we are seeking evidence *for* loss in sweetness. The null hypothesis says "no loss" on the average in a large population of tasters. The alternative hypothesis says "there is a loss." So the hypotheses are

$$H_0: \mu = 0$$
$$H_a: \mu > 0$$

The alternative hypothesis is *one-sided* because we are interested only in whether the cola *lost* sweetness.[3]

EXAMPLE 17.4 Studying Job Satisfaction

Does the job satisfaction of assembly workers differ when their work is machine-paced rather than self-paced? Assign workers either to an assembly line moving at a fixed pace or to a self-paced setting. All subjects work in both settings, in random order. This is a matched pairs design. After two weeks in each work setting, the workers take a test of job satisfaction. The response variable is the difference in satisfaction scores, self-paced minus machine-paced.

The parameter of interest is the mean μ of the differences in scores in the population of all assembly workers. The null hypothesis says that there is no difference between self-paced and machine-paced work—that is,

$$H_0: \mu = 0$$

The authors of the study wanted to know if the two work conditions have different levels of job satisfaction. They did not specify the direction of the difference. The alternative hypothesis is therefore *two-sided:*

$$H_a: \mu \neq 0$$

The hypotheses should express the hopes or suspicions we have *before* we see the data. It is cheating to first look at the data and then frame hypotheses to fit what the data show. For example, the data for the study in Example 17.4 showed that the workers were more satisfied with self-paced work, but this should not influence the choice of H_a. If you do not have a specific direction firmly in mind in advance, use a two-sided alternative.

STATISTICS IN YOUR WORLD

Honest Hypotheses?

Chinese and Japanese people, for whom the number 4 is unlucky, die more often on the fourth day of the month than on other days. The authors of a study did a statistical test of the claim that the fourth day has more deaths than other days and found good evidence in favor of this claim. Can we trust this? Not if the authors looked at all days, picked the one with the most deaths, and then made "this day is different" the claim to be tested. A critic raised that issue, and the authors replied: "No, we had day 4 in mind in advance, so our test was legitimate."

APPLY YOUR KNOWLEDGE

17.3 The GMAT (continued). State the null and alternative hypotheses for the study of the performance on the GMAT of college seniors in the Philippines in Exercise 17.1. Is the alternative hypothesis one-sided or two-sided?

17.4 Inspecting Weights of Boxes of Cookies (continued). State the null and alternative hypotheses for inspecting the net weights of boxes of cookies described in Exercise 17.2. Is the alternative hypothesis one-sided or two-sided?

17.5 Too Early. The examinations in a large multisection statistics class are scaled after grading so that the mean score is 70. The professor thinks that students in the 8:00 A.M. class have trouble paying attention because they are sleepy and suspects that these students have a lower mean score than the class as a whole. The students in the 8:00 A.M. class this semester can be considered a sample from the population of all students in the course, so the professor compares their mean score with 70. State the hypotheses H_0 and H_a.

EricVega/Getty Images

17.6 Women's Incomes. The average income of American women who work full-time and have only a high school diploma is \$37,616. You wonder whether the mean income of female graduates from your local high school who work full-time but have only a high school diploma is different from the national average. You obtain income information from an SRS of 62 female graduates of your high school who work full-time and have only a high school diploma and find that $\bar{x} = \$36,453$. What are your null and alternative hypotheses?

17.7 Stating Hypotheses. In planning a study on the number of days in the past 30 days that high school students texted while driving sometime during the day, a researcher states the hypotheses as

$$H_0: \bar{x} = 15 \text{ days}$$
$$H_a: \bar{x} > 15 \text{ days}$$

What's wrong with this?

17.3 *P*-Value and Statistical Significance

The idea of stating a null hypothesis that we want to find evidence *against* seems odd at first. It may help to think of a criminal trial. The defendant is "innocent until proven guilty." That is, the null hypothesis is innocence, and the prosecution must try to provide convincing evidence against this hypothesis. That's exactly how statistical tests of significance work, though in statistics we deal with evidence provided by data and use a probability to say how strong the evidence is.

The probability that measures the strength of the evidence against a null hypothesis is called a *P-value*. Statistical tests generally work like this:

Test Statistic and *P*-Value

A **test statistic** calculated from the sample data measures how far the data diverge from what we would expect if the null hypothesis H_0 were true. Unusually large values of the statistic show that the data are not consistent with H_0.

The probability, computed assuming that H_0 is true, that the test statistic would take a value as extreme as or more extreme than that actually observed is called the ***P*-value** of the test. The smaller the *P*-value, the stronger the evidence against H_0 provided by the data.

Small *P*-values are evidence against H_0 because they say that the observed result would be unlikely to occur if H_0 were true. Large *P*-values fail to give evidence against H_0. This applies to null hypotheses in general, including those that involve proportions or those that involve comparing the means of two populations (see, for example, Exercises 17.10, 17.32, 17.33, 17.36, and 17.37).

How small a *P*-value is convincing evidence against H_0? We discuss this in detail in Section 18.3 (page 408), and many users of statistics regard values smaller than 0.05 or 0.01 as convincing.

An example of this process of computing a test statistic and the corresponding *P*-value will be given in Section 17.4 (page 392). In practice, people use statistical software to carry out statistics tests. Statistical software will give you the *P*-value of a test when you enter your null and alternative hypotheses and your data. So your most important task is to understand what a *P*-value says.

EXAMPLE 17.5 Sweetening Colas: One-Sided *P*-Value

The study of sweetness loss in Examples 17.2 and 17.3 tests the hypotheses

$$H_0: \mu = 0$$
$$H_a: \mu > 0$$

Because the alternative hypothesis says that $\mu > 0$, values of $\bar{x}$ greater than 0 favor H_a over H_0. The test statistic compares the observed $\bar{x}$ with the hypothesized value $\mu = 0$. For now, let's concentrate on the *P*-value.

The experiment presented in Examples 17.2 and 17.3 actually compared two colas. For the first cola, the 10 tasters found mean sweetness loss $\bar{x} = 0.3$. For the second, the data gave $\bar{x} = 1.02$. *The P-value for each test is the probability of getting an $\bar{x}$ this large when the mean sweetness loss is really $\mu = 0$.*

The shaded area in Figure 17.2 shows the *P*-value when $\bar{x} = 0.3$. The Normal curve is the sampling distribution of $\bar{x}$ when the null hypothesis $H_0: \mu = 0$ is true using the population standard deviation $\sigma = 1$. A Normal probability calculation (Exercise 17.8) shows that the *P*-value $P(\bar{x} \geq 0.3) = 0.1714$.

A value as large as $\bar{x} = 0.3$ would occur just by chance in 17% of all samples when $H_0: \mu = 0$ is true. So observing $\bar{x} = 0.3$ is not *strong* evidence against H_0. On the other hand, you can calculate that the probability that $\bar{x}$ is 1.02 or larger when

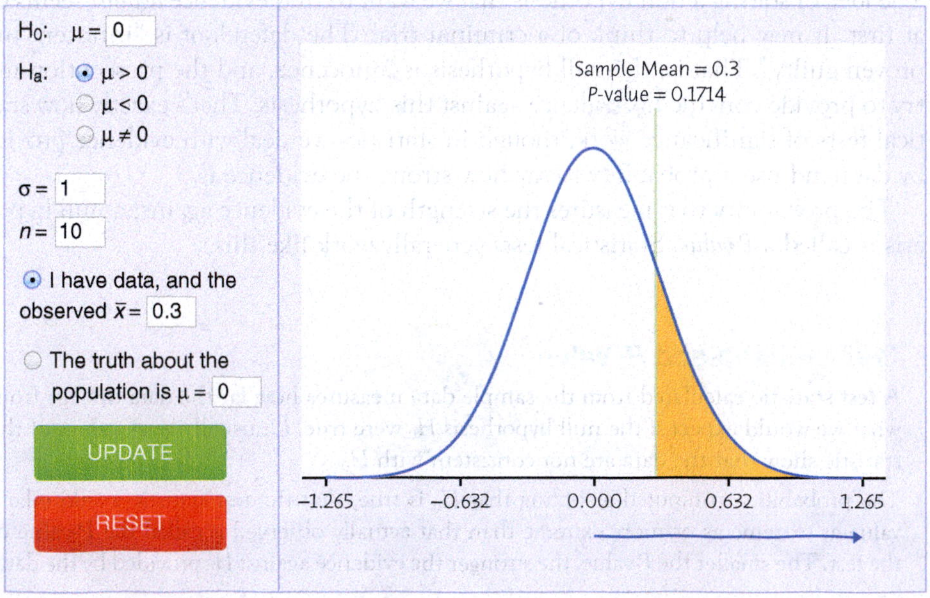

FIGURE 17.2
The one-sided *P*-value for the cola with mean sweetness loss $\bar{x} = 0.3$, for Example 17.5. The figure shows both the input and the output for the *P-Value of a Test of Significance* applet. Note that the *P*-value is the shaded area under the curve, not the unshaded area.

in fact $\mu = 0$ is only 0.0006. We would very rarely observe a mean sweetness loss of 1.02 or larger if H_0 were true. This small *P*-value provides strong evidence against H_0 and in favor of the alternative $H_a: \mu > 0$.

Figure 17.2 is actually the output of the *P-Value of a Test of Significance* applet, along with the information we entered into the applet. This applet automates the work of finding *P*-values for samples of size 50 or smaller under the "simple conditions" for inference about a mean.

The alternative hypothesis sets the direction that counts as evidence against H_0. In Example 17.5, only large positive values count because the alternative is one-sided on the high side. If the alternative is two-sided, both directions count.

EXAMPLE 17.6 Job Satisfaction: Two-Sided *P*-Value

The study of job satisfaction in Example 17.4 (page 386) requires that we test

$$H_0: \mu = 0$$
$$H_a: \mu \neq 0$$

Suppose we know that differences in job satisfaction scores (self-paced minus machine-paced) in the population of all workers follow a Normal distribution with standard deviation $\sigma = 60$.

Data from 18 workers give $\bar{x} = 17$. That is, these workers prefer the self-paced environment on the average. Because the alternative is two-sided, the *P*-value is the probability of getting an $\bar{x}$ at least as far from $\mu = 0$ *in either direction* as the observed $\bar{x} = 17$.

Enter the information for this example into the *P-Value of a Test of Significance* applet and click "Show P." Figure 17.3 shows the applet output as well as the information entered. The *P*-value is the sum of the two shaded areas under the Normal curve. It is $P = 0.2293$. Values as far from 0 as $\bar{x} = 17$ (in either direction) would happen 23% of the time when the true population mean is $\mu = 0$. An outcome that would occur so often when H_0 is true is not good evidence against H_0.

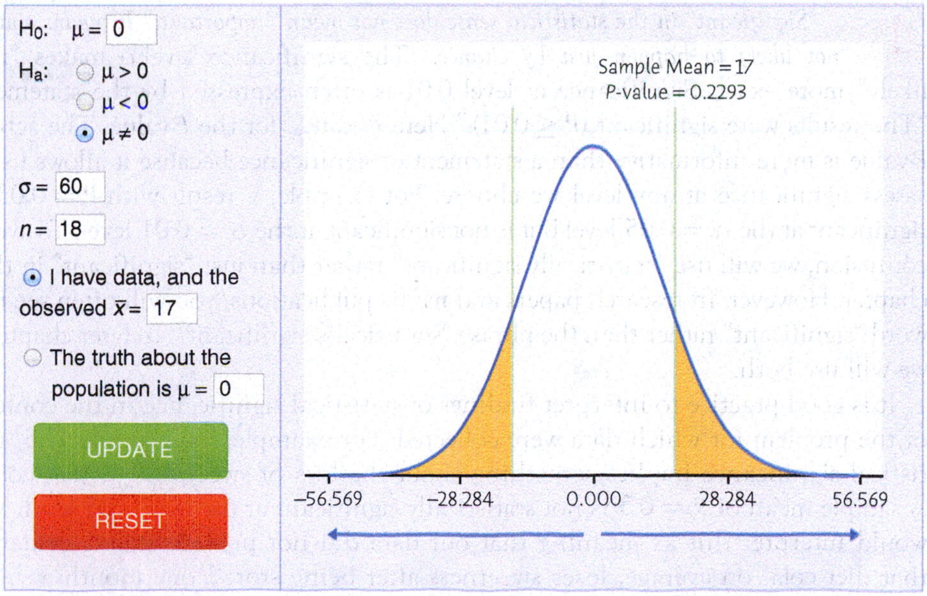

FIGURE 17.3 The two-sided *P*-value, for Example 17.6. The figure shows both the input and the output for the *P-Value of a Test of Significance* applet. Note that the *P*-value is the shaded area under the curve, not the unshaded area.

The conclusion of Example 17.6 is *not* that H_0 is true. The study looked for evidence against $H_0: \mu = 0$ and failed to find strong evidence. That is all we can say. No doubt the mean μ for the population of all assembly workers is not exactly equal to 0. A large enough sample would give evidence of the difference, even if it is very small. Tests of significance assess the evidence against H_0. If the evidence is strong, we can confidently reject H_0 in favor of the alternative. *Failing to find evidence against H_0 means only that the data are not inconsistent with H_0, not that we have clear evidence that H_0 is true.* Only data that are inconsistent with H_0 enable us to make a positive statement that we have strong evidence against H_0.

In Examples 17.5 and 17.6, we decided that *P*-value $P = 0.0006$ was strong evidence against the null hypothesis and that *P*-values $P = 0.1714$ and $P = 0.2293$ did not give convincing evidence. There is no rule for how small a *P*-value we should require to reject H_0; it's a matter of judgment and depends on the specific circumstances.

Nonetheless, we can compare a *P*-value with some fixed values that are in common use as standards for evidence against H_0. The most common fixed values are 0.05 and 0.01. If $P \leq 0.05$, there is no more than one chance in 20 that a sample would give evidence this strong just by chance when H_0 is actually true. If $P \leq 0.01$, we have a result that, in the long run, would happen no more than once per 100 samples if H_0 were true. These fixed standards for *P*-values are called *significance levels*. We use α, the Greek letter alpha, to stand for a significance level. In Section 18.5 we will see that the significance level α is the probability of making a certain type of error. In particular, α is the probability that we would (incorrectly) decide that we have evidence against H_0 and declare H_0 to be false when, in fact, H_0 is true. Setting the significance level small ensures that the chance of making such an error is small.

Statistical Significance

If the *P*-value is as small as or smaller than α, we say that the data are **statistically significant at level α**. The quantity α is called the **significance level,** or the level of significance.

"Significant" in the statistical sense does not mean "important." It means simply "not likely to happen just by chance." The significance level α makes "not likely" more exact. Significance at level 0.01 is often expressed by the statement "The results were significant ($P \leq 0.01$)." Here *P* stands for the *P*-value. The actual *P*-value is more informative than a statement of significance because it allows us to assess significance at any level we choose. For example, a result with $P = 0.03$ is significant at the $\alpha = 0.05$ level but is not significant at the $\alpha = 0.01$ level. To avoid confusion, we will use "statistically significant" rather than just "significant" in this chapter. However, in research papers and media publications, you will often see the word "significant" rather than the phrase "statistically significant." In later chapters, we will use both.

It is good practice to interpret findings of statistical significance in the context of the problem for which data were collected. For example, in Example 17.5, statistical significance implies something about the loss of sweetness in diet colas. A sample mean of $\bar{x} = 0.3$ is not statistically significant at the $\alpha = 0.05$ level. We would interpret this as meaning that our data did not provide strong evidence that diet cola, on average, loses sweetness after being stored one month at high temperature.

STATISTICS IN YOUR WORLD

Significance Strikes Down a New Drug

The pharmaceutical company Pfizer spent $1 billion developing a new cholesterol-fighting drug. The final test for its effectiveness was a clinical trial with 15,000 subjects. To enforce double-blindness, only an independent group of experts saw the data during the trial. Three years into the trial, the monitors declared that there was a statistically significant excess of deaths and of heart problems in the group assigned to the new drug. Pfizer ended the trial.

APPLY YOUR KNOWLEDGE

17.8 Sweetening Colas: Find the *P*-Value. The *P*-value for the first cola in Example 17.5 is the probability (taking the null hypothesis $\mu = 0$ to be true) that $\bar{x}$ takes a value at least as large as 0.3.

(a) What is the sampling distribution of $\bar{x}$ when $\mu = 0$? This distribution appears in Figure 17.2.

(b) Do a Normal probability calculation to find the *P*-value. Your result should agree with Example 17.5 up to roundoff error.

17.9 Job Satisfaction: Find the *P*-Value. The *P*-value in Example 17.6 is the probability (taking the null hypothesis $\mu = 0$ to be true) that $\bar{x}$ takes a value at least as far from 0 as 17.

(a) What is the sampling distribution of $\bar{x}$ when $\mu = 0$? This distribution is shown in Figure 17.3.

(b) Do a Normal probability calculation to find the *P*-value. (Recall that the alternative hypothesis is two-sided.) Your result should agree with Example 17.6 up to roundoff error.

17.10 Lorcaserin and Weight Loss. A double-blind, randomized comparative experiment compared the effect of the drug lorcaserin and a placebo on weight loss in overweight adults. All subjects also underwent diet and exercise counseling. The study reported that, after one year, patients in the lorcaserin group had an average weight loss of 5.8 kilograms (kg), while those on the placebo had an average weight loss of 2.2 kg ($P < 0.001$).[4] Explain to someone who knows no statistics why these results mean that there is good reason to think that lorcaserin works. Include an explanation of what $P < 0.001$ means.

17.11 The GMAT (continued). Exercise 17.1 describes a study of the performance on the GMAT of college seniors in the Philippines. You stated the null and alternative hypotheses in Exercise 17.3. applet

(a) One sample of 250 students had mean GMAT score $\bar{x} = 555$. Enter this $\bar{x}$, along with the other required information, into the *P-Value of a Test of Significance* applet. What is the *P*-value? Is this outcome statistically significant at the $\alpha = 0.05$ level? At the $\alpha = 0.01$ level?

(b) Another sample of 250 students had $\bar{x} = 540$. Use the applet to find the *P*-value for this outcome. Is it statistically significant at the $\alpha = 0.05$ level? At the $\alpha = 0.01$ level?

(c) Explain briefly why these *P*-values tell us that one outcome is strong evidence against the null hypothesis and that the other outcome is not.

17.12 Inspecting Weights of Boxes of Cookies (continued). Exercise 17.2 describes measurements of the net weight of a sample of 12 boxes of cookies. You stated the null and alternative hypotheses in Exercise 17.4. applet

(a) One set of measurements yielded a mean net weight of $\bar{x} = 0.998$. Enter this $\bar{x}$, along with the other required information, into the *P-Value of a Test of Significance* applet. What is the *P*-value? Is this outcome statistically significant at the $\alpha = 0.05$ level? At the $\alpha = 0.01$ level?

(b) Another set of measurements has $\bar{x} = 1.010$. Use the applet to find the *P*-value for this outcome. Is it statistically significant at the $\alpha = 0.05$ level? At the $\alpha = 0.01$ level?

(c) Explain briefly why these *P*-values tell us that one outcome is strong evidence against the null hypothesis and that the other outcome is not.

17.4 Tests for a Population Mean

We have used tests for hypotheses about the mean μ of a population, under the "simple conditions," to introduce tests of significance. The big idea is the reasoning of a test: *sample data that would rarely occur if the null hypothesis H_0 were true provide evidence that H_0 is not true.* The *P*-value gives us a probability to measure "would rarely occur." In practice, the steps in carrying out a test of significance mirror the overall four-step process for organizing realistic statistical problems.

Tests of Significance: The Four-Step Process

STATE: What is the practical question that requires a statistical test?

PLAN: Identify the parameter, state null and alternative hypotheses, and choose the type of test that fits your situation.

SOLVE: Carry out the test in three phases:

1. Check the conditions for the test you plan to use.
2. Calculate the test statistic.
3. Find the *P*-value.

CONCLUDE: Return to the practical question to describe your results in this setting.

Once you have stated your question, formulated hypotheses, and checked the conditions for your test, you or your software can find the test statistic and *P*-value by following a rule. Here is the rule for the test we have used in our examples.

One-Sample *z* Test for a Population Mean

Draw an SRS of size *n* from a Normal population that has unknown mean μ and known standard deviation σ. To test the null hypothesis that μ has a specified value,

$$H_0: \mu = \mu_0$$

calculate the **one-sample *z* test statistic**

$$z = \frac{\bar{x} - \mu_0}{\sigma/\sqrt{n}}$$

In terms of a variable *Z* having the standard Normal distribution, the *P*-value for a test of H_0 against

$H_a: \mu > \mu_0$ is $P(Z \geq z)$

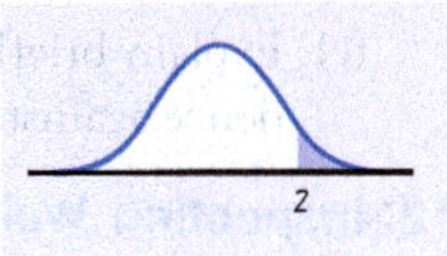

$H_a: \mu < \mu_0$ is $P(Z \leq z)$

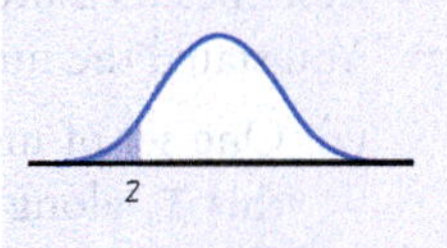

$H_a: \mu \neq \mu_0$ is $P(Z \leq -|z|) + P(Z \geq |z|) = 2P(Z \geq |z|)$

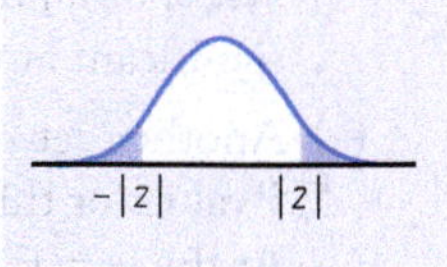

In the graphs shown here, we assume *z* is positive for $H_a: \mu > \mu_0$ because negative *z* would be poor evidence for $\mu > \mu_0$ and similarly for $H_a: \mu < \mu_0$. For a two-sided test, *z* could be positive or negative.

As promised, the test statistic z measures how far the observed sample mean $\bar{x}$ deviates from the hypothesized population value μ_0. The measurement is in the familiar standard scale obtained by dividing by the standard deviation of $\bar{x}$. So we have a common scale for all z tests, and the 68–95–99.7 rule helps us see at once if $\bar{x}$ is far from μ_0. The pictures that illustrate the *P*-value look just like the curves in Figures 17.2 and 17.3, except that they are in the standard scale.

EXAMPLE 17.7 Executives' Cholesterol

4step

STATE: The National Center for Health Statistics reports that the LDL cholesterol for adults has mean 130 and standard deviation 40. The medical director of a large pharmaceutical company looks at the medical records of 72 executives and finds that the mean LDL in this sample is $\bar{x} = 124.86$. Is this evidence that the company's executives have a different mean LDL from the general population?

PLAN: The null hypothesis is "no difference" from the national mean $\mu_0 = 130$. The alternative is two-sided because the medical director did not have a particular direction in mind before examining the data. So the hypotheses about the unknown mean μ of the executive population are

$$H_0: \mu = 130$$
$$H_a: \mu \neq 130$$

We know that the one-sample z test is appropriate for these hypotheses under the "simple conditions."

SOLVE: As part of the "simple conditions," suppose we are willing to assume that executives' LDL follow a Normal distribution with standard deviation $\sigma = 40$. Software can now calculate z and P for you. Going ahead by hand, the test statistic is

$$z = \frac{\bar{x} - \mu_0}{\sigma/\sqrt{n}} = \frac{124.86 - 130}{40/\sqrt{72}}$$
$$= -1.09$$

To help find the *P*-value, sketch the standard Normal curve and mark on it the observed value of z. Figure 17.4 shows that the *P*-value is the probability that a

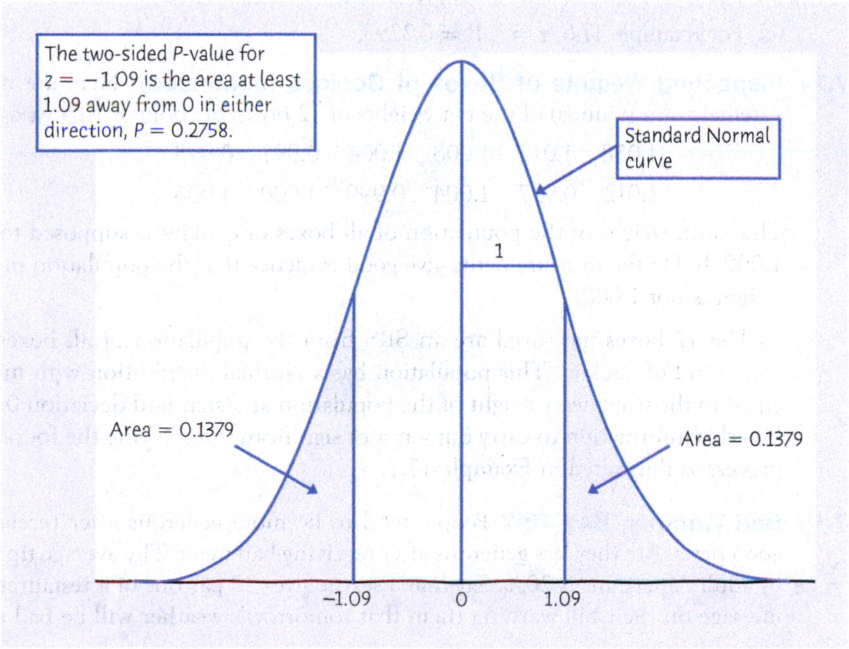
The two-sided P-value for z = −1.09 is the area at least 1.09 away from 0 in either direction, P = 0.2758.
Standard Normal curve
1
Area = 0.1379
Area = 0.1379
−1.09
0
1.09

FIGURE 17.4
The *P*-value for the two-sided test, for Example 17.7. The observed value of the test statistic is $z = -1.09$.

standard Normal variable Z takes a value at least 1.09 away from zero. From Table A or software, this probability is

$$P = 2P(Z > 1.09) = (2)(0.1379) = 0.2758$$

CONCLUDE: More than 27% of the time, an SRS of size 72 from the general adult population would have a mean LDL at least as far from 130 as that of the executive sample. The observed $\bar{x} = 124.86$ is therefore not good evidence that executives differ from other adults.

In this chapter, we are acting as if the "simple conditions" stated on page 367 are true. In practice, you must verify these three conditions.

1. **SRS:** The most important condition is that the 72 executives in the sample are an SRS from the population of all executives in the company. We should check this requirement by asking how the data were produced. If medical records are available only for executives with recent medical problems, for example, the data are of little value for our purpose because of the obvious health bias. It turns out that all executives are given a free annual medical exam and that the medical director selected 72 exam results at random.
2. **Normal distribution:** We should also examine the distribution of the 72 observations to look for signs that the population distribution is not Normal.
3. **Known σ:** It really is unrealistic to suppose that we know that $\sigma = 15$. We will see in Chapter 20 that it is easy to do away with the need to know σ.

APPLY YOUR KNOWLEDGE

17.13 The z Statistic. Published reports of research work are terse. They often report just a test statistic and P-value. For example, the conclusion of Example 17.7 might be stated as "($z = -1.09, P = 0.2758$)." Find the values of the one-sample z statistic needed to complete these conclusions:

(a) For the first cola in Example 17.5 (page 388), $z = ?$, $P = 0.1714$.

(b) For the second cola in Example 17.5, $z = ?$, $P = 0.0006$.

(c) For Example 17.6, $z = ?$, $P = 0.2293$.

4step

17.14 Inspecting Weights of Boxes of Cookies (continued). Here are measurements (in pounds) of the net weights of 12 boxes of cookies: WEIGHTS

1.038	1.012	1.008	1.004	0.997	0.998
1.012	0.997	1.004	0.999	1.000	1.006

The mean weight of the population of all boxes of cookies is supposed to be 1.000 lb. Do the measurements give good evidence that the population mean weight is not 1.000?

The 12 boxes measured are an SRS from the population of all boxes of this brand of cookies. This population has a Normal distribution with mean equal to the true mean weight of the population and standard deviation 0.01. Use this information to carry out a test of significance, following the four-step process as illustrated in Example 17.7.

4step

17.15 Bad Weather, Bad Tip? People tend to be more generous after receiving good news. Are they less generous after receiving bad news? The average tip left by adult Americans is 20%. Say that a server gives 20 patrons of a restaurant a message on their bill warning them that tomorrow's weather will be bad and records the percentage tip they leave. Here are the tips as a percentage of the total bill:[5] TIP3

18.0	19.1	19.2	18.8	18.4	19.0	18.5	16.1	16.8	18.2
14.0	17.0	13.6	17.5	20.0	20.2	18.8	18.0	23.2	19.4

Suppose that percentage tips are Normal with $\sigma = 2$. Is there good evidence that the mean percentage tip left by patrons who have received a bad weather forecast is less than 20%? Follow the four-step process as illustrated in Example 17.7.

luza studios/Getty Images

17.5 Significance from a Table*

Statistics in practice uses technology (graphing calculator or software) to get *P*-values quickly and accurately. In the absence of suitable technology, you can get approximate *P*-values quickly by comparing the value of your test statistic with critical values from a table. For the z statistic, the table is Table C, the same table we used for confidence intervals.

Look at the bottom row of critical values in Table C, labeled z^*. At the top of the table, you see the confidence level C for each z^*. At the bottom of the table, you see both the one-sided and two-sided *P*-values for each z^*. Values of a test statistic z that are farther out than z^* (in the direction given by the alternative hypothesis) are statistically significant at the level that matches z^*.

Significance from a Table of Critical Values

To find the approximate *P*-value for any z statistic, compare z (ignoring its sign) with the critical values z^* at the bottom of Table C. If z falls between two values of z^*, the *P*-value falls between the two corresponding values of *P* in the "One-sided *P*" or the "Two-sided *P*" row of Table C.

EXAMPLE 17.8 Is It Statistically Significant?

The z statistic for a one-sided test is $z = 2.13$. How statistically significant is this result? Compare $z = 2.13$ with the z^* row in Table C.

z^*	2.054	2.326
One-sided *P*	0.02	0.01

It lies between $z^* = 2.054$ and $z^* = 2.326$. So the *P*-value lies between the corresponding entries in the "One-sided P" row, which are level of significance = 0.02 and level of significance = 0.01. This z *is* statistically significant at the $\alpha = 0.02$ level and *is not* statistically significant at the $\alpha = 0.01$ level.

Figure 17.5 illustrates the situation. The area under the Normal curve to the right of $z = 2.13$ is the *P*-value. You can see that *P* falls between the areas to the right of the two critical values, for $P = 0.02$ and $P = 0.01$.

The z statistic in Example 17.7 is $z = -1.09$. The alternative hypothesis is two-sided. Compare $z = -1.09$ (ignoring the minus sign) with the z^* row in Table C.

z^*	1.036	1.282
Two-sided *P*	0.30	0.20

*This material can be skipped if you use software to compute *P*-values.

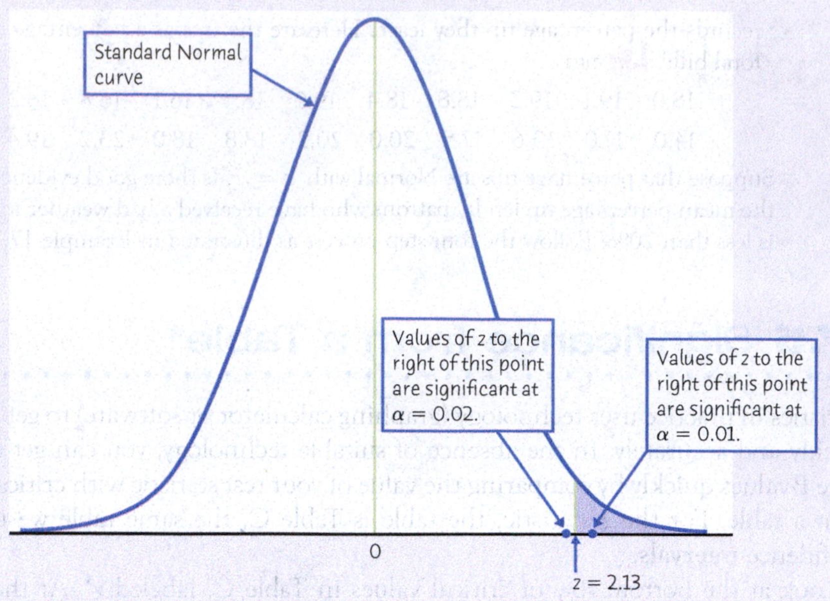
Standard Normal curve
Values of z to the right of this point are significant at α = 0.02.
Values of z to the right of this point are significant at α = 0.01.
0
z = 2.13

FIGURE 17.5 Is it significant? The test statistic value $z = 2.13$ falls between the critical values required for significance at the $\alpha = 0.02$ and $\alpha = 0.01$ levels. So the test *is* significant at $\alpha = 0.02$ and *is not* significant at $\alpha = 0.01$.

It lies between $z^* = 1.036$ and $z^* = 1.282$. So the *P*-value lies between the matching entries in the "Two-sided *P*" row, $P = 0.30$ and $P = 0.20$. This is enough to conclude that the data do not provide good evidence against the null hypothesis.

APPLY YOUR KNOWLEDGE

17.16 Significance from a Table. A test of $H_0: \mu = 0$ against $H_a: \mu > 0$ has test statistic $z = 1.65$. Is this test statistically significant at the 5% level ($\alpha = 0.05$)? Is it statistically significant at the 1% level ($\alpha = 0.01$)?

17.17 Significance from a Table. A test of $H_0: \mu = 0$ against $H_a: \mu \neq 0$ has test statistic $z = 1.65$. Is this test statistically significant at the 5% level ($\alpha = 0.05$)? Is it statistically significant at the 1% level ($\alpha = 0.01$)?

17.18 Testing a Random Number Generator. A random number generator is supposed to produce random numbers that are uniformly distributed on the interval from 0 to 1. If this is true, the numbers generated come from a population with $\mu = 0.5$ and $\sigma = 0.2887$. A command to generate 100 random numbers gives outcomes with mean $\bar{x} = 0.5635$. Assume that the population σ remains fixed. We want to test

$$H_0: \mu = 0.5$$
$$H_a: \mu \neq 0.5$$

(a) Calculate the value of the z test statistic.

(b) Use Table C: Is z statistically significant at the 5% level ($\alpha = 0.05$)?

(c) Use Table C: Is z statistically significant at the 1% level ($\alpha = 0.01$)?

(d) Between which two Normal critical values z^* in the bottom row of Table C does z lie? Between what two numbers does the *P*-value lie? Does the test give good evidence against the null hypothesis?

CHAPTER 17 SUMMARY

- A **test of significance** assesses the evidence provided by data against a **null hypothesis H_0** in favor of an **alternative hypothesis H_a**.
- Hypotheses are always stated in terms of population parameters. Usually H_0 is a statement that no effect is present, and H_a says that a parameter differs from its null value in a specific direction (one-sided alternative) or in either direction (two-sided alternative).
- The essential reasoning of a test of significance is as follows. Suppose for the sake of argument that the null hypothesis is true. If we repeated our data production many times, would we often get data as inconsistent with H_0 as the data we actually have? Data that would rarely occur if H_0 were true provide evidence against H_0.
- A test is based on a **test statistic** that measures how far the sample outcome is from the value stated by H_0.
- The **P-value** of a test is the probability, computed supposing H_0 to be true, that the test statistic will take a value at least as extreme as that actually observed. Small *P*-values indicate strong evidence against H_0. To calculate a *P*-value, we must know the sampling distribution of the test statistic when H_0 is true.
- If the *P*-value is as small as or smaller than a specified value α, the data are **statistically significant at significance level α**.
- **Tests of significance for the null hypothesis $H_0: \mu = \mu_0$** concerning the unknown mean μ of a population are based on the **one-sample z test statistic**

$$z = \frac{\bar{x} - \mu_0}{\sigma/\sqrt{n}}$$

$H_a: \mu > \mu_0$ is $P(Z \geq z)$

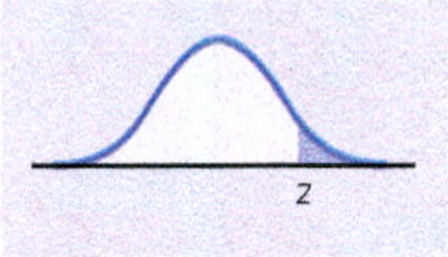

$H_a: \mu < \mu_0$ is $P(Z \leq z)$

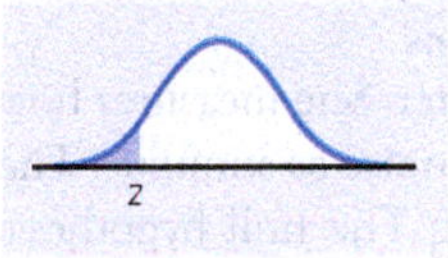

$H_a: \mu \neq \mu_0$ is $P(Z \leq -|z|) + (Z \geq |z|) = 2P(Z \geq |z|)$

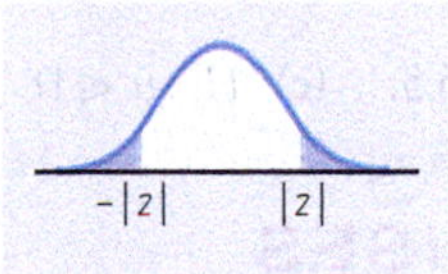

The z test assumes an SRS of size n from a Normal population with known population standard deviation σ. *P*-values can be obtained either with computations from the standard Normal distribution or by using technology (applet, graphing calculator, or software).

CHECK YOUR SKILLS

17.19 You use software to carry out a test of significance. The program tells you that the *P*-value is $P = 0.052$. You conclude that the probability, computed assuming that H_0 is

(a) true, of the test statistic taking a value as extreme as or more extreme than that actually observed is 0.052.

(b) true, of the test statistic taking a value as extreme as or less extreme than that actually observed is 0.052.

(c) false, of the test statistic taking a value as extreme as or more extreme than that actually observed is 0.052.

17.20 You use software to carry out a test of significance. The program tells you that the *P*-value is $P = 0.052$. This result is

(a) not statistically significant at either $\alpha = 0.05$ or $\alpha = 0.01$.

(b) statistically significant at $\alpha = 0.05$ but not at $\alpha = 0.01$.

(c) statistically significant at both $\alpha = 0.05$ and $\alpha = 0.01$.

17.21 The z statistic for a one-sided test is $z = 1.62$. This test is

(a) not statistically significant at either $\alpha = 0.05$ or $\alpha = 0.01$.

(b) statistically significant at $\alpha = 0.05$ but not at $\alpha = 0.01$.

(c) statistically significant at both $\alpha = 0.05$ and $\alpha = 0.01$.

17.22 The gas mileage for a particular model of pickup truck varies but is known to have a standard deviation of $\sigma = 1.0$ mile per gallon in repeated tests in a controlled laboratory environment at a fixed speed of 65 miles per hour. For a fixed speed of 65 miles per hour, gas mileages in repeated tests are Normally distributed. Tests on four trucks of this model at 65 miles per hour give gas mileages of 19.7, 20.1, 19.9, and 19.5 miles per gallon. The z statistic for testing $H_0: \mu = 20$ miles per gallon based on these four measurements is

(a) $z = -0.800$. (b) $z = -0.400$. (c) $z = -0.200$.

17.23 Experiments on learning in animals sometimes measure how long it takes mice to find their way through a maze. The mean time is 18 seconds for one particular maze. A researcher thinks that a loud noise will cause the mice to complete the maze faster. She measures how long each of 10 mice takes with a noise as stimulus. The sample mean is $\bar{x} = 16.5$ seconds. The null hypothesis for the test of significance is

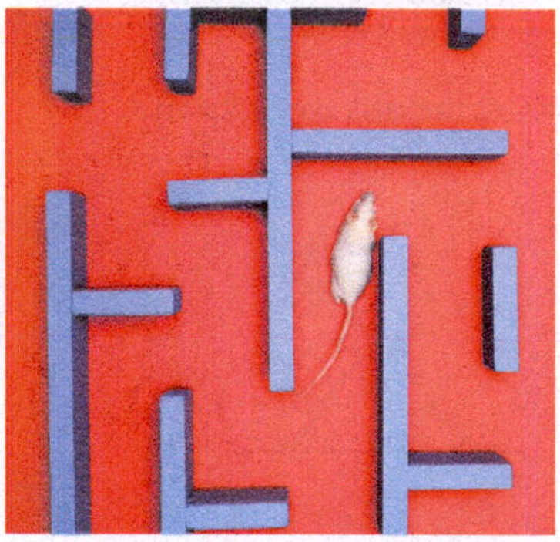

ClassicStock/Alamy Stock Photo

(a) $H_0: \mu = 18$. (b) $H_0: \mu = 16.5$. (c) $H_0: \mu < 18$.

17.24 The alternative hypothesis for the test in Exercise 17.23 is

(a) $H_a: \mu \neq 18$.

(b) $H_a: \mu < 18$.

(c) $H_a: \mu = 16.5$.

17.25 Researchers investigated the effectiveness of oral zinc, as compared to a placebo, in reducing the duration of the common cold when taken within 24 hours of the onset of symptoms. The researchers found that those taking oral zinc had a statistically significantly shorter duration ($P < 0.05$) than those taking the placebo.[6] This means that

(a) the probability that the null hypothesis is true is less than 0.05.

(b) the value of the test statistic, the mean reduction in duration of the cold, is large.

(c) neither of the above is true.

17.26 You are testing H_0: $\mu = 0$ against H_0: $\mu \neq 0$ based on an SRS of 20 observations from a Normal population. What values of the z statistic are statistically significant at the $\alpha = 0.001$ level?

(a) All values for which $|z| > 3.291$

(b) All values for which $z > 3.291$

(c) All values for which $z > 3.091$

17.27 You are testing H_0: $\mu = 0$ against H_0: $\mu > 0$ based on an SRS of 20 observations from a Normal population. What values of the z statistic are statistically significant at the $\alpha = 0.001$ level?

(a) All values for which $|z| > 3.291$

(b) All values for which $z > 3.291$

(c) All values for which $z > 3.091$

CHAPTER 17 EXERCISES

In all exercises that call for P-values, give the actual value if you use software or the P-Value applet. Otherwise, use Table C to give values between which P must fall.

17.28 **Student Study Times.** Exercise 16.19 (page 380) describes a class survey in which students claimed to study an average of $\bar{x} = 13.7$ hours in a typical week. Regard these students as an SRS from the population of all undergraduate students at this university. Does the study give good evidence that students claim to study more than 13 hours per week on the average?

(a) State null and alternative hypotheses in terms of the mean study time, in hours, for the population.

(b) What is the value of the test statistic z?

(c) What is the P-value of the test? Can you conclude that students do claim to study more than 13 hours per week on the average?

17.29 **I Want More Muscle.** If young men thought that their own level of muscle was about what women prefer, the mean "muscle gap" in the study described in Exercise 16.20 (page 380) would be 0. We suspect (before seeing the data) that (most) young men tend to think women prefer more muscle than they themselves have.

(a) State null and alternative hypotheses for testing this suspicion.

(b) What is the value of the test statistic z?

(c) You can tell just from the value of z that the evidence in favor of the alternative is very strong (that is, the P-value is very small). Explain why this is true.

17.30 Hotel Managers' Personalities. Successful hotel managers must have personality characteristics traditionally stereotyped as feminine (such as "compassionate") as well as those traditionally stereotyped as masculine (such as "forceful"). The Bem Sex-Role Inventory (BSRI) is a personality test that gives separate ratings for "female" and "male" stereotypes, both on a scale of 1 to 7. Although the BSRI was developed at a time when these stereotypes were more pronounced, it is still widely used to assess personality types. Unfortunately, the ratings are often referred to as femininity and masculinity scores.

A sample of 148 male general managers of three-star and four-star hotels had mean BSRI femininity score $\bar{x} = 5.29$.[7] The mean score for the general male population is $\mu = 5.19$. Do hotel managers, on the average, differ statistically significantly in femininity score from men in general? Assume that the standard deviation of scores in the population of all male hotel managers is the same as the $\sigma = 0.78$ for the adult male population.

(a) State null and alternative hypotheses in terms of the mean femininity score μ for male hotel managers.

(b) Find the z test statistic.

(c) What is the *P*-value for your z? What do you conclude about male hotel managers?

17.31 Is This What *P* Means? A randomized comparative experiment examined the effect of the attractiveness of an instructor on the performance of students on a quiz given by the instructor. The researchers found a statistically significant difference in quiz scores between students in a class with an instructor rated as attractive and students in a class with an instructor rated as unattractive ($P = 0.005$).[8] When asked to explain the meaning of "$P = 0.005$," a student says, "This means there is only probability of 0.005 that the null hypothesis is true." Explain what $P = 0.005$ really means in a way that makes it clear that the student's explanation is wrong.

17.32 How to Show That You Are Rich. Every society has its own marks of wealth and prestige. In ancient China, it appears that owning pigs was such a mark. Evidence comes from examining burial sites. The skulls of sacrificed pigs tend to appear along with expensive ornaments, which suggests that the pigs, like the ornaments, signal the wealth and prestige of the person buried. A study of burials from around 3500 B.C. concluded that "there are striking differences in grave goods between burials with pig skulls and burials without them . . . . A test indicates that the two samples of total artifacts are statistically significantly different at the 0.01 level."[9] Explain clearly why "statistically significantly different at the 0.01 level" gives good reason to think that there really is a systematic difference between burials that contain pig skulls and those that lack them.

17.33 Alleviating Test Anxiety. Research suggests that pressure to perform well can reduce performance on exams. Are there effective strategies to deal with pressure? In an experiment, researchers had students take a test on mathematical skills. The same students were then asked to take a second test on the same skills. However, for the second test, the researchers added conditions intended to increase the pressure to perform well. Now each student was paired with a partner, and only if both improved their scores would they receive a monetary reward for participating in the experiment. They were also told that their performance would be videotaped and watched by teachers and students.

Students were then randomly divided into two groups. One group served as a control. To help them cope with the pressure, 10 minutes before the second exam, students in the second group were asked to write as candidly as possible about their thoughts and feelings regarding the exam. The difference in the test scores (posttest − pretest score) was computed. "Students who expressed their thoughts before the second high-pressure test showed a statistically significant 5% math accuracy improvement from the pretest to posttest" ($P < 0.03$).[10] A colleague who knows no statistics says that an increase of 5% isn't a lot—and maybe it's just an accident due to natural variation among the students. Explain in simple language how "$P < 0.03$" answers this objection.

17.34 Instructor Gender. In the study described in Exercise 17.33, researchers also examined the effect of the gender of an instructor (assumed to be male or female) on performance of students on a quiz. The researchers found no evidence of a difference in scores ($P = 0.24$). The *P*-value refers to a null hypothesis of "no difference" in quiz scores measured on classes taught by male and female instructors. Explain clearly why this value provides no evidence of a difference.

17.35 5% Versus 1%. Sketch the standard Normal curve for the z test statistic and mark off areas under the curve to show why a value of z that is statistically significant at the 1% level in a one-sided test is always statistically significant at the 5% level. If z is statistically significant at the 5% level, what can you say about its significance at the 1% level?

17.36 The Wrong Alternative. Researchers are interested in the effect of running 30 minutes a day on the performance of undergraduates on the GRE Verbal test. They start with no expectations about whether students who run 30 minutes a day will perform better than students in a control group who follow their usual exercise regimen. After noticing that students running 30 minutes a day tended to have higher GRE scores, the researchers decide to test a one-sided alternative about the mean GRE Verbal test scores,

$$H_0: \mu_{\text{running}} = \mu_{\text{control}}$$
$$H_a: \mu_{\text{running}} > \mu_{\text{control}}$$

The researchers find $z = 1.71$ with one-sided *P*-value $P = 0.0436$.

(a) Explain why the researchers should have used the two-sided alternative hypothesis.

(b) What is the correct *P*-value for $z = 1.71$?

17.37 The Wrong P. The report of a study of seat belt use by drivers says, "Hispanic drivers were not statistically significantly more likely than White/non-Hispanic drivers to overreport safety belt use (27.4% vs. 21.1%, respectively; $z = 1.33$, $P > 1.0$)."[11] How do you know that the *P*-value given is incorrect? What is the correct one-sided *P*-value for test statistic $z = 1.33$?

Exercises 17.38 through 17.41 ask you to answer questions from data. Assume that the "simple conditions" hold in each case. The exercise statements give you the **State** *step of the four-step process. In your work, follow the* **Plan, Solve,** *and* **Conclude** *steps, illustrated in Example 16.3 (page 374) for a confidence interval and in Example 17.7 (page 393) for a test of significance.*

17.38 Pulling Apart Wood. How heavy a load (in pounds) is needed to pull apart pieces of Douglas fir 4 inches long and 1.5 inches square? Here are data from students doing a laboratory exercise: WOOD

33,190	31,860	32,590	26,520	33,280
32,320	33,020	32,030	30,460	32,700
23,040	30,930	32,720	33,650	32,340
24,050	30,170	31,300	28,730	31,920

We are willing to regard the wood pieces prepared for the lab session as an SRS of all similar pieces of Douglas fir. Engineers also commonly assume that characteristics of materials vary Normally. Suppose that the strength of pieces of wood like these follows a Normal distribution, with standard deviation 3000 pounds.

(a) Is there statistically significant evidence at the $\alpha = 0.10$ level against the hypothesis that the mean is 32,500 pounds for the two-sided alternative?

(b) Is there statistically significant evidence at the $\alpha = 0.10$ level against the hypothesis that the mean is 31,500 pounds for the two-sided alternative?

17.39 Bone Loss by Nursing Mothers. As discussed in Exercise 16.26 (page 381), breast-feeding mothers secrete calcium into their milk. Some of the calcium may come from their bones, so mothers may lose bone mineral. Researchers measured the percentage change in mineral content of the spines of 47 mothers during three months of breast-feeding.[12] Here are the data: BONELS

−4.7	−2.5	−4.9	−2.7	−0.8	−5.3	−8.3	−2.1	−6.8	−4.3
2.2	−7.8	−3.1	−1.0	−6.5	−1.8	−5.2	−5.7	−7.0	−2.2
−6.5	−1.0	−3.0	−3.6	−5.2	−2.0	−2.1	−5.6	−4.4	−3.3
−4.0	−4.9	−4.7	−3.8	−5.9	−2.5	−0.3	−6.2	−6.8	1.7
0.3	−2.3	0.4	−5.3	0.2	−2.2	−5.1			

The researchers are willing to consider these 47 women as an SRS from the population of all nursing mothers. Suppose that the percentage change in this population has a Normal distribution with standard deviation $\sigma = 2.5\%$. Do these data give good evidence that, on the average, nursing mothers lose bone mineral?

17.40 This Wine Stinks. Sulfur compounds cause "off odors" in wine, and winemakers want to know the odor threshold—the lowest concentration of a compound that the human nose can detect. The odor threshold for dimethyl sulfide (DMS) in trained wine tasters is about 25 micrograms per liter of wine (μg/L). The untrained noses of consumers may be less sensitive, however. Here are the DMS odor thresholds for 10 untrained students: WINE

30	30	42	35	22	33	31	29	19	23

Assume that the odor threshold for untrained noses is Normally distributed with $\sigma = 7$ μg/L. Is there evidence that the mean threshold for untrained tasters is greater than 25 μg/L?

17.41 Eye Grease. Athletes performing in bright sunlight often smear black eye grease under their eyes to reduce glare. Does eye grease work? In one study, 16 student subjects took a test of visual sensitivity to light-and-dark contrast after three hours facing into bright sun, both with and without eye grease. This is a matched pairs design. If eye grease is effective, subjects will be more sensitive to contrast when they use eye grease. Here are the differences in sensitivity, with eye grease minus without eye grease:[13] EYEGRS

0.07	0.64	−0.12	−0.05	−0.18	0.14	−0.16	0.03
0.05	0.02	0.43	0.24	−0.11	0.28	0.05	0.29

Kathy Willens/AP Images

We want to know whether eye grease increases sensitivity on the average.

(a) What are the null and alternative hypotheses? Say in words what mean μ your hypotheses concern.

(b) Suppose that the subjects are an SRS of all young people with normal vision, that contrast differences follow a Normal distribution in this population, and that the standard deviation of differences is $\sigma = 0.22$. Carry out a test of significance.

17.42 Tests from Confidence Intervals. A confidence interval for the population mean μ tells us which values of μ are plausible (those inside the interval) and which values are not plausible (those outside the interval) at the chosen level of confidence. You can use this idea to carry out a test of any null hypothesis $H_0: \mu = \mu_0$ starting with a confidence interval: *reject H_0 if μ_0 is outside the interval and fail to reject if μ_0 is inside the interval.*

The alternative hypothesis is always two-sided, $H_a: \mu \neq \mu_0$ because the confidence interval extends in both directions from $\bar{x}$. A 95% confidence interval leads to a test at the 5% significance level because the interval is wrong 5% of the time. In general, confidence level C leads to a test at significance level $\alpha = 1 - C$.

(a) In Example 17.7 (page 393), a medical director found LDL $\bar{x} = 124.86$ for an SRS of 72 executives. The standard deviation of the LDL of all adults is $\sigma = 40$. Give a 90% confidence interval for the mean LDL μ of all executives in this company, assuming that the standard deviation is the same as for all adults.

(b) The hypothesized value $\mu_0 = 130$ falls *inside* this confidence interval. Carry out the z test for $H_0: \mu = 130$ against the two-sided alternative. Show that the test is *not statistically significant* at the 10% level.

(c) The hypothesized value $\mu_0 = 134$ falls *outside* this confidence interval. Carry out the z test for $H_0: \mu = 134$ against the two-sided alternative. Show that the test *is statistically significant* at the 10% level.

17.43 Tests from Confidence Intervals. Family caregivers of patients with chronic illness can experience anxiety. Do regular support-group meetings affect these feelings of anxiety? It is possible that they reduce anxiety, perhaps through sharing experiences with other caregivers in similar situations, or increase anxiety, perhaps by reinforcing painful experiences by recounting them to others. To explore the effect of support-group meetings, several family caregivers were enrolled in a support group. After three months, researchers administered a test to measure anxiety, with larger scores indicating greater anxiety. Assume that these caregivers are a random sample from the population of all family caregivers. A 99% confidence interval for the population mean anxiety score μ after participating in a support group is 7.2 ± 0.9.[14]

Use the method described in the previous exercise to answer these questions.

(a) Suppose we know that the mean anxiety score for the population of all family caregivers is 6.2. With a two-sided alternative, can you reject the null hypothesis that $\mu = 6.2$ at the 1% ($\alpha = 0.01$) significance level? Why?

(b) Suppose we know that the mean anxiety score for the population of all family caregivers is 6.4. With a two-sided alternative, can you reject the null hypothesis that $\mu = 6.4$ at the 1% ($\alpha = 0.01$) significance level? Why?

17.44 Tests from Confidence Intervals. Example 16.3 (page 374) computes a 95% confidence interval for the mean percentage tip μ for all patrons of a restaurant when they receive a message on their bill indicating that the next day's weather will be good. Use this confidence interval to test $H_0: \mu = 20$ against the two-sided alternative at the 5% significance level.

17.45 Tests from Confidence Intervals. Exercise 16.6 (page 375) asks you to compute a 90% confidence interval for the true melting point μ of a copper sample purchased from NIST. Here are six measurements on the same copper sample given in Exercise 16.6:

1084.55 1084.89 1085.02 1084.79 1084.69 1084.86

You read online that the melting point of copper is 1084.62°C. If you haven't already done so, compute the confidence interval and use it to test $H_0: \mu = 1084.62$°C against the two-sided alternative at the 10% significance level. Use the four-step process.

CHAPTER

18

Inference in Practice

When you complete this chapter, you will be able to:

18.1 When applying inference procedures, confirm that the required conditions—especially that the sample data were produced using randomization—are met.

18.2 When interpreting confidence intervals, recognize that the margin of error covers only errors occurring by chance in random sampling.

18.3 Understand the factors affecting significance levels in statistical tests and distinguish between statistical significance and practical significance.

18.4 Calculate the required sample size to achieve a desired margin of error in a statistical study.

18.5 Use the concepts of Type I and Type II errors and the power of a test to plan and interpret statistical studies.

To this point, we have met just two procedures for statistical inference. Both concern inference about the mean μ of a population when the "simple conditions" (page 367) are true: the data are an SRS, the population has a Normal distribution, and we know the standard deviation σ of the population. Under these conditions, a confidence interval for the mean μ is

$$\bar{x} \pm z^* \frac{\sigma}{\sqrt{n}}$$

where z^* is the critical value required for a given significance level. To test $H_0: \mu = \mu_0$, we use the one-sample z statistic:

$$z = \frac{\bar{x} - \mu_0}{\sigma/\sqrt{n}}$$

We call such procedures *z procedures* because they both start with the one-sample z statistic and use the standard Normal distribution.

In later chapters, we will modify these procedures for inference about a population mean to make them useful in practice. We will also introduce procedures for confidence intervals and tests in most of the settings we met in learning to explore data. There are libraries—both of books and of software—full of more elaborate statistical techniques. The reasoning with confidence intervals and tests is the same, no matter how elaborate the details of the procedure are.

There is a saying among statisticians that "mathematical theorems are true; statistical methods are effective when used with judgment." That the one-sample z statistic has the standard Normal distribution when the null

Stanislaw Pytel/Getty Images

hypothesis is true is a mathematical theorem. Effective use of statistical methods requires more than knowing such facts. It requires even more than understanding the underlying reasoning. This chapter begins the process of helping you develop the judgment needed to use statistics in practice. That process will continue in examples and exercises through the rest of this book.

18.1 Conditions for Inference in Practice

Any confidence interval or significance test can be trusted only under specific conditions. It's up to you to understand these conditions and judge whether they fit your problem. With that in mind, let's look back at the "simple conditions" for the z procedures for inference about a mean.

Simple Conditions for Inference about a Mean

1. We have a simple random sample (SRS) from the population of interest. There is no nonresponse or other practical difficulty. The population is large compared to the size of the sample.
2. The variable we measure has an exactly Normal distribution $N(\mu, \sigma)$ in the population.
3. We don't know the population mean μ. But we do know the population standard deviation σ.

The final "simple condition"—that we know the standard deviation σ of the population—is rarely satisfied in practice. The z procedures are therefore of little practical use. Fortunately, it's easy to remove the "known σ" condition. Chapter 20 shows how. The condition that the size of the population is large compared to the size of the sample is often easy to verify, and when it is not satisfied, there are special, advanced methods for inference. The other "simple conditions" (SRS, Normal population) are harder to escape. In fact, they represent the kinds of conditions needed if we are to trust almost any statistical inference. As you plan inference, you should always answer the question "Where did the data come from?" and you must often also answer another question: "What is the shape of the population distribution?" This is the point where knowing mathematical facts gives way to the need for judgment.

STATISTICS IN YOUR WORLD

Don't Touch the Plants

We know that confounding can distort inference; however, we don't always recognize how easy it is to confound data. Consider the innocent scientist who visits plants in the field once a week to measure their size. To measure the plants, she has to touch them. A study of six plant species found that one touch a week significantly increased leaf damage by insects in two species and significantly decreased damage in another species.

Where did the data come from? *The most important requirement for any inference procedure is that the data come from a process to which the laws of probability apply.* Inference is most reliable when the data come from a random sample or a randomized comparative experiment. Random samples use chance to choose respondents. Randomized comparative experiments use chance to assign subjects to treatments. The deliberate use of chance ensures that the laws of probability apply to the outcomes, and this in turn ensures that statistical inference makes sense.

Where the Data Come from Matters

When you use statistical inference, you are acting as if your data are a random sample or come from a randomized comparative experiment.

If your data don't come from a random sample or a randomized comparative experiment, your conclusions may be challenged. To answer the challenge, you must usually rely on subject-matter knowledge, not on statistics. It is common to apply statistical inference to data that are not produced by random selection. When you see such a study, consider whether the data can be trusted as a basis for the conclusions of the study.

EXAMPLE 18.1 The Psychologist and the Sociologist

A psychologist is interested in how our visual perception can be fooled by optical illusions. Her subjects are students in Psychology 101 at her university. Most psychologists would agree that it's safe to treat the students as an SRS of all people with normal vision. There is nothing special about being a student that changes visual perception.

A sociologist at the same university uses students in Sociology 101 to examine attitudes toward poor people and antipoverty programs. Students as a group are younger than the adult population as a whole. Even among young people, students as a group come from more prosperous and better-educated homes. Even among students, this university isn't typical of all campuses. Even on this campus, students in a sociology course may have opinions that are quite different from those of engineering students. The sociologist can't reasonably act as if these students are a random sample from any interesting population.

Our first examples of inference, using the z procedures, act as if the data are an SRS from the population of interest. Let's look back at the examples in Chapters 16 and 17.

EXAMPLE 18.2 Is It Really an SRS?

The NHANES survey that produced the BMI data for Example 16.1 (page 368) used a complex multistage sample design, so it's a bit oversimplified to treat the BMI data as coming from an SRS from the population of young men.[1] Although the overall effect of the NHANES sample is close to an SRS, professional statisticians would use more complex inference procedures to match the more complex design of the sample.

The 20 patrons in the tipping study in Example 16.3 (page 374) were chosen from those eating at a particular restaurant to receive one of several treatments being compared in a randomized comparative experiment. Recall that each treatment group in a completely randomized experiment is an SRS of the available subjects. Researchers sometimes act as if the available subjects are an SRS from some population if there is nothing special about where the subjects came from. In some cases, researchers collect demographic data on subjects to help justify the assumption that the subjects are a representative sample from some population. We are willing to regard the subjects as an SRS from the population of patrons of this particular restaurant, but perhaps this needs to be explored further. For example, if the day on which the study was conducted was special in some way (such as Valentine's Day or some other holiday), the patrons may not be representative of those who typically eat at the restaurant.

The cola taste test in Examples 17.2 (page 383) and 17.3 (page 384) uses scores from 10 tasters. All tasters were examined to be sure that they have no medical condition that interferes with normal taste and then carefully trained to score sweetness using a set of standard drinks. We are willing to take their scores as an SRS from the population of trained tasters.

The medical director who examined executives' LDL in Example 17.7 (page 393) actually chose an SRS from the medical records of all executives in this company.

These examples are typical. One is an actual SRS; two are situations in which common practice is to act as if the sample were an SRS; and in the remaining example, procedures that assume an SRS are used for a quick analysis of data from a more complex random sample. *There is no simple rule for deciding when you can act as if a sample is an SRS. Pay attention to these cautions:*

- *Practical problems such as nonresponse in samples or dropouts from an experiment can hinder inference even from a well-designed study.* The NHANES survey has about an 80% response rate. This is much higher than opinion polls and most other national surveys, so by realistic standards, NHANES data are quite trustworthy. (NHANES uses advanced methods to try to correct for nonresponse, but these methods work a lot better when response is high to start with.)
- *Different methods are needed for different designs.* The z procedures aren't correct for random sampling designs more complex than an SRS. Later chapters give methods for some other designs, but we won't discuss inference for really complex designs like that used by NHANES. Always be sure that you (or your statistical consultant) know how to carry out the inference your design calls for.
- *There is no cure for fundamental flaws like voluntary response surveys or uncontrolled experiments.* Look back at the examples of how to sample badly in Sections 8.2 and 9.3 and steel yourself to just ignore data from such studies.

What is the shape of the population distribution? Most statistical inference procedures require some conditions on the shape of the population distribution. Many of the most basic methods of inference are designed for Normal populations. That's the case for the z procedures and also for the more practical procedures for inference about means that we will meet in Chapters 20 and 21. Fortunately, this condition is less essential than where the data come from.

STATISTICS IN YOUR WORLD

Really Wrong Numbers

By now you know that "statistics" that don't come from properly designed studies are often dubious and sometimes just made up. It's rare to find wrong numbers that anyone can see are wrong, but it does happen. A German physicist claimed that 2006 was the first year since 1441 with more than one Friday the 13th. Sorry: Friday the 13th occurred in February and August of 2004, which is a bit more recent than 1441.

This is true because the z procedures and many other procedures designed for Normal distributions are based on Normality of the sample mean $\bar{x}$, not Normality of individual observations. The central limit theorem tells us that the sampling distribution of $\bar{x}$ is more Normal than the distribution of individual observations and that the sampling distribution of $\bar{x}$ becomes more Normal as the size of the sample increases. In practice, the z procedures are reasonably accurate for any roughly symmetric distribution for samples of even moderate size. If the sample is large, the sampling distribution $\bar{x}$ will be close to Normal even if individual measurements are strongly skewed, as Figures 15.4 (page 352) and 15.5 (page 354) illustrate. Later chapters give practical guidelines for specific inference procedures.

There is one important exception to the principle that the shape of the population is less critical than how the data were produced. Outliers can distort the results of inference. *Any inference procedure based on sample statistics like the sample mean $\bar{x}$ that are not resistant to outliers can be strongly influenced by a few extreme observations.*

We rarely know the shape of the population distribution. In practice, we rely on previous studies and on data analysis. Sometimes long experience suggests that our data are likely to come from a roughly Normal distribution, or not. For example, heights of people of the same sex and similar ages are close to Normal, but weights are not. Always explore your data before doing inference. When the data are chosen at random from a population, the shape of the data distribution mirrors the shape of the population distribution. Make a stemplot or histogram of your data and look to see whether the shape is roughly Normal. Remember that small samples have a lot of chance variation, so that Normality is hard to judge from just a few observations. Always look for outliers and try to correct them or justify their removal before performing the z procedures or other inference based on statistics like $\bar{x}$ that are not resistant.

When outliers are present or the data suggest that the population is strongly non-Normal, consider alternative methods that don't require Normality and are not sensitive to outliers. Some of these methods appear in Chapter 28 (available online).

APPLY YOUR KNOWLEDGE

18.1 Rate This Product. An online shopping site asks customers to rate the products they buy on a scale from 1 (strongly dislike) to 5 (strongly like). The invitation to rate a recent purchase is sent by email to customers one week after they purchase a product, and customers can choose to ignore the invitation. Which of the following is the most important reason a confidence interval based on the data from such ratings is of little use for the mean rating by all customers who purchase a particular product. Comment briefly on each reason to explain your answer.

(a) For some products, the number of customers who purchase the product is small, so the margin of error will be large.

(b) Many of the customers may not read their email or may have a spam filter that incorrectly identifies the email requesting a review as spam.

(c) The customers who provide ratings can't be considered a random sample from the population of all customers who purchase a particular product.

© Ilene MacDonald/Alamy

18.2 Running Red Lights. A survey of licensed drivers inquired about running red lights. One question asked, "Of every 10 motorists who run a red light, about how many do you think will be caught?" The mean result for 880 respondents was $\bar{x} = 1.92$, and the standard deviation was $s = 1.83$.[2] For this large sample, s will be close to the population standard deviation σ, so suppose we know that $\sigma = 1.83$.

(a) Give a 95% confidence interval for the mean opinion in the population of all licensed drivers.

(b) The distribution of responses is skewed to the right rather than Normal. This will not strongly affect the z confidence interval for this sample. Why not?

(c) The 880 respondents are an SRS from completed calls among 45,956 calls to randomly chosen residential telephone numbers listed in telephone directories. Only 5029 of the calls were completed. This information gives two reasons to suspect that the sample may not represent all licensed drivers. What are these reasons?

18.3 Sampling Shoppers. A reporter for a local television station visits the city's new upscale shopping mall the day before Christmas to interview shoppers. He questions the first 25 shoppers he meets outside one of department stores at the mall. He asks them whether their overall feelings about Christmas shopping are positive, neutral, or negative. Suggest some reasons why it may be risky to act as if the first 25 shoppers at this particular location are an SRS of all shoppers in the city.

18.2 Cautions about Confidence Intervals

The most important caution about confidence intervals in general is a consequence of the use of a sampling distribution. A sampling distribution shows how a statistic such as $\bar{x}$ varies in repeated random sampling. This variation causes *random sampling error* because the statistic misses the true parameter by a random amount. No other source of variation or bias in the sample data influences the sampling distribution. So *the margin of error in a confidence interval ignores everything except the sample-to-sample variation due to choosing the sample randomly.*

The Margin of Error Doesn't Cover All Errors

The margin of error in a confidence interval covers only random sampling errors. Practical difficulties such as undercoverage and nonresponse are often more serious than random sampling error. The margin of error does not take such difficulties into account.

Recall from Chapter 8 that national opinion polls often have response rates less than 50% and that even small changes in the wording of questions can strongly influence results. In such cases, the announced margin of error is probably unrealistically small. And of course there is no way to assign a meaningful margin of error to results from voluntary response or convenience samples because there is no random selection. Look carefully at the details of a study before you trust a confidence interval.

APPLY YOUR KNOWLEDGE

18.4 What's Your Weight? A 2019 Gallup Poll asked a national random sample of 507 adult men to state their current weight. The mean weight in the sample was $\bar{x} = 196$. We will treat these data as an SRS from a Normally distributed population with standard deviation $\sigma = 35$.

(a) Give a 95% confidence interval for the mean weight of adult men based on these data.

(b) Do you trust the interval you computed in part (a) as a 95% confidence interval for the mean weight of all U.S. adult men? Why or why not?

18.5 Good Weather, Good Tips? Example 16.3 (page 374) described an experiment exploring the size of the tip in a particular restaurant when a message indicating that the next day's weather would be good was written on the bill. You work part-time as a server in a restaurant. You read a newspaper article about the study that reports that, with 95% confidence, the mean percentage tip from restaurant patrons will be between 21.33 and 23.09 when the server writes a message on the bill stating that the next day's weather will be good. Can you conclude that if you begin writing a message on patrons' bills that the next day's weather will be good, approximately 95% of the days you work your mean percentage tip will be between 21.33 and 23.09? Why or why not?

18.6 Sample Size and Margin of Error. Example 16.1 (page 368) described NHANES data on the body mass index (BMI) of 936 young men. The mean BMI in the sample was $\bar{x} = 27.2$ kg/m^2. We treated these data as an SRS from a Normally distributed population with standard deviation $\sigma = 11.6$.

(a) Suppose that we had an SRS of just 100 young men. What would be the margin of error for 95% confidence?

(b) Find the margins of error for 95% confidence based on SRSs of 400 young men and 1600 young men.

(c) Compare the three margins of error. How does increasing the sample size change the margin of error of a confidence interval when the confidence level and population standard deviation remain the same?

18.7 Do You Drink Pop? A July 2015 Gallup Poll asked a national sample of 1009 adults aged 18 and over if they actively avoided drinking soda or pop. Of those sampled, 61% indicated that they do so. Gallup announced the poll's margin of error for 95% confidence as ± 4 percentage points. Which of the following sources of error are included in this margin of error?

(a) Gallup dialed landline telephone numbers at random and so missed all people without landline phones, including people whose only phone is a cell phone.

(b) Some people whose numbers were chosen never answered the phone in several calls or answered but refused to participate in the poll.

(c) There is chance variation in the random selection of telephone numbers.

18.3 Cautions about Significance Tests

Significance tests are widely used in most areas of statistical work. New pharmaceutical products require significant evidence of effectiveness and safety. Courts inquire about statistical significance in hearing class action discrimination cases. Marketers want to know whether a new package design will significantly increase sales. Medical researchers want to know whether a new therapy performs significantly better. In all these uses, statistical significance is valued because it points to an effect that is unlikely to occur simply by chance. Here are some points to keep in mind when you use or interpret significance tests.

How small a *P* is convincing? The purpose of a test of significance is to describe the degree of evidence provided by the sample against the null hypothesis. The *P*-value does this. But how small a *P*-value is convincing evidence against the null hypothesis? This depends mainly on two circumstances:

- *How plausible is H_0?* If H_0 represents an assumption that the people you must convince have believed for years, strong evidence (small *P*) will be needed to persuade them.
- *What are the consequences of rejecting H_0?* If rejecting H_0 in favor of H_a means making an expensive changeover from one type of product packaging to another, you need strong evidence that the new packaging will boost sales.

These criteria are a bit subjective. Different people will often insist on different levels of significance in similar or identical situations. Giving the *P*-value allows each of us to decide individually if the evidence is sufficiently strong.

Users of statistics have often emphasized standard levels of significance such as 10%, 5%, and 1%. This emphasis reflects the time when tables of critical values rather than software dominated statistical practice. The 5% level ($\alpha = 0.05$) is particularly common. *There is no sharp border between "significant" and "not significant," only increasingly strong evidence as the P-value decreases. There is no practical distinction between the P-values 0.049 and 0.051. It makes no sense to treat $P \leq 0.05$ as a universal rule for what is significant.*

Nevertheless, there exist situations where significance at the 5% level is regarded as a strict benchmark. For example, courts have tended to accept 5% as the standard in discrimination cases.[3] Some journals treat 5% as necessary for demonstrating significance of research findings. Regulatory agencies have used 5% as a rule for declaring a finding significant.[4]

Significance depends on the alternative hypothesis You may have noticed that the *P*-value for a one-sided test is one-half the *P*-value for the two-sided test of the same null hypothesis based on the same data. The two-sided *P*-value combines two equal areas, one in each tail of a Normal curve. The one-sided *P*-value is just one of these areas, in the direction specified by the alternative hypothesis. It makes sense that the evidence against H_0 is stronger when the alternative is one-sided because it is based on the data *plus* information about the direction of possible deviations from H_0—the information or reason that led the researcher to choose a one-sided test prior to collecting data. If you lack this added information, always use a two-sided alternative hypothesis.

Significance depends on sample size A sample survey shows that significantly fewer students are heavy drinkers at colleges that ban alcohol on campus. "Significantly fewer" is not enough information to decide whether there is an *important* difference in drinking behavior at schools that ban alcohol. *How important an effect is depends on the size of the effect as well as on its statistical significance.* If the number of heavy drinkers is only 1% less at colleges that ban alcohol than at other colleges, this is probably not an important effect, even if it is statistically significant. (Consider whether you would describe 1% as "significantly" fewer when talking with a friend.) In fact, the sample survey found that 38% of students at colleges that ban alcohol are "heavy episodic drinkers" compared with 48% at other colleges.[5] That difference is large enough to be important. (Of course, this observational study doesn't prove that an alcohol ban directly reduces drinking; it may be that colleges that ban alcohol attract more students who don't want to drink heavily.)

Such examples remind us to always look at the size of an effect (like 38% versus 48%) as well as its significance. They also raise a question: Can a tiny effect really be highly significant? Yes. The behavior of the z test statistic is typical. The statistic is

$$z = \frac{\bar{x} - \mu_0}{\sigma/\sqrt{n}}$$

The numerator measures how far the sample mean deviates from the hypothesized mean μ_0. Larger values of the numerator give stronger evidence against $H_0: \mu = \mu_0$. The denominator is the standard deviation of $\bar{x}$. It measures how much random variation we expect. There is less variation when the number of observations n is large. So z gets larger (more significant) when the estimated effect $\bar{x} - \mu_0$ gets larger *or* when the number of observations n gets larger. Significance depends both on the size of the effect we observe *and* on the size of the sample. Understanding this fact is essential to understanding significance tests.

Sample Size Affects Statistical Significance

Because large random samples have small chance variation, very small population effects can be highly significant if the sample is large.

Because small random samples have a lot of chance variation, even large population effects can fail to be significant if the sample is small.

Statistical significance does not tell us whether an effect is large enough to be important. That is, *statistical significance is not the same thing as practical significance.*

Keep in mind that "statistical significance" means "the sample showed an effect larger than would often occur just by chance." The extent of chance variation changes with the size of the sample, so the size of the sample does matter. Exercise 18.9 demonstrates in detail how increasing the sample size drives down the *P*-value. Here is another example.

STATISTICS IN YOUR WORLD

Should Tests Be Banned?

Significance tests don't tell us how large or how important an effect is. Research in psychology has emphasized these tests, but some think they should be banned from use because of their weaknesses. The American Psychological Association asked a group of experts to consider the issue. They said, "Use anything that sheds light on your study. Use more data analysis and confidence intervals." They also said, "The task force does not support any action that could be interpreted as banning the use of null hypothesis significance testing or *P*-values in psychological research and publication."

EXAMPLE 18.3 It's Significant. Or Not. So What?

We are testing the hypothesis of no correlation between two variables. (We discuss how to do this in Chapter 26.) With 1000 observations, an observed correlation of only $r = 0.08$ is significant evidence at the 1% level that the correlation in the population is not zero but positive. *The small P-value does not mean that there is a strong association, only that there is strong evidence of some association.* The true population correlation is probably quite close to the observed sample value, $r = 0.08$. We might well conclude that for practical purposes we can ignore the association between these variables, even though we are confident (at the 1% level) that the correlation is positive.

On the other hand, if we have only 10 observations, a correlation of $r = 0.5$ is not significantly greater than zero even at the 5% level. Small samples vary so much that a large r is needed if we are to be confident that we aren't just seeing chance variation at work. So a small sample will often fall short of significance even if the true population correlation is quite large.

In normal conversation "significant" and "important" are treated as synonyms. Because the phrase "statistical significance" can be misinterpreted as implying "practical importance," some authors have proposed replacing the term *statistical significance*. Web Exercise 18.58 asks you to investigate some of these proposals.

Beware of multiple analyses Statistical significance ought to mean that you have found an effect you were looking for. The reasoning behind statistical significance works well if you decide what effect you are seeking, design a study to search for it, and use a test of significance to weigh the evidence you get. In other settings, significance may have little meaning.

EXAMPLE 18.4 Cell Phones and Brain Cancer

Bartosz Hadyniak/Getty Images

Might the radiation from cell phones be harmful to users? Many studies have found little or no connection between using cell phones and various illnesses. Here is part of a news account of one study:

> A hospital study that compared brain cancer patients and a similar group without brain cancer found no statistically significant association between cell phone use and a group of brain cancers known as gliomas. But when 20 types of glioma were considered separately an association was found between phone use and one rare form. Puzzlingly, however, this risk appeared to decrease rather than increase with greater mobile phone use.[6]

Suppose that the 20 null hypotheses (no association) for these 20 significance tests are all true. Then each test has a 5% chance of being significant at the 5% level. That's what $\alpha = 0.05$ means: results this extreme occur 5% of the time just by chance when the null hypothesis is true. Because 5% is 1/20, we expect about 1 of 20 tests to give a significant result just by chance. That's what the study observed.

Running one test and reaching the 5% level of significance provides reasonably good evidence that you have found something. Running 20 tests and reaching that level only once does not. The caution about multiple analyses applies to confidence intervals as well. A single 95% confidence interval has probability 0.95 of capturing the true parameter each time you use it. The probability that all of 20 confidence intervals will capture their parameters is much less than 95%. If you think that multiple tests or intervals may have discovered an important effect, you need to gather new data to make inferences about that specific effect.

EXAMPLE 18.5 Publication Bias

A subtle example of multiple analyses is *publication bias.* Suppose 20 researchers are independently studying the effectiveness of a new therapy for treating a disease. To publish their findings, researchers must demonstrate that the new therapy is effective at the 0.05 significance level. One of the researchers obtains statistically significant results, but the other 19 do not. The one researcher who obtained statistically significant results publishes the findings. We do not hear about the 19 researchers who failed to find statistical significance. If we knew that only one of the 20 researchers obtained statistical significance at the 0.05 level, we might suspect that the results of the one researcher are due to chance rather than an actual treatment effect. If we are not aware of the 19 studies that failed to find a treatment effect, we will be "biased" toward treating the findings of the one published study with more importance than it deserves. This is publication bias. The remedy is to encourage replication of the findings with additional studies.

APPLY YOUR KNOWLEDGE

18.8 Is It Significant? In the absence of special preparation, SAT Mathematics (SATM) scores in 2019 varied Normally with mean $\mu = 528$ and $\sigma = 117$. Fifty students go through a rigorous training program designed to raise their SATM scores by improving their mathematics skills. Either by hand or by using the *P-Value of a Test of Significance* applet, carry out a test of

applet

$$H_0: \mu = 528$$
$$H_a: \mu > 528$$

(with $\sigma = 117$) in each of the following situations:

(a) The students' average score is $\bar{x} = 555$. Is this result significant at the 5% level?

(b) The average score is $\bar{x} = 556$. Is this result significant at the 5% level?

The difference between the two outcomes in parts (a) and (b) is of no practical importance. Beware attempts to treat $\alpha = 0.05$ as sacred.

18.9 Detecting Acid Rain. Emissions of sulfur dioxide by industry set off chemical changes in the atmosphere that result in acid rain. The acidity of liquids is measured by pH on a scale of 0 to 14. Distilled water has pH 7.0, and lower pH values indicate acidity. Normal rain is somewhat acidic, so acid rain is sometimes defined as rainfall with a pH below 5.0. Suppose that pH measurements of rainfall on different days in a Canadian forest follow a Normal distribution with standard deviation $\sigma = 0.6$. A sample of n days finds that the mean pH is $\bar{x} = 4.8$. Is this good evidence that the mean pH μ for all rainy days is less than 5.0? The answer depends on the size of the sample.

Either by hand or using the *P-Value of a Test of Significance* applet, carry out four tests of

applet

$$H_0: \mu = 5.0$$
$$H_a: \mu < 5.0$$

Use $\sigma = 0.6$ and $\bar{x} = 4.8$ in all four tests. But use four different sample sizes: $n = 9$, $n = 16$, $n = 36$, and $n = 64$.

(a) What are the *P*-values for the four tests? *The P-value of the same result* $\bar{x} = 4.8$ *gets smaller (more significant) as the sample size increases.*

(b) For each test, sketch the Normal curve for the sampling distribution of $\bar{x}$ when H_0 is true. This curve has mean 5.0 and standard deviation $0.6/\sqrt{n}$. Mark the observed $\bar{x} = 4.8$ on each curve. (If you use the applet, you can just copy the curves displayed by the applet.) *The same result* $\bar{x} = 4.8$ *gets more extreme on the sampling distribution as the sample size increases.*

18.10 Confidence Intervals Help. Give a 95% confidence interval for the mean pH μ for each sample size in the previous exercise. The intervals, unlike the *P*-values, give a clear picture of what mean pH values are plausible for each sample.

18.11 Searching for ESP. A researcher looking for evidence of extrasensory perception (ESP) tests 1000 subjects. Forty-three of these subjects do significantly better ($P < 0.05$) than random guessing.

(a) Forty-three seems like a lot of people, but you can't conclude that these 43 people have ESP. Why not?

(b) What should the researcher now do to test whether any of these 43 subjects have ESP?

18.4 Planning Studies: Sample Size for Confidence Intervals

A wise user of statistics never plans a sample or an experiment without, at the same time, planning the inference. The number of observations is a critical part of planning a study. Larger samples give smaller margins of error in confidence intervals and make significance tests better able to detect effects in the population. But taking observations costs both time and money. How many observations are enough? We will look at this question first for confidence intervals and then for tests. Planning a confidence interval is much simpler than planning a test. It is also more useful because estimation is generally more informative than testing. The section on planning tests is, therefore, optional.

You can arrange to have both high confidence and a small margin of error by taking enough observations. The margin of error of the z confidence interval for the mean of a Normally distributed population is $m = z^* \sigma/\sqrt{n}$. *Notice that it is the size of the sample that determines the margin of error for a given confidence level. The size of the population does not influence the margin of error and hence the sample size we need.* (This is true as long as the population is much larger than the sample.)

To obtain a desired margin of error m, put in the value of z^* for your desired confidence level and solve for the sample size n. Here is the result.

Sample Size for Desired Margin of Error

To estimate the mean of a Normal population using a z confidence interval with given margin of error m and a specified confidence level, the sample size n must be

$$n = \left(\frac{z^*\sigma}{m}\right)^2$$

where z^* is the critical value for the desired level of confidence. Always round n up to the next whole number when you use this formula.

EXAMPLE 18.6 How Many Observations?

In Example 16.3 (page 374), psychologists recorded the size of the tip of 20 patrons in a restaurant when a message indicating that the next day's weather would be good was written on their bill. We know that the population standard deviation is $\sigma = 2$. We want to estimate the mean percentage tip μ for patrons of this restaurant who receive this message on their bill within ± 0.5 with 90% confidence. How many patrons must we observe?

The desired margin of error is $m = 0.5$. For 90% confidence, Table C gives $z^* = 1.645$. Therefore,

$$n = \left(\frac{z^*\sigma}{m}\right)^2 = \left(\frac{1.645 \times 2}{0.5}\right)^2 = 43.3$$

Because 43 patrons will give a slightly larger margin of error than desired, and 44 patrons will give a slightly smaller margin of error, we must observe 44 patrons. *Always round up to the next higher whole number when finding n.*

APPLY YOUR KNOWLEDGE

18.12 Body Mass Index of Young Men. Example 16.1 (page 368) assumed that the body mass index (BMI) of all American young men follows a Normal distribution with standard deviation $\sigma = 11.6$ kg/m^2. How large a sample would be needed to estimate the mean BMI μ in this population to within ± 1 with 95% confidence?

18.13 Number Skills of Eighth-Graders. Suppose that scores on the mathematics part of the National Assessment of Educational Progress (NAEP) test for eighth-grade students follow a Normal distribution with standard deviation $\sigma = 40$. You want to estimate the mean score within ± 1 with 90% confidence. How large an SRS of scores must you choose?

18.5 Planning Studies: The Power of a Statistical Test of Significance*

How large a sample should we take when we plan to carry out a test of significance? We know that if our sample is too small, even large effects in the population will often fail to give statistically significant results. Here are the questions we must answer to decide how many observations we need:

Significance level. How much protection do we want against getting a significant result from our sample when there really is no effect in the population?

Effect size. How large an effect in the population is important in practice?

Effect Size

The **effect size** is the magnitude of the effect in the population.

Power. How confident do we want to be that our study will detect an effect of the size we think is important?

Significance level, effect size, and power are statistical shorthand for three pieces of information. *Power* is a new idea.

*Power calculations are important in planning studies, but this more advanced material is not needed to understand the rest of the book.

EXAMPLE 18.7 Sweetening Colas: Planning a Study

Let's illustrate typical answers to the questions just posed by looking again at the example of testing a new cola for loss of sweetness in storage (Example 17.2, page 383). Ten trained tasters rated the sweetness on a 10-point scale before and after storage. The difference in the before and after scores represents each taster's judgment of the loss of sweetness, with a difference of 0 meaning no loss of sweetness. From experience, we know that sweetness loss scores vary from taster to taster according to a Normal distribution with standard deviation about $\sigma = 1$. To see if the taste test gives reason to think that the cola does lose sweetness, we will test

$$H_0: \mu = 0$$
$$H_a: \mu > 0$$

Are 10 tasters enough, or should we use more?

Significance level. Requiring significance at the 5% level is enough protection against declaring that there is a loss in sweetness when in fact there is no change if we could look at the entire population. This means that when there is no change in sweetness in the population, one out of 20 samples of tasters will incorrectly find a significant loss.

Effect size. A mean sweetness loss of 0.8 point on the 10-point scale will be noticed by consumers and so is important in practice.

Power. We want to be 90% confident that our test will detect a mean loss of 0.8 point in the population of all tasters. We agreed to use significance at the 5% level as our standard for detecting an effect. So we want probability at least 0.9 that a test at the $\alpha = 0.05$ level will reject the null hypothesis $H_0: \mu = 0$ when the true population mean is $\mu = 0.8$.

The probability that the test successfully detects a sweetness loss of the specified size is the *power* of the test. You can think of tests with high power as being highly sensitive to deviations from the null hypothesis. In Example 18.6, we decided that we want power 90% when the truth about the population is that $\mu = 0.8$.

Power

The **power** of a test against a specific alternative is the probability that the test will reject H_0 at a chosen significance level α when the specified alternative value of the parameter is true.

For most statistical tests, calculating power is a job for comprehensive statistical software. Calculating power for the z test is easier than for most statistical tests, but we will not discuss the details. The following examples illustrate two approaches: an applet that shows the meaning of power and statistical software.

EXAMPLE 18.8 Finding Power by Using an Applet

applet

Finding the power of the z test is less challenging than most other power calculations because it requires only a Normal distribution probability calculation. The *Statistical Power* applet does this and illustrates the calculation with Normal curves. Enter the information from Example 18.7 into the applet: hypotheses, significance level $\alpha = 0.05$, alternative value $\mu = 0.8$, standard deviation $\sigma = 1$, and sample size $n = 10$. Click "Update." The applet output appears in Figure 18.1.

The power of the test against the specific alternative $\mu = 0.8$ is 0.812. That is, the test will reject H_0 about 81% of the time when this alternative is true. So 10 observations are too few to give power 90%.

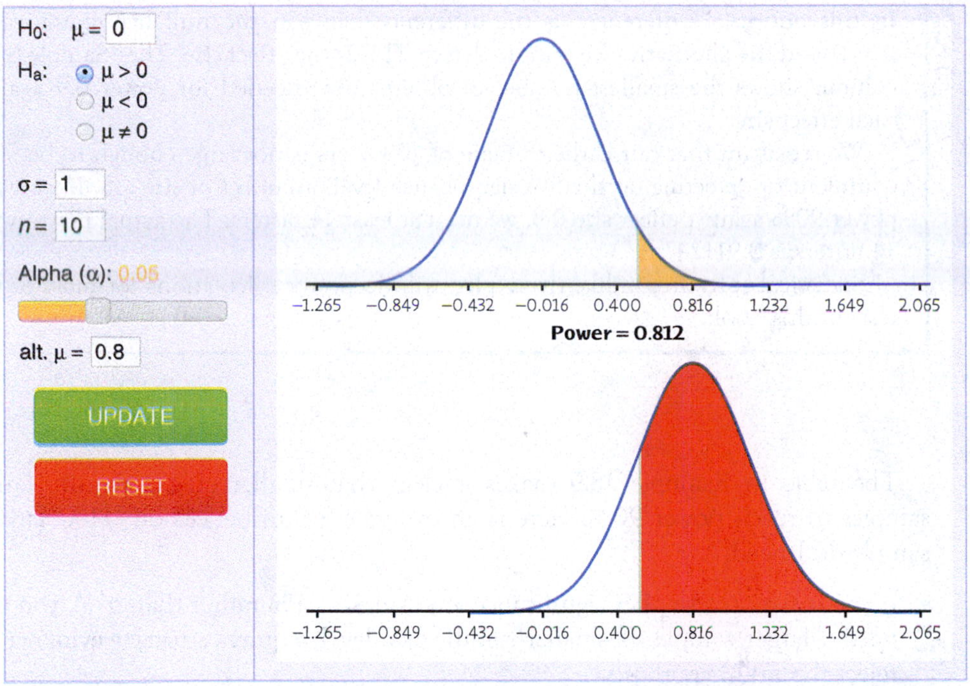

FIGURE 18.1 Output from the *Statistical Power* applet, for Example 18.8, along with the information entered into the applet. The top curve shows the behavior of $\bar{x}$ when the null hypothesis is true ($\mu = 0$). The bottom curve shows the distribution of $\bar{x}$ when $\mu = 0.8$.

The two Normal curves in Figure 18.1 show the sampling distribution of $\bar{x}$ under the null hypothesis $\mu = 0$ (top) and also under the specific alternative $\mu = 0.8$ (bottom). The curves have the same shape because σ does not change. The top curve is centered at $\mu = 0$ and the bottom curve at $\mu = 0.8$. The shaded region at the right of the top curve has area 0.05. It marks off values of $\bar{x}$ that are statistically significant at the $\alpha = 0.05$ level. The lower curve shows the probability of these same values when $\mu = 0.8$. This area is the power, 0.812.

The applet will find the power for any given sample size. It's more helpful in practice to turn the process around and learn what sample size we need to achieve a given power. Statistical software will do this but usually doesn't show the helpful Normal curves that are part of the applet's output.

EXAMPLE 18.9 Finding Power by Using Software

Some software packages (for example, SAS, JMP, Minitab, and R) calculate power. We asked Minitab to find the number of observations needed for the one-sided z test to have power 0.9 against several specific alternatives at the 5% significance level when the population standard deviation is $\sigma = 1$. Here is the table that results:

Difference	Sample Size	Target Power	Actual Power
0.1	857	0.9	0.900184
0.2	215	0.9	0.901079
0.3	96	0.9	0.902259
0.4	54	0.9	0.902259
0.5	35	0.9	0.905440
0.6	24	0.9	0.902259
0.7	18	0.9	0.907414
0.8	14	0.9	0.911247
0.9	11	0.9	0.909895
1.0	9	0.9	0.912315

In this output, "Difference" is the difference between the null hypothesis value $\mu = 0$ and the alternative we want to detect. This is the effect size. The "Sample Size" column shows the smallest number of observations needed for power 0.9 against each effect size.

We see again that our earlier sample of 10 tasters is not large enough to be 90% confident of detecting (at the 5% significance level) an effect of size 0.8. If we want power 90% against effect size 0.8, we need at least 14 tasters. The actual power with 14 tasters is 0.911247.

Statistical software, unlike the applet, will do power calculations for most of the tests in this book.

STATISTICS IN YOUR WORLD

Fish, Fishermen, and Power

Are the stocks of cod in the ocean off eastern Canada declining? Studies over many years failed to find significant evidence of a decline. These studies had low power—that is, they might have failed to find a decline even if one had been present. When it became clear that cod were vanishing, quotas on fishing ravaged the economy in parts of Canada. If the earlier studies had had high power, they would likely have detected the decline. Quick action might have reduced the economic and environmental costs.

The table in Example 18.9 makes it clear that smaller effects require larger samples to reach power 90%. Here is an overview of influences on "How large a sample do I need?"

- If you insist on a smaller significance level (such as 1% rather than 5%), you will need a larger sample. A smaller significance level requires stronger evidence to reject the null hypothesis.
- If you insist on higher power (such as 99% rather than 90%), you will need a larger sample. Higher power gives a better chance of detecting an effect when it is really there.
- At any significance level and desired power, a two-sided alternative requires a larger sample than a one-sided alternative.
- At any significance level and desired power, detecting a small effect requires a larger sample than detecting a large effect.

Planning a serious statistical study always requires an answer to the question "How large a sample do I need?" If you intend to test the hypothesis $H_0: \mu = \mu_0$ about the mean μ of a population, you need at least a rough idea of the size of the population standard deviation σ and of how big a deviation $\mu - \mu_0$ of the population mean from its hypothesized value you want to be able to detect. More elaborate settings, such as comparing the mean effects of several treatments, require more elaborate advance information. You can leave the details to experts, but you should understand the idea of power and the factors that influence how large a sample you need.

To calculate the power of a test, we act as if we are interested in a fixed level of significance such as $\alpha = 0.05$. That's essential to do a power calculation, but remember that in practice, we think in terms of *P*-values rather than a fixed level α. To effectively plan a statistical test, we must find the power for several significance levels and for a range of sample sizes and effect sizes to get a full picture of how the test will behave.

Type I and Type II errors in significance tests We can assess the performance of a test by giving two probabilities: the significance level α and the power for an alternative that we want to be able to detect. The significance level of a test is the probability of reaching the *wrong* conclusion when the null hypothesis is true. The power for a specific alternative is the probability of reaching the *right* conclusion when that alternative is true. We can just as well describe the test by giving the probabilities of being *wrong* under both conditions.

Type I and Type II Errors

If we reject H_0 when in fact H_0 is true, this is a **Type I error.**

If we fail to reject H_0 when in fact H_a is true, this is a **Type II error.**

The **significance level** α of any fixed-level test is the probability of a Type I error.

The **power** of a test against any alternative is the probability of correctly rejecting the null hypothesis for that alternative. It can be calculated as 1 minus the probability of a Type II error for that alternative.

The possibilities are summed up in Figure 18.2. If H_0 is true, our conclusion is correct if we fail to reject H_0 and is a Type I error if we reject H_0. If H_a is true, our conclusion is either correct or a Type II error. Only one error is possible at one time. Figure 18.3 illustrates power, Type I error, and Type II error in the setting of Example 18.7 (page 414).

		Truth about the population	
		H_0 True	H_a True
Conclusion based on sample	Reject H_0	Type I error	Correct conclusion
	Fail to Reject H_0	Correct conclusion	Type II error

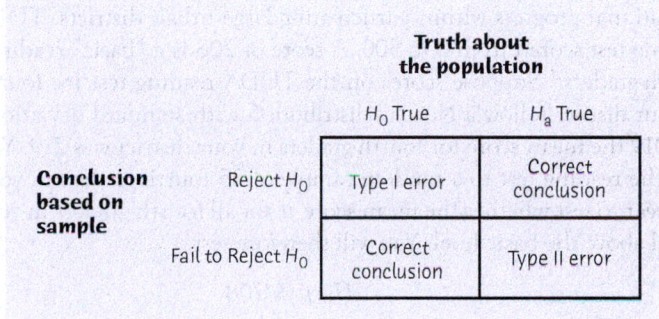

FIGURE 18.2 The two types of error in testing hypotheses.

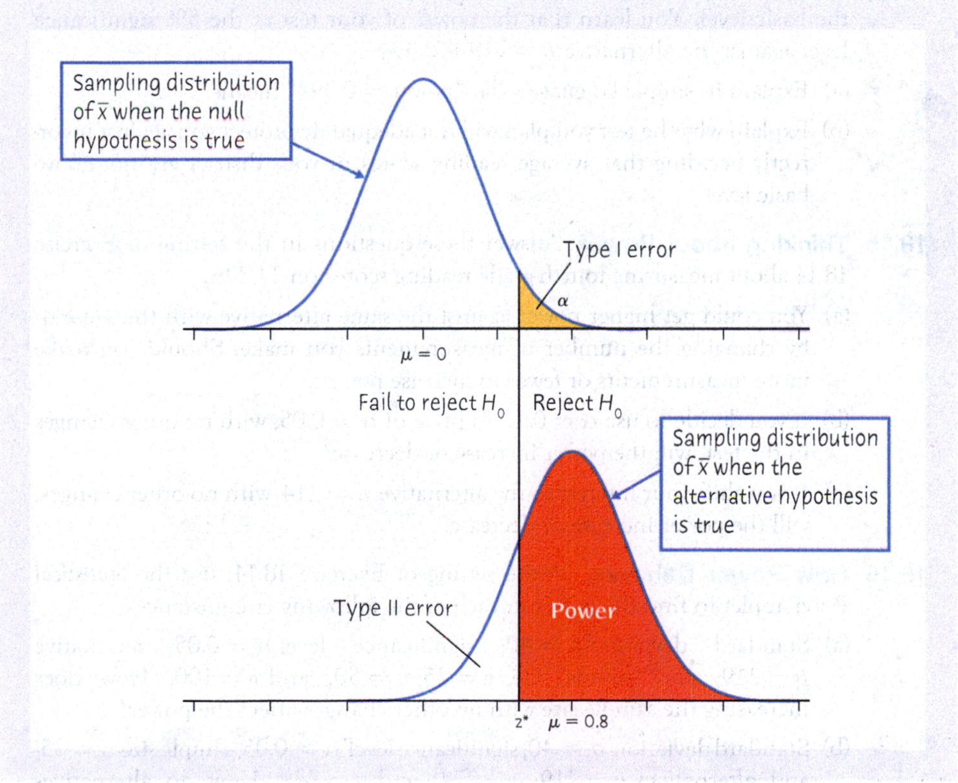

FIGURE 18.3 Illustration of power, Type I error, and Type II error in the setting of Example 18.7. The top Normal curve is the sampling distribution of $\bar{x}$ under the null hypothesis H_0: $\mu = 0$. The area of the yellow shaded region is the significance level α, which is also the Type I error. The bottom Normal curve is the sampling distribution of $\bar{x}$ when $\mu = 0.8$. The area of the red shaded region is the power. The area of the unshaded region is the Type II error. The vertical line is located at the critical value z^* for a level α test. We reject H_0 for values to the right of z^*.

EXAMPLE 18.10 Calculating Error Probabilities

applet

Because the probabilities of the two types of error are just a rewording of significance level and power, we can see from Figure 18.1 (page 415) what the error probabilities are for the test in Example 18.7:

$$P(\text{Type I error}) = P(\text{reject } H_0 \text{ when in fact } \mu = 0)$$
$$= \text{significance level } \alpha = 0.05$$
$$P(\text{Type II error}) = P(\text{fail to reject } H_0 \text{ when in fact } \mu = 0.8)$$
$$= 1 - \text{power} = 1 - 0.812 = 0.188$$

The two Normal curves in Figure 18.1 are used to find the probabilities of a Type I error (top curve, $\mu = 0$) and of a Type II error (bottom curve, $\mu = 0.8$).

APPLY YOUR KNOWLEDGE

18.14 What Is Power? The Trial Urban District Assessment (TUDA) measures educational progress within participating large urban districts. TUDA gives a reading test scored from 0 to 500. A score of 208 is a "basic" reading level for fourth-graders.[7] Suppose scores on the TUDA reading test for fourth-graders in your district follow a Normal distribution with standard deviation $\sigma = 40$. In 2019 the mean score for fourth-graders in your district was 219. You plan to give the reading test to a random sample of 25 fourth-graders in your district this year to test whether the mean score μ for all fourth-graders in your district is still above the basic level. You will therefore test

$$H_0: \mu = 208$$
$$H_a: \mu > 208$$

If the true mean score is again 219, on average, students are performing above the basic level. You learn that the power of your test at the 5% significance level against the alternative $\mu = 219$ is 0.394.

(a) Explain in simple language what "power = 0.394" means.

(b) Explain why the test you plan will not adequately protect you against incorrectly deciding that average reading scores in your district are not above basic level.

18.15 Thinking about Power. Answer these questions in the setting of Exercise 18.14 about measuring fourth-grade reading scores on TUDA.

(a) You could get higher power against the same alternative with the same α by changing the number of measurements you make. Should you make more measurements or fewer to increase power?

(b) If you decide to use $\alpha = 0.10$ in place of $\alpha = 0.05$, with no other changes in the test, will the power increase or decrease?

(c) If you shift your interest to the alternative $\mu = 214$ with no other changes, will the power increase or decrease?

applet

18.16 How Power Behaves. In the setting of Exercise 18.14, use the *Statistical Power* applet to find the power in each of the following circumstances.

(a) Standard deviation $\sigma = 40$; significance level $\alpha = 0.05$; alternative $\mu = 219$; and sample sizes $n = 25$, $n = 50$, and $n = 100$. How does increasing the sample size with no other changes affect the power?

(b) Standard deviation $\sigma = 40$; significance level $\alpha = 0.05$; sample size $n = 25$; and alternatives $\mu = 219$, $\mu = 224$, and $\mu = 229$. How do alternatives more distant from the hypothesis (larger effect sizes) affect the power?

(c) Standard deviation $\sigma = 40$; sample size $n = 25$; alternative $\mu = 219$; and significance levels $\alpha = 0.05$, $\alpha = 0.10$, and $\alpha = 0.20$. How does increasing the desired significance level affect the power?

18.17 How Power Behaves. Use the *Statistical Power* applet to find the power of the test in Exercise 18.14 in each of these circumstances: significance level $\alpha = 0.05$; alternative $\mu = 219$; sample size $n = 25$; and $\sigma = 40$, $\sigma = 30$, and $\sigma = 20$. How does decreasing the variability of the population of measurements affect the power?

applet

18.18 Two Types of Error. Your company markets a computerized medical diagnostic program used to evaluate thousands of people. The program scans the results of routine medical tests (pulse rate, blood tests, etc.) and refers the case to a doctor if there is evidence of a medical problem. The program makes a decision about each person.

(a) What are the two hypotheses and the two types of error that the program can make? Describe the two types of error in terms of false-positive and false-negative test results.

(b) The program can be adjusted to decrease one error probability, at the cost of an increase in the other error probability. Which error probability would you choose to make smaller, and why? (This is a matter of judgment. There is no single correct answer.)

CHAPTER 18 SUMMARY

- A specific confidence interval or test is correct only under specific conditions. The most important conditions concern the method used to produce the data. Other factors such as the shape of the population distribution may also be important.
- Whenever you use statistical inference, you are acting as if your data are a random sample or come from a randomized comparative experiment.
- Always do data analysis before inference to detect outliers or other problems that would make inference untrustworthy.
- The margin of error in a confidence interval accounts for only the chance variation due to random sampling. In practice, errors due to nonresponse or undercoverage are often more serious.
- There is no universal rule for how small a *P*-value in a test of significance is convincing evidence against the null hypothesis. Beware of relying on traditional significance levels such as $\alpha = 0.05$.
- Very small effects can be highly significant (small *P*) when a test is based on a large sample. Always consider whether the **effect size**—the magnitude of the effect in the population—is important in practice. Plot the data to display the effect you are seeking and use confidence intervals to estimate the actual values of parameters.
- On the other hand, lack of significance does not imply that H_0 is true. Even a large effect can fail to be significant when a test is based on a small sample.
- Many tests run at once will probably produce some significant results by chance alone, even if all the null hypotheses are true.
- When you plan a statistical study, plan the inference as well. In particular, determine what sample size you need for successful inference.
- To estimate the mean of a Normal population using a z confidence interval with given margin of error m and a specified confidence level, the sample size n must be

$$n = \left(\frac{z^*\sigma}{m}\right)^2$$

- Here z^* is the critical value for the desired level of confidence. Always round n up to the next whole number when you use this formula.
- The **power** of a significance test measures its ability to detect the truth of an alternative hypothesis. The power against a specific alternative is the probability that the test will reject H_0 at a particular level α when that alternative is true (optional topics).
- Increasing the size of the sample increases the power of a significance test. You can use statistical software to find the sample size needed to achieve a desired power (optional topics).
- If we reject H_0 when in fact H_0 is true, this is a **Type I error.** If we fail to reject H_0 when in fact H_a is true, this is a **Type II error.**

CHECK YOUR SKILLS

18.19 The most important condition for sound conclusions from statistical inference is usually that

(a) the *P*-value we calculate is small.

(b) the population distribution is exactly Normal.

(c) the data can be thought of as a random sample from the population of interest.

18.20 The coach of a Canadian university's women's soccer team records the resting heart rates of the 25 team members. You should not trust a confidence interval for the mean resting heart rate of all female students at this Canadian university based on these data because

(a) the members of the soccer team can't be considered a random sample of all female students at this university.

(b) heart rates may not have a Normal distribution.

(c) with only 25 observations, the margin of error will be large.

18.21 You visit the online Harris Interactive Poll. Based on 2223 responses, the poll reports that 60% of U.S. adults believe that chef is a prestigious occupation.[8] You should refuse to calculate a 95% confidence interval for the proportion of all U.S. adults who believe chef is a prestigious occupation based on this sample because

(a) this percentage is too small.

(b) inference from a voluntary response sample can't be trusted.

(c) the sample is too large.

18.22 Many sample surveys use well-designed random samples, but half or more of the original sample can't be contacted or refuse to take part. Any errors due to this nonresponse

(a) have no effect on the accuracy of confidence intervals.

(b) are included in the announced margin of error.

(c) are in addition to the random variation accounted for by the announced margin of error.

18.23 A writer in a medical journal says: "An uncontrolled experiment in 37 women found a significantly improved mean clinical symptom score after treatment. Methodologic flaws make it difficult to interpret the results of this study." The writer is skeptical about the significant improvement because

(a) there is no control group, so the improvement might be due to the placebo effect or to the fact that many medical conditions improve over time.

(b) the *P*-value given was $P = 0.048$, which is too large to be convincing.

(c) the response variable might not have an exactly Normal distribution in the population.

18.24 Vigorous exercise is associated with an extra several years of life (on the average). Researchers in Denmark found evidence that slow jogging may provide even better life-expectancy benefits than more vigorous running.[9] Suppose that the added life expectancy associated with slow jogging for 30 minutes three times a week is just one month. A statistical test is more likely to find a significant increase in mean life expectancy for those who jog slowly if

(a) it is based on a very large random sample.

(b) it is based on a very small random sample.

(c) The size of the sample has little effect on significance for such a small increase in life expectancy.

18.25 A medical experiment compared zinc supplements with a placebo for reducing the duration of colds. Let μ denote the mean decrease, in days, in the duration of a cold. A decrease to $\mu = 2$ is a practically important decrease. The significance level of a test of $H_0: \mu = 0$ versus $H_a: \mu > 0$ is defined as

(a) the probability that the test fails to reject H_0 when $\mu = 2$ is true.

(b) the probability that the test rejects H_0 when $\mu = 2$ is true.

(c) the probability that the test rejects H_0 when $\mu = 0$ is true.

18.26 **(Optional topic)** The power of the test in Exercise 18.25 against the specific alternative $\mu = 2$ is defined as

(a) the probability that the test fails to reject H_0 when $\mu = 2$ is true.

(b) the probability that the test rejects H_0 when $\mu = 2$ is true.

(c) the probability that the test rejects H_0 when $\mu = 0$ is true.

18.27 **(Optional topic)** The power of a test is important in practice because power

(a) describes how well the test performs when the null hypothesis is actually true.

(b) describes how sensitive the test is to violations of conditions such as Normal population distribution.

(c) describes how well the test performs when the null hypothesis is actually not true.

CHAPTER 18 EXERCISES

18.28 **Hotel Managers.** In Exercise 17.30 (page 399), you carried out a test of significance based on data from 148 general managers of three-star and four-star hotels. Before you trust your results, you would like more information about the data. What facts would you most like to know?

nattrass/Getty Images

18.29 **Attractive Instructors.** A psychologist claims that students are more attentive in classes taught by instructors rated as attractive by students than in classes with instructors rated as unattractive. He surveys students in classes taught by instructors rated as attractive and in classes taught by instructors rated as unattractive. The proportion of students claiming to be highly attentive in class is significantly higher ($P < 0.05$) among students with attractive instructors. What additional information would you want to help you decide whether you believe the pscyhologist's claim?

18.30 **Sampling at the Gourmet Food Store.** A market researcher chooses at random from men entering a gourmet food store. One outcome of the study is a 95% confidence interval for the mean of "the highest price you would pay for a bottle of wine."

(a) Explain why this confidence interval does not give useful information about the population of all men.

(b) Explain why it may give useful information about the population of men who shop at gourmet food stores.

18.31 **Sensitive Questions.** The 2013 Youth Risk Behavior Survey found that 194 individuals in its random sample of 1450 Ohio high school students said that they had carried a weapon such as a gun, knife, or club in the previous 30 days. That's 13.4% of the sample. Why is this estimate likely to be biased? Do you think it is biased high or low? Does the margin of error of a 95% confidence interval for the proportion of all Ohio high school students who had carried a weapon such as a gun, knife, or club in the previous 30 days allow for this bias?

18.32 **College Degrees.** At the Statistics Canada website, `www.statcan.gc.ca`, you can find the percentage of adults in each province or territory who have at least a university certificate, diploma, or degree at bachelor's level or above. It makes no sense to find $\bar{x}$ for these data and use it to get a confidence interval for the mean percentage μ in all 13 provinces or territories. Why not?

18.33 **An Outlier Strikes.** You have data on an SRS of freshmen from your college that shows how long each student spends studying and working on homework. The data contain one high outlier. Will this outlier have a greater effect on a confidence interval for mean completion time if your sample is small or if it is large? Why?

18.34 **Can We Trust This Interval?** Here are data on the percentage change in the total mass (in tons) of wildlife in several West African game preserves in the years 1971 to 1999:[10] WILDMSS

1971	1972	1973	1974	1975	1976	1977	1978	1979	1980
2.9	3.1	−1.2	−1.1	−3.3	3.7	1.9	−0.3	−5.9	−7.9
1981	**1982**	**1983**	**1984**	**1985**	**1986**	**1987**	**1988**	**1989**	**1990**
−5.5	−7.2	−4.1	−8.6	−5.5	−0.7	−5.1	−7.1	−4.2	0.9
1991	**1992**	**1993**	**1994**	**1995**	**1996**	**1997**	**1998**	**1999**	
−6.1	−4.1	−4.8	−11.3	−9.3	−10.7	−1.8	−7.4	−22.9	

Software gives the 95% confidence interval for the mean annual percentage change as −6.66% to −2.55%. There are several reasons we might not trust this interval.

(a) Examine the distribution of the data. What feature of the distribution throws doubt on the validity of statistical inference?

(b) Plot the percentages against year. What trend do you see in this time series? Explain why a trend over time casts doubt on the condition that years 1971 to 1999 can be treated as an SRS from a larger population of years.

18.35 When to Use Pacemakers. A medical panel prepared guidelines for when cardiac pacemakers should be implanted in patients with heart problems. The panel reviewed a large number of medical studies to judge the strength of the evidence supporting each recommendation. For each recommendation, they ranked the evidence as level A (strongest), B, or C (weakest). Here, in scrambled order, are the panel's descriptions of the three levels of evidence.[11] Which is A, which B, and which C? Explain your ranking.

Evidence was ranked as level _____ when data were derived from a limited number of trials involving comparatively small numbers of patients or from well-designed data analysis of nonrandomized studies or observational data registries.

Evidence was ranked as level _____ if the data were derived from multiple randomized clinical trials involving a large number of individuals.

Evidence was ranked as level _____ when consensus of expert opinion was the primary source of recommendation.

18.36 What Is Significance Good For? Which of the following questions does a test of significance answer? Briefly explain your replies.

(a) Is the observed effect large?

(b) Is the observed effect due to chance?

(c) Is the observed effect important?

18.37 Why Are Larger Samples Better? Statisticians prefer large samples. Describe briefly the effect of increasing the size of a sample (or the number of subjects in an experiment) on each of the following:

(a) The *P*-value of a test, when H_0 is false and all facts about the population remain unchanged as n increases

(b) (Optional) The power of a fixed level α test, when α, the alternative hypothesis, and all facts about the population remain unchanged

18.38 Divorces. Divorce rates vary from city to city in the United States. We have lots of data on many U.S. cities. Statistical software makes it easy to perform dozens of significance tests on dozens of variables to see which ones best predict divorce. One interesting finding is that those cities with Major League ballparks tend to have significantly more divorces than other cities. To improve your chances of a lasting marriage, should you use this "significant" variable to decide where to live? Explain your answer.

18.39 A Test Goes Wrong. Software can generate samples from (almost) exactly Normal distributions. Here is a random sample of size 5 from the Normal distribution with mean 20 and standard deviation 2.5: RANDOM

22.94 17.04 17.58 20.96 19.29

These data match the conditions for a z test better than real data will: the population is extremely close to Normal and has known standard deviation $\sigma = 2.5$, and the population mean is $\mu = 20$. Although we know the true value of μ, suppose we pretend that we do not and we test the hypotheses

$$H_0: \mu = 17.5$$
$$H_a: \mu \neq 17.5$$

(a) What are the z statistic and its *P*-value? Is the test significant at the 5% level?

(b) We know that the null hypothesis does not hold, but the test failed to give strong evidence against H_0. Explain why this is not surprising.

18.40 Reducing the Gender Gap. In many science disciplines, women are outperformed by men on test scores. Will "values affirmation training" improve self-confidence and hence performance of women relative to men in science courses? A study conducted at a large university compares the scores of men and women at the end of a large introductory physics course on a nationally normed standardized test of conceptual physics, the Force and Motion Conceptual Evaluation (FMCE). Half the women in the course were randomly assigned to values affirmation training during the course; the other half received no training. The study reports that there was a significant difference ($P < 0.01$) in the gap between men's and women's scores, although the gap for women who received the values affirmation training was much smaller than that for women who did not receive training. As evidence that this gap was reduced for woman who received the training, the study also reports that a 95% confidence interval for the difference in mean scores on the FMCE exam between women who received the training and those who didn't is 13 ± 8 points. You are a faculty member in the physics department, and the provost, who is interested in women in science, asks you about the study.

(a) Explain in simple language what "a significant difference ($P < 0.01$)" means.

(b) Explain clearly and briefly what "95% confidence" means.

(c) Is this study good evidence that requiring values affirmation training of all female students would greatly reduce the gender gap in scores on science tests in college courses?

18.41 How Far Do Rich Parents Take Us? How much education children get is strongly associated with the wealth and social status of their parents. In social science jargon, this is socioeconomic status, or SES. But the SES of parents has little influence on whether children who have graduated from college go on to yet more education. One study looked at whether college graduates took the graduate admissions tests for business, law, and other graduate programs. The effects of the parents' SES on taking the LSAT for law school were "both statistically insignificant and small."

(a) What does "statistically insignificant" mean?

(b) Why is it important that the effects were small in size as well as insignificant?

18.42 This Wine Stinks. How sensitive are the untrained noses of students? Exercise 16.27 (page 381) gives the lowest levels of dimethyl sulfide (DMS) that 10 students could detect. You want to estimate the mean DMS odor threshold among all students, and you would be satisfied to estimate the mean to within ± 0.1 with 99% confidence. The standard deviation of the odor threshold for untrained noses is known to be $\sigma = 7$ micrograms per liter of wine. How large an SRS of untrained students do you need?

18.43 Pulling Apart Wood. You want to estimate the mean load needed to pull apart the pieces of wood in Exercise 16.25 (page 381) to within ± 600 pounds with 95% confidence. How large a sample is needed?

The following exercises concern the optional material on the power of a test.

18.44 The First Child Has Higher IQ. Does the birth order of a family's children influence their IQ scores? A careful study of 241,310 Norwegian 18- and 19-year-olds found that firstborn children scored 2.3 points higher on the average than second children in the same family. This difference was highly significant ($P < 0.001$). A commentator said, "One puzzle highlighted by these latest findings is why certain other within-family studies have failed to show equally consistent results. Some of these previous null findings, which have all been obtained in much smaller samples, may be explained by inadequate statistical power."[12]

(a) Explain in simple language why tests having low power often fail to give evidence against a null hypothesis even when the hypothesis is really false.

(b) Do you think a difference of 2.3 points in IQ scores is an important difference?

18.45 How Valium Works. Valium is a common antidepressant and sedative. A study investigated how valium works by comparing its effect on sleep in seven genetically modified mice and eight normal control mice. There was no significant difference between the two groups. The authors say that this lack of significance "is related to the large inter-individual variability that is also reflected in the low power (20%) of the test."[13]

(a) Explain exactly what power 20% against a specific alternative means.

(b) Explain in simple language why tests having low power often fail to give evidence against a null hypothesis even when the null hypothesis is really false.

(c) What fact about this experiment most likely explains the low power?

18.46 Artery Disease. An article in the *New England Journal of Medicine* describes a randomized controlled trial that compared the effects of using a balloon with a special coating in angioplasty (the repair of blood vessels) compared with a standard balloon. According to the article, the study was designed to have power 90%, with a two-sided Type I error of 0.05, to detect a clinically important difference of approximately 17 percentage points in the presence of certain lesions 12 months after surgery.[14]

(a) What fixed significance level was used in calculating the power?

(b) Explain to someone who knows no statistics why power 90% means that the experiment would probably have been significant if there had been a difference between the use of the balloon with a special coating and the use of the standard balloon.

18.47 Power. applet In Exercise 18.39, a sample from a Normal population with mean $\mu = 20$ and standard deviation $\sigma = 2.5$ failed to reject the null hypothesis $H_0: \mu = 17.5$ at the $\alpha = 0.05$ significance level. Enter the information from this example into the *Statistical Power* applet. (Don't forget that the alternative hypothesis is two-sided.) What is the power of the test against the alternative $\mu \neq 17.5$? Because the power is not high, it isn't surprising that the sample in Exercise 18.39 failed to reject H_0.

18.48 Finding Power by Hand. Even though software is used in practice to calculate power, doing the work by hand builds your understanding. Return to the test in Example 18.7 (page 414). There are $n = 10$ observations from a population with standard deviation $\sigma = 1$ and unknown mean μ. We will test

$$H_0: \mu = 0$$
$$H_a: \mu > 0$$

with fixed significance level $\alpha = 0.05$. Find the power against the alternative $\mu = 0.8$ by following these steps.

(a) The z test statistic is

$$z = \frac{\bar{x} - \mu_0}{\sigma/\sqrt{n}} = \frac{\bar{x} - 0}{1/\sqrt{10}} = 3.162\bar{x}$$

(Remember that you won't know the numerical value of $\bar{x}$ until you have data.) What values of z lead to rejecting H_0 at the 5% significance level?

(b) Starting from your result in part (a), what values of $\bar{x}$ lead to rejecting H_0? The area above these values is shaded under the top curve in Figure 18.1.

(c) The power is the probability that you observe any of these values of $\bar{x}$ when $\mu = 0.8$. This is the shaded area under the bottom curve in Figure 18.1. What is this probability?

18.49 Finding Power by Hand: Two-sided Test. Exercise 18.48 shows how to calculate the power of a one-sided z test. Power calculations for two-sided tests follow the same outline. We will find the power of a test based on a random sample of size 5, discussed in Exercise 18.39. The hypotheses are

$$H_0: \mu = 17.5$$
$$H_a: \mu \neq 17.5$$

The population of all measurements is Normal with standard deviation $\sigma = 2.5$, and the alternative we hope to be able to detect is $\mu = 20$ (the actual population mean). (If you used the *Statistical Power* applet for Exercise 18.47, the two Normal curves for $n = 5$ illustrate parts (a) and (b) below.)

(a) Write the z test statistic in terms of the sample mean $\bar{x}$. For what values of z does this two-sided test reject H_0 at the 5% significance level?

(b) Restate your result from part (a): What values of $\bar{x}$ lead to rejection of H_0?

(c) Now suppose that $\mu = 12$. What is the probability of observing an $\bar{x}$ that leads to rejection of H_0? This is the power of the test.

18.50 Error Probabilities. You read that a statistical test at significance level $\alpha = 0.05$ has power 0.80. What are the probabilities of Type I and Type II errors for this test?

18.51 Power. You read that a statistical test at the $\alpha = 0.01$ level has probability 0.44 of making a Type II error when a specific alternative is true. What is the power of the test against this alternative?

18.52 Find the Error Probabilities. You have an SRS of size $n = 25$ from a Normal distribution with $\sigma = 2.0$. You wish to test

$$H_0: \mu = 0$$
$$H_a: \mu > 0$$

You decide to reject H_0 if $\bar{x} > 0$ and not reject H_0 otherwise.

(a) Find the probability of a Type I error. That is, find the probability that the test rejects H_0 when in fact $\mu = 0$.

(b) Find the probability of a Type II error when $\mu = 0.5$. This is the probability that the test fails to reject H_0 when in fact $\mu = 0.5$.

(c) Find the probability of a Type II error when $\mu = 1.0$.

18.53 Two Types of Error. Go to the *Statistical Significance* applet. This applet carries out tests at a fixed significance level. When you arrive, the applet is set for the cola-tasting test of Example 18.7 (page 414). That is, the hypotheses are

applet

$$H_0: \mu = 0$$
$$H_a: \mu > 0$$

We have an SRS of size 10 from a Normal population with standard deviation $\sigma = 1$, and we will do a test at level $\alpha = 0.05$. At the bottom of the screen, a button allows you to choose a value of the mean μ and then to generate samples from a population with that mean.

(a) Set $\mu = 0$, so that the null hypothesis is true. Each time you click the button, a new sample appears. If the sample $\bar{x}$ lands in the colored region, that sample rejects H_0 at the 5% level. Click 100 times rapidly, keeping track of how many samples reject H_0. Use your results to estimate the probability of a Type I error. If you kept clicking forever, what probability would you get?

(b) Now set $\mu = 0.8$. Example 18.8 shows that the test has power 0.812 against this alternative. Click 100 times rapidly, keeping track of how many samples fail to reject H_0. Use your results to estimate the probability of a Type II error. If you kept clicking forever, what probability would you get?

18.54 Type I and II Errors. Section 13.7 (page 311) discusses the prostate-specific antigen (PSA) test for prostate cancer. The test is not always correct, sometimes indicating prostate cancer (test is positive) when it is not present (a false positive) and often missing prostate cancer (test is negative) that is present (a false negative). Here is a table of the four possibilities.

	Test Result	
	Positive	Negative
Cancer present	Test is correct	False negative
Cancer absent	False positive	Test is correct

If we treat "Cancer absent" as our null hypothesis and the PSA test result as our test statistic, which of the four combinations corresponds to a Type I error? Which corresponds to a Type II error?

18.55 Multiple Testing. *This problem assumes that you have studied optional Chapter 14 on binomial distributions.* If the null hypothesis is true, testing at significance level 0.05 means that the probability is 0.05 of incorrectly rejecting the null hypothesis. Suppose one conducts 20 independent tests at level 0.05 and in each case the null hypothesis is true. Let X denote the number of tests that incorrectly reject the null hypothesis. X can take values from 0 to 20 and will follow a binomial distribution with $n = 20$ observations and probability $p = 0.05$ of success. What is the probability $X \geq 1$? This is the probability that at least one test will incorrectly reject the null hypothesis.

18.56 Publication Bias. Read Example 18.5 (page 411) on publication bias. Use the results of Exercise 18.55 to determine the probability that at least one of the 20 researchers will publish their results as significant at the 0.05 level even if the therapy has no effect on the disease.

CHAPTER

19

From Data Production to Inference: Part III Review

IN THIS CHAPTER, WE COVER . . .

Part III Skills Review

Test Yourself

Supplementary Exercises

In Part I of this book, you mastered *data analysis*, the use of graphs and numerical summaries to organize and explore any set of data. Part II introduced designs for data production. Part III introduced probability and the reasoning of statistical inference. Parts IV and V will deal in detail with practical inference.

Statistical inference draws conclusions about a population on the basis of sample data and uses probability to indicate how reliable the conclusions are. A confidence interval estimates an unknown parameter. A significance test shows how strong the evidence is for some claim about a parameter.

The probabilities in both confidence intervals and tests tell us what would happen if we used the method for the interval or test very many times:

- A confidence level is the success rate of the method for a confidence interval. This is the probability that the method actually produces an interval that captures the unknown parameter. A 95% confidence interval gives a correct result (captures the unknown parameter) 95% of the time when we use it repeatedly.

fotografixx/Getty Images

- A *P*-value tells us how unlikely the observed outcome would be if the null hypothesis were true. That is, *P* is the probability that the test would produce a result at least as extreme as the observed result if the null hypothesis really were true. Very surprising outcomes (small *P*-values) are good evidence against the null hypothesis.

Figures 19.1 and 19.2 use the z procedures introduced in Chapters 16 and 17 to present in picture form the big ideas of confidence intervals and significance tests. These ideas are the foundation for the rest of this book. We will have much to say about many statistical methods and their use in practice. In every case, the basic reasoning of confidence intervals and significance tests remains the same.

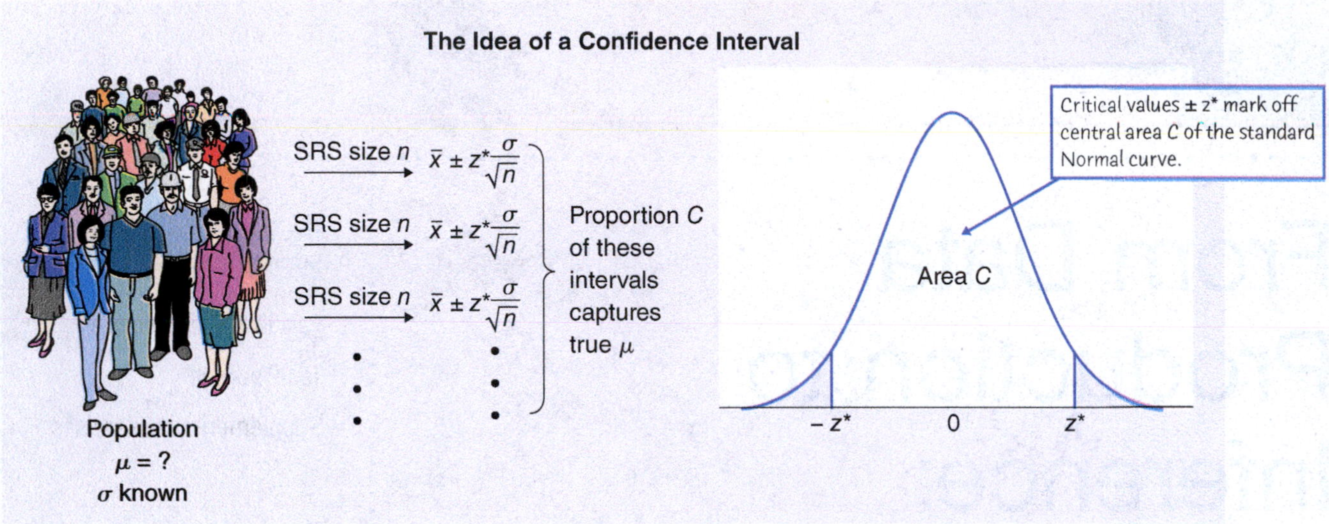

FIGURE 19.1
The idea of a confidence interval.

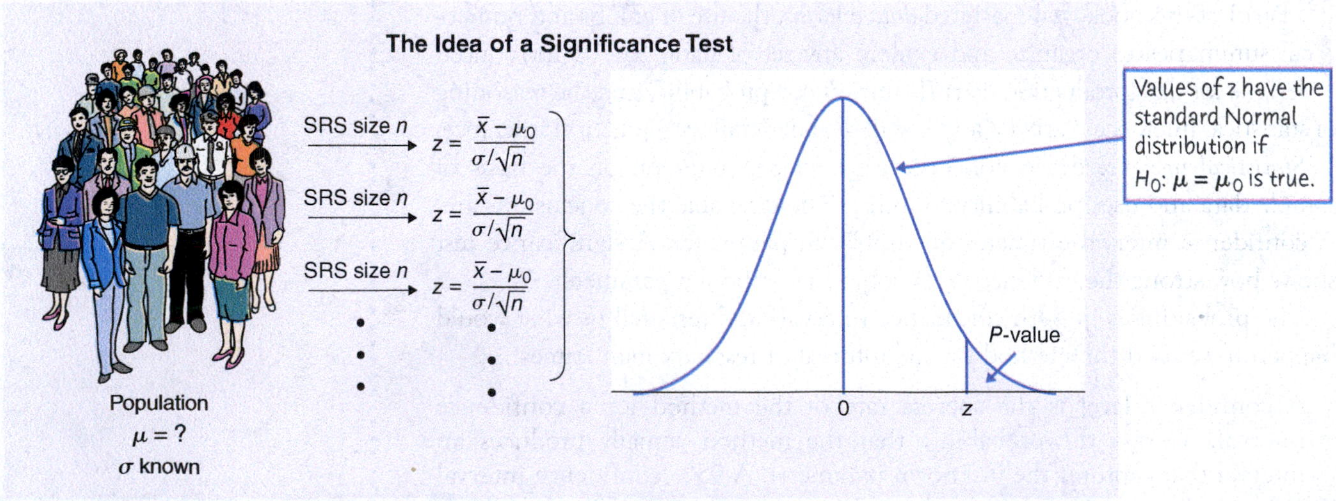

FIGURE 19.2
The idea of a significance test.

Part III Skills Review

Here are the most important skills you should have acquired from reading Chapters 12 through 18. Following each skill in parentheses is the section of the text where the topic is introduced.

A. Probability

1. Recognize that the occurrences of some phenomena are random. Probability describes the long-run regularity of random phenomena. (12.1)
2. Understand that the probability of an event is the proportion of times the event occurs in very many repetitions of a random phenomenon. Use the idea of probability as long-run proportion to think about probability. (12.1)
3. Use basic probability rules to detect illegitimate assignments of probability: any probability must be a number between 0 and 1, and the total probability assigned to all possible outcomes must be 1. (12.4)
4. Use basic probability rules to find the probabilities of events that are formed from other events. The probability that an event does not occur is 1 minus the probability it does occur. If two events are disjoint, the probability that one or the other occurs is the sum of their individual probabilities. (12.4)
5. Find probabilities in a discrete probability model by adding the probabilities of their outcomes. Find probabilities in a continuous probability model as areas under a density curve. (12.5–12.6)
6. Use the notation of random variables to make compact statements about random outcomes, such as $P(\bar{x} \leq 4) = 0.3$. Be able to interpret such statements. (12.7)

B. Sampling Distributions

1. Identify parameters and statistics in a statistical study. (15.1)
2. Recognize the fact of sampling variability: a statistic will take different values when you repeat a sample or an experiment. (15.2)
3. Interpret a sampling distribution as describing the values taken by a statistic in all possible repetitions of a sample or an experiment under the same conditions. (15.3)
4. Interpret the sampling distribution of a statistic as describing the probabilities of its possible values. (15.4)
5. Understand how the sampling distribution of a sample statistic can be used to assess statistical significance. (15.6)

C. General Rules of Probability (not required for later chapters)

1. Use Venn diagrams to picture relationships among several events. (13.1)
2. Use the general addition rule to find probabilities that involve overlapping events. (13.1)
3. Understand the idea of independence. Judge when it is reasonable to assume independence as part of a probability model. (13.2)
4. Use the multiplication rule for independent events to find the probability that all of several independent events occur. (13.2)
5. Use the multiplication rule for independent events in combination with other probability rules to find the probabilities of complex events. (13.2, 13.4)
6. Understand the idea of conditional probability. Find conditional probabilities for individuals chosen at random from a table of counts of possible outcomes. (13.3)

7. Use the general multiplication rule to find *P*(*A* and *B*) from *P*(*A*) and the conditional probability *P*(*B* | *A*). (13.4)
8. Use tree diagrams to organize several-stage probability models. (13.6)

D. Binomial Distributions (not required for later chapters)

1. Recognize the binomial setting: a fixed number n of independent success/failure trials with the same probability p of success on each trial. (14.1)
2. Recognize and use the binomial distribution of the count of successes in a binomial setting. (14.2)
3. Use the binomial probability formula to find probabilities of events involving the count X of successes in a binomial setting for small values of n. (14.3)
4. Find the mean and standard deviation of a binomial count X. (14.5)
5. Recognize when you can use the Normal approximation to a binomial distribution. Use the Normal approximation to calculate probabilities that concern a binomial count X. (14.6)

E. The Sampling Distribution of a Sample Mean

1. Recognize when a problem involves the mean $\bar{x}$ of a sample. Understand that $\bar{x}$ estimates the mean μ of the population from which the sample is drawn. (15.3–15.4)
2. Use the law of large numbers to describe the behavior of $\bar{x}$ as the size of the sample increases. (15.2)
3. Find the mean and standard deviation of a sample mean $\bar{x}$ from an SRS of size n when the mean μ and standard deviation σ of the population are known. (15.4)
4. Understand that $\bar{x}$ is an unbiased estimator of μ and that the variability of $\bar{x}$ about its mean μ gets smaller as the sample size increases. (15.4)
5. Understand that $\bar{x}$ has approximately a Normal distribution when the sample is large (central limit theorem). Use this Normal distribution to calculate probabilities that concern $\bar{x}$. (15.5)

F. Confidence Intervals

1. State in nontechnical language what is meant by "95% confidence" or other statements of confidence in statistical reports. (16.2)
2. Apply the four-step process (page 374) for any confidence interval. This process will be used more extensively in later chapters. Remember to describe your results in the context of the problem. (16.3)
3. Calculate a confidence interval for the mean μ of a Normal population with known standard deviation σ, using the formula $\bar{x} \pm z^* \sigma/\sqrt{n}$. (16.3)
4. Understand how the margin of error of a confidence interval changes with the sample size and the level of confidence C. (16.4)
5. Find the sample size required to obtain a confidence interval of specified margin of error m when the confidence level and other information are given. (18.4)
6. Identify sources of error in a study that are *not* included in the margin of error of a confidence interval, such as undercoverage or nonresponse. (18.2)

G. Significance Tests

1. State the null and alternative hypotheses in a testing situation when the parameter in question is a population mean μ. (17.2)
2. Explain in nontechnical language the meaning of the *P*-value when you are given the numerical value of P for a test. (17.3)

3. Apply the four-step process (page 374) for any significance test. This process will be used more extensively in later chapters. Remember to describe your results in the context of the problem. (17.4)
4. Calculate the one-sample z test statistic and the P-value for both one-sided and two-sided tests about the mean μ of a Normal population. (17.4)
5. Assess statistical significance at standard levels α, either by comparing P with α or by comparing z with standard Normal critical values. (17.3, 17.5)
6. Recognize that significance testing does not measure the size or importance of an effect. Explain why a small effect can be significant in a large sample and why a large effect can fail to be significant in a small sample. (18.3)
7. Recognize that any inference procedure acts as if the data were properly produced. The z confidence interval and test require that the data be an SRS from the population. (18.1)

TEST YOURSELF

The following questions include multiple-choice, calculations, and short-answer questions. They will help you review the basic ideas and skills presented in Chapters 12 through 18.

19.1 A randomly chosen subject arrives for a study of exercise and fitness. Describe a sample space for each of the following. (In some cases, you may have some freedom in your choice of S.)

(a) The subject is either under age 40 or age 40 or over.

(b) After 10 minutes on an exercise bicycle, you ask the subject to rate their effort on the rate of perceived exertion (RPE) scale. RPE ranges in whole-number steps from 6 (no exertion at all) to 20 (maximal exertion).

(c) You measure VO2 max, the maximum volume of oxygen consumed per minute during exercise. VO2 is generally between 2.5 and 6.1 liters per minute.

(d) You measure the maximum heart rate (beats per minute).

Internet Search Engines. *Internet search sites compete for users because they sell advertising space on their sites and can charge more if they are heavily used. Choose an Internet search attempt at random. Here is the probability distribution for the site the search uses:*[1]

Site	Google	Microsoft	Verizon	Ask Network
Probability	0.63	0.25	0.11	?

Use this information to answer Questions 19.2 through 19.4.

19.2 What is the probability of a search attempt being made at Microsoft or Verizon?

(a) 0.25

(b) 0.36

(c) 0.99

(d) Cannot be determined from the information given

19.3 What is the probability of a search attempt being made at Ask Network?

(a) 0.01

(b) 0.36

(c) 0.99

(d) Cannot be determined from the information given

19.4 What is the probability of a search attempt being directed to a site other than Google?

(a) 0.01

(b) 0.25

(c) 0.37

(d) Cannot be determined from the information given

How Many in the House? *In government data, a household consists of all occupants of a dwelling unit. Here is the distribution of household size in the United States:*[2]

Number of persons	1	2	3	4	5	6	7
Probability	0.28	0.35	0.15	0.13	0.06	0.02	0.01

Choose an American household at random and let the random variable Y be the number of persons living in the household. Use this information to answer Questions 19.5 through 19.7.

19.5 Express "more than one person lives in this household" in terms of Y. What is the probability of this event?

19.6 What is $P(2 < Y \leq 4)$?

19.7 What is $P(Y \neq 2)$?

(a) 0.28 (b) 0.35

(c) 0.37 (d) 0.65

How Many Children? *Choose at random an American woman between the ages of 15 and 50. Here is the distribution of the number of children the woman has given birth to:*[3]

X = Number of children	0	1	2	3	4	5 or More
Probability	0.442	0.168	0.217	0.107	0.043	0.023

Use this information to answer Questions 19.8 through 19.11.

19.8 Check that this distribution satisfies the two requirements for a legitimate discrete probability model.

19.9 Describe in words the event $X \leq 2$. What is the probability of this event?

19.10 What is $P(X < 2)$?

(a) 0.168 (b) 0.442

(c) 0.610 (d) 0.827

19.11 Write the event "a woman gives birth to three or more children" in terms of values of X. What is the probability of this event?

Random Number Generators. *Many random number generators allow users to specify the range of the random numbers to be produced. Suppose you specify that the random number Y can take any value between −5 and 5. The density curve of the outcome has a constant height between −5 and 5 and height 0 elsewhere. Use this information to answer Questions 19.12 through 19.15.*

19.12 The random variable Y is

(a) discrete.

(b) continuous but not Normal.

(c) continuous and Normal.

(d) none of the above.

19.13 The height of the density curve between −5 and 5 is

(a) 0.1.
(b) 0.2.
(c) 1.
(d) 5.

19.14 Draw a graph of the density curve and find $P(1 \le Y \le 3)$.

19.15 Find $P(-2 < Y < 2)$.

An IQ Test. *The Wechsler Adult Intelligence Scale (WAIS) is a common IQ test for adults. The distribution of WAIS scores for persons over 16 years of age is approximately Normal with mean 100 and standard deviation 15. Use this information to answer Questions 19.16 through 19.19.*

19.16 What is the probability that a randomly chosen individual has a WAIS score of 105 or higher?

(a) 0.0005
(b) 0.3707
(c) 0.4400
(d) 0.6293

19.17 What are the mean and standard deviation of the average WAIS score $\bar{x}$ for an SRS of 60 people?

(a) Mean = 13.56, standard deviation = 15
(b) Mean = 100, standard deviation = 15
(c) Mean = 100, standard deviation = 1.94
(d) Mean = 100, standard deviation = 0.25

19.18 What is the probability that the average WAIS score of an SRS of 60 people is 105 or higher?

(a) 0.0049
(b) 0.3707
(c) 0.9738
(d) None of the above

19.19 Would your answers to any of Questions 19.16 through 19.18 be affected if the distribution of WAIS scores in the adult population were distinctly non-Normal? Explain.

Reaction Times. *The time that people require to react to a stimulus usually has a right-skewed distribution because lack of attention or tiredness causes some lengthy reaction times. Reaction times for children with attention-deficit/hyperactivity disorder (ADHD) are more skewed because their condition causes more frequent lack of attention. In one study, children with ADHD were asked to press the spacebar on a computer keyboard when any letter other than X appeared on the screen. With two seconds between letters, the mean reaction time was 445 milliseconds (ms), and the standard deviation was 82 ms.[4] Take these values to be the population μ and σ for ADHD children. Use this information to answer Questions 19.20 through 19.22.*

19.20 What are the mean and standard deviation of the mean reaction time $\bar{x}$ for a randomly chosen group of 15 ADHD children? For a group of 150 such children?

19.21 The distribution of reaction time is strongly skewed. Explain briefly why we hesitate to regard $\bar{x}$ as Normally distributed for 15 children but are willing to use a Normal distribution for the mean reaction time of 150 children.

19.22 What is the approximate probability that the mean reaction time in a group of 150 ADHD children is greater than 450 ms?

Michael Nolan/Getty Images

Pesticides in Whale Blubber: Estimation. *The level of pesticides found in the blubber of whales is a measure of pollution of the oceans by runoff from land and can also be used to identify different populations of whales. A sample of eight male minke whales in the West Greenland area of the North Atlantic found the mean concentration of the insecticide dieldrin to be $\bar{x} = 357$ nanograms per gram of blubber (ng/g).[5] Suppose that the concentration in all such whales varies Normally with standard deviation $\sigma = 50$ ng/g. Use this information to answer Questions 19.23 through 19.26.*

19.23 A 95% confidence interval to estimate the mean level is

(a) 344.75 to 369.25.

(b) 339.32 to 374.68.

(c) 322.35 to 391.65.

(d) 259.00 to 455.00.

19.24 A 90% confidence interval to estimate the mean level is

(a) 346.72 to 367.28.

(b) 327.92 to 386.08.

(c) 311.36 to 402.54.

(d) 274.75 to 439.25.

19.25 Find an 80% confidence interval for the mean concentration of dieldrin in the minke whale population.

19.26 What general fact about confidence intervals do the margins of error of your three intervals in the previous problems illustrate?

Estimating LDL Cholesterol. *The distribution of LDL cholesterol level in the population of all adults tested in a large hospital over a 10-year period is close to Normal with standard deviation $\sigma = 40$ milligrams per deciliter (mg/dL). You measure the blood cholesterol of 16 adult patients 20–34 years of age. The mean level is $\bar{x} = 125$ mg/dL. Assume that σ is the same as in the general adult hospital population. Use this information to answer Questions 19.27 through 19.29.*

19.27 A 90% confidence interval for the mean LDL level μ among adult patients 20–34 years of age is

(a) 125 ± 2.50 mg/dL.

(b) 125 ± 10.00 mg/dL.

(c) 125 ± 16.45 mg/dL.

(d) none of the above.

19.28 How large a sample is needed to cut the margin of error in the previous question in half?

(a) 4 (b) 8

(c) 32 (d) 64

19.29 How large a sample is needed to cut the margin of error for a 90% confidence interval to ± 5 mg/dL?

(a) 14

(b) 64

(c) 174

(d) 246

19.30 The Environmental Protection Agency (EPA) fuel economy ratings say that the 2019 Toyota Prius All Wheel Drive hybrid car gets 48 miles per gallon (mpg) on the highway. Deborah wonders whether the actual long-term average highway mileage μ of her new All Wheel Drive Prius is more than 48 mpg. She keeps careful records of gas mileage for 3000 miles of highway driving. Her result is $\bar{x} = 49.2$ mpg. What are her null and alternative hypotheses?

(a) $H_0: \mu = 48, H_a: \mu < 48$

(b) $H_0: \mu = 48, H_a: \mu > 48$

(c) $H_0: \bar{x} = 48, H_a: \bar{x} < 48$

(d) $H_0: \bar{x} = 48, H_a: \bar{x} > 48$

STATISTICS IN YOUR WORLD

How Many Miles Per Gallon?

As gasoline prices rise, more people pay attention to the government's gas mileage ratings of their vehicles. Until recently, these ratings overstated the miles per gallon we can expect in real-world driving. The ratings assumed a top speed of 60 miles per hour, slow acceleration, and no air-conditioning. That doesn't resemble what we see around us on the highway. Maybe it doesn't resemble the way we ourselves drive. Starting with 2008 models, the ratings assume higher speeds (80 miles per hour tops), faster acceleration, and air-conditioning in warm weather. Mileage ratings of the same vehicle dropped by about 12% in the city and 8% on the highway.

19.31 According to the National Survey of Student Engagement (NSSE), the average amount of time that first-year college students spent preparing for class (studying, reading, writing, doing homework or lab work, analyzing data, rehearsing, and other academic activities) in 2019 was 14.44 hours per week. Your college wonders if the average μ for its first-year students in 2019 differed from the national average. A random sample of 500 students who were first-year students in 2019 claims to have spent an average of $\bar{x} = 13.4$ hours per week on homework in their first year. What are the null and alternative hypotheses for a comparison of first-year students at your college with national first-year students in 2019?

(a) $H_0: \bar{x} = 14.44, H_a: \bar{x} \neq 14.44$

(b) $H_0: \bar{x} = 13.4, H_a: \bar{x} > 13.4$

(c) $H_0: \mu = 14.44, H_a: \mu \neq 14.44$

(d) $H_0: \mu = 13.4, H_a: \mu > 13.4$

Testing Blood Cholesterol. *The distribution of blood cholesterol level in the population of all adult patients tested in a large hospital over a 10-year period is close to Normal with mean 130 milligrams per deciliter (mg/dL) and standard deviation 40 mg/dL. You measure the blood cholesterol of 16 adult patients 20–34 years of age. The mean level is $\bar{x} = 125$ mg/dL. Assume that σ is the same as in the general hospital population. Use this information to answer Questions 19.32 through 19.34.*

19.32 We suspect that the mean μ for all young adults aged 20 to 34 years who have been patients in the hospital is lower than that of the population of all adult patients. Thus, we decide to test the hypotheses $H_0: \mu = 130, H_a: \mu < 130$. The z test statistic for testing these hypotheses is

(a) 0.50.
(b) −0.50.
(c) 2.00.
(d) −2.00.

19.33 The result is significant at

(a) $\alpha = 0.01$.

(b) $\alpha = 0.05$ but not at $\alpha = 0.01$.

(c) $\alpha = 0.10$ but not at $\alpha = 0.05$.

(d) none of the above.

19.34 You increase the sample of young adults aged 20–34 years who have been patients in the hospital from 16 subjects to 256. Suppose that this larger sample gives the same mean level, $\bar{x} = 125$ mg/dL. Redo the test in Questions 19.32 and 19.33. The result is significant at

(a) $\alpha = 0.01$.

(b) $\alpha = 0.05$ but not at $\alpha = 0.01$.

(c) $\alpha = 0.10$ but not at $\alpha = 0.05$.

(d) none of the above.

19.35 The Food and Drug Administration regulates the amount of dieldrin in raw food. For some foods, no more than 100 ng/g is allowed. Is there good evidence that the mean concentration μ in whale blubber is above 100 ng/g? Use the information for Exercises 19.23 through 19.26 to carry out a test of the hypotheses $H_0: \mu = 100, H_a: \mu > 100$ assuming that the "simple conditions" (page 367) hold. The *P*-value of your test is

(a) above 0.10.

(b) less than or equal to 0.10 but greater than 0.05.

(c) less than or equal to 0.05 but greater than 0.01.

(d) no more than 0.01.

19.36 Infants weighing less than 1500 grams at birth are classed as "very low birth weight." Low birth weight carries many risks. One study followed 113 male infants with very low birth weight to adulthood. At age 20, the mean IQ score for these men was $\bar{x} = 87.6$.[6] IQ scores vary Normally with standard deviation $\sigma = 15$. Give a 95% confidence interval for the mean IQ score at age 20 for all very-low-birth-weight males.

19.37 IQ tests are scaled so that the mean score in a large population should be $\mu = 100$. We suspect that the very-low-birth-weight population has mean score less than 100. Does the study described in the previous question give good evidence that this is true? State hypotheses, carry out a test assuming that the "simple conditions" (page 367) hold, compute the *P*-value, and give your conclusion in plain language.

19.38 When our brains store information, complicated chemical changes take place. In trying to understand these changes, researchers blocked some processes in brain cells taken from rats and compared these cells with a control group of normal cells. They say that "no differences were seen" between the two groups at significance level 0.05 in four response variables. They give *P*-values 0.45, 0.83, 0.26, and 0.84 for these four comparisons.[7] Which of the following statements is correct?

(a) It is literally true that "no differences were seen." That is, the mean responses were exactly alike in the two groups.

(b) The mean responses were exactly alike in the two groups for at least one of the four response variables measured but not for all of them.

(c) The statement "no differences were seen" means that the observed differences were not statistically significant at the significance level used by the researchers.

(d) The statement "no differences were seen" means that the observed differences were all less than 1 (and were actually 0.45, 0.83, 0.26, and 0.84 for these four comparisons).

19.39 In a 2013 study, researchers compared various measurements on overweight first-born and second-born middle-aged men.[8] They found that first-borns had a significantly higher weight ($P = 0.013$) than second-borns but no significant difference in total cholesterol ($P = 0.74$). Explain carefully why $P = 0.013$ means there is evidence that first-born middle-aged men may have higher weights than second-borns and why $P = 0.74$ provides no evidence that first-born middle-aged men may have different total cholesterol levels than second-borns.

19.40 We often see televised reports of brushfires threatening homes in California. Some people argue that the modern practice of quickly putting out small fires allows fuel to accumulate and so increases the damage done by large fires. A detailed study of historical data suggests that this is wrong, and the damage has risen simply because there are more houses in risky areas.[9] As usual, the study report gives statistical information tersely. Here is the summary of a regression of number of fires on decade (nine data points, for the 1910s to the 1990s): "Collectively, since 1910, there has been a highly significant increase ($r^2 = 0.61$, $P < 0.01$) in the number of fires per decade." How would you explain this statement to someone who knows no statistics? Include an explanation of both the description given by r^2 and its statistical significance.

19.41 **(Optional Topic)** Byron claims that the probability Alabama and Clemson will play in the national championship football game this year is 25%. The number 25% is

(a) the proportion of times Alabama and Clemson have played in the championship game in the past.

(b) Byron's personal probability that Alabama and Clemson will play in the championship football game this year.

(c) the area under a Normal density curve.

(d) all of the above.

19.42 (Optional Topic) Causes of Death. Cancerous tumors, heart disease, and accidents are the leading causes of death for adults. Here are the counts of deaths due to these causes in 2017 for adults in California and New York:

	California	New York
Cancerous tumors	59,516	34,956
Heart disease	62,797	44,092
Accidents	13,840	7,687

(a) Choose at random an adult from California or New York who died from one of these three causes. What is the probability that the adult was from New York?

(b) Find the conditional probability that the victim was from New York, given that the death was accidental.

(c) Use your answers from parts (a) and (b) to explain whether the state the adult was from (California or New York) and these three types of death are independent or not.

(Optional Topic) A Whale of a Time. *Hacksaw's Boats of St. Lucia takes tourists on a daily dolphin/whale watch cruise. Its brochure claims an 85% chance of sighting a whale or a dolphin. Suppose there is a 65% chance of seeing dolphins and a 10% chance of seeing both dolphins and whales. Use this information to answer Questions 19.43 through 19.45. (You may want to make a Venn diagram to help answer Questions 19.43 and 19.44.)*

19.43 The probability of seeing a whale on the cruise is

(a) 0.15.

(b) 0.20.

(c) 0.30.

(d) 0.55.

19.44 The probability of seeing a whale but not a dolphin on the cruise is

(a) 0.15.

(b) 0.20.

(c) 0.30.

(d) 0.55.

19.45 Assume that sightings from day to day are independent. If you take the dolphin/whale watch cruise on two consecutive days, what is the probability you will see a dolphin or a whale on at least one day? (*Hint:* First compute the probability that this event does not occur.)

(a) 0.0225

(b) 0.2775

(c) 0.7225

(d) 0.9775

19.46 (Optional Topic) Alysha makes 40% of her free throws. She takes five free throws in a game. If the shots are independent of each other, the probability that she misses the first two shots but makes the other three is about

(a) 0.230.

(b) 0.115.

(c) 0.023.

(d) 0.600.

19.47 **(Optional Topic)** Alysha makes 40% of her free throws. She takes five free throws in a game. If the shots are independent of each other, the probability that she makes *exactly* one of five shots is about

(a) 0.259. (b) 0.115.

(c) 0.052. (d) 0.200.

(Optional Topic) Tracking Your Health. *Sample surveys show that more people are tracking changes in their health on paper, spreadsheet, mobile device, or just "in their heads." A survey asked a nationwide random sample of 3014 adults, "Now thinking about your health overall, do you currently keep track of your own weight, diet, or exercise routine, or is this not something you currently do?"*[10] *The population that the poll wants to draw conclusions about is all U.S. residents aged 18 and over. Suppose that, in fact, 61% of all adult U.S. residents would say Yes if asked this question. Use this information to answer Questions 19.48 and 19.49.*

19.48 The approximate distribution of the number in the sample of 3014 adults who would say Yes is

(a) $N(61, 26.78)$.

(b) $N(61, 717.03)$.

(c) $N(1838.5, 26.78)$.

(d) $N(1838.5, 717.03)$.

19.49 Find the (approximate) probability that 1900 or more adults in the sample say Yes.

SUPPLEMENTARY EXERCISES

Supplementary exercises apply the skills you have learned in ways that require more thought or more elaborate use of technology.

19.50 The Addition Rule. The addition rule for probabilities, $P(A \text{ or } B) = P(A) + P(B)$, is not always true. Give (in words) an example of real-world events A and B for which this rule is not true.

19.51 Comparing Wine Tasters. Two wine tasters rate each wine they taste on a scale of 1 to 5. From data on their ratings of a large number of wines, we obtain the following probabilities for both tasters' ratings of a randomly chosen wine:

	Taster 2				
Taster 1	1	2	3	4	5
1	0.05	0.02	0.01	0.00	0.00
2	0.02	0.08	0.04	0.02	0.01
3	0.01	0.04	0.25	0.05	0.01
4	0.00	0.02	0.05	0.18	0.02
5	0.00	0.01	0.01	0.02	0.08

(a) Why is this a legitimate discrete probability model?

(b) What is the probability that the tasters agree when rating a wine?

(c) What is the probability that Taster 1 rates a wine higher than Taster 2? What is the probability that Taster 2 rates a wine higher than Taster 1?

19.52 A 14-sided Die. An ancient Korean drinking game involves a 14-sided die. The players roll the die in turn and must submit to whatever humiliation is written on the up-face: something like "Keep still when tickled on face." Six of the 14 faces are squares. Let's call them A, B, C, D, E, and F for short. The other eight faces are triangles, which we will call 1, 2, 3, 4, 5, 6, 7, and 8. Each of the squares is equally likely. Each of the triangles is also equally likely, but the triangle probability differs from the square probability. The probability of getting a triangle is 0.28. Give the probability model for the 14 possible outcomes.

David Moore

19.53 Distributions: Means Versus Individuals. The z confidence interval and test are based on the sampling distribution of the sample mean $\bar{x}$. The National Survey of Student Engagement (NSSE) asks college seniors to rate how much their experience at their institution has contributed to their ability to analyze numerical and statistical information. Ratings are on a scale of 1 to 7, with 7 being the highest (best) rating. Suppose scores for all seniors in 2019 are Normal with mean $\mu = 2.9$ and standard deviation $\sigma = 1.0$. (The ratings take on integer values from 1 to 7, so the population of the ratings of all college seniors could not be normal. However, as discussed in Section 18.1, page 403, it is reasonable to assume that the sample mean $\bar{x}$ is approximately Normal, and it is the Normality of the sample mean that makes the use of z procedures reasonably accurate.)

(a) You take an SRS of 100 seniors. According to the 99.7 part of the 68–95–99.7 rule, about what range of ratings do you expect to see in your sample?

(b) You look at many SRSs of size 100. About what range of sample mean ratings $\bar{x}$ do you expect to see?

19.54 Distributions: Larger Samples. In the setting of the previous exercise, how many seniors must you sample to cut the range of values of $\bar{x}$ in half? This will also cut the margin of error of a confidence interval for μ in half. Do you expect the range of individual scores in the new sample also to be much less than in a sample of size 100? Why?

19.55 Normal Body Temperature. 4step Here are the daily average body temperatures (°F) for 20 healthy adults:[11] BODYTMP

98.74	98.83	96.80	98.12	97.89	98.09	97.87	97.42	97.30	97.84
100.27	97.90	99.64	97.88	98.54	98.33	97.87	97.48	98.92	98.33

(a) Make a stemplot of the data. The distribution is roughly symmetric and single-peaked. There is one mild outlier. We expect the distribution of the sample mean $\bar{x}$ to be close to Normal.

(b) Do these data give evidence that the mean body temperature for all healthy adults is not equal to the "traditional" 98.6°F? (This traditional value has been reevaluated recently. See the reference in the notes.) Follow the four-step process for significance tests (page 374). (Suppose that body temperature varies Normally with standard deviation 0.7°F.)

19.56 Time in a Restaurant. 4step The owner of a pizza restaurant in France knows that the time customers spend in the restaurant on Saturday evening has mean 90 minutes and standard deviation 15 minutes. He has read that pleasant odors can influence customers, so he spreads a lavender odor throughout the restaurant. Here are the times (in minutes) for customers on the next Saturday evening:[12] RSTRNT

92	126	114	106	89	137	93	76	98	108
124	105	129	103	107	109	94	105	102	108
95	121	109	104	116	88	109	97	101	106

(a) Make a stemplot of the times. The distribution is roughly symmetric and single-peaked, so the distribution of $\bar{x}$ should be close to Normal.

(b) Suppose that the standard deviation $\sigma = 15$ minutes is not changed by the odor. Is there reason to think that the lavender odor has increased the mean time customers spend in the restaurant? Follow the four-step process for significance tests (page 374).

19.57 Normal Body Temperature. Use the data in Exercise 19.55 to estimate mean body temperature with 90% confidence. Follow the four-step process for confidence intervals (page 374). BODYTMP

19.58 Time in a Restaurant. 4step Use the data in Exercise 19.56 to estimate the mean time customers spend in this restaurant on Saturday evenings with 95% confidence. Follow the four-step process for confidence intervals (page 374). RSTRNT

19.59 Tests from Confidence Intervals. You read in a U.S. Census Bureau report that a 90% confidence interval for the median income in 2018 of American households was $61,937 ± $94. Based on this interval, can you reject the null hypothesis that the median income in this group is $62,000? What is the alternative hypothesis of the test based on this confidence interval? What is its significance level?

19.60 (Optional Topic) Testing for HIV. Enzyme immunoassay tests are used to screen blood specimens for the presence of antibodies to HIV, the virus that causes AIDS. Antibodies indicate the presence of the virus. The test is quite accurate but is not always correct. Here are approximate probabilities of positive and negative test results when the blood tested does and does not actually contain antibodies to HIV:[13]

	Test Result	
	Positive	Negative
Antibodies present	0.9985	0.0015
Antibodies absent	0.0060	0.9940

Suppose that 1% of a large population carries antibodies to HIV in their blood.

(a) Draw a tree diagram for selecting a person from this population (outcomes: antibodies present or absent) and testing his or her blood (outcomes: test positive or negative).

(b) What is the probability that the test is positive for a randomly chosen person from this population?

19.61 (Optional Topic) Type of High School Attended. Say that you choose a college freshman at random and ask what type of high school he or she attended. Here is the distribution of results:[14]

Type	Regular Public	Public Charter	Public Magnet	Private Religious	Private Independent	Home School
Probability	0.758	0.029	0.035	0.109	0.063	0.006

What is the conditional probability that a college freshman was home schooled, given that he or she did not attend a regular public high school?

19.62 (Optional Topic) False HIV Positives. Continue your work from Exercise 19.60. What is the probability that a person has the antibody, given that the test is positive? (Your result illustrates a fact that is important when considering proposals for widespread testing for HIV, prostate cancer, illegal drugs, or agents of biological warfare: if the condition being tested is uncommon in the population, most positives will be false positives.)

19.63 (Optional Topic) Retention Rates in a Weight-loss Program. Americans spend more than \$30 billion annually on a variety of weight-loss products and services. In a study of retention rates of those using the Rewards Program at Jenny Craig in 2005, it was found that about 18% of those who began the program dropped out in the first four weeks.[15] Assume that we have a random sample of 300 people beginning the program.

(a) Assuming that the results of the 2005 study still characterize the general population, what is the mean number of people who would drop out of the Rewards Program within four weeks in a sample of this size? What is the standard deviation?

(b) What is the probability that at least 235 people in the sample will still be in the Rewards Program after the first four weeks? Check that the Normal approximation is permissible and use it to find this probability. If your software allows, find the exact binomial probability and compare the two results.

19.64 (Optional Topic) Low Power? It appears that eating oat bran lowers cholesterol slightly. At a time when oat bran was popularly considered to promote good health, a paper in the *New England Journal of Medicine* found that it had no significant effect on cholesterol.[16] The paper reported a study with just 20 subjects. Letters to the journal denounced publication of a negative finding from a study with very low power. Explain why lack of significance in a study with low power gives no reason to accept the null hypothesis that oat bran has no effect.

19.65 (Optional Topic) Type I and Type II Errors. Question 19.37 asks for a significance test of the null hypothesis that the mean IQ of very-low-birth-weight male babies is 100 against the alternative hypothesis that the mean is less than 100. State in words what it means to make a Type I error and a Type II error in this setting.

Rawpixel/Getty Images

PART IV

Inference about Variables

With the principles in hand, we proceed to practice—that is, to inference in fully realistic settings. In the remaining chapters of this book, you will meet many of the most commonly used statistical procedures. We have grouped these procedures into two classes, corresponding to our division of data analysis into exploring variables and distributions and exploring relationships. The five chapters of Part IV concern inference about the distribution of a single variable and inference for comparing the distributions of two variables. Part V deals with inference for relationships among variables. In Chapters 20 and 21, we analyze data on quantitative variables. We begin with the familiar Normal distribution for a quantitative variable. Chapters 22 and 23 concern categorical variables so that inference begins with counts and proportions of outcomes. Chapter 24 reviews this part of the text.

The four-step process for approaching a statistical problem can guide much of your work in these chapters. You should review the outlines of the four-step process for a confidence interval (page 374) and for a test of significance (page 392). The statement of an exercise usually does the State step for you, leaving the Plan, Solve, and Conclude steps for you to complete. It is helpful to first summarize the State step in your own words to organize your thinking. Many examples and exercises in these chapters involve both carrying out inference and thinking about inference in practice. Remember that any inference method is useful only under certain conditions and that you must judge these conditions before rushing to inference.

CHAPTER

20

When you complete this chapter, you will be able to:

20.1 Recognize whether conditions for using inferential methods to estimate a population mean are met.

20.2 Given the sample size and desired confidence level for inference based on the t distribution, use software or a table to determine the critical value t^*.

20.3 Use the one-sample t procedure to obtain a confidence interval at a stated level of confidence for the mean μ of a population.

20.4 Carry out a one-sample t test for the hypothesis that a population mean has a specified value and provide the associated P-value

20.5 Use technology to implement one-sample t procedures.

20.6 Recognize matched pairs data and use t procedures to obtain confidence intervals and to perform tests of significance for matched pairs studies.

20.7 Use sample size and an examination of the distribution of sample data to determine whether the robustness of t procedures allows for use of those procedures in specific settings.

Inference about a Population Mean

In Chapters 16 through 18, we began our study of inference. We focused on inference for a population mean based on a sample from a Normal population. We included the unrealistic assumption that we knew the population standard deviation σ. We were able to use what we learned about the Normal distribution in Chapters 3 and 12 and what we learned about the sampling distribution of the sample mean in Chapter 15 to construct confidence intervals for, and to conduct hypothesis tests about, the population mean.

In this chapter we discard the unrealistic condition that we know the population standard deviation σ and present procedures for practical use. We also pay more attention to the real-data setting of our work. The details of confidence intervals and tests change only slightly when you don't know σ. More importantly, you can interpret your results exactly as before. To illustrate this, Example 20.2 repeats an example from Chapter 16.

20.1 Conditions for Inference about a Mean

Confidence intervals and tests of significance for the mean μ of a Normal population are based on the sample mean $\bar{x}$. Confidence intervals and P-values involve probabilities calculated from the sampling distribution of $\bar{x}$. Here are the conditions needed for realistic inference about a population mean.

Richard Nowitz/Getty Images

Conditions for Inference about a Mean

- We can regard our data as a simple random sample (SRS) from the population. This condition is very important.
- Observations from the population have a Normal distribution with mean μ and standard deviation σ. In practice, it is enough that the distribution be symmetric and single-peaked unless the sample is very small. Both μ and σ are unknown parameters.

There is another condition that applies to all the inference methods in this book: *the population must be much larger than the sample–say, at least 20 times as large.*[1] All our examples and exercises satisfy this condition. Practical settings in which the sample is a large part of the population are rather special, and we will not discuss them.

When the conditions for inference are satisfied, the sample mean $\bar{x}$ has the Normal distribution with mean μ and standard deviation $\sigma/\sqrt{n}$. Because we don't know σ, we estimate it by the sample standard deviation s. We then estimate the standard deviation of $\bar{x}$ by $s/\sqrt{n}$. This quantity is called the *standard error* of the sample mean $\bar{x}$.

Standard Error

When the standard deviation of a statistic is estimated from data, the result is called the **standard error** of the statistic. The standard error of the sample mean $\bar{x}$ is $s/\sqrt{n}$.

For example, if a sample of size $n = 20$ from a population has standard deviation $s = 8$, the standard error of the sample mean would be $s/\sqrt{n} = 8/\sqrt{20} = 8/4.472 = 1.789$.

Do not confuse the standard error with the sample standard deviation s. The sample standard deviation s is an estimate of the standard deviation of the population from which the sample is selected. The standard error is not an estimate of the population standard deviation.

APPLY YOUR KNOWLEDGE

20.1 Travel Time to Work. A study of commuting times reports the one-way travel times to work of a random sample of 1000 employed adults in Seattle.[2] The mean is $\bar{x} = 30.1$ minutes, and the standard deviation is $s = 27.2$ minutes. What is the standard error of the mean?

20.2 Standing Height Desks Burn Calories. Researchers at Texas A&M studied the effect of using standing height desks on the energy expended by nine elementary school students. In their paper about the study, they reported descriptive statistics for the nine students. These descriptive statistics were expressed as a mean plus or minus a standard deviation.[3] One such descriptive statistic was the weight of students before using the standing desks, which was reported as 27.0 ± 7.9 kilograms. What are $\bar{x}$ and the standard error of the mean for these students? (This exercise is also a warning to read carefully: that 27.0 ± 7.9 is *not* a confidence interval, yet summaries in this form are common in scientific reports.)

20.2 The *t* Distributions

If we knew the value of σ, we would base confidence intervals and tests for μ on the one-sample z statistic we used in Chapter 17:

$$z = \frac{\bar{x} - \mu}{\sigma/\sqrt{n}}$$

This z statistic has the standard Normal distribution $N(0, 1)$. In practice, we don't know σ, so we substitute the standard error $s/\sqrt{n}$ of $\bar{x}$ for its standard deviation $\sigma/\sqrt{n}$. The statistic that results does not have a Normal distribution. It has a distribution that is new to us, called a *t distribution*.

The One-Sample *t* Statistic and the *t* Distributions

Draw an SRS of size n from a large population that has the Normal distribution with mean μ and standard deviation σ. The **one-sample *t* statistic**

$$t = \frac{\bar{x} - \mu}{s/\sqrt{n}}$$

has the ***t* distribution** with $n - 1$ degrees of freedom.

The t statistic has the same interpretation as any standardized statistic: it says how far $\bar{x}$ is from its mean μ in standard error units. There is a different t distribution for each sample size. We specify a particular t distribution by giving its *degrees of freedom.* The degrees of freedom for the one-sample t statistic come from the sample standard deviation s in the denominator of t. We saw in Chapter 2 (page 58) that s has $n - 1$ degrees of freedom. There are other t statistics with different degrees of freedom, some of which we will meet later. We will write the t distribution with $n - 1$ degrees of freedom as t_{n-1} for short.

Figure 20.1 compares the density curves of the standard Normal distribution and the t distributions with 2 and 9 degrees of freedom. The figure illustrates these facts about the t distributions:

- The density curves of the t distributions are similar in shape to the standard Normal curve. They are symmetric about zero, single-peaked, and bell-shaped.

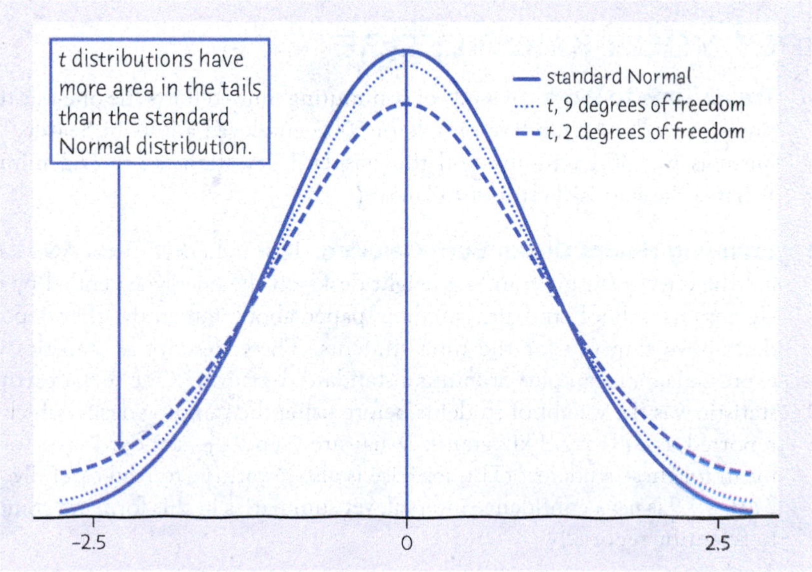

FIGURE 20.1 Density curves for the *t* distributions with 2 and 9 degrees of freedom and for the standard Normal distribution. All are symmetric with center 0. The *t* distributions are somewhat more variable.

- The variability of the t distributions is a bit greater than that of the standard Normal distribution. The t distributions in Figure 20.1 have more probability in the tails and less in the center than does the standard Normal distribution. This is true because substituting the estimate s for the fixed parameter σ introduces more variation into the statistic.
- As the degrees of freedom increase, the t density curve approaches the $N(0, 1)$ curve ever more closely. This happens because s estimates σ more accurately as the sample size increases. So using s in place of σ causes little extra variation in the statistic when the sample is large.

Table C in the back of the book gives critical values for the t distributions. Each row in the table contains critical values for the t distribution whose degrees of freedom appear at the left of the row. For convenience, we label the table entries both by the confidence level C (in percentage) required for confidence intervals and by the one-sided and two-sided P-values for each critical value. You have already used the standard Normal critical values in the z^* row at the bottom of Table C. By looking down any column, you can check that the t critical values approach the Normal values as the degrees of freedom increase. If you use statistical software, you don't need Table C.

EXAMPLE 20.1 *t* Critical Values

Figure 20.1 shows the density curve for the t distribution with 9 degrees of freedom. What point on this distribution has probability 0.05 to its right? In Table C, look in the df = 9 row above one-sided P-value 0.05, and you will find that this critical value is $t^* = 1.833$. To use software, enter the degrees of freedom and the probability you want to the *left*, 0.95 in this case. Here is Minitab's output:

```
Student's t distribution with 9 DF

P(X <= x)      x
   0.95     1.83311
```

For the standard Normal distribution, the point that has probability 0.05 to the right is 1.645. (See the z^* row at the bottom of Table C for the Normal distribution critical values.) The value for the standard Normal distribution is smaller than the value for the t distribution. This is an example of what we mean by the statement that t distributions have more probability in the tails than the standard Normal distribution.

APPLY YOUR KNOWLEDGE

20.3 Critical Values. Use software or Table C to find

(a) the critical value for a one-sided test with level $\alpha = 0.01$ based on the t_1 distribution.

(b) the critical value for a 90% confidence interval based on the t_{30} distribution. How does this compare with the critical value z^* for a 90% confidence interval based on the standard Normal distribution?

20.4 More Critical Values. You have an SRS of size 100 and calculate the one-sample t statistic. What is the critical value t^* such that

(a) t has probability 0.02 to the right of t^*?

(b) t has probability 0.80 to the left of t^*?

(c) How do the values in (b) and (c) compare with the corresponding values of z^* for the standard Normal distribution?

20.3 The One-Sample t Confidence Interval

STATISTICS IN YOUR WORLD

Better Statistics, Better Beer

The t distribution and the t inference procedures were invented by William S. Gosset (1876–1937). Gosset worked for the Guinness brewery, and his goal in life was to make better beer. He used his new t procedures to find the best varieties of barley and hops. Gosset's statistical work helped him become head brewer, a more interesting title than professor of statistics. Because Gosset published under the pen name "Student," you will often see the t distribution called "Student's t" in his honor.

To analyze samples from Normal populations with unknown σ, just replace the standard deviation $\sigma/\sqrt{n}$ of $\bar{x}$ by its standard error $s/\sqrt{n}$ in the z procedures of Chapters 16, 17, and 18. The confidence interval and test that result are *one-sample t procedures*. Critical values and P-values come from the t distribution with $n-1$ degrees of freedom. The one-sample t procedures are similar in both reasoning and computational detail to the z procedures.

The One-Sample t Confidence Interval

Draw an SRS of size n from a large population having unknown mean μ. A level C confidence interval for μ is

$$\bar{x} \pm t^* \frac{s}{\sqrt{n}}$$

where t^* is the critical value for the t_{n-1} density curve with area C between $-t^*$ and t^*. This interval is exact when the population distribution is Normal and is approximately correct for large n in other cases.

We will provide guidelines on how large a sample size is needed for the t confidence interval to be approximately correct in Section 20.7.

EXAMPLE 20.2 Good Weather, Good Tips?

Let's look again at the study of tipping in a restaurant that we met in Example 16.3. We follow the four-step process for a confidence interval, outlined on page 374. TIP2

Ariel Skelley/AGE Fotostock

STATE: Does the expectation of good weather lead to more generous behavior? Psychologists studied the size of the tip in a restaurant when a message indicating that the next day's weather would be good was written on the bill. Here are tips from 20 patrons, measured in percentage of the total bill:[4]

20.8	18.7	19.9	20.6	21.9	23.4	22.8	24.9	22.2	20.3
24.9	22.3	27.0	20.4	22.2	24.0	21.1	22.1	22.0	22.7

This is one of three sets of measurements made, the others being tips received when the message on the bill said that the next day's weather would not be good and tips received when there was no message on the bill. We want to estimate the mean tip for comparison with tips under the other conditions.

PLAN: We will estimate the mean percentage tip μ for all patrons of this restaurant when they receive a message on their bill indicating that the next day's weather will be good by giving a 95% confidence interval.

SOLVE: We must first check the conditions for inference.

- As in Chapter 16 (page 375), we are willing to regard these patrons as an SRS from all patrons of this restaurant.
- The stemplot in Figure 20.2 does not suggest any strong departures from Normality.

We can proceed to calculation. For these data,

$$\bar{x} = 22.21 \quad \text{and} \quad s = 1.963$$

18	7
19	9
20	3468
21	19
22	0122378
23	4
24	099
25	
26	
27	0

FIGURE 20.2 Stemplot of the percentage tips, for Example 20.2.

The degrees of freedom are $n-1=19$. From Table C, we find that for 95% confidence $t^* = 2.093$. The confidence interval is

$$\begin{aligned}\overline{x} \pm t^* \frac{s}{\sqrt{n}} &= 22.21 \pm 2.093 \frac{1.963}{\sqrt{20}} \\ &= 22.21 \pm 0.92 \\ &= 21.29\% \text{ to } 23.13\%\end{aligned}$$

CONCLUDE: We are 95% confident that the mean percentage tip for all patrons of this restaurant when their bill contains a message that the next day's weather will be good is between 21.29 and 23.13.

Our work in Example 20.2 is very similar to what we did in Example 16.3 (page 374). To make the inference realistic, we replaced the assumed $\sigma = 2$ by $s = 1.963$ calculated from the data and replaced the standard Normal critical value $z^* = 1.960$ by the t critical value $t^* = 2.093$. The resulting confidence interval is slightly wider than the one obtained in Example 16.3 (which was 21.33 to 23.09).

The one-sample t confidence interval has the form

$$\text{estimate} \pm t^* \text{SE}_{\text{estimate}}$$

where SE stands for "standard error." We will meet a number of confidence intervals that have this common form. In Example 20.2, the estimate is the sample mean $\overline{x}$, and its standard error is

$$\begin{aligned}\text{SE}_{\overline{x}} &= \frac{s}{\sqrt{n}} \\ &= \frac{1.963}{\sqrt{20}} = 0.439\end{aligned}$$

Software will find $\overline{x}$, s, $\text{SE}_{\overline{x}}$, and the confidence interval from the data. Figure 20.5 (page 450) displays typical software output for Example 20.2.

APPLY YOUR KNOWLEDGE

20.5 Critical Values. What critical value t^* from Table C would you use for a confidence interval for the mean of the population in each of the following situations? (If you have access to software, you can use software to determine the critical values.)

(a) A 99% confidence interval based on $n = 2$ observations

(b) A 95% confidence interval from an SRS of 30 observations

(c) A 90% confidence interval from a sample of size 1001

20.6 How Much Will I Bet? Our decisions depend on how the options are presented to us. Here's an experiment that illustrates this phenomenon. Tell 20 subjects that they have been given \$50 but can't keep it all. Then present them with a long series of choices among bets they can make with the \$50. Scattered among these choices in random order are 64 choices that ask the subject to choose between betting a fixed amount and an all-or-nothing gamble. The odds for all the bets are the same, but in 32 of the 64 choices, the fixed option reads "Keep \$20," and in the other 32 choices, the fixed option reads "Lose \$30." These two fixed options lead to exactly the same outcome, but people are more likely to choose the fixed option that says they lose money. Here are the percentage differences ("Number of times chose 'Lose \$30'" minus "Number of times chose 'Keep \$20'" divided by the number of trials on which the 20 subjects chose the fixed-option gamble rather than the all-or-nothing bet):[5] GAMB1

37.5	30.8	6.2	17.6	14.3	8.3	16.7	20.0	10.5	21.7
30.8	27.3	22.7	38.5	8.3	10.5	8.3	10.5	25.0	7.7

(a) Make a stemplot. Is there any sign of a major deviation from Normality?

(b) All 20 subjects gambled a fixed amount more often when faced with a sure loss than when faced with a sure win. Use the above data to give a 99% confidence interval for the mean percentage difference in gambling a fixed amount when faced with the "Lose $30" option as compared to the "Keep $20" option.

4step

20.7 She Sounds Tall! Presented with recordings of a pair of people of the same sex speaking the same phrase, can a listener determine which speaker is taller simply from the sound of the voice? Twenty-four young adults at Washington University listened to 100 pairs of speakers and, within each pair, were asked to indicate which of the two speakers was taller. Here are the number correct (out of 100) for each of the 24 participants:[6] TALL

65	61	67	59	58	62	56	67	61	67	63	53
68	49	66	58	69	70	65	56	68	56	58	70

Assume that these young adults can be regarded as an SRS of all young adults in the United States. Use a 95% confidence interval to estimate the mean number correct in the population of all young adults in the United States. Follow the four-step process as illustrated in Example 20.2.

20.4 The One-Sample *t* Test

Just as using t^* in place of z^* for confidence intervals is straightforward, so is using the t distribution for tests of significance.

The One-Sample *t* Test

Draw an SRS of size n from a large population having unknown mean μ. To test the hypothesis $H_0: \mu = \mu_0$, compute the **one-sample *t* statistic**

$$t = \frac{\bar{x} - \mu_0}{s/\sqrt{n}}$$

In terms of a variable t having the t_{n-1} distribution, the P-value for a test of H_0 against

$H_a: \mu > \mu_0$ is $P(T \geq t)$

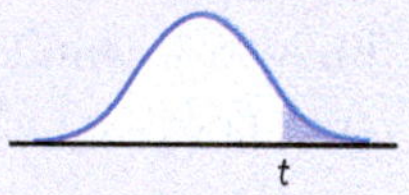

$H_a: \mu < \mu_0$ is $P(T \leq t)$

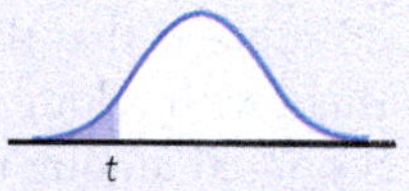

$H_a: \mu \neq \mu_0$ is $2P(T \geq |t|)$

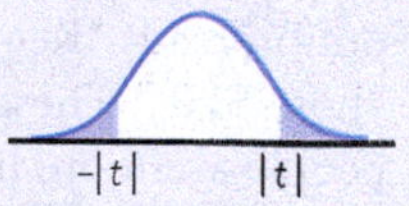

These P-values are exact if the population distribution is Normal and are approximately correct for large n in other cases.

We will provide guidelines on how large a sample size is needed for the *P*-values to be approximately correct in Section 20.7.

EXAMPLE 20.3 Water Quality

We follow the four-step process for a significance test, outlined on page 374.

WQUAL

4step

STATE: To investigate water quality, in late July 2016, the Ohio Department of Health collected vials of water from 12 beaches on Lake Erie in Cuyahoga County. Those vials were tested for fecal coliform, which are *E. coli* bacteria found in human and animal feces. An unsafe level of fecal coliform means there's an increased chance that disease-causing bacteria are present and more risk that a swimmer will become ill if she or he should accidentally ingest some of the water. One website considers it unsafe if a 100-milliliter vial (about 3.3 ounces) of water contains more than 88 coliform bacteria.

PLAN: Fecal coliform levels can change as weather and other conditions change. These beaches had been deemed safe earlier in the summer of 2016, and the reason data were collected was to determine if this continued to be true. So we are looking for evidence that past conditions along this stretch of beaches have deteriorated. We ask the question in terms of the mean fecal coliform level μ for all these beaches. The null hypothesis is "level is safe," and the alternative hypothesis is "level is unsafe":

$$H_0: \mu = 88$$
$$H_a: \mu > 88$$

SOLVE: Here are the fecal coliform levels found by the laboratories:[7]

248 37 146 19 66 236 164 30 13 144 242 20

Do these data provide good evidence that, on average, the fecal coliform levels in these beaches were safe?

First, check the conditions for inference. We are willing to regard these particular 12 samples as an SRS from a large population of possible samples. Figure 20.3 is a histogram of the data. We can't accurately judge Normality from 12 observations; there are no outliers, but the distribution of fecal coliform levels is somewhat skewed. *P*-values for the *t* test may be only approximately accurate.

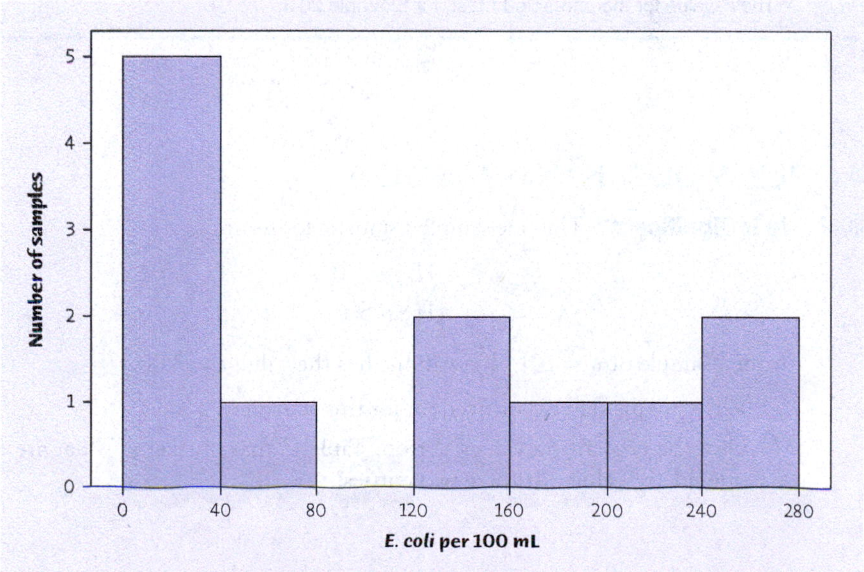

FIGURE 20.3 Histogram of the fecal coliform level (*E. coli* per 100 mL), for Example 20.3.

The basic statistics are

$$\bar{x} = 113.75 \quad \text{and} \quad s = 93.90$$

The one-sample t statistic is

$$t = \frac{\bar{x} - \mu_0}{s/\sqrt{n}} = \frac{113.75 - 88}{93.90/\sqrt{12}} = 0.95$$

The *P*-value for $t = 0.95$ is the area to the right of 0.95 under the t distribution curve with degrees of freedom $n - 1 = 11$. Figure 20.4 shows this area. Software (see Figure 20.6) tells us that $P = 0.1813$.

df = 11

t^*	.876	1.088
One-sided P	.20	.15

Without software, we can pin P between two values by using Table C. Search the df = 11 row of Table C for entries that bracket $t = 0.95$. The observed t lies between the critical values for one-sided *P*-values 0.20 and 0.15.

CONCLUDE: There is not strong evidence ($P = 0.1813$) that, on average, fecal coliform levels in these Ohio beaches are unsafe.

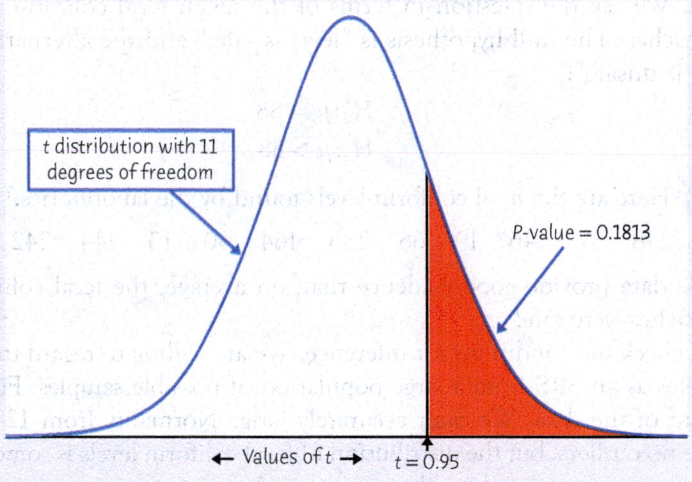

FIGURE 20.4
The *P*-value for the one-sided t test, for Example 20.3.

APPLY YOUR KNOWLEDGE

20.8 Is It Significant? The one-sample t statistic for testing

$$H_0: \mu = 0$$
$$H_a: \mu > 0$$

from a sample of $n = 101$ observations has the value $t = 3.00$.

(a) What are the degrees of freedom for this statistic?

(b) Give the two critical values t^* from Table C that bracket t. What are the one-sided *P*-values for these two entries?

(c) Is the value $t = 3.00$ significant at the 10% level? Is it significant at the 5% level? Is it significant at the 1% level?

(d) (Optional) If you have access to suitable technology, give the exact one-sided *P*-value for $t = 3.00$.

20.9 Is It Significant? The one-sample *t* statistic from a sample of $n = 2$ observations for the two-sided test of

$$H_0: \mu = 50$$
$$H_a: \mu \neq 50$$

has the value $t = 3.00$.

(a) What are the degrees of freedom for *t*?

(b) Locate the two critical values t^* from Table C that bracket *t*. What are the two-sided *P*-values for these two entries?

(c) Is the value $t = 3.00$ statistically significant at the 10% level? At the 5% level? At the 1% level?

(d) (Optional). If you have access to suitable technology, give the exact two-sided *P*-value for $t = 3.00$.

20.10 She Sounds Tall (continued). Do the data of Exercise 20.7 give good reason to think that the mean number of correct identifications in the population of all young adults in the United States is greater than 50 (the expected number correct if one is just guessing)? Carry out a test of significance, following the four-step process as illustrated in Example 20.3. TALL

4step

20.5 Examples of Technology

Any technology suitable for statistics will implement the one-sample *t* procedures. As usual, you can read and use almost any output now that you know what to look for. Figure 20.5 displays output for the 95% confidence interval of Example 20.2 from a graphing calculator, three statistical programs, and a spreadsheet program. The calculator, Minitab, JMP, and CrunchIt! outputs are straightforward. All three give the estimate $\bar{x}$ and the confidence interval plus a clearly labeled selection of other information. The confidence interval agrees with our hand calculation in Example 20.2. In general, software results are more accurate because of the rounding in hand calculations. Excel gives several descriptive measures but does not give the confidence interval. The entry labeled "Confidence Level (95.0%)" is the margin of error. You can use this together with $\bar{x}$ to get the interval using either a calculator or the spreadsheet's formula capability.

Figure 20.6 displays output for the *t* test in Example 20.3. The graphing calculator, Minitab, JMP, and CrunchIt! give the sample mean $\bar{x}$, the *t* statistic, and its *P*-value. Accurate *P*-values are the biggest advantage of software for the *t* procedures. Using Excel is, as usual, more awkward than using software designed for statistics. Excel lacks a one-sample *t* test menu selection but does have a function named TDIST for tail areas under *t* density curves. The Excel output shows functions for the *t* statistic and its *P*-value to the right of the main display, along with their values $t = 0.94990954$ and $P = 0.18128113$.

TEXAS INSTRUMENTS

```
TInterval
 (21.292,23.128)
 x̄=22.2100
 Sx=1.9625
 n=20.0000
```

jmp

Confidence Intervals

Parameter	Estimate	Lower CI	Upper CI	1-Alpha
Mean	22.21	21.29151	23.12849	0.950

Minitab

One-Sample T: percent

Variable	N	Mean	StDev	SE Mean	95% CI
Percent	20	22.210	1.963	0.439	(21.292, 23.128)

CRUNCHIT!

Export

n:	20
Sample Mean:	22.21
Standard Error:	0.4388
df:	19
95% ConfInt:	(21.29, 23.13)

Excel

	B	C	D	E
1	Percent			
2				
3	Mean	22.21		
4	Standard Error	0.4388		
5	Median	22.15		
6	Standard Deviation	1.96		
7	Sample Variance	3.85		
8	Range	8.30		
9	Minimum	18.70		
10	Maximum	27		
11	Count	20		
12	Confidence Level(95.0%)	0.9185		
13				

This is an estimate of μ. (Mean, 22.21)

This is the margin of error $\pm t^*$ SE. (Confidence Level(95.0%), 0.9185)

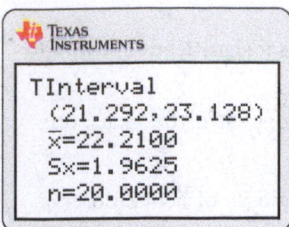

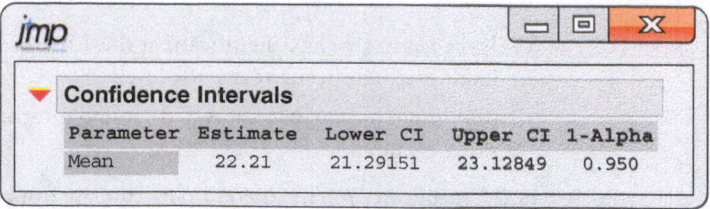

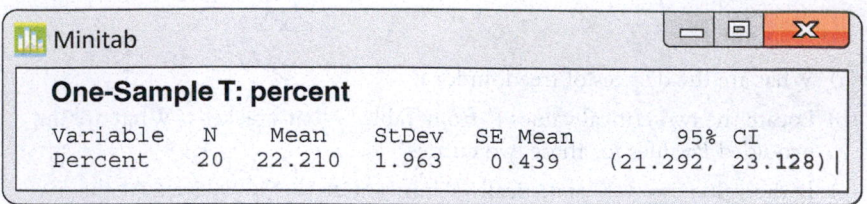

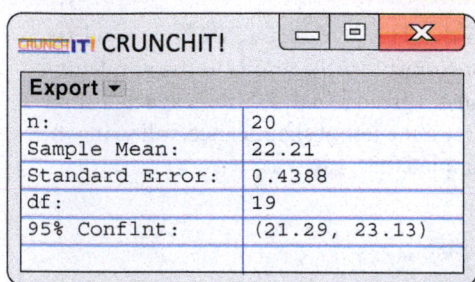

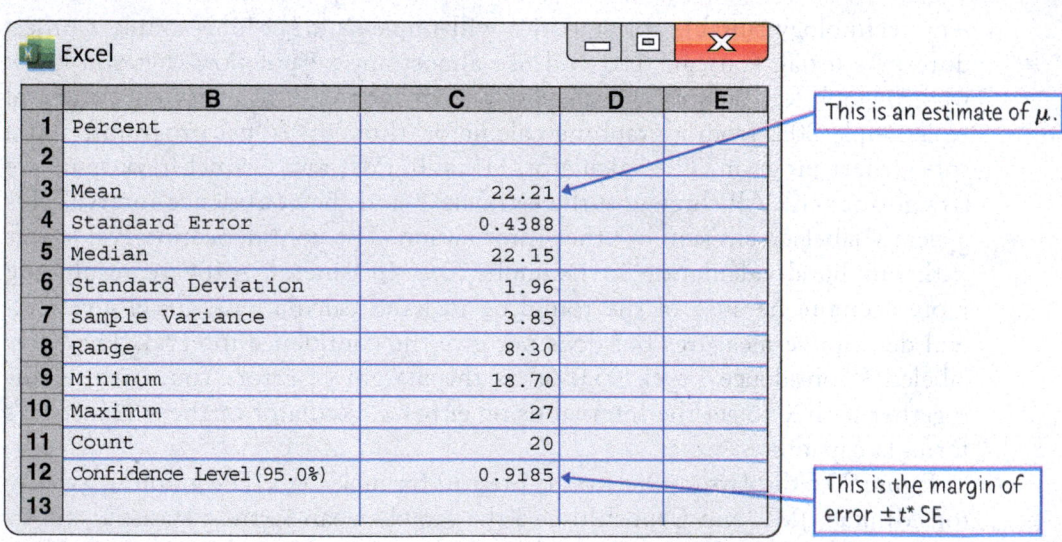

FIGURE 20.5
The *t* confidence interval, for Example 20.2: output from a graphing calculator, three statistical programs, and a spreadsheet program.

Texas Instruments

```
T-Test
 μ>88.0000
 t=0.9499
 p=.1813
 x̄=113.7500
 Sx=93.9043
 n=12.0000
```

jmp

Test Mean

Hypothesized Value	88
Actual Estimate	113.75
DF	11
Std Dev	93.9043

t Test

Test Statistic	0.9499
Prob > |t|	0.3626
Prob > t	0.1813
Prob < t	0.8187

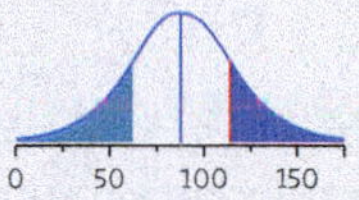

Minitab

One-Sample T: E. coli per 100 ml

```
Test of mu = 88 vs > 88

                                            95%
                                           Lower
Variable    N     Mean   StDev  SE Mean    Bound     T      P
26-Jul     12    113.8    93.9     27.1     65.1  0.95  0.181
```

CRUNCHIT!

Export

Null hypothesis:	Population mean = 88
Alternative hypothesis:	Population mean > 88
n:	12
Sample Mean:	113.8
Standard Error:	27.11
df:	11
t statistic:	0.9499
P-value:	0.1813

Excel

	B	C	D	E	F
1	E. coli per 100 mL				
2					
3	Mean	113.75	(C3-88)/(D4/SQRT(12))	0.94990954	
4	Standard Error	232.0004			
5	Median	1250			
6	Standard Deviation	93.90432752	TDIST (E3,11,1)	0.18128113	
7	Sample Variance	1076483.88			
8	Range	2985			
9	Minimum	15			
10	Maximum	3000			
11	Count	20			
12	Confidence Level (95%)	485.5825			
13					

This is the t statistic.

This is the P-value.

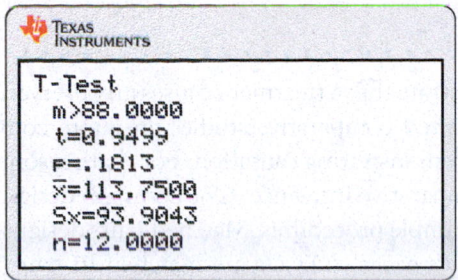

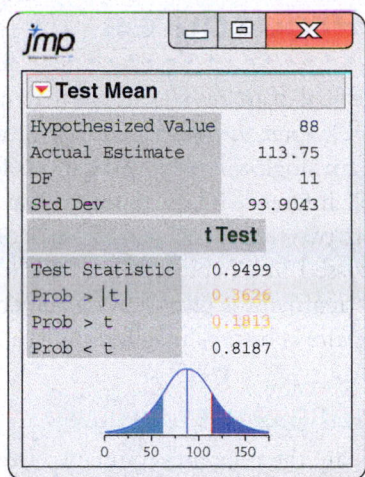

FIGURE 20.6
The t test, for Example 20.3: output from a graphing calculator, three statistical programs, and a spreadsheet program.

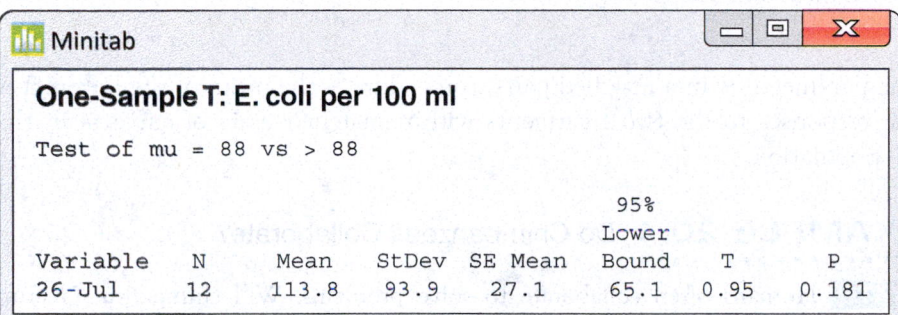

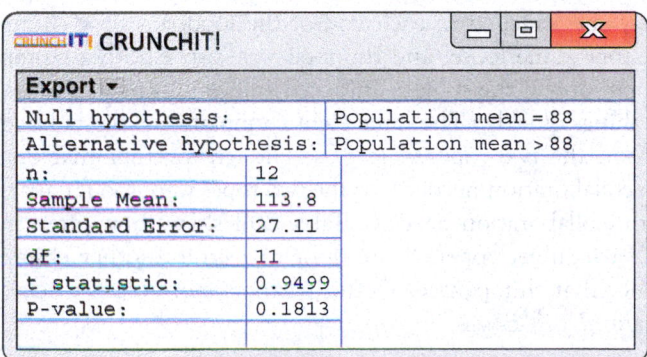

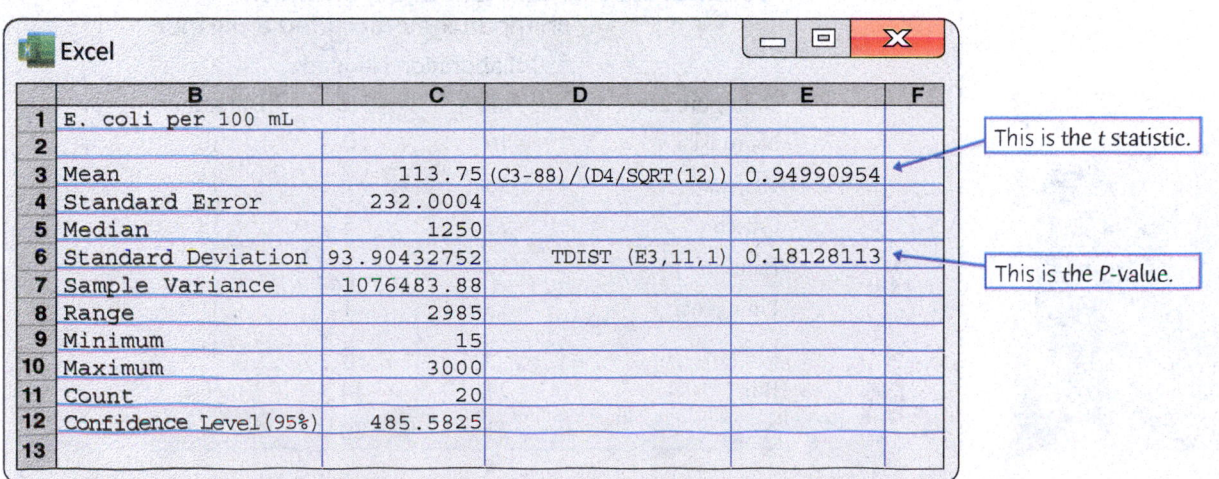

20.6 Matched Pairs *t* Procedures

Often, the goal of an investigation is to demonstrate that a treatment causes an observed effect. In Chapter 9, we learned that randomized comparative studies are more convincing than single-sample investigations for demonstrating causation. For that reason, one-sample inference is less common than comparative inference. One common design to compare two treatments makes use of one-sample procedures. Matched pairs designs were discussed in Chapter 9. In a *matched pairs design*, subjects are matched in pairs, and each treatment is given to one subject in each pair. Another situation calling for matched pairs is before-and-after observations on the same subjects.

Matched Pairs *t* Procedures

To compare the responses to the two treatments in a matched pairs design, find the *difference* between the responses within each pair. Then apply the one-sample *t* procedures to these differences.

The parameter μ in a matched pairs *t* procedure is the mean of the differences in the responses to the two treatments within matched pairs of subjects in the entire population.

EXAMPLE 20.4 Do Chimpanzees Collaborate?

4step

STATE: Humans often collaborate to solve problems. Will chimpanzees recruit another chimp when solving a problem requires collaboration? Researchers presented chimpanzee subjects with food outside their cage that they could bring within reach by pulling two ropes, one attached to each end of the food tray. If a chimp pulled only one rope, the rope came loose, and the food was lost. Another chimp was available as a partner, but only if the subject unlocked a door joining two cages. (Chimpanzees learn these things quickly.) The same eight chimpanzee subjects faced this problem in two versions: the two ropes were close enough together that one chimp could pull both (no collaboration needed), or the two ropes were too far apart for one chimp to pull both (collaboration needed). Table 20.1 shows how often in 24 trials for each version each subject opened the door to recruit another chimp as partner.[8] Is there evidence that chimpanzees recruit partners more often when a problem requires collaboration? CHIMPS

TABLE 20.1 Trials (out of 24) on which chimpanzees recruited a partner

	Collaboration Needed		
Chimpanzee	**Yes**	**No**	**Difference**
Namuiska	16	0	16
Kalema	16	1	15
Okech	23	5	18
Baluku	19	3	16
Umugenzi	15	4	11
Indi	20	9	11
Bili	24	16	8
Asega	24	20	4

Klett/Shutterstock

PLAN: Take μ to be the mean difference (collaboration required minus not) in the number of times a subject recruited a partner. The null hypothesis says that the need for collaboration has no effect, and H_a says that partners are recruited more often when the problem requires collaboration. So we test the hypotheses

$$H_0: \mu = 0$$
$$H_a: \mu > 0$$

SOLVE: The subjects are "semi-free-ranging chimpanzees at Ngamba Island Chimpanzee Sanctuary in Uganda." We are willing to regard them as an SRS from their species. To analyze the data, we examine the difference in the number of times a chimp recruited a partner so subtract the "no collaboration needed" count from the "collaboration needed" count for each subject. The eight differences form a single sample from a population with unknown mean μ . They appear in the "Difference" column in Table 20.1. All the chimpanzees recruited a partner more often when the ropes were too far apart to be pulled by one chimp.

The stemplot in Figure 20.7 creates the impression of a left-skewed distribution. This is a bit misleading as the dotplot in the bottom part of Figure 20.7 shows. As noted in Chapter 1, dotplots give a good picture of distributions with only whole-number values. We know that observations that can take only whole-number values cannot come from a Normal population.[9] In practice, researchers are willing to treat such observations as coming from a Normal population if there are more than just a few possible values and the distribution appears approximately Normal. Of course, we can't assess approximate Normality from just eight observations, but there are no signs of major departures from Normality. The researchers used the matched pairs t test.

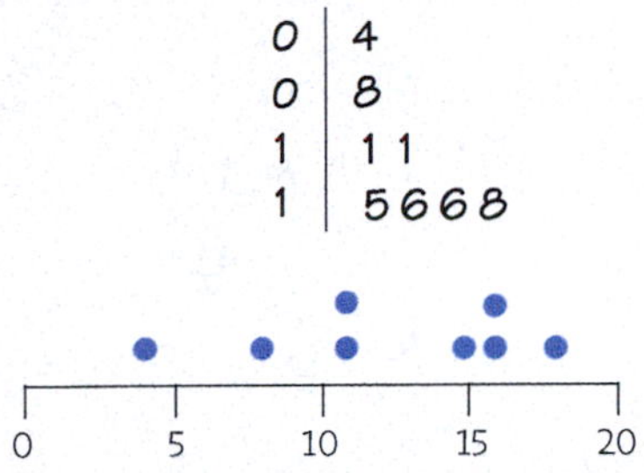

FIGURE 20.7 Stemplot and dotplot of the differences, for Example 20.4.

The eight differences have

$$\bar{x} = 12.375 \quad \text{and} \quad s = 4.749$$

The one-sample t statistic is therefore

$$t = \frac{\bar{x} - 0}{s/\sqrt{n}} = \frac{12.375 - 0}{4.749/\sqrt{8}} = 7.37$$

Find the P-value from the t_7 distribution. (Remember that the degrees of freedom are one less than the sample size.) Table C shows that 7.37 is greater than the critical value for one-sided $P = 0.0005$. The P-value is, therefore, less than 0.0005. Software says that $P = 0.000077$.

df = 7

t^*	4.785	5.408
One-sided P	0.001	0.0005

CONCLUDE: The data give very strong evidence ($P < 0.0005$) that chimpanzees recruit a collaborator more often when faced with a problem that requires a collaborator to solve. That is, chimpanzees recognize when collaboration is necessary, a skill that they share with humans.

Example 20.4 illustrates how to turn matched pairs data into single-sample data by taking differences within each pair. We are making inferences about a single population, the population of all differences within matched pairs. *It is incorrect to ignore the matching and analyze the data as if we had two samples of chimpanzees, one facing ropes close together and the other facing ropes far apart.* Inference procedures for comparing two samples assume that the samples are selected independently of each other. This condition does not hold when the same subjects are measured twice. The proper analysis depends on the design used to produce the data.

APPLY YOUR KNOWLEDGE

Many exercises from this point on ask you to give the P-value of a t test. If you have suitable technology, give the exact P-value. Otherwise, use Table C to give two values between which P lies.

4step

20.11 Eye Grease. Athletes performing in bright sunlight often smear black eye grease under their eyes to reduce glare. Does eye grease work? In one study, 16 student subjects took a test of sensitivity to contrast after three hours facing into bright sun, both with and without eye grease. (Greater sensitivity to contrast improves vision, and glare reduces sensitivity to contrast.) This is a matched pairs design. Here are the differences in sensitivity, with eye grease minus without eye grease:[10] EYEGRS

0.07	0.64	−0.12	−0.05	−0.18	0.14	−0.16	0.03
0.05	0.02	0.43	0.24	−0.11	0.28	0.05	0.29

We want to know whether eye grease increases sensitivity to contrast on the average. Do the data support this idea? Complete the Plan, Solve, and Conclude steps of the four-step process, following the model of Example 20.4.

20.12 Eye Grease (continued). How much more sensitive to contrast are athletes with eye grease than without eye grease? Give a 95% confidence interval to answer this question. EYEGRS

20.7 Robustness of *t* Procedures

The t confidence interval and test are exactly correct when the distribution of the population is exactly Normal. No real data are exactly Normal. At best, the Normal distribution is an excellent approximation to the actual distribution of data from real studies.[11] The usefulness of the t procedures in practice therefore depends on how strongly they are affected by lack of Normality.

Robust Procedures

A confidence interval or significance test is called **robust** if the confidence level or P-value does not change very much when the conditions for use of the procedure are violated.

The condition that the population is Normal effectively rules out outliers in an SRS, so the presence of outliers shows that this condition is not fulfilled. The t

procedures are not robust against outliers unless the sample is large because $\bar{x}$ and s are not resistant to outliers.

Fortunately, the t procedures are quite robust against non-Normality of the population except when outliers or strong skewness are present. (Skewness is more serious than other kinds of non-Normality.) As the size of the sample increases, the central limit theorem ensures that the distribution of the sample mean $\bar{x}$ becomes more nearly Normal and that the t distribution becomes more accurate for critical values and P-values of the t procedures.

Always make a plot to check for skewness and outliers before you use the t procedures for small samples. For most purposes, you can safely use the one-sample t procedures when $n \geq 15$ unless an outlier or quite strong skewness is present. Here are practical guidelines for inference on a single mean.[12]

STATISTICS IN YOUR WORLD

Catching Cheaters

A certification test for surgeons asks 277 multiple-choice questions. Smith and Jones have 193 common right answers and 53 identical wrong choices. The computer flags their 246 identical answers as evidence of possible cheating. They sue. The court wants to know how unlikely it is that exams this similar would occur just by chance. That is, the court wants a P-value. Statisticians offer several P-values based on different models for the exam-taking process. They all say that results this similar would almost never happen just by chance. Smith and Jones fail the exam.

Using the t Procedures

Except in the case of small samples, the condition that the data are an SRS from the population of interest is more important than the condition that the population distribution is Normal.

- *Sample size less than 15*: Use t procedures if the data appear to be close to Normal (roughly symmetric, single-peaked, no outliers). If the data are clearly skewed or if outliers are present, do not use t.
- *Sample size at least 15*: The t procedures can be used except in the presence of outliers or strong skewness.
- *Large samples:* The t procedures can be used even for clearly skewed distributions when the sample is large, roughly $n \geq 40$.

EXAMPLE 20.5 Can We Use t?

Figure 20.8 shows plots of several data sets. For which of these can we safely use the t procedures?[13]

- Figure 20.8(a) is a histogram of the percentage of each state's adult residents who are college graduates. *We have data on the entire population of 50 states, so inference is not needed.* We can calculate the exact mean for the population (of states, not individual people). There is no uncertainty due to having only a sample from the population, and there is no need for a confidence interval or test. *If these data were an SRS from a larger population, t inference would be safe despite the mild skewness because $n = 50$.*
- Figure 20.8(b) is a stemplot of the force required to pull apart 20 pieces of Douglas fir. *The data are strongly skewed to the left, and there may possibly be low outliers, so we cannot trust the t procedures for $n = 20$.*
- Figure 20.8(c) is a stemplot of the lengths of 23 specimens of the red variety of the tropical flower *Heliconia*. *The data are mildly skewed to the right, and there are no outliers. We can use the t distributions for such data.*
- Figure 20.8(d) is a histogram of the heights of the female students in a college class. *This distribution is quite symmetric and appears to be close to Normal. We can use the t procedures for any sample size.*

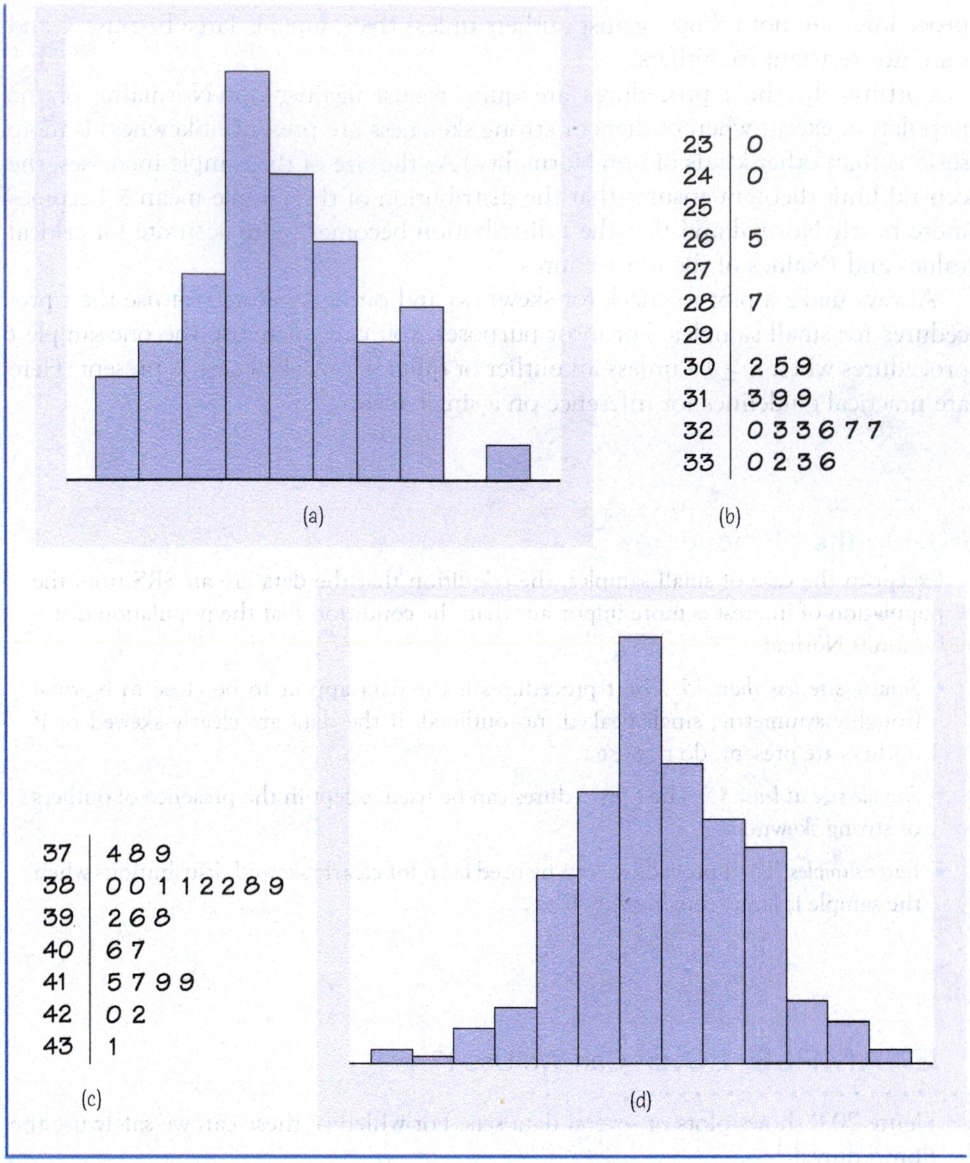

FIGURE 20.8
Can we use *t* procedures for these data? (a) Percentage of adult college graduates in the 50 states. *No*, this is an entire population, not a sample. (b) Force required to pull apart 20 pieces of Douglas fir. *No*, there are just 20 observations and strong skewness. (c) Lengths of 23 tropical flowers of the same variety. *Yes*, the sample is large enough to overcome the mild skewness. (d) Heights of college students. *Yes, for any size sample*, because the distribution is close to Normal.

What can we do if plots suggest that the data are clearly not Normal, especially when we have only a few observations? This is not a simple question. Here are the basic options:

1. If lack of Normality is due to outliers, it may be legitimate to *remove outliers* if you have reason to think that they do not come from the same population as the other observations. Equipment failure that produced a bad measurement, for example, entitles you to remove the outlier and analyze the remaining data. *But if an outlier appears to be "real data," you should not arbitrarily remove it.*

2. In some settings, other standard distributions replace the Normal distributions as models for the overall pattern in the population. The lifetimes in service of equipment or the survival times of cancer patients after treatment usually have right-skewed distributions. Statistical studies in these areas use families of right-skewed distributions rather than Normal distributions. There are inference procedures for the parameters of these distributions that replace the *t* procedures.

3. Modern *bootstrap methods* and *permutation tests* use heavy computing to avoid requiring Normality or any other specific form of sampling distribution. We recommend these methods unless the sample is so small that it may not represent the population well. For an introduction, see Chapter 32, available online as an optional companion chapter.
4. Finally, there are other nonparametric methods, which do not assume any specific form for the distribution of the population. Unlike bootstrap and permutation methods, common nonparametric methods do not make use of the actual values of the observations. For an introduction, see Chapter 28, available online as an optional companion chapter.

APPLY YOUR KNOWLEDGE

20.13 Diamonds. A group of earth scientists studied the small diamonds found in a nodule of rock.[14] Table 20.2 presents data on the nitrogen content (parts per million) and the abundance of carbon-13 in these diamonds. Carbon-12 and carbon-13 are forms of the element carbon, the primary component of diamonds. Carbon-12 makes up almost 99% of natural carbon. The abundance of carbon-13 is measured by the ratio of carbon-13 to carbon-12, in parts per million more or less than a standard. (The minus signs in the data mean that the ratio is smaller in these diamonds than in standard carbon.)

Eric Nathan/Alamy

We would like to estimate the mean abundance of both nitrogen and carbon-13 in the population of diamonds represented by this sample. Examine the data for nitrogen. Can we use a *t* confidence interval for mean nitrogen? Explain your answer. Give a 95% confidence interval if you think the result can be trusted. DMNDS

20.14 Diamonds (continued). Examine the data in Table 20.2 on abundance of carbon-13. Can we use a *t* confidence interval for mean carbon-13? Explain your answer. Give a 95% confidence interval if you think the result can be trusted. DMNDS

TABLE 20.2 Nitrogen and carbon-13 in a sample of diamonds

Diamond	Nitrogen (ppm)	Carbon-13 Ratio	Diamond	Nitrogen (ppm)	Carbon-13 Ratio
1	487	−2.78	13	273	−2.73
2	1430	−1.39	14	94	−2.33
3	60	−4.26	15	69	−3.83
4	244	−1.19	16	262	−2.04
5	196	−2.12	17	120	−2.82
6	274	−2.87	18	302	−0.84
7	41	−3.68	19	75	−3.57
8	54	−3.29	20	242	−2.42
9	473	−3.79	21	115	−3.89
10	30	−4.06	22	65	−3.87
11	98	−1.83	23	311	−1.58
12	41	−4.03	24	61	−3.97

CHAPTER 20 SUMMARY

- Tests and confidence intervals for the mean μ of a Normal population are based on the sample mean $\bar{x}$ of an SRS. Because of the central limit theorem, the resulting procedures are approximately correct for other population distributions when the sample is large.
- The standardized sample mean is the one-sample z statistic

$$z = \frac{\bar{x} - \mu}{\sigma/\sqrt{n}}$$

  If we knew σ, we would use the z statistic and the standard Normal distribution.
- In practice, we do not know σ. Replace the standard deviation $\sigma/\sqrt{n}$ of $\bar{x}$ by the **standard error** $s/\sqrt{n}$ to get the **one-sample *t* statistic**

$$t = \frac{\bar{x} - \mu}{s/\sqrt{n}}$$

  The t statistic has the ***t* distribution** with $n-1$ degrees of freedom.
- There is a t distribution for every positive degree of freedom. All are symmetric distributions similar in shape to the standard Normal distribution. The t distribution approaches the $N(0,1)$ distribution as the degrees of freedom increase.
- A level C confidence interval for the mean μ of a Normal population is

$$\bar{x} \pm t^* \frac{s}{\sqrt{n}}$$

- The critical value t^* is chosen so that the t curve with $n-1$ degrees of freedom has area C between $-t^*$ and t^*.
- Significance tests for $H_0: \mu = \mu_0$ are based on the t statistic. Use P-values or fixed significance levels from the t_{n-1} distribution.
- Use these one-sample procedures to analyze matched pairs data by first taking the difference within each matched pair to produce a single sample.
- The t procedures are quite robust when the population is non-Normal, especially for larger sample sizes. The t procedures are useful for non-Normal data when $n \geq 15$ unless the data show outliers or strong skewness. When $n \geq 40$, the t procedures can be used even for clearly skewed distributions.

CHECK YOUR SKILLS

20.15 We prefer the t procedures to the z procedures for inference about a population mean because

(a) z requires that you know the observations are from a Normal population, while t does not.

(b) z requires that you know the population standard deviation σ, while t does not.

(c) z requires that you can regard your data as an SRS from the population, while t does not.

20.16 You are testing $H_0: \mu = 100$ against $H_a: \mu < 100$ based on an SRS of 25 observations from a Normal population. The data give $\bar{x} = 98.32$ and $s = 4$. The value of the t statistic is

(a) -10.5. (b) -2.1. (c) -0.525.

20.17 You are testing $H_0: \mu = 100$ against $H_a: \mu > 100$ based on an SRS of 25 observations from a Normal population. The t statistic is $t = 2.10$. The degrees of freedom for the t statistic are

(a) 24.

(b) 25.

(c) 26.

20.18 The P-value for the statistic in the previous question

(a) falls between 0.05 and 0.10.

(b) falls between 0.01 and 0.05.

(c) is less than 0.01.

20.19 You have an SRS of nine observations from a Normally distributed population. What critical value would you use to obtain an 80% confidence interval for the mean μ of the population?

(a) 1.397

(b) 1.383

(c) 1.372

20.20 You are testing $H_0: \mu = 100$ against $H_a: \mu \neq 100$ based on an SRS of nine observations from a Normal population. What values of the t statistic are statistically significant at the $\alpha = 0.005$ level?

(a) $t \geq 3.833$

(b) $t \leq -3.833$ or $t \geq 3.833$

(c) $t \leq -3.690$ or $t \geq 3.690$

20.21 Twenty-five adult citizens of the United States were asked to estimate the average income of all U.S. households. The mean estimate was $\bar{x} = \$70,000$ and $s = \$15,000$. (*Note*: The actual average household income at the time of the study was about \$90,000.) Assume the 25 adults in the study can be considered an SRS from the population of all adult citizens of the United States. A 95% confidence interval for the mean estimate of the average income of all U.S. households is

(a) \$67,000 to \$73,000.

(b) \$63,808 to \$76,192.

(c) \$83,808 to \$96,192.

20.22 Which of the following would cause the most worry about the validity of the confidence interval you calculated in the previous exercise?

(a) You notice that there is a clear outlier in the data.

(b) A stemplot of the data is bell-shaped, so a z procedure is the appropriate procedure.

(c) You do not know the population standard deviation σ.

20.23 Which of these settings does *not* allow use of a matched pairs t procedure?

(a) You interview both the instructor and one of the students in each of 20 introductory statistics classes and ask each how many hours per week homework assignments require.

(b) You interview a sample of 15 instructors and another sample of 15 students and ask each how many hours per week homework assignments require.

(c) You interview 40 students in the introductory statistics course at the beginning of the semester and again at the end of the semester and ask how many hours per week homework assignments require.

20.24 Because the t procedures are robust, the most important condition for their safe use is that

(a) the sample size is at least 15.

(b) the population distribution is exactly Normal.

(c) the data can be regarded as an SRS from the population.

CHAPTER 20 EXERCISES

20.25 Read Carefully. You read in the report of a psychology experiment: "Separate analyses for our two groups of 12 participants revealed no overall placebo effect for our student group (mean = 0.08, SD = 0.37, $t_{11} = 0.49$) and a significant effect for our non-student group (mean = 0.35, SD = 0.37, $t_{11} = 3.25$, $p < 0.01$)."[15] The null hypothesis is that the mean effect is zero. What are the correct values of the two t statistics based on the means and standard deviations? Compare each correct t-value with the critical values in Table C. What can you say about the two-sided P-value in each case?

20.26 Body Mass Index of Young Men. In Example 16.1 (page 368), we developed a 95% z confidence interval for the mean body mass index (BMI) of American men aged 20–29 years, based on a national random sample of 936 such men. We assumed there that the population standard deviation was known to be $\sigma = 11.6$. In fact, the sample data had mean BMI $\bar{x} = 27.2$ and standard deviation $s = 11.63$. What is the 95% t confidence interval for the mean BMI of all American young men?

20.27 Mathematics Scores in Dallas. The Trial Urban District Assessment (TUDA) is a government-sponsored study of student achievement in large urban school districts. TUDA gives a mathematics test scored from 0 to 500. A score of 262 is a "basic" mathematics level, and a score of 299 is "proficient." Scores for a random sample of 1100 eighth-graders in Dallas had $\bar{x} = 264$ with standard error 1.3.[16]

(a) We don't have the 1100 individual scores, but use of the t procedures is surely safe. Why?

(b) Give a 99% confidence interval for the mean score of all Dallas eighth-graders. (Be careful: the report gives the standard error of $\bar{x}$, not the standard deviation s.)

(c) Based on your answer in (b), is there good evidence that the mean for all Dallas eighth-graders is different from the basic level? (Just as for z procedures, a level C t confidence interval for a mean μ can be used to test

$$H_0: \mu = \mu_0$$
$$H_a: \mu \neq \mu_0$$

at level $1 - C$ by checking whether μ_0 is in the interval.)

20.28 Color and Cognition. In a randomized comparative experiment on the effect of color on the performance of a cognitive task, researchers randomly divided 69 subjects (27 males and 42 females ranging in age from 17 to 25 years) into three groups. Participants were asked to solve a series of six anagrams. One group was presented with the anagrams on a blue screen, one group saw them on a red screen, and one group had a neutral screen. The time, in seconds, taken to solve the anagrams was recorded. The paper reporting the study gives $\bar{x} = 11.58$ and $s = 4.37$ for the times of the 23 members of the neutral group.[17]

(a) Give a 95% confidence interval for the mean time in the population from which the subjects were recruited.

(b) What conditions for the population and the study design are required by the procedure you used in part (a)? Which of these conditions are important for the validity of the procedure in this case?

20.29 Wearable Technology and Weight Loss. Do wearable devices that monitor diet and physical activity help people lose weight? Researchers had 237 subjects who were already involved in a program of diet and exercise use wearable technology for 24 months. They measured their weight (in kilograms) before using the technology and 24 months after using the technology.[18]

(a) Explain why the proper procedure to compare the mean weight before using the wearable technology and 24 months after using the wearable technology is a matched pairs *t* test.

(b) The 237 differences in weight (weight after 24 months minus weight before using the wearable technology) had $\bar{x} = -3.5$ and $s = 7.8$. Is there significant evidence of a reduction in weight after using the wearable technology?

20.30 Inspecting Ground Chuck. The National Institute of Standards and Technology (NIST) publishes procedures for inspecting the advertised contents of packaged goods.[19] The discussion of these procedures includes an example of inspecting the weights of a sample of packages of ground beef from a particular grocer. An SRS of 12 packages are weighed (in pounds) according to the NIST procedures. The table shows each of the actual weights along with the weight listed on the package label. CHUCK

Sample	1	2	3	4	5	6	7	8	9	10	11	12
Labeled weight	1.85	1.21	1.56	1.98	1.07	1.55	1.02	1.44	1.33	2.03	1.73	1.16
Measured weight	1.67	1.14	1.48	1.84	0.84	1.39	1.00	1.19	1.17	1.83	1.59	1.05

(a) Why is this a matched pairs design? Explain your answer.

(b) Make a stemplot of the differences between labeled and measured weights (measured weight minus labeled weight). Are there outliers or strong skewness that would preclude use of the *t* procedures?

(c) Carry out a *t* test to see if the mean difference in the measured weight minus the labeled weight is less than 0.

20.31 Exhaust from School Buses. In a study of exhaust emissions from school buses, the pollution intake by passengers was determined for a sample of nine school buses used in the Southern California Air Basin. The pollution intake is the amount of exhaust emissions, in grams per person per million grams emitted, that would be inhaled while traveling on the bus during its usual 18-mile trip on congested freeways from South Central LA to a magnet school in West LA. (In comparison, a city of 1 million people will inhale a total of about 12 grams of exhaust per million grams emitted.) Here are the amounts for the nine buses when driven with the windows open:[20] EMIT

1.15 0.33 0.40 0.33 1.35 0.38 0.25 0.40 0.35

(a) Make a stemplot. Are there outliers or strong skewness that would preclude use of the *t* procedures?

(b) A good way to judge the effect of outliers is to do your analysis twice, once with the outliers and a second time without them. Give two 90% confidence intervals, one with all the data and one with the outliers removed, for the mean pollution intake among all school buses used in the Southern California Air Basin that travel the route investigated in the study.

(c) Compare the two intervals in part (b). What is the most important effect of removing the outliers?

20.32 A Big Toe Problem. Hallux abducto valgus (call it HAV) is a deformation of the big toe that often requires surgery. Doctors used X-rays to measure the angle (in degrees) of deformity in 38 consecutive patients under the age of 21 who came to a medical center for surgery to correct HAV. The angle is a measure of the seriousness of the deformity. Here are the data:[21] BIGTOE

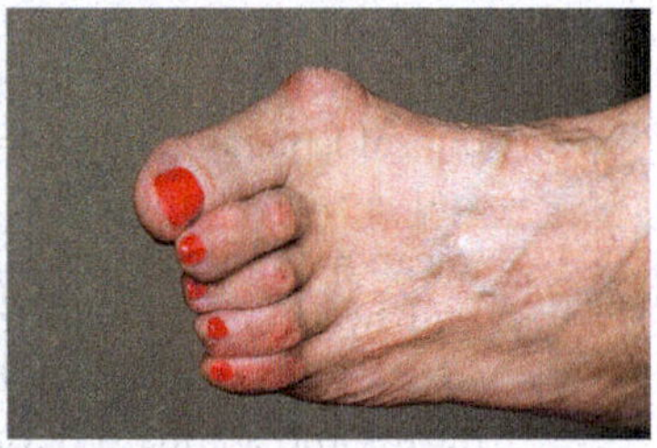

Cristina Lichti/Alamy

28 32 25 34 38 26 25 18 30 26 28 13 20
21 17 16 21 23 14 32 25 21 22 20 18 26
16 30 30 20 50 25 26 28 31 38 32 21

It is reasonable to regard these patients as a random sample of young patients who require HAV surgery? Carry out the Solve and Conclude steps for a 95% confidence interval for the mean HAV angle in the population of all such patients.

20.33 An Outlier's Effect. Our bodies have a natural electrical field that is known to help wounds heal. Does changing the field strength slow healing? A series of experiments with newts investigated this question. In one experiment, the two hind limbs of 12 newts were assigned at random to either experimental or control groups. This is a matched pairs design. The electrical field in the experimental limbs was reduced to zero by applying a voltage. The control limbs were left alone. Here are the rates at which new cells closed a razor cut in each limb, in micrometers per hour:[22] NEWTS

Newt	1	2	3	4	5	6	7	8	9	10	11	12
Control limb	36	41	39	42	44	39	39	56	33	20	49	30
Experimental limb	28	31	27	33	33	38	45	25	28	33	47	23

(a) Why is this a matched pairs design? Explain your answer.

(b) Make a stemplot of the differences between limbs of the same newt (control limb minus experimental limb). There is a high outlier.

(c) A good way to judge the effect of an outlier is to do your analysis twice, once with the outlier and a second time without it. Carry out two t tests to see if the mean healing rate is significantly lower in the experimental limbs, with one test including all 12 newts and another omitting the outlier. What are the test statistics and their P-values? Does the outlier have a strong influence on your conclusion?

(d) (Optional) If you covered optional Section 2.6, make a modified boxplot of the differences between limbs of the same newt (control limb minus experimental limb). You should find two suspected outliers. Carry out two t tests to see if the mean healing rate is significantly lower in the experimental limbs, with one test including all 12 newts (which you already did in part (c)) and another omitting both outliers. What are the test statistics and their P-values? Do the outliers have a strong influence on your conclusion? How do these results compare with what you found in part (c)?

20.34 An Outlier's Effect. A good way to judge the effect of an outlier is to do your analysis twice, once with the outlier and a second time without it. The data in Exercise 20.32 follow a Normal distribution quite closely except for one patient with HAV angle 50 degrees, a high outlier. BIGTOE

(a) Find the 95% confidence interval for the population mean based on the 37 patients who remain after you drop the outlier.

(b) Compare your interval in part (a) with your interval from Exercise 20.32. What is the most important effect of removing the outlier?

20.35 Men of Few Words? Researchers claim that American women speak significantly more words per day than American men. One estimate is that a woman uses about 20,000 words per day, while a man uses about 7,000. To investigate such claims, one study used a special device to record the conversations of male and female university students over a four-day period. From these recordings, the daily word counts of the 20 men in the study were determined. Here are their daily word counts:[23] TALKING

28,408	10,084	15,931	21,688	37,786
10,575	12,880	11,071	17,799	13,182
8,918	6,495	8,153	7,015	4,429
10,054	3,998	12,639	10,974	5,255

(a) Examine the data. Is it reasonable to use the t procedures (assuming that these men are an SRS of all male students at this university)?

(b) If your conclusion in part (a) is Yes, do the data give convincing evidence that the mean number of words per day of men at this university differs from 7,000?

20.36 Genetic Engineering for Cancer Treatment. Here's a new idea for treating advanced melanoma, the most serious kind of skin cancer: genetically engineer white blood cells to better recognize and destroy cancer cells and then infuse these cells into patients. The subjects in a small initial study were 11 patients whose melanoma had not responded to existing treatments. One question was how rapidly the new cells would multiply after infusion, as measured by the doubling time in days. Here are the doubling times:[24] CNCRTRT

1.4	1.0	1.3	1.0	1.3	2.0	0.6	0.8	0.7	0.9	1.9

(a) Examine the data. Is it reasonable to use the t procedures?

(b) Give a 90% confidence interval for the mean doubling time. Are you willing to use this interval to make an inference about the mean doubling time in a population of similar patients? Explain your reasoning.

20.37 Genetic Engineering for Cancer Treatment (continued). Another outcome in the cancer experiment described in Exercise 20.36 is measured by a test for the presence of cells that trigger an immune response in the body and so may help fight cancer. Here are data for the 11 subjects: counts of active immune triggering cells per 100,000 cells before and after infusion of the modified cells. The difference (after minus before) is the response variable. MORECAN

Before	14	0	1	0	0	0	0	20	1	6	0
After	41	7	1	215	20	700	13	530	35	92	108
Difference	27	7	0	215	20	700	13	510	34	86	108

(a) Explain why this is a matched pairs design.

(b) Examine the data. Is it reasonable to use the *t* procedures?

(c) If your conclusion in part (a) is Yes, do the data give convincing evidence that the count of active cells is higher after treatment?

20.38 Kicking a Helium-Filled Football. Does a football filled with helium travel farther than one filled with ordinary air? To test this, the *Columbus Dispatch* conducted a study. Two identical footballs, one filled with helium and one filled with ordinary air, were used. A casual observer was unable to detect a difference in the two footballs. A novice kicker was used to punt the footballs. A trial consisted of kicking both footballs in a random order. The kicker did not know which football (the helium-filled or the air-filled football) he was kicking. The distance of each punt was recorded. Then another trial was conducted. A total of 39 trials were run. Here are the data for the 39 trials, in yards that the footballs traveled. The difference (helium minus air) is the response variable.[25] FTBALL

Helium	25	16	25	14	23	29	25	26	22	26
Air	25	23	18	16	35	15	26	24	24	28
Difference	0	−7	7	−2	−12	14	−1	2	−2	−2
Helium	12	28	28	31	22	29	23	26	35	24
Air	25	19	27	25	34	26	20	22	33	29
Difference	−13	9	1	6	−12	3	3	4	2	−5
Helium	31	34	39	32	14	28	30	27	33	11
Air	31	27	22	29	28	29	22	31	25	20
Difference	0	7	17	3	−14	−1	8	−4	8	−9
Helium	26	32	30	29	30	29	29	30	26	
Air	27	26	28	32	28	25	31	28	28	
Difference	−1	6	2	−3	2	4	−2	2	−2	

(a) Examine the data. Is it reasonable to use the *t* procedures?

(b) If your conclusion in part (a) is Yes, do the data give convincing evidence that the helium-filled football travels farther than the air-filled football?

20.39 Growing Trees Faster. The concentration of carbon dioxide (CO_2) in the atmosphere is increasing rapidly due to our use of fossil fuels. Because plants use CO_2 to fuel photosynthesis, more CO_2 may cause trees and other plants to grow faster. An elaborate apparatus allows researchers to pipe extra CO_2 to a 30-meter circle of forest. They selected two nearby circles in each of three parts of a pine forest and randomly chose one of each pair to receive extra CO_2. The response variable is the mean increase in base area for 30 to 40 trees in a circle during a growing season. We measure this in percentage increase per year. Here are one year's data:[26] TREES

Pair	Control Plot	Treated Plot	Treated − Control
1	9.752	10.587	0.835
2	7.263	9.244	1.981
3	5.742	8.675	2.933

(a) State the null and alternative hypotheses. Explain clearly why the investigators used a one-sided alternative.

(b) Use a *t* procedure to carry out a test and report your conclusion in simple language.

(c) The investigators used the test you just carried out. Any use of the *t* procedures with samples this size is risky. Why?

20.40 Fungus in the Air. The air in poultry-processing plants often contains fungus spores. Inadequate ventilation can affect the health of the workers. The problem is most serious during the summer. To measure the presence of spores, air samples are pumped to an agar plate and "colony-forming units (CFUs)" are counted after an incubation period. Here are data from two locations in a plant that processes 37,000 turkeys per day, taken on four days in the summer. The units are CFUs per cubic meter of air.[27] FUNGUS

	Day 1	Day 2	Day 3	Day 4
Kill room	3175	2526	1763	1090
Processing	529	141	362	224
Kill room − processing	2646	2385	1401	866

(a) Explain carefully why these are matched pairs data.

(b) The spore count is clearly higher in the kill room. Give sample means and a 90% confidence interval to estimate how much higher. Be sure to state your conclusion in plain language.

(c) You will often see the *t* procedures used for data like these. You should regard the results as only rough approximations. Why?

20.41 Weeds Among the Corn. Velvetleaf is a particularly annoying weed in corn fields. It produces lots of seeds, and the seeds wait in the soil for years until conditions are right. How many seeds do velvetleaf plants produce? Here are counts from 28 plants that came up in a corn field when no herbicide was used:[28] WEEDS

REDA &CO srl/Alamy

2450	2504	2114	1110	2137	8015	1623	1531	2008	1716
721	863	1136	2819	1911	2101	1051	218	1711	164
2228	363	5973	1050	1961	1809	130	880		

We would like to give a confidence interval for the mean number of seeds produced by velvetleaf plants in the corn field. Alas, the t interval can't be safely used for these data. Why not?

20.42 Sweetening Colas. Cola makers test recipes for loss of sweetness during storage. Trained tasters rate the sweetness using a "sweetness score" of 1 to 10, with larger scores corresponding to greater sweetness. They rate the sweetness of the cola before and after storage. Here are the sweetness losses (sweetness score before storage minus sweetness score after storage) found by 10 tasters for a cola recipe currently on the market: COLA

1.6	0.4	0.5	−2.0	1.5	−1.1	1.3	−0.1	−0.3	1.2

Take the data from these 10 carefully trained tasters as an SRS from a large population of all trained tasters.

(a) Use these data to see if there is good evidence that the cola lost sweetness.

(b) It is not uncommon to see the t procedures used for data like these. However, you should regard the results as only rough approximations. Why?

20.43 How Much Oil? How much oil will ultimately be produced by wells in a given field is key information in deciding whether to drill more wells. Here are the estimated total amounts of oil recovered from 64 wells in the Devonian Richmond Dolomite area of the Michigan basin, in thousands of barrels:[29] OIL

21.7	53.2	46.4	42.7	50.4	97.7	103.1	51.9
43.4	69.5	156.5	34.6	37.9	12.9	2.5	31.4
79.5	26.9	18.5	14.7	32.9	196	24.9	118.2
82.2	35.1	47.6	54.2	63.1	69.8	57.4	65.6
56.4	49.4	44.9	34.6	92.2	37.0	58.8	21.3
36.6	64.9	14.8	17.6	29.1	61.4	38.6	32.5
12.0	28.3	204.9	44.5	10.3	37.7	33.7	81.1
12.1	20.1	30.5	7.1	10.1	18.0	3.0	2.0

Take these wells to be an SRS of wells in this area.

(a) Give a 95% t confidence interval for the mean amount of oil recovered from all wells in this area.

(b) Make a graph of the data. The distribution is very skewed, with several high outliers. A computer-intensive method that gives accurate confidence intervals without assuming any specific shape for the distribution gives a 95% confidence interval of 40.28 to 60.32. How does the t interval compare with this? Should the t procedures be used with these data?

20.44 *E. coli* in Swimming Areas. To investigate water quality, in early September 2016, the Ohio Department of Health took water samples at 24 beaches on Lake Erie in Erie County. Those samples were tested for fecal coliform, which are *E. coli* bacteria found in human and animal feces. An unsafe level of fecal coliform means there's a higher chance that disease-causing bacteria are present and more risk that a swimmer will become ill if she or he should accidentally ingest some of the water. Ohio considers it unsafe for swimming if a 100-milliliter sample (about 3.3 ounces) of water contains more than 400 coliform bacteria. Here are the *E. coli* levels found by the laboratories:[30] ECOLI

18.7	579.4	1986.3	517.2	98.7	45.7	124.6	201.4
19.9	83.6	365.4	307.6	285.1	152.9	18.7	151.5
365.4	238.2	209.8	290.9	137.6	1046.2	127.4	224.7

Take these water samples to be an SRS of the water in all swimming areas in Erie County.

(a) Are these data good evidence that on average the *E. coli* levels in these swimming areas were safe?

(b) Make a graph of the data. The distribution is very skewed. Another method that gives P-values without assuming any specific shape for the distribution gives a P-value of 0.0043 for the question in part (a). How does the one-sample t test compare with this? Should the t procedures be used with these data?

*The following exercises ask you to answer questions from data without having the details outlined for you. The four-step process is illustrated in Examples 20.2, 20.3, and 20.4. The exercise statements give you the **State** step. Follow the **Plan**, **Solve**, and **Conclude** steps in your work.*

20.45 Natural Weed Control? 4step Fortunately, farmers aren't really interested in the number of seeds velvetleaf plants produce (see Exercise 20.41). The velvetleaf seed beetle feeds on the seeds and might be a natural weed control. Here are the total seeds, seeds consumed by the beetle, and percentage of seeds consumed for 28 velvetleaf plants: WDCTRL

Seeds	2450	2504	2114	1110	2137	8015	1623	1531	2008	1716
Infected	135	101	76	24	121	189	31	44	73	12
Percent	5.5	4.0	3.6	2.2	5.7	2.4	1.9	2.9	3.6	0.7
Seeds	721	863	1136	2819	1911	2101	1051	218	1711	164
Infected	27	40	41	79	82	85	42	0	64	7
Percent	3.7	4.6	3.6	2.8	4.3	4.0	4.0	0.0	3.7	4.3
Seeds	2228	363	5973	1050	1961	1809	130	880		
Infected	156	31	240	91	137	92	5	23		
Percent	7.0	8.5	4.0	8.7	7.0	5.1	3.8	2.6		

Do a complete analysis of the percentage of seeds infected by the beetle. Include a 90% confidence interval for the mean percentage infected in the population of all velvetleaf plants. Do you think that the beetle is very helpful in controlling the weed? Why is analyzing percentage of seeds infected more useful than analyzing number of seeds infected?

20.46 Recruiting T Cells. There is evidence that cytotoxic T lymphocytes (T cells) participate in controlling tumor growth and that they can be harnessed to use the body's immune system to treat cancer. One study investigated the use of a T cell-engaging antibody, blinatumomab, to recruit T cells to control tumor growth. The data below are T cell counts (1000 per microliter) at baseline (beginning of the study) and after 20 days on blinatumomab for six subjects in the study.[31] The difference (after 20 days minus baseline) is the response variable. TCELLS

Baseline	0.04	0.02	0.00	0.02	0.38	0.33
After 20 days	0.28	0.47	1.30	0.25	1.22	0.44
Difference	0.24	0.45	1.30	0.23	0.84	0.11

Do the data give convincing evidence that the mean count of T cells is higher after 20 days on blinatumomab?

20.47 Recruiting T Cells (continued). Give a 95% confidence interval for the mean difference in T cell counts (after 20 days minus baseline) in the previous exercise. TCELLS

20.48 Mutual Funds Performance. Mutual funds often compare their performance with a benchmark provided by an "index" that describes the performance of the class of assets in which the fund invests. For example, the Vanguard International Growth Fund benchmarks its performance against the Spliced International Index. Table 20.3 gives annual returns (percent) for the fund and the index. Does the fund's mean annual percentage return differ significantly from that of its benchmark? MFUND

(a) Explain clearly why the matched pairs t test is the proper choice to answer this question.

(b) Do a complete analysis that answers the question posed.

(c) The paired t test uses the mean of the differences in the annual percentage returns to compare the fund's performance to that of its benchmark over time. What do you think of this as way to compare performance over time? (Consider a simple example over two years. Fund 1 has a 50% annual return in year 1 and a −50% return in year 2. The benchmark has a 0% annual return both years. The mean of the differences in annual percentage returns is 0. However, what is an initial \$100 investment in Fund 1 worth after the two years? What is an initial \$100 investment in the benchmark worth after two years?)

20.49 Right Versus Left. The design of controls and instruments affects how easily people can use them. Timothy Sturm investigated this effect in a course project, asking 25 right-handed students to turn a knob (with their right hands) that moved an indicator by screw action. There were two identical instruments: one with a right-hand thread (the knob turns clockwise) and the other with a left-hand thread (the knob turns

TABLE 20.3 A mutual fund versus its benchmark index

Year	Fund Return (%)	Index Return (%)	Year	Fund Return (%)	Index Return (%)	Year	Fund Return (%)	Index Return (%)	Year	Fund Return (%)	Index Return (%)
1984	−1.02	7.38	1993	44.74	32.56	2002	−17.79	−15.94	2011	−13.68	−13.71
1985	56.94	56.16	1994	0.76	7.78	2003	34.45	38.59	2012	20.01	16.83
1986	56.71	69.44	1995	14.89	11.21	2004	18.95	20.25	2013	22.95	15.29
1987	12.48	24.63	1996	14.65	6.05	2005	15.00	13.54	2014	−5.63	−3.87
1988	11.61	28.27	1997	4.12	1.78	2006	25.92	26.34	2015	−0.67	−5.66
1989	24.76	10.54	1998	16.93	20.00	2007	15.98	11.17	2016	1.71	4.50
1990	−12.05	−23.45	1999	26.34	26.96	2008	−44.94	−43.38	2017	42.96	27.19
1991	4.74	12.13	2000	−8.60	−14.17	2009	41.63	31.78	2018	−12.69	−14.20
1992	−5.79	−12.17	2001	−18.92	−21.44	2010	15.66	8.13	2019	31.56	21.51

counterclockwise). Table 20.4 gives the times, in seconds, each subject took to move the indicator a fixed distance.[32] RTLFT

TABLE 20.4 Performance times (seconds) using right-hand and left-hand threads

Subject	Right Thread	Left Thread	Subject	Right Thread	Left Thread
1	113	137	14	107	87
2	105	105	15	118	166
3	130	133	16	103	146
4	101	108	17	111	123
5	138	115	18	104	135
6	118	170	19	111	112
7	87	103	20	89	93
8	116	145	21	78	76
9	75	78	22	100	116
10	96	107	23	89	78
11	122	84	24	85	101
12	103	148	25	88	123
13	116	147			

(a) Each of the 25 students used both instruments. Explain briefly how you would use randomization in arranging the experiment.

(b) The project hoped to show that right-handed people find right-hand threads easier to use. Do an analysis that leads to a conclusion about this issue.

20.50 Comparing Two Drugs. 4step Makers of generic drugs must show that they do not differ significantly from the "reference" drugs that they imitate. One aspect in which drugs might differ is their extent of absorption in the blood. Table 20.5 gives data taken from 20 healthy nonsmoking male subjects for one pair of drugs.[33] This is a matched pairs design. Numbers 1 through 20 were assigned at random to the subjects. Subjects 1 through 10 received the generic drug first, followed by the reference drug. Subjects 11 through 20 received the reference drug first, followed by the generic drug. In all cases, a washout period separated the administration of the two drugs, so that the first had disappeared from the blood before the subject took the second. By randomizing the order, we eliminate the order in which the drugs were administered from confounding the difference in the absorption in the blood. Do the drugs differ significantly in the amount absorbed in the blood? DRUGS

TABLE 20.5 Absorption extent for two versions of a drug

Subject	Reference Drug	Generic Drug
15	4108	1755
3	2526	1138
9	2779	1613
13	3852	2254
12	1833	1310
8	2463	2120
18	2059	1851
20	1709	1878
17	1829	1682
2	2594	2613
4	2344	2738
16	1864	2302
6	1022	1284
10	2256	3052
5	938	1287
7	1339	1930
14	1262	1964
11	1438	2549
1	1735	3340
19	1020	3050

20.51 Practical Significance? Give a 90% confidence interval for the mean time advantage of right-hand over left-hand threads in the setting of Exercise 20.49. Do you think that the time saved would be of practical importance if the task were performed many times—for example, by an assembly-line worker? To help answer this question, find the mean time for right-hand threads as a percentage of the mean time for left-hand threads. RTLFT

20.52 Bad Weather, Bad Tips? 4step As part of the study of tipping in a restaurant that we met in Example 16.3 (page 374), the psychologists also studied the size of the tip in a restaurant when a message indicating that the next day's weather would be bad was written on the bill. Here are tips from 20 patrons, measured in percentage of the total bill:[34] TIP3

18.0	19.1	19.2	18.8	18.4	19.0	18.5	16.1	16.8	14.0
17.0	13.6	17.5	20.0	20.2	18.8	18.0	23.2	18.2	19.4

Do the data give convincing evidence that the mean percentage tip for all patrons of this restaurant when their bill contains a message that the next day's weather will be bad is less than 20%? (Note that 20% is an often-recommended size for restaurant tips.)

20.53 ***t* vs. *z*.** If you examine Table C, you will notice that critical values of the t distribution get closer and closer to the corresponding critical values of the Normal distribution as the number of degrees of freedom increases. You can see this by comparing the z critical values at the bottom of Table C with the t critical values in the corresponding column. This suggests that, for very large sample sizes, inference based on the Normal probability calculations in Chapters 15 and 16 (supposing σ is known) and inference based on the t distribution as discussed in this chapter (σ is not known) may give essentially the same answer if sample sizes are large and we proceed as though our estimate of σ is the true value of σ. Many statistical software packages will calculate t probabilities. If you have access to such software, answer the following questions.

(a) Use software to determine how large a sample size (or how many degrees of freedom) is needed for the critical value of the t distribution to be within 0.01 of the corresponding critical value of the Normal distribution for a 90%, 95%, and 99% confidence interval for a population mean.

(b) Based on your findings, how large a sample size do you think is needed for inference using the Normal distribution and inference using the t distribution to give very similar results if σ (both the true value and its estimate) is 1? If σ (both the true value and its estimate) is 100?

CHAPTER

21

When you complete this chapter, you will be able to:

21.1 Recognize the difference between two-sample studies and matched pairs studies and distinguish which is appropriate in a given setting.

21.2 Determine whether the conditions for inference comparing two means are met in specific situations.

21.3 Use two-sample *t* procedures to calculate *P*-values for significance tests and to calculate confidence intervals.

21.4 Use technology to compute confidence intervals and *P*-values for two-sample *t* procedures.

21.5 Use sample size and information about the Normality of two populations to determine whether a two-sample *t* procedure is appropriate.

21.6 Use the formula for degrees of freedom to find the approximate distribution of the two-sample *t* statistic when comparing two means.

21.7 State the limitations of the pooled two-sample *t* procedure and why it should not be used if statistical software is available.

21.8 Explain why inferences about population standard deviations should be avoided in basic statistical practice.

Comparing Two Means

In Chapter 20, we studied inference for the mean of a Normal population using procedures based on the *t* distribution. In practice, the most common use of *t* procedures for a single population mean is with matched pairs data because most research studies make comparisons between two or more populations. In this chapter, we discuss *t* procedures for comparing the means of two Normal populations when we have independent samples from these two populations.

Comparing two populations or two treatments is one of the most common situations encountered in statistical practice. We call such situations *two-sample problems*.

Two-Sample Problems

- The goal of inference is to compare the responses to two treatments or to compare the characteristics of two populations.
- We have a separate sample from each treatment or each population.

21.1 Two-Sample Problems

A two-sample problem can arise from a randomized comparative experiment that randomly divides subjects into two groups and exposes each group to a different treatment. Comparing random samples separately selected from two populations is also a two-sample problem. Unlike the matched pairs designs studied earlier, there is no matching of the individuals in the two samples. The two samples are assumed to be independent and can be of different sizes. Inference procedures for two-sample data differ from those for matched pairs. Here are some typical two-sample problems.

Aja Koska/Getty Images

STATISTICS IN YOUR WORLD

Driving While Fasting

Many Muslims fast from sunrise to sunset during the month of Ramadan. Does this affect the rate of traffic accidents? Fasting can improve alertness, reducing accidents. Or it can cause dehydration, increasing accidents. Data from Turkey show a statistically significant increase, starting two weeks into Ramadan. Ah, but because Ramadan follows a lunar calendar, it cycles through the year. Perhaps accidents go down during a winter Ramadan (alertness) but go up during a summer Ramadan (longer fast and dehydration). Ask the statisticians this question and get their favorite answer: we need more data.

EXAMPLE 21.1 Two-Sample Problems

- Does regular physical therapy help lower-back pain? A randomized experiment assigned patients with lower-back pain to two groups: 142 received an examination and advice from a physical therapist; another 144 received regular physical therapy for up to five weeks. After a year, the change in their level of disability (0% to 100%) was assessed by a doctor who did not know which treatment the patients had received.
- An education researcher administers a test of overall subject matter comprehension from assigned readings to a sample of students using tablets in place of traditional print textbooks and to a sample of students using traditional print texbooks. She compares the test scores of the students using tablets with the scores of the students using traditional textbooks.
- A bank wants to know which of two incentive plans will most increase the use of its credit cards. It offers each incentive to independent random samples of credit card customers and compares the amounts charged during the following six months.

APPLY YOUR KNOWLEDGE

Which Data Design? *Each situation described in Exercises 21.1 through 21.4 requires inference about a mean or means. Identify each as involving (1) a single sample, (2) matched pairs, or (3) two independent samples. The procedures of Chapter 20 apply to designs (1) and (2). We are about to learn procedures for design (3).*

21.1 Twitter Posting. Do people post more on Twitter as their numbers of followers increase? To test this, researchers selected a random sample of Twitter users. These users were randomly divided into two groups, a treatment group and a control group. Users did not know to which group they were assigned. Over a period of 100 days, the number of followers was steadily increased by the researchers for subjects in the treatment group. However, those in the treatment group were unaware of the source of this increase. The users in the control group were simply observed by the researchers. The average number of posts by each user over a period of 50 days was recorded. The mean number of posts for the two groups were compared.

21.2 Who Is Smarter? Choose a random sample of 100 households with at least two children of different ages. Measure the IQ scores of the firstborn and the youngest child in each household. Compare the mean IQ score of each firstborn child with that of the youngest child.

21.3 Plant-Based Burgers. A school nutritionist selects a sample of students in a large school district and has them taste and rate a new burger that is being considered as an addition to the school lunch menu. The nutritionist does not tell the students that the new burger is a plant-based burger that is supposed to taste like beef. The ratings are from −5 to 5, with −5 being strongly dislike, 0 being neither like nor dislike, and 5 being strongly like. The nutritionist tests whether the mean rating is greater than 0.

21.4 Plant-Based Burgers (continued). Another nutritionist selects a sample of students in a large school district and randomly divides the sample into two groups. One group rates the taste of a plant-based burger and the other the taste of a traditional beef burger. Neither group is told what their burger is made from. The ratings are from −5 to 5, with −5 being strongly dislike, 0 being neither like nor dislike, and 5 being strongly like. The nutritionist compares the mean ratings of the two groups.

21.2 Comparing Two Population Means

Comparing two populations or the responses to two treatments starts with data analysis: make boxplots, stemplots (for small samples), or histograms (for larger samples) and compare the shapes, centers, and variabilities of the two samples. The most common goal of inference is to compare the average or typical responses in the two populations. When data analysis suggests that both population distributions are symmetric, and especially when they are at least approximately Normal, we want to compare the population means. Here are the conditions for inference about means.

Conditions for Inference Comparing Two Means

- We have two SRSs, from two distinct populations. The samples are *independent*. That is, one sample has no influence on the other. Matching violates independence, for example. We measure the same response variable for both samples.
- Both populations are *Normally distributed*. The means and standard deviations of the populations are unknown. In practice, it is enough that the distributions have similar shapes and that the data have no strong outliers.

The two-sample procedures in this chapter can also be used for studies comparing responses to two treatments. We assume we have data come from a randomized comparative experiment that randomly divides subjects into two groups and exposes each group to a different treatment.

Call the variable we measure x_1 in the first population and x_2 in the second because the variable may have different distributions in the two populations. Here is how we describe the two populations:

Population	Variable	Mean	Standard Deviation
1	x_1	μ_1	σ_1
2	x_2	μ_2	σ_2

There are four unknown parameters: the two means and the two standard deviations. The subscripts remind us which population a parameter describes. We want to compare the two population means, either by giving a confidence interval for their difference, $\mu_1 - \mu_2$, or by testing the hypothesis of no difference, $H_0: \mu_1 = \mu_2$. The null hypothesis of no difference is used when investigating whether a treatment has an effect.

We use the sample means and standard deviations to estimate the unknown parameters. Again, subscripts remind us which sample a statistic comes from. Here is how we describe the samples:

Population	Sample Size	Sample Mean	Sample Standard Deviation
1	n_1	$\bar{x}_1$	s_1
2	n_2	$\bar{x}_2$	s_2

To make inferences about the difference $\mu_1 - \mu_2$ between the means of the two populations, we start from the difference $\bar{x}_1 - \bar{x}_2$ between the means of the two samples.

EXAMPLE 21.2 Daily Activity and Obesity

4step

ACTIVE

STATE: People gain weight when they take in more energy from food than they expend. James Levine and his collaborators at the Mayo Clinic investigated the link between obesity and energy spent on daily activity.[1]

Choose 20 healthy volunteers who don't exercise. Deliberately choose 10 who are lean and 10 who are mildly obese but still healthy. Attach sensors that monitor the subjects' every move for 10 days. Table 21.1 presents data on the time (in minutes per day) that the subjects spent standing or walking, sitting, and lying down. Do lean and obese people differ in the average time they spend standing and walking?

TABLE 21.1 Time (minutes per day) spent in three different postures by lean and obese subjects

Group	Subject	Stand/Walk	Sit	Lie
Lean	1	511.100	370.300	555.500
Lean	2	607.925	374.512	450.650
Lean	3	319.212	582.138	537.362
Lean	4	584.644	357.144	489.269
Lean	5	578.869	348.994	514.081
Lean	6	543.388	385.312	506.500
Lean	7	677.188	268.188	467.700
Lean	8	555.656	322.219	567.006
Lean	9	374.831	537.031	531.431
Lean	10	504.700	528.838	396.962
Obese	11	260.244	646.281	521.044
Obese	12	464.756	456.644	514.931
Obese	13	367.138	578.662	563.300
Obese	14	413.667	463.333	532.208
Obese	15	347.375	567.556	504.931
Obese	16	416.531	567.556	448.856
Obese	17	358.650	621.262	460.550
Obese	18	267.344	646.181	509.981
Obese	19	410.631	572.769	448.706
Obese	20	426.356	591.369	412.919

PLAN: Examine the data and carry out a test of hypotheses. We suspect in advance that lean subjects (Group 1) are more active than obese subjects (Group 2), so we test the hypotheses

$$H_0: \mu_1 = \mu_2$$
$$H_a: \mu_1 > \mu_2$$

SOLVE (first steps): Are the conditions for inference met? The subjects are volunteers, so they are not SRSs from all lean and mildly obese adults. The study tried to recruit comparable groups: all worked in sedentary jobs, none smoked or were taking medication, and so on. Setting clear standards like these helps make up for the fact that we can't reasonably get SRSs for such an invasive study. The subjects were not told that they were chosen from a larger group of volunteers because they did not exercise and were either lean or mildly obese. Because their willingness to volunteer isn't related to the purpose of the experiment, we are willing to treat them as two independent SRSs.

A back-to-back stemplot (see Figure 21.1) displays the "Stand/walk" data in detail. To make the plot, we rounded the data to the nearest 10 minutes and used 100s as stems and 10s as leaves. The distributions are a bit irregular, as we expect with just 10 observations. There are no clear departures from Normality such as extreme outliers or skewness. The lean subjects as a group spend much more time standing and walking than do the obese subjects. Calculating the group means confirms this:

Group	n	Mean $\bar{x}$	Standard Deviation s
Group 1 (lean)	10	525.751	107.121
Group 2 (obese)	10	373.269	67.498

The observed difference in mean time per day spent standing or walking is

$$\bar{x}_1 - \bar{x}_2 = 525.751 - 373.269 = 152.482 \text{ minutes}$$

To complete the Solve step, we must learn the details of inference for comparing two means.

Lean		Obese
	2	67
72	3	567
	4	11236
886410	5	
81	6	

FIGURE 21.1 Back-to-back stemplot of the times spent walking or standing, for Example 21.2. Here, values are in hundreds, so, for example, 2|6 = 260.

21.3 Two-Sample *t* Procedures

To assess the significance of the observed difference between the means of our two samples, we follow a familiar path. Whether an observed difference is surprising depends on the variability of the observations as well as on the two means. Widely different means can arise just by chance if the individual observations vary a great deal. To take variation into account, we would like to standardize the observed difference $\bar{x}_1 - \bar{x}_2$ by dividing by its standard deviation. This standard deviation of the difference in sample means is

$$\sqrt{\frac{\sigma_1^2}{n_1} + \frac{\sigma_2^2}{n_2}}$$

This standard deviation gets larger as either population gets more variable—that is, as σ_1 or σ_2 increases. It gets smaller as the sample sizes n_1 and n_2 increase.

Because we don't know the population standard deviations, we estimate them from the sample standard deviations from our two samples. The result is the standard error, or estimated standard deviation, of the difference in sample means:

$$\text{SE}_{\bar{x}_1 - \bar{x}_2} = \sqrt{\frac{s_1^2}{n_1} + \frac{s_2^2}{n_2}}$$

When we standardize the estimate by dividing it by its standard error, the result is the **two-sample *t* statistic:**

$$t = \frac{(\bar{x}_1 - \bar{x}_2) - (\mu_1 - \mu_2)}{\sqrt{\frac{s_1^2}{n_1} + \frac{s_2^2}{n_2}}}$$

Notice that for hypothesis tests, $H_0\colon \mu_1 = \mu_2$ states that $\mu_1 - \mu_2 = 0$. The statistic t has the same interpretation as any z or t statistic: it has the form

$$\frac{\text{Estimate of parameter} - \text{Null hypothesis value of the parameter}}{\text{Standard error of estimate}}$$

and says how far $\bar{x}_1 - \bar{x}_2 = 0$ is from $\mu_1 - \mu_2 = 0$ in standard error units.

The two-sample t statistic has approximately a t distribution. It does not have exactly a t distribution even if the populations are both exactly Normal. In practice, however, the approximation is very accurate. There are two practical options for using the two-sample t procedures:

Option 1. With software, use the statistic t with accurate critical values from the approximating t distribution. The degrees of freedom are calculated from the data by using a somewhat messy formula. Moreover, the degrees of freedom may not be a whole number.

Option 2. Without software, use the statistic t with critical values from the t distribution with *degrees of freedom equal to the smaller of* $n_1 - 1$ *and* $n_2 - 1$. These procedures are always conservative for any two Normal populations. The confidence interval has a margin of error *as large as or larger than* is needed for the desired confidence level. The significance test gives a *P*-value *equal to or greater than* the true *P*-value.

The two options are exactly the same except for the degrees of freedom used for t critical values and *P*-values. As the sample sizes increase, confidence levels and *P*-values from Option 2 become more accurate. The gap between what Option 2 reports and the truth is quite small unless the sample sizes are both small and unequal.[2]

The Two-Sample *t* Procedures

Draw an SRS of size n_1 from a large Normal population with unknown mean μ_1 and draw an independent SRS of size n_2 from another large Normal population with unknown mean μ_2. A level C confidence interval for $\mu_1 - \mu_2$ is given by

$$(\bar{x}_1 - \bar{x}_2) \pm t^* \sqrt{\frac{s_1^2}{n_1} + \frac{s_2^2}{n_2}}$$

Here t^* is the critical value for confidence level C for the t distribution with degrees of freedom from either Option 1 (software) or Option 2 (the smaller of $n_1 - 1$ and $n_2 - 1$).

To test the hypothesis $H_0: \mu_1 = \mu_2$, calculate the two-sample t statistic

$$t = \frac{\bar{x}_1 - \bar{x}_2}{\sqrt{\frac{s_1^2}{n_1} + \frac{s_2^2}{n_2}}}$$

Find *P*-values from the t distribution with degrees of freedom from either Option 1 (software) or Option 2 (the smaller of $n_1 - 1$ and $n_2 - 1$).

EXAMPLE 21.3 Daily Activity and Obesity

We can now complete Example 21.2.

SOLVE (inference): The two-sample t statistic comparing the average minutes spent standing and walking in Group 1 (lean) and Group 2 (obese) is

$$t = \frac{\bar{x}_1 - \bar{x}_2}{\sqrt{\frac{s_1^2}{n_1} + \frac{s_2^2}{n_2}}}$$

$$= \frac{525.751 - 373.269}{\sqrt{\frac{107.121^2}{10} + \frac{67.498^2}{10}}}$$

$$= \frac{152.482}{40.039} = 3.808$$

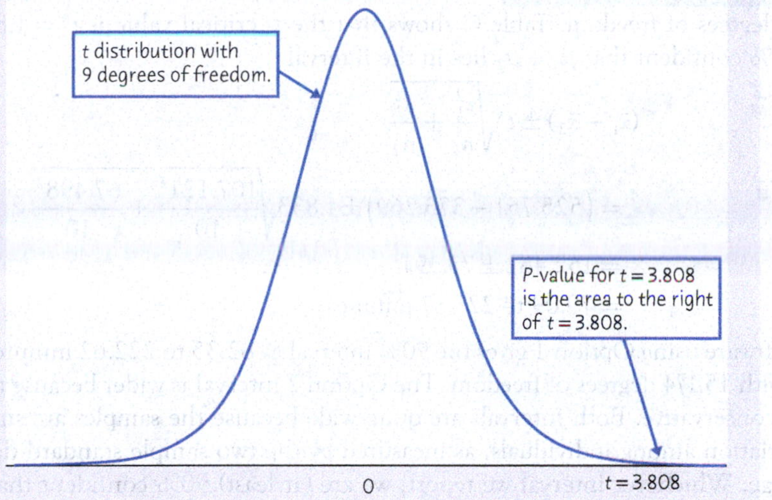

t distribution with 9 degrees of freedom.
P-value for t = 3.808 is the area to the right of t = 3.808.
0
t = 3.808

FIGURE 21.2
Using the conservative Option 2, the *P*-value in Example 21.3 comes from the *t* distribution with 9 degrees of freedom.

Software (Option 1) gives one-sided *P*-value $P = 0.0008$ based on df = 15.174.

Without software, use the conservative Option 2. Because $n_1 - 1 = 9$ and $n_2 - 1 = 9$, there are 9 degrees of freedom. Because H_a is one-sided, the *P*-value is the area to the right of $t = 3.808$ under the t_9 curve. Figure 21.2 illustrates this *P*-value. Table C shows that $t = 3.808$ lies between the critical values t^* for 0.0025 and 0.001. So $0.001 < P < 0.0025$. Option 2 gives a larger (more conservative) *P*-value than Option 1. As usual, the practical conclusion is the same for both versions of the test.

df = 9

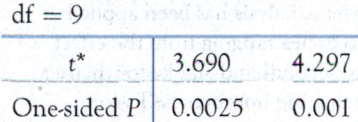

t^*	3.690	4.297
One-sided *P*	0.0025	0.001

CONCLUDE: There is very strong evidence ($P = 0.0008$) that, on average, lean people spend more time walking and standing than do moderately obese people.

Does lack of daily activity *cause* obesity? This is an observational study, and that affects our ability to draw cause-and-effect conclusions. It may be that some people are naturally more active and are therefore less likely to gain weight. Or it may be that people who gain weight reduce their activity level. The study went on to enroll most of the obese subjects in a weight-reduction program and most of the lean subjects in a supervised program of overeating. After eight weeks, the obese subjects had lost weight (mean 8 kg) and the lean subjects had gained weight (mean 4 kg). But both groups kept their original allocation of time to the different postures. This suggests that time allocation may be biological and influences weight rather than the other way around. The authors remark: “It should be emphasized that this was a pilot study and that the results need to be confirmed in larger studies.”

EXAMPLE 21.4 How Much More Active Are Lean People?

PLAN: To estimate how much more active lean people are, we give a 90% confidence interval for $\mu_1 - \mu_2$, the difference in average daily minutes spent standing and walking between lean and mildly obese adults.

4step

SOLVE AND CONCLUDE: As in Example 21.3, the conservative Option 2 uses 9 degrees of freedom. Table C shows that the t_9 critical value is $t^* = 1.833$. We are 90% confident that $\mu_1 - \mu_2$ lies in the interval

$$(\bar{x}_1 - \bar{x}_2) \pm t^* \sqrt{\frac{s_1^2}{n_1} + \frac{s_2^2}{n_2}}$$

$$= (525.751 - 373.269) \pm 1.833\sqrt{\frac{107.121^2}{10} + \frac{67.498^2}{10}}$$

$$= 152.482 \pm 73.391$$

$$= 79.09 \text{ to } 225.87 \text{ minutes}$$

Software using Option 1 gives the 90% interval as 82.35 to 222.62 minutes, based on t with 15.174 degrees of freedom. The Option 2 interval is wider because this method is conservative. Both intervals are quite wide because the samples are small and the variation among individuals, as measured by the two sample standard deviations, is large. Whichever interval we report, we are (at least) 90% confident that the mean difference in average daily minutes spent standing and walking between lean and mildly obese adults lies in this interval.

STATISTICS IN YOUR WORLD

Meta-Analysis

Small samples have large margins of error. Large samples are expensive. Often, we can find several studies of the same issue; if we could combine their results, we would have a large sample with a small margin of error. That is the idea of "meta-analysis." Of course, we can't just lump the studies together because of differences in design and quality. Statisticians have more sophisticated ways of combining the results. Meta-analysis has been applied to issues ranging from the effect of secondhand smoke to whether coaching improves SAT scores.

Note that in Example 21.4 we could have switched the order and computed a 90% confidence interval for $\mu_2 - \mu_1$ using $\bar{x}_2 - \bar{x}_1 = 373.269 - 525.751 = -152.482$. This would have changed the signs in the confidence interval, resulting in the interval -225.87 to -79.09 minutes for $\mu_2 - \mu_1$. However, the final interpretation would be the same as the interval for $\mu_1 - \mu_2$.

EXAMPLE 21.5 Community Service and Attachment to Friends

Hero Images/Getty Images

STATE: Do college students who have volunteered for community service and those who have not differ in how attached they are to their friends? A study obtained data from 57 students who had done service work and 17 who had not. One of the response variables was a measure of attachment to friends, as measured by the Inventory of Parent and Peer Attachment (larger scores indicate greater attachment). In particular, the response is a score based on the responses to 25 questions. Here are the results:[3]

Group	Condition	n	$\bar{x}$	s
1	Service	57	105.32	14.68
2	No service	17	96.82	14.26

PLAN: The investigator had no specific direction in mind for the difference before looking at the data, so the alternative is two-sided. We will test the hypotheses

$$H_0: \mu_1 = \mu_2$$
$$H_a: \mu_1 \neq \mu_2$$

SOLVE: The investigator says that the individual scores, examined separately in the two samples, appear roughly Normal. There is a serious problem with the more important condition that the two samples can be regarded as SRSs from two student populations. We will discuss that after we illustrate the calculations.

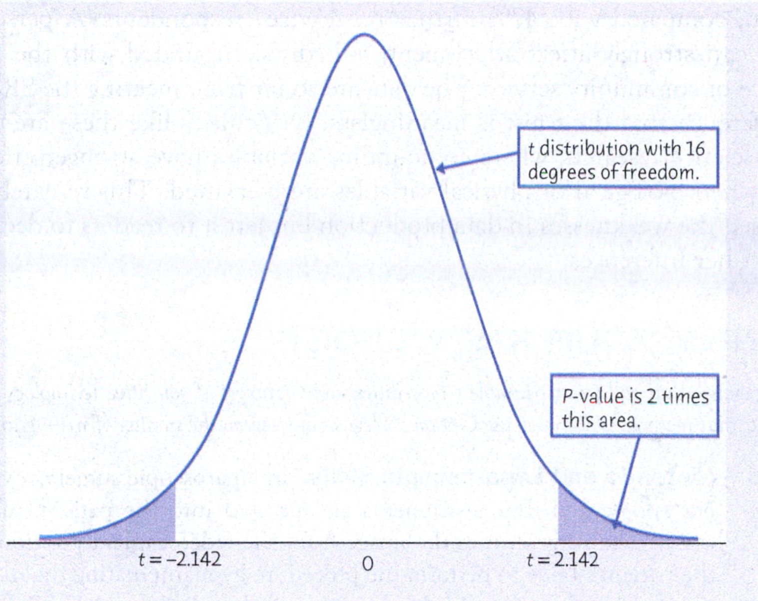

FIGURE 21.3
The *P*-value, for Example 21.5. Because the alternative is two-sided, the *P*-value is double the area to the right of $t = 2.142$.

The two-sample *t* statistic is

$$t = \frac{\bar{x}_1 - \bar{x}_2}{\sqrt{\frac{s_1^2}{n_1} + \frac{s_2^2}{n_2}}}$$

$$= \frac{105.32 - 96.82}{\sqrt{\frac{14.68^2}{57} + \frac{14.26^2}{17}}}$$

$$= \frac{8.5}{3.9677} = 2.142$$

Software (Option 1) says that the two-sided *P*-value is $P = 0.0414$.

Without software, use Option 2 to find a conservative *P*-value. There are 16 degrees of freedom, the smaller of

$$n_1 - 1 = 57 - 1 = 56 \quad \text{and} \quad n_2 - 1 = 17 - 1 = 16$$

Figure 21.3 illustrates the *P*-value. Find it by comparing $t = 2.142$ with the two-sided critical values for the t_{16} distribution. Table C shows that the *P*-value is between 0.05 and 0.04.

df = 16

t^*	2.120	2.235
Two-sided *P*	0.05	0.04

CONCLUDE: The data give moderately strong evidence ($P < 0.05$) that students who have engaged in community service do, on the average, differ from those who have not engaged in community service in how attached they are to their friends (and the data suggest that they are more attached to their friends).

Is the *t* test in Example 21.5 justified? The student subjects were "enrolled in a course on U.S. Diversity at a large midwestern university." Unless this course is required of all students, the subjects cannot be considered a random sample even from this campus. Students were placed in the two groups on the basis of a questionnaire, 39 in the "no service" group and 71 in the "service" group. The data were gathered from a follow-up survey two years later; 17 of the 39 "no service" students responded (44%), compared with 80% response (57 of 71) in the "service" group. Nonresponse is confounded with group: students who had done community service were much more likely to respond. Finally, 75% of the "service" respondents were women, compared with 47% of the "no service" respondents. A person's gender, which can strongly affect attachment, is badly confounded with the presence or absence of community service. The data are so far from meeting the SRS condition for inference that the t test is meaningless. Difficulties like these are common in social science research, where confounding variables have stronger effects than is usual when biological or physical variables are measured. This researcher honestly disclosed the weaknesses in data production but left it to readers to decide whether to trust her inferences.

APPLY YOUR KNOWLEDGE

In exercises that call for two-sample t procedures, use Option 1 if you have technology that implements that method. Otherwise, use Option 2 (degrees of freedom the smaller of $n_1 - 1$ and $n_2 - 1$).

4step

21.5 Nintendo and Laparoscopic Skills. In laparoscopic surgery, a video camera and several thin instruments are inserted into the patient's abdominal cavity. The surgeon uses the image from the video camera positioned inside the patient's body to perform the procedure by manipulating the instruments that have been inserted. It has been found that the Nintendo Wii™, with its motion-seeking interface, replicates the movements required in laparoscopic surgery more closely than other video games. If training with a Nintendo Wii™ can improve laparoscopic skills, it can complement the more expensive training on a laparoscopic simulator. Forty-two medical residents were chosen, and all of them were tested on a set of basic laparoscopic skills. Twenty-one were selected at random to undergo systematic Nintendo Wii™ training for one hour a day, five days a week, for four weeks. The remaining 21 residents were given no Nintendo Wii™ training and were asked to refrain from video games during this period. At the end of four weeks, all 42 residents were tested again on the same set of laparoscopic skills. One of the skills involved a virtual gall bladder removal, with several performance measures, including time to complete the task, recorded. Here are the improvement (before minus after) times in seconds after four weeks for the two groups:[4] NINT

Treatment						Control					
291	134	186	128	84	243	21	66	54	85	229	92
212	121	134	221	59	244	43	27	77	−29	−14	88
79	333	−13	−16	71	−16	145	110	32	90	45	−81
71	77	144				68	61	44			

Does the Nintendo Wii™ training significantly increase the mean improvement time? Follow the four-step process as illustrated in Examples 21.2 and 21.3 (pages 470 and 472).

4step

21.6 Daily Activity and Obesity. We can conclude from Examples 21.2 and 21.3 that mildly obese people spend less time standing and walking (on the average) than lean people. Is there a significant difference between the mean times the two groups spend lying down? Use the four-step process to answer this question from the data in Table 21.1. Follow the model of Examples 21.2 and 21.3. ACTIVE

21.7 Nintendo and Laparoscopic Skills (continued). Use the data in Exercise 21.5 to give a 95% confidence interval for the difference in mean improvement times between the treatment and control groups. NINT

21.4 Examples of Technology

Software should use Option 1 for the degrees of freedom to give accurate confidence intervals and *P*-values. Unfortunately, there is variation in how well software implements Option 1. Figure 21.4 displays output from a graphing calculator, three statistical programs, and a spreadsheet program for the test of Example 21.3 (page 472). All four claim to use Option 1. The two-sample t statistic is exactly as in Example 21.3, $t = 3.808$. You can find this in all four outputs. (Minitab rounds to 3.81; Excel and the graphing calculator give additional decimal places.) The different technologies use different methods to find the *P*-value for $t = 3.808$:

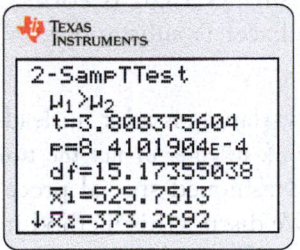

Texas Instruments

```
2-SampTTest
 μ1>μ2
 t=3.808375604
 p=8.4101904E-4
 df=15.17355038
 x̄1=525.7513
↓x̄2=373.2692
```

Minitab

Two-sample T for stand

group	N	Mean	StDev	SE Mean
1	10	526	107	34
2	10	373.3	67.5	21

```
Difference = mu (1) - mu (2)
Estimate for difference: 152.5
T-Test of difference = 0 (vs >): T-Value = 3.81 P-Value = 0.001 DF = 15
```

Excel

	A	B	C
1	t-Test: Two-Sample Assuming Unequal Variances		
2			
3		*Lean*	*Obese*
4	Mean	525.7513	373.2692
5	Variance	11474.8903	4556.019849
6	Observations	10	10
7	Hypothesized Mean	0	
8	df	15	
9	t Stat	3.808375604	
10	P(T<=t) one-tail	0.000856818	
11	t Critical one-tail	1.753051038	
12	P(T<=t) two-tail	0.001713635	
13	P(T<t) two-tail	2.131450856	
14			

CRUNCHIT!

Export

Null hypothesis:	Difference of means = 0
Alternative hypothesis:	Difference of means > 0

Group	Std Dev	Sample Mean	n
1	107.1	525.8	10
2	67.50	373.3	10

df:	15.17
Difference of means:	152.5
t statistic:	3.808
P-value:	0.0008410

jmp

Assuming unequal variances

Difference	152.482	t Ratio	3.808376
Std Err Dif	40.039	DF	15.17355
Upper CL Dif	237.737	Prob > |t|	0.0017*
Lower CL Dif	67.227	Prob > t	0.0008*
Confidence	0.95	Prob < t	0.9992

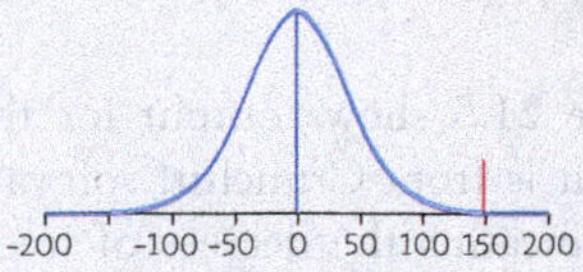

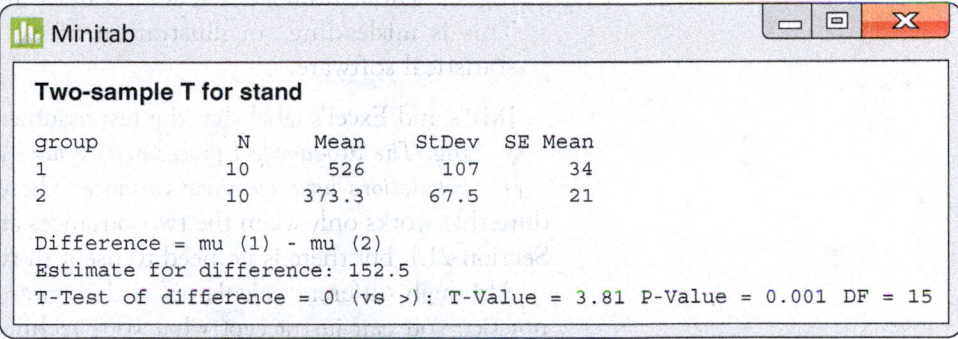

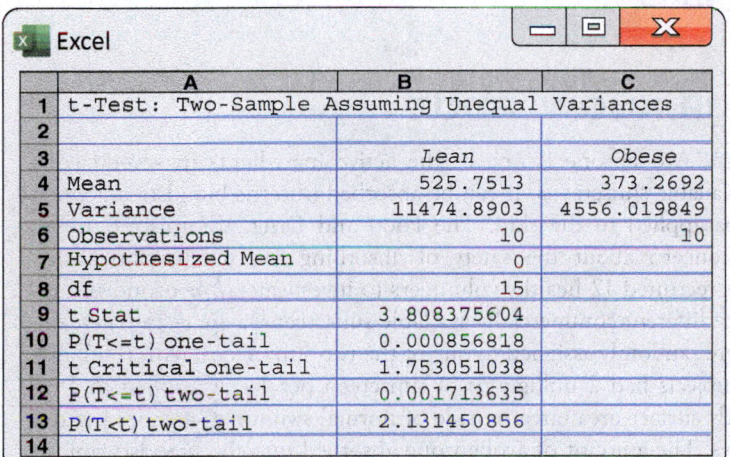

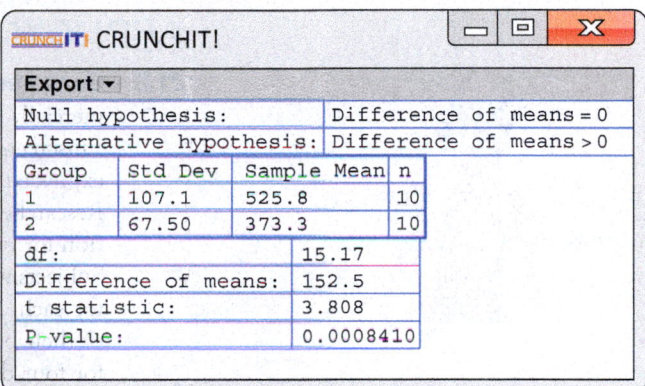

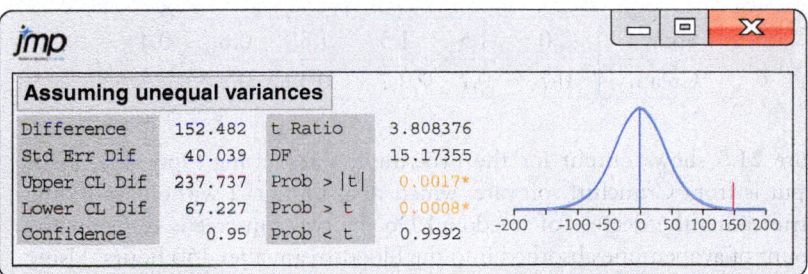

FIGURE 21.4
The two-sample *t* procedures applied to the data on activity and obesity: output from a graphing calculator, three statistical programs, and a spreadsheet program.

- CrunchIt!, JMP, and the calculator get Option 1 completely right. The accurate approximation uses the t distribution with approximately 15.174 (CrunchIt! rounds to 15.17) degrees of freedom. The P-value is $P = 0.0008$.
- Minitab uses Option 1, but it *truncates* the exact degrees of freedom to the next smaller whole number to get critical values and P-values. In this example, the exact df = 15.174 is truncated to df = 15, so that Minitab's results are slightly conservative. That is, Minitab's P-value (rounded to $P = 0.001$ in the output) is slightly larger than the full Option 1 P-value.
- Excel *rounds* the exact degrees of freedom to the nearest whole number so that df = 15.174 becomes df = 15. Excel's method agrees with Minitab's in this example. But when rounding moves the degrees of freedom up to the next higher whole number, Excel's P-values are slightly smaller than is correct. This is misleading, an illustration of the fact that Excel is substandard as statistical software.

JMP's and Excel's label that the test assumes unequal variances is a bit misleading. *The two-sample t procedures we have described work whether or not the two populations have the same variance.* There is an old-fashioned special procedure that works only when the two variances are equal. We discuss this method in Section 21.7, but there is no need to use it in two-sample problems.

Although different calculators and software give slightly different P-values, in practice you can just accept what your technology says. The small differences in P don't affect the conclusion. Even "between 0.001 and 0.0025" from Option 2 (Example 21.3) is close enough for practical purposes.

APPLY YOUR KNOWLEDGE

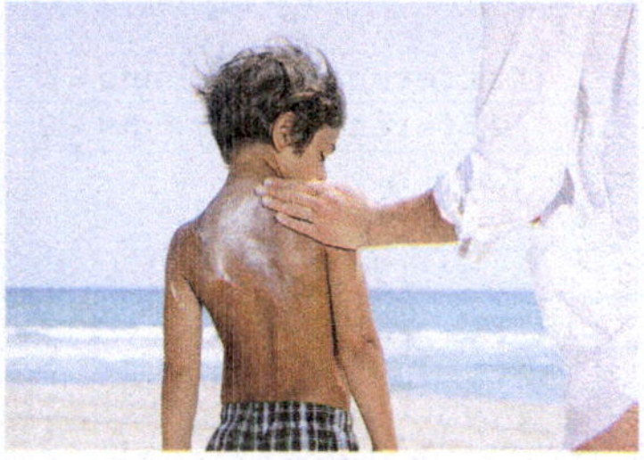

Fertnig/Getty Images

21.8 Sunscreens Avobenzone is one of the active ingredients in several commercially available sunscreens. It can be absorbed into the bloodstream when sunscreen is applied to the skin. The Food and Drug Administration has expressed concern about the safety of absorbing too much avobenzone. Researchers recruited 12 healthy volunteers to investigate avobenzone absorption for two different commercially available sunscreens: a spray and a cream. Subjects were randomly assigned to one of the two sunscreens, with 6 subjects for each. Subjects had 2 milligrams of sunscreen per 1 cm^2 applied to 75% of their body surface area (area outside of normal swimwear) 4 times per day for four days. The amount of avobenzone absorbed into the bloodstream (in nanograms per milliliter applied, ng/mL) after 150 hours was then measured for each subject. Here are the measurements:[5] SUNSCREEN

Spray	2.0	1.5	1.5	1.5	0.6	0.4
Cream	0.7	0.7	0.7	0.3	0.2	0.2

Figure 21.5 shows output for the two-sample t test using Option 1. (This output is from CrunchIt! software, which does Option 1 without rounding or truncating the degrees of freedom.) Do the two sunscreens differ in the amount of avobenzone absorbed into the bloodstream after 150 hours? Using the output in Figure 21.5, write a summary in a sentence or two, including t, df, P, and a conclusion.

CRUNCHIT! CRUNCHIT!

Export

Null hypothesis:	Difference of means = 0
Alternative hypothesis:	Difference of means is not 0

Treatment	n	Sample Mean	Std Dev
Spray	6	1.250	0.6156
Cream	6	0.4667	0.2582

df:	6.706
Difference of means:	0.7833
t statistic:	2.874
P-value:	0.02499

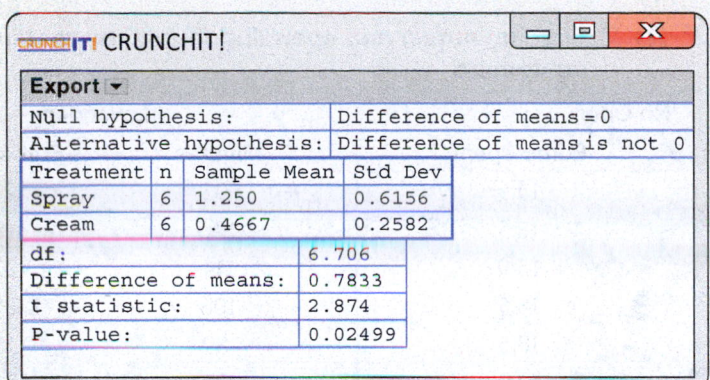

FIGURE 21.5
Two-sample *t* output from CrunchIt!, for Exercise 21.8.

21.5 Robustness Again

The two-sample *t* procedures are more robust (less sensitive to departures from our Normality condition for inference for comparing two means) than the one-sample *t* methods, particularly when the distributions are not symmetric. When the sizes of the two samples are equal and the two populations being compared have distributions with similar shapes, probability values from the *t* table are quite accurate for a broad range of distributions when the sample sizes are as small as $n_1 = n_2 = 5$.[6] When the two population distributions have different shapes, larger samples are needed.

As a guide to practice, adapt the guidelines given on page 455 for the use of one-sample *t* procedures to two-sample procedures by replacing "sample size" with the "sum of the sample sizes," $n_1 + n_2$. These guidelines err on the side of safety, especially when the two samples are of equal size. *In planning a two-sample study, choose equal sample sizes whenever possible. The two-sample t procedures are most robust against non-Normality in this case, and the conservative Option 2 probability values are most accurate.*

APPLY YOUR KNOWLEDGE

21.9 Do Good Smells Bring Good Business? Businesses know that customers often respond to background music. Do they also respond to odors? One study of this question took place in a small pizza restaurant in France on two Saturday evenings in May. On one of these evenings, a relaxing lavender odor was spread through the restaurant. On the other evening, no scent was used. Table 21.2 gives the time (in minutes) that two samples of 30 customers spent in the restaurant and the amount they spent (in euros).[7] The two evenings were comparable in many ways (weather, customer count, and so on), so we are willing to regard the data as independent SRSs from spring Saturday evenings at this restaurant. ODORS2

(a) Does a lavender odor encourage customers to stay longer in the restaurant? Examine the time data and explain why they are suitable for two-sample *t* procedures. Use the two-sample *t* test to answer the question posed.

(b) Does a lavender odor encourage customers to spend more while in the restaurant? Examine the spending data. In what ways do these data deviate from Normality? Explain why, with 30 observations, the *t* procedures are reasonably accurate for these data. Use the two-sample *t* test to answer the question posed.

TABLE 21.2 **Time (minutes) and spending (euros) by restaurant customers**

No Odor		Lavender	
Minutes	Euros Spent	Minutes	Euros Spent
103	15.9	92	21.9
68	18.5	126	18.5
79	15.9	114	22.3
106	18.5	106	21.9
72	18.5	89	18.5
121	21.9	137	24.9
92	15.9	93	18.5
84	15.9	76	22.5
72	15.9	98	21.5
92	15.9	108	21.9
85	15.9	124	21.5
69	18.5	105	18.5
73	18.5	129	25.5
87	18.5	103	18.5
109	20.5	107	18.5
115	18.5	109	21.9
91	18.5	94	18.5
84	15.9	105	18.5
76	15.9	102	24.9
96	15.9	108	21.9
107	18.5	95	25.9
98	18.5	121	21.9
92	15.9	109	18.5
107	18.5	104	18.5
93	15.9	116	22.8
118	18.5	88	18.5
87	15.9	109	21.9
101	25.5	97	20.7
75	12.9	101	21.9
86	15.9	106	22.5

21.10 Travel Times The American Community Survey asks, among much else, about workers' travel times from home to work. Here are the travel times in minutes for 15 workers in North Carolina and 20 workers in New York, chosen at random from U.S. Census Bureau data:[8] TRAVEL

North Carolina

20	35	8	70	5	15	25	30	40	35
10	12	40	15	20					

New York

10	15	55	20	65	50	12	20	10	10
35	50	30	45	15	10	75	40	35	60

(a) Make stemplots to investigate the shapes of the distributions. The travel times for both North Carolina and New York skewed to the right, and the travel times for North Carolina have a high outlier. The high outlier is plausible, but it is also possible that this is an error. Because of the uncertainty, we are not sure whether to remove the outlier.

(b) We suspect that travel times for New York are longer than for North Carolina. Do the data support this suspicion?

(c) Suppose we discover that the outlier in the North Carolina times is an error. Do the data (with the outlier removed) now support the suspicion that travel times for New York are longer than for North Carolina?

21.11 Weeds among the Corn. Lamb's-quarter is a common weed that interferes with the growth of corn. An agriculture researcher planted corn at a uniform rate in 16 small plots of ground and then weeded the plots by hand to allow a fixed number of lamb's-quarter plants to grow in each meter of corn row. No other weeds were allowed to grow. Here are the yields of corn (bushels per acre) for only the experimental plots controlled to have one weed per meter of row and nine weeds per meter of row:[9] WDCORN

One weed/meter	166.2	157.3	166.7	161.1
Nine weeds/meter	162.8	142.4	162.8	162.4

Explain carefully why a two-sample t confidence interval for the difference in mean yields may not be accurate.

21.12 Travel Times (continued). Use the data in Exercise 21.10 to give two 90% confidence intervals for the difference in travel times for New York and North Carolina. One interval should use all the data and the other should use the data with the outlier removed. TRAVEL

21.6 Details of the *t* Approximation*

The exact distribution of the two-sample t statistic is not a t distribution. Moreover, the distribution changes as the unknown population standard deviations σ_1 and σ_2 change. However, an excellent approximation is available. We call this Option 1 for t procedures.

Approximate Distribution of the Two-Sample *t* Statistic

The distribution of the two-sample t statistic is very close to the t distribution with degrees of freedom df given by

$$\text{df} = \frac{\left(\frac{s_1^2}{n_1} + \frac{s_2^2}{n_2}\right)^2}{\frac{1}{n_1 - 1}\left(\frac{s_1^2}{n_1}\right)^2 + \frac{1}{n_2 - 1}\left(\frac{s_2^2}{n_2}\right)^2}$$

This approximation is accurate when both sample sizes n_1 and n_2 are 5 or larger. Notice that the numerator, $\left(\frac{s_1^2}{n_1} + \frac{s_2^2}{n_2}\right)^2$, is equivalent to $\left(SE_{\bar{x}_1}^2 + SE_{\bar{x}_2}^2\right)^2$.

*This section can be omitted unless you are using software and wish to understand what the software does.

EXAMPLE 21.6 Daily Activity and Obesity

ACTIVE

In the experiment of Examples 21.2 and 21.3 (pages 470 and 472), the data on minutes per day spent standing and walking give

Group	n	$\bar{x}$	s
Group 1 (lean)	10	525.751	107.121
Group 2 (obese)	10	373.269	67.498

The two-sample t test statistic calculated from these values is $t = 3.808$.

The one-sided P-value is the area to the right of 3.808 under a t density curve, as in Figure 21.2. The conservative Option 2 uses the t distribution with 9 degrees of freedom. Option 1 finds a very accurate P-value by using the t distribution with degrees of freedom (df) given by

$$\text{df} = \frac{\left(\dfrac{107.121^2}{10} + \dfrac{67.498^2}{10}\right)^2}{\dfrac{1}{9}\left(\dfrac{107.121^2}{10}\right)^2 + \dfrac{1}{9}\left(\dfrac{67.498^2}{10}\right)^2}$$

$$= \frac{2{,}569{,}894}{169{,}367.2} = 15.1735$$

These degrees of freedom appear in the graphing calculator output in Figure 21.4 (page 477). Because the formula is complicated with many variables and roundoff errors are likely, we don't recommend calculating df by hand.

The degrees of freedom is generally not a whole number. It is always at least as large as the smaller of $n_1 - 1$ and $n_2 - 1$. The larger degrees of freedom that result from Option 1 give slightly shorter confidence intervals and slightly smaller P-values than the conservative Option 2 produces. There is a t distribution for any positive degrees of freedom, even though Table C contains entries only for whole-number degrees of freedom. The difference between the t procedures using Options 1 and 2 is rarely of practical importance. That is why we recommend the simpler, conservative Option 2 for inference without software. With software, the more accurate Option 1 procedures are painless.

APPLY YOUR KNOWLEDGE

21.13 Behavioral Intervention on Calorie Intake. Prevention of childhood obesity is important for reducing the risk of future chronic disease. Researchers investigated the effect of a behavioral intervention involving both parents and children on the daily calorie intake (kilocalories/day) of children. The researchers recruited 610 parent/child pairs for the study. They randomly assigned 304 parent/child pairs to the behavioral intervention program and the remaining 306 to a control group that received 6 school readiness sessions over the period of the study. Does the behavioral intervention lead to a lower daily calorie intake compared to that of the control?[10] Software that uses Option 1 gives these summary results:

```
Treatment        n   Mean  Std dev  Std err      t       df      P
Behavioral
intervention   304  1227    363     20.82    -3.117   603.9  <0.002
Control        306  1323    397     22.69
```

Starting from the sample means and standard deviations, verify each of these entries: the standard errors of the means, the degrees of freedom for the two-sample t, and the value of t.

21.14 Sunscreens. Figure 21.5 (page 479) gives output for the sunscreen data in Exercise 21.8 from software that does Option 1 with the correct degrees of freedom. What are $\bar{x}_i$ and s_i for the two treatment groups? Starting from these values, find the t test statistic and its degrees of freedom. Your work should agree with Figure 21.5.

21.15 Behavioral Intervention on Calorie Intake (continued). Write a sentence or two summarizing the comparison of daily calorie intake of children in the behavioral intervention group with that of children in the control group in Exercise 21.13, as if you were preparing a report for publication. Use the output in Exercise 21.13.

21.7 Avoid the Pooled Two-Sample *t* Procedures*

Most software and graphing calculators, including all the ones illustrated in Figure 21.4 (page 477), offer a choice of two-sample *t* statistics. One statistic is often labeled for "unequal" variances and the other for "equal" variances. The "unequal" variance procedure is our two-sample *t*. *This test is valid whether or not the population variances are equal.* The other choice is a special version of the two-sample *t* statistic which assumes that the two populations have the same variance. This procedure averages (or *pools*, in statistical terms) the two sample variances to estimate the common population variance. The resulting statistic is called the *pooled two-sample t statistic.* It is equal to our *t* statistic if the two sample sizes are the same but not otherwise. We could choose to use the pooled *t* for tests and confidence intervals.

The pooled *t* statistic has exactly the *t* distribution with $n_1 + n_2 - 2$ degrees of freedom *if* the two population variances really are equal and the population distributions are exactly Normal. The pooled *t* was in common use before software made it easy to use Option 1 for our two-sample *t* statistic. Of course, in the real world, distributions are not exactly Normal, and population variances are not exactly equal. In practice, the Option 1 two-sample *t* procedures are almost always more accurate than the pooled procedures. Our advice: *never use the pooled t procedures if you have software that will implement Option 1.*

21.8 Avoid Inference about Standard Deviations*

Two basic features of a distribution are its center and variability. In a Normal population, we measure center by the mean and variability by the standard deviation. We use the *t* procedures for inference about population means for Normal populations, and we know that *t* procedures are widely useful for non-Normal populations as well. It is natural to turn next to inference about the standard deviations of Normal populations. Our advice here is short and clear: don't do it without expert advice.

There are methods for inference about the standard deviations of Normal populations. The most common such method is the *F* test for comparing the standard

*This short section offers advice on what not to do. This material is not needed to understand the rest of the book.

deviations of two Normal populations. You will find this test in the menus of most statistical software. *Unlike the t procedures for means, the F test for standard deviations is extremely sensitive to non-Normal distributions.* This lack of robustness does not improve in large samples. When comparing the variability of two populations, it is difficult in practice to tell whether a significant test result is evidence of unequal population variability or simply a sign that the populations are not Normal. Because this test is of little use in practice, we don't give its details.

The deeper difficulty underlying the very poor robustness of Normal population procedures for inference about variability already appeared in our work on describing data. The standard deviation is a natural measure of variability for Normal distributions but not for distributions in general. In fact, because skewed distributions have asymmetric tails, no single numerical measure does a good job of describing the variability of a skewed distribution. In summary, the standard deviation is not always a useful parameter, and even when it is (for symmetric distributions), the results of inference about the standard deviation are not trustworthy. Consequently, *we do not recommend trying to make inferences about population standard deviations in basic statistical practice.*[11]

CHAPTER 21 SUMMARY

- The data in a **two-sample problem** are two independent SRSs, each drawn from a separate population.
- Tests and confidence intervals for the difference between the means μ_1 and μ_2 of two Normal populations start from the difference $\bar{x}_1 - \bar{x}_2$ between the two sample means. Because of the central limit theorem, the resulting procedures are approximately correct for other population distributions when the sample sizes are large.
- Draw independent SRSs of sizes n_1 and n_2 from two Normal populations with parameters μ_1, σ_1, and μ_2, σ_2. The **two-sample *t* statistic** is

$$t = \frac{(\bar{x}_1 - \bar{x}_2) - (\mu_1 - \mu_2)}{\sqrt{\dfrac{s_1^2}{n_1} + \dfrac{s_2^2}{n_2}}}$$

- For hypothesis tests, $H_0: \mu_1 = \mu_2$ means $\mu_1 - \mu_2 = 0$, so the numerator above simplifies to $\bar{x}_1 - \bar{x}_2$. The statistic t has approximately a t distribution.
- There are two choices for the degrees of freedom of the two-sample t statistic. Option 1: software produces accurate probability values using degrees of freedom calculated from the data. Option 2: for conservative inference procedures, use degrees of freedom equal to the smaller of $n_1 - 1$ and $n_2 - 1$.
- The confidence interval for $\mu_1 - \mu_2$ is

$$(\bar{x}_1 - \bar{x}_2) \pm t^* \sqrt{\frac{s_1^2}{n_1} + \frac{s_2^2}{n_2}}$$

- The critical value t^* from Option 1 gives a confidence level very close to the desired level C. Option 2 produces a margin of error at least as wide as is needed for the desired level C.
- Significance tests for $H_0: \mu_1 = \mu_2$ are based on

$$t = \frac{\bar{x}_1 - \bar{x}_2}{\sqrt{\dfrac{s_1^2}{n_1} + \dfrac{s_2^2}{n_2}}}$$

- *P*-values calculated from Option 1 are very accurate. Option 2 *P*-values are always at least as large as the true *P*.

- The two-sample t procedures are quite robust against departures from Normality. Guidelines for practical use are similar to those for one-sample t procedures. Equal sample sizes are recommended.
- Procedures for inference about the standard deviations of Normal populations are very sensitive to departures from Normality. Avoid inference about standard deviations unless you have expert advice.

CHECK YOUR SKILLS

21.16 The 2019 National Assessment of Educational Progress (NAEP) gave a mathematics test to a random sample of eighth-graders in the United States. The mean score was 282 out of 500. To give a confidence interval for the mean score of all eighth-graders in the United States, you would use

(a) the two-sample t interval.

(b) the matched pairs t interval.

(c) the one-sample t interval.

21.17 In the 2019 NAEP sample of eighth-graders in the United States, the mean mathematics scores were 294 for students from Massachusetts and 276 for students from California. To see if this difference is statistically significant, you would use

(a) the two-sample t test.

(b) the matched pairs t test.

(c) the one-sample t test.

21.18 Bananas are thought to provide a high-carbohydrate energy boost before a run and improve performance. To test this, 40 adults from a large city who regularly compete in 10K races are recruited for a study. The runners are randomly divided into two groups of 20. Both groups compete in a 10K race conducted in the city, and their times are recorded. Next, one group is instructed to follow their regular routine for preparing for the next 10K race. The other group is instructed to follow their regular routine and, in addition, eat a banana before running the race. You compare the change in times between the first and second races for the two groups and use

(a) the two-sample t test.

(b) the matched pairs t test.

(c) the one-sample t test.

21.19 One major reason the two-sample t procedures are widely used is that they are quite *robust* This means that

(a) t procedures do not require that we know the standard deviations of the populations.

(b) confidence levels and P-values from the t procedures are quite accurate even if the population distribution is not exactly Normal.

(c) confidence levels and P-values from the t procedures are quite accurate even if the degrees of freedom are not known exactly.

21.20 Children and adolescents with autism spectrum disorders often suffer from obsessive-compulsive behavior. Will the drug fluoxetine help reduce the frequency and severity of this behavior in people with autism spectrum disorders? Researchers randomly assigned subjects, aged 7.7–18 years, to either fluoxetine or a placebo. Of these subjects, 54 were assigned to fluoxetine and 55 to the placebo. Before treatment and after 16 weeks on treatment, subjects were given a test measuring the frequency and severity of obsessive-compulsive behavior. Scores range from 0 to 20, with higher scores indicating more severe symptoms. The change in score (score at outset − score after 16 weeks) was recorded for each participant.[12] To compare the mean change in scores for the two groups using the two-sample t procedures with the conservative Option 2, the correct degrees of freedom is

(a) 55. (b) 54. (c) 53.

21.21 In Question 21.20, are the conditions for two-sample t inference satisfied?

(a) Maybe: the study was randomized, but we need to look at the data to check Normality because the sample sizes are small.

(b) No: scores in a range between 0 and 20 can't be Normal.

(c) Yes: the study was randomized, and large sample sizes make the Normality condition unnecessary.

21.22 In Question 21.20, the researchers were interested in whether fluoxetine decreased symptoms (had a large change in score) compared with a placebo. If we let $\mu_{fluoxetine}$ denote the true mean change in scores after 16 weeks for children and adolescents taking fluoxetine and let $\mu_{placebo}$ denote the true mean change in scores after 16 weeks for children and adolescents taking the placebo, we will test the hypotheses

(a) $H_0: \mu_{fluoxetine} = \mu_{placebo}$ versus $H_a: \mu_{fluoxetine} > \mu_{placebo}$.

(b) $H_0: \mu_{fluoxetine} = \mu_{placebo}$ versus $H_a: \mu_{fluoxetine} \neq \mu_{placebo}$.

(c) $H_0: \mu_{fluoxetine} = \mu_{placebo}$ versus $H_a: \mu_{fluoxetine} < \mu_{placebo}$.

21.23 In Question 21.20, the 54 subjects receiving fluoxetine had a mean change in score of 3.72 with standard deviation 4.22; the 55 subjects receiving the placebo had a mean change in score of 2.53 with standard deviation 5.05. The two-sample t statistic for comparing the population means has value

(a) 1.336.

(b) −1.19.

(c) −1.336.

21.24 The P-value for testing the hypotheses from the previous questions satisfies

(a) $0.01 < P < 0.05$.

(b) $0.05 < P < 0.10$.

(c) $0.10 < P$.

CHAPTER 21 EXERCISES

Exercises 21.25 through 21.34 are based on summary statistics rather than raw data. This information is typically all that is presented in published reports. You can perform inference procedures by hand from the summaries. Use the conservative Option 2 (degrees of freedom the smaller of $n_1 - 1$ and $n_2 - 1$) for two-sample t confidence intervals and P-values. You must trust that the authors understood the conditions for inference and verified that they apply. This isn't always true.

21.25 **Do Women Talk More Than Men?** Equip male and female students with a small device that secretly records sound for a random 30 seconds during each 12.5-minute period over two days. Count the words each subject speaks during each recording period, and from this, estimate how many words per day each subject speaks. The published report includes a table summarizing six such studies.[13] Here are two of the six:

	Sample Size		Estimated Average Number (SD) of Words Spoken per Day	
Study	Women	Men	Women	Men
1	56	56	16,177 (7520)	16,569 (9108)
2	27	20	16,496 (7914)	12,867 (8343)

Readers are supposed to understand that, for example, the 56 women in the first study had $\bar{x} = 16{,}177$ and $s = 7520$. It is commonly thought that women talk more than men. Does either of the two samples support this idea? For each study:

(a) state hypotheses in terms of the population means for men (μ_M) and women (μ_F).

(b) find the two-sample t statistic.

(c) what degrees of freedom does Option 2 use to get a conservative P-value?

(d) compare your value of t with the critical values in Table C. What can you say about the P-value of the test?

(e) What do you conclude from the results of these two studies?

21.26 **Alcohol and Zoning Out.** Healthy men aged 21–35 were randomly assigned to one of two groups: half received 0.82 gram of alcohol per kilogram of body weight; half received a placebo. Participants were then given 30 minutes to read up to 34 pages of Tolstoy's *War and Peace* (beginning at Chapter 1, with each page containing approximately 22 lines of text). Every two to four minutes, participants were prompted to indicate whether they were "zoning out." The proportion of times participants indicated they were zoning out was recorded for each subject. The following table summarizes data on the proportion of episodes of zoning out.[14] (The study report gave the standard error of the mean $s/\sqrt{n}$, abbreviated as SEM, rather than the standard deviation s.)

Teymaz Valley

Group	n	$\bar{x}$	SEM
Alcohol	25	0.25	0.05
Placebo	25	0.12	0.03

(a) What are the two sample standard deviations?

(b) What degrees of freedom does the conservative Option 2 use for two-sample t procedures for these samples?

(c) Using Option 2, give a 95% confidence interval for the mean difference between the two groups.

(d) Would a test of the null hypothesis, using Option 2, of no difference between the two group means against the two-sided alternative be significant at the 0.05 level? Note, as was the case in Exercise 17.42 (page 401), you can carry out a level $\alpha = 1 - C$ two-sided test of the null hypothesis $H_0\colon \mu_1 = \mu_2$ from a level C confidence interval for $\mu_1 - \mu_2$ by rejecting H_0 if $\mu_1 - \mu_2$ is outside the interval.

21.27 **Mood and Food.** In a study of the effects of mood on evaluation of nutritious food, 208 subjects were randomly assigned to read either a happy story (to induce a positive mood) or a control (no story, neutral mood) group. Subjects were then asked to evaluate their attitude toward a certain food on a nine-point scale, with higher numbers indicating a more positive attitude toward the food. The following table summarizes data on the attitude rating:[15]

Group	n	$\bar{x}$	s
Positive mood	104	4.30	2.05
Neutral mood	104	5.50	1.74

(a) What are the standard errors for the sample means of the two groups?

(b) What degrees of freedom does the conservative Option 2 use for two-sample t procedures for these data?

(c) Test the null hypothesis of no difference between the two group means against the two-sided alternative. Use the degrees of freedom from part (b). What is the conclusion of the hypothesis test?

21.28 Are Only Children More Narcissistic? Researchers investigated whether the stereotype that only children are more narcissistic than children with siblings is true. As part of the study, the researchers compared the perceptions of people who were only children with those of people who had siblings. A representative panel of 556 people rated "a typical only child" using the the Narcissistic Admiration and Rivalry Questionnaire Short Scale test. Of the panel of 556 people, 105 were only children and 451 were people with siblings. Here are the summaries of the admiration scores for both groups:[16]

Group	Sample Size	Mean	Standard Deviation
Only child	105	3.50	1.03
Siblings	451	3.55	0.98

Is there evidence that the mean admiration score of a "typical only child" as rated by the population of only children is different from the mean admiration score of a "typical only child" rated by the population of people who have siblings?

21.29 Are Only Children More Narcissistic? Researchers investigated whether the stereotype that only children are more narcissistic than children with siblings is true. A representative panel of 1810 people were given the Narcissistic Admiration and Rivalry Questionnaire Short Scale test. Of these 1810 people, 233 were only children and 1577 had siblings. Here are the summaries of the admiration scores for both groups:[17]

Group	Sample Size	Mean	Standard Deviation
Only child	233	1.95	1.04
Siblings	1577	2.06	1.11

Is there evidence that the mean admiration score of a "typical only child" as rated by the population of only children is lower than the mean admiration score of a "typical only child" rated by the population of people who have siblings?

21.30 Height and the Big Picture. Forty-six college students were randomly divided into two groups of size 23. One group was asked to imagine being on the upper floor of a tall building (where one has a "big-picture view" of the area around the building) and the other on the lowest floor. Participants were then asked to choose between a job that required more detail orientation versus a job that required a more big-picture orientation. They rated their job preferences on an 11-point scale, with higher numbers corresponding to a greater preference for the big-picture job. Here are the summary statistics:[18]

Group	Group Size	Mean	Standard Deviation
Low	23	4.61	3.08
High	23	6.68	3.45

(a) What degrees of freedom would you use in the conservative two-sample t procedures to compare the lower and higher floor groups?

(b) What is the two-sample t test statistic for comparing the mean job preference ratings for the two groups?

(c) Test the null hypothesis of no difference between the two population means against the two-sided alternative. Use your statistic from part (b) with degrees of freedom from part (a).

21.31 Concussions and Brain Size. What is the effect of concussions on the brain? Researchers measured the brain sizes (hippocampal volume, in microliters) of 25 collegiate football players with a history of clinician-diagnosed concussion and 25 collegiate football players without a history of concussion. Here are the summary statistics:[19]

Group	Group Size	Mean	Standard Deviation
Concussion	25	5784	609.3
No concussion	25	6489	815.4

(a) Is there evidence of a difference in mean brain size between football players with a history of concussion and those without concussions?

(b) The researchers in this study stated that participants were "consecutive cases of healthy National Collegiate Athletic Association Football Bowl Subdivision Division I football athletes with ($n = 25$) or without ($n = 25$) a history of clinician-diagnosed concussion . . . between June 2011 and August 2013" at a U.S. psychiatric research institute specializing in neuroimaging among collegiate football players. What effect does this information have on your conclusions in part (a)?

21.32 Coaching and SAT Scores. Coaching companies claim that their courses can raise the SAT scores of high school students. Of course, students who retake the SAT without paying for coaching generally raise their scores, too. A random sample of students who took the SAT twice found 427 who were coached and 2733 who were uncoached.[20] Looking at their Math scores on the first and second tries, we have these summary statistics:

		Try 1		Try 2		Gain	
	n	$\bar{x}$	s	$\bar{x}$	s	$\bar{x}$	s
Coached	427	521	100	561	100	40	58
Uncoached	2733	505	101	527	101	22	50

The summary statistics for Gain are based on the changes in the scores of the individual students. Let's first ask if students who are coached increased their scores significantly.

(a) You could use the information on the Coached line to carry out either a two-sample t test comparing Try 1 with Try 2 for coached students or a matched pairs t test using Gain. Which is the correct test? Why?

(b) Carry out the proper test. What do you conclude?

(c) Give a 99% confidence interval for the mean gain of all students who are coached.

21.33 Coaching and SAT Scores (continued). What we really want to know is whether coached students improve more than uncoached students and whether any advantage is large enough to be worth paying for. Use the information in the previous exercise to answer these questions.

(a) Is there good evidence that coached students improved their scores by more on average than uncoached students?

(b) How much more do coached students gain on average? Give a 99% confidence interval.

(c) Based on your work, what is your opinion: Do you think coaching courses are worth paying for?

21.34 Coaching and SAT Scores: Critique. The data used in the previous two problems came from a random sample of students who took the SAT twice. The response rate was 63%, which is pretty good for nongovernment surveys, so let's accept that the respondents do represent all students who took the exam twice. Nonetheless, we can't be sure that coaching actually *caused* the coached students to gain more than the uncoached students. Explain briefly but clearly why this is so.

21.35 Appliance Sales. A research firm supplies manufacturers with estimates of the sales of their products from samples of stores. Marketing managers often look at the sales estimates and ignore sampling error. An SRS of 50 stores this month shows mean sales of 41 units of a particular appliance with standard deviation of 11 units. During the same month last year, an SRS of 52 stores gave mean sales of 38 units of the same appliance with a standard deviation of 13 units. An increase from 38 to 41 is a rise of 7.9%. The marketing manager is happy because sales are up 7.9%.

(a) Give a 95% confidence interval for the difference in mean number of units of the appliance sold at all retail stores.

(b) Explain in language that management can understand why they cannot be confident that sales rose by 7.9% and, in fact, may have dropped.

21.36 Two Marketing Strategies. Credit card companies earn a percentage of the amount charged on their credit cards, paid by the stores that accept the card. A credit card company compares two proposals for increasing the amount that its customers charge on their credit cards. Proposal 1 offers to eliminate the annual fee for customers who charge $1800 or more during the year on their card. Proposal 2 offers a small percentage of the total amount charged as a cash reward at the end of the year. The credit card company offers each proposal to an SRS of 100 of its existing customers. At the end of the year, the total amount charged by each customer is recorded. Here are the summary statistics.

Group	n	$\bar{x}$	s
Proposal 1	100	$1319	$261
Proposal 2	100	$1372	$274

(a) Do the data show a significant difference between the mean amounts charged by customers offered the two proposed plans? Give the null and alternative hypotheses and calculate the two-sample t statistic. Obtain the P-value, using Option 2. State your practical conclusions.

(b) The distributions of the amounts charged on credit cards are skewed to the right. However, outliers are prevented by the limits that the credit card companies impose on credit balances. Do you think that skewness threatens the validity of the text that you used in part (a)? Explain your answer.

Exercises 21.37 through 21.46 include the actual data. To apply the two-sample t procedures, use Option 1 if you have technology that implements that method. Otherwise, use Option 2.

21.37 Improving Your Tips. Researchers gave 40 index cards to a waitress at an Italian restaurant in New Jersey. Before delivering the bill to each customer, the waitress selected one of the index cards without reading it and wrote on the bill the same message that was printed on the card. Twenty of the cards had the message "The weather is supposed to be really good tomorrow. I hope you enjoy the day!" Another 20 cards contained the message "The weather is supposed to be not so good tomorrow. I hope you enjoy the day anyway!" After the customers left, the waitress recorded the amount of the tip (percentage of bill) before taxes. Here are the tips for those receiving the good-weather message:[21] TIP4

20.8	18.7	19.9	20.6	21.9	23.4	22.8	24.9	22.2	20.3
24.9	22.3	27.0	20.5	22.2	24.0	21.2	22.1	22.0	22.7

The tips for the 20 customers who received the bad weather message are

18.0	19.1	19.2	18.8	18.4	19.0	18.5	16.1	16.8	14.0
17.0	13.6	17.5	20.0	20.2	18.8	18.0	23.2	18.2	19.4

(a) Make stemplots or histograms of both sets of data. Because the distributions are reasonably symmetric with no extreme outliers, the t procedures will work well.

(b) Is there good evidence that the two different messages produce different percentage tips? State the hypotheses, carry out a two-sample t test, and report your conclusions.

21.38 Do Good Smells Bring Good Business? In Exercise 21.9, you examined the effects of a lavender odor on customer behavior in a small restaurant. Lavender is a relaxing odor. The researchers also looked at the effects of lemon, a stimulating odor. The design of the study is described in Exercise 21.9. Here are the times, in minutes, that customers spent in the restaurant when no odor was present: ODORS3

103	68	79	106	72	121	92	84	72	92
85	69	73	87	109	115	91	84	76	96
107	98	92	107	93	118	87	101	75	86

When a lemon odor was present, customers lingered for these times:

78	104	74	75	112	88	105	97	101	89
88	73	94	63	83	108	91	88	83	106
108	60	96	94	56	90	113	97		

(a) Examine both samples. Does it appear that use of two-sample t procedures is justified? Do the sample means suggest that a lemon odor changes the average length of stay?

(b) Does a lemon odor influence the length of time customers stay in the restaurant? State the hypotheses, carry out a t test, and report your conclusions.

21.39 Improving Your Tips (continued). Use the data in Exercise 21.37 to give a 95% confidence interval for the difference between the mean percentage tips for the two different messages. TIP4

21.40 The Power of Positive Thinking? Does the way the press depicts the economic future affect the stock market? To investigate this, researchers analyzed the longest article about economic conditions from the front page of the Money section of *USA Today* from a randomly chosen weekday of each week between August 2007 and June 2009. Articles were rated as to how positive or negative they were about the economic future. For each week, the change in the Dow Jones Industrial Average (average DJIA value the week after the article appeared minus the average the week the article appeared) was computed. Positive values of the change indicate that the DJIA increased.[22] Here are the changes in DJIA corresponding to very positive articles: DJIA

−325	−200	−225	−75	−25	25	50
225	25	−225	−250	200	250	75

Here are the changes in DJIA values corresponding to very negative articles:

150	300	225	125	−175	−225	−375
−175	0	125	175	475		

Is there good evidence that the DJIA performs differently after very positive articles than after very negative articles?

(a) Do the sample means suggest that there is a difference in the change in the DJIA after very positive articles versus after very negative articles?

(b) Make stemplots for both samples. Are there any obvious departures from Normality?

(c) Test the hypothesis $H_0: \mu_1 = \mu_2$ against the two-sided alternative. What do you conclude from part (a) and from the result of your test?

(d) Among a host of different factors that are claimed to have triggered the economic crisis of 2007–2009, one was a "culture of irresponsibility" in the way the future was depicted in the press. Do the data provide any evidence that negative articles in the press contributed to poor performance of the DJIA?

21.41 Paper or Tablet? Two hundred thirty-one students were randomly assigned to read either digital ($n = 119$) or paper ($n = 112$) versions of a leadership paper. Students were then given a 10-item multiple-choice test of recall accuracy.[23] The data set RECALL gives the scores for all the students. RECALL

(a) Do the sample means suggest that there is a difference in the mean recall scores between the two groups?

(b) Make stemplots for both samples. Are there any obvious departures from Normality?

(c) Test the hypothesis $H_0: \mu_{tablet} = \mu_{paper}$ against the one-sided alternative that students who read the paper version have higher recall scores than those who read the tablet version. What do you conclude from part (a) and from the result of your test?

21.42 How Big is the Difference? Continue your work from Exercise 21.40. A researcher wants to know how big a difference there is in the change in the DJIA after reports of a positive economic outlook compared to reports of a negative economic outlook. Give a 90% confidence interval for the difference in mean change in the DJIA. DJIA

21.43 Do Women Talk More Than Men? Another Study. Exercise 21.25 described a series of six studies investigating the number of words women and men speak per day and gives results from two of these studies. Here are the results from another of these studies. The estimated numbers of words spoken per day for 27 women are TALK2

15,357	13,618	9,783	26,451	12,151	8,391	19,763
25,246	8,427	6,998	24,876	6,272	10,047	15,569
39,681	23,079	24,814	19,287	10,351	8,866	10,827
12,584	12,764	19,086	26,852	17,639	16,616	

The estimated numbers of words spoken per day for 20 men are

28,408	10,084	15,931	21,688	37,786	10,575	12,880
11,071	17,799	13,182	8,918	6,495	8,153	7,015
4,429	10,054	3,998	12,639	10,974	5,255	

Does this study provide good evidence that women talk more than men, on average?

(a) Make stemplots for both samples. Are there any obvious deviations from Normality? In spite of these deviations from Normality, it is safe to use the t procedures. Explain.

(b) Test the hypothesis $H_0: \mu_1 = \mu_2$ against the one-sided alternative that the mean number of words per day for women (μ_1) is greater than the mean number of words per day for men (μ_2). What do you conclude?

Do Birds Learn to Time Their Breeding? *Blue titmice eat caterpillars. The birds would like lots of caterpillars around when they have young to feed, but they breed later than peak caterpillar season. Do the birds time when they breed based on the previous year's caterpillar supply? Researchers randomly assigned seven pairs of birds to have the natural caterpillar supply supplemented while feeding their young and another six pairs to serve as a control group relying on natural food supply. The next year, they measured how many days after the caterpillar peak the birds produced their nestlings.*[24] *Exercises 21.44 through 21.46 are based on this experiment.*

Mark L Stanley/Getty Images

21.44 Did the Randomization Produce Similar Groups? The first thing to do is to compare the two groups in the first year. The only difference should be the chance effect of the random assignment. The study report says: "In the experimental year, the degree of synchronization [breeding timed to coincide with caterpillar supply] did not differ between food-supplemented and control females." For this comparison, the report gives $t = -1.05$. What type of t statistic (paired or two-sample) is this? What are the degrees of freedom for this statistic? Show that this t leads to the quoted conclusion.

21.45 Did the Treatment Have an Effect? The investigators expected the control group to adjust their breeding date the next year, whereas the well-fed supplemented group had no reason to change. The report continues: "in the following year food-supplemented females were more out of synchrony with the caterpillar peak than the controls." Here are the data (days after the caterpillar peak): BREED

Control	4.6	2.3	7.7	6.0	4.6	−1.2	
Supplemented	15.5	11.3	5.4	16.5	11.3	11.4	7.7

Carry out a t test and show that it leads to the quoted conclusion.

21.46 Year-to-Year Comparison. Rather than compare the two groups in each year, we could compare the behavior of each group in the first and second years. The study report says: "Our main prediction was that females receiving additional food in the nestling period should not change laying date the next year, whereas controls, which (in our area) breed too late in their first year, were expected to advance their laying date in the second year."

Comparing days behind the caterpillar peak in Years 1 and 2 gave $t = 0.63$ for the control group and $t = -2.63$ for the supplemented group. Are these paired or two-sample t statistics? What are the degrees of freedom for each t? Show that these t-values do *not* agree with the prediction.

The remaining exercises ask you to answer questions from data without having the details outlined for you. The exercise statements give you the ***State*** *step of the four-step process. Follow the* ***Plan, Solve,*** *and* ***Conclude*** *steps as illustrated in Examples 21.2 (page 470) and 21.3 (page 472) for significance tests and Example 21.4 (page 473) for confidence intervals. Remember that examining the data and discussing the conditions for inference are part of the* ***Solve*** *step.*

21.47 Thinking About Money Changes Behavior. 4step Kathleen Vohs of the University of Minnesota and her coworkers carried out several randomized comparative experiments on the effects of thinking about money. Here's part of one such experiment.[25] Ask student subjects to unscramble 30 sets of five words to make a meaningful phrase from four of the five words. The control group unscrambled phrases like "cold it desk outside is" into "it is cold outside." The treatment group unscrambled phrases that led to thinking about money, turning "high a salary desk paying" into "a high-paying salary." Then each subject worked a hard puzzle, knowing that he or she could ask for help. Here are the times, in seconds, until subjects asked for help. For the treatment group: MNYTHNK

609	444	242	199	174	55	251	466	443
531	135	241	476	482	362	69	160	

For the control group:

118	272	413	291	140	104	55	189	126
400	92	64	88	142	141	373	156	

The researchers suspected that money is connected with self-sufficiency, so that the treatment group will ask for help less quickly on the average. Do the data support this idea?

STATISTICS IN YOUR WORLD

Is Money the Root of All Evil?

That may go too far, but Kathleen Vohs and her coworkers show that even thinking about money has strong effects. Exercise 21.47 describes a small part of their work. What does Professor Vohs say about the consequences of having money? "Money makes people feel self-sufficient and behave accordingly." With money, you can achieve your goals with less help from others. You feel less dependent on others and more willing to work toward your own goals. Maybe that's good. You also prefer to be less involved with others, so that self-sufficiency is a barrier to close relationships with others. Maybe that's not good. Scientists don't tell us what's good or not good, just that money increases our sense of self-sufficiency.

21.48 Adolescent Obesity. 4step Adolescent obesity is a serious health risk affecting more than 5 million young people in the United States alone. Laparoscopic adjustable gastric banding has the potential to provide a safe and effective treatment. Fifty adolescents between 14 and 18 years old with a body mass index higher than 35 were recruited from the Melbourne, Australia, community for the study.[26] Twenty-five were randomly selected to undergo gastric banding, and the remaining 25 were assigned to a supervised lifestyle intervention program involving diet, exercise, and behavior modification. All subjects were followed for two years. Here are the weight losses, in kilograms, for the subjects who completed the study. In the gastric banding group: ADOBESE

35.6	81.4	57.6	32.8	31.0	37.6	36.5	−5.4	27.9	49.0	64.8	39.0
43.0	33.9	29.7	20.2	15.2	41.7	53.4	13.4	24.8	19.4	32.3	22.0

In the lifestyle intervention group:

6.0	2.0	−3.0	20.6	11.6	15.5	−17.0	1.4	4.0
−4.6	15.8	34.6	6.0	−3.1	−4.3	−16.7	−1.8	−12.8

Is there good evidence that gastric banding leads to greater weight loss than the lifestyle intervention program?

21.49 Shared Pain and Bonding. 4step Although painful experiences are involved in social rituals in many parts of the world, little is known about the social effects of pain. Will sharing a painful experience in a small group lead to greater bonding of group members than sharing a similar nonpainful experience? Fifty-four university students in South Wales were divided at random into a pain group containing 27 students and a no-pain group containing the remaining 27 students. Pain was induced by two tasks. In the first task, students submerged their hands in freezing water for as long as possible, moving metal balls at the bottom of the vessel into a submerged container. In the second task, students performed a standing wall squat with back straight and knees at 90 degrees for as long as possible. The no-pain group completed the first task using room temperature water for 90 seconds and the second task by balancing on one foot for 60 seconds, changing feet if necessary. In both the pain and no-pain settings, the students completed the tasks in small groups which typically consisted of four students and contained similar levels of group interaction. Afterward, each student completed a questionnaire to create a bonding score based on responses to seven statements such as "I feel the participants in this study have a lot in common" or "I feel I can trust the other participants." Each response was scored on a five-point scale (1 = strongly disagree, 5 = strongly agree), and the scores on the seven statements were averaged to create a bonding score for each subject. Here are the bonding scores for the subjects in the two groups:[27] PAIN

No-pain group:	3.43	4.86	1.71	1.71	3.86	3.14	4.14
	3.14	4.43	3.71	3.00	3.14	4.14	4.29
	2.43	2.71	4.43	3.43	1.29	1.29	3.00
	3.00	2.86	2.14	4.71	1.00	3.71	
Pain group:	4.71	4.86	4.14	1.29	2.29	4.43	3.57
	4.43	3.57	3.43	4.14	3.86	4.57	4.57
	4.29	1.43	4.29	3.57	3.57	3.43	2.29
	4.00	4.43	4.71	4.71	2.14	3.57	

Do the data show that sharing a painful experience in a small group leads to higher bonding scores for group members than sharing a similar nonpainful experience?

21.50 Each Day I Am Getting Better in Math. 4step A "subliminal" message is below our threshold of awareness but may nonetheless influence us. Can subliminal messages help students learn math? A group of students who had failed the mathematics part of the City University of New York Skills Assessment Test agreed to participate in a study to find out.

All received a daily subliminal message, flashed on a screen too rapidly to be consciously read. The treatment group of 10 students (chosen at random) was exposed to "Each day I am getting better in math." The control group of eight students was exposed to a neutral message, "People are walking on the street." All students participated in a summer program designed to raise their math skills, and all took the assessment test again at the end of the program. Table 21.3 gives data on the subjects' scores before and after the program.[28] Is there good evidence that the treatment brought about a greater improvement in math scores than the neutral message? How large is the mean difference in gains between treatment and control? (Use 95% confidence.) SUBLIM

TABLE 21.3 Mathematics skills scores before and after a subliminal message

Treatment Group		Control Group	
Before	After	Before	After
18	24	18	29
18	25	24	29
21	33	20	24
18	29	18	26
18	33	24	38
20	36	22	27
23	34	15	22
23	36	19	31
21	34		
17	27		

21.51 Shared Pain and Bonding (continued).

(a) Use the data in Exercise 21.49 to give a 90% confidence interval for the difference in the mean bonding score for students in the no-pain and pain groups.

(b) Give a 90% confidence interval for the mean bonding score of students in the pain group. PAIN

21.52 Tropical Flowers. Different varieties of the tropical flower *Heliconia* are fertilized by different species of hummingbirds. Over time, the lengths of the flowers and the forms of the hummingbirds' beaks have evolved to match each other. Here are data on the lengths in millimeters of two color varieties of the same species of flower on the island of Dominica:[29] FLOWERS

Art Wolfe/Getty Images

H. Caribaea Red							
41.90	42.01	41.93	43.09	41.47	41.69	39.78	40.57
39.63	42.18	40.66	37.87	39.16	37.40	38.20	38.07
38.10	37.97	38.79	38.23	38.87	37.78	38.01	
H. Caribaea Yellow							
36.78	37.02	36.52	36.11	36.03	35.45	38.13	37.1
35.17	36.82	36.66	35.68	36.03	34.57	34.63	

Is there good evidence that the mean lengths of the two varieties differ? Estimate the difference between the population means. (Use 95% confidence.)

21.53 Student Drinking. A professor asked her sophomore students, "How many drinks do you typically have per session? (A drink is defined as one 12-oz beer, one 4-oz glass of wine, or one 1-oz shot of liquor.)" Some of the students didn't drink. Table 21.4 gives the responses of the female and male students who did drink.[30] It is likely that some of the students exaggerated a bit. The sample is all students in one large sophomore-level class. The class is popular, so we are tentatively willing to regard its members as an SRS of sophomore students at this college. Do a complete analysis that reports on DRINKS

(a) the drinking behavior claimed by sophomore women.

(b) the drinking behavior claimed by sophomore men.

(c) a comparison of the behavior of women and men.

TABLE 21.4 Drinks per session claimed by female and male students

Female Students												
2.5	9	1	3.5	2.5	3	1	3	3	3	3	2.5	2.5
5	3.5	5	1	2	1	7	3	7	4	4	6.5	4
3	6	5	3	8	6	6	3	6	8	3	4	7
4	5	3.5	4	2	1	5	5	3	3	6	4	2
7	7	7	3.5	3	2.5	10	5	4	9	8	1	6
2	5	2.5	3	4.5	9	5	4	4	3	4	6	7
4	5	1	5	3	4	10	7	3	4	4	4	4
2	1	2.5	2.5									
Male Students												
7	7.5	8	15	3	4	1	5	11	4.5	6	4	10
16	4	8	5	9	7	7	3	5	6.5	1	12	4
6	8	8	4.5	10.5	8	6	10	1	9	8	7	8
15	3	10	7	4	6	5	2	10	7	9	5	8
7	3	7	6	4	5	2	5	5.5	9	10	10	4
8	4	2	4	12.5	3	15	2	6	3	4	3	10
6	4.5	5										

CHAPTER

22

When you complete this chapter, you will be able to:

22.1 Solve problems involving the sample proportion $\hat{p}$ and its sampling distribution.

22.2 Use the large-sample z procedure to give a confidence interval for a population proportion p when the conditions are met.

22.3 Calculate the required sample size to estimate a population proportion within a given margin of error.

22.4 Use the z statistic to carry out a test of significance for the hypothesis $H_0: p = p_0$ about a population proportion p against a one-sided or two-sided alternative.

22.5 After confirming that conditions are met, use the plus four confidence interval to estimate a population proportion.

Inference about a Population Proportion

Our discussion of statistical inference to this point has concerned making inferences about population *means*. Estimating the mean response for a population or comparing the means of two populations are common when the response variable is quantitative and takes a numerical value with some unit of measurement. Now we turn to questions about the *proportion* of some outcome in a population. Even when the original response is a quantitative variable, such as total cholesterol, we might be more interested in whether or not someone has cholesterol greater than 200 mg/dL (outcome of interest). Our proportion of interest is then the population proportion of adults who have cholesterol greater than 200 mg/dL, and the methods of this chapter apply. Here are some further examples that call for inference about population proportions.

EXAMPLE 22.1 Risky Behavior in the Age of AIDS

How common is behavior that puts people at risk of AIDS? In the early 1990s, the landmark National AIDS Behavioral Surveys interviewed a random sample of 2673 adult heterosexuals. Of these, 170 had had more than one sexual partner in the past year. That's 6.36% of the sample.[1] Based on these data, what can we say about the percentage of all adult heterosexuals who have multiple partners? We want to *estimate a single population proportion*. This chapter concerns inference about one proportion.

Jennifer Shields/jjs08 images/Getty Images

EXAMPLE 22.2 Cigarette Smoking among High School Seniors

Since the 1990s, daily cigarette smoking among high school seniors has dropped from around 12% to slightly over 3%. In 2017, a random sample of 1725 female and 1564 male high school seniors found that 3.7% of the females and 3.1% of the males had smoked cigarettes daily in the 30 days before the survey.[2] Is this significant evidence that the proportions of daily cigarette smokers differ in the populations of all male and female high school seniors? We want to *compare two population proportions*. This is the topic of Chapter 23.

To make inferences about a population mean μ, we use the mean $\bar{x}$ of a random sample from the population. The reasoning of inference starts with the sampling distribution of $\bar{x}$. Now we follow the same pattern, replacing means with proportions.

22.1 The Sample Proportion $\hat{p}$

We are interested in the unknown proportion p of a population that has some outcome. For convenience, call the outcome we are looking for a "success." In Example 22.1, the population is adult heterosexuals, and the parameter p is the proportion who had had more than one sexual partner in the past year. To estimate p, the National AIDS Behavioral Surveys used random dialing of telephone numbers to contact a sample of 2673 people. Of these, 170 said they had had multiple sexual partners. The statistic that estimates the parameter p is the *sample proportion*:

$$\hat{p} = \frac{\text{number of successes in the sample}}{\text{total number of individuals in the sample}}$$

$$= \frac{170}{2673} = 0.0636$$

Read the sample proportion $\hat{p}$ as "p-hat."

How good is the statistic $\hat{p}$ as an estimate of the parameter p? To find out, we consider the question "What would happen if we took many samples?" The sampling distribution of $\hat{p}$ answers this question. Here are the facts.[3]

Sampling Distribution of a Sample Proportion

Draw an SRS of size n from a large population that contains proportion p of successes. Let $\hat{p}$ be the **sample proportion** of successes,

$$\hat{p} = \frac{\text{number of successes in the sample}}{n}$$

Then:

- The mean of the sampling distribution is p.
- The standard deviation of the sampling distribution is

$$\sqrt{\frac{p(1-p)}{n}}$$

- As the sample size increases, the sampling distribution of $\hat{p}$ becomes approximately Normal. That is, for large n, $\hat{p}$ has approximately the $N(p, \sqrt{p(1-p)/n})$ distribution.

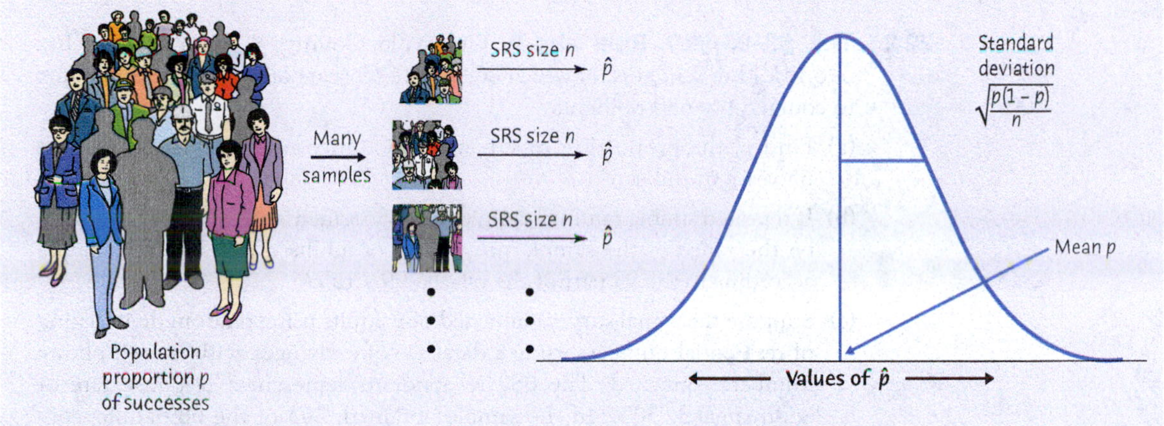

FIGURE 22.1
Select a large SRS from a population in which the proportion p are successes. The sampling distribution of the proportion $\hat{p}$ of successes in the sample is approximately Normal. The mean is p, and the standard deviation is $\sqrt{p(1-p)/n}$.

Figure 22.1 summarizes these facts in a form that helps you recall the big idea of a sampling distribution. The behavior of sample proportions $\hat{p}$ is similar to the behavior of sample means $\bar{x}$, except that the distribution of $\hat{p}$ is only approximately Normal. The mean of the sampling distribution of $\hat{p}$ is the true value of the population proportion p. That is, $\hat{p}$ is an unbiased estimator of p. The standard deviation of $\hat{p}$ gets smaller as the sample size n gets larger, so that estimation is likely to be more accurate when the sample is larger. As is the case for $\bar{x}$, the standard deviation gets smaller only at the rate $\sqrt{n}$. We need four times as many observations to cut the standard deviation in half.

EXAMPLE 22.3 Asking about Risky Behavior

Suppose that, in fact, 6% of all adult heterosexuals had had more than one sexual partner in the past year (and would admit it when asked). The National AIDS Behavioral Surveys interviewed a random sample of 2673 people from this population. In many such samples, the proportion $\hat{p}$ of the 2673 people in the sample who had had more than one partner would vary according to (approximately) the Normal distribution with mean 0.06 and standard deviation

$$\sqrt{\frac{p(1-p)}{n}} = \sqrt{\frac{(0.06)(0.94)}{2673}}$$
$$= \sqrt{0.0000211} = 0.00459$$

APPLY YOUR KNOWLEDGE

22.1 Staph Infections. A study investigated ways to prevent staph infections in surgery patients. In a first step, the researchers examined the nasal secretions of a random sample of 6771 patients admitted to various hospitals for surgery. They found that 1251 of these patients tested positive for *Staphylococcus aureus*, a bacterium responsible for most staph infections.[4]

(a) Describe the population and explain in words what the parameter p is.

(b) Give the numerical value of the statistic $\hat{p}$ that estimates p.

22.2 The 68–95–99.7 Rule and $\hat{p}$. Greenville County, South Carolina, has 396,183 adult residents, of which 80,987 are 65 years or older. A survey wants to contact $n = 689$ residents.[5]

(a) Find p, the proportion of Greenville County adult residents who are 65 years or older.

(b) If repeated simple random samples of 689 residents are taken, what would be the range of the sample proportion of adults over 65 in the sample according to the 95 part of the 68–95–99.7 rule?

(c) Suppose the actual survey contacted 689 adults using random digit dialing of residential numbers using a database of exchanges, with no cell phone numbers contacted. The 689 respondents represent a response rate of approximately 30%. In the sample obtained, 253 of the 689 adults contacted were over 65. Do you have any concerns treating this as a simple random sample from the population of adult residents of Greenville County? Explain briefly.

Stephen Mcsweeny/Shutterstock

22.3 Do You Eat Red Meat? About 60% of American adults include beef and other red meat in their diets.[6] A plant-based meat company contacts an SRS of 1500 American adults and calculates the proportion $\hat{p}$ in this sample who eat red meat.

(a) What is the approximate distribution of $\hat{p}$?

(b) If the sample size were 6000 rather than 1500, what would be the approximate distribution of $\hat{p}$?

22.2 Large-Sample Confidence Intervals for a Proportion

We can follow the same path from sampling distribution to confidence interval as we did for $\bar{x}$ in Chapter 15. To obtain a level C confidence interval for p, we start by capturing the central probability C in the distribution of $\hat{p}$. To do this, go out z^* standard deviations from the mean p, where z^* is the critical value, based on a desired confidence level, that captures the central area C under the standard Normal curve. Figure 22.2 shows the result. The confidence interval is

$$\hat{p} \pm z^* \sqrt{\frac{p(1-p)}{n}}$$

This won't do because we don't know the value of p. So we replace the standard deviation with the **standard error of $\hat{p}$**

$$SE_{\hat{p}} = \sqrt{\frac{\hat{p}(1-\hat{p})}{n}}$$

to get the confidence interval

$$\hat{p} \pm z^* \sqrt{\frac{\hat{p}(1-\hat{p})}{n}}$$

As with previous confidence intervals, this interval has the familiar form

$$\text{estimate} \pm z^* SE_{\text{estimate}}$$

We can trust this confidence interval only for large samples. Because the number of successes must be a whole number, using a continuous Normal distribution to describe the behavior of $\hat{p}$ may not be accurate unless n is large. Because the

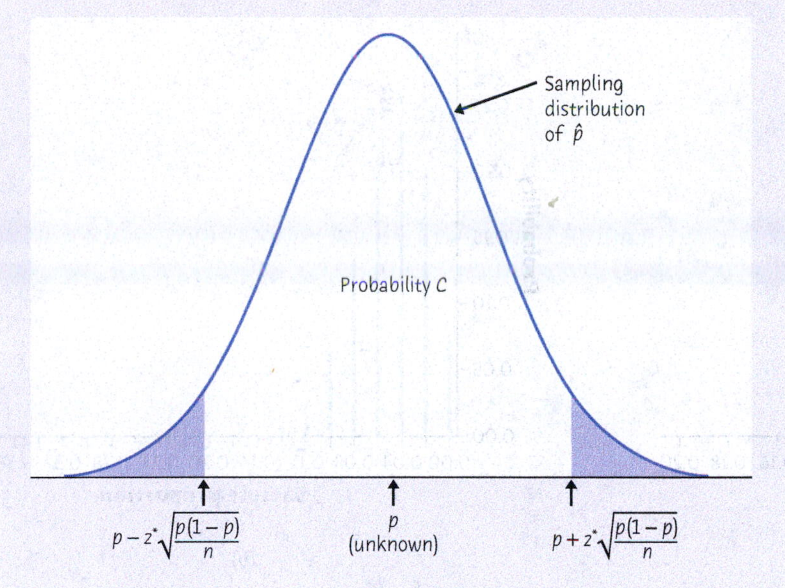

FIGURE 22.2 With probability C, $\hat{p}$ lies within $\pm z^*\sqrt{p(1-p)/n}$ of the unknown population proportion p. That is, in these samples p lies within $\pm z^*\sqrt{p(1-p)/n}$ of $\hat{p}$.

approximation is least accurate for populations that are almost all successes or almost all failures, we require that the sample have both enough successes and enough failures rather than that the overall sample size be large. *Pay attention to both conditions for inference in the following box that summarizes the confidence interval: we must, as usual, be willing to regard the sample as an SRS from the population, and the sample must have both enough successes and enough failures.* The condition on successes and failures ensures that the sample size is large enough to use the Normal approximation without knowing p.

Large-Sample Confidence Interval for a Population Proportion

Draw an SRS of size n from a large population that contains an unknown proportion p of successes. An approximate level C confidence interval for p is

$$\hat{p} \pm z^*\sqrt{\frac{\hat{p}(1-\hat{p})}{n}}$$

where z^* is the critical value for the standard Normal density curve with area C between $-z^*$ and z^*.

Use this interval only when the numbers of successes and failures in the sample are both at least 15.[7]

Figure 22.3 displays the sampling distribution of the sample proportion for a sample of size $n = 50$ and (a) $p = 0.01$, (b) $p = 0.05$, (c) $p = 0.10$, (d) $p = 0.25$, (e) $p = 0.50$, (f) $p = 0.75$, (g) $p = 0.90$, (h) $p = 0.95$, and (i) $p = 0.99$. Our condition that the numbers of successes and failures in the sample are both at least 15 is satisfied in (e) only, and the sampling distribution looks approximately Normal. The condition is nearly satisfied in (d) and (f), and the sampling distribution looks roughly Normal but with a small amount of skewness. As the condition is less and less close to being satisfied, the sampling distribution is increasingly skewed.

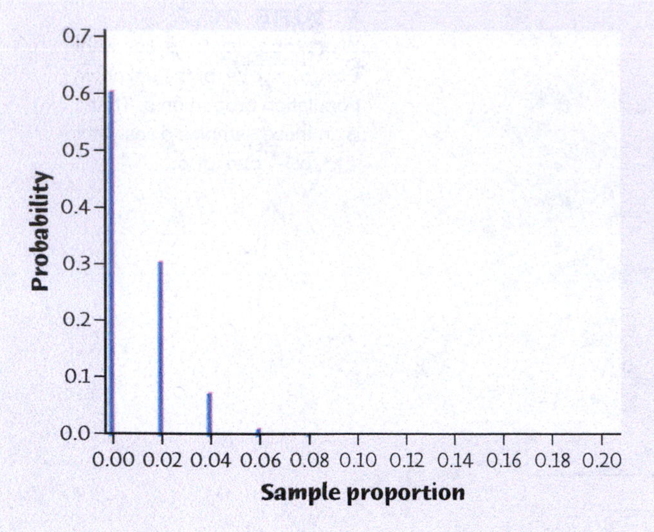
0.7
0.6
0.5
0.4
0.3
0.2
0.1
0.0
Probability
0.00 0.02 0.04 0.06 0.08 0.10 0.12 0.14 0.16 0.18 0.20
Sample proportion

(a)

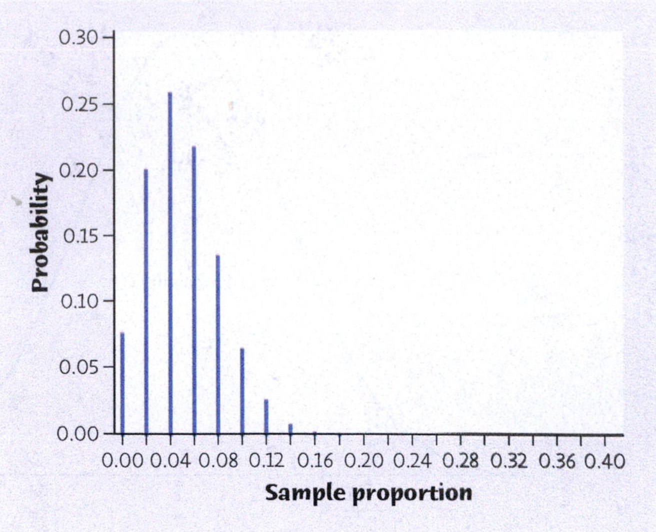
0.30
0.25
0.20
0.15
0.10
0.05
0.00
Probability
0.00 0.04 0.08 0.12 0.16 0.20 0.24 0.28 0.32 0.36 0.40
Sample proportion

(b)

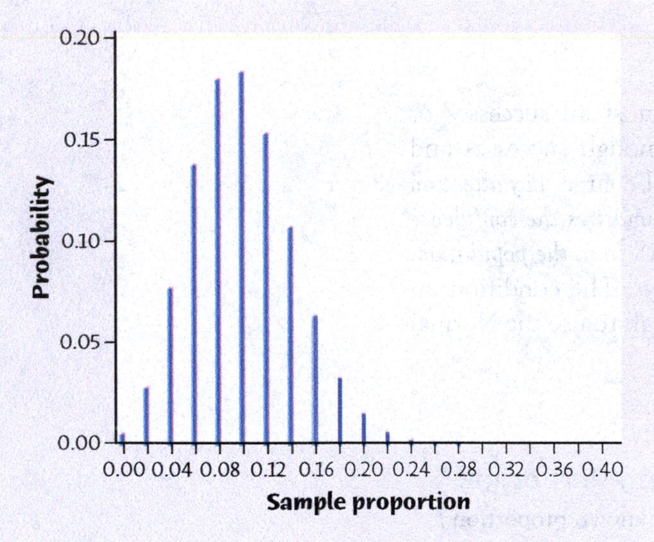
0.20
0.15
0.10
0.05
0.00
Probability
0.00 0.04 0.08 0.12 0.16 0.20 0.24 0.28 0.32 0.36 0.40
Sample proportion

(c)

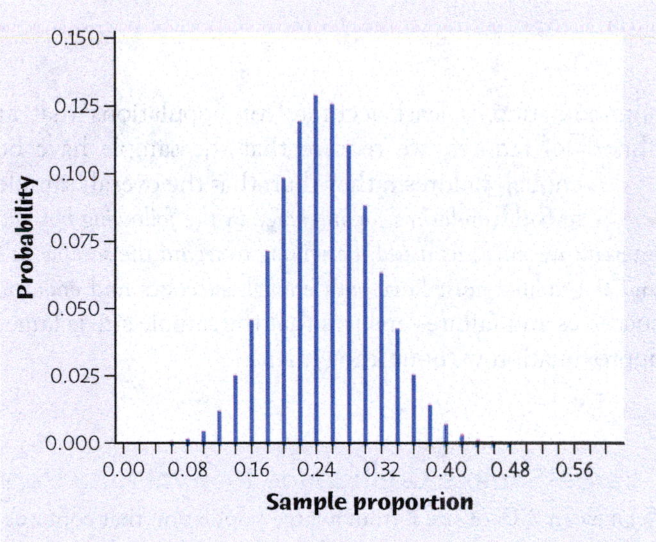
0.150
0.125
0.100
0.075
0.050
0.025
0.000
Probability
0.00 0.08 0.16 0.24 0.32 0.40 0.48 0.56
Sample proportion

(d)

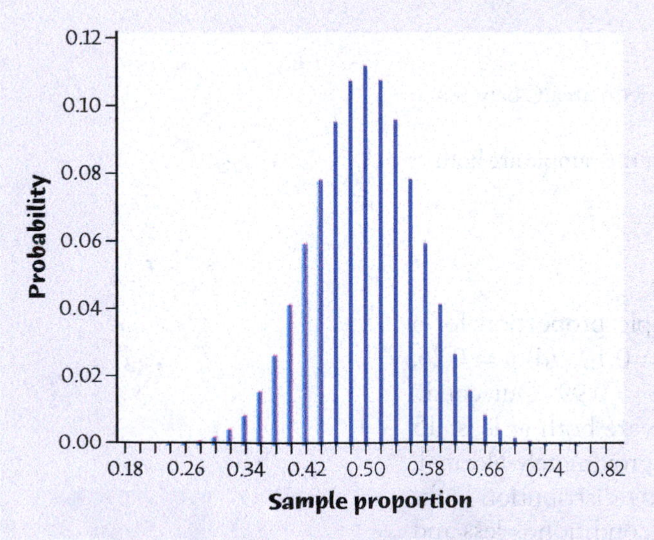
0.12
0.10
0.08
0.06
0.04
0.02
0.00
Probability
0.18 0.26 0.34 0.42 0.50 0.58 0.66 0.74 0.82
Sample proportion

(e)

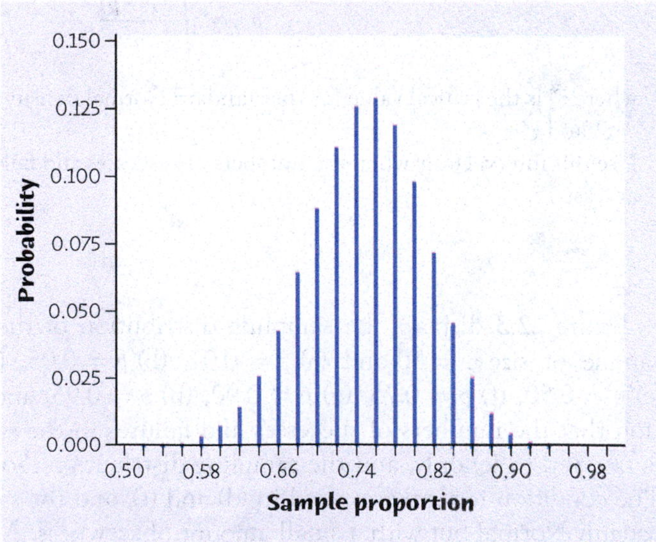
0.150
0.125
0.100
0.075
0.050
0.025
0.000
Probability
0.50 0.58 0.66 0.74 0.82 0.90 0.98
Sample proportion

(f)

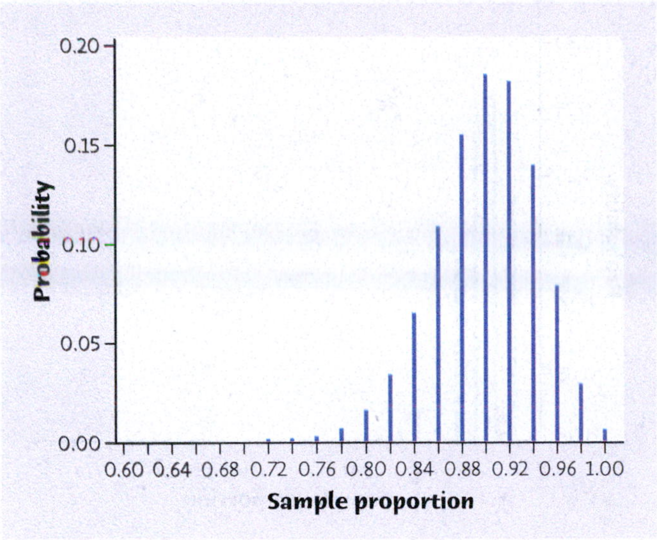

(g)

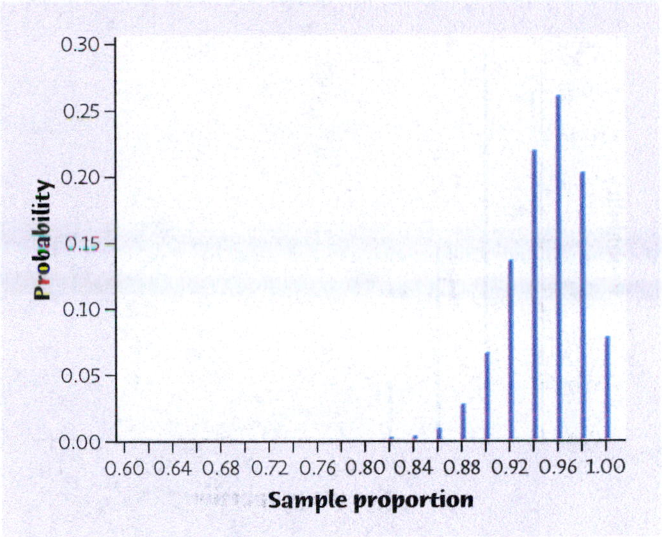

(h)

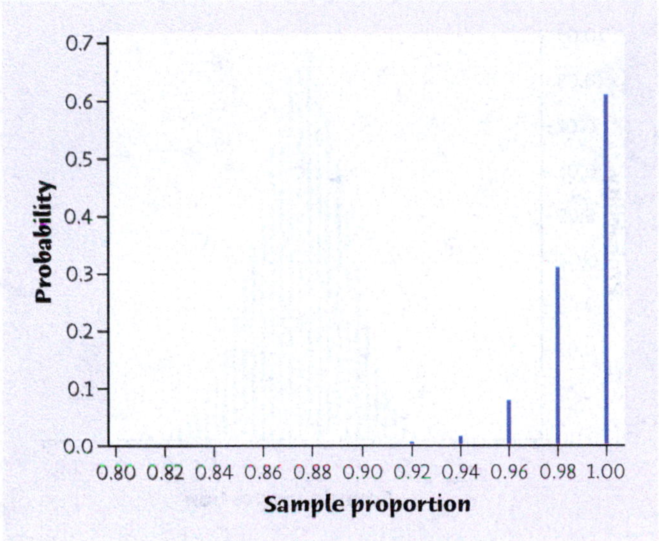

(i)

FIGURE 22.3
The sampling distribution of the sample proportion for $n = 50$ and (a) $p = 0.01$, (b) $p = 0.05$, (c) $p = 0.10$, (d) $p = 0.25$, (e) $p = 0.50$, (f) $p = 0.75$, (g) $p = 0.90$, (h) $p = 0.95$, and (i) $p = 0.99$.

Figure 22.4 displays the sampling distribution of the sample proportion for $p = 0.1$ and sample sizes (a) $n = 10$, (b) $n = 25$, and (c) $n = 250$. Only in (c) is the condition satisfied, and the sampling distribution looks approximately Normal. As the population proportion of successes approaches 0 or 1, and the condition is less and less close to being satisfied, the sampling distribution is increasingly skewed.

Why not *t*? Notice that we *don't* change z^* to t^* when we replace the standard deviation with the standard error. When the sample mean $\bar{x}$ estimates the population mean μ, a separate parameter σ describes the variability of the distribution of $\bar{x}$. We separately estimate σ, and this leads to a t distribution. When the sample proportion $\hat{p}$ estimates the population proportion p, the variability depends on p, not on a separate parameter. There is no t distribution; we just make the Normal approximation a bit less accurate when we replace p in the standard deviation with $\hat{p}$.

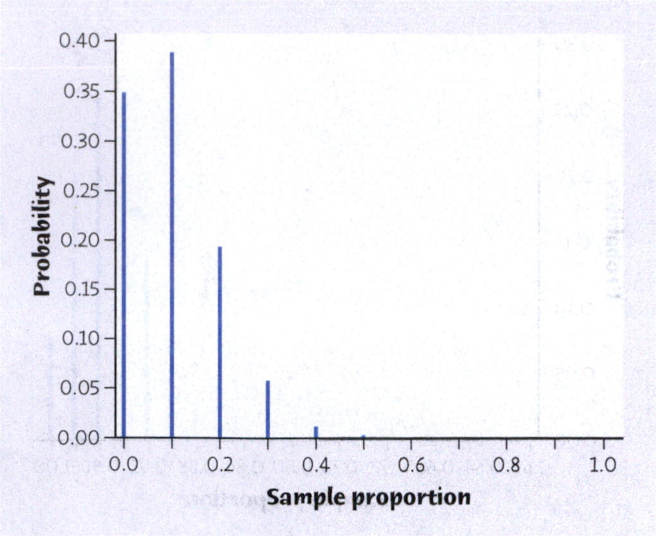

(a)

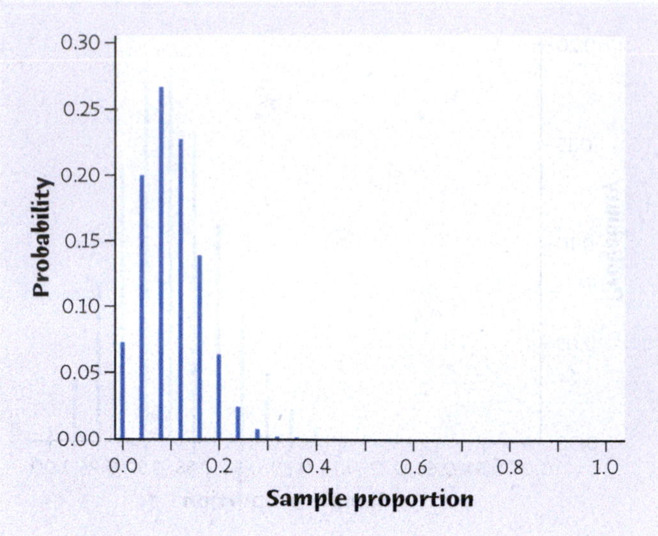

(b)

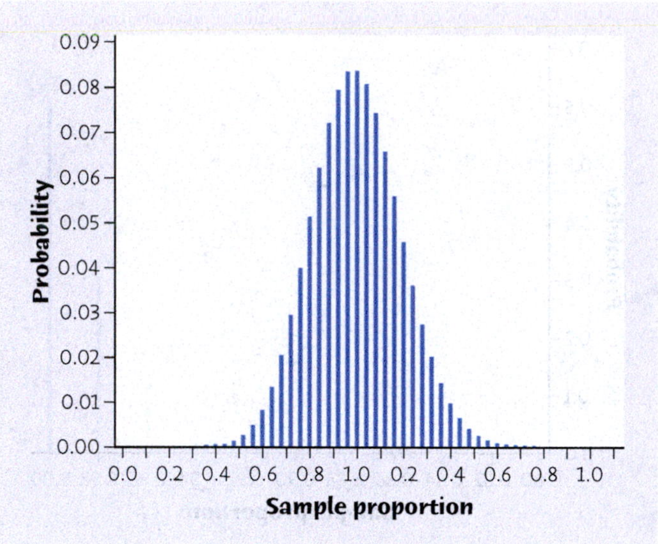

(c)

FIGURE 22.4
The sampling distribution of the sample proportion for $p = 0.1$ and (a) $n = 10$, (b) $n = 25$, and (c) $n = 250$.

EXAMPLE 22.4 Estimating Risky Behavior

4step

The four-step process for any confidence interval is outlined on page 374.

STATE: The National AIDS Behavioral Surveys found that 170 of a sample of 2673 adult heterosexuals had had multiple partners. That is,

$$\hat{p} = \frac{170}{2673} = 0.0636$$

What can we say about the population of all adult heterosexuals?

PLAN: We will give a 99% confidence interval to estimate the proportion p of all adult heterosexuals who have had multiple partners.

SOLVE: First verify the conditions for inference:

- The sampling design was a complex stratified sample, and the survey used inference procedures for that design. The overall effect is close to an SRS, however.

- The sample is large enough: the numbers of successes (170) and failures (2503) in the sample are both much larger than 15.

The sample size condition is easily satisfied. The condition that the sample be an SRS is only approximately met.

A 99% confidence interval for the proportion p of all adult heterosexuals with multiple partners uses the standard Normal critical value $z^* = 2.576$. The confidence interval is

$$\begin{aligned} \hat{p} \pm z^* \sqrt{\frac{\hat{p}(1-\hat{p})}{n}} &= 0.0636 \pm 2.576 \sqrt{\frac{(0.0636)(0.9364)}{2673}} \\ &= 0.0636 \pm 0.0122 \\ &= 0.0514 \text{ to } 0.0758 \end{aligned}$$

CONCLUDE: We are 99% confident that the percentage of adult heterosexuals who had more than one sexual partner in the year prior to the survey lies between about 5.1% and 7.6%.

As usual, the practical problems of a large sample survey weaken our confidence in the AIDS survey's conclusions. Only people in households with landline telephones could be reached. Although at the time of the survey about 89% of American households had landline telephones, as the number of cell phone–only users increases, using a sample of households with landline phones is becoming less acceptable for surveys of the general population (see page 214). Additionally, some groups at high risk for AIDS, such as people who inject illegal drugs, often don't live in settled households and were, therefore, underrepresented in the sample. About 30% of the people reached refused to cooperate. A nonresponse rate of 30% is not unusual in large sample surveys, but it may cause some bias if those who refuse differ systematically from those who cooperate. The survey used statistical methods that adjust for unequal response rates in different groups. Finally, some respondents may not have told the truth when asked about their sexual behavior. The survey team tried to make respondents feel comfortable. For example, Hispanic women were interviewed only by Hispanic women, and Spanish speakers were interviewed by Spanish speakers with the same regional accent (Cuban, Mexican, or Puerto Rican). Nonetheless, the survey report says that some bias is probably present:

> *It is more likely that the present figures are underestimates; some respondents may underreport their numbers of sexual partners and intravenous drug use because of embarrassment and fear of reprisal, or they may forget or not know details of their own or of their partner's HIV risk and their antibody testing history.*[8]

Reading the report of a large study like the National AIDS Behavioral Surveys reminds us that statistics in practice involves much more than formulas for inference.

STATISTICS IN YOUR WORLD

Who Is a Smoker?

When estimating a proportion p, be sure you know what counts as a "success." The news says that 20% of adolescents smoke. Shocking. It turns out that this is the percentage who smoked at least once in the past month. If we say that a smoker is someone who smoked on at least 20 of the past 30 days and smoked at least half a pack on those days, fewer than 4% of adolescents qualify.

APPLY YOUR KNOWLEDGE

22.4 No Confidence Interval. The 2017 Youth Risk Behavioral Survey, in a random sample of 497 high school seniors in Connecticut, found that 0.8% (that's 0.008 as a decimal fraction) smoked cigarettes daily.[9] Explain why we can't use the large-sample confidence interval to estimate the proportion p in the population of all Connecticut high school seniors in 2017 who smoked cigarettes daily.

22.5 Canadian Attitudes toward Guns. Canada has much stronger gun control laws than the United States, and Canadians support gun control more strongly than do Americans. A sample survey asked a random sample of 1525 adult Canadians, "Would you support or oppose Canada having a complete ban on civilian possession of handguns?" Of the 1525 people in the sample, 930 answered either "Strongly support" or "Support."[10]

(a) The survey contacted a randomized sample of large panel of Canadian adults who were invited to the survey process using a wide variety of methods and channels. The randomized sample of panel members was chosen so as to be representative of the Canadian population as a whole. Based on what you know about sample surveys, what is likely to be the biggest weakness in this survey?

(b) Nonetheless, act as if we have an SRS from adults in the Canadian provinces. Give a 95% confidence interval for the proportion who support registration of all firearms.

4step

22.6 Migraines. Migraine is a neurological disease affecting approximately 1 billion people worldwide. Characteristic symptoms of migraine include headache lasting 4 to 72 hours, nausea, and sensitivity to light and sound. Researchers studied the effect of the drug ubrogepant on migraine symptoms. Researchers recruited 1686 subjects, all of whom suffered from migraine, and 464 of them were randomly assigned to receive 50 mg of ubrogepant upon the outset of a migraine attack of moderate or severe pain intensity. Of these, 101 were pain free two hours after taking the medication.[11]

(a) Assuming that the 1686 subjects are a representative sample of migraine sufferers (an assumption often made in clinical trials), give a 90% confidence interval for the proportion of migraine sufferers who will be pain free two hours after taking 50 mg of ubrogepant to treat a moderate to severe migraine. Follow the four-step process as illustrated in Example 22.4.

(b) What concerns do you have about generalizing these results to the population of all migraine sufferers?

22.3 Choosing the Sample Size

In planning a study, we may want to choose a sample size that will allow us to estimate the parameter within a given margin of error. We saw earlier (page 412) how to do this for a population mean. The method is similar for estimating a population proportion.

The margin of error in the large-sample confidence interval for p is

$$m = z^* \sqrt{\frac{\hat{p}(1-\hat{p})}{n}}$$

Here z^* is the standard Normal critical value for the level of confidence we want. Because the margin of error involves the sample proportion of successes $\hat{p}$, we need to guess this value when choosing n. Call our guess p^*. Here are two ways to get p^*:

1. Use a guess p^* based on a pilot study or on past experience with similar studies. You can do several calculations to cover the range of values of $\hat{p}$ you might get.
2. Use $p^* = 0.5$ as the guess. The margin of error m is largest when $\hat{p} = 0.5$, so this guess is conservative in the sense that if we get any other $\hat{p}$ when we do our study, we will get a margin of error smaller than planned.

STATISTICS IN YOUR WORLD

New York, New York

New York City, they say, is bigger, richer, faster, and ruder. Maybe there's something to that. A sample survey firm Zogby International says that, as a national average, it takes five telephone calls to reach a live person. When calling to New York, it takes 12 calls. Survey firms assign their best interviewers to make calls to New York and often pay them bonuses to cope with the stress.

Once you have a guess p^*, the recipe for the margin of error can be solved to give the sample size n needed. Here is the result for the large-sample confidence interval. For simplicity, use this result even if you plan to use the plus four interval discussed in Section 22.5.

Sample Size for a Desired Margin of Error

The level C confidence interval for a population proportion p will have margin of error approximately equal to a specified value m when the sample size is

$$n = \left(\frac{z^*}{m}\right)^2 p^*(1 - p^*)$$

where p^* is a guessed value for the sample proportion. The margin of error will always be less than or equal to m if you take the guess p^* to be 0.5.

Which method for finding the guess p^* should you use? The n you get doesn't change much when you change p^* as long as p^* is not too far from 0.5. You can use the conservative guess $p^* = 0.5$ if you expect the true $\hat{p}$ to be roughly between 0.3 and 0.7. If the true $\hat{p}$ is close to 0 or 1, using $p^* = 0.5$ as your guess will give a sample much larger than you need. Try to use a better guess from a pilot study when you suspect that $\hat{p}$ will be less than 0.3 or greater than 0.7.

EXAMPLE 22.5 Planning a Poll

STATE: A large city is proposing a school tax levy to provide funds to address deteriorating conditions in several older school buildings. You will contact an SRS of registered voters in the city. You want to estimate the proportion p of voters who approve of the levy with 95% confidence and a margin of error no greater than 3%, or 0.03. How large a sample do you need?

PLAN: Find the sample size n needed for margin of error $m = 0.03$ and 95% confidence. Past school tax levies in the city have been decided by close margins, so you decide to use the guess $p^* = 0.5$.

SOLVE: The sample size you need is

$$n = \left(\frac{1.96}{0.03}\right)^2 (0.5)(1 - 0.5) = 1067.1$$

Round the result up to $n = 1068$. (Rounding down would give a margin of error slightly greater than 0.03.)

CONCLUDE: An SRS of 1068 registered voters is adequate for a margin of error ±3%.

If you want a 2.5% margin of error rather than 3%, then (after rounding up)

$$n = \left(\frac{1.96}{0.025}\right)^2 (0.5)(1 - 0.5) = 1537$$

For a 2% margin of error, the sample size you need is

$$n = \left(\frac{1.96}{0.02}\right)^2 (0.5)(1 - 0.5) = 2401$$

Finally, for a 1.5% margin of error (half of 3%), the sample size must be

$$n = \left(\frac{1.96}{0.015}\right)^2 (0.5)(1 - 0.5) = 4268.4$$

which is 4 times that for the margin of error of 3% before rounding up to 4269. One can show that if you want to cut the margin of error in half, you need to multiply the sample size by 4. As usual, smaller margins of error call for larger samples.

APPLY YOUR KNOWLEDGE

22.7 Do You Listen to Podcasts? In January and February 2019, Edison Research conducted a national telephone survey of 1500 Americans aged 12 and older, using random digit dialing techniques to both cell phones and landlines. The survey included questions about the use of mobile devices, Internet audio, podcasting, social media, smart speakers, and more. Of the 1500 people surveyed, 480 said they had listened to a podcast in the past month.[12]

(a) What is the margin of error of the large-sample 95% confidence interval for the proportion of Americans aged 12 and older who had listened to a podcast in the past month?

(b) How large a sample is needed to get the common ±3 percentage point margin of error? Use the January/February survey of 1500 as a pilot study to get p^*.

22.8 Can You Taste PTC? PTC is a substance that has a strong bitter taste for some people and is tasteless for others. The ability to taste PTC is inherited and depends on single gene that codes for a taste receptor on the tongue. Interestingly, although the PTC molecule is not found in nature, the ability to taste it correlates strongly with the ability to taste other naturally occurring bitter substances, many of which are toxins. About 75% of Italians can taste PTC. You want to estimate the proportion of Americans with at least one Italian grandparent who can taste PTC.

(a) Starting with the 75% estimate for Italians, how large a sample must you collect in order to estimate the proportion of PTC tasters within ±0.04 with 90% confidence?

(b) Estimate the sample size required if you made no assumptions about the value of the proportion who could taste PTC. How much has the required sample size changed?

22.4 Significance Tests for a Proportion

The test statistic for the null hypothesis $H_0: p = p_0$ is the sample proportion $\hat{p}$ standardized using the value p_0 specified by H_0,

$$z = \frac{\hat{p} - p_0}{\sqrt{\dfrac{p_0(1 - p_0)}{n}}}$$

This z statistic has approximately the standard Normal distribution when H_0 is true. *P*-values, therefore, come from the standard Normal distribution. Unlike the confidence interval in which p is unknown and must be estimated by $\hat{p}$ when standardizing the estimate, in the test we can replace p by p_0 when standardizing, as p_0

is specified by H_0. Additionally, because H_0 fixes a value of p when standardizing the estimate, the sample size conditions for use of the test are less stringent than for the large-sample confidence interval in which p must be estimated. Here is the procedure for tests.

Significance Tests for a Proportion

Draw an SRS of size n from a large population that contains an unknown proportion p of successes. To test the hypothesis $H_0: p = p_0$, compute the z statistic

$$z = \frac{\hat{p} - p_0}{\sqrt{\frac{p_0(1 - p_0)}{n}}}$$

In terms of a variable Z having the standard Normal distribution, the approximate P-value for a test of H_0 against

$H_a: p > p_0$ is $P(Z \geq z)$

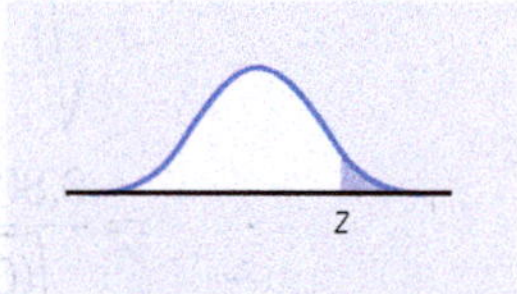

$H_a: p < p_0$ is $P(Z \leq z)$

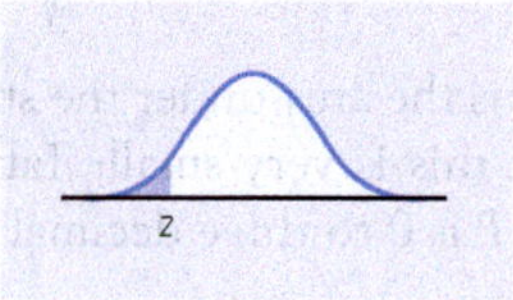

$H_a: p \neq p_0$ is $2P(Z \geq |z|)$

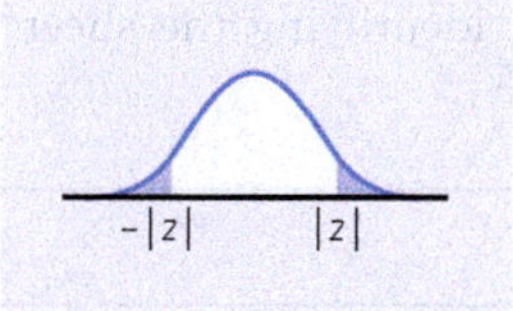

Use this test when the sample size n is so large that both np_0 and $n(1 - p_0)$ are 10 or more.[13]

EXAMPLE 22.6 Can You Match the Dog and Owner?

The four-step process for any significance test is outlined on page 374.

STATE: Researchers presented a random sample of undergraduates with two test sheets, each sheet including 20 photos of the faces of dog-owner pairs taken at a dog lovers' field festival. The sets of dog-owner pairs on the two sheets were equivalent with respect to breed, diversity of appearance, and gender of owners. On one sheet, the dogs were matched with their owners, while on the second sheet, the dogs and owners were deliberately mismatched. Students were asked to "choose the set of dog-owner pairs that resemble each other, Sheet 1 or Sheet 2" and were simply told the aim of the research was a "survey on dog-owner relationships." Of the 61 student judges in this part of the study, 49 chose the sheet with dogs and owners correctly matched.[14] The sample proportion that chose the sheet with the correctly matched dogs and owners was

$$\hat{p} = \frac{49}{61} = 0.8033$$

Darryl Brooks/Shutterstock

Ezzolo/Shutterstock

If students were just guessing, we would expect about 50% of them to correctly identify the sheet with the dogs and owners, while if the dogs and owners do resemble each other, then students should do better than guessing. Although more than half the students correctly matched the dogs and owners, we don't expect a perfect 50–50 split in a random sample. Is this sample evidence that subjects are doing better than guessing?

PLAN: Take p to be the proportion of undergraduates who would choose the sheet with the correctly matched dogs and owners. We want to test the hypotheses

$$H_0: p = 0.5$$
$$H_a: p > 0.5$$

SOLVE: The conditions for inference require that we have a random sample and that $np_0 = (61)(0.5) = 30.5$ and $n(1 - p_0) = (61)(0.5) = 30.5$ are both greater than 10. Because the conditions for inference are met, we can go on to find the z test statistic:

$$z = \frac{\hat{p} - p_0}{\sqrt{\dfrac{p_0(1-p_0)}{n}}} = \frac{0.8033 - 0.5}{\sqrt{\dfrac{(0.5)(0.5)}{61}}} = 4.74$$

The P-value is the area under the standard Normal curve to the right of $z = 4.74$. We know that this is very small; Table C shows that $P < 0.0005$. Minitab (Figure 22.5) says that P is 0 to three decimal places.

CONCLUDE: There is very strong evidence that students are doing better than guessing when identifying the sheet with the correctly matched dog-owner pairs. ($P < 0.001$).

Minitab

Test and CI for One Proportion

```
Test of p = 0.5 p > 0.5

Sample    X    N   Sample p   95% Lower bound   Z-Value   P-Value
1        49   61   0.803279          0.719560      4.74     0.000

Using the normal approximation.
```

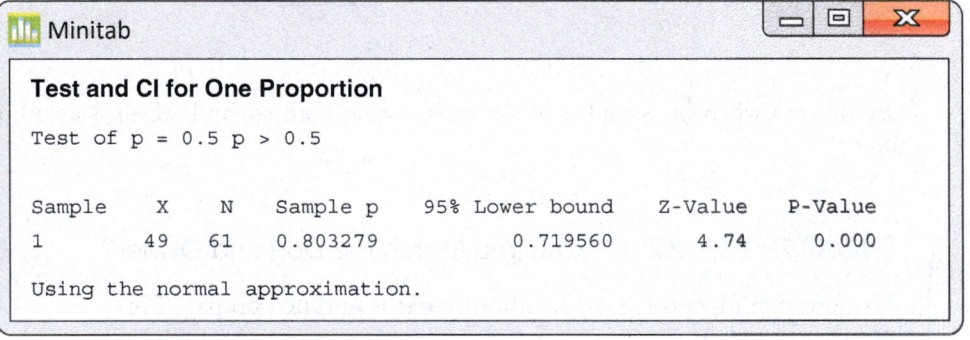

FIGURE 22.5
Minitab output for the significance test, for Example 22.6.

Example 22.6 was the first in a series of experiments that included more than 500 undergraduates from three colleges as subjects. Exercise 22.9 looks at two additional experiments in this series that further examine the resemblance between dogs and owners.

EXAMPLE 22.7 Estimating the Chance of Correctly Matching Dogs and Owners

With 49 successes in 61 trials, we have 49 successes and 12 failures in the sample. The conditions for the large-sample confidence interval are almost met, so we use the confidence interval with some caution (see also Exercise 22.48). The 99% confidence interval is

$$\hat{p} \pm z^* \sqrt{\frac{\hat{p}(1-\hat{p})}{n}} = 0.8033 \pm 2.576\sqrt{\frac{(0.8033)(0.1967)}{61}}$$
$$= 0.8033 \pm 0.1311$$
$$= 0.6722 \text{ to } 0.9344$$

We are 99% confident that between about 67% and 93% of undergraduates can correctly match the dogs and owners.

The confidence interval is more informative than the test in Example 22.6, which tells us only that more than half of undergraduates can correctly match the dogs and owners. The confidence interval tells us that this proportion is substantially larger than 50%.

APPLY YOUR KNOWLEDGE

22.9 More on Matching Dogs and Owners. The results reported in Example 22.6 were part of a larger study that examined the nature of the resemblance between dogs and their owners. In the second experiment, when just the "mouth region" of the owners was blacked out in all the pictures of dogs and owners, 37 of the 51 subjects participating in this part of the study correctly chose the sheet with the dogs and their owners. However, when just the "eye region" of the owners was blacked out in a third experiment, only 30 of the 60 subjects used for this part of the study correctly matched the dogs and the owners. Follow the four-step process as illustrated in Example 22.6 to analyze the results of these two experiments. Summarize your conclusions in the context of the problem. 4step

22.10 Preference for the Middle? When choosing an item from a group, researchers have shown that an important factor influencing choice is the item's location. This occurs in varied situations such as shelf positions when shopping, when filling out a questionnaire, and even when choosing a preferred candidate during a presidential debate. In this experiment, five identical pairs of white socks were displayed in a vertical column on an easel for viewing. One hundred participants from the University of Chester were used as subjects and asked to choose their preferred pair of socks.[15] In choice situations of this type, subjects often exhibit the "center stage effect," which is a tendency to choose the item in the center. In this experiment, 34 subjects chose the pair of socks in the center. Do these data provide evidence of "center stage effect"? Follow the four-step process illustrated in Example 22.6. (*Hint*: If subjects are choosing a pair of socks at random from the five positions, what would be the probability of selecting the pair in the middle? This is the value of p in the null hypothesis.) 4step

22.11 No Test. Explain whether we can use the z test for a proportion in these situations.

(a) You toss a coin 10 times in order to test the hypothesis $H_0: p = 0.5$ that the coin is balanced.

(b) A local congressperson contacts an SRS of 500 of the registered voters in his district to see if there is evidence that more than half support the bill he is sponsoring.

(c) The CEO of a large corporation says, "only 2% of our employees are dissatisfied with our new health insurance plan." You contact an SRS of 150 of the company's 10,000 employees to test the hypothesis $H_0: p = 0.02$.

22.5 Plus Four Confidence Intervals for a Proportion*

The large-sample confidence interval $\hat{p} \pm z^*\sqrt{\hat{p}(1-\hat{p})/n}$ for a sample proportion p is easy to calculate. It is also easy to understand because it rests directly on the approximately Normal distribution of $\hat{p}$. Unfortunately, confidence levels from this interval can be inaccurate, particularly with smaller sample sizes. The actual confidence level is usually *less* than the confidence level you asked for in choosing the critical value z^*. That's bad. What is worse, accuracy does not consistently get better as the sample size n increases. There are "lucky" and "unlucky" combinations of the sample size n and the true population proportion p.

Fortunately, there is a simple modification that is remarkably effective in improving the accuracy of the confidence interval. We call it the "plus four" method because all you need to do is *add four imaginary observations: two successes and two failures*. With the added observations, the *plus four estimate* of p is

$$\tilde{p} = \frac{\text{number of successes in the sample} + 2}{n+4}$$

The formula for the confidence interval is exactly as before, with the new sample size and number of successes.[16] You do not need software that offers the plus four interval: just enter the new sample size (actual size $+4$) and number of successes (actual number $+2$) into the large-sample procedure.

Plus Four Confidence Interval for a Proportion

Draw an SRS of size n from a large population that contains an unknown proportion p of successes. To get the **plus four confidence interval** for p, add four imaginary observations: two successes and two failures. Then use the large-sample confidence interval with the new sample size $(n+4)$ and number of successes (actual number $+2$).

Use this interval when the confidence level is at least 90% and the sample size n is at least 10, with any counts of successes and failures.

EXAMPLE 22.8 Cocaine Traces in Spanish Currency

Andrea Matone/Alamy

STATE: Cocaine users commonly snort the powder up the nose through a rolled-up paper currency bill. Spain has a high rate of cocaine use, so it's not surprising that euro paper currency in Spain often shows traces of cocaine. Researchers collected 20 euro bills in each of several Spanish cities. In Madrid, 17 out of 20 bore traces of cocaine.[17] The researchers note that we can't tell whether the bills had been used to snort cocaine or had been contaminated in currency-sorting machines. Estimate the proportion of all euro bills in Madrid that have traces of cocaine.

PLAN: Take p to be the proportion of bills that show cocaine traces. That is, a "success" is a bill that shows cocaine traces. Give a 95% confidence interval for p.

SOLVE: It is not clear how the bills in the sample were selected, so we don't know if we have an SRS. We will act as though we have an SRS, but we proceed with caution. The conditions for use of the large-sample interval are not met because there are only three failures. To apply the plus four method, add two successes and two failures to the original data. The plus four estimate of p is

$$\tilde{p} = \frac{17+2}{20+4} = \frac{19}{24} = 0.7917$$

*This section is optional.

We calculate the plus four confidence interval in the same way as we do the large-sample interval, but we base it on 19 successes in 24 observations. Here it is:

$$\tilde{p} \pm z^* \sqrt{\frac{\tilde{p}(1-\tilde{p})}{n+4}} = 0.7917 \pm 1.960 \sqrt{\frac{(0.7917)(0.2083)}{24}}$$
$$= 0.7917 \pm 0.1625$$
$$= 0.6292 \text{ to } 0.9542$$

CONCLUDE: Assuming that the sample can be regarded as an SRS, we estimate with 95% confidence that between about 63% and 95% of all euro bills in Madrid bear traces of cocaine.

For comparison, the ordinary sample proportion is

$$\hat{p} = \frac{17}{20} = 0.85$$

The plus four estimate $\tilde{p} = 0.7917$ in Example 22.8 is farther away from 1 than $\hat{p} = 0.85$. The plus four estimate gains its added accuracy by always moving toward 0.5 and away from 1 or 0, whichever is closer. This is particularly helpful when the sample contains only a few successes or a few failures. The numerical difference between a large-sample interval and the corresponding plus four interval is often small. Remember that the confidence level is the probability that the interval will catch the true population proportion *in very many uses*. Small differences every time can add up to more accurate confidence levels from plus four versus the large-sample interval.

How much more accurate is the plus four interval? Computer studies have asked how large n must be to guarantee that the actual probability that a 95% confidence interval covers the true parameter value is at least 0.94 for all samples of size n or larger. If $p = 0.1$, for example, the answer is $n = 646$ for the large-sample interval and $n = 11$ for the plus four interval.[18] The consensus of computational and theoretical studies is that plus four is better than the large-sample interval for many combinations of n and p.

APPLY YOUR KNOWLEDGE

22.12 Black Raspberries and Cancer. Sample surveys usually contact large samples, so we can use the large-sample confidence interval if the sample design is close to an SRS. Scientific studies often use smaller samples that require the plus four method. For example, Familial Adenomatous Polyposis (FAP) is a rare inherited disease characterized by the development of an extreme number of polyps early in life and colon cancer in virtually 100% of patients before the age of 40. A group of 14 people suffering from FAP being treated at the Cleveland Clinic drank black raspberry powder in a slurry of water every day for nine months. The numbers of polyps were reduced in 11 out of 14 of these patients.[19]

(a) Why can't we use the large-sample confidence interval for the proportion p of patients suffering from FAP that will have the number of polyps reduced after nine months of treatment?

(b) The plus four method adds four observations: two successes and two failures. What are the sample size and the number of successes after you do this? What is the plus four estimate $\tilde{p}$ of p?

(c) Give the plus four 90% confidence interval for the proportion of patients suffering from FAP who will have the number of polyps reduced after nine months of treatment.

22.13 Gun Violence and Video Games. People disagree about the impact of video games on gun violence. A survey in 2017 of a random sample of 1501 U.S. adults asked "Does the amount of gun violence in video games contribute a great deal or a fair amount to gun violence?" Of the 1501 sampled, 180 said "Not at all."[20]

(a) Give the 95% large-sample confidence interval for the proportion p of all U.S. adults who said that gun violence in video games does not contribute at all to gun violence. Be sure to verify that the sample size is large enough to use the large-sample confidence interval.

(b) Give the plus four 95% confidence interval for p. If you express the two intervals in percentages, rounded to the nearest 10th of a percent, how do they differ? (The plus four interval always pulls the results toward 50%.)

22.14 Cocaine Traces in Spanish Currency (continued). The plus four method is particularly useful when there are *no* successes or *no* failures in the data. The study of Spanish currency described in Example 22.8 found that in Seville, all 20 of a sample of 20 euro bills had cocaine traces.

(a) What is the sample proportion $\hat{p}$ of contaminated bills? What is the large-sample 95% confidence interval for p? It's not plausible that *every* bill in Seville has cocaine traces, as this interval says.

(b) Find the plus four estimate $\tilde{p}$ and the plus four 95% confidence interval for p. These results are more reasonable in this situation.

CHAPTER 22 SUMMARY

- Tests and confidence intervals for a population proportion p when the data are an SRS of size n are based on the **sample proportion** $\hat{p}$.
- When n is large, the sampling distribution of $\hat{p}$ has approximately the Normal distribution with mean p and standard deviation $\sqrt{p(1-p)/n}$. This standard deviation is the **standard error of** $\hat{p}$.
- The level C **large-sample confidence interval for p** is

$$\hat{p} \pm z^*\sqrt{\frac{\hat{p}(1-\hat{p})}{n}}$$

where z^* is the critical value for the standard Normal curve with area C between $-z^*$ and z^*. Use this interval only when *both* the number of successes and the number of failures in the sample are at least 15.

- The sample size needed to obtain a confidence interval with approximate margin of error m for a population proportion is

$$n = \left(\frac{z^*}{m}\right)^2 p^*(1-p^*)$$

where p^* is a guessed value for the sample proportion $\hat{p}$, and z^* is the standard Normal critical point for the level of confidence you want. If you use $p^* = 0.5$ in this formula, the margin of error of the interval will be less than or equal to m no matter what the value of $\hat{p}$ is.

- **Significance tests for a proportion $H_0: p = p_0$** are based on the z statistic

$$z = \frac{\hat{p} - p_0}{\sqrt{\frac{p_0(1-p_0)}{n}}}$$

with P-values calculated from the standard Normal distribution. Use this test in practice when $np_0 \geq 10$ and $n(1-p_0) \geq 10$.

- **(Optional topic)** To get a more accurate confidence interval for smaller sample sizes, add four imaginary observations—two successes and two failures—to your sample. Then use the same formula for the confidence interval. This is the **plus four confidence interval.** Use this interval in practice for confidence level 90% or higher and sample size n at least 10.

STATISTICS IN YOUR WORLD

Kids on Bikes

In the most recent year for which data are available, 77% of children killed in bicycle accidents were boys. You might take these data as a sample and start from $\hat{p} = 0.77$ to do inference about bicycle deaths in the near future. What you should not do is conclude that boys on bikes are in greater danger than girls. We don't know how many boys and girls ride bikes; it may be that most fatalities are boys because most riders are boys.

CHECK YOUR SKILLS

22.15 Cannabidiol- (CBD-) based products are widely touted for their therapeutic benefits without any psychoactive effects. A 2019 Gallup Poll asked a random sample of 2543 U.S. adults if they personally use CBD products or not. Suppose that, in fact, 15% of all U.S. adults have used CBD products. In repeated samples, the sample proportion $\hat{p}$ would follow approximately a Normal distribution with mean

(a) 381.45.

(b) 0.15.

(c) 0.007.

22.16 The standard deviation of the distribution of $\hat{p}$ in the Question 22.15 is about

(a) 0.007.

(b) 0.357.

(c) 0.1275.

Use the following information for Questions 22.17 through 22.19. A 2018 Gallup survey asked respondents to consider several foods and beverages and to indicate for each whether they actively tried to include it in their diet, actively tried to avoid it, or didn't think about it at all. Of the 1033 adults surveyed, 630 indicated that they actively tried to avoid drinking regular soda or pop. Assume that the sample was an SRS.

22.17 Based on the sample, the large-sample 90% confidence interval for the proportion of all American adults who actively try to avoid drinking regular soda or pop is

(a) 0.61 ± 0.015. (b) 0.61 ± 0.025.

(c) 0.61 ± 0.029.

22.18 In Question 22.17, suppose we computed a large-sample 99% confidence interval for the proportion of all American adults who actively try to avoid drinking regular soda or pop. This 99% confidence interval

(a) would have a smaller margin of error than the 90% confidence interval.

(b) would have a larger margin of error than the 90% confidence interval.

(c) could have either a smaller or a larger margin of error than the 90% confidence interval. This varies from sample to sample.

22.19 How many American adults must be interviewed to estimate the proportion of all American adults who actively try to avoid drinking regular soda or pop within ± 0.01 with 99% confidence using the large-sample confidence interval? Use 0.5 as the conservative guess for p.

(a) $n = 6765$

(b) $n = 9604$

(c) $n = 16590$

22.20 An opinion poll asks an SRS of 100 college seniors how they view their job prospects. In all, 53 say "Good." The large-sample 95% confidence interval for estimating the proportion of all college seniors who think their job prospects are good is

(a) 0.530 ± 0.082.

(b) 0.530 ± 0.098.

(c) 0.530 ± 0.049.

22.21 The sample survey in Question 22.20 actually called 130 seniors, but 30 of the seniors refused to answer. This nonresponse could cause the survey result to be in error. The error due to nonresponse

(a) is in addition to the margin of error found in Question 22.20.

(b) is included in the margin of error found in Question 22.20.

(c) can be ignored because it isn't random.

22.22 Experiments on learning in animals sometimes measure how long it takes mice to find their way through a maze. Only half of all mice complete one particular maze in less than 18 seconds. A researcher thinks that a loud noise will cause the mice to complete the maze faster. She measures the proportion of 40 mice that completed the maze in less than 18 seconds with noise as a stimulus.

The proportion of mice that completed the maze in less than 18 seconds is $\hat{p} = 0.7$. The hypotheses for a test to answer the researcher's question are

(a) $H_0: p = 0.5, H_a: p > 0.5$.

(b) $H_0: p = 0.5, H_a: p < 0.5$.

(c) $H_0: p = 0.5, H_a: p \neq 0.5$.

22.23 The value of the z statistic for the Question 22.22 is 2.53. This test is

(a) not significant at either $\alpha = 0.05$ or $\alpha = 0.01$.

(b) significant at $\alpha = 0.05$ but not at $\alpha = 0.01$.

(c) significant at both $\alpha = 0.05$ and $\alpha = 0.01$.

22.24 A Gallup Poll in November 2019 found that 55% of the people in the sample said they wanted to lose weight. The poll's margin of error for 95% confidence was 4%. This means that

(a) the poll used a method that gets an answer within 4% of the truth about the population 95% of the time.

(b) we can be sure that the percentage of all adults who want to lose weight is between 50% and 58%.

(c) if Gallup takes another poll using the same method, the results of the second poll will lie between 51% and 59%.

CHAPTER 22 EXERCISES

22.25 **Do Smokers Know That Smoking is Bad for Them?** The Harris Poll asked a sample of smokers, "Do you believe that smoking will probably shorten your life, or not?" Of the 1010 people in the sample, 848 said "yes."

alessandro0770/Getty Images

(a) Harris called residential telephone numbers at random in an attempt to contact an SRS of smokers. Based on what you know about national sample surveys, what is likely to be the biggest weakness in the survey?

(b) We will nonetheless act as if the people interviewed are an SRS of smokers. Give a 95% confidence interval for the percent of smokers who agree that smoking will probably shorten their lives.

22.26 **Reporting cheating.** Students are reluctant to report cheating by other students. A student project put this question to an SRS of 172 undergraduates at a large university: "You witness two students cheating on a quiz. Do you go to the professor?" Only 19 answered "yes."[21] Give a 95% confidence interval for the proportion of all undergraduates at this university who would report cheating.

22.27 **Harris Announces a Margin of Error.** Exercise 22.25 describes a Harris Poll survey of smokers in which 848 of a sample of 1010 smokers agreed that smoking would probably shorten their lives. Harris announces a margin of error of ±3 percentage points for all samples of about this size. Opinion polls announce the margin of error for 95% confidence.

(a) What is the actual margin of error (in percent) for the large-sample confidence interval from this sample?

(b) The margin of error is largest when $\hat{p} = 0.5$. What would the margin of error (in percent) be if the sample had resulted in $\hat{p} = 0.5$?

(c) Why do you think that Harris announces a ±3% margin of error for all samples of about this size?

22.28 **Sampling Greenville County.** The Rails to Trails program refers to the conversion of old rail corridors into multipurpose trails for recreation and transportation. Researchers were interested in obtaining information on characteristics of users and nonusers of a 10-mile-long paved greenway trail in Greenville, South Carolina, that connects residential areas to both a university campus and the commercial downtown area of the city. Random digit dialing of residential numbers was done using a database of exchanges. A total of 2461 persons were contacted, of which 726 completed the survey. When a household was reached, surveyors asked to speak to the adult over 18 with the next birthday. No cell phone numbers were included in the sample. In the sample, although 726 completed the survey, only 689 of these respondents provided data on their ages. Among the 689 respondents who provided data on their ages, 253 were over 65 years old.[22]

(a) According to U.S. census data, at the time of the survey 13% of the population of adults in Greenville County were over 65 years old. If the 689 completed surveys that included age data were an SRS of adult Greenville County residents, do you feel that the difference between the proportion of those over 65 that completed the survey and the population proportion of 13% can be easily explained by chance variation? State the hypotheses and give the *P*-value. Give your conclusion in the context of the problem.

(b) Among the 726 surveys that were returned, 24.9% of respondents had used the trail in the six months prior to the survey. If those under 65 were more likely to use the trail, what would be the likely direction of the bias in using 24.9% as the estimate of the proportion of all Greenville County residents who had used the trail in the six months prior to the survey? Explain your answer.

22.29 **Facebook.** Pew Research Center's Internet and American Life Project asked a random sample of 1502 U.S. adults whether they had used Facebook. Of these, 1096 said "yes."[23] Pew dialed landline and cell phone telephone numbers at random in the continental United States in an attempt to contact a random sample of adults. Respondents in the landline sample were selected by randomly asking for the youngest adult male or female who is now at home.

(a) What do you think is likely to be the biggest weakness in the survey?

(b) Act as if the sample is an SRS. Give a large-sample 90% confidence interval for the proportion p of all cell phone users who have used their cell phone to look up health or medical information.

22.30 **Stopping Traffic with a Smile!** Throughout Europe, more than 8000 pedestrians are killed each year in road accidents, with approximately 25% of these dying when using a pedestrian crossing. Although failure to stop for pedestrians at a pedestrian crossing is a serious traffic offense in France, more than half of drivers do not stop when a pedestrian is waiting at a crosswalk. In this experiment, a female research assistant was instructed to stand at a pedestrian crosswalk and stare at the driver's face as a car approached the crosswalk. In 400 trials, the research assistant maintained a neutral expression, and in a second set of 400 trials, the research assistant was instructed to smile. The order of smiling or not smiling was randomized, and several pedestrian crossings were used in a town on the coast in the west of France. The research assistant was dressed in normal attire for her age (jeans, t-shirt, and sneakers).[24]

(a) In the 400 trials in which the assistant maintained a neutral expression, the driver stopped in 229 out of the 400 trials. Find a 95% confidence interval for the proportion of drivers who would stop when a neutral expression is maintained.

(b) In the 400 trials in which the assistant smiled at the driver, the driver stopped in 277 out of the 400 trials. Find a 95% confidence interval for the proportion of drivers who would stop when the assistant is smiling.

(c) What do your results in parts (a) and (b) suggest about the effect of a smile on a driver stopping at a pedestrian crosswalk? Explain briefly. (In Chapter 23, we will consider formal methods for comparing two proportions.)

22.31 **Running Red Lights.** A random digit dialing telephone survey of 880 drivers asked, "Recalling the last 10 traffic lights you drove through, how many of them were red when you entered the intersections?" Of the 880 respondents, 171 admitted that at least one light had been red.[25]

(a) Give a 95% confidence interval for the proportion of all drivers who ran one or more of the last 10 red lights they met.

(b) Nonresponse is a practical problem for this survey: only 21.6% of calls that reached a live person were completed. Another practical problem is that people may not give truthful answers. What is the likely direction of the bias? Do you think more or fewer than 171 of the 880 respondents really ran a red light? Why?

22.32 **Usage of the Olympic National Park.** U.S. National Parks that contain designated wilderness areas are required by law to develop and maintain a wilderness stewardship plan. The Olympic National Park, containing some of the most biologically diverse wilderness in the United States, had a survey conducted in 2012 to collect information relevant to the development of such a plan. National Park Service staff visited 30 wilderness trailheads in moderate- to high-use areas over a 60-day period and asked visitors as they completed their hike to complete a questionnaire. The 1019 completed questionaires, giving a response rate of 50.4%, provided each subject's opinions on the use and management of wilderness. In particular, there were 694 day users and 325 overnight users in the sample.[26]

(a) Why do you think the National Park staff only visited trailheads in moderate- to high-use areas to obtain the sample?

(b) Assuming that the 1019 subjects represent a random sample of users of the wilderness areas in Olympic National Park, give a 90% confidence interval for the proportion of day users.

(c) The response rate was 49% for day users and 52% for overnight users. Does this lessen any concerns you might have regarding the effect of nonresponse on the interval you obtained in part (b)? Explain briefly.

(d) Do you think it would be better to refer to the interval in part (b) as a confidence interval for the proportion of day users or the proportion of day users on the most popular trails in the park? Explain briefly.

22.33 **Vote for the Best Face?** We often judge other people by their faces. It appears that some people judge candidates for elected office by their faces. Psychologists showed head-and-shoulders photos of the two main candidates in 32 races for the U.S. Senate to many subjects (dropping subjects who recognized one of the candidates) to see which candidate was rated "more competent" based on nothing but the photos. On election day, the candidates whose faces looked more competent won 22 of the 32 contests. If faces don't influence voting, half of all two-candidate races in the long run should be won by the candidate with the favored face. Is there evidence that the proportion of times the candidate with the higher-rated face wins is more than 50%?

(a) What are the null and alternative hypotheses H_0 and H_a?

(b) Is the result statistically significant at the 5% level? At the 1% level?

22.34 Do Chemists Have More Girls? Some people think that chemists are more likely than other parents to have female children. (Perhaps chemists are exposed to something in their laboratories that affects the sex of their children.) The Washington State Department of Health lists the parents' occupations on birth certificates. Over a 10-year period, 555 children were born to chemists. Of these births, 273 were girls. During this period, 48.8% of all births in Washington State were girls. Is there evidence that the proportion of girls born to chemists is higher than the statewide proportion? State the null and alternative hypotheses H_0 and H_a and give the *P*-value.

22.35 Coin Tossing. The French naturalist Count Buffon (1707–1788) tossed a coin 4040 times. The result was 2048 heads, with $\frac{2048}{4040} = 0.5069$ the proportion of heads. Is this evidence that the coin was not fair? State the appropriate hypotheses and give the *P*-value.

22.36 ESP. A classic experiment to detect extra-sensory perception (ESP) uses a shuffled deck of cards containing five suits (waves, stars, circles, squares, and crosses). As the experimenter turns over each card and concentrates on it, the subject guesses the suit of the card. A subject who lacks ESP has probability 1-in-5 of being right by luck on each guess. A subject who has ESP will be right more often. Julie is right in 5 of 10 tries. (Actual experiments use much longer series of guesses so that weak or nonexistent ESP can be spotted. No one has ever been right half the time in a long experiment!)

(a) Give H_0 and H_a for a test to see if this result is significant evidence that Julie has ESP.

(b) How convincing was Julie's performance?

22.37 The IRS Plans an SRS. The Internal Revenue Service plans to examine an SRS of individual federal income tax returns from each state. One variable of interest is the proportion of returns claiming itemized deductions. The total number of tax returns in a state varies from almost 30 million in California to approximately 500,000 in Wyoming.

(a) Will the margin of error for estimating the population proportion change from state to state if an SRS of 2000 tax returns is selected in each state? Explain your answer.

(b) Will the margin of error change from state to state if an SRS of 1% of all tax returns is selected in each state? Explain your answer.

22.38 Surveying Students. You are planning a survey of students at a large university to determine what proportion favor an increase in student fees to support an expansion of the student newspaper. Using records provided by the registrar, you can select a random sample of students. You will ask each student in the sample whether he or she is in favor of the proposed increase. Your budget will allow a sample of 100 students.

(a) For a sample of size 100, construct a table of the margins of error for 95% confidence intervals when $\hat{p}$ takes the values 0.1, 0.3, 0.5, 0.7, and 0.9.

(b) A former editor of the student newspaper offers to provide funds for a sample of size 500. Repeat the margin of error calculations in part (a) for the larger sample size. Then write a short thank-you note to the former editor, describing how the larger sample size will improve the results of the survey.

In responding to Exercises 22.35 through 22.41, follow the ***Plan, Solve,*** *and* ***Conclude*** *steps of the four-step process.*

22.39 College-Educated Parents. The National Assessment of Educational Progress (NAEP) includes a "long-term trend" study that tracks reading and mathematics skills over time and obtains demographic information. In the 2012 study (the most recent available as of 2020), a random sample of 9000 17-year-old students was selected.[27] The NAEP sample used a multistage design, but the overall effect is quite similar to an SRS of 17-year-olds who are still in school.

(a) In the sample, 51% of students had at least one parent who was a college graduate. Estimate, with 99% confidence, the proportion of all 17-year-old students in 2012 who had at least one parent graduate from college.

(b) The sample does not include 17-year-olds who dropped out of school, so your estimate is valid only for students. Do you think the proportion of all 17-year-olds with at least one parent who was a college graduate would be higher or lower than 51%? Explain.

22.40 Downloading Music. A husband and wife, Stan and Lucretia, share a digital audio player that has a feature that randomly selects which song to play. A total of 2444 songs have been loaded into the player, some by Stan and the rest by Lucretia. They are interested in determining whether they have loaded different proportions of songs into the player. Suppose that when the player was in the random-selection mode, 26 of the first 40 songs selected were songs loaded by Lucretia. Let p denote the proportion of songs that were loaded by Lucretia.

(a) State the null and alternative hypotheses to be tested. How strong is the evidence that Stan and Lucretia have loaded different proportions of songs into the player? Make sure to check the conditions for the use of this test.

(b) Are the conditions for the use of the large-sample confidence interval met? If so, estimate with 95% confidence the proportion of songs that were loaded by Lucretia.

22.41 **Book Reading.** Although an increasing share of Americans are reading e-books on tablets and smartphones rather than dedicated e-readers, print books continue to be much more popular than books in *digital format* (digital format includes both e-books and audio books). A Pew Research Center survey of 1502 adults nationwide conducted January 8–February 7, 2019, found that 1081 of those surveyed had read a book in either print or digital format in the preceding 12 months.[28]

(You may regard the 1081 adults in the survey who had read a book in the preceding 12 months as a random sample of readers.)

(a) What can you say with 95% confidence about the percentage of all adults who had read a book in either print or digital format in the preceding 12 months?

(b) Of the 1081 surveyed who had read a book in the preceding 12 months, 105 had read only digital books. Among those adults who had read a book in the preceding 12 months, find a 95% confidence interval for the proportion that had read digital books exclusively.

22.42 **Speeding.** It often appears that most drivers on the road are driving faster than the posted speed limit. Situations differ, of course, but here is one set of data. Researchers studied the behavior of drivers on a rural interstate highway in Maryland where the speed limit was 55 miles per hour. They measured speed with an electronic device hidden in the pavement and, to eliminate large trucks, considered only vehicles less than 20 feet long. They found that 5690 out of 12,931 vehicles exceeded the speed limit. Is this good evidence, at a significance level of 0.05, that (at least in this location) fewer than half of all drivers are speeding?

22.43 **Do Cash Incentives Improve Learning?** A high-school teacher in a low-income urban school in Worcester, Massachusetts, used cash incentives to encourage learning in his AP Statistics class.[29] In 2010, 15 of the 61 students enrolled in his class scored a 5 on the AP Statistics exam. Worldwide, the proportion of students who scored a 5 in 2010 was 0.15. Is this evidence that the proportion of students who would score a 5 on the AP Statistics exam when taught by the teacher in Worcester using cash incentives is higher than the worldwide proportion of 0.15?

(a) State the hypotheses, find the *P*-value, and give your conclusions in the context of the problem. Do you have any reservations about using the z test for proportions for this data?

(b) Does this study provide evidence that cash incentives cause an increase in the proportion of 5's on the AP Statistics exam? Explain your answer.

22.44 **Order in Choice.** Does the order in which wine is presented make a difference? In this study, subjects were asked to taste two wine samples in sequence. Both samples given to a subject were the *same* wine, although subjects were expecting to taste two different samples of a particular variety. Of the 32 subjects in the study, 22 selected the wine presented first when presented with two identical wine samples.[30]

(a) Do the data give good reason to conclude that the subjects are not equally likely to choose either of the two positions when presented with two identical wine samples in sequence?

(b) The subjects were recruited in Ontario, Canada, via advertisements to participate in a study of "attitudes and values toward wine." Can we generalize our conclusions to all wine tasters? Explain

22.45 **Fast-Food Accuracy.** Which type of fast-food chain fills orders most accurately at the drive-thru window? The Quick Service Restaurant (QSR) magazine drive-thru study visits restaurants in the largest fast-food chains in all 50 states. Visits occurred throughout the day, starting at 5:00 A.M. and ending at 7:00 P.M. During each visit, the researcher ordered a main item, a side item, and a drink and made a minor special request such as a beverage with no ice. After receiving the order, all food and drink items were checked for complete accuracy. Any food or drink item received that was not exactly as ordered resulted in the order being classified as inaccurate. Also included in the measurement of accuracy were condiments asked for, napkins, straws, and correct change. Any errors in these resulted in the order being classified as inaccurate. In 2019 Arby's, Burger King, Carl's Jr., Chick-fil-A, Dunkin', Hardee's, KFC, McDonald's, Taco Bell, and Wendy's were included in the study. KFC had the most inaccuracies, with 56 of 165 orders classified as inaccurate.[31] What proportion of orders are filled *accurately* by KFC? (Use 95% confidence.)

Ministr-84/Shutterstock

22.46 **Order in Choice: Planning a Study.** How large a sample would be needed to obtain margin of error ± 0.05 in the study of choice order for tasting wine? Use the $\hat{p}$ from Exercise 22.44 as your guess for the unknown p. (Use 95% confidence.)

Exercises 22.47 through 22.50 concern the optional material on the plus four method.

22.47 **Order in Choice (continued).** Does the order in which wine is presented make a difference? In this study, subjects were asked to taste two wine samples in sequence. Both samples given to a subject were the *same* wine, although subjects were expecting to taste two different samples of a particular variety. Of the 32 subjects in the study, 22 selected the wine presented first when presented with two identical wine samples.[32]

(a) Although the conditions for the large-sample significance test were met in Exercise 22.44, show that the conditions for the large-sample confidence interval discussed in this chapter are not met.

(b) Are the conditions for the use of the plus four confidence interval met? If so, use the plus four method to give a 90% confidence interval for the proportion of subjects who would select the first choice presented.

22.48 **Shrubs That Survive Fires.** Some shrubs have the useful ability to resprout from their roots after their tops are destroyed. Fire is a particular threat to shrubs in dry climates because it can injure the roots as well as destroy the aboveground material. One study of resprouting took place in a dry area of Mexico.[33] The investigators clipped the tops of samples of several species of shrubs. In some cases, they also applied a propane torch to the stumps to simulate a fire. Of 12 specimens of the shrub *Krameria cytisoides*, five resprouted after fire. Estimate with 90% confidence the proportion of all shrubs of this species that will resprout after fire.

22.49 **Can You Match the Dog and Owner?** In Example 22.7 (page 506), the large-sample confidence interval for the proportion of undergraduates who can choose the sheet with the dogs and owners correctly matched was computed, although the conditions for the large-sample confidence interval were not quite met.

(a) Are the conditions for the use of the plus four confidence interval met? If so, use the plus four method to give a 99% confidence interval for the proportion of subjects who can choose the sheet with the dogs and owners correctly matched.

(b) If you express the interval as a percentage, round to the nearest 10th of a percent and compare the interval in part (a) with the interval in Example 22.7. (As always, the plus four method pulls results away from 0% or 100%, whichever is closer.)

22.50 **Fast-Food Accuracy (continued).** Which fast-food chain fills orders most accurately at the drive-thru window? The Quick Service Restaurant (QSR) magazine drive-thru study visits restaurants in the largest fast-food chains in all 50 states. Visits occurred throughout the day, starting at 5:00 A.M. and ending at 7:00 P.M. During each visit, the researcher ordered a main item, a side item, and a drink and made a minor special request such as a beverage with no ice. After receiving the order, researchers checked all food and drink items for complete accuracy. Any item that was not exactly as ordered resulted in the order being classified as inaccurate. Also included in the measurement of accuracy were condiments asked for, napkins, straws, and correct change. Any errors in these resulted in the order being classified as inaccurate. In 2019 Arby's, Burger King, Carl's Jr., Chick-fil-A, Dunkin', Hardee's, KFC, McDonald's, Taco Bell, and Wendy's were included in the study. Chick-fil-A had the fewest inaccuracies, with 11 of 183 orders classified as inaccurate. Are the conditions for the use of the plus four confidence interval met? If so, use the plus four method to give a 99% confidence interval for the proportion of orders filled *accurately* by Chick-fil-A.

© Jupiterimages/AgeFotostock

CHAPTER 23

Comparing Two Proportions

When you complete this chapter, you will be able to:

23.1 Recognize two-sample problems involving proportions and understand the steps involved in solving such problems.

23.2 Describe the properties and statistics of the sampling distribution of a difference between two proportions.

23.3 Use the large-sample z procedure to produce confidence intervals for the difference between proportions in two populations, based on samples from the populations, when conditions are met.

23.4 Use technology to compute confidence intervals for the difference between two proportions.

23.5 Use a z statistic to test the hypothesis H_0: $p_1 = p_2$ that proportions in two distinct populations are equal.

23.6 Use plus four confidence intervals for comparing two proportions when conditions are met.

A two-sample problem can arise from a randomized comparative experiment that randomly divides subjects into two groups and exposes each group to a different treatment. Comparing random samples separately selected from two populations is also a two-sample problem. The differences in the types of conclusions that we can reach in comparative experiments versus observational studies were described in Chapter 9.

When a comparison involves the *means* of two populations, we use the two-sample t methods of Chapter 21. In this chapter, we consider two-sample problems in which the measurement on the individual can be categorized as either a success or a failure. Our goal is to compare the *proportions* of successes in the two populations.

23.1 Two-Sample Problems: Proportions

Here are some questions that you will answer in the exercises of this chapter.

EXAMPLE 23.1 Two-Sample Problems for Proportions

- Does the proportion of male high school seniors who smoke cigarettes daily differ from the proportion of female high school seniors who do so? A sample of male and a sample of female high school seniors are obtained, and the proportions in each sample who have smoked cigarettes daily in the 30 days before the survey are to be compared (see Exercise 23.1, page 522).

STATISTICS IN YOUR WORLD

Computer-Assisted Interviewing The days of the interviewer with a clipboard are past. Interviewers now read questions from a computer screen and use the keyboard to enter responses. The computer skips irrelevant items; for example, once a woman says that she has no children, further questions about her children do not appear. The computer can even present questions in random order to avoid bias due to always following the same order. Software keeps records of who has responded and prepares a file of data from the responses. The tedious process of transferring responses from paper to computer, which was once a source of errors, has disappeared.

- Does hand washing with alcohol-based hand sanitizers reduce the risk of infection with the common cold? A randomized experiment assigned 100 subjects to a treatment group that followed a regimen of hand washing with alcohol-based sanitizers and 100 subjects to a control group that used routine hand washing without alcohol-based sanitizers. After 10 weeks, the study compared the proportions in the two groups who were infected with the common cold virus over the study period (see Exercise 23.35, page 535).

We will use notation similar to that used in our study of two-sample t statistics. The groups we want to compare are Population 1 and Population 2. We have a separate SRS from each population or responses form two treatments in a randomized comparative experiment. A subscript shows which group a parameter or statistic describes. Here is our notation:

Population	Population Proportion	Sample Size	Sample Proportion
1	p_1	n_1	$\hat{p}_1$
2	p_2	n_2	$\hat{p}_2$

We compare the populations by doing inference about the difference $p_1 - p_2$ between the population proportions. The statistic that estimates this difference is the difference between the two sample proportions, $\hat{p}_1 - \hat{p}_2$.

EXAMPLE 23.2 Interracial Dating

STATE: "Would you date a person of a different race?" Researchers answered this question by collecting data from the Internet dating site Match.com. When people post profiles on the site, they indicate which races they are willing to date. While several races were studied, we focus on the data collected for Black people dating White people. A random sample of 100 Black males and a random sample of 100 Black females were selected from the dating site, with 75 of the Black males indicating their willingness to date White females and 56 of the Black females indicating their willingness to date White males.[1] Is this good evidence that different proportions of Black males and females on this Internet dating site would be willing to date someone who is White? How large is the difference between the proportions of Black males and females who would be willing to date someone who is White?

PLAN: Take Black males to be Population 1 and Black females to be Population 2. The population proportions who are willing to date someone who is White are p_1 for Black males and p_2 for Black females. We want to test the hypotheses

$$H_0: p_1 = p_2 \text{ (the same as } H_0: p_1 - p_2 = 0)$$
$$H_a: p_1 \neq p_2 \text{ (the same as } H_a: p_1 - p_2 \neq 0)$$

We also want to give a confidence interval for the difference $p_1 - p_2$.

SOLVE: Inference about population proportions is based on the sample proportions

$$\hat{p}_1 = \frac{75}{100} = 0.75 \quad \text{(men)}$$
$$\hat{p}_2 = \frac{56}{100} = 0.56 \quad \text{(women)}$$

We see that 75% of Black males but only 56% of Black females would be willing to date someone who is White. Because the sample sizes are of moderate size and the sample proportions are quite different, we expect that a test will be highly significant (in fact, $P = 0.0046$). So we concentrate on the confidence interval. To estimate $p_1 - p_2$, start from the difference between sample proportions

$$\hat{p}_1 - \hat{p}_2 = 0.75 - 0.56 = 0.19$$

To complete the Solve step, we must know how this difference behaves.

23.2 The Sampling Distribution of a Difference between Proportions

To use $\hat{p}_1 - \hat{p}_2$ for inference, we must know its sampling distribution. Here are the facts we need:

- When the samples are large, the distribution of $\hat{p}_1 - \hat{p}_2$ is approximately Normal.
- The mean of the sampling distribution is $p_1 - p_2$. That is, the difference between sample proportions is an unbiased estimator of the difference between population proportions.
- The standard deviation of the distribution is

$$\sqrt{\frac{p_1(1-p_1)}{n_1} + \frac{p_2(1-p_2)}{n_2}}$$

Notice that we add rather than subtract the terms under the square root in the formula for the standard deviation, despite the fact that this is the standard deviation of a difference. The variability in the difference depends on the *combined* variability of the two samples.

Figure 23.1 displays the distribution of $\hat{p}_1 - \hat{p}_2$. The standard deviation of $\hat{p}_1 - \hat{p}_2$ involves the unknown parameters p_1 and p_2. Just as in Chapter 22, we must replace these with estimates to make inferences. And just as in Chapter 22, we do this a bit differently for confidence intervals and for hypothesis tests.

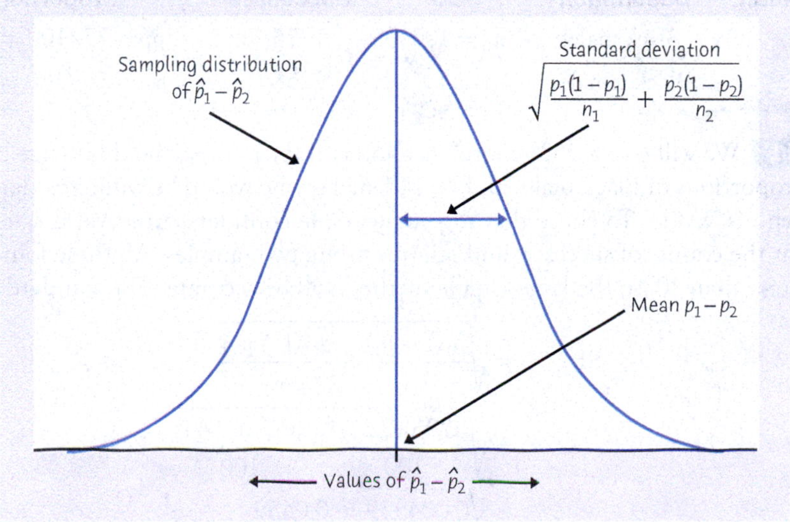

FIGURE 23.1
Select independent SRSs from two populations having proportions of successes p_1 and p_2. The proportions of successes in the two samples are $\hat{p}_1$ and $\hat{p}_2$. When the samples are large, the sampling distribution of the difference $\hat{p}_1 - \hat{p}_2$ is approximately Normal.

23.3 Large-Sample Confidence Intervals for Comparing Proportions

To obtain a confidence interval, replace the population proportions p_1 and p_2 in the standard deviation with the sample proportions. The result is the *standard error* of the statistic $\hat{p}_1 - \hat{p}_2$:

$$SE_{\hat{p}_1-\hat{p}_2} = \sqrt{\frac{\hat{p}_1(1-\hat{p}_1)}{n_1} + \frac{\hat{p}_2(1-\hat{p}_2)}{n_2}}$$

The confidence interval has the same form we met in Chapter 22:

$$\text{estimate} \pm z^* SE_{\text{estimate}}$$

Large-Sample Confidence Intervals for Comparing Two Proportions

Draw an SRS of size n_1 from a large population having proportion p_1 of successes and draw an independent SRS of size n_2 from another large population having proportion p_2 of successes. When n_1 and n_2 are large, an approximate level C confidence interval for $p_1 - p_2$ is

$$(\hat{p}_1 - \hat{p}_2) \pm z^* SE_{\hat{p}_1-\hat{p}_2}$$

In this formula, the standard error $SE_{\hat{p}_1-\hat{p}_2}$ of $\hat{p}_1 - \hat{p}_2$ is

$$SE_{\hat{p}_1-\hat{p}_2} = \sqrt{\frac{\hat{p}_1(1-\hat{p}_1)}{n_1} + \frac{\hat{p}_2(1-\hat{p}_2)}{n_2}}$$

and z^* is the critical value for the standard Normal density curve with area C between $-z^*$ and z^*.

Use this interval only when the numbers of successes and failures are each 10 or more in both samples.

EXAMPLE 23.3 Interracial Dating (continued)

4step

We can now complete Example 23.2. Here is a summary of the basic information:

Population	Population Description	Sample Size	Number of Successes	Sample Proportion
1	Black males	$n_1 = 100$	75	$\hat{p}_1 = 75/100 = 0.75$
2	Black females	$n_2 = 100$	56	$\hat{p}_2 = 56/100 = 0.56$

Phovoir/Shutterstock

SOLVE: We will give a 95% confidence interval for $p_1 - p_2$, the difference between the proportions of Black males and Black females who would be willing to date someone who is White. To check that the large-sample confidence interval is safe to use, look at the counts of successes and failures in the two samples. All these four counts are larger than 10, so the large-sample method will be accurate. The standard error is

$$\begin{aligned} SE_{\hat{p}_1-\hat{p}_2} &= \sqrt{\frac{\hat{p}_1(1-\hat{p}_1)}{n_1} + \frac{\hat{p}_2(1-\hat{p}_2)}{n_2}} \\ &= \sqrt{\frac{(0.75)(0.25)}{100} + \frac{(0.56)(0.44)}{100}} \\ &= \sqrt{0.004339} = 0.0659 \end{aligned}$$

The 95% confidence interval is

$$\begin{aligned}(\hat{p}_1 - \hat{p}_2) \pm z^*\text{SE}_{\hat{p}_1-\hat{p}_2} &= (0.75 - 0.56) \pm (1.960)(0.0659)\\ &= 0.19 \pm 0.13\\ &= 0.06 \text{ to } 0.32\end{aligned}$$

CONCLUDE: We are 95% confident that the percentage of Black males willing to date White women is between 6 and 32 percentage points higher than the percentage of Black females who are willing to date White men on comparable Internet dating sites. Even with sample sizes of 100 in each group, the resulting confidence interval 0.06 to 0.32 is quite wide. As with a single proportion, fairly large sample sizes are required to obtain narrow confidence intervals. In a similar study by the authors, it was found that White men were more willing than White women to date someone who is Black. Note that our conclusions are restricted to comparable Internet dating sites and don't necessarily reflect dating behavior of the general population, as individuals on dating sites may not be representative of the general population.

You will sometimes see confidence intervals for the difference in two proportions used to make inferences about whether the proportions differ. For example, election polls ask voters to indicate which of two candidates they would vote for if the election were held today. If a confidence interval for the difference in the proportion p_1 who would vote for candidate 1 and the proportion p_2 who would vote for candidate 2 includes 0, this may be reported as the election being too close to call. If the confidence interval for the difference $p_1 - p_2$ includes only positive values, this may be reported as evidence that if the election were held today, candidate 1 would win. Likewise, if the confidence interval for the difference $p_1 - p_2$ includes only negative values, candidate 2 would be forecast to win if the election were held today.

Later in this chapter we discuss significance tests for $H_0: p_1 = p_2$, that are designed to answer questions about whether $p_1 \neq p_2$, $p_1 > p_2$, or $p_1 < p_2$. But confidence intervals also provide evidence about this, along with information about the magnitude of any difference. If the confidence interval for the difference $p_1 - p_2$ includes 0, we should not rule out the possibility that the difference is negligible. If all the values in the interval for $p_1 - p_2$ are positive, this provides evidence that $p_1 > p_2$. If all the values in the interval for $p_1 - p_2$ are negative, this provides evidence that $p_1 < p_2$. Wide confidence intervals for $p_1 - p_2$ indicate large uncertainty about the difference. Narrow confidence intervals suggest greater certainty about the magnitude of the difference. However, a narrow confidence interval close to 0, but not including 0, may indicate only that that there is a difference between p_1 and p_2 but that the difference is not of practical importance.

23.4 Examples of Technology

Figure 23.2 displays software output for Example 23.3 from a graphing calculator and two statistical software programs. As usual, you can understand the output even without knowledge of the program that produced it. Minitab gives the test as well as the confidence interval, confirming that the difference between men and women is highly significant. In CrunchIt!, the test and the confidence interval must be requested using separate commands, resulting in the two outputs in the figure.

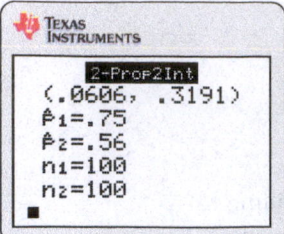

FIGURE 23.2 Output from a graphing calculator, Minitab, and CrunchIt! for the 95% confidence interval of Example 23.3.

TEXAS INSTRUMENTS

```
2-PropZInt
(.0606, .3191)
p̂1=.75
p̂2=.56
n1=100
n2=100
```

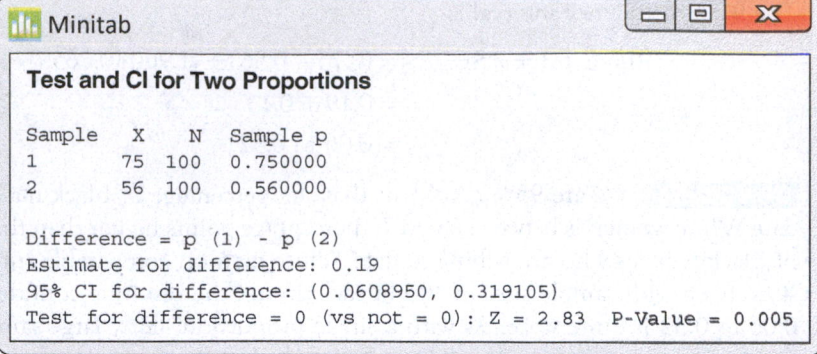

Minitab

Test and CI for Two Proportions

```
Sample   X    N   Sample p
1       75  100   0.750000
2       56  100   0.560000

Difference = p (1) - p (2)
Estimate for difference: 0.19
95% CI for difference: (0.0608950, 0.319105)
Test for difference = 0 (vs not = 0): Z = 2.83  P-Value = 0.005
```

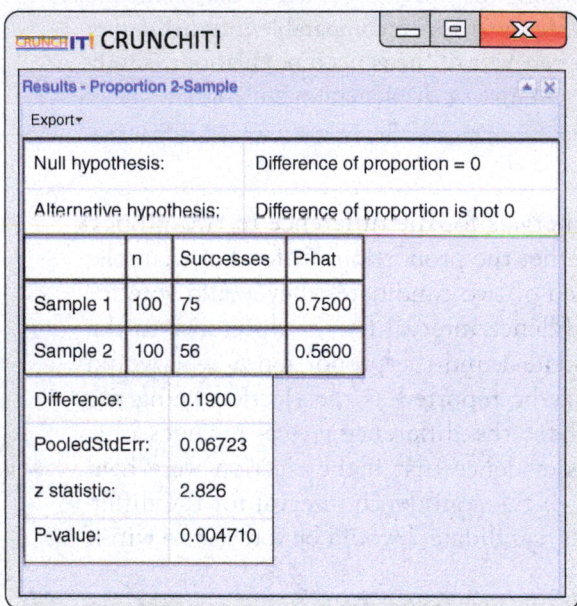

CRUNCHIT! CRUNCHIT!

Results - Proportion 2-Sample

Export▾

Null hypothesis:	Difference of proportion = 0
Alternative hypothesis:	Difference of proportion is not 0

	n	Successes	P-hat
Sample 1	100	75	0.7500
Sample 2	100	56	0.5600

Difference:	0.1900
PooledStdErr:	0.06723
z statistic:	2.826
P-value:	0.004710

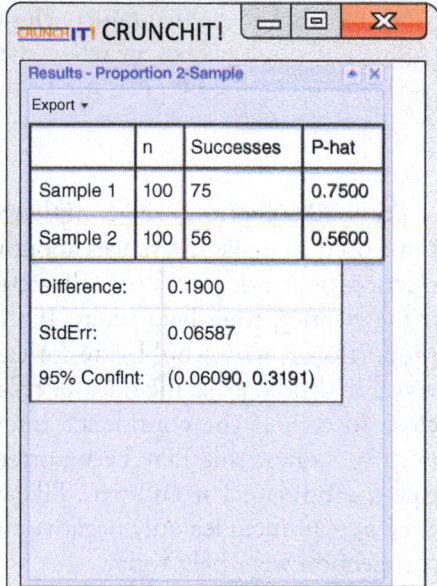

CRUNCHIT! CRUNCHIT!

Results - Proportion 2-Sample

Export▾

	n	Successes	P-hat
Sample 1	100	75	0.7500
Sample 2	100	56	0.5600

Difference:	0.1900
StdErr:	0.06587
95% ConfInt:	(0.06090, 0.3191)

Software packages, including the graphing calculator, allow you to specify the confidence level as part of the commands used to produce the output in Figure 23.2. The default level is 95%, and this is the level displayed in the output from the graphing calculator.

APPLY YOUR KNOWLEDGE

4step

23.1 Smoking among Seniors. Since the 1990s, daily cigarette smoking among high school seniors has dropped from around 12% to slightly over 3%. In 2017, random samples of 1725 female and 1564 male high school seniors found that 3.7% of the females and 3.1% of the males had smoked cigarettes daily in the 30 days before the survey.[2] Give a 95% confidence interval for the difference between the proportions of the populations of male and female high school seniors who had smoked cigarettes daily in the 30 days before the survey. Follow the four-step process as illustrated in Examples 23.2 and 23.3 (pages 518 and 520).

23.2 Aerosolized Vaccine for Measles. An aerosolized vaccine for measles was developed in Mexico and has been used on more than 4 million children since 1980. Aerosolized vaccines have the advantages of being able to be administered by people without clinical training and do not cause injection-associated infections. Despite these advantages, data about efficacy of the aerosolized vaccines against measles compared to subcutaneous injection of the vaccine have been inconsistent. Because of this, a large randomized controlled study was conducted using children in India. The primary outcome was an immune response to measles measured 91 days after the treatments. Among the 785 children receiving the subcutaneous injection, 743 developed an immune response, while among the 775 children receiving the aerosolized vaccine, 662 developed an immune response.[3]

(a) Compute the proportion of subjects experiencing the primary outcome for both the aerosol and injection groups.

(b) Can we safely use the large-sample confidence interval for comparing the proportion of children who developed an immune response to measles in the aerosol and injection groups? Explain.

(c) Give a 95% confidence interval for the difference between the proportion of children in the aerosol and injection groups who experienced the primary outcome.

(d) The study described is an example of a noninferiority clinical trial intended to show that the effect of a new treatment, the aerosolized vaccine, is not worse than that of the standard treatment by more than a specified margin.[4] Specifically, is the percentage of children who developed an immune response for the aerosol treatment more than 5% below the percentage for the subcutaneous injected vaccine? The 5-percentage-point difference was based on previous studies and the fact that with a bigger difference, the aerosolized vaccine would not provide the levels of protection necessary to achieve herd immunity. Using your answer in part (c), do you feel the investigators demonstrated the noninferiority of the aerosolized vaccine? Explain.

23.3 **Euthanasia.** A 2018 Gallup Poll asked two versions of a question concerning euthanasia. 4step

Version A: "When a person has a disease that cannot be cured, do you think doctors should be allowed by law to end the patient's life by some painless means if the patient and his or her family request it?"

Version B: "When a person has a disease that cannot be cured and is living in severe pain, do you think doctors should or should not be allowed by law to assist the patient to commit suicide if the patient requests it?"

A random sample of 542 U.S. adults were asked version A and a random sample of 482 U.S. adults were asked version B. Here are the results:[5]

Question	Sample Size	Number Saying "Yes, Should"
Version A	$n_A = 542$	390
Version B	$n_B = 482$	312

Give a 99% confidence interval for the difference between the proportions of all U.S. adults who would answer "Yes, should" to the question as worded in version A and those who would answer "Yes, should" to the question as worded in version B. Follow the four-step process as illustrated in Examples 23.2 and 23.3 (pages 518 and 520).

23.5 Significance Tests for Comparing Proportions

An observed difference between two sample proportions can reflect an actual difference between the populations, or it may just be due to chance variation in random sampling. Significance tests help us decide if the effect we see in samples is really there in the populations. The null hypothesis says that there is no difference between the two populations:

$$H_0: p_1 = p_2 \text{ (the same as } H_0: p_1 - p_2 = 0\text{)}$$

The alternative hypothesis says what kind of difference we expect.

EXAMPLE 23.4 False Memories

4step

STATE: The political event depicted in Figure 23.3 was remembered by about 31% of those surveyed, despite the fact that it never occurred. In 2010, *Slate*, a current affairs and culture magazine, surveyed more than 5000 readers regarding their perspective on several past political events and their memories of these events. Unbeknown to those surveyed, one of the events shown to each participant was fabricated, making this the largest false memory study conducted to that point.[6] The hypothesis of interest was whether political preferences guided the formation of false memories, as false memories have been shown to be more easily implanted in memory when they are congruent with a person's preexisting attitudes. Figure 23.3 was viewed by 616 participants who categorized themselves as progressive and 49 participants who categorized themselves as conservative. The event was falsely remembered as having occurred by 212 of the progressives surveyed and by seven of the conservatives. How strong is the evidence that a larger proportion of progressives have a false memory of this event than conservatives?

Photo illustration by Holly Allen (Slate Magazine), Roger Clemens photo credit: Al Messerschmidt/Getty Images; George W. Bush credit: Pool/Getty Images

FIGURE 23.3
September 1, 2005: As parts of New Orleans lie underwater in the wake of Hurricane Katrina, President Bush entertains Houston Astros pitcher Roger Clemens at his ranch in Crawford, Texas, for Example 23.4.

PLAN: Call the population proportions p_1 for progressives and p_2 for conservatives. Because the image of a conservative president relaxing at his ranch during a major crisis is more congruent with the preexisting attitude of progressives, our hypothesis gives a direction for the difference before looking at the data, so we have the one-sided alternative:

$$H_0: p_1 = p_2$$
$$H_a: p_1 > p_2$$

SOLVE: Consider those who classify themselves as progressives and conservatives as separate SRSs of progressive and conservative *Slate* readers. The sample proportions who falsely remembered the event in Figure 23.3 are

$$\hat{p}_1 = \frac{212}{616} = 0.344 \quad \text{(progressive)}$$

$$\hat{p}_2 = \frac{7}{49} = 0.143 \quad \text{(conservative)}$$

That is, 34% of those classifying themselves as progressive falsely remembered the event in Figure 23.3, but only 14% of conservatives falsely remembered it. Is this apparent difference statistically significant? To continue the solution, we must learn the proper test.

STATISTICS IN YOUR WORLD

The Cookie Strikes

How many different people clicked on your business website last month? Technology tries to help: when someone visits your site, a little piece of code called a cookie is left on their computer. When the same person clicks again, the cookie says not to count them as a "unique visitor" because this isn't their first visit. But lots of web users delete cookies, either by hand or automatically with software. These people get counted again when they visit your site again. That's bias: your counts of unique visitors are systematically too high. One study found that unique-visitor counts were as much as 50% too high.

To do a hypothesis test, standardize the difference between the sample proportions $\hat{p}_1 - \hat{p}_2$ to get a z statistic. If H_0 is true, both samples come from populations in which the same unknown proportion p have a false memory of the event depicted in Figure 23.3. We take advantage of this by combining the two samples to estimate this single p instead of estimating p_1 and p_2 separately. Call this the **pooled sample proportion.** It is

$$\hat{p} = \frac{\text{number of successes in both samples combined}}{\text{number of individuals in both samples combined}}$$

Substituting $\hat{p}$ in place of both $\hat{p}_1$ and $\hat{p}_2$ in the expression for the standard error $SE_{\hat{p}_1 - \hat{p}_2}$ of $\hat{p}_1 - \hat{p}_2$, we get a z statistic that has the standard Normal distribution when H_0 is true. Here is the test.

Significance Test for Comparing Two Proportions

Draw an SRS of size n_1 from a large population having proportion p_1 of successes and draw an independent SRS of size n_2 from another large population having proportion p_2 of successes. To test the hypothesis $H_0: p_1 = p_2$, first find the pooled proportion $\hat{p}$ of successes in both samples combined. Then compute the z statistic

$$z = \frac{\hat{p}_1 - \hat{p}_2}{\sqrt{\hat{p}(1-\hat{p})\left(\frac{1}{n_1} + \frac{1}{n_2}\right)}}$$

In terms of a variable Z having the standard Normal distribution, the P-value for a test of H_0 against

$H_a: p_1 > p_2$ is $P(Z \geq z)$

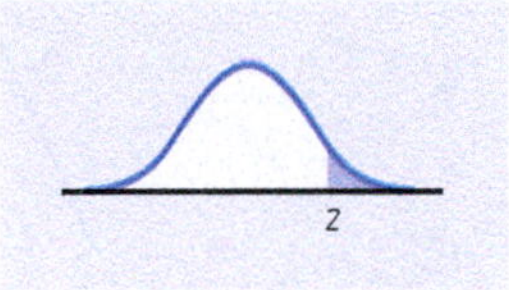

$H_a: p_1 < p_2$ is $P(Z \leq z)$

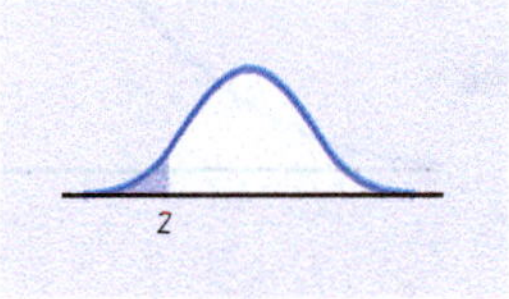

$H_a: p_1 \neq p_2$ is $2P(Z \geq |z|)$

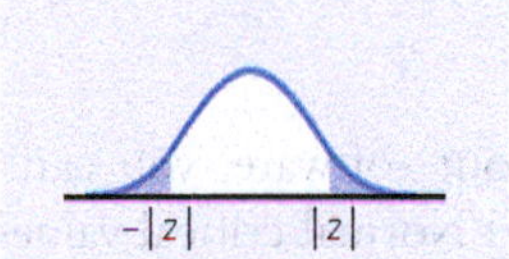

Use this test when the counts of successes and failures are each five or more in both samples.[7]

EXAMPLE 23.5 False Memories (continued)

SOLVE: The data come from an SRS, and the counts of successes and failures are all larger than five. The pooled proportion of progressives and conservatives who falsely remembered this event is

$$\hat{p} = \frac{\text{number “falsely remembered event” among progressives and conservatives combined}}{\text{number of progressives and conservatives combined}}$$

$$= \frac{212 + 7}{616 + 49}$$

$$= \frac{219}{665} = 0.329$$

The z test statistic is

$$z = \frac{\hat{p}_1 - \hat{p}_2}{\sqrt{\hat{p}(1-\hat{p})\left(\frac{1}{n_1} + \frac{1}{n_2}\right)}}$$

$$= \frac{0.344 - 0.143}{\sqrt{(0.329)(0.671)\left(\frac{1}{616} + \frac{1}{49}\right)}}$$

$$= \frac{0.201}{0.0697} = 2.88$$

The one-sided P-value is the area under the standard Normal curve greater than 2.88. Figure 23.4 shows this area. Software tells us that $P = 0.00199$.

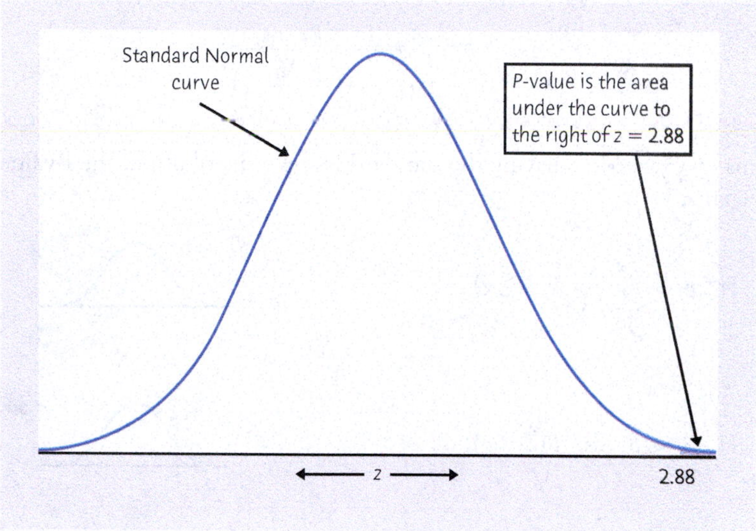
Standard Normal curve
P-value is the area under the curve to the right of z = 2.88
z
2.88

FIGURE 23.4
The P-value for the one-sided test for Example 23.5.

z^*	2.807	3.091
One-sided P	0.0025	0.001

Without software, you can compare $z = 2.88$ with the bottom row of Table C (standard Normal critical values) to approximate P.

It lies between the critical values 2.807 and 3.091 for one-sided P-values 0.0025 and 0.001.

CONCLUDE: There is strong evidence ($P < 0.0025$) that, among *Slate* readers, progressives are more likely than conservatives to have a false memory of Mr. Bush vacationing with Roger Clemens in the wake of the Katrina hurricane. In a second fabricated event by the authors, it was found that conservatives were more likely than progressives to falsely remember President Obama shaking hands with former Iranian President Ahmadinejad at a United Nations conference.

The sample survey in this example selected a single sample of *Slate* readers, not two separate samples of progressives and conservatives. To get two samples, we divided the single sample by political orientation. This means that we did not know the two sample sizes n_1 and n_2 until after the data were in hand. The two-sample z procedures for comparing proportions are valid in such situations. This is an important fact about these methods.

APPLY YOUR KNOWLEDGE

23.4 Matching Dogs and Owners. Researchers constructed two test sheets, each sheet including 20 photos of the faces of dog–owner pairs taken at a dog lovers' field festival. The 20 sets of dog–owner pairs on the two sheets were equivalent with respect to breed, diversity of appearance, and gender of owners. On the first sheet, the dogs were matched with their owners, while on the second sheet, the dogs and owners were deliberately mismatched. Three experiments were conducted, and in all experiments, subjects were asked to "choose the set of dog–owner pairs that resemble each other, Sheet 1 or Sheet 2," and were simply told the aim of the research was a "survey on dog–owner relationships." In the first experiment, the original sheets were shown to subjects; in the second experiment, just the "mouth region" of the owners was blacked out in all the pictures on both sheets; in the third experiment, just the "eye region" of the owners was blacked out. Subjects were assigned at random to the three experimental groups, and in each experiment, the number of subjects who selected the sheet with the dogs and their owners correctly matched was recorded. Experimenters were interested in whether blacking out portions of the face reduced the ability of subjects to correctly match dogs and owners.[8] Here are the results:

Experiment	Number of Subjects	Number Correctly Matched
Experiment 1	61	49
Experiment 2 (mouth blacked out)	51	37
Experiment 3 (eyes blacked out)	60	30

(a) Is there evidence that blacking out the mouth reduces a subject's ability to choose the sheet that correctly matches the dogs and their owners? Follow the four-step process as exhibited in Examples 23.4 and 23.5.

(b) Is there evidence that blacking out the eyes reduces a subject's ability to choose the sheet that correctly matches the dogs and their owners? Follow the four-step process as exhibited in Examples 23.4 and 23.5.

(c) Contrast your conclusions in parts (a) and (b) in the context of the problem, using non-technical language.

23.5 Protecting Skiers and Snowboarders. Most alpine skiers and snowboarders do not use helmets. Do helmets reduce the risk of head injuries? A study in Norway compared skiers and snowboarders who suffered head injuries with a control group who were not injured. Of 578 injured subjects, 96 had worn a helmet. Of the 2992 in the control group, 656 wore helmets.[9] Is helmet use less common among skiers and snowboarders who have head injuries? Follow the four-step process as illustrated in Examples 23.4 and 23.5 (pages 524 and 525). (Note that this is an observational study that compares injured and uninjured subjects. An experiment that assigned subjects to helmet and no-helmet groups would be more convincing.)

23.6 Cardiovascular Disease. Cardiovascular disease is a major cause of death and illness worldwide, with high blood pressure and high LDL cholesterol both being established risk factors. Because most cardiovascular events occur in persons with average risk and no previous cardiovascular disease history, the present research examined the simultaneous use of both blood pressure–reducing drugs and cholesterol-reducing drugs on this population rather than focus on only those at high risk. Subjects included men at least 55 years old and women at least 65 years old without cardiovascular disease who had at least one additional risk factor besides age, such as recent or current smoking, hypertension, or family history of premature coronary heart disease. Those with current cardiovascular disease were excluded from the study. Subjects were randomly assigned to the treatment (cholesterol- and blood pressure–reducing drugs) or a placebo, and the number suffering the primary outcome of a fatal cardiosvascular event or a nonfatal myocardial infarction or a nonfatal stroke were observed. Here are the results for the two groups over the course of the study:[10]

Group	Sample Size	Number Experiencing Primary Outcome
Treatment	3180	113
Placebo	3168	157

How strong is the evidence that the proportion of subjects experiencing the primary outcome in the treatment group differs from that of those who were in the control group? Follow the four-step process as illustrated in Examples 23.4 and 23.5 (pages 524 and 525).

23.6 Plus Four Confidence Intervals for Comparing Proportions*

Like the large-sample confidence interval for a single proportion p, the large-sample interval for $p_1 - p_2$ generally has a true confidence level less than the level you asked for. The inaccuracy is not as serious as in the one-sample case, at least if our guidelines for use are followed. Once again, as discussed in Section 22.5 for the one-sample case, adding imaginary observations can improve the accuracy.[11]

Plus Four Confidence Intervals for Comparing Two Proportions

Draw independent SRSs from two large populations with population proportions of successes p_1 and p_2. To get the **plus four confidence interval for the difference** $p_1 - p_2$, add four imaginary observations: one success and one failure in each of the two samples. Then use the large-sample confidence interval with the new sample sizes (actual sample sizes + 2) and counts of successes (actual counts + 1).

Use this interval when the sample size is at least 5 in each group, with any counts of successes and failures.

If your software does not offer the plus four method, just enter the new plus four sample sizes and success counts into the large-sample procedure.

EXAMPLE 23.6 Abecedarian Early Childhood Education Program: Adult Outcomes

4step

STATE: The Abecedarian Project was a randomized controlled study to assess the effects of intensive early childhood education on children who were at high risk based on several sociodemographic indicators. The project randomly assigned children to a

*This section is optional.

treatment group, which was provided with early educational activities before kindergarten, and the remainder to a control group. A follow-up study interviewed subjects at age 30 and compared the college graduation rates (earning a four-year degree).[12] Here are the data for two groups:

Population	Population Description	Sample Size	Number of Successes	Sample Proportion
1	Treatment	$n_1 = 52$	12	$\hat{p}_1 = 12/52 = 0.2308$
2	Control	$n_2 = 49$	3	$\hat{p}_2 = 3/49 = 0.0612$

How much does early childhood education increase the proportion earning a four-year degree by age 30?

PLAN: Give a 90% confidence interval for the difference between population proportions, $p_1 - p_2$.

SOLVE: The conditions for the large-sample interval are not met because there are only three successes in the control group. However, we can use the plus four method because the sample sizes for the treatment and the control group are both at least five. Add four imaginary observations. The new data summary is

Population	Population Description	Sample Size	Number of Successes	Plus Four Sample Proportion
1	Treatment	$n_1 + 2 = 54$	$12 + 1 = 13$	$\tilde{p}_1 = 13/54 = 0.2407$
2	Control	$n_2 + 2 = 51$	$3 + 1 = 4$	$\tilde{p}_2 = 4/51 = 0.0784$

The standard error based on the new facts is

$$\begin{aligned}\mathrm{SE}_{\tilde{p}_1 - \tilde{p}_2} &= \sqrt{\frac{\tilde{p}_1(1-\tilde{p}_1)}{n_1+2} + \frac{\tilde{p}_2(1-\tilde{p}_2)}{n_2+2}} \\ &= \sqrt{\frac{(0.2407)(0.7593)}{54} + \frac{(0.0784)(0.9216)}{51}} \\ &= \sqrt{0.00480} = 0.0693\end{aligned}$$

The plus four 90% confidence interval is

$$\begin{aligned}(\tilde{p}_1 - \tilde{p}_2) \pm z^*\mathrm{SE}_{\tilde{p}_1 - \tilde{p}_2} &= (0.2407 - 0.0784) \pm (1.645)(0.0693) \\ &= 0.1623 \pm 0.1140 \\ &= 0.048 \text{ to } 0.276\end{aligned}$$

CONCLUDE: We are 90% confident that the early childhood education program increases the proportion of high-risk children who earn a four-year degree by age 30 by between 4.8% and 27.6%.

The plus four interval may be conservative (that is, the true confidence level may be *higher* than you asked for) for very small samples and population p values close to 0 or 1. It is generally much more accurate than the large-sample interval when the samples are small. Nevertheless, the plus four interval in Example 23.6 cannot save us from the fact that sample sizes around 50 will produce a wide confidence interval when comparing two proportions.

APPLY YOUR KNOWLEDGE

23.7 Peanut Allergies. In the past 10 years, the prevalence of peanut allergies has doubled in Western countries. Is consumption or avoidance of peanuts in infants related to the development of peanut allergies in infants at risk? Subjects included infants between four and 11 months with severe eczema, egg allergy, or both but who did not display a preexisting sensitivity to peanuts based on a skin-prick test. The infants were randomly assigned to either a treatment that avoided consuming peanut protein or a treatment in which at least 6 grams of peanut protein were consumed per week. The response was the presence or absence of peanut allergy at 60 months of age. In the avoidance group containing 263 infants, 36 had developed a peanut allergy at 60 months of age, while in the consumption group containing 266 infants, five had developed a peanut allergy.[13]

(a) Despite the large sample sizes in both treatments, why should we not use the large-sample confidence interval for these data?

(b) The plus four method adds one success and one failure in each sample. What are the sample sizes and counts of successes after you do this?

(c) Give the plus four 99% confidence interval for the difference in the probabilities of developing a peanut allergy for the avoidance and consumption treatments. What does your interval say about the comparison of these treatments in the context of the problem?

23.8 Shrubs That Withstand Fire. Fire is a serious threat to shrubs in dry climates. Some shrubs can resprout from their roots after their tops are destroyed. One study of resprouting took place in a dry area of Mexico.[14] The investigators first clipped the tops of all the shrubs in the study. For the treatment, they then applied a propane torch to the stumps of the shrubs to simulate a fire, and for the control, the stumps were left alone. The study included 24 shrubs, of which 12 were randomly assigned to the treatment and the remaining 12 to the control. A shrub is a success if it resprouts. For the shrub *Xerospirea hartwegiana*, all 12 shrubs in the control group resprouted, whereas only eight in the treatment group resprouted. How much does burning reduce the proportion of shrubs of this species that resprout? Give the 95% plus four confidence interval for the amount by which burning reduces the proportion of shrubs of this species that resprout. The plus four method is particularly helpful when, as here, a count of either successes or failures is zero. Follow the four-step process as illustrated in Example 23.6 (page 528).

Leonid Eremeychuk/Getty Images

CHAPTER 23 SUMMARY

- The data in a two-sample problem are two independent SRSs, each drawn from a separate population.
- Tests and confidence intervals to compare the proportions p_1 and p_2 of successes in the two populations are based on the difference $\hat{p}_1 - \hat{p}_2$ between the sample proportions of successes in the two SRSs.
- When the sample sizes n_1 and n_2 are large, the sampling distribution of $\hat{p}_1 - \hat{p}_2$ is close to Normal with mean $p_1 - p_2$.
- The level C large-sample confidence interval for $p_1 - p_2$ is

$$(\hat{p}_1 - \hat{p}_2) \pm z^* \text{SE}_{\hat{p}_1 - \hat{p}_2}$$

where the standard error of $\hat{p}_1 - \hat{p}_2$ is

$$\text{SE}_{\hat{p}_1 - \hat{p}_2} = \sqrt{\frac{\hat{p}_1(1-\hat{p}_1)}{n_1} + \frac{\hat{p}_2(1-\hat{p}_2)}{n_2}}$$

and z^* is a standard Normal critical value.

The true confidence level of the large-sample interval can be substantially less than the planned level C. Use this interval only if the counts of successes and failures in both samples are 10 or greater.

- Significance tests for $H_0: p_1 = p_2$ use the **pooled sample proportion**

$$\hat{p} = \frac{\text{number of successes in both samples combined}}{\text{number of individuals in both samples combined}}$$

and the z statistic

$$z = \frac{\hat{p}_1 - \hat{p}_2}{\sqrt{\hat{p}(1-\hat{p})\left(\frac{1}{n_1} + \frac{1}{n_2}\right)}}$$

P-values come from the standard Normal distribution. Use this test when there are five or more successes and five or more failures in both samples.

- **(Optional topic)** When the conditions for the large-sample confidence interval are not met, to get a more accurate confidence interval, add four imaginary observations: one success and one failure in each sample. Then use the same formula for the confidence interval. This is the **plus four confidence interval.** You can use it whenever both samples have five or more observations.

CHECK YOUR SKILLS

While nearly all toddlers and preschool-age children eat breakfast daily, consumption of breakfast dips as children grow older. The Youth Risk Behavior Surveillance System (YRBSS) monitors health risk behaviors among U.S. high school students. In 2017, the survey randomly selected 4031 ninth-graders and 3396 12th-graders and asked them if they had eaten breakfast on all seven days before the survey.[15] *Of these students, 1536 ninth-graders and 1090 12th-graders said yes. Do these data give evidence that the proportion of ninth-graders who eat breakfast daily is higher than the proportion of 12th-graders eating breakfast daily? Questions 23.9 through 23.13 are based on these results.*

23.9 Take p_9 and p_{12} to be the proportions of all ninth- and 12th-graders who ate breakfast daily. The hypotheses to be tested are

(a) $H_0: p_9 = p_{12}$ versus $H_a: p_9 \neq p_{12}$.

(b) $H_0: p_9 = p_{12}$ versus $H_a: p_9 > p_{12}$.

(c) $H_0: p_9 = p_{12}$ versus $H_a: p_9 < p_{12}$.

23.10 The sample proportions of ninth- and 12th-graders who ate breakfast on all seven days before the survey are

(a) $\hat{p}_9 = 0.381$ and $\hat{p}_{12} = 0.321$.

(b) $\hat{p}_9 = 0.270$ and $\hat{p}_{12} = 0.452$.

(c) $\hat{p}_9 = 0.321$ and $\hat{p}_{12} = 0.381$.

23.11 The pooled sample proportion of respondents who ate breakfast on all seven days before the survey is

(a) $\hat{p} = 0.321$.

(b) $\hat{p} = 0.354$.

(c) $\hat{p} = 0.381$.

23.12 The numerical value of the z statistic for comparing the proportions of ninth- and 12th-graders who ate breakfast daily is

(a) 7.84. (b) 7.49. (c) $z = 5.39$.

23.13 The 90% large-sample confidence interval for the difference $p_9 - p_{12}$ in the proportions of ninth- and 12th-graders who ate breakfast daily is about

(a) 0.060 ± 0.011.

(b) 0.060 ± 0.013.

(c) 0.060 ± 0.018.

23.14 Avobenzone is one of the active ingredients in several commercially available sunscreens. It can be absorbed into the bloodstream when sunscreen is applied to the skin. The Food and Drug Administration has expressed concern about the safety of absorbing too much avobenzone. Amounts less than or equal to 0.5 ng/mL (nanograms absorbed per milliliter applied) are considered acceptable. Researchers recruited 12 healthy volunteers to investigate avobenzone absorption for two different commercially available sunscreens: a spray and a cream. Subjects were randomly assigned to one of the two sunscreens, with 6 subjects for each. Subjects had 2 milligrams of sunscreen per 1 cm^2 applied to 75% of their body surface area (area outside of normal swimwear). The amount of avobenzone absorbed into the bloodstream after 6 hours was then measured for each subject. Four of the six subjects receiving the spray had avobenzone levels exceeding 0.5 ng/mL, and one of the six subjects receiving the cream had levels exceeding 0.5 ng/mL.[16] The z test for "no difference" in the two proportions exceeding 0.5 ng/mL against "the two proportions differ" has

(a) $z = 1.76$, $P < 0.05$.

(b) $z = 1.84$, $P < 0.055$.

(c) $z = 1.76$, $0.05 < P < 0.10$.

23.15 The z test in Question 23.14

(a) may be inaccurate because the populations are too small.

(b) may be inaccurate because the counts of successes and failures are too small.

(c) is reasonably accurate because the conditions for inference are met.

23.16 **(Optional Topic)** The plus four 90% confidence interval for the difference between the proportion of those receiving the spray and those receiving the cream who had avobenzone levels exceeding 0.5 ng/mL is

(a) 0.375 ± 0.451.

(b) 0.375 ± 0.378.

(c) 0.375 ± 0.230.

CHAPTER 23 EXERCISES

When using the large-sample methods of this chapter, be sure to check that the guidelines for their use are met and to state your conclusions in context.

23.17 **Are They Tracking Me?** A survey in 2019 asked a random sample of U.S. adults if they believed the government was tracking all or most of their activities online or on their cell phones. Of the 671 adults aged 18–29, 396 said "yes." Of the 977 adults aged 65 or older, 293 said "yes."[17]

(a) Do these samples satisfy the guidelines for the large-sample confidence interval?

(b) Give a 95% confidence interval for the difference between the proportions of adults aged 18–29 and adults aged 65 or older who believed the government was tracking all or most of their activities online or on their cell phones.

23.18 **Effects of an Appetite Suppressant.** Subjects with preexisting cardiovascular symptoms who received sibutramine, an appetite suppressant, were found to be at increased risk of cardiovascular events while taking the drug. The study included 9804 overweight or obese subjects with preexisting cardiovascular disease and/or type 2 diabetes. The subjects were randomly assigned to sibutramine (4906 subjects) or a placebo (4898 subjects) in a double-blind fashion. The primary outcome measured was the occurrence of any of the following events: nonfatal myocardial infarction or stroke, resuscitation after cardiac arrest, or cardiovascular death. The primary outcome was observed in 561 subjects in the sibutramine group and 490 subjects in the placebo group.[18]

(a) Find the proportion of subjects experiencing the primary outcome for both the sibutramine and placebo groups.

(b) Can we safely use the large-sample confidence interval for comparing the proportions of sibutramine and placebo subjects who experienced the primary outcome? Explain.

(c) Give a 95% confidence interval for the difference between the proportions of sibutramine and placebo subjects who experienced the primary outcome.

23.19 **(Optional Topic) Genetically Altered Mice.** Genetic influences on cancer can be studied by manipulating the genetic makeup of mice. One of the processes that turn genes on or off (so to speak) in particular locations is called "DNA methylation." Do low levels of this process help cause tumors? Compare mice altered to have low levels with normal mice. Of 33 mice with lowered levels of DNA methylation, 23 developed tumors. None of the control group of 18 normal mice developed tumors in the same time period.[19]

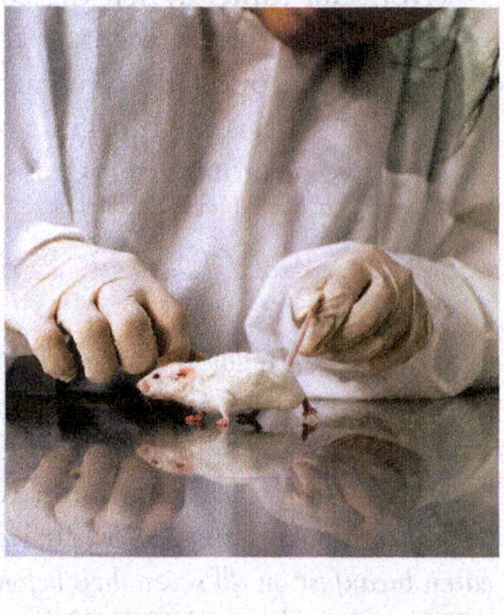

filo/Getty Images

(a) Explain why we cannot safely use either the large-sample confidence interval or the test for comparing the proportions of normal and altered mice that develop tumors.

(b) The plus four method adds two observations, a success and a failure, to each sample. What are the sample sizes and the numbers of mice with tumors after you do this?

(c) Give a 99% confidence interval for the difference in the proportions of the two populations that develop tumors.

23.20 **Effects of an Appetite Suppressant (continued).** Exercise 23.18 describes a study to determine if subjects with preexisting cardiovascular symptoms were at increased risk of cardiovascular events while taking sibutramine. Do the data give good reason to think that there is a difference between the proportions of treatment and placebo subjects who experienced the primary outcome?

(Note that sibutramine has not been available in the United States since the end of 2010 due to its manufacturer's concerns over increased risk of heart attack or stroke, although at the present time, it can still be purchased in other countries.)

(a) State the hypotheses, find the test statistic, and use either software or the bottom row of Table C for the *P*-value. Be sure to state your conclusion.

(b) Explain simply why it was important to have a placebo group in this study.

Tea Drinking and Strokes. *Is drinking tea good for your health? Researchers in China studied a sample of 100,902 adults and classified them as habitual tea drinkers (3 or more cups per week) and never/non-habitual tea drinker (less than 3 cups per week). These subjects were followed for several years, and the number of subjects who had strokes was recorded. Of the 69,017 subjects who were never/non-habitual tea drinkers, 2949 had strokes. Of the 31,885 habitual tea drinkers, 854 had strokes.*[20] *Exercises 23.21 through 23.23 are based on this study.*

23.21 Does Tea Drinking Make a Difference?

(a) Is there a significant difference in the proportions of subjects who were habitual tea drinkers and those who were never/non-habitual tea drinkers who had strokes? State the hypotheses, find the test statistic, and use software or the bottom row of Table C to get a *P*-value.

(b) Is this an observational study or an experiment? Why?

(c) In view of your answer in part (b), carefully state your conclusions about the relationship between tea drinking and strokes.

23.22 How Many had Strokes? Give a 95% confidence interval for the proportion of never/non-habitual tea drinkers who had a stroke.

23.23 How Big a Difference? Give a 95% confidence interval for the difference between the proportions of habitual tea drinkers and never/non-habitual tea drinkers who had strokes. Does your interval support the claim that tea drinking is good for your health? Discuss.

23.24 The Design Matters. Due to concerns about the safety of bariatric weight loss surgery, in 2006 the Centers for Medicare & Medicaid Services (CMS) restricted coverage of bariatric surgery to hospitals designated as Centers for Excellence. Did the CMS restriction improve the outcomes of bariatric surgery for Medicare patients? Among the 1847 Medicare patients in the study having surgery in the 18 months preceding the restriction, 270 experienced overall complications, whereas in the 1639 patients having bariatric surgery in the 18 months following the restriction, 170 experienced overall complications.[21]

(a) What are the sample proportions of Medicare patients who experienced overall complications from bariatric surgery before and after the CMS restriction on coverage? How strong is the evidence that the proportions of overall complications are different before and after the CMS restriction? Use an appropriate hypothesis test to answer this question.

(b) Is this an observational study or an experiment? Can we conclude that the CMS restriction has reduced the proportion of overall complications?

(c) Improved outcomes may be due to several factors, including the use of lower-risk bariatric procedures, increased surgeon experience, or healthier patients receiving the surgery. What types of variables are these, and how do they affect the types of conclusions that you can make?

23.25 Significant Does Not Mean Important. Never forget that even small effects can be statistically significant if the samples are large. To illustrate this fact, consider a sample of 148 small businesses. During a three-year period, 15 of the 106 businesses headed by men and seven of the 42 businesses headed by women failed.[22]

(a) Find the proportions of failures for businesses headed by women and businesses headed by men. These sample proportions are quite close to each other. Give the *P*-value for the z test of the hypothesis that the same proportion of women's and men's businesses fail. (Use the two-sided alternative.) The test is very far from being significant.

(b) Now suppose that the same sample proportions came from a sample 30 times as large. That is, 210 out of 1260 businesses headed by women and 450 out of 3180 businesses headed by men fail. Verify that the proportions of failures are exactly the same as in part (a). Repeat the z test for the new data and show that it is now significant at the $\alpha = 0.05$ level.

(c) It is wise to use a confidence interval to estimate the size of an effect rather than just giving a *P*-value. Give the large sample 95% confidence intervals for the difference between the proportions of women's and men's businesses that fail for the settings of both parts (a) and (b). What is the effect of larger samples on the confidence interval? Do you think the size of the difference between the proportions is an important difference?

In responding to Exercises 23.26 through 23.36, follow the ***Plan, Solve,*** *and* ***Conclude*** *steps of the four-step process.*

23.26 Personal Data and Privacy. There is widespread concern among the general public about how their personal data are used. A Pew Internet survey in 2019 asked a sample of U.S. adults whether they were concerned about how much information law enforcement might know about them. Of the 2887 White non-Hispanics in the survey, 1617 said they were concerned. Of the 445 Black non-Hispanics in the survey, 325 said they were concerned.[23] Is there good evidence that the proportions of White non-Hispanics and Black non-Hispanics who were concerned about how much information law enforcement might know about them differ?

23.27 Personal Data and Privacy. The survey in Exercise 23.26 also looked at possible differences in the proportions of White non-Hispanics and Black non-Hispanics who were concerned about how much information their friends and family might know about them. Of the 2887 White non-Hispanics in the survey, 1010 said they were concerned. Of the 445 Black non-Hispanics in the survey, 271 said they were concerned. Is there evidence of a difference between the proportions of White non-Hispanics and Black non-Hispanics who were concerned about how much information their friends and family might know about them?

23.28 More on Personal Data and Privacy. Continue your work from Exercise 23.26. Estimate the difference between the proportions of White non-Hispanics and Black non-Hispanics who were concerned about how much information law enforcement might know about them. (Use 90% confidence.)

23.29 I'll Stop Smoking—Soon! Chantix is different from most other quit-smoking products in that it targets nicotine receptors in the brain, attaches to them, and blocks nicotine from reaching them. A randomized, double-blind, placebo-controlled clinical trial on Chantix was conducted with a 24-week treatment period. Participants in the study were cigarette smokers who were either unwilling or unable to quit smoking in the next month but were willing to reduce their smoking and make an attempt to quit within the next three months. Subjects received either Chantix or a placebo for 24 weeks, with a target of reducing the number of cigarettes smoked by 50% or more by week 4, 75% or more by week 8, and a quit attempt by 12 weeks. The primary outcome measured was continuous abstinence from smoking during weeks 15 through 24. Of the 760 subjects taking Chantix, 244 abstained from smoking during weeks 15 through 24, whereas 52 of the 750 subjects taking the placebo abstained during this same time period.[24]

(a) Give a 99% confidence interval for the difference (treatment minus placebo) in the proportions of smokers who would abstain from smoking during weeks 15 through 24.

(b) Pfizer, the company that manufactures Chantix, claims that Chatix is proven to help smokers quit. Does your confidence interval support this claim? Discuss.

23.30 Headless Mannequins. The majority of clothing retailers use mannequins to display their merchandise, with approximately one-third displaying mannequins with heads and two-thirds displaying mannequins without heads. Researchers recruited 126 female participants and assigned each to one of the two mannequin styles (head/headless). Participants were asked to imagine that they wanted to buy a new dress and told to go to a named store to make their purchase. They then viewed the dress displayed on a mannequin (head or headless) and were asked whether or not they would buy the dress. Of the 63 participants viewing the dress displayed on a mannequin with a head, 18 indicated that they would buy the dress, while only 10 of the 63 participants viewing the headless mannequin indicated that they would buy the dress.[25]

fancy/Shutterstock

(a) Is there good evidence that the proportion of women who would buy the dress differs between those who viewed the dress displayed on a mannequin with or without a head?

(b) Based on this study, do you think it is a good idea for most manufacturers to use display mannequins without heads?

23.31 Stopping Traffic with a Smile! Throughout Europe, more than 8000 pedestrians are killed each year in road accidents, with approximately 25% of them dying when using a pedestrian crossing. Although failure to stop for pedestrians at a pedestrian crossing is a serious traffic offense in France, more than half of drivers do not stop when a pedestrian is waiting at a crosswalk. In this experiment, a male research assistant was instructed to stand at a pedestrian crosswalk and stare at the driver's face as a car approached the crosswalk. In 400 trials, the research assistant maintained a neutral expression, and in a second set of 400 trials, the research assistant was instructed to smile. The order of smiling or not smiling was randomized, and several pedestrian crossings were used in a town on the coast in the west of France. The research assistants was dressed in normal attire for his age (jeans, t-shirt, and sneakers). In the 400 trials in which the assistant maintained a neutral expression, the driver stopped in 172 out of the 400 trials, while in the 400 trials in which the assistant smiled at the driver, the driver stopped 226 times.[26] Do a test to assess the evidence that a smile increases the proportion of drivers who stopped. (In the portion of the study using female assistants, a smile significantly increased the proportion of drivers who stopped, with the proportion who stopped being significantly higher than for males in both the neutral and smiling conditions.)

23.32 4step **I'll Stop Smoking—Soon! (continued).** The clinical trial described in Exercise 23.29 assigned subjects at random to receive either Chantix or a placebo. To check that the random assignment produced comparable groups, we can compare the two groups at the start of the study.

(a) In the treatment group, 439 of the 760 subjects had made at least two serious attempts to quit smoking by any method since starting smoking, and in the placebo group, 303 of the 750 subjects had made at least two attempts. If the random assignment worked well, there should not be a significant difference in the proportions of subjects in the two groups who had made at least two serious attempts to quit smoking by any method since starting smoking. How significant is the observed difference?

(b) For the subjects in the treatment group, the mean of the number of cigarettes smoked per day in the past month was 20.6, with a standard deviation of 8.5, and in the placebo group, the mean was 20.8, with a standard deviation of 8.2. If the random assignment worked well, there should not be a significant difference in the means of the number of cigarettes smoked per day in the past month for the two groups. How significant is the observed difference?

23.33 4step **(Optional Topic) Lyme Disease.** Lyme disease is spread in the northeastern United States by infected ticks. The ticks are infected mainly by feeding on mice, so more mice result in more infected ticks. The mouse population rises and falls with the abundance of acorns, their favored food. Experimenters studied two similar forest areas in a year when the acorn crop failed. They added hundreds of thousands of acorns to one area to imitate an abundant acorn crop, while leaving the other area untouched. The next spring, 54 of the 72 mice trapped in the first area were in breeding condition, versus 10 of the 17 mice trapped in the second area.[27] Estimate the difference between the proportions of mice ready to breed in good acorn years and bad acorn years. (Use 90% confidence. Be sure to justify your choice of confidence interval.)

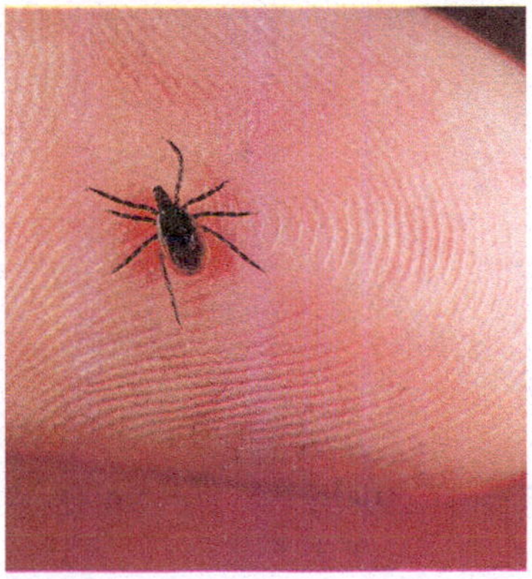
Scott Camazine/Science Source

23.34 4step **Abecedarian Early Childhood Education Program: Adult Outcomes.** The Abecedarian Project is a randomized controlled study to assess the effects of intensive early childhood education on children who were at high risk based on several sociodemographic indicators.[28] The project randomly assigned some children to a treatment group that was provided with early educational activities before kindergarten and the remainder to a control group. A recent follow-up study interviewed subjects at age 30 and evaluated educational, economic, and socioemotional outcomes to learn if the positive effects of the program continued into adulthood. The follow-up study included 52 individuals from the treatment group and 49 from the control group. Out of these, 39 from the treatment group and 26 from the control group were considered "consistently" employed (working 30 + hours per week in at least 18 of the 24 months prior to the interview). Does the study provide significant evidence that subjects who had early childhood education have a higher proportion of consistent employment than those who did not? How large is the difference between the proportions in the two populations that are consistently employed? Do inference to answer both questions. Be sure to explain exactly what inference you choose to do.

23.35 4step **Hand Sanitizers.** Hand disinfection is frequently recommended for prevention of transmission of the rhinovirus that causes the common cold. In particular, hand lotion containing 2% citric acid and 2% malic acid in 70% ethanol (HL+) has been found to have both immediate and persistent ability to inactivate rhinovirus (RV) on the hands in an experimental setting. Is hand disinfection effective in reducing the risk of infection in a natural setting? A total of 212 volunteers were assigned at random to either the HL+ group, which used the hand lotion every three hours or after hand washing, and a control group, which was asked to use routine hand washing but to avoid the use of alcohol-based hand sanitizers. Here are the data on the numbers of subjects with and without RV infection in the two groups over the 10-week study period:[29]

	RV Infection	
	Yes	No
HL+	49	67
Control group	49	47

(a) Is this an experiment or an observational study? Why?

(b) Many manufacturers of hand sanitizers claim that hand sanitizers reduce the chance of RV infection. Do the data give good evidence for such a claim? Discuss.

23.36 Does the Wall Street Culture Corrupt Bankers? Bank employees from a large international bank were recruited, and 67 were assigned at random to a control group and the remaining 61 to a treatment group. All subjects first completed a short online survey. After answering some general filler questions, members of the treatment group were asked seven questions about their professional background, such as "At which bank are you currently employed?" or "What is your function at this bank?" These are referred to as "identity priming" questions. The members of the control group were asked seven innocuous questions unrelated to their profession, such as "How many hours a week on average do you watch television?" After the survey, all subjects performed a coin-tossing task that required tossing any coin 10 times and reporting the results online. They were told they would win $20 for each head tossed, for a maximum payoff of $200. Subjects were unobserved during the task, making it impossible to tell if a particular subject cheated. If the banking culture favors dishonest behavior, it was conjectured that it should be possible to trigger this behavior by reminding subjects of their profession.[30] Here are the results. The first line gives the possible number of heads on 10 tosses, and the next two lines give the number of subjects that reported tossing this number of heads for the control and treatment groups, respectively (for example, 16 control subjects reported getting four heads).

Number of heads	0	1	2	3	4	5	6	7	8	9	10
Control group	0	0	1	8	16	17	14	6	2	1	2
Treatment group	0	0	2	4	8	14	15	7	6	0	5

If a subject is cheating, we would expect them to report doing better than chance, or tossing six or more heads.

(a) Find the proportion of subjects in each group that reported tossing six or more heads.

(b) Test the hypotheses that the proportions reporting tossing six or more heads in the two groups are the same against the appropriate alternative. Explain your conclusions in the context of the problem, being sure to relate this to the researcher's conjecture.

CHAPTER

24

Inference about Variables: Part IV Review

IN THIS CHAPTER, WE COVER . . .

The procedures of Chapters 20 through 23 are among the most common of all statistical inference methods. Now that you have mastered important ideas and practical methods for inference, it's time to review the big ideas of statistics in outline form. Here is a summary of Parts I, II, and III of this book, leading up to Part IV. The outline contains some important warnings: look for the Caution icon.

1. **Data Production**
   - Data basics:
     - Individuals (subjects)
     - Variables: categorical versus quantitative, units of measurement, explanatory versus response
     - Purpose of study
   - Data production basics:
     - Observation versus experiment
     - Simple random samples
     - Completely randomized experiments

Rawpixel/Getty Images

- Beware: weaknesses in data production (for example, sampling students at only one campus) can make generalizing conclusions difficult.
- Beware: really bad data production (voluntary response, confounding) can make interpretation impossible.

2. **Data Analysis**
   - Plot your data. Look for an overall pattern and striking deviations.
   - Add numerical descriptions based on what you see.
   - Beware: averages and other simple descriptions can miss the real story.
   - One quantitative variable:
     Graphs: stemplot, histogram, boxplot
     Pattern: distribution shape, center, variability. Outliers?
     Density curves (such as Normal curves) to describe overall pattern
     Numerical descriptions: five-number summary or $\bar{x}$ and s
   - Relationships between two quantitative variables:
     Graph: scatterplot
     Pattern: relationship form, direction, strength. Outliers? Influential observations?
     Numerical description for linear relationships: correlation, regression line
     Beware the lurking variable: correlation does not imply causation
   - Beware the effects of outliers and influential observations.

3. **The Reasoning of Inference**
   - Inference uses data to infer conclusions about a wider population.
   - When you do inference, you are acting as if your data come from random samples or randomized comparative experiments. Beware: if they don't, you may have "garbage in, garbage out."
   - Always examine your data before doing inference. Inference often requires a regular pattern, such as roughly Normal with no strong outliers.
   - Key idea: "What would happen if we did this many times?"
   - Confidence intervals: estimate a population parameter.
     95% confidence: the method used captures the true parameter 95% of the time in repeated use.
     Beware: the margin of error of a confidence interval does not include the effects of practical errors such as undercoverage and nonresponse.
   - Significance tests: assess evidence against H_0 in favor of H_a.
     P-value: If H_0 were true, how often would we get an outcome favoring the alternative this strongly? Smaller P = stronger evidence against H_0.
     Statistical significance at the 5% level, $P < 0.05$, means that an outcome this extreme would occur less than 5% of the time if H_0 were true.
     Beware: $P < 0.05$ is not sacred. Evidence against H_0 grows stronger as the *P*-value decreases, but there is no sharp border between "significant" and "not significant."
     Beware: statistical significance is not the same as practical significance. Large samples can make small effects statistically significant. Small samples can fail to declare large effects significant.
     Always try to estimate the size of an effect (for example, with a confidence interval), not just its significance.

4. **Methods of Inference**
   - Choose the right inference procedure.
   - Confirm that conditions are met and perform the calculations.
   - State your conclusion.

Part IV of this book introduces the fourth and last part of this outline. To actually do inference, you must choose the right procedure, confirm that conditions are met, and perform the calculations. The Statistics in Summary flowchart offers a compact guide. It is important to do some of the supplementary exercises because now, for the first time, you must decide which of several inference procedures to use. Learning to recognize problem settings in order to choose the right type of inference is a key step in advancing your mastery of statistics. This is the Plan step in the four-step process, in which you translate the real-world problem from the State step into a specific inference procedure.

The flowchart organizes one way of planning inference problems. Let's go through it from left to right.

1. *Do you want to test a claim or estimate an unknown quantity?* That is, will you need a test of significance or a confidence interval?
2. *Are your data a single sample representing one population or two samples chosen to compare two populations or responses to two treatments in an experiment?* Remember

STATISTICS IN SUMMARY

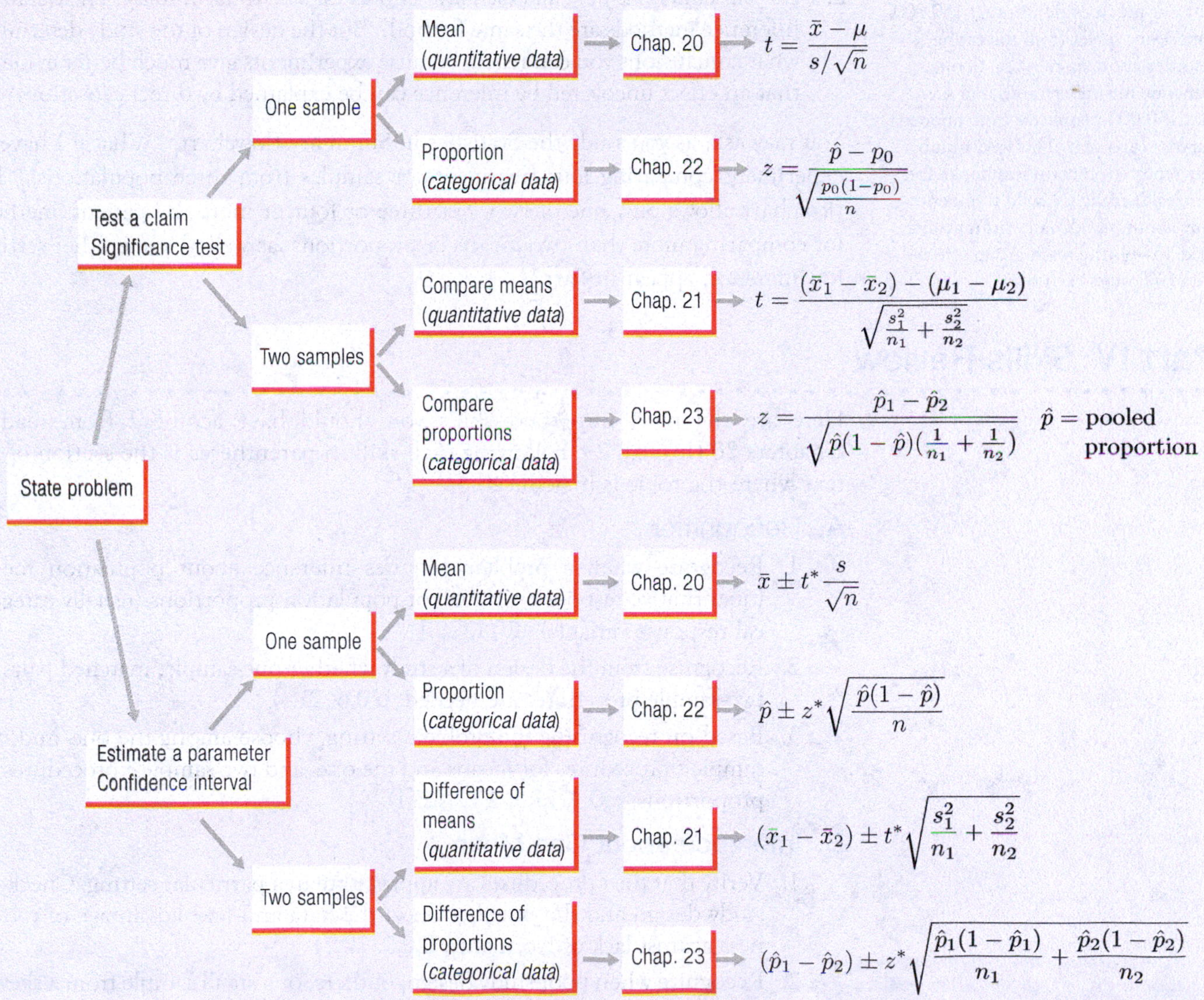

that to work with *matched pairs* data, you form one sample from the differences within pairs.

3. *Is the response variable quantitative or categorical?* Quantitative variables take numerical values with some unit of measurement such as inches or grams. The most common inference questions about quantitative variables concern *mean* responses. If the response variable is categorical, inference most often concerns the *proportion* of some category (call it a "success") among the responses.

The flowchart leads you to a specific test or confidence interval, indicated by a formula at the end of each path. The formula is just an aid to guide you toward the Solve and Conclude steps. You (or your technology) will use the formula as part of the Solve step but don't forget that you must do more.

4. *Are the conditions for this procedure met?* Can you act as if the data come from a random sample or randomized comparative experiment? Does data analysis show extreme outliers or strong skewness that forbids use of inference based on Normality? Do you have enough observations for your intended procedure?

5. *Do your data come from an experiment or from an observational study?* The details of inference methods are the same for both. But the design of the study determines what conclusions you can reach, because experiments give much better evidence that an effect uncovered by inference can be explained by direct causation.

You may ask, as you study the Statistics in Summary flowchart, "What if I have an experiment comparing four treatments or samples from three populations?" The flowchart allows only one or two, not three or four or more. Be patient: methods for comparing more than two means or proportions, as well as some other settings for inference, appear in Part V.

STATISTICS IN YOUR WORLD

How Many Was That?

Good causes often breed bad statistics. An advocacy group claims, without much evidence, that 150,000 Americans suffer from the eating disorder anorexia nervosa. Soon, someone misunderstands and says that 150,000 people *die* from anorexia nervosa each year. This wild number gets repeated in countless books and articles. It really is a wild number: only about 55,000 American women aged 15–44 (the main group affected) die of *all causes* each year.

Part IV Skills Review

Here are the most important skills you should have acquired from reading Chapters 20 through 23. Following each skill in parentheses is the section of the text where the topic is introduced.

A. Recognition

1. Recognize when a problem requires inference about population means (quantitative response variable) or population proportions (usually categorical response variable). (20.1, 22.1)
2. Recognize from the design of a study whether one-sample, matched pairs, or two-sample procedures are needed. (20.6, 21.1)
3. Based on recognizing the problem setting, choose among the one- and two-sample *t* procedures for means and the one- and two-sample *z* procedures for proportions. (20.3, 21.5, 22.2, 23.1)

B. Inference about One Mean

1. Verify that the *t* procedures are appropriate in a particular setting. Check the study design and the distribution of the data and take advantage of robustness against lack of Normality. (20.3, 20.7)
2. Recognize when poor study design, outliers, or a small sample from a skewed distribution make the *t* procedures risky. (20.7)

3. Use the one-sample t procedure to obtain a confidence interval at a stated level of confidence for the mean μ of a population. (20.2, 20.3)
4. Carry out a one-sample t test for the hypothesis that a population mean μ has a specified value against either a one-sided or a two-sided alternative. Use software to find the *P*-value or Table C to get an approximate value. (20.4, 20.5)
5. Recognize matched pairs data and use the t procedures to obtain confidence intervals and to perform tests of significance for such data. (20.6)

C. Comparing Two Means

1. Verify that the two-sample t procedures are appropriate in a particular setting. Check the study design and the distribution of the data and take advantage of robustness against lack of Normality. (21.2, 21.5)
2. Give a confidence interval for the difference between two means. Use software if you have it. Use the two-sample t statistic with conservative degrees of freedom and Table C if you do not have statistical software. (21.3, 21.4)
3. Test the hypothesis that two populations have equal means against either a one-sided or a two-sided alternative. Use software if you have it. Use the two-sample t test with conservative degrees of freedom and Table C if you do not have statistical software. (21.3, 21.4)
4. Know that procedures for comparing the standard deviations of two Normal populations are available but that these procedures are risky because they are not at all robust against non-Normal distributions. (21.8)

D. Inference about One Proportion

1. Verify that you can safely use either the large-sample or the plus four z procedures in a particular setting. Check the study design and the guidelines for sample size. (22.2, 22.5)
2. Use the large-sample z procedure to give a confidence interval for a population proportion p when the conditions are met. Otherwise, the plus four modification of the z procedure can give a confidence interval for p that is accurate even for small samples and for any value of p. (22.2, 22.5)
3. Use the z statistic to carry out a test of significance for the hypothesis $H_0: p = p_0$ about a population proportion p against either a one-sided or a two-sided alternative. Use software or Table A to find the *P*-value or use Table C to get an approximate value. (22.4)

E. Comparing Two Proportions

1. Verify that you can safely use either the large-sample or the plus four z procedures in a particular setting. Check the study design and the guidelines for sample sizes. (23.3, 23.5, 23.6)
2. Use the large-sample z procedure to give a confidence interval for the difference $p_1 - p_2$ between proportions in two populations based on independent samples from the populations when the conditions are met. Otherwise, the plus four modification of the z procedure can give a confidence interval for $p_1 - p_2$ that is accurate even for very small samples and for any values of p_1 and p_2. (23.3, 23.6)
3. Use a z statistic to test the hypothesis $H_0: p_1 = p_2$ that proportions in two distinct populations are equal. Use software or Table A to find the *P*-value or use Table C to get an approximate value. (23.5)

STATISTICS IN YOUR WORLD

How Many Miles per Gallon?

As gasoline prices rise, more people pay attention to the government's gas mileage ratings of their vehicles. Until recently, these ratings overstated the miles per gallon we can expect in real-world driving. The ratings assumed a top speed of 60 miles per hour, slow acceleration, and no air conditioning. That doesn't resemble what we see around us on the highway. Maybe it doesn't resemble the way we ourselves drive. Starting with 2008 models, the ratings assume higher speeds (80 miles per hour tops), faster acceleration, and air conditioning in warm weather. Mileage ratings of the same vehicle dropped by about 12% in the city and 8% on the highway.

TEST YOURSELF

The following questions include both multiple-choice and short-answer questions and calculations. They will help you review the basic ideas and skills presented in Chapters 20 through 23.

Andersen Ross/Getty Images

Multitasking in Class. *Does the presence of a nearby student multitasking on a laptop during lecture create a distraction that inhibits learning during lecture? Researchers randomly assigned 38 undergraduate students to one of two treatment groups, both of size 19. Those in one of the groups were seated in view of someone who was busy multitasking on a laptop during lecture. Those in the other group were seated out of view of anyone multitasking on a laptop. Volunteers were recruited to spend the lecture multitasking on their laptops and were strategically placed around the classroom to provide the distraction. At the end of the lecture, subjects were given a test on the lecture's content, and the scores of the two groups were compared. The mean test score for the students who were not in view of someone multitasking was* $\bar{x} = 73$*, with standard deviation* $s = 12$*.*[1] *Use this information to answer Questions 24.1 and 24.2.*

24.1 A 95% confidence interval for the mean test score in the population that the students who were not in view of someone multitasking are supposed to represent is

(a) 70.25 to 75.75.

(b) 68.23 to 77.77.

(c) 67.60 to 78.40.

(d) 167.22 to 78.78.

24.2 What conditions for the population and the study design are required by the procedure you used to construct your confidence interval? Which of these conditions are important for the validity of the procedure in this case?

Does Nature Heal Better? *Our bodies have a natural electrical field that is known to help wounds heal. Does changing the field strength slow healing? A series of experiments with newts investigated this question. The following table provides the healing rates of cuts (micrometers per hour) in a matched pairs experiment. The pairs are the two hind limbs of the same newt, with the body's natural field in one limb (control) and half the natural value in the other limb (experimental).*[2]

Newt	1	2	3	4	5	6	7	8	9	10	11	12	13	14
Control	25	13	44	45	57	42	50	36	35	38	43	31	26	48
Experimental	24	23	47	42	26	46	38	33	28	28	21	27	25	45
Difference (control - experimental)	1	−10	−3	3	31	−4	12	3	7	10	22	4	1	3

The mean and standard deviation of the differences are 5.71 and 10.56 micrometers per hour, respectively. Use this information to answer Questions 24.3 and 24.4.

24.3 Is there good evidence that changing the electrical field from its natural level slows healing? The *P*-value for your test is

(a) less than 0.01.

(b) between 0.01 and 0.05.

(c) between 0.05 and 0.10.

(d) greater than 0.10.

24.4 Explain why this is a matched pairs experiment. Give a 95% confidence interval for the difference in healing rates (control minus experimental).

Free Speech. *In December 2017, the Gallup Poll asked a random sample of college students: "Do you think the right of freedom of speech is very secure, secure, threatened or very threatened in the country today?"*[3] *Among the 2116 Democrat/Democrat leaning college student respondents, 1248 said secure or very secure, whereas among the 720 Republican/Republican leaning college student respondents, 511 said secure or very secure. Is the difference between the proportion of Democrat/Democrat leaning college* *students and the proportion of Republican/Republican leaning college students who said the right of freedom of speech is secure or very secure statistically significant? Questions 24.5 through 24.8 are based on these results.*

24.5 Take p_D and p_R to be the proportions of all Democrat/Democrat leaning and Republican/Republican leaning college students at the time of the poll who would say the right of freedom of speech is secure or very secure. The hypotheses to be tested are

(a) $H_0: p_D = p_R$ versus $H_a: p_D \neq p_R$.

(b) $H_0: p_D = p_R$ versus $H_a: p_D > p_R$.

(c) $H_0: p_D = p_R$ versus $H_a: p_D < p_R$.

(d) $H_0: p_D < p_R$ versus $H_a: p_D = p_R$.

24.6 The sample proportion of Republican/Republican leaning college students who said the right of freedom of speech is secure or very secure is

(a) 0.59. (b) 0.62.

(c) 0.65. (d) 0.71.

24.7 The pooled sample proportion of Democrat/Democrat leaning and Republican/Republican leaning respondents who said the right of freedom of speech is secure or very secure is

(a) 0.59. (b) 0.62.

(c) 0.65. (d) 0.71.

24.8 The z test for comparing the proportions of Democrat/Democrat leaning college students and Republican/Republican leaning college students who said the right of freedom of speech is secure or very secure has

(a) $P < 0.001$. (b) $0.001 < P < 0.01$.

(c) $0.01 < P < 0.05$. (d) $P > 0.05$.

24.9 **Gastric Bypass Surgery.** How effective is gastric bypass surgery in maintaining weight loss in extremely obese people? Researchers found that 427 of 771 subjects who had received gastric bypass surgery regained at least 25% of their post-surgery weight loss five years after surgery.[4]

(a) Are the conditions for the use of the large-sample confidence interval met? Explain.

(b) Give a 90% confidence interval for the proportion of those receiving gastric bypass surgery who regained at least 25% of their post-surgery weight loss five years after surgery.

(c) Interpret your interval in the context of the problem.

Evil Genius? *Does cheating enhance creativity? Researchers recruited 178 subjects and asked participants to guess whether the outcome of a virtual coin toss would be heads or tails. After indicating their prediction, participants had to press a button to toss the coin virtually. They were asked to press the button only once, but they were given the opportunity to test the button several times before beginning the experiment. Thus, participants could "cheat" by pressing the button before making their prediction and appear to have pressed the button one time for each prediction. Participants then reported whether they had guessed correctly and received a \$1 bonus if they had. The program recorded the outcomes of the initial virtual coin tosses so that researchers could tell whether participants cheated. After the coin-toss task, all participants were given a test that measured creativity. Here are the summary statistics for scores on the creativity test:*[5]

Group	n	$\bar{x}$	s
Cheaters	43	3.60	1.26
Noncheaters	135	2.33	1.00

Use this information to answer Questions 24.10 through 24.13.

24.10 A 90% confidence interval for the mean score on the creativity test for those subjects who cheated is

(a) 3.60 ± 0.19.

(b) 3.60 ± 0.32.

(c) 3.60 ± 2.12.

(d) 1.26 ± 0.43.

24.11 A 90% confidence interval for the mean score on the creativity test for those subjects who did not cheat is

(a) 2.33 ± 0.09.

(b) 2.33 ± 0.14.

(c) 2.33 ± 1.68.

(d) 1.00 ± 0.14.

24.12 Is there a significant difference between the mean scores on the creativity test for those who cheated and those who did not? The value of the t statistic for testing the null hypothesis of no difference in the mean scores on the creativity test is

(a) 0.21. (b) 1.27.

(c) 4.76. (d) 6.03.

24.13 Is there a significant difference between the mean scores on the creativity test for those who cheated and those who did not? The degrees of freedom using the conservative Option 2 for the t statistic for testing the null hypothesis of no difference in the mean scores on the creativity test is

(a) 42. (b) 59.2.

(c) 88. (d) 253.

24.14 Multitasking in Class. Does the presence of a nearby student multitasking on a laptop during lecture create a distraction that inhibits learning during lecture? Researchers randomly assigned 38 undergraduate students to one of two treatments groups, both of size 19. Those in one of the groups were seated in view of someone who was busy multitasking on a laptop during lecture. Those in the other group were seated out of view anyone multitasking on a laptop. Volunteers were recruited to spend the lecture multitasking on a laptops and were strategically placed around the classroom to provide the distraction. At the end of the lecture, subjects were given a test on the lecture's content, and the scores of the two groups were compared. The mean test score for the students who were seated in view of someone multitasking on a laptop was $\bar{x} = 56$, with standard deviation $s = 12$. The mean test score for the students who were not in view of someone multitasking was $\bar{x} = 73$, with standard deviation $s = 12$.[6] Is the observed difference in means statistically significant? Test using the conservative Option 2 for the degrees of freedom. The *P*-value is

(a) less than 0.01.

(b) between 0.01 and 0.05.

(c) between 0.05 and 0.10.

(d) greater than 0.10.

Treating Migraines. *In a randomized clinical trial, 135 youth aged 10 to 17 years diagnosed with chronic migraine were assigned to one of two treatments. One involved 10 cognitive behavioral therapy (CBT) sessions and use of the drug amitriptyline. The other involved 10 headache education sessions plus amitriptyline. Twenty weeks after treatment, the severity of migraines for each subject was assessed using the Pediatric Migraine Disability Assessment Score (PedMIDAS). Here are data on PedMIDAS scores 20 weeks post-treatment for the two groups:*[7]

Group	n	Mean	Standard Deviation
CBT	64	15.5	17.4
Education	71	29.6	42.2

Use this information to answer Questions 24.15 through 24.18.

24.15 Both sets of endurance data are skewed to the right. How do we know this? (Note that PedMIDAS scores are greater than or equal to zero.) Why are t procedures nonetheless reasonably accurate for these data?

24.16 A 95% confidence interval for the mean PedMIDAS score for the CBT group is

(a) 14.97 to 16.03.

(b) 13.32 to 17.68.

(c) 11.87 to 19.13.

(d) 11.15 to 19.85.

24.17 A 95% confidence interval for the mean difference (education minus CBT) in PedMIDAS scores, using the conservative Option 2 for the degrees of freedom, is

(a) 3.18 to 25.02.

(b) 4.98 to 23.22.

(c) 8.64 to 19.56.

(d) 21.24 to 37.96.

24.18 Do the data show that there is a difference in mean PedMIDAS scores for the two treatments? Carry out an appropriate hypothesis test using the conservative Option 2 for the degrees of freedom.

24.19 **Pre-readers in Kindergarten.** A school has two kindergarten classes. There are 21 children in Ms. Hazelcorn's kindergarten class. Of these, 12 are "pre-readers" –children on the verge of reading. There are 19 children in Mr. Shapiro's kindergarten class. Of these, 14 are pre-readers. Is there a statistically significant difference in the proportions of "pre-readers" in the two classes? The z statistic is computed, and the P-value is 0.263.

(a) This test is not reliable because the samples are so small.

(b) This test is of no use because we should be comparing means with a t statistic.

(c) This test is reasonable because the counts of successes and failures are each five or more in both samples.

(d) This test is not appropriate because these samples cannot be viewed as simple random samples taken from a larger population.

24.20 **Healthcare.** If we want to estimate p, the population proportion of likely voters who believe healthcare to be the most urgent national concern, with 95% confidence and a margin of error no greater than 2%, how many likely voters need to be surveyed? Assume that you have no idea of the value of p.

(a) 188

(b) 1691

(c) 2401

(d) 4802

Online Audio. *A 2019 Infinite Dial survey found that 67% of a sample of 1500 Americans aged 12 and older said they listened to online audio (listened to AM/FM radio stations online and/or listened to streamed audio content available only on the Internet) at least once in the month prior to the survey. Assume that the sample was an SRS. Use this information to answer Questions 24.21 through 24.23.*

24.21 Based on the sample, the large-sample 90% confidence interval for the proportion of all Americans aged 12 and older who listened to online audio in the month prior to the survey is

(a) 0.67 ± 0.012. (b) 0.67 ± 0.020. (c) 0.67 ± 0.024.

24.22 In Question 24.21, suppose we computed a large-sample 80% confidence interval for the proportion of all Americans aged 12 and older who listened to online audio in the month prior to the survey. This 80% confidence interval

(a) would have a smaller margin of error than the 90% confidence interval.

(b) would have a larger margin of error than the 90% confidence interval.

(c) could have either a smaller or a larger margin of error than the 90% confidence interval. This varies from sample to sample.

24.23 How many randomly selected Americans aged 12 and older must be interviewed to estimate the proportion who listened to online audio in the month prior to the survey to within ± 0.02 with 99% confidence using the large-sample confidence interval? Use 0.5 as the conservative guess for p.

(a) $n = 1692$ (b) $n = 2401$ (c) $n = 4148$

Very-Low-Birth-Weight Babies. *Starting in the 1970s, medical technology allowed babies with very low birth weight (VLBW, less than 1500 grams, about 3.3 pounds) to survive without major disabilities. It was noticed that these children nonetheless had difficulties in school and as adults. A long-term study has followed 242 VLBW babies to age 20 years, along with a control group of 233 babies from the same population who had normal birth weight.[8] At age 20, 179 of the VLBW group and 193 of the control group had graduated from high school. Use this information to answer Questions 24.24 through 24.29.*

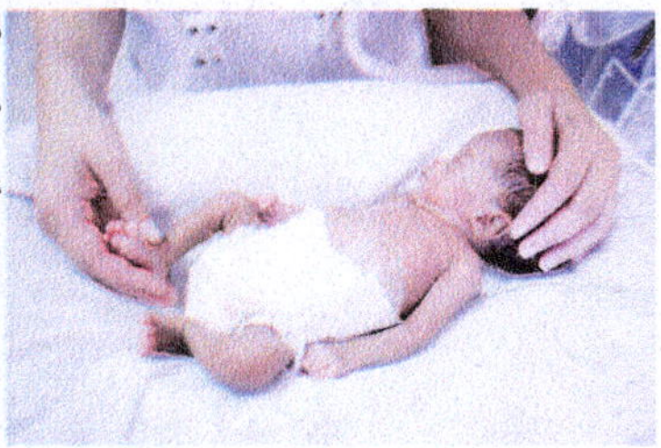
Science Photo Library/Getty Images

24.24 This is an example of

(a) an observational study.

(b) a nonrandomized experiment.

(c) a randomized controlled study.

(d) a matched pairs experiment.

24.25 Take p_{VLBW} and $p_{control}$ to be the proportions of all VLBW and normal-birth-weight (control) babies who would graduate from high school. The hypotheses to be tested are

(a) $H_0: p_{VLBW} = p_{control}$ versus $H_a: p_{VLBW} \neq p_{control}$.

(b) $H_0: p_{VLBW} = p_{control}$ versus $H_a: p_{VLBW} > p_{control}$.

(c) $H_0: p_{VLBW} = p_{control}$ versus $H_a: p_{VLBW} < p_{control}$.

(d) $H_0: p_{VLBW} > p_{control}$ versus $H_a: p_{VLBW} = p_{control}$.

24.26 The pooled sample proportion of subjects who graduated from high school is

(a) $\hat{p} = 0.74$.

(b) $\hat{p} = 0.78$.

(c) $\hat{p} = 0.81$.

(d) $\hat{p} = 0.83$.

24.27 The numerical value of the z test for comparing the proportions of all VLBW and normal-birth-weight (control) babies who would graduate from high school is

(a) $z = -1.65$.

(b) $z = -2.34$.

(c) $z = -2.77$.

(d) $z = -3.14$.

24.28 IQ scores were available for 113 men in the VLBW group and for 106 men in the control group. The mean IQ for the 113 men in the VLBW group was 87.6, and the standard deviation was 15.1. The 106 men in the control group had mean IQ 94.7, with standard deviation 14.9. Is there good evidence that mean IQ is lower among VLBW men than among controls from similar backgrounds? To test this with a two-sample *t* test, the test statistic would be

(a) $t = -1.72$.

(b) $t = -3.50$.

(c) $t = -5.00$.

(d) $t = -7.10$.

24.29 Of the 126 women in the VLBW group, 38 said they had used illegal drugs; 54 of the 124 control group women had done so. The IQ scores for the VLBW women who had used illegal drugs had mean 86.2 (standard deviation 13.4), and the normal-birth-weight controls who had used illegal drugs had mean IQ 89.8 (standard deviation 14.0). Is there a statistically significant difference between the two groups in mean IQ? The *P*-value for this test is

(a) less than 0.01.

(b) between 0.01 and 0.05.

(c) between 0.05 and 0.10.

(d) greater than 0.10.

Spinning Euros. *All euro coins have a national image on the "heads" side and a common design on the "tails" side. Spinning a coin, unlike tossing it, may not give heads and tails equal probabilities. Polish students spun the Belgian euro 250 times, with its portly king, Albert, displayed on the heads side. The result was 140 heads.[9] How significant is this evidence against equal probabilities? Use this information to answer Questions 24.30 and 24.31.*

24.30 Take p to be the probability that a spun euro lands on the heads side. The hypotheses to be tested are

(a) $H_0: p = 0.5$ versus $H_a: p \neq 0.5$.

(b) $H_0: p = 0.5$ versus $H_a: p > 0.5$.

(c) $H_0: p = 0.5$ versus $H_a: p < 0.5$.

24.31 The *P*-value for the hypothesis test is

(a) between 0.15 and 0.20.

(b) between 0.10 and 0.15.

(c) between 0.05 and 0.10.

(d) below 0.05.

24.32 Vote for the Best Face? We often judge other people by their faces. It appears that some people judge candidates for elected office by their faces. Psychologists showed head-and-shoulders photos of the two main candidates in 32 races for the U.S. Senate to many subjects (dropping subjects who recognized either of the candidates) to see which candidate was rated "more competent" based on nothing but the photos. On election day, the candidates whose faces looked more competent won 22 of the 32 contests.[10] If faces don't influence voting, half of all races in the long run should be won by the candidate with the better face. Is there evidence that the candidate with the better face wins more than half the time?

(a) State the null and alternative hypotheses to be tested.

(b) How strong is the evidence that the candidate with the better face wins more than half the time? Make sure to check the conditions for the use of this test.

Who Tweets? *Younger people use Twitter more often than older people do. A random sample of 236 young adults aged 18–29 found that 90 have used Twitter. In a random sample of 391 adults aged 65 and over, 27 have used Twitter.*[11] *Use this information to answer Questions 24.33 and 24.34.*

24.33 Give a 90% confidence interval for the proportion of all young adults aged 18–29 who have used Twitter.

24.34 Give a 95% confidence interval for the difference between the proportions who have used Twitter for these two age groups.

24.35 **I Refuse!** Do our emotions influence economic decisions? One way to examine the issue is to have subjects play an "ultimatum game" against other people and against a computer. Your partner (person or computer) gets \$10, on the condition that it be shared with you. The partner makes you an offer. If you refuse, neither of you gets anything. So it's to your advantage to accept even the unfair offer of \$2 out of the \$10. Some people get mad and refuse unfair offers. Here are data on the responses of 76 subjects randomly assigned to receive an offer of \$2 from either a person they were introduced to or a computer:[12]

	Accept	Reject
Human offers	20	18
Computer offers	32	6

We suspect that emotion will lead to offers from another person being rejected more often than offers from an impersonal computer. Do a test to assess the evidence for this conjecture.

Diet and Bone Density in Cats. *Some dietitians have suggested that highly acidic diets can have an adverse affect on bone density in humans. Alkaline diets have been marketed to avoid or counteract this effect. Is the same thing true for cats, and would an alkaline diet be beneficial? Two groups of four cats were fed diets for 12 months that differed only in acidifying or alkalinizing properties. The bone mineral density* (g/cm^3) *of each cat was measured at the end of 12 months. The summary statistics for bone mineral density appear below:*[13]

Diet	n	$\bar{x}$	s
Acidifying	4	0.63	0.01
Alkalinizing	4	0.64	0.05

Questions 24.36 through 24.39 are based on this study.

24.36 A 90% confidence interval for the mean bone mineral density of cats after 12 months on an acidifying diet is

(a) 0.622 to 0.638.
(b) 0.618 to 0.642.
(c) 0.614 to 0.646.
(d) 0.620 to 0.640.

24.37 Is there strong evidence that cats on an alkalinizing diet have higher mean bone mineral density after 12 months than cats on an acidifying diet? To test this with a two-sample t test, the values of the t statistic and its degrees of freedom using conservative Option 2 are

(a) $t = 0.39$, df $= 3$.

(b) $t = 0.39$, df $= 4$.

(c) $t = 0.01$, df $= 6$.

(d) $t = 0.01$, df $= 7$.

24.38 A 90% confidence interval for the difference in mean bone mineral density after 12 months on an alkalinizing diet and an acidifying diet is (use conservative Option 2 for the degrees of freedom)

(a) 0.01 to 0.05.

(b) −0.03 to 0.05

(c) −0.05 to 0.07.

(d) −0.07 to 0.09.

24.39 What conditions must be satisfied to justify the procedures you used in Question 24.36? In Question 24.37? In Question 24.38?

Choosing an Inference Procedure. *In each of Questions 24.41 through 24.46, say which type of inference procedure from the Statistics in Summary flowchart (page 539) you would use or explain why none of these procedures fits the problem. You do not need to carry out any procedures.*

24.40 Driving too Fast. How seriously do people view speeding in comparison with other annoying behaviors? A large random sample of adults was asked to rate a number of behaviors on a scale of 1 (no problem at all) to 5 (very severe problem). Do speeding drivers get a higher average rating than noisy neighbors?

24.41 Preventing Drowning. Drowning in bathtubs is a major cause of death in children less than five years old. A random sample of parents was asked many questions related to bathtub safety. Overall, 85% of the sample said they used baby bathtubs for infants. Estimate the percentage of all parents of young children who use baby bathtubs.

24.42 Acid Rain? You have data on rainwater collected at 16 locations in the Adirondack Mountains of New York State. One measurement is the acidity of the water, measured by pH on a scale of 0 to 14. (The pH of distilled water is 7.0.) Estimate the average acidity of rainwater in the Adirondacks.

24.43 Athletes' Salaries. Looking online, you find the base salaries of the 25 active players on the roster of the Chicago Cubs as of opening day of the 2019 baseball season. The total salary of these 25 players was $194.1 million, one of the highest in Major League Baseball. Estimate the average salary of the 25 active players on the roster.

24.44 Looking Back on Love. How do young adults look back on adolescent romance? Investigators interviewed 40 couples in their mid-20s. The female and male partners were interviewed separately. Each was asked about his or her current relationship and also about a romantic relationship that lasted at least two months when they were aged 15 or 16. One response variable was a measure on a numerical scale of how much the attractiveness of the adolescent partner mattered. You want to compare the men and women on this measure.

24.45 Preventing AIDS through Education. The Multisite HIV Prevention Trial was a randomized comparative experiment to compare the effects of twice-weekly, small-group AIDS discussion sessions (the treatment) with a single one-hour session (the control). Compare the effects of treatment and control on each of the following response variables:

(a) A subject does or does not use condoms six months after the education sessions.

(b) The number of unprotected intercourse acts by a subject between four and eight months after the sessions.

(c) A subject is or is not infected with a sexually transmitted disease six months after the sessions.

SUPPLEMENTARY EXERCISES

Supplementary exercises apply the skills you have learned in ways that require more thought or more use of technology. Some of these exercises start from actual data rather than from data summaries. Many of these exercises ask you to follow the Plan, Solve, and Conclude steps of the four-step process. Remember that the Solve step includes checking the conditions for the inference you plan.

24.46 Do You Have Confidence? A report of a survey distributed to randomly selected email addresses at a large university says: "We have collected 427 responses from our sample of 2,100 as of April 30, 2004. This number of responses is large enough to achieve a 95% confidence interval with ±5% margin of sampling error in generalizing the results to our study population."[14] Why would you be reluctant to trust a confidence interval based on these data?

24.47 Treating Migraines. Questions 24.15 through 24.18 involve a randomized clinical trial to investigate the effects of two treatments on migraines. One of the treatments, cognitive behavioral therapy (CBT) along with the drug amitriptyline, was expected to be particularly effective. For each of the 64 subjects on this treatment, severity of migraines was assessed using PedMIDAS scores. Here are summary data for these subjects:

Time	$\bar{x}$	s
Before treatment	68.2	31.7
After treatment	15.5	17.4

(a) Which t procedures are correct for comparing the mean PedMIDAS scores before and after treatment: one-sample, matched pairs, or two-sample?

(b) The data summary given is not enough information to carry out the correct t procedures. Explain why not.

24.48 Monkeys and Music. Humans generally prefer music to silence. What about monkeys? In one study, researchers allowed a tamarin monkey to enter a V-shaped cage with food in both arms of the V. After the monkey ate the food, they wondered, which arm would it prefer? The monkey's location determined what it heard: a lullaby played on a flute in one arm and silence in the other. Each of four monkeys was tested six times, on different days and with the music arm alternating between left and right (in case a monkey preferred one direction). The monkeys chose silence for about 65% of their time in the cage. The researchers reported a one-sample t test for the mean percentage of time spent in the music arm, $H_0\colon \mu = 50\%$, against the two-sided alternative, $t = -5.26$, df $= 23$, $P < 0.0001$.[15]

Although the result is interesting, the statistical analysis is not correct. The degrees of freedom df $= 23$ show that the researchers assumed that they had 24 independent observations. Explain why the results of the 24 trials are not independent.

24.49 (Optional Topic) Drug-Detecting Rats? Dogs are big and expensive. Rats are small and cheap. Might rats be trained to replace dogs in sniffing out illegal drugs? A first study of this idea trained rats to rear up on their hind legs when they smelled simulated cocaine. To see how well rats performed after training, they were let loose on a surface with many cups sunk in it, one of which contained simulated cocaine. Four out of six trained rats succeeded in 80 out of 80 trials.[16] How should we estimate the long-term success rate p of a rat that succeeds in every one of 80 trials?

(a) What is the rat's sample proportion $\hat{p}$? What is the large-sample 95% confidence interval for p? It's not plausible that the rat will *always* be successful, as this interval says.

(b) Find the plus four estimate $\tilde{p}$ and the plus four 95% confidence interval for p. These results are more reasonable.

24.50 A New Vaccine. In 2006, the pharmaceutical company Merck released a vaccine named Gardasil for human papilloma virus, the most common cause of cervical cancer in young women. The Merck website gives results from "four placebo-controlled, double-blind, randomized clinical studies" with women 16 to 26 years of age, as follows:[17]

	n	Cervical Cancer	n	Genital Warts
Gardasil	8487	0	7897	1
Placebo	8460	32	7899	91

(a) Give a 99% confidence interval for the difference in the proportions of young women who develop cervical cancer with and without the vaccine.

(b) Do the same for the proportions who develop genital warts.

(c) What do you conclude about the overall effectiveness of the vaccine?

24.51 Starting to Talk. At what age do infants speak their first word of English? Here are data on 20 children (ages in months):[18] 1STWORD

15	26	10	9	15	20	18	11	8	20
7	9	10	11	11	10	12	17	11	10

(In fact, the sample contained one more child, who began to speak at 42 months. Child development experts consider this abnormally late, so the investigators dropped the outlier to get a sample of "typical" children. We are willing to treat these data as an SRS.) Is there good evidence that the mean age at first word among all typical children is greater than one year?

24.52 Fertilizing a Tropical Plant. Bromeliads are tropical flowering plants. Many are epiphytes that attach to trees and obtain moisture and nutrients from air and rain. In an experiment in Costa Rica, Jacqueline Ngai and Diane Srivastava studied whether added nitrogen increases the productivity of bromeliad plants. Bromeliads were randomly assigned to nitrogen or control groups. Here are data on the number of new leaves produced over a seven-month period:[19] FERT

Control	11	13	16	15	15	11	12	
Nitrogen	15	14	15	16	17	18	17	13

Is there evidence that adding nitrogen increases the mean number of new leaves formed?

24.53 Starting to Talk, Continued. Use the data in Exercise 24.51 to give a 90% confidence interval for the mean age at which children speak their first word. 1STWORD

24.54 Dyeing Fabrics. Different fabrics respond differently when dyed. This matters to clothing manufacturers, who want the color of fabric to match their specifications closely. A researcher dyed fabrics made of cotton and of ramie with the same "procion blue" dye applied in the same way. Then she used a colorimeter to measure the lightness of the color on a scale in which black is 0 and white is 100. Here are the data for eight pieces of each fabric:[20] FBCDYE

Cotton	48.82	48.88	48.98	49.04	48.68	49.34	48.75	49.12
Ramie	41.72	41.83	42.05	41.44	41.27	42.27	41.12	41.49

Is there a significant difference between the fabrics? Which fabric is darker when dyed in this way?

24.55 More on Dyeing Fabrics. The color of a fabric depends on the dye used and also on how the dye is applied. The researcher discussed in the previous exercise went on to dye fabric made of ramie with the same procion blue dye applied in two different ways. Here are the lightness scores for eight pieces of identical fabric dyed in each way: FBCDYE2

Method B	40.98	40.88	41.30	41.28	41.66	41.50	41.39	41.27
Method C	42.30	42.20	42.65	42.43	42.50	42.28	43.13	42.45

(a) This is a randomized comparative experiment. Outline the design.

(b) A clothing manufacturer wants to know which method gives the darker color (lower lightness score). Use sample means to answer this question. Is the difference between the two sample means statistically significant? Can you tell from just the *P*-value whether the difference is large enough to be important in practice?

24.56 Do Parents Matter? A professor asked her sophomore students, "Does either of your parents allow you to drink alcohol around him or her?" and "How many drinks do you typically have per session? (A drink is defined as one 12 oz beer, one 4 oz glass of wine, or one 1 oz shot of liquor.)" Table 24.1 contains the responses of the female students who are not abstainers.[21] The sample is all students in one large sophomore-level class. The class is popular, so we are tentatively willing to regard its members as an SRS of sophomore students at this college. Does the behavior of parents make a significant difference in how many drinks female students have on the average? FMDRINK

TABLE 24.1 Drinks per session by female students

Parent Allows Student to Drink												
2.5	1	2.5	3	1	3	3	3	2.5	2.5	3.5	5	2
7	7	6.5	4	8	6	6	3	6	3	4	7	5
3.5	2	1	5	3	3	6	4	2	7	5	8	1
6	5	2.5	3	4.5	9	5	4	4	3	4	6	4
5	1	5	3	10	7	4	4	4	4	2	2.5	2.5
Parent Does Not Allow Student to Drink												
9	3.5	3	5	1	1	3	4	4	3	6	5	3
8	4	4	5	7	7	3.5	3	10	4	9	2	7
4	3	1										

24.57 Parents' Behavior. We wonder what proportion of female students have at least one parent who allows them to drink in their parent's presence. Table 24.1 contains information about a sample of 94 students. Use this sample to give a 95% confidence interval for this proportion. FMDRINK

24.58 Diabetic Mice. The body's natural electrical field helps wounds heal. If diabetes changes this field, that might explain why people with diabetes heal slowly. A study of this idea compared non-diabetic mice and mice bred to develop diabetes. The investigators attached sensors to the right hips and front feet of the mice and measured the difference in electrical potential (millivolts) between these locations. Here are the data:[22] MICE

Diabetic Mice					
14.70	13.60	7.40	1.05	10.55	16.40
10.00	22.60	15.20	19.60	17.25	18.40
9.80	11.70	14.85	14.45	18.25	10.15
10.85	10.30	10.45	8.55	8.85	19.20

Non-diabetic Mice				
13.80	9.10	4.95	7.70	9.40
7.20	10.00	14.55	13.30	6.65
9.50	10.40	7.75	8.70	8.85
8.40	8.55	12.60		

(a) Make a stemplot of each sample of potentials. There is a low outlier in the diabetic group. Does it appear that potentials in the two groups differ in a systematic way?

(b) Is there significant evidence of a difference in mean potentials between the two groups?

(c) Repeat your inference without the outlier. Does the outlier affect your conclusion?

24.59 Keeping Crackers from Breaking. We don't like to find broken crackers when we open a package. How can makers reduce breaking? One idea is to microwave the crackers for 30 seconds right after baking them. Analyze the following results from two experiments intended to examine this idea.[23] Does microwaving significantly improve indicators of future breaking? How large is the improvement? What do you conclude about the idea of microwaving crackers?

(a) The experimenter randomly assigned 65 newly baked crackers to be microwaved and another 65 to a control group that is not microwaved. Fourteen days after baking, three of the 65 microwaved crackers and 57 of the 65 crackers in the control group showed visible checking (small hairline cracks), which is the starting point for breaks.

(b) The experimenter randomly assigned 20 crackers to be microwaved and another 20 to a control group. After 14 days, he broke the crackers. Here are summaries of the pressure needed to break them, in pounds per square inch:

	Microwave	Control
Mean	139.6	77.0
Standard deviation	33.6	22.6

24.60 Falling through the Ice. Table 7.2 (page 194) gives the dates on which a wooden tripod fell through the melting ice of the Tanana River in Alaska, thus deciding the winner of the Nenana Ice Classic contest, for the years 1917 to 2019. Give a 95% confidence interval for the mean date on which the tripod falls through the ice. After calculating the interval in the scale used in the table (days from April 14, which is Day 1), translate your result into calendar dates and hours within the dates. (Each hour is 1/24, or 0.042, of a day.) TANANA

24.61 A Case for the Supreme Court. In 1986, a Texas jury found a Black man guilty of murder. The prosecutors had used "peremptory challenges" to remove 10 of the 11 Blacks and four of the 31 Whites in the pool from which the jury was chosen.[24] The law says that there must be a plausible reason (that is, a reason other than race) for different treatment of Blacks and Whites in the jury pool. When the case reached the Supreme Court 17 years later, the Court said that "happenstance is unlikely to produce this disparity." The inferential methods we have studied cannot safely be used to support the Court's finding that chance is unlikely to produce so large a Black-White difference. Why not?

24.62 Mouse Genes. A study of genetic influences on diabetes compared non-diabetic mice with similar mice genetically altered to remove a gene called *aP2*. Mice of both types were allowed to become obese by eating a high-fat diet. The researchers then measured the levels of insulin and glucose in their blood plasma. Here are some excerpts from their findings.[25] The non-diabetic mice are called "wild type," and the altered mice are called "$aP2^{-/-}$."

Each value is the mean ± SEM of measurements on at least 10 mice. Mean values of each plasma component are compared between $aP2^{-/-}$ mice and wild-type controls by Student's t test ($^{*}P < 0.05$ and $^{**}P < 0.005$).

Parameter	Wild Type	$aP2^{-/-}$
Insulin (ng/mL)	5.9 ± 0.9	$0.75 \pm 0.2^{**}$
Glucose (mg/dL)	230 ± 25	$150 \pm 17^{*}$

Despite much greater circulating amounts of insulin, the wild-type mice had higher blood glucose than the $aP2^{-/-}$ animals. These results indicate that the absence of P2 interferes with the development of dietary obesity-induced insulin resistance.

Other biologists are supposed to understand the statistics reported so tersely.

(a) What does "SEM" mean? What is the expression for SEM, based on n, $\bar{x}$, and s from a sample?

(b) Which of the tests we have studied did the researchers apply?

(c) Explain to a biologist who knows no statistics what $P < 0.05$ and $P < 0.005$ mean. Which is stronger evidence of a difference between the two types of mice?

24.63 Mouse Genes, Continued. The report quoted in the previous exercise says only that the sample sizes were "at least 10." Suppose that the results are based on exactly 10 mice of each type. Use the values in the table to find $\bar{x}$ and s for the insulin concentrations in the two types of mice. Carry out a test to assess the significance of the difference in mean insulin concentration. Does your *P*-value confirm the claim in the report that $P < 0.005$?

24.64 (Optional) Which Typeface? Plain typefaces such as Times New Roman are easier to read than fancy typefaces such as Gigi. A group of 25 volunteer subjects read the same text in both typefaces. (This is a matched pairs design. One-sample procedures for proportions, like those for means, are used to analyze data from matched pairs designs.) Of the 25 subjects, 17 said that they preferred Times New Roman for web use. But 20 said that Gigi was more attractive.[26]

(a) Because the subjects were volunteers, conclusions from this sample can be challenged. Show that the sample size condition for the large-sample confidence interval is not met but that the condition for the plus four interval is met.

(b) Give a 95% confidence interval for the proportion of all adults who prefer Times New Roman for web use. Give a 90% confidence interval for the proportion of all adults who think Gigi is more attractive.

24.65 Parents Doing too Much. A Pew Research Center survey of 9834 U.S. adults nationwide conducted June 25–30, 2019, found that 55% of those surveyed said parents of young adults ages 18 to 29 are doing too much for their adult children. At the time the poll was conducted, what can you say with 95% confidence about the percentage of all adults in the U.S. who said parents of young adults ages 18 to 29 are doing too much for their adult children?

Anthony Mercieca/Science Source

PART V

Inference about Relationships

Statistical inference offers more methods than anyone can know well, as a glance at the offerings of any large statistical software package demonstrates. In an introductory text, we must be selective. Parts I to IV have laid a foundation for understanding statistics:

- The nature and purpose of data analysis
- The central ideas of designs for data production
- The reasoning behind confidence intervals and significance tests
- Experience applying these ideas in practice

Each of the three chapters in Part V offers an introduction to a more advanced topic in statistical inference. You may choose to read any or all of them, in any order.

What makes a statistical method "more advanced"? More complex data, for one thing. In Part IV, we looked only at methods for inference about a single population parameter and for comparing two parameters. All of the chapters in Part V present methods for studying relationships between two variables. In Chapter 25, both variables are categorical, with data given as a two-way table of counts of outcomes. Chapter 26 considers inference in the setting of regressing a response variable on an explanatory variable. This is an important type of relationship between two quantitative variables. In Chapter 27, we meet methods for comparing the mean response in more than two groups. Here, the explanatory variable (group) is categorical, and the response variable is quantitative. These chapters together bring our knowledge of inference to the same point that our study of data analysis reached in Chapters 1 through 7.

With greater complexity comes greater reliance on technology. In these final three chapters, you will more often be interpreting the output of statistical software or using software yourself. With effort, you can do the calculations needed in Chapter 25 with a basic calculator. For the methods covered in Chapters 26 and 27, calculations are arduous, and carrying them out contributes little to learning. Fortunately, you can grasp the ideas without step-by-step arithmetic.

Another aspect of "advanced" methods is new concepts and ideas. This is where we draw the line in deciding what statistical topics we can master in a first course. Part V builds elaborate methods on the foundation we have laid without introducing fundamentally new concepts. You can see that statistical practice does need additional big ideas by reading the sections on "the problem of multiple comparisons" in Chapters 25 and 27. But the ideas you already know place you among the world's statistical sophisticates.

CHAPTER

25

When you complete this chapter, you will be able to:

25.1 Recall and apply the concept of conditional distributions in two-way tables.

25.2 In the context of the relationship between two categorical variables, explain the distinction between multiple individual comparisons and a single comparison of all the relationships taken together.

25.3 Calculate expected counts in two-way tables.

25.4 Describe the chi-square statistic and what it measures.

25.5 Compute chi-square statistics using technology.

25.6 Interpret the chi-square statistic in the context of specific problems involving the relationship between two categorical variables.

25.7 Determine whether the chi-square test is appropriate for a specific data set, based on expected counts.

25.8 Distinguish between the chi-square test of independence and the chi-square test of homogeneity.

25.9 Solve goodness-of-fit problems using the chi-square test.

Two Categorical Variables: The Chi-Square Test

Suppose we want to compare the proportions of White and non-White adult Americans who, when answering the question "are you a have or a have-not," consider themselves to be "haves." Using the two-sample z procedures of Chapter 23, we would treat Whites and non-Whites as two populations. We would then compare the proportion of Whites in our sample who consider themselves haves to the proportion of non-Whites in our sample who consider themselves haves. We can also think of this as a question about the relationship between two categorical variables—race (White or non-White) and haves or have-nots—with each variable having two possible values. However, suppose our data include more than two outcomes for "are you a have or have-not": have, have-not, neither, and refuse to answer. When there are more than two outcomes, or when we want to compare more than two groups, we need a new statistical test. The new test addresses a general question: *Is there a relationship between two categorical variables?*

25.1 Two-Way Tables

We saw in Chapter 6 that we can present data on two categorical variables in a *two-way table* of counts. That's our starting point. Let's begin our exploration of haves and have-nots.

SolStock/Getty Images

EXAMPLE 25.1 Do You Consider Yourself a Have or Have-Not?

HAVES

A sample survey asked a random sample of American adults, "Do you consider yourself a have or have-not?" Table 25.1 is a two-way table of all 1867 people in the sample classified by their political ideology and by whether they considered themselves a have, have-not, neither, or refused to answer.[1] Whether one considered oneself a have, have-not, neither, or refused to answer is a categorical variable. Political ideology is also categorical, and the two-way table divides American adults into three ideological categories: conservative, moderate, and liberal. Table 25.1 gives the counts for all 12 combinations of political ideology and the response to whether one considered oneself a have, have-not, neither, or refused to answer. Each of the 12 counts occupies a *cell* of the table.

TABLE 25.1 American adults by political ideology and whether they consider themselves a have or have-not

	Political Ideology			
Have or Have-Not	**Conservative**	**Moderate**	**Liberal**	**Total**
Have	460	354	248	1062
Have-not	184	242	234	660
Neither	45	27	21	93
Refused to answer	36	6	10	52
Total	725	629	513	1867

As usual, we prepare for inference by first doing data analysis. Because we think that ideology helps explain whether one considered oneself a have or have-not, find the percentages of people in each ideology group who considered themselves a have, have-not, neither, or refused to answer. The percentages appear in Table 25.2. Each column adds to 100% (up to roundoff error) because we are looking at each political ideology separately. In the language of Chapter 6 (page 166), Table 25.2 shows the three *conditional distributions* of have or have-not given a specific ideology.

Figure 25.1 is a bar graph comparing the three conditional distributions. The graph shows a relationship between ideology and have or have-not. As ideology moves from conservative, to moderate, to liberal, the percentage of those who considered themselves haves drops and the percentage who considered themselves have-nots rises. Are these differences among the three ideologies large enough to be statistically significant?

TABLE 25.2 Percentages of each political ideology who identified as a have, have-not, neither, or refused to answer (read down columns).

	Political Ideology		
	Conservative	**Moderate**	**Liberal**
Have	63.4%	56.3%	48.3%
Have-not	25.4%	38.5%	45.6%
Neither	6.2%	4.3%	4.1%
Refused to answer	5.0%	1.0%	1.9%
Total	100.0%	100.1%	99.9%

FIGURE 25.1
Bar graph comparing the three conditional distributions of haves, have-nots, neither, or refused to answer for each political ideology, for Example 25.1.

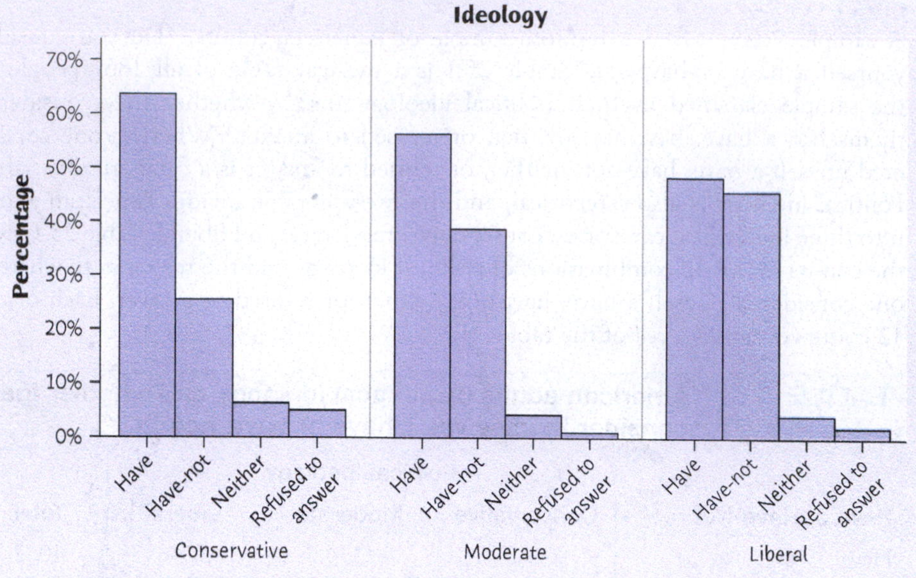

APPLY YOUR KNOWLEDGE

25.1 Would You Give Up Social Media? The Pew Research Center asked a sample of adults whether it would be hard to give up social media. Here is a comparison of the age distributions of those who said it would be hard to give up and those who said it would not be hard to give up:[2] MEDIA

	Age Category			
Hard to Give Up?	**18 to 24**	**25 to 29**	**30 to 49**	**50 or Over**
No	98	91	301	719
Yes	103	60	227	354

(a) For each age category, what percentage said it would be hard to give up social media? What percentage said it would not be hard to give up social media? Each column should add to 100% (up to roundoff error). These are the conditional distributions of whether it would be hard to give up social media for each age category.

(b) Make a bar graph that compares the four conditional distributions. What are the most important differences in whether it would be hard to give up social media for the four age categories?

25.2 Peer Victimization and Depression. The Avon Longitudinal Study of Parents and Children (ALSPAC) included approximately 14,000 children born between 1991 and 1992 in southwestern England and was intended to investigate a wide range of influences on the health and development of children. The data reported here investigate the relationship between being bullied at age 13–which includes incidents such as the taking of personal belongings, being threatened, blackmailed, hit or beat up, being called nasty names or having lies told about them, etc.–and depression at 18 years of age. From the original cohort, 3898 children had data on both the frequency of being bullied and later depression. Here are the results[3]: BULLYING

	Frequency of Being Bullied		
	Never	Occasionally	Frequently
Depressed	97	103	101
Not depressed	1762	1343	582

(a) It appears that 18-year-olds suffering from depression were bullied more often than those not suffering from depression. Give percentages to back up this claim. Make a bar graph that compares your percentages for subjects suffering and not suffering from depression at age 18.

(b) Association does not prove causation. Explain why you can't conclude from this study that being bullied causes depression.

25.2 The Problem of Multiple Comparisons

In Example 25.1 the null hypothesis of interest is that, in the population of all American adults, there is *no difference* among the conditional distributions of have, have-not, neither, or refused to answer for conservatives, moderates, and liberals. If the null hypothesis is true, the differences in the sample are just accidents due to random selection of the sample. Put more generally, the null hypothesis is that there is *no relationship* between two categorical variables:

> H_0: There is no relationship between political ideology and whether a person identified as a have, have-not, neither, or refused to answer for the population of all America adults.

The alternative hypothesis says that there *is* a relationship but does not specify any particular kind of relationship:

> H_a: There is some relationship between political ideology and whether a person identified as a have, have-not, neither, or refused to answer for the population of all American adults.

Any difference among the three distributions of haves, have-nots, neither, or refused to answer in the population of all adults means that the null hypothesis is false and the alternative hypothesis is true. The alternative hypothesis is not one-sided or two-sided. We might call it "many-sided" because it allows any kind of difference.

With only the methods we already know, we might start by comparing the proportions of conservatives and liberals who consider themselves haves. We could similarly compare other pairs of proportions and would end up with many tests and many *P*-values. This is a bad idea. The *P*-values belong to each test separately, not to the collection of all the tests together. Think of the distinction between the probability that a basketball player makes a free throw and the probability that she makes all of her free throws in a game. *When we do many individual tests or confidence intervals, the individual P-values and confidence levels don't tell us how confident we can be in all of the inferences taken together.*

Because of this, it's cheating to pick out one large difference from Table 25.2 and then test its significance as if it were the only comparison we had in mind. For example, the percentages of conservatives and liberals who consider

STATISTICS IN YOUR WORLD

He Started It!

A study of deaths in bar fights showed that, in 90% of the cases, the person who died started the fight. You shouldn't believe this. If you killed someone in a fight, what would you say when the police asked you who started the fight? After all, dead men tell no tales.

themselves haves are significantly different ($z = 5.29$, $P < 0.001$) if we make just this one comparison. But we could also pick a comparison that is not significant—for example, the proportions of moderates and liberals who said neither do not differ significantly ($z = 0.17$, $P = 0.868$). Individual comparisons can't tell us whether the four distributions, each with five outcomes, are significantly different.

The problem of how to do many comparisons at once with an overall measure of confidence in all our conclusions is common in statistics. This is the problem of *multiple comparisons*. Statistical methods for dealing with multiple comparisons usually have two steps:

1. An *overall test* to see if there is good evidence of *any* differences among the parameters that we want to compare.
2. A detailed *follow-up analysis* to decide which of the parameters differ and to estimate how large the differences are.

The overall test, though more complex than the tests we met earlier, is reasonably straightforward. The follow-up analysis can be quite elaborate. We will concentrate on the overall test and use data analysis to describe in detail the nature of the differences.

APPLY YOUR KNOWLEDGE

25.3 Peer Victimization and Bullying. In the setting of Exercise 25.2, we might do several significance tests to compare the frequencies of having been bullied for those 18-year-olds suffering and not suffering from depression. BULLYING

(a) Is there a significant difference between the proportions of being bullied occasionally at age 13 for those 18-year-olds suffering and not suffering from depression? Give the *P*-value.

(b) Is there a significant difference between the proportions of being bullied frequently at age 13 for those 18-year-olds suffering and not suffering from depression? Give the *P*-value.

(c) Explain clearly why *P*-values for individual outcomes like these can't tell us whether the two distributions for all three outcomes for 18-year-olds suffering and not suffering from depression differ significantly.

25.4 Is Astrology Scientific? The University of Chicago's General Social Survey (GSS) is the nation's most important social science sample survey. The GSS asked a random sample of adults their opinion about whether astrology is very scientific, sort of scientific, or not at all scientific. Here is a two-way table of counts for people in the sample who had three levels of higher education degrees:[4] ASTRLGY

	Degree Held		
	Junior College	**Bachelor**	**Graduate**
Not at all scientific	47	181	113
Very or sort of scientific	36	43	13

(a) Give three 95% confidence intervals for the percentages of people with each degree who think that astrology is not at all scientific.

(b) Explain clearly why we are *not* 95% confident that *all three* of these intervals capture their respective population proportions.

25.3 Expected Counts in Two-Way Tables

Our general null hypothesis H_0 is that there is *no relationship* between the two categorical variables that label the rows and columns of a two-way table. To test H_0, we compare the observed counts in the table with the *expected counts*, the counts we would expect—except for random variation—if H_0 were true. If the observed counts are far from the expected counts, that is evidence against H_0. Here is the formula for the expected counts.

Expected Counts

The **expected count** in any cell of a two-way table when H_0 is true is

$$\text{expected count} = \frac{\text{row total} \times \text{column total}}{\text{table total}}$$

EXAMPLE 25.2 Haves and Have-Nots: Expected Counts

HAVES

Let's find the expected counts for the study of haves and have-nots. Look back at the two-way table of counts, Table 25.1. That table includes the row and column totals. The expected count of conservatives who considered themselves haves is

$$\frac{\text{row 1 total} \times \text{column 1 total}}{\text{table total}} = \frac{(1062)(725)}{1867} = 412.40$$

The expected count of liberals who considered themselves haves is

$$\frac{\text{row 1 total} \times \text{column 3 total}}{\text{table total}} = \frac{(1062)(513)}{1867} = 291.81$$

The actual counts were 460 and 248. More conservatives and fewer liberals considered themselves haves than we would expect if there were no relationship between political ideology and whether one considered oneself a have, have-not, neither, or refused to answer. Table 25.3 shows all 12 expected counts.

As this table shows, *the expected counts have exactly the same row and column totals (up to roundoff error) as the observed counts.* That's a good way to check your work. Comparing the actual counts (Table 25.1) and the expected counts (Table 25.3) shows in what ways the data diverge from the null hypothesis.

TABLE 25.3 American adults by political ideology and whether they consider themselves a have or have-not: Expected cell counts

	Political Ideology			
Have or Have-Not	**Conservative**	**Moderate**	**Liberal**	**Total**
Have	412.40	357.79	291.81	1062
Have-not	256.29	222.36	181.35	660
Neither	36.11	31.33	25.55	93
Refused to answer	20.19	17.52	14.29	52
Total	725	629	513	1867

Why the formula works Where does the formula for an expected count come from? Think of a basketball player who makes 70% of her free throws in the long run. If she shoots 10 free throws in a game, we expect her to make 70% of them, or seven of the 10. Of course, she won't make exactly seven every time she shoots 10 free throws in a game. There is chance variation from game to game. But in the long run, seven of 10 is what we expect. In more formal language, if we have n independent tries and the probability of a success on each try is p, we expect np successes.

Now go back to the count of conservatives who considered themselves haves. The proportion of all 1867 adults surveyed who considered themselves haves is

$$\frac{\text{count of successes}}{\text{table total}} = \frac{\text{row 1 total}}{\text{table total}} = \frac{1062}{1867}$$

Think of this as p, the overall proportion of successes. If H_0 is true, we expect (except for random variation) this same proportion of successes in all three ideological groups. So the expected count of successes among the 725 conservatives is

$$np = (725)\left(\frac{1062}{1867}\right) = 412.40$$

That's the formula in the Expected Counts box.

APPLY YOUR KNOWLEDGE

25.5 Would You Give Up Social Media? The two-way table in Exercise 25.1 (page 558) displays data on the relationship between age and whether it would be hard to give up social media for a random sample of adults. The null hypothesis is that there is no relationship between whether it would be hard to give up social media (Yes and No) and age category. MEDIA

(a) If this hypothesis is true, what are the expected counts for the four age categories for those who said it would not be hard to give up social media? This is one row of the two-way table of expected counts. Find the row total and verify that it agrees (up to roundoff error) with the row total for the observed counts.

(b) Those who said it would not be hard to give up social media tend to be older than those who said it would be hard. How is this apparent when comparing the observed and the expected counts for those who said it would not be hard to give up social media?

25.6 Peer Victimization and Depression. Exercise 25.2 (page 558) describes a comparison of the distribution of the frequencies of having been bullied at age 13 for a sample of 18-year-olds in southwestern England suffering and not suffering from depression. The null hypothesis "no relationship" says that in the population of all 18-year-olds in southwestern England, the proportions who have experienced each frequency of bullying at age 13 are the same for those suffering and not suffering from depression. BULLYING

(a) Find the expected cell counts if this hypothesis is true and display them in a two-way table. Add the row and column totals to your table and check that they agree with the totals for the observed counts.

(b) Are there any large deviations between the observed counts and the expected counts? What kind of relationship between the two variables do these deviations point to?

25.4 The Chi-Square Statistic

To test whether the observed differences among the three distributions of have, have-not, neither, or refused to answer given political ideology are statistically significant, we compare the observed and expected counts. The test statistic that makes the comparison is the *chi-square statistic.*

Chi-Square Statistic

The **chi-square statistic** is a measure of how far the observed counts in a two-way table are from the expected counts if H_0 were true. The formula for the statistic is

$$\chi^2 = \sum \frac{(\text{observed count} - \text{expected count})^2}{\text{expected count}}$$

The sum is over all cells in the table.

The symbol χ in the box is the Greek letter chi. Recall from Chapter 2 that Σ (capital Greek sigma) means "add them all up." The chi-square statistic is a sum of terms, one for each cell in the table.

EXAMPLE 25.3 Haves and Have-Nots: The Test Statistic

HAVES

In the study of haves and have-nots, 460 conservatives considered themselves haves. The expected count for this cell is 412.40. So the term of the chi-square statistic from this cell is

$$\frac{(\text{observed count} - \text{expected count})^2}{\text{expected count}} = \frac{(460 - 412.40)^2}{412.40}$$
$$= \frac{2265.76}{412.40} = 5.49$$

The chi-square statistic χ^2 is the sum of 12 terms like this one. Here they are, arranged to match the layout of the two-way table:

$$\begin{aligned} \chi^2 &= 5.49 + 0.04 + 6.58 \\ &\quad + 20.39 + 1.74 + 15.29 \\ &\quad + 2.19 + 0.60 + 0.81 \\ &\quad + 12.37 + 7.57 + 1.29 \\ &= 74.36 \end{aligned}$$

To find the value $\chi^2 = 74.36$, we had to calculate the 12 expected cell counts in Table 25.3 and then the 12 terms of the sum. Each term in the sum represents the "contribution" to the chi-square statistic for one of the cells in the table. *Moreover, even rounding each term to two decimal places as we have done can still create round-off error in the sum. Because of this, software is very handy in finding* χ^2.

Think of χ^2 as a measure of the distance of the observed counts from the expected counts if H_0 were true. Like any other distance, it is always zero or positive, and it is zero only when the observed counts are exactly equal to the expected counts. Large values of χ^2 are evidence against H_0 because they say that the observed counts are far from what we would expect if H_0 were true. *Although the alternative hypothesis* H_a *is many-sided, the chi-square test is one-sided* because any violation of H_0 tends to produce a large value of X^2. Small values of X^2 are not evidence against H_0.

25.5 Examples of Technology

Calculating the expected counts and then the chi-square statistic by hand is time-consuming and can lead to errors due to roundoff. As usual, software saves time and always gets the arithmetic right. Figure 25.2 shows output for the chi-square test for the haves and have-nots data from a graphing calculator and two statistical programs.

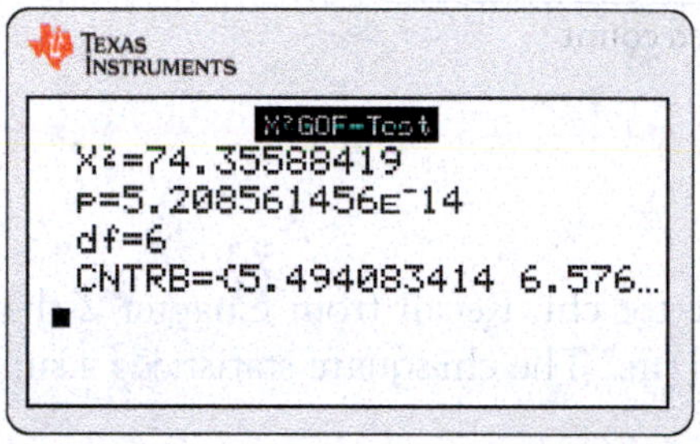

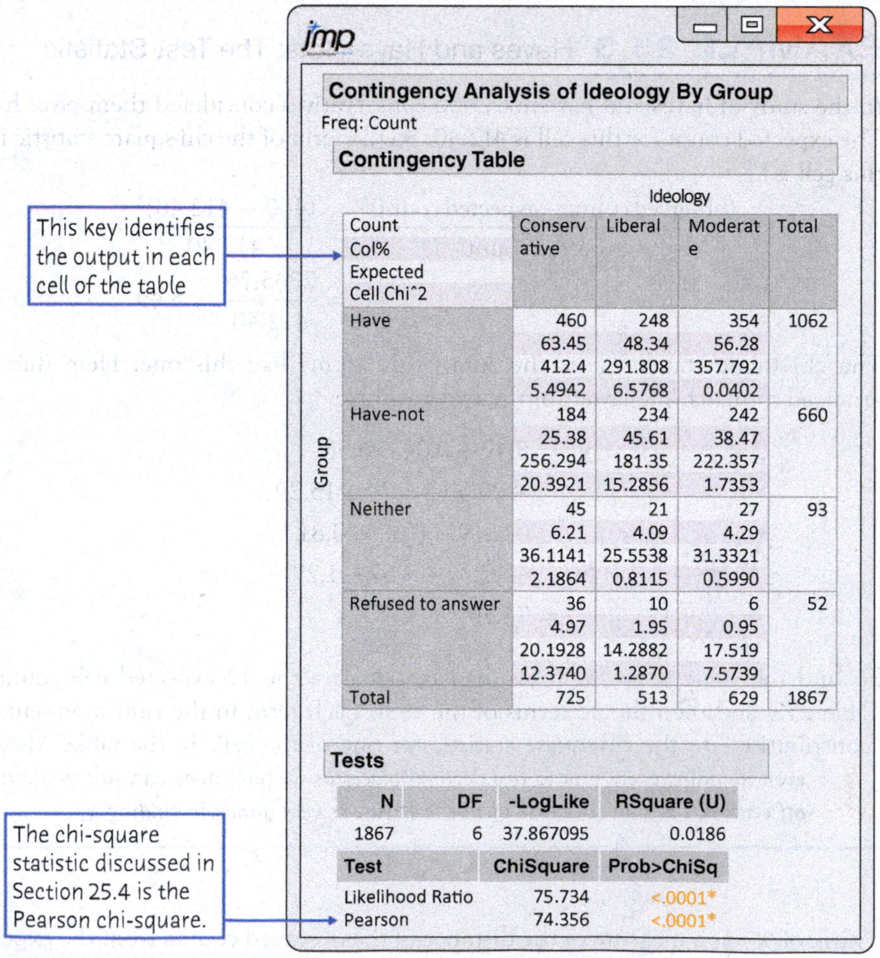

Count / Col% / Expected / Cell Chi^2	Conservative	Liberal	Moderate	Total
Have	460 63.45 412.4 5.4942	248 48.34 291.808 6.5768	354 56.28 357.792 0.0402	1062
Have-not	184 25.38 256.294 20.3921	234 45.61 181.35 15.2856	242 38.47 222.357 1.7353	660
Neither	45 6.21 36.1141 2.1864	21 4.09 25.5538 0.8115	27 4.29 31.3321 0.5990	93
Refused to answer	36 4.97 20.1928 12.3740	10 1.95 14.2882 1.2870	6 0.95 17.519 7.5739	52
Total	725	513	629	1867

Tests

N	DF	-LogLike	RSquare (U)
1867	6	37.867095	0.0186

Test	ChiSquare	Prob>ChiSq
Likelihood Ratio	75.734	<.0001*
Pearson	74.356	<.0001*

FIGURE 25.2
Output from a graphing calculator, JMP, and Minitab for the two-way table in the study of haves and have-nots, for Example 25.4.

FIGURE 25.2
(Continued)

Minitab

Tabulated Statistics: Group, Ideology

Using frequencies in Count

Rows: Group **Columns: Ideology**

	Conservative	Liberal	Moderate	All
Have	460	248	354	1062
	63.45	48.34	56.28	56.88
	412.40	291.81	357.79	
	5.494	6.577	0.040	
Have-not	184	234	242	660
	25.38	45.61	38.47	35.35
	256.29	181.35	222.36	
	20.392	15.286	1.735	
Neither	45	21	27	93
	6.21	4.09	4.29	4.98
	36.11	25.55	31.33	
	2.186	0.812	0.599	
Refused to answer	36	10	6	52
	4.97	1.95	0.95	2.79
	20.19	14.29	17.52	
	12.374	1.287	7.574	
All	725	513	629	1867
	100.00	100.00	100.00	100.00

Cell Contents
Count
% of Column
Expected count
Contribution to Chi-square

Chi-Square Test

	Chi-Square	DF	P-value
Pearson	74.356	6	0.000
Likelihood Ratio	75.734	6	0.000

EXAMPLE 25.4 Haves and Have-Nots: Chi-square Output

All three outputs tell us that the chi-square statistic (rounded to two decimal places) is $\chi^2 = 74.36$, with very small *P*-value. Minitab reports $P = 0.000$. JMP reports $P < 0.0001$. Notice that JMP lists the ideology categories in alphabetical order. The graphing calculator says that $P = 5.2 \times 10^{-14}$, a very small number indeed. This *P*-value comes from an approximation to the sampling distribution of X^2. (We provide more detail on the *P*-value computation in the next section.) The approximation is less accurate far out in the tail than near the center of the distribution, so you should not take 5.2×10^{-14} literally. Just read it as "*P* is very small." The sample gives very strong evidence that whether one considered oneself a have, have-not, neither, or refused to answer is not the same for conservatives, moderates, and liberals.

Statistical software generally offers additional information on request. We instructed JMP and Minitab to show the observed counts and expected counts and also the term in the chi-square statistic for each cell, called the "contribution to chi-square" or "cell chi∧2." The top-left cell has expected count 412.4 and contributes 5.4942 to the chi-square statistic, as we calculated earlier. (Roundoff errors are smaller with software than in hand calculation.) The graphing calculator also displays each cell's term in the chi-square statistic. The first two terms appear next to CNTRB=. To see the chi-square terms of the remaining cells on the calculator's small screen, you must scroll to the right. We also told JMP and Minitab to include the column percentages associated with each count.

HAVES

The chi-square test is an overall test for detecting relationships between two categorical variables. If the test is significant, it is important to look at the data to learn the nature of the relationship. We have three ways to look at the haves and have-nots data:

- **Compare selected percentages:** Which of the categories have, have-not, neither, or refused to answer occur in quite different percentages of the three ideology groups? This is the method we learned in Chapter 6.
- **Look at the terms of the chi-square statistic:** Which cells contribute the most to the value of X^2?
- **Compare observed and expected cell counts:** Which cells have more or fewer observations than we would expect if H_0 were true?

EXAMPLE 25.5 Haves and Have-Nots: Conclusion

HAVES

There is very strong evidence ($X^2 = 74.36$, $P < 0.0001$) that whether one considered oneself a have, have-not, neither, or refused to answer is not the same for conservatives, moderates, and liberals. Comparing selected percentages—specifically, the three conditional distributions of have, have-not, neither, or refused to answer for ideology in Table 25.2 (page 557) and Figure 25.1—shows that the more conservative one's ideology, the more likely one was to consider oneself a have.

The additional information provided by programs like JMP shows what differences among the ideology groups explain the large value of the chi-square statistic. Look at the 12 terms in the chi-square statistic in the JMP output and compare the observed and expected counts in the cells that contribute most to chi-square. Six of the 12 cells, those with values larger than 5 in Figure 25.2, contribute 67.70 of the total chi-square $X^2 = 74.36$. These six cells occur in pairs:

- 5.4942 and 6.5768: fewer liberals than expected and more conservatives than expected considered themselves haves.
- 20.3921 and 15.2856: more liberals than expected and fewer conservatives than expected considered themselves have-nots.
- 12.3740 and 7.5739: more conservatives than expected and fewer moderates than expected refused to answer.

These results show the fact that conservatives were more likely to have considered themselves haves and less likely to have considered themselves have-nots than liberals.

APPLY YOUR KNOWLEDGE

25.7 Would You Give Up Social Media? Figure 25.3 displays JMP output for the two-way table in Exercise 25.1 (page 558). The output includes the two-way table of observed counts, the expected counts, and each cell's contribution to the chi-square statistic. MEDIA

(a) What hypotheses does chi-square test? What are the test statistic and its *P*-value?

(b) Which cells contribute the most to X^2? Compare the observed and expected counts in these cells and comment on the most important differences in the age distributions of those who said it would be hard and those who said it would not be hard to give up social media.

Contingency Table

Age (columns); Hard to give up (rows)

Count / Expected / Cell Chi^2	18-24	25-29	30-49	50+	Total
No	98	91	301	719	1209
	124.429	93.4762	326.857	664.238	
	5.6134	0.0656	2.0455	4.5147	
Yes	103	60	227	354	744
	76.5714	57.5238	201.143	408.762	
	9.1218	0.1066	3.3240	7.3365	
Total	201	151	528	1073	1953

Tests

N	DF	-LogLike	RSquare (U)
1953	3	15.914370	0.0073

Test	ChiSquare	Prob>ChiSq
Likelihood Ratio	31.829	<.0001*
Pearson	32.128	<.0001*

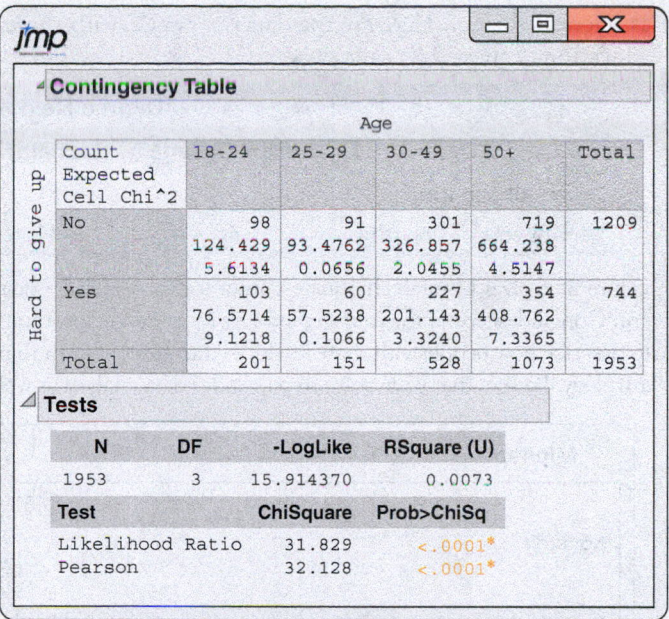

FIGURE 25.3
JMP output for the two-way table of age by whether it would be hard to give up social media, for Exercise 25.7.

25.8 Peer Victimization and Depression. Your data analysis in Exercise 25.2 (page 558) found that 18-year-olds suffering from depression tend to have been bullied more frequently at age 13 than those not suffering from depression. Figure 25.4 gives JMP output for the two-way table in Exercise 25.2. BULLYING

(a) What are the chi-square statistic and its *P*-value? Explain in simple language what it means to reject H_0 in this setting.

(b) Give an overall conclusion that refers to row percentages to describe the nature of the relationship between peer victimization and depression.

Contingency Table

Depressed By Bullied

Count / Row % / Expected / Cell Chi^2	Frequent	Never	Occasional	Total
no	582	1762	1343	3687
	15.79	47.79	36.43	
	631.45	1718.69	1336.86	
	3.8725	1.0914	0.0282	
yes	101	97	103	301
	33.55	32.23	34.22	
	51.5504	140.311	109.139	
	47.4344	13.3690	0.3453	
Total	683	1859	1446	3988

Tests

N	DF	-LogLike	RSquare (U)
3988	2	28.702418	0.0070

Test	ChiSquare	Prob>ChiSq
Pearson	66.141	<.0001*

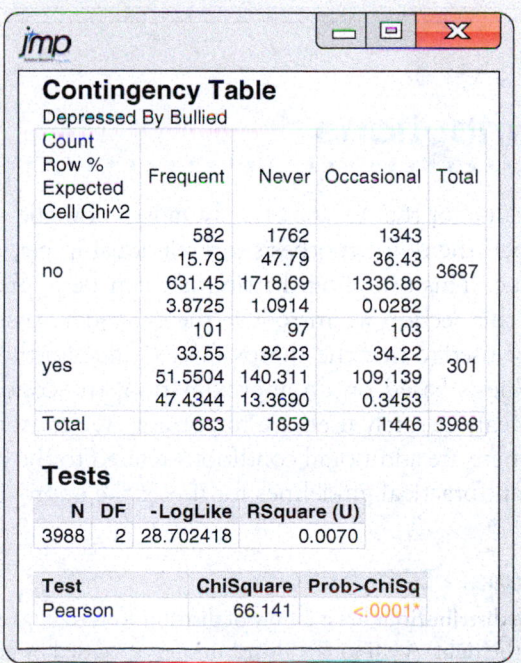

FIGURE 25.4
JMP output for the study of bullying and depression, for Exercise 25.8.

25.9 Is Astrology Scientific? The General Social Survey asked a random sample of adults about their education levels and about their views of astrology as scientific or not. Here are the data for people with three levels of higher education degrees: ASTRLGY

	Degree Held		
	Junior College	**Bachelor**	**Graduate**
Not at all scientific	47	181	113
Very or sort of scientific	36	43	13

Figure 25.5 gives Minitab chi-square output for these data. Follow the Plan, Solve, and Conclude steps of the four-step process in using the information in the output to describe how people with these levels of education differ in their opinions about astrology. Be sure that your Solve step includes data analysis as well as a formal test.

```
Minitab

                    Junior college      Bachelor    Graduate      All

NotScience                      47           181         113      341
                             56.63         80.80       89.68    78.75
                             65.36        176.41       99.23
                             5.160         0.120       1.911

Science                         36            43          13       92
                             43.37         19.20       10.32    21.25
                             17.64         47.59       26.77
                            19.125         0.443       7.084

All                             83           224         126      433
                            100.00        100.00      100.00   100.00

Cell Contents:          Count
                        % of Column
                        Expected count
                        Contribution to Chi-square

Pearson Chi-Square = 33.843,   DF = 2,  P-Value = 0.000
```

FIGURE 25.5
Minitab output for the two-way table of opinion about astrology by degree held, for Exercise 25.9.

25.6 The Chi-Square Distributions

To find the *P*-value for the χ^2 statistic, we must know the sampling distribution of the statistic when the null hypothesis (no relationship between the two categorical variables) is true. This sampling distribution can be approximated by a *chi-square distribution*. In this section we introduce the chi-square distributions and illustrate how they can be used to find the approximate *P*-value for the chi-square test statistic. Since the *P*-value is based on an approximation, the conditions under which the approximation can be safely used are important. We must assume that we have an SRS and that there are additional conditions related to the counts in the cells of the table. We present practical guidelines for use of the approximation in Section 25.7.

The Chi-Square Distributions

The **chi-square distributions** are a family of distributions that take only positive values and are skewed to the right. A chi-square distribution is specified by giving its *degrees of freedom*.

The chi-square test for a two-way table with r rows and c columns uses critical values from the chi-square distribution with $(r-1)\times(c-1)$ degrees of freedom. The *P*-value is the area under the density curve of this chi-square distribution to the right of the value of the test statistic.

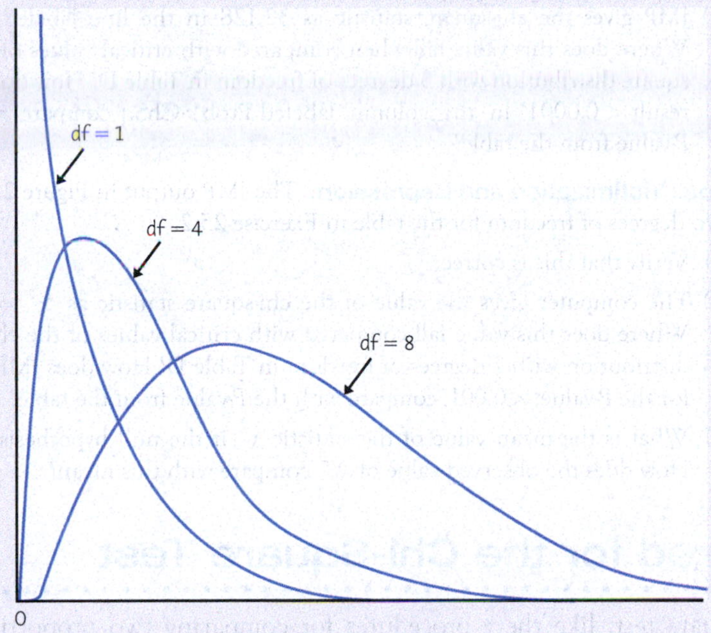

FIGURE 25.6
Density curves for the chi-square distributions with 1, 4, and 8 degrees of freedom. Chi-square distributions take only positive values and are right-skewed.

Figure 25.6 shows the density curves for three members of the chi-square family of distributions. As the degrees of freedom increase, the density curves become less skewed, and larger values become more probable. Table D in the back of the book gives critical values for chi-square distributions. You can use Table D if you do not have software that gives you *P*-values for a chi-square test.

EXAMPLE 25.6 Using the Chi-Square Table

HAVES

The two-way table of five outcomes by three political ideologies for the haves and have-nots study (Table 25.1, page 557) has four rows and three columns. That is, $r = 4$ and $c = 3$. The chi-square statistic therefore has degrees of freedom

$$(r-1)(c-1) = (4-1)(3-1) = (3)(2) = 6$$

All three outputs in Figure 25.2 (page 564) give 6 as the degrees of freedom.

The observed value of the chi-square statistic is $\chi^2 = 74.36$. Look in the df = 6 row of Table D. The value $\chi^2 = 74.36$ falls above the largest critical value in the table, for $P = 0.0005$. Remember that the chi-square test is always one-sided. So the *P*-value of $\chi^2 = 74.36$ is less than 0.0005.

df = 6

p	0.001	0.0005
χ^*	22.46	24.10

We know that all z and t statistics measure the size of an effect in the standard scale centered at zero. We can roughly assess the size of any z or t statistic by the 68–95–99.7 rule, though this is exact only for z. The chi-square statistic does not have any such natural interpretation. But here is a helpful fact: *the mean of any chi-square distribution is equal to its degrees of freedom.* In Example 25.6, χ^2 would have mean 6 if the null hypothesis were true. The observed value $\chi^2 = 74.36$ is so much larger than 6 that we suspect it is significant even before we look at Table D.

APPLY YOUR KNOWLEDGE

25.10 Would You Give Up Social Media? The JMP output in Figure 25.3 gives the degrees of freedom for a table of age category versus the two responses to the question of whether it would be hard to give up social media (No and Yes) as 3 in the column labeled DF.

(a) Show that this is correct for a table with two rows and four columns.

(b) JMP gives the chi-square statistic as 32.128 in the line labeled *Pearson*. Where does this value fall when compared with critical values of the chi-square distribution with 3 degrees of freedom in Table D? How does JMP's result < 0.0001* in the column labeled Prob>ChSq compare with the *P*-value from the table?

25.11 Peer Victimization and Depression. The JMP output in Figure 25.4 gives two degrees of freedom for the table in Exercise 25.2.

(a) Verify that this is correct.

(b) The computer gives the value of the chi-square statistic as $\chi^2 = 66.141$. Where does this value fall compared with critical values of the chi-square distribution with 2 degrees of freedom in Table D? How does JMP's result for the *P*-value, <0.001, compare with the *P*-value from the table?

(c) What is the mean value of the statistic χ^2 if the null hypothesis is true? How does the observed value of χ^2 compare with this mean?

25.7 Cell Counts Required for the Chi-Square Test

The chi-square test, like the z procedures for comparing two proportions, is an approximate method that becomes more accurate as the counts in the cells of the table get larger. We must therefore check that the counts are large enough to allow us to trust the *P*-value. Fortunately, the chi-square approximation is accurate for quite modest counts. Here is a practical guideline.[5]

Cell Counts Required for the Chi-Square Test

You can safely use the chi-square test with critical values from the chi-square distribution when no more than 20% of the expected counts are less than 5 and all individual expected counts are 1 or greater. In particular, all four expected counts in a 2×2 table should be 5 or greater.

Note that the guideline uses *expected* cell counts. The expected counts for the haves and have-nots study of Example 25.1 appear in Table 25.3 (page 561). None of the 12 expected counts are less than 5, so the data meet the guideline for safe use of chi-square. When the expected cell counts are too small for us to trust the computed *P*-value, most computer software will include a warning concerning the use of the approximation as part of the output.

APPLY YOUR KNOWLEDGE

25.12 Peer Victimization and Depression. Your data analysis in Exercise 25.2 found that 18-year-olds suffering from depression tend to have been bullied more frequently at age 13 than those not suffering from depression. Figure 25.4 (page 567) gives JMP output for the two-way table in Exercise 25.2. Verify from the output that the data meet the cell count requirement for use of chi-square.

25.8 Uses of the Chi-Square Test: Independence and Homogeneity

Two-way tables can arise in several ways. Most commonly, the subjects in a single sample are classified by two categorical variables. For example, we classified adults by their political ideology and whether they considered themselves haves, have-nots, neither, or refused to answer. The question of whether or not there is a relationship between these two classification variables can be stated in terms of the independence of these variables, using the definition of independence from our study of probability in Chapter 13. Recall that for two events, A and B, to be independent, we must have $P(A) = P(A|B)$. In this setting, independence of political ideology and whether one considered oneself a have, have-not, neither, or refused to answer would imply that the conditional probability that a randomly selected adult considered themselves a have given their political ideology would be the same for all ideologies. That is, knowing an adult's ideology gives us no information about the probability of them considering themselves a have. The test we have been using to this point is generally referred to as the *chi-square test of independence*, as thus far all the examples and exercises have been questions about whether two classification variables are independent or not.

The next example illustrates a different setting for a two-way table, in which we compare separate samples from two or more populations or from two or more treatments in a randomized controlled experiment. In the setting of a randomized controlled experiment, we think of the different treatment groups as the separate populations. "Which population" is now one of the variables for the two-way table. For each sample, we classify individuals according to one variable (such as whether or not symptoms have subsided in a medical trial), and we are interested in whether or not the probabilities of being classified in each category of this variable are the same for each population. Although our calculations for the chi-square test are unchanged, the method of collecting the data is different. This use of the chi-square test is referred to as the *chi-square test of homogeneity* because we are interested in whether or not the populations from which the samples are selected are homogeneous (the same) with respect to the single classification variable. Notice that tests of homogeneity involve an explanatory variable (membership in different populations) and a response variable, but independence tests do not.

EXAMPLE 25.7 Are Cell-Only Telephone Users Different?

STATE: Random digit dialing (RDD) telephone surveys do not call cell phone numbers. We know that cell-only users tend to be younger (see Exercise 25.32, page 581), which results in younger adults being underrepresented in an RDD sample. Survey organizations can compensate for this by weighting the younger adults in the RDD sample more heavily, but this assumes that the young adults accessible by landline are similar on the survey issue to their cell-only peers, who are excluded. There is growing evidence that this is not always the case. In young adults between 18 and 25 years of age, Pew interviewed separate random samples of cell-only and landline telephone users and compared them on demographic, lifestyle, and attitudinal issues.[6] Here are the results on the living situations for the two groups:

LIVECELL

kyonntra/Getty Images

	Landline Sample	Cell-Only Sample
Live with parents	165	25
Rent	95	74
Own	46	13
Live in a dorm	7	15
Other/refused	16	3
Total	329	130

PLAN: Carry out a chi-square test for

H_0: homogeneity; that is, the distribution of living situation is the same in both populations

H_a: lack of homogeneity; that is, the living situation distribution for the cell-only population differs from that for landline users

Compare column percentages or observed versus expected cell counts or terms of chi-square to see the nature of the differences in the distributions of living situations in the two populations.

SOLVE: The Minitab output in Figure 25.7 includes the column percentages. These give the distribution of living situation for each type of telephone use. Cell-only users are less likely to live with parents (19.23% versus 50.15% of landline users), more likely to rent (56.92% versus 28.88% of landline users), and more likely to live in a dorm (11.54% versus 2.13% of landline users). There is little difference in the percentages for the other two categories. To see if the differences are significant, first check the guidelines for use of chi-square. The samples can be assumed to be SRSs.

Minitab

	Landline sample	Cell-only sample	All
Live with parents	165	25	190
	50.15	19.23	41.39
	136.19	53.81	190.00
	6.096	15.427	*
Rent	95	74	169
	28.88	56.92	36.82
	121.14	47.86	169.00
	5.639	14.270	*
Own	46	13	59
	13.98	10.00	12.85
	42.29	16.71	59.00
	0.326	0.824	*
Live in a dorm	7	15	22
	2.13	11.54	4.79
	15.77	6.23	22.00
	4.876	12.341	*
Other/refused	16	3	19
	4.86	2.31	4.14
	13.62	5.38	19.00
	0.416	1.054	*
All	329	130	459
	100.00	100.00	100.00
	329.00	130.00	459.00
	*	*	*

Cell Contents: Count
% of Column
Expected count
Contribution to Chi-square

Pearson Chi-Square = 61.269, DF = 4, P-Value = 0.000

FIGURE 25.7
Minitab output for the two-way table of living situation and telephone use, for Example 25.7.

The Minitab output shows the expected counts in all of the 10 cells are greater than 5, so the conditions for use of the chi-square (page 570) are satisfied. The chi-square test shows a highly significant difference between the distributions of living situations of the two groups of young adults ($\chi^2 = 61.269$, $P = 0.000$). Comparing observed and expected cell counts again shows that cell-only young adults are less likely to live with parents than would be expected if the null hypothesis of homogeneity were true and more likely to rent or live in a dorm. The contributions to chi-square for the other two categories are quite small.

CONCLUDE: The association between living situation and type of telephone service is statistically significant. Young adults accessible by landline phone are more likely to live with their parents and less likely to rent or live in a dorm. The Pew study also found young adults with landlines tend to be more likely to attend religious services at least once a week and less likely to report drinking alcohol in the past seven days or to say that it is okay for people to smoke marijuana. They also tend to be less technology savvy, using email, texting, and visiting social networking sites less frequently. Because of the rapid increase in cell-only users plus the differences in a number of dimensions between cell-only users and their landline peers, most survey organizations now routinely include a cell-only sample when conducting a survey.[7]

One of the most useful properties of chi-square is that it tests the null hypothesis "the row and column variables are not related to each other" whenever this hypothesis makes sense for a two-way table. It makes sense when we are comparing a categorical response in two or more samples, as when we compared people who have only a cell phone with people who have a landline phone. This is the chi-square test of homogeneity. The hypothesis also makes sense when we have data on two categorical variables for the individuals in a single sample, as when we examined age group and living arrangement for a sample of young adults. This is the chi-square test of independence. Statistical significance has the same intuitive meaning in both settings: "A relationship this strong is not likely to happen just by chance."

STATISTICS IN YOUR WORLD

More Chi-Square Tests

There are other chi-square tests for hypotheses more specific than "no relationship." A sociologist places people in classes by social status, waits 10 years, and classifies the same people again. The row and column variables are the classes at the two times. For example, the row variable might be social status when first measured, and the column variable might be social status 10 years later. She might test the hypothesis that there has been no change in the overall distribution of social status in the group. Or she might ask if moves up in status are balanced by matching moves down. These and other null hypotheses can be tested by variations of the chi-square test.

Uses of the Chi-Square Test

Use the chi-square test to test the null hypothesis

H_0: There is no relationship between two categorical variables.

when you have a two-way table from one of these situations:

- A single SRS, with each individual classified according to both of two categorical variables. In this case, the null hypothesis of no relationship says that the two categorical variables are independent, and the test is called the **chi-square test of independence.**
- Independent SRSs from two or more populations, with each individual classified according to one categorical variable. (The other variable says which sample the individual comes from.) In this case, the null hypothesis of no relationship says the populations are homogeneous, and the test is called the **chi-square test of homogeneity.**

Here are some cautions about chi-square tests:

- Do not apply the chi-square tests when cell counts are too small (see Section 25.7).
- Do not use the chi-square tests when rows or columns are not independent—for example, when each row or column refers to the same subjects.

- Our previous cautions about *P*-values still apply to chi-square tests. A small *P*-value does not mean the association between categorical variables is strong or is of practical importance. A large *P*-value does not mean two categorical variables are independent.

You should also avoid classifying quantitative variables into categories in order to use a chi-square test. Classifying a quantitative variable into categories causes you to lose information. There are more appropriate methods, such as regression (see Chapter 26 and the supplemental Chapter 29), for exploring associations among quantitative variables.

APPLY YOUR KNOWLEDGE

25.13 Preference for the Middle? When choosing an item from a group, researchers have shown that an important factor influencing choice is the item's location. This occurs in varied situations, such as shelf positions when shopping, filling out a questionnaire, and even when choosing a preferred candidate during a presidential debate. In this experiment, five identical pairs of white socks were displayed in a vertical column on an easel with adjustable height settings. The socks were displayed at two heights in the experiment: a high display at approximately head height and a low display at approximately thigh height. One hundred participants from the University of Chester were used in the experiment and divided at random into two groups of 50 each. The first group chose their preferred pair of socks at the high display setting, and the second group viewed the socks at the low display setting. Here are the results, where Loc 1 is the topmost position among the five socks, Loc 3 is the center position, and Loc 5 is the lowest location:[8] MIDDLE1

	Loc 1	Loc 2	Loc 3	Loc 4	Loc 5
High display	14	12	14	6	4
Low display	12	13	20	4	1

Is there a difference in the distribution of preferred location for the two display heights?

(a) Should you use a chi-square test of independence or homogeneity?

(b) Do a complete analysis of these data, following the four-step process.

(c) Does your analysis in (b) answer the question of whether there are any differences among the preferences for the five locations? Explain briefly. (See Exercise 25.16.)

25.14 Is Confidence in Congress Eroding? One of the questions asked by the General Social Survey is, "Would you say you have a great deal of confidence, only some confidence or hardly any confidence in the people running congress?" Here are the data for a random sample of adults from the years 2002, 2006, 2010, 2014, and 2018: CONGRESS

Year	A Great Deal	Only Some	Hardly Any
2018	81	697	745
2014	93	651	900
2010	125	635	587
2006	209	1026	703
2002	117	553	222

(a) Should you use a chi-square test of independence or homogeneity? Explain.

(b) Do a complete analysis of these data, following the four-step process.

25.9 The Chi-Square Test for Goodness of Fit*

The most common and most important use of the chi-square statistic is to test the hypothesis that there is *no relationship between two categorical variables*. A variation of the statistic can be used to test a different kind of null hypothesis: that *a categorical variable has a specified distribution*. Here is an example that illustrates this use of chi-square.

EXAMPLE 25.8 Never on Sunday?

Births are not evenly distributed across the days of the week. Fewer babies are born on Saturday and Sunday than on other days, probably because weekend births are inconvenient for doctors and other medical personnel.

BIRTH700

A random sample of 700 births from local records shows this distribution across the days of the week:

Day	Sun.	Mon.	Tue.	Wed.	Thu.	Fri.	Sat.
Births	84	110	124	104	94	112	72

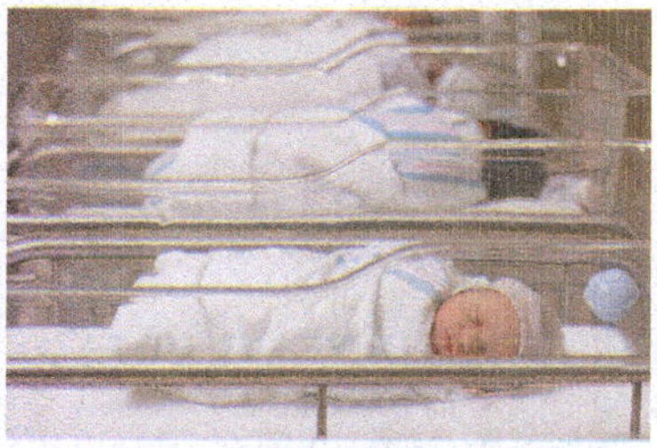

Sure enough, the two smallest counts of births are on Saturday and Sunday. Do these data give significant evidence that local births are not equally likely on all days of the week?

The chi-square test answers the question of Example 25.8 by comparing observed counts with expected counts under the null hypothesis. The null hypothesis for births says that they *are* evenly distributed. To state the hypotheses carefully, write the discrete probability distribution for days of birth:

Day	Sun.	Mon.	Tue.	Wed.	Thu.	Fri.	Sat.
Probability	p_1	p_2	p_3	p_4	p_5	p_6	p_7

The null hypothesis says that the probabilities are the same on all days. In that case, all seven probabilities must be 1/7. So the null hypothesis is

$$H_0: p_1 = p_2 = p_3 = p_4 = p_5 = p_6 = p_7 = \frac{1}{7}$$

The alternative hypothesis says that days are *not* all equally probable:

$$H_a: \text{not all } p_i = \frac{1}{7}$$

As usual in chi-square tests, H_a is a "many-sided" hypothesis that simply says that H_0 is not true. The chi-square statistic is also as usual:

$$\chi^2 = \sum \frac{(\text{observed count} - \text{expected count})^2}{\text{expected count}}$$

The expected count for an outcome with probability p is np, as we saw in the discussion following Example 25.2 (page 561). Under the null hypothesis, all the probabilities p_i are the same, so all seven expected counts are equal to

$$np_i = 700 \times \frac{1}{7} = 100$$

*This special topic is optional.

These expected counts easily satisfy our guidelines for using chi-square. The chi-square statistic is

$$\begin{aligned} X^2 &= \sum \frac{(\text{observed count} - 100)^2}{100} \\ &= \frac{(84-100)^2}{100} + \frac{(110-100)^2}{20} + \cdots + \frac{(72-100)^2}{100} \\ &= 19.12 \end{aligned}$$

df = 6

p	0.005	0.0025
X^*	18.55	20.25

This new use of X^2 requires different degrees of freedom. To find the *P*-value, compare X^2 with critical values from the chi-square distribution with degrees of freedom one less than the number of values the birth day can take. That's $7 - 1 = 6$ degrees of freedom. From Table D, we see that $X^2 = 19.12$ is between the critical values 18.55 and 20.25 in the df = 6 row, which are the critical values for 0.005 and 0.0025, respectively.

The *P*-value is therefore between 0.005 and 0.0025 (software gives the more exact value $P = 0.004$). These 700 births give convincing evidence that births are not equally likely on all days of the week.

STATISTICS IN YOUR WORLD

Chi-Square in the Casino

Gambling devices such as slot machines and roulette wheels are supposed to have a fixed and known distribution of outcomes. Here's a job for the chi-square test of goodness of fit: state gambling regulators use it to verify that casino devices are honest. How much deviation a casino can get away with depends on the state. Nevada cracks down if chi-square is significant at the 5% level. Mississippi gives more leeway, acting only when the 1% level is reached.

The chi-square test applied to the hypothesis that a categorical variable has a specified distribution is called the test for *goodness of fit*. The idea is that the test assesses whether the observed counts "fit" the distribution. The chi-square statistic is the same as for the two-way table test, but the expected counts and degrees of freedom are different. Here are the details.

The Chi-Square Test for Goodness of Fit

A categorical variable has k possible outcomes, with probabilities $p_1, p_2, p_3, \ldots, p_k$. That is, p_i is the probability of the ith outcome. We have n independent observations from this categorical variable.

To test the null hypothesis that the probabilities have specified values

$$H_0: p_1 = p_{10}, p_2 = p_{20}, \ldots, p_k = p_{k0}$$

find the expected count for the ith possible outcome as np_{i0} and use the chi-square statistic

$$X^2 = \sum \frac{(\text{observed count} - \text{expected count})^2}{\text{expected count}}$$

The sum is over all the possible outcomes.

The *P*-value is the area to the right of X^2 under the density curve of the chi-square distribution with $k - 1$ degrees of freedom.

In Example 25.8, the outcomes are days of the week, with $k = 7$. The null hypothesis says that the probability of a birth on the ith day is $p_{i0} = 1/7$ for all days. We observe $n = 700$ births and count how many fall on each day. These are the counts used in the chi-square statistic.

APPLY YOUR KNOWLEDGE

25.15 Saving Birds from Windows. Many birds are injured or killed by flying into windows. It appears that birds don't see windows. Can tilting windows down so that they reflect earth rather than sky reduce bird strikes? Place six windows at the edge of a woods: two vertical, two tilted 20 degrees, and two tilted 40 degrees. During the next four months, there are 53 bird strikes: 31 on the vertical windows, 14 on the 20-degree windows, and 8 on the 40-degree windows.[9] If the tilt has no effect, we expect strikes on windows with all three tilts to have equal probability. Test this null hypothesis. What do you conclude?

25.16 Checking for Survey Errors. The rails-to-trails program refers to the conversion of old rail corridors into multipurpose trails for recreation and transportation. Researchers were interested in obtaining information on residents' usage of a 10-mile-long paved greenway trail in Greenville, South Carolina, that connects residential areas to both a university campus and the commercial downtown area of the city. Random digit dialing of residential numbers was done using a database of exchanges. When a household was reached, surveyors asked to speak to the adult over 18 with the next birthday. No cell phone numbers were included in the sample, and the response rate was approximately 30%.[10] One way of checking the effect of undercoverage, nonresponse, and other sources of error in a sample survey is to compare the sample with known facts about the population. The table below compares the education levels of the sample and the education levels of the 461,299 residents of Greenville County (obtained from the U.S. Census Bureau at the time of the survey).

	Sample Counts	Population Counts
< High school degree	50	69195
High school degree	286	356,123
College degree	320	105,176
Total	656	530,494

(a) Do these data provide evidence that the education levels in the sample differ from those in the population? In what way? Explain in the context of this exercise. (*Hint*: You will first need to compute the population proportions to carry out the chi-square test for goodness of fit.)

(b) Although the proportion in the sample that used the trail in the past six months was approximately 27%, the proportion of trail users among those with less than a high school degree was 16%, among those with a high school degree was 20%, and among those with a college degree was 34%. Given this information and the results from part (a), do you feel that the estimate of 27% trail usage from the sample may be biased? What would be the likely direction of bias? Explain briefly.

25.17 More on Preference for the Middle. When choosing an item from a group, researchers have shown that an important factor influencing choice is the item's location. This occurs in varied situations such as shelf positions when shopping, filling out a questionnaire, and even when choosing a preferred candidate during a presidential debate. In this experiment, five identical pairs of white socks were displayed in a vertical column on an easel with adjustable height settings. The socks were displayed at two heights in the experiment: a high display at approximately head height and a low display at approximately thigh height (see Exercise 25.13, page 574). With the pattern of response found to be similar for the two heights, we now investigate whether there are differences in the preferences for the five locations. Here are the total number of participants among the 100 in the study choosing each of the five locations, where Loc 1 is the topmost position among the five socks, Loc 3 is the center position, and Loc 5 is the lowest location:

4step

Location	Loc 1	Loc 2	Loc 3	Loc 4	Loc 5
Number choosing	26	25	34	10	5

(a) What percentage of the subjects chose each location?

(b) If the subjects were equally likely to select each location, what are the expected counts for each location?

(c) Does the chi-square test for goodness of fit give good evidence that the subjects were not equally likely to choose each location? (State hypotheses, check the guidelines for using chi-square, give the test statistic and its *P*-value, and state your conclusion.)

(d) In these situations, previous research has shown subjects show a preference for the one in the middle, known as the center-stage effect. Is the center-stage effect present in this data?

25.18 Police Harassment? Police may use minor violations such as not wearing a seat belt to stop motorists for other reasons. A large study in Michigan first studied the population of drivers not wearing seat belts during daylight hours by observation at more than 400 locations around the state. Here is the population distribution of seat belt violators by age group:[11]

Age Group	16 to 29	30 to 59	60 or Older
Proportion	0.328	0.594	0.078

The researchers then looked at court records and called a random sample of 803 drivers who had actually been cited by police for not wearing a seat belt. Here are the counts:

Age Group	16 to 29	30 to 59	60 or Older
Count	401	382	20

Does the age distribution of people cited differ significantly from the distribution of ages of all seat belt violators? Which age groups have the larger contributions to chi-square? Are these age groups cited more or less frequently than is justified? (The study found that males, Black drivers, and younger drivers were all overcited.)

4step

25.19 What's Your Sign? For reasons known only to social scientists, the General Social Survey (GSS) regularly asks its subjects their astrological sign. Here are the counts of responses for the 2014 GSS: SIGNS

Sign	Aries	Taurus	Gemini	Cancer	Leo	Virgo
Count	205	174	208	188	227	227
Sign	**Libra**	**Scorpio**	**Sagittarius**	**Capricorn**	**Aquarius**	**Pisces**
Count	212	210	198	182	173	198

If births are spread uniformly across the year, we expect all 12 signs to be equally likely. Are they? Follow the four-step process in your answer.

CHAPTER 25 SUMMARY

- The **chi-square test** for a two-way table tests the null hypothesis H_0 that there is no relationship between the row variable and the column variable. The alternative hypothesis H_a says that there is some relationship but does not say what kind. In a two-way table that classifies individuals from a single SRS into two categorical variables, the absence of a relationship indicates that the two categorical variables are independent, and the test is called the **chi-square test of independence**. If the two-way table is formed from independent SRSs from two or more populations, with each individual classified according to one categorical variable, the absence of a relationship implies that the populations are the same, and the test is called the **chi-square test of homogeneity**.

- The test compares the observed counts of observations in the cells of the table with the counts that would be expected if H_0 were true. The **expected count** in any cell is

$$\text{expected count} = \frac{\text{row total} \times \text{column total}}{\text{table total}}$$

- The **chi-square statistic** is

$$X^2 = \sum \frac{(\text{observed count} - \text{expected count})^2}{\text{expected count}}$$

- The chi-square test compares the value of the statistic X^2 with critical values from the **chi-square distribution** with $(r-1)(c-1)$ degrees of freedom. Large values of X^2 are evidence against H_0, so the *P*-value is the area under the chi-square density curve to the right of X^2.
- The chi-square distribution approximates the probability distribution of the statistic X^2. You can safely use this approximation when all expected cell counts are at least 1 and no more than 20% are less than 5.
- If the chi-square test finds a statistically significant relationship between the row and column variables in a two-way table, do data analysis to describe the nature of the relationship. You can do this by comparing well-chosen percentages, comparing the observed counts with the expected counts, and looking for the largest terms of the chi-square statistic.

CHAPTER 25 SKILLS REVIEW

Here are the most important skills you should have acquired from reading this chapter. Following each skill in parentheses is the section of the text where the topic is introduced.

A. Two-Way Tables

1. Understand that the data for a chi-square test must be presented as a two-way table of counts of outcomes. (25.1)
2. Recognize that a two-way table can occur from a single SRS with each individual classified according to two categorical variables or from independent SRSs from two or more populations with each individual classified according to one categorical variable. (25.6, 25.8)
3. Use percentages to describe the relationship between the row and column variables, starting from the counts in a two-way table. (25.1)

B. Interpreting Chi-Square Tests

1. Locate the chi-square statistic, its *P*-value, and other useful facts (row or column percentages, expected counts, terms of chi-square) in output from your software or calculator. (25.5)
2. Use the expected counts to check whether you can safely use the chi-square test. (25.3, 25.7)
3. Explain what null hypothesis the chi-square statistic tests in a specific two-way table. Understand the difference between tests of independence or homogeneity in a two-way table. (25.2, 25.8)
4. If the test is significant, compare percentages, compare observed with expected cell counts, or look for the largest terms of the chi-square statistic to see what deviations from the null hypothesis are most important. (25.5)

C. Doing Chi-Square Tests by Hand

1. Calculate the expected count for any cell from the observed counts in a two-way table. Check whether you can safely use the chi-square test. (25.3, 25.7)
2. Calculate the term of the chi-square statistic for any cell, as well as the overall statistic. (25.4)
3. Give the degrees of freedom of a chi-square statistic. Make a quick assessment of the significance of the statistic by comparing the observed value with the degrees of freedom. (25.6)
4. Use the chi-square critical values in Table D to approximate the *P*-value of a chi-square test. (25.6)

CHECK YOUR SKILLS

A 2017 Gallup Poll asked a sample of smokers several questions about their smoking habits and beliefs about smoking.[12] *Two of the questions asked on the survey were "In general, how harmful do you feel secondhand smoke is to adults–very harmful, somewhat harmful, not too harmful or not at all harmful?" and "How many packs of cigarettes do you smoke per day?" Here is a two-way table of the results:*

	Cigarettes Per Day		
	Less Than One Pack	One Pack	More Than One Pack
Very harmful	797	228	63
Somewhat harmful	777	446	120
Not too harmful	263	179	69
Not at all harmful	155	139	63

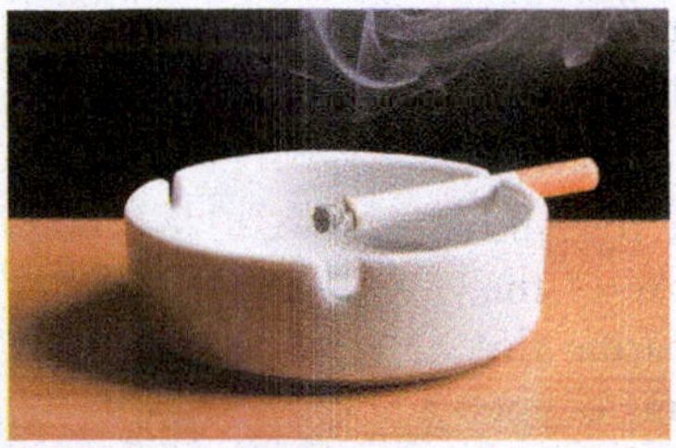
Peter Dazeley/Getty Images

25.20 The percentage of smokers in the sample who smoke more than one pack a day is SECHAND

(a) 32.98%.

(b) 9.55%.

(c) 1.91%.

25.21 The percentage of all the smokers in the sample who feel secondhand smoke is very harmful is

(a) 32.98%.

(b) 9.55%.

(c) 1.91%.

25.22 The percentage of all smokers who smoke more than one pack of cigarettes per day who feel secondhand smoke is very harmful is about

(a) 20.0%.

(b) 5.8%.

(c) 1.9%.

25.23 For this two-way table, we should do a chi-square test of

(a) homogeneity.

(b) independence.

(c) goodness of fit.

25.24 The expected count of smokers who smoke more than one pack of cigarettes per day who feel secondhand smoke is very harmful is about

(a) 103.9. (b) 315.0. (c) 1088.0.

25.25 The term in the chi-square statistic for the cell of smokers who smoke more than one pack of cigarettes per day who feel secondhand smoke is very harmful is about

(a) 103.9.

(b) 152.5.

(c) 16.1.

25.26 The degrees of freedom for the chi-square test for this two-way table are

(a) 6. (b) 8. (c) 12.

25.27 The null hypothesis for the chi-square test for this two-way table is

(a) the distributions of beliefs about the harmfulness of secondhand smoke are the same for less-than-one-pack-a-day smokers, one-pack-a-day smokers, and more-than-one-pack-a-day smokers.

(b) the more you smoke, the less harmful you believe secondhand smoke is.

(c) the distributions of beliefs about the harmfulness of secondhand smoke are different for less-than-one-pack-a-day smokers, one-pack-a-day smokers, and more-than-one-pack-a-day smokers.

25.28 The alternative hypothesis for the chi-square test for this two-way table is

(a) the distributions of beliefs about the harmfulness of secondhand smoke are the same for less-than-one-pack-a-day smokers, one-pack-a-day smokers, and more-than-one-pack-a-day smokers.

(b) the more you smoke, the less harmful you believe secondhand smoke is.

(c) the distributions of beliefs about the harmfulness of secondhand smoke are different for less-than-one-pack-a-day smokers, one-pack-a-day smokers, and more-than-one-pack-a-day smokers.

25.29 Software gives chi-square statistic $\chi^2 = 152.494$ for this table. From the table of critical values, we can say that the *P*-value is

(a) between 0.0025 and 0.001.

(b) between 0.001 and 0.0005.

(c) less than 0.0005.

25.30 We can conclude that

(a) heavy smoking causes you to believe secondhand smoke is not very harmful.

(b) believing that secondhand smoke is harmful causes people to smoke less.

(c) there is a relationship between how much people smoke and what they believe about the effects of secondhand smoke.

CHAPTER 25 EXERCISES

If you have access to software or a graphing calculator, use it to speed your analysis of the data in these exercises. Exercises 25.31 through 25.36 are suitable for hand calculation, if necessary.

25.31 **Smoking Cessation.** A large randomized trial was conducted to assess the efficacy of Chantix for smoking cessation compared with bupropion (more commonly known as Wellbutrin or Zyban) and a placebo. Chantix is different from most other quit-smoking products in that it targets nicotine receptors in the brain, attaches to them, and blocks nicotine from reaching them, whereas bupropion is an antidepressant often used to help people stop smoking. Generally healthy smokers who smoked at least 10 cigarettes per day were assigned at random to take Chantix ($n = 352$), bupropion ($n = 329$), or a placebo ($n = 344$). The study was double-blind, with the response measure being continuous absence from smoking for weeks 9–12 of the study. Here is a two-way table of the results:[13] SMOKE

	Treatment		
	Chantix	Bupropion	Placebo
No smoking in weeks 9–12	155	97	61
Smoked in weeks 9–12	197	232	283

(a) Give a 95% confidence interval for the difference between the proportions of smokers in the bupropion and placebo groups who did not smoke in weeks 9–12 of the study.

(b) What proportion of each of the three groups in the sample did not smoke in weeks 9–12 of the study? Are there statistically significant differences among these proportions? State hypotheses and give a test statistic and its *P*-value.

(c) In part (b), did you use a chi-square test of independence or homogeneity? Explain.

25.32 **Cell-only Versus Landline Users.** We suspect that people who rely entirely on cell phones will as a group be younger than those who have landline telephones. Do data confirm this guess? The Pew Research Center interviewed separate random samples of cell-only and landline telephone users and broke down the samples by age group:[14] CELLAGE

	Landline Sample	Cell-Only Sample
Age 18–29	335	374
Age 30–49	1242	347
Age 50–64	1625	146
Age 65 or older	1481	36
Total	4683	903

(a) Should you use a chi-square test of independence or homogeneity?

(b) Do a complete analysis of these data, following the four-step process.

25.33 **College Students Need Better Sleep!** A random sample of 871 students between the ages of 20 and 24 at a large midwestern university completed a survey including questions about their sleep quality, moods, academic performance, physical health, and psychoactive drug use. Sleep quality was measured using the Pittsburgh Sleep Quality Index (PSQI), with students scoring less than or equal to 5 on the index classified as optimal sleepers, those scoring a 6 or 7 classified as borderline, and those scoring over 7 classified as poor sleepers. The following table looks at the relationship between sleep quality classification and the use of over-the-counter (OTC) or prescription (Rx) stimulant medication more than once a month to help keep awake.[15] SLEEPQ

Use of OTC/Rx Meds to Wake > 1x/month	Sleep Quality on PSQI index		
	Optimal	Borderline	Poor
Yes	37	53	84
No	266	186	245

(a) Regarding this survey as approximately an SRS of college students, give a 95% confidence interval for the proportion of college students who use over-the-counter or prescription stimulant medication more than once a month to keep awake.

(b) Find the conditional distributions of sleep quality for students who use over-the-counter or prescription stimulant medication more than once a month to keep awake and those who don't. Make a graph that compares the two conditional distributions. Use your work to describe the overall relationship between students who use and don't use medications to keep awake and their sleep quality.

(c) Do students who use over-the-counter or prescription stimulant medication more than once a month to keep awake differ significantly in their quality of sleep from those who don't? State hypotheses, give the chi-square statistic and its *P*-value, and state your conclusion.

25.34 **Randomization in Action!** One hundred patients with moderate-to-severe knee osteoarthritis who were eligible for a total knee replacement were enrolled in a randomized controlled study. Fifty patients were assigned to receive a total knee replacement with a 12-week follow-up of nonsurgical treatment consisting of exercise, education, dietary advice, use of insoles, and pain medication. The other 50 patients received only the 12 weeks of nonsurgical treatment. One purpose of the randomization is for the two groups to have similar levels of knee osteoarthritis before undergoing the treatments. After the random assignment, we can compare the two treatment groups to see how well the randomization worked. KNEE_REP

The Kellgren-Lawrence score, which can take the values 0, 1, 2, 3, or 4, is based on a standard X-ray and assesses the severity of knee osteoarthritis, with 0 indicating no osteoarthritis and 4 indicating the most severe symptoms. If the randomization worked, then both treatment groups should have similar levels of osteoarthritis before undergoing the two treatments. Here are the initial Kellgren-Lawrence scores for the two treatment groups:[16]

	Kellgren-Lawrence Score		
Treatment Group	2	3	4
Nonsurgical treament group	5	21	24
Total-knee-replacement group	7	21	22

Do a chi-square test to see if there is a significant difference in the initial distribution of Kellgren-Lawrence scores for the two treatments. What do you find?

25.35 Comparing Two Proportions and the Chi-Square Test. The popularity of computer, video, online, and virtual reality games has raised concerns about their ability to negatively impact youth, although the existing research literature has been inconsistent. A recent survey of 14- to 18-year-olds in Connecticut high schools compared, among other things, aggressive behavior as evidenced by getting into serious fights for girls who have and have not played video games. When comparing two proportions such as these, the data can be presented in a two-way table containing the counts of successes and failures in both samples, with two rows and two columns. Here are the data presented in this manner:[17]

	Serious Fights	
	Yes	No
Played games	36	55
Never played games	578	1436

(a) Is there evidence that the proportions of all 14- to 18-year-old girls who played or have never played video games and have gotten into serious fights differ? Find the two sample proportions, the z statistic, and its P-value.

(b) Is there evidence that the proportions for 14- to 18-year-old girls who have or have not gotten into serious fights differ between those who have played or have never played video games? Find the chi-square statistic X^2 and its P-value.

(c) Show that (up to roundoff error) your X^2 is the same as z^2. The two P-values are also the same. These facts are always true, so you will often see chi-square for 2×2 tables used to compare two proportions.

(d) Suppose we were interested in finding out if the data gave good evidence that video gaming was associated with *increased* aggression in girls, as evidenced by getting into serious fights. Can we use the z test for this hypothesis? What about the X^2 test? What is the important difference between these two procedures?

25.36 Unhappy Rats and Tumors. Some people think that the attitude of cancer patients can influence the progress of their disease. We can't experiment with humans, but here is a rat experiment on this theme. Inject 60 rats with tumor cells and then divide them at random into two groups of 30. All the rats receive electric shocks, but rats in Group 1 can end the shock by pressing a lever. (Rats learn this sort of thing quickly.) The rats in Group 2 cannot control the shocks, which presumably makes them feel helpless and unhappy. We suspect that the rats in Group 1 will develop fewer tumors. The results: 11 of the Group 1 rats and 22 of the Group 2 rats developed tumors.[18]

(a) Make a two-way table of tumors by group. State the null and alternative hypotheses for this investigation.

(b) Although we have a two-way table, the chi-square test can't test a one-sided alternative. Carry out the z test and report your conclusion.

25.37 I Think I'll Be Rich by Age 30. 4step A sample survey asked young adults (females and males, ages 19–25), "What do you think are the chances you will have much more than a middle-class income at age 30?" The JMP output in Figure 25.8 shows the two-way table and related information, omitting a few subjects who refused to respond or who said they were already rich.[19] RICHBY

(a) Should you use a chi-square test of independence or homogeneity?

(b) Use the output as the basis for a discussion of the relationship between sex and a person's assessment of their chances of being rich by age 30.

25.38 Sexy Magazine Ads? 4step Look at full-page ads in magazines with young adult readership. Classify ads that show a model as "not sexual" or "sexual," depending on how the model is dressed (or not dressed). Here are data on 1509 ads in magazines aimed at young men, at young women, or at young adults in general:[20] SEXYADS

	Readers		
	Men	Women	General
Sexual	105	225	66
Not sexual	514	351	248

jmp

Contingency Table

Chance		Sex: Female	Male	
Count / Col% / Expected / Cell Chi^2				
AlmostNo	Count	96	98	194
	Col%	4.06	3.99	
	Expected	95.1508	98.8492	
	Cell Chi^2	0.0076	0.0073	
Some	Count	426	286	712
	Col%	18.00	11.63	
	Expected	349.213	362.787	
	Cell Chi^2	16.8842	16.2525	
5050	Count	696	720	1416
	Col%	29.40	29.28	
	Expected	694.503	721.497	
	Cell Chi^2	0.0032	0.0031	
Good	Count	663	758	1421
	Col%	28.01	30.83	
	Expected	696.955	724.045	
	Cell Chi^2	1.6543	1.5924	
NearCer	Count	486	597	1083
	Col%	20.53	24.28	
	Expected	531.177	551.823	
	Cell Chi^2	3.8424	3.6986	
		2367	2459	4826

Tests

N	DF	-LogLike	RSquare (U)
4826	4	22.067196	0.0066

Test	ChiSquare	Prob>ChiSq
Pearson	43.946	<.0001*

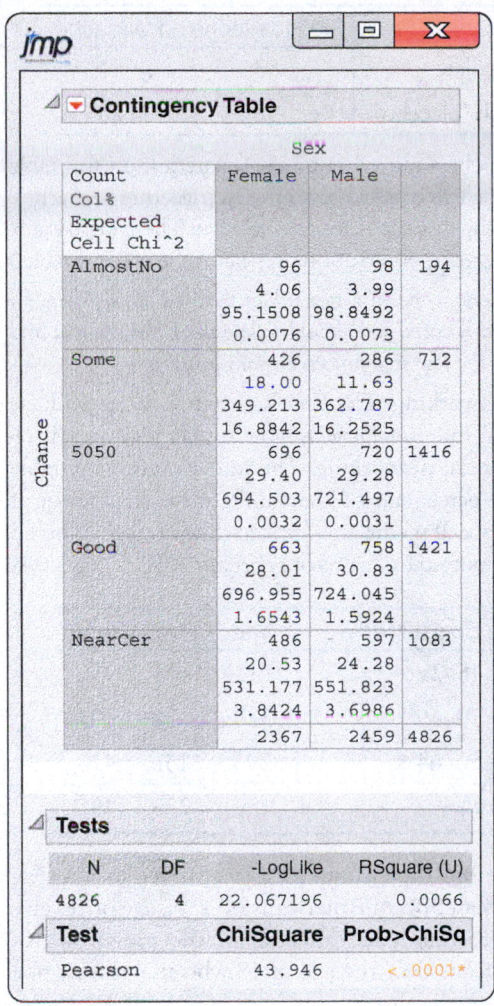

FIGURE 25.8
JMP output for the sample survey responses of Exercise 25.37.

Minitab

	Men	Women	General	All
Sexual	105	225	66	396
	16.96	39.06	21.02	26.24
	162.4	151.2	82.4	396.0
	20.312	36.074	3.265	*
Not sexual	514	351	248	1113
	83.04	60.94	78.98	73.76
	456.6	424.8	231.6	1113.0
	7.227	12.835	1.162	*
All	619	576	314	1509
	100.00	100.00	100.00	100.00
	619.0	576.0	314.0	1509.0
	*	*	*	*

Cell Contents: Count
% of Column
Expected count
Contribution to Chi-square

Pearson Chi-Square = 80.874, DF = 2, P-Value = 0.000

FIGURE 25.9
Minitab output for a study of ads in magazines, for Exercise 25.38.

Figure 25.9 displays Minitab chi-square output. Use the information in the output to describe the relationship between the target audience and the sexual content of ads in magazines for young adults.

Mistakes in using the chi-square test are unusually common. Exercises 25.39 through 25.42 illustrate several kinds of mistakes.

25.39 Sorry, No Chi-Square. An experimenter hid a toy from a dog behind either Screen A or Screen B. In the first phase the toy was always hidden behind Screen A, whereas in the second phase the toy was always hidden behind Screen B. Will the dog continue to look behind Screen A in the second phase? This was tried under three conditions. In the Social-Communicative condition, the experimenter communicated with the dog by establishing eye contact and addressing the dog while hiding the toy; in the Noncommunicative condition, the toy was hidden without communication; in the Nonsocial condition, the toy was dragged by a string to be hidden without any interaction from the experimenter. There were 12 dogs assigned at random to each condition, and the dog had up to three trials to find the toy hidden behind Screen B in phase 2. An error occurred if the dog continued to search behind Screen A, with the number of errors ranging from zero if the dog found the toy behind Screen B on the initial trial up to three if the dog never correctly chose Screen B. Here are the data:[21] HIDETOY

	Number of Errors			
	0	1	2	3
Social-Communicative	0	3	3	6
Noncommunicative	5	3	1	3
Nonsocial	8	2	2	0

(a) The data do show a difference in the number of errors for the different conditions. Show this by comparing suitable percentages.

(b) The researchers used a more complicated but exact procedure rather than chi-square to assess significance for these data. Why can't the chi-square test be trusted in this case?

(c) If you use software, does the chi-square output for these data warn you against using the test?

25.40 Sorry, No Chi-Square. How do U.S. residents who travel overseas for leisure differ from those who travel for business? Here is the breakdown by occupation:[22]

	Leisure Travelers	Business Travelers
Professional/technical	36%	39%
Manager/executive	23%	48%
Retired	14%	3%
Student	7%	3%
Other	20%	7%
Total	100%	100%

Explain why we don't have enough information to use the chi-square test to learn whether these two distributions differ significantly.

25.41 Sorry, No Chi-Square. Using a national random sample from the NEXT Generation Health Study, researchers compared alcohol- and drug-impaired driving between high school seniors with post-high school drivers. The sample included 2407 seniors and 2178 subjects one year post-high school. All subjects in the study completed a questionnaire that included several questions related to recent alcohol and drug use. The following table reports the incidence of several of these behaviors for the two groups. (Binge drinking refers to at least four (for women)/five (for male) drinks within two hours.)[23] Explain why it is not correct to use a chi-square test on this table to compare the "high school seniors" and "post-high school graduate" groups.

	High School Seniors	Post-High School Graduates
Alcohol drinking at least once (last 30 days)	787	1044
Binge drinking at least once (last 30 days)	402	583
Used marijuana at least once (last year)	583	654
Used illicit drugs other than marijuana at least once (last year)	157	162

25.42 Sorry, No Chi-Square. Does eating chocolate trigger headaches? To find out, women with chronic headaches followed the same diet except for eating chocolate bars and carob bars that looked and tasted the same. Each subject ate both chocolate and carob bars in random order with at least three days between. Each woman then reported whether or not she had a headache within 12 hours of eating the bar. Here is a two-way table of the results for the 64 subjects:[24]

	No Headache	Headache
Chocolate	53	11
Carob (placebo)	38	26

The researchers carried out a chi-square test on this table to see if the two types of bar differ in triggering headaches. Explain why this test is incorrect. (*Hint:* There are 64 subjects. How many observations appear in the two-way table?)

The remaining exercises concern larger tables that require software for easy analysis. In many cases, you should follow the **Plan, Solve,** *and* **Conclude** *steps of the four-step process in your answers.*

25.43 Social Networking. The Pew Research Center conducts an annual Internet Project, which includes research related to social networking. The following two-way table about the percentage of U.S. adults surveyed who use at least one social media site, broken down by age, is based on data reported by Pew as of February 2019:[25] MEDIA

Use Social Media	Yes	No	Total
Age 18–29	212	24	236
Age 30–49	324	71	395
Age 50–64	293	131	424
Age 65 +	156	235	391

(a) Regarding the Internet Project survey as approximately an SRS of Americans over the age of 18, give a 99% confidence interval for the proportion of Americans over the age of 18 who use at least one social media site.

(b) Compare the conditional distributions of age for those who use and those who don't use at least one social media site with both a table and a graph. What are the most important differences?

(c) Carry out the chi-square test for the hypothesis of no difference between the age distributions of U.S. adults who use and those who don't use at least one social media site. What would be the mean of the test statistic if the null hypothesis were true? The value of the statistic is so far above this mean that you can see at once that it must be highly significant. What is the approximate *P*-value?

(d) Look at the terms of the chi-square statistic and compare observed and expected counts in the cells that contribute the most to chi-square. Based on this and your findings in part (b), write a short comparison of the age distributions of U.S. adults who use and don't use at least one social media site.

25.44 Who Goes to Religious Services? The General Social Survey (GSS) asked this question: "Have you attended religious services in the last week?" Here are the responses for those whose highest degree was high school or above: SERVICES

	Highest Degree Held			
	High School	Junior College	Bachelor	Graduate
Attended services	400	62	146	76
Did not attend services	880	101	232	105

(a) Carry out the chi-square test for the hypothesis of no relationship between the highest degree attained and attendance at religious services in the last week. What do you conclude?

(b) Make a 2×3 table by omitting the column corresponding to those whose highest degree was high school. Carry out the chi-square test for the hypothesis of no relationship between the type of advanced degree attained and attendance at religious services in the last week. What do you conclude?

(c) Make a 2×2 table by combining the counts in the three columns that have a highest degree beyond high school, so that you are comparing adults whose highest degree was high school against those whose highest degree was beyond high school. Carry out the chi-square test for the hypothesis of no relationship between attaining a degree beyond high school and attendance at religious services for this 2×2 table. What do you conclude?

(d) Using the results from these three chi-square tests, write a short report explaining the relationship between attendance at religious services in the last week and the highest degree attained. As part of your report, you should give the percentages who attended religious services for each of the four highest degrees.

25.45 More on Peer Victimization Are there differences in the frequency of being bullied between males and females? In the ALSPAC study described in Exercise 25.3, data on frequency of peer victimization for 6719 children at age 13 included the subject's sex. Here are the results: BULLYSEX

	Frequency of Being Bullied		
	Never	Occasionally	Frequently
Male	1564	1149	571
Female	1526	1281	629

Describe the most important differences between frequency of being bullied for the two sexes. Is there a significant overall difference in the distribution of frequency of being bullied for males and females?

25.46 How Are Schools Doing? The nonprofit group Public Agenda conducted telephone interviews with a stratified sample of parents of high school children. There were 202 Black parents, 202 Hispanic parents, and 201 White parents. One question asked was, "Are the high schools in your state doing an excellent, good, fair, or poor job, or don't you know enough to say?" Here are the survey results:[26] HIGHSCHL

	Black Parents	Hispanic Parents	White Parents
Excellent	12	34	22
Good	69	55	81
Fair	75	61	60
Poor	24	24	24
Don't know	22	28	14
Total	202	202	201

Are the differences in the distributions of responses for the three groups of parents statistically significant? What departures from the null hypothesis "no relationship between group and response" contribute most to the value of the chi-square statistic? Write a brief conclusion based on your analysis.

25.47 Chronic Kidney Disease and Hearing Loss. Although chronic kidney disease (CKD) has been linked to hearing loss in several specific disease syndromes, the investigation of whether a relationship exists between CKD and hearing loss in the general population has been limited to several small studies. The participants in this larger study consist of a sample of 2564 adults in Australia aged 50 and older. Kidney function was measured using eGFR, which measures the kidney filtration rate and can be determined from a blood test. An eGFR below 60 represents moderate CKD, while values over 90 are in the normal range. Hearing loss was taken to be any measurable hearing loss in an audiometric exam. Here are the results:[27] KIDNEY

Huntstock/Getty Images

	eGFR				
	< 45	45–60	60–75	75–90	≥ 90
Hearing loss	76	203	363	147	71
No hearing loss	27	207	717	458	295

(a) Should you use a chi-square test of independence or homogeneity?

(b) Is there a relationship between CKD and hearing loss? State hypotheses, give the chi-square statistic and its *P*-value, and state your conclusion in context. (*Hint:* You may want to use the proportion of subjects with hearing loss in each eGFR category to describe the relationship when writing your conclusion.)

(c) This is an observational study. Here are the average ages of subjects in each eGFR category:

	eGFR				
	< 45	45–60	60–75	75–90	≥ 90
Average age	78.3	73.3	68.3	64.3	61.0

What type of variable is age? Explain in a simple sentence or two how age could be used to explain the relationship between CKD and hearing loss.

(d) Researchers adjusted for the variable age when analyzing the relationship between CKD and hearing loss, a common practice when analyzing observational data. Although the details are fairly involved, here is a rough idea of the meaning of adjusting for age. Suppose we created a separate two-way table of eGFR and hearing loss for 50-year-olds, 51-year-olds, and so on and found that increased CKD is associated with hearing loss in each table. Explain why you could then no longer use age to explain the relationship between CKD and hearing loss.

25.48 Children at the Monterey Bay Aquarium. The Monterey Bay Aquarium, founded in 1984, is situated on the beautiful coast of Monterey Bay in the historic Cannery Row district. In 1985, the aquarium began a survey program that involved randomly sampling visitors as they exit for the day. The survey included visitor demographic information, use of social media, and opinions on their aquarium visit. For each visitor sampled during 2013–2015, here is the distribution of the number of children in their group[28]: AQUARIUM

	Year		
Number of Children	2013	2014	2015
0	1855	1751	1998
1	585	636	591
2	599	601	483
3 or more	515	506	289

Is there a significant difference in the distribution of the number of children in the group over this three-year period? If so, describe how the distribution has changed.

Political Views. *The General Social Survey (GSS) asked its 2018 sample "Do you usually think of yourself as a liberal, moderate, conservative, or don't know?" The GSS is essentially an SRS of American adults. Here is a large two-way table breaking down the responses by the highest educational degree the subject held:*

	Less Than High School	High School	College or More
Liberal	69	279	317
Moderate	103	477	291
Conservative	68	381	291
Don't know	22	41	9

Exercises 25.49 and 29.50 are based on this table.

25.49 Conservatives. Give a 95% confidence interval for the proportion of adults who said they are conservative. POLIT

25.50 Political Views and Educational Attainment. Use the table to analyze the differences in political views among levels of education. The sample is so large that the differences are bound to be highly significant but give the chi-square statistic and its *P*-value nonetheless. The main challenge is in seeing what the data say. POLIT

CHAPTER

26

Inference for Regression

When you complete this chapter, you will be able to:

26.1 Interpret regression inference as the process of estimating the population parameters α, β, and σ (the standard deviation of y).

26.2 Calculate the regression standard error of a least-squares regression line.

26.3 Interpret computer and calculator outputs of regression statistics.

26.4 Test the hypothesis of no linear relationship between two quantitative variables.

26.5 Use the test for zero regression slope to test the hypothesis of zero correlation between variables.

26.6 Calculate confidence intervals for regression slope.

26.7 Explain the distinction between a confidence interval for the mean response and a prediction interval for an individual response and calculate both intervals.

26.8 Use graphs of residuals to check conditions of regression inference.

In Chapters 4 and 5, we studied scatterplots, correlation, and the least-squares regression line as methods for exploring data. In Chapter 5, we mentioned that we must exercise caution in how we interpret any relationships we observe through such exploratory analyses. We also mentioned that such interpretations rest on the assumption that the relationship is valid in some broader sense. And we promised to explore this more carefully later in this book. In this chapter, we do so by considering inference for regression. Inference for regression allows us to determine whether the relationship we observe in a scatterplot is valid for some larger population.

When a scatterplot shows a linear relationship between a quantitative explanatory variable x and a quantitative response variable y, we can use the least-squares line fitted to the data to predict y for a given value of x. When the data are a sample from a larger population, we need statistical inference to answer questions like these about the population:

- Is there really a linear relationship between x and y in the population, or might the pattern we see in the scatterplot plausibly arise just by chance?
- What is the slope (rate of change) that relates y to x in the population, including a margin of error for our estimate of the slope?
- If we use the least-squares line to predict y for a given value of x, how accurate is our prediction (again, with a margin of error)?

This chapter shows you how to answer these questions. Here is an example we will explore.

Galina Mikhalishina/Alamy

EXAMPLE 26.1 Sea Surface Temperatures and Coral Growth

STATE: Environmental conditions can affect the growth of coral. To study this, a researcher examined a species of coral that is found in the Caribbean Sea and the Gulf of Mexico. At 12 localities, he determined the average annual calcification rate of coral over a period of several years and the average annual maximum sea surface temperature during the same period. Calcification rate affects the growth of coral, with higher rates corresponding to greater growth. Table 26.1 contains data for these 12 localities.[1] How does calcification rate (in grams per square centimeter per year) change with changes in annual maximum sea surface temperature?

TABLE 26.1 Maximum sea surface temperature (°C) and calcification rate (g cm^{-2} yr^{-1})

Maximum Sea Surface Temperature	Calcification Rate	Maximum Sea Surface Temperature	Calcification Rate
29.4	1.48	29.7	1.63
29.4	1.53	29.5	1.53
29.4	1.52	29.4	1.46
29.6	1.48	29.0	1.24
29.1	1.31	29.0	1.29
28.7	1.25	29.0	1.12

PLAN: Make a scatterplot. If the relationship appears linear, use correlation and regression to describe it. Finally, ask whether the linear relationship between annual maximum sea surface temperature and calcification rate is too strong to simply be due to chance. (In other words, ask if there is a *statistically significant* linear relationship.)

SOLVE (first steps): Chapters 4 and 5 introduced the data analysis that must come before inference. The first steps we take are a review of this data analysis. Figure 26.1 is a *scatterplot* of the coral data. Plot the explanatory variable (annual maximum sea surface temperature) horizontally and the response variable (calcification rate) vertically. Look for the form, direction, and strength of the relationship as well as for outliers or other deviations. There is a moderately strong positive linear relationship, with no extreme outliers or potentially influential observations.

Because the scatterplot shows a roughly linear (straight-line) pattern, the *correlation* describes the direction and strength of the relationship. The correlation between annual maximum sea surface temperature and calcification rate is $r = 0.892$. We are

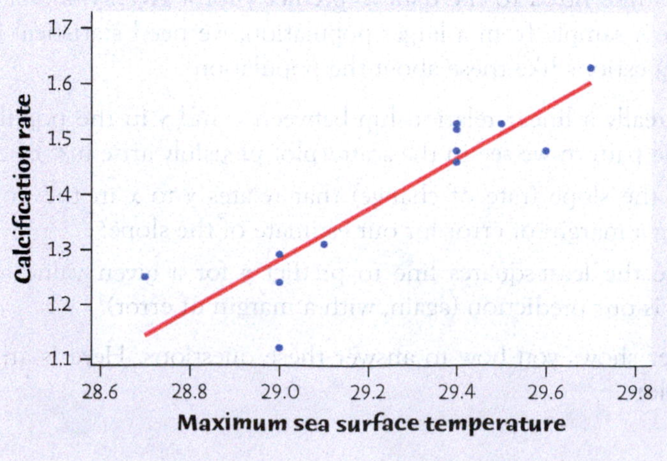

FIGURE 26.1 Scatterplot of the calcification rates of coral against the maximum sea surface temperature, with the least-squares regression line, for Example 26.1.

interested in predicting the response from information about the explanatory variable. So we find the *least-squares regression line* for predicting calcification rate from annual maximum sea surface temperature. The equation of the regression line is

$$\hat{y} = a + bx$$
$$= -12.103 + 0.4615x$$

or,

$$\text{calcification rate} = -12.103 + 0.4615 \times \text{maximum sea surface temperature}$$

CONCLUDE (first steps): Coral in regions with higher annual maximum sea surface temperature tend to have higher rates of calcification. The squared correlation $r^2 = 0.796$ means that 79.6% of the variation in calcification rates is explained by annual maximum sea surface temperature. Thus, prediction of calcification rate will be moderately accurate. Is this observed relationship statistically significant? After all, it is based on only 12 localities. We must now develop tools for inference in the regression setting.

26.1 Conditions for Regression Inference

We can fit a regression line to *any* data relating two quantitative variables, but the results are useful only if the scatterplot shows a linear pattern. Statistical inference requires more detailed conditions. Because the conclusions of inference always concern some *population*, the conditions describe the population and how the data are produced from it. The slope b and intercept a of the least-squares line are *statistics*. That is, we calculated them from the sample data. In the sea surface temperatures and coral growth study, for instance, these statistics would take somewhat different values if we repeated the study with different localities. To do inference, think of a and b as estimates of unknown *parameters* that describe the population of all interest.

Conditions for Regression Inference

We have n observations on an explanatory variable x and a response variable y. Our goal is to study or predict the behavior of y for given values of x.

- For any fixed value of x, the response y varies according to a Normal distribution. Repeated responses y are independent of each other.
- The mean response μ_y has a straight-line relationship with x given by a **population regression line**

$$\mu_y = \alpha + \beta x$$

  The slope β and intercept α are unknown parameters.
- The standard deviation of y (call it σ) is the same for all values of x. The value of σ is unknown.

There are thus three population parameters that we must estimate from the data: α, β, and σ.

These conditions say that in the population there is an "on the average" straight-line relationship between y and x. The population regression line $\mu_y = \alpha + \beta x$ says that the *mean* response μ_y moves along a straight line as the explanatory variable x changes. We can't observe the population regression line. The values of y that we do observe vary about their means according to a Normal distribution. If we hold x fixed and take many observations on y, the Normal pattern will eventually appear

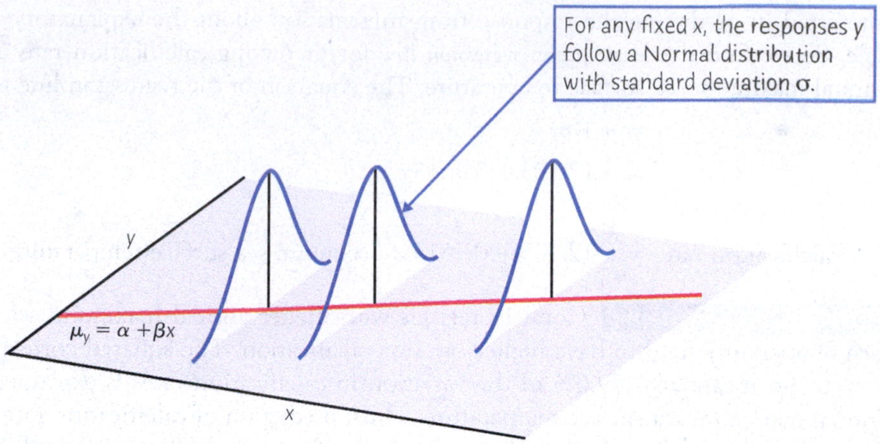
For any fixed x, the responses y follow a Normal distribution with standard deviation σ.
y
$\mu_y = \alpha + \beta x$
x

FIGURE 26.2 The nature of regression data when the conditions for inference are met. The line is the population regression line, which shows how the mean response μ_y changes as the explanatory variable x changes. For any fixed value of x, the observed response y varies according to a Normal distribution having mean μ_y and standard deviation σ.

in a stemplot or histogram. In practice, we observe y for many different values of x, so that we see an overall linear pattern formed by points scattered about the population line. The standard deviation σ determines whether the points fall close to the population regression line (small σ) or are widely scattered (large σ).

Figure 26.2 shows the conditions for regression inference in picture form. The line in the figure is the population regression line. The mean of the response y moves along this line as the explanatory variable x takes different values. The Normal curves show how y will vary when x is held fixed at different values. All the curves have the same σ, so the variability of y is the same for all values of x. You should check the conditions for inference when you do inference about regression. We will see later how to do that.

26.2 Estimating the Parameters

The first step in inference is to estimate the unknown parameters α, β, and σ.

Estimating the Population Regression Line

When the conditions for regression are met and we calculate the least-squares line $\hat{y} = a + bx$, the slope b of the least-squares line is an unbiased estimator of the population slope β, and the intercept a of the least-squares line is an unbiased estimator of the population intercept α.

EXAMPLE 26.2 Sea Surface Temperatures and Coral Growth: Slope and Intercept

NOAA

The data in Figure 26.1 satisfy the condition of scatter about an invisible population regression line reasonably well. The least-squares line is $\hat{y} = -12.103 + 0.4615x$. The slope is particularly important. *A slope is a rate of change.* The population slope β says how much higher the average annual calcification rate is for this species of coral where the average annual maximum sea surface temperature is 1°C higher. Because $b = 0.4615$ estimates the unknown β, we estimate that, on the average, annual calcification rate for this species of coral is about 0.46 g cm^{-2} yr^{-1} higher for each additional 1°C in average annual maximum sea surface temperature.

As discussed in Chapter 5 (page 126), the intercept is the value of average annual calcification rate when average annual maximum sea surface temperature is 0°C.

Here we need the intercept $a = -12.103$ to draw the line, but it has no statistical interest in this example. No average annual maximum sea surface temperature was below 28.7°C, so we have no data near $x = 0$. We know that sea surface temperatures in the Caribbean and Gulf of Mexico will never reach anything approaching 0°C, so that we will never observe $x = 0$.

The remaining parameter is the standard deviation σ, which describes the variability of the response y about the population regression line. The least-squares line estimates the population regression line. So the residuals estimate how much y varies about the population line. Recall from Chapter 5 that the *residuals* are the vertical deviations of the data points from the least-squares line:

$$\text{residual} = \text{observed } y - \text{predicted } y$$
$$= y - \hat{y}$$

There are n residuals, one for each data point. Because σ is the standard deviation of responses about the population regression line, we estimate it by a sample standard deviation of the residuals. We call this sample standard deviation the *regression standard error* to emphasize that it is estimated from data. The residuals from a least-squares line always have mean zero. That simplifies their standard error.

Regression Standard Error

The **regression standard error** is

$$s = \sqrt{\frac{1}{n-2}\sum \text{residual}^2}$$
$$= \sqrt{\frac{1}{n-2}\sum (y - \hat{y})^2}$$

Use s to estimate the standard deviation σ of responses about the means given by the population regression line.

Because we use the regression standard error so often, we just call it s. The quantity $\sum (y - \hat{y})^2$ is the sum of the squared deviations of the data points from the line. We average the squared deviations by dividing by $n - 2$, the number of data points less 2. It turns out that if we know $n - 2$ of the n residuals, the other two are determined. That is, $n - 2$ are the *degrees of freedom* of s. We first met the idea of degrees of freedom in the case of the ordinary sample standard deviation of n observations, which has $n - 1$ degrees of freedom. It is sometimes said that we lose 1 degree of freedom for each parameter we must estimate to compute the standard error. To compute the standard deviation, we must estimate the mean and lose 1 degree of freedom. Now we must estimate two parameters, both the intercept and slope, to compute $\hat{y}$ and then lose 2 degrees of freedom, so the proper degrees of freedom are $n - 2$ rather than $n - 1$.

Calculating s is unpleasant. You must find the predicted response for each x in your data set, and then the residuals, and then s. In practice, you will use software that does this arithmetic instantly. Nonetheless, here is an example to help you understand the standard error s.

STATISTICS IN YOUR WORLD

The Jinx!

Athletes are often jinxed. We read of "the rookie of the year jinx," the "cover of *Sports Illustrated* jinx," and many others. That is, athletes who are recognized for an outstanding performance often fail to do as well in the future. No, nature isn't retaliating against them. It's just random variation about their long-term mean performance. They were recognized because they randomly varied above their typical performance, and in the future, they return to the mean or randomly vary down from it. If they randomly vary down, they can hope for a "comeback" award the next year.

EXAMPLE 26.3 Sea Surface Temperatures and Coral Growth: Residuals and Standard Error

CORALR

Table 26.1 shows that the first locality studied had an average annual maximum sea surface temperature of 29.4°C and a calcification rate of 1.48 g cm^{-2} yr^{-1}. The predicted calcification rate for $x = 29.4$ is

$$\hat{y} = -12.103 + 0.4615x$$
$$= -12.103 + 0.4615(29.4) = 1.4651$$

The residual for this observation is

$$\text{residual} = y - \hat{y}$$
$$= 1.48 - 1.4651 = 0.0149$$

That is, the observed calcification rate for this locality lies 0.0149 g cm^{-2} yr^{-1} above the least-squares line on the scatterplot, and the least-squares line underestimates the calcification rate. In general, a positive residual occurs when the least-squares line underestimates the actual response y.

Repeat this calculation 11 more times, once for each locality. The 12 residuals are

0.01490	0.06490	0.05490	−0.07740	−0.01665	0.10795
0.02645	0.01875	−0.00510	−0.04050	0.00950	−0.16050

Check the calculations by verifying that the sum of the residuals is zero. It is −0.0028—not quite zero—because of roundoff error. Another reason to use software in regression is that roundoff errors in hand calculation can accumulate to make the results inaccurate.

The regression standard error is

$$s = \sqrt{\frac{1}{n-2}\sum \text{residual}^2}$$
$$= \sqrt{\frac{1}{12-2}[(0.01490)^2 + (0.06490)^2 + \cdots + (-0.16050)^2]}$$
$$= \sqrt{\frac{1}{10}(0.05394)}$$
$$= \sqrt{0.005394} = 0.07344$$

We will study several kinds of inference in the regression setting. The regression standard error s is the key measure of the variability of the responses in regression. It is part of the standard error of all the statistics we will use for inference.

APPLY YOUR KNOWLEDGE

© Glowimages RM/age fotostock

26.1 Wine and Cancer in Women. Some studies have suggested that a nightly glass of wine may not only take the edge off a day but also improve health. Is wine good for your health? A study of nearly 1.3 million middle-aged British women examined wine consumption and the relative risk of breast cancer. The relative risk is the proportion of those in the study who drank a given amount of wine and who developed breast cancer divided by the proportion of nondrinkers in the study who developed breast cancer. For example, if 10% of the women in the study who drank 10 grams of wine developed breast cancer and 9% of nondrinkers in the study developed breast cancer, the relative risk of breast cancer for women drinking 10 grams of wine per day would be 10%/9% = 1.11. A relative risk greater than 1 indicates a greater proportion of drinkers in the study developed breast cancer than nondrinkers. Wine intake is the mean wine intake, in grams per day, of all women in the study who drank some wine but less than or equal to two drinks per week, who drank between three and six drinks per week, who drank between seven and 14 drinks per week, and who drank 15 or more drinks per week. Here are the data (for drinkers only):[2] CANCER

Average wine intake (grams per day) (x)	2.5	8.5	15.5	26.5
Relative risk (y)	1.00	1.08	1.15	1.22

(a) Examine the data. Make a scatterplot with wine intake as the explanatory variable and find the correlation. There is a strong linear relationship.

(b) Explain in words what the slope β of the population regression line would tell us if we knew it. Note that these data represent averages over large numbers of women and are an example of an ecological correlation (see page 143), and one must be careful not to interpret the data as applying to individuals. Based on the data, what are the estimates of β and the intercept α of the population regression line?

(c) Calculate by hand the residuals for the four data points. Check that their sum is 0 (up to roundoff error). Use the residuals to estimate the standard deviation σ that measures variation in the responses (relative risk) about the means given by the population regression line. You have now estimated all three parameters.

26.3 Examples of Technology

Basic "two-variable statistics" calculators will find the slope b and intercept a of the least-squares line from keyed-in data. Inference about regression requires in addition the regression standard error s. At this point, software or a graphing calculator that includes procedures for regression inference becomes almost essential for practical work.

Figure 26.3 shows regression output for the data of Table 26.1 from a graphing calculator, three statistical programs, and a spreadsheet program. When we entered the data into the programs, we called the explanatory variable "Maximum sea surface temperature." The software outputs use that label. The graphing calculator just uses "x" and "y" to label the explanatory and response variables. You can locate the basic information in all the outputs. The regression slope is $b = 0.4615$, and the regression intercept is $a = -12.1028$. The equation of the least-squares line is, therefore (after rounding), just as given in Example 26.1. The regression standard error is $s = 0.0734$, and the squared correlation is $r^2 = 0.7957$. Both of these results reflect the rather moderate scatter of the points in Figure 26.1 (page 588) about the least-squares line.

Each output contains other information, some of which we will need shortly and some of which we don't need. In fact, we left out some output to save space. Once you know what to look for, you can find what you want in almost any output and ignore what doesn't interest you.

FIGURE 26.4
JMP output for the introspective ability data, for Exercise 26.2.

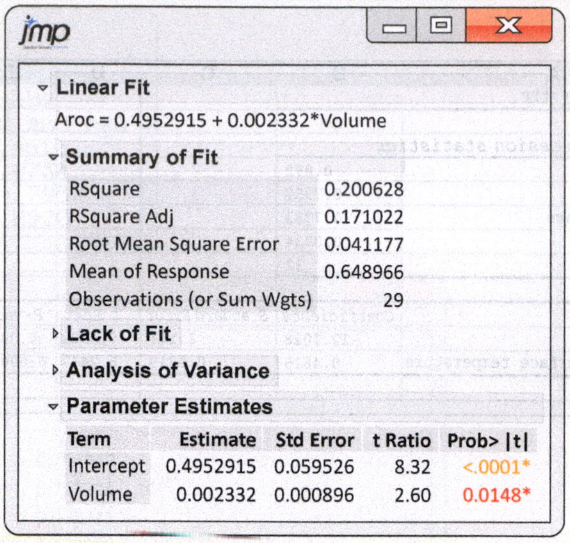
jmp
Linear Fit
Aroc = 0.4952915 + 0.002332*Volume
Summary of Fit
RSquare 0.200628
RSquare Adj 0.171022
Root Mean Square Error 0.041177
Mean of Response 0.648966
Observations (or Sum Wgts) 29
Lack of Fit
Analysis of Variance
Parameter Estimates
Term Estimate Std Error t Ratio Prob>|t|
Intercept 0.4952915 0.059526 8.32 <.0001*
Volume 0.002332 0.000896 2.60 0.0148*

TABLE 26.2 Arctic river discharge (in cubic kilometers), 1936 to 2017

Year	Discharge	Year	Discharge	Year	Discharge	Year	Discharge
1936	1721	1957	1762	1978	2008	1999	1970
1937	1713	1958	1936	1979	1970	2000	1905
1938	1860	1959	1906	1980	1758	2001	1890
1939	1739	1960	1736	1981	1774	2002	2085
1940	1615	1961	1970	1982	1728	2003	1780
1941	1838	1962	1849	1983	1920	2004	1900
1942	1762	1963	1774	1984	1823	2005	1930
1943	1709	1964	1606	1985	1822	2006	1910
1944	1921	1965	1735	1986	1860	2007	2270
1945	1581	1966	1883	1987	1732	2008	2078
1946	1834	1967	1642	1988	1906	2009	1900
1947	1890	1968	1713	1989	1932	2010	1813
1948	1898	1969	1742	1990	1861	2011	1859
1949	1958	1970	1751	1991	1801	2012	1702
1950	1830	1971	1879	1992	1793	2013	1758
1951	1864	1972	1736	1993	1845	2014	1988
1952	1829	1973	1861	1994	1902	2015	2042
1953	1652	1974	2000	1995	1842	2016	1863
1954	1589	1975	1928	1996	1849	2017	1988
1955	1656	1976	1653	1997	2007		
1956	1721	1977	1698	1998	1903		

(a) Make a scatterplot of river discharge against time. Is there a clear increasing trend? Calculate r^2 and briefly interpret its value. There is considerable year-to-year variation, so we wonder if the trend is statistically significant.

(b) Find the least-squares line and draw it on your plot. Then find the regression standard error s, which measures scatter about this line. We will continue the analysis in later exercises.

26.4 Testing the Hypothesis of No Linear Relationship

Example 26.1 asked, "How does calcification rate change with changes in annual maximum sea surface temperature?" Is the association statistically significant? That is, is it too strong to occur just by chance? To answer this question, test hypotheses about the slope β of the population regression line:

$$H_0\colon \beta = 0$$
$$H_a\colon \beta \neq 0$$

A regression line with slope 0 is horizontal. That is, the mean of y does not change at all when x changes. So H_0 says that there is *no linear relationship* between x and y in the population. Put another way, H_0 says that *linear regression of y on x is of no value for predicting y*.

The test statistic is just the standardized version of the least-squares slope b, using the hypothesized value $\beta = 0$ for the mean of b. It is another t statistic. Here are the details.

Significance Test for Regression Slope

To test the hypothesis $H_0\colon \beta = 0$, compute the t statistic

$$t = \frac{b}{\mathrm{SE}_b}$$

In this formula, the standard error of the least-squares slope b is

$$\mathrm{SE}_b = \frac{s}{\sqrt{\sum (x - \bar{x})^2}}$$

The sum runs over all observations on the explanatory variable x. In terms of a random variable T having the t_{n-2} distribution, the P-value for a test of H_0 against

$H_a\colon \beta > 0$ is $P(T \geq t)$

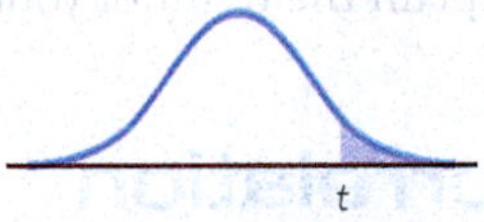

$H_a\colon \beta < 0$ is $P(T \leq t)$

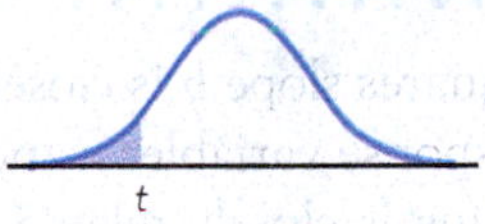

$H_a\colon \beta \neq 0$ is $2P(T \geq |t|)$

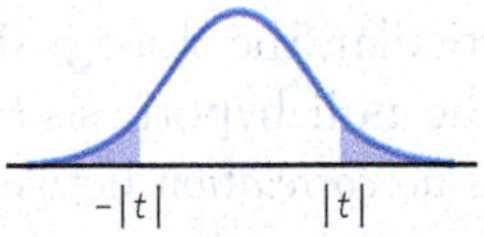

As advertised, the standard error of b is a multiple of the regression standard error s. The degrees of freedom $n - 2$ are the degrees of freedom of s. Although we give the formula for this standard error, you should not try to calculate it by hand. Regression software gives the standard error SE_b along with b itself.

EXAMPLE 26.4 Sea Surface Temperatures and Coral Growth: Is the Relationship Significant?

The hypothesis $H_0\colon \beta = 0$ says that average annual maximum sea surface temperature has no straight-line relationship with calcification rate. We conjecture only that there is a relationship, so we use the two-sided alternative $H_a\colon \beta \neq 0$.

Figure 26.1 shows that there is a positive relationship, and from Figure 26.3 (the software outputs, page 594) we see that $b = 0.46149$ and $SE_b = 0.07394$. Thus,

$$t = \frac{b}{SE_b} = \frac{0.46149}{0.07394} = 6.24$$

so it is not surprising that all the outputs in Figure 26.3 give $t = 6.24$ with two-sided P-value < 0.0001. There is very strong evidence that calcification rate changes as the average annual maximum sea surface temperature increases.

APPLY YOUR KNOWLEDGE

26.4 Wine and Cancer in Women (continued). Exercise 26.1 (page 592) gives data on daily wine consumption and the relative risk of breast cancer in women. Software tells us that the least-squares slope is $b = 0.009012$ with standard error $SE_b = 0.001112$.

(a) What is the t statistic for testing $H_0: \beta = 0$?

(b) How many degrees of freedom does t have? Use Table C to approximate the P-value of t against the one-sided alternative $H_a: \beta > 0$. What do you conclude?

26.5 Great Arctic Rivers: Testing. The most important question we ask of the data in Table 26.2 is this: Is the increasing trend visible in your plot (Exercise 26.3) statistically significant? If so, some climatic trend, such as rising temperatures, may be driving the increase in discharge. Use software to answer this question. Give a test statistic, its P-value, and the conclusion you draw from the test. ARCTIC

26.6 Does Fast Driving Waste Fuel? Exercise 4.8 (page 106) gives data on the fuel consumption of a 2013 Volkswagen Jetta Diesel at various speeds from 20 to 80 miles per hour. Is there statistically significant evidence of straight-line dependence between speed and fuel use? Make a scatterplot and use it to explain the result of your test. FASTDR

26.5 Testing Lack of Correlation

The least-squares slope b is closely related to the correlation r between the explanatory and response variables x and y. In the same way, the slope β of the population regression line is closely related to the correlation between x and y in the population. In particular, the slope is 0 exactly when the correlation is 0.

Testing the null hypothesis $H_0: \beta = 0$ is, therefore, exactly the same as testing that there is *no correlation* between x and y in the population from which we drew our data. You can use the test for zero slope to test the hypothesis of zero correlation between any two quantitative variables. That's a useful trick.

Because correlation also makes sense when there is no explanatory-response distinction, it is handy to be able to test correlation without doing regression. Table E in the back of the book gives critical values of the sample correlation r under the null hypothesis that the correlation is 0 in the population. Use this table when both variables have at least approximately Normal distributions or when the sample size is large.

EXAMPLE 26.5 Testing Lack of Correlation

Figure 26.5 displays two scatterplots that we will use to illustrate testing lack of correlation and also to illustrate once again the need for formal statistical tests. First are data from an experiment on the healing of cuts in the limbs of newts. The data are the healing rates (micrometers per hour) for the two front limbs of 18 newts. The second scatterplot shows the first- and second-round scores for the 87 golfers entered in the 2019 Masters Tournament. (There are fewer than 87 points because of duplicate scores.)

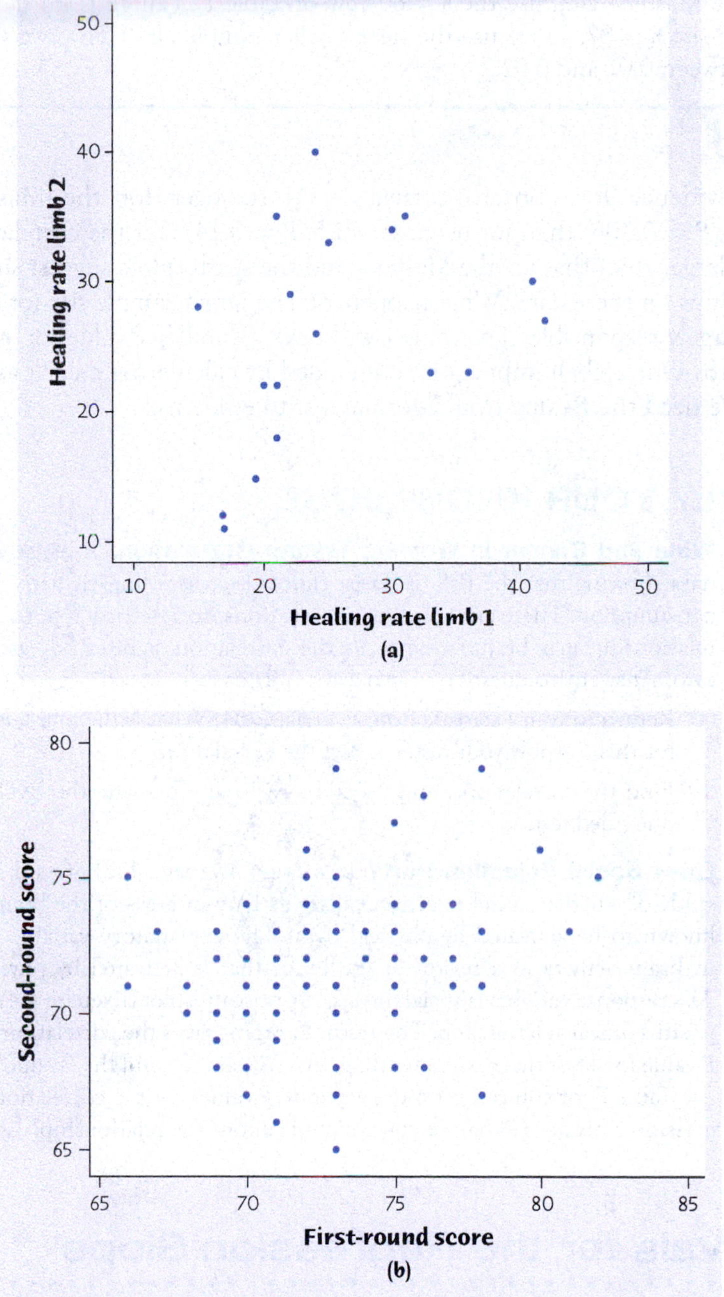

FIGURE 26.5 Two scatterplots for inference about the population correlation, for Example 26.5. (a) Healing rates for the two front limbs of 18 newts. (b) Scores on the first two rounds of the 2019 Masters Tournament.

We will test the hypotheses

$$H_0\text{: population correlation} = 0$$
$$H_a\text{: population correlation} \neq 0$$

for both sets of data. (The Masters scores are all whole numbers, but with n as high as 87, the robustness of t procedures allows their use.) Software gives

```
newts      r = 0.3581   t = 1.5342   P = 0.1445
masters    r = 0.2827   t = 2.72     P = 0.0080
```

The two-sided *P*-values for the *t* statistic for testing slope 0 are also the two-sided *P*-values for testing correlation 0.

Without software, compare the correlation $r = 0.3581$ for newts with the critical values in the $n = 18$ row of Table E. It falls between the table entries for one-tail probabilities 0.05 and 0.10, so the two-sided *P*-value lies between 0.10 and 0.20. For the Masters data, use the $n = 80$ row of Table E. (There is no table entry for sample size $n = 87$, so we use the next smaller sample size.) The two-sided *P*-value lies between 0.01 and 0.02.

The evidence for nonzero correlation is stronger for the Masters scores ($t = 2.72$, $P = 0.008$) than for newts ($t = 1.5$, $P = 0.14$). Yet the correlation for the newts is larger than that for the Masters, and the scatterplots suggest similar linear relationships for these data. What happened? The larger sample size for the Masters data is largely responsible. The same r will have a smaller *P*-value for $n = 87$ than for $n = 18$. Our eyeball impression, even aided by calculating r, can't assess significance. We need the *P*-value from a formal test to guide us.

APPLY YOUR KNOWLEDGE

26.7 Wine and Cancer in Women: Testing Correlation. Exercise 26.1 gives data showing that the risk of breast cancer increases linearly with daily wine consumption. There are only four observations, so we worry that the apparent relationship may be just chance. Is the correlation significantly greater than zero? Answer this question in two ways. CANCER

(a) Return to your *t* statistic from Exercise 26.4. What is the one-sided *P*-value for this *t*? Apply your result to test the correlation.

(b) Find the correlation *r* and use Table E to approximate the *P*-value of the one-sided test.

26.8 Does Social Rejection Hurt? Exercise 4.47 (page 122) gives data from a study of whether social rejection causes activity in areas of the brain that are known to be activated by physical pain. The explanatory variable is change in brain activity in a region of the brain that is activated by physical pain. The response variable is social distress. Your scatterplot (Exercise 4.47) shows a positive linear relationship. The research report gives the correlation *r* and the *P*-value for a test that *r* is greater than zero. What are *r* and the *P*-value? (You can use Table E, or you can get more accurate *P*-values for the correlation from regression software.) What do you conclude about the relationship? REJECT

26.6 Confidence Intervals for the Regression Slope

The slope β of the population regression line is usually the most important parameter in a regression problem. The slope is the rate of change of the mean response as the explanatory variable increases. We often want to estimate β. The slope b of the least-squares line is an unbiased estimator of β. A confidence interval is more useful because it shows how accurate the estimate b is likely to be. The confidence interval for β has the familiar form

$$\text{estimate} \pm t^*\text{SE}_{\text{estimate}}$$

Because b is our estimate, the confidence interval is $b \pm t^*\text{SE}_b$. Here are the details.

Confidence Interval for Regression Slope

A level C **confidence interval for the slope** β of the population regression line is

$$b \pm t^* \mathrm{SE}_b$$

Here t^* is the critical value for the t_{n-2} density curve with area C between $-t^*$ and t^*. The formula for SE_b appears in the box on page 597.

EXAMPLE 26.6 Sea Surface Temperatures and Coral Growth: Estimating the Slope

All the software outputs in Figure 26.3 give the slope $b = 0.46149$ (or $b = 0.4615$) and the standard error $\mathrm{SE}_b = 0.07394$. The outputs use a similar arrangement, a table in which each regression coefficient is followed by its standard error. Excel also gives the lower and upper endpoints of the 95% confidence interval for the population slope β, 0.2967 and 0.6262.

Once we know b and SE_b, it is easy to find the confidence interval. There are 12 data points, so the degrees of freedom are $n - 2 = 10$. In Table C, look in the row for df = 10. We see that the critical value is $t^* = 2.228$. To use software, enter 10 degrees of freedom. For 95% confidence, enter the cumulative proportion 0.975 that corresponds to upper-tail area 0.025. Minitab gives

```
Student's t distribution with 10 DF
P( X <= x )           x
   0.975          2.22814
```

The 95% confidence interval for the population slope β is

$$\begin{aligned} b \pm t^* \mathrm{SE}_b &= 0.46149 \pm (2.22814)(0.07394) \\ &= 0.46149 \pm 0.16475 \\ &= 0.29674 \text{ to } 0.62624 \end{aligned}$$

This agrees with Excel's result. We are 95% confident that mean calcification rate increases by between about 0.297 and 0.626 g cm^{-2} yr^{-1} for each additional 1°C in average annual maximum sea surface temperature.

You can find a confidence interval for the intercept α of the population regression line in the same way, using a and SE_a from the "Constant" line of the Minitab output or the "Intercept" line in Excel, JMP, and CrunchIt!. We rarely need to apply inferential tools such as confidence intervals or tests of significance to α.

APPLY YOUR KNOWLEDGE

26.9 Wine and Cancer in Women: Estimating Slope. Exercise 26.1 (page 592) gives data on wine consumption and the risk of breast cancer. Software tells us that the least-squares slope is $b = 0.009012$ with standard error $\mathrm{SE}_b = 0.001112$. Because there are only four observations, the observed slope b may not be an accurate estimate of the population slope β. Give a 90% confidence interval for β.

26.10 Introspection and Gray Matter: Estimating Slope. Exercise 26.2 (page 595) gives data on introspective ability and gray-matter volume of the brains of subjects. We want a 95% confidence interval for the slope of the population regression line. Starting from the information in the JMP output in Figure 26.4, find this interval. Say in words what the slope of the population regression line tells us about the relationship between Aroc and gray-matter volume.

26.11 Great Arctic Rivers: Estimating Slope. Use the data in Table 26.2 (page 596) to give a 90% confidence interval for the slope of the population regression of Arctic river discharge on year. Does this interval convince you that discharge is actually increasing over time? Explain your answer. ARCTIC

26.7 Inference about Prediction

One of the most common reasons to fit a line to data is to predict the response to a particular value of the explanatory variable. This is another setting for regression inference: we want not simply a prediction but a prediction with a margin of error that describes how accurate the prediction is likely to be.

EXAMPLE 26.7 Beer and Blood Alcohol

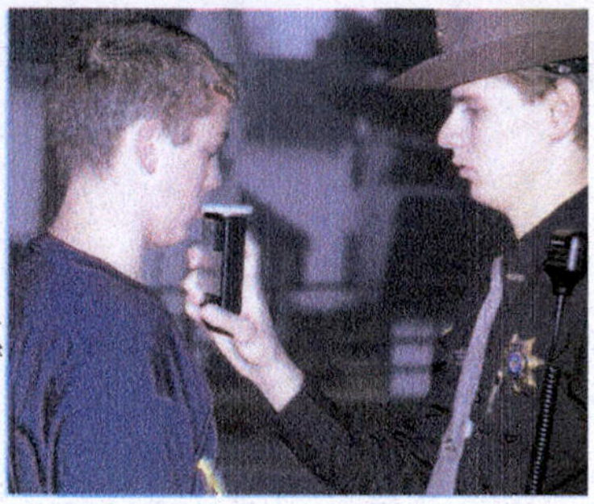

© James Shaffer/PhotoEdit

STATE: The Electronic Encyclopedia of Statistical Exercises and Examples at The Ohio State University includes a study in which 16 student volunteers at the university drank a randomly assigned number of cans of beer. Thirty minutes later, a police officer measured their blood alcohol content (BAC) in grams of alcohol per deciliter of blood. Here are the data:[5]

Student	1	2	3	4	5	6	7	8
Beers	5	2	9	8	3	7	3	5
BAC	0.10	0.03	0.19	0.12	0.04	0.095	0.07	0.06
Student	9	10	11	12	13	14	15	16
Beers	3	5	4	6	5	7	1	4
BAC	0.02	0.05	0.07	0.10	0.085	0.09	0.01	0.05

The students were equally divided between men and women, and they differed in weight and usual drinking habits. Because of this variation, many students don't believe that number of drinks predicts blood alcohol well. Steve thinks he can drive legally 30 minutes after he finishes drinking five beers. The legal limit for driving is BAC 0.08 in all states. We want to predict Steve's blood alcohol content, using no information except that he drinks five beers.

PLAN: Regress BAC on number of beers. Use the regression line to predict Steve's BAC. Give a margin of error that allows us to have 95% confidence in our prediction.

SOLVE: The scatterplot in Figure 26.6 and the regression output in Figure 26.7 show that student opinion is wrong: number of beers predicts blood alcohol content quite well. In fact, $r^2 = 0.80$, so that number of beers explains 80% of the observed variation in BAC. To predict Steve's BAC after five beers, use the equation of the regression line:

$$\begin{aligned}\hat{y} &= -0.0127 + 0.0180x \\ &= -0.0127 + 0.0180(5) = 0.077\end{aligned}$$

That's dangerously close to the legal limit of 0.08. What about 95% confidence? The bottom portion of the output in Figure 26.7 shows *two* 95% intervals. Which should we use?

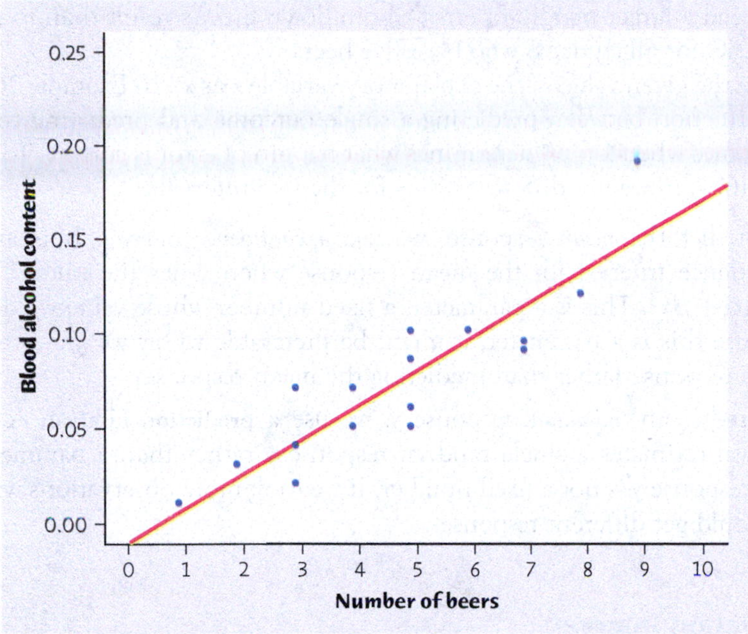

FIGURE 26.6
Scatterplot of students' blood alcohol content against the number of cans of beer consumed, with the least-squares regression line, for Example 26.7.

jmp

Linear Fit

BAC = -0.012701 + 0.0179638*Beers

Summary of Fit

RSquare	0.799841
RSquare Adj	0.785544
Root Mean Square Error	0.020441
Mean of Response	0.07375
Observations (or Sum Wgts)	16

Lack of Fit

Analysis of Variance

Source	DF	Sum of Squares	Mean Square	F Ratio
Model	1	0.02337535	0.023375	55.9443
Error	14	0.00584965	0.000418	Prob > F
C. Total	15	0.02922500		<.0001*

Parameter Estimates

Term	Estimate	Std Error	t Ratio	Prob>|t|
Intercept	-0.012701	0.012638	-1.00	0.3320
Beers	0.0179638	0.002402	7.48	<.0001*

New Observation

Beers	BAC	Predicted BAC	Lower 95% Mean BAC	Upper 95% Mean BAC	Lower 95% Indiv BAC	Upper 95% Indiv BAC
5	•	0.0771182053	0.0661153622	0.0881210485	0.0319171218	0.1223192889

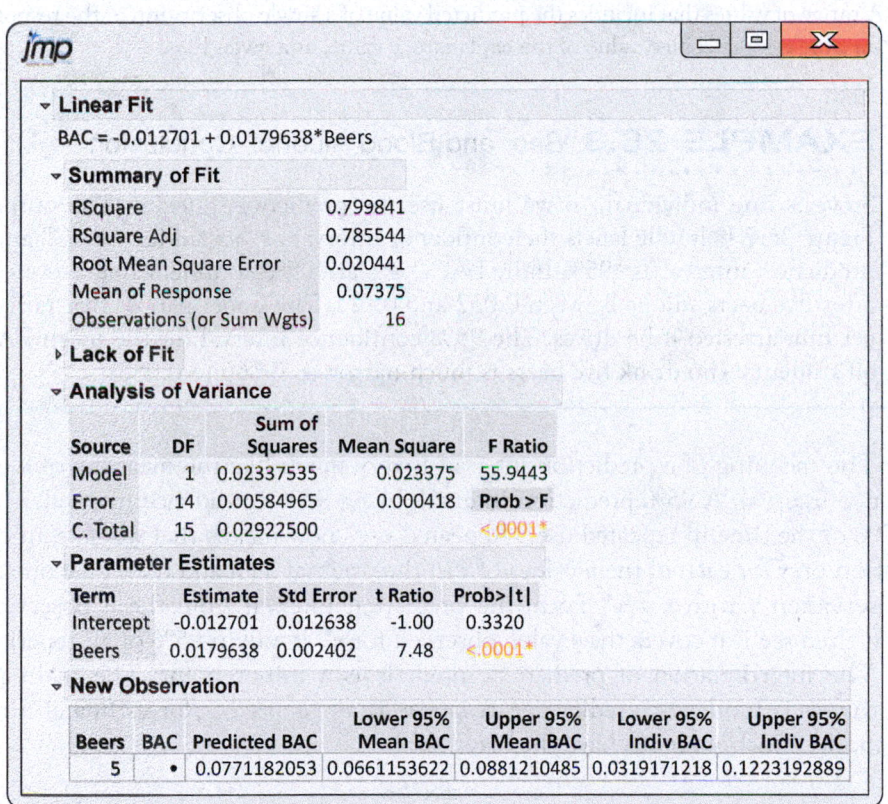

FIGURE 26.7
JMP regression output for the blood alcohol content data, with 5 entered as a new observation, for Example 26.7.

To decide which interval to use, you must answer this question: Do you want to predict the *mean* BAC for *all students* who drink five beers, or do you want to predict the BAC of *one individual student* who drinks five beers? *Both of these predictions may be interesting, but they are two different problems.* The actual prediction is the same, $\hat{y} = 0.077$. But the margin of error is different for the two kinds of prediction. Individual students who drink five beers don't all have the same BAC.

STATISTICS IN YOUR WORLD

Is Regression Garbage?

No—but garbage can be the setting for regression. The U.S. Census Bureau once asked if weighing a neighborhood's garbage would help count its people. So 63 households had their garbage sorted and weighed. It turned out that pounds of plastic in the trash gave the best garbage prediction of the number of people in a neighborhood. The margin of error for a 95% prediction interval in a neighborhood of about 100 households, based on five weeks' worth of garbage, was about ±2.5 people. Alas, that is not accurate enough to help the U.S. Census Bureau.

So we need a larger margin of error to pin down Steve's result than to predict the mean BAC for all students who have five beers.

Write the given value of the explanatory variable x as x^*. In Example 26.7, $x^* = 5$. The distinction between predicting a single outcome and predicting the mean of all outcomes when $x = x^*$ determines what margin of error is correct. To emphasize the distinction, we use different terms for the two intervals.

- To predict the *mean* response, we use a *confidence interval.* It is an ordinary confidence interval for the mean response when x has the value x^*, which is $\mu_y = \alpha + \beta x^*$. This is a parameter, a fixed number whose value we don't know. Because this is a parameter, it might be preferable to say we are estimating the mean response rather than predicting the mean response.
- To predict an *individual* response y, we use a **prediction interval.** A prediction interval estimates a single random response y rather than a parameter like μ_y. The response y is not a fixed number. If we took more observations with $x = x^*$, we would get different responses.

prediction interval

Prediction Interval

A range of values that includes the predicted value of a single observation of the response variable for a specified value of the explanatory value, at a given level.

EXAMPLE 26.8 Beer and Blood Alcohol: Conclusion

Steve is one individual, so we must use the prediction interval. The output in Figure 26.7 helpfully labels the confidence interval as "95% Mean BAC" and the prediction interval as "95% Indiv BAC." We are 95% confident that Steve's BAC after five beers will lie between 0.032 and 0.122. The upper part of that range will get him arrested if he drives. The 95% confidence interval for the mean BAC of all students who drink five beers is much narrower, 0.066 to 0.088.

The meaning of a prediction interval is very much like the meaning of a confidence interval. A 95% prediction interval, like a 95% confidence interval, is right 95% of the time in repeated use. "Repeated use" now means that we take an observation on y for each of the n values of x in the original data and then take one more observation y with $x = x^*$. Form the prediction interval from the n observations and then see if it covers the y value observed for x^*. It will in 95% of all repetitions.

The interpretation of prediction intervals is a minor point. The main point is that it is harder to predict one response than to predict (or estimate) a mean response. Both intervals have the usual form

$$\hat{y} \pm t^* SE$$

but the prediction interval is wider than the confidence interval because individuals are more variable than averages. As a further illustration, consider a professional athlete such as basketball player Kobe Bryant. You can find his career statistics online. During his career as a starter, his season scoring *average* ranges from 7.6 to 35.4 points per game, but individual game performances range from 0 points to a career high of 81 points. Individual game performances of professional athletes are more variable than season average performances. So the margin of error for predictions of individual game performances will be larger than for predictions of season averages. You will rarely need to know the details because software automates the calculation, but we present them here.

STATISTICS IN YOUR WORLD

May the Longer Name Win!

Regression is far from perfect, but it beats most other ways of predicting. A writer in the early 1960s noted a simple method for predicting U.S. presidential elections: just choose the candidate with the longer name. In the 22 elections from 1876 to 1960, this method failed only once. Let's hope that the writer didn't bet the family silver on this idea. The 13 elections from 1964 to 2012 presented 11 tests of the "long name wins" method (the 1980 candidates and the 2000 candidates had names of the same length). The longer name lost eight of the 11.

Confidence and Prediction Intervals for Regression Response

A level C **confidence interval for the mean response** μ_y when x takes the value x^* is

$$\hat{y} \pm t^* \mathrm{SE}_{\hat{\mu}}$$

The standard error $\mathrm{SE}_{\hat{\mu}}$ is

$$\mathrm{SE}_{\hat{\mu}} = s\sqrt{\frac{1}{n} + \frac{(x^* - \bar{x})^2}{\sum (x - \bar{x})^2}}$$

A level C **prediction interval for a single observation** y when x takes the value x^* is

$$\hat{y} \pm t^* \mathrm{SE}_{\hat{y}}$$

The standard error for prediction $\mathrm{SE}_{\hat{y}}$ is

$$\mathrm{SE}_{\hat{y}} = s\sqrt{1 + \frac{1}{n} + \frac{(x^* - \bar{x})^2}{\sum (x - \bar{x})^2}}$$

In both intervals, t^* is the critical value for the t_{n-2} density curve with area C between $-t^*$ and t^*.

There are two standard errors: $\mathrm{SE}_{\hat{\mu}}$ for estimating the mean response μ_y, and $\mathrm{SE}_{\hat{y}}$ for predicting an individual response y. The only difference between the two standard errors is the extra 1 under the square root sign in the standard error for prediction. The extra 1 makes the prediction interval wider. Both standard errors are multiples of the regression standard error s. The degrees of freedom are again $n - 2$, the degrees of freedom of s.

APPLY YOUR KNOWLEDGE

26.12 Wine and Cancer in Women: Prediction. Exercise 26.1 (page 592) gives data on wine consumption and the risk of breast cancer. For a new group of women who drink an average of 10 grams of wine per day, predict their relative risk of breast cancer.

(a) Figure 26.8 is part of the output from Minitab for prediction when $x^* = 10.0$. Which interval in the output is the proper 95% interval for predicting the relative risk?

(b) Minitab gives only one of the two standard errors used in prediction. It is $\mathrm{SE}_{\hat{\mu}}$, the standard error for estimating the mean response. Use this fact along with the output to give a 90% confidence interval for the mean relative risk of breast cancer in all women who drink an average of 10 grams of wine per day.

Minitab

Predictor	Coef	SE Coef	T	P
Constant	0.99309	0.01777	55.88	0.000
Intake	0.009012	0.001112	8.10	0.015

Predicted Values for New Observations

New Obs	Fit	SE Fit	95% CI	95% PI
1	1.08321	0.01057	(1.03775, 1.12868)	(0.98643, 1.18000)

FIGURE 26.8 Partial Minitab output for regressing relative risk of breast cancer on mean daily intake of wine with predicted values for $x^* = 10.0$, for Exercise 26.12.

TABLE 26.3 Winter temperature (°C) and anglerfish latitude, 1977–2001

Year	Temperature	Latitude	Year	Temperature	Latitude
1977	6.26	57.20	1990	6.89	58.13
1978	6.26	57.96	1991	6.90	58.52
1979	6.27	57.65	1992	6.93	58.48
1980	6.31	57.59	1993	6.98	57.89
1981	6.34	58.01	1994	7.02	58.71
1982	6.32	59.06	1995	7.09	58.07
1983	6.37	56.85	1996	7.13	58.49
1984	6.39	56.87	1997	7.15	58.28
1985	6.42	57.43	1998	7.29	58.49
1986	6.52	57.72	1999	7.34	58.01
1987	6.68	57.83	2000	7.57	58.57
1988	6.76	57.87	2001	7.65	58.90
1989	6.78	57.48			

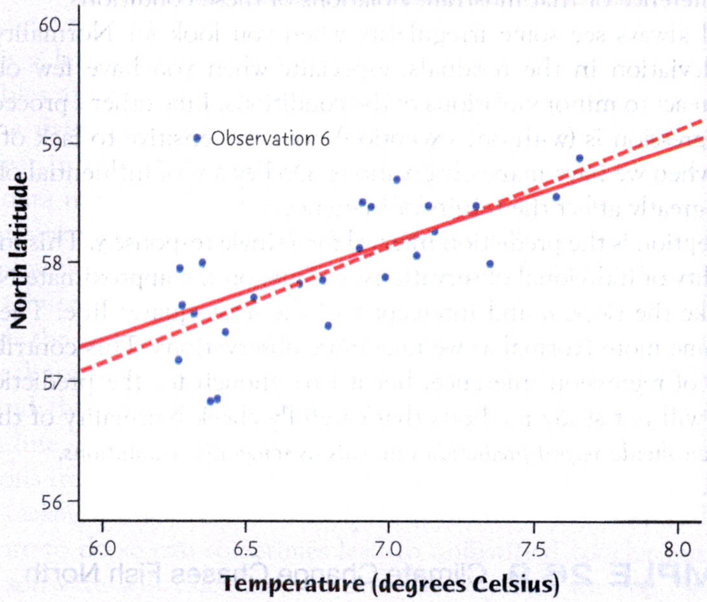

FIGURE 26.10
Scatterplot of the latitude of the center of the distribution of anglerfish in the North Sea against mean winter temperature at the bottom of the sea, for Example 26.9. The two regression lines are for the data with (solid) and without (dashed) Observation 6.

distribution (north latitude) on winter ocean temperature. Software shows that the slope is $b = 0.818$. That is, each degree of ocean warming moves the fish about 0.8 degree of latitude farther north. The t statistic for testing $H_0\colon \beta = 0$ is $t = 3.6287$, with one-sided P-value $P = 0.0007$. There is very strong evidence that the population slope is positive, $\beta > 0$.

CONCLUDE: The data give highly significant evidence that anglerfish have moved north as the ocean has grown warmer. Before relying on this conclusion, we must check the conditions for inference.

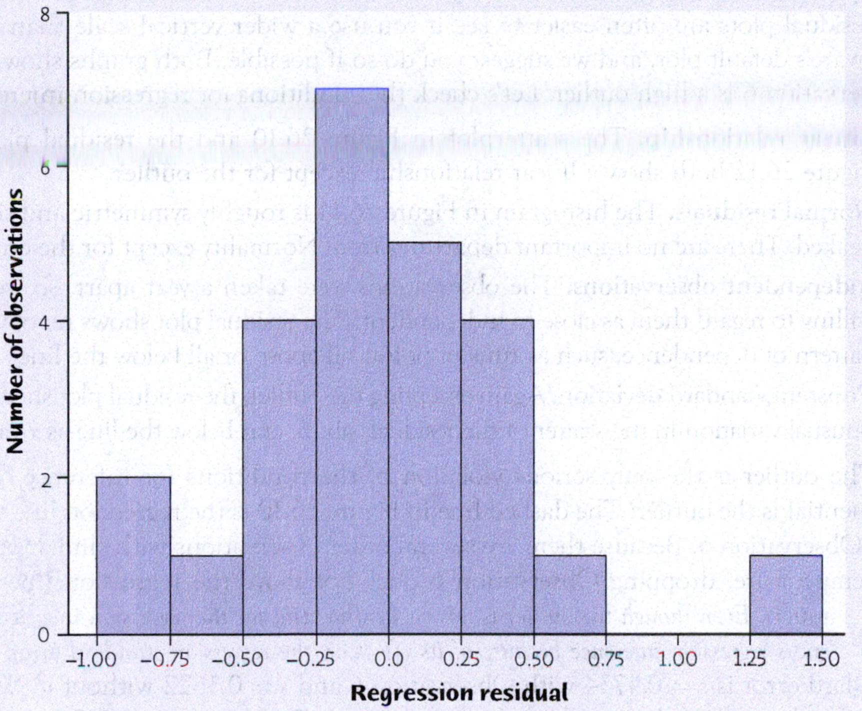

FIGURE 26.11 Histogram of the residuals from the regression of latitude on temperature, for Example 26.9.

The software that did the regression calculations also finds the 25 residuals. In the same order as the observations in Example 26.9, they are

−0.3731	0.3869	0.0687	−0.0240	0.3714	1.4378	−0.8131
−0.8095	−0.2740	−0.0658	−0.0867	−0.1121	−0.5185	0.0415
0.4234	0.3588	−0.2721	0.5152	−0.1821	0.2052	−0.0211
0.0743	−0.4466	−0.0747	0.1899			

Begin by making two graphs of the residuals. Figure 26.11 is a histogram of the residuals. Figure 26.12 is the residual plot—a plot of the residuals against the explanatory variable, sea-bottom temperature. The red "residual = 0" line marks

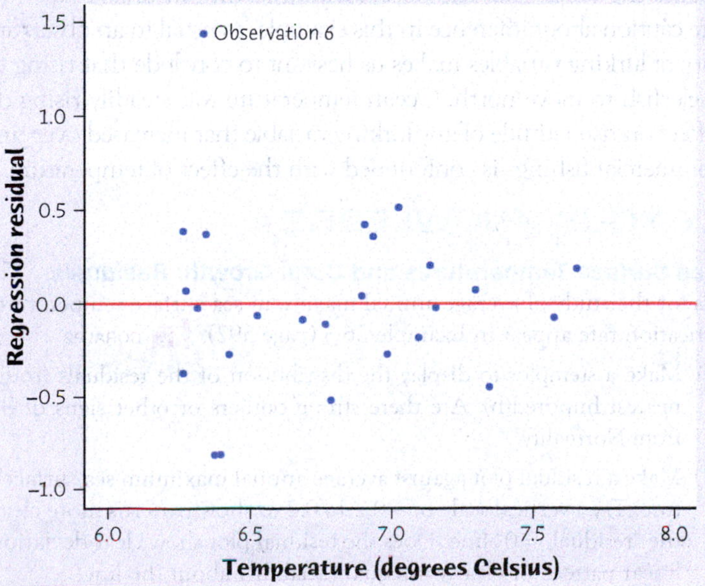

FIGURE 26.12 Residual plot for the regression of latitude on temperature, for Example 26.9.

the position of the regression line relative to each blue-dot observation. Patterns in residual plots are often easier to see if you use a wider vertical scale than your software's default plot, and we suggest you do so if possible. Both graphs show that Observation 6 is a high outlier. Let's check the conditions for regression inference:

- **Linear relationship.** The scatterplot in Figure 26.10 and the residual plot in Figure 26.12 both show a linear relationship except for the outlier.
- **Normal residuals.** The histogram in Figure 26.11 is roughly symmetric and single peaked. There are no important departures from Normality except for the outlier.
- **Independent observations.** The observations were taken a year apart, so we are willing to regard them as close to independent. The residual plot shows no obvious pattern of dependence, such as runs of points all above or all below the line.
- **Constant standard deviation.** Again excepting the outlier, the residual plot shows no unusual variation in the scatter of the residuals above and below the line as x varies.

The outlier is the only serious violation of the conditions for inference. How influential is the outlier? The dashed line in Figure 26.10 is the regression line without Observation 6. Because there are several other observations with similar values of temperature, dropping Observation 6 does not move the regression line very much. *Even though the outlier is not very influential for the regression line, it influences regression inference because of its effect on the regression standard error.* The standard error is $s = 0.4734$ with Observation 6 and $s = 0.3622$ without it. When we omit the outlier, the t statistic changes from $t = 3.6287$ to $t = 5.5599$, and the one-sided P-value changes from $P = 0.0007$ to $P = 0.0000137$. Fortunately, the outlier does not affect the conclusion we drew from the data. Dropping Observation 6 makes the test for the population slope *more* significant and *increases* the percentage of variation in fish location explained by ocean temperature.

Although we see that dropping the outlier, Observation 6, makes the test for the population slope more significant and increases the percentage of variation in fish location explained by ocean temperature, this is not a good reason for excluding it from our analysis. Further investigation is needed before we can justify its removal. If we knew, for example, that the observation is the result of an error or that there are good scientific grounds for regarding it as inconsistent with the other observations, we might reasonably remove it. Otherwise, we note that it is an outlier and perhaps discuss the effect that the observation has on our analysis.

One more caution about inference in this example: as usual in an observational study, the possibility of lurking variables makes us hesitant to conclude that rising temperature is *causing* anglerfish to move north. Ocean temperature was steadily rising during these years. The effect on fish latitude of any lurking variable that increased over time—perhaps increased commercial fishing—is confounded with the effect of temperature.

APPLY YOUR KNOWLEDGE

26.14 Sea Surface Temperatures and Coral Growth: Residuals. The residuals for the study of average annual maximum sea surface temperature and calcification rate appear in Example 26.3 (page 592). CORALR2

(a) Make a stemplot to display the distribution of the residuals (round to the nearest hundredth). Are there strong outliers or other signs of departures from Normality?

(b) Make a residual plot against average annual maximum sea surface temperature. Try a vertical scale of -0.2 to 0.2 to show patterns more clearly. Draw the "residual = 0" line. Does the residual plot show clear deviations from a linear pattern or clearly unequal variability about the line?

26.15 Introspection and Gray Matter: Residuals. Figure 26.4 (page 596) gives part of the JMP output for the data on introspective ability and gray-matter volume in Exercise 26.2. Figure 26.13 is another part of the output. It gives x, y, the predicted response $\hat{y}$, the residual $y - \hat{y}$, and related quantities for each of the 29 observations. Most statistical software provides similar output. Examine the conditions for regression inference one by one. This example illustrates mild violations of the conditions that did not prevent the researchers from doing inference. INTRSPR

(a) **Linear relationship.** Your scatterplot and r^2 from Exercise 26.2 show that the relationship is roughly linear. Plot the residuals against gray-matter volume. Are any deviations from a straight line apparent?

(b) **Normal variation about the line.** Make a stemplot of the residuals. (Round to the third decimal place and don't forget that −0.00 and 0.00 are separate stems.) With only 29 observations, a small amount of skew is not disturbing. Does your plot indicate that any of the observations are outliers? If so, which ones?

(c) **Independent observations.** The data come from 29 subjects who were each measured separately. Can you conclude that the observations are independent?

(d) **Variability about the line stays the same.** Is there any evidence that the variability may be larger at one end than at the other?

jmp

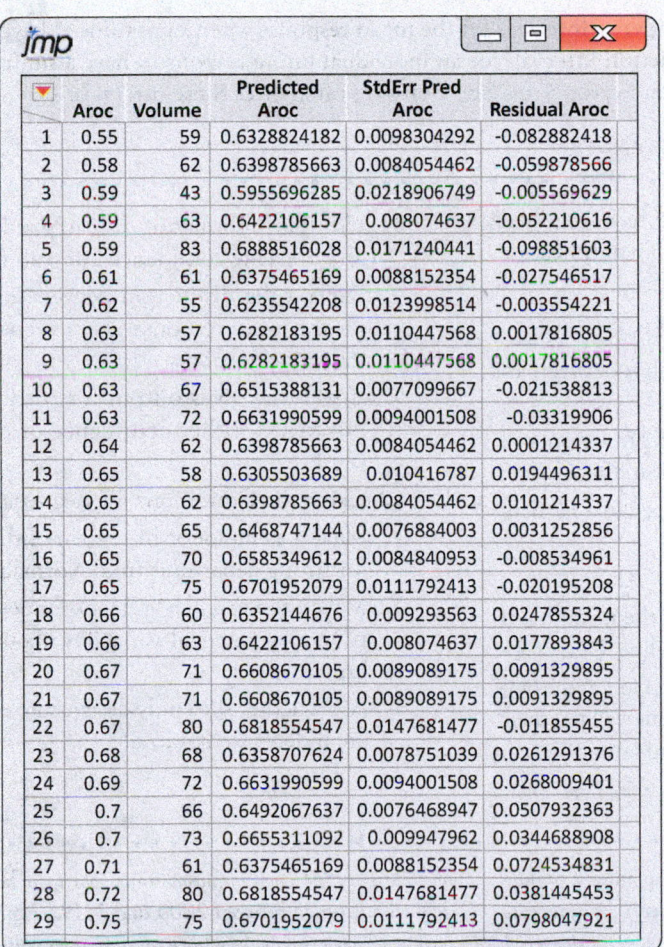

	Aroc	Volume	Predicted Aroc	StdErr Pred Aroc	Residual Aroc
1	0.55	59	0.6328824182	0.0098304292	-0.082882418
2	0.58	62	0.6398785663	0.0084054462	-0.059878566
3	0.59	43	0.5955696285	0.0218906749	-0.005569629
4	0.59	63	0.6422106157	0.008074637	-0.052210616
5	0.59	83	0.6888516028	0.0171240441	-0.098851603
6	0.61	61	0.6375465169	0.0088152354	-0.027546517
7	0.62	55	0.6235542208	0.0123998514	-0.003554221
8	0.63	57	0.6282183195	0.0110447568	0.0017816805
9	0.63	57	0.6282183195	0.0110447568	0.0017816805
10	0.63	67	0.6515388131	0.0077099667	-0.021538813
11	0.63	72	0.6631990599	0.0094001508	-0.03319906
12	0.64	62	0.6398785663	0.0084054462	0.0001214337
13	0.65	58	0.6305503689	0.010416787	0.0194496311
14	0.65	62	0.6398785663	0.0084054462	0.0101214337
15	0.65	65	0.6468747144	0.0076884003	0.0031252856
16	0.65	70	0.6585349612	0.0084840953	-0.008534961
17	0.65	75	0.6701952079	0.0111792413	-0.020195208
18	0.66	60	0.6352144676	0.009293563	0.0247855324
19	0.66	63	0.6422106157	0.008074637	0.0177893843
20	0.67	71	0.6608670105	0.0089089175	0.0091329895
21	0.67	71	0.6608670105	0.0089089175	0.0091329895
22	0.67	80	0.6818554547	0.0147681477	-0.011855455
23	0.68	68	0.6358707624	0.0078751039	0.0261291376
24	0.69	72	0.6631990599	0.0094001508	0.0268009401
25	0.7	66	0.6492067637	0.0076468947	0.0507932363
26	0.7	73	0.6655311092	0.009947962	0.0344688908
27	0.71	61	0.6375465169	0.0088152354	0.0724534831
28	0.72	80	0.6818554547	0.0147681477	0.0381445453
29	0.75	75	0.6701952079	0.0111792413	0.0798047921

FIGURE 26.13
Residuals from JMP, for Exercise 26.15. The table gives the predicted volume and the residual for each observation.

CHAPTER 26 SUMMARY

- *Least-squares regression* fits a straight line to data to predict a response variable y from an explanatory variable x. Inference about regression requires more conditions.
- The **conditions for regression inference** say that there is a **population regression line** $\mu_y = \alpha + \beta x$ that describes how the mean response varies as x changes. The observed response y for any x has a Normal distribution with mean given by the population regression line and with the same standard deviation σ for any value of x. Observations on y are independent.
- The parameters to be estimated are the intercept α and the slope β of the population regression line and also the standard deviation σ. The slope a and intercept b of the least-squares line estimate α and β. Use the **regression standard error s** to estimate σ.
- The regression standard error s has $n-2$ *degrees of freedom*. All t procedures for regression inference have $n-2$ degrees of freedom.
- To test the hypothesis that the slope is zero in the population, use the t statistic $t = b/\text{SE}_b$. This null hypothesis says that straight-line dependence on x has no value for predicting y. In practice, use software to find the slope b of the least-squares line, its standard error SE_b, and the t statistic.
- The t test for regression slope is also a test for the hypothesis that the population correlation between x and y is zero. To do this test without software, use the sample correlation r and Table E.
- **Confidence intervals for the slope** of the population regression line have the form $b \pm t^*\text{SE}_b$.
- **Confidence intervals for the mean response** when x has value x^* have the form $\hat{y} \pm t^*\text{SE}_{\hat{\mu}}$. **Prediction intervals** for an individual future response y have a similar form with a larger standard error, $\hat{y} \pm t^*\text{SE}_{\hat{y}}$. Software often gives these intervals.

CHAPTER 26 SKILLS REVIEW

Here are the most important skills you should have acquired from reading this chapter. Following each skill in parentheses is the section of the text where the topic is introduced.

A. Preliminaries

1. Make a scatterplot to show the relationship between an explanatory and a response variable. (5.1)
2. Use software or a calculator to find the correlation and the equation of the least-squares regression line. (5.3)
3. Recognize which type of inference you need in a particular regression setting. (26.2, 26.4–26.7)

B. Inference Using Software Output

1. Explain in any specific regression setting the meaning of the slope β of the population regression line. (5.1, 26.1)
2. Understand software output for regression. Find in the output the slope and intercept of the least-squares line, their standard errors, and the regression standard error. (26.3)
3. Use that information to carry out tests of H_0: $\beta = 0$ and calculate confidence intervals for β. (26.4, 26.6)
4. Explain the distinction between a confidence interval for the mean response and a prediction interval for an individual response. (26.7)
5. If software gives output for prediction, use that output to give either confidence or prediction intervals. (26.7)

C. Checking the Conditions for Regression Inference

1. Make a stemplot or histogram of the residuals and look for strong departures from Normality. (26.8)
2. Make a residual plot and look for departures from a linear pattern or unequal variability about the "residual = 0" line. (26.8)
3. Ask whether the study design suggests that observations are independent. (26.8)

CHECK YOUR SKILLS

Florida reappraises real estate every year, and a county appraiser's website lists the current "fair market value" of each piece of property. Property usually sells for somewhat more than the appraised market value. Table 26.4 gives the appraised market values and actual selling prices (in thousands of dollars) of 72 condominium units sold in a beachfront building in a 204-month period between 2003 and 2019.[7] *Here is part of the Minitab output for regressing selling price on appraised value, along with prediction for a unit with appraised value $1,000,000:* COND

```
Predictor        Coef      SE Coef         T         P
Constant       325.08       116.77      2.78     0.007
appraised      0.8894       0.1403      6.34     0.000
S = 257.903    R-Sq = 36.5%      R-Sq(adj) = 35.6%
Predicted Values for New Observations
New    Fit   SE Fit       95% CI            95% PI
Obs
 1    1214.5   41.1   (1132.6, 1296.4)  (693.6, 1735.3)
```

Franz Marc Frei/Getty Images

TABLE 26.4 Selling price and appraised market value (thousands of dollars), 2003–2019

Selling Price	Appraised Value	Month	Selling Price	Appraised Value	Month	Selling Price	Appraised Value	Month
825	626	0	1050	774	79	750	584	140
590	492	1	605	470	86	862	639	160
1075	930	3	675	545	88	875	645	160
890	790	9	693	690	93	940	699	164
845	648	13	360	690	94	1000	540	166
1100	942	14	1455	1132	96	420	847	166
715	345	15	1068	867	99	1275	927	167
1325	1032	19	1235	1009	100	1375	1134	171
700	556	21	1083	601	101	910	813	172
1322	879	26	1200	904	109	1020	659	174
1900	1016	26	650	505	110	1360	1231	175
1600	1040	28	1000	877	113	559	1231	176
980	442	34	1100	895	118	1595	602	179
940	771	37	549	481	119	950	819	183
850	715	47	1405	822	120	950	814	184
1100	997	54	1000	870	121	975	1005	185
1164	953	59	1200	752	124	1072	995	196
1425	922	64	1100	678	125	1715	1272	197
1865	1190	64	955	591	129	760	767	198
1450	610	64	1200	888	129	800	689	198
875	806	64	900	850	129	1125	972	199
1510	1241	70	825	516	134	1300	1132	199
1375	813	70	886	626	137	1037	1069	200
560	496	73	885	765	138	1165	706	204

Questions 26.16 through 26.24 are based on this information.

26.16 The equation of the least-squares regression line for predicting selling price from appraised value is

(a) price = 325.08 + 0.8894 × appraised value.

(b) price = 0.8894 + 325.08 × appraised value.

(c) price = 116.77 + 0.1403 × appraised value.

26.17 What is the correlation between selling price and appraised value?

(a) 0.133

(b) 0.365

(c) 0.604

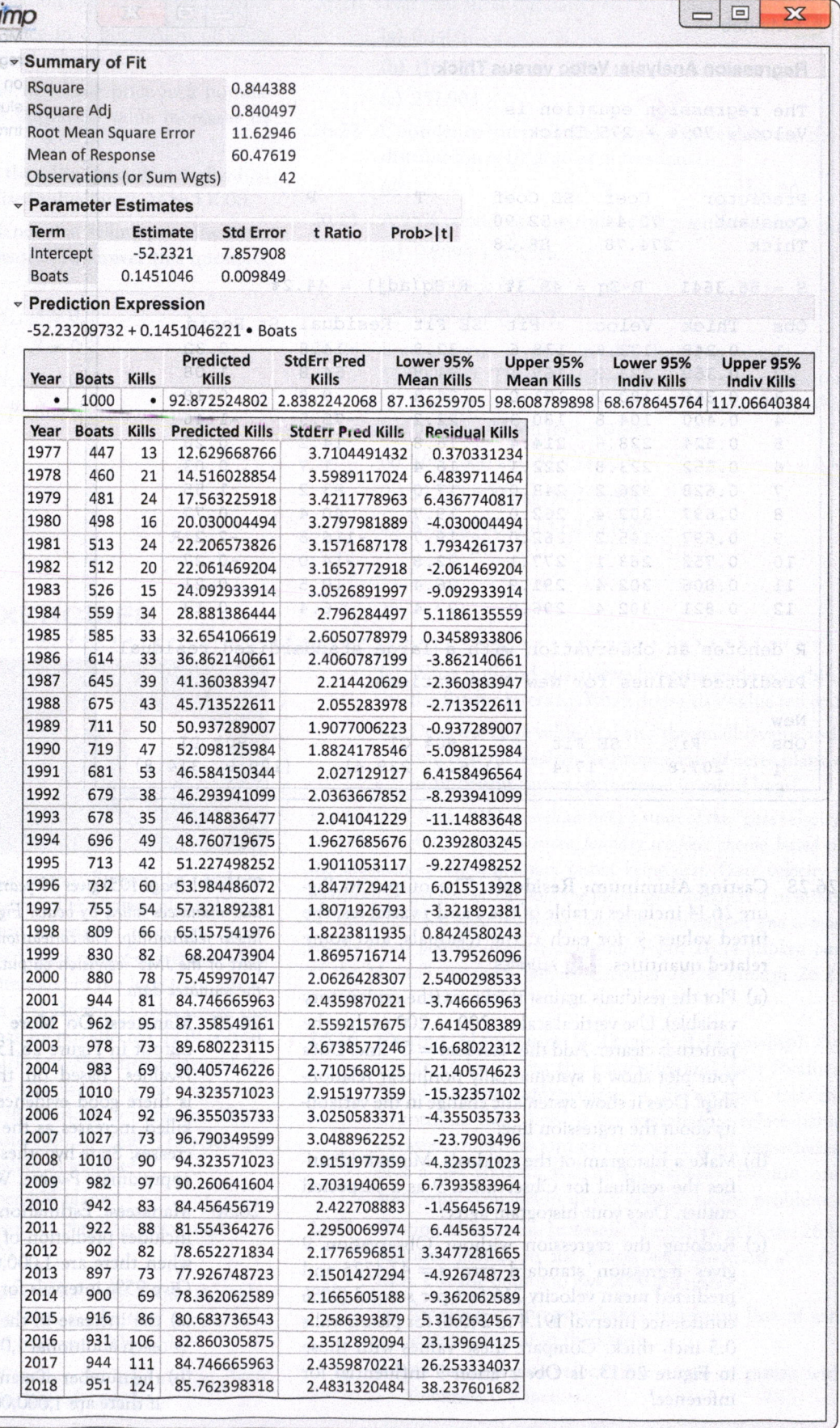

FIGURE 26.15 JMP output for the regression of number of manatees killed by boats on the number of boats (in thousands) registered in Florida, for Exercises 26.29 through 26.31.

jmp

Summary of Fit

RSquare	0.844388
RSquare Adj	0.840497
Root Mean Square Error	11.62946
Mean of Response	60.47619
Observations (or Sum Wgts)	42

Parameter Estimates

Term	Estimate	Std Error	t Ratio	Prob>|t|
Intercept	-52.2321	7.857908		
Boats	0.1451046	0.009849		

Prediction Expression

-52.23209732 + 0.1451046221 • Boats

Year	Boats	Kills	Predicted Kills	StdErr Pred Kills	Lower 95% Mean Kills	Upper 95% Mean Kills	Lower 95% Indiv Kills	Upper 95% Indiv Kills
•	1000	•	92.872524802	2.8382242068	87.136259705	98.608789898	68.678645764	117.06640384

Year	Boats	Kills	Predicted Kills	StdErr Pred Kills	Residual Kills
1977	447	13	12.629668766	3.7104491432	0.370331234
1978	460	21	14.516028854	3.5989117024	6.4839711464
1979	481	24	17.563225918	3.4211778963	6.4367740817
1980	498	16	20.030004494	3.2797981889	-4.030004494
1981	513	24	22.206573826	3.1571687178	1.7934261737
1982	512	20	22.061469204	3.1652772918	-2.061469204
1983	526	15	24.092933914	3.0526891997	-9.092933914
1984	559	34	28.881386444	2.796284497	5.1186135559
1985	585	33	32.654106619	2.6050778979	0.3458933806
1986	614	33	36.862140661	2.4060787199	-3.862140661
1987	645	39	41.360383947	2.214420629	-2.360383947
1988	675	43	45.713522611	2.055283978	-2.713522611
1989	711	50	50.937289007	1.9077006223	-0.937289007
1990	719	47	52.098125984	1.8824178546	-5.098125984
1991	681	53	46.584150344	2.027129127	6.4158496564
1992	679	38	46.293941099	2.0363667852	-8.293941099
1993	678	35	46.148836477	2.041041229	-11.14883648
1994	696	49	48.760719675	1.9627685676	0.2392803245
1995	713	42	51.227498252	1.9011053117	-9.227498252
1996	732	60	53.984486072	1.8477729421	6.015513928
1997	755	54	57.321892381	1.8071926793	-3.321892381
1998	809	66	65.157541976	1.8223811935	0.8424580243
1999	830	82	68.20473904	1.8695716714	13.79526096
2000	880	78	75.459970147	2.0626428307	2.5400298533
2001	944	81	84.746665963	2.4359870221	-3.746665963
2002	962	95	87.358549161	2.5592157675	7.6414508389
2003	978	73	89.680223115	2.6738577246	-16.68022312
2004	983	69	90.405746226	2.7105680125	-21.40574623
2005	1010	79	94.323571023	2.9151977359	-15.32357102
2006	1024	92	96.355035733	3.0250583371	-4.355035733
2007	1027	73	96.790349599	3.0488962252	-23.7903496
2008	1010	90	94.323571023	2.9151977359	-4.323571023
2009	982	97	90.260641604	2.7031940659	6.7393583964
2010	942	83	84.456456719	2.422708883	-1.456456719
2011	922	88	81.554364276	2.2950069974	6.445635724
2012	902	82	78.652271834	2.1776596851	3.3477281665
2013	897	73	77.926748723	2.1501427294	-4.926748723
2014	900	69	78.362062589	2.1665605188	-9.362062589
2015	916	86	80.683736543	2.2586393879	5.3162634567
2016	931	106	82.860305875	2.3512892094	23.139694125
2017	944	111	84.746665963	2.4359870221	26.253334037
2018	951	124	85.762398318	2.4831320484	38.237601682

26.31 Manatees: Conditions for Inference. We know that there is a strong linear relationship between number of boats registered in Florida and number of manatees killed. Let's check the other conditions for inference. Figure 26.15 includes a table of the two variables, the predicted values $\hat{y}$ for each x in the data, the residuals, and related quantities. MANATRS

(a) Round the residuals to the nearest whole number and make a stemplot. The distribution is single peaked and symmetric and appears close to Normal.

(b) Make a residual plot of residuals against boats registered. Use a vertical scale from −40 to 40 to show the pattern more clearly. Add the "residual = 0" line. Is there a clearly nonlinear pattern? Is it reasonable to assume that the standard deviation is the same for all x?

(c) It is reasonable to regard the number of manatees killed by boats in successive years as independent. The number of boats grew over time. Someone says that pollution also grew over time and may explain more manatee deaths. How would you respond to this idea?

(d) Are you willing to trust the inferences in Exercises 26.29 and 26.30? Why or why not?

26.32 Fidgeting Keeps You Slim: Inference. Our first example of regression (Example 5.1, page 125) presented data showing that people who increased their nonexercise activity (NEA) when they were deliberately overfed gained less fat than other people. Use software to add formal inference to the data analysis for these data. FATGAIN

(a) Based on 16 subjects, the correlation between NEA increase and fat gain was $r = -0.7786$. Is this significant evidence that people with higher NEA gain less fat? (Report a t statistic from regression output and give the one-sided P-value.)

(b) The slope of the least-squares regression line was $b = -0.00344$ so that fat gain decreased by 0.00344 kilogram for each added calorie of NEA. Give a 90% confidence interval for the slope of the population regression line. This rate of change is the most important parameter to be estimated.

(c) Sam's NEA increases by 400 calories. His predicted fat gain is 2.13 kilograms. Give a 95% interval for predicting Sam's fat gain.

26.33 Predicting Tropical Storms. Exercise 5.62 (page 157) gives data on William Gray's predictions of the number of named tropical storms in Atlantic hurricane seasons from 1984 to 2018. Use these data for regression inference to answer the following questions. STORMS

(a) Does Professor Gray do better than random guessing? That is, is there a significantly positive correlation between his forecasts and the actual number of storms? (Report a t statistic from regression output and give the one-sided P-value.)

(b) Give a 95% confidence interval for the mean number of storms in years when Professor Gray forecasts 16 storms.

26.34 Coral Growth. How sensitive to changes in water temperature are coral reefs? To find out, scientists examined data on mean sea surface temperatures (in degrees Celsius) and mean coral growth (in centimeters per year) over a several-year period at locations in the Gulf of Mexico and the Caribbean Sea. Here are the data for the Gulf of Mexico:[9] GULF

Mean sea surface temperature	26.7	26.6	26.6	26.5	26.3	26.1
Growth	0.85	0.85	0.79	0.86	0.89	0.92

(a) Do the data indicate that coral growth decreases linearly as mean sea surface temperature increases? Is this change statistically significant?

(b) Use the data to predict with 95% confidence the mean coral growth (centimeters per year) when mean sea surface temperature is 26.4°C.

26.35 Predicting Tropical Storms: Residuals. Make a stemplot of the residuals (round to the nearest whole number) from your regression in Exercise 26.33. Explain why your plot suggests that we should not use these data to get a prediction interval for the number of storms in a single year. STORMS2

26.36 Coral Growth: Residuals. Do the data in Exercise 26.34 on mean sea surface temperatures and coral growth in the Gulf of Mexico and the Caribbean Sea satisfy the conditions for regression inference? To examine this, here are the residuals: GULFR

Mean sea surface temperature	26.7	26.6	26.6	26.5	26.3	26.1
Residual	0.0268	0.0111	−0.0489	0.0053	0.0037	0.0021

(a) **Linear relationship.** A plot of the residuals against the explanatory variable x magnifies the deviations from the least-squares line. Does the plot show any systematic deviation from a roughly linear pattern?

(b) **Normal variation about the line.** Make a histogram of the residuals. With only six observations, no clear shape emerges. Do strong skewness or outliers suggest lack of Normality?

(c) **Independent observations.** Are the six observations independent? Why?

(d) **Variability about the line stays the same.** Does your plot in part (a) show any systematic change in variability as x changes?

26.37 Our Brains Don't Like Losses. Exercise 4.30 (page 118) describes an experiment that showed a linear relationship between how sensitive people are to monetary losses ("behavioral loss aversion") and activity in one part of their brains ("neural loss aversion"). LOSSES

(a) Make a scatterplot with neural loss aversion as x and behavioral loss aversion as y. One point is a high outlier in both the x and y directions. In Exercise 5.41 (page 153), you found that this outlier is not influential for the least-squares line.

(b) The research report says that $r = 0.85$ and that the test for regression slope has $P < 0.001$. Verify these results, using all the observations.

(c) The report recognizes the outlier and says, "However, this regression also remained highly significant ($P = 0.004$) when the extreme data point (top right corner of the scatterplot) was removed from the analysis." Repeat your analysis, this time omitting the outlier. Show that the outlier influences regression inference by comparing the t statistic for testing slope with and without the outlier. Then verify the report's claim about the P-value of this test.

26.38 Time at the Table. Does how long young children remain at the lunch table help predict how much they eat? Here are data on 20 toddlers observed over several months at a nursery school.[10] "Time" is the average number of minutes a child spent at the table when lunch was served. "Calories" is the average number of calories the child consumed during lunch, calculated from careful observation of what the child ate each day. TABTIME

Time	21.4	30.8	37.7	33.5	32.8	39.5	22.8	34.1	33.9	43.8
Calories	472	498	465	456	423	437	508	431	479	454
Time	42.4	43.1	29.2	31.3	28.6	32.9	30.6	35.1	33.0	43.7
Calories	450	410	504	437	489	436	480	439	444	408

(a) Make a scatterplot. Find the correlation and the least-squares regression line. (Be sure to save the regression residuals.) Based on your work, describe the direction, form, and strength of the relationship.

(b) Check the conditions for regression inference. Parts (a) through (d) of Exercise 26.36 provide a handy outline. Use vertical limits −100 to 100 in your plot of the residuals against time to help visualize the pattern. What do you conclude?

(c) Is there significant evidence that more time at the table is associated with more calories consumed? Give a 95% confidence interval to estimate how rapidly calories consumed changes as time at the table increases.

26.39 DNA on the Ocean Floor. We think of DNA as the stuff that stores the genetic code. It turns out that DNA occurs, mainly outside living cells, on the ocean floor. It is important in nourishing seafloor life. Scientists think that this DNA comes from organic matter that settles to the bottom from the top layers of the ocean. "Phytopigments," which come mainly from algae, are a measure of the amount of organic matter that has settled to the bottom. The data contain concentrations of DNA and phytopigments (both in grams per square meter) in 116 ocean locations around the world.[11] Look first at DNA alone. Describe the distribution of DNA concentration and give a confidence interval for the mean concentration. Be sure to explain why your confidence interval is trustworthy in light of the shape of the distribution. The data show surprisingly high DNA concentration, and this by itself was an important finding. DNA

26.40 Time at the Table: Prediction. Rachel attends the nursery school discussed in Exercise 26.38. Over several months, Rachel averages 40 minutes at the lunch table. Give a 95% interval to predict Rachel's average calorie consumption at lunch. TABTIME

Exercises 26.41 through 26.45 ask practical questions involving regression inference without step-by-step instructions. Do complete regression analyses, using the Plan, Solve, and Conclude steps of the four-step process to organize your answers. Follow the model of Example 26.9 (page 607) and the subsequent discussion and check the conditions as part of the Solve step.

26.41 Yukon Squirrels. 4step Table 4.2 for Exercise 4.48 (page 122) gives data on the abundance of the pinecones that red squirrels feed on and the population density of red squirrels over 23 years. How significant is the evidence that more cones leads to higher population density? (Use a vertical scale from −1.5 to 1.5 in your residual plot to show the pattern more clearly.) SQRLCO

26.42 A Big-Toe Problem. 4step Table 7.3 (page 195) and Exercises 7.49 and 7.51 describe the relationship between two deformities of the feet in young patients. Metatarsus adductus (MA) may help predict the severity of hallux abducto valgus (HAV). The paper that reports this study says, "Linear regression analysis, using the hallux abducto angle as the response variable, demonstrated a significant correlation between the metatarsus adductus and hallux abducto angles."[12] Do a suitable analysis to verify this finding. The study authors note that the scatterplot suggests that the variation in y may change as x changes, so they offer a more elaborate analysis as well. BIGTOE

26.43 Beavers and Beetles. 4step Exercise 5.60 (page 157) describes a study that found that the number of stumps from trees felled by beavers predicts the abundance of beetle larvae. Is there good evidence that more beetle larvae clusters are present when beavers have left more tree stumps? Estimate how many more clusters accompany each additional stump, with 95% confidence. BEAVERS

26.44 Sulfur, the Ocean, and the Sun. 4step Sulfur in the atmosphere affects climate by influencing formation of clouds. The main natural source of sulfur is dimethylsulfide (DMS)

produced by small organisms in the upper layers of the oceans. DMS production is in turn influenced by the amount of energy the upper ocean receives from sunlight. Monthly data on solar radiation dose (SRD, in watts per square meter) and surface DMS concentration (in nanomolars) for a region in the Mediterranean are given. Do the data provide convincing evidence that DMS increases as SRD increases? We also want to estimate the rate of increase, with 90% confidence. SULFUR

26.45 **DNA on the Ocean Floor.** Another conclusion of the study introduced in Exercise 26.39 was that organic matter settling down from the top layers of the ocean is the main source of DNA on the seafloor. An important piece of evidence is the relationship between DNA and phytopigments. Do the data give good reason to think that phytopigment concentration helps explain seafloor DNA concentration? (Try vertical limits −0.5 to 0.5 to make the pattern of your residual plot clearer.) DNA

26.46 **A Lurking Variable.** Return to the data on selling price versus appraised value for beachfront condominiums that are the basis for the Exercises 26.16 through 26.24. The data are in order by date of the sale, and the data table includes the number of months from the start of the data period. The data and the residuals are in the data set CONDRES. CONDRES

(a) Plot the residuals against the explanatory variable (appraised value). To make the pattern clearer, use vertical limits −1000 to 1000. Does the pattern you see agree with the conditions of linear relationship and constant standard deviation needed for regression inference?

(b) Make a stemplot of the residuals. Are there strong deviations from Normality that would prevent regression inference?

(c) Next, plot the residuals against month. Are the positive and negative residuals randomly scattered, as would be the case if the conditions for regression inference are satisfied? (Suppose prices for beachfront property rise rapidly during any period. Because property is reassessed just once a year, selling prices might pull away from appraised values over time within a period, creating a pattern of many negative residuals followed by several positive residuals.) As this example illustrates, it is often wise to plot residuals against important lurking variables as well as against the explanatory variable.

26.47 **Standardized Residuals.** Software often calculates *standardized residuals* as well as the actual residuals from regression. Because the standardized residuals have the standard z-score scale, it is easier to judge whether any are extreme. Figure 26.14 and the associated data include the standardized residuals for the regression of gate velocity selected by foundry workers on piston wall thickness, for Exercises 26.26 through 26.28.

(a) Find the mean and standard deviation of the standardized residuals. Why do you expect values close to those you obtain?

(b) Make a stemplot of the standardized residuals. Are there any striking deviations from Normality?

(c) The most extreme standardized residual is $z = -2.21$. Minitab flags this as "large." What is the probability that a standard Normal variable takes a value this extreme (that is, less than −2.21 or greater than 2.21)? Your result suggests that a residual this extreme would be a bit unusual when there are only 12 observations. That's why we examined Observation 9 in Exercise 26.28.

26.48 **Tests for the Intercept.** Figure 26.7 (page 603) gives JMP output for the regression of blood alcohol content (BAC) on number of beers consumed. The t test for the hypothesis that the population regression line has *slope* $\beta = 0$ has $P < 0.001$. The data show a positive linear relationship between BAC and beers. We might expect the *intercept* α of the population regression line to be 0 because no beers ($x = 0$) should produce no alcohol in the blood ($y = 0$). To test

$$H_0\colon \alpha = 0$$
$$H_a\colon \alpha \neq 0$$

we use a t statistic formed by dividing the least-squares intercept a by its standard error SE_a. Locate this statistic in the output of Figure 26.7 and verify that it is, in fact, a divided by its standard error. What is the P-value? Do the data suggest that the intercept is not zero?

26.49 **Confidence Intervals for the Intercept.** The output in Figure 26.7 (page 603) allows you to calculate confidence intervals for both the slope β and the intercept α of the population regression line of BAC on beers in the population of all students. Confidence intervals for the intercept α have the familiar form $a \pm t^* SE_a$ with degrees of freedom $n - 2$. What is the 95% confidence interval for the intercept? Does it contain zero, the value we might guess for α?

CHAPTER

27

When you complete this chapter, you will be able to:

27.1 Explain the advantages of an overall test of differences and follow-up analysis when comparing means of more than two populations over conducting individual pairs of comparisons.

27.2 Identify the null hypothesis in an ANOVA *F* test in a specific setting and reach a conclusion based on the results.

27.3 Locate the *F* statistic and its *P*-value on the output of analysis of variance software.

27.4 Explain how variation within samples and sample size influence the statistical significance of differences among sample means.

27.5 Check the data production, the presence of outliers, and differences in sample standard deviations in order to recognize when you can safely use ANOVA to compare means.

27.6 Find the degrees of freedom for the *F* statistic from the number and sizes of the samples.

27.7 Use software to carry out Tukey pairwise multiple comparisons among means and interpret the results.

27.8 Use the mean square for groups and the mean square error to calculate the ANOVA *F* statistic.

One-Way Analysis of Variance: Comparing Several Means

The two-sample *t* procedures of Chapter 21 compare the means of two populations or the mean responses to two treatments in an experiment. Of course, studies don't always compare just two groups. We need a method for comparing any number of means. Comparing the means of multiple populations allows one to test for a relationship between a categorical variable (the characteristic that makes the populations apparently different from each other) and means.

EXAMPLE 27.1 Comparing Tropical Flowers

STATE: Ethan Temeles and W. John Kress of Amherst College studied the relationship between varieties of the tropical flower *Heliconia* on the island of Dominica and the different species of hummingbirds that fertilize the flowers.[1] Over time, the researchers believe, the lengths of the flowers and the forms of the hummingbirds' beaks have evolved to match each other. If that is true, flower varieties fertilized by different hummingbird species should have distinct distributions of length.

4step

FLOWER

Table 27.1 gives length measurements (in millimeters) for samples of three varieties of *Heliconia*, each fertilized by a different species of hummingbird. Do the three varieties display distinct distributions of length? In particular, are the mean lengths of their flowers different and, if so, by how much do they differ?

inhauscreative/Getty Images

TABLE 27.1 Flower lengths (millimeters) for three *Heliconia* varieties

			H. Bihai				
47.12	46.75	46.81	47.12	46.67	47.43	46.44	46.64
48.07	48.34	48.15	50.26	50.12	46.34	46.94	48.36
			***H. Caribaea* red**				
41.90	42.01	41.93	43.09	41.47	41.69	39.78	40.57
39.63	42.18	40.66	37.87	39.16	37.40	38.20	38.07
38.10	37.97	38.79	38.23	38.87	37.78	38.01	
			***H. Caribaea* yellow**				
36.78	37.02	36.52	36.11	36.03	35.45	38.13	37.10
35.17	36.82	36.66	35.68	36.03	34.57	34.63	

Martin Mecnarowski/Shutterstock

PLAN: Use graphs and numerical descriptions to describe and compare the three distributions of flower length. Then determine whether the differences among the mean lengths of the three varieties are *statistically significant.* The final step, which we return to in Section 27.7, answers the question "Which specific varieties differ from each other, and by how much?"

SOLVE (first steps): Figure 27.1 displays side-by-side stemplots with the stems lined up for easy comparison. The lengths have been rounded to the nearest tenth of a millimeter. Here are the summary measures we will use in further analysis:

Sample	Variety	Sample Size	Mean Length	Standard Deviation
1	*H. bihai*	16	47.60	1.213
2	*H. caribaea* red	23	39.71	1.799
3	*H. caribaea* yellow	15	36.18	0.975

CONCLUDE (first steps): The three varieties differ so much in flower length that there is little overlap among them. In particular, the flowers of *H. bihai* are longer than either *H. caribaea* red or yellow. The mean lengths are 47.6 mm for *H. bihai,* 39.7 mm for *H. caribaea* red, and 36.2 mm for *H. caribaea* yellow. Are these observed differences in sample means statistically significant? We must develop a test for comparing more than two population means.

H. bihai		H. caribaea red		H. caribaea yellow	
34		34		34	66
35		35		35	257
36		36		36	0015788
37		37	489	37	01
38		38	00112289	38	1
39		39	268	39	
40		40	67	40	
41		41	5799	41	
42		42	02	42	
43		43	1	43	
44		44		44	
45		45		45	
46	3467889	46		46	
47	114	47		47	
48	1234	48		48	
49		49		49	
50	13	50		50	

FIGURE 27.1
Side-by-side stemplots comparing the lengths, in millimeters, of samples of flowers from three varieties of *Heliconia,* from Table 27.1.

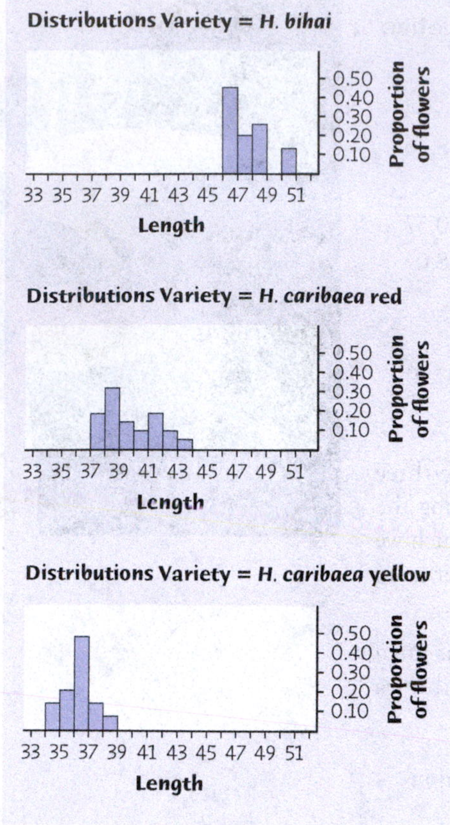

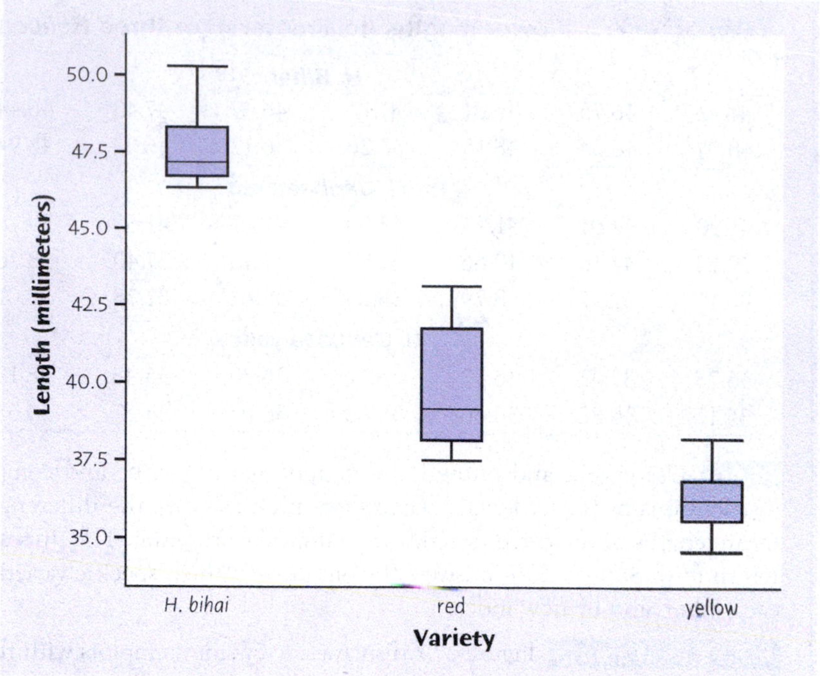

FIGURE 27.2
Histograms (a) and boxplots (b) comparing the lengths, in millimeters, of samples of flowers from three varieties of *Heliconia,* from Table 27.1.

In the Solve step, we have chosen to draw side-by-side stemplots to compare the three distributions. The three distributions can also be compared using histograms or boxplots. Figure 27.2(a) provides comparative histograms, and Figure 27.2(b) provides comparative boxplots. Because we want to compare the size of the responses for the three distributions as well as examine the general shape, it is best to align the response axis for the three graphs for ease of comparison. If the histograms in Figure 27.2(a) were side by side rather than above each other, as in the figure, comparisons of the lengths for the three varieties would be more difficult. For comparison purposes, it is also best to use the proportion or percentage scale for the histograms when the sample sizes differ. Figure 27.2(b) shows the comparative boxplots.

Our impression of the distributions is similar for each of the three graphical displays. The flowers of *H. bihai* are longer than either *H. caribaea* red or yellow, and the variability for the lengths of the red flowers is somewhat larger than for the other two varieties. The choice of an appropriate graph is based on your software capabilities and, to some extent, personal preference. Although boxplots show less detail, they are useful for comparison as long as the sample sizes are not very small.

27.1 Comparing Several Means

Call the mean lengths for the three populations of flowers μ_1 for *H. bihai,* μ_2 for *H. caribaea* red, and μ_3 for *H. caribaea* yellow. The subscript reminds us which group a parameter or statistic describes. To compare these three population means, we might use the two-sample t test several times:

- Test H_0: $\mu_1 = \mu_2$ to see if the mean length for *H. bihai* differs from the mean for *H. caribaea* red.
- Test H_0: $\mu_1 = \mu_3$ to see if the mean length for *H. bihai* differs from the mean for *H. caribaea* yellow.

- Test H_0: $\mu_2 = \mu_3$ to see if the mean length for *H. caribaea* red differs from the mean for *H. caribaea* yellow.

The weakness of doing three tests is that we get three *P*-values, one for each test alone. That doesn't tell us how likely it is that *three* sample means are as different from each other as these are. It may be that $\bar{x}_1 = 47.60$ and $\bar{x}_3 = 36.18$ are significantly different if we look at just two groups but not significantly different if we know that they are the largest and the smallest means in three groups. As we look at more groups, we expect the gap between the largest and smallest sample means to get larger. (Think of comparing the tallest person and shortest person in larger and larger groups of people.) *We can't safely compare many parameters by doing tests or confidence intervals for two parameters at a time.*

The problem of how to do many comparisons at once with an overall measure of confidence in all our conclusions is common in statistics. This is the problem of *multiple comparisons* introduced in Chapter 25. Statistical methods for dealing with multiple comparisons usually have two steps:

1. An *overall test* to see if there is good evidence of *any* differences among the parameters that we want to compare
2. A detailed *follow-up analysis* to decide which of the parameters differ and to estimate how large the differences are

The overall test, though more complex than the tests we have met to this point, is reasonably straightforward. Formal follow-up analysis can be more elaborate. In the next few sections, we will concentrate on the overall test and use data analysis to describe informally the nature of the differences in the means. Section 27.7 returns to the problem of multiple comparisons and provides some of the details of a formal follow-up analysis.

27.2 The Analysis of Variance *F* Test

We want to test the null hypothesis that there are *no differences* among the mean lengths for the three populations of flowers:

$$H_0: \mu_1 = \mu_2 = \mu_3$$

The basic *conditions for inference* (more detail later) are that we have random samples from the three populations and that flower lengths are Normally distributed in each population.

The alternative hypothesis is that there is *some difference*. That is, not all three population means are equal:

$$H_a: \text{not all of } \mu_1, \mu_2, \text{and } \mu_3 \text{ are equal}$$

The alternative hypothesis is no longer one-sided or two-sided. It is "many-sided" because it allows any relationship other than "all three equal." For example, H_a includes the case in which $\mu_2 = \mu_3$ but μ_1 has a different value. The test of H_0 against H_a is called the **analysis of variance *F* test.** Analysis of variance is usually abbreviated as ANOVA. The ANOVA *F* test is almost always carried out by software that reports the test statistic and its *P*-value. **ANOVA *F* test**

The ANOVA *F* Test

Used to test the hypothesis that multiple populations all have the same mean for some quantitative variable.

EXAMPLE 27.2 Comparing Tropical Flowers: ANOVA

4step

SOLVE (inference): Software tells us that, for the flower length data in Table 27.1, the test statistic is $F = 259.12$ with *P*-value $P < 0.0001$. There is very strong evidence that the three varieties of flowers do not all have the same mean length.

The *F* test does not say *which* of the three means are significantly different. It appears from our preliminary data analysis that *H. bihai* flowers are distinctly longer than either *H. caribaea* red or *H. caribaea* yellow. *H. caribaea* red and *H. caribaea* yellow are closer together, but the *H. caribaea* red flowers tend to be longer.

CONCLUDE: There is strong evidence ($P < 0.0001$) that the population means are not all equal. The most important difference among the means is that the *H. bihai* variety has longer flowers than the *H. caribaea* red and yellow varieties.

Example 27.2 illustrates our initial approach to comparing means. The ANOVA *F* test (done by software) assesses the evidence for *some* difference among the population means. We will discuss the meaning of the *F* statistic, in this case 259.12, in Section 27.4. The *P*-value tells us that the probability of obtaining this result (or one more extreme), if the mean flower lengths are equal in the population, is less than 0.0001.

A formal follow-up analysis, as presented in Section 27.7, would allow us to say which means differ and by how much, with (say) 95% confidence that *all* our conclusions are correct. For now, we rely instead on an informal examination of the data using graphs and summary statistics to see what differences are present and whether they are large enough to be interesting.

APPLY YOUR KNOWLEDGE

27.1 Using Tablets in Class. Does the use of technology, such as tablets, in the classroom affect student learning? To explore this, a researcher examined the NAEP mathematics scores for a representative sample of 150,100 eighth-graders. Each student was classified by how frequently he or she had used a tablet in the classroom (never, in fewer that half of all classes, in about half of all classes, in more than half of all classes, or in all classes). Here are the mean NAEP mathematics scores for the five categories of frequency of tablet use in class:[2]

Never	Some	About half	More than half	All
283	279	269	271	279

An analysis of variance finds $F = 10{,}297$, with $P < 0.0001$.

(a) What are the null and alternative hypotheses for the ANOVA *F* test? Be sure to explain what means the test compares.

(b) Based on the sample means and the *F* test, what do you conclude?

27.2 Political Views and Education. The University of Chicago's General Social Survey (GSS) is the nation's most important social science sample survey. The GSS asked a random sample of adults in 2018 their highest degree earned and where they placed themselves on the political spectrum using a 7-point scale from 1 = extremely liberal to 7 = extremely conservative. The analysis reports $F = 9.317$ with *P*-value < 0.0001 and, for each highest degree earned, provides the mean political spectrum score, which can be used to draw a graph like Figure 27.3.[3]

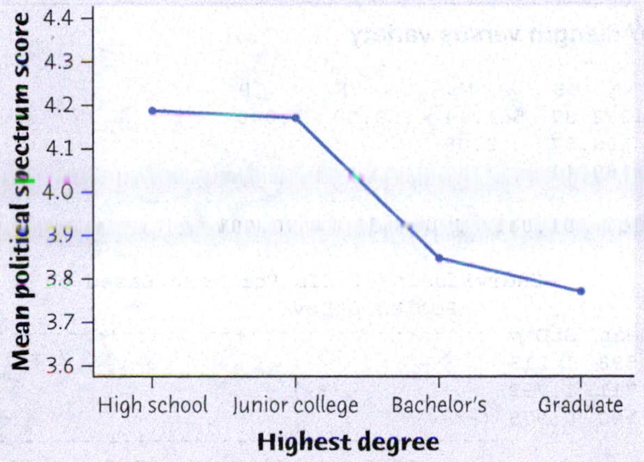

FIGURE 27.3
Line graph comparing the mean political spectrum scores for four levels of education, for Exercise 27.2.

(a) What are the null and alternative hypotheses for the ANOVA *F* test? Be sure to explain what means the test compares.

(b) Based on the graph and the *F* test, what do you conclude?

27.3 Using Technology

Any technology used for statistics should perform analysis of variance. Figure 27.4 displays ANOVA output for the data of Table 27.1 (page 621) from a graphing calculator, two statistical programs, and a spreadsheet program.

Minitab, JMP, and Excel give the sizes of the three samples and their means. These agree with those in Example 27.1. Minitab also gives the standard deviations, Excel gives the variances, and JMP gives the standard errors. All four outputs report the *F* test statistic, $F = 259.12$, and its *P*-value. Minitab reports the *P*-value with zero to three decimal places, while JMP reports that $P < 0.0001$. This is all we need to know about the *P*-value in practice. Excel and the graphing calculator offer the specific value 1.92×10^{-27}. (This would be correct if the population distributions were exactly Normal. In practice, read such values simply as "*P* is very small.") There is very strong evidence that the three varieties of flowers do not all have the same mean length.

All four outputs report degrees of freedom (df), sums of squares (SS), and mean squares (MS). We don't need this information now. Minitab and JMP also give confidence intervals for all three means that help us see which means differ and by how much. None of the intervals overlap, and *H. bihai* is quite far above the other two. These are 95% confidence intervals for each mean separately. We are *not* 95% confident that *all three* intervals cover the three means. This is another example of the peril of multiple comparisons.

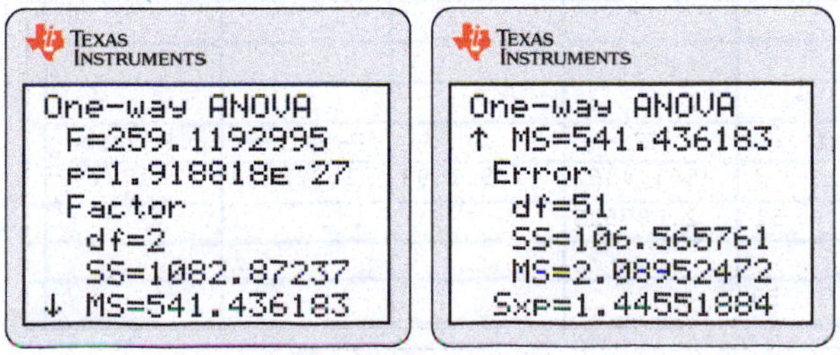

FIGURE 27.4
ANOVA for the flower length data: output from a graphing calculator, two statistical programs, and a spreadsheet program.

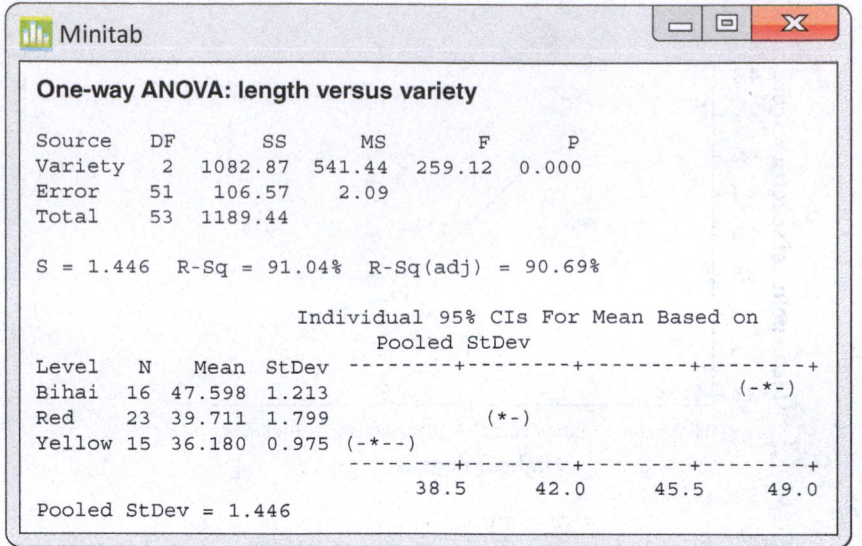

FIGURE 27.4
(Continued)

Minitab

One-way ANOVA: length versus variety

```
Source   DF       SS      MS       F      P
Variety   2  1082.87  541.44  259.12  0.000
Error    51   106.57    2.09
Total    53  1189.44

S = 1.446   R-Sq = 91.04%   R-Sq(adj) = 90.69%

                                 Individual 95% CIs For Mean Based on
                                 Pooled StDev
Level    N    Mean  StDev  --------+---------+---------+---------+
Bihai   16  47.598  1.213                                  (-*-)
Red     23  39.711  1.799                     (*-)
Yellow  15  36.180  0.975  (-*--)
                           --------+---------+---------+---------+
                                38.5      42.0      45.5      49.0

Pooled StDev = 1.446
```

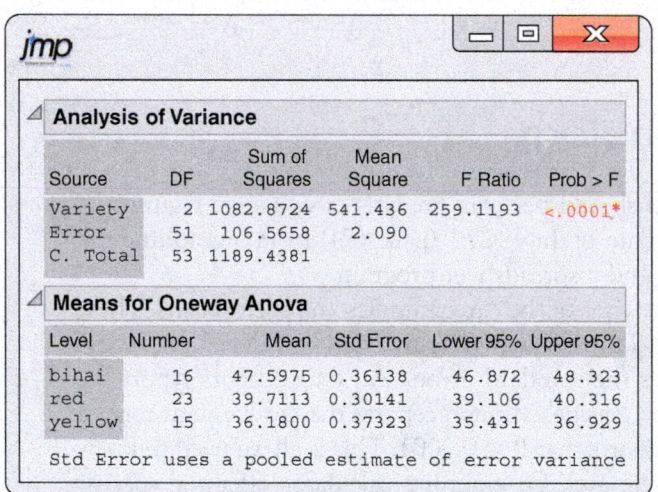

jmp

Analysis of Variance

Source	DF	Sum of Squares	Mean Square	F Ratio	Prob > F
Variety	2	1082.8724	541.436	259.1193	<.0001*
Error	51	106.5658	2.090		
C. Total	53	1189.4381			

Means for Oneway Anova

Level	Number	Mean	Std Error	Lower 95%	Upper 95%
bihai	16	47.5975	0.36138	46.872	48.323
red	23	39.7113	0.30141	39.106	40.316
yellow	15	36.1800	0.37323	35.431	36.929

Std Error uses a pooled estimate of error variance

Excel

	A	B	C	D	E	F	G
1	Anova: Single Factor						
2							
3	SUMMARY						
4	*Groups*	*Count*	*Sum*	*Average*	*Variance*		
5	bihai	16	761.56	47.5975	1.471073		
6	red	23	913.36	39.7113	3.235548		
7	yellow	15	542.7	36.18	0.951257		
8							
9							
10	ANOVA						
11	*Source of variation*	*SS*	*df*	*MS*	*F*	*P-value*	*F crit*
12	Between Groups	1082.872	2	541.4362	259.1193	1.92E-27	3.178799
13	Within Groups	106.5658	51	2.089525			
14							
15	Total	1189.438	53				

APPLY YOUR KNOWLEDGE

27.3 Logging in the Rain Forest. How does logging in a tropical rain forest affect the forest in later years? Researchers compared forest plots in Borneo that had never been logged (Group 1) with similar plots nearby that had been logged one year earlier (Group 2) and eight years earlier (Group 3). Although the study was not an experiment, the authors explained why we can consider the plots to be randomly selected. The data appear in Table 27.2. The variable Trees is the count of trees in a plot; Species is the count of tree species in a plot. The variable Richness is Species/Trees, the number of species divided by the number of individual trees.[4] LOGFULL

TABLE 27.2 Data from a study of logging in Borneo

Group	Trees	Species	Richness	Group	Trees	Species	Richness
1	27	22	0.81481	2	18	15	0.83333
1	22	18	0.81818	2	17	15	0.88235
1	29	22	0.75862	2	14	12	0.85714
1	21	20	0.95238	2	14	13	0.92857
1	19	15	0.78947	2	2	2	1.00000
1	33	21	0.63636	2	17	15	0.88235
1	16	13	0.81250	2	19	8	0.42105
1	20	13	0.65000	3	18	17	0.94444
1	24	19	0.79167	3	4	4	1.00000
1	27	13	0.48148	3	22	18	0.81818
1	28	19	0.67857	3	15	14	0.93333
1	19	15	0.78947	3	18	18	1.00000
2	12	11	0.91667	3	19	15	0.78947
2	12	11	0.91667	3	22	15	0.68182
2	15	14	0.93333	3	12	10	0.83333
2	9	7	0.77778	3	12	12	1.00000
2	20	18	0.90000				

(a) Make an appropriate comparative graph of the variable Trees for the three groups. What effects of logging are visible?

(b) Figure 27.5 shows JMP output for Trees. What do the group means show about the effects of logging?

(c) What are the *F*-value and its *P*-value? What hypotheses does *F* test? What conclusions about the effects of logging on number of trees do the data lead to?

27.4 Stress and Environment in the Prison Population. This study examines the psychological effects of restrictive environments in a medium-security federal correctional institution in Kentucky. A sample of 10 inmates living in the general population (G), 10 inmates living in administrative detention (AD), and 10 inmates living in disciplinary segregation (DS) was selected. When an inmate's presence in the general population poses a threat to the orderly running of the institution, he is placed in administrative detention, where he generally lives with another inmate, while disciplinary segration confines those inmates found guilty of serious violation of regulations to an individual cell removed from the general population. Each sampled individual completed the Brief Symptom Inventory (BSI) consisting of 53 items designed to assess an individual's current level of psychological distress.[5]

(a) Is this an observational study or an experiment? How will this distinction affect your conclusions? Explain briefly.

(b) Figure 27.6 gives the Minitab ANOVA output for these data. What do the mean BSI scores say about the relationship between psychological distress and the living restrictions of inmates?

(c) What are the*F*-value and its *P*-value? What hypotheses does *F* test? Briefly describe the conclusions you can draw from these data.

jmp

Analysis of Variance

Source	DF	Sum of Squares	Mean Square	F Ratio	Prob > F
Group	2	625.1566	312.578	11.4257	0.0002*
Error	30	820.7222	27.357		
C. Total	32	1445.8788			

Means for Oneway Anova

Level	Number	Mean	Std Error	Lower 95%	Upper 95%
1	12	23.7500	1.5099	20.666	26.834
2	12	14.0833	1.5099	11.000	17.167
3	9	15.7778	1.7435	12.217	19.338

Std Error uses a pooled estimate of error variance

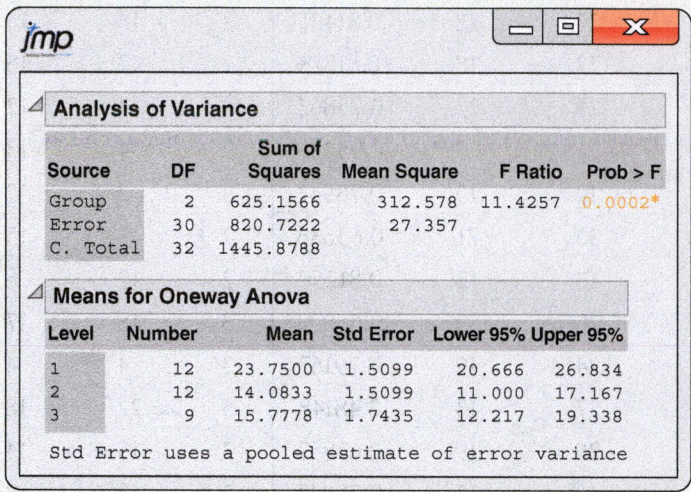

FIGURE 27.5
JMP output for analysis of variance on the number of trees in forest plots, for Exercise 27.3.

Minitab

One-way ANOVA: Stress score versus Environmental restriction

Analysis of Variance

Source	DF	Adj SS	Adj MS	F-Value	P-Value
Environmental restriction	2	2468	1233.8	4.86	0.016
Error	27	6848	253.6		
Total	29	9316			

S	R-Sq	R-Sq(adj)	R-Sq(pred)
15.9259	26.49%	21.04%	9.25%

Environmental restriction	N	Mean	StDev	95% CI
AD	10	66.31	16.00	(55.97, 76.64)
DS	10	73.50	8.00	(63.17, 83.83)
G	10	51.70	21.00	(41.37, 62.03)

Pooled StDev = 15.9259

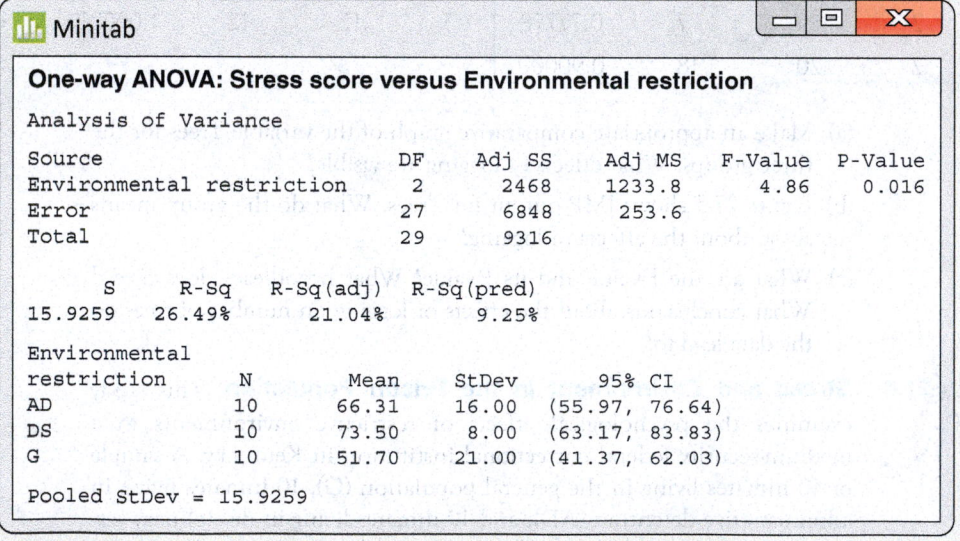

FIGURE 27.6
Minitab output for the data on respondents' BSI scores for three living restrictions for inmates, for Exercise 27.4.

27.4 The Idea of Analysis of Variance

The details of ANOVA are a bit daunting. (They appear in the optional Section 27.8 at the end of this chapter.) The main idea of ANOVA is more accessible and much more important. Here it is: when we ask if a set of sample means gives evidence for differences among the population means, what matters is not how far apart the sample means are but how far apart they are *relative to the variability of individual observations.*

Look at the two sets of boxplots in Figure 27.7. For simplicity, these distributions are all symmetric, so that the mean and median are the same. The center line in each boxplot is therefore the sample mean. The three sample means for the boxplots in Figure 27.7(a) are the same as the three means in 27.7(b). Could differences this large among the means easily arise just due to chance, or are they statistically significant?

- The boxplots in Figure 27.7(a) have tall boxes, which show lots of variation among the individuals in each group. With this much variation among individuals, we would not be surprised if another set of samples gave quite different sample means. The observed differences among the sample means could easily happen just by chance.
- The boxplots in Figure 27.7(b) have the same centers as those in Figure 27.7(a), but the boxes are much shorter. That is, there is much less variation among the individuals in each group. It is unlikely that any sample from the first group would have a mean as small as the mean of the second group. Because means as far apart as those observed would rarely arise just by chance in repeated sampling, they provide good evidence of real differences among the means of the three populations we are sampling from.

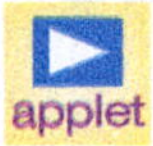

You can use the *One-Way ANOVA* applet to demonstrate the analysis of variance idea for yourself. The applet allows you to change both the group means and the variability within groups. You can watch the value of F and the associated P-value change as you work.

This comparison of the two parts of Figure 27.7 is too simple in one way. It ignores the effect of the sample sizes, an effect that boxplots do not show. *Small differences among sample means can be significant if the samples are large. Large differences among sample means can fail to be significant if the samples are small.* All we can be sure of is that for the same sample size, Figure 27.7(b) will give a much smaller P-value than Figure 27.7(a). Despite this qualification, the big idea remains: if sample means are far apart relative to the variation among individuals in the same groups, that's evidence that something other than chance is at work.

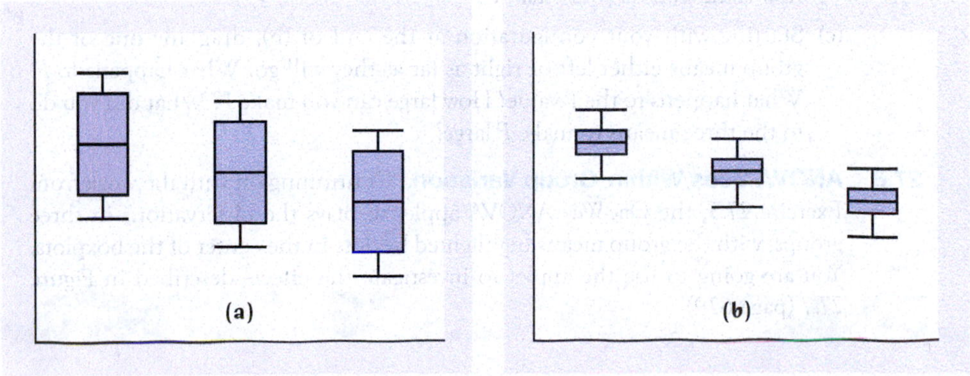

FIGURE 27.7 Boxplots for two sets of three samples each. The sample means are the same in parts (a) and (b). Analysis of variance will find a more significant difference among the means in part (b) because there is less variation among the individuals within those samples.

The Analysis of Variance Idea

ANOVA tests whether more than two populations have the same mean by comparing the variation among sample means with variation within the samples.

It is one of the oddities of statistical language that methods for comparing means are named after the variance. The reason is that the test works by comparing two kinds of variation. Analysis of variance is a general method for studying sources of variation in responses. Comparing several means is the simplest form of ANOVA; this is called **one-way ANOVA.** (The *one-way* in the name one-way ANOVA indicates that this type of analysis involves only one quantitative variable. More advanced ANOVA techniques can test for relationships among more than one variable in multiple populations.)

The ANOVA F Statistic

The **analysis of variance F statistic** for testing the equality of several means has this form:

$$F = \frac{\text{variation among the sample means}}{\text{variation among individuals in the same sample}}$$

For more detail, read the optional Section 27.8 at the end of this chapter. The F statistic can take only values that are zero or positive. It is zero only when all the sample means are identical, and it gets larger as they move farther apart. Large values of F are evidence against the null hypothesis H_0 that all population means are the same. Although the alternative hypothesis H_a is many-sided, the ANOVA F test is one-sided because any violation of H_0 tends to produce a large value of F.

APPLY YOUR KNOWLEDGE

applet

27.5 ANOVA Compares Several Means. The *One-Way ANOVA* applet displays the observations in three groups, and the group means are highlighted by dots in the center of the boxplots.

(a) What are the *F*- and *P*-values for these three samples? (The *P*-value is marked by a red dot that will move along the scale as you modify the samples.)

(b) Which group has the largest mean? If you grab its mean point with the mouse and begin to move it left or right, the *F*-value will change. How small can you make F? What did you do to this mean to make F small? Roughly how significant is your small F?

(c) Starting with your configuration at the end of (b), drag any one of the group means either left or right as far as they will go. What happens to F? What happens to the *P*-value? How large can you make F? What did you do to the three means to make F large?

applet

27.6 ANOVA Uses Within-Group Variation. Continuing on with the work from Exercise 27.5, the *One-Way ANOVA* applet displays the observations in three groups, with the group means highlighted by dots in the center of the boxplots. You are going to use the applet to investigate the effects described in Figure 27.7 (page 629).

(a) Use the mouse to change the standard deviations of each of the groups to 3. You see that the group means do not change, but the variability of the observations in each group increases. What happens to F and P as the variability among the observations in each group increases? What are the values of F and P when the slider is all the way to the right? This is similar to Figure 27.7(a): variation within groups hides the differences among the group means.

(b) Leave the standard deviations at 3. Use the mouse to move the group means apart. What happens to F and P as you do this?

27.5 Conditions for ANOVA

Like all other inference procedures, ANOVA is valid only in some circumstances. Here are the conditions under which we can use ANOVA to compare population means.

Conditions for ANOVA Inference

- We have I independent SRSs, one from each of I populations. We measure the same quantitative response variable for each sample.
- All of the I populations are Normally distributed, and the ith population has unknown mean μ_i. One-way ANOVA tests the null hypothesis that all the population means are the same.
- All the populations have the same standard deviation σ, whose value is unknown.

STATISTICS IN YOUR WORLD

We Weren't Working Anyway

A "consultant" estimated that the annual NCAA men's basketball tournament costs employers $3.8 billion in time wasted by workers participating in office pools, checking game scores, and so on. That's unlikely. Most of the games are played outside work hours, at night and on weekends. More to the point, economists note that workers waste lots of time every workday by talking to other employees, chatting on the phone, shopping online, and so on. The dollar value of time spent on the basketball tournament probably comes in large part from time we were wasting anyway.

The first two conditions are familiar from our study of the two-sample t procedures for comparing two means. As usual, the design of the data production is the most important condition for inference. Biased sampling or confounding can make any inference meaningless. *If we do not actually draw separate SRSs from each population or carry out a randomized comparative experiment, it may be unclear to what population the conclusions of inference apply.* ANOVA, like other inference procedures, is often used when random samples are not available. You must judge each use on its merits, a judgment that usually requires some knowledge of the subject of the study in addition to some knowledge of statistics.

Because no real population has exactly a Normal distribution, the usefulness of inference procedures that assume Normality depends on how sensitive they are to departures from Normality. Fortunately, procedures for comparing means are not very sensitive to lack of Normality. The ANOVA F test, like the t procedures, is *robust*. What matters is Normality of the sample means, so ANOVA becomes safer as the sample sizes get larger because of the central limit theorem. Remember to check for outliers that change the value of sample means and for extreme skewness. When there are no outliers and the distributions are roughly symmetric, you can safely use ANOVA for sample sizes as small as 4 or 5.

The third condition is annoying: ANOVA assumes that the variability of observations, measured by the standard deviation, is the same in all populations. The t test for comparing two means (Chapter 21) does not require equal standard deviations. Unfortunately, the ANOVA F statistic for comparing more than two means is less broadly valid. It is not easy to check the condition that the populations have equal standard deviations. Statistical tests for equality of standard deviations are very sensitive to lack of Normality—so much so that they are of little practical value. You must either seek expert advice or rely on the robustness of ANOVA.

How serious are unequal standard deviations? ANOVA is not too sensitive to violations of the condition, especially when all samples have the same or similar sizes and no sample is very small. When designing a study, try to take samples of about the same size from all the groups you want to compare. The sample standard deviations estimate the population standard deviations, so check before doing ANOVA that the sample standard deviations are similar to each other. We expect some variation among them due to chance. Here is a rule of thumb that is safe in almost all situations.

Checking Standard Deviations in ANOVA

The results of the ANOVA *F* test are approximately correct when the largest sample standard deviation is no more than twice as large as the smallest sample standard deviation.

EXAMPLE 27.3 Comparing Tropical Flowers: Conditions for ANOVA

The study of *Heliconia* blossoms is based on three independent samples that the researchers consider to be random samples from all flowers of these varieties in Dominica. The stemplots in Figure 27.1 show that the *H. bihai* and *H. caribaea* red varieties have slightly skewed distributions, but the sample means of samples of sizes 16 and 23 will have distributions that are close to Normal. The sample standard deviations for the three varieties are

$$s_1 = 1.213 \quad s_2 = 1.799 \quad s_3 = 0.975$$

These standard deviations satisfy our rule of thumb:

$$\frac{\text{largest } s}{\text{smallest } s} = \frac{1.799}{0.975} = 1.85 \quad \text{(less than 2)}$$

We can safely use ANOVA to compare the mean lengths for the three populations.

EXAMPLE 27.4 Policy Justification: Pragmatic Versus Moral

STATE: How does a leader's justification of his or her organization's policy affect support for the policy? This study compared moral, pragmatic, and ambiguous justifications for three public policy proposals: a politician's plan to fund a retirement planning agency, a state governor's plan to repave state highways, and a president's plan to outlaw child labor in a developing country. For example, for the retirement agency proposal, the moral justification was the importance of retirees "to live with dignity and comfort," the pragmatic was "to not drain public funds," and the ambiguous was "to have sufficient funds." Of the 374 volunteer subjects assigned at random to read all three proposals, 122 subjects read the three proposals with a moral justification, 126 subjects read the three proposals with a pragmatic justification, and 126 subjects read the three proposals with an ambiguous justification. Subjects answered several questions measuring support for each policy proposal to create a support score for each proposal, and their scores for the three proposals were then averaged to create an index of policy support, with higher values indicating greater support.[6] Here are the first five observations:

Justification:	**Pragmatic**	**Ambiguous**	**Pragmatic**	**Moral**	**Ambiguous**
Policy support index:	5	7	4.75	7	5.75

The first individual read the proposals with a pragmatic justification with a policy support index of 5, the second with an ambiguous justification and a policy support index of 7, and so forth. On average, are the policy support indices different for the three justifications?

PLAN: Examine the data to compare the effect of the treatments and check that we can safely use ANOVA. If the data allow ANOVA, assess the significance of observed differences in mean indices of policy support.

SOLVE: Figure 27.8 compares the histograms of the data in the three groups. Although there is some irregularity in the histograms, the sample sizes are large, the distributions are not that far from the normal curves superimposed on the histograms, and there are no outliers or strong skewness that would hinder use of ANOVA. The

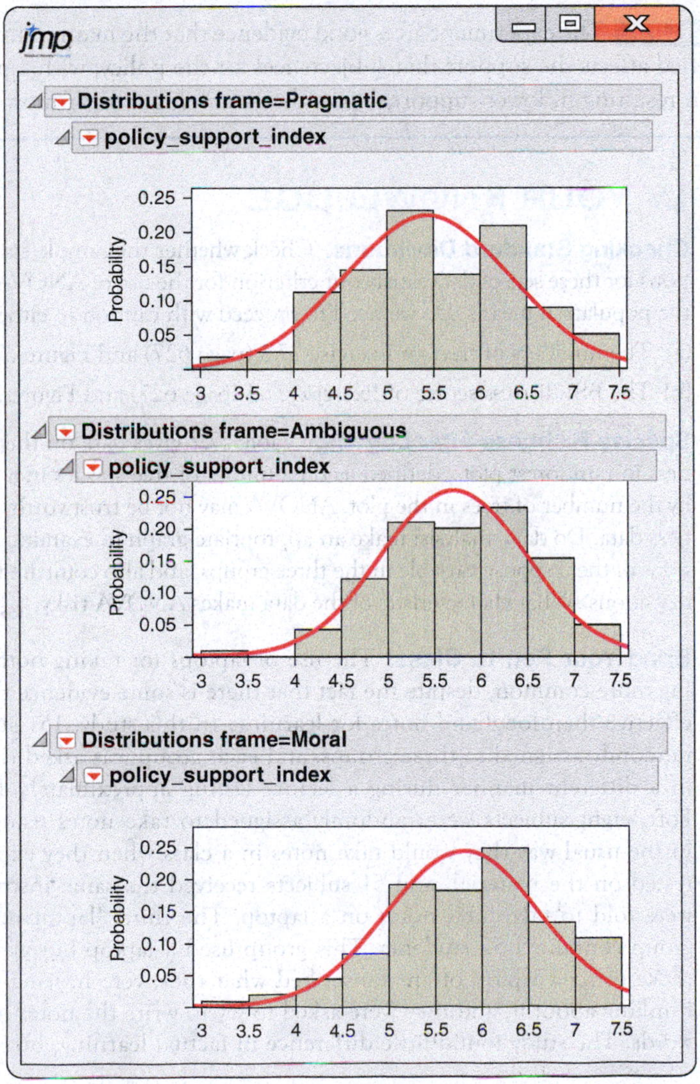

FIGURE 27.8
Histograms comparing the policy support indices for three policy justifications, for Example 27.4.

Minitab ANOVA output in Figure 27.9 shows that the group standard deviations easily satisfy our rule of thumb. The subjects in the pragmatic group had a lower support index (mean 5.44) than did subjects in the ambiguous (mean 5.71) or the the moral (mean 5.81) groups. The three means are significantly different ($F = 6.33$, $P = 0.002$).

```
Minitab

One-way ANOVA: Policy_Support_Index versus Justification

Analysis of Variance

Source    DF        SS       MS  F-Value  P-Value
Frame      2     8.667   4.3337     6.33    0.002
Error    371   253.880   0.6843
Total    373   262.547

       S    R-sq  R-sq(adj)
0.827231   3.30%      2.78%

Justification        N    Mean   StDev
Ambiguous          126  5.7063  0.7722
Moral              122  5.8053  0.8169
Pragmatic          126  5.4444  0.8881
```

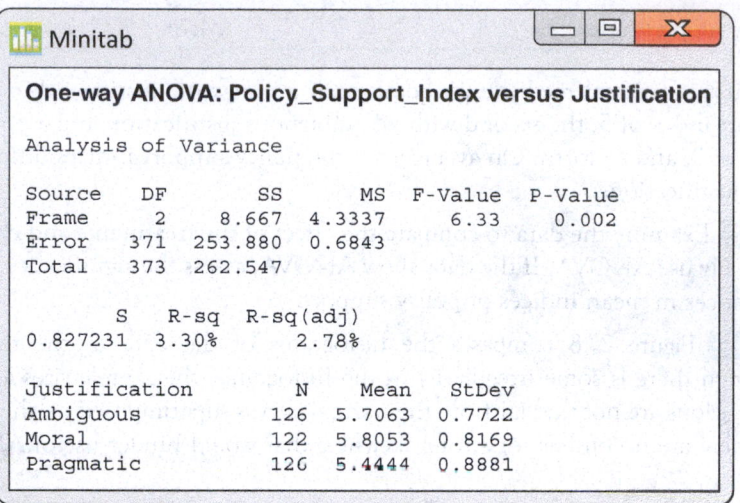

FIGURE 27.9 Minitab ANOVA output for comparing the three treatments in Example 27.4.

CONCLUDE: The experiment gives good evidence that the manner in which a policy is justified affects the support that subjects feel for the policy, with a pragmatic justification resulting in lower support than either an ambiguous or moral justification.

APPLY YOUR KNOWLEDGE

27.7 Checking Standard Deviations. Check whether the sample standard deviations for these sets of data satisfy our criterion for the use of ANOVA to compare the population means. Do we need to proceed with caution in either case?

(a) The numbers of trees in Exercise 27.3 (page 627) and Figure 27.5

(b) The BSI distress scores of Exercise 27.4 (page 627) and Figure 27.6

27.8 Species Richness after Logging. Table 27.2 gives data on the species richness in rain forest plots, defined as the number of tree species in a plot divided by the number of trees in the plot. ANOVA may not be trustworthy for the richness data. Do data analysis: make an appropriate graph to examine the distributions of the response variable in the three groups, and also compare the standard deviations. What characteristic of the data makes ANOVA risky? LOGFULL

4step

27.9 Bring Your Pen to Class! The use of laptops for taking notes is becoming more common, despite the fact that there is some evidence that it is less effective than longhand notes for learning. In this study, 151 subjects were randomly assigned to three groups, and each group was asked to take notes in a different manner during a lecture lasting approximately 15 minutes. Forty-eight subjects were randomly assigned to take notes using longhand in the usual way they would take notes in a class when they expected to be tested on the material, and 51 subjects received the same instructions but were told to take their notes on a laptop. The third "laptop-intervention" group contained 52 students. This group used a laptop but were told that those using a laptop often transcribed what they were hearing rather than thinking about it, and they were asked to try to write the notes in their own words. The study found little difference in factual learning, but those using longhand had greater conceptual learning than either of the other groups. Here we focus on the content of what they wrote—specifically the number of words.[7] Are the differences in the mean word counts for the three groups statistically significant? Analyze these data and discuss the results. Follow the four-step process, as illustrated in Example 27.4. WRDCOUNT

27.6 *F* Distributions and Degrees of Freedom

The ANOVA *F* statistic is

$$F = \frac{\text{variation among the sample means}}{\text{variation among individuals in the same sample}}$$

To find the *P*-value for this statistic, we must know the sampling distribution of *F* when the null hypothesis (all population means equal) is true. This sampling distribution is an *F distribution.*

The *F* distributions are a family of right-skewed distributions that take only values greater than zero. The density curves in Figure 27.10 illustrate their shapes. A specific *F* distribution is determined by the *degrees of freedom* of the numerator and denominator of the *F* statistic. You may have noticed that all our software outputs include degrees of freedom, labeled either "df" or "DF." The box below shows how to find degrees of freedom based on the number of populations compared and the sample sizes. Optional Section 27.8 explains where the numbers come from. When describing an *F* distribution, always give the numerator degrees of freedom first. Our brief notation will be *F*(df1, df2) for the *F* distribution with df1 degrees of freedom in the numerator and df2 in the denominator. *Interchanging the degrees of freedom changes the distribution, so the order is important.*

Tables of *F* critical points are awkward because we need a separate table for every pair of degrees of freedom df1 and df2. Fortunately, software gives you *P*-values for the ANOVA *F* test, without the need for a table.

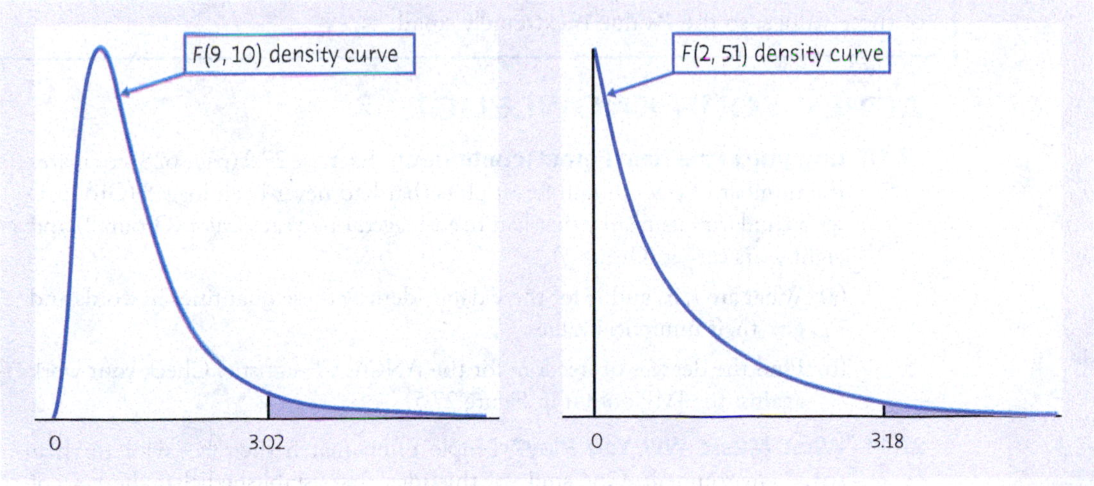

FIGURE 27.10
Density curves for two *F* distributions. Both are right-skewed and take only positive values. The upper 5% critical values are marked under the curves.

The degrees of freedom of the ANOVA F statistic depend on the number of means we are comparing and the number of observations in each sample. That is, the F test takes into account the number of observations. Here are the details.

Degrees of Freedom for the F Test

We want to compare the means of I populations. We have an SRS of size n_i from the ith population, so that the total number of observations in all samples combined is

$$N = n_1 + n_2 + \cdots + n_I$$

If the null hypothesis that all population means are equal is true, the ANOVA F statistic has the F distribution with $I - 1$ degrees of freedom in the numerator and $N - I$ degrees of freedom in the denominator.

EXAMPLE 27.5 Degrees of Freedom for F

In Examples 27.1 and 27.2, we compared the mean lengths for three varieties of flowers, so $I = 3$. Thc three sample sizes are

$$n_1 = 16 \quad n_2 = 23 \quad n_3 = 15$$

The total number of observations is therefore

$$N = 16 + 23 + 15 = 54$$

The ANOVA F test has numerator degrees of freedom

$$I - 1 = 3 - 1 = 2$$

and denominator degrees of freedom

$$N - I = 54 - 3 = 51$$

These are the degrees of freedom given in the outputs in Figure 27.4 (page 625). All three outputs give the degrees of freedom for the F test, labeled "df" or "DF." There are two degrees of freedom in the numerator and 51 in the denominator. P-values for the F test, therefore, come from the F distribution $F(2, 51)$ with two and 51 degrees of freedom. The right-hand curve in Figure 27.10 is the density curve of this distribution. The 5% critical value marked on that curve is 3.18, and the 1% critical value is 5.05. The observed value $F = 259.12$ of the ANOVA F statistic lies far to the right of these values, so the P-value is extremely small.

PhotoAlto sas/Alamy Stock Photo

APPLY YOUR KNOWLEDGE

27.10 Logging in the Rain Forest (continued). Exercise 27.3 (page 628) compares the number of trees in rain forest plots that had never been logged (Group 1) with similar plots nearby that had been logged one year earlier (Group 2) and eight years earlier (Group 3).

(a) What are I, n_i, and N for these data? Identify these quantities in words and give their numerical values.

(b) Find the degrees of freedom for the ANOVA F statistic. Check your work against the JMP output in Figure 27.5.

27.11 What Music Will You Play? People often match their behavior to their social environment. One study of this idea first established that the type of music most preferred by Black college students is R&B and that Whites' most preferred music is rock. Will students hosting a small group of other students choose music that matches the racial composition of the people attending? Assign 90 Black business students at random to three equal-sized groups. Do the same for 96 White students. Each student sees a picture of the people he or she will host. Group 1 sees six Blacks, Group 2 sees three Whites and three Blacks, and Group 3 sees six Whites. Ask how likely the host is to play the type of music preferred by the other race. Use ANOVA to compare the three groups to see whether the racial mix of the gathering affects the choice of music.[8]

(a) For the White subjects, $F = 16.48$. What are the degrees of freedom?

(b) For the Black subjects, $F = 2.47$. What are the degrees of freedom?

27.7 Follow-Up Analysis: Tukey Pairwise Multiple Comparisons

In Example 27.4 (page 632), we saw that there is good evidence that the mean policy support index is not the same for the moral, ambiguous, and pragmatic policy justifications. The sample means in Figure 27.9 suggest that the mean pragmatic justification results in lower policy support than either an ambigous or moral justification.

EXAMPLE 27.6 Comparing Groups: Individual *t* Procedures

JUSTIFY

How much higher is the mean policy support index with a moral justification than with a pragmatic justification? A 95% confidence interval comparing the moral and pragmatic justification groups answers this question. Because the conditions for ANOVA require the population standard deviation to be the same in all three populations of plots, we will use a version of the two-sample *t* confidence interval that also assumes equal standard deviations.

The standard error for the difference of sample means $\bar{x}_{moral} - \bar{x}_{pragmatic}$ estimates the standard deviation of the difference (see Section 21.3), which is

$$\sqrt{\frac{\sigma^2}{n_{moral}} + \frac{\sigma^2}{n_{pragmatic}}} = \sigma\sqrt{\frac{1}{n_{moral}} + \frac{1}{n_{pragmatic}}}$$

because both populations have the same standard deviation σ. The pooled standard deviation s_p is an estimate of σ based on all three samples (see page 644). So the standard error of $\bar{x}_{moral} - \bar{x}_{pragmatic}$ is

$$s_p\sqrt{\frac{1}{n_{moral}} + \frac{1}{n_{pragmatic}}}$$

The Minitab output in Figure 27.9 gives $s_p = 0.8272$. This estimate has 371 degrees of freedom, the degrees of freedom for "Error" in the ANOVA. A 95% confidence interval for $\mu_{moral} - \mu_{pragmatic}$ uses (from software) the critical value 1.966 of the *t* distribution with 371 degrees of freedom:

$$(\bar{x}_{moral} - \bar{x}_{pragmatic}) \pm t^* s_p\sqrt{\frac{1}{n_{moral}} + \frac{1}{n_{pragmatic}}}$$

$$= (5.8053 - 5.4444) \pm (1.966)(0.8272)\sqrt{\frac{1}{122} + \frac{1}{126}}$$

$$= 0.3609 \pm 0.2066$$

$$= 0.1543 \text{ to } 0.5675$$

We are 95% confident that the mean support index for the moral justification is between 0.1543 and 0.5675 units higher than for the pragmatic justification. Because this confidence interval does not contain zero, we can reject the null hypothesis of no difference, H_0: $\mu_{moral} = \mu_{pragmatic}$, in favor of the two-sided alternative at the 5% significance level.

Example 27.6 gives a single 95% confidence interval. We would like to estimate all three *pairwise differences* among the population means,

$$\mu_{moral} - \mu_{pragmatic} \qquad \mu_{moral} - \mu_{ambiguous} \qquad \mu_{ambiguous} - \mu_{pragmatic}$$

Three 95% confidence intervals will not give us 95% confidence that all three *simultaneously capture their true parameter values.* This is the problem of multiple comparisons, discussed on page 623.

In general, we want to give confidence intervals for all pairwise differences among the population means $\mu_1, \mu_2, \ldots, \mu_I$ of I populations. We want an *overall confidence level* of (say) 95%. That is, in very many uses of the method, *all* the intervals will simultaneously capture the true differences 95% of the time. To do this, take the number of comparisons into account by replacing the t critical value t^* in Example 27.6 by another critical value based on the distribution of the difference between the largest and smallest of a set of I sample means. We will call this critical value m^*, for multiple comparisons. Values of m^* depend on the number of populations we are comparing and on the total number of observations in the samples, as well as on the confidence level we want. Tables are therefore long and messy, so in practice we rely on software. This method is named after its inventor, John Tukey (1915–2000), who developed the ideas of modern data analysis.

Tukey Pairwise Multiple Comparisons

In the ANOVA setting, we have independent SRSs of size n_i from each of I populations having Normal distributions with means μ_i and a common standard deviation σ. *Tukey simultaneous confidence intervals* for all pairwise differences $\mu_i - \mu_j$ among the population means have the form

$$(\bar{x}_i - \bar{x}_j) \pm m^* s_p \sqrt{\frac{1}{n_i} + \frac{1}{n_j}}$$

Here $\bar{x}_i$ is the sample mean of the ith sample, and s_p is the pooled estimate of σ. The critical value m^* depends on the confidence level C, the number of populations I, and the total number of observations.

To carry out *simultaneous tests* of the hypotheses

$$H_0: \mu_i = \mu_j$$
$$H_a: \mu_i \neq \mu_j$$

for all pairs of population means at fixed significance level $\alpha = 1 - C$, reject H_0 for any pair whose confidence interval does not contain zero.

If all samples are the same size, Tukey simultaneous confidence intervals provide *overall confidence level* C. That is, C is the probability that *all* of the intervals simultaneously capture the true pairwise differences. If the samples differ in size, the true confidence level is at least as large as C, so that conclusions are conservative. Similarly, if all the samples are the same size, the Tukey simultaneous tests have *overall significance level* 1 − C. That is, 1 − C is the probability that when all the population means are equal, *any* of the tests incorrectly rejects its null hypothesis. If the samples differ in size, the true significance level is smaller than 1 − C, so that conclusions are conservative.

EXAMPLE 27.7 Policy Justification: Multiple Intervals

JUSTIFY

Figures 27.11(a) and (b) contain more Minitab output for the ANOVA comparing the mean policy support indices in three groups of policy justifications. We asked for Tukey multiple comparisons with an overall error rate of 5%. That is, the overall confidence level for the three intervals together is 95%.

The Tukey confidence intervals in Figure 27.11(a) are

$$-0.147 \text{ to } 0.345 \quad \text{for} \quad \mu_{\text{moral}} - \mu_{\text{ambiguous}}$$
$$-0.506 \text{ to } -0.018 \quad \text{for} \quad \mu_{\text{pragmatic}} - \mu_{\text{ambiguous}}$$
$$-0.607 \text{ to } -0.115 \quad \text{for} \quad \mu_{\text{pragmatic}} - \mu_{\text{moral}}$$

Reversing the signs, the interval for $\mu_{\text{moral}} - \mu_{\text{pragmatic}}$ is 0.115 to 0.607, wider than the individual 95% confidence interval in Example 27.6. The wider interval is the price we pay for having 95% confidence not just in one interval but in all three simultaneously.

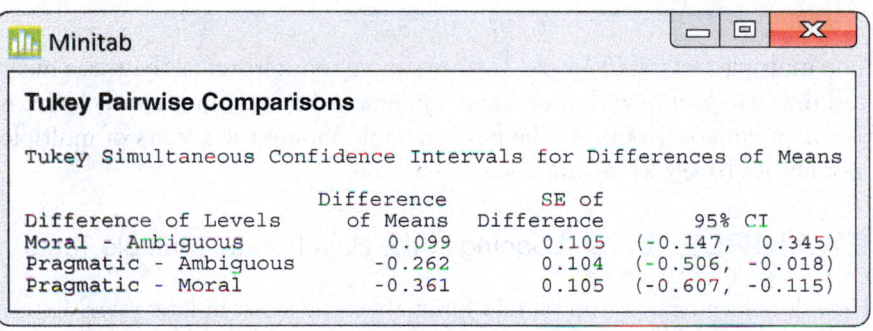

Minitab

Tukey Pairwise Comparisons

```
Tukey Simultaneous Confidence Intervals for Differences of Means

                          Difference       SE of
Difference of Levels        of Means  Difference          95% CI
Moral - Ambiguous              0.099       0.105  (-0.147,  0.345)
Pragmatic - Ambiguous         -0.262       0.104  (-0.506, -0.018)
Pragmatic - Moral             -0.361       0.105  (-0.607, -0.115)
```

(a)

Minitab

```
Grouping Information Using the Tukey Method and 95% Confidence

Justification       N    Mean  Grouping
Moral             122  5.8053  A
Ambiguous         126  5.7063  A
Pragmatic         126  5.4444    B

Means that do not share a letter are significantly different.
```

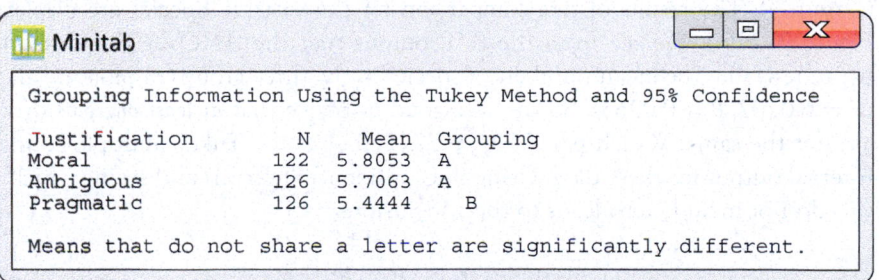

(b)

FIGURE 27.11
Minitab multiple-comparisons output for the policy justification study of Example 27.4. Tukey simultaneous confidence intervals are given in (a), and the results of the multiple tests are summarized in (b).

EXAMPLE 27.8 Policy Justification: Multiple Tests

JUSTIFY

The ANOVA null hypothesis is that all population means are equal,

$$H_0: \mu_{\text{moral}} = \mu_{\text{pragmatic}} = \mu_{\text{ambiguous}}$$

We know from the output in Figure 27.9 that the ANOVA F test rejects this hypothesis ($F = 6.33$, $P = 0.002$). So we have good evidence that *at least one pair* of means are not the same. Which pairs? Look at the simultaneous 95% confidence intervals in Example 27.7. Which of these intervals do not contain zero? *If an interval does not contain zero, we reject the hypothesis that this pair of population means are equal.*

The conclusions are

$$\begin{aligned} &\textit{We can reject} && H_0: \mu_{\text{moral}} = \mu_{\text{pragmatic}} \\ &\textit{We can reject} && H_0: \mu_{\text{ambiguous}} = \mu_{\text{pragmatic}} \\ &\textit{We cannot reject} && H_0: \mu_{\text{moral}} = \mu_{\text{ambiguous}} \end{aligned}$$

This Tukey simultaneous test of three null hypotheses has the property that when all three hypotheses are true, there is only 5% probability that *any* of the three tests wrongly rejects its hypothesis. The output in Figure 27.11(b) summarizes the results of these tests by grouping means for which we do not reject the null hypothesis, a common method of summarizing the results of multiple tests. The idea is that groups sharing the same "Grouping" letter are not significantly different, while groups with a different letter are significantly different from each other. We see that moral and ambiguous are grouped together with the letter A, indicating they are not significantly different from each other, while the pragmatic group with a lower mean has the letter B, indicating that it is in a group which is significantly different than either of the two means in the group associated with the letter A.

The multiple tests in Example 27.8 are simple to interpret as the three means are placed into two groups with moral and ambiguous having significantly larger means than the pragmatic group. As the next example shows, the results of multiple tests do not always have such a simple interpretation.

EXAMPLE 27.9 Logging in the Rain Forest: Multiple Tests

LOGFULL

Fletcher & Baylis/Science Source

How does logging in a tropical rain forest affect the forest in later years? Researchers compared forest plots in Borneo that had never been logged (Group 1) with similar plots nearby that had been logged one year earlier (Group 2) and eight years earlier (Group 3). The results of this comparison for the variable Species are displayed in Figure 27.12(a). We see from the JMP output that the ANOVA F test rejects the hypothesis that the mean numbers of species in the three groups of plots are all equal ($F = 6.0202$, $P = 0.0063$). So we have good evidence that *at least one pair* of means are not the same. Which pairs? Figure 27.12(a) gives the Tukey multiple confidence interval output for these data. Using the confidence intervals and seeing which intervals do not include zero leads to the conclusions

$$\begin{aligned} &\textit{We can reject} && H_0: \mu_1 = \mu_2 \\ &\textit{We cannot reject} && H_0: \mu_1 = \mu_3 \\ &\textit{We cannot reject} && H_0: \mu_2 = \mu_3 \end{aligned}$$

The conclusions for these tests can again be summarized with groupings of the means as in Figure 27.12(b). Groups 1 and 3 both share the letter A so are not significantly different from each other, and Groups 2 and 3 both share the letter B so jmp

Analysis of Variance

Source	DF	Sum of Squares	Mean Square	F Ratio	Prob > F
Group	2	204.38636	102.193	6.0202	0.0063*
Error	30	509.25000	16.975		
C. Total	32	713.63636			

Means for Oneway Anova

Level	Number	Mean	Std Error	Lower 95%	Upper 95%
1	12	17.5000	1.1894	15.071	19.929
2	12	11.7500	1.1894	9.321	14.179
3	9	13.6667	1.3734	10.862	16.471

Std Error uses a pooled estimate of error variance

(a)

jmp

Ordered Differences Report

Level	-Level	Difference	Std ErrorDif	Lower CL	Upper CL	p-Value
1	2	5.750000	1.682013	1.60327	9.896728	0.0051*
1	3	3.833333	1.816781	-0.64564	8.312311	0.1048
3	2	1.916667	1.816781	-2.56231	6.395645	0.5488

Connecting Letters Report

Level			Mean
1	A		17.500000
3	A	B	13.666667
2		B	11.750000

Levels not connected by same letter are significantly different.

(b)

FIGURE 27.12 JMP multiple-comparisons output for the logging study in Example 27.9. Tukey simultaneous confidence intervals are given in (a). The results of the multiple tests are summarized in (b).

are not significantly different from each other. However, since Groups 1 and 2 are assigned different letters, they *are* significantly different from each other. The overlap in the groupings makes the interpretation of the results more difficult.

Think for a moment about the conclusion we have reached. At first glance, it appears to say that μ_1 equals μ_3 and that μ_2 equals μ_3 but that μ_1 and μ_2 are not equal. That sounds like nonsense. Now is the time to recall what a test at a fixed significance level such as 5% tells us: either we *do have enough evidence* to reject the null hypothesis, or the data *do not give enough evidence* to allow rejection. There is no contradiction in saying that

We *do* have enough evidence to conclude that $\mu_1 \neq \mu_2$.

We *do not* have enough evidence to conclude that $\mu_1 \neq \mu_3$.

We *do not* have enough evidence to conclude that $\mu_2 \neq \mu_3$.

That is, $\bar{x}_1 = 17.500$ and $\bar{x}_2 = 11.750$ are far enough apart to conclude that the population means differ, but neither 17.500 nor 11.750 is far enough from $\bar{x}_3 = 13.667$. Notice that the Tukey output provided does not give a *P*-value for the three tests taken together. Rather, we have a set of "reject" or "fail to reject" conclusions with an overall significance level that we fixed in advance: 5% in this example.

There are many other multiple-comparisons procedures that produce various simultaneous confidence intervals with an overall confidence level or simultaneous tests with an overall probability of any false rejection. The Tukey procedures probably are the most useful. If you can interpret results from Tukey, you can understand output from any multiple-comparisons procedure.

APPLY YOUR KNOWLEDGE

27.12 Comparing Flowers. In Example 27.2 (page 624), the ANOVA *F* test had a very small *P*-value, giving good reason to conclude that the three varieties of flowers have different mean lengths. Does the mean length for the *H. bihai* flower differ significantly from the mean length of the *H. caribaea* red or the *H. caribaea* yellow flower, and does the mean length for the *H. caribaea* red flower differ significantly from that of the *H. caribaea* yellow flower? FLOWER

(a) What are the three null hypotheses that formulate these questions?

(b) We want to be 90% confident that we don't wrongly reject any of the three null hypotheses. Tukey pairwise comparisons can give conclusions that meet this condition. What are the conclusions?

27.13 Bring Your Pen to Class! In Exercise 27.9 (page 634), you carried out basic ANOVA to compare the mean number of words for three experimental conditions. WRDCOUNT

(a) Find the Tukey simultaneous 95% confidence intervals for all pairwise differences among the three population means.

(b) Explain in simple language what "95% confidence" means for these intervals.

(c) Which pairs of means differ significantly at the overall 5% significance level?

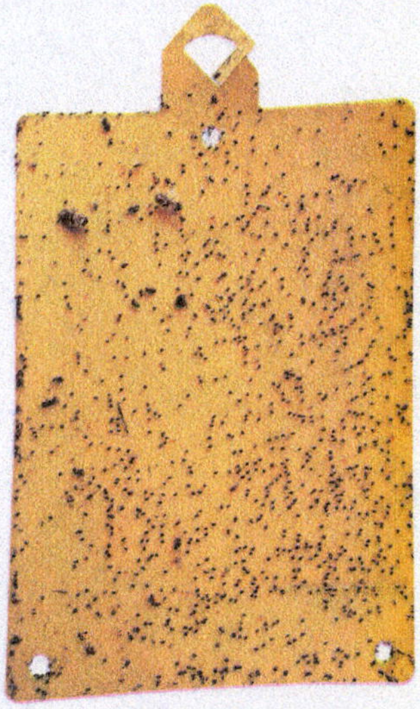

Protasov AN/Shutterstock

27.14 Which Color Attracts Beetles Best? To detect the presence of harmful insects in farm fields, we can put up boards covered with a sticky material and examine the insects trapped on the boards. Which colors attract insects best? Experimenters placed six boards of each of four colors at random locations in a field of oats and measured the number of cereal leaf beetles trapped. Here are the data:[9] BEETLES

Board Color	Beetles Trapped					
Blue	16	11	20	21	14	7
Green	37	32	20	29	37	32
White	21	12	14	17	13	20
Yellow	45	59	48	46	38	47

(a) Is there evidence that the colors differ in their ability to attract beetles? Use ANOVA to answer this question and state carefully the conclusion from this ANOVA.

(b) How many pairwise comparisons are there when we compare four colors?

(c) Which pairs of colors are significantly different when we require significance level 5% for all comparisons as a group? In particular, is yellow significantly better than every other color?

27.8 Some Details of ANOVA*

Now we will give the actual formula for the ANOVA F statistic. We have SRSs from each of I populations. Subscripts from 1 to I tell us to which sample a statistic refers:

Population	Sample Size	Sample Mean	Sample Std. Dev.
1	n_1	$\bar{x}_1$	s_1
2	n_2	$\bar{x}_2$	s_2
⋮	⋮	⋮	⋮
I	n_I	$\bar{x}_I$	s_I

You can find the F statistic from just the sample sizes n_i, the sample means $\bar{x}_i$, and the sample standard deviations s_i. You don't need to go back to the individual observations.

The ANOVA F statistic has the form

$$F = \frac{\text{variation among the sample means}}{\text{variation among individuals in the same sample}}$$

The measures of variation in the numerator and denominator of F are called *mean squares*. A mean square is a more general form of a sample variance. An ordinary sample variance s^2 is an average (or mean) of the squared deviations of observations from their mean, so it qualifies as a "mean square."

Call the overall mean response $\bar{x}$. That is, $\bar{x}$ is the mean of all N observations together. You can find $\bar{x}$ from the I sample means by

$$\bar{x} = \frac{\text{sum of all observations}}{N} = \frac{n_1\bar{x}_1 + n_2\bar{x}_2 + \cdots + n_I\bar{x}_I}{N}$$

(This expression works because multiplying a group mean $\bar{x}_i$ by the number of observations n_i it represents gives the sum of the observations in that group.)

The numerator of F is a mean square that measures variation among the I sample means $\bar{x}_1, \bar{x}_2, \ldots, \bar{x}_I$. To measure this variation, look at the I deviations of the means of the samples from $\bar{x}$,

$$\bar{x}_1 - \bar{x}, \bar{x}_2 - \bar{x}, \ldots, \bar{x}_I - \bar{x}$$

The mean square in the numerator of F is an average of the squares of these deviations. We call it the *mean square for groups*

$$\text{MSG} = \frac{n_1(\bar{x}_1 - \bar{x})^2 + n_2(\bar{x}_2 - \bar{x})^2 + \cdots + n_I(\bar{x}_I - \bar{x})^2}{I - 1}$$

Each squared deviation is weighted by n_i, the number of observations it represents.

The mean square in the denominator of F measures variation among individual observations in the same sample. For any one sample, the sample variance s_i^2 does this job. For all I samples together, we use an average of the individual sample variances. It is another weighted average, in which each s_i^2 is weighted by its degrees of freedom $n_i - 1$. The resulting mean square is called the *mean square error*

$$\text{MSE} = \frac{(n_1 - 1)s_1^2 + (n_2 - 1)s_2^2 + \cdots + (n_I - 1)s_I^2}{N - I}$$

"Error" doesn't mean a mistake has been made. It's a traditional term for chance variation. Here is a summary of the ANOVA test.

*This more advanced section is optional if you are using software to find the F statistic.

The ANOVA *F* Test Formula

Draw an independent SRS from each of I Normal populations that have a common standard deviation but may have different means. The sample from the ith population has size n_i, sample mean $\bar{x}_i$, and sample standard deviation s_i. The total number of observations, $n_1 + n_2 + \cdots + n_I$, is N.

To test the null hypothesis that all I populations have the same mean against the alternative hypothesis that not all the means are equal, calculate the ANOVA F statistic

$$F = \frac{\text{MSG}}{\text{MSE}}$$

The numerator of F is the **mean square for groups**

$$\text{MSG} = \frac{n_1(\bar{x}_1 - \bar{x})^2 + n_2(\bar{x}_2 - \bar{x})^2 + \cdots + n_I(\bar{x}_I - \bar{x})^2}{I - 1}$$

The denominator of F is the **mean square for error**

$$\text{MSE} = \frac{(n_1 - 1)s_1^2 + (n_2 - 1)s_2^2 + \cdots + (n_I - 1)s_I^2}{N - I}$$

When H_0 is true, F has the F distribution with $I - 1$ and $N - I$ degrees of freedom.

The denominators in the formulas for MSG and MSE are the two degrees of freedom $I - 1$ and $N - I$ of the F test. The numerators are called *sums of squares,* from their algebraic form. It is usual to present the results of ANOVA in an *ANOVA table.* Output from software usually includes an ANOVA table.

EXAMPLE 27.10 ANOVA Calculations: Software

Look again at the three outputs in Figure 27.4 (page 625). The two software outputs give the ANOVA table. The calculator, with its small screen, gives the degrees of freedom, sums of squares, and mean squares separately. Each output uses slightly different language to identify the two sources of variation. The basic ANOVA table is

Source of Variation	df	SS	MS	*F* Statistic
Variation among samples	2	1082.87	MSG = 541.44	259.12
Variation within samples	51	106.57	MSE = 2.09	

You can check that each mean square MS is the corresponding sum of squares SS divided by its degrees of freedom df. The F statistic is MSG divided by MSE.

Because MSE is an average of the individual sample variances, it is also called the *pooled sample variance,* written as s_p^2. When all I populations have the same population variance σ^2, as ANOVA assumes that they do, s_p^2 estimates the common variance σ^2. The square root of MSE is the *pooled standard deviation,* s_p. It estimates the common standard deviation σ of the populations. The Minitab and calculator outputs in Figure 27.4 give the value $s_p = 1.446$.

The pooled standard deviation s_p is a better estimator of the common σ than any individual sample standard deviation s_i because it combines (pools) the information in all I samples. We can get a confidence interval for any one of the means μ_i from the usual form

$$\text{estimate} \pm t^*\text{SE}_{\text{estimate}}$$

using s_p to estimate σ. The confidence interval for μ_i is

$$\bar{x}_i \pm t^* \frac{s_p}{\sqrt{n_i}}$$

Use the critical value t^* from the t distribution with $N - I$ degrees of freedom because s_p has $N - I$ degrees of freedom. These are the confidence intervals that appear in Minitab ANOVA output.

EXAMPLE 27.11 ANOVA Calculations: Without Software

We can do the ANOVA test comparing the mean lengths of *H. bihai, H. caribaea* red, and *H. caribaea* yellow flower varieties using only the sample sizes, sample means, and sample standard deviations. These appear in Example 27.1, but it is easy to find them with a calculator. There are $I = 3$ groups with a total of $N = 54$ flowers.

The overall mean of the 54 lengths in Table 27.1 is

$$\begin{aligned}\bar{x} &= \frac{n_1\bar{x}_1 + n_2\bar{x}_2 + n_3\bar{x}_3}{N} \\ &= \frac{(16)(47.598) + (23)(39.711) + (15)(36.180)}{54} \\ &= \frac{2217.621}{54} = 41.067\end{aligned}$$

The mean square for groups is

$$\begin{aligned}\text{MSG} &= \frac{n_1(\bar{x}_1 - \bar{x})^2 + n_2(\bar{x}_2 - \bar{x})^2 + n_3(\bar{x}_3 - \bar{x})^2}{I-1} \\ &= \frac{1}{3-1}\Big[(16)(47.598 - 41.067)^2 + (23)(39.711 - 41.067)^2 \\ &\quad + (15)(36.180 - 41.067)^2\Big] \\ &= \frac{1082.996}{2} = 541.50\end{aligned}$$

The mean square for error is

$$\begin{aligned}\text{MSE} &= \frac{(n_1 - 1)s_1^2 + (n_2 - 1)s_2^2 + (n_3 - 1)s_3^2}{N - I} \\ &= \frac{(15)(1.213^2) + (22)(1.799^2) + (14)(0.975^2)}{51} \\ &= \frac{106.580}{51} = 2.09\end{aligned}$$

Finally, the ANOVA test statistic is

$$F = \frac{\text{MSG}}{\text{MSE}} = \frac{541.50}{2.09} = 259.09$$

Our work differs slightly from the output in Figure 27.4 because of roundoff error. We don't recommend doing these calculations because multiple operations (squaring, dividing, etc.) and roundoff errors frequently cause mistakes.

APPLY YOUR KNOWLEDGE

The calculations of ANOVA use only the sample sizes n_i, the sample means $\bar{x}_i$, and the sample standard deviations s_i. You can therefore re-create the ANOVA calculations when a report gives these summaries but does not give the actual data. These optional exercises ask you to do the ANOVA calculations starting with the summary statistics. P-values require either a table or software for the F distributions.

27.15 Using Tablets in Class. Exercise 27.1 (page 624) describes a study of the effects on learning of the use of tablets in class. Here are the means and standard deviations of the NAEP mathematics scores for a representative sample of eighth-grade students, grouped by frequency of tablet usage:

Frequency in Class	n	$\bar{x}$	s
Never	85,557	283	6.10
Fewer than half	39,026	279	7.02
About half	7,505	269	7.97
More than half	6,004	271	10.60
All	12,008	279	11.90

(a) Check that the standard deviations satisfy the guideline for ANOVA inference.

(b) Calculate the overall mean response $\bar{x}$, the mean squares MSG and MSE, and the ANOVA F statistic.

(c) Which F distribution would you use to find the P-value of the ANOVA F test? Software gives $P < 0.0001$. Write a brief conclusion based on the sample means and the ANOVA F test.

27.16 Political Views and Education. Exercise 27.2 (page 624) describes an observational study that examines the relationship between level of education and political views. Here is a table of the mean and standard deviation of the political spectrum scores for the four levels of education:

Highest Degree	n	$\bar{x}$	s
High school	1123	4.185	1.423
Junior college	193	4.171	1.421
Bachelor's	456	3.846	1.530
Graduate	243	3.770	1.680

(a) Do the standard deviations satisfy the rule of thumb for safe use of ANOVA?

(b) Calculate the overall mean response $\bar{x}$, the mean squares MSG and MSE, and the F statistic.

(c) Which F distribution would you use to find the P-value of the ANOVA F test? Write a brief conclusion based on the sample means and the ANOVA F test.

27.17 Attitudes toward Math. Do high school students from different racial/ethnic groups have different attitudes toward mathematics? Measure the level of interest in mathematics on a 5-point scale (with higher numbers indicating higher interest) for a national random sample of students. Here are summaries for students who were taking math at the time of the survey:[10]

Racial/Ethnic Group	n	$\bar{x}$	s
African American	809	2.57	1.40
White	1860	2.32	1.36
Asian/Pacific Islander	654	2.63	1.32
Hispanic	883	2.51	1.31
Native American	207	2.51	1.28

(a) The conditions for ANOVA are clearly satisfied. Explain why.

(b) Calculate the ANOVA table and the *F* statistic.

(c) Software gives $P < 0.001$. What explains the small *P*-value? Do you think the differences are large enough to be important?

CHAPTER 27 SUMMARY

- **One-way analysis of variance (ANOVA)** compares the means of several populations. The **ANOVA *F* test** tests the null hypothesis that all the populations have the same mean. If the *F* test shows significant differences, examine the data to see where the differences lie and whether they are large enough to be important.
- The conditions for ANOVA state that we have an independent SRS from each population, that each population has a Normal distribution, and that all populations have the same standard deviation.
- In practice, ANOVA inference is relatively robust when the populations are non-Normal, especially when the samples are large. Before doing the *F* test, check the observations in each sample for outliers or strong skewness. Also verify that the largest sample standard deviation is no more than twice as large as the smallest standard deviation.
- When the null hypothesis is true, the **ANOVA *F* statistic** for comparing I means from a total of N observations in all samples combined has the *F* distribution with $I - 1$ and $N - I$ degrees of freedom.
- ANOVA calculations are reported in an ANOVA table that gives sums of squares, mean squares, and degrees of freedom for variation among groups and for variation within groups. In practice, we use software to do the calculations.
- Follow-up analysis is often helpful in a one-way ANOVA setting. **Tukey pairwise multiple comparisons** give confidence intervals for all differences among treatment means with an **overall confidence level.** That is, we can be (say) 95% confident that *all* the intervals simultaneously capture the true population differences between means.

CHAPTER 27 SKILLS REVIEW

Here are the most important skills you should have acquired from reading this chapter. Following each skill in parentheses is the section of the text where the topic is introduced.

A. Recognition

1. Recognize when testing the equality of several means is helpful in understanding data. (27.1)
2. Recognize that the statistical significance of differences among sample means depends on the sizes of the samples and on how much variation there is within the samples. (27.4)
3. Recognize when you can safely use ANOVA to compare means. Check the data production, the presence of outliers, and the sample standard deviations for the groups you want to compare. (27.5)
4. Recognize why a formal follow-up analysis using Tukey's method is more appropriate than doing individual tests or confidence intervals on each pairwise comparison. (27.7)

B. Interpreting ANOVA

1. Explain what null hypothesis *F* tests in a specific setting. (27.2)
2. Locate the *F* statistic and its *P*-value on the output of analysis of variance software. (27.3)
3. Find the degrees of freedom for the *F* statistic from the number and sizes of the samples. (27.6)
4. If the test is significant, use graphs and descriptive statistics to see what differences among the means are most important. (27.2, 27.5)

C. Follow-Up Analysis

1. Decide when it is helpful to know which differences among treatment means are significant. (27.7)
2. Use software to carry out Tukey pairwise multiple comparisons among all the means you want to compare. (27.7)
3. Understand the meaning of the overall confidence level and the overall level of significance provided by Tukey's method for a set of confidence intervals or a set of significance tests. (27.7)

CHECK YOUR SKILLS

27.18 The purpose of analysis of variance is to compare

(a) the variances of several populations.

(b) the standard deviations several populations.

(c) the means of several populations.

27.19 Vitamin D is important for maintaining strong bones by helping the body absorb calcium from food and supplements. The recommended amount of vitamin D is 400 IU (international units) per day. Many people take vitamin D supplements, often at levels greatly exceeding the recommended daily amount. What is the effect of taking too much vitamin D? Researchers investigated this in a double-blind randomized clinical trial. Subjects were divided into three groups. One, consisting of 109 subjects, took 400 IU per day for three years. A second group, consisting of 100 subjects, took 4,000 IU per day for three years. A third group, consisting of 102 subjects, took 10,000 IU per day for three years. The total volumetric bone mass density of the tibia (the large bone located in the lower front portion of the leg, sometimes referred to as the shinbone) was measured in order to assess the effect of the different dosages of vitamin D.[11]

The degrees of freedom for the ANOVA *F* statistic comparing the mean volumetric bone masses are

(a) 2 and 98.

(b) 2 and 308.

(c) 3 and 310.

27.20 The alternative hypothesis for the ANOVA *F* test in the previous question is

(a) either the 4,000 daily IU or the 10,000 daily IU treatment has higher mean volumetric bone masses than the 400 daily IU treatment.

(b) both the 4,000 and 10,000 daily IU treatments have higher mean volumetric bone masses than the 400 daily IU treatment.

(c) the mean volumetric bone masses for the three groups are not all the same.

27.21 Which of the following would be evidence that ANOVA inference is not safe to use?

(a) The sample sizes are less than four.

(b) The largest sample variance is more than twice as large as the smallest sample variance.

(c) The largest sample standard deviation is more than twice as large as the smallest sample standard deviation.

Dogs, Friends, and Stress. *If you are a dog lover, perhaps having your dog along reduces the effect of stress. To examine the effect of pets in stressful situations, researchers recruited 45 women who said they were dog lovers. Fifteen of the subjects were randomly assigned to each of three groups to do a stressful task–alone (the control group), with a good friend present, or with their dog present. The subject's average heart rate during the task is one measure of the effect of stress. Here is partial Minitab output, including the ANOVA table (with several numbers omitted), along with the means and standard deviations of the mean heart rates for the three conditions. Are there significant differences among the mean heart rates for the three conditions?*[12]

Source	DF	SS	MS	F-Value	P-Value
Group	2	2388			0.000
Error	42	3561			
Total	44	5949			

Group	N	Mean	StDev
Control	15	82.52	9.24
Friend	15	91.33	8.34
Pet	15	73.48	9.97

Questions 27.22 to 27.24 are based on this study.

27.22 The conclusion of the ANOVA test is that STRSSPET

(a) there is strong evidence ($P = 0.000$) that the mean heart rates are not the same for all three conditions.

(b) there is strong evidence ($P = 0.000$) that the mean heart rate is different for the control group.

(c) there is strong evidence ($P = 0.000$) that the mean heart rates for the three conditions are all different from each other.

27.23 To compare the treatments, we might use three 90% two-sample t confidence intervals to compare each pair of treatments: control versus friend, control versus pet, and pet versus friend. The weakness of doing this is that

(a) we don't know how confident we can be that all three intervals cover the true differences in means.

(b) 90% confidence is okay for one comparison, but it isn't high enough for three comparisons done at once.

(c) we can't compare the treatments with a control.

27.24 **(Optional topic)** The value of the ANOVA F-statistic for testing equality of the population means of the three treatments is

(a) 0.67.

(b) 2.49.

(c) 14.08.

27.25 A company runs a three-day workshop on strategies for working effectively in teams. On each day, a different strategy is presented. Forty-eight employees of the company attend the workshop. At the outset, all 48 are divided into 12 teams of four. The teams remain the same for the entire workshop. Strategies are presented in the morning. In the afternoon, the teams are presented with a series of small tasks, and the number of these completed successfully using the strategy taught that morning is recorded for each team. The mean number of tasks completed successfully by all teams each day and the standard deviation follow:

Day	n	$\bar{x}$	s
1	12	17.24	7.10
2	12	17.25	14.14
3	12	17.65	14.03

In this example, we notice

(a) the data show very strong evidence of a violation of the assumption that the three populations have the same standard deviation.

(b) ANOVA cannot be used on these data because the sample sizes are less than 20.

(c) the assumption that the data are independent for the three days is unreasonable because the same teams were observed each day.

27.26 In an ANOVA that compares three treatments, how many pairwise comparisons between two of these treatments are there?

(a) two (b) three (c) six

27.27 As part of an ANOVA that compares three treatments, you carry out Tukey pairwise tests at the overall 5% significance level. The Tukey tests find that μ_1 is significantly different from μ_3 but that the other two comparisons show no significant difference. You can be 95% confident that

(a) $\mu_1 \neq \mu_3$ and $\mu_1 = \mu_2$ and $\mu_2 = \mu_3$.

(b) just $\mu_1 \neq \mu_3$; there is not enough evidence to draw conclusions about the other pairs of means.

(c) $\mu_1 = \mu_2$ and $\mu_2 = \mu_3$ and this implies that it must also be true that $\mu_1 = \mu_3$.

CHAPTER 27 EXERCISES

Exercises 27.28 to 27.31 describe situations in which we want to compare the mean responses in several populations. For each setting, identify the populations and the response variable. Then give I, n_i, and N. Finally, give the degrees of freedom of the ANOVA F statistic.

27.28 **Does Art Sell Products?** How does visual art affect the perception and evaluation of consumer products? Subjects were asked to evaluate an advertisement for bathroom fittings that contained an art image, a nonart image, or no image. The art image was Vermeer's painting *Girl with a Pearl Earring*, and the nonart image was a photograph of the actress Scarlett Johansson in the same pose wearing the same garments as the girl in the painting and was taken from the motion picture *Girl with a Pearl Earring*. Thus the art and nonart images were a match on content. College students were divided at random into three groups of 39 each, with each group assigned to one of the three types of advertisements. Students evaluated the product in the advertisement on a scale of 1 to 7, with 1 being the most unfavorable rating and 7 being the most favorable. The paper reported that a one-way ANOVA on the product evaluation index had $F = 6.29$ with $P < 0.05$.[13]

27.29 **Perceived Exertion while Exercising.** Can the introduction of pleasant sensory stimuli lead to a more pleasant exercise environment and decrease perceived exertion during a four-minute stepping task? Forty-three students from a southeastern university were assigned at random to three conditions: "taste," in which participants inserted a lemon-flavored mouth guard during the task; "placebo," in which participants inserted a non-flavored mouth guard; and "control," in which no mouth guard was used. Twelve students were assigned to the taste group, 15 to the placebo group, and 16 to the control group. Ratings of perceived exertion (RPE) scores were measured on a standard 15-point scale ranging from 6 (very, very light) to 20 (exhausted).[14]

27.30 Test Accommodations. Many states require schoolchildren to take regular statewide tests to assess their progress. Children with learning disabilities who read poorly may not do well on mathematics tests because they can't read the problems. Most states allow "accommodations" for learning-disabled children. Randomly assign 100 learning-disabled children in equal numbers to three types of accommodation and a control group: math problems are read by a teacher, math problems are read by a computer, math problems are read by a computer that also shows a video, and standard test conditions. The researcher would like to compare the mean scores on the state mathematics assessment.

27.31 Exercise and Type 2 Diabetes. It is generally accepted that regular exercise provides health benefits to individuals with type 2 diabetes, although the exact exercise regimen (aerobic versus resistance versus both) is unclear. The subjects in this study were sedentary 30- to 75-year-old adults with type 2 diabetes and hemoglobin A1c levels elevated above 6.5%. The level of hemoglobin A1c correlates very well with a person's recent overall blood sugar levels. If the blood sugars have generally been running high during the previous few months, the level of hemoglobin A1c will be high. In a randomized controlled study, 41 subjects were assigned to a nonexercise control group, 73 to resistance training only, 72 to aerobic exercise only, and 76 to combined aerobic and resistance training. The weekly duration of exercise was similar for all three exercise groups, and subjects remained on the exercise regimens for nine months. At the end of nine months, the hemoglobin A1c levels of subjects were measured.[15]

27.32 Don't Handle the Merchandise? Although consumers often want to touch products before purchasing them, they generally prefer that others have not touched products they would like to buy. Can another person touching a product create a positive reaction? Subjects were given instructions to contact a sales associate at a university bookstore who would provide them with a shirt to try on. When meeting the sales associate, subjects were told there was only one shirt left, and it was being tried on by another "customer." The other customer trying on the shirt was a confederate of the experimenter and was either an attractive, well-dressed professional female model or an average-looking female college student wearing jeans and a tee shirt. Subjects, who were either males or females, saw the confederate leaving the dressing room where the shirt was left for them to try on. There was also a control group of subjects who were handed the shirt directly off the rack by the sales associate. Thus there were five treatments: male subjects seeing a model, female subjects seeing a model, male subjects seeing a college student, female subjects seeing a college student, and the control group. Subjects evaluated the product on five dimensions, each dimension on a seven-point scale, with the five scores then averaged to give the subject's evaluation measure, with higher numbers indicating a more positive evaluation. Here are the sample sizes, means, and standard deviations for the five groups:[16]

Treatment Group	n	$\bar{x}$	s
Males seeing a model	22	5.34	0.87
Males seeing a student	23	3.32	1.21
Females seeing a model	24	4.10	1.32
Females seeing a student	23	3.50	1.43
Controls	27	4.17	1.50

(a) Verify that the sample standard deviations allow the use of ANOVA to compare the population means. What do the means suggest about the effect of the subject's sex and the attractiveness of the confederate on the evaluation of the product?

(b) The paper reports the ANOVA $F = 8.30$. What are the degrees of freedom for the ANOVA F statistic and the P-value? State your conclusions.

27.33 Tablets in the Classroom and Learning. Does the use of technology, such as tablets, in the classroom affect student learning? To explore this, a researcher examined the NAEP mathematics scores for a representative sample of 152,300 fourth-graders. Each student was classified by how frequently he or she used a tablet in the classroom (never, in fewer than half of all classes, in about half of all classes, in more than half of all classes, and in all classes). Here are the means and standard deviations of the NAEP mathematics scores for the five categories of frequency of use in class:[17]

Group	Mean Score	Standard Deviation
Never	243	4.38
Fewer than half	240	5.64
About half	231	8.09
More than half	229	9.16
All	232	7.70

(a) Make a graph that compares the mean NAEP scores for the five groups. How do the mean scores change as the frequency of tablet use in class increases?

(b) What hypotheses does ANOVA test in this setting?

(c) Do the standard deviations satisfy our rule of thumb for safe use of ANOVA?

27.34 Can You Hear These Words? To test whether a hearing aid is right for a patient, audiologists play a tape on which words are pronounced at low volume. The patient tries to repeat the words. Some lists of words are supposed to be equally difficult. Are the lists equally difficult when there is background noise? To find out, an experimenter had subjects with normal hearing listen to four lists with a noisy background. The response variable was the percentage of the 50 words in a list that the subject repeated correctly. The researcher collected 96 responses.[18] Here are two study designs that could produce these data:

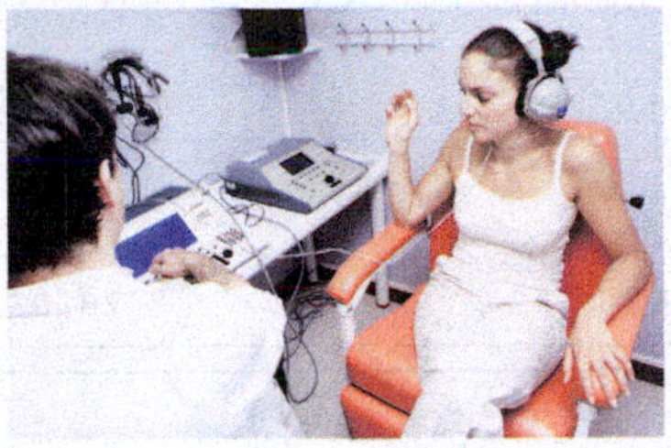

Phanie/Science Source

Design A. The experimenter assigns 96 subjects to four groups at random. Each group of 24 subjects listens to one of the lists. All individuals listen and respond separately.

Design B. The experimenter has 24 subjects. Each subject listens to all four lists in random order. All individuals listen and respond separately.

Does Design A allow use of one-way ANOVA to compare the lists? Does Design B allow use of one-way ANOVA to compare the lists? Briefly explain your answers.

27.35 Sunscreens Avobenzone is one of the active ingredients in several commercially available sunscreens. It can be absorbed into the bloodstream when sunscreen is applied to the skin. The Food and Drug Administration has expressed concern about the safety of absorbing too much avobenzone. Researchers recruited 24 healthy volunteers to investigate avobenzone absorption for four different commercially available sunscreens: two sprays, a lotion, and a cream. Do the types of sunscreens (spray, lotion, and cream) differ in terms of how much avobenzone is absorbed into the blood? Subjects were randomly assigned to one of the four sunscreens, with six subjects for each. Subjects had 2 milligrams of sunscreen per 1 cm^2 applied to 75% of their body surface area (area outside of normal swimwear) four times per day for four days. The amount of avobenzone absorbed into the bloodstream (in nanograms per milliliter applied, ng/mL) after 150 hours was then measured for each subject. For the purpose of analyzing the data using ANOVA, the logarithms of the measurements were used. Thus, for example, a value of 0 corresponds to an absorption of 1 ng/mL and negative values correspond to absorptions less than 1 ng/mL. Larger values of the logarithms correspond to larger levels of absorption. Here are the data:[19] SUNSCREEN1

Spray 1	0.693	0.405	0.405	0.405	−0.511	−0.916
Spray 2	0.588	0.000	0.000	−0.511	−0.693	−1.204
Lotion	1.386	0.693	0.182	0.095	−0.511	−0.511
Cream	−0.357	−0.357	−0.357	−1.204	−1.609	−1.609

Figure 27.13 shows JMP ANOVA output for these data.

(a) Do these data satisfy the conditions for ANOVA?

(b) State H_0 and H_a for the ANOVA F test and explain in words what ANOVA tests in this setting.

(c) Report your overall conclusions about whether the types of sunscreen differ in the mean amount of avobenzone absorbed into the bloodstream after 150 hours.

27.36 Can You Hear These Words? Figure 27.14 displays the Minitab output for one-way ANOVA applied to the hearing data described in Design A in Exercise 27.34. The response variable is "Percent," and "List" identifies the four lists of words. Based on this analysis, is there good reason to think that the four lists are not all equally difficult? Write a brief summary of the study findings.

jmp

Oneway Anova

Analysis of Variance

Source	DF	Sum of Squares	Mean Square	F Ratio	Prob > F
Group	3	4.658577	1.55286	3.5813	0.0320*
Error	20	8.672101	0.43361		
C. Total	23	13.330678			

Means and Std Deviations

Level	Number	Mean	Std Dev	Std Err Mean	Lower 95%	Upper 95%
Cream	6	-0.915479	0.6297893	0.2571104	-1.576402	-0.254556
Lotion	6	0.2225703	0.7306505	0.2982868	-0.5442	0.989341
Spray 1	6	0.0804044	0.6380284	0.260474	-0.589165	0.7499741
Spray 2	6	-0.30336	0.6299645	0.2571819	-0.964467	0.3577474

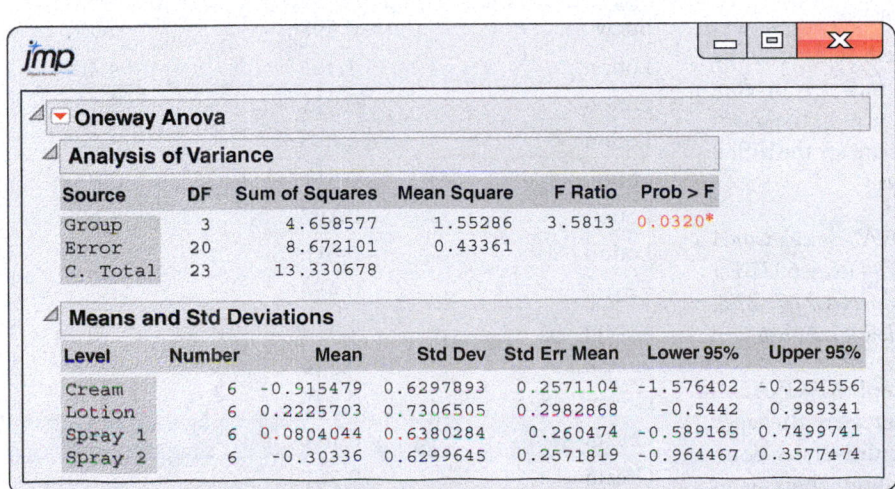

FIGURE 27.13
JMP ANOVA output for comparing avobenzone absorption for four different sunscreens, for Exercise 27.35.

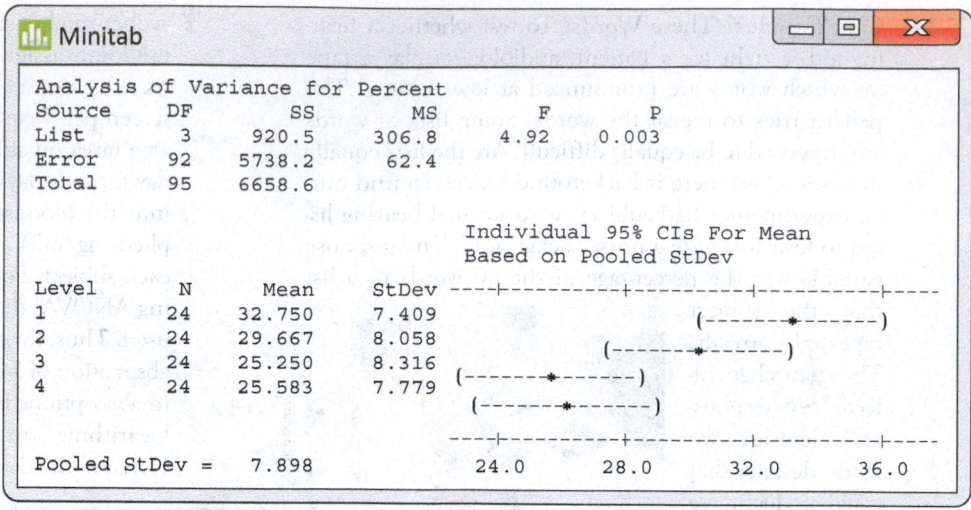

FIGURE 27.14 Minitab ANOVA output for comparing the percents heard correctly in four lists of words, for Exercise 27.36.

Minitab

```
Analysis of Variance for Percent
Source    DF       SS       MS       F       P
List       3    920.5    306.8    4.92   0.003
Error     92   5738.2     62.4
Total     95   6658.6
                                   Individual 95% CIs For Mean
                                   Based on Pooled StDev
Level     N     Mean    StDev  ---+---------+---------+---------+----
1        24   32.750    7.409                          (------*------)
2        24   29.667    8.058                 (------*------)
3        24   25.250    8.316  (------*------)
4        24   25.583    7.779    (------*------)
                               ---+---------+---------+---------+----
Pooled StDev =    7.898          24.0      28.0      32.0      36.0
```

More on Sunscreens *Exercise 27.35 describes a randomized experiment to examine the effects of four different types of sunscreen on the absorption of avobenzone into the bloodstream. The researchers also investigated the absorption of two other compounds, octocrylene and oxybenzone, into the bloodstream after 150 hours. Table 27.3 gives the logarithms of the amounts absorbed. (Amounts absorbed are in nanograms per milliliter applied, ng/mL.) The cream sunscreen did not contain oxybenzone, so there is no oxybenzone absorption data for the cream. Exercises 27.37 to 27.40 are based on this information.*

27.37 Conditions for ANOVA, Octocrylene. Examine the absorption data for octocrylene. Do the conditions for ANOVA appear to be met? SUNSCREEN2

27.38 Results for Octocrylene. (4step) Your work in Exercise 27.35 shows that there were significant differences in the mean absorption of avobenzone after 150 hours for the four types of sunscreen. Is this true for octocrylene? Do a complete analysis of the absorption data for octocrylene and report your conclusions. SUNSCREEN2

27.39 Conditions for ANOVA , Oxybenzone. Examine the absorption data for oxybenzone. Do the conditions for ANOVA appear to be met? SUNSCREEN2

27.40 Results for Oxybenzone. (4step) Do a complete analysis of the absorption data for oxybenzone and report your conclusions. Were there significant differences in the mean absorption of oxybenzone after 150 hours for the different types of sunscreen? SUNSCREEN2

Exercises 27.41 to 27.52 are paired. For the "ANOVA" exercise within each pair, follow the four-step process as in Example 27.4 (page 632). For the "Comparisons" exercise within the pair, use Examples 27.7, 27.8, and 27.9 (pages 639–641) as guides for a formal follow-up analysis.

27.41 College Students Need Better Sleep! ANOVA. (4step) A random sample of 898 students between the ages of 20 and 24 at a large midwestern university completed a survey including questions about their sleep quality, moods, academic performance, physical health, and psychoactive drug use. Sleep quality was measured using the Pittsburgh Sleep Quality Index (PSQI), with students scoring less than or equal to 5 on the index classified as optimal sleepers, those

TABLE 27.3 Sunscreen absorption of two compounds for four sunscreens

Treatment	Octocrylene	Oxybenzone
Spray 1	−0.105	4.007
Spray 1	−0.301	3.912
Spray 1	−0.511	3.689
Spray 1	−0.635	3.497
Spray 1	−0.892	3.332
Spray 1	−0.968	2.773
Spray 2	0.916	3.689
Spray 2	0.788	3.296
Spray 2	0.405	2.996
Spray 2	0.000	2.944
Spray 2	−0.288	2.833
Spray 2	−0.598	2.197
Lotion	1.361	4.407
Lotion	0.916	3.526
Lotion	0.833	3.045
Lotion	0.531	2.833
Lotion	0.262	1.609
Lotion	−0.511	1.099
Cream	0.956	
Cream	0.470	
Cream	0.344	
Cream	0.336	
Cream	−0.105	
Cream	−0.261	

scoring a 6 or 7 classified as borderline, and those scoring over 7 classified as poor sleepers. The depression subscale of the Profile of Moods State (POMS) was used to assess how severely students experienced depression on a typical day, with high scores indicating greater levels of depression. We want to know if there is a significant difference in depression scores among the three classifications of sleep.[20] The full data set is too large to print here, but here are the first seven individuals: DEPRESED

Quality of sleep:	Poor	Poor	Border	Poor	Poor	Optimal	Border
Depression score:	5	8	5	11	7	7	10

(a) Follow the four-step process in data analysis and ANOVA. Be sure to check the conditions for ANOVA and to include an appropriate graph that compares the depression scores for the three qualities of sleep.

(b) Explain why we can trust the ANOVA *F* test to give valid results for these data.

(c) This is an observational study. Explain why. In this study, we can see why association does not prove causation. Explain how poor-quality sleep might lead to higher levels of depression. Then explain how higher depression scores might affect quality of sleep. You see that the cause-and-effect relationship might go in either direction.

27.42 Does Playing Video Games Make Better Surgeons? ANOVA. In laparoscopic surgery, a video camera and several thin instruments are inserted into the patient's abdominal cavity. The surgeon uses the image from the video camera positioned inside the patient's body to perform the procedure by manipulating the instruments that have been inserted. The Top Gun Laparoscopic Skills and Suturing Program was developed to help surgeons develop the skill set necessary for laparoscopic surgery. Because of the similarity in many of the skills involved in video games and laparoscopic surgery, it was hypothesized that surgeons with greater prior video game experience might acquire the skills required in laparoscopic surgery more easily. Thirty-three surgeons participated in the study and were classified into the three categories, never used, under three hours, and more than three hours—depending on the daily number of hours they played video games at the height of their video game use. They also performed Top Gun drills and received a score based on the time to complete the drill and the number of errors made, with lower scores indicating better performance. Here are the Top Gun scores and video game categories for the 33 participants.[21] TOPGUN

Never played:	9379	8302	5489	5334	4605
	4789	9185	7216	9930	4828
	5655	4623	7778	8837	5947
Under three hours:	5540	6259	5163	6149	4398
	3968	7367	4217	5716	
Three or more hours:	7288	4010	4859	4432	4845
	5394	2703	5797	3758	

Give a complete analysis to compare the means of the Top Gun performance scores for the three groups. This is an observational study. Explain why and indicate how this affects the conclusions that can be drawn.

27.43 College Students Need Better Sleep! Comparisons. Exercise 27.41 describes a study of the relationship between sleep quality and depression in college students. Software gives these Tukey 99% simultaneous confidence intervals: DEPRESED

$$-2.712 \text{ to } -0.790 \quad \text{for} \quad \mu_{optimal} - \mu_{borderline}$$
$$0.952 \text{ to } 2.832 \quad \text{for} \quad \mu_{poor} - \mu_{borderline}$$
$$2.761 \text{ to } 4.525 \quad \text{for} \quad \mu_{poor} - \mu_{optimal}$$

(a) How confident are you that all three of these intervals capture the true differences between pairs of population means?

(b) Write a short summary of the results of the ANOVA, including the multiple comparisons.

27.44 Does Playing Video Games Make Better Surgeons? Comparisons. Exercise 27.42 examines the relationship between previous video game experience and a surgeons ability to acquire skills for laparoscopic surgery. Software gives these Tukey 95% simultaneous confidence intervals: TOPGUN

$$-2241 \text{ to } 977 \quad \text{for} \quad \mu_{>3hours} - \mu_{<3hours}$$
$$-66 \text{ to } 2813 \quad \text{for} \quad \mu_{None} - \mu_{<3hours}$$
$$567 \text{ to } 3445 \quad \text{for} \quad \mu_{None} - \mu_{>3hours}$$

(a) How confident are you that all three of these intervals capture the true differences between pairs of population means?

(b) Write a short summary of the results of the ANOVA, including the multiple comparisons.

27.45 Do Good Smells Bring Good Business? ANOVA. Businesses know that customers often respond to background music. Do they also respond to odors? Nicolas Guéguen and his colleagues studied this question in a small pizza restaurant in France on Saturday evenings in May. On one evening, a relaxing lavender odor was spread through the restaurant; on another evening, a stimulating lemon odor; a third evening served as a control, with no odor. The three evenings were comparable in many ways (weather, customer count, and so on), so we are willing to regard the data as independent SRSs from spring Saturday evenings at this restaurant. Table 27.4 contains data on how long (in minutes) customers stayed in the restaurant on each of the three evenings.[22] ODORS

TABLE 27.4 Time (minutes) that customers remain in a restaurant when exposed to odors

Lavender Odor									
92	126	114	106	89	137	93	76	98	108
124	105	129	103	107	109	94	105	102	108
95	121	109	104	116	88	109	97	101	106
Lemon Odor									
78	104	74	75	112	88	105	97	101	89
88	73	94	63	83	108	91	88	83	106
108	60	96	94	56	90	113	97		
No Odor									
103	68	79	106	72	121	92	84	72	92
85	69	73	87	109	115	91	84	76	96
107	98	92	107	93	118	87	101	75	86

(a) Make an appropriate graph comparing the customer times for each evening. Do any of the distributions show outliers, strong skewness, or other clear deviations from Normality?

(b) Do a complete analysis to see whether the groups differ in the average amount of time spent in the restaurant. Follow the four-step process in your work.

27.46 Good Weather and Tipping. ANOVA. Favorable weather has been shown to be associated with increased tipping. Will just the belief that future weather will be favorable lead to higher tips? The researchers gave 60 index cards to a waitress at an Italian restaurant in New Jersey. Before delivering the bill to each customer, the waitress was randomly assigned a card by the researchers and wrote on the bill the same message that was printed on the index card. Twenty of the cards had the message "The weather is supposed to be really good tomorrow. I hope you enjoy the day!" Another 20 cards contained the message, "The weather is supposed to be not so good tomorrow. I hope you enjoy the day anyway!" The remaining 20 cards were blank, indicating that the waitress was not supposed to write any message. Choosing a card at random ensured that there was a random assignment of the diners to the three experimental conditions. Here are the percentage tips for the three messages:[23] TIPPING

Good weather report	20.8	18.7	19.9	20.6	22.0	23.4	22.8
	24.9	22.2	20.3	24.9	22.3	27.0	20.4
	22.2	24.0	21.2	22.1	22.0	22.7	
Bad weather report	18.0	19.0	19.2	18.8	18.4	19.0	18.5
	16.1	16.8	14.0	17.0	13.6	17.5	19.9
	20.2	18.8	18.0	23.2	18.2	19.4	
No weather report	19.9	16.0	15.0	20.1	19.3	19.2	18.0
	19.2	21.2	18.8	18.5	19.3	19.3	19.4
	10.8	19.1	19.7	19.8	21.3	20.6	

Do the data support the hypothesis that there are differences among the tipping percentages for the three experimental conditions? Does a prediction of good weather seem to increase the tip percentage? Follow the four-step process in data analysis and ANOVA. Be sure to check the conditions for ANOVA and to include an appropriate graph which compares the tipping percentages for the three conditions.

27.47 Do Good Smells Bring Good Business? Comparisons. In Exercise 27.45, the effect of a background odor on the time spent in a restaurant was studied. Do the means for the odors lavender and lemon differ significantly from each other, and from the mean for the control group? ODORS

(a) What are the three null hypotheses that formulate these questions?

(b) We want to be 95% confident that we don't wrongly reject any of the three null hypotheses. Tukey pairwise comparisons can give conclusions that meet this condition. Write a short summary of the results of the ANOVA, including the results of the Tukey pairwise comparisons.

27.48 Good Weather and Tipping. Comparisons. In Exercise 27.46, you carried out basic ANOVA to compare the mean tipping percents for the three experimental conditions. TIPPING

(a) Find the Tukey simultaneous 99% confidence intervals for all pairwise differences among the three population means.

(b) Explain in simple language what "99% confidence" means for these intervals.

(c) Which pairs of means differ significantly at the overall 1% significance level?

27.49 Money Changes Behavior. ANOVA. Researchers at the University of Minnesota carried out several randomized comparative experiments on the effects of thinking about money. Here's an outline of one of the experiments. Ask student subjects to unscramble 30 sets of five words to make a meaningful phrase from four of the five. The control group unscrambled phrases like "cold it desk outside is" into "it is cold outside." The "play money" group unscrambled similar sets of words, but a stack of Monopoly money was placed nearby. The "money prime" group unscrambled phrases that led to thinking about money, for instance turning "high a salary desk paying" into "a high-paying salary." Then each subject worked a hard puzzle, knowing that they could ask for help. Table 27.5 shows the time, in seconds, that each subject worked on the puzzle before asking for help.[24] Psychologists think that money tends to make people self-sufficient. If so, the two groups that were encouraged in different ways to think about money should take longer on the average to ask for help. Do the data support this idea? MONEY

TABLE 27.5 Time (seconds) until subjects ask for help with a puzzle

Group	Time	Group	Time	Group	Time
Prime	609	Play	455	Control	118
Prime	444	Play	100	Control	272
Prime	242	Play	238	Control	413
Prime	199	Play	243	Control	291
Prime	174	Play	500	Control	140
Prime	55	Play	570	Control	104
Prime	251	Play	231	Control	55
Prime	466	Play	380	Control	189
Prime	443	Play	222	Control	126
Prime	531	Play	71	Control	400
Prime	135	Play	232	Control	92
Prime	241	Play	219	Control	64
Prime	476	Play	320	Control	88
Prime	482	Play	261	Control	142
Prime	362	Play	290	Control	141
Prime	69	Play	495	Control	373
Prime	160	Play	600	Control	156
		Play	67		

27.50 4step **In Your Own Words, Please. ANOVA.** The data in Exercises 27.9 and 27.13 show that note-takers using longhand write fewer words than those in either of the laptop groups. Do those using laptops have greater verbatim overlap with the lecture? For each set of notes, researchers compared three-word chunks of text in the notes with three-word chunks of text in the lecture transcription and reported a percentage of matches, with a higher percentage suggesting that the note-taker had a greater tendency to transcribe the lecture verbatim rather than write the notes in his or her own words. Give a complete analysis to compare the means of the percentages of matches for the three groups. VERBATIM

27.51 **Money Changes Behavior. Comparisons.** In the situation described in Exercise 27.49, researchers looked at the effects of thinking about money on time until asking for help when working a puzzle. Which of the three conditions are significantly different when we require overall significance level 5% for all comparisons as a group? Write a short summary of the results of the ANOVA, including the multiple comparisons. MONEY

27.52 **In Your Own Words, Please. Comparisons.** The study in Exercise 27.50 compared the verbatim overlap for three methods of taking notes in a lecture. Which of the three methods are significantly different when we require overall significance level 1% for all comparisons as a group. Write a short summary of the results of the ANOVA, including the multiple comparisons. VERBATIM

27.53 **Which Test?** Exercise 27.49 describes one of the experiments done by researchers at the University of Minnesota to demonstrate that even being reminded of money makes people more self-sufficient and less involved with other people. Here are three more of these experiments. For each experiment, which statistical test from Chapters 21, 23, 25, and 27 would you use, and why?

(a) Randomly assign student subjects to money and control groups. The control group unscrambles neutral phrases, and the money group unscrambles money-oriented phrases, as described in Exercise 27.49. Then ask the subjects to volunteer to help the experimenter by coding data sheets, spending about five minutes per sheet. Subjects said how many sheets they would volunteer to code. Participants in the money condition volunteered to help code fewer data sheets than did participants in the control condition.

(b) Randomly assign student subjects to high-money, low-money, and control groups. After playing Monopoly for a short time, the high-money group is left with $4000 in Monopoly money, the low-money group with $200, and the control group with no money. Each subject is asked to imagine a future with lots of money (high-money group), a future with a little money (low-money group), or just their future plans (control group). Another student walks in and spills a box of 27 pencils. How many pencils does the subject pick up? "Participants in the high-money condition gathered fewer pencils" than subjects in the other two groups.

(c) Randomly assign student subjects to three groups. All do paperwork while a computer on the desk shows a screensaver of currency floating underwater (Group 1), a screensaver of fish swimming underwater (Group 2), or a blank screen (Group 3). Each subject must now develop an advertisement and can choose whether to work alone or with a partner. Count how many in each group make each choice. Choosing to perform the task with a coworker was reduced among money condition participants.

NOTES AND DATA SOURCES

Chapter 0 Notes

1. www.nielsen.com/us/en/solutions/?measurement/music-sales-measurement/.
2. Parts of this essay are shared with David S. Moore, "Introduction: Learning from data," in Roxy Peck et al. (eds.), *Statistics: A Guide to the Unknown*, 4th ed., Thomson, 2006.
3. See, for example, Martin Enserink, "The vanishing promises of hormone replacement," *Science*, 297 (2002), pp. 325–326; and Brian Vastag, "Hormone replacement therapy falls out of favor with expert committee," *Journal of the American Medical Association*, 287 (2002), pp. 1923–1924. A National Institutes of Health panel's comprehensive report is *International Position Paper on Women's Health and Menopause*, NIH Publication 02-3284, 2002.
4. A. C. Nielsen, Jr., "Statistics in marketing," in *Making Statistics More Effective in Schools of Business*, Graduate School of Business, University of Chicago, 1986.
5. The data in Figure 0.2 were found online at www.eia.gov/petroleum/gasdiesel/xls/pswrgvwall.xls
6. FUTURE II Study Group, "Quadrivalent vaccine against human papillomavirus to prevent high-grade cervical lesions," *New England Journal of Medicine*, 356 (2007), pp. 1915–1927. We have simplified the conclusions so that students with as yet no statistics background can better follow the essay.
7. The study mentioned in Exercise 0.1 is S. M. Wang, J. H. Fan, P. R. Taylor, et al., "Association of plasma vitamin C concentration to total and cause-specific mortality: A 16-year prospective study in China," *J Epidemiol Community Health*, 72 (2018) pp.1076–1082. The online article mentioned in the question is at https://www.lifeextension.com/magazine/2019/3/vitamin-c-reduces-human-mortality/page-01.
8. http://www.msnbc.com/msnbc/poll-has-europe-let-too-many-refugees.
9. The data in Figure 0.3 were found online at http://energy.gov/eere/vehicles/fact-915-march-7-2016-average-historical-annual-gasoline-pump-price-1929-2015.
10. Gerd Gigerenzer, "Dread risk, September 11, and fatal traffic accidents," *Psychological Science*, 15 (2004), pp. 286–287. The graph in Figure 0.3 was adapted from a graph in the article.

Chapter 1 Notes

1. Higher Education Research Institute 2017 Freshman Survey at www.heri.ucla.edu. The actual data for each field of study consists of the responses for several specialties. For example, under education, the specialties are elementary education, music/art education, physical education/recreation, secondary school teacher in a non-STEM subject, special education, and other education. The percents of first-year students who plan to major in each of these specialties is given and we added these to produce the values for each of the fields of study listed in the example. This may have exacerbated the effect of roundoff error.
2. These data are from *The Infinite Dial 2019* at www.tritondigital.com.
3. See Note 2.
4. See Note 1.
5. Centers for Disease Control and Prevention, National Center for Health Statistics, online at https://www.cdc.gov/nchs/fastats/births.htm.
6. From the National Center for Education Statistics, available online at https://nces.ed.gov/fastfacts/display.asp?id=805.
7. Our eyes do respond to area, but not quite linearly. It appears that we perceive the ratio of two bars to be about the 0.7 power of the ratio of their actual areas. See W. S. Cleveland, *The Elements of Graphing Data*, Wadsworth, 1985, pp. 278–284.
8. From *Young Adults Then and Now* at https://census.socialexplorer.com/young-adults/#/
9. From the Gary Community School Corporation, courtesy of Celeste Foster, Purdue University.
10. The 2018 state information was compiled from the College Board website at https://reports.collegeboard.org/sat-suite-program-results/state-results.
11. From the Centers for Disease Control website at https://www.cdc.gov/lyme/stats/graphs.html. This is the chart of cases by age and sex, 2001–2017. The number of confirmed cases is from the chart of cases by symptom, 2001–2017.
12. The per capita total health expenditures in 2015 were obtained from the Country Data at the Global Health Observatory Data Repository of the World Health Organization at apps.who.int/gho/data/. All amounts are in international dollars at purchasing power parity. That is, the exchange rate between a currency and the dollar is set not at the fluctuating market rate but at the rate that gives a dollar the same buying power in each country.
13. The U.S. Geological Survey maintains data for various water parameters at monitoring sites throughout the United States at waterdata.usgs.gov/nwis. The data can be graphed or downloaded. The data in Figure 1.12 are for USGS 254754080344300 SHARK RIVER SLOUGH NO.1.

14. College Entrance Examination Board, *Trends in College Pricing, 2018*, at https://trends.collegeboard.org/college-pricing. The averages are "enrollment weighted" so that they give average tuition over *students* rather than over *colleges*. The reported averages have been adjusted to constant 2018 dollars.
15. The data are taken from drugabuse.gov/national-survey-drug-use-health.
16. See Note 14. The percentages are given in Figure 23.
17. PPG Industries annual automotive color trend data, https://news.ppg.com/automotive-color-trends/
18. Centers for Disease Control and Prevention publishes yearly reports on "Tobacco Use Among Middle and High School Students—United States," at https://www.cdc.gov/tobacco/data_statistics/fact_sheets/youth_data/tobacco_use/index.htm.
19. Centers for Disease Control and Prevention, National Center for Health Statistics, *Deaths: Final Data for 2017*, 68, No. 9, June 2019, at www.cdc.gov/nchs.
20. Data was taken from "Trends in Student Aid 2018," at https://research.collegeboard.org/pdf/trends-student-aid-2018-full-report.pdf
21. "The Global Mobile Report," September 12, 2017 at www.comscore.com.
22. Tom Lloyd et al., "Fruit consumption, fitness, and cardiovascular health in female adolescents: The Penn State Young Women's Health Study," *American Journal of Clinical Nutrition*, 67 (1998), pp. 624–630.
23. Data provided by Darlene Gordon from her PhD dissertation, "Relationships among academic self-concept, academic achievement, and persistence with self-attribution, study habits, and perceived school environment," Purdue University, 1997.
24. Historical stock returns from the website http://pages.stern.nyu.edu/adamodar/New_Home_Page/datafile/histretSP.html.
25. National Institutes of Health. The data for ratio of omega 3 to omega 6 fatty acids in food oils are no longer available; NIH now uses a new measure to compare amounts of these substances.
26. 2012 *Statistical Abstract of the United States*, Table 165, at www.census.gov.
27. Mortality rates obtained from the Global Health Observatory data repository at apps.who.int/gho/data/node.home.
28. J. Ward Testa (ed.), "Fur Seal Investigations, 2008–2009," *NOAA Technical Memorandum NMFS-AFSC-226*, (2011), p. 74. The most recent data are available at https://www.fisheries.noaa.gov/tags/fur-seal-counts.
29. Domenico Giannotti et al., "Play to become a surgeon: Impact of Nintendo Wii training on laparoscopic skills," *PLOS ONE*, V8, e5272, February 2013 at www.plosone.org.
30. David M. Fergusson and L. John Horwood, "Cannabis use and traffic accidents in a birth cohort of young adults," *Accident Analysis and Prevention*, 33 (2001), pp. 703–711.
31. "The Global Mobile Report," September 12, 2017 (2017), at www.comscore.com.
32. John Morton et al., "Acoustic features mediating height estimation from human speech," abstract from the 166th meeting of the Acoustical Society of America. We would like to thank the authors for supplying the data and additional details of the study.
33. See Note 14.
34. Unemployment data broken down by educational attainment are available from the Bureau of Labor Statistics at www.bls.gov.
35. U.S. Census Bureau, New Residential Construction page at www.census.gov/construction/nrc/. Go to the *Historical data* link to download the data. These are monthly data that are not seasonally adjusted.

Chapter 2 Notes

1. From the 2017 American Community Survey, at the U.S. Census Bureau website, www.census.gov. The data are a subsample of the over 64,000 individuals in the ACS North Carolina sample who had travel times greater than zero.
2. This study is available online at www.dispatch.com/live/content/databases/index.html.
3. This isn't a mathematical theorem. The mean can be less than the median in right-skewed distributions that take only a few values, many of which lie exactly at the median. The rule almost never fails for distributions taking many values, and most counterexamples don't appear clearly skewed in graphs even though they may be slightly skewed according to technical measures of skewness. See Paul T. von Hippel, "Mean, median, and skew: Correcting a textbook rule," *Journal of Statistics Education*, 13, No. 2 (2005), online journal.
4. National Association of College and University Business Officers and TIAA, 2018 NACUBO-TIAA study of endowments, at www.nacubo.org.
5. From the U.S. Census Bureau, www.census.gov/construction/nrs/pdf/uspricemon.pdf.
6. As of the beginning of 2016, yearly data were available through 2014 at the World Bank website, data.worldbank.org/indicator/EN.ATM.CO2E.PC.
7. U.S. Census Bureau, *Historical Income Tables: Households*, at www.census.gov.
8. Brock Bastian et al., "Pain as social glue: Shared pain increases cooperation," *Psychological Science*, 25 (2014), pp. 2079–2085.
9. The U.S. Department of Energy, www.fueleconomy.gov/feg/download.shtml.
10. We would like to thank Patricia Humphrey for supplying the test scores for students at Georgia Southern University.
11. The formal definition of degrees of freedom involves knowledge of a branch of mathematics called matrix algebra. It is associated with properties of the chi-square

distribution, which we will meet in Chapter 25. Roughly speaking, if the formula for a quantity, such as the variance, is the sum of n squared terms, the degrees of freedom is the number of these terms that can be "freely" assigned. For the variance, the n terms are the $x_i - \bar{x}$. You can check that the sum of the $x_i - \bar{x}$ must be 0. We can freely choose the values of $n-1$ of the $x_i - \bar{x}$, but the remaining one must be the value that makes the sum of all n 0. Thus, we say the variance has $n-1$ degrees of freedom. We will encounter degrees of freedom again in later chapters.

12. From the Environmental Protection Agency, www.epa.gov/radon/pubs/consguid.html. For levels in Franklin County, Ohio, see county-radon.info/OH/Franklin.html.
13. C. H. Cannon, D. R. Peart, and M. Leighton, "Tree species diversity in commercially logged Bornean rainforest," *Science*, 281 (1998), pp. 1366–1367. We thank Charles Cannon for providing the data.
14. See Note 27 for Chapter 1.
15. Current Population Survey Tables for Personal Income at www.census.gov/data/tables/time-series/demo/income-poverty/cps-pinc.html.
16. Jesse Bricker et al., "Changes in U.S. family finances from 2013 to 2016: Evidence from the survey of consumer finances," *Federal Reserve Bulletin*, Vol. 103, no. 3 (2017), at www.federalreserve.gov.
17. T. Bjerkedal, "Acquisition of resistance in guinea pigs infected with different doses of virulent tubercle bacilli," *American Journal of Hygiene*, 72 (1960), pp. 130–148.
18. See Note 5 for Chapter 1.
19. Data for 1986 from David Brillinger, University of California, Berkeley. See David R. Brillinger, "Mapping aggregate birth data," in A. C. Singh and P. Whitridge (eds.), *Analysis of Data in Time*, Statistics Canada, 1990, pp. 77–83. A boxplot similar to Figure 2.6 appears in David R. Brillinger, "Some examples of random process environmental data analysis," in P. K. Sen and C. R. Rao (eds.), *Handbook of Statistics*, Vol. 18, *Bioenvironmental and Public Health Statistics*, North Holland, 2000.
20. Paul E. O'Brien et al., "Laparascopic adjustable gastric banding in severely obese adolescents," *Journal of the American Medical Association*, 303 (2010), pp. 519–526. We thank the authors for providing the data.
21. Tim Gamble and Ian Walker, "Wearing a bicycle helmet can increase risk taking and sensation seeking in adults," *Psychological Science*, 27 (2016), pp. 289–294.
22. The salaries were obtained from www.hockey-reference.com/teams/MTL/2019_salary-cap.html.
23. The yearly returns were downloaded from wilshire.com/indexcalculator.
24. Nicolas Guéguen and Christine Petr, "Odors and consumer behavior in a restaurant," *Journal of Hospitality Management*, 25 (2006), pp. 335–339. We thank Nicolas Guéguen for providing the data.
25. Alex B. Van Zant and Don A. Moore, "Leaders' use of moral justifications increases policy support," *Psychological Science*, 26 (2015), pp. 934–943.
26. James C. Rosser et al., "The impact of video games on training surgeons in the 21st century," *Archives of Surgery*, 142 (2007), pp. 181–186. We thank Douglas Gentile for providing the data.
27. Information and data from the NHANES survey can be found at cdc.gov/nchs/nhanes.htm.
28. Revenues for the Global 500 companies can be found at fortune.com/global500/.

Chapter 3 Notes

1. See Note 9 for Chapter 1.
2. Cheryl D. Fryar et al., "Anthropometric reference data for children and adults: United States, 2011–2014," *Vital and Health Statistics*, Series 3, no. 39 (August 2016), at www.cdc.gov/nchs. This report provides the means of various anthropometric measurements. Standard deviations were computed from the sample size and reported standard errors of the mean. Instructions on measuring upper arm length are in the *National Health and Nutrition Examination Survey: Anthropometry Procedures Manual*, January 2007.
3. Monsoon rainfall from B. Parthasarathy, Indian Institute of Tropical Meterology, at www.iges.org. The data cover the years 1871 to 2000.
4. See Note 2.
5. All SAT facts are from the College Board website, www.collegeboard.com, and all ACT facts are from the ACT website, www.act.org.
6. See Note 2.
7. From the NCAA website /fs.ncaa.org/Docs/eligibility_center/Student_Resources/DI_ReqsFactSheet.pdf.
8. All MCAT facts are from the Medical College Admissions Test website, www.aamc.org.
9. Detailed data appear in, "Serum lipids of adults 20–74 Years: United States, 1976-80" *Vital and Health Statistics*, Series 11, no. 242, National Center for Health Statistics, 1993. Standard deviations were computed from the first and third quartiles, assuming Normality.
10. James A. Levine et al., "Inter-individual variation in posture allocation: Possible role in human obesity," *Science*, 307 (2005), pp. 584–586. We thank James Levine for providing the data.
11. Ulric Neisser, "Rising scores on intelligence tests," *American Scientist*, September–October 1997, online edition, www.americanscientist.org.
12. See Note 2.
13. See Note 24 for Chapter 1.
14. The data were provided by Nicolas Fisher.
15. See Note 10 for Chapter 2.
16. See Note 3.
17. See Note 26 for Chapter 2.

Chapter 4 Notes

1. Juan P. Carricart-Ganivet, "Sea surface temperature and the growth of the West Atlantic reef-building coral Montastraea annularis," *Journal of Experimental Marine Biology and Ecology*, 302 (2004), pp. 249–260.
2. The 2018 state information was compiled from the College Board website, at https://reports.collegeboard.org/sat-suite-program-results/state-results.
3. The homicide and suicide rates were found online at https://odh.ohio.gov/wps/wcm/connect/gov/f6130982-7b08-4822-a3bc-2ea5e2fb5696/OHVDRS-2015-Final.pdf?MOD=AJPERES&CONVERT_TO=url&CACHEID=ROOTWORKSPACE.Z18_M1HGGIK0N0JO00QO9DDDDM3000-f6130982-7b08-4822-a3bc-2ea5e2fb5696-mklreeh#:~:text=In%202015%2C%202%2C483%20violent%20deaths,by%20homicide%20(26%20percent). An Internet search of "homicide and suicide rates correlated" turns up several articles that explore this. See, in particular, the article at www.ncbi.nlm.nih.gov/pubmed/10789279 that mentions the implications of this correlation for the delivery of preventive and medical health services.
4. Initial concerns were based on government data for 2005, presented in "An accident waiting to happen?" *Consumer Reports*, March 2007, pp. 16–19. Data for 2018 were found online at web.mit.edu/airlinedata/www/default.html (outsource).
5. The Florida Department of Highway Safety and Motor Vehicles, at www.flhsmv.gov/motor-vehicles-tags-titles/vessels/vessel-owner-statistics/, gives the number of registered vessels. The Florida Wildlife Commission maintains a manatee death database at myfwc.com/research/manatee/.
6. These data were found online at the website https://avt.inl.gov/sites/default/files/pdf/ice/fact2013volkswagenjettatdi.pdf
7. A careful study of this phenomenon is W. S. Cleveland, P. Diaconis, and R. McGill, "Variables on scatterplots look more highly correlated when the scales are increased," *Science*, 216 (1982), pp. 1138–1141.
8. See Note 1.
9. Data for Figure 4.6(b) come from William Gray's website, at hurricane.atmos.colostate.edu. Data for Figure 4.6(c) were provided by Drina Iglesia, Purdue University, from a study reported in D. D. S. Iglesia, E. J. Cragoe Jr., and J. W. Vanable, "Electric field strength and epithelization in the newt (*Notophthalmus viridescens*)," *Journal of Experimental Zoology*, 274 (1996), pp. 56–62. Data for Figure 4.6(d) are from the Wilshire 5000 stock index. As a fine point, plots (b), (c), and (d) are square with the same scales on both axes because both variables measure similar quantities in the same units.
10. The data referred to in this exercise can be found online at www.fueleconomy.gov/feg/download.shtml.
11. These data are for 2019 and were found at the following websites: www.statista.com/statistics/202743/hot-dog-prices-in-major-league-baseball-by-team/ and www.statista.com/statistics/202666/beer-prices-in-major-league-baseball-by-team/.
12. This exercise is based on a study by Shakira F. Suglia et al., "Soft drinks consumption is associated with behavior problems in 5-year-olds," *The Journal of Pediatrics*, 163, No. 5 (2013), pp. 1323–1328. The actual study was more involved than indicated in the exercise. Researchers did find an association between soft drinks consumed and behavior problems but did not compute a correlation.
13. The data are a sample of seven values in a graph in Julie A. Mennella et al., "Preferences for salty and sweet tastes are elevated and related to each other during childhood," *PLOS ONE*, 2014. The paper is available online at www.plosone.org/article/info:doi/10.1371/journal.pone.0092201#pone-0092201-g003. The seven values selected produce a correlation similar to that found for all the data presented in the paper.
14. This exercise is motivated by Scott Berry, "Statistical fallacies in sports," *Chance*, 19, no. 4 (2006), pp. 50–56, where scores from the 2006 Masters are analyzed. Masters scores for 2019 were found online at https://www.pga.com/events/masters/leaderboard/.
15. Andrew J. Oswald et al., "Objective confirmation of subjective measures of human well-being: Evidence from the U.S.A.," *Science*, 327 (2010), pp. 576–579.
16. From a graph in Naomi E. Allen et al., "Moderate alcohol intake and cancer incidence in women," *Journal of the National Cancer Institute*, 101 (2009), pp. 296–305.
17. From a graph in Magdalena Bermejo et al., "Ebola outbreak killed 5000 gorillas," *Science*, 314 (2006), p. 1564.
18. From a graph in Bernt-Erik Saether, Steiner Engen, and Erik Mattysen, "Demographic characteristics and population dynamical patterns of solitary birds," *Science*, 295 (2002), pp. 2070–2073.
19. From a graph in Sabrina M. Tom et al., "The neural basis of loss aversion in decision-making under risk," *Science*, 315 (2007), pp. 515–518.
20. From a graph in J. Curre and H. Schwandt, "Inequality in mortality decreasing among the young while increasing for older adults, 1990–2010," *Science*, 352 (2016), pp. 708–712.
21. From a graph in Camilla A. Hinde et al., "Parent–offspring conflict and coadaptation," *Science*, 327 (2010), pp. 1373–1376.
22. Bruce Rind and David Strohmetz, "Effects of beliefs about future weather conditions on restaurant tipping,"*Journal of Applied Social Psychology*, 31 (2001), pp. 2160–2164. We thank the authors for supplying the original data.
23. The data are available online at https://www.ncdc.noaa.gov/cag/global/time-series/globe/land_ocean/12/12/1880-2019. The website explains that the twentieth-century average global temperature (combined land and ocean surface temperature) was 13.9°C.
24. Brian J. Whipp and Susan A. Ward, "Will women soon outrun men?" *Nature*, 355 (1992), p. 25. An article

in *Scientific American* in 2004 made a similar prediction that women would outrun men in the 100-meter dash in 2156. This article can be found online at www.scientificamerican.com/article.cfm?id=data-trends-suggest-women.

25. From a graph in Glenn J. Tattersall et al., "Heat exchange from the toucan bill reveals a controllable vascular thermal radiator," *Science*, 325 (2009), pp. 468–470.
26. From a graph in Naomi I. Eisenberger, Matthew D. Lieberman, and Kipling D. Williams, "Does rejection hurt? An fMRI study of social exclusion," *Science*, 302 (2003), pp. 290–292.
27. Ben Dantzer et al., "Density triggers maternal hormones that increase adaptive offspring growth in a wild mammal," *Science*, 340 (2013), pp. 1215–1217. The data used here are estimated from a figure in the paper.
28. Data for 2018 college bound seniors SAT scores are from the College Board website, www.collegeboard.org. Average teacher salaries are available at https://nces.ed.gov/programs/digest/d18/tables/dt18_211.60.asp

Chapter 5 Notes

1. From a graph in James A. Levine, Norman L. Eberhardt, and Michael D. Jensen, "Role of nonexercise activity thermogenesis in resistance to fat gain in humans," *Science*, 283 (1999), pp. 212–214.
2. The online article was found at www.livestrong.com/blog/everyone-u-s-exercise-wed-lose-2-billion-pounds/?utm_source=newslette&utm_medium=email&utm_campaign=0312.
3. M. C. Hansen et al., "High-resolution global maps of 21st-century forest cover change," *Science*, 342 (2013), pp. 850–853.
4. See Note 1 for Chapter 4.
5. See Note 3 for Chapter 4.
6. From a graph in Herman Pontzer et al., "Constrained total energy expenditure and metabolic adaption to physical activity in adult humans," *Current Biology*, 26 (2016), pp. 410–417.
7. Contributed by Marigene Arnold, Kalamazoo College.
8. Gannett News Service article appearing in the *Lafayette (Ind.) Journal and Courier*, April 23, 1994.
9. P. Goldblatt (ed.), *Longitudinal Study: Mortality and Social Organization*, Her Majesty's Stationery Office, 1990. At least, so claims Richard Conniff, *The Natural History of the Rich*, Norton, 2002, p. 45. The Goldblatt report is not available to us.
10. Mortality and immunization rates obtained from the Global Health Observatory data repository at apps.who.int/gho/data/node.home.
11. *The Health Consequences of Smoking: 1983*, Public Health Service, Washington DC, 1983.
12. H-L F Eriksen et al., "Effects of Tobacco Smoking in Pregnancy on Offspring Intelligence at the Age of 5," *Journal of Pregnancy*, 2012, Article ID 945196.
13. For more discussion of issues surrounding big data see Tim Harford, "Big data: Are we making a big mistake?" *Significance*, 11 (2014), pp. 14–19.
14. These data are for 2019 and were found at the following websites: www.statista.com/statistics/202743/hot-dog-prices-in-major-league-baseball-by-team/ and www.statista.com/statistics/202666/beer-prices-in-major-league-baseball-by-team/.
15. Data provided by Robert Dale, Purdue University.
16. The data are a sample of seven values in a graph in Julie A. Mennella et al., "Preferences for salty and sweet tastes are elevated and related to each other during childhood," *PLOS ONE*, 2014. The paper is available online at www.plosone.org/article/info:doi/10.1371/journal.pone.0092201#pone-0092201-g003. The seven values selected produce a correlation similar to that found for all the data presented in the paper.
17. G. L. Kooyman et al., "Diving behavior and energetics during foraging cycles in king penguins," *Ecological Monographs*, 62 (1992), pp. 143–163.
18. Found in online advertisements September 26, 2019.
19. The last data pair is the height of one of the authors and his sister. The first 11 data pairs are from Karl Pearson and A. Lee, "On the laws of inheritance in man," *Biometrika*, 2 (1902), p. 357. These first 11 data sets also appear in D. J. Hand et al., *A Handbook of Small Data Sets*, Chapman & Hall, 1994. This book offers more than 500 data sets that can be used in statistical exercises.
20. From a presentation by Charles Knauf, Monroe County (NY) Environmental Health Laboratory.
21. See Note 18 for Chapter 4.
22. Frank J. Anscombe, "Graphs in statistical analysis," *The American Statistician*, 27 (1973), pp. 17–21.
23. Debora L. Arsenau, "Comparison of diet management instruction for patients with non-insulin dependent diabetes mellitus: Learning activity package vs. group instruction," MS thesis, Purdue University, 1993.
24. Found online at www.webmd.com/diet/diet-sodas-and-weight-gain-not-so-fast?page=3.
25. See Note 2 for Chapter 4. Teacher salaries were found online at /www.nea.org/home/2017-2018-average-starting-teacher-salary.html.
26. CO_2 emissions data are available online at www.esrl.noaa.gov/gmd/ccgg/trends/. Global temperature data are available online at www.ncdc.noaa.gov/cag/global/time-series. Historical data on the Dow Jones Industrial Average are available at www.investing.com/indices/us-30-historical-data.
27. Gary Smith, "Do statistics test scores regress toward the mean?" *Chance*, 10 (1997), pp. 42–45.
28. From a graph in G. D. Martinsen, E. M. Driebe, and T. G. Whitham, "Indirect interactions mediated by changing plant chemistry: Beaver browsing benefits beetles," *Ecology*, 79 (1998), pp. 192–200.
29. P. Velleman, *ActivStats 2.0*, Addison Wesley Interactive, 1997.
30. From William Gray's website, hurricane.atmos.colostate.edu. Forecasts are those made each June.

31. Data for 1936–1999 are from a graph in Bruce J. Peterson et al., "Increasing river discharge to the Arctic Ocean," *Science*, 298 (2002), pp. 2171–2173. Data for 2000–2008 are from a graph in I. Ashik et al., "Arctic report card: Update for 2010," available online at https://arctic.noaa.gov/Portals/7/ArcticReportCard/Documents/ArcticReportCard_full_report2010.pdf. The graph is on page 41 of the report. Data for 2009–2010 are from a graph in K. R. Arrigo et al., "Arctic report card: Update for 2011," available online at https://arctic.noaa.gov/Portals/7/ArcticReportCard/Documents/ArcticReportCard_full_report2011.pdf. The graph is on page 157 of the report. The data for 2011–2014 are from a table in R. M. Holmes et al., "Arctic report card: Update for 2015," available online at https://arctic.noaa.gov/Portals/7/ArcticReportCard/Documents/ArcticReportCard_full_report2015.pdf. The data for 2015–2017 are from a table in R. M. Holmes et al., "Arctic report card: Update for 2018," https://arctic.noaa.gov/Portals/7/ArcticReportCard/Documents/ArcticReportCard_full_report2018.pdf.

32. See Note 23 for Chapter 4.

Chapter 6 Notes

1. The data are from the 2017 *Digest of Education Statistics* at the website of the National Center for Education Statistics, nces.ed.gov.

2. Rani A. Desai et al., "Video-gaming among high school students: Health correlates, gender differences, and problematic gaming," *Pediatrics*, 126 (2010), pp. 1416–1424.

3. From the October 2017 Current Population Survey, at www.census.gov.

4. Siem Oppe and Frank De Charro, "The effect of medical care by a helicopter trauma team on the probability of survival and the quality of life of hospitalized victims," *Accident Analysis and Prevention*, 33 (2001), pp. 129–138. The authors give the data in Example 6.4 as a "theoretical example" to illustrate the need for their more elaborate analysis of actual data using severity scores for each victim.

5. Found online at www.sports-reference.com/cbb/schools/michigan/2018.html.

6. I. Westbrooke, "Simpson's paradox: An example in a New Zealand survey of jury composition," *Chance*, 11 (1998), pp. 40–42.

7. The data came from the Pew Social Networking Fact Sheet, obtained at www.pewinternet.org/fact-sheets/social-networking-fact-sheet/.

8. This General Social Survey exercise presents a table constructed using the search function at the GSS archive, sda.berkeley.edu/archive.htm. These data are from the 2012 GSS.

9. Gregory D. Myer et al., "Youth versus adult weightlifting injuries presenting to United States emergency rooms: Accidental versus nonaccidental injury mechanisms," *Journal of Strength and Conditioning Research*, 23 (2009), pp. 2054–2060.

10. These data are available at the U.S. Census Bureau website at, www.census.gov/data/tables/time-series/demo/income-poverty/cps-pinc/pinc-02.html.

11. M. Radelet, "Racial characteristics and imposition of the death penalty," *American Sociological Review*, 46 (1981), pp. 918–927.

12. D. Gonzales et al., "Varenicline, an α4β2 nicotinic acetylcholine receptor partial agonist, vs sustained-release bupropion and placebo for smoking cessation: A randomized controlled trial," *Journal of the American Medical Association*, 340 (1999), pp. 685–691.

13. Michael Gurian, "Where have the men gone? No place good," *Washington Post*, December 4, 2005, at www.washingtonpost.com. The data are from the 2017 *Digest of Education Statistics* at the website of the National Center for Education Statistics, nces.ed.gov.

14. Nancy J. O. Birkmeyer, "Hospital complication rates with bariatric surgery in Michigan," *Journal of the American Medical Association*, 304 (2010), pp. 435–442.

15. The data for the University of Michigan Health and Retirement Study (HRS) can be downloaded from the website, ssl.isr.umich.edu/hrs/start.php.

16. Data on Phil's predictions can be found online at www.groundhog.org. Historical data on March temperatures can be found online at www.ncdc.noaa.gov/cag/.

17. Hannah Lund et al., "Sleep patterns and predictors of disturbed sleep in a large population of college students," *Journal of Adolescent Health*, 46 (2010), 124–132. We would like to thank the authors for supplying the data.

Chapter 7 Notes

1. Harry B. Meyers, "Investigations of the life history of the velvetleaf seed beetle, *Althaeus folkertsi Kingsolver*," master's thesis, Purdue University, 1996.

2. Educational attainment data for OECD countries were obtained at https://www.oecd-ilibrary.org/education/educational-attainment-of-25-64-year-olds-2019_75230926-en.

3. From a plot in K. Krishna Kumar et al., "Unraveling the mystery of Indian monsoon failure during El Niño," *Science*, 314 (2006), pp. 115–119.

4. *Audio Today 2019: How America Listens*, at www.arbitron.com.

5. See Note 5 for Chapter 3.

6. See Note 2 for Chapter 3.

7. Data provided by Brigitte Baldi, University of California at Irvine.

8. Data on spending were found online at www.capology.com/nhl/payrolls/2018-19/. Points earned by each team were found online at https://www.hockey-reference.com/leagues/franchise_points.html.

9. Juan P. Carricart-Ganivet, "Sea surface temperature and the growth of the West Atlantic reef-building coral Montastraea annularis," *Journal of Experimental Marine Biology and Ecology*, 302 (2004), pp. 249–260.

10. J. T. Dwyer et al., "Memory of food intake in the distant past," *American Journal of Epidemiology*, 130 (1989), pp. 1033–1046.
11. Data from a plot in Josef P. Rauschecker, Biao Tian, and Marc Hauser, "Processing of complex sounds in the macaque nonprimary auditory cortex," *Science*, 268 (1995), pp. 111–114. The paper states that there are $n = 41$ observations, but only $n = 37$ can be read accurately from the plot.
12. These and other data are available online at `inys.indiana.edu/docs/survey/indianaYouthSurvey_2018.pdf`.
13. "Dancing in step," *Economist*, March 22, 2001.
14. From a plot in Jon J. Ramsey et al., "Energy expenditure, body composition, and glucose metabolism in lean and obese rhesus monkeys treated with ephedrine and caffeine," *American Journal of Clinical Nutrition*, 68 (1998), pp. 42–51.
15. These data were found online at `www.cdc.gov/tobacco/data_statistics/fact_sheets/adult_data/cig_smoking/index.htm`.
16. Janice E. Williams et al., "Anger proneness predicts coronary heart disease risk," *Circulation*, 101 (2000), pp. 2034–2039.
17. From `https://ozonewatch.gsfc.nasa.gov/statistics/annual_data.html`.
18. From the Nenana Ice Classic website, `www.nenanaakiceclassic.com`. See Raphael Sagarin and Fiorenza Micheli, "Climate change in nontraditional data sets," *Science*, 294 (2001), p. 811, for a careful discussion.
19. Raymond Fisman and Edward Miguel, "Cultures of corruption: Evidence from diplomatic parking tickets," National Bureau of Economic Research Working Paper 12312, June 2006, at `www.nber.org`.
20. Louie H. Yang, "Periodical cicadas as resource pulses in North American forests," *Science*, 306 (2004), pp. 1565–1567. The data are simulated Normal values that match the means and standard deviations reported in this article.
21. Alan S. Banks et al., "Juvenile hallux abducto valgus association with metatarsus adductus," *Journal of the American Podiatric Medical Association*, 84 (1994), pp. 219–224.
22. Todd W. Anderson, "Predator responses, prey refuges, and density-dependent mortality of a marine fish," *Ecology*, 81 (2001), pp. 245–257.
23. From a graph in Craig Packer et al., "Ecological change, group territoriality, and population dynamics in Serengeti lions," *Science*, 307 (2005), pp. 390–393.
24. Peter H. Chen, Neftali Herrera, and Darren Christiansen, "Relationships between gate velocity and casting features among aluminum round castings," no date. Provided by Darren Christiansen.
25. Stefan Wojcik and Adam Hughes, "Sizing up Twitter users," at `www.pewresearch.org/internet/2019/04/24/sizing-up-twitter-users/`.
26. R. Shine, T. R. L. Madsen, M. J. Elphick, and P. S. Harlow, "The influence of nest temperatures and maternal brooding on hatchling phenotypes in water pythons," *Ecology*, 78 (1997), pp. 1713–1721.
27. See Note 23 for Chapter 1.

Chapter 8 Notes

1. Information about the online poll was found at `kgab.com/online-poll-majorities-favor-legalizing-weed-keeping-superday/`.
2. Poll results are from `www.pollingreport.com/guns.htm`. The methodological statement can be found at `www.nytimes.com` and is similar for most polls listed.
3. Gary S. Foster and Craig M. Eckert, "Up from the grave: A sociohistorical reconstruction of an African American community from cemetary data in the rural Midwest," *Journal of Black Studies*, 33 (2003), pp. 468–489.
4. Monica Anderson and Jingjing Jiang, "Teens, social media & technology 2018" (2018). The Pew full report can be found at `www.pewresearch.org/internet/2018/05/31/teens-social-media-technology-2018`.
5. The regulations that govern seat belt survey design can be found at `www-nrd.nhtsa.dot.gov`. Details on the Alaska survey are in *Alaska 2016 survey of seat belt use*, at `/www.dot.state.ak.us/stwdplng/hwysafety/assets/pdf/2016_Observational_Seatbelt_Survey.pdf`.
6. Donald L. McCabe, Linda Klebe Trevino, and Kenneth D. Butterfield, "Dishonesty in academic environments," *Journal of Higher Education*, 72 (2001), pp. 29–45.
7. For information about the response rates for the American Community Survey of households (there is a separate sample of group quarters), go to `www.census.gov/acs/www/methodology/sample-size-and-data-quality/response-rates`.
8. Katherine Zickuhr and Lee Rainie, "E-reading rises as device ownership jumps," (2014). The full report is at `www.pewinternet.org/2014/01/16/e-reading-rises-as-device-ownership-jumps/`.
9. For more detail on the limits of memory in surveys, see N. M. Bradburn, L. J. Rips, and S. K. Shevell, "Answering autobiographical questions: The impact of memory and inference on surveys," *Science*, 236 (1987), pp. 157–161.
10. The immigration detention center questions are from the Quinnipiac University Poll taken July 25 to 28, 2019, found at `www.pollingreport.com`. The responses on electing a gay president are from a Quinnipiac University Poll taken April 26 to 29, 2019, found at `www.pollingreport.com`. The example on the effect of question order is cited in Daniel Kahnemann et al., "Would you be happier if you were richer? A focusing illusion," *Science*, 312 (2006), pp. 1908–1910.
11. The research study "Health reform and the decline of physician private practice" (2010), found at `www.physiciansfoundation.org`.

12. You can go to www.pollingreport.com to see the results of many polling agencies compiled on a variety of issues.
13. Information from various articles in the special issue on cell phone surveys, *Public Opinion Quarterly*, 71, no. 5 (2007). The 2014 cell phone use numbers were obtained from the CDC website, www.cdc.gov.
14. See Mick P. Couper, "Web surveys: A review of issues and approaches," *Public Opinion Quarterly*, 64 (2000), pp. 464–494.
15. The Harris Poll, "Back-to-school shopping: Kids influenced by social media push parents to overspend," at theharrispoll.com/back-to-school-shopping-kids-influenced-by-social-media-push-parents-to-overspend/.
16. The American Association for Public Opinion Research, "AAPOR report on online panels," 2010, at www.aapor.org.
17. Rachel Sherman and John Hickner, "Academic physicians use placebos in clinical practice and believe in the mind–body connection," *Journal of General Internal Medicine*, 23 (2008), pp. 7–10.
18. The *New York Times* SurveyMonkey poll can be found at www.surveymonkey.com/curiosity/nyt-october-2019-cci/?ut_source=content_center&ut_source2=topic.
19. Stephen J. Blumberg and Julian V. Luke, "Wireless substitution: Early release of estimates from the national health interview survey, January–June 2017," at www.cdc.gov.
20. Euridice M. Steele et al., "Ultra-processed foods and added sugars in the US diet: Evidence from a nationally representative cross-sectional study," *British Medical Journal Open*, 6 (2016), at bmjopen.bmj.com.
21. Andrew Rzepa and Julie Ray, "Is there an outbreak of doubt about vaccines in the U.S.?" (2019). Available online at www.gallup.com.
22. The poll results, including a link to the methodology, are available at news.gallup.com/poll/246899/blacks-rate-race-relations-whites-bad.aspx.
23. Anna E. Price and Julian A. Reed, "Use and nonuse of a rail trail conversion for physical activity: Implications for promoting trail use," *American Journal of Health Education*, 45 (2014), pp. 249–256.
24. The questions are available from a Gallup poll, April 1–9, 2019, at https://news.gallup.com/poll/1714/taxes.aspx.
25. The online poll results were found at poll.fm/10434126/results.
26. The poll results were found at news.gallup.com/poll/1645/guns.aspx.
27. See Note 4.
28. Bryan E. Porter and Thomas D. Berry, "A nationwide survey of self-reported red light running: Measuring prevalence, predictors, and perceived consequences," *Accident Analysis and Prevention*, 33 (2001), pp. 735–741.
29. Mario A. Parada et al., "The validity of self-reported seatbelt use: Hispanic and non-Hispanic drivers in El Paso," *Accident Analysis and Prevention*, 33 (2001), pp. 139–143.
30. Giuliana Coccia, "An overview of non-response in Italian telephone surveys," *Proceedings of the 99th Session of the International Statistical Institute*, 1993, Book 3, pp. 271–272.
31. "Amplifiers study: The Twitters users who are most likely to retweet and how to engage them," January, 2013, posted by Taylor Schreiner at blog.twitter.com/2013/.
32. Information about the Ontario College of Pharmacists obtained from www.ocpinfo.com. The numbers reported are from the 2018 Annual Report.
33. The Health Care in Canada Survey can be found by going to survey reports and presentations at the website www.hcic-sssc.ca.
34. Clyde O. McDaniel, Jr., "Dating roles and reasons for dating," *Journal of Marriage and the Family*, 31 (1969), pp. 97–107.
35. Robert C. Parker and Patrick A. Glass, "Preliminary results of double-sample forest inventory of pine and mixed stands with high- and low-density LiDAR," in Kristina F. Connoe (ed.), *Proceedings of the 12th Biennial Southern Silvicultural Research Conference*, U.S. Department of Agriculture, Forest Service, Southern Research Station, 2004.
36. The article can be found at www2.macleans.ca/2010/07/16/sometimes-a-gaffe-is-more-than-a-gaffe/.
37. The results and description of the survey were found at www.nytimes.com/interactive/2016/us/elections/polls.html.

Chapter 9 Notes

1. James C. Rosser et al., "The impact of video games on training surgeons in the 21st century," *Archives of Surgery*, 142 (2007), pp. 181–186. We thank Douglas Gentile for providing the data.
2. Sam Harper et al., "Trends in socioeconomic inequalities in motor vehicle accident deaths in the United States, 1995–2010," *American Journal of Epidemiology*, 182 (2015), pp. 606–614.
3. Hyunjin Song and Norbert Schwarz,"If it's hard to read, it's hard to do: Processing fluency affects effort prediction and motivation," *Psychological Science*, 19 (2008), pp. 986–988.
4. Hsin-Chieh Yeh et al.,"Smoking, smoking cessation, and risk for type 2 diabetes mellitus: A cohort study," *Annals of Internal Medicine*, 152 (2010), pp. 10–17.
5. Charles A. Nelson III et al., "Cognitive recovery in socially deprived young children: The Bucharest Early Intervention Project," *Science*, 318 (2007), pp. 1937–1940.
6. See Note 24 for Chapter 2. The description of the factors and the response are based on the study, although the public vs. private factor was within subjects.

7. See Note 19 for Chapter 2.
8. Michael L. Lowe and Kelly L. Haws,"Sounds big: The effects of acoustic pitch on product perceptions," *Journal of Marketing Research*, 54 (2017), pp. 332–346.
9. The description of the factors and the response is based on a study by M. K. Baloch and F. Bibi, "Effect of harvesting and storage conditions on the post harvest fruit quality and shelf life of mango, (*Mangifera indica* L.) fruit," *South African Journal of Botany*, 83 (2012), pp. 109–116.
10. Janice K. Kiecolt-Glaser et al., "Marital distress, depression, and a leaky gut: Translocation of bacterial endotoxin as a pathway to inflammation," *Psychoneuroendocrinology*, 58 (2018), pp. 52–60.
11. See Note 28 for Chapter 1.
12. See Note 8 for Chapter 2.
13. Julie Mares et al., "Healthy diets and the subsequent prevalence of nuclear cataract in women," *Archives of Opthalmology*, 128 (2010), pp. 738–749.
14. Raine Sihvonen et al., "Arthroscopic partial meniscectomy versus sham surgery for a degenerative meniscal tear," *New England Journal of Medicine*, 369 (2013), pp. 2515–2524.
15. Libby Moulton, "The active placebo effect: Patent eligible subject matter," in the *The Columbia Science and Technology Law Review*, December 2010.
16. Chittaranhan Andrade, "Prayer and healing: A medical and scientific perspective on randomized controlled trials," *Indian Journal of Psychiatry*, 51 (2009), pp. 247–253.
17. Christopher N. Blesso et al., "Whole egg consumption improves lipoprotein profiles and insulin sensitivity to a greater extent than yolk-free egg substitute in individuals with metabolic syndrome," *Metabolism*, 62 (2013), pp. 400–410.
18. The description of this experiment is based on the methodology described in "How to win the battle of the bugs," *Consumer Reports*, July 2015, pp. 34–37.
19. D. G. Jakovljevic and A. K. McConnell, "Influence of different breathing frequencies on the severity of inspiratory muscle fatigue induced by high-intensity front crawl swimming," *Journal of Strength and Conditioning Research*, 23, no. 4 (2009), pp. 1169–1174.
20. Sterling C. Hilton et al., "A randomized controlled experiment to assess technological innovations in the classroom on student outcomes: An overview of a clinical trial in education," manuscript, no date. A brief report is Sterling C. Hilton and Howard B. Christensen, "Evaluating the impact of multimedia lectures on student learning and attitudes," *Proceedings of the 6th International Conference on the Teaching of Statistics*, at `www.stat.aukland.ac.nz`.
21. Lucy Bowes et al., "Peer victimisation during adolescence and its impact on depression in early adulthood: Prospective cohort study in the United Kingdom," *British Medical Journal*, 350 (2015), h2469.
22. Morgan K. Ward and Darren W. Dahl, "Should the devil sell Prada? Retail rejection increases aspiring consumers' desire for the brand," *Journal of Consumer Research*, 41 (2014), pp. 590–609.
23. An Pan et al., "Red meat consumption and mortality," *Archives of Internal Medicine*, 172 (2012), pp. 555–563.
24. Mirva Rottensteiner et al., "Physical activity, fitness, glucose homeostasis, and brain morphology in twins," *Medicine and Science in Sports and Exercise*, 47 (2015), pp. 509–518.
25. N. Kalak et al., "Daily morning running for 3 weeks improved sleep and psychological functioning in healthy adolescents compared with controls," *Journal of Adolescent Health*, 51 (2012), pp. 615–622.
26. Marc Neuman and Rachel Werner, "Marital status and postoperative functional recovery," *Journal of the American Medical Association Surgery*, 151 (2016), pp. 194–196.
27. Jo Phelan et al., "The stigma of homelessness: The impact of the label 'homeless' on attitudes towards poor persons," *Social Psychology Quarterly*, 60 (1997), pp. 323–337.
28. Heather I Katcher et al., "The effects of a whole grain enriched hypocaloric diet on cardiovascular disease risk factors in men and women with metabolic syndrome13," *American Journal of Clinical Nutrition* 87 (2008), pp. 79–90.
29. Tina Poon and Bianca Grohmann, "Spatial density and ambient scent: Effects on consumer anxiety," *American Journal of Business*, 29 (2014), pp. 76–94.
30. This problem is based on the COcoa Supplement and Multivitamin Outcomes Study (COSMOS). Information about the study is available online at `www.whi.org/studies/COSMOS/SitePages/Home.aspx`.
31. Anne Mangen et al., "Handwriting versus keyboard writing: Effect on word recall," *Journal of Writing Research*, 7 (2015), pp. 227–247. In the actual experiment, a third method for writing down words (using an iPad) was also used.
32. The description of the factors and the response is based on a portion of the study by Brian Wansink and Pierre Chandon,"Can 'low-fat' nutrition labels lead to obesity," *Journal of Marketing Research*, 43 (2006), pp. 605–617.
33. Fredrik Lange et al., "Store-window creativity's impact on shopper behavior," *Journal of Business Research*, 69 (2016), pp. 1014–1021.
34. Ian G. Williamson et al., "Antibiotics and topical nasal steroid for treatment of acute maxillary sinusitis," *Journal of the American Medical Association*, 298 (2007), pp. 2487–2496.
35. Found at `www.forbes.com`, January 14, 2014.
36. The description of the factors and the response is based on a study by Daniel Schwartz et al., "Advertising energy saving programs: The potential environmental cost of emphasizing monetary savings," *Journal of Experimental Psychology: Applied*, 21 (2015), pp. 153–165.
37. Christopher A. Bail et al., "Exposure to opposing views on social media can increase political polarization," *Proceedings of the National Academy of Sciences of the United States of America*, online at `www.pnas.org/content/early/2018/08/27/1804840115`.

38. The study is described in Gina Kolata, "New study finds vitamins are not cancer preventers," *New York Times*, July 21, 1994. Look in the *Journal of the American Medical Association* of the same date for the details.
39. George I. Papakostas et al., "*S*-Adenosyl methionine (SAMe) augmentation of serotonin reuptake inhibitors for antidepressant nonresponders with major depressive disorder: A double-blind, randomized clinical trial," *American Journal of Psychiatry*, 167 (2010), pp. 942–948.

Chapter 10 Notes

1. The U.S. Department of Health and Human Services maintains a website that reports cases of misconduct. The web address is `ori.hhs.gov`. If you visit the site, you will see reports of actual cases of misconduct. Many of these involve handling of data.
2. John C. Bailar III, "The real threats to the integrity of science," *Chronicle of Higher Education*, April 21, 1995, pp. B1–B2.
3. For more on the problems with open-access publishing, see Gina Kolata, "Scientific articles accepted (personal checks, too)," *The New York Times*, April 7, 2013, and the March 28, 2013, issue of *Nature*.
4. This statement can be found at the *Journal of the American Medical Association* website: `jama.jamanetwork.com/public/instructionsForAuthors.aspx`. Search under "Requirements for Reporting" and look for the link "Ethical Approval of Studies and Informed Consent."
5. See the details on the website of the Office for Human Research Protections of the Department of Health and Human Services, `www.hhs.gov/ohrp`.
6. The quote is from the Texas A&M International University website, `www.tamiu.edu/irb/structure\_IRB.shtml`.
7. The difficulties of interpreting guidelines for informed consent and for the work of institutional review boards in medical research are a main theme of Beverly Woodward, "Challenges to human subject protections in U.S. medical research," *Journal of the American Medical Association*, 282 (1999), pp. 1947–1952. The references in this paper point to other discussions. Updated regulations and guidelines appear on the OHRP website (see Note 5).
8. See Note 5.
9. See, for example, `www.lc.org/hotissues/2001/aba_1-18/public_records_laws_by_state.htm`.
10. The Heartbleed Bug, which came to public attention in early 2014, is a security bug that affects open-source OpenSSL cryptographic software. This bug renders many websites thought to be secure, vulnerable to attacks by hackers. You can read more about this online at `heartbleed.com`.
11. For more discussion of the dangers of observational studies in clinical trials, see David Madigan et al., "A systematic statistical approach to evaluating evidence from observational studies," *Annual Review of Statistics and Its Application*, 1 (2014), pp. 11–39.
12. Quotation from the *Report of the Tuskegee Syphilis Study Legacy Committee*, May 20, 1996. A detailed history is James H. Jones, *Bad Blood: The Tuskegee Syphilis Experiment*, Free Press, 1993.
13. "More than Tuskegee: Understanding Mistrust about Research Participation," *Journal of Health Care for the Poor and Underserved*, 21 (2010), pp. 879–897, found at `https://www.ncbi.nlm.nih.gov/pmc/articles/PMC4354806/`.
14. Dr. Hennekens's words are from an interview in the Annenberg/Corporation for Public Broadcasting video series *Against All Odds: Inside Statistics*. The lack of certainty that Dr. Hennekens refers to is now called "clinical equipoise" in discussions of ethics.
15. For more discussion of the ethics of the use of placebos in clinical trials, see `www.ama-assn.org/ama/pub/physician-resources/medical-ethics/code-medical-ethics/opinion8083.page` on the American Medical Association website.
16. Ezekial J. Emanuel, David Wendler, and Christine Grady, "What makes clinical research ethical?" *Journal of the American Medical Association*, 283 (2000), pp. 2701–2711.
17. R. D. Middlemist, E. S. Knowles, and C. F. Matter, "Personal space invasions in the lavatory: Suggestive evidence for arousal," *Journal of Personality and Social Psychology*, 33 (1976), pp. 541–546.
18. For a review of domestic violence experiments, see C. D. Maxwell et al., *The Effects of Arrest on Intimate Partner Violence: New Evidence from the Spouse Assault Replication Program*, U.S. Department of Justice, NCH188199, 2001. Available online at `www.ojp.usdoj.gov/nij/pubs-sum/188199.htm`.
19. See `www.pnas.org/content/111/24/8788.full` for the article and a description of the Facebook experiment.
20. Joseph Millum and Ezekial J. Emanuel, "The ethics of international research with abandoned children," *Science*, 318 (2007), pp. 1874–1875. This paper has some useful comments on international research in general.

Chapter 11 Notes

1. Monic Sun and Remi Trudel, "The effect of recycling versus trashing on consumption: Theory and experimental evidence," *Journal of Marketing Research*, 54 (April 2017), pp. 293–305.
2. Trine Madsen et al., "Association between traumatic brain injury and risk of suicide," *Journal of the American Medical Association*, 320 (2018), pp. 580–588.
3. The description of the factors and the response is based on a portion of the study by Alice Healy et al.,"Terrorism after 9/11: Reactions to simulated news reports," *American Journal of Psychology*, 122 (2009), pp. 153–165.

4. A. Swanson, "Study: Birth month might affect health," *Columbus Dispatch*, June 20, 2015.
5. Simplified from Sanjay K. Dhar, Claudia González-Vallejo, and Dilip Soman, "Modeling the effects of advertised price claims: tensile versus precise pricing," *Marketing Science*, 18 (1999), pp. 154–177.
6. Julia Belluz, "Is one drink per day really unsafe? That new alcohol study explained," *Vox*, August 29, 2018. Online at `www.vox.com/science-and-health/2018/8/29/17790118/alcohol-lancet-health-study`. The research paper referred to in the article is Max G. Griswold et al., "Alcohol use and burden for 195 countries and territories, 1990–2016: A systematic analysis for the Global Burden of Disease Study 2016," *Lancet*, August 23, 2018, published online at `dx.doi.org/10.1016/S0140-6736(18)31310-2`.
7. Alysha Fligner and Xiaoyan Deng, "The effect of font naturalness on perceived healthiness of food products," Senior thesis, The Ohio State University.
8. Margot Roosevelt, "Study finds traffic pollution can speed hardening of arteries," *LA Times*, February 14, 2010.
9. Donald G. McNeil Jr., "Flu shots in children can help community," *New York Times*, March 10, 2010, p. A16.

Chapter 12 Notes

1. The *Washington Post*/ABC News poll was conducted September 2–5, 2019. Found online at `www.pollingreport.com/guns.htm`.
2. The *Washington Post*/ABC News poll was conducted by telephone from Sept. 2–5, 2019 among a random national sample of 1003 adults, with 65% reached on cell phones and 35% on landlines. The margin of sampling error is plus or minus 3.5 percentage points overall and is larger among subgroups. `games-cdn.washingtonpost.com/notes/prod/default/documents/c9b1e401-36b4-420e-953c-741837849c5d/note/bfa3f530-447d-4df9-b55c-90a8ad641ef8.pdf#page=1`.
3. Note that pennies have rims that make spinning more stable. The probability of a head in spinning a coin depends on the type of coin and also on the surface. See Exercise 21.3 for an account of 56% of heads in spinning a Belgian 1-euro coin. *Chance News 11.02* at `www.dartmouth.edu/~chance` reports about 45% heads in more than 20,000 spins of American pennies by Robin Lock's students at Saint Lawrence University.
4. The percentages were found at the GMAT website, `www.gmac.com/-/media/files/gmac/research/gmat-test-taker-data/profile-of-gmat-testing-citizenship-ty2014-ty2018.pdf`.
5. Data from the website of Statistics Canada, `www.statcan.gc.ca`.
6. Some statisticians prefer the use of the terminology "discrete probability model" because it is more general than the terminology "finite probability model." Discrete probability models include finite sample spaces as well as sample spaces that are infinite, such as the set of all positive integers. An example of such an infinite discrete sample space would be the sample space for the number of free-throw attempts until a basketball player makes her first free throw. This could occur on her first attempt, her second attempt, her third attempt, and so on. All positive integers are possible outcomes. Assigning probabilities to individual outcomes in an infinite discrete sample space is more complicated than for a finite discrete sample space. To avoid these complexities, we restrict our attention to finite sample spaces.
7. You can find a mathematical explanation of Benford's law in Ted Hill, "The first-digit phenomenon," *American Scientist*, 86 (1996), pp. 358–363; and Ted Hill, "The difficulty of faking data," *Chance*, 12, No. 3 (1999), pp. 27–31. Applications in fraud detection are discussed in the second paper by Hill and in Mark A. Nigrini, "I've got your number," *Journal of Accountancy*, May 1999, available online at `www.aicpa.org/pubs/jofa/joaiss.htm`.
8. Based on a July 2015 Gallup Poll, found at `gallup.com/poll/6424/Nutrition-Food.aspx`.
9. Software random number generators don't really generate any number between 0 and 1. The numbers generated must be rounded to a finite number of decimal places and technically must be described by a discrete probability model. Continuous probability models are thus an abstraction. We imagine that, in theory, outcomes could be any number to an arbitrary number of decimal places, but for practical reasons, they must be truncated to a finite number of decimal places.
10. All MCAT facts are from the Medical College Admission Test website, `www.aamc.org`.
11. Information from `http://gradedistribution.registrar.indiana.edu/reports.php`.
12. Thomas K. Cureton et al., "Endurance of young men," *Monographs of the Society for Research in Child Development*, 10, (1945).
13. Based on data from 2018 available at `https://crashstats.nhtsa.dot.gov/#/`.
14. Based on a November 2019 Gallup Poll, found at `news.gallup.com/poll/15370/party-affiliation.aspx`.
15. Data from 2017 found online at `https://nhts.ornl.gov/vehicle-trips`.
16. See Note 17 for Chapter 1.
17. National population estimates for July 1, 2018, at the Census Bureau website, at `www.census.gov`. The table omits people who consider themselves to belong to more than one race.
18. Based on data from the Census Bureau website, at `www.census.gov`. The website treats gender as binary.
19. This exercise is based on a famous example discussed in Part II of *The Design of Experiments* by Sir R. A. Fisher. The example involves a lady tasting tea. Fisher used this

example to illustrate the logic of hypothesis testing. Presumably, if milk is added to hot tea, the milk can curdle and produce a slightly sour taste. This is avoided by adding the milk first.

20. Found online November 2019 at climate.nasa.gov/evidence/.

Chapter 13 Notes

1. Andrew Perrin, "One-in-five Americans now listen to audiobooks," September 25, 2019 at www.pewresearch.org/fact-tank/2019/09/25/one-in-five-americans-now-listen-to-ft audiobooks/_19-09-25_bookreadingformats_print-books-more-popular-e-books-audiobooks/.
2. Patrick Van Kessle et al., "A week in the life of popular YouTube channels," July 25, 2019, from the Pew Research Center, Internet & Technology, at www.pewresearch.org/internet/2019/07/25/a-week-in-the-life-of-popular-youtube-channels/.
3. Data are for 2015–2016 and found online at nces.ed.gov/programs/raceindicators/indicator_REE.asp#info.
4. This is one of several tests discussed in Bernard M. Branson, "Rapid HIV testing: 2005 update," a presentation by the Centers for Disease Control and Prevention, at www.cdc.gov. The Malawi clinic result is reported by Bernard M. Branson, "Point-of-care rapid tests for HIV antibody," *Journal of Laboratory Medicine*, 27 (2003), pp. 288–295.
5. P. Habibzadeh, "Decay of references to web sites in articles published in general medical journals: Mainstream vs small journals," *Applied Clinical Informatics*, 4 2013, pp. 455–464.
6. From the International Trade Administration website, at https://www.trade.gov/td/otm/autostats.asp.
7. Sales in 2018 from the website of the Entertainment Software Association, at www.theesa.com.
8. Emily A. Vogels, "Millennials stand out for their technology use, but older generations also embrace digital life," September 19, 2019, online at www.pewresearch.org/fact-tank/2019/09/09/us-generations-technology-use/.
9. Aaron Smith, "15% of American adults have used online dating sites or mobile dating apps," February 11, 2016, from the Pew Research Center, at www.pewinternet.org. The distribution of ages was found at www.census.gov.
10. The probabilities for this exercise were obtained from pets.thenest.com/blindness-white-cats-9034.html and cats.about.com/od/coatcolorpatternstypes/ss/White-Cats-Profile-All-You-Need-To-Know-About-White-Cats.htm. For probabilities of events that were reported as an interval of values, we used the center of the interval as the probability of the event for the exercise.
11. Hal R. Arkes and Wolfgang Gaissmaier, "Psychological research and the prostate-cancer screening controversy," *Journal of Psychological Science*, 23 (2012), pp. 547–553.
12. F. H. Schroder et al., "Screening and prostate-cancer mortality in a randomized European study," *New England Journal of Medicine*, 360 (2009) pp. 1320–1328.
13. See Note 8.
14. W. Uter et al., "Inter-relation between variables determining constitutional UV sensitivity in Caucasian children," *Photodermatology, Photoimmunology and Photmeicine*, 20 (2004), pp. 9–13.
15. Digest of Education Statistics, 1960–2017 at nces.ed.gov. The data are contained in Table 302.10.
16. M. McCulloch et al., "Diagnostic accuracy of canine scent detection in early- and late-stage lung and breast cancers," *Integrative Cancer Therapies*, 5 (2006), pp. 30–39.
17. See Note 20 for Chapter 1.
18. About one in 1500 births exhibit characteristics of both male and female. According to the World Health Organization website, "Gender describes those characteristics of women and men that are largely socially created, while sex encompasses those that are biologically determined. However, these terms are often mistakenly used interchangeably in scientific literature, health policy, and legislation." https://www.who.int/genomics/gender/en/.
19. Projections from U.S. Department of Education, *Projections of Education Statistics to 2027*, February 2019, at nces.ed.gov.
20. Data provided by Patricia Heithaus and the Department of Biology at Kenyon College.
21. The probabilities given are realistic according to the 2018 fact sheet on tobacco, at www.cdc.gov.
22. F. J. G. M. Klaassen and J. R. Magnus, "How to reduce the service dominance in tennis? Empirical results from four years at Wimbledon," in S. J. Haake and A. O. Coe (eds.), *Tennis Science and Technology*, Blackwell, 2000, pp. 277–284.
23. R. S. Gupta et al., "The prevalence, severity, and distribution of childhood food allergy in the United States," *Journal of Pediatrics*, 128 (2011), pp. e9–e17.
24. From the National Institutes of Health's National Digestive Diseases Information Clearinghouse, at wrongdiagnosis.com.
25. The data were obtained from the GSS Cumulative Datafile 1972–2014, at sda.berkeley.edu/archive.htm. The data were restricted to the year 2014.
26. B. Budowle et al., "Population data on the thirteen CODIS core short tandem repeat loci in African Americans, U.S. Caucasians, Hispanics, Bahamians, Jamaicans, and Trinidadians,"*Journal of Forensic Sciences*, (1999), pp. 1277–1286.
27. Marilyn vos Savant, "Ask Marilyn column," *Parade Magazine*, September 9, 1990, p. 16.

Chapter 14 Notes

1. Matthew A. Carlton and William D. Stansfield, "Making babies by the flip of a coin?" *American Statistician*, 59 (2005), pp. 180–182.
2. Richard F. MacLeod et al., "Damage to fresh tomatoes can be reduced" at ucce.ucdavis.edu/files/repositoryfiles/ca3012p10-72079.pdf.
3. Pew Internet at www.pewinternet.org provides a full disposition of all sampled numbers in many of its reports. The response rates vary somewhat, ranging from 5% to 15%.
4. From the Canadian Survey on Disability, 2017 at www.statcan.gc.ca.
5. Richard Fry, "For first time in modern era, living with parents edges out other living arrangements for 18- to 34-year olds," May 2016 at pewsocialtrends.org. The data given are for 2019 from www.census.gov/data/tables/time-series/demo/families/adults.html.
6. Data are from 2016, found at www.statcan.gc.ca.
7. Information obtained from the Centers for Disease Control and Prevention website, at www.cdc.gov.
8. Nicola Low et al., "A randomized, controlled trial of an aerosolized vaccine against measles," *New England Journal of Medicine*, 372 (2015), pp.1519–1529.
9. Alain Cohn et al., "Business culture and dishonesty in the banking industry," *Nature*, 516 (2014), pp. 86–89.
10. Sales data from www.hyundai.com.
11. Paul Rodway et. al., "Preferring the one in the middle: Further evidence for the centre-stage effect," *Applied Cognitive Psychology*, 26 (2012), pp. 215–222.
12. Associated Press news item dated December 9, 2007, found at www.msnbc.msn.com.
13. See demonstrations.wolfram.com/MonteCarloEstimateForPi/ for an online demonstration of this idea.

Chapter 15 Notes

1. U.S. Census Bureau data available online at https://www.census.gov/data/tables/time-series/demo/income-poverty/cps-hinc.html.
2. Technically, the probability that $\bar{x} - \mu$ differs in absolute value by less than any fixed arbitrary small amount goes to 1 as the number of observations increases.
3. The data can be found at the Bureau of Labor Statistics website and are available for download at http://www.bls.gov/tus/home.htm.
4. Strictly speaking, the formula $\sigma/\sqrt{n}$ for the standard deviation of $\bar{x}$ assumes that we draw an SRS of size n from an *infinite* population. If the population has finite size N, this standard deviation is multiplied by $\sqrt{\frac{N-n}{N-1}} = \sqrt{1-\frac{n-1}{N-1}}$. This "finite population correction" approaches 1 as N increases. When the population is at least 20 times as large as the sample, the correction factor is between about 0.97 and 1. It is reasonable to use the simpler form $\sigma/\sqrt{n}$ in these settings.
5. More precisely, if we take many, many samples, calculate the unbiased estimator for each sample, and then average these many, many values of the unbiased estimator, the average will be very close to (equal to as the number of samples becomes infinite) the value of the parameter.
6. Personal income for all 141,251 individuals were downloaded from the Integrated Public Use Microdata Series (IPUMS), which provides data from the Current Population Survey. Data can be accessed at cps.ipums.org/cps/. The histograms in Figure 15.5 were produced from the downloaded data.
7. See Note 10 for Chapter 3.
8. Found online at pages.stern.nyu.edu/adamodar/New_Home_Page/datafile/histret.html. Sophisticates will note that for compounding over several years, we want the geometric mean return, which was 9.73%.

Chapter 16 Notes

1. Note 4 for Chapter 15 explains the reason for this condition in the case of inference about a population mean.
2. Fryar C. D., et al., Anthropometric reference data for children and adults: United States, 2011–2014. National Center for Health Statistics. *Vital Health Statistics* 3(2016), at https://www.cdc.gov/nchs/data/series/sr_03/sr03_039.pdf.
3. Information about the NAEP test can be found online at https://www.nationsreportcard.gov.
4. B. Rind and D. Strohmetz, "Effect of beliefs about future weather conditions on restaurant tipping," *Journal of Applied Social Psychology*, 31 (2001), pp. 2160–2164.
5. See Note 23 for Chapter 1.
6. The National Survey of Student Engagement conducts an annual survey that includes hours spent preparing for class each week. The numbers in this exercise are based on the results of the 2015 survey. You can find survey results at nsse.indiana.edu/html/findings.cfm.
7. Chi-Fu Jeffrey Yang, Peter Gray, and Harrison G. Pope, Jr., "Male body image in Taiwan versus the West," *American Journal of Psychiatry*, 162 (2005), pp. 263–269.
8. M. Ann Laskey et al., "Bone changes after 3 mo of lactation: Influence of calcium intake, breast-milk output, and vitamin D–receptor genotype," *American Journal of Clinical Nutrition*, 67 (1998), pp. 685–692.

Chapter 17 Notes

1. Data on scores on the GMAT can be found at www.gmac.com/gmat/learn-about-the-gmat-exam/, www.gmac.com/gmat-other-assessments/accessing-gmat-exam-scores-and-reports/gmat-

scoring-by-exam-section-normal-view, and www.gmac.com/market-intelligence-and-research/assessment-data/about-the-geographic-trend-reports. The mean and standard deviation for the GMAT scores is based on data from 2015–2018 and is based on the results of 739,752 tests.

2. For detailed information about the procedures, see nvlpubs.nist.gov/nistpubs/hb/2019/NIST.HB.133-2019.pdf.
3. For one-sided alternatives, some statisticians prefer to write the null hypothesis so that it includes all possible values of μ that are not specified by the alternative. They would write either

$$H_0: \mu \leq 0$$
$$H_a: \mu > 0$$

or

$$H_0: \mu \geq 0$$
$$H_a: \mu < 0$$

and interpret a null hypothesis such as $H_0: \mu = 0$ as implying either $H_0: \mu \leq 0$ or $H_0: \mu \geq 0$ for one-sided alternatives.

Sir R. A. Fisher, one of the greatest statisticians and scientists of the twenty-first century, argued that "the null hypothesis must be exact, that is free from vagueness or ambiguity . . . " (see his book *The Design of Experiments*, Section 8 of Part II). His reason was that the null hypothesis must determine an exact distribution from which probability calculations can be made. He would advocate the use of $H_0: \mu = 0$, and we have followed Fisher's formulation in this textbook.

4. Steven R. Smith et al., "Multicenter, placebo-controlled trial of lorcaserin for weight management," *The New England Journal of Medicine*, 363 (2010), pp. 245–256.
5. See Note 3 for Chapter 16.
6. Rashmi Ranjan Das and Meenu Singh, "Oral zinc for the common cold," *Journal of the American Medical Association*, 311 (2014), pp. 1440–1441.
7. Ajay Ghei, "An empirical analysis of psychological androgeny in the personality profile of the successful hotel manager," master's thesis, Purdue University, 1992.
8. Richard Westfall et al., "Effects of instructor attractiveness on learning," *The Journal of General Psychology*, 143 (2016), pp. 161–171.
9. Seung-Ok Kim, "Burials, pigs, and political prestige in Neolithic China," *Current Anthropology*, 35 (1994), pp. 119–141.
10. Gerardo Ramirez and Sian L. Bellock, "Writing about testing worries boosts exam performance in the classroom," *Science*, 331 (2011), pp. 211–213.
11. Mario A. Parada et al., "The validity of self-reported seatbelt use: Hispanic and non-Hispanic drivers in El Paso," *Accident Analysis and Prevention*, 33 (2001), pp. 139–143.
12. See Note 7 for Chapter 16.
13. Data simulated from a Normal distribution based on information in Brian M. DeBroff and Patricia J. Pahk, "The ability of periorbitally applied antiglare products to improve contrast sensitivity in conditions of sunlight exposure," *Archives of Ophthamology*, 121 (2003), pp. 997–1001.
14. This exercise is based on Shannon S. Carson, et al., "Effect of pallative care-led meetings for families of patients with chronic illness: A randomized clinical trial," *Journal of the American Medical Association*, 316 (2016), pp. 51–62.

Chapter 18 Notes

1. See www.cdc.gov/nchs/tutorials/currentnhanes/SurveyDesign/SampleDesign/intro.htm.
2. Bryan E. Porter and Thomas D. Berry, "A nationwide survey of self-reported red light running: Measuring prevalence, predictors, and perceived consequences," *Accident Analysis and Prevention*, 33 (2001), pp. 735–741.
3. For a discussion of statistical significance in the legal setting, see D. H. Kaye, "Is proof of statistical significance relevant?" *Washington Law Review*, 61 (1986), pp. 1333–1365. Kaye argues: "Presenting the *P*-value without characterizing the evidence by a significance test is a step in the right direction. Interval estimation, in turn, is an improvement over *P*-values."
4. See, for example, the regulatory guide at www.nrc.gov/docs/ML0037/ML003739948.pdf.
5. From a press release from the Harvard School of Public Health College Alcohol Study, April 12, 2001, at www.hsph.harvard.edu/cas/.
6. Warren E. Leary, "Cell phones: questions but no answers," *New York Times*, October 26, 1999.
7. Information about TUDA can be found online at www.nationsreportcard.gov/highlights/reading/2019/.
8. Poll published August 17, 2016, at http://www.theharrispoll.com/health-and-life/Chef-Prestigious-Occupation.html. A note at the bottom of the page says, "Because the sample is based on those who agreed to participate in the Harris Interactive panel, no estimates of theoretical sampling error can be calculated."
9. Alice G. Walton, "Why jogging may be better for your health than running." Found online at http://www.forbes.com/sites/alicegwalton/2015/02/03/why-jogging-may-be-better-for-your-health-than-running/#112bf03d36f6.
10. Justin S. Brashares et al., "Bushmeat hunting, wildlife declines, and fish supply in West Africa," *Science*, 306 (2004), pp. 1180–1183. The data used here (and in Figure 1B of the article) are found in the online supplementary material.

11. Gabriel Gregoratos et al., "ACC/AHA guidelines for implantation of cardiac pacemakers and antiarrhythmia devices: executive summary," *Circulation*, 97 (1998), pp. 1325–1335.
12. From the commentary by Frank J. Sulloway, "Birth order and intelligence," *Science*, 316 (2007), pp. 1711–1712. The study report appears in the same issue, Petter Kristensen and Tor Bjerkedal, "Explaining the relation between birth order and intelligence," *Science*, 316 (2007), p. 1717.
13. C. Kopp et al., "Modulation of rhythmic brain activity by diazepam: GABAA receptor subtype and state specificity," *Proceedings of the National Academy of Sciences*, 101 (2004), pp. 3674–3679.
14. Kenneth Rosenfield et al., "Trial of a paclitaxel-coated ballon for femoropopliteal artery disease," *New England Journal of Medicine*, 373, No. 2 (2015), pp. 145–153.

Chapter 19 Notes

1. Data for U.S. searches (desktop only) in October 2019 from `https://www.comscore.com/Insights/Rankings#tab_search_share/`.
2. Data for U.S, household size for 2019 from `www.census.gov/data/tables/time-series/demo/families/households.html`. We have grouped households with size 7 or more into the single category, size 7.
3. U.S. Census Bureau, *Fertility of American Women: June 2014*, at `www.census.gov`.
4. Aaron S. Hervey et al., "Reaction time distribution analysis of neuropsychological performance in an ADHD sample," *Child Neuropsychology*, 12 (2006), pp. 125–140.
5. K. E. Hobbs et al., "Levels and patterns of persistent organochlorines in minke whale (*Balaenoptera acutorostrata*) stocks from the North Atlantic and European Arctic," *Environmental Pollution*, 121 (2003), pp. 239–252.
6. Maureen Hack et al., "Outcomes in young adulthood for very-low-birth-weight infants," *New England Journal of Medicine*, 346 (2002), pp. 149–157.
7. Mikyoung Park et al., "Recycling endosomes supply AMPA receptors for LTP," *Science*, 305 (2004), pp. 1972–1975.
8. Benjamin B. Albert et al., "Among overweight middle-aged men, first-borns have lower insulin sensitivity than second-borns," *Scientific Reports*, 4 (2014), article number 3906. Available online at `www.nature.com/srep/index.html`.
9. Jon E. Keeley, C. J. Fotheringham, and Marco Morais, "Reexamining fire suppression impacts on brushland fire regimes," *Science*, 284 (1999), pp. 1829–1831.
10. The survey question is reported in Susanah Fox and Maeve Duncan, "Tracking for health" January, 2013 at `pewinternet.org`. In fact, the estimate in the survey was 60% saying "Yes."
11. Data simulated from a Normal distribution with $\mu = 98.2$ and $\sigma = 0.7$. These values are based on P. A. Mackowiak, S. S. Wasserman, and M. M. Levine, "A critical appraisal of 98.6° F, the upper limit of the normal body temperature, and other legacies of Carl Reinhold August Wunderlich," *Journal of the American Medical Association*, 268 (1992), pp. 1578–1580.
12. Nicolas Guéguen and Christine Petr, "Odors and consumer behavior in a restaurant," *International Journal of Hospitality Management*, 25 (2006), pp. 335–339. We thank Nicolas Guéguen for providing the data.
13. Probabilities from trials with 2897 people known to be free of HIV antibodies and 673 people known to be infected, reported in J. Richard George, "Alternative specimen sources: methods for confirming positives," 1998 Conference on the Laboratory Science of HIV, found online at the Centers for Disease Control and Prevention, `www.cdc.gov`.
14. Higher Education Research Institute 2013 Freshman Survey, at `www.heri.ucla.edu`.
15. Corby K. Martin et al., "Weight loss and retention rates and weight loss in a commercial weight loss program and the effect of corporate partnership," *International Journal of Obesity*, 34 (2010), pp. 742–750. The retention rate after 4 weeks was read from a graph in the paper.
16. J. F. Swain et al., "Comparison of the effects of oat bran and low-fiber wheat on serum lipoprotein levels and blood pressure," *New England Journal of Medicine*, 322 (1990), pp. 147–152.

Chapter 20 Notes

1. Note 4 for Chapter 15 explains the reason for this condition in the case of inference about a population mean.
2. The American Community Survey collects data on one-way commute times for cities in the United States. Data can be found at `www.census.gov/newsroom/press-releases/2017/acs-5yr.htm`.
3. Mark E. Benden et al., "Within-subjects analysis of the effects of a stand-biased classroom intervention on energy expenditure," *Journal of Exercise Physiology*, 15 (2012). Available online at `www.asep.org/asep/asep/JEPonline_April_2012.html`.
4. See Note 4 for Chapter 16.
5. From a graph in Benedetto De Martino et al., "Frames, biases, and rational decision-making in the human brain," *Science*, 313 (2006), pp. 684–687. We simplified the design a bit for easier comprehension: the starting amounts and gambles offered differed from trial to trial, though still matched in pairs; 32 very unbalanced "catch trials" were mixed with the 64 experimental trials to be sure subjects were paying attention; and all money amounts were in British pounds, not dollars.
6. John Morton, et al., "Acoustic features mediating height estimation from human speech," Abstract from the 166th meeting of the Acoustical Society of America. We would

like to thank the authors for supplying the data and additional details of the study.

7. These data were found online at `http://publicapps.odh.ohio.gov/BeachGuardPublic/SearchResults.aspx?instatepark=False&beachtype=PUB_PUB_ACC`.
8. Alice P. Melis, Brian Hare, and Michael Tomasello, "Chimpanzees recruit the best collaborators," *Science*, 311 (2006), pp. 1297–1300. A Normal quantile plot does not show major lack of Normality, and a saddlepoint approximation that allows for skew gives $P = 0.0039$. So the t test is reasonably accurate despite the skew and small sample size.
9. Data that take only whole-number values are discrete. Thus, technically, they cannot come from a population with a continuous distribution, such as the Normal distribution.
10. Data simulated from a Normal distribution based on information in Brian M. DeBroff and Patricia J. Pahk, "The ability of periorbitally applied antiglare products to improve contrast sensitivity in conditions of sunlight exposure," *Archives of Ophthalmology*, 121 (2003), pp. 997–1001.
11. The fact that all data must be truncated to a finite number of decimal places implies that real data can never exactly follow any continuous distribution. For example, integer-valued data are discrete so that, at best, such data can only be roughly approximated by a continuous distribution.
12. For a qualitative discussion explaining why skewness is the most serious violation of the Normal shape condition, see Dennis D. Boos and Jacqueline M. Hughes-Oliver, "How large does n have to be for the Z and t intervals?" *American Statistician*, 54 (2000), pp. 121–128. Our recommendations are based on extensive computer work. See, for example, Harry O. Posten, "The robustness of the one-sample t-test over the Pearson system," *Journal of Statistical Computation and Simulation*, 9 (1979), pp. 133–149; and E. S. Pearson and N. W. Please, "Relation between the shape of population distribution and the robustness of four simple test statistics," *Biometrika*, 62 (1975), pp. 223–241.
13. For more advanced users, a good way to ascertain if the t procedures are safe is to compare the 95% confidence interval produced by t with the BCa interval from a bootstrap with at least 1000 resamples. For part (b), the t interval is 29,428 to 32,254 and a BCa interval is 29,106 to 31,894. For part (c), on the other hand, t gives 38.93 to 40.49 and BCa gives 38.97 to 40.44. These results confirm the judgment that t is safe for part (c) but not for part (b).
14. Table 1 of E. Thomassot et al., "Methane-related diamond crystallization in the earth's mantle: Stable isotopes evidence from a single diamond-bearing xenolith," *Earth and Planetary Science Letters*, 257 (2007), pp. 362–371.
15. From the online supplement to Tor D. Wager et al., "Placebo-induced changes in fMRI in the anticipation and experience of pain," *Science*, 303 (2004), pp. 1162–1167.
16. TUDA results for 2019 from the National Center for Education Statistics, at `nationsreportcard.gov`.
17. Ravi Mehta and Rui Zhu, "Blue or red? Exploring the effect of color on cognitive task performances," *Science*, 323 (2009), pp. 1226–1229.
18. John M. Jakicic et al., "Effect of wearable technology combined with a lifestyle intervention on long-term weight loss, the IDEA randomized clinical trial," *JAMA*, 316 (2016), pp. 1161–1171.
19. See Note 2 for Chapter 17.
20. J. D. Marshall et al., "Vehicle self-pollution intake fraction: Children's exposure to school bus emissions," *Environmental Science and Technologhy*, 39 (2005), pp. 2559–2563.
21. See Note 19 for Chapter 7.
22. Data provided by Drina Iglesia, Purdue University. The data are part of a larger study reported in D. D. S. Iglesia, E. J. Cragoe, Jr., and J. W. Vanable, "Electric field strength and epithelization in the newt (*Notophthalmus viridescens*)," *Journal of Experimental Zoology*, 274 (1996), pp. 56–62.
23. Matthias R. Mehl et al., "Are women really more talkative than men?" *Science* 317 (2007), p. 82
24. Data for Cohort 2 in Richard A. Morgan et al., "Cancer regression in patients after transfer of genetically engineered lymphocytes," *Science*, 314 (2006), pp. 126–129. The doubling time data are given in the paper, and the immune response data appear in the supplementary online material.
25. M. B. Laferty, "OSU scientist gets a kick out of sports controversy," *The Columbus Dispatch*, November 21, 1993.
26. I thank Jason Hamilton, University of Illinois, for providing the data. The study is reported in Evan H. DeLucia et al., "Net primary production of a forest ecosystem with experimental CO_2 enhancement," *Science*, 284 (1999), pp. 1177–1179. No method for inference can be trusted with $n = 3$. In this study, each observation is very costly, so the small n is inevitable.
27. Michael W. Peugh, "Field investigation of ventilation and air quality in duck and turkey slaughter plants," MS thesis, Purdue University, 1996.
28. Harry B. Meyers, "Investigations of the life history of the velvetleaf seed beetle, *Althaeus folkertsi* Kingsolver," MS thesis, Purdue University, 1996. The 95% t interval is 1227.9 to 2507.6. A 95% bootstrap BCa interval is 1444 to 2718, confirming that t inference is inaccurate for these data.
29. J. Marcus Jobe and Hutch Jobe, "A statistical approach for additional infill development," *Energy Exploration and Exploitation*, 18 (2000), pp. 89–103. The comparison interval is the BCa interval based on 1000 bootstrap resamples.

30. These data were found online at http://publicapps.odh.ohio.gov/BeachGuardPublic/SearchResults.aspx?instatepark=False&beachtype=PUB_PUB_ACC.
31. Ralf Bargou et al., "Tumor regression in cancer patients by very low doses of a T cell engaging antibody," *Science*, 321 (2008), pp. 974–977.
32. Data provided by Timothy Sturm.
33. Lianng Yuh, "A biopharmaceutical example for undergraduate students," manuscript, no date.
34. See Note 3 for Chapter 16.

Chapter 21 Notes

1. James A. Levine et al., "Inter-individual variation in posture allocation: Possible role in human obesity," *Science*, 307 (2005), pp. 584–586. We thank James Levine for providing the data.
2. Detailed information about the conservative *t* procedures can be found in Paul Leaverton and John J. Birch, "Small sample power curves for the two sample location problem," *Technometrics*, 11 (1969), pp. 299–307; Henry Scheffé, "Practical solutions of the Behrens-Fisher problem," *Journal of the American Statistical Association*, 65 (1970), pp. 1501–1508; and D. J. Best and J. C. W. Rayner, "Welch's approximate solution for the Behrens-Fisher problem," *Technometrics*, 29 (1987), pp. 205–210.
3. Kathleen G. McKinney, "Engagement in community service among college students: Is it affected by significant attachment relationships?" *Journal of Adolescence*, 25 (2002), pp. 139–154. To see the questions in the Inventory of Parent and Peer Attachments, go to chipts.cch.ucla.edu/assessment/IB/List_Scales/inventory%20parent%20and%20peer%20attachment.htm.
4. Domenico Giannotti et al., "Play to become a surgeon: Impact of Nintendo Wii training on laparoscopic skills," *PLOS ONE*, 8 (2013), p. e5272.
5. Murali K. Matta et al., "Effect of sunscreen application under maximal use conditions on plasma concentration of sunscreen active ingredients a randomized clinical trial," *JAMA*, 321 (2019), pp. 2082–2091. The data in the problem are estimated from a graph in the paper and are rounded to the first decimal.
6. See the extensive simulation studies in Harry O. Posten, "The robustness of the two-sample *t*-test over the Pearson system," *Journal of Statistical Computation and Simulation*, 6 (1978), pp. 295–311; and Harry O. Posten, H. Yeh, and Donald B. Owen, "Robustness of the two-sample *t*-test under violations of the homogeneity assumption," *Communications in Statistics*, 11 (1982), pp. 109–126.
7. Nicolas Guéguen and Christine Petr, "Odors and consumer behavior in a restaurant," *Journal of Hospitality Management*, 25 (2006), pp. 335–339. I thank Nicolas Guéguen for providing the data. Although the spending data are quite discrete, a bootstrap BCa 95% confidence interval for the difference in means based on 1000 resamples is 2.394 to 4.826, close to the Option 1 95% interval 2.209 to 4.736. So the sample means are sufficiently Normal to allow use of *t* procedures.
8. See Note 1 for Chapter 2.
9. Data provided by Samuel Phillips, Purdue University.
10. Shari L. Barkin et al., "Effect of a behavioral intervention for underserved preschool-age children on change in body mass index a randomized clinical trial," *JAMA*, 320 (2018), pp. 450–460.
11. The problem of comparing variabilities is difficult even with advanced methods. Common distribution-free procedures do not offer a satisfactory alternative to the *F* test because they are sensitive to unequal shapes when comparing two distributions. A recent survey of possible approaches is Dennis D. Boos and Cavell Brownie, "Comparing variances and other measures of dispersion," *Statistical Science*, 19 (2005), pp. 571–578.
12. Dinah S. Reddihough et al., "Effect of fluoxetine on obsessive-compulsive behaviors in children and adolescents with autism spectrum disorders," *JAMA*, 322 (2019), pp. 1561–1569.
13. Matthias R. Mehl et al., "Are women really more talkative than men?" *Science*, 317 (2007), p. 82.
14. Michael A. Sayette et al., "Lost in the sauce, the effects of alcohol on mind wandering," *Psychological Science*, 20 (2009), pp. 747–752.
15. Meryl P. Gardner et al., "Better moods for better eating? How mood influences food choice," *Journal of Consumer Psychology*, 24 (2013), pp. 320–335.
16. Michael Dufner et al., "The end of a stereotype: Only children are not more narcissistic than people with siblings," *Social Psychological and Personality Science* (2019), available online at journals.sagepub.com/doi/10.1177/1948550619870785.
17. See Note 15.
18. Pankaj Aggarwal and Min Zhao, "Seeing the big picture: The effect of height on the level of construal," *Journal of Marketing Research*, 52 (2015), pp. 120–133.
19. Rashmi Singh et al., "Relationship of collegiate football experience and concussion with hippocampal volume and cognitive outcomes," *JAMA*, 311 (2014), pp. 1883–1888.
20. Wayne J. Camera and Donald Powers, "Coaching and the SAT I," *TIP* (1999), available online at www.siop.org/tip.
21. Bruce Rind and David B. Strohmetz, "Effect of beliefs about future weather conditions on restaurant tipping," *Journal of Applied Social Psychology*, 31 (2001), pp. 2160–2164.
22. A. Timur Sevincer et al., "Positive thinking about the future in newspaper reports and presidential addresses predicts economic downturn," *Psychological Science*, 25 (2014). Available online at pss.sagepub.com. Data

are estimates from a plot in the article, with positive articles corresponding to articles with a positive thinking z score above 2 and negative articles corresponding to a positive thinking z score below −2.

23. Anne Niccoli, "Paper or tablet? Reading recall and comprehension," *EDUCAUSE Review* (2015), available online at `er.educause.edu/articles/2015/9/paper-or-tablet-reading-recall-and-comprehension`.
24. Fabrizio Grieco, Arie J. van Noordwijk, and Marcel E. Visser, "Evidence for the effect of learning on timing of reproduction in blue tits," *Science*, 296 (2002), pp. 136–138. The data in Exercise 21.45 are from a graph in this paper.
25. Kathleen D. Vohs, Nicole L. Mead, and Miranda R. Goode, "The psychological consequences of money," *Science*, 314 (2006), pp. 1154–1156. I thank Kathleen Vohs for supplying the data.
26. Paul E. O'Brien et al., "Laparascopic adjustable gastric banding in severely obese adolescents," *Journal of the American Medical Association*, 303 (2010), pp. 519–526. I thank the authors for providing the data.
27. Brock Bastian et al., "Pain as social glue: Shared pain increases cooperation," *Psychological Science*, 25 (2014), pp. 2079–2085.
28. Data provided by Warren Page, New York City Technical College, from a study done by John Hudesman.
29. Ethan J. Temeles and W. John Kress, "Adaptation in a plant-hummingbird association," *Science*, 300 (2003), pp. 630–633. We thank Ethan J. Temeles for providing the data.
30. Data provided by Marigene Arnold, Kalamazoo College.

Chapter 22 Notes

1. Joseph H. Catania et al., "Prevalence of AIDS-related risk factors and condom use in the United States," *Science*, 258 (1992), pp. 1101–1106.
2. From the 2017 Youth Risk Behavior Surveillance System at `nccd.cdc.gov/Youthonline/App/Default.aspx`. The data are from a complex multistage sample, so that acting as if we have SRSs is oversimplified.
3. Strictly speaking, the formula $\sqrt{p(1-p)/n}$ for the standard deviation of $\hat{p}$ assumes that we draw an SRS of size n from an *infinite* population. If the population has finite size N, this standard deviation is multiplied by $\sqrt{1-(n-1)/(N-1)}$. This "finite population correction" approaches 1 as N increases. When the population is at least 20 times as large as the sample, the correction factor is between about 0.97 and 1. It is reasonable to use the simpler form $\sqrt{p(1-p)/n}$ in these settings. See also Note 3 for Chapter 15.
4. L. G. M. Bode, "Preventing surgical-site infections in nasal carriers of *Staphylococcus aureus*," *New England Journal of Medicine*, 362 (2010), pp. 9–17.
5. These are 2018 population counts, found online at `datausa.io/profile/geo/greenville-county-sc`.
6. This proportion has remained steady for nearly 20 years. See `news.gallup.com/poll/6424/nutrition-food.aspx`.
7. This rule of thumb is based on study of computational results in the papers cited in Note 15 and discussion with Alan Agresti.
8. The quotation is from page 1104 of the article cited in Note 1.
9. See Note 2.
10. From a 2019 poll by the Angus Reid Institute at `http://angusreid.org/gun-control-handgun-ban/`.
11. Richard B. Lipton et al., "Effect of ubrogepant vs placebo on pain and the most bothersome associated symptom in the acute treatment of migraine the ACHIEVE II randomized clinical trial," *JAMA* 322 (2019), pp. 1887–1898.
12. These data are from *The Infinite Dial 2019* at `www.tritondigital.com`.
13. In fact, P-values for two-sided tests are more accurate than those for one-sided tests. Our rule of thumb is a compromise to avoid the confusion of too many rules.
14. Sadahiko Nakajima, "Dogs and owners resemble each other in the eye region," *Anthrozoös*, 26(4) (2013), pp. 551–556.
15. See Note 11 for Chapter 14.
16. This interval is proposed by Alan Agresti and Brent A. Coull, "Approximate is better than 'exact' for interval estimation of binomial proportions," *The American Statistician*, 52 (1998), pp. 119–126. Note in particular that the plus four interval is often more accurate than the Clopper-Pearson "exact interval" based on the binomial distribution of the sample count and implemented by, for example, Minitab.

    There are several even more accurate but considerably more complex intervals for p that might be used in professional practice. See Lawrence D. Brown, Tony Cai, and Anirban DasGupta, "Interval estimation for a binomial proportion," *Statistical Science*, 16 (2001), pp. 101–133. A detailed theoretical study that uncovers the reason the large-sample interval is inaccurate is Lawrence D. Brown, Tony Cai, and Anirban DasGupta, "Confidence intervals for a binomial proportion and asymptotic expansions," *Annals of Statistics*, 30 (2002), pp. 160–201.
17. BBC News, December 25, 2006, at `news.bbc.co.uk`.
18. From Alan Agresti and Brian Caffo, "Simple and effective confidence intervals for proportions and differences of proportions result from adding two successes and two failures," *The American Statistician*, 45 (2000), pp. 280–288. When can the plus four interval be safely used? The answer depends on just how much accuracy you insist on. Brown and coauthors (see Note 15) recommend $n \geq 40$. Agresti and Coull (see Note 15)

demonstrate that performance is almost always satisfactory in their eyes when $n \geq 5$. Our rule of thumb $n \geq 10$ allows for confidence levels C other than 95% and fits our philosophy of not insisting on more exact results than practice requires. The big point is that plus four is very much more accurate than the standard interval for most values of p and all but very large n.

19. Gary Stoner et al., "Regression of rectal polyps in familial adenomatous polyposis patients with freeze dried black raspberries," *Cancer Prevention Research*, 1 (2008), p. PR-14.
20. Kim Parker et al., "America's complex relationship with guns an in-depth look at attitudes and experiences of U.S, adults," June 2017, at `www.pewinternet.org`. The sampling scheme was more complex than an SRS, so the computation of the number in the sample describing themselves as "gamers" and acting as if it were an SRS is oversimplified.
21. Michele L. Head, "Examining college students' ethical values," Consumer Science and Retailing honors project, Purdue University, 2003.
22. See Note 23 for Chapter 8.
23. Andrew Perrin and Monica Anderson, "Share of U.S. adults using social media, including Facebook, is mostly unchanged since 2018," April 2019, at `https://www.pewresearch.org/fact-tank/2019/04/10/share-of-u-s-adults-using-social-media-including-facebook-is-mostly-unchanged-since-2018/`.
24. Nicolas Guéguen et al., "A pedestrian's smile and drivers' behavior: When a smile increases careful driving," *Journal of Safety Research*, 56 (2016), pp. 83–88.
25. See Note 28 for Chapter 8.
26. W. Vinson Pierce and Robert E. Manning, "Day and overnight visitors to the Olympic wilderness," *Journal of Outdoor Recreation and Tourism*, 12 (2015), pp. 14–24.
27. *The Nation's Report Card: Trends in Academic Progress 2012 (NCES 2013 456)* can be found on the website `nces.ed.gov`.
28. See Note 1 for Chapter 13. The sampling scheme was more complex than an SRS, so the computation of the number in the sample who have read in the preceding 12 months and acting as if it were an SRS is oversimplified.
29. Sam Dillon, "Incentives for advanced work let pupils and teachers cash in," *New York Times*, October 2, 2011.
30. A. Mantonakis et al., "Order in Choice: Effects of Serial Position on Preferences," *Psychological Science*, 20 (2009), pp. 1309–1312.
31. Sam Oches and Rachel Pittman, "The 2019 QSR drive-thru study," October 2019, `www.qsrmagazine.com/reports/2019-qsr-drive-thru-study`.
32. See Note 28.
33. Francisco Lloret et al., "Fire and resprouting in Mediterranean ecosystems: Insights from an external biogeographical region, the Mexican shrubland," *American Journal of Botany*, 88 (1999), pp. 1655–1661.

Chapter 23 Notes

1. Shauna B. Wilson et al., "Dating across race: An examination of African American Internet personal advertisements," *Journal of Black Studies*, 37 (2007), pp. 964–982.
2. From the 2017 Youth Risk Behavior Surveillance System, at `nccd.cdc.gov/Youthonline/App/Default.aspx`. The data are from a complex multistage sample so that acting as if we have SRSs is oversimplified.
3. See Note 8 for Chapter 14.
4. Committee for Proprietary Medicinal Products, "Points to consider on switching between superiority and non-inferiority," *British Journal of Clinical Pharmacology*, 52 (2001), pp. 223–228.
5. Megan Brenan, "Americans' strong support for euthanasia persists," Gallup Poll, May 2018, online at `news.gallup.com/poll/235145/americans-strong-support-euthanasia-persists.aspx`.
6. Steven J. Frenda et al., "False memories of fabricated political events," *Journal of Experimental Social Psychology*, 49 (2013), pp. 280–286. We would like to thank the authors for providing the data and the photo in Figure 23.3.
7. This rule of thumb is quite conservative. It is in fact safe to arrange the data as a 2×2 table and apply the rule of thumb from Chapter 25 that all four *expected* counts must be five or greater. We give the conservative rule here because expected counts are messy to explain in the present context.
8. See Note 13 for Chapter 22.
9. Steiner Sulheim et al., "Helmet use and risk of head injuries in alpine skiers and snowboarders," *Journal of the American Medical Association*, 295 (2006), pp. 919–924.
10. Salim Yusuf et al., "Blood-pressure and cholesterol lowering in persons without cardiovascular disease," *New England Journal of Medicine*, 374 (2016), pp. 2032–2043.
11. The plus four method is due to Alan Agresti and Brian Caffo. See Note 15 for Chapter 22.
12. Frances A. Campbell et al., "Adult outcomes as a function of an early childhood educational program: An Abecedarian Project follow up," *Developmental Psychology*, 48 (2012), pp. 1033–1043.
13. George Du Toit et al., "Randomized trial of peanut consumption in infants at risk for peanut allergy," *New England Journal of Medicine*, 372 (2015), pp. 803–813.
14. See Note 33 for Chapter 22.
15. See Note 2 for Chapter 22.
16. See Note 5 for Chapter 21.
17. See Brooke Auxier et al., "Americans and privacy: Concerned, confused and feeling lack of control over personal information," Pew Research Center, November 2019, at `www.pewresearch.org/internet/2019/11/15/how-americans-think-about-privacy-and-the-vulnerability-of-their-personal-data/`.

18. W. P. T. James et al.,"Effect of Sibutramine on cardiovascular outcomes in overweight and obese subjects," *New England Journal of Medicine*, 363 (2010), pp. 905–917.
19. François Gaudet et al., "Induction of tumors in mice by genomic hypomethylation," *Science*, 300 (2003), pp. 489–492.
20. Xinyan Wang et al., "Tea consumption and the risk of atherosclerotic cardiovascular disease and all-cause mortality: The China-PAR project," *European Journal of Preventive Cardiology* (2020), available online at journals.sagepub.com/doi/10.1177/2047487319894685.
21. Ninh T. Nguyen et al., "Improved bariatric surgery outcomes for medicare beneficiaries after implementation of the medicare national coverage determination," *Archives of Surgery*, 145 (2010), pp. 72–78.
22. Arne L. Kalleberg and Kevin T. Leicht, "Gender and organizational performance: Determinants of small business survival and success," *The Academy of Management Journal*, 34 (1991), pp. 136–161.
23. See Note 17.
24. Jon O. Ebbert et al., "Effect of varenicline on smoking cessation through smoking reduction," *Journal of the American Medical Association*, 313, (2015), pp. 687–694.
25. Anika Lindstrom et al., "Does the presence of a mannequin head change shopping behavior?" *Journal of Business Research*, 69 (2016), pp. 517–524.
26. See Note 23 for Chapter 22.
27. Clive G. Jones et al., "Chain reactions linking acorns to gypsy moth outbreaks and Lyme disease risk," *Science*, 279 (1998), pp. 1023–1026.
28. See Note 12.
29. R. B. Turner et al., "A randomized trial of the efficacy of hand disinfection for prevention of rhinovirus infection," *Clinical Infectious Diseases*, 54 (2012), pp. 1422–1426.
30. See Note 9 for Chapter 14.

Chapter 24 Notes

1. Faria Sana, Tina Weston, and Nicholas J. Cepeda, "Laptop multitasking hinders classroom learning for both users and nearby peers," *Computers and Education*, 62 (2013), pp. 24–31.
2. Data provided by Drina Iglesia, Purdue University. The data are part of a larger study reported in D. D. S. Iglesia, E. J. Cragoe, Jr., and J. W. Vanable, "Electric field strength and epithelization in the newt (*Notophthalmus viridescens*)," *Journal of Experimental Zoology*, 274 (1996), pp. 56–62.
3. "Free expression on campus: What college students think about first amendment issues," December 2017. The full report is available online at https://knightfoundation.org/reports/free-expression-on-campus-what-college-students-think-about-first-amendment-issues/.
4. Wendy C. King et al., "Comparison of the performance of common measures of weight regain after bariatric surgery for association with clinical outcomes," *JAMA*, 320 (2018), pp. 1560–1569.
5. Francesca Gino and Scott S. Wilermuth, "Evil genius? How dishonesty can lead to greater creativity," *Psychological Science*, 25 (2014), pp. 973–981.
6. See Note 1.
7. Scott W. Powers et al., "Cognitive behavioral therapy plus amitriptyline for chronic migraine in children and adolescents," *JAMA*, 310 (2013), pp. 2622–2630.
8. Maureen Hack et al., "Outcomes in young adulthood for very-low-birth-weight infants," *New England Journal of Medicine*, 346 (2002), pp. 149–157. The exercises are simplified, in that the measures reported in this paper have been statistically adjusted for "sociodemographic status."
9. Data found on the New Scientist website, at www.newscientist.com/article/dn1748-euro-coin-accused.
10. Alexander Todorov et al., "Inferences of competence from faces predict election outcomes," *Science*, 308 (2005), pp. 1623–1626.
11. Andrew Perrin and Monica Anderson, "Share of U.S. adults using social media, including Facebook, is mostly unchanged since 2018," Pew Research Center, April 2019, at www.pewresearch.org/fact-tank/2019/04/10/share-of-u-s-adults-using-social-media-including-facebook-is-mostly-unchanged-since-2018/.
12. Based on Alan G. Sanfey et al., "The neural basis of economic decision-making in the ultimatum game," *Science*, 300 (2003), pp. 1755–1758. The paper reports a chi-square test (equivalent to a two-sided z test). This analysis is incorrect for the paper's data, as there were in fact only 19 participants, each appearing twice in each row of the table given in the exercise. Question 23.31 therefore amends the data, assuming 76 participants, so that the elementary analysis is correct.
13. Joseph W. Bartges et al., "Influence of acidifying or alkalizing diets on bone mineral density and urine relative supersaturation with calcium oxalate and struvite in healthy cats," *American Journal of Veterinary Research*, 74 (2013), pp. 1347–1352.
14. Jin Ha Lee and J. Stephen Downie, "Survey of music information needs, uses, and seeking behaviors: Preliminary findings," online Proceedings of the 5th International Conference on Music Information Retrieval, 2004, at ismir2004.ismir.net.
15. Josh McDermott and Marc D. Hauser, "Nonhuman primates prefer slow tempos but dislike music overall," *Cognition*, 104 (2007), pp. 654–668. Failure to take account of repeated measures on the same subjects is one of the most common errors observed in statistical analysis.
16. James Otto, Michael F. Brown, and William Long III, "Training rats to search and alert on contraband odors," *Applied Animal Behaviour Science*, 77 (2002), pp. 217–232.

17. From the Merck website, www.merckvaccines.com/gardasilProductPage_frmst.html.
18. These data were originally collected by L. M. Linde of UCLA but were first published by M. R. Mickey, O. J. Dunn, and V. Clark, "Note on the use of stepwise regression in detecting outliers," *Computers and Biomedical Research*, 1 (1967), pp. 105–111. The data have been used by several authors. We found them in N. R. Draper and J. A. John, "Influential observations and outliers in regression," *Technometrics*, 23 (1981), pp. 21–26.
19. Jacqueline T. Ngai and Diane S. Srivastava, "Predators accelerate nutrient cycling in a bromeliad ecosystem," *Science*, 314 (2006), p. 963. I thank Jacqueline Ngai for providing the data.
20. Yvan R. Germain, "The dyeing of ramie with fiber reactive dyes using the cold pad-batch method," MS thesis, Purdue University, 1988.
21. Data provided by Marigene Arnold, Kalamazoo College.
22. Data provided by Corinne Lim, Purdue University, from a student project supervised by Professor Joseph Vanable.
23. Saiyad S. Ahmed, "Effect of microwave drying on checking and mechanical strength of low-moisture baked products," MS thesis, Purdue University, 1994.
24. Michael O. Finkelstein and Bruce Levin, "Statistical proof of discrimination in peremptory challenges," *Chance*, 17 (2004), pp. 35–38.
25. G. S. Hotamisligil et al., "Uncoupling of obesity from insulin resistance through a targeted mutation in *aP2*, the adipocyte fatty acid binding protein," *Science*, 274 (1996), pp. 1377–1379.
26. Data simulated from a Normal distribution with the mean and standard deviation reported by Sarah Morrison and Jan Noyes, "A comparison of two computer fonts: Serif versus ornate sans serif," *Usability News*, 5.2 (2003), at psychology.wichita.edu/surl/usability news.html.

Chapter 25 Notes

1. Jeffrey M. Jones, "Majority rejects idea of haves, have-nots divide in U.S.," 2019, at news.gallup.com/poll/248216/majority-rejects-idea-haves-nots-divide.aspx. Details about the poll are at news.gallup.com/file/poll/248237/190403HaveHave-Nots.pdf.
2. Aaron Smith and Monica Anderson, "Social media use in 2018," at www.pewresearch.org/internet/2018/03/01/social-media-use-in-2018/.
3. See Note 20 for Chapter 9.
4. All General Social Survey exercises in this chapter present tables constructed using the search function at the GSS archive, sda.berkeley.edu/archive.htm or at gssdataexplorer.norc.org/trends. Most concern data from the 2014 GSS.
5. There are many computer studies of the accuracy of chi-square critical values for χ^2. Our guideline goes back to W. G. Cochran (1954). Later work has shown that it is often conservative in the sense that, if the expected cell counts are all similar and the degrees of freedom exceed 1, the chi-square approximation works well for an average expected count as small as 1 or 2. Our guideline protects against dissimilar expected counts. It has the added advantage that it is safe in the 2×2 case, where the chi-square approximation is least good. So our guideline is helpful for beginners: there is no single condition that is not conservative and that applies to 2×2 and larger tables with similar and dissimilar expected cell counts. There are exact procedures that (with software) should be used for tables that do not satisfy our guideline. For a survey, see Alan Agresti, "A survey of exact inference for contingency tables," *Statistical Science*, 7 (1992), pp. 131–177.
6. Scott Keeter et al., "What's missing from national RDD surveys? The impact of the growing cell-only population," 2007, at www.pewresearch.org.
7. Leah Christian, et al., "Assessing the cell phone challenge to survey research in 2010," at www.pewresearch.org.
8. See Note 11 for Chapter 14.
9. Based on a news item in *Science*, 305 (2004), p. 1560. The study, by Daniel Klem, appeared in the *Wilson Journal*.
10. See Note 23 for Chapter 8.
11. David W. Eby et al., "The effect of changing from secondary to primary safety belt enforcement on police harassment," *Accident Analysis and Prevention*, 36 (2000), pp. 819–828.
12. Justin McCarthy, "Smaller majority of heavy smokers say they want to quit," 2017, at news.gallup.com/poll/214925/smaller-majority-heavy-smokers-say-quit.aspx.
13. See Note 12 for Chapter 6.
14. See Note 7.
15. See Note 17 for Chapter 6.
16. Saren T. Skou et al., "Randomized, controlled trial of total knee replacement," *The New England Journal of Medicine*, 373 (2015), pp. 1597–1606.
17. See Note 2 for Chapter 6.
18. Adapted from M. A. Visintainer, J. R. Volpicelli, and M. E. P. Seligman, "Tumor rejection in rats after inescapable or escapable shock," *Science*, 216 (1982), pp. 437–439.
19. See Note 1.
20. Tom Reichert, "The prevalence of sexual imagery in ads targeted to young adults," *Journal of Consumer Affairs*, 37 (2003), pp. 403–412.
21. József Topál et al., "Differential sensitivity to human communication in dogs, wolves and human infants," *Science*, 325 (2009), pp. 1269–1272. Many statistical software packages offer "exact tests" that are valid even when there are small expected counts.

22. U.S. Department of Commerce, Office of Travel and Tourism Industries, in-flight survey, 2007, at `tinet.ita.doc.gov`.
23. Kaigang Li et al., "Marijuana-, alchohol-, and drug-impaired driving among emerging young adults: Changes from high school to one-year post-high school," *Journal of Safety Research*, 58 (2016), pp. 15–20.
24. D. A. Marcus et al., "A double-blind provocative study of chocolate as a trigger of headache," *Cephalalgia*, 17 (1997), pp. 855–862. We have simplified slightly: the table in the paper is exactly as in the exercise but contains data for 63 subjects plus data from one type of bar for 3 subjects who dropped out. Although the authors say that their chi-square refers to this table, they give a nonsignificant value that contradicts what the table shows.
25. See Note 7 for Chapter 6.
26. Data compiled from a table of percentages in "Americans view higher education as key to the American dream," press release by the National Center for Public Policy and Higher Education, May 3, 2000, at `www.highereducation.org`.
27. Eswari Vilayur et al., "The association between reduced GFR and hearing loss: A cross-sectional population-based study," *American Journal of Kidney Diseases*, 56 (2010), pp. 661–669.
28. See Note 10 for Chapter 22.

Chapter 26 Notes

1. Juan P. Carricart-Ganivet, "Sea surface temperature and the growth of the West Atlantic reef-building coral Montastraea annularis," *Journal of Experimental Marine Biology and Ecology*, 302 (2004), pp. 249–260.
2. From a graph in Naomi E. Allen et al., "Moderate alcohol intake and cancer incidence in women," *Journal of the National Cancer Institute*, 101 (2009), pp. 296–305. These data represent averages over large numbers of women and are an example of an ecological correlation (see page 143 in Chapter 5); one must be careful not to interpret the data as applying to individuals.
3. From a graph in Stephen M. Fleming et al., "Relating introspective accuracy to individual differences in brain structure," *Science*, 329 (2010), pp. 1541–1543.
4. See Note 31 for Chapter 5.
5. Electronic Encyclopedia of Statistical Examples and Exercises (EESEE) at the text website, `www.saplinglearning.com/ibiscms/login/`. Students will need a pass-word to access the website.
6. From a graph in Allison L. Perry et al., "Climate change and distribution shifts in marine fishes," *Science*, 308 (2005), pp. 1912–1915. The explanatory variable is the five-year running mean of winter (December to March) sea-bottom temperature.
7. Data for the building at 1800 Ben Franklin Drive, Sarasota, Florida, starting in February 2003. From the website of the Sarasota County Property Appraiser, `www.sc-pa.com`.
8. Yanhui Lu et al., "Mirid bug outbreaks in multiple crops correlated with wide-scale adoption of Bt cotton in China," *Science*, 328 (2010), pp. 1151–1154.
9. See Note 1.
10. Based on Marion E. Dunshee, "A study of factors affecting the amount and kind of food eaten by nursery school children," *Child Development*, 2 (1931), pp. 163–183. This article gives the means, standard deviations, and correlation for 37 children, from which the data in the exercise are simulated.
11. From Table S2 in the online supplement to Antonio Dell'Anno and Roberto Danovaro, "Extracellular DNA plays a key role in deep-sea ecosystem functioning," *Science*, 309 (2005), p. 2179.
12. See Note 19 for Chapter 7.

Chapter 27 Notes

1. Ethan J. Temeles and W. John Kress, "Adaptation in a plant–hummingbird association," *Science*, 300 (2003), pp. 630–633. We thank Ethan J. Temeles for providing the data.
2. Helen Lee Bouygues, "Addendum: The 2019 NAEP data on technology and achievement outcomes," November 22, 2019, The Reboot Foundation. Online at `reboot-foundation.org/does-educational-technology-help-students-learn/`.
3. The data from the General Social Survey for this exercise were constructed using the search function and download capabilities at the GSS archive, `sda.berkeley.edu/archive.htm`.
4. See Note 13 for Chapter 2.
5. Holly A. Miller, "Reexamining psychological distress in the current conditions of segregation," *Journal of Correctional Health Care*, 1 (1994), pp. 39–53.
6. See Note 25 for Chapter 2.
7. Pam A. Mueller and Daniel M. Oppenheimer, "The pen is mightier than the keyboard: Advantages of longhand over laptop note taking," *Psychological Science*, 25, no. 6 (2014), pp. 1159–1168.
8. David B. Wooten, "One-of-a-kind in a full house: Some consequences of ethnic and gender distinctiveness," *Journal of Consumer Psychology*, 4 (1995), 205–224.
9. Modified from M. C. Wilson and R. E. Shade, "Relative attractiveness of various luminescent colors to the cereal leaf beetle and the meadow spittlebug," *Journal of Economic Entomology*, 60 (1967), pp. 578–580.
10. John P. Thomas, "Influences on mathematics learning and attitudes among African American high school students," *Journal of Negro Education*, 69 (2000), pp. 165–183.
11. Lauren A. Burt et al., "Effect of high-does vitamin D supplementation on volumetric bone density and bone strength a randomized clinical trial," *Journal of the American Medical Association*, 322 (2019), pp. 736–745.

12. Data from the EESEE story "Stress among pets and friends."
13. Henrik Hagvedt and Vanessa M. Patrick, "Art infusion: The influence of visual art on the perception and evaluation of consumer products," *Journal of Marketing Research*, 45 (2008), pp. 379–389.
14. Jason Ritchie et al., "The effects of lemon taste on attention, perceived exertion, and affect during a stepping task," *Psychology of Sport and Exercise*, 25 (2016), pp. 9–16.
15. Timothy Church et al., "Effects of aerobic and resistance training on hemoglobin A1c levels in patients with type 2 diabetes: A randomized controlled trial," *Journal of the American Medical Association*, 304 (2010), pp. 2253–2262.
16. Jennifer J. Argo et al., "Positive consumer contagion: Responses to attractive others in a retail context," *Journal of Marketing Research*, 45 (2008), pp. 690–701.
17. See Note 2.
18. The data and the full story can be found in the Data and Story Library at `lib.stat.cmu.edu`. The original study is by Faith Loven, "A study of interlist equivalency of the CID W-22 word list presented in quiet and in noise," MS thesis, University of Iowa, 1981.
19. Murali K. Matta, et al., "Effect of sunscreen application under maximal use conditions on plasma concentration of sunscreen active ingredients a randomized clinical trial," *Journal of the American Medical Association*, 321 (2019), pp. 2082–2091. The data in the problem were estimated from a graph in the paper and were rounded to the first decimal. The natural logarithms of these values were then calculated. The graphs in the paper plotted the data on a logarithmic scale, suggesting the use of the logarithms of the data in the analysis.
20. See Note 17 for Chapter 6.
21. See Note 26 for Chapter 2.
22. See Note 24 for Chapter 2.
23. See Note 21 for Chapter 21.
24. See Note 25 for Chapter 21.

TABLES

Table entry for *z* is the area under the standard Normal curve to the left of *z*.

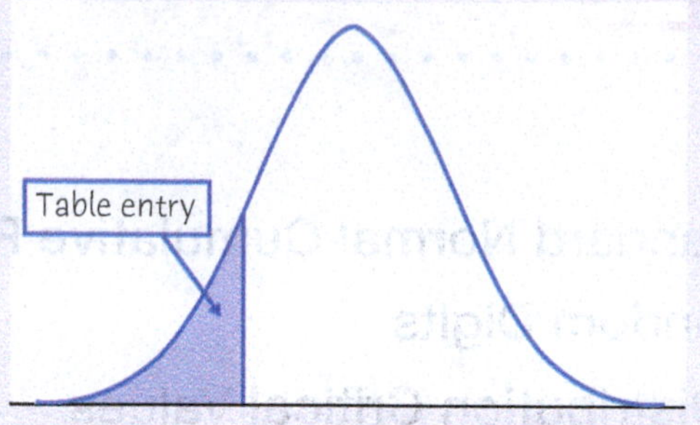

TABLE A Standard Normal cumulative proportions

z	.00	.01	.02	.03	.04	.05	.06	.07	.08	.09
−3.4	.0003	.0003	.0003	.0003	.0003	.0003	.0003	.0003	.0003	.0002
−3.3	.0005	.0005	.0005	.0004	.0004	.0004	.0004	.0004	.0004	.0003
−3.2	.0007	.0007	.0006	.0006	.0006	.0006	.0006	.0005	.0005	.0005
−3.1	.0010	.0009	.0009	.0009	.0008	.0008	.0008	.0008	.0007	.0007
−3.0	.0013	.0013	.0013	.0012	.0012	.0011	.0011	.0011	.0010	.0010
−2.9	.0019	.0018	.0018	.0017	.0016	.0016	.0015	.0015	.0014	.0014
−2.8	.0026	.0025	.0024	.0023	.0023	.0022	.0021	.0021	.0020	.0019
−2.7	.0035	.0034	.0033	.0032	.0031	.0030	.0029	.0028	.0027	.0026
−2.6	.0047	.0045	.0044	.0043	.0041	.0040	.0039	.0038	.0037	.0036
−2.5	.0062	.0060	.0059	.0057	.0055	.0054	.0052	.0051	.0049	.0048
−2.4	.0082	.0080	.0078	.0075	.0073	.0071	.0069	.0068	.0066	.0064
−2.3	.0107	.0104	.0102	.0099	.0096	.0094	.0091	.0089	.0087	.0084
−2.2	.0139	.0136	.0132	.0129	.0125	.0122	.0119	.0116	.0113	.0110
−2.1	.0179	.0174	.0170	.0166	.0162	.0158	.0154	.0150	.0146	.0143
−2.0	.0228	.0222	.0217	.0212	.0207	.0202	.0197	.0192	.0188	.0183
−1.9	.0287	.0281	.0274	.0268	.0262	.0256	.0250	.0244	.0239	.0233
−1.8	.0359	.0351	.0344	.0336	.0329	.0322	.0314	.0307	.0301	.0294
−1.7	.0446	.0436	.0427	.0418	.0409	.0401	.0392	.0384	.0375	.0367
−1.6	.0548	.0537	.0526	.0516	.0505	.0495	.0485	.0475	.0465	.0455
−1.5	.0668	.0655	.0643	.0630	.0618	.0606	.0594	.0582	.0571	.0559
−1.4	.0808	.0793	.0778	.0764	.0749	.0735	.0721	.0708	.0694	.0681
−1.3	.0968	.0951	.0934	.0918	.0901	.0885	.0869	.0853	.0838	.0823
−1.2	.1151	.1131	.1112	.1093	.1075	.1056	.1038	.1020	.1003	.0985
−1.1	.1357	.1335	.1314	.1292	.1271	.1251	.1230	.1210	.1190	.1170
−1.0	.1587	.1562	.1539	.1515	.1492	.1469	.1446	.1423	.1401	.1379
−0.9	.1841	.1814	.1788	.1762	.1736	.1711	.1685	.1660	.1635	.1611
−0.8	.2119	.2090	.2061	.2033	.2005	.1977	.1949	.1922	.1894	.1867
−0.7	.2420	.2389	.2358	.2327	.2296	.2266	.2236	.2206	.2177	.2148
−0.6	.2743	.2709	.2676	.2643	.2611	.2578	.2546	.2514	.2483	.2451
−0.5	.3085	.3050	.3015	.2981	.2946	.2912	.2877	.2843	.2810	.2776
−0.4	.3446	.3409	.3372	.3336	.3300	.3264	.3228	.3192	.3156	.3121
−0.3	.3821	.3783	.3745	.3707	.3669	.3632	.3594	.3557	.3520	.3483
−0.2	.4207	.4168	.4129	.4090	.4052	.4013	.3974	.3936	.3897	.3859
−0.1	.4602	.4562	.4522	.4483	.4443	.4404	.4364	.4325	.4286	.4247
−0.0	.5000	.4960	.4920	.4880	.4840	.4801	.4761	.4721	.4681	.4641

Table entry for z is the area under the standard Normal curve to the left of z.

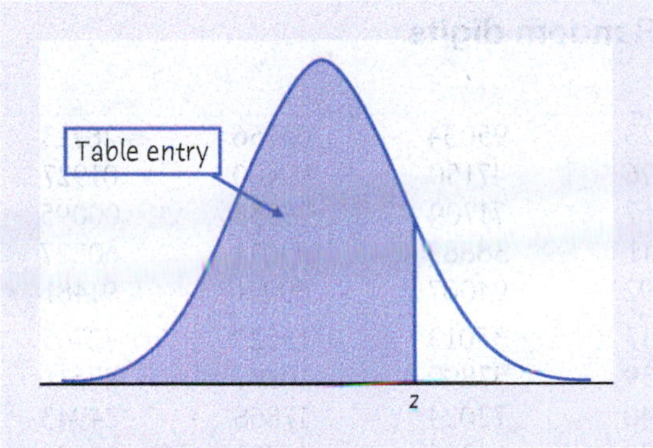

TABLE A Standard Normal cumulative proportions (*continued*)

z	.00	.01	.02	.03	.04	.05	.06	.07	.08	.09
0.0	.5000	.5040	.5080	.5120	.5160	.5199	.5239	.5279	.5319	.5359
0.1	.5398	.5438	.5478	.5517	.5557	.5596	.5636	.5675	.5714	.5753
0.2	.5793	.5832	.5871	.5910	.5948	.5987	.6026	.6064	.6103	.6141
0.3	.6179	.6217	.6255	.6293	.6331	.6368	.6406	.6443	.6480	.6517
0.4	.6554	.6591	.6628	.6664	.6700	.6736	.6772	.6808	.6844	.6879
0.5	.6915	.6950	.6985	.7019	.7054	.7088	.7123	.7157	.7190	.7224
0.6	.7257	.7291	.7324	.7357	.7389	.7422	.7454	.7486	.7517	.7549
0.7	.7580	.7611	.7642	.7673	.7704	.7734	.7764	.7794	.7823	.7852
0.8	.7881	.7910	.7939	.7967	.7995	.8023	.8051	.8078	.8106	.8133
0.9	.8159	.8186	.8212	.8238	.8264	.8289	.8315	.8340	.8365	.8389
1.0	.8413	.8438	.8461	.8485	.8508	.8531	.8554	.8577	.8599	.8621
1.1	.8643	.8665	.8686	.8708	.8729	.8749	.8770	.8790	.8810	.8830
1.2	.8849	.8869	.8888	.8907	.8925	.8944	.8962	.8980	.8997	.9015
1.3	.9032	.9049	.9066	.9082	.9099	.9115	.9131	.9147	.9162	.9177
1.4	.9192	.9207	.9222	.9236	.9251	.9265	.9279	.9292	.9306	.9319
1.5	.9332	.9345	.9357	.9370	.9382	.9394	.9406	.9418	.9429	.9441
1.6	.9452	.9463	.9474	.9484	.9495	.9505	.9515	.9525	.9535	.9545
1.7	.9554	.9564	.9573	.9582	.9591	.9599	.9608	.9616	.9625	.9633
1.8	.9641	.9649	.9656	.9664	.9671	.9678	.9686	.9693	.9699	.9706
1.9	.9713	.9719	.9726	.9732	.9738	.9744	.9750	.9756	.9761	.9767
2.0	.9772	.9778	.9783	.9788	.9793	.9798	.9803	.9808	.9812	.9817
2.1	.9821	.9826	.9830	.9834	.9838	.9842	.9846	.9850	.9854	.9857
2.2	.9861	.9864	.9868	.9871	.9875	.9878	.9881	.9884	.9887	.9890
2.3	.9893	.9896	.9898	.9901	.9904	.9906	.9909	.9911	.9913	.9916
2.4	.9918	.9920	.9922	.9925	.9927	.9929	.9931	.9932	.9934	.9936
2.5	.9938	.9940	.9941	.9943	.9945	.9946	.9948	.9949	.9951	.9952
2.6	.9953	.9955	.9956	.9957	.9959	.9960	.9961	.9962	.9963	.9964
2.7	.9965	.9966	.9967	.9968	.9969	.9970	.9971	.9972	.9973	.9974
2.8	.9974	.9975	.9976	.9977	.9977	.9978	.9979	.9979	.9980	.9981
2.9	.9981	.9982	.9982	.9983	.9984	.9984	.9985	.9985	.9986	.9986
3.0	.9987	.9987	.9987	.9988	.9988	.9989	.9989	.9989	.9990	.9990
3.1	.9990	.9991	.9991	.9991	.9992	.9992	.9992	.9992	.9993	.9993
3.2	.9993	.9993	.9994	.9994	.9994	.9994	.9994	.9995	.9995	.9995
3.3	.9995	.9995	.9995	.9996	.9996	.9996	.9996	.9996	.9996	.9997
3.4	.9997	.9997	.9997	.9997	.9997	.9997	.9997	.9997	.9997	.9998

TABLE B Random digits

Line								
101	19223	95034	05756	28713	96409	12531	42544	82853
102	73676	47150	99400	01927	27754	42648	82425	36290
103	45467	71709	77558	00095	32863	29485	82226	90056
104	52711	38889	93074	60227	40011	85848	48767	52573
105	95592	94007	69971	91481	60779	53791	17297	59335
106	68417	35013	15529	72765	85089	57067	50211	47487
107	82739	57890	20807	47511	81676	55300	94383	14893
108	60940	72024	17868	24943	61790	90656	87964	18883
109	36009	19365	15412	39638	85453	46816	83485	41979
110	38448	48789	18338	24697	39364	42006	76688	08708
111	81486	69487	60513	09297	00412	71238	27649	39950
112	59636	88804	04634	71197	19352	73089	84898	45785
113	62568	70206	40325	03699	71080	22553	11486	11776
114	45149	32992	75730	66280	03819	56202	02938	70915
115	61041	77684	94322	24709	73698	14526	31893	32592
116	14459	26056	31424	80371	65103	62253	50490	61181
117	38167	98532	62183	70632	23417	26185	41448	75532
118	73190	32533	04470	29669	84407	90785	65956	86382
119	95857	07118	87664	92099	58806	66979	98624	84826
120	35476	55972	39421	65850	04266	35435	43742	11937
121	71487	09984	29077	14863	61683	47052	62224	51025
122	13873	81598	95052	90908	73592	75186	87136	95761
123	54580	81507	27102	56027	55892	33063	41842	81868
124	71035	09001	43367	49497	72719	96758	27611	91596
125	96746	12149	37823	71868	18442	35119	62103	39244
126	96927	19931	36809	74192	77567	88741	48409	41903
127	43909	99477	25330	64359	40085	16925	85117	36071
128	15689	14227	06565	14374	13352	49367	81982	87209
129	36759	58984	68288	22913	18638	54303	00795	08727
130	69051	64817	87174	09517	84534	06489	87201	97245
131	05007	16632	81194	14873	04197	85576	45195	96565
132	68732	55259	84292	08796	43165	93739	31685	97150
133	45740	41807	65561	33302	07051	93623	18132	09547
134	27816	78416	18329	21337	35213	37741	04312	68508
135	66925	55658	39100	78458	11206	19876	87151	31260
136	08421	44753	77377	28744	75592	08563	79140	92454
137	53645	66812	61421	47836	12609	15373	98481	14592
138	66831	68908	40772	21558	47781	33586	79177	06928
139	55588	99404	70708	41098	43563	56934	48394	51719
140	12975	13258	13048	45144	72321	81940	00360	02428
141	96767	35964	23822	96012	94591	65194	50842	53372
142	72829	50232	97892	63408	77919	44575	24870	04178
143	88565	42628	17797	49376	61762	16953	88604	12724
144	62964	88145	83083	69453	46109	59505	69680	00900
145	19687	12633	57857	95806	09931	02150	43163	58636
146	37609	59057	66967	83401	60705	02384	90597	93600
147	54973	86278	88737	74351	47500	84552	19909	67181
148	00694	05977	19664	65441	20903	62371	22725	53340
149	71546	05233	53946	68743	72460	27601	45403	88692
150	07511	88915	41267	16853	84569	79367	32337	03316

Table entry for *C* is the critical value t^* required for confidence level *C*. To approximate one- and two-sided *P*-values, compare the value of the *t* statistic with the critical values of t^* that match the *P*-values given at the bottom of the table.

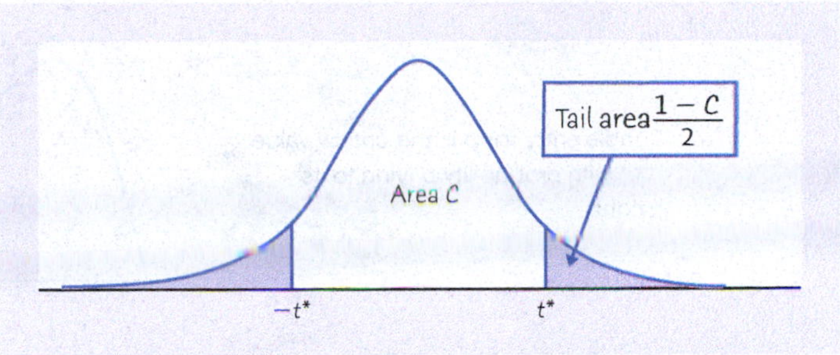

TABLE C *t* distribution critical values

Degrees of freedom	Confidence Level C											
	50%	60%	70%	80%	90%	95%	96%	98%	99%	99.5%	99.8%	99.9%
1	1.000	1.376	1.963	3.078	6.314	12.71	15.89	31.82	63.66	127.3	318.3	636.6
2	0.816	1.061	1.386	1.886	2.920	4.303	4.849	6.965	9.925	14.09	22.33	31.60
3	0.765	0.978	1.250	1.638	2.353	3.182	3.482	4.541	5.841	7.453	10.21	12.92
4	0.741	0.941	1.190	1.533	2.132	2.776	2.999	3.747	4.604	5.598	7.173	8.610
5	0.727	0.920	1.156	1.476	2.015	2.571	2.757	3.365	4.032	4.773	5.893	6.869
6	0.718	0.906	1.134	1.440	1.943	2.447	2.612	3.143	3.707	4.317	5.208	5.959
7	0.711	0.896	1.119	1.415	1.895	2.365	2.517	2.998	3.499	4.029	4.785	5.408
8	0.706	0.889	1.108	1.397	1.860	2.306	2.449	2.896	3.355	3.833	4.501	5.041
9	0.703	0.883	1.100	1.383	1.833	2.262	2.398	2.821	3.250	3.690	4.297	4.781
10	0.700	0.879	1.093	1.372	1.812	2.228	2.359	2.764	3.169	3.581	4.144	4.587
11	0.697	0.876	1.088	1.363	1.796	2.201	2.328	2.718	3.106	3.497	4.025	4.437
12	0.695	0.873	1.083	1.356	1.782	2.179	2.303	2.681	3.055	3.428	3.930	4.318
13	0.694	0.870	1.079	1.350	1.771	2.160	2.282	2.650	3.012	3.372	3.852	4.221
14	0.692	0.868	1.076	1.345	1.761	2.145	2.264	2.624	2.977	3.326	3.787	4.140
15	0.691	0.866	1.074	1.341	1.753	2.131	2.249	2.602	2.947	3.286	3.733	4.073
16	0.690	0.865	1.071	1.337	1.746	2.120	2.235	2.583	2.921	3.252	3.686	4.015
17	0.689	0.863	1.069	1.333	1.740	2.110	2.224	2.567	2.898	3.222	3.646	3.965
18	0.688	0.862	1.067	1.330	1.734	2.101	2.214	2.552	2.878	3.197	3.611	3.922
19	0.688	0.861	1.066	1.328	1.729	2.093	2.205	2.539	2.861	3.174	3.579	3.883
20	0.687	0.860	1.064	1.325	1.725	2.086	2.197	2.528	2.845	3.153	3.552	3.850
21	0.686	0.859	1.063	1.323	1.721	2.080	2.189	2.518	2.831	3.135	3.527	3.819
22	0.686	0.858	1.061	1.321	1.717	2.074	2.183	2.508	2.819	3.119	3.505	3.792
23	0.685	0.858	1.060	1.319	1.714	2.069	2.177	2.500	2.807	3.104	3.485	3.768
24	0.685	0.857	1.059	1.318	1.711	2.064	2.172	2.492	2.797	3.091	3.467	3.745
25	0.684	0.856	1.058	1.316	1.708	2.060	2.167	2.485	2.787	3.078	3.450	3.725
26	0.684	0.856	1.058	1.315	1.706	2.056	2.162	2.479	2.779	3.067	3.435	3.707
27	0.684	0.855	1.057	1.314	1.703	2.052	2.158	2.473	2.771	3.057	3.421	3.690
28	0.683	0.855	1.056	1.313	1.701	2.048	2.154	2.467	2.763	3.047	3.408	3.674
29	0.683	0.854	1.055	1.311	1.699	2.045	2.150	2.462	2.756	3.038	3.396	3.659
30	0.683	0.854	1.055	1.310	1.697	2.042	2.147	2.457	2.750	3.030	3.385	3.646
40	0.681	0.851	1.050	1.303	1.684	2.021	2.123	2.423	2.704	2.971	3.307	3.551
50	0.679	0.849	1.047	1.299	1.676	2.009	2.109	2.403	2.678	2.937	3.261	3.496
60	0.679	0.848	1.045	1.296	1.671	2.000	2.099	2.390	2.660	2.915	3.232	3.460
80	0.678	0.846	1.043	1.292	1.664	1.990	2.088	2.374	2.639	2.887	3.195	3.416
100	0.677	0.845	1.042	1.290	1.660	1.984	2.081	2.364	2.626	2.871	3.174	3.390
1000	0.675	0.842	1.037	1.282	1.646	1.962	2.056	2.330	2.581	2.813	3.098	3.300
z^*	0.674	0.841	1.036	1.282	1.645	1.960	2.054	2.326	2.576	2.807	3.091	3.291
One-sided *P*	.25	.20	.15	.10	.05	.025	.02	.01	.005	.0025	.001	.0005
Two-sided *P*	.50	.40	.30	.20	.10	.05	.04	.02	.01	.005	.002	.001

Table entry for p is the critical value χ^* with probability p lying to its right.

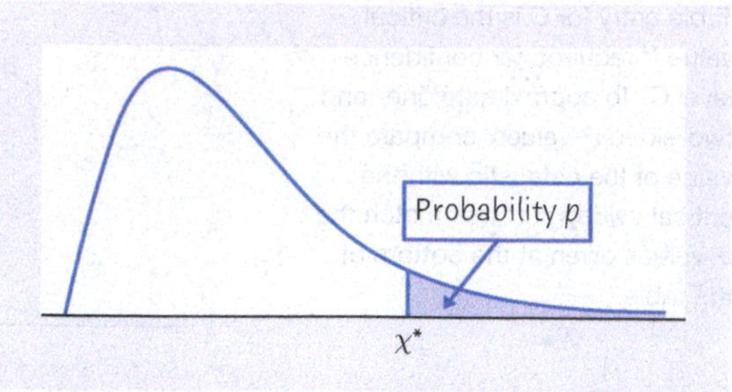

TABLE D Chi-square distribution critical values

	p											
df	**.25**	**.20**	**.15**	**.10**	**.05**	**.025**	**.02**	**.01**	**.005**	**.0025**	**.001**	**.0005**
1	1.32	1.64	2.07	2.71	3.84	5.02	5.41	6.63	7.88	9.14	10.83	12.12
2	2.77	3.22	3.79	4.61	5.99	7.38	7.82	9.21	10.60	11.98	13.82	15.20
3	4.11	4.64	5.32	6.25	7.81	9.35	9.84	11.34	12.84	14.32	16.27	17.73
4	5.39	5.99	6.74	7.78	9.49	11.14	11.67	13.28	14.86	16.42	18.47	20.00
5	6.63	7.29	8.12	9.24	11.07	12.83	13.39	15.09	16.75	18.39	20.51	22.11
6	7.84	8.56	9.45	10.64	12.59	14.45	15.03	16.81	18.55	20.25	22.46	24.10
7	9.04	9.80	10.75	12.02	14.07	16.01	16.62	18.48	20.28	22.04	24.32	26.02
8	10.22	11.03	12.03	13.36	15.51	17.53	18.17	20.09	21.95	23.77	26.12	27.87
9	11.39	12.24	13.29	14.68	16.92	19.02	19.68	21.67	23.59	25.46	27.88	29.67
10	12.55	13.44	14.53	15.99	18.31	20.48	21.16	23.21	25.19	27.11	29.59	31.42
11	13.70	14.63	15.77	17.28	19.68	21.92	22.62	24.72	26.76	28.73	31.26	33.14
12	14.85	15.81	16.99	18.55	21.03	23.34	24.05	26.22	28.30	30.32	32.91	34.82
13	15.98	16.98	18.20	19.81	22.36	24.74	25.47	27.69	29.82	31.88	34.53	36.48
14	17.12	18.15	19.41	21.06	23.68	26.12	26.87	29.14	31.32	33.43	36.12	38.11
15	18.25	19.31	20.60	22.31	25.00	27.49	28.26	30.58	32.80	34.95	37.70	39.72
16	19.37	20.47	21.79	23.54	26.30	28.85	29.63	32.00	34.27	36.46	39.25	41.31
17	20.49	21.61	22.98	24.77	27.59	30.19	31.00	33.41	35.72	37.95	40.79	42.88
18	21.60	22.76	24.16	25.99	28.87	31.53	32.35	34.81	37.16	39.42	42.31	44.43
19	22.72	23.90	25.33	27.20	30.14	32.85	33.69	36.19	38.58	40.88	43.82	45.97
20	23.83	25.04	26.50	28.41	31.41	34.17	35.02	37.57	40.00	42.34	45.31	47.50
21	24.93	26.17	27.66	29.62	32.67	35.48	36.34	38.93	41.40	43.78	46.80	49.01
22	26.04	27.30	28.82	30.81	33.92	36.78	37.66	40.29	42.80	45.20	48.27	50.51
23	27.14	28.43	29.98	32.01	35.17	38.08	38.97	41.64	44.18	46.62	49.73	52.00
24	28.24	29.55	31.13	33.20	36.42	39.36	40.27	42.98	45.56	48.03	51.18	53.48
25	29.34	30.68	32.28	34.38	37.65	40.65	41.57	44.31	46.93	49.44	52.62	54.95
26	30.43	31.79	33.43	35.56	38.89	41.92	42.86	45.64	48.29	50.83	54.05	56.41
27	31.53	32.91	34.57	36.74	40.11	43.19	44.14	46.96	49.64	52.22	55.48	57.86
28	32.62	34.03	35.71	37.92	41.34	44.46	45.42	48.28	50.99	53.59	56.89	59.30
29	33.71	35.14	36.85	39.09	42.56	45.72	46.69	49.59	52.34	54.97	58.30	60.73
30	34.80	36.25	37.99	40.26	43.77	46.98	47.96	50.89	53.67	56.33	59.70	62.16
40	45.62	47.27	49.24	51.81	55.76	59.34	60.44	63.69	66.77	69.70	73.40	76.09
50	56.33	58.16	60.35	63.17	67.50	71.42	72.61	76.15	79.49	82.66	86.66	89.56
60	66.98	68.97	71.34	74.40	79.08	83.30	84.58	88.38	91.95	95.34	99.61	102.7
80	88.13	90.41	93.11	96.58	101.9	106.6	108.1	112.3	116.3	120.1	124.8	128.3
100	109.1	111.7	114.7	118.5	124.3	129.6	131.1	135.8	140.2	144.3	149.4	153.2

Table entry for p is the critical value r^* of the correlation coefficient r with probability p lying to its right.

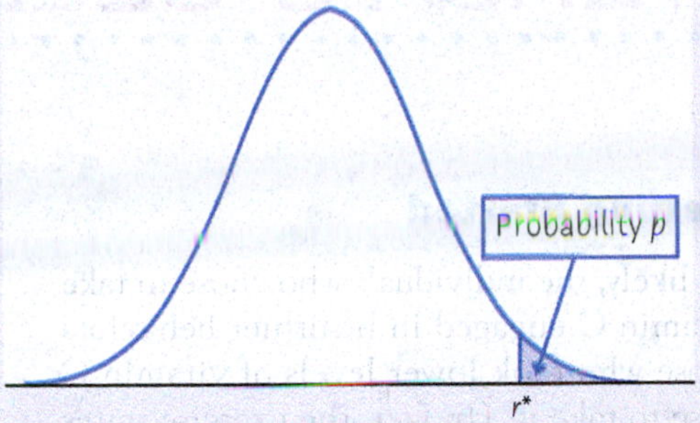

TABLE E Critical values of the correlation *r*

	Upper Tail Probability p									
n	.20	.10	.05	.025	.02	.01	.005	.0025	.001	.0005
3	0.8090	0.9511	0.9877	0.9969	0.9980	0.9995	0.9999	1.0000	1.0000	1.0000
4	0.6000	0.8000	0.9000	0.9500	0.9600	0.9800	0.9900	0.9950	0.9980	0.9990
5	0.4919	0.6870	0.8054	0.8783	0.8953	0.9343	0.9587	0.9740	0.9859	0.9911
6	0.4257	0.6084	0.7293	0.8114	0.8319	0.8822	0.9172	0.9417	0.9633	0.9741
7	0.3803	0.5509	0.6694	0.7545	0.7766	0.8329	0.8745	0.9056	0.9350	0.9509
8	0.3468	0.5067	0.6215	0.7067	0.7295	0.7887	0.8343	0.8697	0.9049	0.9249
9	0.3208	0.4716	0.5822	0.6664	0.6892	0.7498	0.7977	0.8359	0.8751	0.8983
10	0.2998	0.4428	0.5494	0.6319	0.6546	0.7155	0.7646	0.8046	0.8467	0.8721
11	0.2825	0.4187	0.5214	0.6021	0.6244	0.6851	0.7348	0.7759	0.8199	0.8470
12	0.2678	0.3981	0.4973	0.5760	0.5980	0.6581	0.7079	0.7496	0.7950	0.8233
13	0.2552	0.3802	0.4762	0.5529	0.5745	0.6339	0.6835	0.7255	0.7717	0.8010
14	0.2443	0.3646	0.4575	0.5324	0.5536	0.6120	0.6614	0.7034	0.7501	0.7800
15	0.2346	0.3507	0.4409	0.5140	0.5347	0.5923	0.6411	0.6831	0.7301	0.7604
16	0.2260	0.3383	0.4259	0.4973	0.5177	0.5742	0.6226	0.6643	0.7114	0.7419
17	0.2183	0.3271	0.4124	0.4821	0.5021	0.5577	0.6055	0.6470	0.6940	0.7247
18	0.2113	0.3170	0.4000	0.4683	0.4878	0.5425	0.5897	0.6308	0.6777	0.7084
19	0.2049	0.3077	0.3887	0.4555	0.4747	0.5285	0.5751	0.6158	0.6624	0.6932
20	0.1991	0.2992	0.3783	0.4438	0.4626	0.5155	0.5614	0.6018	0.6481	0.6788
21	0.1938	0.2914	0.3687	0.4329	0.4513	0.5034	0.5487	0.5886	0.6346	0.6652
22	0.1888	0.2841	0.3598	0.4227	0.4409	0.4921	0.5368	0.5763	0.6219	0.6524
23	0.1843	0.2774	0.3515	0.4132	0.4311	0.4815	0.5256	0.5647	0.6099	0.6402
24	0.1800	0.2711	0.3438	0.4044	0.4219	0.4716	0.5151	0.5537	0.5986	0.6287
25	0.1760	0.2653	0.3365	0.3961	0.4133	0.4622	0.5052	0.5434	0.5879	0.6178
26	0.1723	0.2598	0.3297	0.3882	0.4052	0.4534	0.4958	0.5336	0.5776	0.6074
27	0.1688	0.2546	0.3233	0.3809	0.3976	0.4451	0.4869	0.5243	0.5679	0.5974
28	0.1655	0.2497	0.3172	0.3739	0.3904	0.4372	0.4785	0.5154	0.5587	0.5880
29	0.1624	0.2451	0.3115	0.3673	0.3835	0.4297	0.4705	0.5070	0.5499	0.5790
30	0.1594	0.2407	0.3061	0.3610	0.3770	0.4226	0.4629	0.4990	0.5415	0.5703
40	0.1368	0.2070	0.2638	0.3120	0.3261	0.3665	0.4026	0.4353	0.4741	0.5007
50	0.1217	0.1843	0.2353	0.2787	0.2915	0.3281	0.3610	0.3909	0.4267	0.4514
60	0.1106	0.1678	0.2144	0.2542	0.2659	0.2997	0.3301	0.3578	0.3912	0.4143
80	0.0954	0.1448	0.1852	0.2199	0.2301	0.2597	0.2864	0.3109	0.3405	0.3611
100	0.0851	0.1292	0.1654	0.1966	0.2058	0.2324	0.2565	0.2786	0.3054	0.3242
1000	0.0266	0.0406	0.0520	0.0620	0.0650	0.0736	0.0814	0.0887	0.0976	0.1039

ANSWERS TO SELECTED EXERCISES

Chapter 0 – Getting Started

0.1 **(a)** More than likely, the individuals who chose to take higher levels of vitamin C engaged in healthier behaviors in general than those who took lower levels of vitamin C or who didn't choose to take it. (In fact, the exercise states that "higher Vitamin C levels are associated with people who practice healthier behavior patterns.") Those taking higher levels of vitamin C may also have been more affluent (i.e., had money available to purchase the vitamins and, possibly, had better health care generally). **(b)** In a randomized experiment, people of all types are randomly assigned to the treatments, equalizing out the effects of influences other than vitamin C. If vitamin C in the bloodstream does not reduce the risk of dying, then we would not expect to see a difference in mortality rates between those with high levels of vitamin C and those with lower levels in the experiment.

0.3 **(a)** The proportion of respondents who do not feel that Europe has let in too many refugees is likely to be different from 32% (although it is difficult to determine in which direction). The online poll used voluntary response: people were not randomly selected but made their own decision to participate or not. (*Note:* The "poll" is associated with an article which mentions that the Dalai Lama thinks Europe has let in too many refugees, so respondents might be primed to agree with the Dalai Lama.) **(b)** As long as the poll was one of voluntary response, the sample size (3,000 or 30,000) doesn't matter; it is a biased sample and will not reflect the views of people nationwide.

Chapter 1 – Picturing Distributions with Graphs

1.1 **(a)** The individuals are the car makes and models. **(b)** For each individual, the variables recorded are vehicle class (categorical), transmission type (categorical), number of cylinders (usually treated as quantitative), city mpg (quantitative), highway mpg (quantitative), and annual fuel cost in dollars (quantitative).

1.3 **(a)** 90% use these top social media sites; 10% use other sites most often. **(b)** A bar graph is provided.

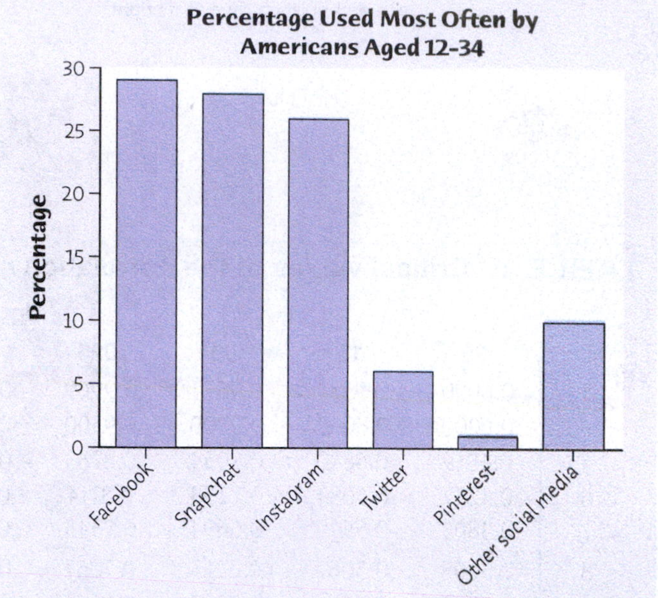

(c) If you include an "Other" category, then, yes, a pie chart is appropriate. **(d)** If a company is considering advertising to Americans aged 12–34, these data would provide information about where to advertise to the target audience.

1.5 A pie chart can be made because the days are nonoverlapping and make up the whole. Some births are scheduled (e.g., induced labor) and, probably, most are scheduled for weekdays.

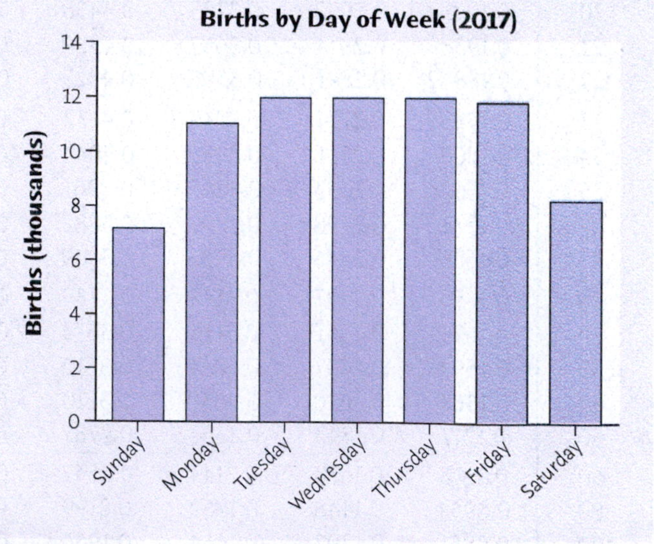

1.7 Use the applet to answer these questions.

1.9 **(a)** There are two clear peaks in the distribution. If we gave only one center, it would most likely be between these and would not be truly representative. **(b)** Young boys might spend a lot of time outdoors playing and engaged in sports; their time outside in places where they would encounter ticks might well be less as they go through the college age and younger adulthood. With families and yard work, their time outside might increase beginning in their thirties. **(c)** This is incorrect. Hiking in the woods at any age will increase a person's likelihood of encountering the ticks that spread Lyme disease. There are fewer cases for people aged 65 and older because fewer people hike at those ages. **(d)** The histograms have the same shapes, but females have a slightly lower incidence rate until age 75, at which point females have a slightly higher rate. Females under age 75 possibly spend less time outdoors in areas where they would encounter ticks.

1.11 A stemplot for health expenditures per capita (in PPP) is given. (*Note*: The stemplot was produced by JMP, which puts the stems in decreasing order. It is also correct to put the stems in increasing order as shown in Chapter 1.) Data are rounded to units of hundreds. Stems are thousands and are split. This distribution is right-skewed with a single high outlier (United States). There seem to be two clusters of countries. The center of this distribution is around 26 ($2600 spent per capita). The distribution varies from about 0 | 1 (about $100 spent per capita) to about 9 | 5 (about $9500 spent per capita).

Stem	Leaf
9	5
8	
7	6
6	2
5	11334
4	145568
3	124
2	46
1	001134447
0	1224689

0|1 represents 100

1.13 (a)

1.15 (b)

1.17 (c)

1.19 (b)

1.21 (c)

1.23 **(a)** Individuals are students who have finished medical school. **(b)** Five, in addition to "Name," "Age" (in years), and "USMLE" (score points) are quantitative. The others are categorical.

1.25 Green accounts for 1%. A bar graph is provided. A pie chart could also be made.

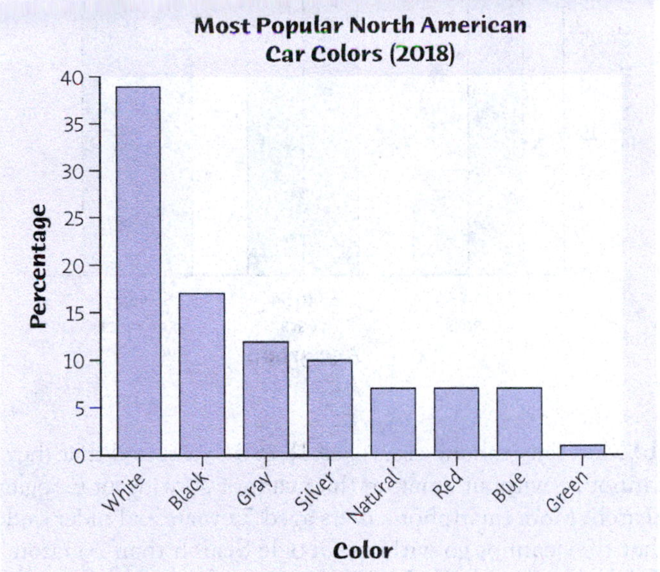

1.27 **(a)** A bar graph is given.

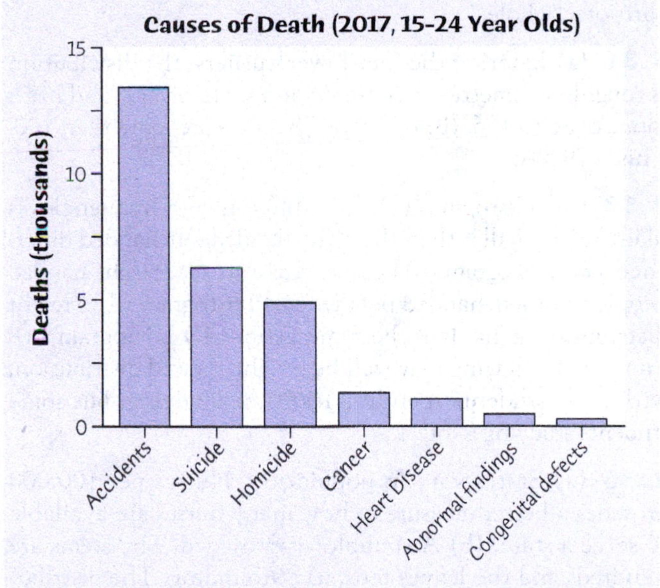

(b) Yes, we can construct a pie chart if we provide an "Other" category, where the total number of deaths in the "Other" category is $32{,}025 - 13{,}441 - 6252 - 4905 - 1374 - 1126 - 501 - 362 = 4064$. The creation of an "Other" category is required for a pie chart so that the number of deaths in each subcategory sum to the total number of deaths. Without an "Other" category, we cannot construct a pie chart.

1.29 (a) A bar graph is provided.

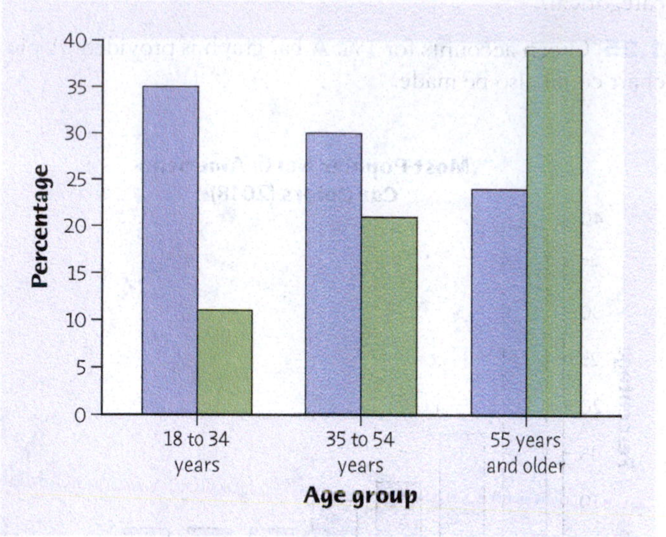

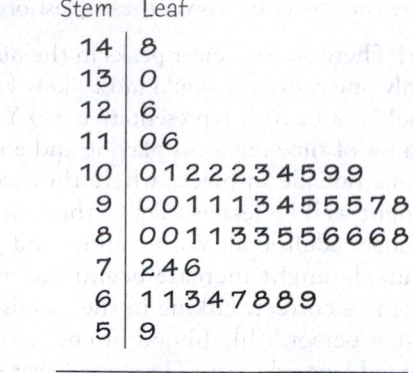

(b) More smartphone users aged 18 to 34 years said that they cannot go without Amazon than cannot go without Google Search. More smartphone users aged 55 years and older said that they cannot go without Google Search than Amazon. Smartphone users aged 35 to 54 years are more even with respect to which app they cannot go without, although there is still a preference for Amazon over Google Search. (c) A pie chart is not appropriate because these data do not represent parts of a "whole."

1.31 (a) Ignoring the four lower outliers, the distribution is roughly symmetric, is centered at a score of 111, and has a range of 86 to 136. (b) 62 of the 78 scores are more than 100. This is 79.5%.

1.33 (1.) Histogram (c). The difference in frequencies is likely to be smaller than the right-handed/left-handed difference. (2.) Histogram (b) because there are more right-handed people than left-handed people. (3.) Histogram (d). Height distribution is likely to be symmetric. (4.) Histogram (a). Time spent studying may well be a right-skewed distribution, with most students spending less time studying, but some students studying a lot.

1.35 (a) States vary in population. Nurses per 100,000 provides a better measure of how many nurses are available to serve a state. (b) A stemplot is provided. The stems are hundreds and the leaves tens, after rounding. The distribution is slightly left-skewed, with a center around 900 and a range from 585 to 1483 nurses per 100,000. The observation with 1483 nurses per 100,000 is an outlier. This corresponds to Washington, DC; many people live in states surrounding DC but commute to DC to health care.

Stem	Leaf
14	8
13	0
12	6
11	06
10	0122234599
9	001113455578
8	001133556668
7	246
6	11347889
5	9

5|9 represents 590

(c) Splitting the stems make the right tail more visible and allows you to see the variability between the large number of states with between 800 and 1100 nurses per 100,000.

1.37 The shape of the distribution is bimodal and (perhaps) left-skewed. There is a high outlier around 246 thousand. The center is about 170 thousand. The data range from 75.7 thousand to 245.93 thousand.

Stem	Leaf
24	6
23	
22	
21	
20	1244
19	2
18	22
17	022399
16	68
15	9
14	6
13	
12	3
11	0
10	3
9	257
8	1
7	6

7|6 represents 76

1.39 The decline in the population is not seen in the stemplot made in Exercise 1.37.

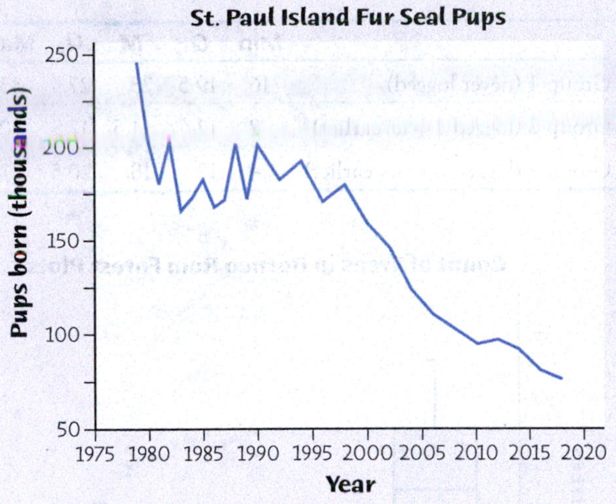

1.41 (a) A bar graph is provided.

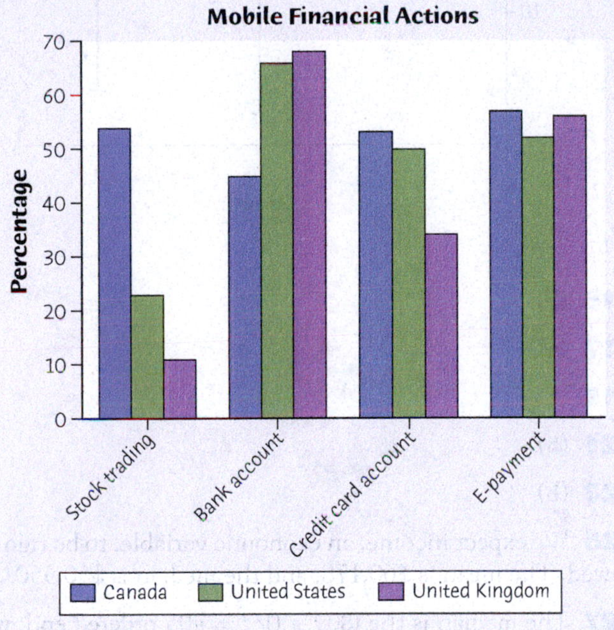

(b) Approximately the same percent of mobile shoppers use e-payments in Canada, the United States, and the United Kingdom. (The bars are similar in height.) A markedly higher percent of mobile shoppers in Canada than in the United States or in the United Kingdom do stock trading. (The bar for Canada is much higher than for the United States or the United Kingdom.) (c) A pie chart is not appropriate because these data do not represent parts of a "whole."

1.43 (a) Graph (a) appears to show the greatest increase. Vertical scaling can impact the perception of the data. (b) In 2000, tuition was about \$5000, and it rose to about \$10,000 in 2018; this is an increase of approximately \$5000. Both plots describe the same data.

1.45 (a) It seems as though winter quarters are typically associated with lower housing starts. (b) and (c) Over the long run, housing starts have risen, except in crisis years, which are shown by the sharp decrease between 2006 and 2008. (d) Since 2011, it appears that housing starts are again increasing from year to year.

Chapter 2 – Describing Distributions with Numbers

2.1 $\bar{x} = \dfrac{291.0 + 10.9 + \cdots + 9.6}{16} = 56.28$ per 100 ml. Only three (86, 190.4, 291) lakes have *E. coli* levels greater than the mean. The mean is greater than most of the observations because of the three outliers (86, 190.4, 291).

2.3 $\bar{x} = 33.1$ minutes. The median is 32.5 minutes. The mean is larger than the median, which is expected with a right-skewed distribution.

2.5 A histogram is given. The mean is larger than the median because of the right-skew. $\bar{x} = 4.9676$, and the median is 2.753.

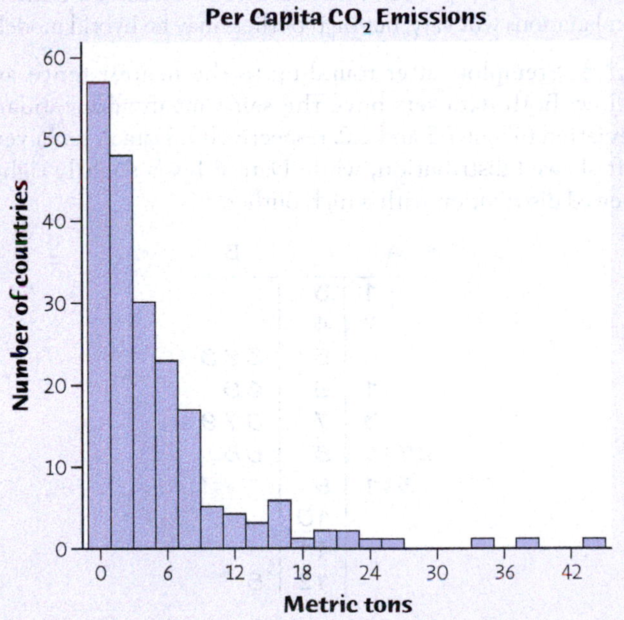

2.7 (a) Minimum = 12, $Q_1 = 23$, Median = 26, $Q_3 = 30.5$, Maximum = 56. **(b)** The boxplot shows right skew in the distribution of MPG values. There are high outliers (which are most likely hybrid cars).

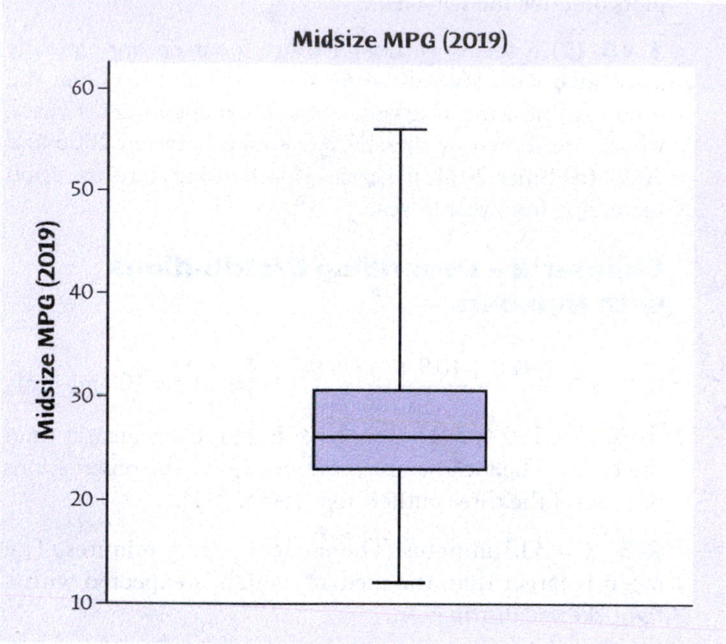

2.9 $IQR = 30 - 23 = 7$, so $Q_3 + 1.5 \times IQR = 30 + 15 \times 7 = 40.5$. Suspected outliers are 41, 41, 41, 41, 42, 42, 43, 44, 44, 46, 46, 48, 50, 52, 52, 52, 56. Because $Q_1 - 1.5 \times IQR = 23 - 1.5 \times 7 = 12.5$, there are no potential low outliers. Explanations will vary, but high outliers may be hybrid models.

2.11 Stemplots after rounding to the nearest tenth are below. Both data sets have the same mean and standard deviation (about 7.5 and 2.0, respectively). Data A has a very left-skewed distribution, while Data B has a slightly right-skewed distribution with a high outlier.

A		B
1	3	
7	4	
	5	368
1	6	69
3	7	079
8711	8	58
311	9	
	10	
	11	
	12	5

2.13 STATE: We'd like to know how logging impacts the number of trees. PLAN: Create side-by-side boxplots for the three types of plots and compute summary statistics. SOLVE: None of the distributions are symmetric; Groups 2 and 3 have a low outlier and are left-skewed, while Group 1 is right-skewed. The five-number summaries are computed. CONCLUDE: Plots that have never been logged have more trees than either group of logged plots. If we compare the distributions for the two groups of logged plots, it takes a long time for the rain forest to recover from logging.

	Min	Q_1	M	Q_3	Max
Group 1 (never logged)	16	19.5	23	27.5	33
Group 2 (logged 1 year earlier)	2	12	14.5	17.5	20
Group 3 (logged 8 years earlier)	4	12	18	20.5	22

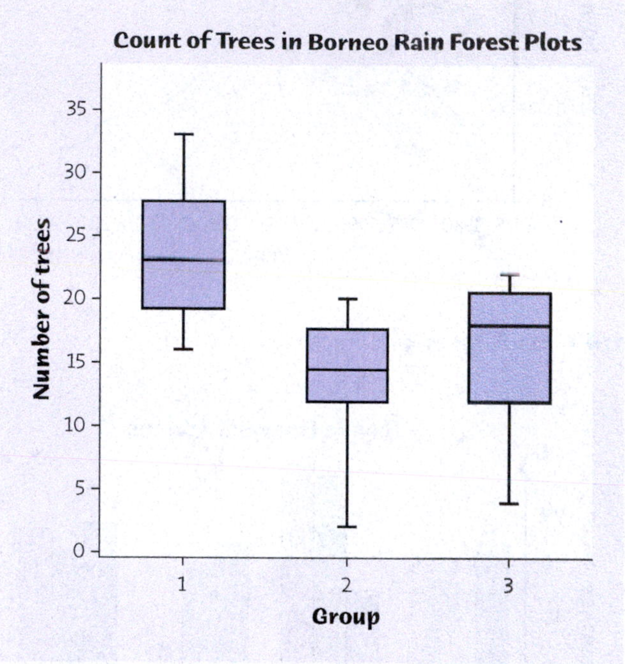

2.15 (b)

2.17 (a)

2.19 (c)

2.21 (b)

2.23 (b)

2.25 We expect income, an economic variable, to be right-skewed. The mean is $60,178, and the median is $50,350.

2.27 The median is the $(809 + 1)/2 = 405$ ordered endowment. The first quartile, Q_1, is the $(404 + 1)/2 = 202.5$ endowment. Q_3, is the $405 + 202.5 = 607.5$ endowment.

2.29 In this case, the boxplots do not fail to reveal any important information because there are no gaps in the stemplots.

	Min	Q_1	M	Q_3	Max
MW	80.2	83.75	86.75	88.45	91
NE	81.8	85.35	87.9	89	90.5
S	73.2	82.45	86.9	89.35	89.8
W	71.1	78.1	79.7	84.25	86.2

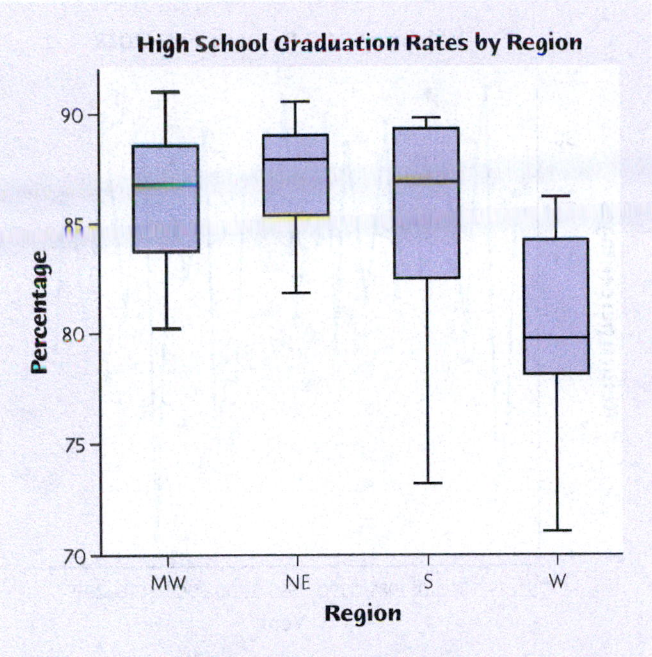

2.31 (a) A histogram is given. The distribution is strongly right-skewed, with center around 100 days and range from about 0 to about 600 days. (b) We should use the five-number summary: 43, 82.5, 102.5, 151.5, 598 days. The median is closer to Q_1 than to Q_3.

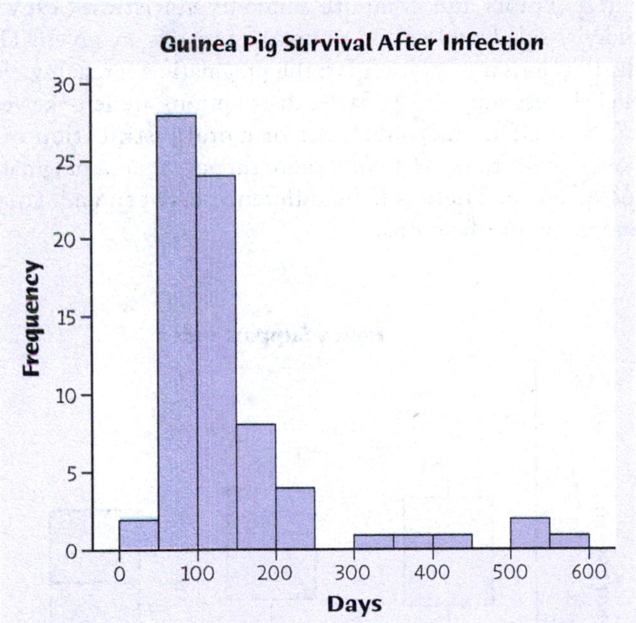

2.33 (a) Symmetric distributions are best summarized by $\bar{x}$ and s. The distribution for the treatment group was right-skewed. The control distribution could be called rather symmetric, but it has a high outlier. (b) The mean decreased by 8.45 seconds; the standard deviation decreased by 12.03. (c) The median is less affected by the outlier than the mean.

	With Outlier	Without Outlier
Mean	59.7	51.25
Standard deviation	63.0	50.97
Median	61	57.5

2.35 (a) The sixth observation must be placed at the median for the original five observations. (b) No matter where you put the seventh observation, the median is one of the two (repeated) values above because it will be the fourth (ordered) observation. In the output, the seventh point is the one on the far left.

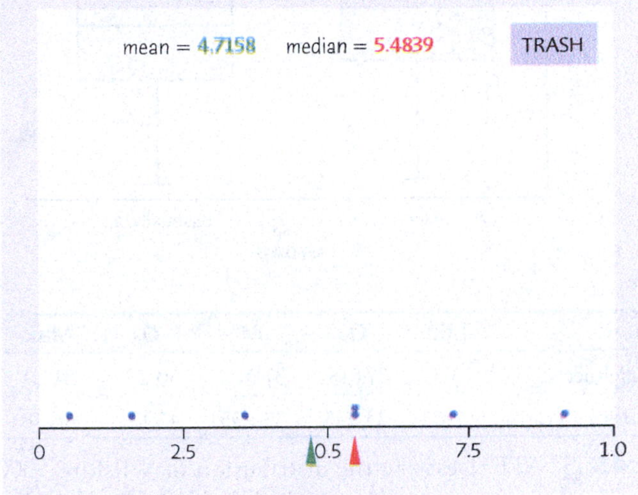

2.37 $\bar{x} = 33.9\%$. More populated states carry more weight in the national percentage.

2.39 (a) The smallest possible standard deviation will come from choosing all four numbers to be the same; for example, choose the numbers (2, 2, 2, 2). (b) The largest possible standard deviation is with the four numbers (0, 0, 10, 10). (c) There is more than one choice in part (a) but not in part (b).

2.41 Many answers are possible. One solution: (1, 2, 3, 4, 5, 100). In general, a "large" high outlier will guarantee that the mean is greater than the median.

2.43 STATE: We want to determine how wearing a helmet relates to the measure of risk behavior. PLAN: We will make side-by-side boxplots and compute five-number summaries of the scores for each group. We will compare the distributions for each group to make a conclusion on how wearing a helmet relates to the average number of pumps (risk taking). SOLVE: The boxplots are provided. The minimum, first quartile, and median of the helmet group is only slightly larger than of the baseball cap group. There is a greater discrepancy between the third quartiles and maximums, with those being much larger for the helmet group than the baseball cap group. There is greater variability in the helmet group. CONCLUDE: Wearing a helmet appears to be related to risky behavior. Although it didn't increase the behavior by much for most individuals, some helmet-wearing individuals displayed much riskier behavior.

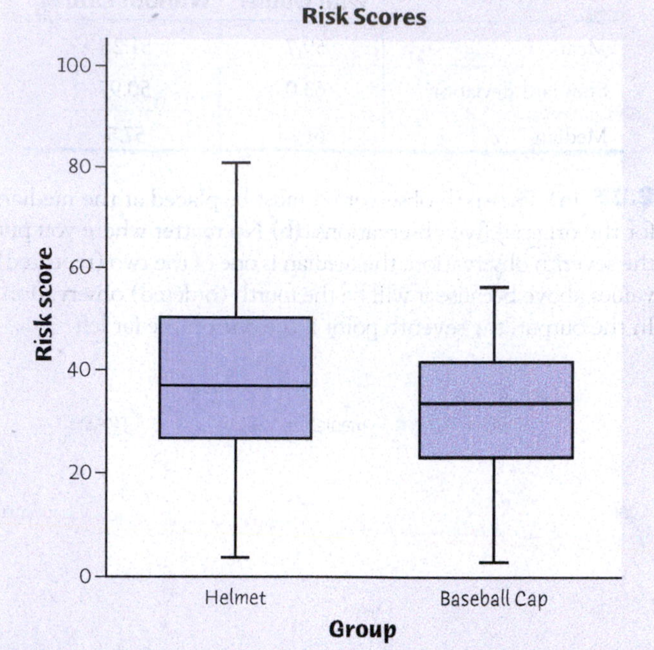

	Min	Q_1	M	Q_3	Max
Helmet	3.67	27.015	37.0	50.2	81.29
Baseball cap	2.68	23.935	33.635	42.17	56.58

2.45 STATE: Describe the distribution of Wilshire 5000 stock index returns over the period 1971–2018. PLAN: Graph the return with a histogram and a time plot. Compute and report appropriate summary statistics. SOLVE: The histogram, time plot, and summary statistics are given. CONCLUDE: The distribution of average returns is left-skewed. Most years, the average return is positive. Returns range from about −40% to about 40%, with the median return about 16%.

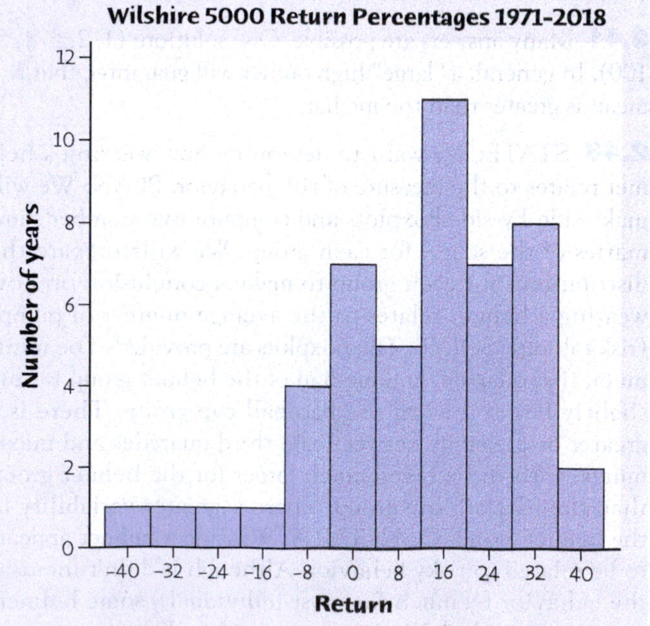

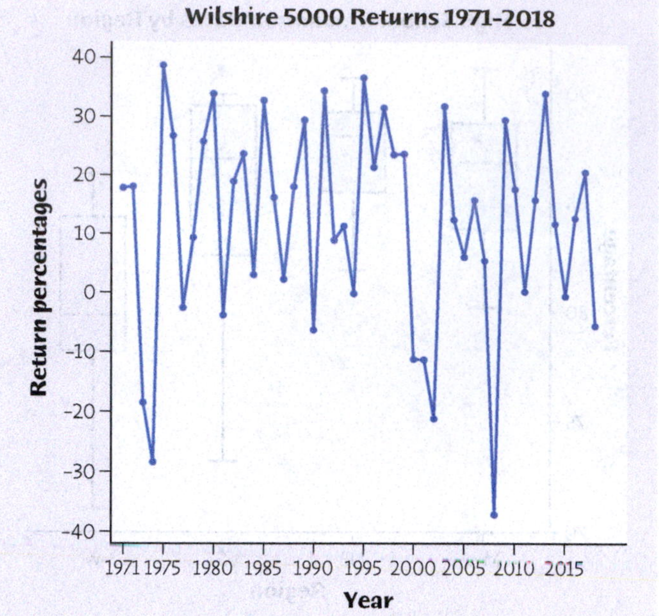

Mean	St. Dev.	Min	Q_1	Median	Q_3	Max
11.96	17.64	−37.34	0.265	15.99	24.56	38.47

2.47 STATE: We want to know how a leader's justification affects support for the policy. PLAN: Create side-by-side boxplots and compute summary statistics. SOLVE: Side-by-side boxplots and summary statistics are given. The distributions are very similar; the pragmatic approach yields slightly less support. All three distributions are left-skewed. CONCLUDE: An ambiguous or moral justification of a policy tends to have slightly more support than a pragmatic justification. There is little difference between ambiguous and moral justifications.

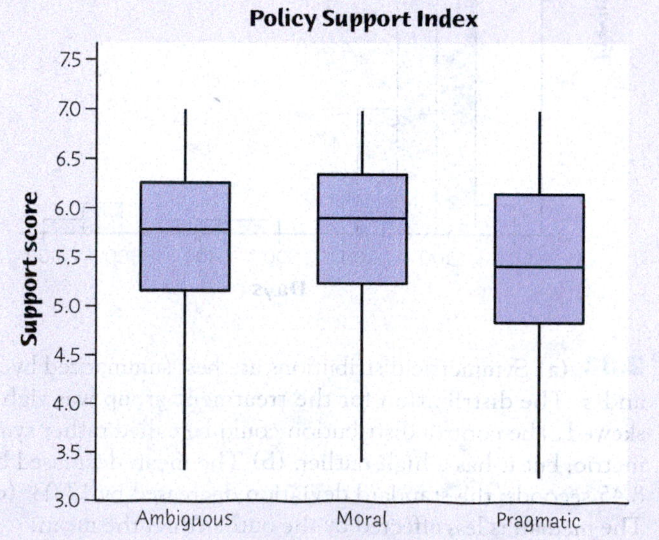

	Mean	St. Dev.	Minimum	Q_1	Median	Q_3	Maximum
Ambiguous	5.71	0.77	3.33	5.17	5.79	6.27	7
Moral	5.81	0.82	3.42	5.23	5.92	6.35	7
Pragmatic	5.44	0.89	3.25	4.83	5.42	6.17	7

2.49 **(a)** Side-by-side boxplots are provided. All three distributions are right-skewed with high outliers. The median increases slightly with increasing age. We also see an increase in variability as people age, although the IQRs are relatively the same. **(b)** Unless their original cholesterol levels were *extremely* high, the 4 or 24 people on medication in their 20s and 30s, respectively, probably wouldn't affect these distributions much. However, more than 10% of people in their 40s are on medication. If those 117 had not been on medication, the distribution would likely show more variability and higher cholesterol readings.

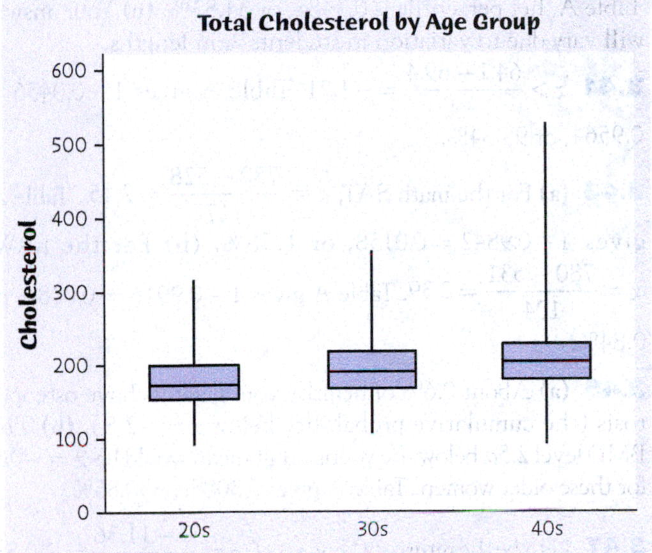

2.51 **(a)** The mean and median are 3.71 and 4, respectively. With the two smallest observations omitted, the mean and median are 3.9 and 4.14. Omitting the two smallest observations had a greater impact on the mean than the median. The median is more robust to a small number of unusually small or large values. **(b)** Yes, the rule does identify these two scores as suspected outliers. **(c)** It is possible that after randomization, subjects who experience little bonding regardless of the grouping were assigned to the pain group by chance.

2.53 The five-number summary is 92, 154, 173, 199, 318. We have $IQR = 199 - 154 = 45$. Outliers would be values smaller than $154 - 1.5 \times 45 = 86.5$ or larger than $199 + 1.5 \times 45 = 266.5$. There are no low outliers, but there are 18 high-end outliers.

Chapter 3 – The Normal Distributions

3.1 Sketches will vary. **(a)** Symmetric distributions are mirror images on either side of the center. **(b)** A distribution that is skewed to the right has a long right tail.

3.3 **(a)** $\mu = 5$, the balance point. The median is also 5 because the distribution is symmetric. **(b)** The first quartile is 2.5, and the third quartile is 7.5.

3.5 A sketch of the distribution is given. The tick marks are placed at the mean and at 1, 2, and 3 standard deviations above and below the mean.

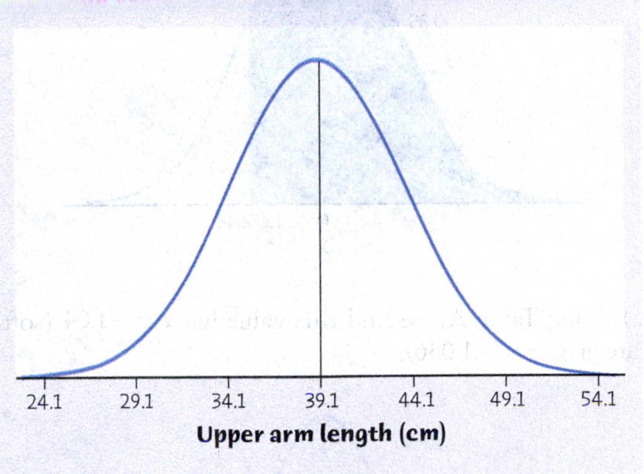

3.7 **(a)** In the middle 95% of all years, monsoon rain levels are between $852 \pm 2(82) = 688$ and 1016 mm. **(b)** The driest 2.5% of monsoon rainfalls are less than 688 mm (more than 2σ below μ).

3.9 A woman 5.5 feet tall has $z = \frac{66 - 64.1}{3.7} = 0.51$. A man 5.5 feet tall has $z = \frac{66 - 69.4}{3.1} = -1.10$. A woman 5.5 feet tall is 0.51 standard deviation taller than average for women. A man 5.5 feet tall is 1.10 standard deviations *below* average for men.

3.11 Let x be the monsoon rainfall in a given year. **(a)** $x \leq 697$ mm corresponds to $z \leq \frac{697 - 852}{82} = -1.89$. Table A gives 0.0294, or 2.94%. **(b)** $682 < x < 1022$ corresponds to $\frac{682 - 852}{82} < z < \frac{1022 - 852}{82}$, or $-2.07 < z < 2.07$. Table A gives $0.9808 - 0.0192 = 0.9616$, or 96.16%.

3.13 **(a)** Using Table A, we find this value has $z = 0.67$ (software gives $z = 0.6745$).

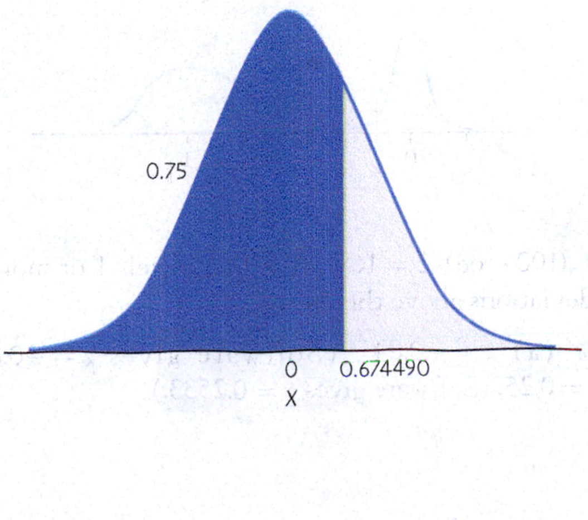

(b) We want a proportion of 0.85 below. Using Table A, $z = 1.04$ (software gives $z = 1.036$).

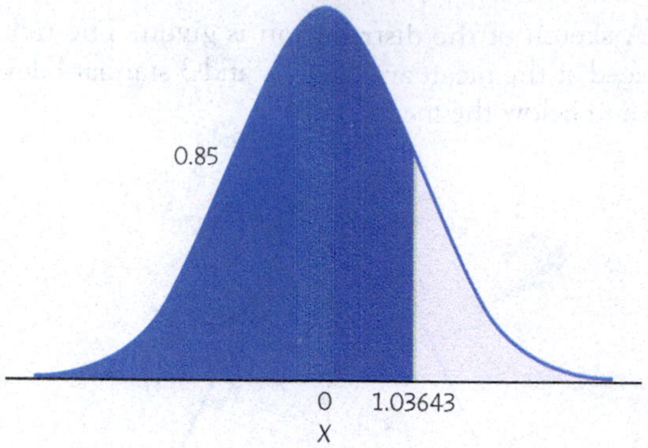

(c) Using Table A, we find this value has $z = -1.04$ (software gives $z = -1.036$).

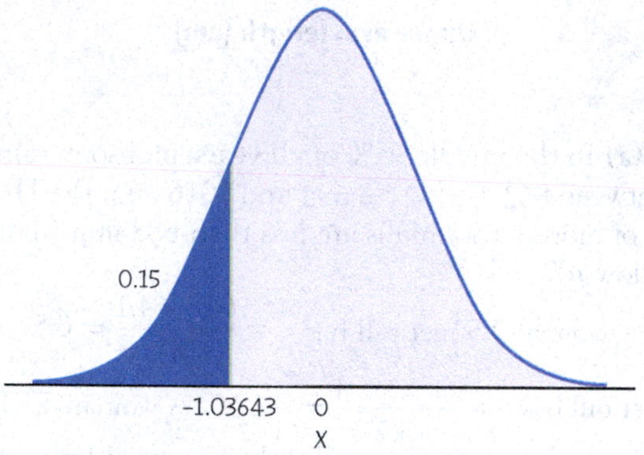

3.15 (c)

3.17 (b)

3.19 (c)

3.21 (b)

3.23 (b)

3.25 Sketches will vary, but these should be some variation on the one shown here; the peak at 0 should be "tall and skinny," while near 1, the curve should be "short and fat."

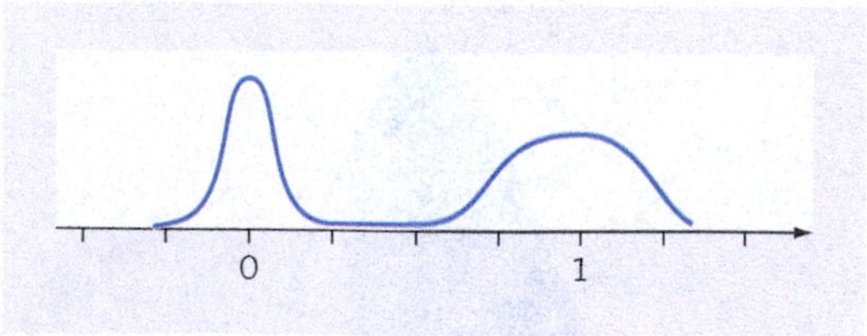

3.27 $(100 - 68)/2 = 16\%$ have LDL levels 1 or more standard deviations above the mean.

3.29 **(a)** $z = -0.84$. (Software gives $z = -0.8416$.) **(b)** $z = 0.25$. (Software gives $z = 0.2533$.)

3.31 $x < 5.0$ corresponds to $z < \frac{5.0 - 5.43}{0.54} = -0.80$; Table A gives 0.2119.

3.33 **(a)** $z > \frac{130 - 100}{15} = 2$. Table A gives $1 - 0.9772 = 0.0228$, or 2.28%. **(b)** $z > \frac{130 - 120}{15} = 0.67$. Table A gives $1 - 0.7486 = 0.2514$, or 25.14%.

3.35 $z > \frac{29 - 22.8}{4.8} = 1.29$. Table A gives $1 - 0.9015 = 0.0985$, or 9.85%.

3.37 Q_1 and Q_3 have $z = -0.67$ and 0.67, respectively. $Q_1 = 22.8 - (0.67)(4.8) = 19.58$ mpg and $Q_3 = 22.8 + (0.67)(4.8) = 26.02$ mpg.

3.39 **(a)** Cecile's arm has $z = \frac{33.9 - 35.9}{5.1} = -0.39$. Using Table A, her percentile is 0.3483, or 34.83%. **(b)** Your answer will vary due to variation in students' arm lengths.

3.41 $z > \frac{64.1 - 69.4}{3.1} = -1.71$. Table A gives $1 - 0.0436 = 0.9564$, or 95.64%.

3.43 **(a)** For the math SAT, $z = \frac{780 - 528}{117} = 2.15$. Table A gives $1 - 0.9842 = 0.0158$, or 1.58%. **(b)** For the ERW, $z = \frac{780 - 531}{104} = 2.39$. Table A gives $1 - 0.9916 = 0.0084$, or 0.84%.

3.45 **(a)** About 0.6% of healthy young adults have osteoporosis (the cumulative probability below $z = -2.5$). **(b)** The BMD level 2.5σ below the young adult mean would be $z = -0.5$ for these older women. Table A gives 0.3085, or 30.85%.

3.47 Let x be the return. **(a)** For $x > 0$, $z = \frac{0 - 11.36}{19.58} = -0.58$. Table A gives $1 - 0.2810 = 0.7190$, or 71.9% of returns are greater than 0. For $x > 30$, $z = \frac{30 - 11.36}{19.58} = 0.95$. Table A gives $1 - 0.8289 = 0.1711$, or 17.11% of returns are greater than 30. **(b)** About 72.5% and of the actual returns are greater than 0, and about 18.7% of the actual returns are greater than 30. The value for returns greater than 0 is close to what we expect from the $N(11.36, 19.58)$ distribution, as is the value for returns greater than 30. This Normal distribution is a good approximation.

3.49 **(a)** A histogram is provided and is roughly symmetric. **(b)** Mean = 536.95, Median = 540, Standard deviation = 69.88, $Q_1 = 490$, $Q_3 = 580$. The mean and median are close. The distances between the median and the quartiles, 50 (for Q_1) and 40 (for Q_3), are similar. This is consistent with a Normal distribution. **(c)** Assume GSU freshman scores have a $N(536.95, 69.88)$ distribution. The proportion higher than 511 corresponds to $z > \frac{511 - 536.95}{69.88} = -0.37$. Table A gives this proportion to be $1-0.3557=0.6443$, or 64.43%. **(d)** 63.6% of GSU freshmen scored higher than 511. This is close to the probability computed in part (c).

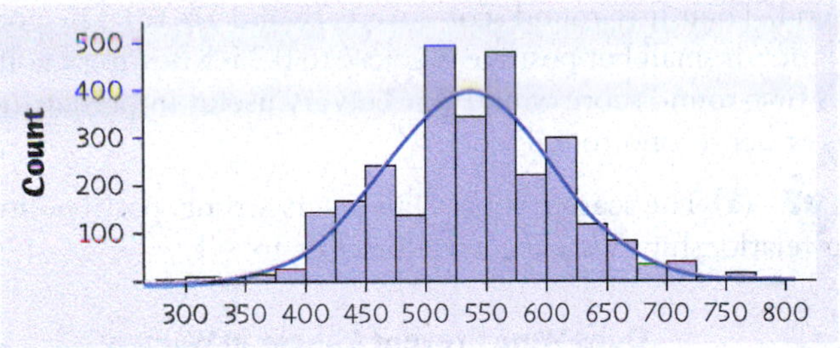

3.51 **(a)** $14/548=0.0255$ (2.55%) weighed less than 100 pounds. $x<100$ corresponds to $z<\dfrac{100-161.58}{48.96}=-1.26$. Using Table A, the area is 0.1038 (10.38%). **(b)** $33/548=0.0602$ (6.02%) weighed more than 250 pounds. $x>250$ corresponds to $z>\dfrac{250-161.58}{48.96}=1.81$. Using Table A, about $1-0.9649=0.0351(3.51\%)$ would weigh more than 250 pounds. **(c)** The Normal distribution model predicts 10.38% of women weigh less than 100 pounds, while actually about 2.55% do. The Normal model also predicts 3.51% of women weigh more than 250 pounds, while we actually observed 6.02% of women weigh more than 250 pounds. This is a substantial error.

3.53 Because the quartiles of any distribution have 50% of observations between them, place the flags so that the reported area is 0.5. The closest the applet gets is an area of 0.4978, between −0.671 and 0.671. The quartiles of any Normal distribution are about 0.67σ above and below μ.

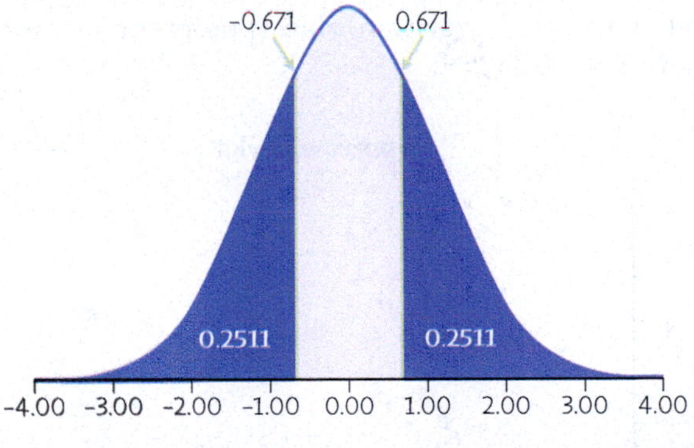

Chapter 4 – Scatterplots and Correlation

4.1 **(a)** Explanatory: number of times a student accessed the course website for your statistics course; response: grade on the final exam for the course. **(b)** Explanatory: number of hours per week spent exercising; response: calories burned per week. **(c)** Explanatory: hours per week spent online using social media; response: grade point average. **(d)** Explore the relationship.

4.3 Your answer will vary from those of your peers and your instructor. Examples: weight, sex, blood pressure, country of origin, etc.

4.5 Outsource percent is the explanatory variable. The data do not support concerns of the critics.

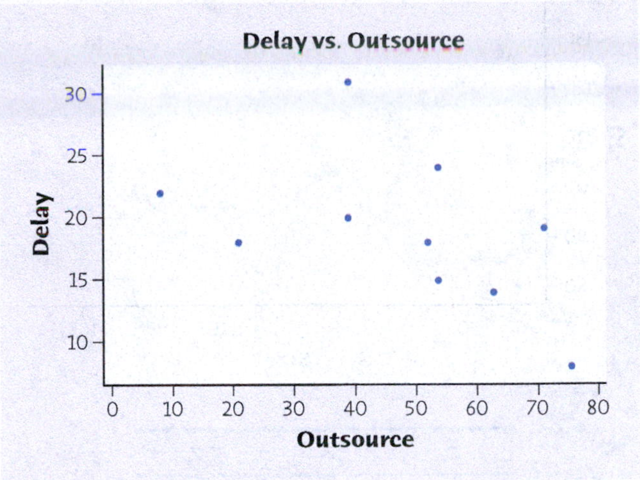

4.7 One could consider Allegiant and Spirit to be outliers. Both have low outsourcing percentages and moderate delay percentages. Without these two airlines, there is a stronger negative relationship; with the two airlines, the relationship is weakly negative.

4.9 **(a)** Counties with very large populations (above 800,000) are marked with red dots; the others are marked with blue dots.

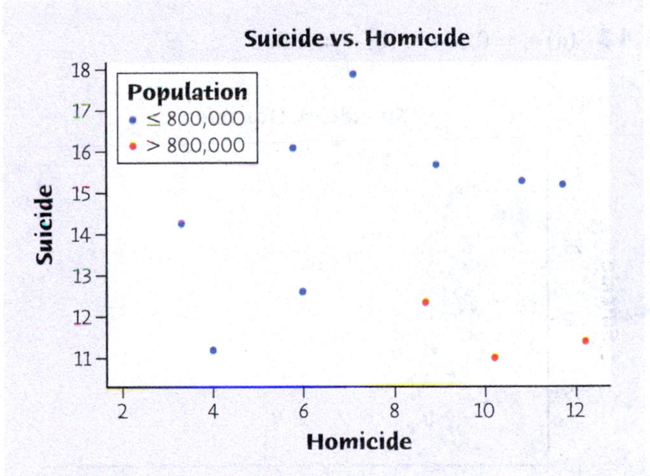

(b) For the counties with very large populations, there appears to be no relationship between homicide and suicide rates. For the smaller counties, there is a weak positive relationship between homicide and suicide rates.

4.11 **(a)** See the scatterplot on the next page. Brain size is the explanatory variable. **(b)** $\bar{x}=95.17$ (10,000 pixels), $s_x=6.77$(10,000 pixels), $\bar{y}=108$ points, $s_y=24.29$ points. See the accompanying table for the standardized scores. The correlation is $r=0.374$. This is consistent with the weak, positive association depicted in the scatterplot. **(c)** Software gives $r=0.377$. The answer in part (c) is erroneous at the thousandths place due to rounding.

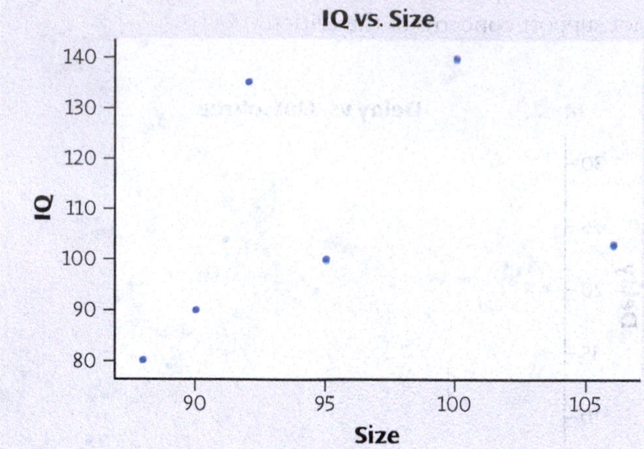

z_x	z_y	$z_x z_y$
0.71	1.32	0.94
−0.76	−0.74	0.56
−0.03	−0.33	0.01
−0.47	1.11	−0.52
−1.06	−1.15	1.22
1.60	−0.21	−0.34
	Sum	1.87

4.13 (a) $r = 0.0645$. (b) $r = 0.8085$.

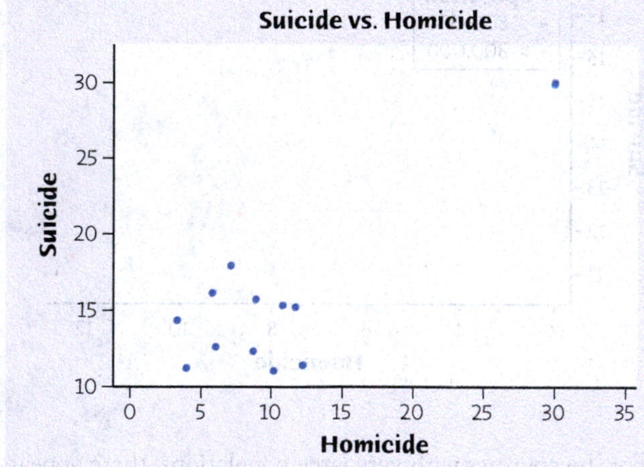

(c) Point *A* strengthens the positive linear association, because, when *A* is included the points of the scatterplot seem to actually have a linear relationship (your eye is drawn to that point in the upper right).

4.15 (b)

4.17 (b)

4.19 (c)

4.21 (c)

4.23 (b)

4.25 (a) The lowest first-round score was 66, scored by two golfers. One of the golfers scored 71 in the second round, while the other score 75 in the second round. (b) Jose Maria Olazabal and Jovan Rebula both scored 79 in the second round. Their first-round scores were 73 and 78. (c) The correlation is small but positive, so close to 0.25. Knowing a golfer's first-round score would not be very useful in predicting his or her second-round score.

4.27 (a) The scatterplot reveals a very strong, positive linear relationship. We expect r to be close to $+1$.

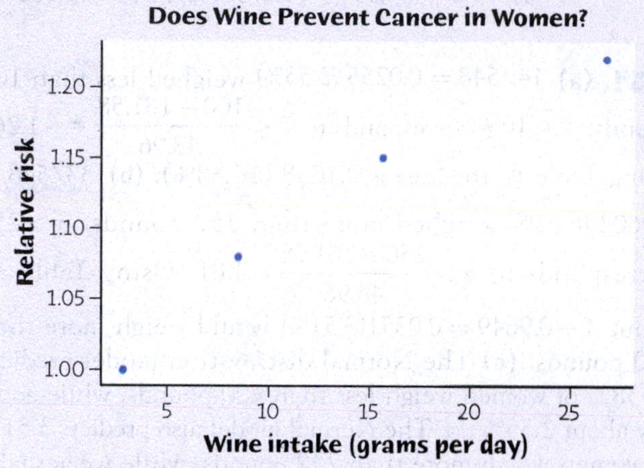

(b) Using software, $r = 0.9851$. The data suggest that women who consume more wine tend to have higher risk of breast cancer. (c) We cannot conclude a causal relationship between wine consumption and breast cancer in women because this is an observational study. Women who drink more wine may differ in many respects from women who drink less wine.

4.29 (a) The scatterplot shows a negative, somewhat linear relationship. Correlation is an appropriate measure of strength; $r = -0.7485$.

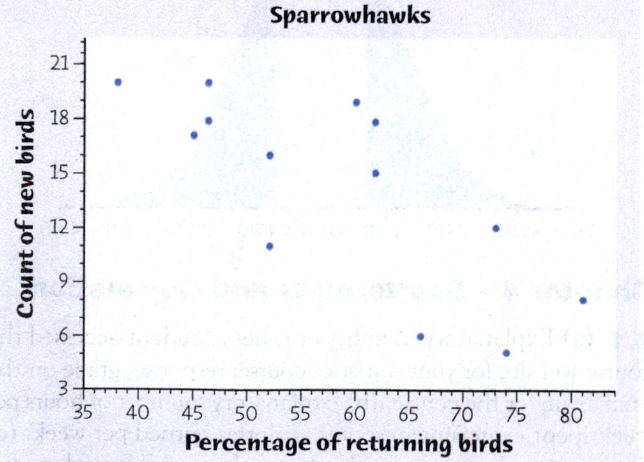

(b) Because this association is negative, we conclude that the sparrowhawk is a long-lived territorial species.

4.31 (a) The scatterplot is shown; note that poverty percentile rank is explanatory (and so should be on the horizontal axis)

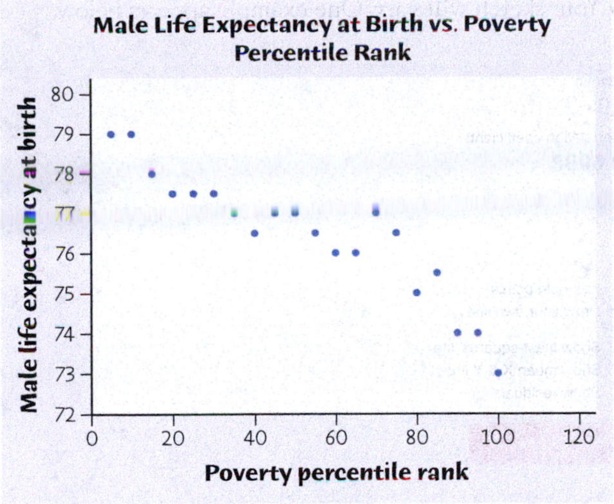

(b) The association is strong, negative, and linear; $r = -0.924$.

4.33 (a) The scatterplot is provided.

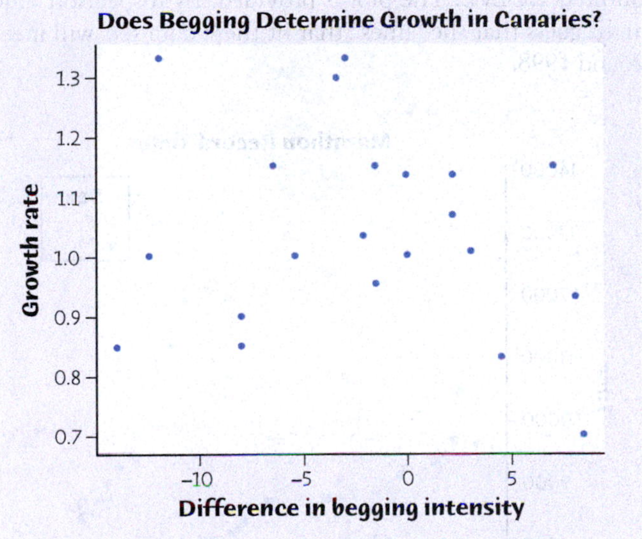

(b) The scatterplot suggests no linear relationship (we can see a bit of a curve). $r = -0.1749$. r is not helpful here because the relationship is not linear. (c) Neither theory is strongly supported, but growth rate increases initially as begging intensity increases and then levels off or decreases as parents begin to ignore increases in begging by the foster babies.

4.35 (a) Correlation would not change (correlation does not depend on units). (b) Correlation would not change. Subtracting 0.25 from all risks moves the plot "down" by 0.25, but the strength and direction of the linear relationship do not change. (c) There would be a perfect positive relationship; $r = +1$.

4.37 Explanations and sketches will vary but should note that correlation measures the strength of the linear association, not the slope of the line. The hypothetical Funds A and B mentioned in the report, for example, might have a linear relationship having line of slope 2 or 1/2.

4.39 The person who wrote the article interpreted a correlation close to 0 as if it were a correlation close to -1 (implying a negative association between teaching rating and research productivity). Professor McDaniel's findings mean there is little linear association between research productivity and teaching rating.

4.41 (a) Two scatterplots are given below. To the naked eye, the two plots look identical. (b) For data set A, $r = 0.664$. For data set B, $r = 0.834$. The increase in r is due to more points at (1, 1) and (4, 4). You would not expect the difference in r if you simply looked at the plots.

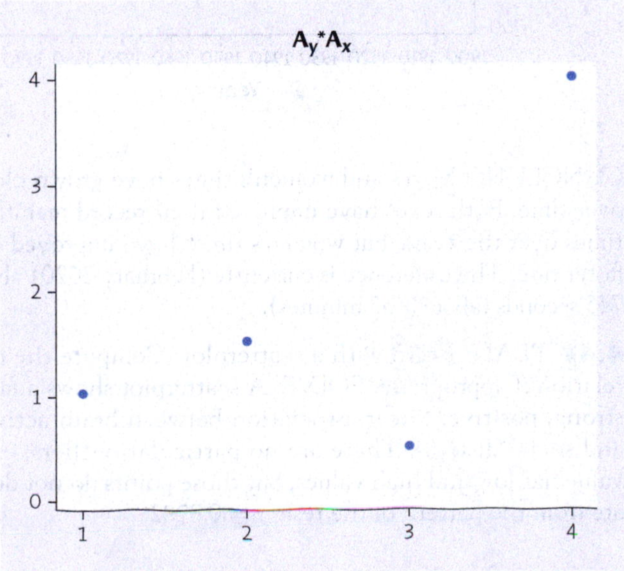

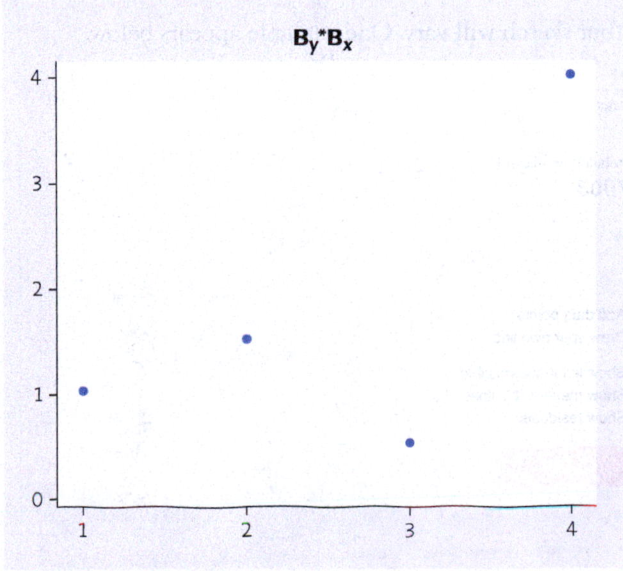

4.43 **(a)** Because two points determine a line, the correlation is always −1 or +1.

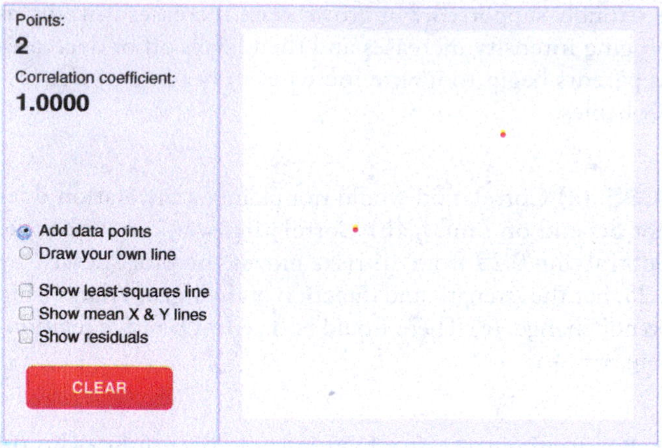

(b) Your sketch will vary. One example appears below. Note that the scatterplot must be positively sloped, but r is affected only by the scatter about the line, not by the steepness of the slope of that line.

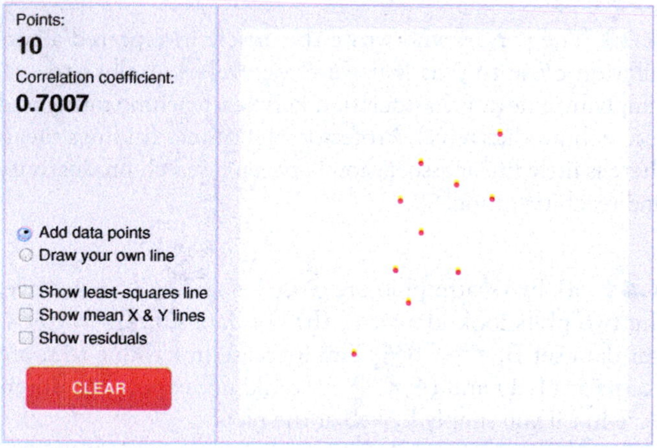

(c) Your sketch will vary. One example appears below.

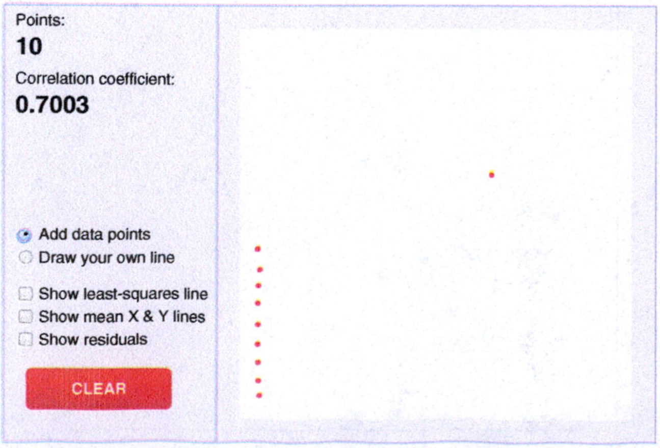

(d) Your sketch will vary. One example appears below.

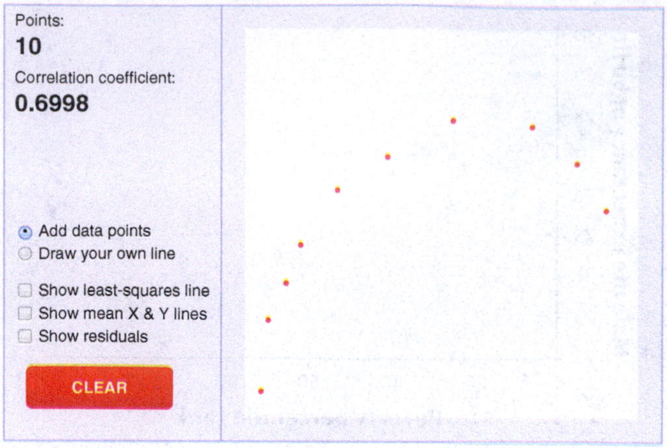

4.45 **(a and b)** PLAN: Plot the data on the same scatterplot. Examine the plot to see if women are beginning to outrun men. SOLVE: The plot is provided. By inspection, one might guess that the "lines" that fit these data sets will meet around 1998.

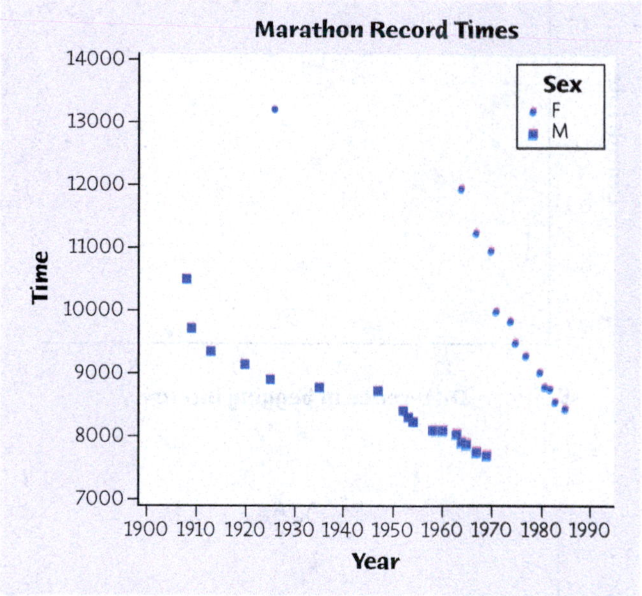

CONCLUDE: Men's and women's times have grown closer over time. Both sexes have improved their record marathon times over the years, but women's times have improved at a faster rate. The difference is currently (February 2020) about 745 seconds (about 12.5 minutes).

4.47 PLAN: Begin with a scatterplot. Compute the correlation if appropriate. SOLVE: A scatterplot shows a fairly strong, positive, linear association between brain activity and social distress. There are no particular outliers; each value has low and high values, but those points do not deviate from the pattern of the rest. $r = 0.8782$.

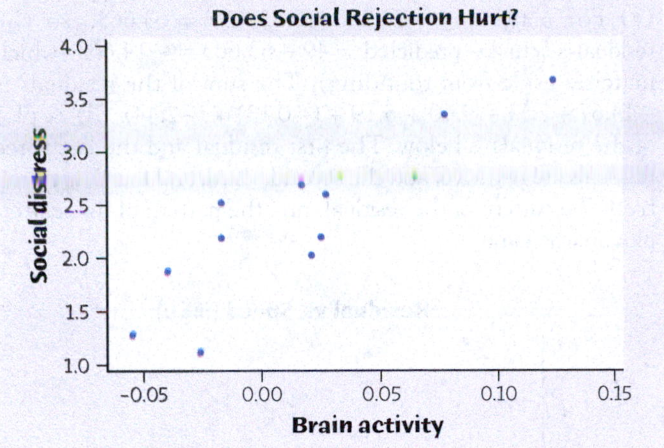

CONCLUDE: Social exclusion does appear to trigger a pain response: higher social distress measurements are associated with increased activity in the pain-sensing area of the brain. However, no cause-and-effect conclusion is possible because this was not a designed experiment.

4.49 PLAN: Salary is the explanatory variable, and SAT score is the response. Create a scatterplot and compute the correlation between the two variables. SOLVE: The scatterplot is shown. If there is a linear relationship, it is very weak. Most notable is that states with the highest average teacher salaries seem to have low SAT Math scores. $r = -0.266$.

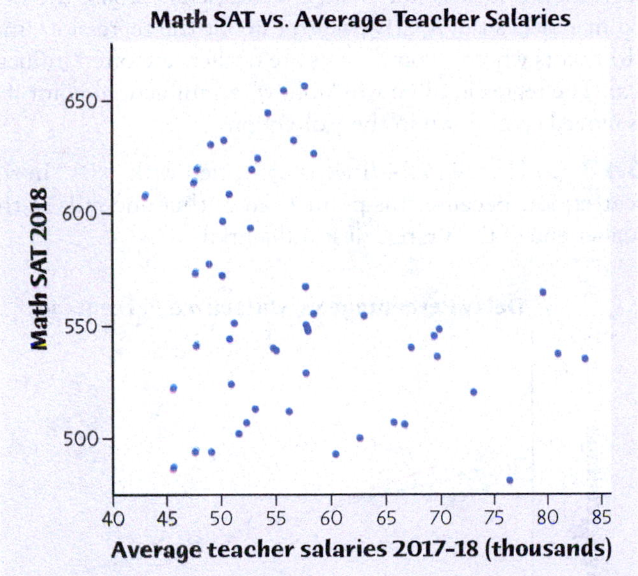

CONCLUDE: The relationship between teacher salaries and SAT Math scores (based on averages by state) is weakly linear and decreasing; these data do *not* support the idea that higher teacher salaries lead to greater student accomplishment (as measured by SAT Math scores).

Chapter 5 – Regression

5.1 **(a)** The slope is 0.914. On average, highway mileage increases by 0.914 mpg for each additional 1 mpg increase in city mileage. **(b)** The intercept is 8.720 mpg. This is the highway mileage for a nonexistent car that gets 0 mpg in the city. **(c)** For a car that gets 16 mpg in the city, we predict highway mileage to be $8.720 + (0.914)(16) = 23.344$ mpg. For a car that gets 28 mpg in the city, we predict highway mileage to be $8.720 + (0.914)(28) = 34.312$ mpg. **(d)** The scatterplot is shown. The regression line passes through both points of prediction from part (c).

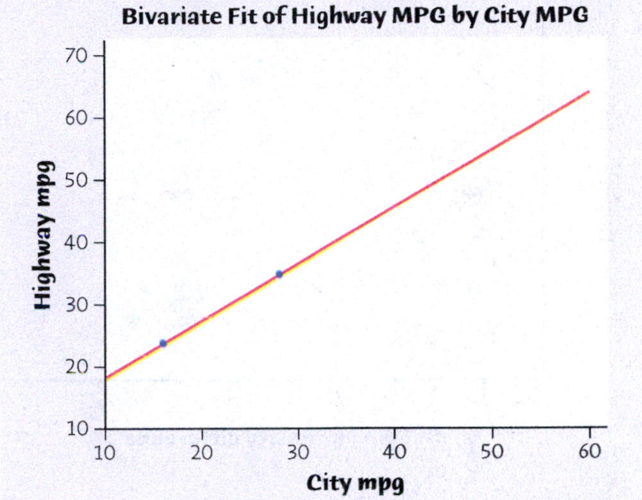

5.3 **(a)** The slope is 1021. That is, for each year since 2000, forest loss averages about 1021 km^2. **(b)** In square meters, the slope would be $1021 \times 10^6 = 1{,}021{,}000{,}000$, a loss of 1 billion square meters per year (on average). **(c)** In thousands of km^2, the slope would be 1.021, a loss of a bit more than 1000 km^2 per year (on average). Units matter in regression.

5.5 **(a)** The scatterplot (with the regression line) is shown.

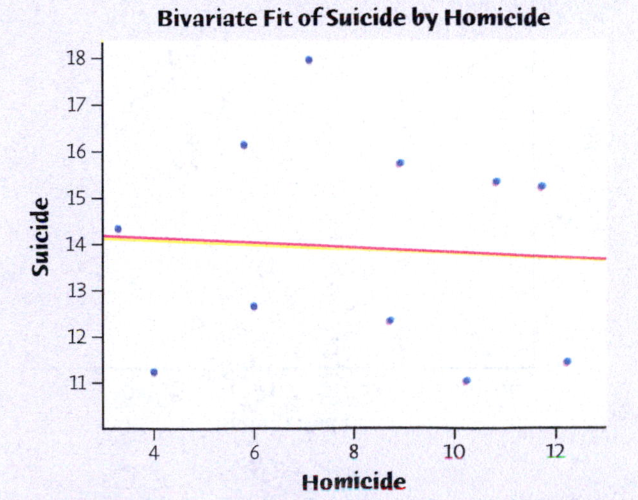

(b) From software, Suicide = 14.306 – 0.0492(Homicide). **(c)** The slope means that, for every additional homicide (per 100,000 people), there is an average decrease of 0.0492 suicide (per 100,000 people) in these Ohio counties. **(d)** The predicted suicide rate is 14.306 – (0.0492)(8.0) = 13.91 suicides (per 100,000 people).

5.7 (a) The scatterplot (with the regression line) is shown. The plot suggests a slightly curved pattern, not a strong linear pattern. A regression line is not useful for making predictions.

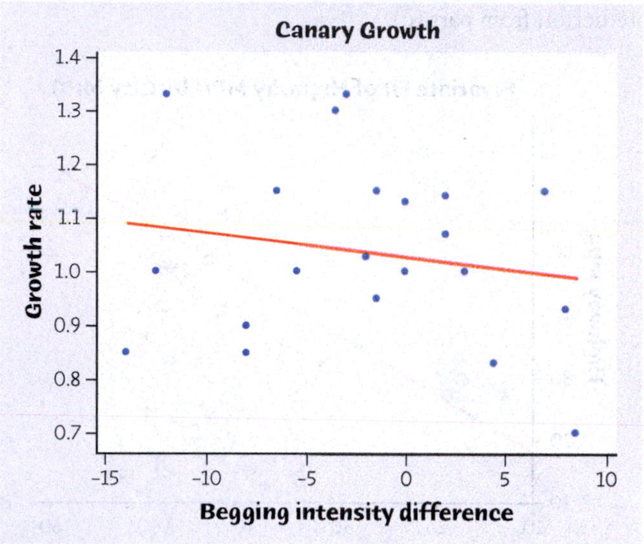

(b) $r^2 = 0.031$. This confirms what we see in the graph: the regression line does a poor job of summarizing the relationship. Only about 3% of the variation in growth rate is explained by the least-squares regression on difference in begging intensity.

5.9 (a) The scatterplot (with the regression line) is shown.

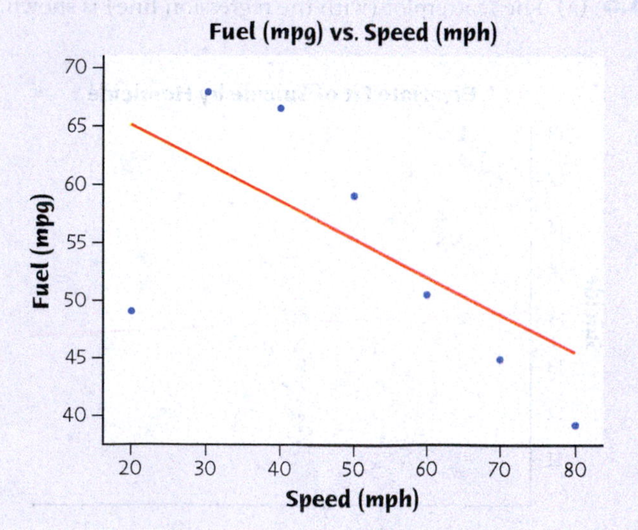

(b) The pattern is curved; linear regression is not appropriate. **(c)** For $x = 20$, $\hat{y} = 70.243 - 0.329(20) = 63.663$, so the residual is actual – predicted = 49 – 63.663 = –14.663 (which matches aside from rounding). The sum of the residuals is –14.67 + 7.51 + 9.40 + 5.19 + (–0.13) + (–2.44). **(d)** A plot of the residuals is below. The first residual and the last three residuals are negative, and the second, third, and fourth are positive. The pattern of the residuals and the pattern of the scatterplot are the same.

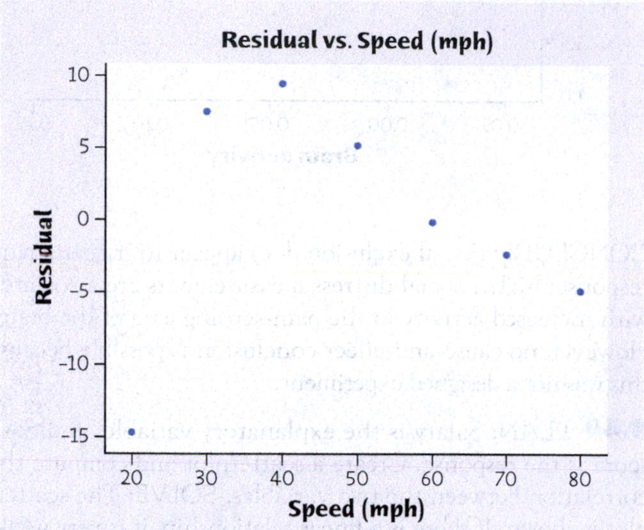

5.11 (a) Any point that falls exactly on the regression line will not increase the sum of squared vertical distances, so the regression line does not change. r increases because the new point reduces the relative scatter about the regression line. **(b)** Points whose x coordinates are outliers are often influential. The regression line will "follow" an influential point if it is moved up or down in the y direction.

5.13 (a) Hawaiian Airlines is identified with "HA" in the scatterplot. Because this point is an outlier and falls at the higher end of the x range, it is influential.

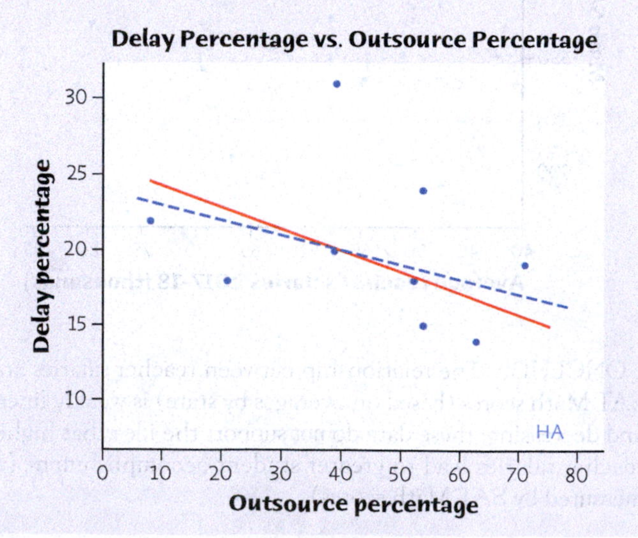

(b) With the outlier, $r = -0.4943$. If the outlier is deleted, $r = -0.2990$. The outlier is somewhat influential for correlation. (c) The two regression lines are plotted. The line based on the full data set has been pulled down toward the outlier, indicating that the outlier is influential. The regression line based on the complete data set is $\hat{y} = 25.7120 - 0.1433x$; when $x = 78.4$, we predict 14.48% delays. Omitting the outlier, $\hat{y} = 23.5150 - 0.0766x$, so our prediction would be 17.51% delays. The outlier influences prediction because it influences the regression line.

5.15 (a) $\hat{y} = -52.2321 + 0.1451x$ (or, *Predicted Kills* $= -52.2321 + 0.1451$ *Boats*). (b) If 950,000 boats are registered, $x = 950$, and $\hat{y} = -52.2321 + 0.1451(950) = 85.61$ manatees killed. The prediction seems reasonable, as long as conditions remain the same, because 950 is within the range of observed values of x. (c) If $x = 0$ (corresponding to no registered boats), then we would "predict" -52.2321 manatees to be killed by boats. This is absurd because it is clearly impossible. This illustrates the folly of extrapolation; $x = 0$ is well outside the range of observed values of x.

5.17 Possible lurking variables include the IQ and socioeconomic status of the mother. These variables are associated with smoking in various ways and are also predictive of a child's IQ.

5.19 One example might be that men who are married, widowed, or divorced may be more "invested" in their careers than men who are single. There may still be a feeling of societal pressure for a man to "provide" for his family.

5.21 (c)

5.23 (a)

5.25 (c)

5.27 (a)

5.29 (a)

5.31 (a) The least-squares regression line says that increasing the size of a diamond by 1 carat increases its price by \$11,975.14, on average. (b) A diamond of size 0 carats would have a predicted price of $-$\$6047.75. This is extrapolation because the data set on which the line was constructed almost certainly had no rings with diamonds of size 0 carats. (And it does not make sense that a retailer would pay you to buy a ring.)

5.33 (a) $\hat{y} = 0.919 + 2.0647x$. For every degree Celsius, the toucan will lose about 2.06% more heat through its beak, on average. (b) $\hat{y} = 0.919 + 2.0647(25) = 52.5$. At a temperature of 25 degrees Celsius, we predict a toucan to lose 52.5% more heat through its beak, on average. (c) $r^2 = 83.6$. (d) $r = \sqrt{r^2} = \sqrt{0.836} = 0.914$. Correlation is positive because the regression line has a positive slope.

5.35 (a) $b = r\frac{s_y}{s_x} = 0.5\left(\frac{8}{40}\right) = 0.1$, and $a = \bar{y} - b\bar{x} = 75 - (0.1)(280) = 47$. The regression equation is $\hat{y} = 47 + 0.1x$. Each point of pre-exam total score means an additional 0.1 point on the final exam, on average. (b) $\hat{y} = 47 + 0.1(300) = 77$. (c) With $r = 0.5$, $r^2 = (0.5)^2 = 0.25$, so the regression line accounts for only 25% of the variability in student final exam scores; the regression line doesn't predict final exam scores very well.

5.37 (a) $\hat{y} = 28.037 + 0.521x$. $r = 0.555$. The plot is provided below.

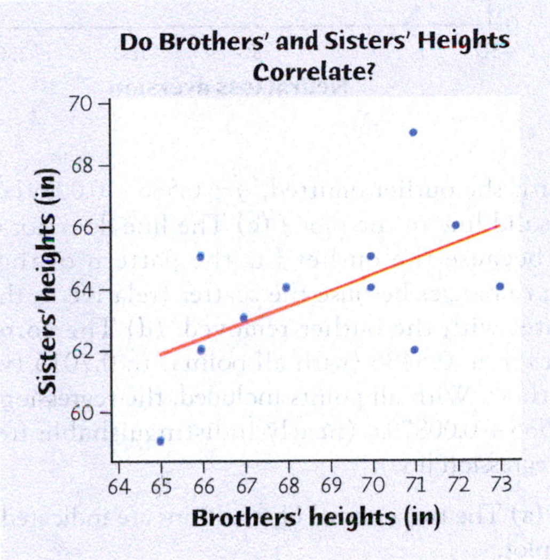

(b) $\hat{y} = 28.037 + 0.521(70) = 64.5$ inches (rounded). This prediction isn't expected to be very accurate because the correlation isn't very large; $r^2 = (0.555)^2 = 0.308$.

5.39 (a) $\hat{y} = 31.934 - 0.304x$.

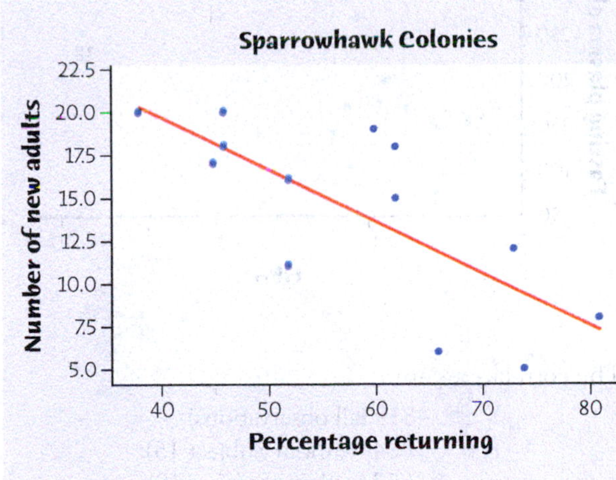

(b) On the average, for each additional 1% increase in returning birds, the number of new birds joining the colony decreases by 0.304. (c) When $x = 60$, we predict $\hat{y} = 31.934 - 0.304(60) = 13.69$ new birds.

5.41 (a) The outlier is in the upper-right corner.

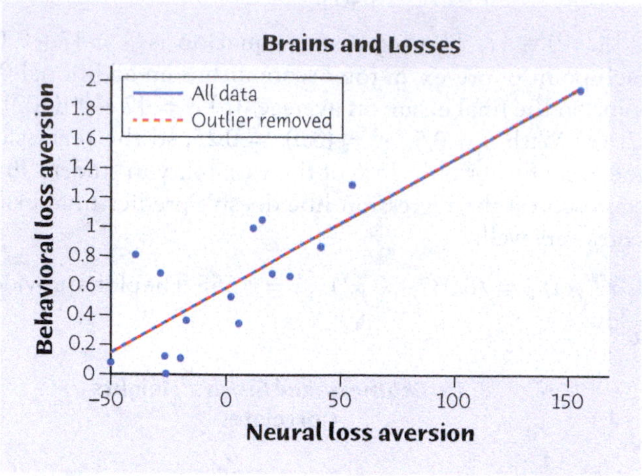

(b) With the outlier omitted, $\hat{y} = 0.586 + 0.00891x$. (This is the solid line in the plot.) (c) The line does not change much because the outlier fits the pattern of the other points; r changes because the scatter (relative to the line) is greater with the outlier removed. (d) The correlation changes from 0.8486 (with all points) to 0.7015 (without the outlier). With all points included, the regression line is $\hat{y} = 0.585 + 0.00879x$ (nearly indistinguishable from the other regression line).

5.43 (a) The two unusual observations are indicated on the scatterplot.

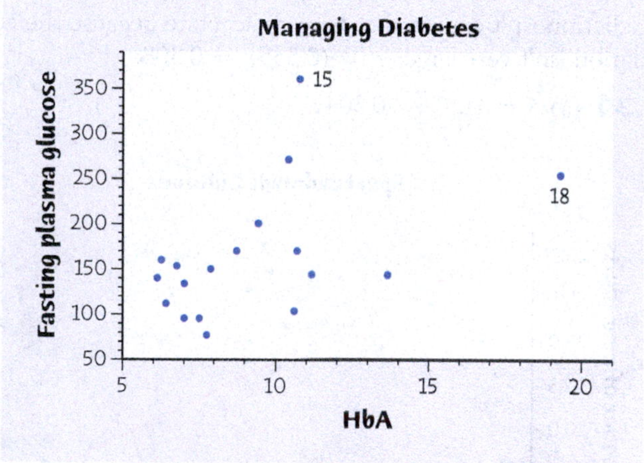

(b) The correlations are

$r_1 = 0.4819$ (all observations)
$r_2 = 0.5684$ (without Subject 15)
$r_3 = 0.3837$ (without Subject 18)

Both outliers change the correlation. Removing Subject 15 increases r because its presence makes the scatterplot less linear. Removing Subject 18 decreases r because its presence decreases the relative scatter about the linear pattern.

5.45 The scatterplot with regression lines added is given.

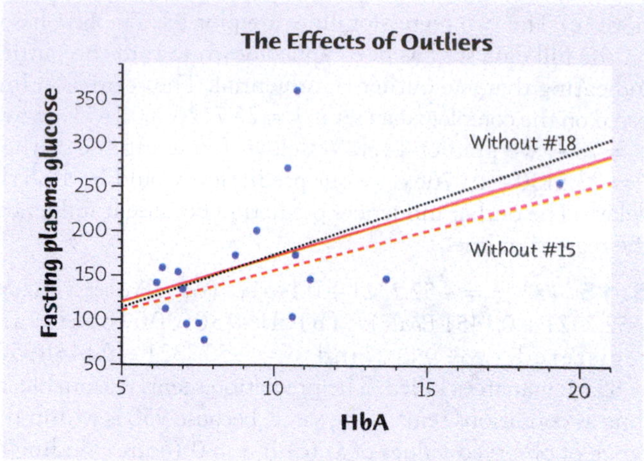

The equations are

$\hat{y} = 66.4 + 10.4x$ (all observations)
$\hat{y} = 69.5 + 8.92x$ (without Subject 15)
$\hat{y} = 52.3 + 12.1x$ (without Subject 18)

While the equation changes in response to removing either subject, one could argue that neither one is particularly influential because the line moves very little over the range of x (HbA) values. Subject 15 is an outlier in terms of its y value; such points are typically not influential. Subject 18 is an outlier in terms of its x value, but it is not particularly influential because it is consistent with the linear pattern suggested by the other points.

5.47 The correlation would be smaller. Individual income will vary much more than the average income for a given age.

5.49 Responses will vary. For example, students who choose the online course might have more self-motivation or might have better computer skills.

5.51 (a) $\hat{y} = 495.524 + 0.0005x$. Increasing the average teacher salary by \$1 increases the mean Math SAT score by 0.0005 point, on average. (b) $\hat{y} = 384.233 + 0.0041x$. Increasing the average teacher salary by \$1 increases the mean Math SAT score by 0.0041 point, on average. (c) The slopes here have opposite signs from the slope found in Exercise 5.50. Consideration of a third (lurking) variable changed the relationship.

5.53 For example, a student who, in the past, might have received a grade of B (and a lower SAT score) now receives an A (but has a lower SAT score than an A student in the past). While this is a bit of an oversimplification, this means that today's A students are yesterday's A and B students, today's B students are yesterday's C students, and so on. Because of the grade inflation, we are not comparing students with equal abilities in the past and today.

5.55 We have slope $b = r\frac{s_y}{s_x}$, intercept $a = \bar{y} - b\bar{x}$, and $\hat{y} = a + bx$. When $x = \bar{x}$, $\hat{y} = a + b\bar{x} = (\bar{y} - b\bar{x}) + b\bar{x} = \bar{y}$.

5.57 Note that $\bar{y} = 46.6 + 0.41\bar{x}$. We predict that Octavio will score 4.1 points above the mean on the final exam: $\bar{y} = 46.6 + 0.41(\bar{x} + 10) = 46.6 + 0.41\bar{x} + 4.1 = \bar{y} + 4.1$. (Alternatively, because the slope is 0.41, we can observe that an increase of 10 points on the midterm yields an average of 4.1 on the predicted final exam score.)

5.59 **(a)** Drawing the "best line" by eye is an inaccurate process; few people choose the best line. **(b)** Most people tend to overestimate the slope for a scatterplot with $r = 0.7$; that is, the least-squares line usually is less steep than an estimated line drawn through the center of the points.

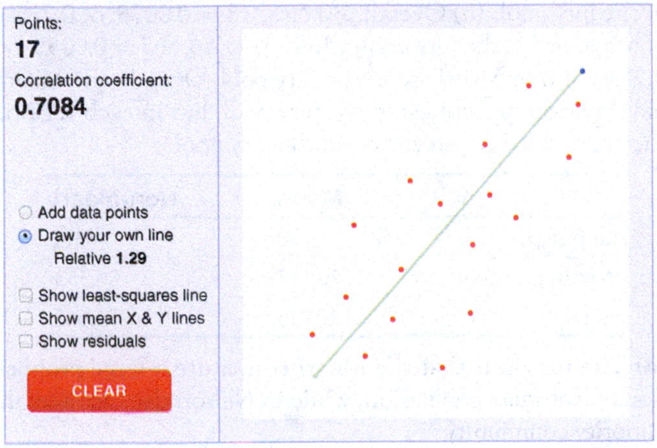

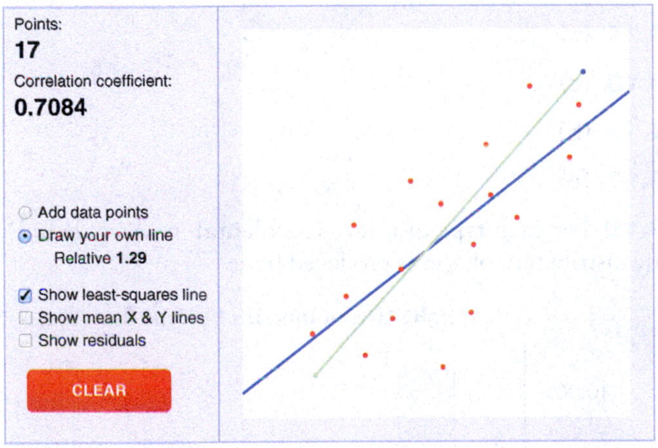

5.61 PLAN: We construct a scatterplot, with distance as the explanatory variable, using different symbols for the left and right hands, and (if appropriate) find separate regression lines for each hand. SOLVE: In the scatterplot, right-hand points are red dots, and left-hand points are blue. In general, the right-hand points lie below the left-hand points, meaning the right-hand times are shorter, so the subject is likely right-handed. There is no striking pattern for the left-hand points; the patterns for right-hand points is obscured because they are "squeezed" at the bottom of the plot. While neither relationship looks particularly linear, we might nonetheless find the two regression lines: for the right hand, $\hat{y} = 99.364 + 0.0283x$ $(r = 0.305,\ r^2 = 9.3\%)$, and, for the left hand, $\hat{y} = 171.5 + 0.2619x$ $(r = 0.318,\ r^2 = 10.1\%)$. CONCLUDE: Neither regression is particularly useful for prediction. Distance accounts for only 9.3% (right) and 10.1% (left) of the variation in time.

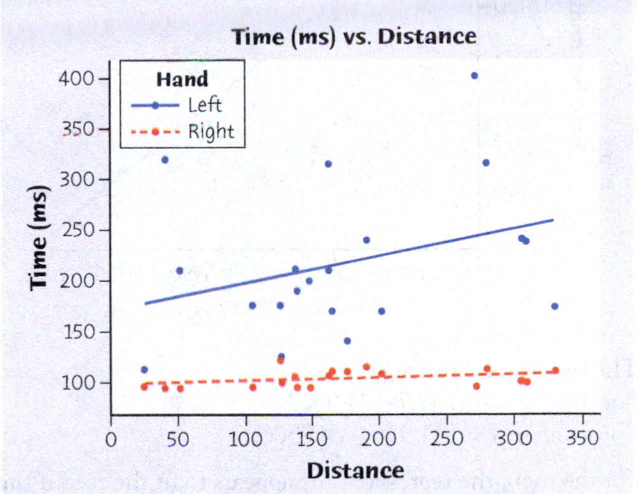

5.63 PLAN: We plot the data, producing a time-series plot. If appropriate, we consider fitting a regression line. SOLVE: The plot is provided. We see that during the recent 15 to 20 years, the volume of discharge has become highly variable, but, before then, the rate increased slowly.

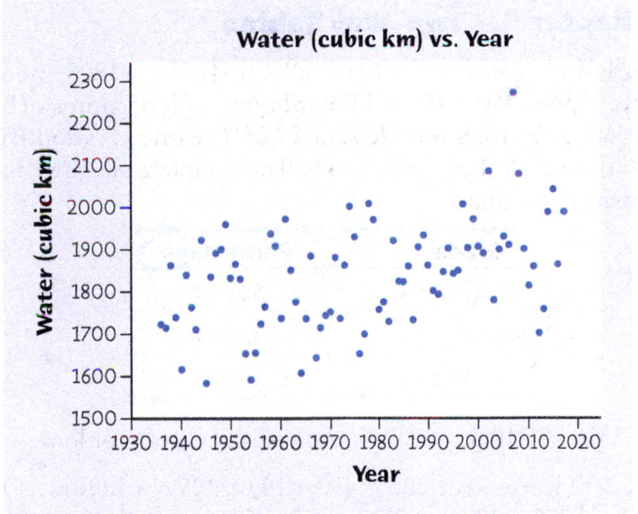

CONCLUDE: If there is a relationship between year and discharge, it isn't strongly linear, and use of a regression line would not be useful to predict discharge from year to year.

5.65 PLAN: We plot marathon times by year for each sex, using different symbols. If appropriate, we fit least-squares regression lines for predicting time from year for each sex. We then use these lines to guess when the times will agree. SOLVE: The scatterplot is provided on the next page, with regression lines plotted.

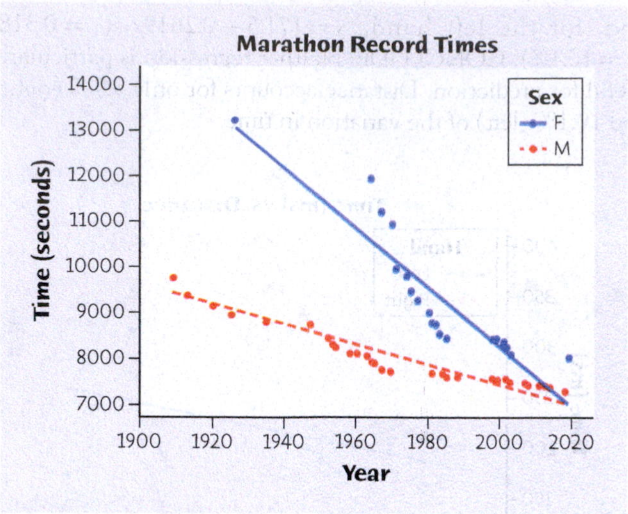

The regression lines are

For men: $\hat{y} = 51{,}247.7 - 21.9013x$

For women: $\hat{y} = 137{,}936 - 64.8306x$

For the men, the regression line appears to fit the record times reasonably well. For the women, the regression line would fit better if we omitted the outlier associated with year 1926 and if the relationship between record time and year were more linear (it appears to be curved). CONCLUDE: Using the regression lines plotted, we might expect women to "outrun" men by the year 2019. Since the record time for women in 2019 was higher than the record time for men in 2019, this did not happen.

Chapter 6 – Two-Way Tables

6.1 (a) $736 + 450 + 193 + 205 + 144 + 80 = 1808$ people. $736 + 450 + 193 = 1379$ played video games. (b) $(736 + 205)/1808 = 0.5205$, or 52.05%, earned A's and B's. Do this for all three grade levels. The complete marginal distribution for grades is

Grade	Percentage
A's and B's	52.05%
C's	32.85%
D's and F's	15.10%

$32.85\% + 15.10\% = 47.95\%$ received a grade of C or lower.

6.3 There are $736 + 450 + 193 = 1379$ players. Of these, $736/1379 = 0.5337$, or 53.37%, earned A's or B's. There are $205 + 144 + 80 = 429$ nonplayers. Of these, $205/429 = 0.4779$, or 47.79%, earned A's or B's. Continuing in like manner, the conditional distributions of grades follows:

Grades	Players	Nonplayers
A's and B's	53.37%	47.79%
C's	32.63%	33.57%
D's and F's	14.00%	18.65%

If anything, players have slightly higher grades than nonplayers, but this could be due to chance.

6.5 Two examples are shown. In general, choose a to be any number from 0 to 50, and then all of the other entries can be determined.

30	20
30	20

50	0
10	40

6.7 (a) For Rotorua district, $79/8889 = 0.0089$, or 0.9%, of Maori are in the jury pool, while $258/24{,}009 = 0.0107$, or 1.07%, of the non-Maori are in the jury pool. For Nelson district, the corresponding percentages are 0.08% for Maori and 0.17% for non-Maori. In each district, the percentage of non-Maori in the jury pool exceeds the percentage of Maori in the jury pool. (b) Overall, $80/10{,}218 = 0.0078$, or 0.78%, of Maori are in the jury pool, while $314/56{,}667 = 0.0055$, or 0.55%, of non-Maori are in the jury pool. Overall, the Maori have a larger percentage in the jury pool, but in each region, they have a lower percentage in the jury pool.

	Maori	Non-Maori
In jury pool	80	314
Not in jury pool	10,138	56,353
Total	10,218	56,667

(c) The reason is that the Maori constitute a large proportion of Rotorua's population, while in Nelson they are a small minority community.

6.9 (b)

6.11 (a)

6.13 (c)

6.15 (b)

6.17 (b)

6.19 For each type of injury (accidental, not accidental), the distribution of ages is produced here:

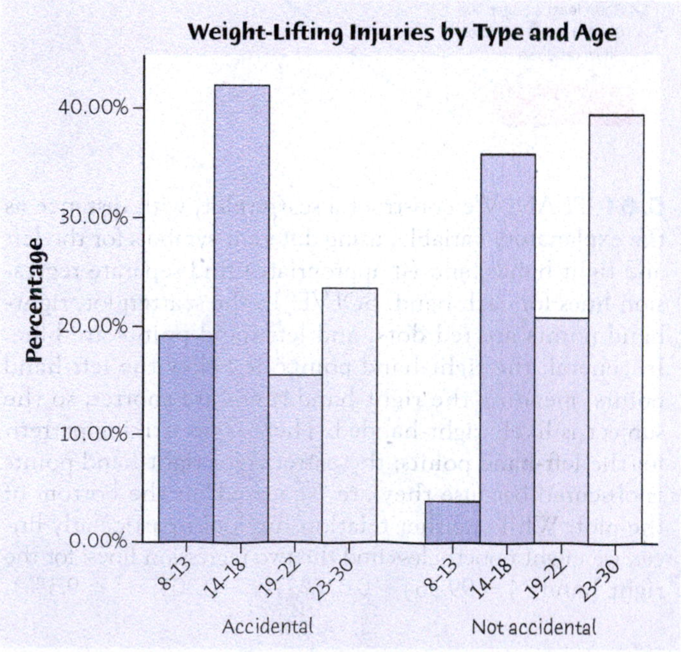

	Accidental	Not Accidental
8–13	19.0%	4.0%
14–18	42.2%	35.8%
19–22	15.4%	20.8%
23–30	23.4%	39.4%

Among accidental weight-lifting injuries, the percentage of younger lifters is larger, while among the injuries that are not accidental, the percentage of older lifters is larger.

6.21 The percentage of single men with no income is $5514/39{,}776 = 0.1386$, or 13.86%. The percentage of men with no income who are single is $5514/8135 = 0.6778$, or 67.78%.

6.23 (a) We need to compute percentages rather than rely on counts because more married men than single men were counted in the census. (b) A table is provided below. Divorced and widowed men had similar percentages between the two income groups studied. Single men had much higher percentages of no income; married men had much higher percentages of incomes at and over $100,000.

	Single	Married	Divorced	Widowed
No income	67.78%	23.76%	6.79%	1.67%
$100,000 and up	11.57%	79.85%	7.11%	1.48%

6.25 (a) The two-way table of race (White, Black) versus death penalty (death penalty, no death penalty) follows:

	White Defendant	Black Defendant
Death penalty	19	17
No death penalty	141	149

(b) For Black victims, percentage of White defendants given the death penalty is $0/9 = 0$, or 0%, and the percentage of Black defendants given the death penalty is $6/103 = 0.058$, or 5.8%. For White victims, White defendants were given the death penalty in $19/151 = 0.126$, or 12.6%, of cases, and Black defendants were given the death penalty in $11/63 = 0.175$, or 17.5%, of cases. For both Black and White victims, Black defendants were given the death penalty relatively more often than White defendants. However, overall, $19/160 = 0.119$, or 11.9%, of White defendants got the death penalty, while $17/166 = 0.102$, or 10.2%, of Black defendants got the death penalty. (c) For White defendants, $(19 + 132)/(19 + 132 + 0 + 9) = 0.9438$, or 94.4%, of victims were White. For Black defendants, $(11 + 52)/(11 + 52 + 6 + 97) = 0.3795$, or 37.95%, of victims were White. The death penalty was predominantly assigned to cases involving White victims: 14.0% of all cases with a White victim and only 5.4% of all cases with a Black victim had a death penalty assigned. Because most White defendants' victims were White and cases with White victims carried additional risk of a death penalty, White defendants were assigned the death penalty more often overall.

6.27 PLAN: Find and compare the conditional distributions of outcome for each treatment. SOLVE: The percentages for each column are provided. For example, for Chantix, the percentage of successes is $155/(155 + 197) = 0.4403$, or 44.0%.

	Chantix	Bupropion	Placebo
Percentage no smoking in weeks 9–12	44.0%	29.5%	17.7%

CONCLUDE: Clearly, a larger percentage of subjects using Chantix were not smoking during weeks 9–12, compared with results for either of the other treatments.

6.29 PLAN: Calculate and compare the conditional distributions of sex for each degree level. SOLVE: For example, $644/(644 + 405) = 0.6139$, or 61.39%, of associate's degrees went to women. The table shows the percentage of women at each degree level, which is all we need.

Degree	% Female
Associate's	61.39%
Bachelor's	57.53%
Master's	58.23%
Professional or doctorate	52.66%

CONCLUDE: Women constitute a substantial majority of associate's, bachelor's, and master's degrees and a small majority of doctorate and professional degrees.

6.31 PLAN: Find and compare the conditional distributions for health for smokers and for nonsmokers. SOLVE: The table provides the percentage of subjects with various health outlooks for each group. CONCLUDE: Clearly, the outlooks of current smokers are generally bleaker than those of current nonsmokers. Much larger percentages of nonsmokers reported being in "excellent" or "very good" health, while much larger percentages of smokers reported being in "fair" or "poor" health.

	Health Outlook				
	Excellent	Very Good	Good	Fair	Poor
Current smoker	6.2%	28.5%	35.9%	22.3%	7.2%
Current nonsmoker	12.4%	39.9%	33.5%	14.0%	0.3%

6.33 PLAN: Because the numbers of students who use (or do not use) medications are different, we find the conditional distributions of those who do and do not use medications. SOLVE: The table provides the percentages of subjects with various levels of sleep quality for the drug use groups. CONCLUDE: Those who use medications are less likely to have optimal sleep. This is a case where we cannot ascribe causation: We don't know whether those who use medications to stay awake have poor sleep quality because they use the medication or because they had poor sleep quality before using them.

	Sleep Quality		
	Optimal	Borderline	Poor
Use medications	21.3%	30.5%	48.3%
Do not use medications	38.2%	26.7%	35.2%

Chapter 7 – Exploring Data: Part I Review

7.1 (c)

7.3 (d)

7.5 (b)

7.7 (c)

7.9 (c)

7.11 (a) grams. (b) centimeters. (c) centimeters. (d) grams^2.

7.13 (c)

7.15 (a)

7.17 (a) $40 < x < 50$ corresponds to $\frac{40-44.8}{2.1} < z < \frac{50-44.8}{2.1}$, or $-2.29 < z < 2.48$. This proportion is $0.9934 - 0.0110 = 0.9824$, or 98.24%. (b) Approximately 95% of all values are within 2σ of μ in a Normal distribution; this becomes $44.8 - 2(2.1) = 40.6$ inches to $44.8 + 2(2.1) = 49.0$ inches. (c) The 70th percentile corresponds to $z = 0.52$ because the area to the left of 0.52 under the standard Normal curve is 0.6985. Then $x = 44.8 + 0.52(2.1) = 45.892$ inches.

7.19 About 3.76%. Slots meeting specifications correspond to $0.8725 < x < 0.8775$, which for the $N(0.8750, 0.0012)$ distribution corresponds to $-2.08 < z < 2.08$, for which Table A gives $0.9812 - 0.0188 = 0.9624$. The proportion of slots that do not meet these specifications is $1 - 0.9624 = 0.0376$.

7.21 (a) Minimum = 7.2, $Q_1 = 8.5$, $M = 9.3$, $Q_3 = 10.9$, Maximum = 12.8. (b) $M = 27$. (c) 25% of values exceed $Q_3 = 30$. (d) Yes. All Torrey pine needles are longer than all Aleppo pine needles.

7.23 (a)

7.25 (c)

7.27 (d)

7.29 (d)

7.31 (d)

7.33 (c)

7.35 (b)

7.37 (a) No. (b) $r^2 = 0.64$, or 64%.

7.39 (a) 8.683 kg. (b) 10.517 kg. (c) Such a comparison is unreasonable because the lean group has less lean body mass, and therefore would be expected to burn less energy on average. (d) See the scatterplot below.

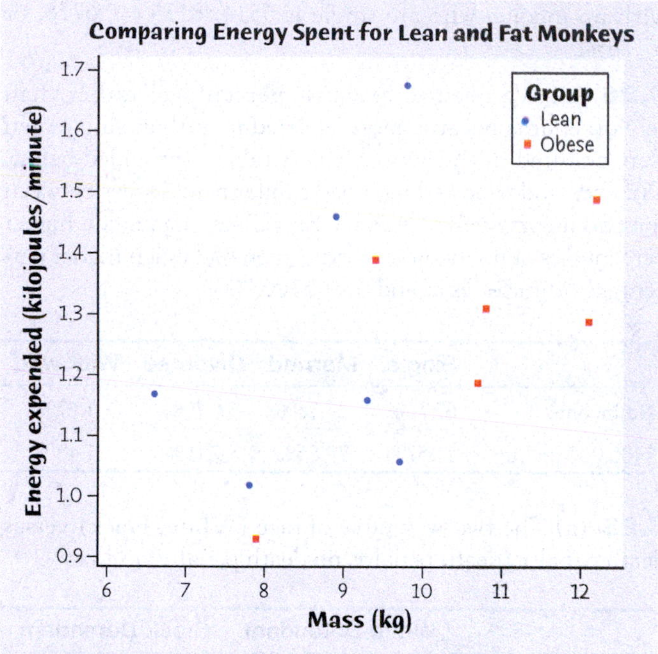

(e) It appears that the rate of increase in energy burned per kilogram of mass is about the same for both groups. The obese monkeys burn less energy than the lean monkeys because their points tend to be below the others.

7.41 (a) $190/8474 = 0.0224$, or 2.24%. (b) $633/8474 = 0.0747$, or 7.47%. (c) $27/633 = 0.0427$, or 4.27%. (d) $4621/8284 = 0.5578$, or 55.78%. (e) The conditional distribution of CHD for each level of anger is tabulated below.

Low anger	Moderate anger	High anger
1.70%	2.33%	4.27%

The result for the high anger group was computed in part (c). Clearly, angrier people are at greater risk of CHD.

7.43 PLAN: Make a time plot to show how the size of the ozone hole changed between 1979 and 2019. SOLVE: The time plot is provided below.

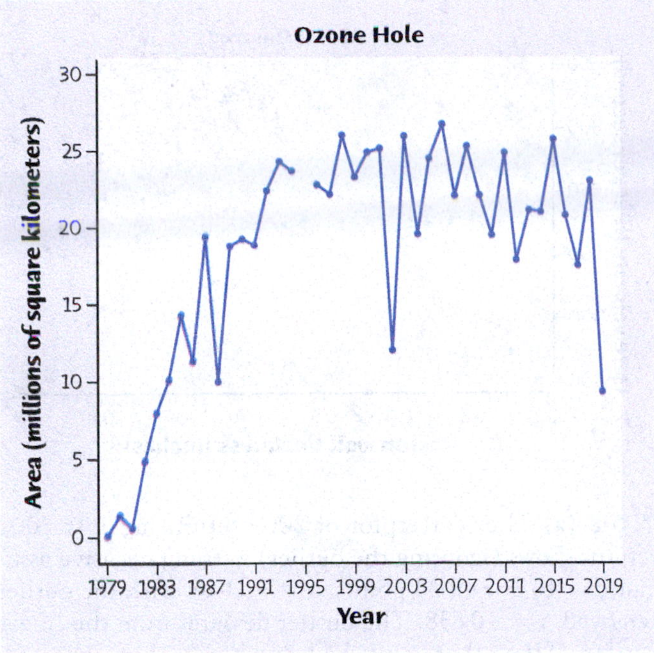

In addition to year-to-year variation, the time plot shows two distinct trends between 1979 and 2019. From 1979 to the mid-1990s, there is a very strong, positive linear relationship between year and ozone hole area. The slope of a regression line fit to this portion of the data would be quite large. From 1993 to 2019, the relationship between year and ozone hole area is quite different; here the linear relationship is much weaker and is negative. In addition, the data points for 2002 and 2019 are low outliers. No cyclical fluctuation is present.

7.45 (a) The plot is provided below.

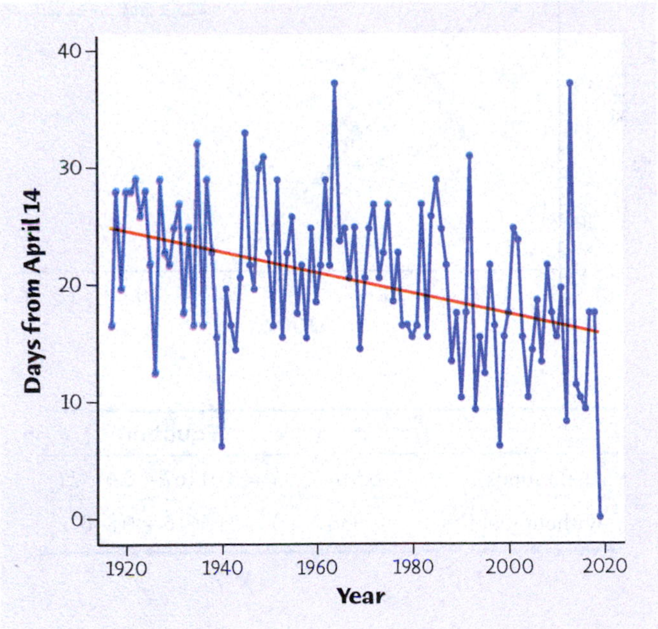

(b) $\hat{y} = 189.985 - 0.0860x$. The slope is negative, suggesting that the ice breakup day is occurring earlier (by about 0.086 day per year). (c) The regression line is not very useful for prediction because it accounts for only about 15.4% ($r^2 = 0.1538$) of the variation in the ice breakup date.

7.47 PLAN: Create side-by-side boxplots to compare the distributions for countries identified as "developing" and "developed" and compute appropriate summary statistics. SOLVE: Both groups (developing countries and developed countries) have right-skewed distribution. Developing countries have more outliers than developed countries, but the most unpaid tickets belong to a developed country (Kuwait). Because of the outliers, the five-number summaries are appropriate.

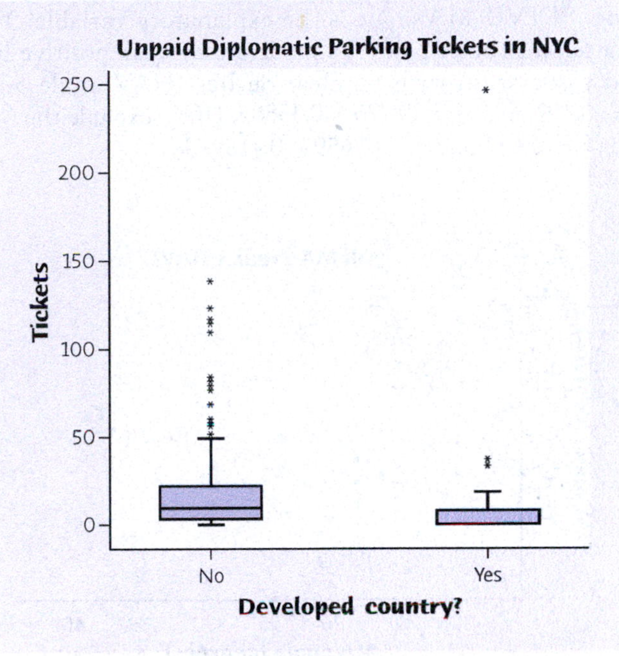

	Min	Q_1	M	Q_3	Max
Developed	0	0	0.7	8.13	246.2
Not developed	0	3.2	9.5	22.8	139.6

CONCLUDE: Comparing the distribution, developing countries' diplomats tend to have more unpaid tickets. National income alone does not explain countries whose diplomats have more or fewer unpaid tickets; the country with the largest number of unpaid tickets is classified as "developed," but it is an Arab emirate; perhaps the culture there has an impact.

7.49 PLAN: Graph the data and compute appropriate numerical summaries. SOLVE: A stemplot is given below:

1	34
1	66788
2	000111123
2	55556666888
3	00012224
3	88
4	
4	
5	0

The distribution seems fairly Normal, apart from a high outlier of 50°. The five-number summary is preferred because of the outlier: Minimum = 13°, Q_1 = 20°, M = 25°, Q_3 = 30°, Maximum = 50°. If we can discard the outlier as an error, we can use the mean and standard deviation to describe the center spread: $\bar{x} = 24.8°$, $s = 6.3°$. CONCLUDE: Descriptions of the distribution will vary. Most patients have a deformity angle in the range of 15° to 35°.

7.51 PLAN: We examine the relationship with a scatterplot and (if appropriate) correlation and the regression line. SOLVE: MA angle is the explanatory variable. The scatterplot (below) shows a moderate to weak positive linear association with on clear outlier (HAV angle 50°). $r = 0.302$, and $\hat{y} = 19.723 + 0.3388x$. (If we exclude the outlier, $r = 0.443$ and $\hat{y} = 17.659 + 0.4189x$.)

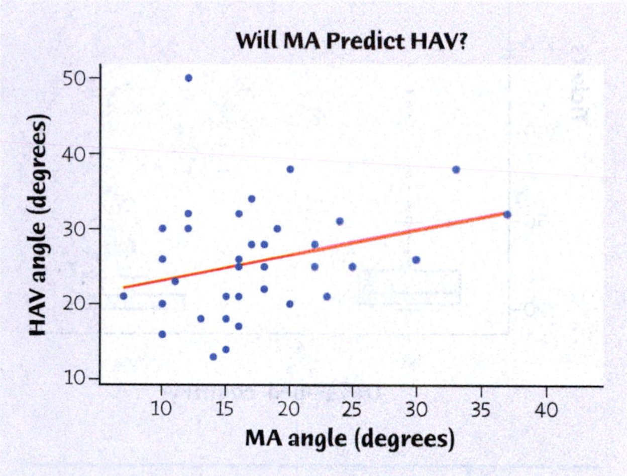

CONCLUDE: MA angle can be used to give estimates of HAV angle, but the variability is so large that estimates would not be very reliable. The linear relationship explains only $r^2 = 9.1\%$ of the variation in HAV angle with the entire data set (only $r^2 = 19.6\%$ of the variation in HAV angle with the outlier removed).

7.53 PLAN: We will examine the relationship with a scatterplot and (if appropriate) correlation and regression line. SOLVE: The scatterplot, shown with the line $\hat{y} = 70.44 + 274.78x$, shows a moderate, positive linear relationship. The linear relationship explains about $r^2 = 49.3\%$ of the variation in gate velocity. CONCLUDE: The regression formula might be used as a rule of thumb for new workers to follow, but the large variability in the scatterplot suggests that there may be other factors that should be taken into account.

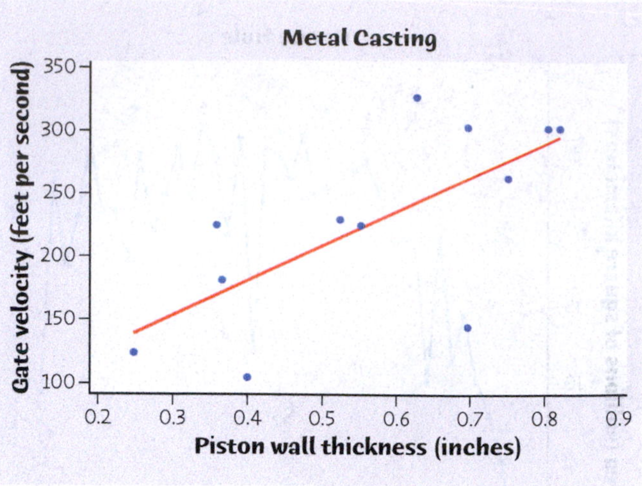

7.55 (a) The scatterplot of 2003 returns against 2002 returns shows (ignoring the outlier) a strong negative association. (b) For all 23 points, $r = -0.616$; with the outlier removed, $r = -0.838$. The outlier deviates from the linear pattern of the other points by being higher than expected; removing it makes the negative association stronger, and so r moves closer to -1. (c) Regression formulas are given below. The dashed regression line for 22 funds other than Gold must pivot up toward Fidelity Gold in order to minimize the sum of squares for all 23 deviations. Fidelity Gold is very influential.

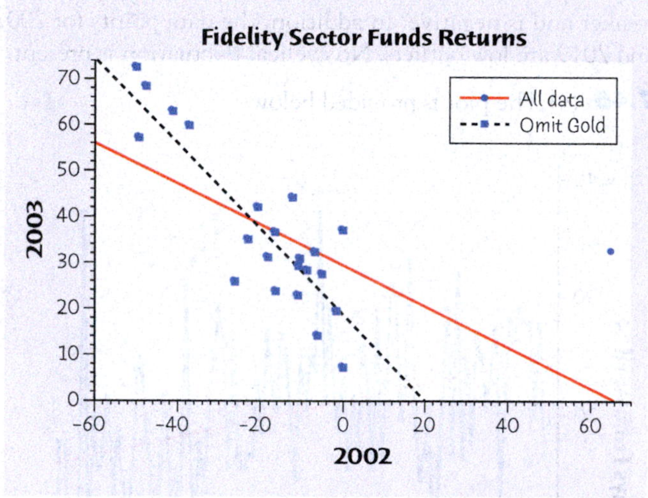

	r	Equation
All 23 funds	−0.616	$\hat{y} = 31.1167 - 0.4132x$
Without Gold	−0.838	$\hat{y} = 21.4616 - 0.8403x$

7.57 (a) Fish catch is the explanatory variable. The point for 1999 is at the bottom of the plot.

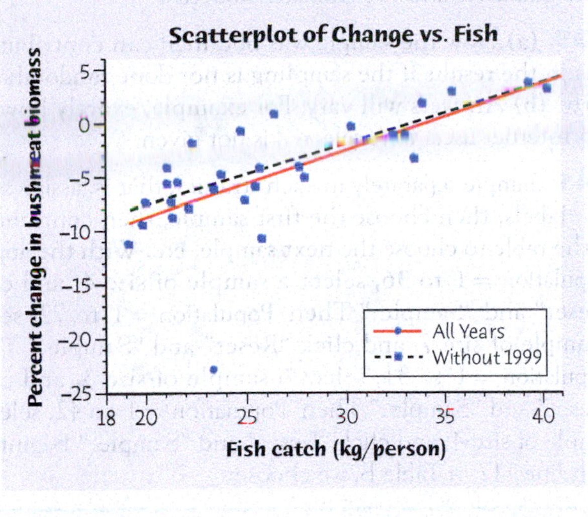

(b) For all 41 points, $r = 0.672$; with the point for 1999 removed, $r = 0.804$. The outlier decreases r because it weakens the strength of the linear association. (c) Regression formulas are given below. The effect of the outlier on the line is small: the tight clustering of the other points weakens the role of the outlier in altering the regression line. Also, 1999 was not particularly extreme in the amount of fish caught, so the regression line does not need to pivot down toward the point for 1999 in order to minimize the sums of squares for all 41 observations.

	r	Equation
All points	0.672	$\hat{y} = -21.09 + 0.6345x$
Without 1999	0.804	$\hat{y} = -19.05 + 0.5788x$

7.59 (a) There are two somewhat low IQs—72 qualifies as an outlier by the 1.5 × *IQR* rule, while 74 is on the boundary. However, for a small sample, this stemplot looks reasonably Normal.

7	24
7	
8	
8	69
9	13
9	68
10	023334
10	578
11	11222444
11	89
12	0
12	8
13	02

(b) We compute $\bar{x} = 105.84$ and $s = 14.27$ and find: $23/31 = 0.7419$, or about 74.2%, within $\bar{x} \pm 1s$, or 91.6 to 120.1, and $29/31 = 0.9355$, or about 93.6%, within $\bar{x} \pm 2s$, or 77.3 to 134.4.

For an exactly Normal distribution, we would expect these proportions to be 68% and 95%. Given the small sample, this is reasonably close agreement.

Chapter 8 – Producing Data: Sampling

8.1 (a) The population is (all) students at the university who live in off-campus housing. (b) The sample is the 78 students who live in off-campus housing and returned the questionnaire.

8.3 (a) The population is all users of the software. Unless the company's market is primarily educational, the 1100 individuals (who are mostly faculty and who received the software free of charge) will not represent the population. (b) The sample is the 186 people who completed the survey.

8.5 This is a biased sampling method. Those who take the time to write an online review are likely to do so because they are upset with the service they received. The direction of the bias is likely to overestimate the proportion of customers who have a negative opinion of the service.

8.7 Number from 01 to 24 alphabetically (down the columns). With the applet: Population = 1 to 24, select a sample of size 3, and click "Reset" and "Sample." With Table B, enter at line 127 and choose 06 = Deis, 08 = Fernandez, and 11 = Gemayel.

8.9 With the election close at hand, the polling organization wants to increase the accuracy of its results. Larger samples provide better information about the population.

8.11 Label the suburban townships from 01 to 30 alphabetically (down the columns). With the applet: Population = 1 to 30, select a sample of size 4, and click "Reset" and "Sample." With Table B, enter at line 118 and choose 19 = Orland, 03 = Bloom, 25 = Riverside, and 04 = Bremen. Next, label the Chicago townships from 1 to 8 alphabetically (down the columns). With the applet: Population = 1 to 8, select a sample of size 2, and click "Reset" and "Sample." With Table B, enter at line 127 and choose 4 = Lake View and 3 = Lake.

8.13 (a) The population is all physicians practicing in the United States. The sample size is $n = 2379$. If the 2379 were randomly selected, we could draw conclusions. However, they were not randomly selected, and there was too much nonresponse in this voluntary response sample. (b) The nonresponse rate is $\frac{100{,}000 - 2379}{100{,}000} = 0.9762$, or 97.62%. We don't know the attitudes of the nonrespondents about health care reform, so the results may not be credible. (c) They only received 2379 respondents.

8.15 (a) Inference from voluntary response samples is misleading because the method of choosing the sample is biased. In particular, we need to consider how the more than 2 million people who take SurveyMonkey surveys differ from the rest of the U.S. adults who do not take surveys on SurveyMonkey. (b) Individuals who do not have regular access to the Internet are likely to be missed in this survey. (In addition, the reason that they do not have regular Internet access is likely related to their financial situation.)

8.17 (a)

8.19 (c)

8.21 (b)

8.23 (c)

8.25 (b)

8.27 (a) The population is all Hispanic residents of Denver. The sample is the 200 adults who answer the questions at the selected mailing addresses. (b) This survey may suffer from response bias because the officer doing the questioning is also Hispanic or because the sample excludes Hispanic people who do not live in Hispanic neighborhoods.

8.29 (a) Population = residents of Greenville, South Carolina. Response rate $= 726/2461 = 0.2950$, or 29.50%. (b) $436/689 = 0.6328$ or 63.28% of the sample is between 18 and 64. $356{,}123/461{,}299 = 0.7720$, or 77.2% of the population is between 18 and 64. This is not surprising because cell phone numbers were not included in the sample, and younger adults are more likely to have only a cell phone and no landline. (c) The bias may underestimate because younger adults, who are more likely to use the trail, are underrepresented in the survey.

8.31 Question A drew the 48%. The "too high" option in Question B may have prompted respondents with an option that more clearly reflected their views.

8.33 The questions are worded very differently. The Second Amendment in the U.S. Constitution allows guns, so some respondents might feel that the first question, which specifically mentions a "law," challenges the Second Amendment. The second question addresses a "ban," which some respondents might (even unconsiously) interpret as being different from a new law.

8.35 (a) Assign labels 0001 through 5024. With the applet: Population = 1 to 5024, select a sample of size 5, and click "Reset" and "Sample." Using Table B and starting at line 118, select 0325, 3304, 4702, 1887, and 2099. (b) More than 171 respondents have run red lights. We would not expect very many people to claim they *have* run red lights when they have not, but some people will deny running red lights when they have.

8.37 (a) $300/30{,}000 = 0.01$; $100/10{,}000 = 0.01$. (b) To be a simple random sample, *every* possible sample of size 400 must have the same chance of being selected. This is not the case because the only possible samples are those with 300 undergraduates and 100 graduate students.

8.39 (a) How the sample was obtained can contribute to bias in the results if the sampling is not done randomly and fairly. (b) Answers will vary. For example, exactly how the 655 Internet users were selected is not given.

8.41 Sample separately in each stratum; that is, assign separate labels, then choose the first sample, then continue on in the table to choose the next sample, etc. With the applet: Population = 1 to 36, select a sample of size 4, and click "Reset" and "Sample." Then Population = 1 to 72, select a sample of size 7, and click "Reset" and "Sample." Then Population = 1 to 31, select a sample of size 3, and click "Reset" and "Sample." Then Population = 1 to 42, select a sample of size 4, and click "Reset" and "Sample." Beginning with line 112 in Table B, we choose:

Forest Type	Labels	Parcels Selected
Climax 1	01 to 36	04, 11, 19, 35
Climax 2	01 to 72	27, 30, 57, 62, 56, 02, 06
Climax 3	01 to 31	08, 02, 25
Secondary	01 to 42	11, 17, 14, 29

8.43 (a) Because $200/5 = 40$, we will choose 1 of the first 40 names at random. With the applet: Population = 1 to 40, select a sample of size 1, and click "Reset" and "Sample." Beginning on line 128 of Table B, the addresses are 15, 55, 95, 135, and 175. (Only the first number is chosen from the table; each subsequent number is 40 more than the previous number.) (b) All addresses are equally likely; each has chance $1/40$ of being selected. This is not an SRS because the only possible samples have exactly 1 address from the first 40, one address from the second 40, and so on. An SRS could contain any 5 of the 200 addresses in the population.

8.45 (a) This design omits households without telephones, those with only cell phones, and those with unlisted numbers. Such households would likely be made up of individuals who cannot afford a phone, those who choose not to have landline phones, and those who do not wish to have their phone numbers published. (b) Those with unlisted landline numbers would be included in the sampling frame when a random digit dialer is used. If the exchange selected is one that can be used for cell phones, some of those individuals would be included in the sampling frame as well.

8.47 (a) The wording is clear but will almost certainly be slanted toward high agreement, because of the reference to climate change. (b) The wording is clear, and it makes the case for a national health care system and so will slant responses toward Yes. (c) This survey question is not clear.

8.49 The minister is wrong. The sample may not be representative of the population because people who fill out the optional long-form questions will be systematically different from those who don't. Larger samples do not address such problems of bias.

8.51 (a) Assign labels 01 through 25. With the applet: Population = 1 to 25, select a sample of size 5, and click "Reset" and "Sample." Using Table B and starting at line 121, select 07, 22, 10, 25, and 13. (b) To sample 100 students, you need to sample four dorms because three dorms would not be enough.

Chapter 9 – Producing Data: Experiments

9.1 (a) Explanatory: level of education; response: rate of motor vehicle crash deaths. (b) Age of car, crash test rating, and presence of safety features are lurking variables as they are neither the primary explanatory nor response variables. (c) No, because a lurking variable could be the true cause.

9.3 This is an observational study, so it is not reasonable to conclude any cause-and-effect relationship. Furthermore, even if quitting smoking does increase risk of diabetes, that risk might be offset by the health risks of smoking.

9.5 Subjects: the students. Factor: pitch of voice. Treatments: high pitch and low pitch. Response: rating of perceived size of sandwich.

9.7 There was no comparison group for the study. The researchers were not able to compare the results with results from couples who were not discussing touchy subjects, so "discussing touchy subjects" is a lurking variable. Perhaps just being involved in the experiment contributed to the higher levels of LPS-binding protein. In addition, there is some confounding because the experimenters only used healthy couples, and there could be a difference in LPS-binding protein behavior in healthy and unhealthy couples.

9.9 (a) Diagram follows.

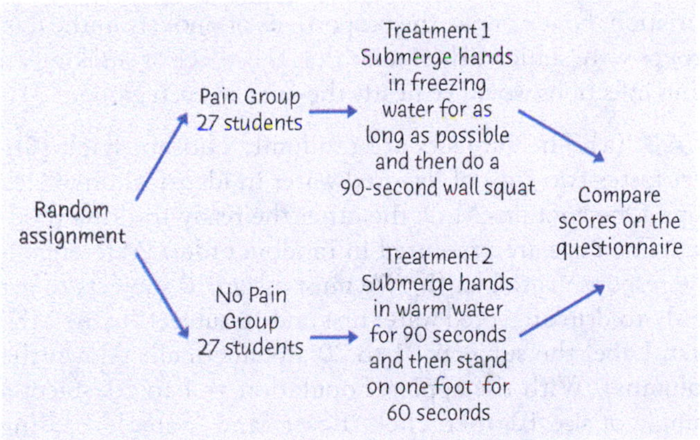

(b) Label the subjects 01 to 54. With the applet: Population = 1 to 54, select a sample of size 27, and click "Reset" and "Sample." Using Table B, beginning at line 125, the first few subjects selected are 21, 49, 37, 18, 44, 23, 51, and so on. (c) The experiment used similar tasks to control for bonding that may result from certain tasks regardless of whether there is pain. This is important because, if the tasks are dissimilar, we cannot conclude whether a difference in bond is attributed to pain or to the particular activity.

9.11 In a controlled scientific study, the effects of factors other than the nonphysical treatment (e.g., the place effect, differences in the prior health of the subjects) can be eliminated or accounted for so that the differences in improvement between the subjects can be attributed to the differences in treatments.

9.13 (a) The researchers did not alter the subjects' diets. (b) Such language is reasonable because, with observational studies, no cause-and-effect conclusion would be reasonable.

9.15 Here "single-blinded" means that only the evaluators of the subjects' cholesterol levels did not know which treatment the subjects were receiving. There is no way to blind the subjects to which treatment they are receiving.

9.17 Use a completely randomized design like the one outlined below, and compare times on the final run.

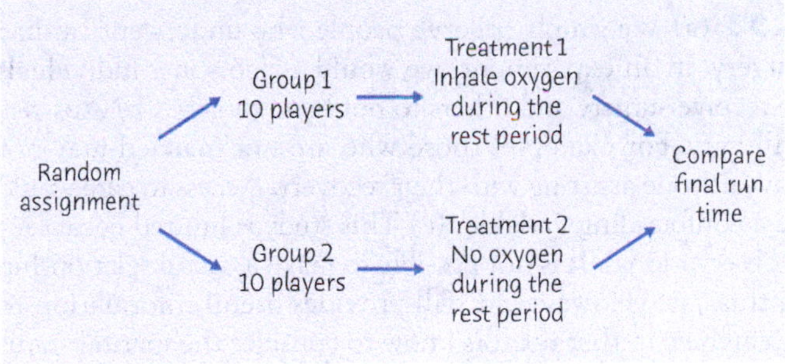

This experiment could alternatively be done as matched pairs, where all 20 players inhale oxygen during the rest period and during a separate trial do not inhale oxygen (the control) during the rest period. Sufficient time would have to pass between the two trials to allow for recovery. The order of the oxygen versus no-oxygen trials would be randomly assigned to each player.

To do the randomization, label the players 01 to 20. With the *Simple Random Sample* applet: Population = 1 to 20, select a sample of size 10, and click "Reset" and "Sample." Using Table B, and beginning at line 142, the first 10 players selected are 02, 08, 17, 10, 05, 09, 19, 06, 16, and 01. In a completely randomized design, these 10 players would be in the "inhale oxygen during the rest period" group, and the remaining 10 players would be in the "no oxygen during the rest period" group. In the matched pairs design, these 10 players would complete the "inhale oxygen during the rest period" trial first and, after a sufficient amount of time, the "no oxygen during the rest period" trial. The remaining 10 players would reverse the order of these two trials.

9.19 (a)

9.21 (b)

9.23 (c)

9.25 (a)

9.27 (a)

9.29 (a) Observational study; the subjects chose how much red meat to eat. Explanatory variable: red meat consumption; response: whether or not a subject dies. (b) Many answers are possible. For example, smoking is known to increase the risk of cancer. Variables like physical activity, smoking status, drinking behavior, and body mass index are called lurking variables. (c) Many answers are possible. For example, how many servings of fruit and vegetables were consumed along with the red meat?

9.31 (a) Matched pairs design. Explanatory variable: activity level; response variables: body fat percentage, endurance level, and insulin sensitivity. (b) This is an observational study because no treatment is assigned. (c) Blind means the person taking the measurements did not know whether he or she was measuring the active or inactive twin. This is important because the person recording this information will not be able to influence the results.

9.33 (a) We simply observe people who underwent cardiac surgery. In an experiment, we would assign some individuals to receive surgery and others to not have surgery. (b) Answers will vary. For example, those who are not married may not have anyone assisting with their recovery. Access to care would be a confounding variable. (c) This study is limited because it is observational. It is not possible to make a causal relationship in this case. However, it still provides useful information to researchers in that doctors know to consider the marital status of a patient when developing a recovery plan.

9.35 (a) Diagram follows.

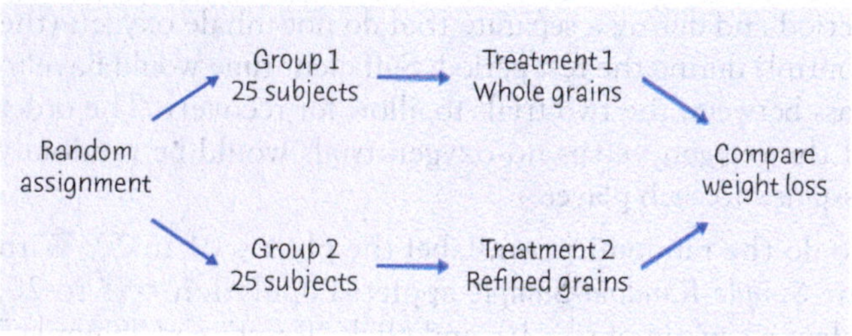

(b) Label the subjects 01 to 50. With the *Simple Random Sample* applet: Population = 1 to 50, select a sample of size 10, and click "Reset" and "Sample." Using Table B, beginning at line 107, the first 10 adults are 20, 11, 38, 31, 48, 07, 20, 24, 17, and 49.

9.37 (a) Diagram follows.

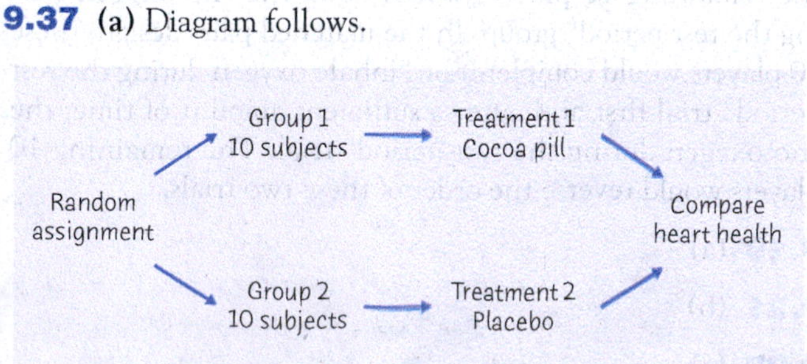

(b) Label the subjects 01 to 20 alphabetically (down the columns). With the *Simple Random Sample* applet: Population = 1 to 20, select a sample of size 10, and click "Reset" and "Sample." Using Table B, beginning at line 129 and choose 13 = Reichert, 18 = Williams, 03 = Birkel, 05 = DeVore, 16 = Scannell, 17 = Stout, 20 = Worbis, 19 = Wilson, 04 = Bower, 07 = Fritz. (c) This could be run as a double-blind experiment, assuming that subjects can't distinguish between the cocoa pill and the placebo, and the persons evaluating the subjects' heart health don't know which treatment the subjects received.

9.39 (a) There are two factors. The first is type of granola (regular and low-fat). The second factor is serving size label (two servings, one serving, and no label). There are six treatment combinations (regular granola at two servings, regular granola at one serving, regular granola with no serving label, low-fat granola at two servings, low-fat granola at one serving, low-fat granola with no serving label). At 20 subjects per treatment, there were 120 subjects in the experiment.

	Granola Type	
Serving Size Label	**Regular**	**Low-Fat**
2 servings	20 subjects	20 subjects
1 serving	20 subjects	20 subjects
No label	20 subjects	20 subjects

9.41 (a) The factors are pill type and spray type. "Double-blind" means the treatment assigned to a patient was unknown to both the patient and those responsible for assessing the effectiveness of that treatment. "Placebo-controlled" means that some of the subjects were given placebos. Even though placebos possess no medical properties, some subjects may show improvement or benefits just as a result of participating in the experiment. (b) "No significant difference" does *not* mean the groups are identical. It means that the actual observed differences are easily explained by chance variation. For example, the proportions of smokers in the four groups were sufficiently similar that the effect of smoking on sinus infections would be nearly the same in each group.

9.43 (a) The subjects are randomly chosen. Each subject tastes two cups of flavored water in identical unlabeled cups. One contains MiO, the other the ready-to-drink product. The cups are presented in random order. Preference is the response variable. (b) We must assign 10 subjects to get ready-to-drink flavored water first and 10 subjects to get MiO first. Label the subjects 01 to 20 alphabetically (down the columns). With the applet: Population = 1 to 20, select a sample of size 10, then click "Reset" and "Sample." Using Table B, beginning at line 138, the "MiO first" group is 16, 08, 15, 13, 17, 04, 10, 19, 12, and 18.

9.45 (a) The subjects are all consumers who are recruited and assigned a combination of program and advertisement. The factors are the energy-saving programs (Conservation or Peak Saving) and advertisement (save money, save energy, or save both). The treatments are the six combinations of program and advertisement (see part (b)). The response variable was whether the consumer decided to enroll in the program after reading the advertisement. **(b)** Diagram follows.

		Advertisement		
		Save Money	Save Energy	Save Both
Program	Conservation	1	2	3
	Peak shaving	4	5	6

9.47 There are two explanatory variables, each with two levels, so there will be four treatments: wine and snack, wine and no snack, no wine and snack, no wine and no snack. The response variable is how many times the subject wakes up during the night. To offset the effect of lurking variables such as amount of sleep the previous night, randomize the subjects into each of the four treatments.

9.49 Age is completely confounded with treatment. If adults under age 65 and adults age 65 or older respond differently to the treatment, or if they tend to differ in the response variable regardless of treatment, then the experiment will be strongly biased.

9.51 **(a)** "Randomized" means that the subjects in the study were assigned at random (by chance) to the treatments. This reduces the chance that the two groups will differ in some way that influences the outcome, other than the fact that one group receives the treatment and the other does not. "Double-blind" means that neither the subjects nor the persons evaluating their symptoms of depression knew whether the subjects were in the treatment group (took SAMe) or the control group (took a placebo). Even though placebos possess no medical properties, some subjects may show improvement or benefits just as a result of participating in the experiment; the placebos allow those doing the study to observe this effect. **(b)** Statistical significance means that the SAMe group had a greater difference in response (more had a positive response) than could be attributed to chance. This means that it appears SAMe helps reduce depression when used with standard treatment. **(c)** Diagram follows.

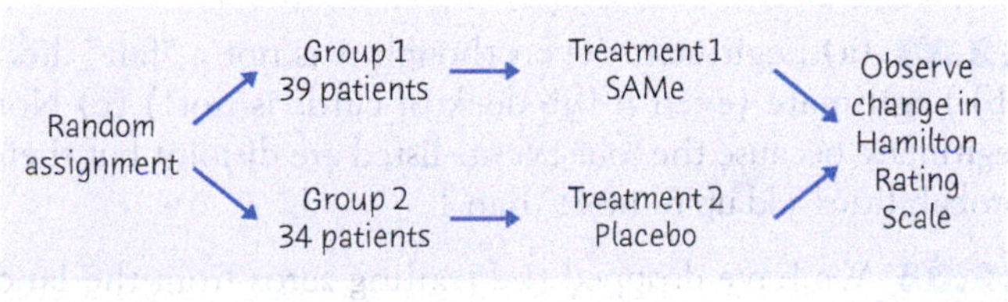

Chapter 10 – Data Ethics

As the text states, "Most of these exercises pose issues for discussion. There are no right or wrong answers, but there are more and less thoughtful answers." We have not tried to supply answers for exercises that are largely matters of opinion. For that reason, only answers to a few exercises are provided.

10.1 Answers will vary. Many answers will indicate that option (a) qualifies as minimal risk, and most will agree that option (e) goes beyond minimal risk.

10.3 **(a)** Losing a job and feeling as if there is no alternative to earn money would induce most subjects to agree to participate. **(b)** Pressuring a new employee to participate may be viewed as a threat to continued employment.

10.15 The responses are confidential if survey responses are reported only in aggregate—that is, if a subject's name or other identifying information is not attached to his or her response. The subject is not anonymous to the interviewer.

10.19 The ethical issue here is that prisoners may consent under undue influence. In addition, the nonrandom assignment of prisoners to the treatments could result in confounding: compliant prisoners may differ from noncompliant prisoners in ways that affect the outcome on the study.

Chapter 11 – Producing Data: Part II Review

11.1 (c)

11.3 **(a)** Label the students from 01 to 30 alphabetically (down the columns). **(b)** With the *Simple Random Sample* applet: Population = 1 to 30, select a sample of size 5, and click "Reset" and "Sample." With Table B, enter at line 122 and choose 13 = Hans, 15 = Jeter, 05 = Collins, 29 = Verducci, and 09 = Drake. **(c)** The response variable is how much the subject trusts information about politics from the Internet.

11.5 (d)

11.7 (a)

11.9 Many answers are possible. One possible lurking variable is "student attitude about the purpose of college." (Students with a view that college is about partying, rather than studying, may be more likely to binge drink and more likely to have lower grades.)

11.11 (c)

11.13 People who follow C-SPAN on Twitter don't represent American adults broadly. Those people responding to the survey went out of their way to respond to the poll, and, as C-SPAN Twitter followers, they have some information about politics from C-SPAN. (We do not know if they follow other political Twitter feeds.) It seems reasonable to believe that the respondents in this voluntary response sample differ in their political views from other American adults.

11.15 (b)

11.17 (c)

11.19 This is an experiment because the introduction type (the explanatory variable) was assigned for each call made. The response variable is whether or not the interview was completed.

11.21 (a) increase. (b) decrease. (c) increase. (d) decrease.

11.23 (a) The explanatory variable is the amount of alcohol one drinks. The response is whether or not one has cancer. (b) Even if taking zero drinks does not ensure that a person will not get cancer, it still might be a good recommendation. (Taking zero drinks will not cause you not to have a car accident, but taking zero drinks before driving is a good recommendation.) Health risks from drinking small amounts of alcohol may be negligible and offset by potential benefits.

11.25 (a) Subjects were not assigned where to live, so this study is not an experiment. (b) Answers will vary. For example, those who live near highways might have less money and might be attracted to the lower housing costs near a highway. Money would be a confounding variable. (c) No, researchers cannot randomly assign a person to live near a highway or not.

Chapter 12 – Introducing Probability

12.1 In the long run of a *large* number of five-card poker hands, the fraction in which you will be dealt a flush is 1/508. It *does not* mean that exactly 1 out of 508 such five-card poker hands would yield a flush. As you are dealt more hands, the proportion of hands that are a flush *approaches* 1 in 508, but that proportion may not be exact even for many thousands of hands.

12.3 (a) There are 21 zeros among the first 200 digits of the table (rows 101–105), for a proportion of 0.105. (b) Answers will vary, but more than 99% of all results should fall between 9 and 33 heads out of 200 flips when $p = 0.1$.

12.5 (a) S = {has a pet, does not have a pet}. (b) S = {All numbers between _______ and _____}. (Choices of lower and upper limits will vary.) (c) $S = \{000, 001, 002, \ldots, 999\}$. (d) S = {January, February, . . . , December}.

12.7 Add 2 to each pair-total: $S = \{4, 5, 6, 7, 8, 9, 10\}$. Each of the 16 possible pairings is equally likely, so (for example) the probability of a total of 6 is 3/16 because 3 pairings add to 4 (and then we add 2). The complete set of probabilities follows.

Total	Probability
4	1/16 = 0.0625
5	2/16 = 0.125
6	3/16 = 0.1875
7	4/16 = 0.25
8	3/16 = 0.1875
9	2/16 = 0.125
10	1/16 = 0.0625

12.9 (a) Event B specifically rules out obese subjects, so there is no overlap with event A. (b) A or B is the event "The person chosen is overweight or obese." $P(A \text{ or } B) = P(A) + P(B) = 0.40 + 0.32 = 0.72$. (c) $P(C) = 1 - P(A \text{ or } B) = 1 - 0.72 = 0.28$.

12.11 (a) Disjoint. (b) Not disjoint. For example, $300,000 is more than $100,000 and more than $250,000. (c) Disjoint. If one die is a 3, then the sum of both dice must be more than 3.

12.13 (a) $A = \{4, 5, 6, 7, 8, 9\}$, $P(A) = 0.097 + 0.079 + 0.067 + 0.058 + 0.051 + 0.046 = 0.398$. (b) $B = \{2, 4, 6, 8\}$, $P(B) = 0.176 + 0.097 + 0.067 + 0.051 = 0.391$. (c) A or B $= \{2, 4, 5, 6, 7, 8, 9\}$, $P(A \text{ or } B) = 0.176 + 0.097 + 0.079 + 0.067 + 0.058 + 0.051 + 0.046 = 0.574$. This is different from $P(A) + P(B)$ because A and B are not disjoint.

12.15 (a) $P(Y \le 0.6) = 0.6$. (b) $P(Y < 0.6) = 0.6$. (c) $P(0.4 \le Y \le 0.8) = 0.4$. (d) $P(0.4 < Y \le 0.8) = 0.4$

12.17 (a) $\{X \ge 510\}$ (b) $P(X \ge 510) = P\left(Z \ge \frac{510 - 500.9}{10.6}\right) = P(Z \ge 0.86) = 1 - 0.8051 = 0.1949$ (using Table A).

12.19 (a) Continuous random variable: It can take on any value within an interval. (b) $P(Y \ge 8)$ is "the probability that a randomly selected able-bodied male student runs a mile in 8 minutes or more." $P(Y \ge 8) = P\left(Z \ge \frac{8 - 7.11}{0.74}\right) = P(Z \ge 1.20) = 1 - 0.8849 = 0.1151$ (using Table A). (c) $\{Y < 6\}$. $P(Y < 6) = P\left(Z < \frac{6 - 7.11}{0.74}\right) = P(Z < -1.50) = 0.0668$ (using Table A).

12.21 (a) If Joe says P(Louisville wins) = 0.05, then Joe believes P(North Carolina wins) = 0.10 and P(Duke wins) = 0.20. (b) Joe's probabilities for Louisville, North Carolina, and Duke add up to 0.35, so that leaves probability 0.65 total for the other 15 teams.

12.23 (b)

12.25 (b)

12.27 (b)

12.29 (c)

12.31 (c)

12.33 (a) Legitimate (even though it is not a "fair" die.) (b) Legitimate (even if the deck of cards is not!) (c) Not legitimate because the four events listed are disjoint but their probabilities add up to more than 1.

12.35 We have dropped the trailing zeros from the land area figure. (a) P(area is forested) = 4176/9094 = 0.4592. (b) P(area is not forested) = 1 − 0.4592 = 0.5408.

12.37 (a) The given probabilities add to 0.99, so other colors must account for the remaining 0.01. (b) P(neither white nor silver) = 1 − P(white or silver) = 1 − (0.39 + 0.10) = 1 − 0.49 = 0.51.

12.39 The probabilities of 2, 3, 4, and 5 are unchanged (1/6), so $P(1 \text{ or } 6)$ must still be 1/3. If $P(6) = 0.2$, then $P(1) = 1/3 - 0.2 = 2/15$ (or about 0.1333).

Face	⚀	⚁	⚂	⚃	⚄	⚅
Probability	0.13	1/6	1/6	1/6	1/6	0.2

12.41 (a) Legitimate: Each person must fall into exactly one category, the probabilities are all between 0 and 1, and they add up to 1. **(b)** $P(\text{Hispanic}) = 0.003 + 0.011 + 0.161 + 0.009 = 0.184$. **(c)** $P(\text{not non-Hispanic White}) = 1 - P(\text{non-Hispanic White}) = 1 - 0.605 = 0.395$.

12.43 (a) A = {never married woman, married woman, divorced woman, widowed woman, married man}. $P(A) = 0.152 + 0.261 + 0.057 + 0.045 + 0.259 = 0.774$. This is different from the sum of the probabilities in parts (c) and (d) of Exercise 12.42 because that sum counts the probability of a married woman (0.261) twice; "woman" and "married" are not disjoint events.

12.45 (a) X is discrete because it has a finite sample space. **(b)** "At least one nonword error" is the event $\{X \geq 1\}$ or $\{X > 0\}$. $P(X \geq 1) = 1 - P(X = 0) = 1 - 0.1 = 0.9$. **(c)** $\{X \leq 2\}$ is "no more than two nonword errors," or "fewer than 3 nonword errors." $P(X \leq 2) = P(X = 0) + P(X = 1) + P(X = 2) = 0.1 + 0.2 + 0.3 = 0.6$. $P(X < 2) = P(X = 0) + P(X = 1) = 0.1 + 0.2 = 0.3$.

12.47 (a) Just using initials: {(A, D), (A, M), (A, S), (A, R), (D, M), (D, S), (D, R), (M, S), (M, R), (S, R)}. **(b)** Each has probability 1/10 = 0.1, or 10%. **(c)** Mei-Ling is chosen in 4 of the 10 possible outcomes: 4/10 = 0.4, or 40%. **(d)** There are three pairs with neither Sam nor Roberto: 3/10 = 0.3, or 30%.

12.49 The possible values of Y are 1, 2, 3, . . . , 12, each with probability 1/12.

12.51 (a) This is a continuous random variable because the set of possible values is an interval. **(b)** The height should be 0.5 because the area under the curve must be 1. (For a rectangle, area = base × height.) The density curve is illustrated below.

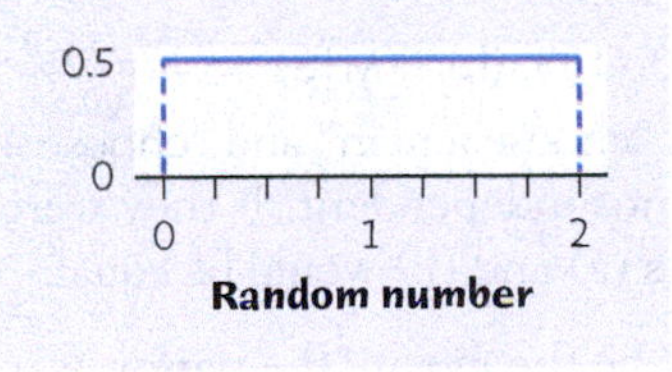

(c) $P(Y \leq 1) = 0.5$.

12.53 (a) $P(0.44 \leq V \leq 0.48) = P\left(\frac{0.44 - 0.46}{0.011} \leq Z \leq \frac{0.48 - 0.46}{0.011}\right) = P(-1.82 \leq Z \leq 1.82) = 0.9656 - 0.0344 = 0.9312$.

(b) $P(V \geq 0.43) = P\left(Z \geq \frac{0.43 - 0.46}{0.011}\right) = P(Z \geq -2.73) = 1 - 0.0032 = 0.9968$.

12.55 (a) Because there are 10,000 equally likely four-digit numbers (0000 through 9999), the probability of an exact match is 1/10,000 = 0.0001. **(b)** There is a total of 24 arrangements of the digits 5, 9, 7, and 4, so the probability of a match in any order is 24/10,000 = 0.0024.

12.57 (a)–(c) Results will vary, but after n tosses, the distribution of the proportion (call it $\hat{p}$) is approximately Normal with mean 0.5 and standard deviation $1/(2\sqrt{n})$, while the distribution of the count of heads is approximately Normal with mean 0.5n and standard deviation $\sqrt{n}/2$. Therefore, using the 68–95–99.7 rule, we have the results shown in the table below. Note that the range for the proportion $\hat{p}$ gets narrower, while the range for the count gets wider.

n	99.7% Range for $\hat{p}$	99.7% Range for Count
50	0.5 ± 0.212	25 ± 10.6
150	0.5 ± 0.122	75 ± 18.4
250	0.5 ± 0.095	125 ± 23.7
500	0.5 ± 0.0.67	250 ± 33.5

12.59 (a) With $n = 50$, the variability in the proportion (call it $\hat{p}$) is larger. With $n = 100$, nearly all answers will be between 0.20 and 0.48. With $n = 400$, nearly all answers will be between 0.27 and 0.41. **(b)** Results will vary.

Chapter 13 – General Rules of Probability

13.1 (a) Because $P(A \text{ or } B) = 0.34 = P(A) + P(B) - P(A \text{ and } B) = 0.30 + 0.13 - P(A \text{ and } B)$, we must have that $P(A \text{ and } B) = 0.30 + 0.13 - 0.34 = 0.09$. A Venn diagram is provided.

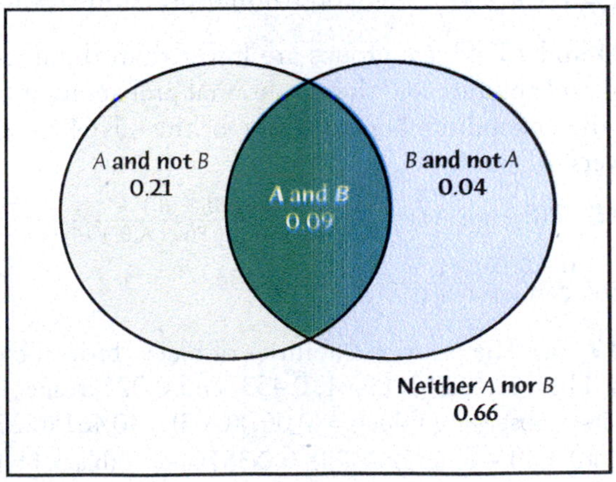

(b) Event (A and B) is an English-language video that involves games and hobbies & skills. Event (A and not B) is an English-language video that involves games but not hobbies & skills. Event (B and not A) is an English-language video that involves hobbies & skills but not games. Event (neither A nor B) is an English-language video that does not involve games and also does not involve hobbies & skills. **(c)** $P(A \text{ and } B) = 0.09$; $P(A \text{ and not } B) = 0.30 - 0.09 = 0.21$; $P(B \text{ and not } A) = 0.13 - 0.09 = 0.04$; $P(\text{neither } A \text{ nor } B) = 1 - 0.34 = 0.66$. Probabilities are shown in the Venn diagram.

13.3 It is unlikely that these events are independent. In particular, it is reasonable to expect that younger adults are more likely than older adults to be college students.

13.5 If we assume that each site reference remaining valid is independent of the others, $P(\text{all seven are still good}) = (1-0.47)^7 = (0.53)^7 = 0.0117$.

13.7 Using the Venn diagram from Exercise 13.1 above: $P(A \mid B) = \frac{P(A \text{ and } B)}{P(B)} = \frac{0.09}{0.13} = 0.6923$.

13.9 Let H be the event that an adult belongs to a health club and T be the event that the adult goes at least twice a week. Note that $P(T \text{ and } H) = P(T)$ because one has to be a member of a health club to attend. $P(T) = P(H)P(T \mid H) = (0.45)(0.08) = 0.036$.

13.11 **(a)** Let W be the event that a professor is a woman. $P(W) = \frac{94+134+244}{253+314+949} = \frac{472}{1516} = 0.3113$. **(b)** Let F be the event that the professor is a full professor. $P(W \mid F) = \frac{P(W \text{ and } F)}{P(F)} = \frac{244}{949} = 0.257$. **(c)** No, rank and sex are not independent because $P(W \mid F) \neq P(W)$.

13.13 **(a)** $P(\text{age 18–24} \mid \text{online date no}) = \frac{P(\text{Age 18–24 and online date no})}{P(\text{online date no})} = \frac{0.1241}{0.1241+0.3198+0.3654} = \frac{0.1241}{0.8093} = 0.1533$; $P(\text{age 25–44} \mid \text{online date no}) = \frac{0.3198}{0.8093} = 0.3952$; $P(\text{age 45–64} \mid \text{online date no}) = \frac{0.3654}{0.8093} = 0.4515$. **(b)** The conditional probabilities for the 18–24 and 25–44 age groups are lower than the unconditional probabilities, and the conditional probability is higher than the unconditional probability for the 45–64 age group. Answers will vary.

13.15 $P(B_1 \mid \text{not } A) = \frac{P(\text{not } A \mid B_1)P(B_1)}{P(\text{not } A \mid B_1)P(B_1) + P(\text{not } A \mid B_2)P(B_2)} = \frac{(1-0.21)(0.063)}{(1-0.21)(0.063)+(1-0.06)(1-0.063)} = 0.0535$.

13.17 **(a)** The prior probabilities of black, brown, blonde, and red hair are 0.061, 0.461, 0.453, and 0.025, respectively. **(b)** First, observe $P(\text{blue}) = (0.061)(0.03) + (0.461)(0.259) + (0.453)(0.562) + (0.025)(0.473) = 0.3876$. $P(\text{black} \mid \text{blue}) = \frac{P(\text{black and blue})}{P(\text{blue})} = \frac{(0.061)(0.03)}{0.3876} = 0.0047$; $P(\text{brown} \mid \text{blue}) = \frac{(0.461)(0.259)}{0.3876} = 0.3080$; $P(\text{blonde} \mid \text{blue}) = \frac{(0.453)(0.562)}{0.3876} = 0.6568$; $P(\text{red} \mid \text{blue}) = \frac{(0.025)(0.473)}{0.3876} = 0.0305$. The probabilities are what one would expect. People with blonde and red hair have the highest probability of having blue eyes, so we would expect the conditional probabilities for those two groups to be larger than the prior probabilities.

13.19 (b)

13.21 (b)

13.23 (c)

13.25 (b)

13.27 (b)

13.29 $P(\text{none are O-negative}) = (1-0.072)^{10} = 0.4737$, so $P(\text{at least one is O-negative}) = 1 - 0.4737 = 0.5263$.

13.31 PLAN: Let I be the event "infection occurs" and let F be the event "the repair fails." We have been given $P(I) = 0.03$, $P(F) = 0.14$ and $P(I \text{ and } F) = 0.01$. We want $P(\text{not } I \text{ and not } F)$. SOLVE: First, $P(I \text{ or } F) = P(I) + P(F) - P(I \text{ and } F) = 0.03 + 0.14 - 0.01 = 0.16$. $P(\text{not } I \text{ and not } F) = 1 - P(I \text{ or } F) = 0.84$. CONCLUDE: 84% of operations succeed and are free from infection.

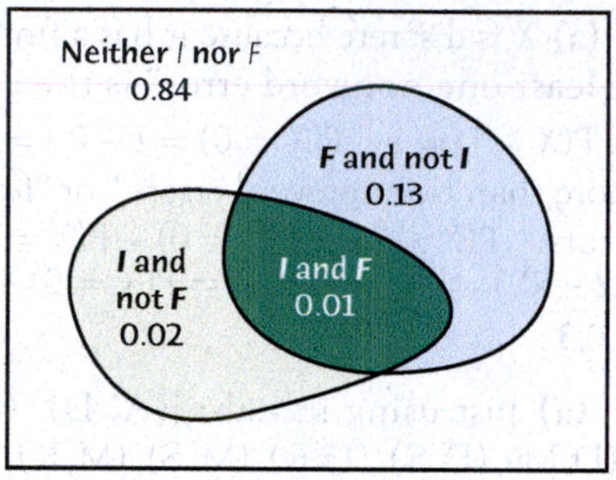

13.33 PLAN: Let I be the event "infection occurs" and let F be the event "the repair fails." Refer to the Venn diagram in Exercise 13.31. We want $P(I \mid \text{not } F)$. SOLVE: $P(I \mid \text{not } F) = \frac{P(I \text{ and not } F)}{P(\text{not } F)} = \frac{0.02}{0.86} = 0.0233$. CONCLUDE: The probably of infection given successful surgery is 0.0233.

13.35 **(a)** $P(\text{debt is at least } \$20{,}000) = 0.17 + 0.12 + 0.08 + 0.10 = 0.47$. **(b)** $P(\text{debt is at least } \$50{,}000 \mid \text{debt is at least } \$20{,}000) = \frac{P(\text{debt is at least } \$50{,}000)}{P(\text{debt is at least } \$20{,}000)} = \frac{0.10}{0.47} = 0.2128$.

13.37 Let M be the event "the person is a man" and S be the event "the person earned a master's degree." **(a)** $P(M) = 1657/3990 = 0.4153$. **(b)** $P(M \mid S) = \frac{P(M \text{ and } S)}{P(S)} = \frac{340}{813} = 0.4182$ **(c)** The events "choose a man" and "choose a master's degree recipient" are not independent. If they were, the two probabilities in parts (a) and (b) would be equal.

13.39 Let W be the event "the person is a woman" and A be the event "the person earned an associate's degree." **(a)** $P(W) = 2333/3990 = 0.5847$. **(b)** $P(A \mid W) = \frac{P(A \text{ and } W)}{P(W)} = \frac{653}{2333} = 0.2799$. **(c)** Using the multiplication rule, $P(W \text{ and } A) = P(W)P(A \mid W) = (0.5847)(0.2799) = 0.1637$; using the table, $P(W \text{ and } A) = 653/3990 = 0.1637$.

13.41 (a) and (b) These probabilities are provided below.

P(1st card is ♠)	$13/52 = 0.25$
P(2nd card is ♠ | first card is ♠)	$12/51 = 0.2353$
P(3rd card is ♠ | first two are ♠)	$11/50 = 0.22$
P(4th card is ♠ | first three are ♠)	$10/49 = 0.2041$
P(5th card is ♠ | first four are ♠)	$9/48 = 0.1875$

(c) The product of these conditional probabilities gives the probability of a flush in spades. The product of the five probabilities is 0.0004952. (d) Since there are four possible suits in which to have a flush, the probability of a flush is four times that found in part (c), or about 0.001981.

13.43 This conditional probability is $P\left(\text{cover} > \frac{2}{3} \mid \text{not } D\right)$ $= \frac{176}{151+158+177+176} = \frac{176}{662} = 0.2659$, or 26.59%.

13.45 The Venn diagram for Exercises 13.45 through 13.47 is provided. P(no tobacco products) = 0.65.

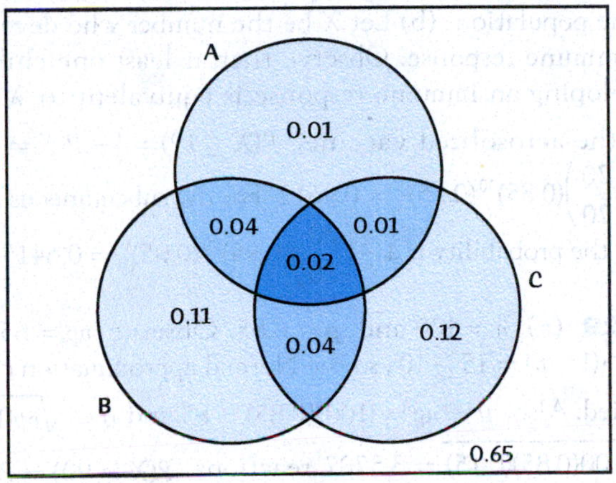

13.47 $P(A \mid B) = \frac{P(A \text{ and } B)}{P(B)} = \frac{0.06}{0.21} = 0.2857$; $P(B \mid A) = \frac{P(A \text{ and } B)}{P(A)} = \frac{0.06}{0.08} = 0.75$.

13.49 The tree diagram provided organizes this information. The total probability of the serving player winning a point is $0.4307 + 0.208034 = 0.638734$.

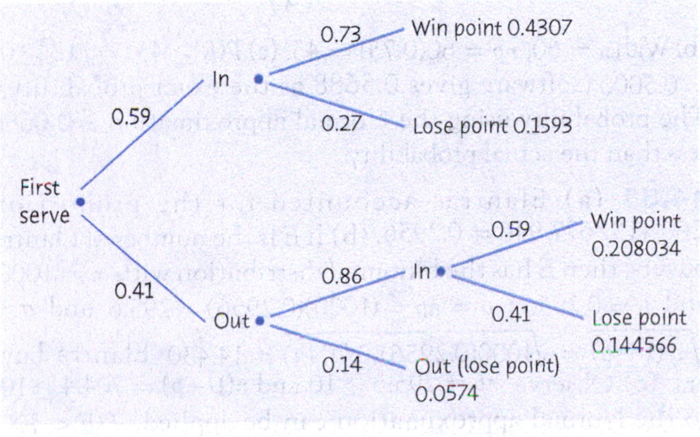

13.51 P(first serve in | server won point) = $\frac{P(\text{first serve in } \textit{and} \text{ server won point})}{P(\text{server won point})} = \frac{0.4307}{0.638734} = 0.6743$.

13.53 For a randomly selected resident of the United States, let W, B, A, and L be (respectively) the events that the resident is White, Black, Asian, and lactose intolerant. (a) $P(L) = P(W \text{ and } L) + P(B \text{ and } L) + P(A \text{ and } L) = (0.82)(0.15) + (0.14)(0.70) + (0.04)(0.90) = 0.257$. (b) $P(A \mid L) = \frac{P(A \text{ and } L)}{P(L)} = \frac{(0.04)(0.90)}{0.257} = 0.1401$.

13.55 For a randomly selected degree recipient in the United States, let A, B, M, and D be (respectively) the events that the recipient earned an associate's degree, a bachelor's degree, a master's degree, and a doctorate or other advanced degree. Let I be the event that the recipient is Asian or a Pacific Islander. (a) $P(A) = 0.26$, $P(B) = 0.49$, $P(M) = 0.2$, and $P(D) = 0.05$.

(b) $P(A \mid I) = \frac{P(A \text{ and } I)}{P(I)} = \frac{P(I|A)P(A)}{P(I|A)P(A)+P(I|B)P(B)+P(I|M)P(M)+P(I|D)P(D)}$
$= \frac{(0.07)(0.26)}{(0.07)(0.26)+(0.06)(0.49)+(0.06)(0.2)+(0.11)(0.05)} = \frac{0.0182}{0.0651} = 0.2796$;

$P(B \mid I) = \frac{(0.06)(0.49)}{(0.07)(0.26)+(0.06)(0.49)+(0.06)(0.2)+(0.11)(0.05)} = 0.4516$;

$P(M \mid I) = \frac{(0.06)(0.20)}{(0.07)(0.26)+(0.06)(0.49)+(0.06)(0.2)+(0.11)(0.05)} = 0.1843$;

$P(D \mid I) = \frac{(0.11)(0.05)}{(0.07)(0.26)+(0.06)(0.49)+(0.06)(0.2)+(0.11)(0.05)} = 0.0845$.

Explanations will vary.

13.57 The proportion having combination (16, 17) is $(2)(0.232)(0.213) = 0.098832$.

13.59 If the DNA profile found on the hair is possessed by one in 1.6 million individuals, then we would expect about three individuals in the database of 4.5 million convicted felons to demonstrate a match. The proportion of people with the DNA profile is 1/(1.6 million). The expected number of matches is 1/(1.6 million) × (4.5 million) = 2.8125, which was rounded up to 3.

Chapter 14 – Binomial Distributions

14.1 Binomial. (1) We have a fixed number of observations ($n = 20$). (2) It is reasonable to believe that each call is independent of the others. (3) "Success" means a useable response to the poll is provided; "failure" is any other outcome. (4) Each randomly dialed number has chance $p = 0.10$ of getting a usable response to the poll.

14.3 Not binomial. The trials aren't independent. If one tile in a box is cracked, there are likely more cracked tiles.

14.5 (a) C, the number caught, is binomial with $n = 10$ and $p = 0.7$. M, the number missed, is binomial with $n = 10$ and $p = 0.3$.

(b) $P(M = 3) = \binom{10}{3}(0.3)^3(0.7)^7 = (120)(0.027)(0.08235) = 0.2668$. With software, $P(M \geq 3) = 1 - P(M \leq 2) = 0.6172$.

14.7 (a) `5 choose 2` returns 10. (b) `500 choose 2` returns 124,750, and `500 choose 100` returns 2.041694×10^{107}. (c) `(10 choose 1)*0.11 * 0.89 ^ 9` returns 0.38539204407.

14.9 (a) X is binomial with $n = 10$ and $p = 0.3$; Y is binomial with $n = 10$ and $p = 0.7$. (b) The mean of Y is $(10)(0.7) = 7$ errors caught, and the mean of X is $(10)(0.3) = 3$ errors missed. (c) The standard deviation of Y (or X) is $\sigma = \sqrt{10(0.7)(0.3)} = 1.4491$ errors.

14.11 (a) Let X be the number of students who accept the admissions offer. Then X is binomial with $n = 1250$ and $p = 0.4$. The mean is $\mu = np = (1250)(0.4) = 500$, and $\sigma = \sqrt{np(1-p)} = \sqrt{1250(0.4)(0.6)} = 17.32$ students. (b) We observe that $np = (1250)(0.4) = 500 \geq 10$ and $n(1-p) = (1250)(0.6) = 750 \geq 10$, so n is large enough for the Normal approximation. $P(X \geq 501) = P\left(Z \geq \frac{501 - 500}{17.32}\right) = P(Z \geq 0.06)$ $= 1 - 0.5239 = 0.4761$ (c) The exact binomial probability is 0.4877, so the Normal approximation is 0.0116 too low. (d) To decrease the chance of more students than desired, the college needs to decrease the number admitted. If $n = 1181$, we have $P(X \geq 501) = 0.0479$. If $n = 1182$, we have $P(X \geq 501) = 0.0503$. The college should admit, at most, 1181 students.

14.13 (a)

14.15 (a)

14.17 (c)

14.19 (c)

14.21 (a)

14.23 (a) A binomial distribution is *not* an appropriate choice for field goals made because, given the different situations the kicker faces (wind, distance, etc.), his probability of success is likely to change from one attempt to another. In addition, conditions affecting field goal success may vary from season to season, so 20 attempts from one particular season may not have the same p as in the past. (b) It would be reasonable to use a binomial distribution for free throws made because each is from the same position with respect to the basket, with no interference allowed for a shot, and, presumably, each shot is independent of any other shot.

14.25 (a) $n = 5$ and $p = 0.256$. (b) The possible values are the integers 0, 1, 2, 3, 4, and 5. (c) $P(X = 0) = \binom{5}{0}(0.256)^0 (0.744)^5 = 0.2280$; $P(X = 1) = \binom{5}{1}(0.256)^1(0.744)^4 = 0.3922$; $P(X = 2) = \binom{5}{2}(0.256)^2(0.744)^3 = 0.2699$; $P(X = 3) = \binom{5}{3}(0.256)^3(0.744)^2 = 0.0929$; $P(X = 4) = \binom{5}{4}(0.256)^4(0.744)^1 = 0.0160$; $P(X = 5) = \binom{5}{5}(0.256)^5(0.744)^0 = 0.0011$.

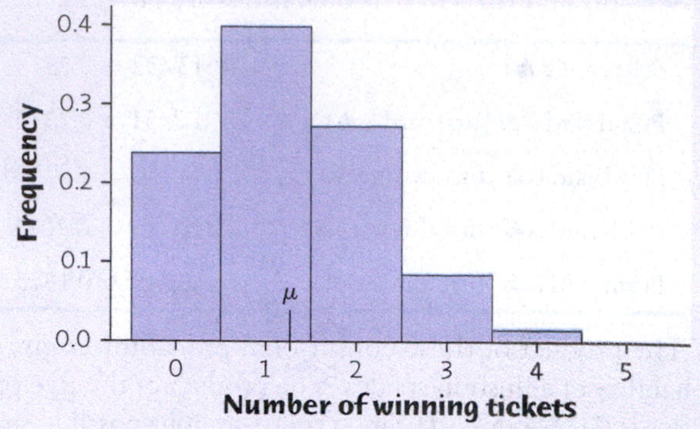

(d) $\mu = np = (5)(0.256) = 1.28$ and $\sigma = \sqrt{np(1-p)} = \sqrt{5(0.256)(0.744)} = 0.9759$ winning tickets.

14.27 (a) Provided that none of the children are related, all children should be independent in terms of immune response. There are a fixed number of children to be observed, and (we assume) each has the same probability of having an immune response. Our sample size is presumably much less than 5% of the population. (b) Let X be the number who developed an immune response. Observe that at least one child *not* developing an immune response is equivalent to $X \leq 19$. For the aerosolized vaccine, $P(X \leq 19) = 1 - P(X = 20) = 1 - \binom{20}{20}(0.85)^{20}(0.15)^0 = 0.9612$. For the subcutaneous injection, the probability is $1 - \binom{20}{20}(0.95)^{20}(0.05)^0 = 0.6415$.

14.29 (a) $n = 100$ and $p = 0.85$. Observe $np = 85 \geq 10$ and $n(1-p) = 15 \geq 10$, so the Normal approximation can be applied. Also, $\mu = np = (100)(0.85) = 85$ and $\sigma = \sqrt{np(1-p)}$ $= \sqrt{100(0.85)(0.15)} = 3.5707$ reactions. $P(X \geq 90) = P\left(Z \geq \frac{90-85}{3.5707}\right) = P(Z \geq 1.40) = 1 - 0.9192 = 0.0808$. The exact probability is 0.0994, which is 0.0186 larger than the approximate probability. (b) If the subcutaneous injection is used, then $np = 95 \geq 10$, but $n(1-p) = 5$, which is not at least 10, so the Normal approximation cannot be used.

14.31 (a) If R is the number of red-blossomed plants out of a sample of 4, then $P(R = 3) = \binom{4}{3}(0.75)^3(0.25)^1 = 0.4219$. (b) With $n = 60$, $np = 60(0.75) = 45$. (c) $P(R \geq 45) = P(Z \geq 0)$ $= 0.5000$ (software gives 0.5688 as the exact probability). The probability using the Normal approximation is 0.0688 less than the actual probability.

14.33 (a) Elantras accounted for the proportion $200{,}415/677{,}946 = 0.2956$. (b) If E is the number of Elantra buyers, then E has the binomial distribution with $n = 1000$ and $p = 0.2956$; $\mu = np = (1000)(0.2956) = 295.6$ and $\sigma = \sqrt{np(1-p)} = \sqrt{1000(0.2956)(0.7044)} = 14.430$ Elantra buyers. (c) Observe $np = 295.6 \geq 10$ and $n(1-p) = 704.4 \geq 10$, so the Normal approximation can be applied. $P(E < 300)$ $= P(E \leq 299) = P\left(Z \leq \frac{299-295.6}{14.430}\right) = P(Z \leq 0.24) = 0.5948$.

14.35 **(a)** With $n = 100$ and $p = 0.75$, $\mu = np = (100)(0.75) = 75$ and $\sigma = \sqrt{np(1-p)} = \sqrt{100(0.75)(0.25)} = 4.3301$ questions. $P(70 \le X \le 80) = P\left(\frac{70-75}{4.3301} \le Z \le \frac{80-75}{4.3301}\right) = P(-1.15 \le Z \le 1.15) = 0.8749 - 0.1251 = 0.7498$ (software gives 0.7518). **(b)** With $n = 250$ and $p = 0.75$, $\mu = np = (250)(0.75) = 187.5$ and $\sigma = \sqrt{np(1-p)} = \sqrt{250(0.75)(0.25)} = 6.8465$ questions. $P(175 \le X \le 200) = P\left(\frac{175-187.5}{6.8465} \le Z \le \frac{200-187.5}{6.8465}\right) = P(-1.83 \le Z \le 1.83) = 0.9664 - 0.0336 = 0.9328$ (software gives 0.9428).

14.37 **(a)** Answers will vary, but more than 99.7% of samples should have zero to four bad tomatoes. **(b)** Each time we choose a sample of size 10, the probability that we have one bad tomato is 0.3854; out of 20 samples, the number of times that we have exactly one bad tomato has a binomial distribution with $n = 20$ and $p = 0.3854$. Usually, between 2 and 14 out of 20 samples have exactly one bad tomato.

14.39 The number N of infections untreated Bob Jones University students is binomial with $n = 1400$ and $p = 0.80$, so $\mu = np = (1400)(0.80) = 1120$ and $\sigma = \sqrt{np(1-p)} = \sqrt{1400(0.80)(0.20)} = 14.9666$ students. We can use the Normal approximation because $np = 1120 \ge 10$ and $n(1-p) = 280 \ge 10$. Of the 1400 Bob Jones University students, 75% is 1050. $P(N \ge 1050) = P\left(Z \ge \frac{1050-1120}{14.9666}\right) = P(Z \ge -4.68)$, which is very near to 1. (Exact binomial computation gives 0.999998.)

14.41 Let V and U be the number, respectively, of new infections among the vaccinated and unvaccinated children. **(a)** $P(V = 1) = 0.3741$ and $P(U = 1) = 0.0960$. Because the events are (assumed) independent, $P(V = 1 \text{ and } U = 1) = P(V = 1)P(U = 1) = (0.3741)(0.0960) = 0.0359$. **(b)** We have $P(2 \text{ infections}) = P(V = 0 \text{ and } U = 2) + P(V = 1 \text{ and } U = 1) + P(V = 2 \text{ and } U = 0) = P(V = 0)P(U = 2) + P(V = 1)P(U = 1) + P(V = 2)P(U = 0) = (0.4181)(0.3840) + (0.3741)(0.0960) + (0.1575)(0.0080) = 0.1977$.

14.43 **(a)** $np = (20)(0.5) = 10 \ge 10$ and $n(1-p) = (20)(0.5) = 10 \ge 10$. **(b)** $\mu = np = 10$ and $\sigma = \sqrt{np(1-p)} = \sqrt{20(0.5)(0.5)} = 2.236$ cases. $P(X \ge 13) = P\left(Z \ge \frac{13-10}{2.236}\right) = P(Z \ge 1.34) = 1 - 0.9099 = 0.0901$. **(c)** $P(X \ge 13) = P(X \ge 12.5) = P\left(Z \ge \frac{12.5-10}{2.236}\right) = P(Z \ge 1.12) = 1 - 0.8686 = 0.1314$.

Chapter 15 – Sampling Distributions

15.1 Both are statistics; they came from the 11 subjects in the experiment.

15.3 27.3% and 4.4% are statistics; they are based on the 14,765 American high school students. 26.5% is a parameter; it is based on all American high school students.

15.5 Although the probability of having to pay for a total loss for one or more of the 10 policies is very small, if this were to happen, it would be financially disastrous. For thousands of policies, the law of large numbers says that the average claim on many policies will be close to the mean, so the insurance company can be assured that the premiums it collects will (almost certainly) cover the claims.

15.7 (a) The histogram is provided.

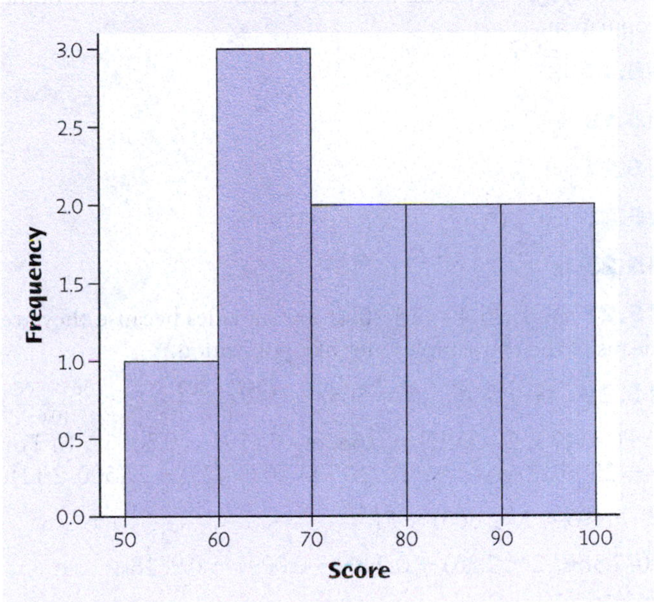

(b) The mean is $\mu = 75.6$. (c) and (d) Results will vary. The result for one sample is shown.

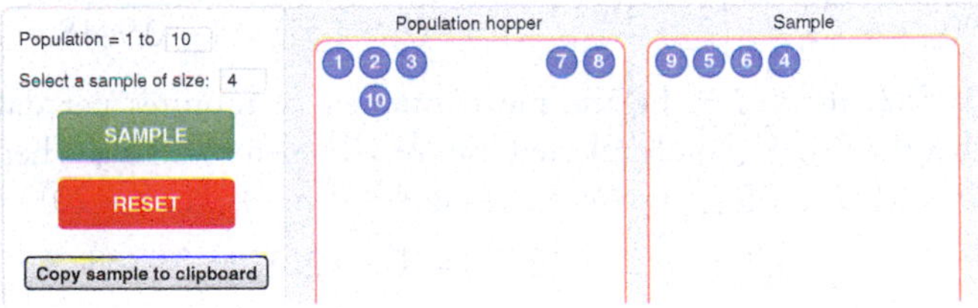

This sample selects students 9, 5, 6, and 4, with scores of 88, 72, 84, and 75, respectively. Their mean is $\bar{x} = \frac{88+72+84+75}{4} = 79.75$. Results from all 10 samples of size $n = 4$ will vary from this result and from each other.

15.9 (a) The sampling distribution of $\bar{x}$ is $N(182 \text{ mg/dL}, 3.7 \text{ mg/dL})$. Therefore, using Table A, $P(180 < \bar{x} < 184) = P\left(\frac{180-182}{3.7} < Z < \frac{184-182}{3.7}\right) = P(-0.54 < Z < 0.54) = 0.7054 - 0.2946 = 0.4108$. (b) With $n = 1000$, the sampling distribution of $\bar{x}$ is $N(182 \text{ mg/dL}, 1.17 \text{ mg/dL})$, so, using Table A, $(180 < \bar{x} < 184) = P(-1.71 < Z < 1.71) = 0.9564 - 0.0436 = 0.9128$.

15.11 Targets (a) and (c) show bias; targets (b) and (c) show large sampling variability.

15.13 (a) The mean is 2.2, and the standard deviation is $\frac{\sigma}{\sqrt{n}} = \frac{3.9}{\sqrt{50}} = 0.5515$. (b) The sampling distribution of $\bar{x}$ is $N(2.2,\ 0.5515)$. Therefore, using Table A, $P(\bar{x} > 3.0) = P\left(Z > \frac{3.0-2.2}{0.5515}\right) = P(Z > 1.45) = 1 - 0.9265 = 0.0735$.

15.15 A sample variance of 9.2 ($s = 3.033$) is on the low end of the histogram shown in Figure 15.8(b) of the text. However, because it is not unusual, we cannot say that the variability of adults aged 65 and older is different from that of the rest of the population.

15.17 (b)

15.19 (b)

15.21 (c)

15.23 (c)

15.25 (a)

15.27 Both 25.40 and 20.41 are statistics because they are means of the two samples, not of a population.

15.29 (a) $P(495 < X < 505) = P\left(\frac{495-500}{10.6} < Z < \frac{505-500}{10.6}\right) = P(-0.47 < Z < 0.47) = 0.6808 - 0.3192 = 0.3616$. (b) For $n = 25$ students, $\bar{x}$ is $N(500, 10.6/\sqrt{25}) = N(500, 2.12)$. (c) $P(495 < \bar{X} < 505) = P\left(\frac{495-500}{2.12} < Z < \frac{505-500}{2.12}\right) = P(-2.36 < Z < 2.36) = 0.9909 - 0.0091 = 0.9818$.

15.31 (a) Let $\bar{x}$ be the mean number of minutes per day that the five randomly selected mildly obese people spend walking. Then $\bar{x}$ is $N(373, 67/\sqrt{5}) = N(373 \text{ min}, 29.96 \text{ min})$. $P(\bar{x} > 420) = P\left(Z > \frac{420-373}{29.96}\right) = P(Z > 1.57) = 1 - 0.9418 = 0.0582$. (b) Let $\bar{x}$ be the mean number of minutes per day that the five randomly selected lean people spend walking. Then $\bar{x}$ is $N(526, 107/\sqrt{5}) = N(526 \text{ min}, 47.85 \text{ min})$. $P(\bar{x} > 420) = P\left(Z > \frac{420-526}{47.85}\right) = P(Z > -2.22) = 1 - 0.0132 = 0.9868$.

15.33 (a) For the emissions X of a single car, $P(X > 86) = P\left(Z > \frac{86-80}{4}\right) = P(Z > 1.5) = 1 - 0.9332 = 0.0668$. (b) The average $\bar{x}$ has the $N(80, 4/\sqrt{25}) = N(80 \text{ mg/mi}, 0.8 \text{ mg/mi})$. Therefore, $P(\bar{X} > 86) = P\left(Z > \frac{86-80}{0.8}\right) = P(Z > 7.5)$, which is essentially 0.

15.35 The mean NOX level for 25 cars has an $N(80 \text{ mg/mi}, 0.8 \text{ mg/mi})$ distribution, and $P(Z > 2.33) = 0.01$ if Z is $N(0, 1)$, so $L = 80 + (2.33)(0.8) = 81.864$ mg/mi.

15.37 STATE: What is the probability that the total weight of the 22 passengers exceeds 4500 pounds? PLAN: Use the central limit theorem to approximate this probability. SOLVE: If W is total weight, then the sample mean weight is $\bar{x} = W/22$. The event that the total weight exceeds 4500 pounds is equivalent to the event that $\bar{x}$ exceeds $4500/22 = 204.55$ pounds. Note that $\bar{x}$ is approximately Normal with mean 195 pounds and standard deviation $35/\sqrt{22} = 7.462$ pounds, so $P(W > 4500) = P(\bar{x} > 204.55) = P\left(Z > \frac{204.55-195}{7.462}\right) = P(Z > 1.28) = 1 - 0.8997 = 0.1003$. CONCLUDE: There is about a 10% chance that the total weight exceeds 4500 pounds.

15.39 (a) 99.7% of all observations fall within 3 standard deviations, so we want $3\sigma/\sqrt{n} = 1$. The standard deviation of $\bar{x}$ (i.e., $\sigma/\sqrt{n}$) must therefore be 1/3 point. (b) We need to choose n so that $10.6/\sqrt{n} = 1/3$. This means $\sqrt{n} = (10.6)(3) = 31.8$, so $n = 1011.24$. Because n must be a whole number, take $n = 1012$.

15.41 (a) With $n = 14{,}000$, the mean and standard deviation of $\bar{x}$ are \$0.60 and $\$18.96/\sqrt{14{,}000} = \$0.1602$, respectively. (b) $P(\$0.50 < \bar{X} < \$0.70) = P\left(\frac{0.50-0.60}{0.1602} < Z < \frac{0.70-0.60}{0.1602}\right) = P(-0.62 < Z < 0.62) = 0.7324 - 0.2676 = 0.4648$.

15.43 (a) The estimate in Exercise 15.41(b) was 0.4648, so the Normal approximation underestimates the exact answer by 0.0313. (b) With $n = 3500$, the Normal approximation gives $P(\$0.50 < \bar{X} < \$0.70) = P\left(\frac{0.50-0.60}{0.3205} < Z < \frac{0.70-0.60}{0.3205}\right) = P(-0.31 < Z < 0.31) = 0.6217 - 0.3783 = 0.2434$ (using Table A). This is smaller than the exact answer by 0.1614. (c) With $n = 150{,}000$, the Normal approximation gives $P(\$0.50 < \bar{X} < \$0.70) = P\left(\frac{0.50-0.60}{0.0490} < Z < \frac{0.70-0.60}{0.0490}\right) = P(-2.04 < Z < 2.04) = 0.9793 - 0.0207 = 0.9586$ (using Table A). This is smaller than the exact answer by only 0.0043.

15.45 The probability found in Exercise 12.50 was $1/20 = 0.05$. This outcome is very unlikely if we assume that our Canadian friend is selecting by chance alone, so the result is statistically significant.

Chapter 16 – Confidence Intervals: The Basics

16.1 (a) The sampling distribution of $\bar{x}$ has standard deviation $\frac{\sigma}{\sqrt{n}} = \frac{40}{\sqrt{147{,}400}} = 0.1042$. (b) 95% of all values of $\bar{x}$ fall within 2 standard deviations of the mean μ (within $2(0.1042) = 0.2084$ point). (c) 282 ± 0.2084, or between 281.7916 and 282.2084 points.

16.3 Answers will vary due to randomness. In 99.7% of all repetitions in part (a), the applet should produce between 5 and 10 hits. Out of 1000 80% confidence intervals, nearly all hit rates will fall between 76% and 84%.

16.5 Search Table A for the probability that is closest to 0.1250. The value $z = -1.15$ has area 0.1251 to its left, so the critical value is $z^* = 1.15$.

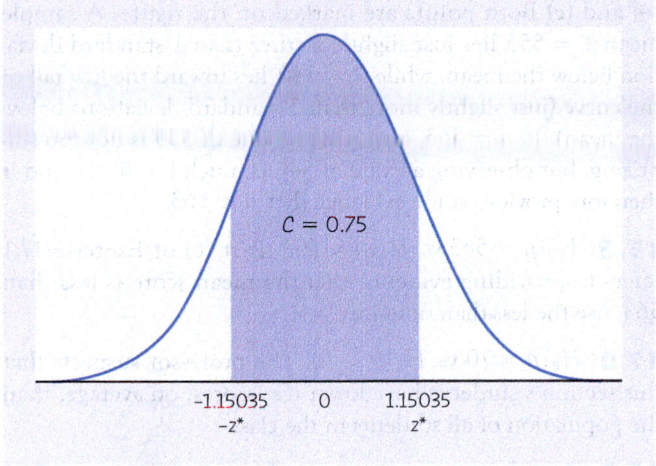

16.7 (a) A stemplot is provided. There are no apparent deviations from Normality.

13	6
13	02
12	677888
12	003344
11	55688999
11	0000111122223333334444
10	555666777789
10	0022333344
9	6778
9	0133
8	69

(b) STATE: What is the mean IQ μ of all seventh-grade girls in this school district? PLAN: We estimate μ by giving a 99% confidence interval. SOLVE: The problem states that these girls are an SRS of the population, which is assumed to be very large. We saw that the scores are consistent with having come from a Normal population, and we know σ, so the conditions are met. With $\bar{x} = 110.73$ points and $z^* = 2.576$, our 99% confidence interval for μ is given by $110.73 \pm 2.576 \frac{11}{\sqrt{74}} = 110.73 \pm 3.29$ points. CONCLUDE: We are 99% confident that the mean IQ of seventh-grade girls in this district is between 107.44 and 114.02 points. (c) No. The phrase "99% confidence" is about the process of constructing a confidence interval, not about the percentage of individual IQ test scores that would fall in the interval. "99% confidence" means that, if we repeat the sampling process many times and construct a 99% confidence interval each time, 99% of the intervals will include the mean IQ for all seventh-grade girls.

16.9 With $z^* = 1.96$ and $\sigma = 11.6$, the margin of error is $z^* \frac{\sigma}{\sqrt{n}} = \frac{22.736}{\sqrt{n}}$. (a) and (b) The margins of error are given in the table.

n	m.e.
100	2.2736
400	1.1368
1600	0.5684

(c) Margin of error decreases by one-half as n increases fourfold.

16.11 (c)

16.13 (b)

16.15 (b)

16.17 (a)

16.19 (a) We use $\bar{x} \pm z^* \frac{\sigma}{\sqrt{n}} = 13.7 \pm 2.576 \frac{7.4}{\sqrt{463}} = 13.7 \pm 0.886 = 12.814$ to 14.586 hours. (b) The 463 students in this class must be a random sample of all of the first-year students at this university, and the entire population of first-year students must be at least 9260 to satisfy conditions for inference.

16.21 The margin of error is now $2.576 \frac{7.4}{\sqrt{464}} = 0.885$, so the extra observation has minimal impact on the margin of error. However, if $\bar{x} = 35.2$, then the 99% confidence interval for the average amount of time spent studying becomes $35.2 \pm 0.885 = 34.315$ to 36.085 hours, which is very different from the result in Exercise 16.19.

16.23 The student is confused. If we repeatedly took samples, then 95% of all future sample means would be within 1.96 standard deviations of μ (that is, within $1.96 \frac{\sigma}{\sqrt{n}}$) of the true, unknown value of μ. Future samples will have no memory of this sample.

16.25 (a) A stemplot of the data is provided. Notice that the distribution is noticeably skewed to the left. The data do not appear to follow a Normal distribution.

33	0237
32	033677
31	399
30	259
29	
28	7
27	
26	5
25	
24	1
23	0

(b) PLAN: We will estimate μ by giving a 95% confidence interval. SOLVE: The problem states that we are willing to take this sample to be an SRS of the population. In spite of the shape of the stemplot, we are told to assume that this distribution is Normal, with standard deviation $\sigma = 3000$ lb. We have $\bar{x} = 30{,}841$ lb, so the 95% confidence interval for μ is given by $30{,}841 \pm 1.96 \frac{3000}{\sqrt{20}} = 30{,}841 \pm 1314.8 = 29{,}526.2$ to $32{,}155.8$. CONCLUDE: With 95% confidence, the mean load μ

required to break apart pieces of Douglas fir is between 29,526.2 and 32,155.8 lb; however, given the shape of the distribution of the data, we cannot put much reliance in this interval. **(c)** The student is not correct. The confidence interval is about the population mean μ, not about future measurements.

16.27 **(a)** A stemplot of the data is provided. It seems reasonable that the sample comes from a Normal distribution. For inference, we must assume that the 10 students with untrained noses were selected randomly from the population of all people with untrained noses.

4	2
3	5
3	0013
2	9
2	23
1	9

(b) PLAN: We will estimate μ with a 95% confidence interval. SOLVE: We have assumed that we have a random sample and that the population we are sampling from is Normal. We obtain $\bar{x} = 29.4\ \mu g/L$. Our 95% confidence interval for μ is given by $29.4 \pm 1.96\frac{7}{\sqrt{10}} = 29.4 \pm 4.34 = 25.06$ to $33.74\ \mu g/L$. CONCLUDE: With 95% confidence, the mean sensitivity for all people with untrained noses is between 25.06 and 33.74 $\mu g/L$.

16.29 No, we cannot conclude that. The 4% margin of error is based on the entire sample of 1108 adults. Since the sample size for Independents is smaller than the sample size for the whole sample, the margin of error will be larger than 4%.

Chapter 17 – Tests of Significance: The Basics

17.1 **(a)** If $\mu = 563$, the sampling distribution of $\bar{x}$ is approximately Normal with mean $\mu = 563$ and standard deviation $\frac{\sigma}{\sqrt{n}} = \frac{118}{\sqrt{250}} = 7.463$. The density curve is provided.

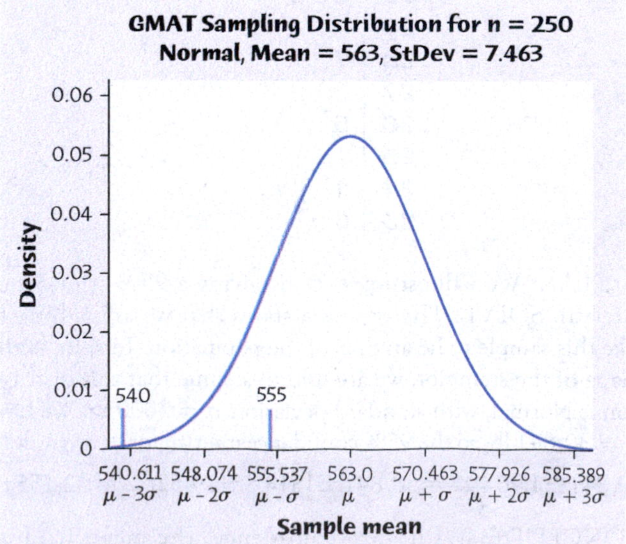

(b) and **(c)** Both points are marked on the figure. A sample mean $\bar{x} = 555$ lies just slightly farther than 1 standard deviation below the mean, while $\bar{x} = 540$ lies toward the low tail of the curve (just slightly more than 3 standard deviations below the mean). If $\mu = 563$, observing a value of 555 is not too surprising, but observing a value of 540 is much less likely, and it therefore provides some evidence that $\mu < 563$.

17.3 $H_0: \mu = 563$ vs. $H_a: \mu < 563$. Part (c) of Exercise 17.1 refers to providing evidence that the mean score is less than 563; use the less-than one-sided test.

17.5 $H_0: \mu = 70$ vs. $H_a: \mu < 70$. The professor suspects that this section's students have lower test scores, on average, than the population of all students in the class.

17.7 Hypotheses are statements about parameters, not statistics. The research question should not be about the sample mean, $\bar{x}$, but should be about the population mean, μ.

17.9 **(a)** With $\sigma = 60$ and $n = 18$, the standard deviation of the sampling distribution of $\bar{x}$ is $\frac{\sigma}{\sqrt{n}} = \frac{60}{\sqrt{18}} = 14.1421$, so when $\mu = 0$, the distribution of $\bar{x}$ is $N(0, 14.1421)$. **(b)** The P-value is $P = 2P\left(z \geq \left|\frac{17-0}{14.1421}\right|\right) = 2P(Z \geq 1.20) = 2(0.1151) = 0.2302$.

17.11 **(a)** The applet using $\bar{x} = 555$ provides a P-value of 0.1419. This is not significant at either $\alpha = 0.05$ or $\alpha = 0.01$.

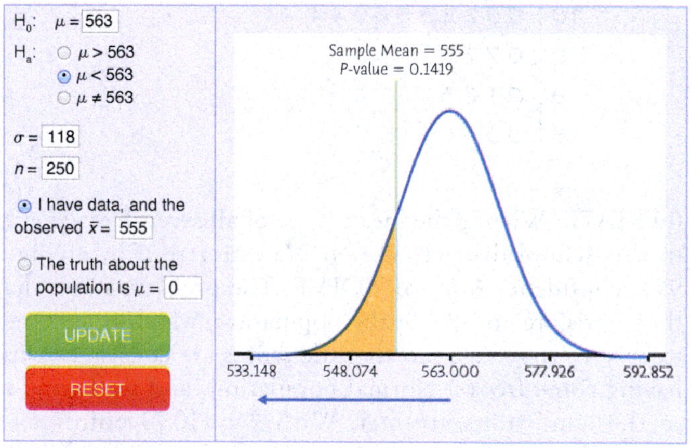

(b) The applet using $\bar{x} = 540$ provides a P-value of 0.0010. This is significant at both $\alpha = 0.05$ and $\alpha = 0.01$.

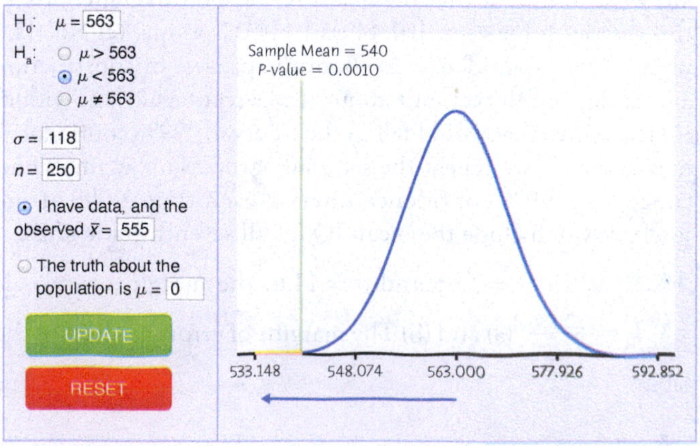

(c) If $\mu = 563$ (that is, if H_0 were true), observing a value of 555 would not be very unlikely, but observing a value of 540 is not likely at all, and it therefore provides strong evidence that $\mu < 563$.

17.13 (a) $z = \frac{0.3-0}{1/\sqrt{10}} = \frac{0.3}{0.3162} = 0.9488$. (b) $z = \frac{1.02-0}{1/\sqrt{10}} = \frac{1.02}{0.3162} = 3.226$. (c) $z = \frac{17-0}{60/\sqrt{18}} = \frac{17}{14.1421} = 1.2021$. Note that in part (c), the test is two-sided, while in parts (a) and (b), it is one-sided.

17.15 STATE: Is there evidence that the average percentage tip when bad news is received (a bad weather prediction) is less than 20%? PLAN: Let μ be the average percentage tip for all customers receiving bad news. We test $H_0\!: \mu = 20$ against $H_a\!: \mu < 20$. SOLVE: We have a random sample of $n = 20$ customers and were told to assume that tips have a Normal distribution. We observe $\bar{x} = 18.19\%$. The standard deviation of $\bar{x}$ is $\frac{\sigma}{\sqrt{n}} = \frac{2}{\sqrt{20}} = 0.4472$, so $z = \frac{18.19-20}{0.4472} = -4.05$, and the *P*-value is $P(Z \le -4.05) \approx 0$. CONCLUDE: There is overwhelming evidence that the average tip percentage when bad news is delivered is lower than the overall average tip percentage (20%).

17.17 Using the z^* row of Table C, $z = 1.65$ is not significant at $\alpha = 0.05$ because it is not larger than 1.960 or smaller than -1.960. It is also not significant at $\alpha = 0.01$ because $|z|$ is smaller than 2.576.

17.19 (a)

17.21 (a)

17.23 (a)

17.25 (c)

17.27 (c)

17.29 (a) $H_0\!: \mu = 0$ vs. $H_a\!: \mu > 0$. (b) $z = \frac{2.35-0}{2.5/\sqrt{200}} = 13.29$. (c) The *P*-value is essentially 0. Under H_0, it would be virtually impossible to observe a sample mean as large as 2.35 based on a sample of 200 men. This sample mean cannot be explained by random chance, as we would easily reject H_0.

17.31 "$P = 0.005$" means H_0 is not likely to be correct–but only in the sense that it provides a poor explanation of the data observed. It means that if H_0 were true, a sample as contrary to H_0 as our sample would occur by chance alone about 0.5% of the time if the experiment were repeated over and over. However, it does not mean that there is a 0.5% chance that H_0 is true.

17.33 The person is confusing practical significance with statistical significance. In fact, a 5% increase isn't a lot in a pragmatic sense. However, $P = 0.03$ means that random chance probably does not explain the difference observed.

17.35 In the sketch, the "significant at 1%" region includes only the dark shading ($Z \ge 2.326$). The "significant at 5%" region of the sketch includes both the light and dark shading ($Z \ge 1.645$). If $P < 0.01$, we must have $P < 0.05$. The converse is false; something that occurs fewer than 5 times in 100 repetitions is not necessarily as rare as something that happens less than once in 100 repetitions, so a test that is significant at the 5% level is not necessarily significant at the 1% level.

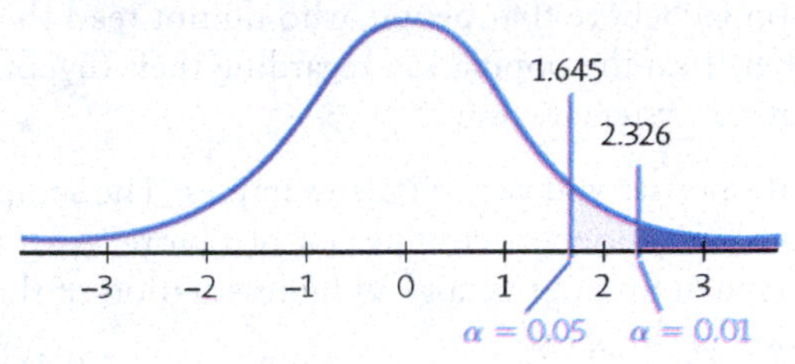

17.37 Because a *P*-value is a probability, it can never be greater than 1. The correct *P*-value is $P(Z \ge 1.33) = 0.0918$.

17.39 PLAN: Test the hypotheses $H_0\!: \mu = 0\%$ against $H_a\!: \mu < 0\%$. SOLVE: $\bar{x} = -3.587$, $z = \frac{-3.587-0}{2.5/\sqrt{47}} = -9.84$, and $P = P(Z \le -9.84) \approx 0$. CONCLUDE: There is overwhelming evidence that, on average, nursing mothers lose bone mineral.

17.41 (a) We test $H_0\!: \mu = 0$ vs. $H_a\!: \mu > 0$, where μ is the mean sensitivity difference in the population of people wearing eye grease. (b) PLAN: We test the hypotheses stated in part (a). SOLVE: $\bar{x} = 0.10125$, $z = \frac{0.10125-0}{0.22/\sqrt{16}} = 1.84$, and $P = P(Z \ge 1.84) = 0.0329$. CONCLUDE: The sample gives significant evidence (at $\alpha = 0.05$) that eye grease increases sensitivity, on average.

17.43 (a) Yes, because 6.2 falls outside of the 99% confidence interval, which is (6.3, 8.1). (b) No, because 6.4 falls in the 99% confidence interval.

17.45 STATE: We will use a 90% confidence interval to test the mean melting point of copper. PLAN: We test $H_0\!: \mu = 1084.62°\text{C}$ vs. $H_a\!: \mu \ne 1084.62°\text{C}$, where μ is the mean melting point of copper. SOLVE: $\bar{x} = 1084.8$. From Table C, $z^* = 1.645$. The 90% confidence interval for μ is given by $1084.8 \pm 1.645\frac{0.25}{\sqrt{6}} = 1084.8 \pm 167.89 = 916.9$ to $1252.7°\text{C}$. Notice that the true melting point of copper (1084.62°C) is in the interval. CONCLUDE: Since 1084.62°C is in the 90% confidence interval, we cannot conclude that the population mean melting point of copper differs from 1084.8°C.

Chapter 18 – Inference in Practice

18.1 Reason (c) is most important; this is a voluntary response survey consisting only of those customers that choose to respond to the email. This is not an SRS. Anything we learn from this sample will not extend to the larger population. The other two reasons are valid but less important. Reason (a)—the size of the sample and large margin of error—would make the interval less informative, even if the sample were representative of the population. Reason (b)—nonresponse—is a potential problem with every survey, but there is no reason to believe that people who do not read the invitation differ from the population regarding their (hypothetical) rating in any systematic way.

18.3 Responses will vary. Some examples: The sample isn't random. Also, shoppers coming out of a store in an upscale shopping mall may not be a good representation of the entire population of shoppers.

18.5 You cannot conclude this. The restaurant you work at is most likely different in ways that could affect tip amounts from the one where the experiment took place.

18.7 The only source of error included in the margin of error is that due to random sampling variability, so (c).

18.9 (a) and (b) The results and the curve for $n = 9$ are shown here. We see that as the sample size increases, the same difference between μ_0 and $\bar{x}$ goes from being not at all significant to highly significant.

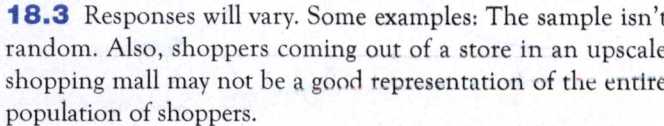

n	P-value
9	0.1587
16	0.0912
36	0.0228
64	0.0038

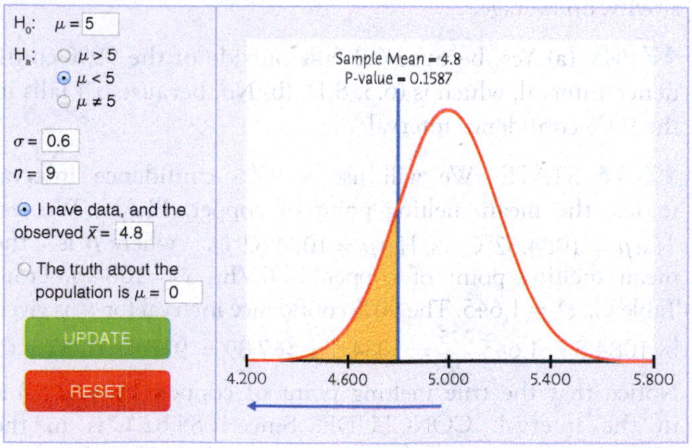

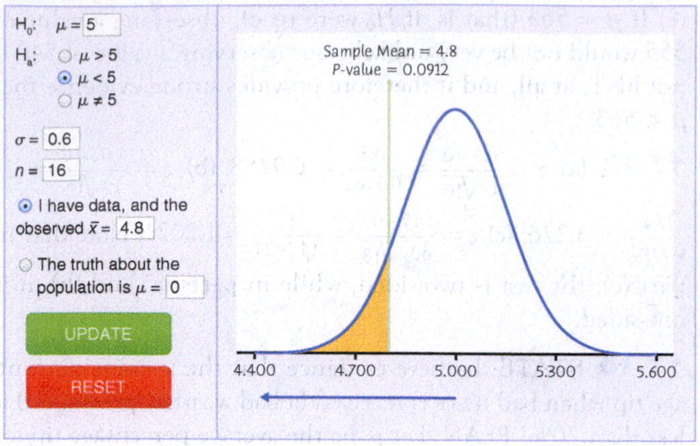

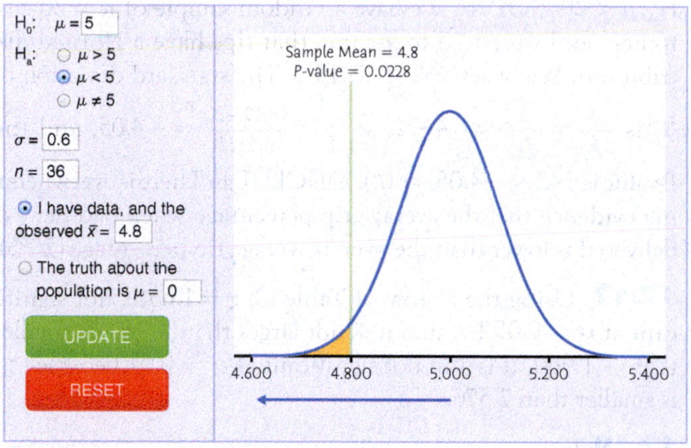

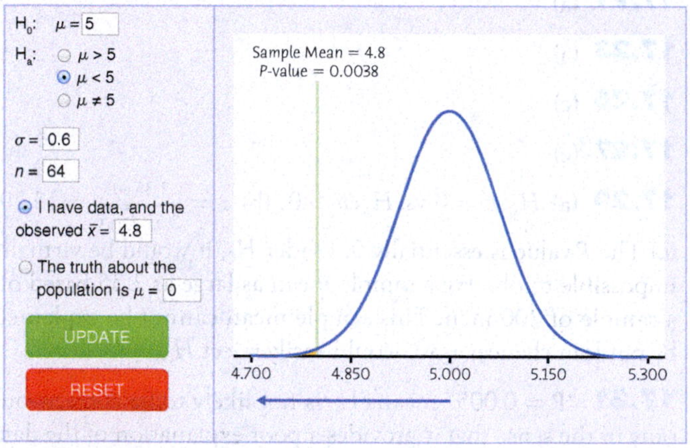

18.11 (a) Each test (subject) has a 5% chance of being deemed "significant" at the 5% level when the null hypothesis (no ESP) is true. With 1000 tests, we would expect about 50 such occurrences just by chance. (b) Retest the 43 promising subjects with a different version of the test.

18.13 For margin of error ±10, we need at least $n = \left(\frac{1.645 \times 40}{1}\right)^2 = 4329.64$, so $n = 4330$.

18.15 (a) Increase power by taking more measurements. (b) If you increase α, you make it easier to reject H_0, thus increasing power. (c) A value of $\mu = 214$ is closer to the stated value of $\mu = 208$ under H_0, so power decreases.

18.17

σ	40	30	20
Power	0.394	0.575	0.865

As σ decreases, power increases. Less variability in the population increases the researcher's ability to recognize a false null hypothesis.

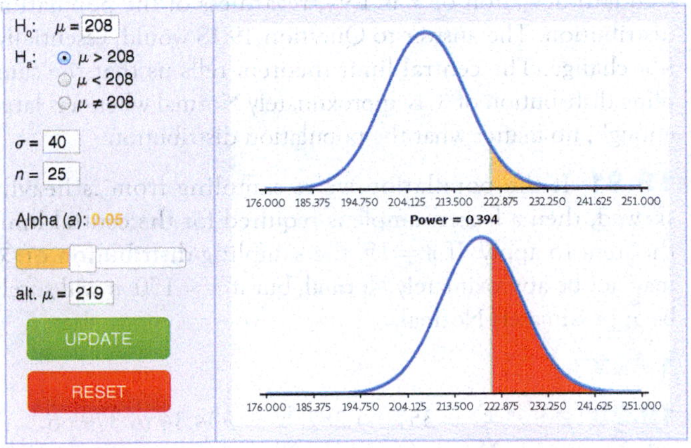

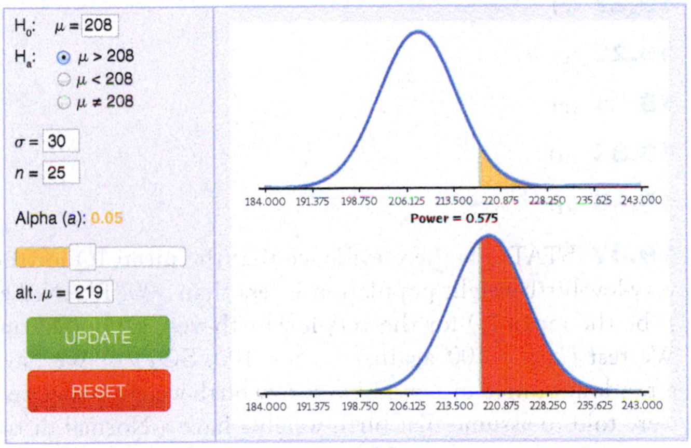

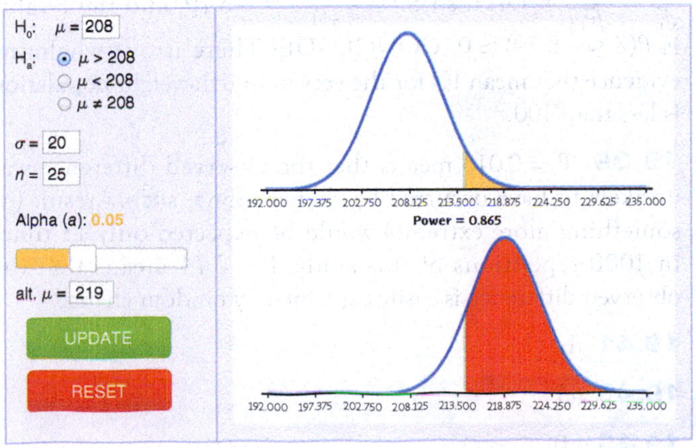

18.19 (c)

18.21 (b)

18.23 (a)

18.25 (c)

18.27 (c)

18.29 We need to know that the samples taken from both populations are random. It would also be helpful to know the size of each sample, since large sample sizes could result in statistically significant results even without practical significance. To this end, it would be helpful to know the effect size (a measure of the magnitude of the difference without accounting for the sample sizes).

18.31 Many high school students might be reluctant to admit that they had carried a weapon in the past 30 days. Thus, this response is likely to be biased low. The margin of error only covers random sampling errors and does not allow for this response bias.

18.33 The effect size is greater if the sample is small. With a larger sample, the impact of any one value is small.

18.35 Opinion—even expert opinion—unsupported by data is the weakest type of evidence, so the third description is level C. The second description refers to experiments (clinical trials) and large samples; that is the strongest evidence (level A).

18.37 (a) The *P*-value decreases. (b) Power increases.

18.39 (a) $\bar{x} = 19.562, z = \dfrac{19.562 - 17.5}{2.5/\sqrt{5}} = 1.84, P = 2P(Z \geq 1.84) = 2(0.0329) = 0.0658$. This is not significant at the 5% level. We would not reject 17.5 as a plausible value of μ, even though $\mu = 20$. (b) The small sample size makes it difficult to detect a difference that is really there, unless the difference is much greater than 2.5.

18.41 (a) "Statistically insignificant" means that the differences observed were no more than might have been expected to occur by chance even is SES had no effect on LSAT applications. (b) If the results are based on a small sample, then even if the null hypothesis were not true, the test might not be sensitive enough to detect the effect (i.e., have low power). Knowing the effects were small tells us that the test was not insignificant merely because of a small sample size. Furthermore, a small effect may be of no *practical* significance.

18.43 $n = \left(\dfrac{1.96 \times 3000}{600}\right)^2 = 96.04$; take $n = 97$.

18.45 (a) This test has a 20% chance of rejecting H_0 when the alternative is true. (b) If the test has 20% power, then when the alternative is true, it will fail to reject H_0 80% of the time. (c) The sample sizes are very small, which typically leads to low power. This is exacerbated by the large σ, as noted by the study author.

18.47 From the applet, against the alternative $\mu = 20$, power $= 0.609$.

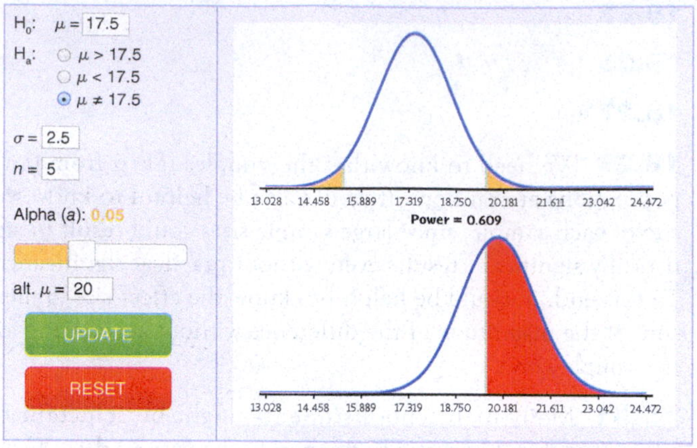

18.49 (a) $z = \frac{\bar{x} - \mu_0}{\sigma/\sqrt{n}} = \frac{\bar{x} - 17.5}{2.5/\sqrt{5}} = 0.894\bar{x} - 15.652$. Because the alternative is $\mu \neq 17.5$, we reject H_0 at the 5% level when $z \geq 1.96$ or $z \leq -1.96$. (b) We reject H_0 at the 5% level when $z = 0.894\bar{x} - 15.652 \geq 1.96$ (i.e., when $\bar{x} \geq 19.70$) or $z = 0.894\bar{x} - 15.652 \leq -1.96$ (i.e., when $\bar{x} \leq 15.315$). (c) When $\mu = 20$, the power is $P(\bar{x} \geq 19.70) + P(\bar{x} \leq 15.315) = P\left(Z \geq \frac{19.70 - 20}{2.5/\sqrt{5}}\right) + P\left(Z \leq \frac{15.315 - 20}{2.5/\sqrt{5}}\right) = P(Z \geq -0.27) + P(Z \leq -4.19) \approx 0.6064 + 0.0000 = 0.6064$.

18.51 Power $= 1 - P(\text{Type II error}) = 1 - 0.44 = 0.56$.

18.53 (a) In the long run, this probability should be 0.05. Out of 100 simulated tests, the number of false rejections will have a binomial distribution with $n = 100$ and $p = 0.05$. Most results will be between 0 and 10 rejections (inclusive). (b) If the power is 0.812, the probability of a Type II error is 0.188. Out of 100 simulated tests, the number of false nonrejections will have a binomial with $n = 100$ and $p = 0.188$. Most results will be between 10 and 29 nonrejections (inclusive).

18.55 X is binomial with $n = 20$ and $p = 0.05$. $P(X \geq 1) = 1 - P(X < 1) = 1 - P(X = 0) = 1 - \binom{20}{0}(0.05)^0(0.95)^{20} = 1 - 0.3585 = 0.6415$.

Chapter 19 – From Data Production to Inference: Part III Review

19.1 (a) $S = \{\text{Under age 40, Age 40 or over}\}$ (b) $S = \{6, 7, 8, \ldots, 19, 20\}$. (c) $S = \{\text{All values } 2.5 \leq \text{VO2} \leq 6.1 \text{ liters per minute}\}$. (d) $S = \{\text{All heart rates such that } x \text{ bpm} < \text{heart rate} < y \text{ bpm}\}$. Values chosen for x and y will vary.

19.3 (a)

19.5 $\{Y > 1\}$ or $\{Y \geq 2\}$; $P(Y > 1) = 1 - 0.28 = 0.72$.

19.7 (d)

19.9 $\{X \leq 2\}$ is the event that a woman between the ages of 15 and 50 has given birth to two or fewer children. $P(X \leq 2) = 0.442 + 0.168 + 0.217 = 0.827$.

19.11 $\{X \geq 3\}$; $P(X \geq 3) = 0.107 + 0.043 + 0.023 = 0.173$.

19.13 (a)

19.15 $P(-2 < Y < 2) = [2 - (-2)](0.1) = 4(0.1) = 0.4$.

19.17 (c)

19.19 The answer to Question 19.16 would change because this refers to the population distribution, which is now non-Normal. The answer to Question 19.17 would not change; the standard deviation of $\bar{x}$ is 1.94, regardless of the population distribution. The answer to Question 19.18 would, essentially, not change. The central limit theorem tells us that the sampling distribution of $\bar{x}$ is approximately Normal when n is large enough, no matter what the population distribution.

19.21 If the population we're sampling from is heavily skewed, then a larger sample is required for the central limit theorem to apply. If $n = 15$, the sampling distribution of $\bar{x}$ may not be approximately Normal, but if $n = 150$, it will surely be approximately Normal.

19.23 (c)

19.25 $\bar{x} \pm z^* \frac{\sigma}{\sqrt{n}} = 357 \pm 1.282 \frac{50}{\sqrt{8}} = 334.34$ to 379.66.

19.27 (c)

19.29 (c)

19.31 (c)

19.33 (d)

19.35 (d)

19.37 STATE: Is there evidence that the mean IQ for the very-low-birth-weight population is less than 100? PLAN: Let μ be the mean IQ for the very-low-birth-weight population. We test $H_0: \mu = 100$ against $H_a: \mu < 100$. SOLVE: We have a random sample of $n = 113$ very-low-birth-weight males and were told to assume that birth weights have a Normal distribution. We observe $\bar{x} = 87.6$. The standard deviation of $\bar{x}$ is $\frac{\sigma}{\sqrt{n}} = \frac{15}{\sqrt{113}} = 1.411$, so $z = \frac{87.6 - 100}{1.411} = -8.79$, and the P-value is $P(Z \leq -8.79) \approx 0$. CONCLUDE: There is overwhelming evidence that mean IQ for the very-low-birth-weight population is less than 100.

19.39 $P = 0.013$ means that the observed difference was unlikely to have occurred by chance alone; such a result (or something more extreme) would be expected only 13 times in 1000 repetitions of this study. $P = 0.74$ means that the observed difference is easily explained by random chance.

19.41 (b)

19.43 (c)

19.45 (d)

19.47 (a) This is a binomial with $n = 5$ and $p = 0.4$.

19.49 $P(X \geq 1900) = P\left(Z \geq \frac{1900 - 1838.5}{26.78}\right) = P(Z \geq 2.30) = 1 - 0.9893 = 0.0107$.

19.51 (a) All probabilities are between 0 and 1, and their sum is 1. (b) Let R_1 be Taster 1's rating and R_2 be Taster 2's rating. Add the probabilities on the diagonal (upper left to lower right): $P(R_1 = R_2) = 0.05 + 0.08 + 0.25 + 0.18 + 0.08 = 0.64$. (c) $P(R_1 > R_2) = 0.18$. This is the sum of the 10 numbers in the "lower left" part of the table. $P(R_2 > R_1) = 0.18$. This is the sum of the 10 numbers in the "upper right" part of the table.

19.53 (a) Nearly all the *individual students* should be in the range $\mu \pm 3\sigma = 2.9 \pm 3(1.0) = 2.9 \pm 3.0 = -0.1$ to 5.9. Since NSSE scores cannot be negative, nearly all the *individual students* should be in the range 0 to 5.9. (b) The sample mean $\bar{x}$ has a $N(\mu, \sigma/\sqrt{n}) = N(2.9, 0.1)$ distribution, so nearly all such *means* should be in the range $2.9 \pm 3(0.1) = 2.9 \pm 0.3 = 2.6$ to 3.2.

19.55 (a) The stemplot (below) confirms the description given in the text. (Arguably, there are two high "mild outliers," although the $1.5 \times IQR$ criterion only flags the highest as an outlier.)

96	8
97	344
97	888889
98	0133
98	5789
99	
99	6
100	2

(b) STATE: Is there evidence that the mean body temperature for healthy adults is not equal to the "traditional" 98.6°F? PLAN: Let μ be the mean body temperature. $H_0: \mu = 98.6°$ versus $H_a: \mu \neq 98.6°$. SOLVE: Assume that we have a Normal distribution and an SRS. $\bar{x} = 98.203°$, so $z = \frac{98.203 - 98.6}{0.7/\sqrt{20}} = -2.54$. $P = 2P(Z < -2.54) = 0.0110$. CONCLUDE: We have fairly strong evidence—significant at $\alpha = 0.05$ but not at $\alpha = 0.01$—that mean body temperature is not equal to 98.6°F. (Specifically, the data suggest that mean body temperature is lower.)

19.57 STATE: What is the mean body temperature for healthy adults? PLAN: WE will estimate μ by giving a 90% confidence interval. SOLVE: Assume that we have a Normal distribution and an SRS (conditions were checked in Exercise 19.55). With $\bar{x} = 98.203°$, our 90% confidence interval for μ is $\bar{x} \pm z^* \frac{\sigma}{\sqrt{n}} = 98.203 \pm 1.645 \frac{0.7}{\sqrt{20}} = 98.203 \pm 0.257 = 97.95$ to 98.46. CONCLUDE: We have 90% confidence that the mean body temperature for healthy adults is between 97.95°F and 98.46°F.

19.59 For the two-sided test $H_0: M = \$62,000$ versus $H_a: M \neq \$62,000$ with significance level $\alpha = 0.10$, we fail to reject H_0 because \$62,000 falls inside the 90% confidence interval.

19.61 Let H be the event the student was home schooled. Let R be the event the student attended a regular public school. We want $P(H \mid \text{not } R)$. Note that the event "H and not R" = "H." Then $P(H \mid \text{not } R) = \frac{P(H)}{P(\text{not } R)} = \frac{0.006}{1 - 0.758} = 0.025$.

19.63 (a) $\mu = np = 300(0.18) = 54$ people and $\sigma = \sqrt{np(1-p)} = \sqrt{300(0.18)(1-0.18)} = \sqrt{44.28} = 6.65$ people. (b) $np = 54$ and $n(1-p) = 246$; both are more than 10, so the Normal approximation holds. Let X be the number of people that dropped out of the program. Then, using the Normal approximation: $P(X \leq 65) = P\left(Z \leq \frac{65 - 54}{6.65}\right) = P(Z \leq 1.65) = 0.9505$. Using software to find the exact binomial probability, $P(X \leq 65) = 0.9554$. The Normal approximation 0.9505 underestimates the true probability by about 0.0049.

19.65 A Type I error means that we conclude the mean IQ of very-low-birth-weight male babies is less than 100 when it really is 100 (or more). A Type II error means that we conclude the mean IQ of very-low-birth-weight male babies is 100 (or more) when it really is less than 100.

Chapter 20 – Inference about a Population Mean

20.1 $SE = \frac{s}{\sqrt{n}} = \frac{27.2}{\sqrt{1000}} = 0.8601$.

20.3 (a) $t^* = 31.82$. (b) $t^* = 1.697$. This is greater than $z^* = 1.645$.

20.5 (a) $df = 2 - 1 = 1$, so $t^* = 63.66$. (b) $df = 30 - 1 = 29$, so $t^* = 2.045$. (c) $df = 1001 - 1 = 1000$, so $t^* = 1.646$.

20.7 STATE: Among young adults in the United States, what is the mean number (out of 100 attempts) of correct answers to identifying the taller of two people by voice? PLAN: We will estimate μ with a 95% confidence interval. SOLVE: We are told to view the observations as an SRS. A stemplot (below) shows some possible bimodality but no outliers.

4	9
5	
5	3
5	
5	666
5	8889
6	11
6	23
6	55
6	6777
6	889
7	00

With $\bar{x} = 62.1667$ and $s = 5.806$ correct, $df = 24 - 1 = 23$, and $t^* = 2.069$, the 95% confidence interval for μ is $\bar{x} \pm 2.069 \frac{s}{\sqrt{n}} = 62.1667 \pm 2.069 \frac{5.806}{\sqrt{24}} = 62.1667 \pm 2.4521 =$

59.71 to 64.62. CONCLUDE: We are 95% confident that the mean number (out of 100) of correct answers to identifying the taller of two people by voice is between 59.71 and 64.62.

20.9 (a) df $= 2 - 1 = 1$. (b) $t = 3.00$ is bracketed by $t^* = 1.963$ (with two-tail probability 0.30) and $t^* = 3.078$ (with two-tail probability 0.20), so $0.20 < P < 0.30$. (c) This test is not significant at the 10%, 5%, or 1% levels because $P > 0.20$. (d) From software, $P = 0.2048$.

20.11 STATE: Is there evidence that eye grease increases sensitivity to contrast on average? PLAN: Take μ to be the mean difference (with eye grease minus without) in sensitivity. We test $H_0\colon \mu = 0$ against $H_a\colon \mu > 0$, using a one-sided alternative because if the eye grease works, it should increase sensitivity. SOLVE: We must assume that the students in the experiment can be regarded as an SRS of all students, that the treatments were randomized, and that athletes would experience a similar effect as the students. We were provided the differences for each student; a stemplot of these differences (below) seems to show two outliers; in this plot $-1\,|\,8$ represents -0.18.

-1	8621
-0	5
0	23557
1	4
2	489
3	
4	3
5	
6	4

Checking with the $1.5 \times IQR$ rule, these are not outliers. (Using JMP, $Q_1 = -0.095$, $Q_3 = 0.27$, and $Q_3 + 1.5 \times IQR = 0.27 + 1.5(0.365) = 0.8175$; calculating these values by hand, $Q_1 = -0.08$, $Q_3 = 0.26$, and $Q_3 + 1.5 \times IQR = 0.26 + 1.5(0.34) = 0.77$.) However P-values will only be approximate due to the skew and relatively small sample size. With $\bar{x} = 0.1013$ and $s = 0.2263$, $t = \frac{0.1013 - 0}{0.2263/\sqrt{16}} = 1.79$ with df $= 15$. Using Table C, $0.025 < P < 0.05$ (software gives 0.0469). CONCLUDE: We have evidence that eye grease does increase sensitivity to contrast, on average. Due to the skew in the data, we may not want to place a lot of emphasis on this result.

20.13 The stemplot below suggests that the distribution of nitrogen contents is heavily skewed with a strong outlier, 1430. Although t procedures are robust to violations of normality, they should not be used if the population being sampled is this heavily skewed.

0	00000000000111
0	2222233
0	44
0	
0	
1	
1	
1	4

20.15 (b)

20.17 (a)

20.19 (a)

20.21 (b)

20.23 (b)

20.25 For the student group: $t = \frac{0.08 - 0}{0.37/\sqrt{12}} = 0.749$. For the nonstudent group: $t = \frac{0.35 - 0}{0.37/\sqrt{12}} = 3.277$. From Table C, the first P-value (assuming a two-sided alternative hypothesis) is between 0.4 and 0.5 (software gives 0.47), and the second P-value is between 0.005 and 0.01 (software gives 0.007). These P-values support the researchers' findings of no strong evidence of an effect for students but a significant effect for nonstudents.

20.27 (a) With $n = 1100$, it is safe to use the t procedures (which can be used even for clearly skewed distributions when $n \geq 40$). (b) From Table C, $t^* = 2.581$(df $= 1000$); or; using software, $t^* = 2.580$(df $= 1099$). For either value, the 99% confidence interval is $\bar{x} \pm t^*(1.3) = 264 \pm 3.4 = 260.6$ to 267.4, rounded to one decimal place. (c) Because the 99% confidence interval for μ contains 262, the data do not provide evidence that the mean for all Dallas eighth-graders is different from the basic level.

20.29 (a) This is a matched pairs t test because each person's weight is measured twice. (b) Let μ be the mean weight difference (weight after wearing technology minus weight before). We test $H_0\colon \mu = 0$ against $H_a\colon \mu < 0$. We are given $\bar{x} = -3.5$, $s = 7.8$ and $n = 237$, so $t = \frac{-3.5 - 0}{7.8/\sqrt{237}} = -6.908$. From Table C, with df $= 236$, $P < 0.0005$ ($P < 0.0001$ using software). There is an extreme amount of evidence that there is a decrease in average weight after wearing the device.

20.31 (a) A stemplot (below) suggests the presence of outliers. The sample is small, and the stemplot is skewed, so use of t procedures is not appropriate.

2	5
3	3358
4	00
5	
6	
7	
8	
9	
10	
11	5
12	
13	5

(b) In the first interval, using all nine observations, we have df $= 8$ and $t^* = 1.860$. For the second interval, removing the two outliers (1.15 and 1.35), df $= 6$ and $t^* = 1.943$. The two 90% confidence intervals are

$$0.549 \pm 1.860\left(\frac{0.403}{\sqrt{9}}\right) = 0.299 \text{ to } 0.799 \text{ gram}$$

$$0.349 \pm 1.943\left(\frac{0.053}{\sqrt{7}}\right) = 0.310 \text{ to } 0.388 \text{ gram}$$

(c) The confidence interval computed without the two outliers is much narrower and has a lower center.

20.33 (a) This is a matched pairs design because each newt is measured twice. (b) The following stemplot clearly shows the high outlier (31).

```
-1 | 3
-0 | 6
-0 |
 0 | 1 2
 0 | 5 7 8 9
 1 | 0 1 2
 1 |
 2 |
 2 |
 3 | 1
```

(c) Let μ be the population mean difference (control minus experimental) in healing rates. Test $H_0: \mu = 0$ versus $H_a: \mu > 0$. With all 12 differences, $\bar{x} = 6.417$ and $s = 10.7065$, so $t = \frac{6.417 - 0}{10.7065/\sqrt{12}} = 2.08$. With df = 11, $P = 0.0311$ (using software). Omit the outlier: $\bar{x} = 4.182$ and $s = 7.7565$, so $t = \frac{4.182 - 0}{7.7565/\sqrt{11}} = 1.79$. With df = 10, $P = 0.0520$ (using software). With all 12 differences, there is more evidence that the population mean healing time is greater for the control limb. When we omit the outlier, the evidence is weaker. (d) The modified boxplot (below) shows the anticipated two outliers (−13 and 31). Omit both outliers: $\bar{x} = 5.9$ and $s = 5.5468$, so $t = \frac{5.9 - 0}{5.5468/\sqrt{10}} = 3.36$. With df = 9, $P = 0.0042$. By removing both outliers, there is very strong evidence that the population mean healing time is greater for the control limb (much stronger evidence than was found in part (c) when using the data from all 12 newts).

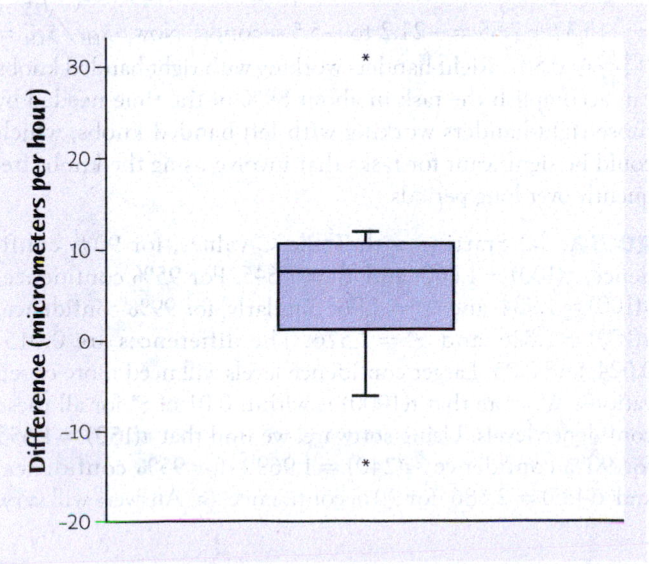

20.35 (a) A histogram (below) shows a significant outlier and indicates skew. We might consider applying t procedures to the sample after removing the most extreme observation (37,786).

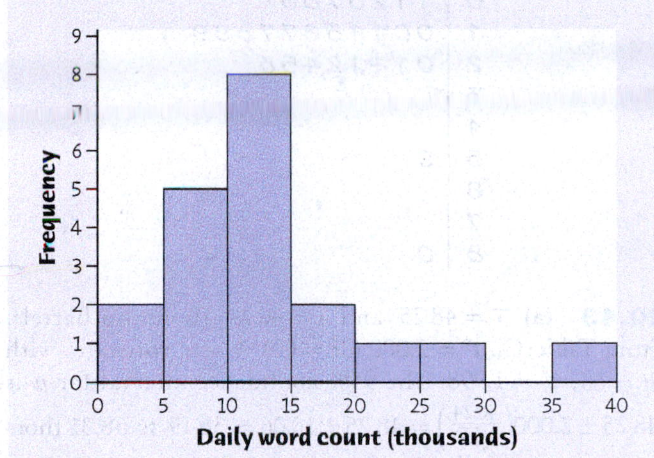

(b) If we remove the largest observation, the remaining sample is not heavily skewed and has no outliers. Now we test $H_0: \mu = 7{,}000$ versus $H_a: \mu \neq 7{,}000$. With the outlier removed, $\bar{x} = 11{,}555.16$ and $s = 6{,}095.015$, so $t = \frac{11{,}555.16 - 7000}{6{,}095.015/\sqrt{19}} = 3.258$. With df = 18, $P = 0.0044$ (using software). There is overwhelming evidence that the population mean number of words per day for men at this university differs from 7,000.

20.37 (a) Each patient was measured before and after treatment. (b) The stemplot of the differences (below) shows an extreme right skew and one or two high outliers. The t procedures should not be used.

```
0 | 0 0 1 2 2 3 8
1 | 0
2 | 1
3 |
4 |
5 | 1
6 |
7 | 0
```

(c) Suppose researchers perform a test ($H_0: \mu = 0$ versus $H_a: \mu > 0$) using t procedures, despite the presence of strong skew and outliers in the sample. If so, they would find $\bar{x} = 156.36$, $s = 234.2952$, and $t = 2.213$, yielding $P = 0.0256$ (using software).

20.39 (a) We test $H_0: \mu = 0$ versus $H_a: \mu > 0$, where μ is the population mean difference (treated minus control). This test is one-sided because the researchers have reason to believe that CO_2 will increase growth rate. (b) $\bar{x} = 1.916$ and $s = 1.050$, so $t = \frac{1.916 - 0}{1.050/\sqrt{3}} = 3.16$. With df = 2, $P = 0.0436$ (using software). This is significant at the 5% significance level. (c) For very small samples, t procedures should be used only when we can assume that the population is Normal. We have no way to assess the Normality of the population based on these three differences.

20.41 A stemplot (below) reveals that these data contain two extreme high outliers (5793 and 8015). Thus, t procedures are not appropriate.

0	1123788
1	00115677899
2	01112458
3	
4	
5	9
6	
7	
8	0

20.43 (a) $\bar{x} = 48.25$ and $s = 40.24$ thousand barrels. From Table C, $t^* = 2.000$ (df = 60). Using software, with df = 63, $t^* = 1.998$. The 95% confidence interval for μ is $48.25 \pm 2.000\left(\frac{40.24}{\sqrt{64}}\right) = 48.25 \pm 10.06 = 38.19$ to 58.31 thousand barrels. (Using software, the confidence interval is 38.20 to 58.30 thousand barrels.) (b) A stemplot (below) confirms the skewness and outliers (156.5, 196, 204.0) described. The two intervals have similar widths, but the new interval is shifted higher by about 2000 barrels. Although t procedures are fairly robust, we should be cautious about trusting the result in part (a) because of the strong skew and outliers. The computer-intensive method may produce a more reliable interval.

0	00001111111111
0	22222223333333333333
0	44444445555555
0	6666667
0	8899
1	01
1	
1	5
1	
1	9
2	0

20.45 Analyzing the percentage of seeds infected is more useful than analyzing the number of seeds infected is better because it allows us to compare plants with different seed counts. PLAN: We will construct a 90% confidence interval for μ the population mean percentage of beetle-infected seeds. SOLVE: A stemplot (below) shows a single-peaked and roughly symmetric distribution.

0	07
1	9
2	24689
3	666778
4	0000336
5	157
6	
7	00
8	57

We assume that the 28 plants can be viewed as an SRS of the population, so t procedures are appropriate. $\bar{x} = 4.0786$ and $s = 2.0135$. Using $t^* = 1.703$ (df = 27), the 90% confidence interval μ is $4.0786 \pm 1.703\left(\frac{2.0135}{\sqrt{28}}\right) = 4.0786 \pm 0.648 = 3.43\%$ to 4.73%. CONCLUDE: We estimate that the beetle infects less than 5% of seeds, so it is unlikely to be effective in controlling velvetleaf.

20.47 $\bar{x} = 0.5283$ and $s = 0.4574$. Using $t^* = 2.571$ (df = 5), the 95% confidence interval μ is $0.5283 \pm 2.571\left(\frac{0.4574}{\sqrt{6}}\right) = 0.5283 \pm 0.4801 = 0.0482$ to 1.0084 thousand cells.

20.49 (a) For each subject, randomly select which knob (right or left) that subject should use first. (b) PLAN: We test $H_0: \mu = 0$ versus $H_a: \mu < 0$, where μ denotes the population mean difference in time (right-thread time − left-thread time). SOLVE: A stemplot of the differences (below) gives no reasons that t procedures are not appropriate.

−5	2
−4	853
−3	511
−2	94
−1	66621
−0	74331
0	02
1	1
2	03
3	8

We assume our sample can be viewed as an SRS. $\bar{x} = -13.32$ seconds and $s = 22.936$ seconds, so $t = \frac{-13.32 - 0}{22.936/\sqrt{25}} = -2.90$. With df = 24, we find $P = 0.0039$ (using software). CONCLUDE: We have good evidence (significant at the 1% level) that the population mean difference really is negative—that is, the population mean time for right-hand-thread knobs is less than the population mean time for left-hand-thread knobs.

20.51 Refer to Exercise 20.49. With df = 24, $t^* = 1.711$, so the 90% confidence interval for μ is $-13.32 \pm 1.711\left(\frac{22.936}{\sqrt{25}}\right) = -13.32 \pm 7.85 = -21.2$ to -5.5 seconds. Now, $\bar{x}_{RH} / \bar{x}_{LH} = \frac{104.12}{117.44} = 0.887$. Right-handers working with right-handed knobs can accomplish the task in about 89% of the time needed by those right-handers working with left-handed knobs, which could be significant for tasks that involve using the knobs frequently over long periods.

20.53 (a) Starting with Table C values, for 90% confidence, $t(100) = 1.660$ and $z^* = 1.645$. For 95% confidence, $t(100) = 1.984$ and $z^* = 1.96$. Similarly, for 99% confidence, $t(100) = 2.626$ and $z^* = 2.576$. The differences are 0.015, 0.024, and 0.05. Larger confidence levels will need more observations. We note that $t(1000)$ is within 0.01 of z^* for all these confidence levels. Using software, we find that $t(150) = 1.655$ for 90% confidence, $t(240) = 1.9699$ for 95% confidence, and $t(485) = 2.586$ for 99% confidence. (b) Answers will vary.

We'll note that the effect of the standard deviation increasing from 1 to 100 multiplies the margin of error in the calculation by 100, which implies that getting similar results requires more observations with $\sigma = 100$ than for $\sigma = 1$. Using $n = 485$ with $\sigma = 100$, the 99% margin of error using t is 11.74, compared to a 99% margin of error using z of 11.70. Using $\sigma = 1$, the margins of error are both 0.117, rounding to three decimal places.

Chapter 21 – Comparing Two Means

21.1 This situation involves two independent samples.

21.3 This situation involves a single sample.

21.5 STATE: Is there evidence that Nintendo Wii training significantly increases the mean improvement time? PLAN: Test $H_0: \mu_{Wii} = \mu_{NoWii}$ versus $H_a: \mu_{Wii} > \mu_{NoWii}$. We use a one-sided alternative because practice with the Wii should result in more improvement than performing the same operation again without Wii training. SOLVE: These data come from participants in a randomized experiment, so the two groups are independent. Stemplots suggest some deviation from Normality and a possible high outlier for the No Wii group. Boxplots indicate no outliers and a relatively symmetric distribution for the Wii group, but both the −81 and 229 are outliers in the No Wii group. We proceed with the t test for two samples appealing to robustness (especially good with equal sample sizes; see Section 21.5). We have $\bar{x}_{Wii} = 132.71$, $\bar{x}_{NoWii} = 59.67$, $s_{Wii} = 98.44$, $s_{NoWii} = 63.04$, $n_{Wii} = 21$, and $n_{NoWii} = 21$. So, $SE = \sqrt{\frac{s_{Wii}^2}{n_{Wii}} + \frac{s_{NoWii}^2}{n_{NoWii}}} = 25.509$ and $t = \frac{\bar{x}_{Wii} - \bar{x}_{NoWii}}{SE} = 2.863$. Since both sample sizes are 21, we have $df = 20$, and $0.0025 < P < 0.005$. Using software, $df = 34.04$ and $P = 0.0036$. CONCLUDE: There is very strong evidence that playing with a Nintendo Wii does help improve the skills of student doctors, at least in terms of the mean time to complete a virtual gall bladder operation.

21.7 From Exercise 21.5, we have $\bar{x}_{Wii} = 132.71$, $\bar{x}_{NoWii} = 59.67$, $n_{Wii} = n_{NoWii} = 21$ (so $df = 20$ for Option 2), and $SE = 25.509$. A 95% confidence interval for the population mean difference in improvement in time to complete the virtual gall bladder operation is $(\bar{x}_{Wii} - \bar{x}_{NoWii}) \pm t^*SE = 73.04 \pm 2.086$ $(25.509) = 19.828$ to 126.252 seconds. Software uses $df = 34.04$ and gives an interval of 21.203 to 124.88 seconds.

21.9 (a) Back-to-back stemplots appear to be reasonably Normal, and the discussion in the exercise justifies our treating the data as independent SRSs, so we can use t procedures. We test $H_0: \mu_1 = \mu_2$ versus $H_a: \mu_1 < \mu_2$, where μ_1 is the population mean time in the restaurant with no scent and μ_2 is the population mean time in the restaurant with a lavender odor. Here with $\bar{x}_1 = 91.27$, $\bar{x}_2 = 105.700$, $s_1 = 14.930$, $s_2 = 13.105$, and $n_1 = n_2 = 30$: $SE = \sqrt{\frac{s_1^2}{n1} + \frac{s_2^2}{n2}} = 3.627$ and $t = \frac{\bar{x}_1 - \bar{x}_2}{SE} = -3.98$. Using software, $df = 57.041$, and $P = 0.0001$. Using the more conservative $df = 29$ (both samples are of size 30, so $df = 30 - 1$) and Table C, $P < 0.0005$. There is very strong evidence that customers spend more time on average in the restaurant when the lavender scent is present. (b) Back-to-back stemplots of the spending data are skewed and have many gaps (perhaps due to pricing). Even so, the central limit theorem tells us that the t procedures are approximately correct for other population distributions when the sample sizes are large. (Both samples are size 30.) We test $H_0: \mu_1 = \mu_2$ versus $H_a: \mu_1 < \mu_2$, where μ_1 is the population mean amount spent in the restaurant with no scent and μ_2 is the population amount spent in the restaurant with a lavender odor. Here, with $\bar{x}_1 =$ €17.5133, $\bar{x}_2 =$ €21.1233, $s_1 =$ €2.3588, $s_2 =$ €2.3450, and $n_1 = n_2 = 30$: $SE = \sqrt{\frac{s_1^2}{n1} + \frac{s_2^2}{n2}}$ = €0.6073 and $t = \frac{\bar{x}_1 - \bar{x}_2}{SE} = -5.94$. Using software, $df = 57.998$, and $P < 0.0001$. Using the more conservative $df = 29$ and Table C, $P < 0.0005$. There is very strong evidence that, on average, customers spend more money when the lavender scent is present.

21.11 We have two small samples ($n_1 = n_2 = 4$), so the t procedures are not reliable unless both distributions are Normal.

21.13 Let BI represent "Behavioral Intervention" and C represent "Control." Here are the details of the calculations:

$$SE_{BI} = \frac{s_{BI}}{\sqrt{n_{BI}}} = \frac{363}{\sqrt{304}} = 20.82$$

$$SE_C = \frac{s_C}{\sqrt{n_C}} = \frac{397}{\sqrt{306}} = 22.69$$

$$df = \frac{(SE_{BI}^2 + SE_C^2)^2}{\frac{1}{(n_{BI}-1)}(SE_{BI}^2)^2 + \frac{1}{(n_C-1)}(SE_C^2)^2}$$

$$= \frac{\left(\frac{363^2}{304} + \frac{397^2}{306}\right)^2}{\frac{1}{(303)}\left(\frac{363^2}{304}\right)^2 + \frac{1}{(305)}\left(\frac{397^2}{306}\right)^2} = 603.9$$

$$t = \frac{\bar{x}_{BI} - \bar{x}_C}{\sqrt{SE_{BI}^2 + SE_C^2}} = \frac{1227 - 1323}{\sqrt{\frac{363^2}{304} + \frac{397^2}{306}}} = -3.117$$

21.15 Reading from the software output shown in Exercise 21.13, we find that there is a significant difference between the population mean daily calorie intake of children in the behavioral intervention group and the population mean daily calorie intake of children in the control group ($t = -3.117$, $df = 603.9$, $P < 0.002$). Because the average daily caloric intake is lower for children in the behavioral intervention group, it appears that behavioral intervention leads to a lower daily calorie intake when compared to the control.

21.17 (a)

21.19 (b)

21.21 (c)

21.23 (a)

21.25 (a) To test the belief that women talk more than men, test $H_0: \mu_F = \mu_M$ versus $H_a: \mu_F > \mu_M$. (b)–(d) The small table that follows provides a summary of t statistics, degrees of freedom, and P-values for both studies. We take the conservative approach for computing df as the smaller sample size minus 1.

Study	t	df	Table C Values	P-value
1	−0.248	55	$\lvert t\rvert < 0.679$	$P > 0.25$
2	1.507	19	$1.328 < \lvert t\rvert < 1.729$	$0.05 < P < 0.10$

Note that for Study 1, we reference df = 50 in Table C. (e) The first study gives no support to the belief that women talk more than men; the second study gives weak support, and it is significant only at a relatively high significance level (say, $\alpha = 0.10$).

21.27 (a) Standard errors are $SE_P = \frac{s_P}{\sqrt{n_P}} = \frac{2.05}{\sqrt{104}} = 0.201$ and $SE_N = \frac{s_N}{\sqrt{n_N}} = \frac{1.74}{\sqrt{101}} = 0.171$ for the positive mood and neutral mood groups, respectively. (b) Using the conservative approach for computing df as the smaller sample size minus 1, df = 104 − 1 = 103. (Both samples are of size 104.) (c) We test $H_0: \mu_P = \mu_N$ versus $H_a: \mu_P \neq \mu_N$, where μ_P is the population mean attitude score for the positive mood group and μ_N is the population mean attitude score for the neutral mood group. The test statistic is $t = \frac{\bar{x}_P - \bar{x}_N}{\sqrt{SE_P^2 + SE_N^2}} = \frac{4.30 - 5.50}{\sqrt{\frac{2.05^2}{104} + \frac{1.74^2}{104}}} = -4.551$, and with df = 103 (rounded down to 100), Table C shows $P < 0.001$. There is overwhelming evidence that the population mean attitude toward the indulgent food was different for those who read the happy story (positive mood) than for those who did not read the story (neutral mood).

21.29 We test $H_0: \mu_O = \mu_S$ versus $H_a: \mu_O < \mu_S$, where μ_O is the mean admiration score of a "typical only child" as rated by the population of only children and μ_S is the mean admiration score of a "typical only child" as rated by the population of people who have siblings. The test statistic is $t = \frac{\bar{x}_O - \bar{x}_S}{\sqrt{SE_O^2 + SE_S^2}} = \frac{1.95 - 2.06}{\sqrt{\frac{1.04^2}{233} + \frac{1.11^2}{1577}}} = -1.494$, and with df = 232 (rounded down to 100), Table C shows $0.05 < P < 0.10$ (using df = 315.35 and software, $P = 0.0681$). We do not have evidence at the $\alpha = 0.05$ level that the mean admiration score of a "typical only child" as rated by the population of only children is lower than the mean admiration score of a "typical only child" rated by the population of people who have siblings.

21.31 (a) Let μ_C be the population mean brain size for football players who have had concussions and let μ_{NC} be the population mean brain size for football players who have not had concussions. We test $H_0: \mu_C = \mu_{NC}$ versus $H_a: \mu_C \neq \mu_{NC}$. This is a two-sided test because we simply want to know if there is a difference in mean brain size. The test statistic is $t = \frac{\bar{x}_C - \bar{x}_{NC}}{\sqrt{SE_C^2 + SE_{NC}^2}} = \frac{5784 - 6489}{\sqrt{609.3^2/25 + 815.4^2/25}} = -3.463$. Using the conservative approach for computing df as the smaller sample size minus 1, df = 25 − 1 = 24. (Both samples are of size 25.) Table C shows $0.002 < P < 0.005$. (Using software, df = 44.43 and $P = 0.0012$.) There is strong evidence that the population mean brain size is different for football players who have had concussions as opposed to those who have not had concussions. (b) The fact that the subjects were referrals to a psychiatric institute indicates that they are not a random sample of all football players who have or have not had concussions. That could weaken or negate the results of the test. We would need more information about how and why these players were referred to the institute.

21.33 (a) Let μ_C be the population mean gain among coached students and let μ_{NC} be the population mean gain among all uncoached students. The hypotheses are $H_0: \mu_C = \mu_{NC}$ versus $H_a: \mu_C > \mu_{NC}$. We find $SE = \sqrt{\frac{s_C^2}{n_C} + \frac{s_{NC}^2}{n_{NC}}} = \sqrt{\frac{58^2}{427} + \frac{50^2}{2733}} = 2.9653$ and $t = \frac{\bar{x}_C - \bar{x}_{NC}}{SE} = \frac{40 - 22}{2.9653} = 6.070$. Using the conservative approach for computing df as the smaller sample size minus 1, df = 427 − 1 = 426. (The smaller sample size is 427.) Using Table C and rounding df down to 100, $P < 0.0005$. (Using software, df = 529.56 and $P < 0.0001$.) There is strong evidence that coached students had a greater average increase than uncoached students. (b) The 99% confidence interval for the difference in score improvement for coached versus uncoached students is $(\bar{x}_C - \bar{x}_{NC}) \pm t^*(SE) = 18 \pm t^*(2.9653)$, where $t^* = 2.626$ (using df = 100 with Table C) or 2.585 (df = 529.56 with software). This gives 10.21 to 25.79, or 10.33 to 25.67, respectively. (c) Increasing one's score by about 10 to 26 points (see results of part (b)) is not likely to make a difference in being granted admission to or receiving scholarships from any colleges, so coaching may not be worth the cost.

21.35 (a) Let μ_1 be the population mean number of units of the appliance sold this month and let μ_2 be the population mean number of units of the appliance sold this same month last year. Then $\bar{x}_1 = 41$, $\bar{x}_2 = 38$, $s_1 = 11$, $s_2 = 13$, $n_1 = 50$, $n_2 = 52$, and $SE = \sqrt{\frac{s_1^2}{n_1} + \frac{s_2^2}{n_2}} = 2.381$. The 95% confidence interval is $(\bar{x}_1 - \bar{x}_2) \pm t^*(SE) = 3 \pm t^*(2.381)$, where $t^* = 2.021$ (using the Table C conservative approach with df = 40) or 1.984 (df = 98.427, with software). This gives either −1.812 to 7.812 units or −1.724 to 7.724 units, respectively. (b) Both 95% confidence intervals contain 0; it is possible that there is not a significant difference in the mean number of units of the appliance sold in all stores this month to the same month last year. Because the intervals contain negative numbers, it is possible that the mean number of units of the appliance sold in all stores this month is less than those sold in the same month last year.

21.37 (a) The stemplot for the good-weather message group (on the next page) shows a potential outlier (27.0%). The stemplot for the bad-weather message group shows three potential outliers (13.6%, 14.0%, and 23.2%). The boxplots show the same four potential outliers.

Good-weather message

18	7
19	9
20	3568
21	29
22	0122378
23	4
24	099
25	
26	
27	0

Bad-weather message

13	6
14	0
15	
16	18
17	05
18	0024588
19	0124
20	02
21	
22	
23	2

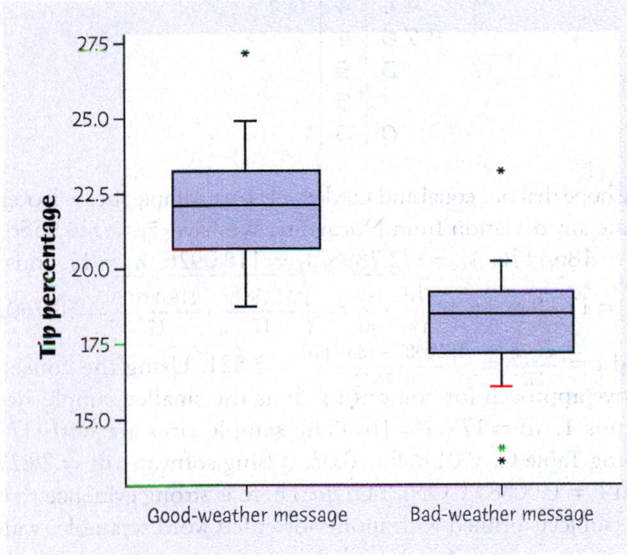

Since t procedures are robust to violations of Normality, it is reasonable to proceed. (b) Let μ_1 be the population mean tip percentage when the forecast is good and let μ_2 be the population mean tip percentage when the forecast is bad. We test $H_0: \mu_1 = \mu_2$ versus $H_a: \mu_1 \neq \mu_2$. We have $\bar{x}_1 = 22.22$, $\bar{x}_2 = 18.19$, $s_1 = 1.955$, $s_2 = 2.105$, $n_1 = 20$, and $n_2 = 20$. Here, $SE = \sqrt{\frac{s_1^2}{n_1} + \frac{s_2^2}{n_2}} = 0.642$ and $t = \frac{\bar{x}_1 - \bar{x}_2}{SE} = \frac{22.22 - 18.19}{0.642} = 6.274$. Using the conservative approach for computing df as the smaller sample size minus 1, df $= 20 - 1 = 19$. (Both samples are of size 20.) Using Table C, $P < 0.001$. (Using software, df $= 37.8$ and $P < 0.00001$.) There is overwhelming evidence that the population mean tip percentage differs between the two types of forecasts presented to patrons.

21.39 Refer to the results in Exercise 21.37. The 95% confidence interval for the difference in mean tip percentages between these two populations is $(\bar{x}_1 - \bar{x}_2) \pm t^*(SE) = (22.22 - 18.19) \pm t^*(0.642)$, where $t^* = 2.093$ (using the Table C conservative approach with df $= 19$) or 2.025 (df $= 37.8$, with software). This gives either 2.69% to 5.37% or 2.73% to 5.33%, respectively.

21.41 (a) The means for the two samples are $\bar{x}_{tablet} = 7.059$ and $\bar{x}_{paper} = 7.232$. These do not appear different enough to suggest that there is a difference in the mean recall scores between the two groups. (b) The stemplots (below) suggest that the population distributions are both moderately left-skewed (not Normal).

Tablet

2	00
3	000
4	00000
5	00000000000
6	0000000000000000000
7	000000000000000000000
8	0000000000000000000000000000000000000
9	000000000000000
10	000000

Paper

3	0000000
4	0000
5	0000000000
6	0000000000000
7	0000000000000000000000
8	0000000000000000000000000
9	0000000000000000000
10	000000000000

(c) Recall that t procedures are robust to non-Normal populations unless there are outliers or strong skewness. There are no outliers in either data set, and the skewness is not that extreme, so we can proceed using t procedures. We test $H_0: \mu_{tablet} = \mu_{paper}$ versus $H_a: \mu_{tablet} < \mu_{paper}$. We have $\bar{x}_{tablet} = 7.059$, $\bar{x}_{paper} = 7.232$, $s_{tablet} = 1.738$, $s_{paper} = 1.908$, $n_{tablet} = 119$, and $n_{paper} = 112$. Here, $SE = \sqrt{\frac{s_{tablet}^2}{n_{tablet}} + \frac{s_{paper}^2}{n_{paper}}} = \sqrt{\frac{1.738^2}{119} + \frac{1.908^2}{112}} = 0.2406$ and $t = \frac{\bar{x}_{tablet} - \bar{x}_{paper}}{SE} = \frac{7.059 - 7.232}{0.2406} = -0.719$. Using the conservative approach for computing df as the smaller sample size minus 1, df $= 112 - 1 = 111$. (The smallest sample is of size 112.) Using Table C (and df rounded down to 100), $0.20 < P < 0.25$. (Using software, df $= 223.74$ and $P = 0.2360$.) The results of our hypothesis test agree with our thoughts from part (a). It does not appear that students who read the paper version have higher recall scores, on average, than those who read the tablet version.

21.43 (a) The stemplot for the number of words women speak per day (below) shows a potential outlier (39,681). The stemplot for the number of words men speak per day (below) shows a potential outlier (37,786). Both stemplots suggest that there is some skew in both populations, but the sample sizes should be large enough to overcome this problem.

Words/day for women

0	668889
1	0002223
1	5567999
2	344
2	566
3	
3	9

Words/day for men

0	34
0	56788
1	00001223
1	57
2	1
2	8
3	
3	7

(b) Let μ_1 be the population mean number of words per day spoken by women and let μ_2 be the population mean number of words per day spoken by men. We test $H_0: \mu_1 = \mu_2$ versus $H_a: \mu_1 > \mu_2$. We have $\bar{x}_1 = 16{,}496.1$, $\bar{x}_2 = 12{,}866.7$, $s_1 = 7{,}914.35$, $s_2 = 8{,}342.47$, $n_1 = 27$, and $n_2 = 20$. Here, $SE = \sqrt{\frac{s_1^2}{n_1} + \frac{s_2^2}{n_2}} = \sqrt{\frac{7914.35^2}{27} + \frac{8342.47^2}{20}} = 2{,}408.26$ and $t = \frac{\bar{x}_1 - \bar{x}_2}{SE} = \frac{16{,}496.1 - 12{,}866.7}{2{,}408.26} = 1.51$. Using the conservative approach for computing df as the smaller sample size minus 1, df $= 20 - 1 = 19$. (The smallest sample size is 20.) Using Table C, $0.05 < P < 0.10$. (Using software, df $= 39.8$ and $P = 0.0698$.) There is some evidence that, on average, women say more words per day than men, but the evidence is not particularly strong.

21.45 Let μ_1 be the population mean days behind caterpillar peak for the control group and let μ_2 be the population mean days behind caterpillar peak for the supplemented group. We test $H_0: \mu_1 = \mu_2$ versus $H_a: \mu_1 \neq \mu_2$. We have $\bar{x}_1 = 4.0$, $\bar{x}_2 = 11.3$, $s_1 = 3.10934$, $s_2 = 3.92556$, $n_1 = 6$, and $n_2 = 7$. Here, $SE = \sqrt{\frac{s_1^2}{n_1} + \frac{s_2^2}{n_2}} = \sqrt{\frac{3.10934^2}{6} + \frac{3.92556^2}{7}} = 1.95263$ and $t = \frac{\bar{x}_1 - \bar{x}_2}{SE} = \frac{4.0 - 11.3}{1.95263} = -3.74$. Using the conservative approach for computing df as the smaller sample size minus 1, df $= 6 - 1 = 5$. (The smallest sample size is 6.) Using Table C, $0.01 < P < 0.02$. (Using software, df $= 10.96$ and $P = 0.0033$.) These results agree with the stated conclusion that there is a significant difference.

21.47 STATE: Does the treatment group ask for help less quickly on the average? PLAN: Let μ_1 be the population mean time in seconds until subjects in the treatment group ask for help and let μ_2 be the population mean time in seconds until subjects in the control group ask for help. We test $H_0: \mu_1 = \mu_2$ versus $H_a: \mu_1 > \mu_2$. The alternative hypothesis is one-sided because the researcher suspects that the treatment group will wait longer before asking for help. SOLVE: We must assume that the data come from an SRS of the intended population; we cannot check this with the data. The back-to-back stemplot (below) shows some irregularity in the treatment times and skewness in the control times.

Treatment		Control
65	0	5689
3	1	012444
976	1	58
44	2	
5	2	79
	3	
6	3	7
44	4	01
876	4	
3	5	
	5	
0	6	

We hope that our equal and moderately large sample sizes will overcome any deviation from Normality. We have $\bar{x}_1 = 314.0588$, $\bar{x}_2 = 186.1176$, $s_1 = 172.7898$, $s_2 = 118.0926$, $n_1 = 17$, and $n_2 = 17$. Here, $SE = \sqrt{\frac{s_1^2}{n_1} + \frac{s_2^2}{n_2}} = \sqrt{\frac{172.7898^2}{17} + \frac{118.0926^2}{17}} = 50.7602$ and $t = \frac{\bar{x}_1 - \bar{x}_2}{SE} = \frac{314.0588 - 186.1176}{50.7602} = 2.521$. Using the conservative approach for computing df as the smaller sample size minus 1, df $= 17 - 1 = 16$. (The sample sizes are both 17.) Using Table C, $0.01 < P < 0.02$. (Using software, df $= 28.27$ and $P = 0.0088$.) CONCLUDE: There is strong evidence that the subjects primed with money-oriented word scrambles wait longer, on average, to ask for help.

21.49 STATE: Does sharing a painful experience in a small group lead to higher bonding scores for group members than sharing a similar nonpainful experience? PLAN: Let μ_1 be the population mean bonding score for subjects in the pain group and let μ_2 be the population mean bonding score for subjects in the no-pain group. We test $H_0: \mu_1 = \mu_2$ versus $H_a: \mu_1 > \mu_2$. SOLVE: We must assume that the data come from an SRS of the intended population; we cannot check this with the data. The back-to-back stemplot (below) shows that both groups are slightly skewed left. Also, using the $1.5 \times IQR$ criterion, the pain group has two low outliers (1.29 and 1.43).

Pain		No pain
42	1	022
	1	77
221	2	14
	2	78
44	3	00011144
855555	3	778
44422110	4	11244
877755		78

We will remove these outliers and hope that our moderately large sample sizes will overcome any deviation. We have $\bar{x}_1 = 3.903$, $\bar{x}_2 = 3.138$, $s_1 = 0.7734$, $s_2 = 1.0876$, $n_1 = 25$, and $n_2 = 27$. Here, $SE = \sqrt{\frac{s_1^2}{n_1} + \frac{s_2^2}{n_2}} = \sqrt{\frac{0.7734^2}{25} + \frac{1.0876^2}{27}} = 0.2603$ and $t = \frac{\bar{x}_1 - \bar{x}_2}{SE} = \frac{3.903 - 3.138}{0.2603} = 2.940$. Using the conservative approach for computing df as the smaller sample size minus 1, df $= 25 - 1 = 24$. (The smallest sample sizes is 25.) Using Table C, $0.0025 < P < 0.005$. (Using software, df $= 46.98$ and $P = 0.0025$.) There is strong evidence that a painful experience in a small group leads to higher average bonding scores for group members than sharing a similar nonpainful experience.

21.51 (a) Refer to Exercise 21.49 for details. The 90% confidence interval for the difference in the population mean bonding score for students in the pain and no-pain groups is $(\bar{x}_1 - \bar{x}_2) \pm t^*(SE) = (3.903 - 3.138) \pm t^*(0.2603)$, where $t^* = 1.711$ (using the Table C conservative approach with df $= 24$) or 1.678 (df $= 46.98$, with software). This gives either 0.32 to 1.21 or 0.33 to 1.20, respectively. (b) Using the notation from Exercise 21.49, the 90% confidence interval for the population mean bonding score for students in the pain group is $\bar{x}_1 \pm t^*\left(\frac{s_1}{\sqrt{n_1}}\right) = 3.903 \pm 1.711\left(\frac{0.7734}{\sqrt{25}}\right) = 3.638$ to 4.168 (where df $= 25 - 1 = 24$).

21.53 Responses may vary because specific hypotheses and confidence levels were not stated. This solution gives 95% confidence intervals for the means in parts (a) and (b) and performs a hypothesis test and gives a 95% confidence interval for part (c). Note that the first two problems call for single-sample *t* procedures (Chapter 20), and the last uses the Chapter 21 procedures. Answers should be formatted according to the "four-step process" of the text; these answers are not formatted as such but can be used to check any result. We begin with summary statistics.

	n	$\bar{x}$	s
Women	95	4.2737	2.1472
Men	81	6.5185	3.3471

A back-to-back stemplot of responses for men and women reveals that the distribution of claimed drinks per day for women is slightly skewed but has no outliers. For men, the distribution is only slightly skewed but contains four outliers. However, these outliers are not too extreme. In all parts of the exercise, it seems that use of *t* procedures is reasonable. Let μ_w be the population mean number of claimed drinks per session for women and let μ_m be the population mean number of claimed drinks per session for men.

(a) We construct a 95% confidence interval for μ_w. Here $t^* = 1.990$ (df $= 80$ in Table C) or $t^* = 1.9855$ (df $= 94$, with software) and $SE = \frac{2.1472}{\sqrt{95}} = 0.2203$. A 95% confidence interval for μ_w is $4.2737 \pm 1.990(0.2203) = 3.84$ to 4.71 drinks per session. The interval using software is virtually the same. With 95% confidence, the population mean number of claimed drinks per session for women is between 3.84 and 4.71 drinks. (b) We construct a 95% confidence interval for μ_m. Here $t^* = 1.990$. (df $= 80$ in Table C or with software) and $SE = \frac{3.3471}{\sqrt{81}} = 0.3719$. A 95% confidence interval for μ_m is $6.5185 \pm 1.990(0.3719) = 5.78$ to 7.26 drinks per session. With 95% confidence, the population mean number of claimed drinks per session for men is between 5.78 and 7.26 drinks. (c) We test $H_0\!: \mu_w = \mu_m$ versus $H_a\!: \mu_w \neq \mu_m$. We have $SE = \sqrt{\frac{2.1472^2}{95} + \frac{3.3471^2}{81}} = 0.4322$ and $t = \frac{4.2737 - 6.5185}{0.4322} = -5.193$. Regardless of the choice of df (80 or 132.14), the results are highly significant ($P < 0.001$). We have very strong evidence that the claimed number of drinks per session is different for women and men. To construct a 95% confidence interval for $\mu_w - \mu_m$, we use $t^* = 1.990$(df $= 80$) or 1.9781 (df $= 132.15$, with software). Using $(\bar{x}_w - \bar{x}_m) \pm t^*(SE)$, we obtain either -2.2448 ± 0.8601 or -2.2448 ± 0.8549. After rounding either interval, we report that, with 95% confidence, on average sophomore men at the college who drink claim an additional 1.4 to 3.1 drinks per session compared with sophomore women who drink.

Chapter 22 – Inference about a Population Proportion

22.1 (a)The population is surgical patients. p is the proportion of all surgical patients who will test positive for *Staphylococcus aureus*. (b) $\hat{p} = \frac{1251}{6771} = 0.185$, or 18.5%.

22.3 (a) Because n is large, the approximate distribution of $\hat{p}$ is Normal, with mean $p = 0.6$ and standard deviation $\sqrt{\frac{p(1-p)}{n}} = \sqrt{\frac{(0.6)(0.4)}{1500}} = 0.0126$. (b) If the sample size were 6000, the approximate distribution of $\hat{p}$ would still be Normal with mean $p = 0.6$, but the standard deviation would decrease to $\sqrt{\frac{p(1-p)}{n}} = \sqrt{\frac{(0.6)(0.4)}{6000}} = 0.0063$.

22.5 (a) The biggest weakness would be under coverage if the survey missed soliciting some groups of Canadian adults who were not reached by the "wide variety of methods and channels" that were used to solicit respondents. (b) $\hat{p} = \frac{930}{1525} =$

0.6098, so $SE = \sqrt{\frac{\hat{p}(1-\hat{p})}{n}} = \sqrt{\frac{(0.6098)(0.3902)}{1525}} = 0.0125$. The 95% confidence interval is $\hat{p} \pm 1.96SE = 0.6098 \pm 1.96(0.0125) =$ 0.5853 to 0.6343, or 58.5% to 63.4%.

22.7 **(a)** $\hat{p} = \frac{480}{1500} = 0.320$, so the margin of error is $1.96\sqrt{\frac{0.320(0.680)}{1500}} = 0.0236$. **(b)** We need a sample of size $n = \left(\frac{z^*}{m}\right)^2 p^*(1-p^*) = \left(\frac{1.96}{0.03}\right)^2 (0.32)(0.68) = 928.8$, so 929.

22.9 STATE: Is the proportion of judges who correctly identify the dog-owner pairs the same as the proportion who do not? PLAN: Let p_1 be the population proportion of undergraduates who would correctly match the sheet when the mouth region is blacked out. Let p_2 be the population proportion of undergraduates who would correctly match the sheet when the eye region is blacked out. We want to test the hypotheses $H_0: p_1 = 0.5$ versus $H_a: p_1 > 0.5$. We also want to test the hypotheses $H_0: p_2 = 0.5$ versus $H_a: p_2 > 0.5$. SOLVE: When the mouth region is blacked out: We expect $51(0.5) = 25.5$ successes and 25.5 failures. Both are larger than 10, so assuming this sample can be treated like an SRS, the conditions are met. $\hat{p} = \frac{37}{51} = 0.7255$, $z = \frac{\hat{p}-p_0}{\sqrt{\frac{p_0(1-p_0)}{n}}} = \frac{0.7255-0.5}{\sqrt{\frac{0.5(0.5)}{51}}} = 3.22$, and the P-value is 0.0006. When the eye region is blacked out: We expect $60(0.5) = 30$ successes and 30 failures. Both are larger than 10, so assuming this sample can be treated like an SRS, the conditions are met. $\hat{p} = \frac{30}{60} = 0.5$, $z = \frac{\hat{p}-p_0}{\sqrt{\frac{p_0(1-p_0)}{n}}} = \frac{0.5-0.5}{\sqrt{\frac{0.5(0.5)}{60}}} = 0$, and the P-value is 0.5. CONCLUDE: When the mouth region is blacked out, there is strong evidence that the population proportion of undergraduates who can correctly match the dog-owner pairs is greater than 0.5. When the eye region is blacked out, there is *not* evidence that the population proportion of undergraduates who can correctly match the dog-owner pairs is greater than 0.5.

22.11 **(a)** Since we expect $10(0.5) = 5 < 10$, the number of trials is not large enough. **(b)** As long the sample can be viewed as an SRS, a z test for a proportion can be used. **(c)** Under the null hypothesis, we expect only $150(0.02) = 3$ successes, which is smaller than the 10 necessary to use a z test for a proportion.

22.13 **(a)** We are told that we have a random sample. There are 180 successes and $1501 - 180 = 1321$ failures, both of which are at least 15. $\hat{p} = \frac{180}{1501} = 0.1199$. A large-sample 95% confidence interval for p is $\hat{p} \pm 1.96\sqrt{\frac{\hat{p}(1-\hat{p})}{n}} = 0.1199 \pm 1.96\sqrt{\frac{(0.1199)(0.8801)}{1501}} = 0.1035$ to 0.1363, or 10.4% to 13.6%. **(b)** The confidence level (95%) is larger than 90%, and the sample size is at least 10, so the plus-four interval can be used. $\tilde{p} = \frac{180+2}{1501+4} = 0.1209$. The 95% plus four confidence interval is $\tilde{p} \pm 1.96\sqrt{\frac{\tilde{p}(1-\tilde{p})}{n+4}} = 0.1209 \pm 1.96\sqrt{\frac{(0.1201)(0.8799)}{1505}} = 0.1044$, to 0.1374, or 10.4% to 13.7%. The plus-four interval is approximately the same width as the large-sample confidence interval because the sample size is so large.

22.15 (b)

22.17 (b)

22.19 (c)

22.21 (a)

22.23 (c)

22.25 **(a)** The survey excludes those who have no phone or who have only cell phone service. **(b)** We have 848 "yes" answers and 162 "no" answers; both of these are at least 15. With the sample proportion $\hat{p} = \frac{848}{1010} = 0.8396$, the large-sample 95% confidence interval is $\hat{p} \pm 1.96(\text{SE}) = 0.8396 \pm 1.96\sqrt{\frac{(0.8396)(0.1604)}{1010}} = 0.8170$ to 0.8622.

22.27 **(a)** $\hat{p} = \frac{848}{1010} = 0.8396$, SE $= 0.01155$, so the margin of error for 95% confidence is $1.96(\text{SE}) = 0.02263$, or 2.26%. **(b)** If $\hat{p} = 0.50$, then SE $= 0.01573$, and the margin of error for 95% confidence is $1.96(\text{SE}) = 0.03084$, or 3.08%. **(c)** For samples of about this size, the margin of error is no more than about $\pm 3\%$, no matter what $\hat{p}$ is.

22.29 **(a)** The biggest weakness is likely to be that the survey excludes U.S. adults who live in Alaska or Hawaii or who do not have a cellular or landline phone. **(b)** We have 1096 successes and 406 failures; both of these are at least 15. With the sample proportion $\hat{p} = \frac{1096}{1502} = 0.7297$, the large-sample 95% confidence interval is $\hat{p} \pm 1.645(\text{SE}) = 0.7297 \pm 1.645\sqrt{\frac{(0.7297)(0.2703)}{1502}} = 0.7108$ to 0.7485, or 71.1% to 74.9%.

22.31 **(a)** We can construct a large-sample confidence interval because we have 171 successes and 709 failures, both of which are greater than 15. With the sample proportion $\hat{p} = \frac{171}{880} = 0.1943$, the large-sample 95% confidence interval is $\hat{p} \pm 1.96(\text{SE}) = 0.1943 \pm 1.96\sqrt{\frac{(0.1943)(0.8057)}{880}} = 0.1682$ to 0.2204, or 16.8% to 22.0%. **(b)** It is likely that more than 171 respondents ran red lights. We would not expect very many people to claim that they have run red lights when they have not, but some people will deny running red lights when they have.

22.33 **(a)** Let p be the population proportion of times that the candidate with the higher-rated face wins. We want to test the hypotheses $H_0: p = 0.5$ versus $H_a: p > 0.5$. **(b)** We expect $32(0.5) = 16$ successes and 16 failures. Both are at least 10, so assuming that this sample can be treated like an SRS, the conditions are met. $\hat{p} = \frac{22}{32} = 0.6875$, $z = \frac{\hat{p}-p_0}{\sqrt{\frac{p_0(1-p_0)}{n}}} =$

$\frac{0.6875-0.5}{\sqrt{\frac{0.5(0.5)}{32}}} = 2.12$, and the *P*-value is 0.0170. Since $0.01 < p < 0.05$, the results are statistically significant at the 5% level but not at the 1% level.

22.35 Let p be the population proportion of times that the coin would come up heads. We want to test the hypotheses $H_0: p = 0.5$ versus $H_a: p \neq 0.5$. We expect $4040(0.5) = 2020$ successes and 2020 failures. Both are larger than 10, and based on the simple method of a coin-flipping experiment we can treat the data as an SRS, so the conditions are met. The exercise gives us $\hat{p} = 0.5069$, so $z = \frac{\hat{p}-p_0}{\sqrt{\frac{p_0(1-p_0)}{n}}} = \frac{0.5069-0.5}{\sqrt{\frac{0.5(0.5)}{4040}}} = 0.88$. This is a two-tailed test, and the *P*-value is $2P(Z > 0.88) = 0.3788$. We do not have evidence to suggest that the coin was not fair.

22.37 **(a)** The margin of error will slightly change because the sample proportion of returns claiming itemized reductions will change from state to state. These are likely similar from one state to another, so the margin of error will be similar for the states. **(b)** Yes, it will change because the sample size used for each margin of error will be very different.

22.39 **(a)** STATE: What is the proportion of all 17-year-old students still in school in 2012 who had at least one parent graduate from college? PLAN: We construct a 99% confidence interval. SOLVE: We are told to treat this sample as an SRS of 17-year-olds still in school. The sample size is $n = 9000$, and we are told that $\hat{p} = 0.51$, so we have $9000(0.51) = 4590$ students with a parent who graduated from college and 4410 students whose parents did not graduate from college. Both are larger than 15, so the large-sample confidence interval is appropriate. The large-sample 99% confidence interval is $\hat{p} \pm 2.576(SE) = 0.51 \pm 2.576\sqrt{\frac{(0.51)(0.49)}{9000}} = 0.4964$ to 0.5236, or 49.6% to 52.4%. CONCLUDE: With 99% confidence, the proportion of all 17-year-old students still in school in 2012 who had at least one parent graduate from college is between about 0.496 and 0.524. **(b)** Answers will vary, but 17-year-olds who dropped out of school are more likely to have neither parent graduate from college. If so, then the proportion for the entire population is likely to be lower.

22.41 **(a)** STATE: What percentage of all American adults had read a book in either print or digital format in the preceding 12 months? PLAN: We construct a 95% confidence interval. SOLVE: We are told to treat this sample as an SRS of readers. There are 1081 successes and 421 failures. Both are larger than 15, so the large-sample confidence interval is appropriate. $\hat{p} = \frac{1081}{1502} = 0.7197$. The large-sample 95% confidence interval is $\hat{p} \pm 1.96(SE) = 0.7197 \pm 1.96\sqrt{\frac{(0.7197)(0.2803)}{1502}} = 0.6970$ to 0.7424, or 69.7% to 74.2%. CONCLUDE: We are 95% confident that the percentage of all American adults who had read a book in either print or digital format in the preceding 12 months is between 69.7% and 74.2%. **(b)** STATE: What percentage of all adults who had read a book in the preceding 12 months had read only digital books? PLAN: We construct a 95% confidence interval. SOLVE: We are told to treat this sample as an SRS of readers. There are 105 successes and 976 failures. Both are larger than 15, so the large-sample confidence interval is appropriate. $\hat{p} = \frac{105}{1081} = 0.0971$. The large-sample 95% confidence interval is $\hat{p} \pm 1.96(SE) = 0.0971 \pm 1.96\sqrt{\frac{(0.0971)(0.9029)}{1081}} = 0.0794$ to 0.1148, or 7.94% to 11.48%. CONCLUDE: We are 95% confident that among all adults who had read a book in the preceding 12 months, the percentage that had read only digital books is between 7.94% and 11.48%.

22.43 **(a)** Let p be the population proportion of students who scored a 5 in 2010 on the AP Statistics exam. We want to test the hypotheses $H_0: p = 0.15$ versus $H_a: p > 0.15$. We expect $61(0.15) = 9.15$ successes and $61(0.85) = 51.85$ failures. The number of successes is less than 10, so we should have some concerns about using the z test for proportions. Additionally, these students are probably not an SRS of any meaningful population, so we should be cautious about any conclusions we make. We have $\hat{p} = \frac{15}{61} = 0.2459$, so $z = \frac{\hat{p}-p_0}{\sqrt{\frac{p_0(1-p_0)}{n}}} = \frac{0.2459-0.15}{\sqrt{\frac{0.15(0.85)}{61}}} = 2.10$. This is a right-tailed test, and the *P*-value is $P(Z > 2.10) = 0.0179$. There is evidence to suggest that the population proportion of students who would score a 5 on the AP Statistics exam when given a cash incentive is greater than 0.15. **(b)** This is an observational study, not a randomized experiment, so we cannot conclude that cash incentives *cause* an increase in the proportion of 5's on the AP Statistics exam.

22.45 STATE: What is the proportion of all drive-thru orders at KFC that are filled accurately? PLAN: We construct a 95% confidence interval. SOLVE: We need to assume that we can treat this sample as an SRS of all drive-thru orders at KFC. There are 109 successes (accurately-filled orders) and 56 failures. Both are larger than 15, so the large-sample confidence interval is appropriate. $\hat{p} = \frac{109}{165} = 0.6606$. The large-sample 95% confidence interval is $\hat{p} \pm 1.96(SE) = 0.6606 \pm 1.96\sqrt{\frac{(0.6606)(0.3394)}{165}} = 0.5883$ to 0.7329, or 58.8% to 73.3%. CONCLUDE: We are 95% confident that the proportion of all drive-thru orders at KFC that are filled accurately is between 58.8% and 73.3%.

22.47 **(a)** We have 22 people who preferred the first presented wine and 10 who preferred the second. Because 10 is less than 15, the conditions for the large-sample confidence interval are not met. **(b)** STATE: We want to estimate the population proportion of subjects who would select the first choice presented. PLAN: We construct a 90% confidence interval. SOLVE: The confidence level (90%) is equal to 90%, and the sample size is at least 10, so the plus-four interval can be used. $\tilde{p} = \frac{22+2}{32+4} = 0.6667$. The 95% plus four confidence interval is

$\tilde{p} \pm 1.645\sqrt{\frac{\tilde{p}(1-\tilde{p})}{n+4}} = 0.6667 \pm 1.645\sqrt{\frac{(0.6667)(0.3333)}{36}} = 0.5374$ to 0.7959, or 53.74% to 79.59%. CONCLUDE: We are 90% confidence that the population proportion of subjects who would select the first choice presented is between 0.5374 and 0.7959.

22.49 (a) Provided that we have an SRS, the conditions are met since the total number of observations is at least 10 and the confidence level is greater than 90%. $\tilde{p} = \frac{49+2}{61+4} = 0.7846$. The 99% plus four confidence interval is $\tilde{p} \pm 2.576\sqrt{\frac{\tilde{p}(1-\tilde{p})}{n+4}} = 0.7846 \pm 2.576\sqrt{\frac{(0.7846)(0.2154)}{65}} = 0.6532$ to 0.9160. (b) As a percentage, the interval is from 65.3% to 91.6%. This interval is shifted slightly; it is less than the interval in Example 22.7 by about 1.9%.

Chapter 23 – Comparing Two Proportions

23.1 (a) STATE: We want to estimate the difference between the proportions of male and female high school seniors who had smoked cigarettes daily in the 30 days before the survey. PLAN: Let p_M be the proportion of all male high school seniors who had smoked cigarettes daily in the 30 days before the survey and p_F be the proportion of all female high school seniors who had smoked cigarettes daily in the 30 days before the survey. We want a 95% confidence interval for the difference in these proportions. SOLVE: The samples were large, with clearly more than 10 "successes" and 10 "failures" in each sample. Assume that the observations from each group can be thought of as an SRS from their respective populations. $\hat{p}_F = 0.037$, $\hat{p}_M = 0.031$, $SE = \sqrt{\frac{\hat{p}_F(1-\hat{p}_F)}{n_F} + \frac{\hat{p}_M(1-\hat{p}_M)}{n_M}} = \sqrt{\frac{0.037(0.963)}{1725} + \frac{0.031(0.969)}{1564}} = 0.0063$. The 95% confidence interval is $(\hat{p}_F - \hat{p}_M) \pm 1.96(SE) = (0.037 - 0.031) \pm 1.96(0.0063) = -0.0063$ to 0.0183. CONCLUDE: We are 95% confident that between 0.6% less and 1.8% more female than male high school seniors smoked cigarettes daily in the 30 days before the survey.

23.3 STATE: We want to estimate the difference between the proportions of all U.S. adults who would answer "Yes, should" to the question as worded in version A and those who would answer "Yes, should" to the question as worded in version B. PLAN: Let p_A be the proportion of all U.S. adults who would answer "Yes, should" to the question as worded in version A and p_B be the proportion of all U.S. adults who would answer "Yes, should" to the question as worded in version B. We want a 99% confidence interval for the difference in these proportions. SOLVE: There are 390 successes and 152 failures in the version A sample and 312 successes and 170 failures in the version B sample, all of which are greater than 10. Assume that the observations from each group can be thought of as an SRS from their respective populations. $\hat{p}_A = \frac{390}{542} = 0.7196$, $\hat{p}_B = \frac{312}{482} = 0.6473$, $SE = \sqrt{\frac{\hat{p}_A(1-\hat{p}_A)}{n_A} + \frac{\hat{p}_B(1-\hat{p}_B)}{n_B}} = \sqrt{\frac{0.7196(0.2804)}{542} + \frac{0.6473(0.3527)}{482}} = 0.0291$. The 99% confidence interval is $(\hat{p}_F - \hat{p}_M) \pm 2.576(SE) = (0.7196 - 0.6473) \pm 2.576(0.0291) = -0.0027$ to 0.1473. CONCLUDE: We are 99% confident that the proportion of U.S. adults who would answer "Yes, should" to version A is between 0.003 lower and 0.147 higher than the proportion who would answer "Yes, should" to version B.

23.5 STATE: Is helmet use less common among skiers and snowboarders who have suffered head injuries when compared to those who have not suffered head injuries? PLAN: Let p_1 be the proportion of all skiers and snowboarders who have suffered head injuries and p_2 be the proportion of all skiers and snowboarders who have not suffered head injuries. We test $H_0: p_1 = p_2$ versus $H_a: p_1 < p_2$. SOLVE: Assume that the observations from each group can be thought of as an SRS. The smallest count is 96, which is greater than 5, so the significance testing procedure is safe to use. $\hat{p}_1 = \frac{96}{578} = 0.1661$, $\hat{p}_2 = \frac{656}{2992} = 0.2193$. The pooled proportion is $\hat{p} = \frac{96 + 656}{578 + 2992} = 0.2106$, and $SE = \sqrt{\hat{p}(1-\hat{p})\left(\frac{1}{n_1} + \frac{1}{n_2}\right)} = \sqrt{0.2106(0.7894)\left(\frac{1}{578} + \frac{1}{2992}\right)} = 0.01853$. $z = \frac{\hat{p}_1 - \hat{p}_2}{SE} = \frac{0.1661 - 0.2193}{0.01853} = -2.87$, and $P = P(Z < -2.87) = 0.0021$. CONCLUDE: We have strong evidence that skiers and snowboarders who have had head injuries are less likely to use helmets than skiers and snowboarders who have not had head injuries.

23.7 (a) We cannot calculate a large-sample confidence interval because only five infants in the peanut consumption group developed a peanut allergy, and the method requires at least 10 successes and at least 10 failures. (b) The sample sizes become 265 and 268, with the counts of successes now equal to 37 and 6, respectively. (c) Let p_A be the population proportion of infants who develop a peanut allergy after avoiding peanuts and p_C be the population proportion of infants who develop a peanut allergy after consuming peanuts. Assume the observations from each group can be thought of as an SRS. $\tilde{p}_A = \frac{36+1}{263+2} = 0.1396$, $\tilde{p}_C = \frac{5+1}{266+2} = 0.0224$, $SE = \sqrt{\frac{\tilde{p}_A(1-\tilde{p}_A)}{n_A+2} + \frac{\tilde{p}_C(1-\tilde{p}_C)}{n_C+2}} = \sqrt{\frac{0.1396(0.8604)}{265} + \frac{0.0224(0.9776)}{268}} = 0.0231$. The 99% plus four confidence interval is $(\tilde{p}_A - \tilde{p}_C) \pm 2.576(SE) = (0.1396 - 0.0224) \pm 2.576(0.0231) = 0.0577$ to 0.1767. We are 99% confident that between 5.77% and 17.67% more infants with severe eczema, egg allergy, or both will develop a peanut allergy by 60 months after avoiding peanuts compared to those who consume them.

23.9 (b)

23.11 (b)

23.13 (c)

23.15 (b)

23.17 (a) The smallest count is $671 - 396 = 275$, which is more than 10, so a large-sample confidence interval can be used. **(b)** Let p_1 be the proportion of all adults aged 18–29 who believed the government was tracking all or most of their activities online or on their cell phones and p_2 be the proportion of all adults aged 65 or older who believed the government was tracking all or most of their activities online or on their cell phones. We want a 95% confidence interval for the difference in these proportions. $\hat{p}_1 = \frac{396}{671} = 0.5902$, $\hat{p}_2 = \frac{293}{977} = 0.2999$, $SE = \sqrt{\frac{\hat{p}_1(1-\hat{p}_1)}{n_1} + \frac{\hat{p}_2(1-\hat{p}_2)}{n_2}} = \sqrt{\frac{0.5902(0.4098)}{671} + \frac{0.2999(0.7001)}{977}} = 0.0240.$ The 99% confidence interval for the difference in proportions is $(\hat{p}_1 - \hat{p}_2) \pm 1.96(SE) = (0.5902 - 0.2999) \pm 1.96(0.0240) = 0.2433$ to 0.3373. We are 95% confident that the difference between the population proportions of adults aged 18–29 and adults aged 65 or older who believed the government was tracking all or most of their activities online or on their cell phones is between 0.2433 and 0.3373.

23.19 (a) We cannot calculate a large-sample confidence interval nor conduct a hypothesis test, because one of the counts is zero and the method requires at least 10 successes and at least 10 failures. **(b)** The sample size for the treatment group is 35, 24 of which have tumors; the sample size for the control group is 20, one of which has a tumor. **(c)** Assume that the mice were randomly assigned to the treatment. $\tilde{p}_1 = \frac{23+1}{33+2} = 0.6857$, $\tilde{p}_C = \frac{0+1}{18+2} = 0.05$, $SE = \sqrt{\frac{\tilde{p}_1(1-\tilde{p}_1)}{n_1+2} + \frac{\tilde{p}_2(1-\tilde{p}_2)}{n_2+2}} = \sqrt{\frac{0.6857(0.3143)}{35} + \frac{0.05(0.95)}{20}} = 0.0924.$ The 99% plus four confidence interval is $(\tilde{p}_1 - \tilde{p}_1) \pm 2.576$ $(SE) = (0.6857 - 0.05) \pm 2.576(0.0924) = 0.3977$ to 0.8737. CONCLUDE: We are 99% confident that lowering DNA methylation increases the incidence of tumors by between about 40% and 87%.

23.21 (a) Let p_1 be the proportion of all habitual tea drinkers who had a stroke and p_2 be the proportion of all never/non-habitual tea drinkers who had a stroke. We test $H_0: p_1 = p_2$ versus $H_a: p_1 \neq p_2$. Assume that the observations from each group can be thought of as an SRS. The smallest count is 854, which is greater than 5, so the significance testing procedure is safe to use. $\hat{p}_1 = \frac{854}{31{,}885} = 0.0268$, $\hat{p}_2 = \frac{2949}{69{,}017} = 0.0427$. The pooled proportion is $\hat{p} = \frac{854+2949}{31{,}885+69{,}017} = 0.0377$, and $SE = \sqrt{\hat{p}(1-\hat{p})\left(\frac{1}{n_1}+\frac{1}{n_2}\right)} = \sqrt{0.0377(0.9623)\left(\frac{1}{31{,}885}+\frac{1}{69{,}017}\right)} = 0.00129.$ $z = \frac{\hat{p}_1 - \hat{p}_2}{SE} = \frac{0.0268-0.0427}{0.00129} = -12.33$, and $P = 2P(Z < -12.33) \approx 0$. We have very strong evidence that there is a difference in the proportions of all habitual tea drinkers who had a stroke and of all never/non-habitual tea drinkers who had a stroke. **(b)** This is an observational study because the researchers did not assign the subjects to the treatments. **(c)** We cannot conclude cause and effect unless we have a randomized experiment. Additionally, the extremely large sample sizes would have resulted in a tiny P-value even if there were very little difference in the proportions.

23.23 The smallest count is 854, which is greater than 10, so a large-sample confidence interval can be constructed. $\hat{p}_1 = \frac{854}{31{,}885} = 0.0268$, $\hat{p}_2 = \frac{2949}{69{,}017} = 0.0427$. $SE = \sqrt{\frac{\hat{p}_1(1-\hat{p}_1)}{n_1+2} + \frac{\hat{p}_2(1-\hat{p}_2)}{n_2+2}}$ $= \sqrt{\left(\frac{0.0268(0.9732)}{31{,}885} + \frac{0.0427(0.9573)}{69{,}017}\right)} = 0.00119$. The large-sample 95% confidence interval is $(\hat{p}_1 - \hat{p}_2) \pm 1.96(SE) = (0.0268 - 0.0427) \pm 1.96(0.00119) = -0.0183$ to -0.01345. Since the interval is entirely below 0, it appears that the stroke rate is lower for those who are habitual tea drinkers. This does support the claim that tea drinking is good for your health. We need to keep in mind that this is an observational study, so we cannot conclude causation.

23.25 (a) Let p_W be the proportion of all businesses headed by women that fail and p_M be the proportion of all businesses headed by men that fail. We test $H_0: p_W = p_M$ versus $H_a: p_W \neq p_M$. $\hat{p}_W = \frac{7}{42} = 0.1667$, $\hat{p}_M = \frac{15}{106} = 0.1415$. The pooled proportion is $\hat{p} = \frac{7+15}{42+106} = 0.1486$, and $SE = \sqrt{\hat{p}(1-\hat{p})\left(\frac{1}{n_1}+\frac{1}{n_2}\right)} = \sqrt{0.1486(0.8514)\left(\frac{1}{42}+\frac{1}{106}\right)} = 0.06485$. $z = \frac{\hat{p}_1 - \hat{p}_2}{SE} = \frac{0.1667-0.1415}{0.06485} = 0.39$, and $P = 2P(Z > 0.39) = 0.6966$, which provides virtually no evidence of a difference in failure rates. **(b)** $\hat{p}_W = \frac{210}{1260} = 0.1667$, $\hat{p}_M = \frac{450}{3180} = 0.1415$, and $\hat{p} = 0.1486$, but now $SE = \sqrt{\hat{p}(1-\hat{p})\left(\frac{1}{n_1}+\frac{1}{n_2}\right)} = \sqrt{0.1486(0.8514)\left(\frac{1}{1260}+\frac{1}{3180}\right)} =$ 0.01184, so $z = \frac{0.1667-0.1415}{0.01184} = 2.13$, and $P = 0.0332$. **(c)** For case (a), the 95% confidence interval for the difference is $(\hat{p}_W - \hat{p}_M) \pm 1.96\sqrt{\frac{\hat{p}_W(1-\hat{p}_W)}{n_W} + \frac{\hat{p}_M(1-\hat{p}_M)}{n_M}} = (0.1667 - 0.1415) \pm 1.96(0.0667) = -0.1055$ to 0.1559. For case (b), the resulting confidence interval is $= (0.1667 - 0.1415) \pm 1.96(0.0122) = 0.0013$ to 0.0491. The confidence interval is narrower with larger sample sizes and shows that the absolute difference between the proportions is less than 0.05.

23.27 STATE: Is there a difference between the population proportions of White non-Hispanics and Black non-Hispanics who were concerned about how much information their friends and family might know about them? PLAN: Let p_W be the population proportion of White non-Hispanics who were concerned and p_B be the population proportion of Black non-Hispanics who were concerned. We test $H_0: p_W = p_B$ versus $H_a: p_W \neq p_B$. SOLVE: Assume that the observations from each group can be thought of as an SRS. The smallest count is $445 - 271 = 174$, which is greater than 5, so the significance testing procedure is safe to use. $\hat{p}_W = \frac{1010}{2887} = 0.3498$, $\hat{p}_B = \frac{271}{445} = 0.6090$. The pooled proportion is $\hat{p} = \frac{1010+271}{2887+445} = 0.38445$, and $SE = \sqrt{\hat{p}(1-\hat{p})\left(\frac{1}{n_1}+\frac{1}{n_2}\right)} = \sqrt{0.38445(0.61555)\left(\frac{1}{2887}+\frac{1}{445}\right)} = 0.02477.$

$z = \frac{\hat{p}_1 - \hat{p}_2}{SE} = \frac{0.3498 - 0.6090}{0.02477} = -10.46$, and $P = 2P(Z < -10.46) \approx 0$. CONCLUDE: We have strong evidence that there is a difference between the population proportions of White non-Hispanics and Black non-Hispanics who were concerned about how much information their friends and family might know about them.

23.29 STATE: We want to estimate the difference (treatment minus placebo) in the proportions of smokers who would abstain from smoking during weeks 15 through 24. PLAN: Let p_1 be the proportion of all adults who would abstain from smoking with Chantix and p_2 be the proportion of all adults who would abstain from smoking without Chantix. We want a 99% confidence interval for the difference in these proportions. SOLVE: The trial was a randomized, double-blind, placebo-controlled experiment. The smallest count is 52, which is more than 10, so we can calculate a large-sample confidence interval. $\hat{p}_1 = \frac{244}{760} = 0.3211$, $\hat{p}_2 = \frac{52}{750} = 0.0693$, $SE = \sqrt{\frac{\hat{p}_1(1-\hat{p}_1)}{n_1} + \frac{\hat{p}_2(1-\hat{p}_2)}{n_2}} = \sqrt{\frac{0.3211(0.6789)}{760} + \frac{0.0693(0.9307)}{750}} = 0.0193$. The 99% confidence interval for the difference in proportions is $(\hat{p}_1 - \hat{p}_2) \pm 2.576(SE) = (0.3211 - 0.0693) \pm 2.675(0.0193) =$ 0.2021 to 0.3015. CONCLUDE: We are 99% confident that the population proportion who abstain from smoking on Chantix is between 20.21% and 30.15% higher than those without Chantix.

23.31 STATE: Is there evidence that a smile increases the proportion of drivers who stop? PLAN: Let p_1 be the population proportion who stopped when a pedestrian had a neutral expression and p_2 be the population proportion who stopped when a pedestrian was smiling. We test $H_0: p_1 = p_2$ versus $H_a: p_1 < p_2$. SOLVE: Assume that the observations from each group can be thought of as an SRS. The smallest count is 172, which is greater than 5, so the significance testing procedure is safe to use. $\hat{p}_1 = \frac{172}{400} = 0.430$, $\hat{p}_2 = \frac{226}{400} = 0.5650$. The pooled proportion is $\hat{p} = \frac{172+226}{400+400} = 0.4975$, and $SE = \sqrt{\hat{p}(1-\hat{p})\left(\frac{1}{n_1} + \frac{1}{n_2}\right)} = \sqrt{0.4975(0.5025)\left(\frac{1}{400} + \frac{1}{400}\right)} = 0.03535$. $z = \frac{\hat{p}_1 - \hat{p}_2}{SE} = \frac{0.43 - 0.565}{0.03535} = -3.82$, and $P = P(Z < -3.82) < 0.0002$. CONCLUDE: There is evidence that a smile increases the population proportion of drivers who stop for a pedestrian at a pedestrian crossing. We can conclude a cause-and-effect relationship because this was a randomized experiment.

23.33 STATE: We want to estimate the difference between the proportions of mice ready to breed in good acorn years and in bad acorn years. PLAN: Let p_G be the population proportion of mice ready to breed in good acorn years and p_B be the population proportion of mice ready to breed in bad acorn years. We want a 90% confidence interval for the difference in these proportions. SOLVE: Assume that the mice represent an SRS of the population of mice. The smallest count is 7, and the guidelines for using the large-sample method require all counts to be at least 10. We use the plus four method. $\tilde{p}_G = \frac{54+1}{72+2} = 0.7432$, $\tilde{p}_C = \frac{10+1}{17+2} = 0.5789$, $SE = \sqrt{\frac{\tilde{p}_G(1-\tilde{p}_G)}{n_G+2} + \frac{\tilde{p}_B(1-\tilde{p}_B)}{n_B+2}} = \sqrt{\frac{0.7432(0.2568)}{74} + \frac{0.5789(0.4211)}{19}} = 0.12413$. The 90% plus four confidence interval is $(\tilde{p}_G - \tilde{p}_B) \pm 1.645(SE) = (0.7432 - 0.5789) \pm 1.645(0.12413) = -0.0399$ to0.3685. CONCLUDE: We are 90% confident that the population proportion of mice ready to breed in good acorn years is between 0.04 lower and 0.3685 higher than the population proportion in bad acorn years.

23.35 (a) This is an experiment because the researchers assigned subjects to the groups being compared. (b) STATE: Do hand sanitizers reduce the chance of rhinovirus (RV) infection? PLAN: Let p_1 be the population proportion of subjects who have an RV infection for the HL+ group and p_2 be the population proportion of subjects who have an RV infection for the control group. We test $H_0: p_1 = p_2$ versus $H_a: p_1 < p_2$. SOLVE: Assume the observations from each group can be thought of as an SRS. The smallest count is 47, which is greater than 5, so the significance testing procedure is safe to use. $\hat{p}_1 = \frac{49}{49+67} = 0.4224$, $\hat{p}_2 = \frac{49}{49+47} = 0.5104$. The pooled proportion is $\hat{p} = \frac{49+49}{116+96} = 0.46226$, and $SE = \sqrt{\hat{p}(1-\hat{p})\left(\frac{1}{n_1} + \frac{1}{n_2}\right)} = \sqrt{0.46226(0.53774)\left(\frac{1}{116} + \frac{1}{96}\right)} = 0.06879$. $z = \frac{\hat{p}_1 - \hat{p}_2}{SE} = \frac{0.4224 - 0.5104}{0.06879} = -1.28$, and $P = P(Z < -1.28) =$ 0.1003. CONCLUDE: We do not have enough evidence to reject the null hypothesis. There is little evidence to conclude that the population proportion of HL+ users with a rhinovirus infection is less than that for non-HL+ users.

Chapter 24 – Inference about Variables: Part IV Review

24.1 (c)

24.3 (b)

24.5 (a)

24.7 (b)

24.9 (a) There are 427 successes and 344 failures, both of which are larger than 15. Assuming that the observations can be treated as an SRS of the population, the large-sample confidence interval is appropriate. (b) With $\hat{p} = \frac{427}{771} = 0.5538$, the 90% confidence interval is $\hat{p} \pm 1.645\sqrt{\frac{\hat{p}(1-\hat{p})}{n}} = 0.5538 \pm 1.645\sqrt{\frac{(0.5538)(0.4462)}{771}} =$ 0.5244 to 0.5832, or 52.4% to 58.3%. (c) We are 90% confident that between 52.4% and 58.3% of extremely obese people who had gastric bypass surgery regained at least 25% of their post-surgery weight loss five years after surgery.

24.11 (b)

24.13 (a)

24.15 Notice that the standard deviations are larger than the mean. Because PedMIDAS scores must be greater than or equal to zero, all of the scores that are less than the mean differ from the mean by less than one standard deviation. This means several scores that are greater than the mean must be greater by more than one standard deviation (otherwise, the standard deviation would not be so large), "pulling" the distribution to the right. The sample sizes are fairly large ($n = 64$ and 71), so the sample means should be approximately Normal by the central limit theorem.

24.17 (a)

24.19 (c)

24.21 (b)

24.23 (c)

24.25 (c)

24.27 (b)

24.29 (d)

24.31 (c)

24.33 There are 90 successes and 146 failures, both of which are larger than 15. We are told that we have a random sample, so the large-sample confidence interval is appropriate. With $\hat{p} = \frac{90}{236} = 0.3814$, the 90% confidence interval is $\hat{p} \pm 1.645\sqrt{\frac{\hat{p}(1-\hat{p})}{n}} = 0.3814 \pm 1.645\sqrt{\frac{(0.3814)(0.6186)}{236}} = 0.3294$ to 0.4334. We are 90% confident that between 32.9% and 43.3% of young adults aged 18-29 have used Twitter.

24.35 Let p_H be the population proportion of human offers rejected and p_C be the population proportion of computer offers rejected. Assume the observations from each group can be thought of as an SRS. The smallest count is 6, which is greater than 5, so the significance testing procedure is safe to use. We test $H_0: p_H = p_C$ versus $H_a: p_H > p_C$. $\hat{p}_H = \frac{18}{38} = 0.4737$, $\hat{p}_C = \frac{6}{38} = 0.1579$. The pooled proportion is $\hat{p} = \frac{18+6}{38+38} = 0.3158$, and

$$\text{SE} = \sqrt{\hat{p}(1-\hat{p})\left(\frac{1}{n_1}+\frac{1}{n_2}\right)} = \sqrt{0.3158(0.6842)\left(\frac{1}{38}+\frac{1}{38}\right)} = 0.10664.$$

$z = \frac{\hat{p}_1 - \hat{p}_2}{\text{SE}} = \frac{0.4737 - 0.1579}{0.10664} = 2.96$, and $P = P(Z > 2.96) = 0.0015$ (with software; using Table C, $0.001 < P < 0.0025$). There is strong evidence that offers from another person are rejected more often than offers from a computer.

24.37 (a)

24.39 In all three cases, the observations must be able to be seen as random samples of both types of diets. Also, the populations must be Normally distributed.

24.41 A large-sample (or plus four) confidence interval for a population proportion.

24.43 This is the entire population of Chicago Cubs players on the active roster. Statistical inference is not appropriate.

24.45 (a) Two-sample test or confidence interval for difference in proportions. (b) Two-sample test or confidence interval for difference in means. (c) Two-sample test or confidence interval for difference in proportions.

24.47 (a) This is a matched pairs situation; the responses of each subject before and after treatment are not independent. (b) We need to know the standard deviation of the differences, not the two individual sample standard deviations. (Note that the sample mean difference is equal to the difference in the two sample means, which is why we only need to know the standard deviation of the differences.)

24.49 (a) $\hat{p} = \frac{80}{80} = 1$. The large-sample margin of error for 95% confidence (or any level of confidence) is 0 because $z^*\sqrt{\frac{1(1-1)}{n}} = 0$. Almost certainly, if more trials were performed, a rat would eventually make a mistake, so the actual success rate is less than 1. (b) The confidence level (95%) is greater than 90%, and the sample size (80) is at least 10, so the plus four interval can be used. $\tilde{p} = \frac{80+2}{80+4} = 0.9762$, and the 95% plus four confidence interval is $\tilde{p} \pm 1.96\sqrt{\frac{\tilde{p}(1-\tilde{p})}{n+4}} = 0.9762 \pm 1.96\sqrt{\frac{(0.9762)(0.0238)}{84}} = 0.9436$ to 1.0088. Ignoring the upper limit, we are 95% confident that the actual success rate is 0.9436 or greater.

24.51 STATE: Is there good evidence that the mean age at first word among all typical children is greater than one year? PLAN: We test $H_0: \mu = 12$ versus $H_a: \mu > 12$, where μ denotes the population mean age at first word. SOLVE: We regard our sample as an SRS. A stemplot (below) shows that the data are right skewed with an outlier (26).

0	7
0	899
1	00001111
1	2
1	55
1	7
1	8
2	00
2	
2	
2	6

If we proceed with the t procedures despite this, we find $\bar{x} = 13$ and $s = 4.9311$ months. $t = \frac{13-12}{4.9311/\sqrt{20}} = 0.907$. With df = 19, we find $P = 0.1879$ (using software). If we delete the outlier mentioned above, $\bar{x} = 12.3158$ and $s = 3.9729$ months. $t = \frac{12.3158-12}{3.9729/\sqrt{19}} = 0.346$. With df = 18, we find $P = 0.3665$ (using software). CONCLUDE: We cannot conclude that the mean age at first word is greater than one year.

24.53 STATE: What is the mean age at first word among all typical children? PLAN: We construct a 90% confidence interval for μ. SOLVE: We checked the conditions in Exercise 24.51. Ignoring the outlier: df = 19 and $t^* = 1.729$. The 90% confidence interval for μ is $\bar{x} \pm 1.729\frac{s}{\sqrt{n}} = 13.0 \pm 1.729 \frac{4.9311}{\sqrt{20}} = 11.09$ to 14.91. CONCLUDE: We are 90% confident that the mean age at first word for typical children is between 11.1 and 14.9 months.

24.55 (a) The design is shown below.

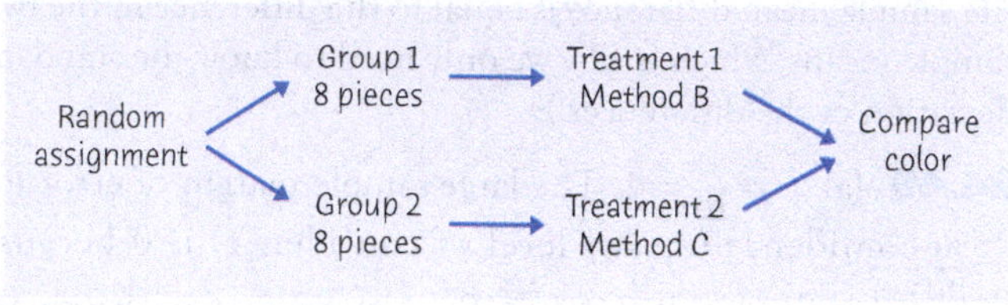

(b) STATE: Which method (B or C) gives the darker color (lower lightness score), on average? PLAN: Let μ_B be the population mean time lightness score for Method B and let μ_C be the population mean time lightness score for Method C. We test $H_0: \mu_B = \mu_C$ versus $H_a: \mu_B \neq \mu_C$. SOLVE: Stemplots (below) do not appear to violate Normality. The samples are independent because we have a randomized experiment. We can use t procedures.

408	8
409	8
410	
411	
412	78
413	09
414	
415	0
416	6

42	223
42	445
42	6
42	
43	1

Here with $\bar{x}_B = 41.2825$, $\bar{x}_C = 42.4925$, $S_B = 0.2550$, $S_C = 0.2939$, and $n_1 = n_2 = 8$: $\text{SE} = \sqrt{\frac{s_B^2}{n_B} + \frac{s_C^2}{n_C}} = 0.1376$ and $t = \frac{\bar{x}_B - \bar{x}_C}{\text{SE}} = -8.79$. Using software, df = 13.73, $P < 0.0001$. Using the more conservative df = 7 (both samples are of size 8, so df = 8 − 1) and Table C, $P < 0.001$. CONCLUDE: There is overwhelming evidence that the two dying methods do, on average, differ in lightness scores, and the data suggest that Method B gives darker color. However, the *P*-value tells us nothing about the magnitude of the difference, and it may be too small to be important in practice.

24.57 STATE: What proportion of female students have at least one parent who allows them to drink in their parent's presence? PLAN: We construct a 95% confidence interval. SOLVE: We are told that the sample represents an SRS. We have 65 successes and 29 failures, which are both larger than 15, so large-sample methods may be used. With $\hat{p} = \frac{65}{94} = 0.6915$, the 95% confidence interval is $\hat{p} \pm 1.96\sqrt{\frac{\hat{p}(1-\hat{p})}{n}} = 0.6915 \pm 1.96 \sqrt{\frac{(0.6915)(0.3085)}{94}} = 0.5981$ to 0.7849. CONCLUDE: With 95% confidence, the proportion of female students who have at least one parent who allows them to drink in their presence is between 0.598 and 0.785.

24.59 (a) STATE: Does microwaving reduce the percentage of cracked crackers? PLAN: Let p_1 be the proportion of all microwaved crackers that show checking and p_2 be the proportion of all control crackers that show checking. We want a 95% confidence interval for the difference in these proportions. SOLVE: Assume that the crackers represent an SRS of the population of crackers. The smallest count is three, and the guidelines for using the large-sample method require all counts to be at least 10. We use the plus four method. $\tilde{p}_1 = \frac{3+1}{65+2} = 0.0597$, $\tilde{p}_2 = \frac{57+1}{65+2} = 0.8657$, $\text{SE} = \sqrt{\frac{\tilde{p}_1(1-\tilde{p}_1)}{n_1+2} + \frac{\tilde{p}_2(1-\tilde{p}_2)}{n_2+2}} = \sqrt{\frac{0.0597(0.9403)}{67} + \frac{0.8657(0.1343)}{67}} = 0.05073$. The 95% plus four confidence interval is $(\tilde{p}_1 - \tilde{p}_2) \pm 1.96(\text{SE}) = (0.0597 - 0.8657) \pm 1.96(0.05073) = -0.9054$ to -0.7066. CONCLUDE: We are 95% confident that microwaving reduces the percentage of cracked crackers by between 70.7% and 90.5%. (b) STATE: Does microwaving crackers change their mean breaking strength? PLAN: Let μ_1 be the population mean breaking strength of microwaved crackers and let μ_2 be the population mean breaking strength of control crackers. We test $H_0: \mu_1 = \mu_2$ versus $H_a: \mu_1 \neq \mu_2$ and construct a 95% confidence interval for the difference between the two means. SOLVE: We assume the data can be considered SRSs from the populations of crackers. We do not have the data to examine, but our sample sizes are large enough to use t procedures. We have $\bar{x}_1 = 139.6$, $\bar{x}_2 = 77.0$, $s_1 = 33.6$, $s_2 = 22.6$, $n_1 = n_2 = 20$. Here, $SE = \sqrt{\frac{s_1^2}{n_1} + \frac{s_2^2}{n_2}} = \sqrt{\frac{33.6^2}{20} + \frac{22.6^2}{20}} = 9.0546$ and $t = \frac{\bar{x}_1 - \bar{x}_2}{\text{SE}} = \frac{139.6 - 77.0}{9.0546} = 6.914$. Using the conservative approach for computing df as the smaller sample size minus 1, df = 20 − 1 = 19 (the sample sizes are both 20). Using Table C, $P < 0.001$ (using software, df = 33.27 and $P \approx 0$). The 95% confidence interval is $(\bar{x}_1 - \bar{x}_2) \pm t^*(\text{SE}) = (139.6 - 77.0) \pm t^*(9.0546)$, where $t^* = 2.093$ (using the Table C conservative approach with df = 19) or 2.0339 (df = 33.27, with software). This gives either 43.65 to 81.55 psi or 44.18 to 81.02 psi, respectively. CONCLUDE: There is very strong evidence that microwaving crackers changes their mean breaking strength. We are 95% confident that microwaving crackers increases their mean breaking strength by between 43.65 and 81.55 psi.

24.61 Two of the counts (one and four) are too small to safely perform a significance test.

24.63 Let μ_1 be the population mean blood glucose for the wild-type mice and let μ_2 be the population mean blood glucose for the $aP2^{-/-}$ mice. We test $H_0: \mu_1 = \mu_2$ versus $H_a: \mu_1 \neq \mu_2$. The group means are $\bar{x}_1 = 5.9$ (wild type) and $\bar{x}_2 = 0.75$ ($aP2^{-/-}$) ng/ml, and the standard deviations are $\text{SEM}_1 = 0.9$ and $\text{SEM}_2 = 0.2$ ng/ml. The sample sizes are $n_1 = n_2 = 10$. We assume that the data can be considered SRSs from the populations of mice. We will use t procedures as was done in the paper. We have $\bar{x}_1 = 5.9, \bar{x}_2 = 0.75$, $s_1 = 0.9\sqrt{10} = 2.846$, $s_2 = 0.2\sqrt{10} = 0.632$, $n_1 = n_2 = 10$. Here, $SE = \sqrt{0.9^2 + 0.2^2} = 0.9220$ and $t = \frac{\bar{x}_1 - \bar{x}_2}{\text{SE}} = \frac{5.9 - 0.75}{0.9220} = 5.59$. Using the conservative approach for computing df as the smaller sample size minus 1, df $= 10 - 1 = 9$ (the sample sizes are both 10) or df $= 9.89$ (using software) and $P < 0.001$. The evidence is even stronger than the paper claimed.

24.65 STATE: What percentage of all adults in the United States would say parents of young adults ages 18 to 29 are doing too much for their adult children? PLAN: We construct a 95% confidence interval. SOLVE: We assume that we can treat this sample as an SRS of all adults in the United States. There are $9834(0.55) = 5408.7$ successes and $9834(0.45) = 4425.3$ failures. Both are larger than 15, so the large-sample confidence interval is appropriate. $\hat{p} = 0.55$. The large-sample 95% confidence interval is $\hat{p} \pm 1.96(\text{SE}) = 0.55 \pm 1.96\sqrt{\frac{(0.55)(0.45)}{9834}} = 0.5402$ to 0.5598, or 54.0% to 56.0%. CONCLUDE: We are 95% confident that between 54.0% and 56.0% adults in the United States would say parents of young adults ages 18 to 29 are doing too much for their adult children.

Chapter 25 – Two Categorical Variables: The Chi-Square Test

25.1 (a) The table provided gives percentages in each category:

	Age Category			
Hard to give up	18–24	25–29	30–49	50 or over
No	48.8%	60.3%	57.0%	67.0%
Yes	51.2%	39.7%	43.0%	33.0%

(b) The bar graph reveals that only the 18–24 age group has a higher percentage who think it would be hard to give up social media. The largest difference is for those aged 50 or older; a much higher proportion of those respondents do not think it would be hard to give up social media.

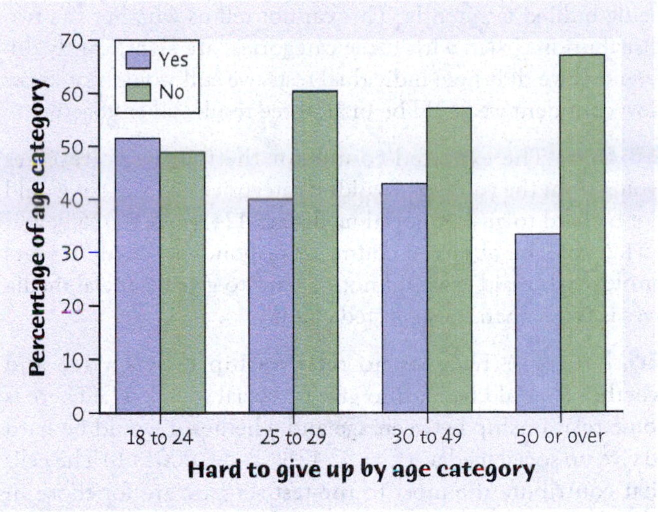

25.3 (a) Let p_1 be the proportion of those suffering from depression who were bullied occasionally and p_2 be the proportion of those not suffering from depression who were bullied occasionally. We test $H_0: p_1 = p_2$ versus $H_a: p_1 \neq p_2$. Assume that the observations from each group can be thought of as an SRS. The smallest count is 103, which is greater than 5, so we can conduct the significance test. $\hat{p}_1 = \frac{103}{97+103+101} = 0.3422$, $\hat{p}_2 = \frac{1343}{1762+1343+582} = 0.36425$. The pooled proportion is $\hat{p} = \frac{103+1343}{97+103+101+1762+1343+582} = 0.3626$, and $\text{SE} = \sqrt{\hat{p}(1-\hat{p})\left(\frac{1}{n_1}+\frac{1}{n_2}\right)} = \sqrt{0.3626(0.6374)\left(\frac{1}{301}+\frac{1}{3687}\right)} = 0.02882$. $z = \frac{\hat{p}_1 - \hat{p}_2}{\text{SE}} = \frac{0.3422 - 0.36425}{0.02882} = -0.765$, and $P = 2P(Z < -0.765) = 0.4440$. There is not evidence of a significant difference between the proportions bullied occasionally for those suffering and not suffering from depression. (b) Let p_1 be the proportion of those suffering from depression who were bullied frequently and p_2 be the proportion of those not suffering from depression who were bullied frequently. We test $H_0: p_1 = p_2$ versus $H_a: p_1 \neq p_2$. Assume that the observations from each group can be thought of as an SRS. The smallest count is 101, which is greater than 5, so we can conduct the significance test. $\hat{p}_1 = \frac{101}{97+103+101} = 0.3355$, $\hat{p}_2 = \frac{582}{1762+1343+582} = 0.1579$. The pooled proportion is $\hat{p} = \frac{101+582}{97+103+101+1762+1343+582} = 0.1713$, and $\text{SE} = \sqrt{\hat{p}(1-\hat{p})\left(\frac{1}{n_1}+\frac{1}{n_2}\right)} = \sqrt{0.1713(0.8287)\left(\frac{1}{301}+\frac{1}{3687}\right)} = 0.0226$. $z = \frac{\hat{p}_1 - \hat{p}_2}{\text{SE}} = \frac{0.3355 - 0.1579}{0.0226} = 7.86$, and $P = 2P(Z > 7.86) \approx 0$. There is strong evidence that the proportions who were bullied frequently are different for those suffering and not suffering from depression. (c) The P-values only indicate the strength of evidence for a difference between two particular categories of being bullied frequently. This cannot tell us whether the two distributions, each with three categories, are significantly different. If we did three individual tests, we still would not know how confident we could be in all three results taken together.

25.5 (a) The expected counts for the four age categories (going from the youngest to oldest categories) who said it would not be hard to give up social media are 124.4, 93.5, 326.9, and 664.2. (b) The observed count for respondents aged 50 years or older who said it would not be hard to give up social media ages is larger than the expected count.

25.7 (a) H_0: There is no relationship between age and whether it would be hard to give up social media. H_a: There is some relationship between age and whether it would be hard to give up social media. $\chi^2 = 32.128$; $P < 0.0001$. (b) The cells that contribute the most to the test statistic are for those in the 18–24 years and 50 years or older age groups who said it would be hard to give up social media.The count for those in the 18–24 years age group is larger than the expected count, while the count for those in the 50 years or older age group is smaller than the expected count.

25.9 STATE: Is there a relationship between the degree held and the view of astrology? PLAN: We want to test H_0: There is no relationship between the degree held and the view of astrology. H_a: There is some relationship between the degree held and the view of astrology. SOLVE: The observed percentage who view astrology as a science is much lower for bachelor's degree and graduate degree than for junior college degree. $\chi^2 = 33.843$. $P < 0.0001$. (It is reported to be 0, but we cannot have $P = 0$.) CONCLUDE: There is strong evidence that the distribution of views on astrology is related to the degree held.

25.11 (a) $df = (2-1)(3-1) = 2$. (b) The value of the test statistic is much larger than the largest value in Table D for $df = 2$, so we can conclude that $P < 0.0005$. JMP is more precise than the table, giving $P < 0.0001$. (c) If the null hypothesis is true, the mean of the test statistic is $df = 2$. The observed value is much larger than the mean.

25.13 (a) This would be a chi-square test of homogeneity. (b) STATE: Determine if there is a relationship between which sock is chosen and whether the display was high or low. PLAN: Conduct a chi-square test of the hypotheses H_0: There is no relationship between sock location and display height.H_a: There is some relationship between sock location and display height. SOLVE: The conditional distributions for the high and low display are provided in the table below. The values in parentheses are the expected counts.

	Loc 1	Loc 2	Loc 3	Loc 4	Loc 5
High display	0.28 (13)	0.24 (12.5)	0.28 (17)	0.12 (5)	0.08 (2.5)
Low display	0.24 (13)	0.26 (12.5)	0.40 (17)	0.08 (5)	0.02 (2.5)

The observed proportion who chose the middle location is a bit different for the high and low display. Note that 8 of 10, or 80%, of cells have expected counts that are at least 5, and all cells have expected counts of at least 1. It is safe to conduct a chi-square test. $\chi^2 = \frac{(14-13)^2}{13} + \frac{(12-12.5)^2}{12.5} + \cdots + \frac{(1-2.5)^2}{2.5} = 3.453$. $df = (2-1)(5-1) = 4$. $P > 0.25$. CONCLUDE: We fail to reject the null hypothesis. There is not evidence of a relationship between the location and display height. (c) No, this test does not answer that question. This test answers the question of whether there is a relationship between preferred location and display height. Whether or not there are any differences among the preferences for the five locations is separate from the relationship between preference and display height, so the chi-square test of homogeneity is not appropriate.

25.15 We test $H_0: p_1 = p_2 = p_3 = \frac{1}{3}$ versus H_a: The probabilities are not all $\frac{1}{3}$. The expected counts are each $53 \times \frac{1}{3} = 17.67$. $\chi^2 = \sum \frac{(\text{observed count} - 17.67)^2}{17.67} = \frac{(31-17.67)^2}{17.67} + \frac{(14 \; 17.67)^2}{17.67} + \frac{(8-17.67)^2}{17.67} = 16.11$. $df = (3-1) = 2$, $P < 0.0005$. There is very strong evidence that the three tilts differ. The data show that more birds than expected strike the vertical window and fewer than expected strike the 40-degree window.

25.17 (a) The percentage of subjects who chose each location (ordered from location 1 to location 5) are 26%, 25%, 34%, 10%, and 5%. (b) If all locations are equally likely, we would expect 100/5, or 20%, to choose each location. Thus, the expected count is 20 for each location. (c) STATE: Determine if the subjects were equally likely to choose each location. PLAN: Conduct a chi-square test of the hypotheses $H_0: p_1 = p_2 = \cdots = p_5 = 0.2$. H_a: The probabilities are not all 0.2. SOLVE: We can perform a chi-square test because the expected count is greater than 5 for each group. $\chi^2 = \sum \frac{(\text{observed count} - 20)^2}{20} = \frac{(26-20)^2}{20} + \cdots + \frac{(5-20)^2}{20} = 29.1$. $df = 5 - 1 = 4$, $P < 0.0005$. CONCLUDE: There is strong evidence that the locations are not all equally likely to be chosen. (d) STATE: Is there evidence of the center-stage effect? PLAN: Let p denote the proportion that choose the center location. Test the hypotheses $H_0: p = 0.2$ versus $H_a: p > 0.2$. SOLVE: We expect $100(0.2) = 20$ successes and 80 failures. Both are larger than 10, so assuming that this sample can be treated like an SRS, the conditions are met. $\hat{p} = \frac{34}{100} = 0.34$, $z = \frac{\hat{p} - p_0}{\sqrt{\frac{p_0(1-p_0)}{n}}} = \frac{0.34 - 0.2}{\sqrt{\frac{0.2(0.8)}{100}}} = 3.50$, and, using Table C, $P < 0.0002$. CONCLUDE: There is strong evidence that the item in the center is chosen more often than those in other locations.

25.19 STATE: Are all 12 astrological signs equally likely? PLAN: Conduct a chi-square test of the hypotheses $H_0: p_1 = p_2 = \cdots = p_{12} = \frac{1}{12}$. H_a: The 12 astrological birth sign probabilities are not equally likely. SOLVE: There are 2402 subjects in this sample. Under H_0, we expect $2402/12 = 200.17$ per sign, so all cells have expected counts greater than 5, and a chi-square test is appropriate. $\chi^2 = \frac{(205-200.17)^2}{200.17} + \frac{(174-200.17)^2}{200.17} + \cdots + \frac{(198-200.17)^2}{200.17} = 18.34$. df $= 12 - 1 = 11$. Using Table C, $0.05 < P < 0.10$ ($P = 0.0740$ using software). CONCLUDE: There is little support for the notion that astrological signs are not equally likely.

25.21 (a)

25.23 (b)

25.25 (c)

25.27 (a)

25.29 (c)

25.31 (a) STATE: We compare rates of success at smoking cessation in the three treatment groups. PLAN: Let p_c be the proportion of smokers taking Chantix who did not smoke in weeks 9–12 of the study, p_B be the proportion of smokers taking bupropion who did not smoke in weeks 9–12 of the study, and p_P be the proportion of smokers taking the placebo who did not smoke in weeks 9–12 of the study. The smallest count is 61, which is greater than 10, so a large-sample confidence interval can be constructed. SOLVE: We want a 95% confidence interval for $p_B - p_P$. $\hat{p}_B = \frac{97}{329} = 0.2948$, $\hat{p}_P = \frac{61}{344} = 0.1773$, SE $= \sqrt{\frac{\hat{p}_B(1-\hat{p}_B)}{n_B} + \frac{\hat{p}_P(1-\hat{p}_P)}{n_P}} = \sqrt{\frac{0.2948(0.7052)}{329} + \frac{0.1773(0.8227)}{344}} = 0.0325.$ The 95% confidence interval for the difference in proportions is $(\hat{p}_B - \hat{p}_P) \pm 1.96(SE) = (0.2948 - 0.1773) \pm 1.96(0.0325) =$ 0.0538 to 0.1812. CONCLUDE: We are 95% confident that the proportion of smokers taking bupropion who did not smoke in weeks 9–12 of the study is between 0.0538 and 0.1812 greater than the proportion for smokers taking the placebo.
(b) The sample proportion for Chantix is $\hat{p}_C = \frac{155}{352} = 0.4403$. To test H_0: $p_C = p_B = p_P$ versus H_a: not all proportions are equal. The smallest expected count is 100.47 (see table below), which is greater than 5, and the trials are independent, so we can perform a chi-square test of homogeneity.

	Chantix	Bupropion	Placebo	Total
No	155	97	61	313
	107.49	100.47	105.05	
Yes	197	232	283	712
	244.51	228.53	238.95	

Chi-square = 56.992, df = 2, P-value = 0.000

$\chi^2 = 56.992$. df $= 3 - 1 = 2$, $P < 0.0001$. (It is reported to be 0, but we cannot have $P = 0$.) The treatments are not equally successful at helping people to quit smoking. The Chantix group, in particular, has a higher success rate. (c) This is a test of homogeneity because the subjects were randomly assigned to three different treatments.

25.33 (a) Let p be the proportion of college students who use over-the-counter or prescription stimulants more than once a month to keep awake. We are told to consider this to be an SRS. There are 174 successes and 697 failures, both of which are greater than 15. $\hat{p} = \frac{174}{871} = 0.1998$. A large-sample 95% confidence interval for p is $\hat{p} \pm 1.96\sqrt{\frac{\hat{p}(1-\hat{p})}{n}} = 0.1998 \pm 1.96\sqrt{\frac{(0.1998)(0.8002)}{871}} = 0.1732$ to 0.2264. (b) It appears that students who use OTC/Rx stimulants tend to have borderline and poor sleep quality, while students who do not use OTC/Rx stimulants have similar proportions for each sleep quality.

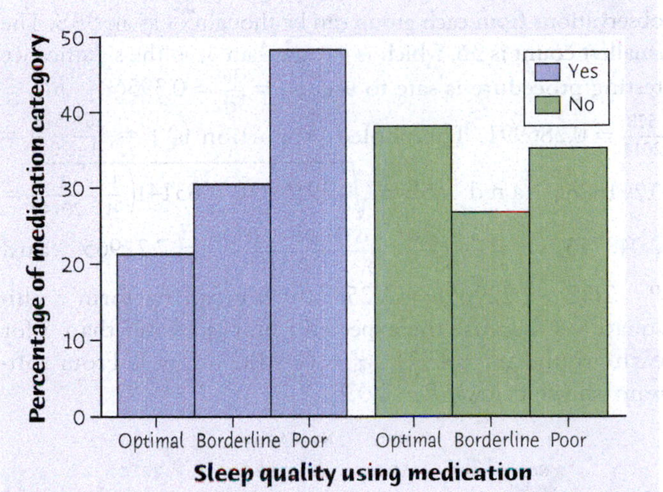

(c) PLAN: Conduct a chi-square test of the hypotheses H_0: There is no relationship between taking OTC/Rx stimulants and sleep quality. H_a: There is some relationship between taking OTC/Rx stimulants and sleep quality. SOLVE: The smallest expected count is 47.75 (see Minitab output below), which is greater than 5, and the trials are independent, so we can perform a chi-square test of independence.

	Yes	No	All
Optimal	37	266	303
	21.26	38.16	34.79
	60.53	242.47	
	9.15	2.28	
Borderline	53	186	239
	30.46	26.69	27.44
	47.75	191.25	
	0.58	0.14	
Poor	84	245	329
	48.28	35.15	37.77
	65.72	263.28	
	5.08	1.27	
All	174	697	871
	100.00	100.00	100.00

Cell Contents: Count
% of Column
Expected count
Contribution to Chi-square

Chi-Square Test

	Chi-Square	DF	P-value
Pearson	18.50	2	<0.0001
Likelihood Ratio	19.52	2	<0.0001

$\chi^2 = 18.50$. df $= 2$. $P < 0.0001$. CONCLUDE: We reject the null hypothesis. There is evidence of an association between taking OTC/Rx stimulants to stay awake and sleep quality.

25.35 (a) Let p_G be the proportion of all 14- to 18-year-old girls who played video games and have gotten into serious fights and p_{NG} be the proportion of all 14- to 18-year-old girls who have never played video games and have gotten into serious fights. $H_0: p_G = p_{NG}$ versus $H_a: p_G \neq p_{NG}$. Assume that the observations from each group can be thought of as an SRS. The smallest count is 36, which is greater than 5, so the significance testing procedure is safe to use. $\hat{p}_G = \frac{36}{91} = 0.395604$, $\hat{p}_{NG} = \frac{578}{2014} = 0.286991$. The pooled proportion is $\hat{p} = \frac{36+578}{91+2014} = 0.291686$, and $SE = \sqrt{0.291686(0.708314)\left(\frac{1}{91}+\frac{1}{2014}\right)} = 0.048713$. $z = \frac{\hat{p}_G - \hat{p}_{NG}}{SE} = \frac{0.395604-0.286991}{0.048713} = 2.22965$, and $P = 2P(Z > 2.22965) = 0.0258$. (b) We can perform a chi-square test because the expected count is greater than 5 for each group. $\chi^2 = 4.971$. df $= (2-1)(2-1) = 1$. From software (shown below), $P = 0.026$.

```
          Fight    No Fight      All
No          578        1436     2014
          587.5      1426.5   2014.0
         0.1522      0.0627        *

Yes          36          55       91
           26.5        64.5     91.0
         3.3690      1.3874        *
Cell Contents:       Count
                     Expected count
                     Contribution to
                     chi-square

Pearson chi-square = 4.971, df = 1,
P-value = 0.026
```

(c) $z^2 = (2.22965)^2 = 4.971$, which is equal to X^2. Obviously, P-values also agree. (d) We would use a one-sided z test. The chi-square test is inherently two-sided because it tests for association generally instead of for a particular direction of association.

25.37 (a) This was a single sample, and individuals are summarized in the table by two categorical variables, so this is a test of independence. (b) STATE: Is there a difference between how men and women assess their chances of being rich by age 30? PLAN: We want to test H_0: There is no relationship between sex and self-assessment of chances of being rich. H_a: There is some relationship between sex and self-assessment of chances of being rich. SOLVE: All expected cell counts exceed 5. $\chi^2 = 43.946$. df $= 4$. $P < 0.0001$. CONCLUDE: Overall, men give themselves a better chance of being rich. This difference shows up most noticeably in the second and fifth rows of the table: women were more likely to say "some chance, but probably not," whereas men more often responded "almost certain." There was virtually no difference between men and women in the "almost no chance" and "a 50-50 chance" responses and little difference in the "a good chance" response.

25.39 (a) We compare the percentage of dogs in each "condition type" that make a specified number of errors. The table below summarizes the data:

	Number of Errors			
	0	1	2	3
Social-Communicative	0%	25%	25%	50%
Noncommunicative	41.7%	25%	8.3%	25%
Nonsocial	66.7%	16.7%	16.7%	0%

We see that under the Social-Communicative condition, dogs tend to make more errors, whereas under the Nonsocial condition, dogs tend to make fewer errors. (b) The expected counts are in the table below.

	Number of Errors			
	0	1	2	3
Social-Communicative	4.33	2.67	2	3
Noncommunicative	4.33	2.67	2	3
Nonsocial	4.33	2.67	2	3

We can safely use a chi-square test when all expected cell counts are at least 1 and no more than 20% are less than 5. Here, all of the cells have expected counts of at least 1, but 100% are less than 5, so we cannot use a chi-square test. Also, each dog had up to three trials, so observations are not independent. (c) Software should warn users against using the chi-square test. The software package Minitab does provide such a warning.

25.41 Presumably, many of the individuals were included in more than one category of alcohol and drug use. This must be true for the post-high school group because the values in that column add up to more than the sample size. The chi-square test should be used only when each individual is classified in one specific category.

25.43 (a) Let p be the population proportion of Americans over age 18 who use at least one social media site. We are told to regard this as an SRS. There are 985 successes and 461 failures, both of which are greater than 15. $\hat{p} = \frac{985}{1446} = 0.6812$. A large-sample 99% confidence interval for p is $\hat{p} \pm 2.576\sqrt{\frac{\hat{p}(1-\hat{p})}{n}} = 0.6812 \pm 2.576\sqrt{\frac{(0.6812)(0.3188)}{1446}} = 0.6496$ to 0.7128. (b) The conditional distributions are shown in the second row of each cell in the Minitab output below. A bar graph is also provided. The table and graph show that younger individuals (under age 65) are much more likely to use at least one social media site than older individuals (age 65 or over).

	Yes	No	All
18-29	212	24	236
	21.52	5.21	16.32
	160.76	75.24	
	16.33	34.89	
30-49	324	71	395
	32.89	15.40	27.32
	269.07	125.93	
	11.21	23.96	
50-64	293	131	424
	29.75	28.42	29.32
	288.82	135.18	
	0.06	0.13	
65+	156	235	391
	15.84	50.98	27.04
	266.35	124.65	
	45.72	97.68	
All	985	461	1446
	100.00	100.00	100.00

Cell Contents: Count
% of Column
Expected count
Contribution to Chi-square

Chi-Square Test

	Chi-Square	DF	P-value
Pearson	229.98	3	<0.0001
Likelihood Ratio	232.76	3	<0.0001

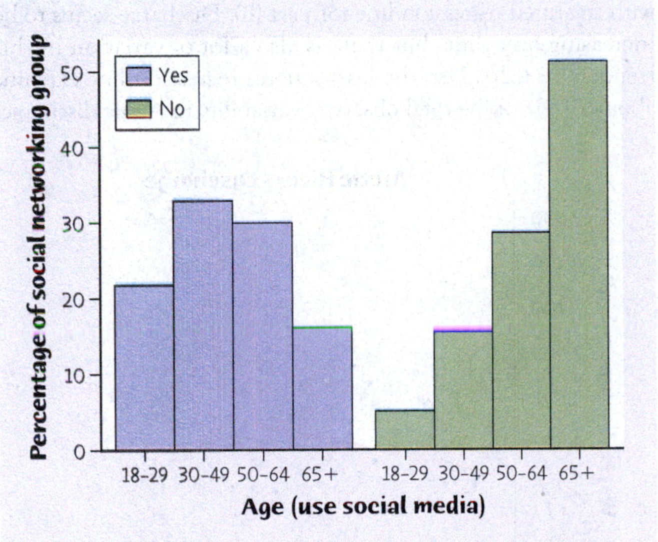

(c) Conduct a chi-square test of the hypotheses H_0: The distribution of ages for adults who use at least one social media site is the same as that distribution for adults who do not use at least one social media site. H_a: The distribution of ages for adults who use at least one social media site differs from that for adults who do not use at least one social media site. The conditional distributions and expected counts are in the output above. Note that all cells have expected counts that are at least 5. It is safe to conduct a chi-square test. $\chi^2 = 229.98$. df $= (4-1)(2-1) = 3$. If the null hypothesis were true, the mean of the test statistic would be 3. Our observed test statistic is much larger than 3. $P < 0.0001$. There is strong evidence that the age distributions of those who use and don't use at least one social networking site differ. **(d)** The cells that contribute most to the chi-square statistic are the 18–29 and "No" cell, the 65+ and "Yes" cell, and the 65+ and "No" cell. For 18–29, we observe a much lower number of non-social-media users than expected if the null hypothesis were true. For the 65+ group, we observe a much lower number of "Yes" (social media users) and a much higher number of "No" (non-users) than expected if the null hypothesis were true. It appears that younger individuals are much more likely to use at least one social media site than are individuals age 65 or over.

25.45 STATE: Determine if there is a difference in the distribution of frequency of being bullied for males and females. PLAN: Conduct a chi-square test of the hypotheses H_0: There is no relationship between sex and frequency of bullying. H_a: There is some relationship between sex and frequency of bullying. SOLVE: The conditional distributions and expected counts are in the output below. The frequencies are quite similar for males and females. There are slightly more females who have experienced either occasional or frequent bullying, whereas males have a slightly larger percentage of never having been bullied. Note that all cells have expected counts that are at least 5. It is safe to conduct a chi-square test. $\chi^2 = 7.01$. df $= (2-1)(3-1) = 2$. $P = 0.0301$. CONCLUDE: There is some evidence that the distribution of frequency of being bullied is not the same for males and females.

	Never	Occasionally	Frequently	All
Male	1564	1149	571	3284
	47.62	34.99	17.39	100.00
	1510.05	1187.52	586.43	
	1.93	1.25	0.41	
Female	1526	1281	629	3436
	44.41	37.28	18.31	100.00
	1579.95	1242.48	613.57	
	1.84	1.19	0.39	
All	3090	2430	1200	6720
	45.98	36.16	17.86	100.00

Cell Contents: Count
% of Row
Expected count
Contribution to Chi-square

Chi-Square Test

	Chi-Square	DF	P-value
Pearson	7.01	2	0.0301
Likelihood Ratio	7.01	2	0.0301

25.47 (a) We should use a test of independence because we have one SRS, and each individual was classified according to the eGFR and hearing loss. (b) H_0: There is no relationship between eGFR and hearing loss. H_a: There is some relationship between CKD and hearing loss. The conditional distributions and expected counts are in the output below. All cells have expected counts that are at least 5. It is safe to conduct a chi-square test. $\chi^2 = 177.787$. df $= (2-1)(5-1) = 4$. $P < 0.0001$. The low P-value is strong evidence of a relationship between eGFR and hearing loss. It appears those under age 45 are less likely to have hearing loss while those age 60 and over are more likely to have hearing loss.

Hearing loss	Count Col % Expected Cell Chi^2	eGFR <45	45-60	60-75	75-90	>90	Total
	No	27 26.21 68.4524 25.1022	207 50.49 272.48 15.7358	717 66.39 717.754 0.0008	458 75.70 402.075 7.7787	295 80.60 243.239 11.0148	1704
	Yes	76 73.79 34.5476 49.7373	203 49.51 137.52 31.1788	363 33.61 362.246 0.0016	147 24.30 202.925 15.4127	71 19.40 122.761 21.8247	860
	Total	103	410	1080	605	366	2564

Test	ChiSquare	Prob>ChiSq
Likelihood Ratio	174.508	<.0001*
Pearson	177.787	<.0001*

(c) Age is a quantitative variable. It could also be a lurking variable because it may be related to both eGFR and hearing loss. (d) If the relationship between eGFR and hearing loss is the same for each age group, then age is not contributing to that relationship.

25.49 Let p be the proportion of all American adults who said they are conservative. We are told to consider this to be an SRS. There are 740 successes and 1608 failures, both of which are greater than 15. $\hat{p} = \frac{740}{2348} = 0.3152$. A large-sample 95% confidence interval for p is $\hat{p} \pm 1.96\sqrt{\frac{\hat{p}(1-\hat{p})}{n}} = 0.3152 \pm 1.96\sqrt{\frac{(0.3152)(0.6848)}{2348}} = .2964$ to 0.3340. We are 95% confident that the proportion of all American adults who said they are conservative is between 0.2964 and 0.3340.

Chapter 26 – Inference for Regression

26.1 (a) A scatterplot of the data is provided, along with the least-squares regression line. From software, $r = 0.985$.

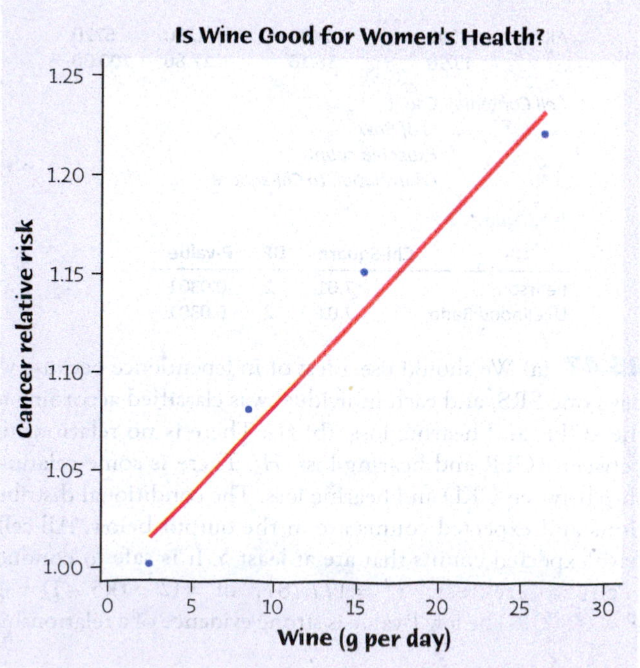

(b) If we knew it, the slope β would tell us how much relative risk of breast cancer changes in women for each increase in intake of 1 gram of wine per day (on average). If these data were on individual women, we would estimate that, on average, an increase in intake of 1 gram per day increases relative risk of breast cancer by about 0.009 (this is b in the least-squares regression). We would also estimate that wine intake of 0 grams per day is associated with a relative risk of breast cancer of 0.9931 (about 1; this is a in the least-squares regression). However, these data are based on averages, not individuals, so we should not assume that these estimates hold for individual women. (c) The residuals, calculated as $y - \hat{y} = y - (0.9931 + 0.0090x)$, are in the table below.

x	y	$\hat{y}$	Residual $(y-\hat{y})$	$(y-\hat{y})^2$
2.5	1.00	1.0156	−0.0156	0.00024
8.5	1.08	1.0697	0.0103	0.00011
15.5	1.15	1.1328	0.0172	0.00030
26.5	1.22	1.2319	−0.0119	0.00014
		Sum	0	0.00079

We estimate the standard deviation σ with $s = \sqrt{\frac{1}{n-2}\Sigma(\text{residual})^2} = \sqrt{\frac{0.00079}{4-2}} = 0.01987$.

26.3 (a) A scatterplot of discharge by year is provided, along with the fitted regression line for part (b). Discharge seems to be increasing over time, but there is also a lot of variation in this trend. $r^2 = 0.2017$, so the least-squares regression line explains about 20.2% of the total observed variability in Arctic discharge.

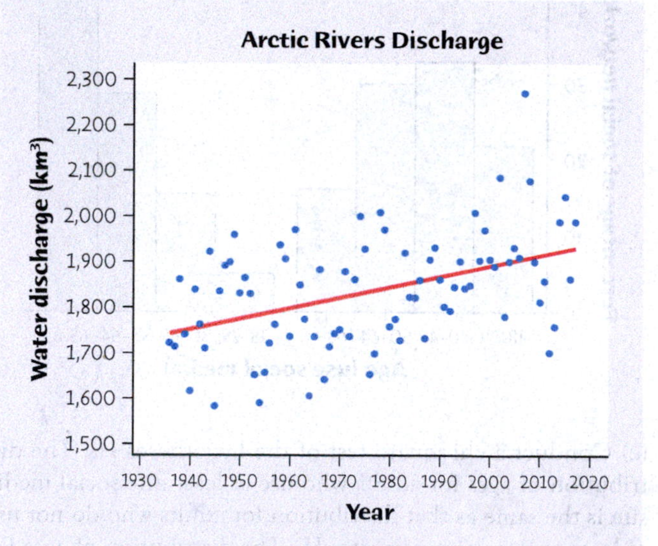

(b) $\hat{y} = -2792.93 + 2.3423x$, $s = 111.669$.

26.5 H_0: $\beta = 0$ versus H_a: $\beta > 0$. We compute $t = \frac{b}{SEb} = \frac{2.3423}{0.5210} = 4.50$. Here, df $= n - 2 = 81 - 2 = 79$. In referring to Table C, we round df down to df $= 60$; $P < 0.0005$. Using software, $P < 0.0001$. There is strong evidence of an increase in Arctic discharge over time.

26.7 (a) $H_0: \beta = 0$ versus $H_a: \beta > 0$, $t = 8.10$ with $df = 2$ (all from software), $0.005 < P < 0.01$. This test is equivalent to testing H_0: population correlation = 0 versus H_a: population correlation > 0. (b) $r = 0.985$. From Table E with $n = 4$, $0.005 < P < 0.01$.

26.9 $t^* = 2.920$, $df = 4 - 2 = 2$. A 90% confidence interval is $b \pm t^* SE_b = 0.009012 \pm 2.920(0.001112) = 0.00576$ to 0.01226. With 90% confidence, the expected increase in relative risk of breast cancer associated with a 1 gram per day increase in alcohol consumption is between 0.00576 and 0.01226.

26.11 $b = 2.3423$ and $SE_b = 0.5210$. With $n = 81$, $df = 79$. Using Table C and $df = 60$, $t^* = 1.671$ ($t^* = 1.664$ from software). A 90% confidence interval is $b \pm t^* SE_b = 2.3423 \pm 1.671(0.5210) = 1.4717$ to 3.2129 cubic kilometers per year (software: 1.4754 to 3.2092). With 90% confidence, the yearly increase in Arctic discharge is between 1.47 and 3.21 cubic kilometers. Zero is not in this confidence interval, and the entire interval is positive, so there is evidence that Arctic discharge is increasing over time.

26.13 (a) If $x^* = 70$, $\hat{\mu} = 0.4952915 + 0.002332(70) = 0.6585$. (b) $SE_{\hat{\mu}} = 0.008484$. For $df = 29 - 2 = 27$ and 95% confidence, $t^* = 2.052$. A 95% confidence interval for mean Aroc in people with mean gray-matter volume of 70 is given by $\hat{\mu} \pm t^* SE_{\hat{\mu}} = 0.6585 \pm 2.052(0.008484) = 0.6411$ to 0.6759.

26.15 (a) The residual plot provided does not suggest any deviation from a straight-line relationship between Aroc score and brain volume. The point on the left side of the plot shows that the volume 43 is an outlier in the x direction. There is also a large (in absolute value) residual in the lower right (residual −0.0989).

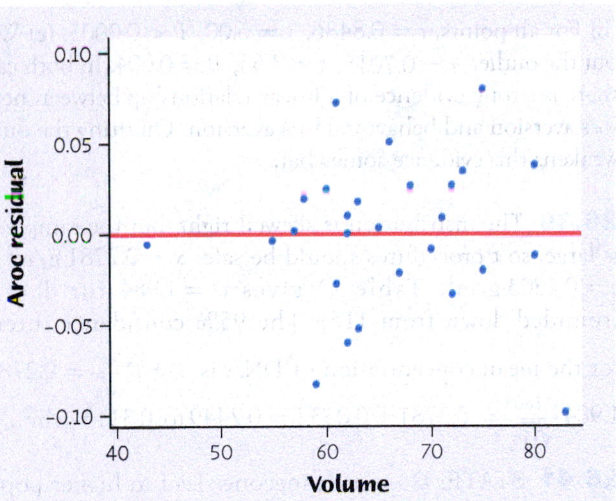

(b) The provided stemplot of residuals does not suggest that the distribution of residuals departs strongly from Normality. The values −0.08 and −0.09 from Observations 1 and 5 may be outliers. Other than that, the residuals are relatively symmetric and mound shaped.

```
-0 | 98
-0 |
-0 | 55
-0 | 3222
-0 | 1000
 0 | 000000111
 0 | 22233
 0 | 5
 0 | 77
```

(c) It is reasonable to assume that observations are independent because we have 29 different subjects, measured separately. (d) Other than the large residuals noted in part (a), there is no indication that variability changes.

26.17 (c)

26.19 (a)

26.21 (c)

26.23 (b)

26.25 (a) Scientists estimate that each additional increase of 1% in the proportion of Bt cotton plants results in an average increase of 6.81 mirid bugs per 100 plants. (b) The proportion of Bt plants explains 90% of the variability in mirid bug density. (c) $H_0: \beta = 0$ versus $H_a: \beta > 0$ (H_0: population correlation = 0 versus H_a: population correlation > 0). There is strong evidence of a linear relationship between the proportion of Bt cotton plants and the density of mirid bugs. (d) We cannot conclude a causal relationship; this was not a designed experiment.

26.27 For 90% confidence intervals with $df = 10$, use $t^* = 1.812$. (a) $b \pm t^* SE_b = 274.78 \pm 1.812(88.18) = 115.0$ to 434.6 fps/inch. (b) This is the "90% CI" given in Figure 26.14: 176.2 to 239.4 fps. To confirm this, use $\hat{\mu} = 207.8$ and $SE_{\hat{\mu}} = 17.4$. $\hat{\mu} \pm t^* SE_{\hat{\mu}} = 207.8 \pm 1.812(17.4) = 176.3$ to 239.3 fps, which agrees with the output up to roundoff error.

26.29 We test $H_0: \beta = 0$ versus $H_a: \beta > 0$. With $df = 42 - 2 = 40$, we have $t = \frac{b}{SEb} = \frac{0.1451046}{0.009849} = 14.733$ and $P < 0.0005$. There is overwhelming evidence that the number of manatees killed increases as the number of boats registered increases.

26.31 (a) The provided stemplot of residuals does not suggest that the distribution of residuals departs strongly from Normality. The values 23, 26, and 38 may be outliers. Other than that, the residuals are relatively symmetric and mound shaped.

-2	41
-1	75
-1	1
-0	999855
-0	44444332211
0	0001233
0	556666678
1	4
1	
2	3
2	6
3	
3	8

(b) There is no clear pattern, but the variability about the "residual = 0" line increases as the number of boats increases.

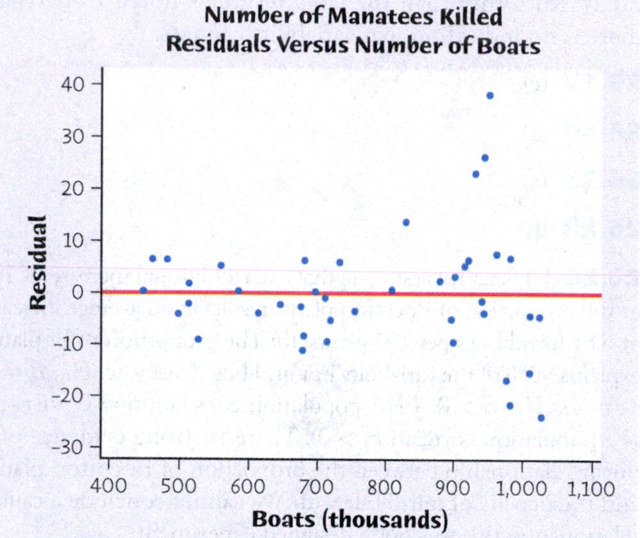

(c) $r^2 = 0.844$, so the least-squares regression line on number of boats explains about 84.4% of the total observed variability in manatee deaths. Only 15.6% of the observed variability is left over to be explained by other variables, including pollution. **(d)** It seems reasonable for us to trust the inferences in Exercises 26.29 and 26.30 since the assumptions for inference for regression appear to be satisfied.

26.33 (a) H_0: population correlation $= 0$ versus H_a: population correlation > 0. $t = 4.64$ with df $= 35 - 2 = 33$, $P < 0.0001/2 = 0.00005$. There is very strong evidence of a positive correlation between Gray's forecasted number of storms and the number of storms that actually occur. **(b)** $\hat{\mu} = 1.6376 + 0.9495(16) = 16.8296$ storms, and software gives the 95% confidence interval for the mean as 16.4128 to 17.2464 storms.

26.35 The stemplot suggests that the residuals may not follow a Normal distribution. Specifically, there are both a low and a high outlier that seem extreme. This makes regression inference and interval procedures unreliable.

-0	8
-0	
-0	544
-0	33333322
-0	0011
0	000001
0	2222222233
0	5
0	6
0	
1	
1	2

26.37 (a) The scatterplot shows two regression lines, indistinguishable because they nearly overlap. The outlier is at (155.2, 1.94).

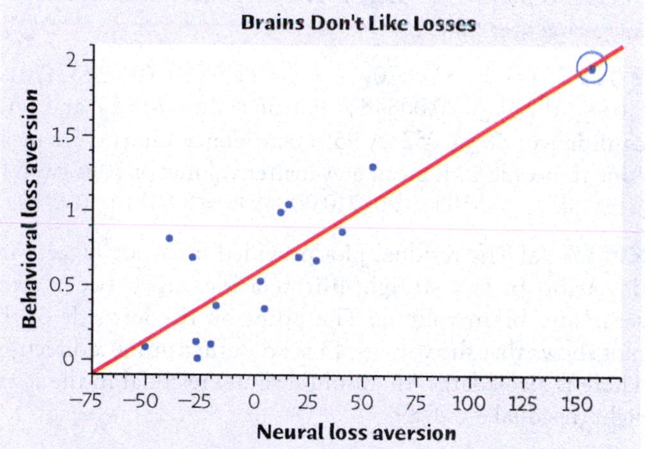

(b) For all points, $r = 0.8486$, $t = 6.00$, $P < 0.0005$. **(c)** Without the outlier, $r = 0.7015$, $t = 3.55$, $P = 0.004$. In both cases, there is strong evidence of a linear relationship between neural loss aversion and behavioral loss aversion. Omitting the outlier weakens this evidence somewhat.

26.39 The distribution is skewed right, but the sample size is large, so t procedures should be safe. $\bar{x} = 0.2781$ g/m^2 and $s = 0.1803$ g/m^2. Table C gives $t^* = 1.984$ for df $= 100$ (rounded down from 115). The 95% confidence interval for the mean concentration of DNA is $\bar{x} \pm t^* \frac{s}{\sqrt{n}} = 0.2781 \pm 1.984 \frac{0.1803}{\sqrt{116}} = 0.2781 \pm 0.0332 = 0.2449$ to 0.3113 g/m^2.

26.41 STATE: Do more pinecones lead to higher population density of red squirrels? PLAN: We examine the relationship between pinecone abundance and squirrel density, seeking a positive correlation using a scatterplot and regression. SOLVE: A scatterplot (on the next page) indicates a positive relationship that is roughly linear with what appears to be an outlier at the upper right of the graph. Regression gives predicted squirrel density as $\hat{y} = 0.961 + 0.205x$. We test H_0: $\beta = 0$ versus H_a: $\beta > 0$. With df $= 23 - 2 = 21$, we have $t = 3.13$ and $P = 0.0025$. Conditions for inference seem to be violated. The residual plot shows increasing variability with increasing cone values. The stemplot of the residuals shows two large positive outliers; the distribution may be right-skewed.

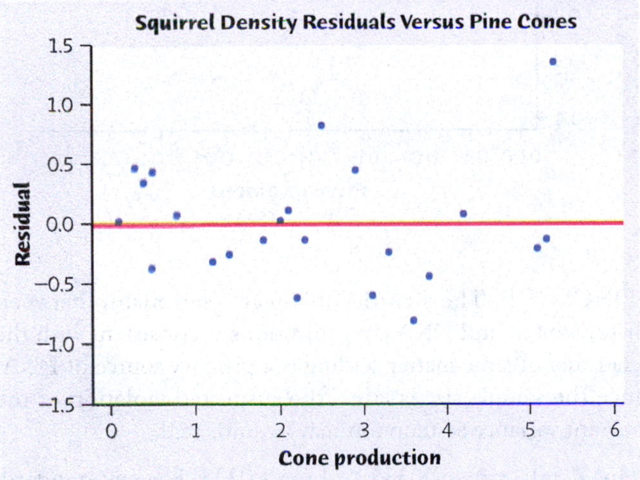

-0	8
-0	66
-0	44
-0	3222
-0	111
0	00111
0	
0	4445
0	
0	8
1	
1	
1	4

CONCLUDE: We have strong evidence of a positive linear relationship between pinecone abundance and squirrel density; however, conditions for inference may not be satisfied.

26.43 STATE: Are more beetle larvae clusters present when beavers have left more tree stumps? PLAN: We will examine the relationship between beaver stumps and beetle larvae using a scatterplot and regression. We specifically test for a positive slope β and find a confidence interval for β. SOLVE: The scatterplot shows a positive linear association. $\hat{y} = -1.286 + 11.894x$. A stemplot of the residuals does not suggest non-Normality, even with the small gap around zero. There is an unusual pattern in the residual plot that should be investigated as there may be a lack of independence.

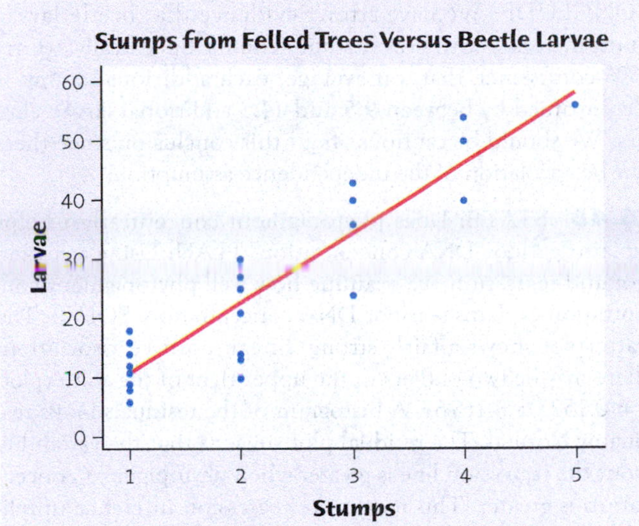

-1	3
-1	00
-0	9
-0	6
-0	5
-0	3222
-0	
0	01
0	223
0	445
0	677
0	89

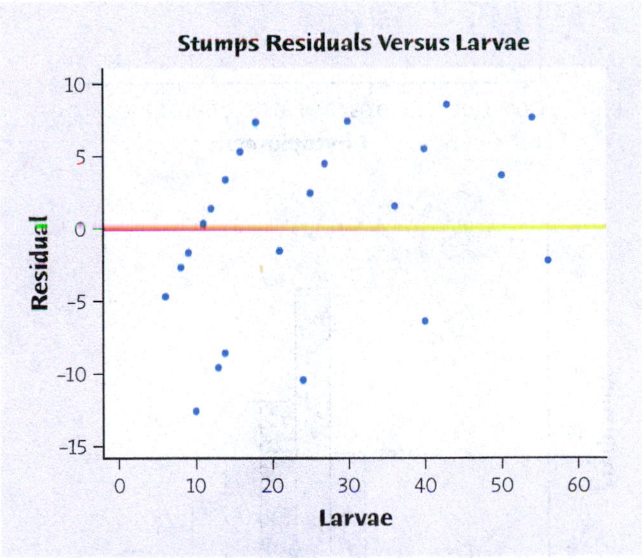

For H_0: $\beta = 0$ versus H_a: $\beta > 0$, df $= 21$, $t = 10.47$, and $P < 0.0005$. For 95% confidence, $t^* = 2.080$. $b \pm t^* SE_b = 11.894 \pm 2.080(1.136) = 9.531$ to 14.257.

CONCLUDE: We have strong evidence that beetle larvae count increases with beaver stump counts. Specifically, we are 95% confidence that, on average, each additional stump is accompanied by between 9.5 and 14.3 additional larvae clusters. We should be cautious about this conclusion since there may be a violation of the independence assumption.

26.45 STATE: Does phytopigment concentration helps explain seafloor DNA concentration? PLAN: Using a scatterplot and regression, we examine how well phytopigment concentration explains seafloor DNA concentration. SOLVE: The scatterplot shows a fairly strong, linear, positive association. There may be two outliers in the upper right of the scatterplot. $\hat{y} = 0.1523 + 8.1676x$. A histogram of the residuals looks reasonably Normal. The residual plot suggests that the variability about the regression line is greater when phytopigment concentration is greater. This may make regression inference unreliable. It is reasonable that the observations are independent. For H_0: $\beta = 0$ versus H_a: $\beta \neq 0$, df = 114, $t = 13.25$, and $P < 0.0001$. For 95% confidence, $t^* = 1.981$. $b \pm t^* SE_b = 8.1676 \pm 1.981(0.6163) = 6.95$ to 9.39.

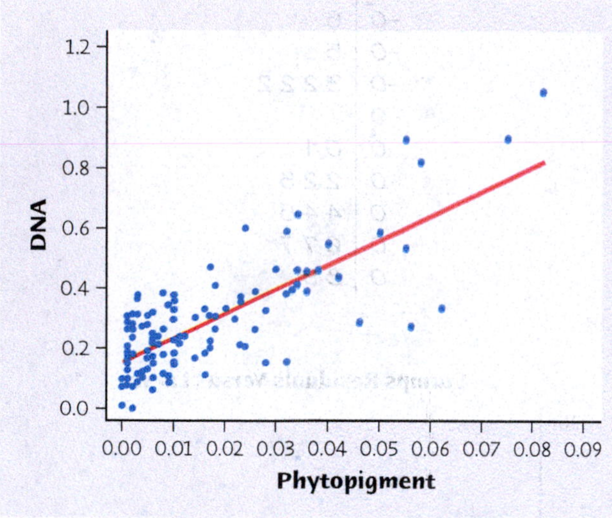

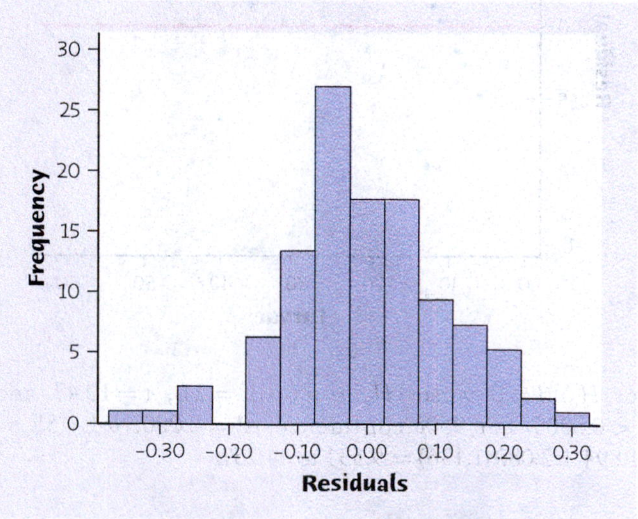

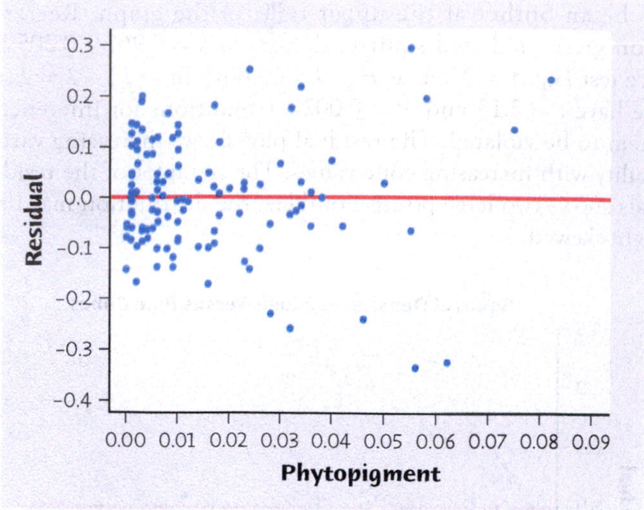

CONCLUDE: The significant linear relationship between phytopigment and DNA concentrations is consistent with the belief that organic matter settling is a primary source of DNA. Since the sample size is large, the suspected violation of the constant variance assumption may be mitigated.

26.47 (a) $\bar{x} = -0.00333$ and $s = 1.0233$. For any standardized set of values, we expect the mean and standard deviation to be 0 and 1, respectively. (b) The stemplot does not look particularly symmetric, but it is not strikingly non-Normal for such a small sample. (c) The probability is about 0.0272.

26.49 For df = 14 and a 95% confidence interval, $t^* = 2.145$. $a \pm t^* SE_a = -0.01270 \pm 2.145(0.01264) = -0.0398$ to 0.0144. This interval does contain 0.

Chapter 27 – One-Way Analysis of Variance: Comparing Several Means

27.1 (a) The null hypothesis is "all five categories have the same (population) mean NAEP mathematics score," and the alternative hypothesis is "at least one category has a different mean." (b) The results of the F test are quite significant, giving strong evidence that the means are different.

27.3 (a) Side-by-side boxplots are provided. The boxes for Groups 2 and 3 overlap and are both lower than the box for Group 1. There are more trees in the plots that have never been logged (Group 1).

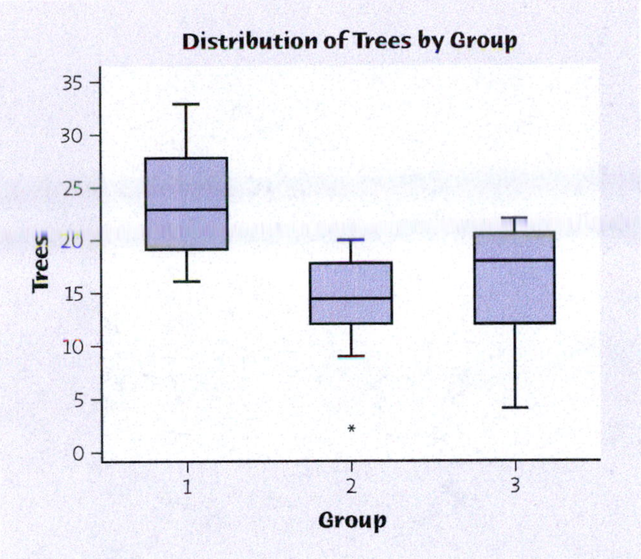

(b) The mean number of trees for the group that has never been logged (Group 1) is the largest. The mean number of trees for the group that was logged eight years ago (Group 3) is slightly larger than for the group logged one year ago (Group 2). (c) The null hypothesis is "all three groups have the same (population) mean number of trees," and the alternative hypothesis is "at least one category has a different mean." The F statistic is $F = 11.4257$, and the P-value is $P = 0.0002$. There is strong evidence that logging is related to differences in the mean number of trees for a forest plot.

27.5 (a) Answers will vary due to randomness. (b) The smallest F-values will vary. F is made small by moving the largest mean to midway between the other two means. The smaller F has a higher P-value and is therefore less significant. (c) The largest F-values will vary. F increases and the P-value decreases as any mean is pulled toward the top or bottom. F is largest when two of the means are very high or very low and the third is the opposite.

27.7 (a) $s_1 = 5.0655$, $s_2 = 4.9810$, and $s_3 = 5.7615$. The ratio of largest to smallest is $5.7615/4.9810 = 1.16$, which is less than 2. Conditions are satisfied. (b) $s_G = 21$, $s_{DS} = 8$, $s_{AD} = 16$. The ratio of largest to smallest $21/8 = 2.625$, which is not less than 2. Conditions are not satisfied. Proceed with caution in the study on the psychological effects of restrictive environments.

27.9 STATE: Are there differences in the mean word counts for the three groups? PLAN: We will examine the data by first looking at side-by-side boxplots. We will assess whether it is safe to use ANOVA and do the analysis, if appropriate. SOLVE: Those who took longhand notes wrote fewer words on average than those in either laptop condition. The laptop group averaged the most words. The ratio of largest to smallest standard deviations is $118.5/59.64 = 1.99$, which is just slightly less than 2. The laptop group seems right-skewed (at least more than the other two conditions), and both laptop groups show outliers. However, with the smallest sample size being 48, the central limit theorem says the sample means should be approximately Normal. So it is safe to use the F-test. The output shows $F = 17.04$ and $P < 0.0005$.

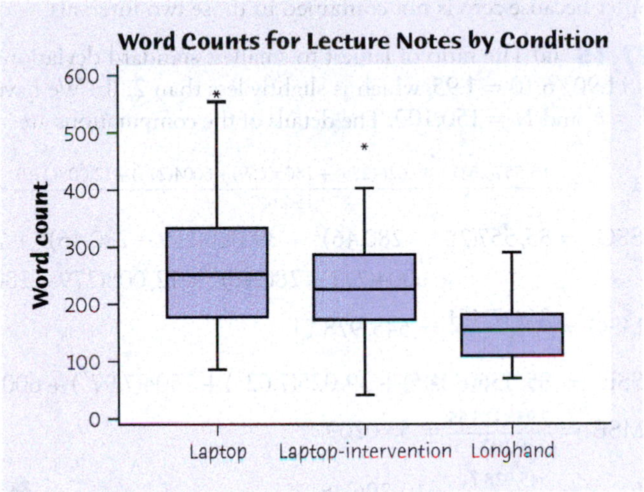

Variable	Condition	N	N*	Mean	SE Mean	StDev	Minimum	Q1
wordcount	Laptop	51	0	260.9	16.6	118.5	85.0	177.0
	Laptop-intervention	52	0	229.0	11.8	84.8	40.0	171.8
	Longhand	48	0	155.98	8.61	59.64	68.00	108.25

Variable	Condition	Median	Q3	Maximum
wordcount	Laptop	252.0	332.0	566.0
	Laptop-intervention	226.0	285.0	473.0
	Longhand	153.50	181.00	289.00

One-way ANOVA: wordcount versus condition

Source	DF	SS	MS	F	P
Condition	2	284599	142300	17.04	0.000
Error	148	1235978	8351		
Total	150	1520577			

S = 91.38 R-Sq = 18.72% R-Sq(adj) = 17.62%

CONCLUDE: There is clearly a difference in the number of words written while taking notes with the three methods. It seems fairly obvious from the graphs that those who take notes on a laptop write more; those taking longhand notes write the least.

27.11 (a) $I = 3$ and $N = 96$, so df = 2 and 93. (b) $I = 3$ and $N = 90$, so df = 2 and 87.

27.13 (a) Let μ_1 be the mean word count for the laptop group, μ_2 be the mean word count for the laptop-intervention group, and μ_3 be the mean word count for the longhand group. The intervals are

$\mu_1 - \mu_2$	61.43 to 148.45
$\mu_1 - \mu_3$	29.73 to 116.35
$\mu_2 - \mu_3$	−10.74 to 74.54

(b) "95% confidence" means that there is probability 0.95 that all of the intervals simultaneously capture the true pairwise differences in mean word count. (c) The laptop and laptop-intervention group and the laptop and longhand group significantly differ because zero is not contained in those two intervals.

27.15 (a) The ratio of largest to smallest standard deviations is $11.90/6.10 = 1.95$, which is slightly less than 2. (b) We have $I = 5$ and $N = 150{,}100$. The details of the computations are:

$$\bar{x} = \frac{85{,}557(283)+39{,}026(279)+7505(269)+6004(271)+12{,}008(279)}{150{,}100} = 280.46$$

$$\text{SSG} = 85{,}557(283-280.46)^2 + 39{,}026(279-280.46)^2 + 7505(269-280.46)^2 + 6004(271-280.46)^2 + 12{,}008(279-280.46)^2 = 2{,}183{,}714.84$$

$$\text{MSG} = \frac{2{,}183{,}714.84}{5-1} = 545{,}928.71$$

$$\text{SSE} = 85{,}556(6.10^2) + 39{,}025(7.02^2) + 7504(7.97^2) + 6003(10.60^2) + 12{,}007(11.90^2) = 7{,}958{,}175.55$$

$$\text{MSE} = \frac{7{,}958{,}175.55}{150{,}100-5} = 53.0209$$

$$F = \frac{545{,}928.71}{53.0209} = 10{,}296.48$$

(c) We use an F distribution with df = 4 and 150,095. We have very strong evidence that the mean NAEP mathematics score differs among the tablet use groups. These results are not surprising because the sample sizes were very large. The differences might not have practical importance. (The largest difference in mean scores is 14, which is relatively small on a 500-point scale.)

27.17 (a) We have independent samples from the five groups, and the standard deviations easily satisfy our rule of thumb ($1.40/1.28 = 1.09 < 2$). (b) The details of the computations, with $I = 5$ and $N = 4413$, are:

$$\bar{x} = \frac{809(2.57)+1860(2.32)+654(2.63)+883(2.51)+207(2.51)}{4413} = 2.459$$

$$\text{SSG} = 809(2.57-2.459)^2 + 1860(2.32-2.459)^2 + 654(2.63-2.459)^2 + 883(2.51-2.459)^2 + 207(2.51-2.459)^2 = 67.86$$

$$\text{MSG} = \frac{67.86}{5-1} = 16.97$$

$$\text{SSE} = 808(1.40^2) + 1859(1.36^2) + 653(1.32^2) + 882(1.31^2) + 206(1.28^2) = 8010.98$$

$$\text{MSE} = \frac{8010.98}{4413-5} = 1.82$$

$$F = \frac{16.97}{1.82} = 9.34$$

(c) The ANOVA is very significant ($P < 0.001$), but this is not surprising because the sample sizes were very large. The differences might not have practical importance. (The largest difference in mean scores is 0.31, which is relatively small on a 5-point scale.)

27.19 (b)

27.21 (c)

27.23 (a)

27.25 (c)

27.27 (b)

27.29 The populations are students who wear a lemon-flavored mouth guard, students who wear a non-flavored mouth guard, and students who wear no mouth guard. The response is the rating of perceived exertion. $I = 3$, $N = 43$, $n_1 = 12$, $n_2 = 15$, and $n_3 = 16$. The degrees of freedom are 2 and $43 - 3 = 40$.

27.31 The response variable is hemoglobin A1c level. We have $I = 4$ populations: a control (sedentary) population, an aerobic exercise population, a resistance training population, and a combined aerobic and resistance training population. Sample sizes are $n_1 = 41$, $n_2 = 73$, $n_3 = 72$, and $n_4 = 76$. Our total sample size is $N = 41 + 73 + 72 + 76 = 262$. We have $I - 1 = 4 - 1 = 3$ and $N - I = 262 - 4 = 258$. So, the degrees of freedom for F are 3 and 258.

27.33 (a) The graph suggests that NAEP mathematics scores decrease as tablet use increases because the mean NAEP mathematics scores for groups with more than some tablet use are all much lower than the mean.

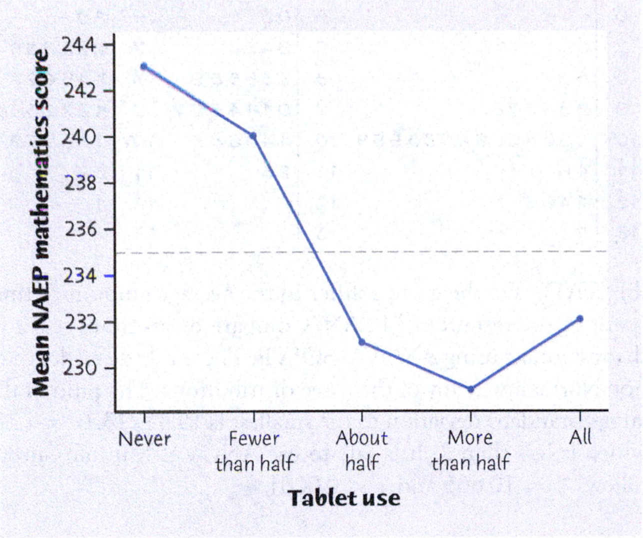

(b) The null hypothesis is "all five groups have the same mean NAEP mathematics score." The alternative is "at least one group has a different mean NAEP mathematics score." (c) No. The ratio of largest to smallest standard deviations is $9.16/4.38 = 2.09$, which exceeds 2.

27.35 (a) The stemplots and table are provided below. The ratio of the largest to smallest standard deviations is $0.7307/0.6298 = 1.16$, which is less than 2. Even so, ANOVA is risky with these data because the distribution of avobenzone absorption for the cream is rather irregular, with outliers (although it is difficult to judge with such small samples).

Spray 1		Spray 2		Lotion		Cream	
-0	95	-1	2	-0	55	-1	66
-0		-0	65	-0		-1	
0	444	-0		0	01	-1	2
0	6	0	00	0	6	-1	
		0	5	1	3	-0	
						-0	
						-0	
						-0	333

Treatment	n	$\bar{x}$	s
Spray 1	6	0.0802	0.6378
Spray 2	6	−0.3033	0.6300
Lotion	6	0.2223	0.7306
Cream	6	−0.9155	0.6294

(b) We wish to test $H_0: \mu_1 = \mu_2 = \mu_3 = \mu_4$ (the mean amount of avobenzone absorbed into the bloodstream after 150 hours is the same for all four treatments) versus H_a: at least one mean is different. (c) The means shown in the table suggest that avobenzone is absorbed into the blood least with the cream and most with the lotion. ANOVA gives a statistically significant result ($F = 3.5809$, df = 3, and 20, $P = 0.0321$), but as noted in part (a), the conditions for ANOVA may not be satisfied.

Analysis of Variance

Source	DF	Sum of Squares	Mean Square	F Ratio	Prob > F
Treatment	3	4.656561	1.55219	3.5812	0.0320*
Error	20	8.668531	0.43343		
C. Total	23	13.325092			

27.37 The stemplots and table are provided. The means suggest that octocrylene is absorbed into the bloodstream most with lotion and least with Spray 1. The standard deviations satisfy the rule of thumb ($0.6449/.3342 = 1.93 < 2$). However, ANOVA is risky with these data because the distributions of octocrylene absorption for the lotion and cream are rather irregular, with outliers (although it is difficult to judge with such small samples).

Spray 1		Spray 2		Lotion		Cream	
-0	98	-0	5	-0	5	-0	2
-0	6	-0		-0		-0	1
-0	5	-0	2	-0		0	
-0	3	-0		0		0	33
-0	1	0	0	0	2	0	4
		0		0	5	0	
		0	4	0		0	9
		0	7	0	89		
		0	9	1			
				1	3		

Treatment	n	$\bar{x}$	s
Spray 1	6	−0.5687	0.3342
Spray 2	6	0.2038	0.6025
Lotion	6	0.5653	0.6449
Cream	6	0.2900	0.4337

27.39 The stemplots (with values rounded rather than truncated) and table are provided. The means suggest that oxybenzone is absorbed into the bloodstream most with lotion and least with Spray 1. However, the standard deviations do not satisfy the rule of thumb ($1.2222/0.4499 = 2.72 > 2$). ANOVA is risky with these data because the distribution of oxybenzone absorption for the lotion is not unimodal and is not mound-shaped (although it is difficult to judge with such small samples).

Spray 1		Spray 2		Lotion	
2	8	2	2	1	1
3	3	2	89	1	6
3	579	3	03	2	
4	0	3	7	2	8
				3	0
				3	5
				4	4

Treatment	n	$\bar{x}$	s
Spray 1	6	3.5350	0.4499
Spray 2	6	2.9925	0.4976
Lotion	6	2.7532	1.2222

27.41 (a) STATE: Is there a significant difference in depression scores among the three classifications of sleep? PLAN: Examine the data to determine if ANOVA can be used. Assume that these students are close to an SRS and that students are independent. SOLVE: Side-by-side boxplots show outliers, but with such large samples, the sample means will be approximately Normal. The rule of thumb for standard deviations is satisfied because $4.719/2.560 = 1.84 < 2$. $F = 75.52$ and $P < 0.0001$.

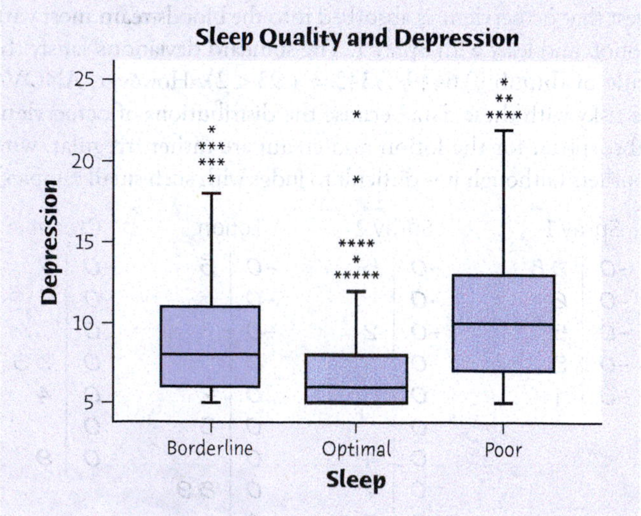


```
Source    DF       SS       MS      F      P
Sleep      2   2162.3   1081.1  72.52  0.000
Error    895  13343.7     14.9
Total    897  15506.0

Level        N     Mean   StDev
Borderline  246    8.764  3.892
Optimal     309    7.013  2.560
Poor        343   10.656  4.719
```

CONCLUDE: The mean depression scores for the three levels of sleep quality are not the same. It appears that the mean depression score for poor sleepers is highest, and the mean depression score for optimal sleepers is lowest. (b) Assuming that the students were randomly selected, the large sample size would leave us to believe that these students are most likely representative of other college students. (c) Students were not randomly assigned to sleep conditions. Explanations about causation will vary, but this might well be a case of one condition (poor sleep) feeding the other (depression) in a "vicious cycle."

27.43 (a) We can be 99% confident that all three of these intervals capture the true difference between pairs of population means. (b) Combining the results from the F-test and the multiple comparisons, we can conclude that, on average, depression is greatest for those with poor sleep quality and lowest for those with optimal sleep quality.

27.45 (a) Stemplots are provided. There is not strong evidence of non-Normality in any of the distributions, even though the data corresponding to lemon odor has a somewhat left-skewed distribution. There are no outliers.

	Lavender		Lemon		No Odor
5		5	6	5	
6		6	03	6	89
7	6	7	3458	7	223569
8	89	8	338889	8	445677
9	234578	9	0144677	9	1222368
10	1234556678899	10	145688	10	136779
11	46	11	23	11	58
12	1469	12		12	1
13	7	13		13	

(b) STATE: Do the groups differ in the average amount of time spent in the restaurant? PLAN: Compare mean times spent in the restaurant using ANOVA. SOLVE: There is little evidence of non-Normality in any of the three distributions. The ratio of the largest standard deviation to the smallest is $15.44/13.10 = 1.18$, which is less than 2. It is safe to use ANOVA. Minitab output follows. $F = 10.861$ and $P < 0.0001$.

One-way ANOVA: Time versus Odor

Source	DF	SS	MS	F	P
Odor	2	4569	2285	10.86	0.000
Error	85	17879	210		
Total	87	22448			

S = 14.50 R-Sq = 20.35% R-Sq(adj) = 18.48%

Level	N	Mean	StDev
Lavendar	30	105.70	13.10
Lemon	28	89.79	15.44
No odor	30	91.27	14.93

CONCLUDE: There is overwhelming evidence of a difference in the mean amount of time that customers spend in the restaurant, depending on the odor present. Lavender odor yields the longest mean time, while lemon odor reduces time spent on average, compare with no odor at all.

27.47 (a) Let μ_L denote the mean time spent with lavender scent, μ_{Le} denote the mean time spent with lemon scent, and μ_N denote the mean time spent with no odor. Test the three hypotheses $H_0\colon \mu_L = \mu_{Le}$, $H_0\colon \mu_L = \mu_N$, and $H_0\colon \mu_{Le} = \mu_N$. (b) JMP output for the Tukey pairwise comparisons at a 95% overall confidence level is provided.

Level	- Level	Difference	Std Err Dif	Lower CL	Upper CL
lavender	lemon	15.91429	3.810966	6.82334	25.00523
lavender	noodor	14.43333	3.744683	5.50051	23.36616
noodor	lemon	1.48095	3.810966	-7.60999	10.57189

In Exercise 27.45, we concluded that the means of the three groups are not all the same. Based on the pairwise comparisons, we can conclude with 95% confidence that the smell of lavender is related to a higher mean time than the other two smells, and lemon and no odor are similar with respect to the mean time.

27.49 STATE: Do the groups differ in the average length of time they wat to ask for help? PLAN: Examine the data to compare the effect of the treatments and determine if we can use ANOVA to test the significance of the observed differences in mean times to ask for help. SOLVE: Histograms of the data for each of the three groups and JMP output are provided. The histograms are not symmetric, but with 17 or 18 observations in each group, the sample sizes are likely large enough for Normality to hold because there is not strong skewness, nor are there outliers. The standard deviations satisfy our rule of thumb $(172.79/118.09 = 1.46 < 2)$, so it is safe to use ANOVA. Using ANOVA to test the null hypothesis that the mean time for the three groups is the same against the alternative that at least one mean is different, we get $F = 3.73$ and $P = 0.031$.

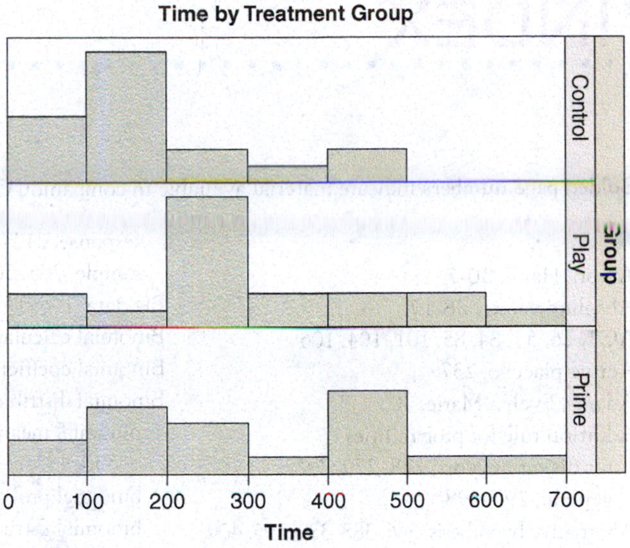

Analysis of Variance

Source	DF	Sum of Squares	Mean Square	F Ratio	Prob > F
Group	2	174911.6	87455.8	3.7278	0.0311*
Error	49	1149567.8	23460.6		
C. Total	51	1324479.4			

Level	Number	Mean	Std Dev	Std Err Mean	Lower 95%	Upper 95%
control	17	186.11765	118.09259	28.641661	125.40004	246.83526
play	18	305.22222	162.46866	38.29423	224.42846	386.01599
prime	17	314.05882	172.78978	41.907678	225.21852	402.89913

CONCLUDE: There is good evidence that the mean time to ask for help is not the same for the three groups. Based on the histograms and summary statistics, it appears that being reminded about money is related to people taking longer to ask for help.

27.51 The 95% Tukey pairwise comparison intervals are in the provided JMP output. It appears that there is only a significant difference in mean time between the prime group and the control group and that the other two comparisons are not significantly different at the 0.05 significance level.

Level	- Level	Difference	Std Err Dif	Lower CL	Upper CL
prime	control	127.9412	52.53634	0.965	254.9716
play	control	119.1046	51.80153	-6.096	244.3050
prime	play	8.8366	51.80153	-116.364	134.0370

27.53 (a) Two-sample *t* test. (b) ANOVA. (c) Chi-square test of homogeneity.

INDEX

Bolded page numbers indicate material available in companion Chapters 28–32; available online

D

I

J

K

L

M